THE
BACTERIAL
CHROMOSOME

THE BACTERIAL CHROMOSOME

EDITED BY **N. PATRICK HIGGINS**

Department of Biochemistry and Molecular Genetics and
Howell Heflin Center for Human Genetics
University of Alabama at Birmingham
Birmingham, Alabama

ASM
PRESS

Washington, D.C.

Address editorial correspondence to ASM Press, 1752 N St. NW, Washington, DC 20036-2904, USA

Send orders to ASM Press, P.O. Box 605, Herndon, VA 20172, USA
Phone: (800) 546-2416 or (703) 661-1593
Fax: (703) 661-1501
E-mail: books@asmusa.org
Online: www.asmpress.org

Library of Congress Cataloging-in-Publication Data

The bacterial chromosome / edited by N. Patrick Higgins.
     p. ; cm.
   Includes bibliographical references and index.
   ISBN 1-55581-232-5 (hardcover)
   1. Bacterial chromosomes. 2. Bacterial genetics.
   [DNLM: 1. Bacteria—genetics. 2. Chromosomes, Bacterial. QW 51 B13035 2004] I. Higgins, N. Patrick.

QH434.B332 2004
572.8′7293—dc22   2004008702

10 9 8 7 6 5 4 3 2 1

Cover images: (Upper left) *Escherichia coli* colonies containing a stable green fluorescent protein expressed from a low-copy-number plasmid in which transcription originates in a fragment from the *dnaA* region of the *E. coli* K-12 chromosome (nucleotides 3877671 to 3877850). The colonies grew for 41 h at 32°C on minimal salts-glucose-Casamino acids agar before digital photography under fluorescence illumination with a Zeiss Axiomat microscope with a 2.5× uncorrected objective. The colonies, which were photographed by Jim Shapiro at the University of Chicago, measured 2 to 3 mm in diameter. (Lower right) A microarray illustrating the transposition preference of bacteriohage Mu for the protein-coding genes of the *E. coli* genome. Cy-5-labeled probe (red) was derived from purified phage DNA, whereas Cy-3-labeled probe (green) was derived from random primed *E. coli* genomic DNA (D. Manna, A. M. Breier, and N. P. Higgins, *Proc. Natl. Acad. Sci. USA* **101**:9780–9785, 2004).

CONTENTS

CONTRIBUTORS

Irina Artsimovitch
Department of Microbiology, Ohio State University, 484 West 12th Ave., Columbus, OH 43210

Stuart Austin
Gene Regulation and Chromosome Biology Laboratory, Division of Basic Sciences, NCI-Frederick, Frederick, MD 21702-1201

François-Xavier Barre
Laboratoire de Microbiologie et de Génétique moléculaire du CNRS, 118 route de Narbonne, 31062 Toulouse Cedex, France

Therese Brendler
Gene Regulation and Chromosome Biology Laboratory, Division of Basic Sciences, NCI-Frederick, Frederick, MD 21702-1201

Johanna Eltz Camara
Department of Biochemistry and Molecular Biology, Georgetown University Medical Center, 3900 Reservoir Rd. NW, Washington, DC 20007

Sherwood Casjens
Department of Pathology, University of Utah Medical School, Salt Lake City, UT 84132, and Pittsburgh Bacteriophage Institute, University of Pittsburgh, Pittsburgh, PA 15260

George Chaconas
Department of Biochemistry and Molecular Biology and Department of Microbiology and Infectious Diseases, University of Calgary, Calgary, Alberta T2N 4N1, Canada

Keith Champion
Department of Biochemistry and Molecular Genetics, University of Alabama at Birmingham, 720 20th St. South, Birmingham, AL 35294-2170

Carton W. Chen
Institute of Genetics, National Yang-Ming University, Shih-Pai, Taipei 112, Taiwan

François Cornet
Laboratoire de Microbiologie et de Génétique moléculaire du CNRS, 118 route de Narbonne, 31062 Toulouse Cedex, France

Michael M. Cox
Department of Biochemistry, University of Wisconsin—Madison, 433 Babcock Dr., Madison, WI 53706-1544

Elliott Crooke
Department of Biochemistry and Molecular Biology, Georgetown University Medical Center, 3900 Reservoir Rd. NW, Washington, DC 20007

Shuang Deng
Department of Biochemistry and Molecular Genetics, University of Alabama at Birmingham, 720 20th St. South, Birmingham, AL 35294-2170

Keith M. Derbyshire
Division of Infectious Disease, Wadsworth Center, New York State Department of Health, and Department of Biomedical Sciences, State University of New York at Albany, Albany, NY 12201-2002

Simon L. Dove
Division of Infectious Diseases, Children's Hospital, Harvard Medical School, Boston, MA 02115

Jeffrey F. Gardner
Department of Microbiology, University of Illinois at Urbana-Champaign, Urbana, IL 61801

Kati Geszvain
Department of Bacteriology, University of
Wisconsin—Madison, Madison, WI 53706

Nigel D. F. Grindley
Department of Molecular Biophysics and
Biochemistry, Yale University, New Haven, CT
06520-8114

Roger W. Hendrix
Department of Biological Sciences and Pittsburgh
Bacteriophage Institute, University of Pittsburgh,
Pittsburgh, PA 15260

N. Patrick Higgins
Department of Biochemistry and Molecular
Genetics, University of Alabama at Birmingham, 720
20th St. South, Birmingham, AL 35294-2170

Manju M. Hingorani
Wesleyan University, 205 Hall-Atwater
Laboratories, Middletown, CT 06459

Ann Hochschild
Department of Microbiology and Molecular
Genetics, Harvard Medical School, Boston, MA
02115

Lianna M. Johnson
Life Science Core Curriculum, University
of California, Los Angeles, Los Angeles,
CA 90095-1606

Reid C. Johnson
Department of Biological Chemistry, David Geffen
School of Medicine at UCLA, Los Angeles, CA
90095-1737

Roberto Kolter
Department of Microbiology and Molecular
Genetics, Harvard Medical School, 200 Longwood
Ave., Boston, MA 02115

Stephen C. Kowalczykowski
Sections of Microbiology and of Molecular
and Cellular Biology, Center for Genetics and
Development, University of California, Davis, CA
95616-8665

Kenneth N. Kreuzer
Department of Biochemistry, Duke University
Medical Center, Box 3711, Durham, NC 27710

Peter Kuempel
Department of Molecular, Cellular and
Developmental Biology, University of Colorado,
Boulder, CO 80309

Sidney R. Kushner
Department of Genetics, University of Georgia,
Athens, GA 30602

Andrei Kuzminov
Department of Microbiology, University of Illinois,
Urbana-Champaign, B103 C&LSL, 601 S. Goodwin
Ave., Urbana, IL 61801-3709

Robert C. Landick
Department of Bacteriology, University of
Wisconsin—Madison, Madison, WI 53706

Michael T. Laub
Department of Developmental Biology, Stanford
University Medical Center, B300 Beckman
Center, 279 Campus Dr., Palo Alto,
CA 94304-5329

Jeffrey G. Lawrence
Pittsburgh Bacteriophage Institute and Department
of Biological Sciences, University of Pittsburgh,
Pittsburgh, PA 15260

Jean-Michel Louarn
Laboratoire de Microbiologie et de Génétique
moléculaire du CNRS, 118 route de Narbonne,
31062 Toulouse Cedex, France

Susan T. Lovett
Rosenstiel Center, Brandeis University, Waltham,
MA 02454-9110

Dipankar Manna
Department of Biochemistry and Molecular
Genetics, University of Alabama at Birmingham,
720 20th St. South, Birmingham, AL 35294-2170

Kenneth J. Marians
Molecular Biology Program, Memorial
Sloan-Kettering Cancer Center, 1275 York Ave.,
New York, NY 10021

M. G. Marinus
Biochemistry and Molecular Pharmacology,
University of Massachusetts Medical School, 55
Lake Ave., Worcester, MA 01655

Harley H. McAdams
Department of Developmental Biology, Stanford University Medical Center, B300 Beckman Center, 279 Campus Dr., Palo Alto, CA 94304-5329

Bénédicte Michel
Génétique Microbienne, Institut National de la Recherche Agronomique, 78352 Jouy-en-Josas, France

Abraham Minsky
Department of Organic Chemistry, The Weizmann Institute of Science, Rehovot 76100, Israel

Hiroshi Nakai
Department of Biochemistry and Molecular Biology, Georgetown University Medical Center, 3900 Reservoir Rd. NW, Washington, DC 20007

Stella H. North
Department of Biochemistry and Molecular Biology, Georgetown University Medical Center, 3900 Reservoir Rd. NW, Washington, DC 20007

Mike O'Donnell
Rockefeller University and Howard Hughes Medical Institute, 1230 York Ave., New York, NY 10021

Zhenhua Pang
Department of Biochemistry and Molecular Genetics, University of Alabama at Birmingham, 720 20th St. South, Birmingham, AL 35294-2170

Jeffrey Roberts
Department of Molecular Biology and Genetics, 349 Biotechnology Building, Cornell University, Ithaca, NY 14853-2703

John R. Roth
Microbiology Section, Division of Biological Sciences, Center for Genetics and Development, University of California, Davis, Davis, CA 95616

John W. Schmidt
Department of Microbiology, University of Illinois at Urbana-Champaign, Urbana, IL 61801

Lucy Shapiro
Department of Developmental Biology, Stanford University Medical Center, B300 Beckman Center, 279 Campus Dr., Palo Alto, CA 94304-5329

David J. Sherratt
Division of Molecular Genetics, Department of Biochemistry, University of Oxford, South Parks Rd., Oxford OX1 3QU, United Kingdom

Maria Spies
Sections of Microbiology and of Molecular and Cellular Biology, Center for Genetics and Development, University of California, Davis, CA 95616-8665

Franklin W. Stahl
Institute of Molecular Biology, University of Oregon, Eugene, OR 97403-1229

Richard A. Stein
Department of Biochemistry and Molecular Genetics, University of Alabama at Birmingham, 720 20th St. South, Birmingham, AL 35294-2170

Bernard S. Strauss
Department of Molecular Genetics and Cell Biology, The University of Chicago, 920 E. 58th St., Chicago, IL 60637

Malcolm F. White
Centre for Biomolecular Sciences, University of St. Andrews, North Haugh, St. Andrews, Fife KY16 9ST, United Kingdom

PREFACE

In 1997, the genome sequence of one *Escherichia coli* K-12 strain was published (1). A colleague of mine who was very fond of computational protein science predicted in a lecture to our incoming students that experimental science with *E. coli* as a model organism would soon be obsolete. According to this scholar, computers would solve the three-dimensional structures of all proteins from first principles, all biochemical pathways would be obvious from linking the encoded enzymes in the proper order, and all that any student would need to know about *E. coli* would come on a computer disk inserted in a sleeve on the last page of the textbook *Lehninger Principles of Biochemistry*. My reaction to this astounding announcement of the demise of experimental prokaryotic biology was, "I wouldn't bet the farm on that."

Well, *E. coli* continues to be a robust experimental system today, and this book updates an ASM book with the same title published in 1990 (3) and *Organization of the Prokaryotic Genome* printed in 1999 (2). Computational science has certainly provided some dramatic advances. For example, clever analyses of numerous complete genomes revealed an intriguing secret of the evolutionary mechanism. Horizontal transfer of genetic material among bacterial groups is now firmly established as the major engine of bacterial evolution. Chapter 2, by Lawrence, addresses this process and the paradox of how bacterial genomes change and how they remain the same. Chapter 1, Roth's overview of section I, provides a geneticist's view of the hunt for elements that might contribute to chromosome compaction, movement, and segregation.

Experimental bacterial science also benefited enormously from learning the sequence of many bacteriophage genomes. Bacteriophages provide one mechanism for horizontal DNA transfer; chapter 3, by Casjens and Hendrix, and chapter 27, by North and Nakai, explore ways in which bacteriophages interact with and contribute to the evolution of their bacterial hosts. Chapter 26, by Derbyshire and Grindley, addresses a way to think about different types of transposons that populate all genomes.

New experimental technology coupled to computer analyses was developed soon after the publication of the *E. coli* sequence. Through a fusion of automated robots, novel DNA synthesis methods, and micromanipulators, it became possible to study genome function of organisms as disparate as bacteria and humans by using the knowledge base present in a complete sequence. Chapter 4, by Laub et al., illustrates how high-density microarrays containing complete sets of organismic genes have been used to identify new pathways and study the control elements that coordinate the cell cycle.

New approaches were also developed to study chromosome dynamics in living cells. Protein fusions between DNA-binding proteins and green fluorescent protein made it possible to analyze the cellular distribution of proteins in vivo on a real-time basis. Chapter 11, by Brendler and Austin, illustrates the dynamic properties of replication forks in dividing cells by using the hemimethyl-binding protein SeqA. The beautifully comprehensive chapter 5, by Johnson et al., focuses on the abundant small DNA-binding proteins and illustrates how DNA-shaping ensembles are used in a wide range of chromosomal processes. Chapters 6, by Higgins et al., and 7, by Minsky and Kolter, explain new genetic and physical techniques designed to study basic chromosome structure during exponential growth and stationary phase.

DNA replication is a central problem of biology. The section II overview by Marians (chapter 8) introduces the current ideas on the replisome, which is the central engine of DNA synthesis. The process of initiation is discussed in chapter 9, by Camara and Crooke. Chapter 10, by Hingorani and O'Donnell, focuses on elongation mechanisms; the authors show how the three-dimensional X-ray crystal structures of individual subunits provide the fabric for chemical and physical models of this high-fidelity process. Termination of DNA replication is not understood as completely as initiation. However, chapter 13, by Louarn et al., discusses mechanisms for segregation, and chapters 28, by Barre and Sherratt, and 29, by Chaconas and Chen, discuss the new enzymes associated with the resolution of chromosomes with circular and linear genetic structures, respectively.

After replication, the second major problem of chromosome biology is the mechanism of transcription: how cells make the structural RNAs for translation and mRNAs that encode proteins. *E. coli* and the phages like T4, λ, and T7 that alter the bacterial transcription pattern have provided the interlacing systems for understanding transcription. Initiation, elongation, and termination are the stages of transcription that represent major prizes sought in the pursuit of understanding gene regulation throughout biology. The section III overview by Roberts (chapter 14) addresses how the stages are linked and what type of traffic on DNA can influence each phase of transcription. Initiation control mechanisms are covered by Dove and Hochschild in chapter 16. The detailed X-ray crystal structures of RNA polymerases are discussed by Geszvain and Landick in chapter 15. RNA polymerase pausing and transcript termination are covered by Artsimovitch in chapter 17. In bacteria, most mRNAs have a half-life of only several minutes, and the machinery responsible for mRNA decay is discussed by Kushner in chapter 18.

Homologous DNA recombination has been called the third rail of genome science. It seemed that everyone in the field of recombination had one set of observations that they considered facts, but the blueprints explaining why and how recombination started and stopped seemed overwhelming. My own personal failure was remembering how a *recB recC recD sbcB sbcC sbcD* mutant behaves. The nuances of various mutations became clearer when the connections between recombination, replication, and repair were revealed through genetic and biochemical discoveries made over the past 10 years. In chapter 19, Kuzminov and Stahl provide an overview for this fusion. Chapter 20, by Cox, explains how the RecA protein serves a central role in managing the traffic between promoting homologous chromosome exchange, saving collapsed replication forks, and promoting chromosomal recovery. The interface between RecA protein and a complex RecBCD helicase/nuclease is described by Spies and Kowalczykowski in chapter 21. In chapter 22, White covers enzymes designed to resolve the Holliday intermediate in homologous exchange.

Recombination and repair are really stories about how chromosomes are shielded from the ravages of life in a cell that is often loaded with DNA-reactive chemicals. In chapter 12, Kreuzer and Michel discuss the cross-wiring of replication and recombination machinery in bacterial and phage systems. Chapter 24, by Strauss, explains how multiple systems have evolved to monitor DNA damage, catalyze excision repair, and allow replication machines to bypass DNA damage. Chapter 25, by Lovett, explains special cases of recombination at unique short sequences. A Dr. Jekyll-and-Mr. Hyde aspect of the MutSLH mismatch repair system is the subject of chapter 23, by Marinus. These chapters are particularly relevant to the field of human biology because mutations in repair and recombination enzymes that are homologous to bacterial proteins are root causes of many cancer-causing genetic syndromes.

Thus, contrary to one fearless prognosticator's vision, work on organisms like *E. coli, Salmonella, Bacillus subtilis, Borrelia burgdorferi,* and *Caulobacter crescentus* has thrived since their sequences were published. A combination of powerful genetic and biochemical tools made our perspective broader, generated new excitement, and propelled the exploration of evolutionary biology to a significantly higher plane. Changing patterns of information transfer from the printed page to electronic media raise doubts about how books will be written and distributed in the future. This could be the last ASM book devoted to the subject of bacterial chromosomes. But I wouldn't bet the farm on that.

N. Patrick Higgins
Department of Biochemistry and Molecular Genetics
University of Alabama at Birmingham

REFERENCES

1. Blattner, F. R., G. Plunkett, C. A. Bloch, N. T. Perna, V. Burland, M. Riley, J. Collado-Vides, J. D. Glasner, C. K. Rode, G. F. Mayhew, J. Gregor, N. W. Davis, H. A. Kirkpatrick, M. A. Goeden, D. J. Rose, B. Mau, and Y. Shao. 1997. The complete genome sequence of *Escherichia coli* K-12. *Science* 277:1453–1474.
2. Charlebois, R. L. (ed.). 1999. *Organization of the Prokaryotic Genome.* ASM Press, Washington, D.C.
3. Drlica, K., and M. Riley (ed.). 1990. *The Bacterial Chromosome.* American Society for Microbiology, Washington, D.C.

INTRODUCTION

To introduce a volume for which my own knowledge of the contents is confined mainly to the chapter titles, it seems appropriate to discuss a few problems related to the volume's topic. These are mostly questions that have been with us for decades but whose definition has been made more precise by recent knowledge, especially by whole-genome sequencing. A long-standing interest in unanswered questions maintains the passion that some of us feel for our own branch of science. I hope that both the reader and I will detect some progress toward the answers in the chapters of this volume.

That many bacteria (*Escherichia coli* in particular) have most of their genes on a single, circular DNA structure is accepted by all but a few diehard skeptics. (For a contrarian view, see reference 1.) Most of my questions relate in some way to that fact.

TRAFFIC PATTERNS

The chromosome replicates once per division cycle. In rapidly growing cells, one round of replication is not complete until after the next one starts, so the chromosome is always replicating. Also, the chromosome is continually being transcribed, and the transcripts begin to be translated before detaching from the DNA. How is all this traffic managed without a traffic jam? Related questions were raised long ago at early molecular biology meetings—especially, as I remember, by the late Ole Maaløe. This was in the days before PowerPoint, when the audience could perceive that the speaker's brain was functioning even as he spoke.

RNA polymerase moves along DNA (or, if you prefer, the DNA moves past RNA polymerase) at a rate of about 45 bp/s. Its direction is either clockwise or counterclockwise, depending on the orientation of the operon. The big replication machine containing DNA polymerase moves at about 1,000 bp/s, from origin to terminus. Does DNA polymerase sweep RNA polymerase out of its way, or can it be slowed down by opposing transcription? French (8) found that insertion close to the replication origin of a strong promoter oriented oppositely to the direction of replication detectably delayed replication fork movement. A slight retardation of the replication fork advance was also seen in vitro with T4 DNA polymerase (11).

There are other intriguing questions as well. Transcriptional initiation is frequently controlled by repressors bound to operator sites on DNA. When a replication fork passes, not only does the repressor risk temporary dislodgement, but now there are two operator sites to be covered. Is access of RNA polymerase restricted behind a replication fork so as to allow recruitment of fresh repressor?

DYNAMICS AND BOOKKEEPING

One long-standing desire is that one day we may have a reliable snapshot of the typical chromosome during steady-state growth: how many molecules of RNA polymerase are concurrently transcribing, how many transcripts are concurrently being translated, how many molecules of other types are bound, and what their on-and-off rates are. This is part of a more general goal that concerns not just the chromosome but also the bacterial cell. How close are we to an accurate diagram of material flow through the growing cell? Every minute, some number of ammonia molecules traverse the cell membrane on their way to the cell interior. The nitrogen atoms are then assimilated into various macromolecular fractions, each composed of diverse molecular species. Some molecules are very stable; some decay quickly. Each molecular species ends up with a steady-state molar abundance and turnover rate.

Some values are available and have been for years. Two-dimensional gels can assess the molar abundance of many proteins and, with pulse labeling, can indicate synthesis rates as well (13). Microchips give a global view of transcription rates, but as currently used they are more informative as to how a given RNA species changes with time or ambient conditions than on the relative abundance of different RNA species. There has been major progress, and what is being done is perhaps a more urgent priority, but a complete flow sheet is still to come.

DIRECTIONALITY

One early observation that seemed relevant to chromosome trafficking (3) is that most of those *E. coli* genes believed to be highly expressed in rapidly growing cells are oriented in the direction of replication fork movement. This should minimize collisions of RNA polymerase with DNA polymerase. The 16S rRNA operons are a classical case in point. Their products are needed in abundance, so the operons are present in several copies, all oriented with the direction of replication and all near the origin. Proximity to the origin provides the added dividend that, in rapidly growing cells, the copy number of a gene close to the origin is as much as four times higher than that of one near the terminus.

More-complete information emerged from whole-genome sequencing (2). In the first 100 kb in either direction from the origin, most genes are oriented in the direction of replication. In the rest of the chromosome, orientation is closer to random.

There are other correlates of replication direction. These include "GC skew," where $G > C$ in the leading strand and $G < C$ in the lagging strand (9, 12), and preferential orientation of oligonucleotide or longer elements. The rather striking case of lambdoid prophage orientation has held up amazingly well with an expanding database (4, 5). There is also the orientation of octameric Chi sites, which probably reflects gene orientation, codon usage, and amino acid usage (6) but is not entirely explained by those factors.

The explanations for GC skew can be classified as mutation pressure (as elaborated by Francino and Ochman [7]) or selective pressure. The existence of strand biases in various elements of different lengths raises the possibility that, whatever the original driving force may have been, at this stage the most critical force may be coadaptation of the orientation of different elements (5). A corollary is that, with time, it should become increasingly difficult for gene order to change.

EVOLUTION

Such a picture fits nicely with some known examples of conservation of order, as in *E. coli* and

Salmonella enterica. But order has not always been conserved so well. *Escherichia, Haemophilus, Pseudomonas*, and *Bacillus* must all have had a common ancestor, but their genetic maps have little in common. When and how did order change?

With increasing sequence information, other conserved attributes of genomes (some of them recognized for many years) have become more conspicuous. For example, some DNA statistics, e.g., $G + C$ content, nearest-neighbor relationships, and patterns of codon usage, show little variation within a genome (9). Through a combination of known and unknown factors, the genome as a whole experiences selective and/or mutational pressures to generate properties that are genomewide and genome specific. These pressures must also have diversified over time to produce the variety of organisms now extant.

Segments of alien DNA that entered genomes by lateral transfer are presumed to have ameliorated to the genome-specific values, where full amelioration may require upwards of 100 million years (10). This calculation assumes that the parameters determining those genome-specific values have not changed within the past 100 million years. If that is true, when and how have they changed? If not now, when?

The analysis of gene phylogenies and divergence times has developed into a sophisticated science, although resting on dubious assumptions about the constancy of molecular clocks. For whole-genome properties (gene order, $G + C$ content, etc.), there is no accepted metric. There are three a priori possibilities:

1. These properties are changing slowly all the time. According to a uniformitarian view, in 1 billion years there should be 10 times as much divergence as in 100 million years. This seems unlikely to be the whole story. If we compare *E. coli* and *S. enterica*, increasing the divergence 10-fold will not give us *Haemophilus* or *Pseudomonas*.

2. Periods of change alternate with periods of stasis (a kind of punctuated equilibrium). Some species are currently changing rapidly, while others are hardly changing at all.

3. These properties were much more variable in an earlier epoch and now have become largely fixed.

With regard to the evolution of gene orders, one fact made more conspicuous through genomics is that not every bacterial species has a single, circular molecule of chromosomal DNA. Both linear DNAs and multiple DNAs have been encountered. It is noteworthy that structural changes at this level (which might seem fundamental) have no deep-seated taxonomic or phylogenetic basis. *Borrelia* and its plasmids have linear DNAs, whereas other spirochetes have circular DNA. Variations in number occur within genera (e.g., *Rhodobacter*) or families (e.g., *Rhizobium* versus *Bradyrhizobium*). Such apparently easy transitions may temper enthusiasm for the notion that gene order and orientation are highly coevolved.

However, we do not know how different the alternative configurations really are in terms of location in the bacterial cell or replication times of different genes. The discovery that the ends of the linear *Streptomyces* chromosome colocalize in vivo (14) is intriguing. If a bacterial chromosomal is defined as it was 25 years ago, as a DNA-containing structure within the cell, it is reasonable to imagine that some bacterial species may have circular chromosomes with linear DNA. Without belaboring the semantics, one can say that the structure clearly may be circular even if the DNA is discontinuous at one point.

At any rate, I for one look forward to scrutinizing the contents of this volume for insights relevant to some of the issues I have touched on.

Allan Campbell
Department of Biological Sciences
Stanford University

REFERENCES

1. Bendich, A. J. 2001. The form of chromosomal DNA molecules in bacterial cells. *Biochimie* 83:177–186.
2. Blattner, F. R., G. Plunkett, C. A. Bloch, N. T. Perna, V. Burland, M. Riley, J. Collado-Vides, J. D. Glasner, C. K. Rode, G. F. Mayhew, J. Gregor, N. W. Davis, H. A. Kirkpatrick, M. A. Goeden, D. J. Rose, B. Mau, and Y. Shao. 1997. The complete genome sequence of *Escherichia coli* K-12. *Science* 277:1453–1474.
3. Brewer, B. J. 1990. Replication and the transcriptional organization of the *Escherichia coli* chromosome, p. 61–84. *In* K. Drlica and M. Riley (ed.), *The Bacterial Chromosome*. American Society for Microbiology, Washington, D.C.
4. Campbell, A. M. 1992. Chromosomal insertion sites for phages and plasmids. *J. Bacteriol.* 174:7495–7499.
5. Campbell, A. M. 2002. Preferential orientation of natural lambdoid prophages and bacterial chromosome organization. *Theor. Popul. Biol.* 61:503–507.
6. Colbert, T., A. F. Taylor, and G. P. Smith. 1998. Genomics, Chi sites and codons: 'islands of preferred DNA pairing' are oceans of ORFs. *Trends Genet.* 14:485–488.
7. Francino, M. P., and H. Ochman. 1997. Strand asymmetries in DNA evolution. *Trends Genet.* 13:240–245.
8. French, S. 1992. Consequences of replication fork movement through transcription units in vivo. *Science* 258:1362–1365.
9. Karlin, S., A. M. Campbell, and J. Mrázek. 1998. Comparative DNA analysis across diverse genomes. *Annu. Rev. Genet.* 32:186–226.
10. Lawrence, J. G., and H. Ochman. 1997. Amelioration of bacterial genomes: rates of change and exchange. *J. Mol. Evol.* 44:383–397.
11. Liu, B., and B. M. Alberts. 1995. Head-on collision between a DNA replication apparatus and RNA polymerase transcription complex. *Science* 267:1131–1137.
12. Lobry, J. R. 1996. Asymmetric substitution patterns in the two DNA strands of bacteria. *Mol. Biol. Evol.* 13:660–665.
13. VanBogelen, R. A., K. Z. Abshire, A. Pertsemlidis, R. L. Clark, and F. C. Neidhardt. 1996. Gene-protein database of *Escherichia coli* K-12, edition 6, p. 2067–2117. *In* F. C. Neidhardt, R. Curtiss III, J. L. Ingraham, E. C. C. Lin, K. B. Low, B. Magasanik, W. S. Reznikoff, M. Riley, M. Schaechter, and H. E. Umbarger (ed.), Escherichia coli *and* Salmonella: *Cellular and Molecular Biology*, vol. 2. ASM Press, Washington D.C.
14. Yang, M. C., and R. Losick. 2001. Cytological evidence for association at the ends of the linear chromosome in *Streptomyces coelicolor*. *J. Bacteriol.* 183:5180–5186.

I. GENETIC AND PHYSICAL STRUCTURE

The Bacterial Chromosome
Edited by N. Patrick Higgins
© 2005 ASM Press, Washington, D.C.

Chapter 1

Where's the Beef? Looking for Information in Bacterial Chromosomes

This chapter is a collection of thoughts, questions, and problems to consider while reading this book. It is not intended as a comprehensive review or a unifying theory for chromosome structure. In looking at the structure or sequence of a genome, one sees all aspects of its information content, regardless of how they got there or how they are maintained. Many of these features are difficult to understand as functions contributing to the fitness of a single cell. Some aspects of chromosome structure that are shaped by selfish behavior with few conventionally maintained features that may be understandable only at a population level are pointed out. The principles described here are well known but are often forgotten because we are so accustomed to identifying functions genetically (mutations with phenotypes) or biochemically (proteins with interesting properties). Genetics is good at revealing genes and other sequence elements that affect the phenotype of a single cell and that are likely to be maintained by conventional selection. Genome sequences reveal all information, even that maintained by aspects of population biology that are unlikely to help us understand chromosome behavior. Some conserved features may be positively noxious. The challenge to those trying to explain chromosome behavior is to identify the critical sequence elements (i.e., "the beef").

THE QUESTIONS

1. How do bacteria compact their chromosomes?
2. How do bacterial chromosomes move? Do chromosomal sites, perhaps recognizable as sequence elements, serve as attachment points for cytoskeletal motor proteins? Where in the cell are these fibers attached?
3. How is gene expression and its regulation integrated with chromosome replication, compaction, and movement?

APPROACHING THE QUESTIONS

An advantage of chromosome work in bacteria is the large number of complete genome sequences that are available for comparison. This body of data should help identify aspects of sequence that are universal, overrepresented, and perhaps critical for compaction and movement of chromosomes. However, in extracting this information, we must learn to distinguish the critical features from features that are unlikely to be mechanistically informative but may be interesting in their own right.

FORCES THAT DICTATE THE SIZE OF CHROMOSOMES

One might expect that a genome would be under constant selective pressure to expand, adding new genes that broaden the capabilities of the organism. Three aspects of population biology and lifestyle limit this expansion: mutation rate, population size, and recombination frequency. These principles were suggested by Muller (15, 35, 48, 49). First, if all other factors remain constant, mutation rate limits genome size; beyond a critical point, the genome can expand only if the mutation rate drops. This is because more information can be kept under selection if fewer mutations occur. (With no mutations, a huge genome

---

John R. Roth • Microbiology Section, Division of Biological Sciences, Center for Genetics and Development, University of California, Davis, Davis, CA 95616.

could be maintained without selection.) Second, larger populations allow genome expansion (with other factors constant) because selection is more effective (drift is less significant) and genes with a smaller fitness contribution can be maintained. Third, sexual recombination allows genome expansion by permitting assembly of intact information sets from those damaged by mutation. Expressed differently, selection works better on individual genes if recombination continuously rearranges allele combinations. Together these considerations suggest that (if other factors are held constant) genome size will be inversely related to mutation rate and directly related to population size and recombination frequency (35). Each organism or population may satisfy these relationships in different ways, but the values remain interdependent. (Organisms with a very high mutation rate generally have small genomes and large population sizes.) While the exact mathematical function is unknown, these general rules of Muller reflect the problems of maintaining genes in the face of continued mutation pressure. However, these forces may work differently on different genes.

A basic assumption made here is that each gene is individually subject to loss by mutation and can counteract that loss by individual strategies. That is, different genes compete with each other for maintenance in a chromosome (35). A gene remains in the chromosome (and avoids loss by mutation and drift) conventionally if its function enhances fitness; mutations that impair such a function reduce fitness and are removed from the population by selection. Some genes enhance their ability to compete with other genes by private (selfish) strategies that do not enhance fitness and may actually be detrimental to the organism.

INTRODUCTION TO SELFISH BEHAVIOR

Much of sequence analysis is based on the expectation that conservation implies a selectively valuable function that can be unraveled and explained. Most chromosome features do contribute to cell fitness and provide a conventionally selectable phenotype; these are likely to be reliable clues to mechanism. However, other sequence features may provide no increase in fitness to the organism or may actually be deleterious, but are conserved by private (selfish) selection schemes.

Selfish features can be "addictive"—they do little or nothing when present, but cause a fitness loss when removed by mutation. Other selfish features are "infectious"; that is, they spread (by replication or horizontal transfer) faster than they can be removed from the genome (and population) by counterselection or simple mutation and drift. It is suggested below that high copy number may be an additional distinct form of selfish behavior. Distinguishing selfish from conventional sequence elements can be difficult because elements can be maintained by a combination of conventional and selfish selective forces. Some conventional genes may enhance their ability to stay in the chromosome by minor use of a selfish mechanism. Some primarily selfish elements may enhance their survival by modestly enhancing cell fitness. In extreme cases, elements may be maintained by selfish mechanisms alone.

A point to keep in mind is that selfish behavior improves a gene's ability to compete with others for a position in the genome, but it does not make that gene more beneficial. Genes maintained by selfish behavior still impose a cost on aggregate fitness of the organism. Selfish behavior allows a gene to remain despite its inadequate or even negative contributions to fitness. Selfish elements take a free ride, but ultimately their fare has to be paid by conventional genes—you could not have a genome made up entirely of selfish elements.

CONSERVATION OF CHROMOSOME GENE ORDER: GENE POSITION MAY BE ADDICTIVE

One might expect that a particular gene would work equally well regardless of its chromosomal position. This idea is supported by the observation that chromosome map order is not well conserved over extensive evolutionary distances (50), suggesting that order is not under strong selection. However, order seems to be conserved on a shorter time scale. *Salmonella* and *Escherichia coli* have maintained essentially the same map since their divergence about 140 million years ago. Since rearrangements are mechanistically possible in these organisms, this conservation suggests that gene order might be under some kind of selection. However, gene order could also be an addictive feature of chromosome structure.

Genes can, in principle, work anywhere, but once their promoters are tuned for any particular location, their transplantation causes a loss of fitness. We presume that the strength of each promoter is selectively adjusted to optimize gene expression and that this optimization necessarily takes into consideration the local influences on expression. One local characteristic may be the degree of supercoiling that is characteristic of the site (see below); another may be the gene dosage status imposed by the position a gene occupies vis-à-vis the origin of replication.

When chromosome replication requires a significant fraction of the cell cycle, as is frequently the case for bacteria, average dosage of a particular gene varies with its position in the chromosome (distance from the origin). Genes replicated early are present in multiple doses for much of the cell cycle, while those replicated late have two copies only for the period between their replication and cell division. Dichotomous replication during rapid growth (when new forks are initiated before completion of the previous round) can increase this dosage gradient. Optimization of promoters and regulatory elements will necessarily take into consideration the time-averaged copy number of each gene.

Expression of a gene could be optimized for any location in the chromosome, but once that optimization is achieved, moving the gene to a new site would render suboptimal the tuning of its promoter. In essence, gene order might be an addictive property of chromosome structure restraining rearrangement. Even if tuning of an individual gene is only modestly affected by an inversion or translocation, the accumulated small fitness costs incurred by the many genes involved in a rearrangement could allow selection to efficiently eliminate individuals with an altered gene order.

Evidence that these changes occur is that expression of a particular gene changes as it is moved farther from the replication origin (61). Furthermore, evolution appears to have exploited this fact because genes with high expression levels (judged by a high codon adaptation index) tend to be located near the replication origin (64, 65).

Conservation of gene order is often attributed to lack of mechanisms for rearrangements or to the need to maintain the proper spacing of sequence features that are critical to the packing or movement of chromosomes. These could be factors, but the above arguments suggest that conservation could equally well be attributed to the "addictive" effects of well-tuned promoters.

OPERONS AND OTHER GENE CLUSTERS

A notable feature of bacterial chromosomes is the clustering of genes that contribute to a single selectable phenotype (11). Some of these clusters are operons (30), groups of genes cotranscribed from a single promoter, while others are made up of closely linked but independently expressed genes. Initially the evolution of operons was ascribed to economies of coregulation (29) or to setting the molecular ratio of proteins that acted together (1). More recently, this clustering has been ascribed to a selfish property

of gene clusters (36) that gives them an advantage (over identical unclustered versions) in the larger context of horizontal transfer between genomes of different bacteria. This model suggests that clustering gives genes a selfish property of the infectious sort. Previously, Wheelis and Stanier suggested that horizontal transfer might drive clustering of different operons controlling parts of the same pathway (74).

The earliest regulatory explanations of operon evolution were called into question because the benefits proposed to explain operon formation would accrue only after genes were successfully brought together. It was hard to visualize how two genes could be brought together by slow steps with progressive increases in selective value, as expected for most evolutionary processes. Furthermore, the energy or material saved by coregulating genes in an operon seemed very small, once it was realized that a regulatory binding site of only a few bases is sufficient to allow effective coregulation of distant genes. Finally, the regulatory models did not explain clustering of independently transcribed genes with related functions.

The selfish operon model suggests that clustering provides no selective benefit to the cell (36). Instead the clustered state is a selfish property of genes that contributes to the ease with which they can spread horizontally between conspecific individuals or species. If clustering improves the likelihood that a set of genes moves horizontally, it may provide an advantage (to the genes themselves, not to their hosts) vis-à-vis identical genes in an unclustered state. The advantage conferred on genes by clustering accrues without any required improvement in the fitness of the organism (outlined below in more detail). If this is true, then operons and gene clusters may be examples of a conserved chromosome feature that does not tell us much about the physiological functions within an individual bacterial cell. The origin and maintenance of this feature is rather explained by horizontal gene transfer and population biology. Occasional loss of the genes by mutation can be corrected by horizontal reacquisition, and naïve organisms are more likely to acquire a clustered version than a collection of unclustered genes. The genes persist on a global scale because they can move horizontally fast enough to avoid elimination by insufficient (or episodic) selection.

The conventional (and reasonable) idea is that genes are maintained in a genome only if they provide a selectable phenotype. Selection continually purges a population of mutant individuals that have lost a valuable gene. The purging of mutants works best for genes that make a big fitness contribution (whose mutants are very deleterious). Conversely, genes that are only weakly (or only occasionally) selected are most likely to be lost from a population by mutation and drift (35). Thus, all but the most essential genes are subject to loss by mutation in cells that are still encountered in the population. Genes that are weakly selected can enhance their survival by nonconventional (selfish) gimmicks, which become more important as their selective value drops.

From the point of view of a gene, the likelihood of persistence in the world is enhanced by horizontal transfer to new hosts, which provides a way of staying ahead of loss by mutation and counterselection (or drift). Transferred bacterial genes will be held in their new host only if they provide a phenotype of sufficient selective value. Many phenotypes require multiple proteins (and thus multiple genes). For a gene to take advantage of horizontal transfer as a way of maintaining itself, that gene must be transferred with all the other genes needed for its particular phenotype. This presents a problem in that only small DNA fragments are transferred from one bacterium to another. Thus a gene gains the advantage of horizontal transferability in direct proportion to the probability that it can be transferred simultaneously with all of the genes that allow it to confer a selectable phenotype—that is, in proportion to the proximity of the cooperating genes.

Two sets of genes, each contributing a single phenotype, may make the same fitness contribution regardless of their position in the genome. One set can gain the added selfish advantage of transferability if its genes are sufficiently close together to allow occasional cotransfer. The closer they are together, the more probable is this cotransfer. This provides a positive selection for clustering that becomes stronger as genes get closer together. Therefore, this model allows gene clustering to be achieved by a series of small steps; the selfish advantage increases progressively as the genes are brought closer together (36). Fusion of clustered genes into transcription units or operons is proposed to further enhance transmissibility by minimizing the problems of promoter adaptation.

This selfish operon model predicts that clustering will be a property of genes that are dispensable and subject to loss and reacquisition, and this is the general observation. For example, biosynthetic operons escape selection when their end product is present in the environment, as do genes for degradative pathways when cells grow on an alternative carbon and energy source. In contrast, clustering is not predicted for essential genes, which cannot be lost and thus might not be expected to be under selection for horizontal reacquisition. While essential genes are usually not clustered, there are exceptions.

(The large operon encoding ribosomal proteins is especially notable.)

Most clusters of essential genes encode proteins that interact intimately. If a mutation arises that improves the function (e.g., protein synthesis), that new mutation would be under selection to spread globally if its gene could transfer horizontally. However, the mutant gene can move horizontally only if it brings along its companion proteins that work well together. The advantageous mutant protein may not work well when inserted in unfamiliar protein complexes. The transferred ribosomal gene clusters may supersede existing homologues rather than adding a novel function to a naive cell. Thus, intimate interactions between proteins may make it possible for horizontal transfer to drive clustering of essential genes. Because these clusters are essential, they are less prone to mutational loss or disruption. Evidence has been presented that clusters of this sort are stable over longer evolutionary periods (75).

A powerful tool has been developed based on the strong tendency of functionally related genes to cluster. While this has been implemented in several ways, a program we have found extremely powerful is the functional coupling program within the WIT (Integrated Genomics) program package (54). This program takes a submitted gene sequence and tabulates its neighbors (genes often found nearby in the chromosome) in a large number of complete bacterial genomes. The result is a list of genes that are frequently located near the submitted sequence. The ability of this program to identify known genes with related functions is astounding. This ability generates confidence that surprising examples of frequent neighbors are likely to reflect unexpected functional relatedness. In view of the selfish operon model, it seems likely that this powerful tool for identifying genes of related functions may owe its power to the selfish tendency of genes to enhance their survival by frequent horizontal transfer.

TRANSPOSONS AS SELFISH ELEMENTS

The above argument for selfish gene clusters is a less extreme example of an argument made long ago to explain the success of transposable elements. Insertion sequences (IS elements) are usually regarded as selfish elements that are dangerous to their hosts because they generate mutations and rearrangements that are almost all deleterious (see digression below). When considered as part of the bacterial genome, transposable elements maintain themselves despite counterselection, because replicative transposition allows them to stay ahead of purifying selection by generating new copies faster than selection can remove them from the population. Replication can occur by transposition within the genome in an asexual lineage or by horizontal transfer to new lineages. By moving fast enough, these elements can stay ahead of counterselection without providing a beneficial function and can even overcome their substantial inherent cost in mutagenesis. They replicate more often per unit time than a conventional gene—a formula for success.

Some transposable elements contain drug resistance determinants that confer a selectable phenotype; this suggests that they might be maintained by conventional selection. This evidence for selective value does not contradict the initial idea that these elements are basically selfish. Multiple forms of selection can act simultaneously on any element. Transposons may depend primarily on selfish mechanisms for their survival, but enhance it by accumulating genes that provide a positively selectable function. This positive selection is demonstrable but insufficient to allow them to remain without their selfish behavior (see below). If the included genes could maintain themselves by conventional selection (e.g., for drug resistance), one would expect them to lose the costly ability to transpose and mutagenize the genome. Their continued association with transposition activity (and its deleterious effects on the host) suggests that the nonselfish aspects of their phenotype are not sufficient to ensure maintenance. It is proposed below that these nonselfish selective phenotypes are weak ones in a natural setting; i.e., drug resistance may only rarely be of selective importance.

Additional evidence suggesting a selective value of transposons has come from growth competition experiments in which strains with the elements do better than isogenic strains without them (10, 24). This advantage was seen for elements that encode no protein other than their own transposase. In the competition experiments, the advantage appeared to reflect a mutator effect in that the element caused mutations that allowed for improved growth (9, 23, 47). In the digression below, it is argued that these experiments reveal short-term advantages and do not assess the true long-term costs of carrying these elements. That is, the mutator phenotype is not likely to explain the persistence of transposable elements.

Transposons seem likely to be maintained primarily by a selfish advantage derived from transposition and horizontal transfer. In addition to being small enough to be transferred frequently, transposable elements actively contribute to their horizontal transfer by providing mechanisms for integration into the recipient genome; some also contribute to cell-cell transfer (63). Some of these elements enhance their maintenance by acquiring genes with selective value for the host, but the selfish behavior is still required for their maintenance. Transposable elements and some phages may also have "addictive" properties (see below).

A DIGRESSION ON THE COSTS OF MUTAGENESIS

While competition experiments suggest that transposable elements confer a selective advantage by their mutator effect, it seems more attractive that this is a short-term advantage seen under restricted conditions and is unlikely to explain the natural long-term persistence of these essentially selfish elements. The vast majority of random mutations are expected to be deleterious. This was shown directly for transposons (14, 62), but seems inescapable on first principles. There is a very large target for mutations that cause loss of valuable information (information maintained by selection). In contrast, there is likely to be a very tiny target for mutations that provide a fitness increase in any particular circumstance. For many conditions, the ratio of beneficial to deleterious mutations could easily approach 10^{-6} (20, 59). In short-term growth experiments, rare valuable mutations may arise and sweep the population selectively; adaptation may be aided by an increased mutation rate if the supply of mutations is limiting and if the fitness increase is large enough to outweigh the cost of the associated mutations. This can occur in small laboratory experiments that demand rare mutations with very large fitness increases (e.g., drug resistance) under a very specific set of conditions. Under these circumstances, mutation supply limits adaptation, and short-term benefit masks the long-term cost of the preponderant deleterious mutations.

Work done by Miller and coworkers (46) demonstrates clearly that mutators (strains with an increased mutation rate) are favored under conditions of strong selection (sequential acquisition of resistance to several drugs). This observation is often cited as evidence that increased mutation rate speeds genetic adaptation. The results demonstrate beautifully that mutators can be selected in short-term experiments, but do not address long-term cost. Later experiments showed that the cost of mutators is very high (18), making it unlikely that transposons are maintained in bacteria because they increase the mutation rate. There are easier ways to do this with mutation types that are more likely to be valuable, and it probably is not a good idea even then (59). The interacting parameters described by Muller (mutation rate, population size, recombination frequency) can be set at a variety of compatible equilibrium values. However, once these values are set, they are likely to be addictive in the sense that increasing one parameter (e.g., mutation rate) can be costly. In general, mutagenesis brings a prohibitive long-term cost because deleterious mutations are so frequent. The numerical problem of increasing mutation rates has been discussed for one particular genetic system with pointers to some of the extensive previous literature (59).

PHAGE, COMPENSATION, AND ADDICTION

Phages are usually visualized as independent entities that happen to grow in bacteria. However, when viewed from a bacterial perspective (as an extension of the bacterial genome), they are the ultimate in selfish elements. They prosper by replicating and having a free-living form that provides for all aspects of horizontal transfer (exit, persistence outside the host, entrance and incorporation into the new genome). While virulent phages seem purely selfish, temperate phages clearly spend considerable time as part of a host chromosome. It is therefore not surprising that some temperate phages (like some transposable elements) have acquired genes that are not central to phage growth but confer a conventional selective advantage on their host (25). As for transposable elements, prophages appear to confer a selective advantage in mixed cultures, when lysogens and nonlysogens compete (7, 12, 37). This could reflect phage genes that provide a useful host phenotype. Regardless of these advantages, phages are likely to have strong costs to the bacterium in terms of lysis, and their persistence in the face of these costs suggests that they are substantially selfish. That is, their long-term maintenance still depends on occasional lytic growth (before mutational inactivation) and consequent horizontal transfer to new hosts. Survival of phage and transposons may also be enhanced by addiction mechanisms put in place by compensatory mutations in the host (see below).

Work on drug resistance has shown that when a strain becomes resistant (for example) to streptomycin by mutational ribosomal alteration, its resistance is associated with a growth disadvantage (in the absence of drug) due to impaired ribosome function (2). This impairment can be corrected (compensated) by secondary mutations that restore more rapid growth without loss of drug resistance (5). These compensatory mutations reduce fitness when present in strains without an initial streptomycin resistance mutation (44). It seems likely that many genetic situations may provide similar conditions.

New mutations may provide a short-term benefit with a long-term cost. Compensatory mutations reduce the long-term cost without removing the benefit provided by the original mutation. However, if the original beneficial mutation is removed, the compensatory mutations impose a new cost. This phenomenon might contribute to some of the apparent selective benefit conferred by prophages and transposons.

If a wild-type bacterium (e.g., *E. coli* K-12 carrying phage lambda) spends considerable time growing with its lambda prophage, it may accumulate multiple mutations that minimize the cost of carrying that prophage. These mutations might, for example, minimize the probability of spontaneous prophage induction. When such a compensated strain is brought into the laboratory and cured of its prophage, the compensatory mutations remain and may be slightly deleterious in the absence of the prophage. Normal growth could be restored by replacement of the prophage and restoration of the conditions under which compensation was selected.

Similar events might explain growth advantages provided by certain transposable elements. That is, cells may have spent recent evolutionary time with such elements and acquired mutations that compensate for the cost of the transposon (e.g., reducing transposition frequency); such mutations may be slightly deleterious when the transposon is removed (especially under the rich conditions that force fast growth in the laboratory). Note that this argument suggests that some of the apparent selective value of phages and transposons could be due to host mutations that minimize the cost of carrying the element; the compensatory mutations serve to addict the host to its load of infective elements. This makes the elements appear valuable.

Many genetic features of a genome may be addiction modules in the sense that they do not contribute a useful phenotype but are tolerated better if certain compensatory genetic changes occur. (An example is the suggestion made above for addictive maintenance of gene order.) This raises serious problems for those trying to understand the physiological basis of the growth advantage provided by a phage or other genetic element. Such elements may in fact contribute nothing to normal cell function, even though a competitive disadvantage results from their removal.

Virulent phages (viewed as an extension of the bacterial genome) represent an extreme that may help to make these arguments clearer. From a bacterial point of view, these phages are solidly deleterious. One would not expect them to include genes that improve host fitness. Despite their deleterious effects, virulent phages persist because they can replicate, accumulate mutations that circumvent resistance, and set up a standoff situation of balanced warfare in which both combatants (cells and phage) remain stably on battlefield. In essence, they are replicating and changing faster than they can be removed by selection. While it seems easier to view virulent phages as independent entities competing with bacteria, their underlying strategy is an extreme example of that used by temperate phages, transposons, and gene clusters: horizontal transfer is the fundamental underpinning of their survival.

THE PROBLEM OF PLASMIDS

Plasmids are important features of bacterial genomes that may represent a different aspect of selfish behavior. Conjugative plasmids (and mobilizable plasmids) are capable of intracellular transfer and thus are likely to be maintained in large part by selfish mechanisms (like phages and transposable elements), despite the fact that they often include genes that provide selectable cellular phenotypes. Nonconjugative plasmids present a problem.

Nonconjugative plasmids (in common with IS elements) are not obviously transmissible, but may show enhanced horizontal transfer because of their status as an independent replicon. This may enhance their ability to establish themselves in a new genome following transmission (by transduction or transformation); this is the same transmission benefit that transposition provides to an IS element. In this sense, nonconjugative plasmids may still be considered infectious selfish elements.

A second consideration in understanding plasmids is their high copy number. This might be considered an analogue of transposition (for IS elements), allowing genes to replicate faster than they can be removed by selection. However, multiple copies and genetic versions of a transposon can be held for extensive time in a single genome. High-copy-number plasmids (subject to a single-copy control mechanism) quickly clone themselves and do not stably maintain a heterozygous condition when different versions are carried by the same cell (53). Therefore, it seems likely that plasmid copy number may contribute a subtly different sort of selfish behavior; genes in high copy have a larger effective population size than chromosomal genes and thus may be maintainable by weaker selection (see Muller's rules above). In a sense, the higher copy number gives a measure of defense against mutation when an element is maintained by weak selective forces. Copy increase may be a selfish strategy that is distinct from infectious behavior.

STANDARD-LOOKING GENES CAN BENEFIT FROM SELFISH ASPECTS OF SELECTION

Selfish mechanisms used by phage, transposons, and plasmids can also contribute to maintenance of genes that are respectable chromosome residents. Below are examples of standard sequences with addictive or infectious behavior.

Restriction/modification systems are thought to confer a conventional selective advantage in that they protect against invading foreign (unmodified) DNA. However, the mutational loss of just the modification function is lethal if the restriction activity remains; this addicts cells (partially) to the modification function. While this may seem axiomatic, given the nature of the functions involved, it is interesting that simultaneous loss of both modification and restriction functions is also lethal for some systems, because the half-life of the existing restriction enzyme is longer than that of the modification enzyme (32, 51). It should be noted that modification genes can still be lost without cost if the restriction activity is removed first. This does not eliminate a contribution of selfish behavior to maintenance, because the situation still counterselects one means of functional loss.

Not all restriction enzyme systems show this addictive resistance to loss of both functions. This suggests that some bacterial lifestyles include a sufficient influx of harmful DNA that the restriction system is maintained solely by its conventional protective function. In contrast, environments that provide little influx of harmful DNA may reduce the selection on restriction systems sufficiently that they cannot be maintained in the genome by this mechanism alone. Under these conditions, a contribution from addictive behavior may be required for maintenance. The fact that *Helicobacter pylori* genomes include over 50 restriction modification systems (52) seems unlikely to reflect frequent exposure to invasive DNA (since these organisms inhabit sequestered sites) and may suggest (paradoxically) that the lifestyle of this bacterium includes so little infective DNA that only strongly addictive restriction systems have remained and that this addictive behavior has been sufficient to provide for both maintenance and horizontal spread.

An example of "infectious" behavior may be provided by the uptake signal sequences (USSs) found in *Haemophilus* and *Neisseria*. These sequences were originally considered part of a sexual exchange system that allows preferential uptake of conspecific DNA and is therefore highly overrepresented in the genome (70). However, it has been convincingly argued that bacteria take up DNA for use as a food source and that the concomitant sexual exchange is an unavoidable (and unintended) side effect (55). If such a nutritional uptake system gained efficiency (lowered its K_m) by recognition of a specific sequence, it could take up extensive DNA sequences that include this sequence, even when the concentration of DNA is low. If this occurred, the genomic abundance of the recognized sequence might increase secondarily due to a selfish property—frequent horizontal transfer. The nature of the uptake system dictates that all incoming DNA sequences include the USS; therefore, to the extent that DNA is ever added to the chromosome from an outside source, the USS element copy number will increase. Stated differently, every time a USS arises by chance in a *Haemophilus* genome, it is likely to sweep all *Haemophilus* cells simply because it is efficiently taken up by all organisms that share the uptake mechanism and therefore has a chance of recombining into their genome with no required selection. In conclusion, a sequence element may be heavily overrepresented for selfish rather than functional reasons. Encountered without any knowledge of DNA transport, the overrepresented USS might well be a candidate for a role in chromosome packaging or movement, when in fact it may have evolved to improve uptake of a nutrient. A possible unifying idea would be that exogenous DNA is pulled into the cell following attachment of some multifunctional chromosome-moving mechanism at the specific USS.

USING GENE POSITION TO MAKE PRELIMINARY INFERENCES REGARDING SELECTIVE VALUE: A HIERARCHY OF SELFISHNESS

Above are described some genome features that depend to varying extents on their selfish behavior for their maintenance in the genome. These assumptions can be used in reverse to draw conclusions regarding the fitness contribution of genes based on their genomic position. The basic idea is the following: the more heavily an element depends on selfish behavior, the smaller is the likely contribution of its information to conventional fitness.

Consider a hierarchy of genes placed in order based on decreasing dependence on conventional selective value for their maintenance—and increasing dependence on selfish mechanisms (horizontal transmission or addiction). A shamelessly simplistic suggestion is shown in Table 1. At the top of the list are standard genes with important physiological roles and no need for selfish behavior. At the bottom are virulent phages that seem purely deleterious (from a bacterial point of view); they have no selective value and are maintained solely by their own selfish (in-

fectious) means. Between these extremes is a series of elements ordered (very speculatively) according to perceived increasing dependence on selfish behavior. Ranking higher in the list than transmissible elements are standard chromosomal genes that are under conventional selection but enhance their (perhaps only moderate) selective value by having selfish properties that improve their horizontal transferability or cause a measure of addiction. Examples are functionally related gene clusters, genes flanked by repeated sequences, addictive genes, and sequences that enhance their own horizontal transfer by serving as transformation recognition signals (USSs mentioned above).

In the genome of one organism, a certain set of genes is unclustered and is therefore presumed to be maintained by conventional selection alone. In a different organism the same genes are driven to cluster, arguably by the added selfish benefit derived by enhanced horizontal transmissibility. The need for this extra measure of selection in the latter case suggests that these genes are of lower selective value to the second organism (in the case of a biosynthetic pathway, the end product may be more abundant in the environment of the second organism). The pathway is dispensable (at least temporarily) and its genes are therefore subject to mutational loss during periods of weak selection. Most clustered genes are dispensable (under some growth conditions) and are thus subject to mutational loss during periods of relaxed selection. Essential genes are rarely clustered. Exceptions to this (discussed above) are proposed to involve proteins that interact closely (36).

Genes with flanking repeats (described below) may enhance their chances of horizontal acquisition by having a means of excising by circularization and adding themselves to a new genome by homologous recombination following transfer. These repeat-flanked genes may also have an enhanced chance of amplifying during growth under selection (see below).

Next in Table 1 come conjugative plasmids and transposons that include conventionally selected genes but still maintain sophisticated mechanisms for infectious spread. These rank higher than simple IS elements, whose maintenance depends entirely on transposition. Temperate bacteriophages, which spend some time as part of the bacterial genome and may include genes with conventionally selectable functions, rank above IS elements with no included genes. These phages are maintained mainly by selfish transmissibility, but enhance their ability to remain in the genome by standard selection for included functions (and perhaps by host addiction). At the bottom of the hierarchy are elements that confer no fitness

Table 1. Genetic elements listed in order of decreasing fitness contribution (and increasing selfish behavior)

General behavior	Genetic feature	Means of selective maintenance	
		Conventional fitness contribution	Selfish behavior
Purely conventional selection	Essential chromosomal gene	High	None
	Chromosomal gene with important or frequent fitness contribution	Quite high	None
Conventional with slight contribution of selfish behavior	Clustered chromosomal genes of related function	Moderate	Clustering enhances horizontal transmissibility.
	Chromosomal genes flanked by repeats	Moderate	Flanking repeats aid in recombinational integration following transfer.
	Chromosomal genes with addictive property	Moderate	Addictive behavior gives resistance to mutational loss.
Mainly selfish with some conventional fitness contribution	Conjugative plasmid with a useful phenotype	Weak	Sophisticated machinery for horizontal transfer and maintenance as replicon
	Transposons conferring a selectable phenotype	Weak	Transposition ability allows escape from elimination and enhances inheritance following horizontal transfer.
	Temperate phages with host phenotypes	Very weak	Sophisticated mechanism for horizontal transfer and integration into genome of new host
Purely selfish	Insertion sequence without host phenotype	None or deleterious	Transposition minimizes loss by mutation and drift. Enhances integration following horizontal transfer.
	Plasmids without selectable phenotypes	None or deleterious	Controlled high copy number gives resistance to mutational loss. Replicon status is maintained after horizontal transfer.
	Temperate phages without phenotypes	None or deleterious	Sophisticated mechanism for horizontal transfer and integration into genome of new host Possible addiction
	Virulent phages	Deleterious	Rapid infectious transfer

gain on their host and may be neutral or deleterious. These elements depend entirely on selfish mechanisms to maintain a position in the genome.

This suggests a rule of thumb: the most strongly selected genes are likely to be located at standard positions in the chromosomes rather than clustered with genes contributing to the same function. The genes carried by highly developed selfish elements (plasmids, transposons, and phages) are probably under the weakest conventional selection. While some genes within a prophage, transposon, or plasmid may contribute a measurable fitness increase, one does not expect a cell's important or essential genes to be found there. In considering the validity of this suggestion, it is important to remember that a weakly selected gene may either make a continuous very small contribution to fitness or, alternatively, could be absolutely

essential under conditions that are only rarely encountered (e.g., drug resistance).

DO BACTERIAL CHROMOSOMES EXPLOIT A "WIND OF SUPERCOILING"?

The replication origin and terminus of *E. coli* define the ends of two equal-sized domains replicated divergently from a single origin. This feature is combined with a preferential orientation of the most heavily transcribed genes and operons so that their transcription proceeds in the same direction as their replication. It seems attractive to imagine that these two features reflect interacting effects of supercoiling. Both transcription and replication introduce positive supercoiling ahead of their direction of travel and negative supercoils in their wake. The positive supercoiling introduced by one terminus-directed transcription unit would be taken up by the next (more terminus-proximal) promoter oriented in the same direction. Thus, heavy transcripts would avoid interfering with each other. Similarly, a replication fork would be expected to inhibit activity of promoters ahead of its path and stimulate those it has passed. One might imagine that supercoils introduced in this way are pushed inexorably toward the terminus, where serious problems seem likely to arise.

THE PROBLEM OF CONVERGING REPLICATION FORKS

While minor local supercoiling problems might be solved by topoisomerases, it seems possible that something special is needed to solve the more intense problems that arise as replication forks converge. Each fork introduces positive coils ahead of its path, and these are focused on a smaller and smaller region as the forks come closer together. If it becomes difficult to open DNA (or to relieve this supercoiling), one might expect a blockage of fork movement. Solving this problem might require some special features of the termination region (e.g., action of XerCD, temporary strand interruptions at or near *dif*) or alternatively by ends of linear chromosomes (below).

To the extent that fork convergence requires specialized sequence features present in the terminus region, it poses a problem for sister-strand recombination events. It seems increasingly attractive to imagine that recombination involves initiation of new replication forks (38). Two replication forks generated by recombinational repair of a double-strand break could converge in a region not designed to accommodate such forks. Ter and Chi sites may contribute to a system for minimizing this problem (see below).

CIRCULAR VERSUS LINEAR CHROMOSOMES

The form of a chromosome seems profoundly important to many models for compaction and movement. Yet many bacteria support both linear and circular replicons, suggesting that the difference is not as profound as one might imagine (8). Despite the difference in chromosome structure, however, the distribution and orientation of genes (and the consequent skew in base composition) is biased in *Borrelia*, just as it is in *E. coli* (19; also see below). It will be interesting to see if the *E. coli* chromosome can segregate normally if its circular chromosome is

disrupted (perhaps at the *dif* site) and the two ends are supplied with the sort of telomeres found in chromosomes and plasmids of *Borrelia*. The switch might be managed by the circle-linear interconverting function found in the lambdoid phage N15 (33). It seems attractive to imagine that the difference between linearity and circularity is not central to the behavior of bacterial chromosomes.

The supercoiling problem associated with converging replication forks near the terminus of a circular chromosome, discussed above, seems to be solved for linear chromosomes, whose ends might be expected to rotate freely. Compaction of bacterial chromosomes by supercoiling might be harder to visualize for linear chromosomes if their ends are free to rotate. However, if supercoiled domains are defined as regions between points at which the chromosome is either cross-linked or anchored (perhaps to the membrane), then the difference between linear and circular structures seems less critical.

SHORT ELEMENTS THAT SEEM IMPORTANT TO CHROMOSOME BEHAVIOR: Ter AND Chi

Two short sequence elements have been extensively studied with regard to their effects on chromosome replication (Ter) and recombination (Chi). Replication forks terminate when they encounter a properly oriented Ter sequence with a bound Tus protein (27). Following chromosome breakage, degradation of a double-strand end by RecBCD is terminated at a properly oriented Chi sequence, producing a single-strand 3' end that can be used by RecA to initiate recombination, possibly by reestablishing a replication fork (34, 69, 73). While Ter and Chi have been extensively analyzed mechanistically, their biological roles seem less clear. Both elements show a nonrandom distribution and orientation in the genome of *E. coli* and *Salmonella*.

Ter sequences are located in the half of the chromosome farthest from the origin and are symmetrically distributed around the *dif* site, where replication normally terminates and where the XerCD functions act to ensure partition of completed chromosome copies to daughter cells. These Ter sites are oriented so as to allow passage of forks headed toward *dif* and to block rare replication forks that manage to pass the *dif* site. It has been proposed that Ter is a "fail-safe" device to ensure that opposed replication forks meet and are terminated near *dif*; yet some Ter sites are located far from *dif* and would only affect forks that had passed multiple similarly oriented Ter sites.

Chi sequences are overrepresented in all parts of the genome and are preferentially oriented such that RecBCD acting at a double-strand break would favor recombinational activation of the end nearest the origin and degradation of the end nearest the terminus; 75% of Chi sequences are in this orientation. The reasons for this are unclear.

Perhaps Chi and Ter elements are parts of a system that acts to ensure that replication forks formed during DNA repair recombination are directed toward the terminus region (*dif*) and not toward each other or toward new forks progressing from the origin. This may help avoid the supercoiling problem generated when forks converge, especially when they converge in regions that are not equipped to deal with the problem (Fig. 1; also see below).

At a double-strand break, the two ends are subject to resection and can then invade a sister chromosome to initiate forks. Depending on the timing of these invasions, forks that diverge or converge could be produced, or one fork could be lost (Fig. 1). The distribution of Chi sequences would favor prompt activation of the end nearest the origin and extensive degradation of the end nearest the terminus. If degradation proceeds to the old fork, the origin-proximal fork (directed toward the terminus) supersedes the old fork; this result is favored by the action of Chi and RecBCD. If the terminus-proximal end (following degradation) encounters a rare reversed Chi and initiates a fork, that fork is likely to converge with the origin-proximal fork and, if problems result, might have a chance to abort and try again. The third possibility is that one of the forks starts quickly and passes the other free end before it can initiate a fork, resulting in divergent forks, with one of them opposed to any new fork coming from the origin. An important function of Ter may be to terminate these "escaped" recombinational forks. Thus, the combination of Chi and Ter systems may act to preferentially direct new recombinational forks toward the terminus. Ter sites may be less essential near the origin, since forks that escape in this region can proceed across the origin.

THE GENOME OF *SALMONELLA ENTERICA* SEROVAR TYPHI: A TEST CASE FOR MAINTAINING GENE ORDER

It was suggested above that chromosomal gene order is under selection (perhaps by conventional functional selection to maintain a wind of supercoiling, perhaps by some selfish mechanism). It is striking that the basic gene order in *S. enterica* and *E. coli* has been maintained for perhaps 200 million years—the aggregate time elapsed as each diverged from their common ancestor. However, isolates of *S. enterica* serovar Typhi (40) and other host-specific

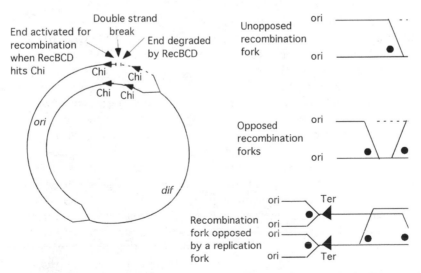

Figure 1. A proposal for the role of Chi and Ter sites in managing recombinational replication. (Left) Diagram of a double-strand break and the positions of nearby Chi sequences in their predominant orientation. The left end is likely to be promptly activated for recombination and lead to a fork moving toward the terminus. The right end is likely to be extensively degraded before engaging in recombination and could form a fork moving toward the origin. (Right) Diagram of the possible outcomes if the two broken ends establish independent recombinational forks. Depending on the timing of fork initiation, the two forks may diverge, or converge, or one end may be destroyed, leaving a single fork. Forks bound toward the origin might be terminated at Ter sites before converging with approaching replication forks. The concerted action of Chi and Ter might act to minimize fork collision and preferentially direct all recombinational replication toward the terminus.

Salmonella strains (39) provide exceptions that may help test rules for chromosome conservation.

In *E. coli* and *S. enterica*, the positions and orientation of seven repeated rRNA operons (*rrn*) are conserved. There are two widely separated, direct-order copies on one side of the origin and five similarly separated direct-order copies on the other side; all are oriented so that transcription proceeds away from the origin of replication. Although these extensive repeated sequences provide targets for recombination that could lead to inversions and transpositions of blocks of sequence, these rearrangements are not found in most natural isolates. In contrast, isolates of *S. enterica* serovar Typhi (a human-specific pathogen) show rearrangements that result from exchanges between these prominent *rrn* repeats. The behavior of serovar Typhi does not seem to indicate that the organism is more proficient at recombination between internal repeats; inversions form readily between inverse-order repeats in *E. coli* and in *S. enterica* (13).

The serovar Typhi rearrangements do not disturb the "wind of supercoiling" or the bias in GC content (described below). If all rearrangements are achieved by recombination between *rrn* repeats, none of the resulting rearrangements will reverse the orientation of any gene vis-à-vis the direction of replication. This is a necessary consequence of the fact that all *rrn* loci within a single replicore are located in the same orientation. That is, inversions made by recombination between *rrn* sequences must all include the origin (or terminus) of replication. These rearrangements could, however, alter the distance between a particular gene and the replication origin, but most observed rearrangements seem to conserve this spacing. This may suggest that the observed rearrangements are not random, but conserve the features that are most critical to chromosome function. Alternatively, serovar Typhi may grow under conditions (or at a rate) that relax some of the selective constraints on gene order. The rearrangements observed in serovar Typhi may be mechanistically permitted for all enteric bacteria, but the frequency distribution of permitted arrangements is broader (thus the selective cost of rearrangement is lower) in this particular subgroup.

INTERPRETING SMALL SEQUENCE ELEMENTS THAT ARE OVERREPRESENTED IN THE CHROMOSOME

Several small sequence elements (not obviously transposable) are found in the genome of *E. coli* (31). The REP (or PU) elements are imperfect palindromes, 30 to 40 bp in length (21, 26, 72), that have been placed in several general sequence classes (3, 22). These elements appear frequently at intergenic sites within transcribed regions as arrays in direct or inverse order. The palindromic character of these elements appears better conserved toward the outside ends, reminiscent of some transposable elements. While these sequences are frequently discussed as remnants of transposons, they have never been shown to transpose, and the conservation of their sequences suggests strongly that they may now be under selection and may play some role in cell physiology (or may be maintained by some unrecognized selfish mechanism).

Evidence that REP elements modulate the relative amounts of proteins encoded in a single message suggested a role in message half-life or translation (72). However, some of these units are located outside regions known to be transcribed, and evidence has been presented for interaction between these sequences and DNA-binding enzymes, including gyrase (76), IHF factor (6), and PolA. One of the largest effects of REP elements is their participation in formation of duplications and deletions (68). It seems possible that these REP sequences will prove important to bacterial chromosome behavior.

ERIC or IRU elements are larger and have a more complex secondary structure (28). While functions have not been demonstrated for these elements, they may stimulate recombination and be sites of frequent replication fork stalling or chromosome breakage (67; R. Dawson and J. R. Roth, unpublished).

STRAND-BIASED BASE SUBSTITUTION RATES

The base composition of the two strands of many bacterial chromosomes is skewed (the two strands show different overall base compositions). This is most striking for G/C pairs, which are positioned such that an excess of G residues appears in the leading strand (the untranscribed, sense strand for most highly expressed genes, which use the lagging strand as transcription template). Thus, this skew shows striking reversal at the origin and terminus of replication (42, 43). Evidence has been presented that this skew is the result of differential mutational pressure on the two strands.

Much of this bias is now thought to result from more frequent cytosine deamination (C → T changes) in single-stranded versus double-stranded DNA. It is suggested that transcription displaces the nontemplate strand and thereby exposes its C residues to deamination; over time this reduces the C level and increases the T level in this strand. This becomes a genome-wide bias because heavily expressed genes tend to be oriented so their transcription proceeds away from the origin; i.e., their transcription template is the lagging strand (17). In principle, this strand bias should reverse for each gene in the opposite orientation. However, the observed strand bias is heavier for genes transcribed away from the origin. This may reflect the fact that genes transcribed away from the origin are more heavily transcribed and consequently subject the leading strand to more frequent deamination of C residues; alternatively, the leading strand may be subject to superimposed additional deamination because it is exposed as a single strand when it serves as template for lagging strand replication (56).

Regardless of the underlying reasons for strand bias in base composition, the leading strand is generally richer in G residues (and slightly richer in T residues) than the lagging strand in many bacteria (45, 57). Because of this bias, any G-rich sequence element is more likely to appear by chance in the leading strand and will show a strand preference that reverses sharply at the terminus and origin. This systematic skew is seen in many bacterial genomes and has been used to infer origins and termini of replication.

ORIENTATION-BIASED OLIGONUCLEOTIDE SEQUENCES

Inspection of genome sequences has revealed short, nonpalindromic oligonucleotides that are preferentially oriented vis-à-vis the direction of chromosome replication (60). The abundance of some of these sequences increases in the region of the terminus, and their orientation preference shifts sharply at the *dif* site. These elements (perhaps justifiably) have irresistible appeal for those investigating the nature of the terminus or looking for the attachment sites of proteins responsible for chromosome movement. However, the considerations in the preceding section predict that G-rich motifs will be overrepresented in the leading strand simply because of the higher G content of this strand. Thus, overrepresented or strand-biased sequences being considered for a role in cell physiology must be well analyzed to be sure they reflect more than secondary consequences of mutationally generated skew in base composition (see above). Increasing frequency near the terminus may not be sufficient, since mutation rates appear to increase with distance from the origin (66). Two such G-rich elements (Chi and Rag) show a directional bias (75 and 82% in the leading strand) that is clearly in excess of that predicted by GC skew for sequences of the same composition (56%) (41).

CONVENTIONALLY SELECTED SEQUENCE ELEMENTS CAN HAVE FUNCTIONS THAT ARE SEEN ONLY AT THE POPULATION LEVEL

There is a formal possibility that some conventionally maintained (nonselfish) sequence elements have functions that can only be understood at the population level. For example, in *Haemophilus* and *Neisseria*, base runs appear in genes (contingency loci) that encode proteins involved in the interactions between the bacterium and its host (4). At these particular sequences, there is a high probability of frameshift mutations that inactivate (and reactivate) the gene, making a sort of on-off toggle switch. Each of the multiple contingency loci is thereby switched on and off stochastically, generating a variety of individuals in the population with different patterns of gene expression. The basic idea is that any appreciable population will include at least one individual that will show a pattern that allows it to face the latest challenge thrown up by the host. This behavior should be distinguished from that of mutators (discussed above), whose high frequency of randomly distributed deleterious mutations seems unlikely to be beneficial (59). In the case of contingency loci, the mutations affect a limited gene set, and the (on-off) reversibility prevents long-term destruction of information.

Another example might be directly repeated sequence elements flanking genes or including genes whose amplification is frequently selected (16, 58, 71). These repeats would have no immediate function within a cell, but might enhance the frequency of a particular duplication and allow amplification under selection. The role of these repeats would be difficult to analyze functionally using conventional genetic analysis. This sort of mechanism has been suggested for genes involved in nitrogen fixation which are repeated in a plasmid. These repeats support amplification of the genes for nitrogen fixation when the cells are placed under selective conditions. Repeats of this sort might be thought of as contributing to programmed amplification or as parts of a regulatory mechanism whose specificity is provided by natural selection.

In both contingency loci and the hypothetical programmed amplification, the selected event is reversible and the conserved sequence elements support interconversion of a few states for a small number of particular genes. In this sense, the examples resemble phase variation mechanisms, which might be regarded as another case in which identifiable sequence elements (genes and noncoding sequences) have a biological function that can be understood only at a population level.

SUMMARY

Whole chromosomes show us all the information that they carry. Some of this information is highly functional and intensely informative for those seeking a mechanistic understanding of chromosome behavior and cell function. However, much information owes its presence to events that occur at a population level and are unlikely to contribute to single-cell physiology. These features include selfish elements and conventional sequences whose importance is realized only in a population. In looking for sequence elements that are directly relevant to chromosome packaging, movement, and expression, the selfish and population-based functions may be a fascinating distraction.

Acknowledgments. I thank Dan Andersson, Dan Dykhuizen, Glenn Herrick, Pat Higgins, Ichizo Kobayashi, Jeff Lawrence, Jon Seger, Valley Stewart, and David Witherspoon for stimulating discussions and useful suggestions. They are not responsible for any inadequacies in presentation or errors in fact.

REFERENCES

1. **Ames, B. N., and R. G. Martin.** 1964. Biochemical aspects of genetics: the operon. *Annu. Rev. Biochem.* **33:**235–258.
2. **Andersson, D. I., and B. R. Levin.** 1999. The biological cost of antibiotic resistance. *Curr. Opin. Microbiol.* **2:**489–493.
3. **Bachellier, S., J. M. Clement, M. Hofnung, and E. Gilson.** 1997. Bacterial interspersed mosaic elements (BIMEs) are a major source of sequence polymorphism in *Escherichia coli* intergenic regions including specific associations with a new insertion sequence. *Genetics* **145:**551–662.
4. **Bayliss, C. D., D. Field, and E. R. Moxon.** 2001. The simple sequence contingency loci of *Haemophilus influenzae* and *Neisseria meningitidis. J. Clin. Investig.* **107:**657–662.
5. **Björkman, J., I. Nagaev, O. G. Berg, D. Hughes, and D. I. Andersson.** 2000. Effects of environment on compensatory mutations to ameliorate costs of antibiotic resistance. *Science* **287:**1479–1482.
6. **Boccard, F., and P. Prentki.** 1993. Specific interaction of IHF with RIBs, a class of bacterial repetitive DNA elements located at the 3′ end of transcription units. *EMBO J.* **12:**5019–5027.
7. **Campbell, J. H., D. Dykhuizen, and B. G. Rolfe.** 1978. Effects of the rex gene of phage lambda on lysogeny. *Genet. Res.* **32:**257–263.
8. **Casjens, S., N. Palmer, R. van Vugt, W. M. Huang, B. Stevenson, P. Rosa, R. Lathigra, G. Sutton, J. Peterson, R. J. Dodson, D. Haft, E. Hickey, M. Gwinn, O. White, and C. M. Fraser.** 2000. A bacterial genome in flux: the twelve linear and nine circular extrachromosomal DNAs in an infectious isolate of the Lyme disease spirochete *Borrelia burgdorferi. Mol. Microbiol.* **35:**490–516.
9. **Chao, L., and S. M. McBroom.** 1985. Evolution of transposable elements: an IS*10* insertion increases fitness in *Escherichia coli. Mol. Biol. Evol.* **2:**359–369.
10. **Chao, L., C. Vargas, B. B. Spear, and E. C. Cox.** 1983. Transposable elements as mutator genes in evolution. *Nature* **303:**633–635.
11. **Demerec, M., and P. Hartman.** 1959. Complex loci in microorganisms. *Annu. Rev. Microbiol.* **13:**377–406.
12. **Edlin, G., L. Lin, and R. Bitner.** 1977. Reproductive fitness of P1, P2, and Mu lysogens of *Escherichia coli. J. Virol.* **21:**560–564.
13. **Edwards, R. A., G. J. Olsen, and S. R. Maloy.** 2002. Comparative genomics of closely related Salmonellae. *Trends Microbiol.* **10:**94–99.
14. **Elena, S. F., L. Ekunwe, N. Hajela, S. A. Oden, and R. E. Lenski.** 1998. Distribution of fitness effects caused by random insertion mutations in *Escherichia coli. Genetica* **102-103:**349–358.
15. **Felsenstein, J.** 1974. The evolutionary advantage of recombination. *Genetics* **78:**737–756.
16. **Flores, M., P. Mavingui, X. Perret, W. J. Broughton, D. Romero, G. Hernandez, G. Davila, and R. Palacios.** 2000. Prediction, identification, and artificial selection of DNA rearrangements in Rhizobium: toward a natural genomic design. *Proc. Natl. Acad. Sci. USA* **97:**9138–9143.
17. **Francino, M. P., L. Chao, M. A. Riley, and H. Ochman.** 1996. Asymmetries generated by transcription-coupled repair in enterobacterial genes. *Science* **272:**107–109.
18. **Funchain, P., A. Yeung, J. L. Stewart, R. Lin, M. M. Slupska, and J. H. Miller.** 2000. The consequences of growth of a mutator strain of *Escherichia coli* as measured by loss of function among multiple gene targets and loss of fitness. *Genetics* **154:**959–970.
19. **Garcia-Lara, J., M. Picardeau, B. J. Hinnebusch, W. M. Huang, and S. Casjens.** 2000. The role of genomics in approaching the study of *Borrelia* DNA replication. *J. Mol. Microbiol. Biotechnol.* **2:**447–454.
20. **Gerrish, P. J., and R. E. Lenski.** 1998. The fate of competing beneficial mutations in an asexual population. *Genetica* **102-103:**127–144.
21. **Gilson, E., J. M. Clement, D. Brutlag, and M. Hofnung.** 1984. A family of dispersed repetitive extragenic palindromic DNA sequences in *E. coli. EMBO J.* **3:**1417–1421.
22. **Gilson, E., W. Saurin, D. Perrin, S. Bachellier, and M. Hofnung.** 1991. The BIME family of bacterial highly repetitive sequences. *Res. Microbiol.* **142:**217–222.
23. **Hartl, D. L., D. E. Dykhuizen, and D. E. Berg.** 1984. Accessory DNAs in the bacterial gene pool: playground for coevolution. *Ciba Found. Symp.* **102:**233–245.
24. **Hartl, D. L., D. E. Dykhuizen, R. D. Miller, L. Green, and J. de Framond.** 1983. Transposable element IS*50* improves growth rate of *E. coli* cells without transposition. *Cell* **35:**503–510.
25. **Hendrix, R. W., J. G. Lawrence, G. F. Hatfull, and S. Casjens.** 2000. The origins and ongoing evolution of viruses. *Trends Microbiol.* **8:**504–508.
26. **Higgins, C. F., G. F. Ames, W. M. Barnes, J. M. Clement, and M. Hofnung.** 1982. A novel intercistronic regulatory element of prokaryotic operons. *Nature* **298:**760–762.
27. **Hill, T. M.** 1996. Features of the chromosomal terminus region, p. 1602–1614. *In* F. C. Neidhardt, R. Curtiss III, J. L. Ingraham, E. C. C. Lin, K. B. Low, B. Magasanik, W. S. Reznikoff, M. Riley, M. Schaechter, and H. E. Umbarger (ed.), Escherichia coli *and* Salmonella typhimurium: *Cellular and Molecular Biology*, 2nd ed., vol. 2. ASM Press, Washington, D.C.
28. **Hulton, C. S., C. F. Higgins, and P. M. Sharp.** 1991. ERIC sequences: a novel family of repetitive elements in the genomes of *Escherichia coli, Salmonella typhimurium* and other enterobacteria. *Mol. Microbiol.* **5:**825–834.
29. **Jacob, F., and J. Monod.** 1962. On the regulation of gene activity. *Cold Spring Harbor Symp. Quant. Biol.* **26:**193–211.
30. **Jacob, F., D. Perrin, C. Sanchez, and J. Monod.** 1960. L'opéron: groupe de gènes à expression coordinée par un opérateur. *Crit. Rev. Acad. Sci.* **250:**1727–1729.

31. Kawano, M., S. Kanaya, T. Oshima, Y. Masuda, T. Ara, and H. Mori. 2002. Distribution of repetitive sequences on the leading and lagging strands of the *Escherichia coli* genome: comparative study of Long Direct Repeat (LDR) sequences. *DNA Res.* 9:1–10.

32. Kobayashi, I. 2001. Behavior of restriction-modification systems as selfish mobile elements and their impact on genome evolution. *Nucleic Acids Res.* 29:3742–3756.

33. Kobryn, K., and G. Chaconas. 2001. The circle is broken: telomere resolution in linear replicons. *Curr. Opin. Microbiol.* 4:558–564.

34. Kowalczykowski, S. C. 2000. Initiation of genetic recombination and recombination-dependent replication. *Trends Biochem. Sci.* 25:156–165.

35. Lawrence, J. G., and J. R. Roth. 1999. Genomic flux: genome evolution by gene loss and acquisition, p. 263–289. *In* R. L. Charlebois (ed.), *Organization of the Prokaryotic Genome.* ASM Press, Washington, D.C.

36. Lawrence, J. G., and J. R. Roth. 1996. Selfish operons: horizontal transfer may drive the evolution of gene clusters. *Genetics* 143:1843–1860.

37. Lin, L., R. Bitner, and G. Edlin. 1977. Increased reproductive fitness of *Escherichia coli* lambda lysogens. *J. Virol.* 21:554–559.

38. Liu, J., L. Xu, S. J. Sandler, and K. J. Marians. 1999. Replication fork assembly at recombination intermediates is required for bacterial growth. *Proc. Natl. Acad. Sci. USA* 96:3552–3555.

39. Liu, S. L., and K. E. Sanderson. 1998. Homologous recombination between *rrn* operons rearranges the chromosome in host-specialized species of *Salmonella. FEMS Microbiol. Lett.* 164:275–281.

40. Liu, S. L., and K. E. Sanderson. 1995. Rearrangements in the genome of the bacterium *Salmonella typhi. Proc. Natl. Acad. Sci. USA* 92:1018–1022.

41. Lobry, C., and J. Louarn. 2003. Polarization of prokaryotic chromosomes. *Curr. Opin. Microbiol.* 6:101–108.

42. Lobry, J. R. 1996. Asymmetric substitution patterns in the two DNA strands of bacteria. *Mol. Biol. Evol.* 13:660–665.

43. Lobry, J. R., and C. Lobry. 1999. Evolution of DNA base composition under no-strand-bias conditions when the substitution rates are not constant. *Mol. Biol. Evol.* 16:719–723.

44. Maisnier-Patin, S., O. G. Berg, L. Liljas, and D. I. Andersson. 2002. Compensatory adaptation to the deleterious effect of antibiotic resistance in *Salmonella typhimurium. Mol. Microbiol.* 46:355–366.

45. McLean, M. J., K. H. Wolfe, and K. M. Devine. 1998. Base composition skews, replication orientation, and gene orientation in 12 prokaryote genomes. *J. Mol. Evol.* 47:691–696.

46. Miller, J. H., A. Suthar, J. Tai, A. Yeung, C. Truong, and J. L. Stewart. 1999. Direct selection for mutators in *Escherichia coli. J. Bacteriol.* 181:1576–1584.

47. Miller, R. D., D. E. Dykhuizen, and D. L. Hartl. 1988. Fitness effects of a deletion mutation increasing transcription of the 6-phosphogluconate dehydrogenase gene in *Escherichia coli. Mol. Biol. Evol.* 5:691–703.

48. Muller, H. 1932. Some genetic aspects of sex. *Am. Nat.* 66:118–138.

49. Muller, H. J. 1964. The relation of recombination to mutational advance. *Mutat. Res.* 1:2–9.

50. Mushegian, A., and E. Koonin. 1996. Gene order is not conserved in bacterial evolution. *Trends Genet.* 12:289–290.

51. Naito, T., K. Kusano, and I. Kobayashi. 1995. Selfish behavior of restriction-modification systems. *Science* 267:897–899.

52. Nobusato, A., I. Uchiyama, and I. Kobayashi. 2000. Diversity of restriction-modification gene homologues in *Helicobacter pylori. Gene* 259:89–98.

53. Novick, R. P., and F. C. Hoppensteadt. 1978. On plasmid incompatibility. *Plasmid* 1:421–434.

54. Overbeek, R., M. Fonstein, M. D'Souza, G. D. Pusch, and N. Maltsev. 1999. Use of contiguity on the chromosome to predict functional coupling. *In Silico Biol.* 1:93–108.

55. Redfield, R. J. 2001. Do bacteria have sex? *Nat. Rev. Genet.* 2:634–639.

56. Rocha, E. P., and A. Danchin. 2001. Ongoing evolution of strand composition in bacterial genomes. *Mol. Biol. Evol.* 18:1789–1799.

57. Rocha, E. P., A. Danchin, and A. Viari. 1999. Universal replication biases in bacteria. *Mol. Microbiol.* 32:11–16.

58. Romero, D., and R. Palacios. 1997. Gene amplification and genomic plasticity in prokaryotes. *Annu. Rev. Genet.* 31:91–111.

59. Roth, J. R., E. Kofoid, F. P. Roth, O. G. Berg, J. Seger, and D. I. Andersson. 2003. Regulating general mutation rates: examination of the hypermutable state model for Cairnsian adaptive mutation. *Genetics* 163:1483–1496.

60. Salzberg, S. L., A. J. Salzberg, A. R. Kerlavage, and J. F. Tomb. 1998. Skewed oligomers and origins of replication. *Gene* 217:57–67.

61. Schmid, M. B., and J. R. Roth. 1987. Gene location affects expression level in *Salmonella typhimurium. J. Bacteriol.* 169:2872–2875.

62. Schneider, D., E. Duperchy, E. Coursange, R. E. Lenski, and M. Blot. 2000. Long-term experimental evolution in *Escherichia coli.* IX. Characterization of insertion sequence-mediated mutations and rearrangements. *Genetics* 156:477–488.

63. Scott, J. R., and G. G. Churchward. 1995. Conjugative transposition. *Annu. Rev. Microbiol.* 49:367–397.

64. Sharp, P. M., and W. H. Li. 1987. The Codon Adaptation Index—a measure of directional synonymous codon usage bias, and its potential applications. *Nucleic Acids Res.* 15:1281–1295.

65. Sharp, P. M., and W. H. Li. 1986. An evolutionary perspective on synonymous codon usage in unicellular organisms. *J. Mol. Evol.* 24:28–38.

66. Sharp, P. M., D. C. Shields, K. H. Wolfe, and W. H. Li. 1989. Chromosomal location and evolutionary rate variation in enterobacterial genes. *Science* 246:808–810.

67. Sharples, G. J., and R. G. Lloyd. 1990. A novel repeated DNA sequence located in the intergenic regions of bacterial chromosomes. *Nucleic Acids Res.* 18:6503–6508.

68. Shyamala, V., E. Schneider, and G. F. Ames. 1990. Tandem chromosomal duplications: role of REP sequences in the recombination event at the join-point. *EMBO J.* 9:939–946.

69. Smith, G. R., S. M. Kunes, D. W. Schultz, A. Taylor, and K. L. Triman. 1981. Structure of chi hotspots of generalized recombination. *Cell* 24:429–436.

70. Smith, H. O., M. L. Gwinn, and S. L. Salzberg. 1999. DNA uptake signal sequences in naturally transformable bacteria. *Res. Microbiol.* 150:603–616.

71. Sonti, R., and J. R. Roth. 1988. Role of gene duplications in the adaptation of *Salmonella typhimurium* to growth limiting carbon sources. *Genetics* 123:19–28.

72. Stern, M. J., G. F. Ames, N. H. Smith, E. C. Robinson, and C. F. Higgins. 1984. Repetitive extragenic palindromic sequences: a major component of the bacterial genome. *Cell* 37:1015–1026.

73. Thaler, D. S., and F. W. Stahl. 1988. DNA double-chain breaks in recombination of phage lambda and of yeast. *Annu. Rev. Genet.* 22:169–197.

74. **Wheelis, M., and R. Stanier.** 1970. The genetic control of dissimilatory pathways in *Pseudomonas putida*. *Genetics* 66:245–266.

75. **Wolf, Y. I., I. B. Rogozin, A. S. Kondrashov, and E. V. Koonin.** 2001. Genome alignment, evolution of prokaryotic genome organization, and prediction of gene function using genomic context. *Genome Res.* 11:356–372.

76. **Yang, Y., and G. F. Ames.** 1988. DNA gyrase binds to the family of prokaryotic repetitive extragenic palindromic sequences. *Proc. Natl. Acad. Sci. USA* 85:8850–8854.

Chapter 2

The Dynamic Bacterial Genome

JEFFREY G. LAWRENCE

When asked to describe the dynamic nature of bacterial chromosomes, one might start by outlining what properties of chromosomes would make them appear to be static, that is, relatively uniform among many organisms in space and only slowly changing over time. This exercise, not being entirely pedantic, then serves to highlight the large numbers of evolutionary processes and molecular biological mechanisms that act continually to alter the content and composition of bacterial chromosomes, all the while leaving untouched classes of information which have lent an air of perpetuity to these very same molecules. Since many of these mechanisms are discussed in detail in other chapters in this volume, here I will survey briefly the many ways by which chromosomes' content, context, and composition can be altered and yet act so as to leave a cursory impression that chromosomes of closely related bacteria lack significant spatial or temporal heterogeneity.

GENOMES AT REST

The model of a static bacterial chromosome arose from early comparisons of the genetic maps of *Escherichia coli* and *Salmonella typhimurium* (now termed *Salmonella enterica* serovar Typhimurium). The accumulation of mutations affecting primarily pathways for nutrient biosynthesis (e.g., amino acids, cofactors, and nucleosides) culminated in the construction of continually refined circular genetic maps for both species in the mid-1960s (164–166, 197–199); these maps were remarkably congruent (Fig. 1), with even the famous terminus-centered inversion not described until several years later (167). The striking colinearity of gene order on these two maps strongly suggested that the chromosomes of *E. coli* and *S. enterica* serovar Typhimurium had not changed very dramatically since the separation of their lineages, eventually estimated to have occurred about 100 million to 150 million years ago (146, 147). This view was entirely consistent with the understanding that bacterial cells reproduce primarily by binary fission, that is, by the production of two daughter cells without any requisite genetic exchange among individuals. Therefore, one could visualize a bacterial chromosome's being passed down faithfully through generations of cell divisions, untouched except by point mutational processes and other mechanisms that made only modest changes to chromosome content. With such large population sizes and many generations over which selection could act, these chromosomes perhaps represented the culmination of a long evolutionary process by which each gene has been shaped and placed on the chromosome exactly where it performs best.

At the time the maps were being constructed, even nucleotide substitutions were thought to be somewhat rare events. Contemporary population genetic theory centered on the idea that alterations in genetic material would be strongly counterselected, since most mutations undoubtedly conferred detrimental effects and the resulting variants would be removed from the population. The only cells to survive would be those bearing presumably rare beneficial mutations (therefore, any detectable variation must reflect some adaptive role). The two organisms were known to differ in important ways—*E. coli* fermented lactose (note the *lac* operon on the *E. coli* genetic map [Fig. 1]), while *Salmonella* evolved H_2S from thiosulfate, grew on citrate, and was notably pathogenic; these differences were viewed as likely arising from the alteration of ancestral genetic material by duplication and divergence, thus preserving the synteny of the surrounding genetic material. This model had been invoked for the evolution of

Jeffrey G. Lawrence • Pittsburgh Bacteriophage Institute and Department of Biological Sciences, University of Pittsburgh, Pittsburgh, PA 15260.

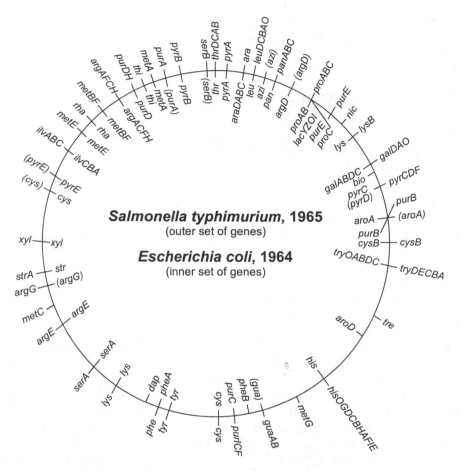

Figure 1. Correspondence between the first genetic maps of *E. coli* (197), whose loci are denoted on the inside of the circle, and *S. enterica* serovar Typhimurium (166), whose loci are denoted on the outside of the circle. Genes whose positions were less defined are depicted in parentheses; spacing between genes was adjusted to allow for facile alignment of the two maps. The loci shared between the maps show remarkable conservation of order.

eukaryotic genes, starting with the globin gene family (217, 218).

The next 20 years brought little to change this view of a docile bacterial chromosome. While there were facets of microbial evolution that hinted at other possibilities—such as wide-ranging horizontal transfer of antibiotic resistance genes on plasmids (34), the existence of conjugative plasmids and transducing bacteriophages (104, 105, 216), and the great levels of heterozygosity among bacterial genes (121) reflecting the accumulation of neutral mutations (80)—they offered no comprehensive model of a dynamic chromosome that could compete with the powerful picture of stability offered by the congruence of genetic maps between related bacteria.

This view was effectively codified in the excellent periodic selection model offered by Levin (106), whereby an occasional beneficial mutation could arise in a bacterial population and carry the originating chromosome to high frequency by virtue of the ben-

efits it conferred. Surveys of numerous bacterial isolates (2, 3, 24, 25, 144, 145, 206–209) showed large clones of strains deemed identical (or nearly so) by multilocus enzyme electrophoresis (MLEE), with variant alleles exhibiting strong linkage disequilibrium. These data were all consistent with the periodic selection model. More important, the predictions of this model were strongly supported by extensive DNA hybridization surveys which showed that bacteria clustered into groups of closely related strains (15–20). Major contemporary reviews of *E. coli* population genetics offered a compellingly complete picture of a slowly changing bacterial chromosome (62).

GENOMES IN ACTION

This view changed somewhat in the late 1980s and early 1990s, when an idea was introduced that would significantly refine the periodic selection model for overall bacterial evolution. The standard periodic selection model predicted that genes within a bacterial chromosome would coevolve; as a result, phylogenies inferred from different genes from the same set of variant strains would be congruent. As it turned out, this was not the case. From the same subset of strains of *E. coli*, phylogenies were first drawn for genes of the *trp* locus (123) and then for alleles of the *phoA* (40) and *gnd* (42) genes. Comparison of the three phylogenies illustrated two important points. First, exchange of genes among strains of the same "species" must have occurred, since the relationships among the strains inferred for each of the three genes were not congruent; this result has since been upheld in the analyses of numerous genes, including the *sppA, gapA, pabB,* and *zwf* (57), *mdh* (14), and *putP* (135) loci. Second, periodic selection events were evident, but recombination did not allow an entire chromosome to "hitchhike" to high frequency as the result of carrying a single beneficial mutation. Rather, the "selective sweep," as it was termed, likely carried to high frequency only those genes closely linked to a beneficial mutation; as a result, the overall level of divergence was different for the *trp, phoA,* and *gnd* loci.

This patchwork view of bacterial chromosomes, described as clonal segments embedded within a clonal frame (122, 124), broke the notion that bacterial genomes were static entities that slowly accumulated neutral point mutations until the chance creation of a beneficial allele allowed a lucky organism to rise to high frequency. Rather, bacterial species themselves could be defined as a set of strains in which the driving force behind the phenotypic and genotypic cohesion of its lineage was the exchange of genetic material by homologous recombination (42) (Fig. 2), rather than periodic selection events carrying entire chromosomes to high frequency within the population. Variants of this model have been offered which either highlight (90) or downplay (30) the role of homologous recombination in defining bacterial "species" (if this is even possible). Regardless of the approach taken, homologous recombination—at some frequency, although it clearly varied among lineages (117)—must be integrated into any view of bacterial evolution.

The idea that bacteria could exchange genes was certainly not new. The mechanisms for gene exchange—plasmid-mediated conjugation, bacteriophage-mediated transduction, and direct transformation by naked DNA (142)—were elucidated decades earlier, and used routinely in the construction of bacterial strains in the laboratory. However, the actual rates of gene exchange by homologous recombination were largely

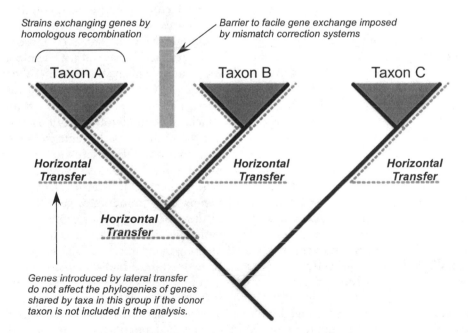

Figure 2. Mechanisms of gene transfer and their effects on inferring phylogeny. Homologous recombination serves to unify strains within bacterial taxa; as a result, phylogenies of different genes within these groups will not be congruent, but phylogenies of the same genes found in different lineages—that is, those which do not exchange genes because of the imposition of mismatch correction systems (111–113, 160, 190, 201, 202, 215)—will be congruent. This system has been invoked to define bacterial species (42). Gene exchange across large phylogenetic distances does not disrupt these patterns as long as the donor taxa are not included in the analyses. Limitations of this approach are discussed elsewhere (90).

unknown, but not thought to be tremendously frequent, at least in *E. coli* or *Salmonella*, in light of the strong linkage disequilibrium that was observed by MLEE. The incongruent phylogenies of the *trp*, *phoA*, and *gnd* loci pointed to the effects of recombination, but the rate of recombination among closely related strains—and variation in this rate among different species—could not be estimated from these data alone. Later experiments, using different approaches and much larger data sets, showed that the rate of recombination among *E. coli* strains was on the order of the mutation rate (57, 58); that is, the probability of gaining a variant allele by mutation was similar to the probability of gaining a variant allele by recombination. More refined estimates, using multilocus sequence typing, showed these rates to be underestimates by 20- to 50-fold (46, 47). In addition, recombination rates varied remarkably among lineages, with some (e.g., *Mycobacterium* and *Buchnera*), showing little evidence for recombination (130, 189) and others (e.g., *Neisseria* and *Helicobacter*) showing evidence for massive rates of gene exchange (1, 45, 46, 66, 85, 168, 188, 195).

Like periodic selection, homologous recombination offers a plausible mechanism for the clustering of bacterial strains as observed by DNA hybridization (15–20). In addition, the high rates of recombination among strains of *E. coli* inferred by multilocus sequence typing can also be reconciled with the observations of clonal lineages and strong linkage disequilibrium revealed by MLEE. Unlike sexual exchange in typical diploid eukaryotes, which exchange entire haploid genomes during meiosis and syngamy, genetic exchange in prokaryotes typically involves the transfer of only fragments of DNA. The size of these DNA fragments is limited both by the mechanisms introducing DNA into the cytoplasm (142) and by the action of restriction endonuclease systems within the host cell (119). As a result, recombination is necessarily limited to a small portion of the bacterial chromosome. Hence, localized recombination (184, 186, 187) could exchange genetic material at one locus, while leaving intact linkage disequilibrium at distantly placed loci that were uninvolved in the recombination event; therefore, selective sweeps could take place locally within a chromosome while allowing coadapted gene complexes to persist.

These substantial rates of gene exchange by homologous recombination reveal a dynamic nature of bacterial chromosomes that is often not appreciated when the complete genome sequence of a bacterial genome is determined. The genome sequence of a single strain is often viewed as the genetic blueprint of a bacterial species; this view parallels the idea that a "type strain" is the physiological representative of a bacterial species. Yet this view of a bacterial "species" can be improved, in most cases, by recognizing that a single organism cannot serve as the sole representative of its group's genetic variation, and a collection of organisms exchanging variant DNA by homologous recombination offers a more complete picture. The groups of strains participating in this type of facile gene exchange may be delineated by the barriers imposed by mismatch correction systems, which prevent recombination between DNA heteroduplexes that bear significant numbers of mismatched bases (111–113, 160, 190, 201, 202, 215). This variability among strains does not detract from the "static genome" model arising from comparison of genetic maps, yet has strong implications for what genomic diversity may lie within accepted "species" boundaries when other sources of sequence variation—aside from alleles generated by point mutation—are considered.

GENOMES IN FLUX

The acquisition of complete genome sequences from more than one strain of a bacterial species did not highlight the allelic differences among these isolates and showcase gene exchange among closely related strains by homologous recombination as a major factor in microbial evolution. Rather, these differences were overshadowed by a more striking result, which was that even closely related strains varied tremendously in gene content. For example, the genome sequence of the enterohemorrhagic strain *E. coli* O157:H7 revealed about 1,000 kb of DNA not found in the commensal strain *E. coli* K-12, which, in turn, contained half as much unique DNA not found in the pathogenic isolate (155). Similarly, the genomes of two serovars of *S. enterica*, Typhimurium and Typhi, also showed large differences in gene content (43, 118, 153), with each strain containing large amounts of genetic material not found in the other. In light of these data, we see that homologous recombination can serve not only to reassort variant alleles of shared genes among closely related strains, but it can also disseminate newly acquired genes as well by allowing for recombination at sequences flanking the variable segments. The action of these two processes makes the gene content of bacterial populations far more fluid than previously imagined.

Analyses of complete genome sequences by several methods revealed that the differences in gene content were the result of two complementary processes: the gain of new genes by horizontal gene transfer from distantly related organisms (here the new DNA is not integrated by homologous recombination; rather, illegitimate or site-specific recombination mechanisms are employed to integrate the new DNA fragments into the existing chromosome), and the loss of ancestral genes from descendent lineages. Here, differences between lineages have not arisen by the modification of genetic material originating in the parental lineage; rather, genes and gene clusters which confer distinct physiological activities have been acquired from an outside source, and the genes for other capabilities have been lost entirely. Appreciation of the impact of these processes necessarily relies on their accurate quantification, which has proved to be difficult, as it relies on both the identification of genes novel to a genome and some estimate of their times of arrival. Two general approaches have been used to identify potentially new genes in bacterial genomes; each approach has its advantages and its limitations (97).

First, phylogenetic methods can detect both gene acquisition and gene loss events by (i) detecting an unduly high level of similarity between genes found in otherwise unrelated organisms (e.g., see references 29, 71, and 120) or (ii) finding that homologues of a gene are present in a wide range of phylogenetically unrelated organisms, but they are absent from closely related organisms (e.g., see references 29, 37–39, 81, 89, 150, and 157). In both cases, the most parsimonious explanation for the unusual gene distribution invokes lateral gene transfer. These methods are powerful in being able to detect ancient, even unsuspected gene transfer events (e.g., the transfer of the GAPDH gene from eukaryotes into proteobacteria [36] or the transfer of a proteobacterium to eukaryotes to form the mitochondrion [9]), but can be muddled by variation in rates of evolution between lineages and by the expansion of gene families which confound the identification of orthologues. In addition, their power lies entirely in the depth and breadth of the sequence database, and they work best when several closely related organisms can be analyzed simultaneously. So while these methods have been used to make strong claims—e.g., that a large portion of the *Thermotoga* genome arose from transfer from archaeal lineages (136), results that are supported by other analyses (212)—these results can be interpreted in other ways (109).

Alternatively, genomes may be analyzed to detect sequences introduced by horizontal transfer by employing the one feature that all acquired genes share: they arrived in that genome from an outside source. These methods assume that genes native to—or long-term residents of—a bacterial genome will appear similar owing to the action of directional mutation pressures (191–194), which are discussed further below. Directional mutation pressures provide a distinct "fingerprint" to a bacterial genome owing to the differential mutational proclivities of DNA polymerases, the nature and number of mismatch correction systems, the numbers and abundances of tRNA species, and even relative concentrations of precursor nucleotide pools. Thus, genes which appear atypical in their current genomic context may reflect the direction pressures of a donor genome.

A number of measures may be used to detect atypical genes, including nucleotide and/or dinucleotide composition (73–75, 95, 96), biases in relative codon usage (76, 77, 128), or the patterns found by Markov chain analyses (63). One advantage of these parametric approaches is that genomes may be examined for acquired genes without reliance on the sequence database or interpretations of comparisons with homologous genes. However, parametric methods are limited to detecting recent transfer events; over time, the directional mutation pressures of the recipient genome will act to ameliorate the aberrant sequence features so that the genes no longer appear atypical (95, 96, 141). Additional limitations of these methods include difficulties in establishing the criteria which describe what a "typical" gene is, and the fact that genes of foreign origin may not look atypical upon arrival (81, 204). Moreover, due to their different approaches in identifying foreign genes, phylogenetic methods and parametric methods may define different subsets of genes as being horizontally transferred (97, 158), since genes introduced over different time scales are being identified.

Both classes of detection methods have been successful in identifying impressive arrays of diverse genes which have been introduced into genomes by horizontal gene transfer, including genes encoding central metabolic functions such as glyceraldehyde-3-phosphate dehydrogenase (GAPDH), β-hydroxy-β-methylglutaryl-coenzyme A (HMG-CoA) reductase, glutamine synthases, and phosphoglucose isomerase (13, 36, 78, 137, 150, 156); complete biosynthetic pathways such as cytochrome *c* biogenesis (84) or coenzyme B_{12} biosynthesis (99, 100); and components of the transcription and translation machinery (68, 210, 211), including elongation factors (79), ribosomal proteins L32, L33, S14, and S18 (21, 114), and even rRNA (132, 214). No gene appears to be immune to the possibility of transfer between distantly related organisms. As detailed below, the organization of bacterial genes into clusters and operons (whose genes are cotranscribed) allows horizontal gene transfer to introduce in a single step all of the genes required to perform a complex physiological function (91, 92, 103), making gene transfer a powerful mechanism in promoting niche invasion and eventual lineage diversification (that is, speciation [90]). Virtually all bacterial genome sequences show signs of recent horizontal transfer events (as in Fig. 3, inferred from the presence of atypical genes; see also references 89 and 140), and all genomes show signs of ancient horizontal transfer events detected by phylogenetic approaches (e.g., see reference 210).

Yet it is also clear that not all genes are transferred with equal proficiency. Issues of clustering into operons aside, one should expect that genomes would be more recalcitrant in accepting some classes of genes for two reasons. First, some genes may provide a selective benefit to only very few organisms, thereby limiting the pool of potential recipients. For example, genes required for the synthesis of cofactors specific to methanogenesis would have little utility in organisms that do not perform methanogenesis or methane degradation; indeed, the one known case of transfer involving these genes is from a likely

methanogen donor into a (currently) methyltrophic recipient (27). Similarly, genes for photosystem II would be of little utility outside of photosynthetic organisms possessing photosystem I (213). Alternatively, a gene might be quite useful, but have a widespread or ubiquitous distribution; in these cases, recipient lineages will already have a gene performing the function in question, and the acquired version will have a disadvantage in not being attuned to the transcription and translation machinery of the cell or to other proteins with which it must interact (70). Examples of such genes include ribosomal proteins or aminoacyl-tRNA synthetases, although cases of transfer of both have been reported (21, 210). In these cases, either the native and acquired genes were equally proficient, and the acquired gene survived by chance, or the acquired gene conferred a selective benefit (e.g., antibiotic resistance) which promoted its replacement of the native gene. This prediction is upheld by inspection of genes recently transferred into bacterial genomes, identified by the atypical sequence

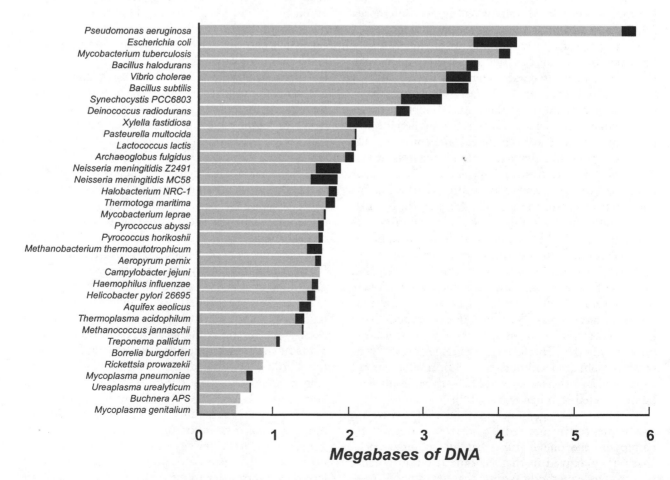

Figure 3. The distribution of recently acquired DNA, inferred from the numbers of atypical genes, in various bacterial genomes. Gray bars denote amounts of typical protein-coding sequences, while black bars denote atypical protein-coding sequences, identified as having aberrant composition, dinucleotide fingerprints, and patterns of codon usage bias.

features. Rather than representing all classes of genes equally, certain classes of genes—such as those encoding transport apparatuses, catabolic pathways, or biosynthesis of fimbrae—are overrepresented.

Variation in the amount of recently acquired DNA is evident upon inspection of Fig. 3, but is not readily predicted from genome size (e.g., *E. coli* and *Mycobacterium tuberculosis* have similar genome sizes but differ greatly in the amount of recently transferred DNA), lifestyle (the two methanogens, *Methanococcus jannaschii* and *Methanobacterium thermautotrophicum*, vary greatly in their amounts of foreign DNA), or the ability to take up DNA by transformation (e.g., the genome of naturally transformable *Haemophilus influenzae* has little recently acquired DNA, while those of *Neisseria* species show a great deal more). One trend is clear, which is that genomes of intracellular parasites generally have little recently acquired DNA; this dearth may reflect a lowered exposure to the agents of gene transfer (94, 142, 205). As discussed below, much of the variation in amounts of recently acquired DNA can be predicted from the ability of a genome to utilize genes which encode function of low selective value (89).

Translating the amounts of acquired DNA found in a bacterial genome into a rate of acquisition is a bit more difficult. Initial studies (95, 96) focused on estimating the ages of acquired genes within their new host genome, relying on the observation that directional mutation pressures lead to some universal patterns that can be inferred from the analysis of many different genomes (131). This idea was extended to posit that newly arriving genes would conform to these relationships since their sequence features would resemble those of their donor genome. Over time, however, these features would ameliorate under the directional mutation pressures of the recipient genome; during the process of amelioration, genes would resemble neither their donor nor recipient genomes, and their time of arrival could be estimated from the disparity of their sequence features from the global patterns previously inferred (95, 96, 115). Yet this method provides only a ballpark estimate for the general ages of acquired genes which, when taken en masse, provide an estimate for the rate of lateral gene transfer. Estimates for the ages of individual acquisition events can be traced more readily by phylogenetic analyses (140).

GENOMES RESHUFFLING

One of the facets of bacterial chromosome organization that facilitates the transfer of DNA among distantly related lineages is the aggregation of genes which cooperate in a single metabolic pathway into clusters or cotranscribed operons. It has been postulated that this organization arose by virtue of gene transfer, since following the transfer of a chromosome segment bearing all genes required for a selectable function, those genes not contributing toward a beneficial function would be deleted (103). As a result, the remaining genes would form a cluster composed only of genes under selection. At this point, cotranscription of the genes within a cluster may be selected by a number of mechanisms, including the transfer of the newly formed gene cluster (facilitated by its now smaller size) into an organism with a significantly different transcription apparatus, requiring transcription from a promoter at the site of insertion and selecting strongly for cotranscription of all genes in the cluster, leading to the reduction of intergenic spaces. Alternatively, cotranscription may be selected in situ if it is advantageous. One advantage could include benefits incurred from the coregulation of the clustered genes; however, this advantage is unlikely to select for the formation of a gene cluster since coregulation can be readily attained for the multiple promoters of unclustered genes, thereby eliminating the need for multiple, fortuitous chromosomal rearrangements that precisely juxtapose the appropriate genes. After its formation, the bacterial operon provides a promiscuous package of DNA that can be expressed at the site of insertion by a native promoter, and allow for the expression of multiple protein products via translational coupling. For these reasons, the transfer of operons may allow for efficient and effective niche invasion, since all of the genes required for the expression of novel functions may be introduced in a single event (91, 92).

Yet operons are not permanent fixtures of chromosomes once they are created. While selection for coexpression may provide a benefit after an operon has been formed, and perhaps counterselect operon disruption, surveys of bacterial genomes have shown that operon organization can be transient (69). Examples of operon disruption include the likely split of the *cysHIJ* and *cysDNC* operons in *E. coli* (12) (the intervening sequence has signatures of being foreign in origin [96]), the split of the *trpEDF* and *trpBA* operons by a prophage in *H. influenzae* (49), and the numerous cases of operon disruption in *Neisseria meningitidis* (152). Figure 4 shows the degradation of the *nuo* operon in *N. meningitidis*; here, the genes interrupting the *nuo* operon were likely acquired by horizontal transfer, since they are significantly more AT-rich than typical *Neisseria* genes. While these surveys show that most gene clusters are transient, some genes are quite likely to be found together across multiple genomes; in these cases, the encoded protein

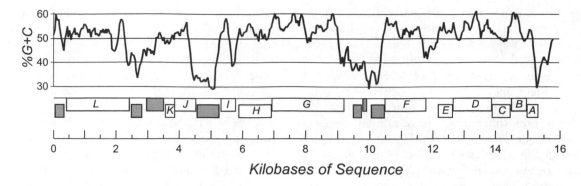

Figure 4. Disruption of the *nuo* operon in *N. meningitidis* serogroup A strain Z2491 (152). Letters indicate *nuo* genes; non-*nuo* genes are indicated by the gray boxes. The nucleotide composition plot shows the %G + C for a 200-base window starting at the position indicated.

products physically interact (33). This perhaps provides an additional selective benefit for translation from the same mRNA, since the local molar concentration of each product would be highest at the same location, facilitating the formation of the protein complex. Additional benefits of cotranscription could include the delivery of newly translated products to particular cellular locations; recent advances in the localization of gene products to specific cellular locations (110, 176) has made this possibility more compelling.

GENOMES AT A LOSS

Not only do genomes acquire new genes by lateral gene transfer, but genomes lose content when genes are deleted. While the selective forces which allow for the maintenance of relatively high deletion rates among most bacteria are still under debate (94), it is clear that genes not under selection are rapidly lost from most genomes (e.g., the *phoA* gene was lost from *S. enterica* [41], and transposons are lost rapidly from strains of *E. coli* [93, 98, 133]). An exception to this pattern is the long-term retention of pseudogenes in *Buchnera* species (126, 183, 196), which reflects a decrease in the overall deletion rate. However, gene loss may reflect more than a gene's lack of utility; in some cases, gene expression may actively interfere with the adoption of a new lifestyle, placing positive selection on gene loss. Examples of such cases include the loss of the *cadA* (116) or *ompT* (134) genes from *Shigella*, where their expression attenuates the virulence in this pathogen. Loss of genetic material reflects constraints on genome size: not so much physical size, but the information contained therein that must be retained by natural selection (89). These constraints are discussed further below.

Gene loss is especially evident among intracellular parasites, whose genomes are often smaller than 1 Mb (28, 125, 127, 143, 205). Their small size is likely the result of at least two forces: the lack of incoming DNA, and the low information content afforded by such small populations of organisms which rarely engage in gene exchange by homologous recombination. Models for genome reduction (6–9, 94) usually cite a period of relaxed selection on gene function—often as a result of reduced population sizes and lower rates of gene exchange by homologous recombination, which raises the value of an effectively neutral mutation (148, 149)—as leading to the transient accumulation of pseudogenes. These fragments of noncoding DNA are eventually deleted, leading to genomes of reduced physical size. The active process of genome reduction is evident in the genomes of *Mycobacterium leprae* (31), where more than 1,100 pseudogenes have been identified, and *Rickettsia prowazekii* (7, 9), where many large pseudogenes are evident even after a long period of genome reduction. Pseudogenes have persisted for millions of years in the genomes of *Buchnera* species (126, 183, 196) and are becoming evident in the newly created pathogenic lineage of *S. enterica* serovar Typhi (153). These organisms all represent steps along the pathway of genome reduction.

GENOMES ADRIFT

A change in the bacterial genome more subtle than the mechanisms of genome evolution illustrated above—the exchange of variant alleles by homologous recombination, gene loss, and gene acquisition by horizontal gene transfer—is the slow alteration of genetic material by directional mutation pressure (191–194). This phenomenon is evidenced by the

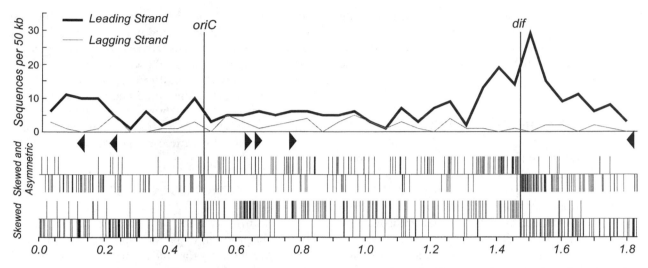

Figure 5. Distribution of oligomers in the *H. influenzae* genome. Sequence is from Fleischmann et al. (49); origin and terminus of replication are inferred from strand asymmetry analysis; triangles represent positions as direction of transcription of rRNA operons. The lower panel shows the effects of mutation biases. Strand asymmetry in the genome of *Haemophilus* is manifested by 14 different octameric oligonucleotides, which are drawn on either the forward or reverse strand. The middle panel shows sequences with both strand asymmetry and a biased distribution with respect to the terminus of replication; the distribution of 23 octamers is shown. The top panel shows a histogram of the distribution of the octamers shown in the middle panel on the leading and lagging strands for 50-kb intervals; intervals were chosen so that the terminus of replication fell between two intervals.

change in nucleotide composition, codon usage bias, and other measures of genome composition that vary among lineages. While the tempo and mode of these changes are not completely understood, some basic tenets are clear. First, chromosomes can change in overall composition rather slowly; for example, genes in *S. enterica* are slightly more GC-rich than their *E. coli* homologues, meaning that changes in composition are not entirely episodic. Second, rapid directional change is possible, as seen in the tendency for parasitic lineages to attain high AT contents over short periods of time (126). Directional mutation pressures also result in compositional asymmetries between the DNA strands (163), since the nature of mutations that occur on leading and lagging strands differ (50–54). As a result, some sequences are found preferentially on one strand. These sequences may be coopted into performing a useful role in cellular metabolism, such as the χ recombination-stimulation factor for the RecBCD complex in *E. coli* (82, 83, 185), which may aid in the reestablishment of collapsed replication forks (87, 88). An example of such strand asymmetry is depicted in Fig. 5, where skewed octamers resulting from differential substitution rates on the leading and lagging strands in *H. influenzae* neatly define the origin and terminus of replication. Additional roles for strand-specific sequences in creating chromosome stability are discussed below.

Directional mutation pressure may be opposed by selection for other sequence features, such as preference for certain codons (reflected in codon usage bias [177–182]). Since degree of codon usage bias is variable among lineages—and the subset of genes that may experience strong selection for codon usage may be different in different lineages (76, 179)—the results of directional mutation pressures may vary among lineages. These alterations may change genomes very subtly over time but have large long-term effects; for example, changes in nucleotide composition would alter the likelihood of facile DNA bending, thereby changing the roles for DNA binding proteins such as IHF, HN-S, and HU in gene expression, DNA supercoiling, and nucleoid organization. Again, these dynamic changes are not immediately obvious upon genome comparison, leading to a view of a static chromosome over time, especially if gene content has not largely changed. Yet the effects of directional mutation pressure are quite evident in the strong AT-richness of most intracellular parasites and the great variation in nucleotide composition among other bacterial lineages. Such changes also impact other mechanisms by which chromosomes change over time. For example, the action of the *E. coli* RecA protein is compromised when it encounters GC-rich DNA (56), which may lead to preferential incorporation of AT-rich sequences into the genome; recently acquired genes in *E. coli* are notably AT-rich (96), and such effects demonstrate how changes in nucleotide composition can have a broad impact, such as altering the potential for organisms to act as efficient donors for horizontal gene exchange.

GENOMES UNDER ATTACK

Bacterial genomes have large numbers of prophages (94) (see chapter 3), insertion sequences (169), conjugative transposons (162), integrons (60, 61), and other mobile genetic elements. Surveys have shown that that more than 98% of natural isolates of *S. enterica* are lysogens (170); even more striking, 95% of these lysogens carry generalized transducing phages, thereby providing a plausible mechanism for frequent gene exchange via homologous recombination among enteric bacteria. While it may seem that these genetic parasites would do little to disrupt the appearance of a static genome, their roles in genome evolution can be multifold. First, phages may carry genes which, when expressed from the integrated prophage, improve the fitness of their hosts; examples include the toxin genes in the enterohemorrhagic strain *E. coli* O157:H7 located on phage 933W (155), the RecE and RecT recombinase proteins expressed from the *E. coli rac* prophage (26, 86), toxin genes on *Shigella* phage Sfv (4), or the *Vibrio cholerae* toxin genes on ϕCtx (203). Therefore, integrated phages may provide a source of new genes for bacterial chromosomes. Second, insertion (IS) elements have been implicated in the capture of exogenous cytoplasmic DNA, offering a mechanism by which foreign DNA injected into a cell's cytoplasm may become integrated into a stable replicon (96, 142). Genetic parasites also offer a large block of genes presumably under little or no selection, which can offer sites for the integration of other fragments of foreign DNA via illegitimate recombination without disruption of native genes. Last, repeated sequences such as IS elements offer regions of sequence identity that can promote inversions, translocations, and other genomic rearrangements (59, 161).

Aside from providing substrates for further genome dynamics, genetic parasites themselves may promote an increase in a chromosome's natural rate of deletion (94). That is, since prophages and IS elements incur long-term detrimental effects for their hosts, the benefits of their removal may allow for high rates of deletion, which would preferentially remove these dangerous genetic elements (by virtue of their terminal repeats as well as other factors). The detrimental effects of such a high rate of deletion—that is, the occasional removal of native chromosomal genes

with important functions—would be offset by the benefits of removing genetic parasites. As predicted by this model, genomes sheltered from genetic parasites have larger numbers of pseudogenes (94), which may reflect a lower overall rate of chromosomal deletion. For example, the aphid endosymbiont *Buchnera aphidicola* has harbored pseudogenes for 50 million years (196), long after the time they would have been deleted by a free-living organism.

GENOMES IN MOTION

Aside from changes in gene content, gene order has also been found to be more plastic than once assumed. Mechanisms for DNA rearrangement are well known and have been well measured in the laboratory (55, 59, 161, 171–175). Yet despite the opportunities for chromosomal rearrangement, the genetic maps of *E. coli* and *S. enterica* seemed to be largely congruent (Fig. 1), save the inversion about the terminus of replication. Genome sequences have led to three conclusions regarding genome rearrangements. First, inversions which are symmetric with respect to the origin and terminus of replication are frequent among many bacterial lineages (44). Second, inversion and "backflipping" (that is, a second inversion returning the chromosome to its original configuration) can lead to the appearance of stability when the chromosome is actually quite dynamic; this is especially evident when the backflipping events do not use the exact same sequences as were used for inversion, thereby leading to "ragged" ends in the regions of inversion (e.g., the ragged inversion about the terminus is evident upon comparison of the *E. coli*, *Salmonella*, and *Klebsiella* genetic maps [11]). While these ragged ends are often not detected on crude genetic maps, leading to an illusion of chromosome stability over time, these seeming translocations are easily detected upon comparison of complete genome sequences (44). Last, some lineages can accumulate large numbers of inversions that are otherwise counterselected in closely related organisms (107, 108, 139), meaning that the selection against chromosome rearrangements is not uniform among organisms. In these cases, the inversions may be insufficiently detrimental to be counterselected, perhaps due to the small information content afforded by the population (89) (see below).

Since sequences that would allow for high-frequency inversion (e.g., rRNA and tRNA loci, IS elements [98, 133, 169], repetitive extragenic palindromic sequences [35], enterobacterial repetitive intergenic consensus sequences [67], bacterial interspersed mosaic elements [10], and other repeat sequences) are found throughout the chromosome, what prevents most inversions except those that are more or less symmetrical with respect to the origin and/or terminus of replication? This bias toward some inversions' being difficult to observe in nature has been verified in the laboratory, where they are difficult to select (173, 175). One explanation may be that some sequences act to stall replication forks after they pass the terminus of replication (22, 154), defined as the position of the *dif* sequence, the site of action for the XerCD recombinase (65); some of these sequences (e.g., *ter*) bind to specific proteins (Tus) that may mediate this interaction (5, 32, 138). As seen in Fig. 5 for *H. influenzae*, there are other oligonucleotide sequences which cluster significantly about the terminus of replication (see also reference 22) and may play a role in the efficient termination of replication or segregation of chromosomes; similarly distributed sequences are evident in all other genomes analyzed (H. Hendrickson and J. G. Lawrence, unpublished data). While the appropriate placement and orientation of these sequences would provide a selective detriment to chromosomal rearrangements in many bacteria, populations with low information content (as defined below) may not be able to counterselect strains which have undergone genome rearrangement (107, 108, 139).

GENOMES UP AGAINST THE WALL

What constrains genomes in terms of their gene content and gene arrangement? After all, if genomes can acquire selectively beneficial functions by horizontal transfer, then what has prevented organisms from acquiring numerous genes that would allow for an impressive repertoire of physiological capabilities and an equally impressive range of environments in which they could survive? Examination of genome sequences reveals fundamental fallibilities in all organisms, either limitations in what compounds can serve as sources of carbon or energy (not all organisms are autotrophs), or which nutrients they require (not all organisms are prototrophs). Each organism appears to have a more or less plausible set of genes that would be required to live in the environmental space in which it is found.

The reason behind this pattern of gene distribution among organisms is relatively simple: populations of organisms can maintain only a finite amount of information under selection at the same time (101, 102). For a physiological function to be maintained, the genes encoding the necessary proteins must not suffer from the accumulation of deleterious mutations, or be deleted altogether. Yet mutations and deletions occur in bacterial chromosomes

during every cell division. Since, to a first approximation, bacterial populations maintain a constant size, only the fittest organisms (those with the capabilities best suited to exploit an environment) will survive. Mutations that eliminate important functions will normally be eliminated from a population as organisms bearing those mutations are outcompeted by organisms bearing mutations that cause less detrimental effects. Organisms bearing highly deleterious mutations can rise to high frequency only in populations of small size through the action of genetic drift; populations large in numbers can stave off the effects of genetic drift and maintain more information (meaning more genes and more capabilities). In addition, homologous recombination, seen above as reassorting alleles among conspecific individuals, can defeat the action of genetic drift by recreating mutation-free cells by combining sets of mutation-free genes from two differentially impaired parents, thus defeating the crippling effects of Muller's ratchet (48, 129). Therefore, the information that can be maintained in a bacterial genome is finite, and is a function of population size and rate of recombination.

As a result, microbial genomes have a maximal information content given a population size and rate of recombination (89, 101). If information is gained by horizontal transfer, it can be maintained only at the expense of losing information that was once under selection. That is, if mutations in the newly acquired DNA are to be counterselected (so that individuals without mutations in these genes contribute to the next generation), mutations will necessarily accumulate in other genes in the genome. In this way, the limitations of genome information content lead to cycles of gene loss and gene acquisition as organisms constantly refine their gene content to best exploit an environment. Since organisms vary in their population sizes and rates of recombination, we would expect them to vary in the total amount of information they can carry. If a population cannot maintain a large amount of information, it is unlikely that genes acquired by horizontal transfer will confer sufficiently beneficial functions that they will be maintained. In contrast, if a population can maintain a very large amount of information, acquired genes are more likely to be retained, since they need only confer a modest benefit to do so.

Indeed, when genomic information content is quantitated (89), a robust relationship between overall information content and the amount of recently acquired DNA is detected. In Fig. 6, low-value information, quantitated as the corrected values of the codon usage bias measurement χ^2/L, serves as a surrogate measure for the threshold of an effectively neutral mutation. Aberrant points in Fig. 6, shown as open circles, represent the genomes of *Synechocystis* strain PCC6803 (72) and *Mycoplasma pneumoniae* (64). In these cases, we postulate that the genomes contain an excess of newly acquired DNA, either as unidentified genetic parasites or as acquired DNA tolerated due to a recent expansion in population size or recombination rate, thereby allowing for the maintenance of a greater genomic information content. Consistent with the latter idea, the disparity of these points from the overall relationship depicted in Fig. 6 vanishes if very recently acquired DNA—aberrant DNA that otherwise conforms completely to the Muto and Osawa relationships (131), therefore showing no signs of amelioration (95)—is not included in the analysis.

These data support a model whereby bacterial genomes constantly acquire new genes by horizontal transfer, and lose preexisting genes under the pressure of the maximal information content they can maintain. As a result, gene content will be in constant flux, as new genes arrive and other genes—those which presumably make the lowest overall contribution to organismal fitness—are lost. Genes which provide for strongly selected functions would rarely be deleted, since only replacement by a gene performing a similar function would allow their loss.

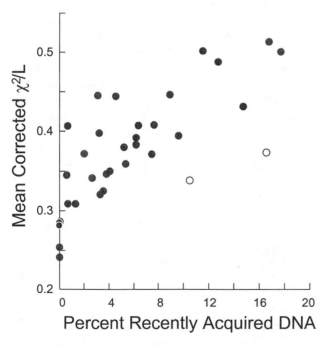

Figure 6. Relationship between the amount of recently acquired DNA and information content in bacterial genomes. Genomes used for analysis are shown in Fig. 3; information content is measured as corrected, length-normalized average χ^2 of codon usage as described elsewhere (89).

For this reason, linkage maps of closely related organisms that depict rather important genes, such as those shown in Fig. 1, will be quite congruent. The genes which have been gained and lost by horizontal transfer will not be shared among these chromosomes (e.g., the *E. coli lac* operon), and therefore will not disrupt the appearance of chromosome stability implied by the large numbers of shared loci. However, if a bacterial "species" occupies several distinct ecological niches, each best exploited by a different set of genes, one may expect that the gene content of subpopulations adapted to these different environments would be different. Here, the genome of the "species" may include many genes that are stably maintained in only a subset of strains, and a true "genome" sequence would include the cumulative gene content of many varied isolates.

GENOMES IN THE DRIVER'S SEAT

If microbial genomes are so fluid, what can microbiologists hope to learn from individual genome sequences? While it is true that each genome sequence provides only a snapshot of the tumultuous sea of genetic material flowing between isolated bits of cytoplasm, the pattern of gene loss and acquisition detailed above paints an intriguing picture of how organisms adapt to their environment. That is, organisms appear to make major adaptations by gaining new functions via lateral gene transfer; often, all of the genes required for a selectable function are found clustered together in an operon, facilitating this process (91, 92, 102, 103). Additional adaptations may entail active gene loss or the alteration of gene expression by point mutation. Traditionally, a microbiologist would view these series of changes as allowing an organism to adapt to a particular environment or ecological niche.

This interpretation is reinforced by laboratory experiments wherein bacteria grown for thousands of generations under constant selective conditions adapt to those conditions, becoming fitter than their parents (151). Yet this situation, while quite instructive in uncovering the tempo and mode of evolution as organisms adapt to an environment, may not accurately reflect the process of organismal adaptation in natural environments. First, it is clear that even the most rigorously controlled environment can be partitioned by its residents into multiple environments, allowing for the persistence of different variants arising from the same parental stock (159, 200). In these cases, multiple genotypically distinct lineages arose and were maintained in a "constant" environment, either a glucose-limited chemostat or a standing broth culture.

Second, if variant lineages readily colonize available niches in laboratory settings, even in seemingly homogeneous environments, one may ask how organisms effectively invade novel niches in natural environments where potential competitors must be overcome. Here, one can think about the model of a bacterial genome in flux in a different way. Rather than envisioning a cell which betters its performance in a target niche by the refinement of preexisting capabilities (either by point mutation or by the acquisition of novel genes by lateral transfer), we view the niche as the environmental space best exploited by a cell given its unique complement of genes. By this model, selection does not act to hone a cell's capabilities so that it becomes increasingly more adept at exploiting one particular niche, as is the case in laboratory settings; here new genes would be retained only if they contributed toward fitness in a predefined set of environmental conditions. On the contrary, we may postulate that new genes will be retained if they allow for an increase in organismal fitness regardless of which niche is being occupied. That is, there is active interplay between the organism, its genome, and the ecological domain it can best exploit. The genome, with all its dynamic parts, steers the organism into the environmental space it is best suited to exploit. Rather than a stale collection of genes having reached optimal performance after billions of years of evolution, one may view a bacterial genome as an ever-changing consortium of genes which cooperate in perpetuating their host organism.

Acknowledgments. This work was supported by a grant from the David and Lucile Packard Foundation.

REFERENCES

1. Achtman, M., T. Azuma, D. E. Berg, Y. Ito, G. Morelli, Z. J. Pan, S. Suerbaum, S. A. Thompson, A. van der Ende, and L. J. van Doorn. 1999. Recombination and clonal groupings within *Helicobacter pylori* from different geographical regions. *Mol. Microbiol.* 32:459–470.

2. Achtman, M., M. Heuzenroeder, B. Kusecek, H. Ochman, D. Caugant, R. K. Selander, V. Vaisanen-Rhen, T. K. Korhonen, S. Stuart, F. Orskov, and I. Orskov. 1986. Clonal analysis of *Escherichia coli* O2:K1 isolated from diseased humans and animals. *Infect. Immun.* 51:268–276.

3. Achtman, M., A. Mercer, B. Kusecek, A. Pohl, M. Heuzenroeder, W. Aaronson, and R. Silver. 1983. Six widespread bacterial clones among *E. coli* K1 isolates. *Infect. Immun.* 39:315–335.

4. Allison, G. E., D. Angeles, N. Tran-Dinh, and N. K. Verma. 2002. Complete genomic sequence of SfV, a serotype-converting temperate bacteriophage of *Shigella flexneri*. *J. Bacteriol.* 184:1974–1987.

5. Andersen, P. A., A. A. Griffiths, I. G. Duggin, and R. G. Wake. 2000. Functional specificity of the replication fork-arrest complexes of *Bacillus subtilis* and *Escherichia coli*:

significant specificity for Tus-Ter functioning in *E. coli. Mol. Microbiol.* 36:1327–1335.

6. Andersson, J. O. 2000. Evolutionary genomics: is *Buchnera* a bacterium or an organelle? *Curr. Biol.* 10:R866–R868.

7. Andersson, J. O., and S. G. Andersson. 1999. Genome degradation is an ongoing process in *Rickettsia. Mol. Biol. Evol.* 16:1178–1191.

8. Andersson, J. O., and S. G. Andersson. 1999. Insights into the evolutionary process of genome degradation. *Curr. Opin. Genet. Dev.* 9:664–671.

9. Andersson, S. G., A. Zomorodipour, J. O. Andersson, T. Sicheritz-Ponten, U. C. Alsmark, R. M. Podowski, A. K. Naslund, A. S. Eriksson, H. H. Winkler, and C. G. Kurland. 1998. The genome sequence of *Rickettsia prowazekii* and the origin of mitochondria. *Nature* 396:133–140.

10. Bachellier, S., J. M. Clement, M. Hofnung, and E. Gilson. 1997. Bacterial interspersed mosaic elements (BIMEs) are a major source of sequence polymorphism in *Escherichia coli* intergenic regions including specific associations with a new insertion sequence. *Genetics* 145:551–562.

11. Bender, R. A. 1996. Variations on a theme by *Escherichia*, p. 4–9. *In* F. C. Neidhardt, R. Curtiss III, J. L. Ingraham, E. C. C. Lin, K. B. Low, B. Magasanik, W. S. Reznikoff, M. Riley, M. Schaechter, and H. E. Umbarger (ed.), Escherichia coli *and* Salmonella: *Cellular and Molecular Biology*, 2nd ed., vol. 1. ASM Press, Washington, D.C.

12. Blattner, F. R., G. R. Plunkett, C. A. Bloch, N. T. Perna, V. Burland, M. Riley, J. Collado-Vides, J. D. Glasner, C. K. Rode, G. F. Mayhew, J. Gregor, N. W. Davis, H. A. Kirkpatrick, M. A. Goeden, D. J. Rose, B. Mau, and Y. Shao. 1997. The complete genome sequence of *Escherichia coli* K-12. *Science* 277:1453–1474.

13. Boucher, Y., H. Huber, S. L'Haridon, K. O. Stetter, and W. F. Doolittle. 2001. Bacterial origin for the isoprenoid biosynthesis enzyme HMG-CoA reductase of the archaeal orders thermoplasmatales and archaeoglobales. *Mol. Biol. Evol.* 18:1378–1388.

14. Boyd, E. F., K. Nelson, F. S. Wang, T. S. Whittam, and R. K. Selander. 1994. Molecular genetic basis of allelic polymorphism in malate dehydrogenase (*mdh*) in natural populations of *Escherichia coli* and *Salmonella enterica. Proc. Natl. Acad. Sci. USA* 91:1280–1284.

15. Brenner, D. J. 1978. Characterization and clinical identification of *Enterobacteriaceae* by DNA hybridization. *Prog. Clin. Pathol.* 7:71–117.

16. Brenner, D. J., and D. B. Cowie. 1968. Thermal stability of *Escherichia coli-Salmonella typhimurium* deoxyribonucleic acid duplexes. *J. Bacteriol.* 95:2258–2262.

17. Brenner, D. J., and S. Falkow. 1971. Molecular relationships among members of the Enterobacteriaceae. *Adv. Genet.* 16:81–118.

18. Brenner, D. J., G. R. Fanning, K. E. Johnson, R. V. Citarella, and S. Falkow. 1969. Polynucleotide sequence relationships among members of the *Enterobacteriaceae. J. Bacteriol.* 98:637–650.

19. Brenner, D. J., G. R. Fanning, F. J. Skerman, and S. Falkow. 1972. Polynucleotide sequence divergence among strains of *Escherichia coli* and closely related organisms. *J. Bacteriol.* 109:953–965.

20. Brenner, D. J., M. A. Martin, and B. H. Hoyer. 1967. Deoxyribonucleic acid homologies among some bacteria. *J. Bacteriol.* 94:486–487.

21. Brochier, C., H. Philippe, and D. Moreira. 2000. The evolutionary history of ribosomal protein RpS14: horizontal gene transfer at the heart of the ribosome. *Trends Genet.* 16:529–533.

22. Capiaux, H., F. Cornet, J. Corre, M. Guijo, K. Perals, J. E. Rebollo, and J. Louarn. 2001. Polarization of the *Escherichia coli* chromosome. A view from the terminus. *Biochimie* 83:161–170.

23. Reference deleted.

24. Caugant, D. A., B. R. Levin, and R. K. Selander. 1984. Distribution of multilocus genotypes of *Escherichia coli* within and between host families. *J. Hyg. Camb.* 92:377–384.

25. Caugant, D. A., B. R. Levin, and R. K. Selander. 1981. Genetic diversity and temporal variation in the *E. coli* population of a human host. *Genetics* 98:467–490.

26. Chang, H. W., and D. A. Julin. 2001. Structure and function of the *Escherichia coli* RecE protein, a member of the RecB nuclease domain family. *J. Biol. Chem.* 276:46004–46010.

27. Chistoserdova, L., J. A. Vorholt, R. K. Thauer, and M. E. Lidstrom. 1998. C1 transfer enzymes and coenzymes linking methylotrophic bacteria and methanogenic Archaea. *Science* 281:99–102.

28. Clark, M. A., L. Baumann, M. L. Thao, N. A. Moran, and P. Baumann. 2001. Degenerative minimalism in the genome of a psyllid endosymbiont. *J. Bacteriol.* 183:1853–1861.

29. Clarke, G. D., R. G. Beiko, M. A. Ragan, and R. L. Charlebois. 2002. Inferring genome trees by using a filter to eliminate phylogenetically discordant sequences and a distance matrix based on mean normalized BLASTP scores. *J. Bacteriol.* 184:2072–2080.

30. Cohan, F. M. 2001. Bacterial species and speciation. *Syst. Biol.* 50:513–524.

31. Cole, S. T., K. Eiglmeier, J. Parkhill, K. D. James, N. R. Thomson, P. R. Wheeler, N. Honore, T. Garnier, C. Churcher, D. Harris, K. Mungall, D. Basham, D. Brown, T. Chillingworth, R. Connor, R. M. Davies, K. Devlin, S. Duthoy, T. Feltwell, A. Fraser, N. Hamlin, S. Holroyd, T. Hornsby, K. Jagels, C. Lacroix, J. Maclean, S. Moule, L. Murphy, K. Oliver, M. A. Quail, M. A. Rajandream, K. M. Rutherford, S. Rutter, K. Seeger, S. Simon, M. Simmonds, J. Skelton, R. Squares, S. Squares, K. Stevens, K. Taylor, S. Whitehead, J. R. Woodward, and B. G. Barrell. 2001. Massive gene decay in the leprosy bacillus. *Nature* 409:1007–1011.

32. Coskun-Ari, F. F., and T. M. Hill. 1997. Sequence-specific interactions in the Tus-Ter complex and the effect of base pair substitutions on arrest of DNA replication in *Escherichia coli*. *J. Biol. Chem.* 272:26448–26456.

33. Dandekar, T., B. Snel, M. Huynen, and P. Bork. 1998. Conservation of gene order: a fingerprint of proteins that physically interact. *Trends Biochem. Sci.* 23:324–328.

34. Davies, J. 1996. Origins and evolution of antibiotic resistance. *Microbiologia* 12:9–16.

35. Dimri, G. P., K. E. Rudd, M. K. Morgan, H. Bayat, and G. F. Ames. 1992. Physical mapping of repetitive extragenic palindromic sequences in *Escherichia coli* and phylogenetic distribution among *Escherichia coli* strains and other enteric bacteria. *J. Bacteriol.* 174:4583–4593.

36. Doolittle, R. F., D. F. Feng, K. L. Anderson, and M. R. Alberro. 1990. A naturally occurring horizontal gene transfer from a eukaryote to a prokaryote. *J. Mol. Evol.* 31:383–388.

37. Doolittle, W. F. 1999. Lateral genomics. *Trends Cell Biol.* 9:M5–M8.

38. Doolittle, W. F. 2000. The nature of the universal ancestor and the evolution of the proteome. *Curr. Opin. Struct. Biol.* 10:355–358.

39. Doolittle, W. F. 1999. Phylogenetic classification and the universal tree. *Science* 284:2124–2129.

40. DuBose, R. F., D. E. Dykhuizen, and D. L. Hartl. 1988. Genetic exchange among natural isolates of bacteria: recombination within the *phoA* gene of *Escherichia coli*. *Proc. Natl. Acad. Sci. USA* 85:7036–7040.

41. DuBose, R. F., and D. L. Hartl. 1990. The molecular evolution of alkaline phosphatase: correlating variation among enteric bacteria to experimental manipulations of the protein. *Mol. Biol. Evol.* 7:547–577.

42. Dykhuizen, D. E., and L. Green. 1991. Recombination in *Escherichia coli* and the definition of biological species. *J. Bacteriol.* 173:7257–7268.

43. Edwards, R. A., G. J. Olsen, and S. R. Maloy. 2002. Comparative genomics of closely related Salmonellae. *Trends Microbiol.* 10:94–99.

44. Eisen, J. A., J. F. Heidelberg, O. White, and S. L. Salzberg. 2000. Evidence for symmetric chromosomal inversions around the replication origin in bacteria. *Genome Biol.* 1:1–11.

45. Falush, D., C. Kraft, N. S. Taylor, P. Correa, J. G. Fox, M. Achtman, and S. Suerbaum. 2001. Recombination and mutation during long-term gastric colonization by *Helicobacter pylori*: estimates of clock rates, recombination size, and minimal age. *Proc. Natl. Acad. Sci. USA* 98:15056–15061.

46. Feil, E. J., E. C. Holmes, D. E. Bessen, M. S. Chan, N. P. Day, M. C. Enright, R. Goldstein, D. W. Hood, A. Kalia, C. E. Moore, J. Zhou, and B. G. Spratt. 2001. Recombination within natural populations of pathogenic bacteria: short-term empirical estimates and long-term phylogenetic consequences. *Proc. Natl. Acad. Sci. USA* 98:182–187.

47. Feil, E. J., J. M. Smith, M. C. Enright, and B. G. Spratt. 2000. Estimating recombinational parameters in *Streptococcus pneumoniae* from multilocus sequence typing data. *Genetics* 154:1439–1450.

48. Felsenstein, J. 1974. The evolutionary advantage of recombination. *Genetics* 78:737–756.

49. Fleischmann, R. D., M. D. Adams, O. White, R. A. Clayton, E. F. Kirkness, A. R. Kerlavage, C. J. Bult, J. F. Tomb, B. A. Dougherty, J. M. Merrick, et al. 1995. Whole-genome random sequencing and assembly of *Haemophilus influenzae* Rd. *Science* 269:496–512.

50. Francino, M. P., L. Chao, M. A. Riley, and H. Ochman. 1996. Asymmetries generated by transcription-coupled repair in enterobacterial genes. *Science* 272:107–109.

51. Francino, M. P., and H. Ochman. 1999. A comparative genomics approach to DNA asymmetry. *Ann. N. Y. Acad. Sci.* 870:428–431.

52. Francino, M. P., and H. Ochman. 2001. Deamination as the basis of strand-asymmetric evolution in transcribed Escherichia coli sequences. *Mol. Biol. Evol.* 18:1147–1150.

53. Francino, M. P., and H. Ochman. 1997. Strand asymmetries in DNA evolution. *Trends Genet.* 13:240–245.

54. Francino, M. P., and H. Ochman. 2000. Strand symmetry around the beta-globin origin of replication in primates. *Mol. Biol. Evol.* 17:416–422.

55. Galitski, T., and J. R. Roth. 1997. Pathways for homologous recombination between chromosomal direct repeats in *Salmonella typhimurium*. *Genetics* 146:751–767.

56. Gruss, A., V. Moretto, S. D. Ehrlich, P. Duwat, and P. Dabert. 1991. GC-rich DNA sequences block homologous recombination. *J. Biol. Chem.* 266:6667–6669.

57. Guttman, D. S., and D. E. Dykhuizen. 1994. Clonal divergence in *Escherichia coli* as a result of recombination, not mutation. *Science* 266:1380–1383.

58. Guttman, D. S., and D. E. Dykhuizen. 1994. Detecting selective sweeps in naturally occurring *Escherichia coli*. *Genetics* 138:993–1003.

59. Haack, K. R., and J. R. Roth. 1995. Recombination between chromosomal IS200 elements supports frequent duplication formation in *Salmonella typhimurium*. *Genetics* 141:1245–1252.

60. Hall, R. M., and C. M. Collis. 1995. Mobile gene cassettes and integrons: capture and spread of genes by site-specific recombination. *Mol. Microbiol.* 15:593–600.

61. Hall, R. M., C. M. Collis, M. J. Kim, S. R. Partridge, G. D. Recchia, and H. W. Stokes. 1999. Mobile gene cassettes and integrons in evolution. *Ann. N. Y. Acad. Sci.* 870:68–80.

62. Hartl, D. L., and D. E. Dykhuizen. 1984. The population genetics of *Escherichia coli*. *Annu. Rev. Genet.* 18:31–68.

63. Hayes, W. S., and M. Borodovsky. 1998. How to interpret an anonymous bacterial genome: machine learning approach to gene identification. *Genome Res.* 8:1154–1171.

64. Himmelreich, R., H. Hilbert, H. Plagens, E. Pirkl, B. C. Li, and R. Herrmann. 1996. Complete sequence analysis of the genome of the bacterium *Mycoplasma pneumoniae*. *Nucleic Acids Res.* 24:4420–4449.

65. Hiraga, S. 1993. Chromosome partition in *Escherichia coli*. *Curr. Opin. Genet. Dev.* 3:789–801.

66. Holmes, E. C., R. Urwin, and M. C. J. Maiden. 1999. The influence of recombination on the population structure and evolution of the human pathogen *Neisseria meningitidis*. *Mol. Biol. Evol.* 16:741–749.

67. Hulton, C. S., C. F. Higgins, and P. M. Sharp. 1991. ERIC sequences: a novel family of repetitive elements in the genomes of *Escherichia coli*, *Salmonella typhimurium* and other enterobacteria. *Mol. Microbiol.* 5:825–834.

68. Ibba, M., S. Morgan, A. W. Curnow, D. R. Pridmore, U. C. Vothknecht, W. Gardner, W. Lin, C. R. Woese, and D. Soll. 1997. A euryarchaeal lysyl-tRNA synthetase: resemblance to class I synthetases. *Science* 278:1119–1122.

69. Itoh, T., K. Takemoto, H. Mori, and T. Gojobori. 1999. Evolutionary instability of operon structures disclosed by sequence comparisons of complete microbial genomes. *Mol. Biol. Evol.* 16:332–346.

70. Jain, R., M. C. Rivera, and J. A. Lake. 1999. Horizontal gene transfer among genomes: the complexity hypothesis. *Proc. Natl. Acad. Sci. USA* 96:3801–3806.

71. Jiang, W., W. W. Metcalf, K. S. Lee, and B. L. Wanner. 1995. Molecular cloning, mapping, and regulation of Pho regulon genes for phosphonate breakdown by the phosphonatase pathway of *Salmonella typhimurium* LT2. *J. Bacteriol.* 177:6411–6421.

72. Kaneko, T., S. Sato, H. Kotani, A. Tanaka, E. Asamizu, Y. Nakamura, N. Miyajima, M. Hirosawa, M. Sugiura, S. Sasamoto, T. Kimura, T. Hosouchi, A. Matsuno, A. Muraki, N. Nakazaki, K. Naruo, S. Okumura, S. Shimpo, C. Takeuchi, T. Wada, A. Watanabe, M. Yamada, M. Yasuda, and S. Tabata. 1996. Sequence analysis of the genome of the unicellular cyanobacterium *Synechocystis* sp. strain PCC6803. II. Sequence determination of the entire genome and assignment of potential protein-coding regions. *DNA Res.* 3:109–136.

73. Karlin, S. 1998. Global dinucleotide signatures and analysis of genomic heterogeneity. *Curr. Opin. Microbiol.* 1:598–610.

74. Karlin, S., and C. Burge. 1995. Dinucleotide relative abundance extremes: a genomic signature. *Trends Genet.* 11:283–290.

75. Karlin, S., A. M. Campbell, and J. Mrázek. 1998. Comparative DNA analysis across diverse genomes. *Annu. Rev. Genet.* 32:185–225.

76. Karlin, S., and J. Mrazek. 2000. Predicted highly expressed genes of diverse prokaryotic genomes. *J. Bacteriol.* 182:5238–5250.

77. Karlin, S., J. Mrazek, and A. M. Campbell. 1998. Codon usages in different gene classes of the *Escherichia coli* genome. *Mol. Microbiol.* 29:1341–1355.

78. Katz, L. A. 1996. Transkingdom transfer of the phosphoglucose isomerase gene. *J. Mol. Evol.* 43:453–459.

79. Ke, D., M. Boissinot, A. Huletsky, F. J. Picard, J. Frenette, M. Ouellette, P. H. Roy, and M. G. Bergeron. 2000. Evidence for horizontal gene transfer in evolution of elongation factor Tu in enterococci. *J. Bacteriol.* 182:6913–6920.

80. Kimura, M. 1983. *The Neutral Theory of Molecular Evolution*. Cambridge University Press, Cambridge, United Kingdom.

81. Koski, L. B., R. A. Morton, and G. B. Golding. 2001. Codon bias and base composition are poor indicators of horizontally transferred genes. *Mol. Biol. Evol.* 18:404–412.

82. Kowalczykowski, S. C. 2000. Initiation of genetic recombination and recombination-dependent replication. *Trends Biochem. Sci.* 25:156–165.

83. Kowalczykowski, S. C., D. A. Dixon, A. K. Eggleston, S. D. Lauder, and W. M. Rehrauer. 1994. Biochemistry of homologous recombination in *Escherichia coli*. *Microbiol. Rev.* 58:401–465.

84. Kranz, R. G., and B. S. Goldman. 1998. Evolution and horizontal transfer of an entire biosynthetic pathway for cytochrome *c* biogenesis: *Helicobacter*, *Deinococcus*, Archae and more. *Mol. Microbiol.* 27:871–874.

85. Kroll, J. S., K. E. Wilks, J. L. Farrant, and P. R. Langford. 1998. Natural genetic exchange between *Haemophilus* and *Neisseria*: intergeneric transfer of chromosomal genes between major human pathogens. *Proc. Natl. Acad. Sci. USA* 95:12381–12385.

86. Kusano, K., N. K. Takahashi, H. Yoshikura, and I. Kobayashi. 1994. Involvement of RecE exonuclease and RecT annealing protein in DNA double-strand break repair by homologous recombination. *Gene* 138:17–25.

87. Kuzminov, A. 1995. Collapse and repair of replication forks in *Escherichia coli*. *Mol. Microbiol.* 16:373–384.

88. Kuzminov, A. 1999. Recombinational repair of DNA damage in *Escherichia coli* and bacteriophage lambda. *Microbiol. Mol. Biol. Rev.* 63:751–813.

89. Lawrence, J. G. 2001. Catalyzing bacterial speciation: correlating lateral transfer with genetic headroom. *Syst. Biol.* 50:479–496.

90. Lawrence, J. G. 2002. Gene transfer in bacteria: speciation without species? *Theor. Popul. Biol.* 61:449–460.

91. Lawrence, J. G. 1997. Selfish operons and speciation by gene transfer. *Trends Microbiol.* 5:355–359.

92. Lawrence, J. G. 1999. Selfish operons: the evolutionary impact of gene clustering in the prokaryotes and eukaryotes. *Curr. Opin. Genet. Dev.* 9:642–648.

93. Lawrence, J. G., D. E. Dykhuizen, R. F. DuBose, and D. L. Hartl. 1989. Phylogenetic analysis using insertion sequence fingerprinting in *Escherichia coli*. *Mol. Biol. Evol.* 6:1–14.

94. Lawrence, J. G., R. W. Hendrix, and S. Casjens. 2001. Where are the pseudogenes in bacterial genomes? *Trends Microbiol.* 9:535–540.

95. Lawrence, J. G., and H. Ochman. 1997. Amelioration of bacterial genomes: rates of change and exchange. *J. Mol. Evol.* 44:383–397.

96. Lawrence, J. G., and H. Ochman. 1998. Molecular archaeology of the *Escherichia coli* genome. *Proc. Natl. Acad. Sci. USA* 95:9413–9417.

97. Lawrence, J. G., and H. Ochman. 2002. Reconciling the many faces of gene transfer. *Trends Microbiol.* 10:1–4.

98. Lawrence, J. G., H. Ochman, and D. L. Hartl. 1992. The evolution of insertion sequences within enteric bacteria. *Genetics* 131:9–20.

99. Lawrence, J. G., and J. R. Roth. 1995. The cobalamin (coenzyme B_{12}) biosynthetic genes of *Escherichia coli*. *J. Bacteriol.* 177:6371–6380.

100. Lawrence, J. G., and J. R. Roth. 1996. Evolution of coenzyme B_{12} synthesis among enteric bacteria: evidence for loss and reacquisition of a multigene complex. *Genetics* 142:11–24.

101. Lawrence, J. G., and J. R. Roth. 1999. Genomic flux: genome evolution by gene loss and acquisition, p. 263–289. *In* R. L. Charlebois (ed.), *Organization of the Prokaryotic Genome*. ASM Press, Washington, D.C.

102. Lawrence, J. G., and J. R. Roth. 1998. Roles of horizontal transfer in bacterial evolution, p. 208–225. *In* M. Syvanen and C. I. Kado (ed.), *Horizontal Transfer*. Chapman and Hall, London, England.

103. Lawrence, J. G., and J. R. Roth. 1996. Selfish operons: horizontal transfer may drive the evolution of gene clusters. *Genetics* 143:1843–1860.

104. Lederberg, J. 1947. Gene recombination and linked segregations in *Escherichia coli*. *Genetics* 32:505–525.

105. Lederberg, J., and E. L. Tatum. 1946. Gene recombination in *Escherichia coli*. *Nature* 158:558.

106. Levin, B. 1981. Periodic selection, infectious gene exchange, and the genetic structure of *E. coli* populations. *Genetics* 99:1–23.

107. Liu, S. L., and K. E. Sanderson. 1996. Highly plastic chromosomal organization in *Salmonella typhi*. *Proc. Natl. Acad. Sci. USA* 93:10303–10308.

108. Liu, S. L., and K. E. Sanderson. 1995. Rearrangements in the genome of the bacterium *Salmonella typhi*. *Proc. Natl. Acad. Sci. USA* 92:1018–1022.

109. Logsdon, J. M., and D. M. Fuguy. 1999. *Thermotoga* heats up lateral gene transfer. *Curr. Biol.* 9:R747–R751.

110. Losick, R., and L. Shapiro. 1999. Changing views on the nature of the bacterial cell: from biochemistry to cytology. *J. Bacteriol.* 181:4143–4145.

111. Majewski, J., and F. M. Cohan. 1999. DNA sequence similarity requirements for interspecific recombination in *Bacillus*. *Genetics* 153:1525–1533.

112. Majewski, J., and F. M. Cohan. 1998. The effect of mismatch repair and heteroduplex formation on sexual isolation in *Bacillus*. *Genetics* 148:13–18.

113. Majewski, J., P. Zawadzki, P. Pickerill, F. M. Cohan, and C. G. Dowson. 2000. Barriers to genetic exchange between bacterial species: *Streptococcus pneumoniae* transformation. *J. Bacteriol.* 182:1016–1023.

114. Makarova, K. S., L. Aravind, M. Y. Galperin, N. V. Grishin, R. L. Tatusov, Y. I. Wolf, and E. V. Koonin. 1999. Comparative genomics of the Archaea (Euryarchaeota): evolution of conserved protein families, the stable core, and the variable shell. *Genome Res.* 9:608–628.

115. Martin, W. 1999. Mosaic bacterial chromosomes: a challenge en route to a tree of genomes. *Bioessays* 21:99–104.

116. Maurelli, A. T. 1994. Virulence protein export systems in *Salmonella* and *Shigella*: a new family or lost relatives. *Trends Cell Biol.* 4:240–242.

117. Maynard Smith, J., N. H. Smith, M. O'Rourke, and B. G. Spratt. 1993. How clonal are bacteria? *Proc. Natl. Acad. Sci. USA* 90:4384–4388.

118. McClelland, M., K. E. Sanderson, J. Spieth, S. W. Clifton, P. Latreille, L. Courtney, S. Porwollik, J. Ali, M. Dante, F. Du, S. Hou, D. Layman, S. Leonard, C. Nguyen, K. Scott, A. Holmes, N. Grewal, E. Mulvaney, E. Ryan, H. Sun, L. Florea, W. Miller, T. Stoneking, M. Nhan, R. Waterston,

and R. K. Wilson. 2001. Complete genome sequence of *Salmonella enterica* serovar Typhimurium LT2. *Nature* 413:852–856.

119. McKane, M., and R. Milkman. 1995. Transduction, restriction and recombination patterns in *Escherichia coli*. *Genetics* 139:35–43.

120. Metcalf, W. W., and B. L. Wanner. 1993. Evidence for a fourteen-gene, *phnC* to *phnP* locus for phosphonate metabolism in *Escherichia coli*. *Gene* 129:27–32.

121. Milkman, R. 1973. Electrophoretic variation in *Escherichia coli* from natural sources. *Science* 182:1024–1026.

122. Milkman, R., and M. M. Bridges. 1990. Molecular evolution of the *E. coli* chromosome. III. Clonal frames. *Genetics* 126:505–517.

123. Milkman, R., and I. P. Crawford. 1983. Clustered third-base substitutions among wild strains of *Escherichia coli*. *Science* 221:378–379.

124. Milkman, R., and A. Stoltzfus. 1988. Molecular evolution of the *Escherichia coli* chromosome. II. Clonal segments. *Genetics* 120:359–366.

125. Mira, A., H. Ochman, and N. A. Moran. 2001. Deletional bias and the evolution of bacterial genomes. *Trends Genet.* 17:589–596.

126. Moran, N. A. 2002. Microbial minimalism: genome reduction in bacterial pathogens. *Cell* 108:583–586.

127. Moran, N. A., and A. Mira. 2001. The process of genome shrinkage in the obligate symbiont *Buchnera aphidicola*. *Genome Biol.* 2(12):research0054.1–research0054.12. [Online.]

128. Moszer, I., E. P. Rocha, and A. Danchin. 1999. Codon usage and lateral gene transfer in *Bacillus subtilis*. *Curr. Opin. Microbiol.* 2:524–528.

129. Müller, H. 1932. Some genetic aspects of sex. *Am. Nat.* 66:118–138.

130. Musser, J. M., A. Amin, and S. Ramaswamy. 2000. Negligible genetic diversity of *Mycobacterium tuberculosis* host immune system protein targets: evidence of limited selective pressure. *Genetics* 155:7–16.

131. Muto, A., and S. Osawa. 1987. The guanine and cytosine content of genomic DNA and bacterial evolution. *Proc. Natl. Acad. Sci. USA* 84:166–169.

132. Mylvaganam, S., and P. P. Dennis. 1992. Sequence heterogeneity between the two genes encoding 16S rRNA from the halophilic archaebacterium *Haloarcula marismortui*. *Genetics* 130:399–410.

133. Naas, T., M. Blot, W. M. Fitch, and W. Arber. 1994. Insertion sequence-related genetic variation in resting *Escherichia coli* K-12. *Genetics* 136:721–730.

134. Nakata, N., T. Tobe, I. Fukuda, T. Suzuki, K. Komatsu, M. Yoshikawa, and C. Sasakawa. 1993. The absence of a surface protease, OmpT, determines the intercellular spreading ability of *Shigella*: the relationship between *ompT* and *kcpA* loci. *Mol. Microbiol.* 9:459–468.

135. Nelson, K., and R. K. Selander. 1992. Evolutionary genetics of the proline permease gene (*putP*) and the control region of the proline utilization operon in populations of *Salmonella* and *Escherichia coli*. *J. Bacteriol.* 174:6886–6895.

136. Nelson, K. E., R. A. Clayton, S. R. Gill, M. L. Gwinn, R. J. Dodson, D. H. Haft, E. K. Hickey, J. D. Peterson, W. C. Nelson, K. A. Ketchum, L. McDonald, T. R. Utterback, J. A. Malek, K. D. Linher, M. M. Garrett, A. M. Stewart, M. D. Cotton, M. S. Pratt, C. A. Phillips, D. Richardson, J. Heidelberg, G. G. Sutton, R. D. Fleischmann, J. A. Eisen, and C. M. Fraser. 1999. Evidence for lateral gene transfer between Archaea and bacteria from genome sequence of *Thermotoga maritima*. *Nature* 399:323–329.

137. Nesbo, C. L., S. L'Haridon, K. O. Stetter, and W. F. Doolittle. 2001. Phylogenetic analyses of two "archaeal" genes in *Thermotoga maritima* reveal multiple transfers between Archaea and Bacteria. *Mol. Biol. Evol.* **18**:362–375.

138. Neylon, C., S. E. Brown, A. V. Kralicek, C. S. Miles, C. A. Love, and N. E. Dixon. 2000. Interaction of the *Escherichia coli* replication terminator protein (Tus) with DNA: a model derived from DNA-binding studies of mutant proteins by surface plasmon resonance. *Biochemistry* **39**:11989–11999.

139. Ng, I., S.-L. Liu, and K. Sanderson. 1999. Role of genomic rearrangements in producing new ribotypes of *Salmonella typhi. J. Bacteriol.* **181**:3536–3541.

140. Ochman, H., and I. B. Jones. 2000. Evolutionary dynamics of full genome content in *Escherichia coli. EMBO J.* **19**: 6637–6643.

141. Ochman, H., and J. G. Lawrence. 1996. Phylogenetics and the amelioration of bacterial genomes, p. 2627–2637. *In* F. C. Neidhardt, R. Curtiss III, J. L. Ingraham, E. C. C. Lin, K. B. Low, B. Magasanik, W. S. Reznikoff, M. Riley, M. Schaechter, and H. E. Umbarger (ed.), Escherichia coli *and* Salmonella: *Cellular and Molecular Biology*, 2nd ed., vol. 2. ASM Press, Washington, D.C.

142. Ochman, H., J. G. Lawrence, and E. Groisman. 2000. Lateral gene transfer and the nature of bacterial innovation. *Nature* **405**:299–304.

143. Ochman, H., and N. A. Moran. 2001. Genes lost and genes found: evolution of bacterial pathogenesis and symbiosis. *Science* **292**:1096–1099.

144. Ochman, H., and R. K. Selander. 1984. Evidence for clonal population structure in *Escherichia coli. Proc. Natl. Acad. Sci. USA* **81**:198–201.

145. Ochman, H., T. S. Whittam, D. A. Caugant, and R. K. Selander. 1983. Enzyme polymorphism and genetic population structure in *Escherichia coli* and *Shigella. J. Gen. Microbiol.* **129**:2715–2726.

146. Ochman, H., and A. C. Wilson. 1988. Evolution in bacteria: evidence for a universal substitution rate in cellular genomes. *J. Mol. Evol.* **26**:74–86.

147. Ochman, H., and A. C. Wilson. 1987. Evolutionary history of enteric bacteria, p. 1649–1654. *In* F. C. Neidhardt, J. L. Ingraham, K. B. Low, B. Magasanik, M. Schaechter, and H. E. Umbarger (ed.), Escherichia coli *and* Salmonella typhimurium: *Cellular and Molecular Biology*. American Society for Microbiology, Washington, D.C.

148. Ohta, T. 1976. Role of very slightly deleterious mutations in molecular evolution and polymorphism. *Theor. Popul. Biol.* **10**:254–275.

149. Ohta, T. 1973. Slightly deleterious mutant substitutions in evolution. *Nature* **264**:96–98.

150. Olendzenski, L., L. Liu, O. Zhaxybayeva, R. Murphey, D. G. Shin, and J. P. Gogarten. 2000. Horizontal transfer of archaeal genes into the deinococcaceae: detection by molecular and computer-based approaches. *J. Mol. Evol.* **51**:587–599.

151. Papadopoulos, D., D. Schneider, J. Meier-Eiss, W. Arber, R. E. Lenski, and M. Blot. 1999. Genomic evolution during a 10,000-generation experiment with bacteria. *Proc. Natl. Acad. Sci. USA* **96**:3807–3812.

152. Parkhill, J., M. Achtman, K. D. James, S. D. Bentley, C. Churcher, S. R. Klee, G. Morelli, D. Basham, D. Brown, T. Chillingworth, R. M. Davies, P. Davis, K. Devlin, T. Feltwell, N. Hamlin, S. Holroyd, K. Jagels, S. Leather, S. Moule, K. Mungall, M. A. Quail, M. A. Rajandream, K. M. Rutherford, M. Simmonds, J. Skelton, S. Whitehead, B. G. Spratt, and B. G. Barrell. 2000. Complete DNA sequence of a serogroup A strain of *Neisseria meningitidis* Z2491. *Nature* **404**:502–506.

153. Parkhill, J., G. Dougan, K. D. James, N. R. Thomson, D. Pickard, J. Wain, C. Churcher, K. L. Mungall, S. D. Bentley, M. T. Holden, M. Sebaihia, S. Baker, D. Basham, K. Brooks, T. Chillingworth, P. Connerton, A. Cronin, P. Davis, R. M. Davies, L. Dowd, N. White, J. Farrar, T. Feltwell, N. Hamlin, A. Haque, T. T. Hien, S. Holroyd, K. Jagels, A. Krogh, T. S. Larsen, S. Leather, S. Moule, P. O'Gaora, C. Parry, M. Quail, K. Rutherford, M. Simmonds, J. Skelton, K. Stevens, S. Whitehead, and B. G. Barrell. 2001. Complete genome sequence of a multiple drug resistant *Salmonella enterica* serovar Typhi CT18. *Nature* **413**:848–852.

154. Perals, K., F. Cornet, Y. Merlet, I. Delon, and J. M. Louarn. 2000. Functional polarization of the *Escherichia coli* chromosome terminus: the *dif* site acts in chromosome dimer resolution only when located between long stretches of opposite polarity. *Mol. Microbiol.* **36**:33–43.

155. Perna, N. T., G. Plunkett, V. Burland, B. Mau, J. D. Glasner, D. J. Rose, G. F. Mayhew, P. S. Evans, J. Gregor, H. A. Kirkpatrick, G. Posfai, J. Hackett, S. Klink, A. Boutin, Y. Shao, L. Miller, E. J. Grotbeck, N. W. Davis, A. Lim, E. T. Dimalanta, K. D. Potamousis, J. Apodaca, T. S. Anantharaman, J. Lin, G. Yen, D. C. Schwartz, R. A. Welch, and F. R. Blattner. 2001. Genome sequence of enterohaemorrhagic *Escherichia coli* O157:H7. *Nature* **409**:529–533.

156. Pesole, G., C. Gissi, C. Lanave, and C. Saccone. 1995. Glutamine synthetase gene evolution in bacteria. *Mol. Biol. Evol.* **12**:189–197.

157. Ragan, M. A. 2001. Detection of lateral gene transfer among microbial genomes. *Curr. Opin. Genet. Dev.* **11**:620–626.

158. Ragan, M. A. 2001. On surrogate methods for detecting lateral gene transfer. *FEMS Microbiol. Lett.* **201**:187–191.

159. Rainey, P. B., and M. Travisano. 1998. Adaptive radiation in a heterogeneous environment. *Nature* **394**:69–72.

160. Rayssiguier, C., D. S. Thaler, and M. Radman. 1989. The barrier to recombination between *Escherichia coli* and *Salmonella typhimurium* is disrupted in mismatch-repair mutants. *Nature* **342**:396–401.

161. Roth, J., N. Benson, T. Galitski, K. Haack, J. G. Lawrence, and L. Miesel. 1996. Rearrangements of the bacterial chromosome: formation and applications, p. 2256–2276. *In* F. C. Neidhardt, R. Curtiss III, J. L. Ingraham, E. C. C. Lin, K. B. Low, B. Magasanik, W. S. Reznikoff, M. Riley, M. Schaechter, and H. E. Umbarger (ed.), Escherichia coli *and* Salmonella: *Cellular and Molecular Biology*, 2nd ed., vol. 2. ASM Press, Washington, D.C.

162. Salyers, A. A., N. B. Shoemaker, A. M. Stevens, and L. Y. Li. 1995. Conjugative transposons: an unusual and diverse set of integrated gene transfer elements. *Microbiol. Rev.* **59**:579–590.

163. Salzberg, S. L., A. J. Salzberg, A. R. Kerlavage, and J. F. Tomb. 1998. Skewed oligomers and origins of replication. *Gene* **217**:57–67.

164. Sanderson, K. E. 1970. Current linkage map of *Salmonella typhimurium. Bacteriol. Rev.* **34**:176–193.

165. Sanderson, K. E. 1967. Revised linkage map of *Salmonella typhimurium. Bacteriol. Rev.* **31**:354–372.

166. Sanderson, K. E., and M. Demerec. 1965. The linkage map of *Salmonella typhimurium. Genetics* **51**:897–913.

167. Sanderson, K. E., and C. A. Hall. 1970. F-prime factors of *Salmonella typhimurium* and an inversion between *S. typhimurium* and *Escherichia coli. Genetics* **64**:215–228.

168. Saunders, N. J., D. W. Hood, and E. R. Moxon. 1999. Bacterial evolution: bacteria play pass the gene. *Curr. Biol.* **11**:R180–R183.

169. Sawyer, S. A., D. E. Dykhuizen, R. F. DuBose, L. Green, T. Mutangadura-Mhlanga, D. F. Wolczyk, and D. L. Hartl.

1987. Distribution and abundance of insertion sequences among natural isolates of *Escherichia coli*. *Genetics* **115**: 51–63.

170. **Schicklmaier, P., E. Moser, T. Wieland, W. Rabsch, and H. Schmieger.** 1998. A comparative study on the frequency of prophages among natural isolates of *Salmonella* and *Escherichia coli* with emphasis on generalized transducers. *Antonie Leeuwenhoek* **73**:49–54.

171. **Schmid, M. B., and J. R. Roth.** 1983. Genetic methods for analysis and manipulation of inversion mutations in bacteria. *Genetics* **105**:517–537.

172. **Schmid, M. B., and J. R. Roth.** 1983. Selection and endpoint distribution of bacterial inversion mutations. *Genetics* **105**:539–557.

173. **Segall, A., M. J. Mahan, and J. R. Roth.** 1988. Rearrangement of the bacterial chromosome: forbidden inversions. *Science* **241**:1314–1318.

174. **Segall, A. M., and J. R. Roth.** 1994. Approaches to half-tetrad analysis in bacteria: recombination between repeated, inverse-order chromosomal sequences. *Genetics* **136**:27–39.

175. **Segall, A. M., and J. R. Roth.** 1989. Recombination between homologies in direct and inverse orientation in the chromosome of *Salmonella*: intervals which are nonpermissive for inversion formation. *Genetics* **122**:737–747.

176. **Shapiro, L., and R. Losick.** 1997. Protein localization and cell fate in bacteria. *Science* **276**:712–718.

177. **Sharp, P. M.** 1991. Determinants of DNA sequence divergence between *Escherichia coli* and *Salmonella typhimurium*: codon usage, map position, and concerted evolution. *J. Mol. Evol.* **33**:23–33.

178. **Sharp, P. M., M. Averof, A. T. Lloyd, G. Matassi, and J. F. Peden.** 1995. DNA sequence evolution: the sounds of silence. *Philos. Trans. R. Soc. Lond. B* **349**:241–247.

179. **Sharp, P. M., E. Cowe, D. G. Higgins, D. C. Shields, K. H. Wolfe, and F. Wright.** 1988. Codon usage patterns in *Escherichia coli*, *Bacillus subtilis*, *Saccharomyces cerevisiae*, *Schizosaccharomyces pombe*, *Drosophila melanogaster*, and *Homo sapiens*; a review of the considerable within-species diversity. *Nucleic Acids Res.* **16**:8207–8211.

180. **Sharp, P. M., and W.-H. Li.** 1987. The codon adaptation index—a measure of directional synonymous codon usage bias, and its potential applications. *Nucleic Acids Res.* **15**: 1281–1295.

181. **Sharp, P. M., and W.-H. Li.** 1986. Codon usage in regulatory genes in *Escherichia coli* does not reflect selection for "rare" codons. *Nucleic Acids Res.* **14**:7737–7749.

182. **Sharp, P. M., and W.-H. Li.** 1987. The rate of synonymous substitution in enterobacterial genes is inversely related to codon usage bias. *Mol. Biol. Evol.* **4**:222–230.

183. **Shimomura, S., S. Shigenobu, M. Morioka, and H. Ishikawa.** 2002. An experimental validation of orphan genes of *Buchnera*, a symbiont of aphids. *Biochem. Biophys. Res. Commun.* **292**:263–267.

184. **Smith, G. R.** 1994. Hotspots of homologous recombination. *Experientia* **50**:234–241.

185. **Smith, G. R., S. K. Amundsen, P. Dabert, and A. F. Taylor.** 1995. The initiation and control of homologous recombination in *Escherichia coli*. *Philos. Trans. R. Soc. Lond. B* **347**:13–20.

186. **Smith, J. M., C. G. Dowson, and B. G. Spratt.** 1991. Localized sex in bacteria. *Nature* **349**:29–31.

187. **Smith, M. W., D.-W. Feng, and R. F. Doolittle.** 1992. Evolution by acquisition: the case for horizontal gene transfers. *Trends Biochem. Sci.* **17**:489–493.

188. **Smith, N. H., E. C. Holmes, G. M. Donovan, G. A. Carpenter, and B. G. Spratt.** 1999. Networks and groups within the genus *Neisseria*: analysis of *argF*, *recA*, *rho*, and 16S rRNA sequences from human *Neisseria* species. *Mol. Biol. Evol.* **16**:773–783.

189. **Sreevatsan, S., X. Pan, K. E. Stockbauer, N. D. Connell, B. N. Kreiswirth, T. S. Whittam, and J. M. Musser.** 1997. Restricted structural gene polymorphism in the Mycobacterium tuberculosis complex indicates evolutionarily recent global dissemination. *Proc. Natl. Acad. Sci. USA* **94**:9869–9874.

190. **Stambuk, S., and M. Radman.** 1998. Mechanism and control of interspecies recombination in *Escherichia coli*. I. Mismatch repair, methylation, recombination and replication functions. *Genetics* **150**:553–542.

191. **Sueoka, N.** 1988. Directional mutation pressure and neutral molecular evolution. *Proc. Natl. Acad. Sci. USA* **85**:2653–2657.

192. **Sueoka, N.** 1993. Directional mutation pressure, mutator mutations, and dynamics of molecular evolution. *J. Mol. Evol.* **37**:137–153.

193. **Sueoka, N.** 1992. Directional mutation pressure, selective constraints, and genetic equilibria. *J. Mol. Evol.* **34**:95–114.

194. **Sueoka, N.** 1962. On the genetic basis of variation and heterogeneity in base composition. *Proc. Natl. Acad. Sci. USA* **48**:582–592.

195. **Suerbaum, S., J. M. Smith, K. Bapumia, G. Morelli, N. H. Smith, E. Kunstmann, I. Dyrek, and M. Achtman.** 1998. Free recombination within *Helicobacter pylori*. *Proc. Natl. Acad. Sci. USA* **95**:12619–12624.

196. **Tamas, I., L. Klasson, B. Canback, A. K. Näslund, A.-S. Eriksson, J. J. Wernegreen, J. P. Sandström, N. A. Moran, and S. G. E. Andersson.** 2002. 50 million years of genomic stasis in endosymbiotic bacteria. *Science* **296**:2376–2379.

197. **Taylor, A. L., and M. S. Thoman.** 1964. The genetic map of *Escherichia coli* K-12. *Genetics* **50**:659–677.

198. **Taylor, A. L.** 1970. Current linkage map of *Escherichia coli*. *Bacteriol. Rev.* **34**:155–175.

199. **Taylor, A. L., and C. D. Trotter.** 1967. Revised linkage map of *Escherichia coli*. *Bacteriol. Rev.* **31**:332–353.

200. **Treves, D. S., S. Manning, and J. Adams.** 1998. Repeated evolution of an acetate-crossfeeding polymorphism in long-term populations of *Escherichia coli*. *Mol. Biol. Evol.* **15**: 789–797.

201. **Vulic, M., F. Dionisio, F. Taddei, and M. Radman.** 1997. Molecular keys to speciation: DNA polymorphism and the control of genetic exchange in Enterobacteria. *Proc. Natl. Acad. Sci. USA* **94**:9763–9767.

202. **Vulic, M., R. E. Lenski, and M. Radman.** 1999. Mutation, recombination, and incipient speciation of bacteria in the laboratory. *Proc. Natl. Acad. Sci. USA* **96**:7348–7351.

203. **Waldor, M. K., and J. J. Mekalanos.** 1996. Lysogenic conversion by a filamentous phage encoding cholera toxin. *Science* **272**:1910–1914.

204. **Wang, B.** 2001. Limitations of compositional approach to identifying horizontally transferred genes. *J. Mol. Evol.* **53**: 244–250.

205. **Wernegreen, J. J., H. Ochman, I. B. Jones, and N. A. Moran.** 2000. Decoupling of genome size and sequence divergence in a symbiotic bacterium. *J. Bacteriol.* **182**:3867–3869.

206. **Whittam, T. S.** 1996. Genetic variation and evolutionary processes in natural populations of *Escherichia coli*, p. 2708–2720. *In* F. C. Neidhardt, R. Curtiss III, J. L. Ingraham, E. C. C. Lin, K. B. Low, B. Magasanik, W. S. Reznikoff, M. Riley, M. Schaechter, and H. E. Umbarger (ed.), Escherichia coli *and* Salmonella typhimurium: *Cellular and Molecular Biology*, 2nd ed., vol. 2. ASM Press, Washington, D.C.

207. **Whittam, T. S., and S. Ake.** 1992. Genetic polymorphisms and recombination in natural populations of *Escherichia*

coli, p. 223–246. *In* N. Takahata and A. G. Clark (ed.), *Mechanisms of Molecular Evolution.* Japan Scientific Society Press, Tokyo, Japan.

208. **Whittam, T. S., H. Ochman, and R. K. Selander.** 1984. Geographical components of linkage disequilibrium in natural populations of *Escherichia coli. Mol. Biol. Evol.* **1:** 67–83.

209. **Whittam, T. S., H. Ochman, and R. K. Selander.** 1983. Multilocus genetic structure in natural populations of *Escherichia coli. Proc. Natl. Acad. Sci. USA* **80:**1751–1755.

210. **Woese, C. R., G. J. Olsen, M. Ibba, and D. Soll.** 2000. Aminoacyl-tRNA synthetases, the genetic code, and the evolutionary process. *Microbiol. Mol. Biol. Rev.* **64:**202–236.

211. **Wolf, Y. I., L. Aravind, N. V. Grishin, and E. V. Koonin.** 1999. Evolution of aminoacyl-tRNA synthetases—analysis of unique domain architectures and phylogenetic trees reveals a complex history of horizontal gene transfer events. *Genome Res.* **9:**689–710.

212. **Worning, P., L. J. Jensen, K. E. Nelson, S. Brunak, and D. W. Ussery.** 2000. Structural analysis of DNA sequence: evidence for lateral gene transfer in *Thermotoga maritima. Nucleic Acids Res.* **28:**706–709.

213. **Xiong, J., K. Inoue, and C. E. Bauer.** 1998. Tracking molecular evolution of photosynthesis by characterization of a major photosynthesis gene cluster from *Heliobacillus mobilis. Proc. Natl. Acad. Sci. USA* **95:**14851–14856.

214. **Yap, W. H., Z. Zhang, and Y. Wang.** 1999. Distinct types of rRNA operons exist in the genome of the actinomycete *Thermomonospora chromogena* and evidence for horizontal transfer of an entire rRNA operon. *J. Bacteriol.* **181:** 5201–5209.

215. **Zawadzki, P., M. S. Roberts, and F. M. Cohan.** 1995. The log-linear relationship between sexual isolation and sequence divergence in *Bacillus* transformation is robust. *Genetics* **140:**917–932.

216. **Zinder, N. D., and J. Lederberg.** 1952. Genetic exchange in *Salmonella. J. Bacteriol.* **64:**679–697.

217. **Zuckerkandl, E.** 1965. The evolution of hemoglobin. *Sci. Am.* **212:**110–118.

218. **Zuckerkandl, E., and L. Pauling.** 1965. Molecules as documents of evolutionary history. *J. Theor. Biol.* **8:**357–366.

Chapter 3

Bacteriophages and the Bacterial Genome

SHERWOOD CASJENS AND ROGER W. HENDRIX

Bacteriophages have had and continue to have an enormous effect on bacteria and their chromosomes. Their effects range from exerting evolutionary pressures to carrying genes that are expressed in the everyday life of a bacterium. Bacteriophages are the most numerous type of organism on Earth, and their presence is likely a constant influence on their bacterial and archael hosts. Current measurements suggest that there are between 10^{29} and 10^{31} free double-stranded DNA (dsDNA) phage virions in Earth's biosphere, which is about 10 times the number of bacterial cells (7, 127, 129). We will focus our discussion on the dsDNA phages, and especially on the temperate phages. While virulent phages certainly perform transduction and engage in evolutionary sparring with their hosts and so influence their evolution, this chapter focuses mainly on the complex interactions of temperate phages with their hosts. Temperate phage genomes can exist benignly within their hosts (either integrated into the host chromosome or as a plasmid) for indefinitely long times, becoming in effect part of the cell's genome.

BACTERIOPHAGES AND BACTERIAL GENOME EVOLUTION

Horizontal Transfer of Bacterial Genetic Information

Analysis of the nucleotide sequences of bacterial genomes has made it abundantly clear that most bacteria have engaged frequently in the horizontal transfer of genetic information both within and between phyla (e.g., references 65, 74, and 86; see also chapter 2). Bacteriophages appear very likely to have been the vehicles by which at least some bacterial horizontal transfer was mediated, but have they been present throughout bacterial evolution or are they recently arrived? Bacteriophages appear to be ancient, since their virion structural proteins are very highly diverged. For example, phage portal proteins, given their very specific function and similar (low resolution) structures (reference 107 and references therein), seem unlikely to have been "invented" more than once. Portal proteins form the hole through which DNA enters and part of the motor that drives DNA into the phage head; they are universally present in dsDNA tailed phages (as well as in the almost certainly evolutionarily related herpesviruses), and homologues that have other roles have never been found. The different *Escherichia coli* phage portal proteins, which have identical functions and are very unlikely to be the result of convergent evolution, are often not recognizably similar in amino acid sequence to one another; for example, the portal proteins of the relatively well studied *E. coli* phages λ, P2, Mu, 933W, HK620, T4, and T7 have no convincing sequence similarity to one another (and that of HK97 is barely recognizably similar only to the P2 portal protein). Likewise, major capsid and other virion structural proteins with arguably common ancestry have diverged to the very limits of current detection methods and probably beyond (our unpublished observations). Although we lack an accurate calibration for the rate of the mutational clock in phage genes, what information is available argues that it cannot be dramatically different from that of bacterial genes. We interpret these observations to mean that phages are almost certainly very ancient organisms that have likely been coevolving with bacteria since relatively soon after the beginning. Phages are clearly not new arrivals on the bacterial evolutionary scene.

Although it may be impossible to decipher in detail the mechanisms by which horizontal transfers of genetic information between individual bacteria

Sherwood Casjens • Department of Pathology, University of Utah Medical School, Salt Lake City, UT 84132, and Pittsburgh Bacteriophage Institute, University of Pittsburgh, Pittsburgh, PA 15260.   **Roger W. Hendrix** • Department of Biological Sciences and Pittsburgh Bacteriophage Institute, University of Pittsburgh, Pittsburgh, PA 15260.

occurred in nature in the past, our current knowledge of phage biology argues persuasively that phages must have played a significant role. Such transfer could happen by phage-mediated generalized or specialized transduction (reviewed in reference 70). Generalized transduction has usually been studied in temperate phages but can also be carried out by virulent phages, whereas specialized transduction is unique to integrating temperate phages. In the former, phages such as P22 and P1 are known to "accidentally" encapsidate a piece of host DNA instead of their own phage chromosome, and this host DNA can then be injected into a susceptible host as if it were a phage chromosome. This can lead to recombinational replacement or insertion of resident host sequences by new or related sequences. In the case of the Mu-like phages, each virion carries several kilobase pairs of DNA from its previous host fused to its chromosome's termini, which is available for recombination with resident host sequences. Generalized transductional mechanisms by which phage carry bacterial chromosomal sequences between cells rely largely on homologous recombination for permanent acquisition by the recipient bacterium, and so could be in part limited to transfer between fairly close relatives. However, transduction of plasmids or transposons by phage could lead to introduction of the plasmid or transposon into any host that receives the DNA. Specialized transduction can occur without the aid of homologous recombination through lysogenization and integration of a phage that "accidentally" carries a section of its previous host's DNA integrated into its own chromosome (reviewed in reference 126). Thus, if injection and integration were achieved by the transducing phage DNA even in a more distantly related bacterium, horizontal transfer could be achieved.

In generalized transduction by the well-characterized phages P22 and P1, a small percentage of the virions made in a lytic infection contain, respectively, a 43.5- or 90-kbp section of the previous host's genome (27, 49). Not every type of dsDNA phage is normally capable of generalized transduction; some have a DNA packaging apparatus that is so specific that essentially no host DNA fragments are successfully packaged. Specialized transducing phage genomes are also generated quite rarely ($\leq 10^{-4}$ per induced cell); however, sometimes these are not defective and the added DNA might even be useful to the phage, so genes from a host can become a part of the phage genome (and there are other mechanistically less well understood pathways for phages to incorporate genes from their host into their own genomes [44]). These numbers suggest that

bacteriophage-mediated transduction as carried out in the laboratory is typically a relatively rare event compared with normal phage infection, and specificity of adsorption of virions to bacterial cells limits utilization of this mechanism in transfer of genetic material between distant phyla. Nonetheless, unlike conjugation, transduction does not require physical contact between the donor and recipient cells, and DNA in transducing particles might be protected from environmental insult for long periods of time. Thus, given the enormous numbers of phages on Earth, and the apparent genetic contact between phages with different hosts (45), it seems impossible to imagine that transduction did not play a significant role in the evolution of many if not most bacterial genomes (see also references 78 and 90).

Deletion Pressure?

Bacteriophages no doubt also have other, less direct effects on the evolution of bacterial chromosomes. For example, it has been suggested that they contribute to a kind of "deletion pressure" that is exerted on bacterial genomes (64). This idea derives in part from the observation that spontaneous deletions (which are usually mediated by homologous recombination between very short sequence repeats [reviewed in reference 100]) occur at a noticeable frequency in bacteria. This presents a paradox: one would imagine that such more or less random deletions would usually have deleterious effects, so why has evolution not "fixed" this bad situation by requiring homologous recombination machinery to utilize longer regions of sequence identity, thus making such deletions much less likely? It is argued that a prophage represents a significant threat to a bacterial lineage, in that prophage induction kills the cell. Consequently, a lineage carrying a prophage is at a selective disadvantage relative to a lineage not carrying such a molecular time bomb. In this view, the high deletion rate exhibited by bacteria is the result of an equilibrium that balances the selective advantage of a high deletion rate in ridding the cell of genetic parasites (prophages, transposons, and others) with the selective disadvantage of too frequent deletion of important bacterial genes. If correct, this provides an attractive explanation for one of the most striking properties of bacterial genomes, namely, that bacteria typically have very compact genomes with few pseudogenes and little noncoding space between genes (64). That is, an active deletion mechanism will delete, among other things, pseudogenes and noncoding DNA, and since such deletions should carry no selective penalty, they will persist in the population.

Bacteriophages may thus have contributed to the current compact nature of bacterial genomes. This of course is a very speculative idea, but it does serve to point out that bacterial and bacteriophage evolution are likely to be intertwined in many and devious ways.

Current Bacterial Genome Content and the Ubiquity of Prophages

We know that phage virions are common in the environment, but are prophages also common? Before complete genome sequencing, a number of the temperate phages that were intensively studied in the laboratory were originally found as prophages (e.g., enteric bacteriophages λ, P1, P2, P22, and Sf6), and it is anecdotally possible to induce resident prophages into lytic growth from many bacterial isolates. For example, Schmieger and colleagues found that 92.5% of 173 *Salmonella enterica* serovar Typhimurium strains tested and 78.5% of 107 *E. coli* strains tested released at least one functional phage type (102, 103). Fifty-one different phages were induced from 27 enterohemorrhagic *E. coli* strains by Osawa et al. (89), and the well-studied serovar Typhimurium laboratory strain LT2 harbors no fewer than four apparently fully functional prophages (33, 132). In addition, particular prophage attachment sites in bacterial chromosomes may be occupied by a prophage in a large fraction of individuals. For example, the *E. coli* phage 21 attachment site was occupied in 28 of 77 independently isolated strains that were analyzed by Wang et al. (125). Finally, various defective (also called "cryptic") prophages have been identified and studied in some detail, for example, qsr′ (53), e14 (96), Rac (55), and DLP12 (67) in *E. coli* strain K-12, and SKIN (77) and PBSX (59) in *Bacillus subtilis* strain 168.

However, only with the advent of whole phage and bacterial genome sequencing did it become possible to attempt to recognize all the prophages in a genome and so address this question more systematically. Although phage genes are notoriously variable in sequence (because of their ancient origins and ability to exchange horizontally), we now believe that it is possible to identify most prophages in genomes of bacteria in those phylogenetic groups where bacteriophages have been studied in some detail, in particular in the genomes of proteobacteria (especially the γ subdivision) and the low-G + C gram-positive bacteria. Some prophage-like sequences are found in sequenced genomes of more distantly related bacteria where no phages have been studied, but others could well be missed in such genomes. For example, only

recently, after the sequencing of new prophage genomes in the proteobacterium *Xylella fastidiosa* (108), was it convincingly recognized (through new sequence similarity connections) that the sequenced genome of the spirochete *Borrelia burgdorferi* may harbor as many as 12 or 14 prophages (22, 28). We have examined the approximately 100 currently published bacterial genome complete nucleotide sequences, and have been able to recognize about 285 prophages that are related to known bacteriophages (19, 21). Table 1 lists a sampling of these prophage-containing bacterial genomes. Prophages (we use the term to include defective prophages) are clearly common in bacterial genomes. In fact, in some cases they occupy a significant fraction of the genome: in two *E. coli* O157:H7 isolates (41, 88, 95) and two *Streptococcus pyogenes* strains, MGAS315 and SSI-1 (6, 82), recognizable prophages make up at least 12% of the genome (about 0.67 of 5.5 Mbp in the former case), and plasmid prophages may make up as much as 20% of the 1.5-Mbp *B. burgdorferi* genome (22, 28, 29). We also note that not all bacterial genomes contain recognizable prophages: most of the small-genome bacteria (genomes less than 1.5 Mbp) and a few large-genome bacteria appear to have none (21, 64).

The prophages identified in bacterial genome sequences are not all functional; in fact, many appear to be defective and in a state of partial decay. Of the more than 280 prophages in the currently sequenced bacterial genomes, only a few are known to be fully functional bacteriophages (but a substantial number appear to be largely intact and could potentially be viable phages). Genes on both intact and decaying prophages can be expressed to modify the host in some way. It is not possible to know whether some of the genes in the defective prophages might at some time in the future be left intact to perform a valuable function for the bacteria after the rest of the phage sequences are gone, but it appears that such events have happened in the past (see below). In one studied case, the important *Shigella dysenteriae* toxin, Shiga toxin, is encoded by two genes that lie on a remnant of a λ-like prophage (72) (also see below). Two lines of evidence suggest that, as defective prophages decay, deletions of parts of the prophage may often occur as fast or faster than point mutational inactivation of the remaining genes. First, there are a number of laboratory cases in which mutational activation of the expression of functional but unexpressed genes on defective prophage genes have given rise to bacteria with altered properties; for example, in *E. coli* K-12, RecE recombination exonuclease (54), MvrC methyl viologen resistance (80), RusA

Table 1. Prophages in some completely sequenced bacterial genomes

Bacterial host[a]	No. of prophages[b]	Name if functional[c]	Comment[d]
Bacillus anthracis Ames	4		All largely intact
Bacillus cereus ATCC14579	7		Several largely intact, one is a linear plasmid
Bacillus halodurans C-125	4		1 largely intact
Bacillus subtilis 168	3	SPβ	Others defective
Borrelia burgdorferi B31	12–14		5–7 largely intact, all plasmids
Caulobacter crescentus CB215	1		Defective
Chlamydia pneumoniae AR39	1		φX174-like circle
Clostridium acetobutylicum ATCC824	2		Both largely intact
Clostridium tetani TLS	3		2 largely intact
Deinococcus radiodurans R1	2		1 largely intact
Escherichia coli K-12	11	λ	10 defective
Escherichia coli O157:H7 EDL933	18–20[e]	933W	Several largely intact
Escherichia coli O157:H7 Sakai	18		Several largely intact
Haemophilus influenzae RD KW20	3		1 largely intact
Lactococcus lactis IL1403	6		3 largely intact
Listeria monocytogenes CLP1162	6		Several largely intact
Mesorhizobium loti MAFF303099	3		2 largely intact
Mycobacterium tuberculosis H37Rv	2		Both defective
Neisseria meningitidis MC58	2		Both defective
Neisseria meningitidis Z2491	3		1 largely intact
Pasteurella multocida Pm70	1		Defective
Pseudomonas aeruginosa PAO1	3	F, R pyocins	1 integrated filamentous phage?
Pseudomonas putida KT2440	4		Several largely intact
Ralstonia solanacearum GMI1000	8		Several largely intact
Salmonella enterica CT18	12		Several largely intact
Salmonella enterica LT2	7	Fels-1, Fels-2, Gifsy-1, Gifsy-2	Others defective
Shewanella oneidensis MR-1	3		All largely intact
Shigella flexneri 301	11		Several largely intact
Staphylococcus aureus N315	1		Largely intact
Staphylococcus aureus Mu50	2		Both largely intact
Streptococcus pyogenes M1SF370	4		Largely intact
Streptococcus pyogenes MGAS315	6		Most largely intact
Vibrio cholerae N16961	2	CTXφ	1 defective, 1 filamentous
Xanthomonas axonopodis 903	2		1 largely intact
Xylella fastidiosa 9a5c	10		4 largely intact, 1 filamentous
Yersinia pestis CO92	7		Several largely intact
Yersina pestis KIM10 +	5		Several largely intact

[a]This is not a complete list of the bacterial genomes with completed sequences, but it is meant to give the flavor of genomes that contain recognizable prophages. GenBank annotations and genome sequences for these bacteria can be obtained on the Internet at http://www.ncbi.nlm.nih.gov/PMGifs/Genomes/bact.html.

[b]The number includes all recognizable prophages whether intact or obviously defective. This list does not include "prophagelike elements" that have only an integrase gene or other single, isolated phagelike genes. There may well be additional unrecognizable prophages present, especially in those phyla in which phages have not previously been studied in detail.

[c]Phages that have been induced to grow lytically and have been subsequently propagated, i.e., are fully functional bacteriophages.

[d]By sequence analysis alone, some prophages appear to be "largely intact" in that they contain enough open reading frames to be a potentially functional phage, and there is no evidence that suggests that they are not functional. These have not been demonstrated to be fully functional phages, and many may well not be functional.

[e]It is not clear exactly how many prophages are in strain EDL933. At the time of this writing there is still a gap in the sequence at prophage CP-933X, and in several cases prophages appear to have integrated inside preexisting prophages and the number of phages that originally integrated is different depending upon the particular rearrangements one envisions happened during and since the original integration events.

Holliday junction resolvase (68), DicB cell division inhibitor (4), and NmpC outer membrane porin (14) are all genes in defective prophages that are normally not expressed but can be mutationally activated. Second, unexpressed prophage genes can recombine into the genome of an infecting related phage and successfully perform their functions there (reviewed in reference 17). Thus, many genes in decaying pro-phages remain potentially functional, and it seems likely that some might be mutationally switched on to a bacterium's advantage in a wild setting, in which case they could be retained by selection even as the rest of the prophage disappears over time.

It has recently become clear that bacterial genomes (unlike those of humans and other eukaryotic organisms) are in general extremely plastic. Different individuals within a species have a "core genome" that is very similar, but substantial portions of their genomes may be very different (reviewed in reference 20). In *E. coli*, chromosome sizes range from 4.5 to 5.5 Mbp (8), and prophages appear to be responsible for a substantial fraction of this variation (40, 95). This is also the case in the gram-positive species *S. pyogenes* (2). It is not yet known if this is true in other bacterial phyla, but it seems reasonable from current knowledge to expect that it will be so in many of them. It has also been noted that lambdoid prophages are invariably oriented so that bacterial chromosome replication passes over them in the same relative direction (18); we have noted that this is true in several gram-positive species as well (e.g., *Listeria* [36], *Streptococcus* [82], and *Lactobacillus* [58] spp.). The reasons for this are not known, but one consequence is that single recombination events between homologous prophages in the two bacterial replichores will cause an inversion rather than deletion of the DNA between them. When multiple genomes of the same or similar species are compared, in several cases clear examples of large inversions mediated by recombination between prophages have been identified (e.g., *E. coli* [94], *X. fastidiosa* [120], and *S. pyogenes* [82]). Thus, prophages can also provide a vehicle for bacterial genome rearrangements.

Bacteriocins, Gene Transfer Agents, and Prophage Gene Incorporation into the Host Genome

It seems very possible that as prophages decay, some prophage genes might be useful to the host and eventually be left on the chromosome after the remainder of the prophage has completely disappeared (e.g., Shiga toxin genes; see above). There are two rather complex types of genetic entity in which this appears to have happened: the phage taillike bacteriocins and the gene transfer agents. Some bacteria can release structures that are very much like phage tails. Such "bacteriocins" are known in a number of bacterial phyla (e.g., the genera *Erwinia* [87], *Bacillus* [104], and *Xenorhabdus* [116]). These structures can bind to other bacteria, presumably in a manner perfectly analogous to attachment of tailed phages to susceptible cells, and then in some way kill the target bacterium. The ability to kill one's competitors can be very advantageous. In *Pseudomonas aeruginosa*, in which these structures have been studied in the most detail, two types of taillike *Pseudomonas*-specific bacteriocins (R- and F-type "pyocins") are specific for killing other strains of *Pseudomonas*. One resembles a phage λ tail (F type), and the other resembles a phage P2 tail (R type) (84). The recent sequence of the portion of the *P. aeruginosa* PAO1 chromosome

harboring the pyocin genes has shown that they are encoded by two contiguous parts of a single gene cluster (84), and that the F and R genes are strikingly homologous to λ and P2 tail genes, respectively. These pyocin genes are arranged in two adjacent clusters, each of which encodes one pyocin, and whose gene order is nearly identical to their homologues on the two phages. In addition, there are cell lysis genes present at the promoter proximal end of the R cluster that mediate the release of the pyocins, and there are two genes adjacent to the pyocin cluster that regulate expression of both pyocin clusters. Finally, very similar clusters of genes that encode R- and F-type pyocins are present in many *P. aeruginosa* isolates (84). The convincing similarity in both sequence and genetic organization between the pyocins and known phage tail genes shows that they must share a common ancestry. Although it is not possible to be positive, it seems likely to us that the pyocins are derived from two "unrelated" and now highly deleted prophages and less likely that the pyocins evolved first and were later taken over by phages to be phage tails. The facts that there do not appear to be associated decaying pseudogenes, that they have gained a novel regulatory system, and that they are present in many isolates make it extremely likely that these are portions of phage genomes that have been completely appropriated for the host's own purposes and are not simply deleted prophages that accidentally have bactericidal properties.

The gene transfer agents (GTAs) present a parallel example of apparent appropriation of a whole tailed-phage "virion." GTAs are encoded by a cluster of phage-like genes that can build a virionlike particle which can package fragments of the bacterial genome and deliver them to a new bacterium; they have been described in two very different bacterial species and in one archaeon (9, 48, 98, 134). The best characterized of these GTAs is that of *Rhodobacter capsulatus* (62). The *Rhodobacter* GTA genes have been shown to be regulated by genes that respond to environmental signals and which are not known to be associated with other phagelike genes in any other context. Upon induction, *Rhodobacter* GTA "virions" are made that contain apparently random, headful-sized 4.5-kbp fragments of the bacterial genome. Curiously, the 4.5-kbp packaged fragments are much smaller than the block of genes that encode the particles (about 17 kbp), so not only is there no detectable bias in favor of packaging the genes encoding the proteins of the particles, but also the packaging capacity of the particles is much too small to accommodate all of those genes in a single particle. These viruslike particles are in effect generalized transducing particles that can deliver their genetic payloads to other cells that have the right surface receptors. Like the pyocins, the *R. capsulatus* GTA genes which encode the particle do not appear to be accompanied by other phagelike genes or pseudogenes, there are nearby regulatory genes that control the expression of the GTA genes, and GTAs are present in most *Rhodobacter* isolates (reference 63 and references therein). These facts suggest that this is not just a decaying, partially deleted prophage that happens by chance to have these properties, but is rather a set of phagelike genes that are part of the host genome and serve some purpose (63). Are GTAs "domesticated" prophages serving a purpose for their host, or might the tailed bacteriophages have evolved from cell-derived GTAs (and taillike bacteriocins)? Given the rarity of GTAs compared to phages, the former seems a more parsimonious explanation, but this is speculation. What (if any) evolutionary purpose the GTAs have also remains unclear. Might the cell be putting its genes into "escape pods" and sending them out into the universe to fend for themselves upon impending disaster? Do GTAs play a significant role in maintaining the genetic diversity or stability of the species? What selective force maintains the GTA genes in the population, given that their expression will destroy the cell in which they reside? Clearly, many fascinating questions remain.

Other prophagelike entities are known which produce virionlike particles that contain only random fragments of host DNA in their heads, but which kill the bacteria to which they adsorb (e.g., references 110 and 137). These entities function as bacteriocins and so are less likely to successfully deliver their DNA into a new, live host. PBSX of *B. subtilis* is an example of such a device that has been studied in some detail (59), but it is unclear whether it is a defective prophage in the early stages of decay that happens by accident to have these properties, whether it is being used by its host as a bacteriocin to kill related bacteria that do not carry PBSX, or whether it is perhaps able to transduce its DNA under some conditions, like the GTAs.

The genes for pyocins and GTAs are two cases of phage genes that appear to have been appropriated for bacterial purposes. In both of these cases, the apparently appropriated phage genes have important and direct evolutionary consequences: the pyocins can affect competition between bacterial relatives, and the GTAs can affect gene transfer among individual members of a species. There are numerous other examples of individual phagelike genes in isolation on bacterial chromosomes, and in these cases we do not know if they might represent the original bacterial source of a phage gene or, perhaps more likely, may themselves have been procured from a phage through phage integration followed by deletion of phage genes that did not provide a selective benefit to the bacterial chromosome.

PROPHAGES DIRECTLY ALTER PROPERTIES OF THE HOST BACTERIUM

Prophages also have much more direct and immediate effects on their hosts, which in turn will have an effect on the evolution of the host's genome. Essentially every temperate phage that has been examined in detail has been found to express genes that affect its host in some way. Even the laboratory workhorse, phage λ, originally chosen for study without knowledge of such things, is now known to express at least six genes (*cI* repressor, *lom*, *rexA*, *rexB*, *sieB*, and *bor*) that inhibit attack by other phages or modify the bacterial surface in such a way that pathogenicity of the host is affected. In addition, other phages that are members of the lambdoid phage family carry genes that express the important Shiga-like toxins and enzymes that alter the structure and antigenicity of surface polysaccharides of their hosts, as well as other host-modifying factors (see below). Alteration of host properties by a prophage has been termed "lysogenic conversion." To date, protection from other phages (or programmed death caused by plasmid loss) and disease virulence factors are the lysogenic conversion genes that have been discovered and studied in the laboratory, but this likely reflects their ease of study and the lifestyles of the hosts studied. It seems quite possible that prophage lysogenic conversion genes in nonparasitic, free-living host bacteria might affect those bacteria in ways that enhance their ability to survive in particular niches, just as virulence factors do in parasitic hosts (e.g., phages are known that enhance sporulation in *B. subtilis*, *Bacillus pumilus*, *Bacillus thuringiensis*, and *Clostridium perfringens* [93, 106, 112]). The short- and long-term evolutionary values of a dangerous device like a prophage, a "booby trap" that is valuable in the short term but could kill the host that carries it at any time, are poorly understood. Is this a temporary accident of horizontal transfer, and if we were to look again in the distant future will the valuable (to the bacteria) phage genes have become an integral part of the bacterial chromosome (i.e., become so like other bacterial genes in their relationship to the rest of the bacterial genome that their origin as phage genes is unidentifiable)? Or is there some inherent value in keeping these genes on mobile elements even though the elements that carry them have attendant dangers? We discuss several cases of lysogenic conversion in more detail below.

Protection from Other Phages and Plasmids

Temperate phages often carry lysogenic conversion genes that protect the host from attack by other phages. The lambdoid phages express several such systems, for example, the immunity, alteration of surface receptor polysaccharides, *sieA*, *sieB*, and *rex* systems. "Immunity" refers to the fact that prophage repressors repress gene expression from phages with similar early operon operators, thereby blocking the growth of phages with closely related repressors. Changes in the chemical nature of the surface polysaccharide can affect adsorption of phages that use these polymers as receptors (e.g., see references 25, 46, 47, 71, and 136). Expression of the *sieA* gene prevents injection of DNA from related phages by an unknown mechanism (113). SieB blocks growth of at least some other lambdoid phages by a different mechanism that acts after injection of the superinfecting phage DNA (114). The phage λ *rexA* and *rexB* gene proteins, which at least partially exclude a number of other phages (105), occupy a hallowed place in the history of molecular genetics, since their expression from the λ prophage causes exclusion of the virulent phage T4 *rII* mutants but not wild-type T4. This in turn allowed detailed analysis of T4 *rII* mutants, which contributed greatly to our understanding of the nature of genes (5). The precise mechanism of this exclusion remains unclear to this day, but RexB is now known to be an inhibitor of *E. coli* ClpP proteases (109), and this inhibition appears to protect lysogens from death after plasmid loss due to some plasmid-borne "addiction" systems (31).

Such contributions to the host's susceptibility to phage attack are not limited to the lambdoid phages. For example, *E. coli* phage P1 carries a DNA restriction modification system that can degrade unmodified incoming phage DNA and a gene whose expression interferes with lambdoid phage growth (reviewed in reference 133). The defective *E. coli* prophage e14 carries a tRNA "restriction" system that could help protect against those phages that encode some of their own tRNAs (92). Other prophages may give themselves (and their host) an advantage by producing substances that kill nonlysogenic bacteria of the same species as their host. For example, *B. subtilis* phage SPβ produces an uncharacterized, secreted protein that kills nonlysogens but not SPβ lysogens of its host species (43). Such varied, apparently protective functions are very common in temperate phages, and each seems likely to contribute to the well-being of the host by conferring protection from an outside threat, but in so doing each protective function also has the "selfish" role of allowing the prophage DNA to prosper.

Bacterial Virulence Factor Genes on Phage Chromosomes

It has been known for half a century that the diphtheria toxin gene resides on the converting β phage chromosome and is expressed from the prophage after it establishes a lysogenic state (35, 37), but only much more recently has it become known how pervasive this arrangement is for bacterial "virulence genes" (13, 15, 24, 73, 123). Table 2 lists a sampling of the currently known genes that reside on phage genomes and affect the host bacterium's interaction with its host. It is very clear that unrelated toxins reside on numerous phage chromosomes, so this strategy appears to have been independently "invented" many times, and it appears to have general value to both the phage and the host. Most phage-carried virulence factors, for example, exfoliating toxin on the *Staphylococcus* phage ϕETA (131), have been shown to be expressed from the prophage, but recently this has been found not to be universally true for all such genes. The Shiga-like toxin phage genes that make *E. coli* O157 strains so virulent lie in what appears to be an operon that is expressed only at late times during lytic growth; at least in some cases experimental evidence suggests that in fact lytic growth of the phage is required for the toxin to be present in high levels in the infected animal host (122). Thus, there appears to be a second, less well understood level of interaction between phages and their bacterial hosts in which genes expressed during lytic infection may be advantageous to the bacterial population. This notion at first appears to be self-contradictory, but perhaps lysis of a few cells is advantageous to siblings that make up the local population.

The numerous parallel but nonhomologous cases of different bacterial virulence genes on phage chromosomes, and the apparent functional and regulated incorporation of these genes into the phage genome, suggest that it is much more than an accident of history that such genes are phage borne. There must be some evolutionary advantage to having these genes reside on mobile elements. Whether that advantage accrues to the bacteria, the phages, the genes themselves, or to some combination of these remains to be seen.

New Genes May Enter the Bacterial Genome as "Morons"

How do phages acquire lysogenic conversion genes? We do not know the mechanism by which such genes become part of the phage genome, but phages that are otherwise quite closely related in sequence and gene organization often differ in the

Table 2. Some bacteriophages with genes that affect the interaction of pathogenic bacteria with their eukaryote hosts

Bacteriophage	Virulence factor[a]	Bacterial host	Reference(s)[b]
φBB-1 family	Erp outer surface proteins, inhibition of complement system	*Borrelia burgdorferi*	42, 111
C1	Botulism neurotoxin	*Clostridium botulinum*	39
NA1	Gas gangrene toxin	*Clostridium novyi*	30
Converting β	Diphtheria toxin	*Corynebacterium diphtheriae*	35, 37
H-19B, VT1-Sa[c] (and others)	Shiga-like toxin type 1, Lom	*Escherichia coli*	85, 117, 135
933W, VT2-Sa (and others)	Shiga-like toxin type 2, Lom, Bor	*Escherichia coli*	76, 97
ΦP27	Shiga-like toxin type 2e	*Escherichia coli*	81
λ	Bor, serum resistance; Lom, epithelial cell adhesion	*Escherichia coli*	3, 118
Atlas	Immunoglobulin (IgG) binding protein	*Escherichia coli*	101
ΦC3208	Enterohemolysin	*Escherichia coli*	12
oβ1	Heat-labile toxin	*Escherichia coli*	115
Defective prophage?[c]	Ovine foot rot virulence	*Dichelobacter nodosus*	23, 128
MAV1	Mouse arthritis	*Mycoplasma arthritidis*	121
Pnm1, MuMenB[c]	Bacterial surface-exposed proteins	*Neisseria meningitidis*	69
ΦCTX	Cytotoxin	*Pseudomonas aeruginosa*	83
D3	Surface polysaccharide acetylation	*Pseudomonas aeruginosa*	60
FIZ15	Surface polysaccharide alteration	*Pseudomonas aeruginosa*	119
ε34, P22	Surface polysaccharide glucosylation	*Salmonella enterica*	130, 136
ε15, φ27	Surface polysaccharide main chain alteration	*Salmonella enterica*	1, 16
SopEΦ	SopE, TTSS[d] protein, cell membrane ruffling and bacterial uptake	*Salmonella enterica*	38, 75
Gifsy-1	GipA, intestinal survival; GogB, possible TTSS[d] protein	*Salmonella enterica*	33
Gifsy-2	Superoxide dismutase; MsgA, macrophage survival; Lom	*Salmonella enterica*	33
Gifsy-3	SspH1, TTSS[d] protein; PagJ virulence protein	*Salmonella enterica*	33
Fels-1	Neuraminidase; superoxide dismutase; Lom	*Salmonella enterica*	33
Defective prophage[c]	Shiga toxin (dysentery enterotoxin)	*Shigella dysenteriae*	72
SfII, SfV, SfX	Surface polysaccharide glucosylation	*Shigella flexneri*	46, 47, 71
Sf6	Surface polysaccharide acetylation	*Shigella flexneri*	25
7888	Shiga-like toxin type 1	*Shigella sonnei*	11
PS42D	Enterotoxin A	*Staphylococcus aureus*	10
3GL16	Enterotoxin C	*Staphylococcus aureus*	51
Φ13	Staphylokinase, enterotoxin A	*Staphylococcus aureus*	26
ΦETA	Exfoliative toxin A	*Staphylococcus aureus*	26
L2043	Hemolysin[e]	*Staphylococcus aureus*	131
ΦPVL	Leukocidin toxin	*Staphylococcus aureus*	56
SaPAI1[f]	Toxic shock toxin	*Staphylococcus aureus*	34, 66
T12, T270	Scarlet fever toxin	*Streptococcus pyogenes*	50
CTXΦ	Cholera toxin	*Vibrio cholerae*	124
VPIΦ	Phage CTXΦ receptor	*Vibrio cholerae*	57
K139	Glo virulence protein	*Vibrio cholerae*	99

[a]Numerous other prophage genes discovered in genome sequencing projects are predicted to have such roles (e.g., references 32, 41, 61, and 95).
[b]In general, references were chosen to give the reader access to current literature, rather than to credit initial discoveries.
[c]Defective prophage, or perhaps a satellite phage.
[d]TTSS, protein secreted by a type III secretion system. This machinery is often responsible for moving bacterially encoded proteins from the bacteria that made them into cells of their hosts.
[e]Virulence factor gene not yet identified.
[f]The *Staphylococcus* toxic shock toxin element, pathogenicity island SaPAI1, is a satellite phage that is efficiently "transduced" by phage α80 (34, 66). The element does not encode a fully functional phage by itself, but it is induced to excise and replicate, and is efficiently packaged by phage α80 proteins.

complement of lysogenic conversion genes they carry. A more general manifestation of this phenomenon is seen quite dramatically in genomic comparisons between related phages (44, 52, 79, 91). In these comparisons it is frequently found that a gene is present in one phage genome in a surrounding context that has no such gene in the comparison phage. Sequence analysis typically indicates that these "extra" genes have entered the phage genome in recent evolutionary time (44, 52). Such genes are also frequently flanked by transcription signals—an upstream promoter and a downstream terminator—that allow their expression from an otherwise repressed prophage. Together with their transcription signals, these genes have been termed "morons" to emphasize the fact that their insertion into the phage genome results in "more" DNA in the genome than in its ancestor (52). Morons that carry the necessary associated transcription signals can and do function as lysogenic conversion genes, and to the extent that their expression provides a selective advantage to the host, their presence should mitigate against the deletion of the prophage. (Perhaps other morons have been retained in the phage genome, not because they benefit the bacterial host but because they enhance lytic growth of the phage [44, 79].) These observations led to the following speculative picture of what might be termed the moron cycle: a moron enters a phage genome, and it is retained if it provides a selective benefit to the phage, either directly by enhancing the lytic growth of the phage or indirectly by providing a selective benefit to cells that do not delete the prophage. In the latter case, although there is a selective penalty for deleting the beneficial moron DNA, there is still a selective benefit in deleting those parts of the prophage that will kill the cell upon prophage induction. Thus the long-term expectation is that in a lysogen there will be preferential retention of the morons that benefit the host and preferential loss of the prophage genes that are potentially lethal, with the consequent conversion of the moron from a "phage" gene into a "bacterial" gene. The observation that many prophages are defective, and that some important virulence factors lie in defective prophage genomes (e.g., Table 1 and references 72 and 135), suggests that real-life defective prophages may follow these expectations, but more detailed analyses are needed. It is at this point completely obscure by what biochemical mechanism a moron "jumps" into a new phage genome, nor is it clear what the ultimate source of morons is.

SUMMARY

Our ideas about how bacteriophages have affected the nature of the bacterial chromosome are necessarily based on extrapolations from things we know about bacteriophage biology and from inferences based on the current structure of the bacterial genomes, and not on direct observation of those processes over evolutionary time. It is particularly difficult to know the relative quantitative importance of the different bacteriophage-based mechanisms that we postulate to have a role in shaping the bacterial genome. Nonetheless, it is hard to escape the conclusion that phages have indeed had a major role in this process. Some of the mechanisms we describe—for example, generalized transduction within a closely related population, mediated either by conventional phages or the phagelike GTAs—might be expected to maintain the existing structure of the genome by providing a mechanism for replacing mutated genes with nonmutant copies from other members of the population. Other processes have a more creative role in that they generate novel rearrangements of the genome through addition, deletion, or transposition. Such novel rearrangements also can be caused by transduction, especially transduction between phylogenetically distant bacteria, but most dramatically, and perhaps most important, they result from the addition of a prophage to the bacterial genome through lysogenization.

When an infecting phage enters the lysogenic cycle and becomes a prophage, the cell is presented with both challenges and opportunities. On the one hand, the potential of the prophage to kill the cell through prophage induction reduces the long-term fitness of the cell's lineage, and this fact may account for the high deletion rates in bacteria and the consequent compact genetic organization that characterizes bacterial (and archaeal) genomes. On the other hand, phages typically have genes that are expressed from the prophage and confer a selective benefit on the host—for example, by protecting the host from infection by other phages or by encoding a factor that enhances the cell's ability to replicate or survive in a particular environment (e.g., virulence factor genes in bacterial pathogens); such a benefit to the host mitigates against the gene's deletion, even when some or all of the other prophage genes may be lost. It is not clear at this point how many "bacterial" genes may have gained entry into the bacterial chromosome by such a mechanism. Finally, the phage taillike bacteriocins and the GTAs appear to provide examples in which phage genes whose initial function was strictly in lytic growth of the phage have been captured by the host cell, placed under regulation appropriate to their new cellular role, and can now no longer be considered to be phage genes at all except in the matter of their ancestry.

REFERENCES

1. **Bagdian, G., O. Luderitz, and A. M. Staub.** 1966. Immunochemical studies on *Salmonella*. XI. Chemical modification correlated with conversion of group B *Salmonella* by bacteriophage 27. *Ann. N. Y. Acad. Sci.* **133:**405–424.
2. **Banks, D. J., S. B. Beres, and J. M. Musser.** 2002. The fundamental contribution of phages to GAS evolution, genome diversification and strain emergence. *Trends Microbiol.* **10:**515–521.
3. **Barondess, J. J., and J. Beckwith.** 1995. bor gene of phage lambda, involved in serum resistance, encodes a widely conserved outer membrane lipoprotein. *J. Bacteriol.* **177:**1247–1253.

4. Bejar, S., F. Bouche, and J. P. Bouche. 1988. Cell division inhibition gene *dicB* is regulated by a locus similar to lambdoid bacteriophage immunity loci. *Mol. Gen. Genet.* 212:11–19.

5. Benzer, S. 1959. On the topology of genetic fine structure. *Proc. Natl. Acad. Sci. USA* 45:1607–1620.

6. Beres, S. B., G. L. Sylva, K. D. Barbian, B. Lei, J. S. Hoff, N. D. Mammarella, M. Y. Liu, J. C. Smoot, S. F. Porcella, L. D. Parkins, D. S. Campbell, T. M. Smith, J. K. McCormick, D. Y. Leung, P. M. Schlievert, and J. M. Musser. 2002. Genome sequence of a serotype M3 strain of group A *Streptococcus*: phage-encoded toxins, the high-virulence phenotype, and clone emergence. *Proc. Natl. Acad. Sci. USA* 99: 10078–10083.

7. Bergh, O., K. Y. Borsheim, G. Bratbak, and M. Heldal. 1989. High abundance of viruses found in aquatic environments. *Nature* 340:467–468.

8. Bergthorsson, U., and H. Ochman. 1998. Distribution of chromosome length variation in natural isolates of *Escherichia coli*. *Mol. Biol. Evol.* 15:6–16.

9. Bertani, G. 1999. Transduction-like gene transfer in the methanogen *Methanococcus voltae*. *J. Bacteriol.* 181:2992–3002.

10. Betley, M. J., and J. J. Mekalanos. 1985. Staphylococcal enterotoxin A is encoded by phage. *Science* 229:185–187.

11. Beutin, L., E. Strauch, and I. Fischer. 1999. Isolation of *Shigella sonnei* lysogenic for a bacteriophage encoding gene for production of Shiga toxin. *Lancet* 353:1498.

12. Beutin, L., U. H. Stroeher, and P. A. Manning. 1993. Isolation of enterohemolysin (Ehly2)-associated sequences encoded on temperate phages of *Escherichia coli*. *Gene* 132:95–99.

13. Bishai, W., and J. Murphy. 1988. Bacteriophage gene products that cause human disease, p. 683–724. *In* R. Calendar (ed.), *The Bacteriophages*, vol. 2. Plenum Press, New York, N.Y.

14. Blasband, A. J., W. R. Marcotte, Jr., and C. A. Schnaitman. 1986. Structure of the *lc* and *nmpC* outer membrane porin protein genes of lambdoid bacteriophage. *J. Biol. Chem.* 261:12723–12732.

15. Boyd, E. F., and H. Brussow. 2002. Common themes among bacteriophage-encoded virulence factors and diversity among the bacteriophages involved. *Trends Microbiol.* 10: 521–529.

16. Bray, D., and P. Robbins. 1967. Mechanism of ε15 conversion studied with bacteriophage mutants. *J. Mol. Biol.* 30:457–475.

17. Campbell, A. 1994. Comparative molecular biology of lambdoid phages. *Annu. Rev. Microbiol.* 48:193–222.

18. Campbell, A. M. 2002. Preferential orientation of natural lambdoid prophages and bacterial chromosome organization. *Theor. Popul. Biol.* 61:503–507.

19. Canchaya, C., C. Proux, G. Fournous, A. Bruttin, and H. Brussow. 2003. Prophage genomics. *Microbiol. Mol. Biol. Rev.* 67:238–276.

20. Casjens, S. 1998. The diverse and dynamic structure of bacterial genomes. *Annu. Rev. Genet.* 32:339–377.

21. Casjens, S. 2003. Prophages and bacterial genomics: what have we learned so far? *Mol. Microbiol.* 49:277–300.

22. Casjens, S., N. Palmer, R. van Vugt, W. M. Huang, B. Stevenson, P. Rosa, R. Lathigra, G. Sutton, J. Peterson, R. J. Dodson, D. Haft, E. Hickey, M. Gwinn, O. White, and C. M. Fraser. 2000. A bacterial genome in flux: the twelve linear and nine circular extrachromosomal DNAs in an infectious isolate of the Lyme disease spirochete *Borrelia burgdorferi*. *Mol. Microbiol.* 35:490–516.

23. Cheetham, B. F., D. B. Tattersall, G. A. Bloomfield, J. I. Rood, and M. E. Katz. 1995. Identification of a gene encoding a bacteriophage-related integrase in a vap region of the *Dichelobacter nodosus* genome. *Gene* 162:53–58.

24. Cheetham, F., and M. Katz. 1995. A role for bacteriophages in the evolution and transfer of bacterial virulence determinants. *Mol. Microbiol.* 18:201–208.

25. Clark, C. A., J. Beltrame, and P. A. Manning. 1991. The oac gene encoding a lipopolysaccharide O-antigen acetylase maps adjacent to the integrase-encoding gene on the genome of *Shigella flexneri* bacteriophage Sf6. *Gene* 107:43–52.

26. Coleman, D. C., D. J. Sullivan, R. J. Russell, J. P. Arbuthnott, B. F. Carey, and H. M. Pomeroy. 1989. *Staphylococcus aureus* bacteriophages mediating the simultaneous lysogenic conversion of beta-lysin, staphylokinase and enterotoxin A: molecular mechanism of triple conversion. *J. Gen. Microbiol.* 135:1679–1697.

27. Ebel-Tsipis, J., and D. Botstein. 1971. Superinfection exclusion by P22 prophage in lysogens of *Salmonella typhimurium*. 1. Exclusion of generalized transducing particles. *Virology* 45:629–637.

28. Eggers, C. H., S. Casjens, S. F. Hayes, C. F. Garon, C. J. Damman, D. B. Oliver, and D. S. Samuels. 2000. Bacteriophages of spirochetes. *J. Mol. Microbiol. Biotechnol.* 2:365–373.

29. Eggers, C. H., S. Casjens, and D. S. Samuels. 2001. Bacteriophages of *Borrelia burgdorferi* and other spirochetes, p. 35–44. *In* M. Saier and J. Garcia-Lara (ed.), *The Spirochetes. Molecular and Cellular Biology.* Horizon Scientific Press, Wymondham, United Kingdom.

30. Eklund, M. W., F. T. Poysky, J. A. Meyers, and G. A. Pelroy. 1974. Interspecies conversion of *Clostridium botulinum* type C to *Clostridium novyi* type A by bacteriophage. *Science* 186:456–458.

31. Engelberg-Kulka, H., M. Reches, S. Narasimhan, R. Schoulaker-Schwarz, Y. Klemes, E. Aizenman, and G. Glaser. 1998. rexB of bacteriophage lambda is an anti-cell death gene. *Proc. Natl. Acad. Sci. USA* 95:15481–15486.

32. Ferretti, J. J., W. M. McShan, D. Ajdic, D. J. Savic, G. Savic, K. Lyon, C. Primeaux, S. Sezate, A. N. Suvorov, S. Kenton, H. S. Lai, S. P. Lin, Y. Qian, H. G. Jia, F. Z. Najar, Q. Ren, H. Zhu, L. Song, J. White, X. Yuan, S. W. Clifton, B. A. Roe, and R. McLaughlin. 2001. Complete genome sequence of an M1 strain of *Streptococcus pyogenes*. *Proc. Natl. Acad. Sci. USA* 98:4658–4663.

33. Figueroa-Bossi, N., S. Uzzau, D. Maloriol, and L. Bossi. 2001. Variable assortment of prophages provides a transferable repertoire of pathogenic determinants in *Salmonella*. *Mol. Microbiol.* 39:260–271.

34. Fitzgerald, J. R., S. R. Monday, T. J. Foster, G. A. Bohach, P. J. Hartigan, W. J. Meaney, and C. J. Smyth. 2001. Characterization of a putative pathogenicity island from bovine *Staphylococcus aureus* encoding multiple superantigens. *J. Bacteriol.* 183:63–70.

35. Freeman, V. 1951. Studies on the virulence of bacteriophage-infected strains of *Corynebacterium diphtheriae*. *J. Bacteriol.* 61:675–688.

36. Glaser, P., L. Frangeul, C. Buchrieser, C. Rusniok, A. Amend, F. Baquero, P. Berche, H. Bloecker, P. Brandt, T. Chakraborty, A. Charbit, F. Chetouani, E. Couve, A. de Daruvar, P. Dehoux, E. Domann, G. Dominguez-Bernal, E. Duchaud, L. Durant, O. Dussurget, K. D. Entian, H. Fsihi, F. G. Portillo, P. Garrido, L. Gautier, W. Goebel, N. Gomez-Lopez, T. Hain, J. Hauf, D. Jackson, L. M. Jones, U. Kaerst, J. Kreft, M. Kuhn, F. Kunst, G. Kurapkat, E. Madueno, A. Maitournam, J. M. Vicente, E. Ng, H. Nedjari, G. Nordsiek,

S. Novella, B. de Pablos, J. C. Perez-Diaz, R. Purcell, B. Remmel, M. Rose, T. Schlueter, N. Simoes, A. Tierrez, J. A. Vazquez-Boland, H. Voss, J. Wehland, and P. Cossart. 2001. Comparative genomics of *Listeria* species. *Science* **294**:849–852.

37. Groman, N. B. 1984. Conversion by corynephages and its role in the natural history of diphtheria. *J. Hyg.* **93**:405–417.

38. Hardt, W. D., L. M. Chen, K. E. Schuebel, X. R. Bustelo, and J. E. Galan. 1998. S. typhimurium encodes an activator of Rho GTPases that induces membrane ruffling and nuclear responses in host cells. *Cell* **93**:815–826.

39. Hauser, D., M. W. Eklund, P. Boquet, and M. R. Popoff. 1994. Organization of the botulinum neurotoxin C1 gene and its associated non-toxic protein genes in *Clostridium botulinum* C 468. *Mol. Gen. Genet.* **243**:631–640.

40. Hayashi, T., K. Makino, M. Ohnishi, K. Kurokawa, K. Ishii, K. Yokoyama, C. G. Han, E. Ohtsubo, K. Nakayama, T. Murata, M. Tanaka, T. Tobe, T. Iida, H. Takami, T. Honda, C. Sasakawa, N. Ogasawara, T. Yasunaga, S. Kuhara, T. Shiba, M. Hattori, and H. Shinagawa. 2001. Complete genome sequence of enterohemorrhagic *Escherichia coli* O157:H7 and genomic comparison with a laboratory strain K-12. *DNA Res.* **8**:11–22.

41. Hayashi, T., K. Makino, M. Ohnishi, K. Kurokawa, K. Ishii, K. Yokoyama, C. G. Han, E. Ohtsubo, K. Nakayama, T. Murata, M. Tanaka, T. Tobe, T. Iida, H. Takami, T. Honda, C. Sasakawa, N. Ogasawara, T. Yasunaga, S. Kuhara, T. Shiba, M. Hattori, and H. Shinagawa. 2001. Complete genome sequence of enterohemorrhagic *Escherichia coli* O157:H7 and genomic comparison with a laboratory strain K-12 (Supplement). *DNA Res.* **8**(Suppl.):47–52.

42. Hellwage, J., T. Meri, T. Heikkila, A. Alitalo, J. Panelius, P. Lahdenne, I. J. Seppala, and S. Meri. 2001. The complement regulator factor H binds to the surface protein OspE of *Borrelia burgdorferi*. *J. Biol. Chem.* **276**:8427–8435.

43. Hemphill, H. E., I. Gage, S. A. Zahler, and R. Z. Korman. 1980. Prophage-mediated production of a bacteriocinlike substance by SP beta lysogens of *Bacillus subtilis*. *Can. J. Microbiol.* **26**:1328–1333.

44. Hendrix, R. W., J. G. Lawrence, G. F. Hatfull, and S. Casjens. 2000. The origins and ongoing evolution of viruses. *Trends Microbiol.* **8**:504–508.

45. Hendrix, R. W., M. C. Smith, R. N. Burns, M. E. Ford, and G. F. Hatfull. 1999. Evolutionary relationships among diverse bacteriophages and prophages: all the world's a phage. *Proc. Natl. Acad. Sci. USA* **96**:2192–2197.

46. Huan, P. T., R. Taylor, A. A. Lindberg, and N. K. Verma. 1995. Immunogenicity of the *Shigella flexneri* serotype Y (SFL 124) vaccine strain expressing cloned glucosyl transferase gene of converting bacteriophage SfX. *Microbiol. Immunol.* **39**:467–472.

47. Huan, P. T., B. L. Whittle, D. A. Bastin, A. A. Lindberg, and N. K. Verma. 1997. *Shigella flexneri* type-specific antigen V: cloning, sequencing and characterization of the glucosyl transferase gene of temperate bacteriophage SfV. *Gene* **195**:207–216.

48. Humphrey, S. B., T. B. Stanton, N. S. Jensen, and R. L. Zuerner. 1997. Purification and characterization of VSH-1, a generalized transducing bacteriophage of *Serpulina hyodysenteriae*. *J. Bacteriol.* **179**:323–329.

49. Ikeda, H., and J. Tomizawa. 1965. Transducing particles in generalized transduction by phage P1. I. Molecular origins of the fragments. *J. Mol. Biol.* **14**:85–109.

50. Johnson, L. P., and P. M. Schlievert. 1984. Group A streptococcal phage T12 carries the structural gene for pyrogenic exotoxin type A. *Mol. Gen. Genet.* **194**:52–56.

51. Johnson, L. P., P. M. Schlievert, and D. W. Watson. 1980. Transfer of group A streptococcal pyrogenic exotoxin production to nontoxigenic strains of lysogenic conversion. *Infect. Immun.* **28**:254–257.

52. Juhala, R. J., M. E. Ford, R. L. Duda, A. Youlton, G. F. Hatfull, and R. W. Hendrix. 2000. Genomic sequences of bacteriophages HK97 and HK022: pervasive genetic mosaicism in the lambdoid bacteriophages. *J. Mol. Biol.* **299**:27–51.

53. Kaiser, K. 1980. The origin of Q-independent derivatives of phage lambda. *Mol. Gen. Genet.* **179**:547–554.

54. Kaiser, K., and N. Murray. 1980. On the nature of SbcA mutations in *E. coli* K-12. *Mol. Gen. Genet.* **179**:555–563.

55. Kaiser, K., and N. E. Murray. 1979. Physical characterisation of the "Rac prophage" in *E. coli* K12. *Mol. Gen. Genet.* **175**:159–174.

56. Kaneko, J., T. Kimura, S. Narita, T. Tomita, and Y. Kamio. 1998. Complete nucleotide sequence and molecular characterization of the temperate staphylococcal bacteriophage φPVL carrying Panton-Valentine leukocidin genes. *Gene* **215**:57–67.

57. Karaolis, D. K., S. Somara, D. R. Maneval, Jr., J. A. Johnson, and J. B. Kaper. 1999. A bacteriophage encoding a pathogenicity island, a type-IV pilus and a phage receptor in cholera bacteria. *Nature* **399**:375–379.

58. Kleerebezem, M., J. Boekhorst, R. van Kranenburg, D. Molenaar, O. P. Kuipers, R. Leer, R. Tarchini, S. A. Peters, H. M. Sandbrink, M. W. Fiers, W. Stiekema, R. M. Lankhorst, P. A. Bron, S. M. Hoffer, M. N. Groot, R. Kerkhoven, M. de Vries, B. Ursing, W. M. de Vos, and R. J. Siezen. 2003. Complete genome sequence of *Lactobacillus plantarum* WCFS1. *Proc. Natl. Acad. Sci. USA* **100**:1990–1995.

59. Krogh, S., M. O'Reilly, N. Nolan, and K. M. Devine. 1996. The phage-like element PBSX and part of the skin element, which are resident at different locations on the *Bacillus subtilis* chromosome, are highly homologous. *Microbiology* **142**:2031–2040.

60. Kropinski, A. M. 2000. Sequence of the genome of the temperate, serotype-converting, *Pseudomonas aeruginosa* bacteriophage D3. *J. Bacteriol.* **182**:6066–6074.

61. Kuroda, M., T. Ohta, I. Uchiyama, T. Baba, H. Yuzawa, I. Kobayashi, L. Cui, A. Oguchi, K. Aoki, Y. Nagai, J. Lian, T. Ito, M. Kanamori, H. Matsumaru, A. Maruyama, H. Murakami, A. Hosoyama, Y. Mizutani-Ui, N. K. Takahashi, T. Sawano, R. Inoue, C. Kaito, K. Sekimizu, H. Hirakawa, S. Kuhara, S. Goto, J. Yabuzaki, M. Kanehisa, A. Yamashita, K. Oshima, K. Furuya, C. Yoshino, T. Shiba, M. Hattori, N. Ogasawara, H. Hayashi, and K. Hiramatsu. 2001. Whole genome sequencing of methicillin-resistant *Staphylococcus aureus*. *Lancet* **357**:1225–1240.

62. Lang, A. S., and J. T. Beatty. 2000. Genetic analysis of a bacterial genetic exchange element: the gene transfer agent of *Rhodobacter capsulatus*. *Proc. Natl. Acad. Sci. USA* **97**:859–864.

63. Lang, A. S., and J. T. Beatty. 2001. The gene transfer agent of *Rhodobacter capsulatus* and "constitutive transduction" in prokaryotes. *Arch. Microbiol.* **175**:241–249.

64. Lawrence, J. G., R. W. Hendrix, and S. Casjens. 2001. Where are the bacterial pseudogenes? *Trends Microbiol.* **9**:535–540.

65. Lawrence, J. G., and J. R. Roth. 1996. Selfish operons: horizontal transfer may drive the evolution of gene clusters. *Genetics* **143**:1843–1860.

66. Lindsay, J. A., A. Ruzin, H. F. Ross, N. Kurepina, and R. P. Novick. 1998. The gene for toxic shock toxin is carried by

a family of mobile pathogenicity islands in *Staphylococcus aureus*. *Mol. Microbiol.* **29**:527–543.

67. Lindsey, D. F., D. A. Mullin, and J. R. Walker. 1989. Characterization of the cryptic lambdoid prophage DLP12 of *Escherichia coli* and overlap of the DLP12 integrase gene with the tRNA gene *argU*. *J. Bacteriol.* **171**:6197–6205.

68. Mahdi, A. A., G. J. Sharples, T. N. Mandal, and R. G. Lloyd. 1996. Holliday junction resolvases encoded by homologous *rusA* genes in *Escherichia coli* K-12 and phage 82. *J. Mol. Biol.* **257**:561–573.

69. Masignani, V., M. M. Giuliani, H. Tettelin, M. Comanducci, R. Rappuoli, and V. Scarlato. 2001. Mu-like prophage in serogroup B *Neisseria meningitidis* coding for surface-exposed antigens. *Infect. Immun.* **69**:2580–2588.

70. Masters, M. 1996. Generalized transduction, p. 2421–2441. *In* F. C. Neidhardt, R. Curtiss III, J. L. Ingraham, E. C. C. Lin, K. B. Low, B. Magasanik, W. S. Reznikoff, M. Riley, M. Schaechter, and H. E. Umbarger (ed.), Escherichia coli *and* Salmonella: *Cellular and Molecular Biology*, 2nd ed., vol. 2. ASM Press, Washington, D.C.

71. Mavris, M., P. A. Manning, and R. Morona. 1997. Mechanism of bacteriophage SfII-mediated serotype conversion in *Shigella flexneri*. *Mol. Microbiol.* **26**:939–950.

72. McDonough, M. A., and J. R. Butterton. 1999. Spontaneous tandem amplification and deletion of the shiga toxin operon in *Shigella dysenteriae* 1. *Mol. Microbiol.* **34**:1058–1069.

73. Miao, E. A., and S. I. Miller. 1999. Bacteriophages in the evolution of pathogen-host interactions. *Proc. Natl. Acad. Sci. USA* **96**:9452–9454.

74. Milkman, R., E. A. Raleigh, M. McKane, D. Cryderman, P. Bilodeau, and K. McWeeny. 1999. Molecular evolution of the Escherichia coli chromosome. V. Recombination patterns among strains of diverse origin. *Genetics* **153**:539–554.

75. Mirold, S., W. Rabsch, M. Rohde, S. Stender, H. Tschape, H. Russmann, E. Igwe, and W. D. Hardt. 1999. Isolation of a temperate bacteriophage encoding the type III effector protein SopE from an epidemic Salmonella typhimurium strain. *Proc. Natl. Acad. Sci. USA* **96**:9845–9850.

76. Miyamoto, H., W. Nakai, N. Yajima, A. Fujibayashi, T. Higuchi, K. Sato, and A. Matsushiro. 1999. Sequence analysis of Stx2-converting phage VT2-Sa shows a great divergence in early regulation and replication regions. *DNA Res.* **6**:235–240.

77. Mizuno, M., S. Masuda, K. Takemaru, S. Hosono, T. Sato, M. Takeuchi, and Y. Kobayashi. 1996. Systematic sequencing of the 283 kb 210 degrees–232 degrees region of the *Bacillus subtilis* genome containing the SKIN element and many sporulation genes. *Microbiology* **142**:3103–3111.

78. Moreira, D. 2000. Multiple independent horizontal transfers of informational genes from bacteria to plasmids and phages: implications for the origin of bacterial replication machinery. *Mol. Microbiol.* **35**:1–5.

79. Morgan, G., G. Hatfull, S. Casjens, and R. Hendrix. 2002. Bacteriophage Mu genome sequence: analysis and comparison with Mu-like prophages in *Haemophilus*, *Neisseria* and *Deinococcus*. *J. Mol. Biol.* **317**:337–359.

80. Morimyo, M., E. Hongo, H. Hama-Inaba, and I. Machida. 1992. Cloning and characterization of the *mvrC* gene of *Escherichia coli* K-12 which confers resistance against methyl viologen toxicity. *Nucleic Acids Res.* **20**:3159–3165.

81. Muniesa, M., J. Recktenwald, M. Bielaszewska, H. Karch, and H. Schmidt. 2000. Characterization of a shiga toxin 2e-converting bacteriophage from an *Escherichia coli* strain of human origin. *Infect. Immun.* **68**:4850–4855.

82. Nakagawa, I., K. Kurokawa, A. Yamashita, M. Nakata, Y. Tomiyasu, N. Okahashi, S. Kawabata, K. Yamazaki,

T. Shiba, T. Yasunaga, H. Hayashi, M. Hattori, and S. Hamada. 2003. Genome sequence of an M3 strain of *Streptococcus pyogenes* reveals a large-scale genomic rearrangement in invasive strains and new insights into phage evolution. *Genome Res.* **13**:1042–1055.

83. Nakayama, K., S. Kanaya, M. Ohnishi, Y. Terawaki, and T. Hayashi. 1999. The complete nucleotide sequence of φCTX, a cytotoxin-converting phage of *Pseudomonas aeruginosa*: implications for phage evolution and horizontal gene transfer via bacteriophages. *Mol. Microbiol.* **31**:399–419.

84. Nakayama, K., K. Takashima, H. Ishihara, T. Shinomiya, M. Kageyama, S. Kanaya, M. Ohnishi, T. Murata, H. Mori, and T. Hayashi. 2000. The R-type pyocin of *Pseudomonas aeruginosa* is related to P2 phage, and the F-type is related to lambda phage. *Mol. Microbiol.* **38**:213–231.

85. Neely, M. N., and D. I. Friedman. 1998. Arrangement and functional identification of genes in the regulatory region of lambdoid phage H-19B, a carrier of a Shiga-like toxin. *Gene* **223**:105–113.

86. Nelson, K. E., R. A. Clayton, S. R. Gill, M. L. Gwinn, R. J. Dodson, D. H. Haft, E. K. Hickey, J. D. Peterson, W. C. Nelson, K. A. Ketchum, L. McDonald, T. R. Utterback, J. A. Malek, K. D. Linher, M. M. Garrett, A. M. Stewart, M. D. Cotton, M. S. Pratt, C. A. Phillips, D. Richardson, J. Heidelberg, G. G. Sutton, R. D. Fleischmann, J. A. Eisen, and C. M. Fraser. 1999. Evidence for lateral gene transfer between Archaea and bacteria from genome sequence of *Thermotoga maritima*. *Nature* **399**:323–329.

87. Nguyen, A. H., T. Tomita, M. Hirota, T. Sato, and Y. Kamio. 1999. A simple purification method and morphology and component analyses for carotovoricin Er, a phage-tail-like bacteriocin from the plant pathogen *Erwinia carotovora* Er. *Biosci. Biotechnol. Biochem.* **63**:1360–1369.

88. Ohnishi, M., K. Kurokawa, and T. Hayashi. 2001. Diversification of *Escherichia coli* genomes: are bacteriophages the major contributors? *Trends Microbiol.* **9**:481–485.

89. Osawa, R., S. Iyoda, S. I. Nakayama, A. Wada, S. Yamai, and H. Watanabe. 2000. Genotypic variations of Shiga toxin-converting phages from enterohaemorrhagic *Escherichia coli* O157:H7 isolates. *J. Med. Microbiol.* **49**:565–574.

90. Paul, J. H. 1999. Microbial gene transfer: an ecological perspective. *J. Mol. Microbiol. Biotechnol.* **1**:45–50.

91. Pedulla, M. L., M. E. Ford, J. M. Houtz, T. Karthikeyan, C. Wadsworth, J. A. Lewis, D. Jacobs-Sera, J. Falbo, J. Gross, N. R. Pannunzio, W. Brucker, V. Kumar, J. Kandasamy, L. Keenan, S. Bardarov, J. Kriakov, J. G. Lawrence, W. R. Jacobs, Jr., R. W. Hendrix, and G. F. Hatfull. 2003. Origins of highly mosaic mycobacteriophage genomes. *Cell* **113**:171–182.

92. Penner, M., I. Morad, L. Snyder, and G. Kaufmann. 1995. Phage T4-coded Stp: double-edged effector of coupled DNA and tRNA-restriction systems. *J. Mol. Biol.* **249**:857–868.

93. Perlak, F. J., C. L. Mendelsohn, and C. B. Thorne. 1979. Converting bacteriophage for sporulation and crystal formation in *Bacillus thuringiensis*. *J. Bacteriol.* **140**:699–706.

94. Perna, N., J. Glasner, V. Burland, and G. Plunkett III. 2002. The genomes of *Escherichia coli* K-12 and pathogenic *E. coli*, p. 3–53. *In* M. Donnenberg (ed.), Escherichia coli. *Virulence Mechanisms of a Versatile Pathogen*. Academic Press, New York, N.Y.

95. Perna, N. T., G. Plunkett III, V. Burland, B. Mau, J. D. Glasner, D. J. Rose, G. F. Mayhew, P. S. Evans, J. Gregor, H. A. Kirkpatrick, G. Posfai, J. Hackett, S. Klink, A. Boutin, Y. Shao, L. Miller, E. J. Grotbeck, N. W. Davis, A. Lim, E. T. Dimalanta, K. D. Potamousis, J. Apodaca, T. S. Anantharaman, J. Lin, G. Yen, D. C. Schwartz,

R. A. Welch, and F. R. Blattner. 2001. Genome sequence of enterohaemorrhagic *Escherichia coli* O157:H7. *Nature* **409:**529–533.

96. Plasterk, R. H., and P. van de Putte. 1985. The invertible P-DNA segment in the chromosome of *Escherichia coli*. *EMBO J.* **4:**237–242.

97. Plunkett, G., III, D. J. Rose, T. J. Durfee, and F. R. Blattner. 1999. Sequence of Shiga toxin 2 phage 933W from *Escherichia coli* O157:H7: Shiga toxin as a phage late-gene product. *J. Bacteriol.* **181:**1767–1778.

98. Rapp, B., and J. Wall. 1987. Genetic transfer in *Desulfovibrio desulfuricans. Proc. Natl. Acad. Sci. USA* **84:**9128–9130.

99. Reidl, J., and J. J. Mekalanos. 1995. Characterization of *Vibrio cholerae* bacteriophage K139 and use of a novel mini-transposon to identify a phage-encoded virulence factor. *Mol. Microbiol.* **18:**685–701.

100. Roth, J. R., N. Benson, T. Galitski, K. Haack, J. G. Lawrence, and L. Miesel. 1996. Rearrangements of the bacterial chromosome: formation and applications, p. 2256–2276. *In* F. C. Neidhardt, R. Curtiss III, J. L. Ingraham, E. C. C. Lin, K. B. Low, B. Magasanik, W. S. Reznikoff, M. Riley, M. Schaechter, and H. E. Umbarger (ed.), Escherichia coli *and* Salmonella: *Cellular and Molecular Biology*, 2nd ed., vol. 2. ASM Press, Washington, D.C.

101. Sandt, C. H., and C. W. Hill. 2000. Four different genes responsible for nonimmune immunoglobulin-binding activities within a single strain of *Escherichia coli. Infect. Immun.* **68:**2205–2214.

102. Schicklmaier, P., E. Moser, T. Wieland, W. Rabsch, and H. Schmieger. 1998. A comparative study on the frequency of prophages among natural isolates of *Salmonella* and *Escherichia coli* with emphasis on generalized transducers. *Antonie Leeuwenhoek* **73:**49–54.

103. Schmieger, H. 1999. Molecular survey of the *Salmonella* phage typing system of Anderson. *J. Bacteriol.* **181:**1630–1635.

104. Seldin, L., and E. G. Penido. 1990. Production of a bacteriophage, a phage tail-like bacteriocin and an antibiotic by *Bacillus azotofixans. An. Acad. Bras. Cienc.* **62:**85–94.

105. Shinedling, S., D. Parma, and L. Gold. 1987. Wild-type bacteriophage T4 is restricted by the lambda *rex* genes. *J. Virol.* **61:**3790–3794.

106. Silver-Mysliwiec, T. H., and M. G. Bramucci. 1990. Bacteriophage-enhanced sporulation: comparison of spore-converting bacteriophages PMB12 and SP10. *J. Bacteriol.* **172:**1948–1953.

107. Simpson, A. A., Y. Tao, P. G. Leiman, M. O. Badasso, Y. He, P. J. Jardine, N. H. Olson, M. C. Morais, S. Grimes, D. L. Anderson, T. S. Baker, and M. G. Rossmann. 2000. Structure of the bacteriophage φ29 DNA packaging motor. *Nature* **408:**745–750.

108. Simpson, A. J., F. C. Reinach, P. Arruda, F. A. Abreu, M. Acencio, R. Alvarenga, L. M. Alves, J. E. Araya, G. S. Baia, C. S. Baptista, M. H. Barros, E. D. Bonaccorsi, S. Bordin, J. M. Bove, M. R. Briones, M. R. Bueno, A. A. Camargo, L. E. Camargo, D. M. Carraro, H. Carrer, N. B. Colauto, C. Colombo, F. F. Costa, M. C. Costa, C. M. Costa-Neto, L. L. Coutinho, M. Cristofani, E. Dias-Neto, C. Docena, H. El-Dorry, A. P. Facincani, A. J. Ferreira, V. C. Ferreira, J. A. Ferro, J. S. Fraga, S. C. Franca, M. C. Franco, M. Frohme, L. R. Furlan, M. Garnier, G. H. Goldman, M. H. Goldman, S. L. Gomes, A. Gruber, P. L. Ho, J. D. Hoheisel, M. L. Junqueira, E. L. Kemper, J. P. Kitajima, J. E. Krieger, E. E. Kuramae, F. Laigret, M. R. Lambais, L. C. Leite, E. G. Lemos, M. V. Lemos, S. A. Lopes,

C. R. Lopes, J. A. Machado, M. A. Machado, A. M. Madeira, H. M. Madeira, C. L. Marino, M. V. Marques, E. A. Martins, E. M. Martins, A. Y. Matsukuma, C. F. Menck, E. C. Miracca, C. Y. Miyaki, C. B. Monteiro-Vitorello, D. H. Moon, M. A. Nagai, A. L. Nascimento, L. E. Netto, A. Nhani, Jr., F. G. Nobrega, L. R. Nunes, M. A. Oliveira, M. C. de Oliveira, R. C. de Oliveira, D. A. Palmieri, A. Paris, B. R. Peixoto, G. A. Pereira, H. A. Pereira, Jr., J. B. Pesquero, R. B. Quaggio, P. G. Roberto, V. Rodrigues, M. R. A. J. de, V. E. de Rosa, Jr., R. G. de Sa, R. V. Santelli, H. E. Sawasaki, A. C. da Silva, A. M. da Silva, F. R. da Silva, W. A. da Silva, Jr., J. F. da Silveira, et al. 2000. The genome sequence of the plant pathogen *Xylella fastidiosa. Nature* **406:**151–157.

109. Snyder, L., and G. Kaufmann. 1994. T4 phage exclusion mechansims, p. 391–396. *In* J. Karam (ed.), *Molecular Biology of Bacteriophage T4.* ASM Press, Washington, D.C.

110. Steensma, H. Y. 1981. Effect of defective phages on the cell membrane of *Bacillus subtilis* and partial characterization of the phage protein involved in killing. *J. Gen. Virol.* **56:**275–286.

111. Stevenson, B., N. El-Hage, M. E. Hines, J. H. Miller, and K. Babb. 2002. Differential binding of host complement inhibitor factor H by *Borrelia burgdoferi* Erp surface proteins: a possible mechanism behind the expansive host range of Lyme disease spirochetes. *Infect. Immun.* **70:**491–497.

112. Stewart, A. W., and M. G. Johnson. 1977. Increased numbers of heat-resistant spores produced by two strains of *Clostridium perfringens* bearing temperate phage s9. *J. Gen. Microbiol.* **103:**45–50.

113. Susskind, M. M., D. Botstein, and A. Wright. 1974. Superinfection exclusion by P22 prophage in lysogens of *Salmonella typhimurium.* III. Failure of superinfecting phage DNA to enter sieA+ lysogens. *Virology* **62:**350–366.

114. Susskind, M. M., A. Wright, and D. Botstein. 1974. Superinfection exclusion by P22 prophage in lysogens of *Salmonella typhimurium.* IV. Genetics and physiology of *sieB* exclusion. *Virology* **62:**367–384.

115. Takeda, Y., and J. R. Murphy. 1978. Bacteriophage conversion of heat-labile enterotoxin in *Escherichia coli. J. Bacteriol.* **133:**172–177.

116. Thaler, J. O., S. Baghdiguian, and N. Boemare. 1995. Purification and characterization of xenorhabdicin, a phage tail-like bacteriocin, from the lysogenic strain F1 of *Xenorhabdus nematophilus. Appl. Environ. Microbiol.* **61:**2049–2052.

117. Unkmeir, A., and H. Schmidt. 2000. Structural analysis of phage-borne stx genes and their flanking sequences in shiga toxin-producing *Escherichia coli* and *Shigella dysenteriae* type 1 strains. *Infect. Immun.* **68:**4856–4864.

118. Vaca Pacheco, S., O. Garcia Gonzalez, and G. L. Paniagua Contreras. 1997. The *lom* gene of bacteriophage lambda is involved in *Escherichia coli* K12 adhesion to human buccal epithelial cells. *FEMS Microbiol. Lett.* **156:**129–132.

119. Vaca-Pacheco, S., G. L. Paniagua-Contreras, O. Garcia-Gonzalez, and M. de la Garza. 1999. The clinically isolated FIZ15 bacteriophage causes lysogenic conversion in Pseudomonas aeruginosa PAO1. *Curr. Microbiol.* **38:**239–243.

120. Van Sluys, M. A., M. C. de Oliveira, C. B. Monteiro-Vitorello, C. Y. Miyaki, L. R. Furlan, L. E. A. Camargo, A. C. R. da Silva, D. H. Moon, M. A. Takita, E. G. M. Lemos, M. A. Machado, M. I. T. Ferro, F. R. da Silva, M. H. S. Goldman, G. H. Goldman, M. V. F. Lemos, H. El-Dorry, S. M. Tsai, H. Carrer, D. M. Carraro, R. C. de Oliveira, L. R. Nunes, W. J. Siqueira, L. L. Coutinho, E. T. Kimura, E. S. Ferro, R. Harakava, E. E. Kuramae,

C. L. Marino, E. Giglioti, I. L. Abreu, L. M. C. Alves, A. M. do Amaral, G. S. Baia, S. R. Blanco, M. S. Brito, F. S. Cannavan, A. V. Celestino, A. F. da Cunha, R. C. Fenille, J. A. Ferro, E. F. Formighieri, L. T. Kishi, S. G. Leoni, A. R. Oliveira, V. E. Rosa, Jr., F. T. Sassaki, J. A. D. Sena, A. A. de Souza, D. Truffi, F. Tsukumo, G. M. Yanai, L. G. Zaros, E. L. Civerolo, A. J. G. Simpson, N. F. Almeida, Jr., J. C. Setubal, and J. P. Kitajima. 2003. Comparative analyses of the complete genome sequences of Pierce's disease and citrus variegated chlorosis strains of *Xylella fastidiosa*. *J. Bacteriol.* **185**:1018–1026.

121. Voelker, L. L., and K. Dybvig. 1999. Sequence analysis of the *Mycoplasma arthritidis* bacteriophage MAV1 genome identifies the putative virulence factor. *Gene* **233**:101–107.

122. Wagner, P. L., M. N. Neely, X. Zhang, D. W. Acheson, M. K. Waldor, and D. I. Friedman. 2001. Role for a phage promoter in Shiga toxin 2 expression from a pathogenic *Escherichia coli* strain. *J. Bacteriol.* **183**:2081–2085.

123. Wagner, P. L., and M. K. Waldor. 2002. Bacteriophage control of bacterial virulence. *Infect. Immun.* **70**:3985–3993.

124. Waldor, M. K., and J. J. Mekalanos. 1996. Lysogenic conversion by a filamentous phage encoding cholera toxin. *Science* **272**:1910–1914.

125. Wang, F. S., T. S. Whittam, and R. K. Selander. 1997. Evolutionary genetics of the isocitrate dehydrogenase gene (icd) in *Escherichia coli* and *Salmonella enterica*. *J. Bacteriol.* **179**:6551–6559.

126. Weisberg, R. A. 1996. Specialized transduction, p. 2442–2448. *In* F. C. Neidhardt, R. Curtiss III, J. L. Ingraham, E. C. C. Lin, K. B. Low, B. Magasanik, W. S. Reznikoff, M. Riley, M. Schaechter, and H. E. Umbarger (ed.), Escherichia coli *and* Salmonella: *Cellular and Molecular Biology*, 2nd ed., vol. 2. ASM Press, Washington, D.C.

127. Whitman, W. B., D. C. Coleman, and W. J. Wiebe. 1998. Prokaryotes: the unseen majority. *Proc. Natl. Acad. Sci. USA* **95**:6578–6583.

128. Whittle, G., G. A. Bloomfield, M. E. Katz, and B. F. Cheetham. 1999. The site-specific integration of genetic elements may modulate thermostable protease production, a virulence factor in *Dichelobacter nodosus*, the causative agent of ovine footrot. *Microbiology* **145**:2845–2855.

129. Wommack, K. E., and R. R. Colwell. 2000. Virioplankton: viruses in aquatic ecosystems. *Microbiol. Mol. Biol. Rev.* **64**:69–114.

130. Wright, A. 1971. Mechanism of conversion of *Salmonella* O antigen by bacteriophage epsilon 34. *J. Bacteriol.* **105**:927–936.

131. Yamaguchi, T., T. Hayashi, H. Takami, K. Nakasone, M. Ohnishi, K. Nakayama, S. Yamada, H. Komatsuzawa, and M. Sugai. 2000. Phage conversion of exfoliative toxin A production in *Staphylococcus aureus*. *Mol. Microbiol.* **38**:694–705.

132. Yamamoto, K. 1967. The origin of bacteriophage P221. *Virology* **33**:545–547.

133. Yarmolinsky, M. B., and N. Sternberg. 1988. Bacteriophage P1, p. 291–438. *In* R. Calendar (ed.), *The Bacteriophages*, vol. 2. Plenum Press, New York, N.Y.

134. Yen, H. C., N. T. Hu, and B. L. Marrs. 1979. Characterization of the gene transfer agent made by an overproducer mutant of *Rhodopseudomonas capsulata*. *J. Mol. Biol.* **131**:157–168.

135. Yokoyama, K., K. Makino, Y. Kubota, C. H. Yutsudo, S. Kimura, K. Kurokawa, K. Ishii, M. Hattori, I. Tatsuno, H. Abe, T. Iida, K. Yamamoto, M. Onishi, T. Hayashi, T. Yasunaga, T. Honda, C. Sasakawa, and H. Shinagawa. 1999. Complete nucleotide sequence of the prophage VT2-Sakai carrying the verotoxin 2 genes of the enterohemorrhagic *Escherichia coli* O157:H7 derived from the Sakai outbreak. *Genes Genet. Syst.* **74**:227–239.

136. Young, B., Y. Fukazawa, and P. Hartman. 1964. A P22 bacteriophage mutant defective in antigen conversion. *Virology* **23**:279–283.

137. Zink, R., M. J. Loessner, and S. Scherer. 1995. Characterization of cryptic prophages (monocins) in *Listeria* and sequence analysis of a holin/endolysin gene. *Microbiology* **141**:2577–2584.

Chapter 4

Global Approaches to the Bacterial Cell as an Integrated System

MICHAEL T. LAUB, LUCY SHAPIRO, AND HARLEY H. MCADAMS

The past several years have seen an explosion in the number of fully sequenced bacterial genomes. More than 100 will have been published by the end of 2004, and several hundred more are to be finished in the next few years (5). New technologies made possible by this sequence data, such as DNA microarrays, in combination with the small size and ease of genetic manipulation of bacteria, now make it possible to identify the complete genetic regulatory circuitry that controls the bacterial cell. Analysis of the global gene expression profile of the bacterial cell during its cell cycle, under conditions of environmental challenge, and during pathogen invasion of host organisms will provide an unprecedented understanding of the bacterial cell as an integrated system. Tools and techniques for genetic circuit analysis perfected by application to microbial models will provide the paradigm for analysis of the regulatory circuitry of more complex organisms.

This chapter addresses the use of microarrays for study of a wide range of microbiological problems with emphasis on the profoundly different results that this genome-wide technique provides relative to the analysis of single genes and conventional forward genetics. By assaying the response of all genes to a given genetic or environmental perturbation in parallel and simultaneously, the microarray results identify whole pathways or subroutines of the organism's genetic regulatory circuitry. Examples below illustrate the versatility of microarrays for application to problems in microbiology well beyond RNA expression profiling. For any of these applications, microarray-based experiments are most powerful in combination with an experimentally tractable system so that the microarray results can be placed in context with standard genetic, biochemical, and cell biological analyses.

The following sections describe how arrays can provide rapid global, high-resolution characteriza-tion of bacterial gene expression or DNA content under different conditions (Table 1). The ability to characterize how the gene expression profile of the cell changes with time provides an exceptionally powerful tool. The progression of normal cell processes, such as the cell cycle, can be tracked with high resolution. In addition, the adaptation of the cellular physiology and morphology to changing environmental circumstances can also be tracked as needed genes are switched on and unneeded genes are switched off.

Two leading variants of the technology are discussed in the next section followed by discussion of several applications. The challenges and emerging solutions for managing and analyzing the large data sets produced by expression arrays are discussed in the final sections.

OVERVIEW OF MICROARRAY TECHNOLOGY

PCR Amplicon Arrays

Microarrays were initially used to assay genome-wide RNA expression patterns; this has been the widest application of microarrays in bacteria as well as eukaryotes. The most common type of microarray experiment is diagrammed in Fig. 1A. In this DNA microarray technique, fluorescently labeled single-stranded cDNAs are hybridized to cDNA that has been arrayed and immobilized on a solid substrate. The immobilized arrays, or spots, of DNA are typically the products of PCR that generate amplicons ranging from a few hundred base pairs to several kilobases in length. PCR amplicons can be generated for each predicted gene in a bacterium's genome and then robotically arrayed on a glass microscope slide so that each spot of immobilized DNA corresponds to a single gene. For an RNA expression analysis

Michael T. Laub, Lucy Shapiro, and Harley H. McAdams • Department of Developmental Biology, Stanford University Medical Center, B300 Beckman Center, 279 Campus Dr., Palo Alto, CA 94304-5329.

Table 1. Applications of microarrays for analysis of bacterial systems

Array type	Sample type	Application	Examples of bacterial systems (reference[s])
Oligonucleotide	RNA	High-resolution mapping of transcriptional start and stop sites, secondary structure, and antisense expression	21
PCR amplicons	RNA	Analysis of gene expression	12, 14, 24, 26
PCR amplicons	DNA	Genotyping	1, 20
PCR amplicons	DNA	Monitoring DNA replication	13

experiment, RNA is harvested from a sample culture and from the selected reference culture. Then, cDNAs are made from each of these RNA samples by reverse transcription with fluorescently labeled nucleotides incorporated into the cDNAs: conventionally, Cy3 (green) and Cy5 (red) are incorporated into the reference and sample populations, respectively. These labeled cDNAs are hybridized to a microarray and then scanned to measure the red and green fluorescent signal level for each spot on the array. Resulting ratios of red to green signal in each spot then represent the ratio of the RNA level for the corresponding gene in the sample compared with the reference. However, for the fluorescence signal ratio to be a valid indicator of respective mRNA levels, differences in labeling and detection efficiencies for the fluorescent labels and for differing amounts of RNA in the initial samples must be taken into account. There are several approaches of varying complexity for this essential normalization process (see reference 8). Ratios of red to green signal greater than 1 for a spot indicate that more RNA was expressed from the corresponding gene in the sample culture than in the reference culture, and vice versa for ratios less than 1. For technical aspects of spotted microarray fabrication and additional variations of the technique, see reference 4.

Oligonucleotide Arrays

A second array technology, diagrammed in Fig. 1B, uses arrays of oligonucleotides, typically only 20 to 30 nucleotides in length, that are directly synthesized on a solid substrate by a modified photolithographic technique (15). For expression analysis, an RNA sample is converted to cDNA by reverse transcription and biotinylated. The sample's cDNA is hybridized alone (not competitively, as with the spotted microarrays) to an oligonucleotide array and stained with streptavidin-phycoerythrin. A fluorescence readout gives a direct, quantitative measure of each probe's representation in the sample assayed. The oligonucleotide arrays have several advantages

relative to printed arrays (Table 2). Most important, oligonucleotide arrays can be manufactured with much higher spot densities so that a single array can contain more than 200,000 unique probes. As a result, the oligonucleotide probes on the array can be "tiled" across a genome such that each gene is sampled by a whole set of these short probes. Since each individual probe is complementary to a different region of a gene, the final data can give a higher-resolution picture of gene expression. In principle, at least, the tiling of the probes across a gene or operon region should allow precise determination of transcriptional start and stop sites based on which probes are enhanced in a sample. Probes against nontranscribed regions should give low or no signal relative to transcribed regions. If a signal is detected for a probe specific to a predicted intergenic region, it may serve to refine the predicted transcription start and stop sites, or, in some cases, it may be the result of an expressed, but untranslated, small RNA. Because the oligonucleotides, unlike the printed PCR amplicons, are specific to one strand or the other, tiling of both strands of the chromosome can also identify regions of the genome subject to antisense expression.

Each of these various applications has been demonstrated using 25-mer probes for the *Escherichia coli* genome, with an average of one probe per 30 bp over the entire chromosome (21). The array included sets of probes specific for the expressed strand of every open reading frame (ORF), tRNA, and rRNA as well as both strands of every intergenic region larger than 40 bp. A second array contained probes specific for the antisense strand of each of the 4,290 predicted ORFs. These arrays were used to analyze RNA prepared from cells grown in rich media to either mid-log phase or to late stationary phase. Overall, 1,529 RNAs (including mRNAs representing ORFs, tRNAs, and rRNAs) were found to be significantly increased or decreased in stationary-phase cells. The results included mRNAs from genes previously known to be decreased in stationary phase, such

A

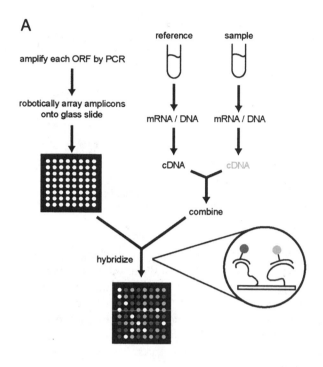

B

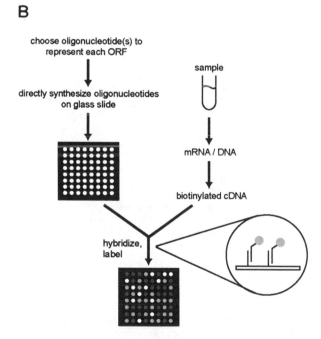

Figure 1. Overview of the microarray technique. (A) With arrays made by robotically spotting PCR amplicons, either RNA or DNA levels in a reference culture sample can be compared with RNA or DNA levels in a sample culture. The RNA or DNA samples are converted to fluorescently labeled cDNA samples which are competitively hybridized on a spotted microarray. The reference and sample are labeled with different colors. Comparison of the relative fluorescent levels from each label in a spot indicates relative ratios of the corresponding gene in the genome. (B) Oligonucleotide arrays are made by direct synthesis of short probes on a solid substrate. A population of cDNAs, derived from either RNA or DNA, is biotinylated and hybridized to an oligonucleotide array. Staining with streptavidin-phycoerythrin then provides a quantitative fluo-

as protein synthesis genes, and from others previously known to be increased, such as the stress and starvation genes.

Besides simply identifying differentially expressed genes, the use of high-resolution oligonucleotide arrays in this study yielded additional RNA-level data. By using probes for all of the predicted intergenic regions of the *E. coli* genome, sections of transcribed but not translated DNA were detected, thereby helping to identify candidate small RNAs. For example, the results verified the stationary-phase-induced expression of *csrB*, a small "intergenic," untranslated RNA not identified in the annotation of the *E. coli* genome (21). In another experiment, a stationary-phase mRNA sample was hybridized to an "antisense" array, leading to detection of expression of antisense transcripts within more than 3,000 ORFs (21). This surprising result suggests that antisense transcription may be a more common, genomewide phenomenon than previously thought, although this observation remains to be verified. Additionally, patterns of probe hybridizations showed a strong correlation between changes in signal intensity and the location of known transcription start sites, stop sites, and secondary structure elements. These results indicate that arrays may be generally useful for genomewide mapping of transcript start and stop sites, operon structures, and RNA secondary structures. Development of this information has always required study of a single gene or operon at a time using laborious and time-consuming assays. As a result, only a relatively small proportion of genes were characterized, and the results for different genes were far less conducive to comparison owing to differing experimental conditions. High-density oligonucleotide arrays can be used to study these properties of every gene in a genome simultaneously and in parallel.

MICROARRAY ANALYSIS OF CELL CYCLE REGULATION

Caulobacter crescentus: a Model System for Study of the Bacterial Cell Cycle

Application of microarray-based genomic analysis to study the cell cycle of *C. crescentus* (referred to simply as *Caulobacter* hereafter), a system with well-established genetics, biochemistry, and cell biology, has recently led to a dramatic increase in our

rescent signal with the signal strength (shaded spots) for each oligonucleotide probe, indicating the level of the cRNA or cDNA in the sample culture.

Table 2. Comparison of microarray technologies

Feature	PCR amplicon/spotted DNA arrays	Oligonucleotide arrays
Cost	Relatively inexpensive	Expensive
Flexibility	Easy to modify design	Difficult to modify after original design
Density of probes	$<\sim 30{,}000$ elements	$>200{,}000$ elements
Storage requirements	Large number of cloned DNA fragments or PCR amplicons	None
Type of measurement	Relative ratios	Direct, quantitative
Transcript start/stop site resolution?	Low	High
Detection of strand-specific expression?	No	Yes
Genotyping?	Detect relatively large deletions and insertions	Detect deletions, insertions, and individual base changes

understanding of regulation of the bacterial cell cycle (14). The cell cycle of *Caulobacter*, diagrammed in Fig. 2, has well-defined G_1, S, and G_2 phases. The *Caulobacter* chromosome is replicated only once per cell cycle, with no overlapping replication phases as in *E. coli*. The inherently asymmetric division of *Caulobacter* (Fig. 2) allows easy isolation of newly divided swarmer progeny cells for cell cycle studies involving synchronized cell populations. Swarmer cells do not have a polar stalk, leading to swarmer cell hydrodynamics that enables their separation from other *Caulobacter* cell types by differential centrifugation. The ability to isolate pure populations of synchronized, G_1-phase swarmer cells makes *Caulobacter* an extremely tractable system for cell cycle analysis.

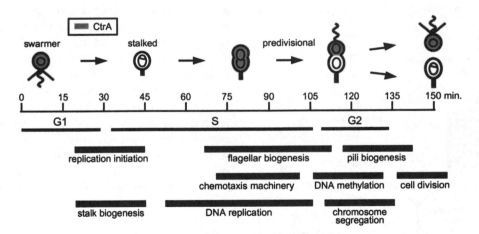

Figure 2. Temporally coordinated events of the *Caulobacter* cell cycle. Cells in the G_1 phase have a single polar flagellum and several polar pili. These motile "swarmer" cells are unable to initiate replication of their single, circular chromosome. In response to signals that are not yet understood, swarmer cells differentiate by shedding their polar flagellum and pili and subsequently synthesizing a stalk at that same pole. This sessile "stalked" cell has a tubular extension of the cell envelope, the stalk, with a holdfast substance at the tip allowing the cell to adhere to various surfaces. Coincident with the morphological transition of swarmer to stalked cell, DNA replication is initiated (G_1-to-S transition). As the stalked cell proceeds through S phase, it establishes a cell division site by constructing a FtsZ ring and appears pinched. Soon thereafter, these pinched predivisional cells begin constructing a new polar flagellum at the pole opposite the stalk. Upon completion of DNA replication, the daughter cell chromosomes segregate to opposite ends of the cell followed by an asymmetric cell division at a site slightly closer to the new flagellar pole. This asymmetric (in both size and morphology) division produces daughter cells with different morphologies and distinct cell fates. The smaller progeny swarmer cells are equivalent to G_1-phase cells and cannot replicate their DNA until after the obligate swarmer-to-stalked cell transition. The progeny stalked cell, on the other hand, immediately reinitiates replication of its chromosome without an intervening G_1 phase. Progeny stalked cells thus function as "stem cells" that produce new swarmer cells at each division. Black bars indicate the approximate time of execution of cell cycle events. Gray shading indicates the cell types in which the master regulator CtrA is present and activated (by phosphorylation). CtrA is present in swarmer cells where it can repress DNA replication initiation by binding to the origin of replication. During the swarmer-to-stalked cell transition, CtrA protein is rapidly degraded so that chromosome replication can initiate (2, 3). After replication is initiated, synthesis of CtrA is restarted in the stalked cell.

Premicroarray Studies of the *Caulobacter* Cell Cycle

The construction of the polar flagellum in the *Caulobacter* predivisional cell involves more than 40 genes organized in a transcriptional hierarchy of four classes (7). The flagellum is built from the interior of the cell toward the exterior, with the order of transcription of the flagellar genes reflecting the order of assembly of their gene products. The membrane-embedded motor and basal body are built first, followed by export and assembly of the hook subunits, and finally export and polymerization of the flagellar filament (Color Plate 1A [see color insert]). The basal body genes (class II and III) are transcribed and expressed first, while genes whose products are required later in assembly, such as genes for the hook (class III) and filament (class IV) proteins, are expressed last. Each class of flagellar genes includes *trans*-acting factors that activate the next class of genes, thereby coupling order of assembly to order of expression. The class I designation was initially reserved for the unknown factor that set the entire cascade in motion. The flagellar assembly order and most of the class II, III, and IV genes were known before the master regulator was finally identified in 1996 (18). The master regulator gene, *ctrA*, was found by use of a temperature-sensitive screen for genes affecting flagellar class II transcription. CtrA proved to be a two-component response regulator protein with a central role in regulation of the *Caulobacter* cell cycle, including initiation of DNA replication, DNA methylation, cell division, and flagellum synthesis. Two-component signal transduction genes, encompassing response regulators and histidine kinases, are a dominant form of signal transduction in the bacterial kingdom as well as being found in plants and some eukaryotes. The two-component proteins are primarily known for their function as signaling factors involved in adaptive responses to environmental changes, but it is now demonstrated that they also can play essential roles in control of internal cell processes such as cell cycle progression.

Analysis of Wild-Type *Caulobacter* Cells

Recognition that a complex genetic regulatory network governs *Caulobacter*'s cell cycle progression and morphogenesis motivated the use of microarrays to undertake a global assay of cell cycle-regulated transcription in wild-type *Caulobacter* cells. DNA microarrays containing nearly all of *Caulobacter*'s 3,767 genes were designed and constructed. A population of swarmer cells was isolated and allowed to progress synchronously through the 150-min cell cycle, with RNA samples collected every 15 min at 11 time points. Each sample was then competitively hybridized on a microarray against a common reference RNA pool from unsynchronized *Caulobacter* cells (14). The result was a simultaneous profile of the expression level of each of nearly 3,000 genes as a function of time in the cell cycle. Analysis of the resulting expression profiles identified 553 cell cycle-regulated transcripts, which included the 73 previously identified as cell cycle regulated by other methods.

Examination of the known or predicted functions for these 553 cell cycle-regulated genes led to three general conclusions: (i) genes are expressed immediately before or coincident with the time at which they are needed for function during the cell cycle, (ii) genes encoding proteins that function together as large molecular complexes are coexpressed, and (iii) biogenesis of large, multiprotein complexes is often temporally controlled by staged, hierarchical gene expression.

The flagellar biogenesis genes illustrate all three of these principles (Color Plate 1A). The cell cycle expression profiles for the flagellar genes show that they are expressed during the second half of the cell cycle, precisely when the predivisional *Caulobacter* cell assembles a new flagellum at the swarmer cell pole (Fig. 2). The flagellar genes are coexpressed, but at a detailed level of time resolution they are seen to be expressed in a staged fashion corresponding to their order of assembly (14).

Functionally related sets of genes associated with essentially every major event in the *Caulobacter* cell cycle were also found to be coordinately expressed. Genes involved in DNA replication initiation are maximally expressed during G_1 in the swarmer cells (Color Plate 1B). At the G_1-S transition, genes required for replication elongation, nucleotide synthesis, and DNA repair are expressed. These are followed in turn by peak expression of genes known or predicted to play roles in chromosome decatenation and segregation. In summary, these patterns of gene expression show that transcriptional regulation plays a major role in controlling the timing and execution of DNA replication functions in *Caulobacter* (Color Plate 1B).

A breakdown of cell cycle-regulated transcripts by functional category also points to other, previously unrecognized processes that may be executed in a cell cycle-dependent fashion. Most intriguing among these are three sets of genes encoding subunits of the ribosome, RNA polymerase, and the NADH dehydrogenase complex of oxidative respiration (14). The upregulation of these three sets of genes at approximately the same time, in stalked or S-phase cells, suggests that stalked and predivisional cells may have a stronger metabolic capacity than swarmer cells. Whether the G_1-S transition includes a shift in general metabolic state remains to be shown, but is a tantalizing hypothesis made possible by this global, non-hypothesis-driven analysis.

In these experiments, the time series of microarray expression profiles in the synchronized population was analyzed to distinguish cell cycle- from non-cell-cycle-regulated genes. The results showed that detailed analysis of the temporal ordering of peak expression times of genes known to have related functions can suggest how their regulatory control network is organized. The expression peaks of genes known to be regulated by a simple cascade of transcription regulation are seen to follow one after the other, as in the ordering of the expression peaks of class I, II, and III flagellar genes (Color Plate 1A). With advances in the technology, one can expect higher and more definitive time resolution along with noise reduction in the measurement of expression levels. Then, it will be possible to identify other potential transcriptional cascade regulatory structures by analysis of temporal time series of expression profiles.

Analysis of Mutant *Caulobacter* Cells

The microarray studies described above identified global patterns of transcription during the *Caulobacter* cell cycle and laid the foundation for a determination of the complete cell cycle genetic regulatory network. Determining the regulatory network requires additional information beyond the wild-type mRNA expression profiles: (i) What regulatory molecules control the temporal patterns of expression? (ii) Is the cell cycle-dependent expression of these genes necessary for proper execution of the cell cycle? Microarrays were also used to address these questions.

A set of 40 known or predicted regulatory molecules were identified within the 553 cell cycle-regulated genes, of which only 13 had been previously identified and characterized (14). This set included 35 two-component signal transduction genes (19 response regulators and 16 histidine kinases) and 5 sigma factors. This list of candidate regulators includes the master regulator CtrA whose mRNA peaks in expression in late-stalked, early-predivisional cells. The cell cycle-dependent expression of these regulatory genes suggests that they may control the periodic expression of other sets of genes.

To determine the full CtrA regulon, the expression profiles of cells bearing temperature-sensitive, loss-of-function mutant alleles of *ctrA* were analyzed using microarrays. RNA was isolated from the *ctrA*^ts strain at the permissive temperature of 30°C and after a 4-h shift to the restrictive temperature, immediately before the strain began losing viability. Comparison of these RNA samples on a microarray, after elimination of genes responding merely to the temperature change, identified 144 genes whose expression was both dependent on CtrA and cell cycle regulated in wild-type cells (14). To distinguish between the genes directly regulated and those indirectly regulated by CtrA, the predicted, upstream regulatory region of each of these 144 genes was examined for a consensus CtrA-binding site, which had been defined previously (16). Thirty-eight genes were found to have canonical CtrA-binding sites. Further verification of CtrA's direct control of these 38 genes will require either in vitro footprinting of individual promoters or application of new, high-throughput techniques for in vivo footprinting of DNA-binding proteins (11, 19).

The 144 directly and indirectly controlled CtrA-dependent genes included all genes previously known to be regulated by CtrA, such as *ccrM*, *ftsZ*, the origin of DNA replication, and many of the flagellar biogenesis genes. However, the microarray-based, genome-wide analysis dramatically expanded the set of genes known to be directly or indirectly regulated by CtrA (Fig. 3). It now is known that CtrA regulates structural and morphological processes such as chemotaxis and pili biogenesis. In addition, CtrA appears to play a role in regulating the expression of metabolic genes, including ribosomal subunits, RNA polymerase subunits, and the NADH dehydrogenase complex required for oxidative respiration. Thus, CtrA is a major hub of the genetic network driving cell cycle progression in *Caulobacter* (Fig. 3).

This work on *ctrA* demonstrates the power of combining microarrays with a genetically tractable system for purposes of dissecting the regulatory circuitry of bacterial cells. There are expected to be a small number of additional top-level master regulatory proteins that coordinate the remaining cell cycle-regulated functions not controlled by CtrA. The timing of these additional master regulators is expected to be coordinated with CtrA so that they together orchestrate the entire cell cycle.

OTHER MICROARRAY STUDIES OF BACTERIAL PHYSIOLOGY

Analysis of Bacterial Metabolism

Microarrays have also been used, in combination with sophisticated genetics and biochemistry, to

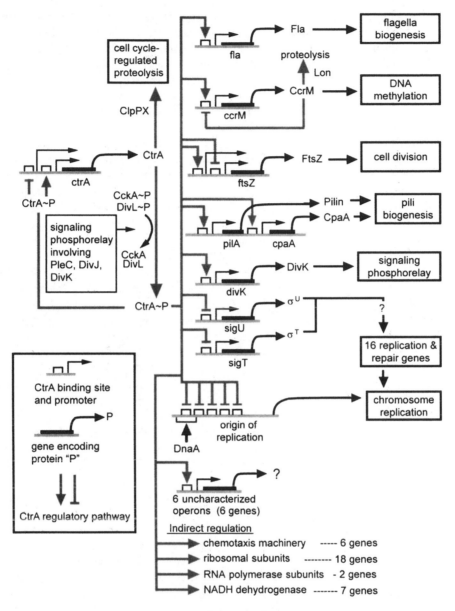

Figure 3. Regulatory network governing cell cycle progression in *Caulobacter*. Phosphorylated CtrA acts as a transcription factor to control a wide range of cell cycle-dependent events, as described in the text. Gene expression microarrays now provide a critical tool for identification of such large-scale regulons. (Modified from reference 14 with permission of the publisher.)

study tryptophan and nitrogen metabolism in *E. coli*. For tryptophan metabolism, the global transcriptional responses to tryptophan excess and starvation were assayed using whole-genome *E. coli* microarrays (12). The starvation response was examined by addition of a tryptophan analog that mimics intracellular tryptophan starvation and by use of a *trp* mutant in which the *trpR* repressor is inactivated by a well-characterized mutation. Genes that are members of the *trpR* regulon were expected to be decreased in *trp* excess and increased in *trp* starvation, whether environmentally or genetically induced. Only

nine genes, in five operons, met these criteria. Notably, all of these genes had been previously identified, by standard methodologies, as members of the *trp* regulon and were thus known to be directly involved in tryptophan metabolism. Since these genes and their protein products were well characterized, the changes in mRNA levels could be correlated with the protein changes reported previously under similar conditions. The correlation was high, with all mRNA and protein changes in agreement, within a factor of 2. Thus it appears that, at least in the case of tryptophan metabolism, changes in final, relevant enzymatic activities are controlled predominantly at the transcriptional level. If other metabolic adaptive responses are similar, microarrays will provide a powerful tool for the rapid dissection of metabolic regulons.

In addition to the nine genes with patterns conforming to the *trpR* regulon, 160 other genes were found to change significantly in one or more of the conditions tested. This set of 160 included many genes involved in nitrogen metabolism and aromatic amino acid biosynthesis. It was particularly interesting that the arginine biosynthesis genes, which are predominantly members of the Arg/Art regulon, were strongly downregulated under conditions of tryptophan starvation. While amino acid starvation might be expected to cause a general decrease in amino acid synthesis, the enhanced response of arginine biosynthetic genes relative to all others is thought to suggest a significant connection or synergy between the arginine and tryptophan biosynthetic pathways. These results illustrate how using microarrays to examine expression changes of every gene in an unbiased fashion can uncover novel, and often subtle, physiological changes. Classical genetic approaches have usually focused on identification of genes with strong, direct effects. In contrast, microarray studies are able to identify more subtle, indirect effects. One of the major challenges of the genomic era is to understand the physiological relevance and selective advantage of these complex interactions.

Another recent microarray-based analysis of metabolism examined the global expression changes in response to perturbations of the nitrogen regulatory cascade of *E. coli* (26). NtrC (nitrogen regulatory protein C) is a transcription factor that activates the expression of a wide range of genes in response to nitrogen starvation, typically sensed as an intracellular depletion of glutamine. Microarray comparison of a strain in which the NtrC gene is constitutively activated (*glnL*[Up] or *ntrB*[Con]) and a strain in which *ntrC* is inactivated (*glnG*) led to the identification of approximately 75 NtrC-regulated genes. Among this set of 75 were 27 genes from 9 operons known from prior studies to be in the NtrC regulon and involved in ammonium assimilation into the intermediates glutamate and glutamine, catabolic production of these intermediates, and relief of dependence on these intermediates or regulatory genes. The other 48 genes identified as directly or indirectly controlled by NtrC were dominated by those encoding a variety of transport systems, specific for putrescine, histidine, nucleosides, oligopeptides, and dipeptides generated during murein synthesis. The preponderance of transport systems specific for nitrogen-containing compounds points to nitrogen scavenging as one of the major physiological responses to nitrogen starvation. The microarray-based global analysis was essential for identifying this broad web of regulatory interactions which had not been identified after years of conventional single-gene experiments.

Bacterial Response to Antibiotics

Microarrays have also been used to investigate the effects of antibiotics on pathogenic bacteria (6, 10, 23, 24). By assaying changes in the global patterns of gene expression following antibiotic treatment, the mode of action of the drug can be determined by considering the sets of genes and pathways activated or repressed. This has two immediate benefits. First, if the affected pathway is known, then pathway complements are strong candidates for new drug targets. Second, the expression profile in response to the drug provides a fingerprint or signature for that inhibitor or even a whole class of related inhibitors. Such signatures may assist in analysis of other uncharacterized or poorly characterized inhibitors. As one example of this use of microarrays, the effects of isoniazid (INH) on growth of *Mycobacterium tuberculosis* was assessed by using microarrays (24). INH, an antituberculosis drug, targets biosynthesis of cell wall lipids, preventing synthesis of mycolic acids. Exposure to INH or the related drug ethionamide was shown to cause upregulation of 14 genes. Six of these upregulated genes are directly involved in mycolic acid synthesis, and two are involved in general lipid metabolism. Presumably, the cell senses a block in synthesis of the mycolate lipid and consequently upregulates the genes encoding the critical enzymes. The set of upregulated genes included the previously known drug targets, suggesting that other members of this critical pathway in *M. tuberculosis* should also be investigated as potential targets. Microarrays, as exemplified in the INH tuberculosis study, hold promise as a powerful new tool for accelerating the identification of novel drug targets by helping us to understand how bacteria respond to, adapt to, and resist antibiotics.

Tryptophan and nitrogen metabolism in *E. coli*, as well as the response of *M. tuberculosis* to INH, are processes that have been studied for many years. Microarray analysis of these systems served largely as a verification tool, with some novel genes and pathways identified. Previous genetic and biochemical studies had identified and characterized the major genes of relevance in each of these adaptive responses. Now microarrays can be used on problems that are not nearly as well characterized or that may involve large numbers of genes. For example, microarray analyses were used to identify hundreds of genes whose expression levels differed in rich versus minimal media and in log phase versus stationary phase in *E. coli* (21, 22). Similarly, differences in gene expression patterns in *Bacillus subtilis* have been characterized in aerobic versus anaerobic growth (25). The lists of affected genes generated by these experiments are providing new insights into the complex physiological changes that accompany these dramatic changes in growth state and into the ways the cell is organized and controlled. Only a few of these global expression profiles of bacteria under differing conditions have been completed, but an enormous number of such experiments will be available for analysis in the very near future.

USE OF MICROARRAYS TO EXAMINE DNA CONTENT DIRECTLY

Bacterial Genotyping

Although microarrays were initially developed to analyze RNA levels, they can also be used to examine DNA samples (Table 2). Fragments of DNA, often sheared pieces of chromosomal DNA, can be labeled by incorporation of fluorescent nucleotides during PCR or any other reaction using a DNA template. These labeled DNA fragments can then be hybridized to a DNA microarray as described above for cDNA derived from RNA samples (Fig. 1). Typically, the reference sample, or red channel, contains all of the DNA that is represented on the array. Thus, after competitive hybridization, red spots indicate genes that are not present in the sample, and yellow spots (i.e., both green and red signals [Fig. 1]) indicate genes that are present. The results of such a hybridization are usually not quantitative and simply indicate, in a yes-or-no fashion, whether or not the DNA sample being tested contains each of the arrayed DNA fragments. DNA microarrays have been used in this way to assay the DNA content, that is, the genotype of several bacterial species (see below).

PCR amplicon microarrays constructed from all of the predicted ORFs of the *M. tuberculosis* H37Rv reference strain sequence have been used to examine the DNA content of the closely related bacteria *M. tuberculosis, Mycobacterium bovis,* and several live attenuated strains of *M. bovis* comprising the *M. bovis* bacillus Calmette-Guérin (BCG) vaccines (1). Genomic DNA was prepared from each of these closely related strains, fluorescently labeled, and then hybridized to an array containing spots of DNA for each *M. tuberculosis* H37Rv ORF. Competitive hybridization of labeled genomic DNA fragments from *M. tuberculosis* and *M. bovis* allows identification of genes present in *M. tuberculosis* but not in *M. bovis.* In total, this analysis identified 90 genes in 11 genomic regions that are absent in *M. bovis* relative to H37Rv. Similar comparisons with the BCG vaccine strains identified 38 H37Rv ORFs in five other regions that were absent in at least one of the BCG strains. While the comparison was unidirectional, i.e., genes absent in H37Rv but present in *M. bovis* or in a BCG strain cannot be detected, this identification of strain-specific genes provides valuable new information about tuberculosis pathogens. The identification of genes specific to *M. tuberculosis*, *M. bovis,* and the BCG strains could potentially enable better diagnostics for infection or immunization by these strains. In addition, more detailed follow-up experiments on these strain-specific genes can produce an understanding of the phenotypic and clinical differences between the different strains at the molecular level. Finally, these data, combined with the historical record of BCG strain dispersal around the world, may provide a unique look at the recent evolution of these bacterial pathogens.

Falkow and colleagues took a similar approach to analyze the genetic diversity of 15 clinically isolated strains of *Helicobacter pylori*, a bacterium that causes a wide range of stomach diseases (9, 20). Presence or absence of specific sets of genes has been shown to correlate with severity of the disease state; in particular, strains with a 40-kb pathogenicity island and the *cagA* gene show stronger virulence. Genomic DNA from each of the 15 clinical isolates was labeled and hybridized to a microarray containing every predicted ORF for two fully sequenced *H. pylori* strains, J99 and 26695. The results identified 362 genes that were present on the array (and hence present in either strain J99, 26695, or both) but not present in one of the 15 clinical isolates analyzed, and are thus strain-specific genes. These 362 genes, comprising nearly 22% of the *H. pylori* genome, are thus dispensable for general growth of *H. pylori*, but may encode products that are responsible for the differences in virulence and survival capabilities of these strains (20). Nearly a third of the strain-specific genes in *H. pylori* are in one of two locations on the chromosome, the pathogenicity island or the so-called plasticity zone. The rest are found throughout the chromosomes, in clusters ranging in size from one to eight genes. Phylogenetic profiles for each strain-specific gene were generated where a given gene's profile indicates whether the gene is present or absent in each of the 15 strains analyzed. Cluster analysis was then used to identify which genes had similar or identical profiles, which can indicate sets of genes that are always present together and so possibly coinherited. Of course, strain-specific genes adjacent in the genome showed strong correlation in this cluster analysis and are likely to be coinherited. Interestingly, though, genes at disparate locations were also often correlated in their presence or absence, suggesting that the coinheritance of these distinct genes may be maintained by selective pressure. The critical issue now is to determine how the pattern of coinheritance correlates with clinical phenotypes, particularly with respect to the presence of the pathogenicity islands. These two reports, analyzing the DNA content of *M. tuberculosis* and *H. pylori* strains, illustrate the use of microarrays as a tool for genotyping strains and thereby gaining molecular-level insight into the basis of pathogenicity.

Analysis of DNA Replication Progression

Microarrays were used to study the role of topoisomerases in DNA replication (13). A combined analysis of topoisomerase inhibitors and mutant strains showed that both gyrase and topoisomerase IV (topo IV) were required for some aspect of normal DNA replication elongation. Gyrase has long been known to play such a role, but topo IV was previously thought to function only in the decatenation of fully replicated daughter chromosomes. Addition of a gyrase inhibitor, novobiocin, to *E. coli* cells caused complete arrest of DNA replication within 30 min. However, during the 30 min before complete arrest, replication continued, suggesting that replication is not dependent entirely on gyrase for relaxing supercoils created by replication fork movement. To test whether topo IV is a factor capable of preventing the accumulation of positive supercoils, the same concentration of novobiocin was added to a *parE*ts strain (a temperature-sensitive allele of a topo IV subunit gene) at the restrictive temperature. The result was a precipitous decline in replication with nearly complete arrest in just 10 min. This suggests that topo IV normally aids in DNA replication elongation.

To test whether topo IV has a direct role in DNA replication, *E. coli* microarrays were used to directly analyze replication fork progression in these experimental conditions (13). This experiment exploited the inherent twofold increase in DNA dosage for the portions of the chromosome that have been replicated to assay the rate of replication progress (Color Plate 2 [see color insert]). By comparing DNA fragments from partially replicated chromosomes with a reference pool of completely replicated DNA, ORFs that had been replicated at the time of the sampling could be identified based on their signal level's being twofold higher than that of the reference DNA. Mapping this information back to the chromosomal location of the ORFs allowed relatively precise delineation of replication fork location on the *E. coli* chromosome. The experiment was repeated for various levels of novobiocin. In the absence of novobiocin, the assay monitors normal replication fork progression. The rate of fork progression was estimated to be 45 kb per min, or 51 min to replicate the entire chromosome, values that are consistent with previous estimates by other methods. Low levels of novobiocin decreased the fork progression rate to 14 kb per min. In combination with data showing that similar levels of novobiocin cause an inhibition of gyrase but not topo IV, the results support the conclusion that topo IV can support replication elongation, but at a much slower rate than gyrase. The ability of topo IV to prevent the accumulation of positive supercoils was verified by a plasmid-based in vivo supercoiling assay (13). In total, these results demonstrate how microarrays, combined with a genetically and biochemically tractable system, can produce new biological insights.

BIOINFORMATICS AND DATA MANAGEMENT TOOLS

The need for resources and standards for archiving experimental data on gene expression array in a public repository is widely appreciated, and there are several active efforts to address this problem. A central consideration in design of expression data archives is that understanding the context of the experiment is essential; without it, the expression levels alone are meaningless. This problem arises because, in contrast to sequence data that document a reproducible physical property of the organism, essentially all current microarray expression assays are relative. That is, the data show which genes are expressed differently in an experiment in comparison with another experiment, or in relation to another gene in the same experiment. The Microarray Gene Expression Database Group (http://www.mged.org) working group on Microarray Data Annotations is developing a specification for the minimum information that must be reported about an array-based gene expression monitoring experiment to ensure the interpretability of the results and to enable verification by third parties. The specification calls for detailed information in six categories (Table 3) (see also http://www.mged.org/Annotations-wg/MGEDmiame Apr2001.1.0.doc).

In addition to emerging data management and archiving standards, there has been rapid growth in the number of computational tools and techniques available to analyze the data. Widely used statistical

Table 3. Information categories for documenting microarray experiments

Category	Description
Experimental design	The set of hybridization experiments as a whole
Array design	Each array used and each element (spot) on the array
Samples	Samples used, extract preparation and labeling
Hybridizations	Procedures and parameters
Measurements	Images, quantification, specifications
Normalization controls	Types, values, specifications

techniques include hierarchical clustering and self-organizing maps (17). Various clustering algorithms can be used to identify subsets of genes in the genome that have correlated responses under diverse conditions. These genes are frequently found to be coregulated, and examination of their predicted function may suggest that they belong to a regulon related to a specific physiological function.

DEVELOPMENT OF A GLOBAL MODEL OF BACTERIAL REGULATORY CIRCUITRY

Regardless of the mathematical techniques applied to the data, any analysis must be based on a strong appreciation of the biology of the organism, particularly of the mechanisms of gene regulation. Analysis of the predicted operons, regulators, and transcription factors, combined with clustering analyses and other analytical methods applied to the gene expression data sets for experiments on wild-type strains, produces an initial prediction of the organism's regulatory network. As noted above, analysis of temporal gene expression profiles has shown that genes whose products function together in genetic cascades or as multiprotein complexes are frequently expressed at about the same time.

Further analysis of coregulated gene subsets is required to verify that the correlations indeed result from coregulation and to continuously test and refine the investigator's emerging picture of the organism's genetic regulatory network. For example, when a cell population is switched to conditions of phosphate starvation, the set of genes whose expression changes significantly would be considered candidate participants in phosphate metabolism or in regulation of phosphate metabolism. But verification that coordinately expressed genes are involved in activation or repression of a coherent and complete physiological

response still requires construction and analysis of mutant strains and demonstration that their phenotype corresponds to the predicted function.

Testing and validating the global regulatory network model also requires a level of genetic analysis that goes well beyond simply seeking patterns and correlations in gene expression data sets. Maintaining the results of microarray expression studies in a relational database provides a valuable data browsing capability for comparing gene expression microarray data sets with the predicted functionality of the genetic networks. In a relational database environment, queries such as "Find all genes whose expression went up more than 2× in experiment A, but was unchanged in experiment B and experiment C" are easy to formulate and implement. Thus the investigator can readily do a "sanity check" of the regulatory circuit model. Additional cross-checks of data with model can include (i) identification of genes whose expression levels are not consistent with the hypothesized operon organization of the genome (for example, genes A and B are supposedly in the same operon, but in a given experiment the expression level of A increases and the expression level of B decreases) and (ii) selection of genes whose expression levels conflict with the expression levels of their presumed controlling transcription factor, such as genes whose expression levels increase at the same time that the expression level of a transcription factor thought to induce the expression of those genes has decreased.

In summary, identification of the organism's global regulatory circuitry requires applying complementary analytical approaches to the experimental data sets. The objective of this analysis is to develop a detailed description of the different levels of regulation (e.g., master regulators and regulatory proteins lower in the hierarchy) that control the cell. Iterative refinement of the developing regulatory circuitry map will identify the regulatory networks in the cell that control cell growth, cell cycle, and adaptive responses in natural changing environments and in stress conditions.

FUTURE DIRECTIONS FOR GLOBAL ANALYSES OF BACTERIAL REGULATION

As with sequence data, interorganism comparison of the regulatory circuit architectures and design details will provide deeper levels of biological insight. How many different regulatory strategies will we find for responding to common stress conditions? How diverse are the expression profiles of individual cells in populations in "natural" environments, i.e., biofilms, in competitive situations with other bacterial species, and in differentiating colonies. The capability to observe the detailed regulatory dynamics of infection processes offers an enormous opportunity. Phage infections of host bacteria are one branch of these studies. Tracking the progress of bacterial infections of eukaryotes is much more experimentally complex, but offers the promise of identifying entirely new strategies for medical intervention. For all of these problems, microarrays provide a powerful new tool as described in this chapter.

REFERENCES

1. Behr, M. A., M. A. Wilson, W. P. Gill, H. Salamon, G. K. Schoolnik, S. Rane, and P. M. Small. 1999. Comparative genomics of BCG vaccines by whole-genome DNA microarray. *Science* **284**:1520–1523.

2. Domian, I. J., K. C. Quon, and L. Shapiro. 1997. Cell type-specific phosphorylation and proteolysis of a transcriptional regulator controls the G1-to-S transition in a bacterial cell cycle. *Cell* **90**:415–424.

3. Domian, I. J., A. Reisenauer, and L. Shapiro. 1999. Feedback control of a master bacterial cell-cycle regulator. *Proc. Natl. Acad. Sci. USA* **96**:6648–6653.

4. Eisen, M. B., and P. O. Brown. 1999. DNA arrays for analysis of gene expression. *Methods Enzymol.* **303**:179–205.

5. Fraser, C. M., J. A. Eisen, and S. L. Salzberg. 2000. Microbial genome sequencing. *Nature* **406**:799–803.

6. Gmuender, H., K. Kuratli, K. Di Padova, C. P. Gray, W. Keck, and S. Evers. 2001. Gene expression changes triggered by exposure of Haemophilus influenzae to novobiocin or ciprofloxacin: combined transcription and translation analysis. *Genome Res.* **11**:28–42.

7. Gober, J. W., and J. C. England. 2000. Regulation of flagellum biosynthesis and motility in *Caulobacter*, p. 319–339. *In* Y. V. Brun and L. J. Shimkets (ed.), *Prokaryotic Development.* ASM Press, Washington, D.C.

8. Hegde, P., R. Qi, K. Abernathy, C. Gay, S. Dharap, R. Gaspard, J. E. Hughes, E. Snesrud, N. Lee, and J. Quackenbush. 2000. A concise guide to cDNA microarray analysis. *BioTechniques* **29**:548–550, 552–554, 556 passim.

9. Israel, D. A., N. Salama, C. N. Arnold, S. F. Moss, T. Ando, H. P. Wirth, K. T. Tham, M. Camorlinga, M. J. Blaser, S. Falkow, and R. M. Peek, Jr. 2001. Helicobacter pylori strain-specific differences in genetic content, identified by microarray, influence host inflammatory responses. *J. Clin. Investig.* **107**:611–620.

10. Ivanov, I., C. Schaab, S. Planitzer, U. Teichmann, A. Machl, S. Theml, S. Meier-Ewert, B. Seizinger, and H. Loferer. 2000. DNA microarray technology and antimicrobial drug discovery. *Pharmacogenomics* **1**:169–178.

11. Iyer, V. R., C. E. Horak, C. S. Scafe, D. Botstein, M. Snyder, and P. O. Brown. 2001. Genomic binding sites of the yeast cell-cycle transcription factors SBF and MBF. *Nature* **409**:533–538.

12. Khodursky, A. B., B. J. Peter, N. R. Cozzarelli, D. Botstein, P. O. Brown, and C. Yanofsky. 2000. DNA microarray analysis of gene expression in response to physiological and genetic changes that affect tryptophan metabolism in Escherichia coli. *Proc. Natl. Acad. Sci. USA* **97**:12170–12175.

13. Khodursky, A. B., B. J. Peter, M. B. Schmid, J. DeRisi, D. Botstein, P. O. Brown, and N. R. Cozzarelli. 2000. Analysis of topoisomerase function in bacterial replication fork movement: use of DNA microarrays. *Proc. Natl. Acad. Sci. USA* **97**:9419–9424.

14. Laub, M. T., H. H. McAdams, T. Feldblyum, C. M. Fraser, and L. Shapiro. 2000. Global analysis of the genetic network controlling a bacterial cell cycle. *Science* **290**:2144–2148.

15. Lockhart, D. J., H. Dong, M. C. Byrne, M. T. Follettie, M. V. Gallo, M. S. Chee, M. Mittmann, C. Wang, M. Kobayashi, H. Horton, and E. L. Brown. 1996. Expression monitoring by hybridization to high-density oligonucleotide arrays. *Nat. Biotechnol.* **14**:1675–1680.

16. Ouimet, M. C., and G. T. Marczynski. 2000. Analysis of a cell-cycle promoter bound by a response regulator. *J. Mol. Biol.* **302**:761–775.

17. Quackenbush, J. 2001. Computational analysis of microarray data. *Nat. Rev. Genet.* **2**:418–427.

18. Quon, K. C., G. T. Marczynski, and L. Shapiro. 1996. Cell cycle control by an essential bacterial two-component signal transduction protein. *Cell* **84**:83–93.

19. Ren, B., F. Robert, J. J. Wyrick, O. Aparicio, E. G. Jennings, I. Simon, J. Zeitlinger, J. Schreiber, N. Hannett, E. Kanin, T. L. Volkert, C. J. Wilson, S. P. Bell, and R. A. Young. 2000. Genome-wide location and function of DNA binding proteins. *Science* **290**:2306–2309.

20. Salama, N., K. Guillemin, T. K. McDaniel, G. Sherlock, L. Tompkins, and S. Falkow. 2000. A whole-genome microarray reveals genetic diversity among Helicobacter pylori strains. *Proc. Natl. Acad. Sci. USA* **97**:14668–14673.

21. Selinger, D. W., K. J. Cheung, R. Mei, E. M. Johansson, C. S. Richmond, F. R. Blattner, D. J. Lockhart, and G. M. Church. 2000. RNA expression analysis using a 30 base pair resolution Escherichia coli genome array. *Nat. Biotechnol.* **18**:1262–1268.

22. Tao, H., C. Bausch, C. Richmond, F. R. Blattner, and T. Conway. 1999. Functional genomics: expression analysis of Escherichia coli growing on minimal and rich media. *J. Bacteriol.* **181**:6425–6440.

23. Troesch, A., H. Nguyen, C. G. Miyada, S. Desvarenne, T. R. Gingeras, P. M. Kaplan, P. Cros, and C. Mabilat. 1999. Mycobacterium species identification and rifampin resistance testing with high-density DNA probe arrays. *J. Clin. Microbiol.* **37**:49–55.

24. Wilson, M., J. DeRisi, H. H. Kristensen, P. Imboden, S. Rane, P. O. Brown, and G. K. Schoolnik. 1999. Exploring drug-induced alterations in gene expression in Mycobacterium tuberculosis by microarray hybridization. *Proc. Natl. Acad. Sci. USA* **96**:12833–12838.

25. Ye, R. W., W. Tao, L. Bedzyk, T. Young, M. Chen, and L. Li. 2000. Global gene expression profiles of Bacillus subtilis grown under anaerobic conditions. *J. Bacteriol.* **182**:4458–4465.

26. Zimmer, D. P., E. Soupene, H. L. Lee, V. F. Wendisch, A. B. Khodursky, B. J. Peter, R. A. Bender, and S. Kustu. 2000. Nitrogen regulatory protein C-controlled genes of Escherichia coli: scavenging as a defense against nitrogen limitation. *Proc. Natl. Acad. Sci. USA* **97**:14674–14679.

The Bacterial Chromosome
Edited by N. Patrick Higgins
© 2005 ASM Press, Washington, D.C.

Chapter 5

Major Nucleoid Proteins in the Structure and Function of the *Escherichia coli* Chromosome

REID C. JOHNSON, LIANNA M. JOHNSON, JOHN W. SCHMIDT, AND JEFFREY F. GARDNER

The bacterial nucleoid is distinguished from the eukaryotic nucleus in a number of fundamental ways, not the least of which are the lack of a membrane separating the chromosomal DNA from the cytosol and the lack of a stable and ordered nucleosomal structure to the DNA. In the absence of histones, a set of abundant "nucleoid-associated" proteins coats a significant fraction of the bacterial chromosome. While these proteins have often been referred to as "histone-like," their molecular structures and DNA binding properties are very different, and thus the analogy is not particularly appropriate. Bacterial nucleoid-associated proteins function in multiple capacities; they contribute to varying extents to the packaging of the bacterial chromosome in the nucleoid and perform specific regulatory functions in a wide spectrum of DNA transactions. An important aspect is that the levels of some of these proteins vary dramatically in relation to cell physiology. This feature, along with their ability to contort DNA structure, figures prominently in their roles as regulators of DNA transactions. This chapter focuses on the major nucleoid-associated proteins and summarizes our current knowledge of how these proteins contribute to the structure of the nucleoid and function in specific reactions involving the chromosome. Earlier reviews discussing these proteins can be found in references 101, 257, 323, 365, and 367.

The 4.6-Mb circular chromosome of *Escherichia coli* would have an unfolded circumference of about 1.6 mm. Given that the nucleoid constitutes about 25 to 50% of the internal volume of the cell (Fig. 1A), or about 1 μm in diameter, over 1,000-fold condensation of the chromosome is required, resulting in a DNA concentration calculated to be between 20 and 50 mg/ml (206). Even greater condensation may be required under fast growth rates when multiple chromosomes exist in a single cell. The physical requirement for DNA compaction must be balanced by the high DNA activity demanded by the streamlined genome and rapid growth rates exhibited by a bacterium like *E. coli*. Under optimal growth conditions, *E. coli* replicates, repackages, and segregates the daughter chromosomes every 20 min. Any damaged DNA or replication errors must be efficiently repaired, and if necessary, be available for recombination-promoted replication reinitiation at random positions. Moreover, essentially the entire chromosome is competent for transcription. Transcription within regions in the chromosome can vary from continual transcription fork passage, such as within genes whose products are involved in ribosome biogenesis and function, to less than one transcript per generation. Even quiescent genes can be rapidly activated to respond to sudden environmental changes. These properties demand a dynamic and accessible chromosome structure. The comparatively static structure of eukaryotic chromosomal DNA, which is packaged in regular nucleosomal arrays, is less suited for such active DNA metabolism.

ORGANIZATION OF THE CONDENSED CHROMOSOME IN THE NUCLEOID

A combination of factors is believed to coordinate the compaction of the chromosome into the nucleoid. The relative contributions of these factors are still under debate and are discussed in greater depth in chapter 6 (also see references 324, 366, and 426). The first level of organization involves the formation of 40 to 200 topologically constrained

Reid C. Johnson • Department of Biological Chemistry, David Geffen School of Medicine at UCLA, Los Angeles, CA 90095-1737.
Lianna M. Johnson • Life Science Core Curriculum, University of California, Los Angeles, Los Angeles, CA 90095-1606. John W. Schmidt and Jeffrey F. Gardner • Department of Microbiology, University of Illinois at Urbana-Champaign, Urbana, IL 61801.

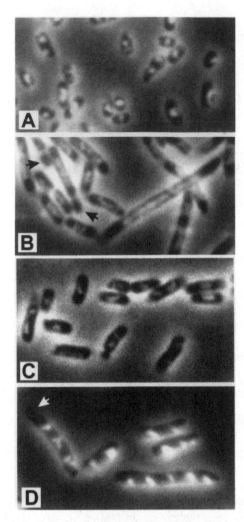

Figure 1. DAPI-stained *E. coli* cells highlighting the nucleoids. (A) MG1655 wild-type cells. (B) MG1655 *hupA hupB* mutants. Cells are often filamented with decondensed nucleoids (100, 170). Anucleated cells, which constitute about 10% of the population, are denoted with arrows. (C) MG1655 *hupA hupB* mutants expressing the *Saccharomyces cerevisiae* HMGB protein NHP6A (318). Cell and nucleoid morphology is similar to that of wild type. (D) MG1655 *fis* mutants. Cells are often elongated with multiple condensed nucleoids. The arrow points to a budding minicell, which has also been observed in *fis* mutant cell populations by Spaeny-Dekking et al. (402). Cells grown in Luria broth were stained with DAPI by the method of Hiraga (160, 318). Similar preparations of *ihf* or *hns* mutant cells do not show significant differences from wild type.

DNA loops per chromosome of sizes that probably range from 10 to 100 kb in size. Visual evidence for these loops comes from the electron micrographs of *E. coli* nucleoids by Kavenoff and Bowen generated over 25 years ago (204). Compelling biochemical evidence has been obtained by measuring unfolding of the chromosome by increasing γ-irradiation or DNase treatment within isolated nucleoids (465) or in vivo by trimethylpsoralen cross-linking (325, 394).

Current evidence is most consistent with a stochastic organization of chromosome domains without defined boundaries to supercoil diffusion (157, 408) (see chapter 6). Trun and Marko have calculated that if the *E. coli* chromosome is organized into 100 equal-sized relaxed loops, it will have an effective diameter of 2 μm (426).

The DNA within the topologically isolated domains is negatively supercoiled, with about half of the linking number deficit present as free (plectonemic) supercoils, which can be released upon DNA nicking (326). The interwound and highly branched structure of plectonemic supercoiling contributes to DNA compaction within the nucleoid (444). The remaining half of the linking number deficit is believed to be primarily stabilized by interactions with the major nucleoid-associated proteins, probably in conjunction with polyamines such as spermidine. In vivo studies employing trimethylpsoralen photocross-linking of *E. coli* DNA and phage λ Int-mediated site-specific recombination are also consistent with chromosomal DNA containing a roughly equivalent proportion of restrained and unrestrained supercoils (36, 393).

Recent studies have highlighted the importance of macromolecular crowding forces exerted by the cytosol in constraining the nucleoid (81, 280–282, 492). The protein plus RNA concentration within the cytoplasm depends somewhat on the external osmolarity but has been estimated to be in the range of 200 to 340 mg/ml (61, 493), and thus is greater than the DNA concentration in the nucleoid. Evidence for the role of cytoplasmic crowding forces on constraining nucleoids comes from observing permeabilized cells under conditions in which the cell envelope remains intact (280). As the cytoplasm leaks out, the nucleoids expand to the entire internal volume of the cell. In vitro studies with isolated nucleoids have clearly documented the importance of macromolecular crowding agents such as polyethylene glycol in stabilizing the condensed nucleoid structure (279, 281). Indeed, ≥4% polyethylene glycol can stabilize the compact form of salt-extracted nucleoids that are deficient in the major nucleoid-associated proteins.

MAJOR PROTEINS ASSOCIATED WITH THE NUCLEOID

Intact nucleoids containing much of their associated proteins are typically isolated under low-salt conditions in the presence of counterions such as spermidine or, more recently, polylysine to stabilize their structure (223, 279, 282, 442). The nucleoid particles are separated from cytoplasmic and loosely associated proteins using sucrose gradients. These preparations contain attached membrane or cell envelope material, which may reflect a physiologically relevant membrane association. In addition, a substantial amount of RNA and even 70S ribosomes copurify. The isolated nucleoids are rich in protein, containing a weight ratio of protein to DNA of about 5:1 (279, 282). The DNA within the isolated nucleoids is highly sensitive to digestion with DNase, implying it is readily accessible to the external environment. The proteins that are specifically associated with the DNA in these preparations are perhaps best assessed by mild digestion of the nucleoids with DNase I. Under these conditions a subset of proteins is released, with HU, H-NS, and Fis constituting the most abundant species. Smaller amounts of integration host factor (IHF) as well as RNA polymerase are also released. Similar incubations with RNase did not release these proteins. Earlier studies also consistently observed HU and H-NS as being dominant constituents released from denatured nucleoids or native plasmid preparations (53, 351, 442, 477). Recent work has indicated that another protein, StpA, which is a close homologue of H-NS and not distinguishable from H-NS by standard sodium dodecyl sulfate-gel electrophoresis, may also be a major constituent of the nucleoid (411, 412, 491). A common component of earlier preparations was a 27-kDa protein that cross-reacts with antibody raised against histone H2A, but this protein has been identified as ribosomal protein S3 (50).

The current view is that the *E. coli* nucleoid is dominated by five different proteins under nutrient-rich exponential growth conditions: HU and its paralog IHF, H-NS and its paralog StpA, and Fis (Table 1). Immunological analysis of these proteins indicates that HU and Fis are the most abundant nucleoid-associated proteins, while IHF, H-NS, and StpA are present at 30 to 50% of the levels of HU or Fis. Each of these proteins exhibits physiologically significant sequence-independent affinity for DNA and bends or condenses DNA in vitro. Assuming three chromosome equivalents per exponentially growing cell in rich media, along with their DNA binding site sizes and cellular levels, one can estimate that about 20% of the chromosome is bound by these proteins. These numbers agree remarkably well with in vivo formaldehyde cross-linking experiments performed by Varshavsky et al. in the mid-1970s, in which one protomer (at the time considered to be primarily HU and H-NS) was estimated to be distributed every 150 to 200 bp throughout the chromosome (441, 442). As elaborated below, biochemical and mutant analyses implicate HU as the most important protein with respect to condensing DNA within the nucleoid, but the other proteins perform

at least partially overlapping roles. HU (*hupAB*) mutants are viable but contain disordered nucleoids (Fig. 1B), reduced plasmid supercoiling, and a severe chromosome segregation phenotype. Whereas *ihf* and *hns* mutants display only a modest nucleoid phenotype, mutants deficient in both HU and IHF or HU and H-NS display synthetic phenotypes, and mutants simultaneously defective in HU, IHF, and H-NS cannot be obtained (196, 483).

Significant changes occur in the structures and associated proteins of nucleoids during stationary phase (120, 174, 412). Fis is absent and the protein Dps becomes the major constituent (Table 1). HU and IHF levels change to a lesser extent, and a significant fraction of H-NS becomes modified (233, 403). Association of the chromosome with Dps is believed to confer protection from exposure to hydroxyl radicals generated by iron-promoted Fenton reactions (139). The properties of Dps and the bacterial nucleoid in stationary phase are discussed in chapter 7.

OTHER PROTEINS CONTRIBUTING TO THE NUCLEOID STRUCTURE

Although present in much smaller amounts than the major nucleoid-associated proteins that are the subject of this chapter, other structural proteins and enzymes have profound, and in some cases essential, functions in organizing and packaging DNA in the nucleoid. The most important of these are DNA gyrase, topoisomerases I and IV, and bacterial SMC (e.g., MukBEF in *E. coli*). These proteins appear to act in concert to recondense newly replicated DNA as well as to maintain the steady-state levels of DNA supercoiling and perhaps chromosomal domains within the nucleoid (164, 363, 408, 488). Other DNA binding proteins that have been found to be present in large amounts in the cell or extracted from isolated nucleoids include Rob (CbpB [396]), CbpA (stationary phase [431]), CspE (168), YejK, and RdgC (278). In most cases, little is currently known about their roles in the structure and function of the nucleoid.

HU

Identification of HU

HU protein was discovered in 1975 as a host factor present in *E. coli* extracts that stimulated transcription from phage λ DNA in vitro (352). It was originally called factor U because it was isolated from *E. coli* strain U93. Early work showed that the host factor was an abundant low-molecular-weight protein

Table 1. The major nucleoid-associated proteins

Protein	Structure[a]	Preferred DNA target[b]	Binding site size (bp)[c]	Exponential phase Copies/cell[d]	% of chromosomal DNA bound[e]	Early stationary phase Copies/cell[d]	% of chromosomal DNA bound[e]	Late stationary phase Copies/cell[d]	% of chromosomal DNA bound[e]
HU	Heterodimer (X ray, NMR)	Kinked, gapped, 3- or 4-way junctions	36	30,000	8	16,000	6	7,500	6
IHF	Heterodimer (X ray)	WATCAANNNNTTR	36	6,000–17,000	4	27,500	11	15,000	12
H-NS	Dimer-oligomer (NMR-partial)	Curved	10	10,000	1	7,500	1	5,000	1
StpA	Dimer-oligomer (X ray)	Curved	10	12,500	1	5,000	<1	4,000	1
Fis	Homodimer (X ray)	GNtYAaWWWtTRaNC	21–27	30,000	6	<1,000	<1	<100	<1
Dps	Dodecamer (X ray)	None	90?	500	<1	10,000	10	15,000	30

[a]Quaternary structure and whether the atomic structure has been determined.

[b]Each of the proteins binds DNA nonspecifically with physiologically significant affinities. A preferred binding sequence or DNA structure is noted. Y = C or T, R = G or A, W = A or T, and N = any base.

[c]Binding site size is estimated from atomic structures of the DNA complex (HU and IHF), footprinting data (H-NS, StpA, and Fis), or modeling based on the atomic structures of the unbound protein (Fis and Dps). For HU, 14 to 19 bp are actually contacted in the crystal structures, but contacts covering up to 36 bp are supported by the structures (see text). The size and stoichiometry of the Dps-DNA complex is highly speculative (139).

[d]Fis, HU, and IHF are expressed in dimers per cell, H-NS and StpA are expressed as dimers per cell, but the functional binding form may be a tetramer or higher order. Data are expressed in dimers per cell, and Dps is expressed in dodecamers per cell. H-NS and StpA are expressed as dimers per cell. Data are collated from individual studies and reference 412 from *E. coli* cells cultured in rich media. In the case of StpA, some studies imply a lower abundance (see text).

[e]Percentage of chromosomal DNA bound by each protein. Calculations are based on three, two, and one chromosome equivalents per cell in exponential, early-stationary, and late-stationary phases, respectively, where the *E. coli* chromosome is 4.6×10^6 bp. These values are approximate because of the assumptions used in the calculations.

that has an unusual basic amino acid content for an *E. coli* protein. Indeed, its amino acid content is very similar to that of the eukaryotic histone H2B. Subsequently, HU was classified as a histonelike protein (hence HU, for histonelike protein from strain U93) because of its amino acid composition, abundance, small size, basic pI, and the appearance of beadlike structures along double-stranded DNA in early electron microscopy experiments (101, 354). Later work showed that HU is associated with the bacterial nucleoid in vivo and when cell lysates are prepared under low-salt conditions. These properties suggested that HU could play a role in compacting DNA and were consistent with later observations that HU participates in a variety of DNA transactions (see below).

Cross-linking studies of HU from *E. coli* produced a dimeric species with an apparent molecular mass of ~20,000 kDa (352). Phosphocellulose chromatography resolved purified HU into three fractions. The majority of HU is a heterodimer composed of two closely related subunits of approximately 9.5 kDa called HU-α and HU-β. The two minor fractions are homodimers composed of HU-α or HU-β (70, 71, 353). Both the heterodimeric and homodimeric forms are active in vivo (see below).

Sequences homologous to HU are found in each major division of bacteria (Fig. 2). Sequences homologous to HU have also been found in both archaea and eukaryotes, but they do not appear to be widely distributed in these kingdoms. Whereas in several enterobacteria, such as *E. coli*, HU is a heterodimer, it is present as a homodimer in most bacteria. In *E. coli*, the HU heterodimer is encoded by the *hupA* and *hupB* genes that map at 10 and 90 min, respectively (198, 199, 201). An alignment of HU proteins is shown in Fig. 2, and its structure is described below.

A few years after the discovery of HU, IHF, a relative of HU, was identified as a host factor required for site-specific recombination by bacteriophage λ (208). HU and IHF show significant homology, and alignment of the four *E. coli* subunits encoding these proteins showed that they have more than 45% identity or similarity (101). While HU binds double-stranded DNA without sequence specificity, IHF was found to bind with enhanced affinity to specific sites (76). IHF is discussed in the following section.

General Properties of *hup* Mutants

E. coli and *Salmonella enterica* serovar Typhimurium strains with mutations that inactivate both subunits of HU are viable, but the cells are sick and tend to accumulate suppressors that restore more

rapid growth. HU mutant cells display the most severe and varied phenotypes in comparison to strains containing mutations in the other major nucleoid proteins. IHF is probably substituting for some of the HU functions in *hupAB* mutants, since *hupAB ihf* mutants are difficult to construct and are severely compromised for growth and survival (196; R. Isaksson and R. Johnson, unpublished data). HU-deficient cells are cold sensitive, have a lethal phenotype following cold or heat shock, have long doubling times at temperatures below 30°C, exhibit altered responses in DNA supercoiling after cold or heat shock, and also exhibit a long lag phase when diluted from stationary-phase cultures into fresh medium (170, 271, 304). Populations of *hupAB* cells contain large numbers of filamented as well as anucleated cells (Fig. 1B). *hupAB* mutants are sensitive to gamma and UV irradiation (42), exhibit increased illegitimate recombination (379), are defective in recombinational repair of UV damage (245), and show partial relaxation of chromosomal and plasmid supercoiling (158, 167). Mutants lacking HU are defective in maintaining *oriC* minichromosomes and also have alterations in other cellular functions, including defects in initiation of DNA synthesis, chromosome partitioning, and cell division (see below) (100, 176, 305).

Structure of HU Dimers

The first structure of an HU/IHF family member was an X-ray crystallographic analysis of HU protein from *Geobacillus stearothermophilus* resolved to 3 Å (Fig. 3) (414, 457). The structure showed that the two subunits wrap around each other and make extensive intermolecular contacts. Eight conserved hydrophobic (Leu, Phe, Val, or Ile) residues at positions 6, 29, 32, 36, 44, 47, 50, and 79 (Fig. 2) contribute to the hydrophobic core. The N-terminal region of the protein contains two α-helices (α1 and α2) that are connected by a turn. The turn places the two helices in a V-shaped structure, which is stabilized by van der Waals interactions between conserved alanines at positions 11 and 21. The α2 helix is connected to the body of the protein by the highly conserved Gly-39 residue. The remainder of the protein contains a three-stranded antiparallel β-sheet structure (β1, β2, and β3) followed by a short α3 helix. β-strands 1 and 2 are connected by a sharp turn, which contains a highly conserved Gly-Phe-Gly sequence. Most of the other HU and IHF proteins that lack this triad contain a change at only one of these residues.

β-strands 2 and 3 are connected by an extended arm region containing 26 amino acids from residues 52 through 77. This arm contains a conserved

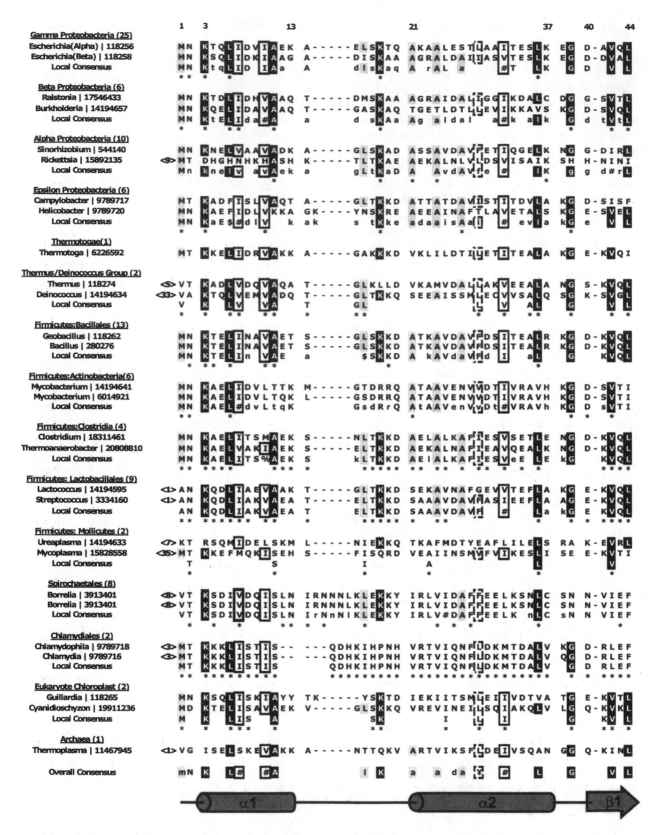

Figure 2. Amino acid sequence alignment of members of the HU family. The sequences are identified by the species from whose genome they originated and their National Center for Biotechnology Information (NCBI) GenInfo Identifier (GI) number. The HU sequences were obtained by searching the SWISSPROT database for "DBH." The NCBI PSI-BLAST program was used with *E. coli* HUα and HUβ query sequences to obtain additional HU sequences not contained in the SWISSPROT database. Sequences not described as HU, histonelike DNA binding protein, or DNA binding protein were discarded. The alignment is numbered according to the *Geobacillus stearothermophilus* HU sequence. The sequences were aligned by using Blosum62-12-2 (75) at the website http://prodes.toulouse.inra.fr/multalin/multalin.html. The following settings were used: the symbol comparison

(continued)

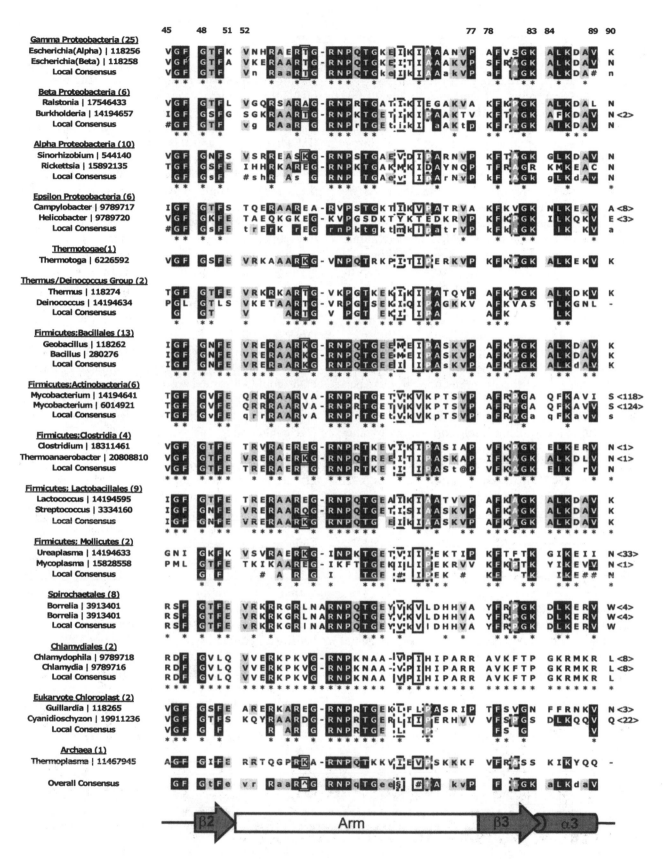

Figure 2 (*continued*) table was blosum 62, the gap weight was 12, and the gap length weight was 2. Overall consensus notation: a lowercase letter indicates the residue is found in 50 to 74% of sequences and is shaded gray where conserved. An uppercase letter indicates the residue is found in 75 to 100% of sequences overall, and the background is black where conserved. # indicates I or V residues are found in 75 to 100% of the sequences; the conserved sequences are boxed by a solid line. ¥ indicates that F, L, I, or V residues are found at this position 90 out of 97 times; the conserved sequences are boxed by a broken line. ^ indicates that T or K residues are found 60 out of 97 times at this position; the conserved sequences are boxed

Arg-X-Gly-Arg-Asn-Pro sequence between positions 58 and 63. The proline is found in virtually all HU proteins and is the residue that intercalates into the DNA (see below). The region of the arm closest to the main body of the protein has a β-ribbon conformation, but 12 of the residues at the tip of the arm were not visible in the electron density map. Tanaka et al. (414) proposed that the two arms are involved in binding DNA because the visible regions of the arms formed a concave surface that is complementary to right-handed B-DNA. The HU dimer could interact with the sugar-phosphate backbone of the DNA through electrostatic interactions, consistent with the fact that the highly conserved Arg-58, Lys-59, and Arg-61 residues are in a region of the protein that would be appropriate for DNA interactions (414, 457).

The X-ray structure did not reveal whether the arms interact with the minor or major grooves of DNA. However, chemical protection studies performed on the related IHF protein-DNA complex (478) showed that IHF binds DNA through the minor groove. Subsequent higher-resolution X-ray crystallographic and nuclear magnetic resonance (NMR) studies on HU indicated that the arm contains a hinge region that is rich in alanine and proline residues (443, 458). This flexible hinge could allow the β-ribbon to twist and reposition the side chains when binding DNA. Based on the comparison with the IHF structure (see "Structure of the IHF Heterodimer and IHF Interactions with DNA"), the highly conserved hydrophobic residues Met-69 and Ile-71 could be involved in opening the minor groove and bending the DNA (458).

Recently, Rice and Swinger solved a crystal structure of the homodimeric *Anabaena* HU bound to DNA (Fig. 4 and 5) (P. Rice and K. Swinger, personal communication). The DNA oligonucleotides present in the crystals formed a symmetric sequence that contained a T-T mismatch and two extra Ts on each strand. The DNA duplexes pack to form a serpentine or double "U" structure with an HU dimer located at the apex of each bend (Fig. 5A). Both the crystal

packing arrangement and the overall structure of the HU-DNA complex is very similar to the IHF-DNA crystal structure determined earlier by Rice and coworkers (341, 342) (Fig. 5B) (see "Structure of the IHF Heterodimer and IHF Interactions with DNA"). The β-ribbon arms wrap around the minor groove of the DNA, and the conserved prolines at the tips of the arms intercalate into the DNA to induce two pronounced kinks. Although both the protein and the DNA site are symmetric, the structure of the complex is asymmetric because there is a sharper bend on one side of the DNA. The IHF-DNA complex is also asymmetric, but that is expected because both the protein and its binding site are asymmetric. Due to crystal packing, an unpaired T residue on each strand of the DNA flips out of the duplex. However, the HU-DNA and IHF-DNA structures superimpose very well, despite the distortion in the DNA substrate. The *Anabaena* HU crystal structure indicates that the DNA could be curved around the protein by as much as 105 to 140°. However, the basic electrostatic surface of HU extends down the sides of the protein so HU could induce or stabilize a bend approaching 180°, similar to IHF (see below).

Likewise, the number of base pairs contacted by HU may vary under different conditions. Bonnefoy et al. calculated a minimal number of 9 bp per HU dimer bound in solution, which fits with the number of base pairs between the intercalated prolines at the kinks in the X-ray structure (41). However, the X-ray structure indicates that an additional 5 bp on each side, or a total of 19 bp, are interacting with HU within a single complex (Fig. 5). The amount of DNA contacted is likely to be even greater: the DNA extending down the sides of the protein is bent away because of contacts with the neighboring HU proteins in the crystal. Thus, the amount of DNA contacted could be as high as 25 to 35 bp, similar to the amount of DNA contacted by IHF. Factors such as flexibility of the DNA are likely to influence the amount of DNA contacted (Rice and Swinger, personal communication). As described below, in vivo and in vitro data are

---

Figure 2 (*continued*) by a double line. § indicates that I, L, M, or V residues are found 94 out of 97 times at this position; the conserved sequence is boxed by a broken line. † indicates that P or A residues are found 72 out of 97 times at position 72, and 81 out of 97 times at position 81. The conserved sequences are boxed by a broken line and are white with a gray background. Local consensus notation: a lowercase letter indicates the residue is found in 50 to 74% of sequences locally. An uppercase letter indicates the residue is found in 75 to 100% of sequences locally. # indicates I or V residues are found in 75 to 100% of the sequences. $ indicates L or F residues are found in 75 to 100% of the sequences. % indicates I or M residues are found in 75 to 100% of the sequences. @ indicates V or A residues are found in 75 to 100% of the sequences. An asterisk below the residue in the local consensus indicates that residue is absolutely conserved in that local grouping. Secondary structure: the secondary structure cartoon shown below the alignments is based on the crystal structure of the *G. stearothermophilus* HU homodimer performed by Tanaka et al. (414) (Protein Data Bank [PDB] accession code 1HUU). The secondary structure is as follows: α1 helix residues 3 to 13, α2 helix residues 21 to 37, β1-strand residues 40 to 44, β2-strand residues 48 to 51, "arm" region residues 52 to 77, β3-strand residues 78 to 83, and α3 helix residues 84 to 89.

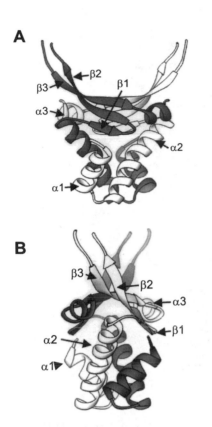

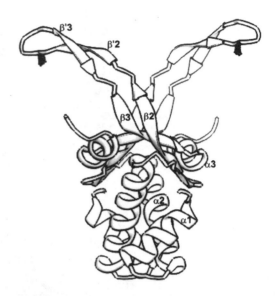

Figure 4. The structure of the homodimeric HU from *Anabaena* (K. Swinger and P. Rice, personal communication). The positions of the prolines that intercalate into the DNA are shown near the ends of the two β-ribbon arms. The α-helices of one subunit (α1, α2, and α3) are labeled. The β2'- and β3'-strands that are not visible in the *G. stearothermophilus* HU structure are revealed in the *Anabaena* HU-DNA complex. (Courtesy of P. Rice.)

Figure 3. The crystal structure of HU from *G. stearothermophilus* (414, 457). The α-helices (α1, α2, and α3) are indicated for the subunit shown in white, and the antiparallel β-sheet (β1-, β2-, and β3-strands) are indicated for the dark subunit. Residues 59 to 69 in the arms of the monomers are disordered and are not shown. (A and B) Two different views of the structure with the arms extended on each side. (Courtesy of P. Rice.)

consistent with strong bends being introduced into DNA upon binding of moderate concentrations of HU.

DNA Binding by HU

HU binds to supercoiled DNA, linear duplex DNA, RNA, and single-stranded DNA (reviewed in reference 101). In addition, HU interacts with double-stranded RNA and RNA-DNA hybrids with affinities similar to that of double-stranded DNA (17, 327, 387). Binding of HU to double-stranded DNA is nonspecific and is salt sensitive (39, 49). The nonspecific binding to linear DNA is weak. Depending upon the dimeric form of HU and the salt conditions, the K_d can range between 400 nM and 30 μM (327), which is similar to that of IHF for nonspecific sites. Nonspecific binding to DNA is stimulated by increasing levels of negative supercoiling, and HU can also be recruited to specific sites in DNA in a supercoiling-dependent manner (216).

HU binds to a variety of bent DNA structures, such as kinked, gapped, and cruciform structures

with affinities as low as 2 to 20 nM. HU also selectively binds to a variety of intermediates of recombination and repair, such as DNA with 3′ overhangs, various incomplete four-way junctions, three-way junctions, double-stranded forks, and invasion structures. Footprinting of these complexes suggests that HU is specifically recognizing the bent or flexible junctions that are common to these DNA motifs (41, 60, 192, 327, 333, 387).

The composition of HU dimers influences the affinity of the protein for different substrates. The DNA binding properties of the heterodimeric HU-αβ and the homodimeric HU-α2 are similar. They bind to linear, nicked, gapped, and cruciform DNA. The HU-β2 homodimer binds cruciform DNA but shows decreased affinity for gapped or nicked DNA and binds poorly to linear DNA (327).

DNA bending by HU can be assayed by measuring the stimulation by HU on microcircle formation. The intrinsic stiffness of DNA normally inhibits the circularization of fragments less than 150 bp in length. However, in the presence of HU, circles as small as 78 bp can ligate, and circles of 88 and 99 bp ligate efficiently (163, 316, 317). Previous studies have estimated that a circular complex of 80 to 100 bp would contain 8 to 10 HU dimers and that 12 to 13 HU dimers would be required to produce a superhelical turn (413, 458). However, the stoichiometry of HU binding to circular DNA is likely to be

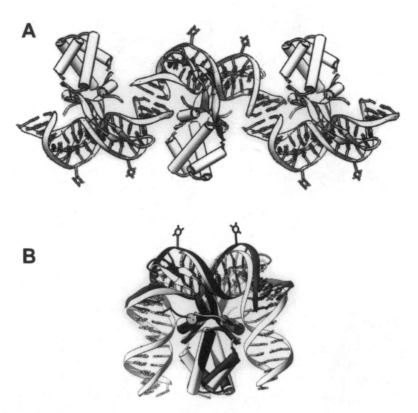

Figure 5. The structure of *Anabaena* HU bound to DNA (Swinger and Rice, personal communication). (A) The structure of three HU-DNA complexes as they are aligned in the crystal. Two thymines in each complex are flipped out. (B) A single HU-DNA complex (dark shading) is superimposed over the IHF-DNA crystal structure (light shading) (342). The T residues that are flipped out of the HU-DNA complex are shown. (Courtesy of P. Rice.)

much lower, since only two HU dimers are able to bind to a 75-bp microcircle to form a very stable complex (B. Wong and R. Johnson, unpublished data). In vitro studies of the Mu transpososome complex and the GalR repression loop indicate that a single HU dimer can form a substantial bend in DNA (216, 244). Under these low-density conditions, HU dimers may be phased to form a DNA circle or supercoil. At high densities, which could be present at localized positions within the chromosome, it is possible that serpentine structures such as those found in the crystals (Fig. 5A) may form.

HU in DNA Supercoiling, Condensation, and Nucleoid Structure

Experiments employing fluorescently tagged HU or indirect immunofluorescence all indicate that HU is distributed randomly throughout the nucleoid (382, 411, 455). The abundance of HU, combined with its nonspecific binding and potent DNA bending activities, is consistent with its having an important role in compacting the chromosome. This role is supported by the changed appearance of 4',6'-diamidino-2-

phenylindole (DAPI)-stained nucleoids in *hupAB* mutants (Fig. 1B) (100, 170). For example, in one study, nucleoid areas from *hupAB* mutant cells were found to be very heterogeneous and averaged twice the size of their *hupAB*+ parent (318). Since expression of an unrelated eukaryotic HMGB protein can largely suppress the nucleoid decondensation phenotype of *hupAB* mutants (Fig. 1C), the effect on nucleoid structure by HU is most likely due to its nonspecific binding and bending properties (261, 318).

HU efficiently promotes formation of negative supercoils in DNA when it is incubated with DNA in the presence of eukaryotic topoisomerase I (49, 354). Supercoiling of chromosomal and plasmid DNA is partially relaxed in mutants lacking HU (158, 167), supporting an in vivo role for HU in regulating the degree of supercoiling. Indeed, changes in the activities of topoisomerase I appear to compensate for the loss of HU, suggesting that HU and topoisomerase I collaborate to regulate intracellular supercoiling (26). Using HU-nuclease chimeras, Kobryn et al. (216) showed that nonspecific binding of DNA by HU is stimulated by increasing levels of supercoiling. The sensitivity of HU binding to DNA as a function of supercoiling suggests that HU could indirectly regulate a variety of cellular processes depending upon the growth phase and the environment. Furthermore, the supercoiling-sensing property of HU suggests that it could contribute to the homeostatic control of free and restrained chromosomal supercoiling (216).

As mentioned above, *hupAB* mutants grow slowly and accumulate suppressors. Small colonies give a mixture of small and large colonies when streaked on plates. The large colonies are also more resistant to novobiocin than the parental *hupA-hupB* strain. Further genetic analyses indicated that some of these suppressors map in the *gyrB* gene. In addition, a plasmid-encoded *gyrB* gene suppressed an HU-deficient strain, indicating that overexpression of GyrB itself can overcome an HU deficiency. To account for these observations, Malik et al. (251) proposed that HU normally facilitates the action of DNA gyrase. HU stimulates DNA gyrase and lowers the activity of topoisomerase I. When cells lack HU, gyrase activity is lowered, leading to a decrease in DNA supercoiling and poor growth. *gyrB*-encoded suppressors increase DNA gyrase activity in the absence of HU by increasing the concentration or activity of DNA gyrase (251). The exact molecular mechanisms by which HU facilitates gyrase activity and the mechanism of suppressor action are unknown. Interestingly, HU-deficient mutants show an increase in illegitimate recombination that is likely mediated by DNA gyrase. Some of the *gyrB*-encoded suppressors, as well as overproduction of *gyrA*, depress the higher rate of illegitimate recombination exhibited by *hupAB* strains (379).

HU in DNA Replication, Chromosome Partitioning, and Cell Division

As described above, *hup* mutants have abnormal nucleoids and generate up to 10% anucleated cells, implying that they have defects related to chromosome structure and partitioning. HU plays several interconnected roles in DNA replication and appears to be involved in several steps involving initiation of chromosomal DNA synthesis, chromosomal partitioning, and cell division. Strains deficient in HU protein accumulate cells with three origins at a significant frequency, indicating that initiation of DNA synthesis is altered. This observation suggests that HU is required for proper synchrony of DNA synthesis in vivo, although it is also possible that the defect is due to the accumulation of secondary mutations that are known to occur in cells deficient in HU protein (176). In vitro studies of *oriC* replication have demonstrated a role for HU in the DnaA-promoted unwinding step leading to open complex formation

(44, 172, 331). Only five HU dimers per *oriC* are required to stimulate open complex formation on a supercoiled substrate, and high amounts of HU are inhibitory (14, 395). IHF is capable of efficiently substituting for HU in the unwinding step (44, 172).

HU protein may also be directly or indirectly involved in chromosome partitioning and septum formation. Cells lacking HU accumulate minicells, and some *hupAB* suppressors are in the *minB* and *minC* genes, supporting a role for HU in septum formation (176). Like *hupAB* strains, mutants with deletions of the *mukF*, *mukE*, or *mukB* genes, whose products are involved in supercoiling and chromosome partitioning (453), generate anucleate cells at high frequency. It is not possible to construct synthetic *hupA mukB* double mutants, suggesting that cells need at least one of the two proteins for proper partitioning of chromosomes, but the precise mechanism behind the synthetic lethality is not known (176).

HU plays a negative regulatory role in activation of phage λ DNA replication. Initiation of λ DNA replication in vivo and in crude systems in vitro is dependent upon transcription of the λ replication origin (*ori* λ). In an in vitro system containing purified replication proteins, HU blocks the initiation of phage λ DNA replication. It appears that HU interferes with the assembly or function of complexes at *ori* λ containing the λ O and P proteins and the host DnaB helicase. If the *ori* λ O-P-DnaB complex is formed before addition of HU, DNA replication proceeds. When the complex is assembled in the presence of HU, DNA replication is inhibited, but if the template is transcribed by RNA polymerase, the inhibition by HU is relieved. Thus, HU probably inhibits the assembly or function of complexes at *ori* λ, and active transcription by RNA polymerase is required to form a complex proficient in initiation of DNA synthesis (264).

Regulation of HU Synthesis

The intracellular levels of HU do not vary dramatically as a function of growth phase. In exponentially growing cells there are approximately 30,000 dimers per cell, and the concentration of HU decreases to about 16,000 dimers per cell in early stationary phase (Table 1) (94, 101, 412). The amount of HU in exponential cells is sufficient to bind approximately 8% of the *E. coli* chromosome, assuming three chromosome equivalents per cell (Table 1).

The subunit composition of HU does vary with growth phase, and different promoters are utilized at different times in the growth cycle. Regulation of HU synthesis is complex, and the genes are regulated by different mechanisms (Fig. 6). As a result, synthesis

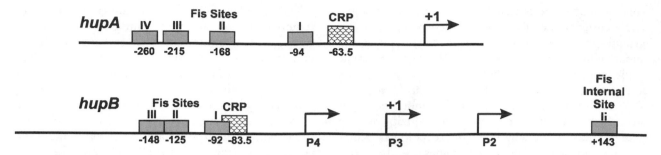

Figure 6. Organization of the upstream regulatory sequences of the *hupA* (top) and *hupB* (bottom) genes (70). The *hupB* gene has three promoters designated P2, P3, and P4; the numbering is relative to the P3 start site (70, 221). An additional promoter (P1) initiating transcription at about −10 relative to the ATG of *hupB* was described by Kohno et al. (221) but not observed by Claret and Rouviere-Yaniv (70). The four Fis binding sites (I, II, III, and IV [gray boxes]) upstream of the start of transcription of *hupA*, the start site (indicated by an arrow), and the CRP site are shown. Transcription of *hupB* is stimulated by CRP and repressed by Fis (sites I, II, III, and Ii). The Fis site I and the downstream site Ii are required for repression. Binding of CRP prevents Fis binding to site I and relieves repression of promoter P3.

of the two subunits is not coordinated. In exponential phase, the αβ and the α2 forms predominate and there is very little β2, whereas in late exponential and stationary phase only the αβ form predominates (71). The physiological reasons for why the *hupA* and *hupB* genes are regulated by such different mechanisms are not well understood. The reason that the β2 form of HU is not present during exponential growth could be because it is deleterious to the cell. For example, the β2 form does not constrain DNA in vitro so that the superhelical density of the chromosome could be altered. Alternatively, synthesis of the β subunit could be used to change the distribution of the α2 and αβ forms of HU. The αβ form of HU is required for maintenance of viability during prolonged culturing in stationary phase (71).

The monocistronic *hupA* gene is transcribed from a single promoter that is activated by both cyclic AMP (cAMP) receptor protein (CRP) and Fis (Fig. 6). Transcription of *hupA* is activated during exponential growth and decreases rapidly as cells enter stationary phase. The region upstream of the promoter contains a single CRP site where CRP binding could activate transcription. There are four upstream Fis binding sites where Fis could activate transcription by contacts between RNA polymerase and Fis in a manner similar to Fis activation of stable RNA operons (70) (see "Mechanism of Transcriptional Activation by Fis").

The region upstream of *hupB* has three promoters—P2, P3, and P4—with P2 and P3 being the most active (Fig. 6). The P2 promoter is active at mid- to late-exponential growth, and the P3 promoter is activated as cells enter stationary phase. The P4 promoter is most active during the transition from stationary phase to exponential phase. Transcription

of the *hupB* gene is regulated by multiple mechanisms. It is stimulated by CRP and repressed by Fis. Repression of the *hupB* gene by Fis requires both the downstream Fis binding site in the *hupB* gene and a sequence called site I upstream of the multiple promoters. When CRP binds to its site, Fis binding to the overlapping site I is prevented, and repression of promoter P3, but not P2, is relieved. The *hupB* gene is also under stringent control that is characteristic of ribosomal operons (70). The regulatory mechanisms that are responsible for the sequential activation of the *hupB* promoters during exponential phase and stationary phase are not known (71).

The *hupA* and *hupB* genes are also regulated differentially upon induction of cold shock. *hupA* and *hupB* transcripts were analyzed after a temperature downshift in mid-exponential growth from 37 to 10°C. Transcription of *hupA* mRNA was decreased to about 20% of the initial amount. It is not known whether Fis or CRP are involved in the cold shock response. Conversely, expression of *hupB* was induced by cold shock. The *hupB* P2 and P3 promoters remained active during cold shock, and transcription from P4 was stimulated. In addition, the stability of *hupB* mRNA was increased after cold shock. Thus, it appears that a shift favoring the accumulation of the αβ and β2 forms of HU plays a role in cold adaptation (126).

HU-Mediated Interactions of DNA Binding Proteins

HU can modulate binding of some sequence-specific DNA binding proteins. Flashner and Gralla (115) studied the effects of HU on binding of the sequence-specific DNA binding proteins Lac repressor, CRP protein, and Trp repressor. Footprinting

experiments showed that HU cooperatively promoted binding of Lac repressor and CRP-cAMP
complex to their DNA binding sites. HU had a small
inhibitory effect on binding of Trp repressor, indicating that HU is not a general stimulator for DNA
binding proteins. BaCl$_2$, which has been shown to
promote DNA flexibility, mimicked the HU effect,
suggesting that HU induces a DNA structure recognized by DNA binding proteins. Flashner and Gralla
(115) proposed a model in which the dynamic association with HU occasionally distorts DNA to a form
that is preferred by some DNA binding proteins
such as Lac repressor and CRP. Interestingly, HU can
also assist binding of IHF to DNA that does not
contain a consensus sequence, in addition to stimulating the formation of specific IHF-DNA complexes
(39, 40).

HU can also inhibit binding by sequence-specific
regulatory proteins. Preobrajenskaya et al. have proposed that the distortion of DNA by HU binding
through interactions via the minor groove disrupt the
major groove interactions by LexA (335). This mechanism is supported by the observation that the minor
groove binding drug distamycin also disrupted LexA-
operator complexes.

Regulation of Transcription by HU

A well-characterized example of HU participation in a transcription repression complex is in the *gal*
operon regulatory region. The *gal* operon is repressed
by the GalR repressor in the absence of the sugar
galactose (Fig. 7). Transcription of the *gal* operon can
be initiated from two overlapping promoters, P1 and
P2. There are two GalR operator sites, called O$_E$ and
O$_I$, that are bound by GalR. The O$_E$ and O$_I$ sites are
positioned approximately 50 bp upstream and downstream of the promoters and must be in the proper
helical phase with each other. Repression of the
P2 promoter requires HU, the two GalR operators,
and DNA supercoiling. Repression occurs when GalR
proteins bind to the operators and form a looped
DNA structure held together by protein-protein interactions between the GalR repressors bound to the
two sites. HU binds in an orientation-dependent
fashion to a 34- to 40-bp site, called *hbs*, between the
GalR operators and induces a bend that facilitates
protein-protein interactions between the distantly
bound GalR proteins (3, 243, 244). Recent work by
Kar and Adhya (202) indicates that GalR directs
binding of HU to DNA by a protein-protein interaction. The association may be transient, and HU may
not contact GalR in the final repression structure. It is
possible that similar multiprotein looped complexes
promoted by HU occur in other systems.

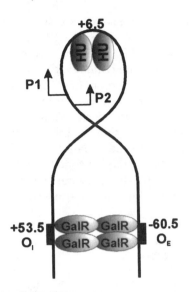

Figure 7. The HU-dependent repression loop complex of the *gal* operon (3, 244). In the presence of supercoiled DNA, GalR dimers bind to the O$_E$ and O$_I$ operators and form a DNA loop. Specific protein-protein interactions between HU and GalR facilitate recruitment of HU to the DNA complex. This association may be transient, or GalR and HU may continue to contact each other in the repression complex (202).

Regulation of Translation by HU

As cells enter the transition from exponential
growth into starvation phase, the RpoS sigma factor
content of the cell increases. RpoS is a sigma factor
that is involved in transcription of genes that promote cell survival under starvation and stress conditions. Several proteins, such as the RNA binding
protein Hfq, the *dsrA* regulatory RNA, the H-NS
protein, and the regulatory RNA OxyS, are involved
in regulation of *rpoS* translation (152). Balandina
et al. showed that HU also stimulates translation of
the RpoS protein (16). They found that a synthetic
fragment of *rpoS* mRNA containing the ribosome
recognition sequence and the AUG translation initiation codon was selectively bound by HU in a gel-
shift assay even under stringent, high-salt conditions.
A computer model of the *rpoS* RNA indicated that
the RNA secondary structure might form a three-
way RNA junction that binds HU. Precisely how HU
binding to the *rpoS* RNA leads to increased translation remains to be determined.

HU in Site-Specific Recombination

HU can also participate in the formation of
looped DNA complexes that support site-specific
DNA recombination. The flagellar phase variation
system of *S. enterica* Typhimurium is an example of

a site-specific DNA inversion system that is stimulated by HU (Fig. 8A) (see also "Regulation of DNA Recombination by Fis"). During formation of the activated recombination complex, the Hin recombinase binds to two *hix* recombination sites at the boundaries of the invertible segment. Fis binds to two nearby sites within an "enhancer" located about 100 bp from one of the *hix* sites within the invertible segment. Fis bound to the enhancer and Hin bound to the *hix* sites then assemble together into a tripartite complex called the invertasome (Fig. 8B). HU promotes assembly of the invertasome by facilitating the formation of the short DNA loop between the Hin bound to the nearby *hix* site and the Fis proteins bound to the enhancer (147, 184). If the enhancer is moved such that the size of the DNA loops in the invertasome are larger, the requirement for HU is much less. HMGB proteins are effective in substituting for HU in the Hin reaction, confirming that HU is functioning strictly as a DNA architectural factor without the need for direct interactions with the other protein components (317). The role of Fis in forming the invertasome and regulating recombination is discussed below.

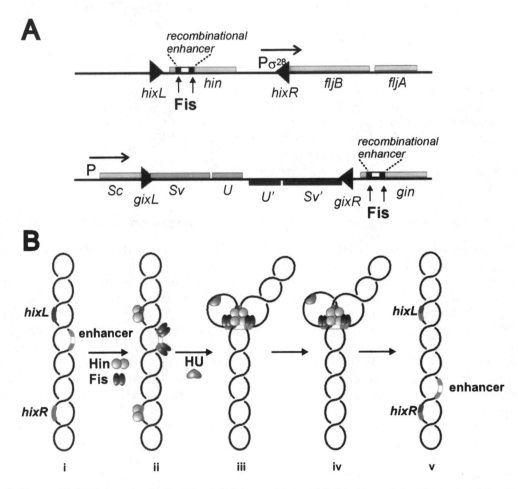

Figure 8. Site-specific DNA inversion by the Hin and Gin recombinases. (A) Structures of the Hin-catalyzed flagellar phase variation control region from *S. enterica* serovar Typhimurium (392) and the Gin-catalyzed G inversion segment from phage Mu that controls tail fiber gene expression (218). The recombination sites (*hixR, hixL; gixR, gixL*) are designated by triangles, and the coding regions are designated by gray bars. A recombinational enhancer sequence is located within the N-terminal segments of the *hin* and *gin* coding sequences. Fis binds to two essential sites within the enhancers. In the Hin system, *fljB* and *fljA* code for the H2 form of flagellin and a repressor of the unlinked *fliC* flagellin gene, respectively. In this orientation of the invertible segment, the cell expresses the H2 flagellin, which is transcribed from the σ28-dependent promoter located within the 996-bp invertible segment. In the Gin system, a constant 5′ region is linked to the alternate forms of the 3′ regions of the S gene, depending on the orientation of the 3,015-bp invertible segment. (B) Pathway leading to Hin-catalyzed site-specific DNA inversion (182). Hin and Fis bind (ii) to their respective binding sites in the starting supercoiled plasmid substrate (i). (iii) An invertasome complex is assembled at the base of a supercoiled DNA branch. Within this complex Fis activates Hin to coordinately cleave both DNA strands within each recombination site. HU facilitates the looping of DNA between the *hixL* recombination site and the enhancer, which are about 100 bp apart in their native context. (iv) DNA exchange is accompanied by a 180° clockwise rotation of duplex DNA strands to invert the internal segment (v).

The *Bacillus subtilis* Hbsu protein, which is equivalent to HU, promotes site-specific DNA resolution and inversion reactions performed by the β recombinase found on plasmids pSM19035 and pAMβ1 and the Sin recombinase from *Staphylococcus aureus*. Its role in these recombination systems is probably to introduce appropriate bends into the DNA to allow the assembly of synaptic complexes in a manner analogous to the role of HU in the Hin system (6, 322, 355).

HU in Transposition

HU stimulates an early step in phage Mu transposition by facilitating the assembly of the transpososome (15, 77, 234, 449). Each Mu end contains three binding sites for the MuA transposase, and a tetramer of MuA plus the end-proximal binding sites assemble into the final stable transpososome structure (63). Chaconas and colleagues coupled an iron-EDTA cleavage reagent to HU and generated a footprint of HU bound within the Mu transpososome (235). They found that a single HU dimer binds in the spacer region between the L1 and L2 *att* sites (Fig. 9). DNA in the transpososome becomes wrapped around HU in a supercoiling-dependent manner to shorten the distance between the two DNA-bound MuA proteins. HU bound to the spacer region between the sites and produced a cleavage pattern suggesting that the DNA is bent 155° (236), consistent with the recent Swinger and Rice crystal structure. Further analysis suggested that HU recognizes a DNA structure that is induced by supercoiling rather than being targeted to a specific sequence between the L1 and L2 sites. Since the level of supercoiling affected HU binding, Kobryn et al. proposed that HU acts as a "supercoiling sensor" to regulate transposition as a function of the level of supercoiling in the donor cell (216). This could be a way to coordinate Mu transposition with DNA superhelicity and growth phase.

HU has also been shown to be required for Tn916 transposition (73). Tn916 is a conjugative transposon with a wide host range that includes both gram-positive and gram-negative bacteria (69). Experiments in *E. coli* have shown that HU is required for the initial excision reaction from the donor DNA, which proceeds by a mechanism that resembles excision by phage λ. Since HU homologues are present throughout bacteria, a nonspecific architectural protein like HU is ideally suited for Tn916 transposition into diverse hosts.

HU in DNA Recombination and Repair

Cells deficient in HU are extremely sensitive to γ-radiation, which causes double-strand breaks in DNA (42). Li and Waters showed that *hupAB* cells are also sensitive to UV irradiation and defective in recombinational DNA repair (245). Further genetic analysis suggests that the defect appears to be at the step of Holliday junction resolution. As described above, HU binds to and potentially stabilizes a variety of DNA structures that could be intermediates in DNA recombination and repair. HU could also be directly enhancing the interactions of repair proteins with DNA in a manner similar to its stimulation of other DNA regulatory proteins discussed above. HU has been shown to protect DNA from radiation-induced breaks in vitro (42). Thus, the coating of DNA by HU could have an indirect protecting role as well as function to inhibit the nucleolytic degradation of damaged DNAs.

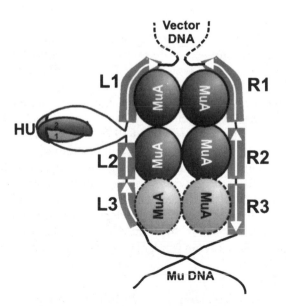

Figure 9. Structure of the Mu transpososome (63, 216). MuA monomers bind to *attL* (L1, L2, and L3) sites and *attR* (R1, R2, and R3) sites to initiate transpososome assembly. The stabilization of a loop between the L1 and L2 sites by HU is believed to be required for assembly of the catalytically competent transpososome. The relative positions of the HU-α and HU-β subunits are shown. The MuA protomers at the L3 and R3 binding sites are loosely associated and not required for transpososome function.

IHF

Identification of IHF

IHF was discovered as a host-encoded protein that was required for bacteriophage λ site-specific recombination in vitro (208, 287). Early work also showed that lytic growth of bacteriophage Mu was depressed in strains that lacked IHF (269). Over the ensuing years the roles of IHF in a variety of cellular processes, including transposition, gene expression, and DNA replication, have been discovered (reviewed in reference 286).

IHF homologues have been identified in several proteobacteria, and Fig. 10 shows an alignment of several IHF proteins. The structure of IHF is similar to that of HU protein, and the subunits contain all of the secondary structure elements found in HU (see below). In *E. coli* and *S. enterica* serovar Typhimurium, IHF is a heterodimeric protein composed of α and β subunits that are approximately 11 and 9.5 kDa in size, respectively (287). The amino acid sequences of the subunits of IHF and HU proteins show significant homology, with approximately 45% identical or similar residues. In *E. coli* the unlinked *ihfA* (formerly *himA*) (267, 268) and *ihfB* (formerly *hip* or *himD*) (114, 207, 268) genes that encode IHF were originally identified by mutants that affected phage λ site-specific recombination in vivo. The *ihfA* gene is located at min 38 of the *E. coli* genome and is part of the *pheST* operon. Its expression is repressed by IHF and is also under control of the SOS network and tRNA-Phe-dependent attenuation (12, 260, 268). The *ihfB* gene maps at min 20 of the *E. coli* genome. It lies downstream of the *rpsA* gene and is weakly repressed by IHF (12, 114). The autogenous regulation of IHF synthesis is possibly mediated by binding of IHF to sites upstream of the *ihfA* and *ihfB* genes. The amount of IHF increases in cells in late stationary phase (see below), and this increase is dependent upon the *rpoS* gene and ppGpp (12, 93).

General Properties of *ihf* Mutants

Mutants deficient in IHF are viable, but they have subtle phenotypes. A comparison of protein profiles from strains deficient for IHF revealed many differences relative to wild type (121). When the expressions of most genes known to be affected by IHF are examined in IHF-deficient cells or in cells that have mutated IHF binding sites, the effects are usually modest, with only two- to fivefold differences. However, site-specific recombination systems such as λ and HK022, as well as operons activated by the NifA, NarL, and TdcR regulators, show large reductions in activity in *ihf* backgrounds (114, 166, 207, 268, 270, 336, 373, 467). As noted by Nash, the proteins induced by these regulators synthesize ammonia from alternative nitrogen sources and are only expressed under anaerobic or microaerobic conditions (286).

This suggests that IHF might be essential for responses to limited nitrogen under anaerobic conditions.

Arfin et al. (7) measured gene expression profiles in isogenic *ihf*[+] and *ihf* mutant *E. coli* strains. The cells were grown in glucose-minimal MOPS (morpholinepropanesulfonic acid) medium. The study showed that IHF directly or indirectly affects expression of over 100 genes. The genes that were differentially expressed were examined for putative IHF binding sites in the 500 bp upstream of the transcription initiation site. The criteria for identifying high-affinity sites were that the sites must have a 12 out of 13 bp match with the core consensus sequence and that at least 10 of the 15 bp 5′ (upstream) to the core consensus must be dA-dT base pairs. The study identified 46 candidates that matched the criteria, some of which were documented in other studies. The criteria used were quite stringent because IHF binding sites can vary significantly from the consensus (see below), and it is likely that several functional sites upstream of genes were missed. Among the genes identified were *arcA*, a global regulatory protein, as well as several putative regulatory proteins. Other genes with documented IHF sites upstream of the promoter include the *ilvGMEDA* operon, *sodA* and *ilvP_a*, *ihfB*, *ompF*, *dps*, and *ndh*. The study also identified seven genes, including *ihfA*, that were expressed only when IHF was present and eight genes that were expressed only in IHF-deficient cells. Mechanisms of repression and activation mediated by IHF are discussed below.

IHF is also required for starvation survival, and 14 glucose survival proteins are induced in wild-type cells but not in an *ihfA* mutant after glucose depletion (302). Several of the proteins were identified as belonging to different starvation and stress stimulons, including ones regulated by RpoS, Lrp, OxyR, and OmpR. Overproduction of IHF in exponentially growing cells was not sufficient to induce any of the glucose survival proteins that are induced as wild-type cells enter stationary phase. Interestingly, overexpression of IHF in exponential phase did result in increased synthesis of *rpoH*-dependent heat shock proteins. However, it is not known whether this stimulation occurs under physiologically relevant conditions.

Genetic Analyses of IHF Binding to DNA

Mutagenesis and footprinting studies have shown that IHF binds to sequences 30 to 35 bp in length. These sites are asymmetric and contain a core element with the consensus sequence WATCAAN-NNNTTR (where W is A or T, R is purine, and N is any base) (135). Genetic analyses of mutant IHF proteins agree well with the IHF-DNA cocrystal structure (see below). In one study, Lee et al. (240) isolated substitution mutants in IHF with altered, expanded DNA binding specificities. The mutants were the first evidence that suggested that the α subunit interacts with the WATCAA element, and the β subunit interacts with the TTR element. Two of the mutant proteins recognized the wild-type λ H′ site in addition to a mutant site with a change of the second A:T of the WATCAA sequence to a C:G. The mutants changed the proline-64 residue to leucine or the lysine-65 residue to a serine in the α subunit. Pro-64 of the α subunit intercalates into the DNA so the change to leucine must still allow DNA binding. The exact mechanism by which these mutants expand the recognition specificity is not known.

Three other relaxed-specificity mutants change Glu-44 of the β subunit to lysine, valine, or glycine. These mutants bound to the wild-type H′ site and a variant site with a change of the middle T:A of the TTR element to an A:T. These mutants appear to act indirectly, because Glu-44 does not contact the DNA. The crystal structure shows that Glu-44 and Arg-42 form a network of interactions that place Arg-46 in a position where it can interact within the T44 residue in the DNA. The altered specificity mutants do not have side chains that form the network that anchors Arg-46 in a position where it can contact the T44 residue. This appears to increase the conformational freedom of the Arg-46 side chain so that it does not form a disruptive contact with the mutant TTR element. As a result, both the wild-type and mutant sites are bound by the mutant proteins (338, 342).

Mengeritsky et al. (263) analyzed β-subunit mutants with substitutions of residues Arg-42, Glu-44, and Arg-46 and found that the mutant proteins had reduced IHF activity in vivo. The Arg-42 to alanine and Glu-44 to alanine substitutions had small effects on IHF function, consistent with the results above that showed that disruption of the Arg-42 and Glu-44 interaction expanded recognition but did not drastically decrease DNA binding. The Arg-46 to cysteine change had the most dramatic loss of function phenotype in a variety of assays. However, the same mutation had marginal phenotypes in other assays (138). Interestingly, Zulianello et al. (496) showed that a mutant with Arg-46 changed to histidine decreased DNA binding dramatically. The results are generally consistent with the crystal structure work indicating that Arg-46 is the only β-subunit residue that makes a specific contact with DNA. However, loss of this contact can have a vari-

able effect, depending upon the IHF binding site and functional assay used.

Granston and Nash (138) isolated loss-of-function mutants that were still able to form dimers. Such mutants would be expected to affect residues that contact the DNA. These results were consistent with the crystal structure and suggested that the arms and flanks of the two subunits play different roles in DNA binding (see below). Granston and Nash found that all of their mutants with amino acid substitutions in the α subunit changed residues in the β-ribbon structure that interact with the WATCAA element. However, the α-subunit flank is also important for binding, because a protein containing a deletion of the 15 C-terminal amino acids shows no detectable DNA binding. The eight C-terminal residues in the α3 helix are not required for DNA binding, because deletion of these residues has no effect (496). A mutant with a serine replacement of the proline at the tip of the β-subunit arm (residue 64) was defective in DNA binding. However, most of the substitutions isolated in the β subunit were in the flank, suggesting that this region also plays a role in DNA binding (138). In addition, deletion of the last four C-terminal residues of the β-subunit α3 helix has no effect on DNA binding, while deletion of the last 10 residues decreases the affinity but not specificity of IHF for DNA (496).

Homodimers of IHF, which also bind DNA sequence-specifically, can be formed in vitro (454, 495). However, 10 to 50 times more α or β homodimer is required to achieve the same amount of in vitro λ recombination supported by the native heterodimeric protein (454). It is possible that a small number of homodimers assemble in vivo, but the majority of IHF activity must come from heterodimers, because deletions of the genes encoding either *ihfA* or *ihfB* usually produce the *ihf* null phenotype (138, 454, 495). There is no evidence for heterodimeric proteins composed of subunits of IHF and HU.

Structure of the IHF Heterodimer and IHF Interactions with DNA

An early model of IHF was based on the *G. stearothermophilus* HU crystal structure solved in the absence of DNA (see above) (414, 457). A model was built using extensive DNA footprinting information derived from the phage λ H′ site complexed with IHF (478). Using the HU structure as a reference, the model proposed that the arms of IHF wrap around the minor groove of DNA. The DNA was bent as it wrapped around the body of the protein, creating a bend in the DNA that approached

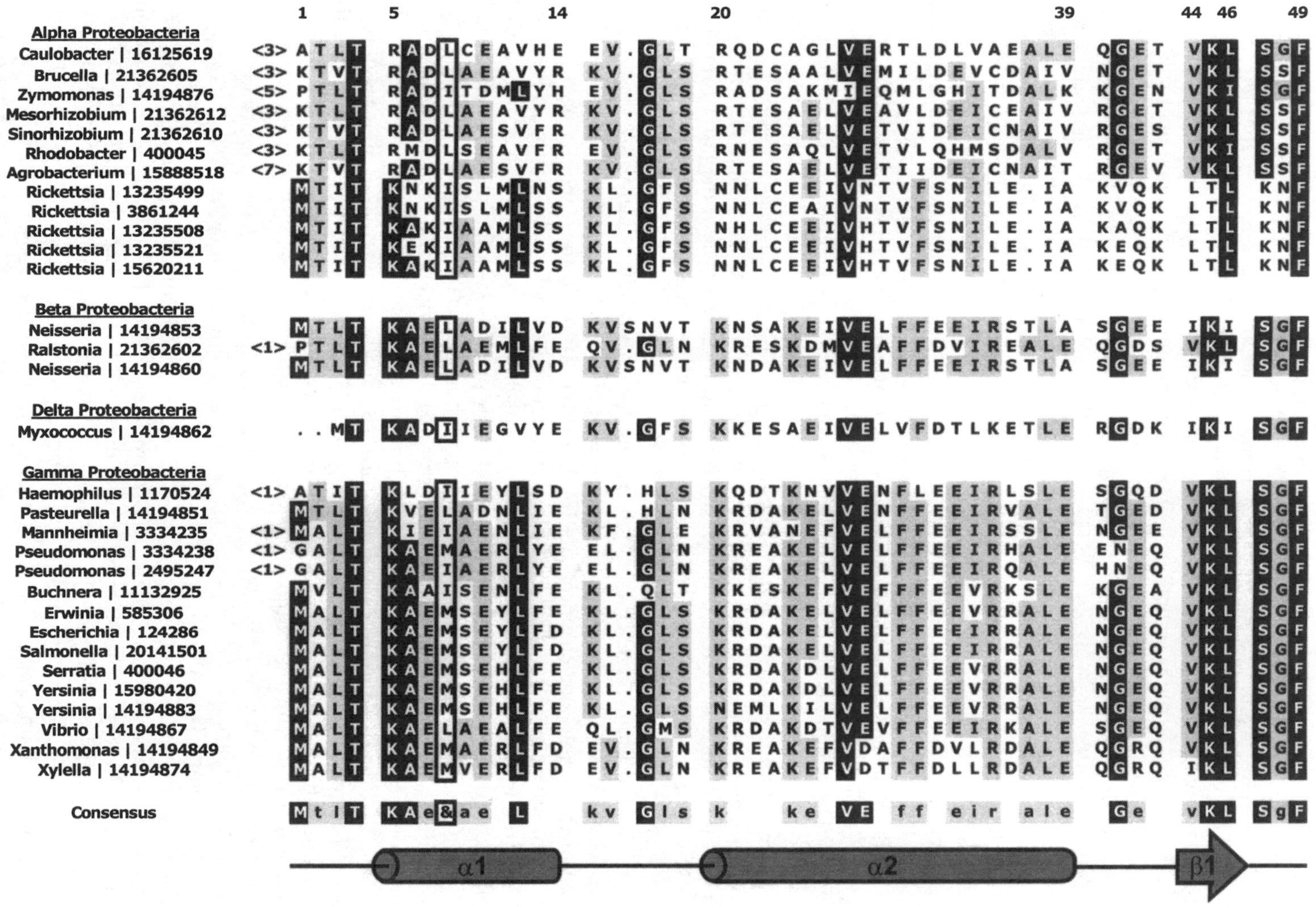

Figure 10. Alignment of the α and β subunits, members of the IHF family. The sequences are identified by the species from whose genome they originated and their NCBI GI number. The IHF sequences were obtained by searching the SWISSPROT database for "IHF" and for "integration host factor." The NCBI PSI-BLAST program was used with *E. coli* IHFα and IHFβ query sequences to obtain additional IHF sequences not contained in the SWISSPROT database. Sequences not described as integration host factor were discarded. The alignments were performed as described in the legend to Fig. 2 and numbered according to the *E. coli* IHF sequences. Consensus notation: a lowercase letter indicates the residue is found in 75 to 100% of sequences overall and is shaded gray where conserved. An uppercase letter indicates the residue is found in 75 to 100% of sequences overall, and the background is black where conserved. The # indicates that I or V residues are found in 75 to 100% of the sequences and are boxed by a solid line. The & indicates that I, L, or M is always found at this position, and the residues

(continued)

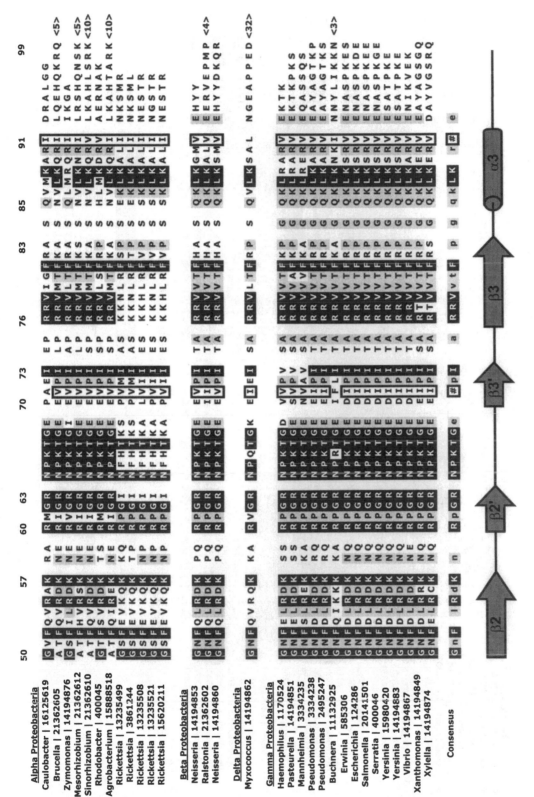

Figure 10 (*continued*) are boxed by a solid line. Secondary structure: the secondary structure cartoon shown below the alignments is based on the crystal structure of the *E. coli* IHF-DNA complex (342) (PDB accession code, 1IHF). The secondary structure elements for the α subunit are as follows: α1 helix residues 5 to 14, α2 helix residues 20 to 39, β1-sheet residues 44 to 46, β2-sheet residues 50 to 57, β2′-sheet residues 60 to 63, β3′-sheet residues 70 to 73, β3-sheet residues 76 to 83, and α3 helix residues 85 to 91. The secondary structure elements for the β subunit are as follows: α1 helix residues 3 to 13, α2 helix residues 19 to 38, β1-sheet residues 43 to 45, β2-sheet residues 49 to 56, β2′-sheet residues 59 to 62, β3′-sheet residues 69 to 72, β3-sheet residues 75 to 82, and α3 helix residues 84 to 90.

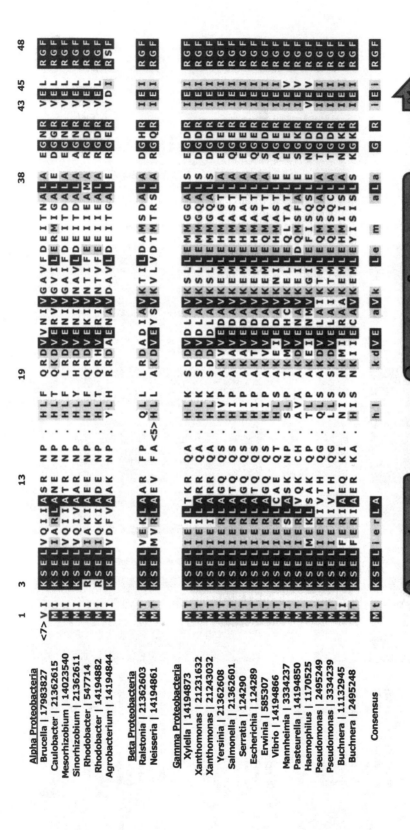

Figure 10 (*continued*)

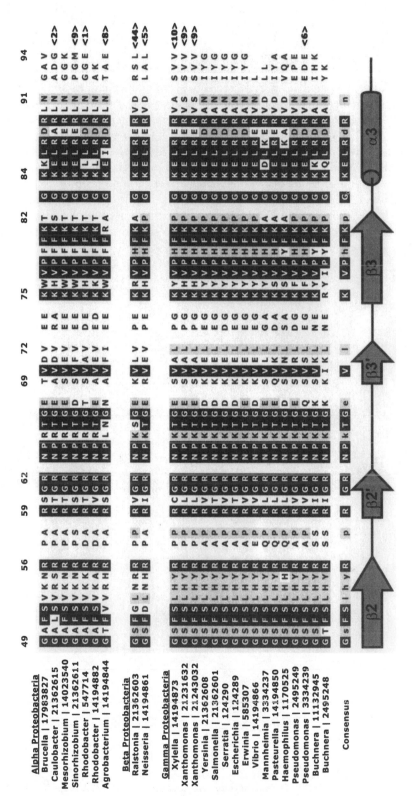

Figure 10 (*continued*)

180° (478). Subsequent genetic studies that identified residues of IHF likely to be near or in contact with DNA (see above) (138, 240, 495) and cross-linking studies (480) were consistent with the model.

Crystal structures of the *E. coli* IHF protein bound to the phage λ H′ IHF binding site have been reported by Rice et al. (341, 342) (Fig. 11). In one structure, both DNA strands contained a nick with an 8-bp overlap in the center, whereas a second structure contained only a single nick in one of the oligonucleotides. Solution studies have shown that the nicks have a negligible effect on the IHF-induced bend angle (247). The DNA in the crystals formed a U-turn around the protein, and the ends of the DNA formed a pseudocontinuous helix (Fig. 11A). The overall bend of the DNA was between 160 and 180° depending upon whether contacts with neighboring

DNA are considered. The β-ribbon arms of IHF wrap around the DNA and interact with the minor groove. All of the contacts with the DNA are either in the minor groove or involve electrostatic interactions with the phosphate backbone.

The fold of IHF is almost identical to that of HU. The N terminus of each subunit of the IHF heterodimer contains two amphipathic α-helices (α1 and α2) that form a helix-turn-helix structure. The N-terminal amino acids of the α1 helix of each subunit interact with the phosphate backbone of DNA. The α2 helices are amphipathic, and the hydrophobic residues form the major portion of the dimerization surface. The C termini contain a third amphipathic α-helix (α3). The N termini of both the α2 and α3 helices contact the phosphate backbone of the DNA. Each subunit also has a β-ribbon arm composed of

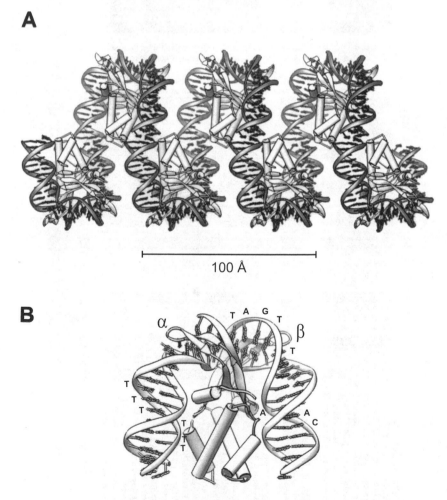

Figure 11. Structure of IHF bound to the H′ site (341, 342). (A) Crystal packing of IHF and DNA with seven asymmetric units shown (342). The DNA forms a serpentinelike helix that zigzags through the crystal. (B) A single IHF-H′ DNA complex. The α subunit (dark gray) and the β subunit (light gray) are shown. The six consecutive dT residues of the dA-dT-rich element are shown on the left. The WATCAA element (shown as the opposite strand: ATAGTT) is on the top of the complex and the TTR element (shown as AAC) is on the right. The arms of the α and β subunits interact with the minor groove of the DNA. The DNA is bent around the protein to generate a bend angle of more than 160°. (Courtesy of P. Rice.)

two antiparallel β-strands (β2 and β3) that protrude from the body of the protein. The C-terminal end of the β2-strand has the highly conserved Arg-X-Gly-Arg-Asn-Pro motif that is also observed in HU.

The bend in the DNA is generated in part by two IHF-induced kinks in the DNA spaced 9 bp apart. The kinks are formed by intercalation of the conserved proline residues (residue 65 in the α subunit and 64 in the β subunit) that are located at the tip of the β-ribbon arm. This stabilizes the disruption of the base stacking required to produce the kinks. The bend is also stabilized by multiple interactions of positively charged residues in the body of the protein with the phosphate backbone of the DNA. Residues in the α-helices and β1-β2 turn on the sides of the protein form a "tripartite clamp" that lies on the minor groove of the DNA.

IHF Binding Sites

As described above, IHF binds sequences 30 to 35 bp in length. The adjacent C and A residues of the WATCAA element are the most highly conserved. Some sites with perfect 3' core element consensus sequences bind DNA relatively poorly, while other sites with weak consensus sequences bind relatively well. This lack of correlation between the similarities of sites to the consensus sequence and their affinities for DNA could be due to differential contributions of the 5' flanking domain to DNA binding. The 5' domain, composed of approximately 20 bp, has no obvious consensus sequence and differs for every site. Many IHF binding sites, including the λ H' site, have an A tract containing three to six consecutive adenines in the 5' domain. A tracts tend to be straight and contain a narrow minor groove (92, 291). It is likely that recognition of the A tract occurs through DNA structure rather than by specific base interactions, because the sequences of this region vary greatly among IHF binding sites (342).

Domain-swapping experiments have shown that removal of a dA + dT-rich element from a relatively high affinity site (λ H' site) can decrease the affinity of the new site for IHF. Addition of the same dA + dT-rich element to a lower-affinity site that lacks a 5' dA + dT element (λ H1 site) increases the affinity of the hybrid site for IHF (144, 145). The footprint of the H' site is larger than that of the H1 site, also indicating that the dA + dT element contributes to binding (76). Interestingly, high binding affinity per se does not require a 5' dA + dT-rich sequence; Goodman and Kay found no bias for dA + dT-rich 5' sequences among high-affinity IHF binding sites selected in vitro (133).

Chemical protection and interference studies showed that IHF makes contacts with the minor grooves of these consensus elements (478). The minor groove of the WATCAA element is contacted by the α subunit. The highly conserved C:G and A:T base pairs (WAT<u>C</u><u>A</u>A) interact with two arginine residues (Arg-60 and Arg-63) that make the only two specific contacts with the bases in the element. Rice et al. (342) proposed that these arginine residues select a DNA conformation that is accommodated by the base pairs at these two positions. The intercalation of Pro-65 between the A residues of the DNA buckles these residues into a conformation that allows interactions with Arg-60 and Arg-63 residues 2 and 5 bp away, respectively. These arginines per se do not fully explain IHF's sequence specificity, as they are also highly conserved in HU proteins that bind DNA nonspecifically. The remainder of the recognition is mediated by a large series of interactions through sequence-dependent distortion of the DNA, or indirect readout (310). The TTR element is recognized by the body of the β subunit. Only one arginine (Arg-46) makes a specific contact with the DNA. Interestingly, this arginine is not conserved among HU proteins that bind DNA nonspecifically.

Natural IHF binding sites are asymmetric, although IHF can recognize symmetric sites. Werner et al. (454) proposed that the predominance of asymmetric sequences indicates that they were selected for during evolution. Because the sizes of IHF binding sites are so large, IHF cannot interact with the whole site in a single step. IHF could recognize asymmetric sites in sequential stages rather than in a single step. Werner et al. argue that IHF initially interacts with 10 bp, while the size of the entire site is 30 bp. If one of the two arms of IHF initially recognizes the DNA, the remaining arm could be involved in stabilizing the complex. This model predicts that one of the arms and its recognition determinant would coevolve to optimize the initial interactions. The remaining arm would coevolve with its target sequence in a way that optimizes IHF binding and bending to function at the particular site. Recent synchrotron X-ray footprinting studies also suggest that IHF binds DNA through a stepwise mechanism. The first step is proposed to be diffusion along the DNA followed by binding and bending of the DNA (91).

IHF in DNA Supercoiling, Condensation, and Nucleoid Structure

IHF binds to specific sites with K_d values ranging from 0.3 to 20 nM (283, 448, 479). The specificity ratio of natural specific sites to nonspecific sites is 1,000 to 10,000, so its nonspecific binding affinity for DNA is similar to that of HU for linear DNA (283, 327, 448, 479). Indeed, the majority of IHF is believed to be bound nonspecifically to DNA (see below). IHF and HU are estimated to cover between 10 and 20% of the DNA in the cell, depending on growth conditions (Table 1), and thus they likely have a profound effect on chromosome structure.

The intracellular concentration of IHF varies with growth phase. Ditto et al. estimated by immunoblotting that exponentially growing cells contain 8,500 (7 μM) to 17,000 (14 μM) dimers per cell, and stationary-phase cells contain about fivefold-greater amounts (93). In a later study, Talukder et al. estimated that exponentially growing cells contain 6,000 dimers per cell (412). As cells progress into early stationary phase, the number of dimers rises to about 27,500 per cell, and the number drops to about 15,000 dimers per cell in late-stationary-phase cells.

Dimethyl sulfate footprinting in vivo performed by Yang and Nash estimated that the free intracellular concentration of IHF is 15 nM and 35 nM in exponentially growing and stationary-phase cells, respectively (479). Their calculations indicated that several of the natural sites tested have affinities only slightly higher than that required for occupancy. More recent UV laser footprinting studies by Murtin et al. suggest that the amount of free IHF in cells is even lower, varying between 0.7 nM for exponential-phase cells and 5 nM for stationary-phase cells (283). The same study also measured occupancy of three sites—*yjbE* ORF, BIME-1 *gyrB*, and IS*1*), under exponential and stationary phases of growth. The higher-affinity *yjbE* ORF and BIME-1 *gyrB* sites were nearly saturated in stationary-phase cells and 50 to 70% saturated in exponential-phase cells. The weaker IS*1* site was about 70% saturated in stationary-phase cells and 25 to 30% saturated in exponential-phase cells. These results argue that only high-affinity sites would be efficiently occupied in exponential-phase cells. Thus, growth-dependent variations in intracellular IHF levels could have important consequences for regulation of gene expression and in the formation of architectural structures that regulate gene expression and other cellular functions.

Since the total concentration of IHF in exponential- and stationary-phase cells varies between 6,000 and 27,500 dimers per cell, only a small fraction of the dimers can be bound to specific sites. If one assumes that the number of specific IHF binding sites is probably less than 1,000 per chromosome (7, 435), most of the IHF is bound nonspecifically to chromosomal DNA.

The importance of IHF binding to nonspecific sites is supported by results of a study done by Ali et al. (4). They analyzed compaction of single λ DNA molecules by IHF and concluded that DNA molecules have a more compact, coiled structure in the presence of IHF. They argue that the compaction is likely due to interactions with multiple low-affinity sites because of the large range of concentrations of IHF that produce the effects, the observation that the same degree of compaction is observed with short DNA molecules containing or lacking a specific site, and the observation that compaction can be competed with nonspecific competitor DNA (4).

IHF in Site-Specific Recombination

Studies on λ site-specific recombination have demonstrated that IHF functions as a sequence-specific architectural element in assembling nucleoprotein structures called intasomes (reviewed in reference 13). Footprinting and mutation studies show that IHF binds to three sites (H1, H2, and H′) in the phage *attP* site (Fig. 12). IHF binding to all three sites is required for integrative recombination. In addition, *attP* has five arm-type sites (P1, P2, P′1, P′2, and P′3) that bind Int through its N-terminal domain. The C-terminal domain of Int also interacts with two core sites (C and C′) that flank the region where the strand cleavages and ligations occur during recombination. Int and IHF cooperate to form the *attP* intasome that is required for synapsis of partner DNA sites and then strand exchange to generate recombinant products. After integrative recombination, the *attL* site at the left phage-chromosome junction contains the H′ site, and the *attR* site at the right phage-chromosome junction contains the H1 and H2 sites (76, 250, 272, 344, 346, 347).

The reverse reaction, excisive recombination, also involves Int binding to a subset of the arm binding sites and IHF binding to the H2 and H′ sites (415, 420). In addition, the phage-encoded Xis protein is required, and the reaction is stimulated by Fis. Xis binds to two sites in direct repeat orientation, X1 and X2, and Fis binds to the F site that overlaps X2. Binding of Xis to X2 and binding of Fis to the F site are mutually exclusive (417). During excision, intasomes are formed on *attR* (Int, IHF, and Xis-Fis) and *attL* (Int and IHF), which undergo synapsis to initiate the excision reaction (reviewed in reference 13). The roles of Fis and Xis in excisive recombination are described below (see "Phage λ site-specific recombination" below).

Because the sizes of IHF footprints were so large, it was initially proposed that the DNA could be wrapped around the protein to form bends in the DNA at the H1, H2, and H′ sites (344, 419). The bends could be used to help form intasomes containing IHF and Int during integration and containing IHF, Int, and Xis during excision. The first

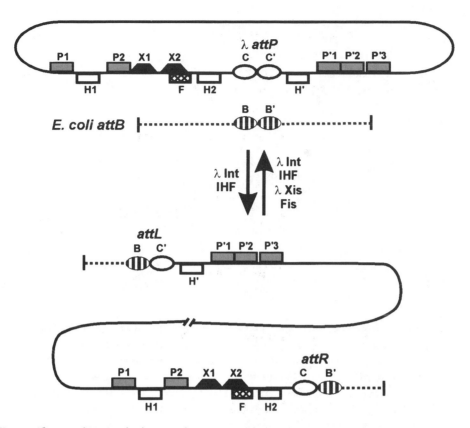

Figure 12. Site-specific recombination by bacteriophage λ (13). The 250-bp phage attachment site (*attP*) contains multiple binding sites for Int and accessory proteins. Int binds to five arm sites (designated P1, P2, P'1, P'2, and P'3) and two core-type sites (designated C and C'). IHF binds to three sites (designated H1, H2, and H'). Xis binds to two sites (designated X1 and X2), and Fis binds to a single site (designated F). Binding of Fis to the F site and binding of Xis to the X2 site are mutually exclusive. Recombination between *attP* and the bacterial attachment site (*attB*) generates the recombinant *attL* and *attR* sites. Int and IHF are required for integration, and Int, IHF, and Xis are required for excision. Excision is also stimulated by Fis.

experimental evidence for an architectural role of IHF was the "bend-swap" experiments of Goodman and Nash (134). They replaced the H1 IHF binding site in the *attP* site of λ (Fig. 12) with a CRP site or intrinsically bent DNA sequences and found that Int could promote recombination in the presence of CRP with the CRP site-substituted DNA or with the intrinsically bent DNA. More recently, Goodman and Kay (133) made substrates with regions of single-stranded DNA to provide a flexible tether to mimic bending of the DNA. They found that the substrates enhanced recombination in the absence of IHF, albeit they were not fully active.

Other studies have demonstrated that an IHF-induced bend is necessary to form an *attL* intasome. The core sites of *attL* are arranged as inverted repeats and interact with the C-terminal domain of Int (Fig. 13). The P'1, P'2, and P'3 sites are arranged in direct-repeat orientation and interact with the N-terminal domain of Int. The H' IHF binding site lies between the core-type and arm-type sites. Biochemical and genetic studies showed that both IHF and Int

are required to form the *attL* intasome. An Int monomer binds simultaneously to the core C' site through its C-terminal domain and to the arm P'1 site through its N-terminal domain. This bridging interaction between the two sites requires IHF, which induces a bend in the intervening DNA to allow the two domains of Int to bind simultaneously to the C' and P'1 sites (Fig. 13). Mutations in the B' site, the P'1 site, or the H' site reduce formation of the complex dramatically. Mutations in the C site or the P'2 and P'3 arm sites have a much less dramatic effect on formation of the complex (211, 250, 377).

HU can substitute for IHF in excisive recombination between *attL* and *attR* sites (377). Presumably, HU substitutes for IHF at the H2 and H' sites and collaborates with Int to form active *attL* and *attR* intasomes. In support of this model, HU and Int formed complexes on *attL* DNA that displayed electrophoretic activity similar to that of complexes containing Int and IHF. The ability of IHF and HU to replace each other is consistent with the proposal that they function by bending DNA and not

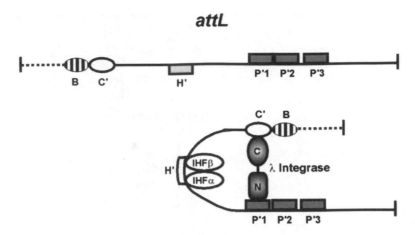

Figure 13. Structure of the λ *attL* intasome (209, 211, 250). IHF binds to the H′ site located between the core C′ site and the arm P′1 site. An Int monomer forms a bridge, where its N-terminal domain binds the P′1 site and the C-terminal domain binds the C′ core site. An Int monomer bound to the P′2 site forms a *trans* interaction with the partner *attR* site (not shown) (211).

by making protein-protein contacts with Int. HU also substituted for IHF with Int and Xis on *attR* DNA, although formation of the *attR* complex was not as efficient as formation of the *attL* complex. HU cannot substitute for IHF in the integration reaction, probably because it is not possible to assemble an active intasome with three HU dimers and four Int monomers simultaneously on an *attP* substrate.

The changes in cellular IHF concentrations as a function of growth phase may contribute to regulation of the directionality of λ site-specific recombination. Work done in vitro by Thompson et al. showed that the moderate-affinity H1 site must be occupied by IHF for integration but must be unbound for excision (415). Thus, when the intracellular levels of IHF are high (early stationary phase), the H1 site will be occupied, and the integration reaction will be favored. Excision would be undesirable in stationary phase because the cellular resources will be insufficient to support efficient lytic growth after excision (230).

The conjugative transposon CTnDOT, which carries resistance to tetracycline and erythromycin, has been found in isolates of anaerobic bacteria of the genus *Bacteroides* (390, 459). Its host range is broad, and it can integrate at low frequency into the *E. coli* chromosome (66). Integration of CTnDOT into its chromosomal site requires a CTnDOT-encoded tyrosine recombinase that is related to λ integrase and at least one protein encoded by the host. The reaction has been reconstituted in vitro and shown to require CTnDOT integrase and a host extract for optimal activity. Interestingly, purified *E. coli* IHF can substitute for the host factor (66). The mechanism of IHF function has not been established, but it is likely that

it binds CTnDOT joined ends and induces a bend in the DNA that is required to form the integrative intasome. Since IHF homologues are found in a variety of bacteria, it could be utilized by other conjugative transposons to form intasomes for integration of the element into new host chromosomes, as has been suggested for HU and Tn*916* (see above).

IHF in Transposition

IHF binding sites have been found near the ends of several transposable elements, such as IS*10*/Tn*10*, IS*1*, and γδ (125, 169, 460). In the case of IS*10*/Tn*10*, IHF plays a complex role in the transposition reaction, as it influences both the rates of the reaction and the types of transposition products produced (64, 391). In vitro, IHF strongly stimulates the reaction under suboptimal conditions, such as reduced supercoiling, and is absolutely required when linear transposition substrates are employed (64, 358).

Staged in vitro reactions employing linear substrates have demonstrated that IHF is required early during the transpososome assembly step. Transposase initially interacts with DNA on both sides of the IHF site, whose core is positioned 30 to 42 bp from the transposon outer end, to form a tightly wrapped complex (78). IHF, together with the upstream nonspecific DNA contacts by transposase, then appears to be released coincident with a conformational change triggered by the addition of metal (5, 78). This initial IHF-dependent assembly followed by its release has been referred to as the loading and then firing of a molecular spring (64, 358). The unfolding of the complex is believed to be a prerequisite for capture of the target sequence prior to the strand-transfer reaction.

In the presence of high levels of IHF, strand-transfer products are "channeled" away from intermolecular transpositions, which are generated by random collisions between the transpososome and target DNA, to intramolecular products. Most of the latter products are the result of a reaction into a target close to the transposon end that results in an inversion circle. Under these conditions, IHF is thought to rebind to the transpososome and orient the ends such that strand transfer with a nearby target site within the same DNA is strongly favored (64, 358).

Regulation of Transcription by IHF

IHF has been found to directly inhibit transcription in several systems, including the *ompB* operon and the λ *Pcin* and $P_{R'}$ promoters. In these cases, IHF functions as a repressor by binding to specific sites that overlap the −10 or −35 regions of the promoter (141, 227, 427). It is likely that some of the genes identified in the microarray analysis by Arfin et al. are also repressed directly by IHF (7).

IHF can activate some promoters by stimulating interactions between the C-terminal domain of the α subunit (αCTD) of RNA polymerase and DNA upstream of the promoter. The P_L promoter of phage λ (128) (Fig. 14), the Pe promoter of phage Mu (226, 439, 440), and the Pu promoter of *Pseudomonas putida* (31) are activated by IHF. The mechanism of IHF activation in these systems is through an IHF-induced bend that promotes contacts between the αCTD of RNA polymerase and DNA upstream of the promoters. Giladi et al. used molecular modeling to examine the spatial positions of residues in IHF, αCTD, and the IHF binding site in the P_L promoter region (128). Because IHF binds DNA asymmetrically (240, 342, 454), they positioned IHF on the DNA in the orientation in which residues in the β

subunit interact with the A/T-rich UP element. In support of this configuration, amino acid substitutions of some of these residues reduce bending of the DNA and decrease activation of P_L. IHF is proposed to generate a DNA structure that enables interaction by one or both αCTD elements, with the αCTD and IHF occupying the same region but on opposite sides of the DNA (Fig. 14).

Engelhorn and Geiselmann isolated IHF mutants that showed increased activation of a modified *malT* promoter by IHF binding to an upstream site (104). Most mutants had multiple substitutions, but the mutants tended to remove positive charges or add negative ones. Many amino acid substitutions were at residues on the surface of the α and β subunits. Mutants with substitutions of Lys-45 of the α subunit and Lys-27 of the β subunit showed the most activation. The authors propose that the substitutions alter the planarity of the bend induced by the mutant proteins to enable a more favorable interaction between RNA polymerase and upstream DNA.

The *E. coli ilvP*$_G$ promoter is stimulated by IHF by a novel mechanism that is proposed to involve transmission of superhelical stress (Fig. 15). The upstream region of the *ilvP*$_G$ promoter has an IHF site at position −92 and a dA + dT-rich region upstream of the IHF site. Hatfield and coworkers have shown that the duplex DNA in the dA + dT-rich region is destabilized by superhelical stress, but when IHF binds to its site, the DNA structure is stabilized. As a consequence, the superhelical energy normally absorbed by the upstream region is transferred to the downstream −10 element, which becomes destabilized, leading to enhanced open complex formation (383).

Studies in a number of systems show that complex nucleoprotein structures involving IHF, together with other nucleoid and specialized regulatory proteins, can inhibit or activate gene expression. The

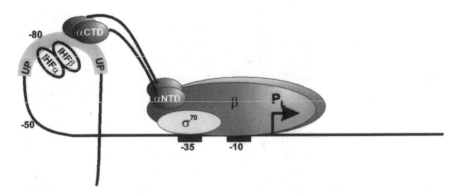

Figure 14. Activation of the λ P_L promoter by IHF (128). Binding of IHF induces a bend in the DNA that brings the UP (thick gray line) sequence closer to the promoter. This allows the αCTD of RNA polymerase to make contact with the UP element and stimulate transcription.

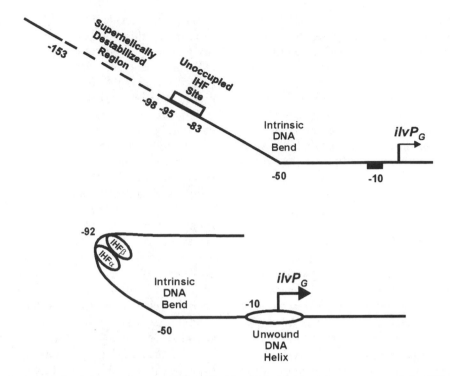

Figure 15. Activation of the *ilvP_G* promoter by IHF (383). The dA + dT-rich region upstream of the IHF binding site is destabilized by superhelical stress. When IHF binds, the superhelical stress is transferred downstream to the *ilvP_G* promoter to facilitate destabilization of the duplex in the −10 region, thereby increasing the rate of open complex formation by RNA polymerase.

bends induced by IHF modify the geometry of nucleoprotein structures and thereby affect gene expression. Essentially all studies indicate that IHF acts indirectly by bending DNA and does not participate in additional protein-protein interactions, similar to its role in λ site-specific recombination. A comprehensive description of these IHF-regulated transcription systems is beyond the scope of this chapter, but a few well-documented examples that illustrate basic mechanisms are summarized below.

Many sigma 54-dependent promoters provide a relatively simple example of IHF acting strictly as an architectural factor to stimulate transcription. In the case of the *Klebsiella pneumoniae glnHp2* promoter (166), IHF binds between the NtrA activator protein, which is centered at about 130 bp upstream of the transcription start site, and the promoter. The IHF-induced bend allows NtrA to interact with RNA polymerase bound at the promoter to trigger open complex formation.

Control of *ompF* expression is an example in which IHF participates in forming looped structures that promote repression. In *E. coli* the EnvZ and OmpR proteins constitute a two-component regulatory system that responds to osmolarity (reviewed in reference 333a). EnvZ phosphorylates OmpR and may

also be involved in dephosphorylation of phospho-OmpR (OmpR-P) (87, 173). These proteins control the levels of outer membrane proteins OmpF and OmpC. At low osmolarity, OmpF predominates in the outer membrane but is replaced by OmpC at high osmolarity. The *ompF* regulatory region contains three contiguous OmpR binding sites (F1, F2, and F3) in the region between the −100 and −35 sites upstream of the transcription start site. There is also an additional OmpR binding site (F4) located at position −360. Under low osmolarity, OmpR-P binds to the F1, F2, and F3 sites and is in a conformation that activates expression of the *ompF* promoter. Under conditions of high osmolarity, OmpR-P undergoes a conformational change that facilitates repression. IHF bound at a site between the distal (F4) and proximal OmpR-P sites makes a bend in the DNA that promotes repression by protein-protein contacts between OmpR proteins (254).

The *nir* promoter is an even more complex example in which multiple proteins form a structure that inhibits transcription (Fig. 16). FNR-dependent transcription of the *nir* promoter of *E. coli* is repressed by Fis, IHF, and H-NS (47). These proteins, in conjunction with FNR, form a nucleoprotein structure that contains several bends induced by Fis

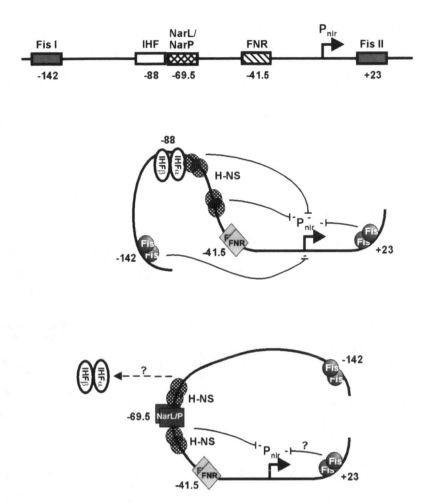

Figure 16. Regulation of *nir* by IHF, Fis, and H-NS (47). (Top) Organization of the Fis (I and II), IHF, NarL/P, and FNR binding sites within the *nir* promoter region. (Middle) Binding of IHF, H-NS, and Fis forms a complex that inhibits transcription initiation. (Bottom) It has been proposed that binding of phospho-NarL/NarP displaces IHF from the complex. This facilitates interactions between RNA polymerase and upstream sequences, resulting in activation of transcription.

and IHF. Binding of IHF within the complex forms an architecture that inhibits binding of RNA polymerase by orienting the upstream sequences away from the promoter. The binding site for NarL/NarP overlaps the IHF binding site. Browning et al. proposed that NarL or NarP binding displaces IHF from the complex. This changes the conformation of the complex into one in which RNA polymerase can make activation contacts with sequences upstream of the promoter, presumably through its αCTD (47).

Control of DNA Replication at *oriC* by IHF

ihf mutant strains exhibit an asynchronous replication phenotype in flow cytometry experiments, which is more severe than that observed in *hupAB* strains (43, 356, 445). Polaczek first noted that an IHF binding site is present in the left half of *oriC*

(Fig. 17) (330), and mutations within this site cause the loss of *oriC* plasmids and moderate replication asynchrony when recombined into the chromosome (350, 450). Leonard and coworkers have shown by in vivo footprinting that IHF undergoes cell cycle-dependent binding to this site (58, 356). At the time of initiation of DNA synthesis, IHF is transiently bound to *oriC*. Interestingly, Fis seems to be bound throughout most of the remaining part of the cell cycle, indicating that the IHF-bound and Fis-bound states may be mutually exclusive (58) (see "Control of DNA Replication at *oriC* by Fis"). DnaA is also bound to high-affinity sites R1, R2, and R4 within *oriC* throughout most of the cell cycle, but the weaker binding sites at R5(M), R3, and I1 to I3 become occupied coincidentally with IHF binding and open complex formation (58, 356). Enhancement of DnaA-ATP binding to its weaker sites by IHF can be

A) Most of cell cycle:

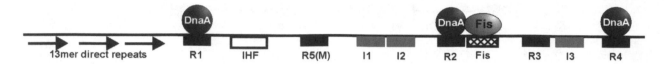

B) Pre-replication complex:

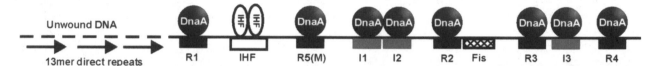

Figure 17. Organization of the *E. coli* origin of replication. The black boxes indicate the DnaA binding sites R1, R2, R3, R4, and R5(M). The gray boxes indicate the weaker DnaA-ATP binding sites I1, I2, and I3 that are detected by dimethyl sulfate footprinting in the presence of IHF (142). (A) During the majority of the cell cycle DnaA is bound to the R1, R2, and R4 sites, and the Fis protein is also bound. (B) During prereplication a nucleoprotein complex is formed in which all the R sites and weaker I sites are bound by DnaA. At this time, IHF is bound and Fis is absent. HU or IHF facilitates the DnaA-ATP-dependent unwinding of the DNA duplex in the region of the 13-bp direct repeats, which then enables loading of the DnaBC helicase plus other replication proteins to form the replication competent complex (356).

reproduced in vitro on either supercoiled or linear *oriC* substrates (142, 356). DnaA-ATP loading to the weaker sites is not observed when the *oriC* IHF or the strong DnaA site R4 is mutated. These findings led Leonard and coworkers to propose that IHF may be playing an active role in assembling the replication-competent preinitiation complex. By inducing a defined bend between DnaA sites R1 and R4, IHF may promote the formation of an interwrapped structure that recruits multiple DnaA-ATP protomers to the weak intervening sites. At this stage, DnaA-ATP, with the help of either HU or IHF, promotes unwinding within the 13-bp repeat region (405) (see above), and the DnaB helicase is loaded onto the open complex.

H-NS

Identification of H-NS

H-NS (histonelike nucleoid structuring protein) was first identified biochemically as a DNA binding protein and originally referred to as protein 1 or H1 (80, 177, 442). The *hns* gene is located at 27 min on the *E. coli* chromosome and encodes a 137-amino-acid (15.4-kDa) neutral protein (108, 136, 171, 332). Mutants in *hns* have been identified in many genetic screens and found to grow more slowly than wild-

type strains, display increased osmosensitivity, exhibit prolonged lags and slow growth when shifted to low temperature, and have severely decreased motility (22, 30, 89, 159, 476). Although recent data suggest that the nonmotile phenotype may be due in part to direct interactions with the flagellar motor protein FliG (95, 215, 253), most of these phenotypes can be explained by changes in gene expression, identifying H-NS as a global regulator of transcription (reviewed in references 10, 99, 137, 156, 436, and 461).

H-NS homologues are found in many different gram-negative bacteria, with sequence similarities present throughout the primary sequences of *hns* genes from gamma proteobacteria (Fig. 18). The most diverged segment, between residues 77 and 86, is believed to correspond to an unstructured region linking an N-terminal oligomerization domain to a C-terminal DNA binding domain (82) (see below). *hns*-like genes in more distantly related bacteria have diverged more extensively, but considerable conservation remains within their C-terminal DNA binding domains. Some of these weak homologues, such as *Rhodobacter capsulatus* HvrA, which displays only 30% amino acid similarity to *E. coli* H-NS, have been shown to preferentially bind to curved DNA, a hallmark of H-NS activity (see below), and to support partial complementation of some *E. coli hns* mutant phenotypes (27, 28).

Physical Properties of H-NS

Isoforms

H-NS is found in three isoforms: H1a, b, and c (233, 403). Each isoform is distinguished by one charge difference, with the most basic form (H1a at a pI of 7.5) predominating, especially in stationary phase. Although the posttranslational modifications causing these charge differences have not been determined, a deletion containing only the first 67 amino acids of *hns* is found in two isoforms (96). In addition, an A18E or L26P mutation in the full-length protein results in the disappearance of the most basic form of the protein. These results imply that at least one of these posttranslational modifications occurs in the N-terminal 67 amino acids, possibly near amino acids 18 to 26. As this N-terminal region is involved in oligomerization (see below), it is possible that the posttranslational modifications regulate protein-protein interactions, thereby affecting both gene regulation and chromatin condensation.

Oligomerization

Although the quaternary structure of H-NS appears to be critical for both its affinity and ability to bend DNA (388, 407), its oligomeric state has been highly disputed (10, 461). In vitro cross-linking studies have revealed that a high percentage of H-NS exists as at least a dimer (108, 462). Small-pore gel filtration chromatography (Sephadex G-75) suggests that H-NS partitions into both dimeric and tetrameric forms (407). However, large-pore gel filtration (Sephadex G-100) indicates that H-NS exists primarily as a tetramer in solution, although dimers are found in small amounts depending on the temperature, salt, and protein concentrations (62). Ceschini et al. argue that the larger pore size is less disruptive and thus more accurately represents the equilibrium state of the protein (62). To account for differential sensitivities to ion concentrations, they proposed that H-NS forms a nonsymmetrical tetramer in which the subunits forming dimers are held together primarily by hydrophobic interactions, and dimers associate to form a tetramer via ionic interactions. Smyth et al. were unable to find distinct dimer or tetramer forms using either gel filtration or analytical ultracentrifugation, but instead found heterodisperse oligomeric states ranging up to 20-mers (398).

Domain structure

A number of studies have indicated that H-NS is organized into two distinct domains: an N-terminal dimerization-oligomerization domain (residues 1 to 63) and the C-terminal DNA binding domain (residues 90 to 137), with a protease-sensitive flexible hinge linking these domains (Fig. 19) (82, 105, 340, 374, 388). The dimerization domain is composed of two short and one long (25 residues) α-helices. The long helix contains a heptad repeat of leucine or valine and forms a coiled-coil structure upon dimerization (105, 340). The surface of the coiled-coil region is highly negatively charged and thus would not be expected to be associated with DNA. Addition of residues up to amino acid 89 leads to the formation of higher-order oligomers (398). Mutational analysis supports the N-terminal domain as being critical for multimerization since truncated proteins consisting of at least the first 46 amino acids are capable of multimerizing but incapable of binding DNA (398, 432, 462), and mutations located in the long α-helix, such as at Leu-29 within the heptad repeat, result in loss of multimerization (433). Esposito et al. proposed that the formation of H-NS oligomers may occur by a head-to-tail arrangement involving residues within the N terminus of the dimerization domain and the linker region (105). The DNA binding domains extending out from this oligomer could contact separate DNA duplexes resulting in the bridged DNA structures discussed below (Fig. 19).

A folded monomeric domain consisting of the C-terminal 47 residues can be isolated after incubation with trypsin (388). This region contains the DNA binding domain, albeit its affinity for DNA is poor relative to that of the intact protein. The solution structure of this monomeric domain (residues 90 to 137) reveals an unusual fold for a DNA binding domain (Fig. 19), consisting of an anti-parallel β-sheet followed by an α-helix and a 3_{10}-helix (388). The overall fold is stabilized by one or two hydrophobic residues within each of the secondary structure elements whose side chains define the core. ^{1}H and ^{15}N chemical shift mapping using an H-NS derivative containing residues 60 to 137 and a 14-bp A/T-rich oligonucleotide strongly implicates two adjacent but flexible regions as being close to DNA (389). These are the basic regions prior to β1 (residues Arg-90, Gln-92, Arg-93, and Lys-96) and the loop between β2 and the helix (residues Thr-110 to Ala-117) (Fig. 19). Mutations in H-NS that alter regulation of *proU* (see below) cluster in precisely these regions (432, 462). In addition, tryptophan fluorescence experiments implicate Trp-109 as being close to DNA (422).

DNA Binding Properties of H-NS

Binding to DNA is nonspecific, but H-NS has a preference for curved DNA or A/T-rich sequences

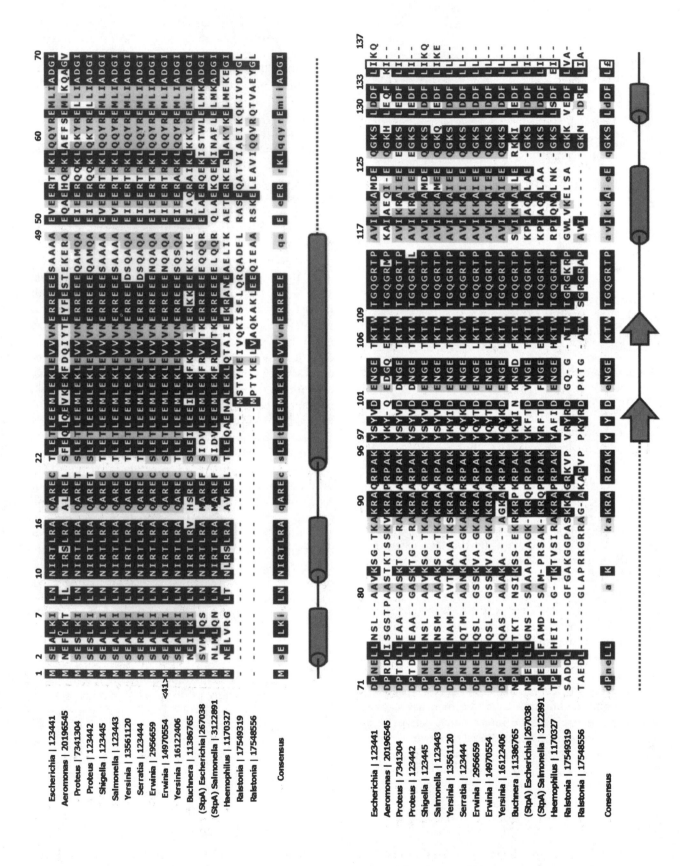

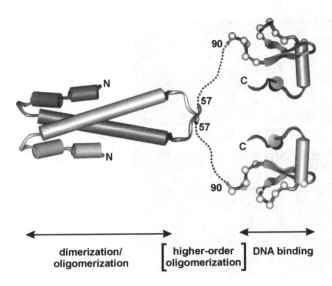

**dimerization/
oligomerization**   [**higher-order
oligomerization**]   **DNA binding**

Figure 19. H-NS domain structures. The structure of the N-terminal domain that mediates dimerization and is involved in the formation of higher-order H-NS oligomers is from the NMR study of residues 1 to 57 (105). A recent NMR-based structure of residues 1 to 46 of the *E. coli* H-NS dimer peptide revealed that the long α3-helices are associated in an antiparallel orientation; a similar configuration was also observed in a crystal structure of the N-terminal domain of an H-NS homologue from *Vibrio cholerae* (see Addendum in Proof). The structure of the C-terminal trypsin-resistant domain (residues 90 to 137) that mediates DNA binding is from Shindo et al. (388). The locations of residues implicated as being close to DNA by NMR chemical shift mapping (389) or by the properties of mutant proteins (434) are highlighted with spheres. The structurally undefined region between residues 57 and 90 that links the two domains is important for higher-order oligomerization (398). Computer-based analysis predicts that this linker region is mostly α-helical with a centrally located unstructured region or β-turn.

(108, 122, 249, 312, 428, 435, 475). In particular, fragments containing in-phase repeats of A_5 and A_6 tracts, which induce a bend with a planar orientation, are preferred, with K_d values as low as 3 nM measured by DNase I footprinting (343). Selective binding has also been demonstrated by competitive gel-shift studies (96, 109, 249, 312, 421, 485). However, K_d

values for curved DNA measured using this technique are in the 1- to 3-μM range (187, 400, 494). In all cases, the preference for curved DNA was about 1 to 2 orders of magnitude greater than that for noncurved sequences (187, 343, 428). The minimal binding site determined by DNase I footprinting appears to be 8 to 10 bp (249, 343), but H-NS most often binds cooperatively to extended regions. Binding within the minor groove is implied from the inhibitory effect of the minor groove binding drug distamycin (474).

In addition to selective binding to curved DNA at low concentrations, H-NS binding can become extensive with increasing concentrations of protein. Spreading via cooperative binding to nonspecific sites as well as increased affinity for remote specific sites has been observed using both DNA footprinting and gel-shift assays (1, 107, 109, 249, 343, 491). The initial binding sites thus appear to serve as nucleation points. The spreading of H-NS from these nucleation points was visualized recently by using atomic force microscopy: hairpinlike structures were found extending from the curved A_5 and A_6 tracts (83). The observed preferential binding to DNA in the vicinity of curved sequences was proposed to be due to the higher probability of forming oligomers between DNA-bound H-NS dimers on either side of the curved DNA (Fig. 20) (see "Regulation of Transcription by H-NS"). Similar H-NS–DNA filaments have been observed by transmission electron microscopy (83, 84, 368, 428). This DNA bridging model is consistent with earlier observations that mutations which affect oligomerization lead to loss of recognition of curved DNA sequences (407).

H-NS can promote the ligation of 155-bp noncurved fragments into minicircles, implying an ability to introduce bends into DNA in addition to preferentially binding to curved DNA (407). However, if a fragment containing curved DNA is used in this assay, the ligation frequency is not further stimulated, suggesting that H-NS does not further bend an already curved fragment. These results are compatible

Figure 18. Alignment of H-NS family members. The sequences are identified by the species from whose genome they originated and their NCBI GI number. The H-NS sequences were obtained by searching the SWISSPROT database, excluding eukaryotic sequences, for "hns." Also, the NCBI PSI-BLAST program was used with *E. coli* H-NS and StpA query sequences to obtain additional H-NS sequences not present in the SWISSPROT database. Sequences not described as H-NS, StpA, histonelike DNA binding protein, or DNA binding protein were discarded. Sequences identified as StpA are denoted as such. The sequences were aligned as described in the legend to Fig. 2 and numbered according to the *E. coli* H-NS sequence. Consensus notation: a lowercase letter indicates the residue is found in 50 to 74% of sequences overall and is shaded gray where conserved. An uppercase letter indicates the residue is found in 75 to 100% of sequences overall, and the background is black where conserved. £ indicates that I or L residues are found 16 out of 17 times at this position, and the conserved residues are boxed by a solid line. Secondary structure: the secondary structure cartoon shown below the alignments is based on the structures determined by Esposito et al. (105) (PDB accession code 1LR1) and Shindo et al. (388) (PDB accession codes 1HNR and 1HNS). The secondary structure elements are as follows: α-helices at residues 2 to 7, 10 to 16, 22 to 49, 117 to 125, and 130 to 133 (3_{10}-helix), and β-strands at residues 97 to 101 and 106 to 109. The structure of the region between residues 50 and 96 is not known and is marked with a dotted line.

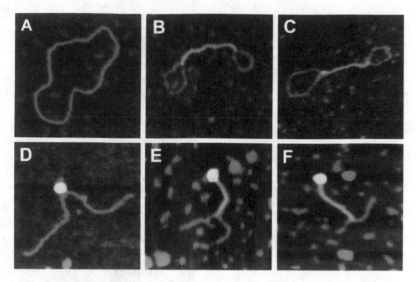

Figure 20. Bridging of DNA segments by H-NS. (A to C) Atomic force microscopy images of nicked pUC19 without (A) and with (B and C) addition of H-NS showing H-NS-mediated filaments associating two DNA segments (83). (D to F) Atomic force microscopy images of a 1.2-kb DNA fragment containing the *rrnB* promoter region and RNA polymerase (84). H-NS was added to the reactions in panels E and F. In the presence of H-NS, RNA polymerase is able to form open complexes that are competent for abortive initiation but not elongation (375). (Images kindly provided by Remus Dame, Nora Goosen, and Claire Wyman.)

with the DNA bridging model presented above since the increased ligation rates can be explained by the binding of H-NS to disparate regions of the DNA molecule, thereby bringing them together and effectively decreasing the persistence length.

H-NS in DNA Supercoiling, Condensation, and Nucleoid Structure

H-NS effectively constrains negative supercoils in vitro, as assayed by incubation with either supercoiled or relaxed plasmid DNA, followed by treatment with topoisomerase I (403, 428, 491). The ability to constrain supercoils is highly sensitive to K^+ ion concentration, presumably reflecting the sensitivity of oligomerization and/or DNA binding to salt concentrations (62, 421, 428).

The role of H-NS in modulating DNA superhelicity in vivo is less clear. Isolation of plasmid DNA from different *hns* mutants has resulted in either an increase, a decrease, or no effect on superhelical density depending on the particular *hns* mutation or even when same *hns* allele is examined in *E. coli* versus *Shigella flexneri* or *S. enterica* serovar Typhimurium (98, 155, 159, 171, 200, 294, 312). When directly compared, the difference from wild type in plasmid linking number from *hns* mutants is much less than that observed with *hupAB* mutants, and *hns hupAB* triple mutants exhibit only a small further reduction of plasmid supercoiling over that of *hupAB* mutants (483; Isaksson and Johnson, unpublished).

Further complicating our understanding of the role of H-NS in modulating global DNA topology, rates of trimethylpsoralen-mediated DNA cross-linking at several chromosomal loci in vivo were increased in *hns* mutants, implying greater superhelical densities without H-NS present (273).

As noted above, atomic force microscopy and electron microscopy have revealed DNA complexes in which H-NS is observed to coat the DNA, often forming lateral tracts in which two regions are associated (e.g., Fig. 20B and C) (83, 84, 428). Apparent DNA lengths within the bridged DNA structures were found to be reduced by only about 3 to 6% (83, 428). Atomic force microscopy analysis of H-NS–DNA complexes also revealed discrete foci at which apparent DNA lengths were reduced by up to 25% and very high ratios of H-NS to circular DNA (1 dimer per 6 bp) generated condensed but unstable rodlike structures (83). The formation of these structures could explain earlier observations implicating DNA compaction within H-NS complexes obtained by gel electrophoresis and sedimentation in sucrose gradients (403).

H-NS does not appear to be essential for maintaining the global structure of nucleoids in vivo, since the appearance of DAPI-stained nucleoids from *hns* mutant cells is largely indistinguishable from that of wild-type cells (191; S. McLeod and R. Johnson, unpublished data). The absence of an obvious cytological difference between *hns* mutant and wild-type nucleoids, together with the minimal effect on plasmid linking number by *hns* mutants, is consistent with the relatively small proportion of the chromosome that is predicted to be bound by H-NS (Table 1; see also below). On the other hand, H-NS almost certainly influences DNA structure locally. Indeed, when H-NS is massively overproduced, nucleoids become highly compact with an accompanying decrease in global transcription and cell viability (406). Similarly overproduced H-NS mutants that are defective in oligomerization displayed no adverse effects on cell viability, reinforcing the importance of higher-order oligomerization for H-NS activity in vivo.

Regulation of H-NS Synthesis

While earlier reports suggested that H-NS levels increase in stationary phase (90, 109, 403, 430), recent studies agree that H-NS levels remain relatively constant from log phase (20,000 to 25,000 molecules of H-NS per cell growing in nutrient-rich conditions) through stationary phase (10,000 to 15,000 molecules of H-NS per cell) (116, 281, 412, 483). The approximately twofold decrease in H-NS levels observed in stationary phase relative to exponential phase is consistent with the decrease in DNA content of the cell. In fact, *hns* mRNA levels are directly coupled to DNA synthesis, since they have been observed to decline when DNA synthesis is blocked (116). Assuming that H-NS is at least a dimer (yielding about 10,000 H-NS dimers per rapidly growing cell) and that the size of the minimal H-NS footprint is 8 to 10 bp (see above), only a small percentage of the chromosome is predicted to be bound by H-NS. These numbers are consistent with the large effects of H-NS on gene regulation in localized regions but only a marginal global effect on nucleoid structure.

The relatively constant ratio of H-NS to DNA at different stages of growth is in part mediated through autoregulation, whereby H-NS (or its paralog StpA; see below) represses transcription of the *hns* gene (90, 109, 430). This repression is antagonized by Fis, leading to somewhat higher levels of H-NS during log phase, when Fis is abundant, and decreased levels in stationary phase, when Fis is absent (106). In addition, regulation of H-NS synthesis is finely tuned in stationary phase via DsrA, an 87-nucleotide regulatory RNA (239). H-NS is downregulated in the presence of high levels of DsrA due to a decrease in *hns* mRNA half-life, from 4 min to less than 0.5 min (238). Sequence-specific interactions of the 5′ and 3′ *hns* mRNA ends with DsrA have been demonstrated and are thought to enhance turnover, either by providing a direct target for RNase action or by inhibiting translation, and thereby indirectly increasing turnover (238).

In addition, H-NS is one of the major cold shock-induced proteins, and its expression is activated by the cold-shock regulator CspA (232). Mutational analysis indicates that a CG clamp centered at +10 in the *hns* mRNA is the site of action of CspA, which facilitates unwinding of RNA secondary structures both in exponentially growing cells (in which CspA concentrations are high [45]) and under cold shock conditions (127). In addition, H-NS has been found to be induced during phosphate starvation (437); a PhoB box overlapping the −35 region of the *hns* promoter may facilitate this regulation (127). A G/C-rich discriminator-like sequence is located between −5 and +1, but stringent control of the *hns* promoter has not yet been examined.

Regulation of Transcription by H-NS

Two-dimensional gel analysis comparing protein levels from *hns* mutant and wild-type strains has implicated up to 100 genes with altered expression patterns (29, 165, 475, 485, 491). In addition, H-NS was identified in many mutant screens that looked for a loss of repression, again implicating it as a global regulator (e.g., *drdX* [136], *bglY* [29], *osmZ* [171], *pilG* [404], and *virR* [98]). In general, H-NS acts as a transcriptional repressor, with possibly an indirect function in activating some genes (179, 288, 401). Although the types of genes regulated by H-NS appear to be quite diverse, there seems to be an overrepresentation of environmentally regulated or stress-related genes (10). In particular, H-NS often plays a role in repressing genes that are required during starvation, conditions of high osmolarity, and high temperatures. Many, but not all, of these H-NS–repressed genes are dependent on the stress and stationary-phase σ factor RpoS (22). Global regulators such as H-NS and Fis appear to modulate the specificity and level of gene expression of many RpoS-dependent genes (reviewed in reference 151).

Repression, or gene silencing, by H-NS has been most intensively studied at the molecular level in the *hns* and *virF* genes and the *rrnB*, *bgl*, and *proU* operons (Fig. 21). In each case the promoter region is associated with both a region of intrinsically curved DNA and one or more H-NS preferred binding sites. Footprinting studies have generally revealed binding at these nucleation sites followed by a cooperative polymerization of H-NS along the DNA (1, 107, 109, 249, 343, 428, 491). In the *proU* operon, this coating of the DNA is proposed to inhibit open complex formation (188). However, at *rrnB* P1, KMnO₄-sensitive open complexes are formed in the presence of H-NS that are capable of synthesizing 2- to 3-nucleotides-long aborted transcripts (375). These

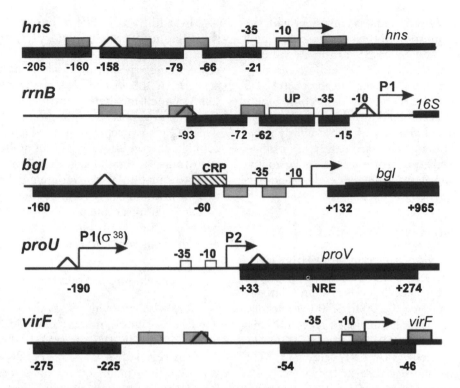

Figure 21. Examples of H-NS-regulated genes. Regulatory regions for the *hns* gene, *rrnB* operon, *bgl* operon, *proU* operon, and *virF* gene are depicted; the transcription initiation sites are indicated with arrows. Centers of curved DNA are indicated with a caret (^). Primary H-NS control regions are shown with black boxes, and Fis binding sites are depicted with gray boxes. Binding sites were determined by DNase I footprinting in the *hns* (five of the seven Fis sites are shown) (106, 109), *rrnB* P1 (1, 349, 422), and *virF* promoter regions (107, 110). The upstream and downstream regulatory regions required for H-NS regulation of *bgl* were determined by deletion analysis (277, 372). The negative regulatory element (NRE) in the *proU* operon control region is depicted, as determined by deletion analysis (85, 311, 312). High-affinity H-NS binding occurs within multiple discrete regions throughout the negative regulatory element, particularly between about +60 and +180, and coating extends to cover the entire promoter region at higher concentrations of H-NS (249, 428).

properties, along with scanning force microscopy images of these complexes, led to the model of H-NS bridging two regions of DNA, with an RNA polymerase either trapped in the loop or occluded from the DNA (84). This model is particularly attractive because it can explain the dimer-tetramer-oligomeric states of H-NS, whereby a dimer can bind to a preferred site on one side of the curved DNA and then associate with another dimer at a second preferred site to bridge the DNA. Alternatively, a tetramer could bind at a preferred site and search for the second preferred site. Once established, this bridged DNA would quickly lead to cooperative binding of additional H-NS dimers (or tetramers), causing either RNA polymerase trapping or occlusion.

How could this transcriptional repression or gene silencing be relieved? Many examples exist in which binding of a site-specific activator (or repressor) leads to derepression. In the model described above, this could be envisioned by disruption of the DNA bridging leading to an opening of the DNA

structure. For example, Fis acts to antagonize H-NS binding at the *hns*, *rrnB*, and *virF* promoters (106, 110, 421). In competitive gel-shift assays, Fis can inhibit H-NS binding, but H-NS cannot prevent Fis binding (1, 106, 110, 421). These properties would lead to derepression during exponential growth, when Fis concentrations are high, but to repression in stationary phase, when Fis concentrations are low. Other DNA binding proteins have also been found to alleviate H-NS repression. H-NS repression of the Mu Pe promoter is relieved by IHF binding (439), and the cytolysin A operon is derepressed by CRP-cAMP (456). In the cryptic *bgl* operon, which is kept transcriptionally silent by H-NS, derepression can occur by mutations in the CRP binding site, which increase the similarity to the consensus sequence (277). Caramel and Schnetz (56) have further shown that site-specific DNA binding proteins such as CRP promote expression of *bgl* not by directly activating transcription through RNA polymerase interactions but by simply binding within the H-NS silencing

DNA region. They inserted either Lac or λ repressor binding sites into the upstream region of the *bgl* promoter and observed derepression in the presence of the appropriate repressor (56).

High osmolarity can also lead to derepression of H-NS-controlled genes. Cells grown at high osmolarity accumulate potassium glutamate (137, 339). High K^+ concentrations in turn have been shown to affect the dimer-tetramer equilibrium (62, 422) and decrease H-NS binding in vitro (421). These data suggest a model whereby DNA bridging and/or DNA binding could be disrupted by K^+ ion concentration alone via transition from tetramer to dimer. In fact, Rajkumari et al. showed that an increase in K^+ ion concentration directly reverses H-NS repression of *proU* P2 transcription in vitro (337).

High temperature is another mechanism which counteracts transcriptional silencing by H-NS. Although temperature affects the dimer-tetramer equilibrium, it does so at temperatures lower than appear to be relevant for in vivo regulation (62). Instead, temperature may play a direct role in modulating DNA curvature, as shown at the *virF* promoter (107). In these studies, binding of H-NS to two sites centered around −250 and −1 is separated by an intrinsic DNA bend. H-NS binds cooperatively to these two sites below 32 but not at 37°C (possibly forming a bridged DNA structure, as described above). The *virF* promoter fragment containing the two H-NS sites undergoes a temperature-dependent conformational transition at approximately 32°C (107). This alteration of the DNA target conformation may modulate the cooperative interaction between H-NS molecules bound at two distant sites in the *virF* promoter region, and thus may be the physical basis for the H-NS-dependent thermoregulation of virulence genes.

H-NS in DNA Recombination and Transposition

H-NS has been implicated in regulating the site-specific DNA inversion reaction that controls phase variation of type 1 fimbriae in *E. coli*. Transcription of *fimA*, encoding the structural gene for type 1 pilin, is directed from a promoter within a 314-bp invertible DNA segment (102). In wild-type strains there is a switching bias favoring the orientation of transcription away from the *fimA* gene (the off position). This orientation preference is modulated by environmental cues such as temperature and changes in growth media. *hns* mutations have been shown to alleviate the temperature control and accelerate the inversion catalyzed by the FimB recombinase up to 100-fold (205). One way in which H-NS is known to modulate switching is by directly repressing the *fimB*

promoter, thus reducing the levels of the FimB recombinase (96, 306, 307, 376). Expression of a second recombinase, FimE, which is required for inversion in the on-to-off orientation, is affected by H-NS to a much smaller degree. Mutations in a predicted curved DNA region within the invertible segment or the *fimA* promoter itself also disrupt the orientation bias of switching (303, 364). H-NS binding to this region and the resulting changes in P_{fimA} transcription are proposed to indirectly modulate the rate of inversion by affecting DNA topology or by interfering with formation of recombination complexes.

H-NS, but not the other major nucleoid-associated proteins, is required for IS*1* transposition (385). The effect on transposition by H-NS could not be attributed to changes in transcription from the IS*1* transposase promoter or to preferential binding of H-NS to regions within or surrounding IS*1*. Induction of IS*1* transposase induces the cellular SOS response, presumably due to the formation of 3' breaks at the transposon termini. However, SOS induction does not occur in *hns* mutants, implying that the transposition reaction is blocked at a step prior to DNA cleavage. IS*1* can form transposon circles containing a spacer sequence of 6 to 9 bp that can undergo further reaction leading to intermolecular transposition (386). Surprisingly, whereas the formation of IS*1* circles absolutely requires H-NS, some transposition of IS*1* circles occurs in its absence. Ohtsubo and coworkers suggest that H-NS may be required to assemble an IS*1* transpososome complex but is somewhat dispensable when the ends are already close to each other in the transposon circle (385). H-NS may also influence IS*1* target site selection, since a broader spectrum of insertion sites, which were not limited to A/T-rich regions, were observed from transposition of IS*1* circles in *hns* mutants (386).

H-NS has not been reported to play a role in homologous recombination; however, H-NS appears to inhibit RecA-independent illegitimate recombination. Lejeune and Danchin found a 10- to 100-fold increase in the formation of spontaneous deletions at different loci in *hns* mutants (241). Shanado et al. found that *hns* mutants led to a small increase in the spontaneous formation of λ*bio* transducing phage; this effect was enhanced to about a 10-fold increase after γ irradiation but not UV irradiation (378). The distribution of recombination endpoints in the λ*bio* transducing phage after γ irradiation was not significantly altered in *hns* mutants, however. Viability of *hns* mutants is not adversely affected by γ irradiation, and *hns* mutants do not display an enhanced frequency of rifampin resistance mutations as do most conventional mutator genes (241, 378).

H-NS and the Cell Division Cycle

Whereas the morphology of *hns* mutant cells is relatively indistinguishable from that of *hns*⁺ cells, a marked increase in the number of anucleated cells under slow growth rates in minimal media has been reported (191). The mechanism(s) responsible for this apparent chromosome partitioning defect is currently unknown. Replication run-out experiments employing flow cytometry of cells after inhibition of cell division and new rounds of chromosomal replication have shown that *hns* mutant cells display a normal synchronous pattern of replication control (9, 191). Thus, H-NS does not appear to have a primary role in controlling replication at *oriC*, although complex effects when *hns* mutations are combined with *dnaA* mutations have been reported (203). The replication run-out experiments do show that *hns*-deficient mutants have a reduced number of origins per cell (ploidy) under both rich and minimal media growth conditions (9, 191). Cells expressing different levels of H-NS suggest that low H-NS primarily leads to faster replication periods, whereas aberrantly high H-NS inhibits cell division, probably through indirect mechanisms (9).

H-NS Homologues and Interacting Proteins: StpA, Hha/Ymo, and Phage T7 Gene 5.5

StpA, an H-NS homologue, was first identified based on its ability to suppress a splicing defect in phage T4 thymidylate synthase mRNA (489), and later found to be a multicopy suppressor of *hns* mutants (384, 399). The *stpA* gene was subsequently cloned and found to be unlinked to *hns* (located at 60.24 min [384]). Sequence analysis revealed it to be 133 amino acids in length, sharing 58% identity and 67% similarity to H-NS, but considerably more basic than H-NS (Fig. 18) (489).

There exists some uncertainty as to the levels of StpA in the cell and how large a role it plays in regulating gene expression and nucleoid structure. Reported levels of *stpA* mRNA have varied with assay conditions but are generally considered to be quite low, particularly when compared with *hns* mRNA levels (117, 399, 462, 491). *stpA* mRNA is transiently induced during mid-log phase and is much higher throughout the growth curve in minimal media, due to activation by Lrp (117, 399). Expression of *stpA* was also found to be repressed by both H-NS and StpA but independent of Fis. Carbon starvation resulted in repression of transcription, while activation occurred upon osmotic shock and upon shift to higher temperatures.

Determination of the levels of StpA protein has been confounded by the fact that most antibodies for StpA also cross-react with H-NS. Despite this complication, levels of StpA have been estimated in a wild-type strain grown in rich media and found to be very similar to the levels of H-NS (Table 1) (412). This is much higher than predicted from the RNA levels and suggests that StpA is either a major player in nucleoid structure and gene regulation together with H-NS, or that it has other cellular functions. Immunolocalization of StpA indicates that it is uniformly distributed within the nucleoid as is H-NS (411).

Genetic analysis suggests a minor role, or at least a different one, for StpA compared with H-NS, as *stpA* mutants do not mimic *hns* mutants (86, 489), and mutations in *hns* have strong effects on gene expression even when wild-type StpA is present (29, 86, 165, 475, 485, 491). If the functions of the two proteins were completely redundant, one would not expect to detect any phenotype unless both genes were inactivated. The growth properties of the different mutant strains also indicate that StpA is dispensable in rich media, whereas H-NS mutants exhibit moderately reduced growth rates. However, mutants defective in both *hns* and *stpA* have a slower growth rate than *hns* mutants alone, suggesting that StpA can provide partially for H-NS function (399, 491). StpA can also suppress *hns* mutants when provided in multicopy or by overproduction, suggesting that at its normal level it is unable to substitute for H-NS (384).

Comparison of StpA and H-NS DNA binding properties indicates a similar preference for curved DNA (with StpA having a slightly higher affinity than H-NS) and a similar ability to constrain supercoils in vitro (400, 491). However, an *stpA* or even an *stpA hns* double mutant has little effect on plasmid supercoiling densities in vivo (Isaksson and Johnson, unpublished). Both proteins are involved in transcriptional repression, although StpA seldom acts independently of H-NS (491). In addition, both proteins autoregulate their own genes and cross-regulate the expression of each other's genes (399, 462, 491). One way in which they significantly differ is that StpA is a much more effective RNA chaperone than H-NS (82, 490, 491). StpA RNA chaperone activity has recently been found to play a physiological role in regulating *micF* antisense RNA stability (86, 88). Therefore, although the two proteins clearly descend from a common ancestor, it appears that StpA has evolved independent activities and is no longer functionally redundant with H-NS.

With the similarity in sequence, one might expect StpA to interact directly with H-NS to form heterodimers or hetero-oligomers. Indeed, evidence supporting the formation of mixed oligomers in vivo, as well as in vitro, has come from several different lines of investigation. Recent studies have shown that StpA is subject to Lon-mediated degradation in vivo in the absence of H-NS (181). This degradation is completely inhibited by H-NS, and further analysis has revealed two regions, which are required for multimerization, that mediate this stabilization (180). In vivo cross-linking data indicate that most StpA found in a wild-type strain is in the form of a heterodimer with H-NS, whereas in an *hns* mutant, StpA forms tetramers and higher-order oligomers that are more susceptible to Lon degradation (180). Analysis of dominant-negative mutations revealed that *stpA* mutations could interfere with the functioning of the wild-type H-NS and, likewise, dominant-negative mutations in *hns* could interfere with StpA function (462). Finally, mutations in the DNA binding domain of H-NS that effect silencing of the *bgl* operon can be suppressed by StpA, presumably by forming heterodimers with the truncated H-NS and directing its binding to the *bgl* operon (118, 119). It is interesting to speculate as to whether one role of StpA may be to modulate the oligomeric state of H-NS and thus regulate its gene silencing activity.

A second class of weakly related H-NS homologues has been uncovered which are similar to H-NS and StpA only in the oligomerization domain (293). In *E. coli*, Hha is an 8.5-kDa protein involved in the regulation of the toxin α-hemolysin (57, 276), and in *Yersinia enterocolitica*, YmoA is a regulator of the *yop* virulence regulon (74, 228). Using His-tagged proteins, a direct interaction between H-NS and Hha or Ymo was detected that was independent of DNA (293). Although Hha and Ymo are able to bind DNA, they have no homology with H-NS in their DNA binding domains but do have amino acid identities scattered throughout their oligomerization domains. It was proposed that Hha regulates α-hemolysin expression through interactions with H-NS to modulate gene silencing at that operon.

The gene 5.5 protein of phage T7 has been found to interact directly with H-NS but does not display any homology to H-NS (246). Gene 5.5 encodes a small (11 kDa) highly expressed protein, which is required for normal plaque size and burst yield (409). Purification of gene 5.5 protein in *hns*+ cells yielded a 1:1 complex with H-NS, and gel-shift experiments revealed a supershifted complex with a fragment containing the *proU* H-NS binding site (246). T7 RNA polymerase was found to be more sensitive to inhibition by H-NS than was *E. coli* RNA polymerase in vitro, and gene 5.5 protein was able to prevent this inhibition both in vitro and in vivo. The physiological role of gene 5.5 in T7 phage growth was postulated to sequester H-NS released after the host genome is degraded by T7 enzymes. In this way, T7 transcription would be unheeded by H-NS binding.

FIS

Identification of Fis

The Fis protein was first identified in 1985 as a heat-stable activity from *E. coli* required for site-specific DNA inversion catalyzed by the *S. enterica* serovar Typhimurium Hin and phage Mu Gin recombinases, hence the name factor for inversion stimulation, Fis (186, 189). A year later it was purified to homogeneity and shown to be a homodimer of 11.2-kDa subunits (184, 217). The subsequent cloning of the gene revealed a coding sequence of 98 amino acids with a calculated pI of 9.4 located at min 73.5 on the *E. coli* chromosome (183, 220). *fis* is the second gene of an operon; the gene immediately upstream, *yhdG* (renamed *dusB*), encodes a dihydrouridine synthase, which catalyzes the formation of dihydrouridine within a subset of tRNAs (35). Fis is present in the currently sequenced genomes of the γ proteobacteria, and N-terminally truncated forms are found in β proteobacteria (Fig. 22), but it has not been found thus far in gram-positive bacteria. The *fis* coding sequences in the different genomes where it exists exhibit considerably greater conservation than the overall genomic sequence. For example, the amino acid sequence from *S. enterica* serovar Typhimurium and other closely related enteric bacteria such as *K. pneumoniae* and *Serratia marcescens* are identical to that of *E. coli* (25, 309). The *dusB* coding sequence is almost always upstream of *fis*, although it diverges to a somewhat greater extent.

Fis contains a helix-turn-helix (HTH) DNA binding motif at its C terminus and therefore is not related to HU/IHF or H-NS/StpA. In addition to recognizable similarities to other HTH proteins over the minimal HTH region, Fis bears significant homology over the 70-amino-acid C-terminal dimerization and DNA binding domain to members of the NtrC family of transcriptional regulators (183, 214, 274). This relationship is substantiated by a comparison between the atomic structures of Fis and *S. enterica* serovar Typhimurium NtrC (319). The *yhdG/dusB* gene of the *Enterobacteriaceae* is most similar to *nifR3* in α proteobacteria, which is often associated in an operon with *ntrC*. Additionally, the

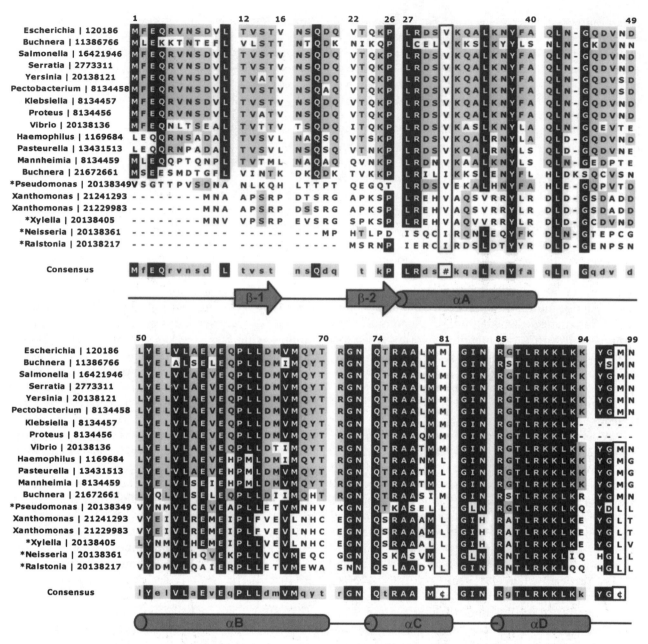

Figure 22. Alignment of Fis homologues. The sequences are identified by the species from whose genome they originated and their NCBI GI number. The Fis sequences were obtained by searching the SWISSPROT database, excluding eukaryotic sequences, for "Fis" or "factor for inversion stimulation." Also, the NCBI PSI-BLAST program was used with the *E. coli* Fis query sequence to obtain additional Fis sequences not contained in the SWISSPROT database. Sequences described as "Fis-like" DNA binding protein are marked with an asterisk. Sequences not described as Fis or Fis-like DNA binding protein or DNA binding protein were discarded. The alignment was performed as described in the legend to Fig. 2 and is numbered according to the *E. coli* Fis sequence. Consensus notation: a lowercase letter indicates the residue is found in 50 to 74% of sequences overall and is shaded gray where conserved. An uppercase letter indicates the residue is found in 75 to 100% of sequences overall, and the background is black where conserved. # indicates that I or V residues are found in 75 to 100% of the sequences and are boxed by a solid line. ¢ indicates that L or M residues are found in 75 to 100% of the sequences and are boxed by a solid line. Secondary structure: the secondary structure designation shown below the alignment is based on the crystal structures of *E. coli* Fis by Kostrewa et al. (224) (PDB accession code 1FIA), Yuan et al. (486) (PDB accession code 3FIS), and Cheng et al. (67) (PDB accession code 1ETY) and of Fis mutant K36E by Safo et al. (357) (PDB accession code 1F36). The secondary structure elements are follows: β1-strand residues 12 to 16, β2-strand residues 22 to 26, αA-helix residues 27 to 40, αB-helix residues 50 to 70, αC-helix residues 74 to 81, and αD-helix residues 85 to 94.

C-terminal domain of α-proteobacterial NtrC is more closely related to γ-proteobacterial *fis* than are the respective segments from γ-proteobacterial *ntrC* genes. These observations have led Morett and Bork to propose that the *dusB-fis* operon has been acquired by members of the gamma subdivision from an ancestral α-proteobacterial lineage by horizontal gene transfer (274).

General Properties of *fis* Mutants

Fis null mutants of *E. coli* and *S. enterica* serovar Typhimurium are viable but display pleiotropic properties. Growth rates of *fis* mutants under nutrient-rich environments can be up to 20 to 30% slower than those of their otherwise isogenic parents, depending on the strain background and conditions (295, 297, 309). In addition, longer lag phases are often observed in *fis* mutants, particularly after prolonged periods in stationary phase. The growth properties of *fis*⁺ and *fis* mutant cells cultured in minimal media with a unique carbon source are largely indistinguishable. As discussed below, the differences observed under fast-growth conditions are believed to be related to the role of Fis on the expression of genes involved in protein synthesis. It has been reported that *fis* mutants are unable to grow anaerobically in minimal glucose medium due to a deficiency in expression of *adhE*, which encodes

ethanol oxidoreductase (262). Fis mutants also appear to have altered membrane properties, as recognized by their propensity to clump and their hypersensitivity to hydrophobic compounds such as drugs, dyes, and detergents (473). The pleiotropic effects on membrane properties may be related to their reduced motility (309). Fis mutants do not appear hypersensitive to UV radiation, and SOS functions are not induced under standard growth conditions. However, *fis* mutant cells have a tendency to become elongated or moderately filamented, particularly at high temperatures (112, 309). This phenotype may be related to the aberrant regulation of initiation of DNA replication or cell division (see below).

Structure of the Fis Dimer

The first crystal structures of Fis were reported by two groups in 1991 and revealed an ellipsoidal structure with dimensions of 25 by 35 by 50 Å (224, 225, 486). In both cases the wild-type Fis crystallized in an orthorhombic form which failed to resolve the N-terminal 24 residues of each subunit of the dimer. The lack of electron density for residues 1 to 25 was presumed to be due to disorder or flexibility of this segment. Residues 26 to 98, which are sufficient for dimerization and DNA binding, are assembled into four α-helices connected by β-turns (Fig. 23).

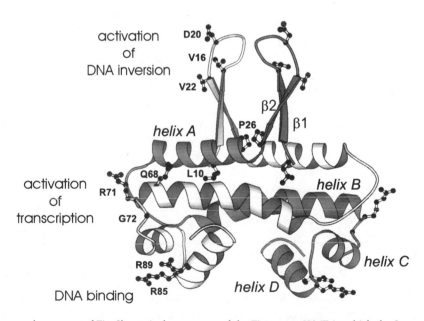

Figure 23. X-ray crystal structure of Fis. Shown is the structure of the Fis mutant K36E in which the β-arms (β1 and β2 plus connecting turn) are resolved (357). The polypeptide chain for both subunits starts at residue 10. The locations of some of the functionally or structurally important side chains are shown and labeled for one of the subunits. The secondary structure elements are labeled for the other subunit.

A hydrophobic core stabilizing the dimer is formed predominantly from interdigitating side chains from residues emanating from helices A, A' and B, B'; additional forces contributing to dimer stability involve several intersubunit hydrogen bonding networks involving residues in helices B to D. Fis subunits have been found to exchange rapidly between dimers with a half-life of 2 to 4 min in solution, but the presence of DNA reduces this exchange over 100-fold (266).

The C-terminal half of helix B, together with helices C and D, forms a prototypical HTH DNA binding motif within each subunit (Fig. 23), and mutations within this region, in particular helix D, have a severe effect on DNA binding (219, 308). In nearly all dimeric HTH proteins, the separation between recognition helices is 32 to 34 Å, which corresponds to the separation between equivalent positions within adjacent major grooves on the same side of the DNA helix. However, the distances between Cα atoms of the first seven residues of the D and D' recognition helices range from 23 to 25 Å. This short separation is found in all five crystal forms that have been solved to date and is believed to be responsible for a large part of the Fis-induced bending of DNA (67, 224, 357, 486) (see below).

The structure of the N-terminal segment has been resolved from the X-ray analyses of four different mutants: K36E, R71L, R71Y, and G72D (67, 357). Each of these mutants crystallized in a different form and revealed all or parts of the region. Residues 12 to 16 and 22 to 26 in each subunit form antiparallel β-strands separated by a five-residue turn (Fig. 23). The β-arms are connected to the core of the protein by insertion of Leu-11 into a hydrophobic pocket primarily consisting of residues from the other subunit. Consequently, mutations at Leu-11 result in a disruption of the β-arm structure (357). The switch from β-strand 2 to α-helix A occurs at proline-26. Amino acid substitutions at Pro-26 also destabilize the β-arm structure, with the X-ray structure of a proline-to-alanine mutation showing that residues 22 to 26 from one subunit were converted from a β-strand structure to an α-helix, extending helix A by two turns (482). The N-terminal 10 residues assume a random coil that is loosely associated with the surface of the core (67). No biological function has been assigned to this segment, and it is not present in some of the more distantly related *fis* homologues (Fig. 22).

The β-arms, which protrude over 20 Å from the dimer core, are flexible. Superimposition of the visible regions of the β-arms from the four mutants that exhibit electron density for the N terminus shows that the trajectories of the arms are different in each of the mutants (67). Moreover, even though the two arms of the dimer do not contact each other in any of the crystal forms, pairs of cysteines introduced at several positions that face toward each other in the β-arms rapidly form disulfide cross-links upon oxidation (357). Although the Fis β-arms bear some structural similarity to the β-strand region that mediates DNA binding in IHF/HU, there is no evidence that they perform an analogous function in Fis. Rather, residues at the ends of the arms directly contact and regulate the activity of DNA invertases (see below).

DNA Binding Properties: Sequence-Specific and Nonspecific

Whereas Fis is often portrayed as a sequence-specific DNA binding and bending protein, Fis associates with DNA of random sequence with relatively high affinity (32, 314). Nonspecific binding is most easily recognized by gel-shift assays performed without competitor DNA. A series of discrete complexes are formed on DNA segments of mixed sequence with increasing concentrations of Fis, with the apparent K_d for the first complex typically being ≤ 20 nM. Binding of additional Fis dimers is largely noncooperative and is saturated at 100 nM with a Fis dimer bound every 25 to 30 bp. The formation of contiguous arrays of Fis dimers on DNA can also be observed by DNA scission patterns obtained using Fis–1,10-phenanthroline–copper chimeras (255). When the chemical nuclease is tethered at the Fis C terminus, a regular pattern of cleavages is generated every 18 to 22 bp over end-labeled DNA fragments.

Fis forms complexes with specific DNA sites, with K_d values measured in the presence of nonspecific DNA typically in the 2 to 5 nM range. The distinguishing features of a specific Fis-DNA complex are that the protein remains stably bound in the presence of an excess of nonspecific DNA and that a defined footprint is generated. Whereas nonspecific Fis-DNA complexes dissociate in the presence of excess DNA within seconds, specific Fis-DNA complexes exchange slowly between binding sites. The rate of exchange is dependent upon the concentration of specific sites, suggesting direct transfer between DNA segments (314). DNase I footprints of specific Fis-DNA complexes typically extend over 27 to 29 bp and exhibit a characteristic pattern of hypersensitive sites spaced about 12 bp apart on each strand (Fig. 24). These sites are believed to reflect positions of structural distortions within the DNA helix, presumably expanded minor grooves, which are induced by binding and bending of DNA by Fis. Hydroxyl radical protections can extend up to 25 bp depending on the specific binding site (38).

tatattcggg**G**NTY**A**ww**W**ww**T**R**a**N**C**aaaatctaat

Figure 24. Fis consensus binding sequence. The 15-bp core recognition sequence is in bold capital letters, with the size reflecting the degree of conservation (150, 314, 435). Y indicates C or T, R indicates A or G, W indicates A or T, and N is no obvious preference. Ten base pairs of flanking sequences denoted in lowercase letters are shown; the left flanking sequence is from the *hin* enhancer proximal site, where little bending is observed, and the right flanking sequence is from the λ *attR* F binding site, which displays a large amount of bending (see reference 314). DNase I typically generates an interrupted protection pattern over the Fis binding site; the positions designated with an arrow are often hypersensitive to cleavage. Protections from dimethyl sulfate reactivity on guanines, when present in the sequence, are denoted by asterisks. Positions of oxidative cleavage by 1,10-phenanthroline–copper when tethered to Fis residues 71 or 73 are designated by solid triangles; on these flanking sequences, poor cleavage is obtained on the left side whereas strong cleavage is obtained on the right side (314). Positions designated from below refer to locations in the bottom DNA strand, which is not shown.

A comparison of the DNA sequences at over 50 specific binding sites that have been footprinted reveals a degenerate 15-bp "consensus" palindromic sequence (Fig. 24) (113, 314). Information theory and hidden Markov models have been used to generate sequence logos that match the consensus (150, 435). Most, but not all, sequences that match this consensus form stable complexes with Fis. However, many biologically relevant binding sites deviate at one or two of the more highly conserved positions. Moreover, single base pair substitutions, even at conserved positions within this 15-bp sequence, often do not significantly disrupt Fis binding. A synthetically derived 15-bp core sequence is sufficient to specify a high-affinity Fis binding site when inserted into a DNA fragment (51). However, binding experiments with oligonucleotides demonstrate that at least 3 bp on either side of the 15-bp core is required to form a complex and that high-affinity binding requires 5 to 6 bp of flanking DNA (314).

The amount of curvature introduced into the DNA upon Fis binding to a specific site has been estimated from the mobility of complexes in polyacrylamide gels to vary from 45 to 90° (129, 258, 314, 321, 416). The variation is largely a function of the identity of the DNA sequences flanking the core and presumably reflects their flexibility and thus their ability to dynamically wrap around the sides of Fis. For example, Fis-DNA complexes containing the *hin-D* 15-bp core site with the phage λ F site flanking sequences have an estimated curvature of 90°, but the same core with a GC-rich flanking sequence displays 45° of curvature, even though similar equilibrium binding constants are measured at both sites (314). Experiments with Fis derivatives containing the chemical nuclease 1,10-phenanthroline-copper (OP-Cu) tethered at specific sites also support the variable amounts of bending induced by Fis within the flanking sequences (313–315). On one hand, Fis chimeras with OP-Cu at residue 98 near the C-terminal end of the recognition helix cleave DNA in different Fis complexes with approximately equal efficiencies 3 to 4 bp outside the 15-bp core, consistent with a 50° curvature of the intervening DNA. However, Fis derivatives with the OP-Cu linked to position 71 or 73 generate widely variable amounts of scission located 5 to 6 nucleotides from the core at different binding sites, indicating that the DNA must be wrapped around the sides of Fis to different extents depending on the flanking sequence. Scission at both flanking sequences implies at least 70° of overall DNA curvature (313). This wrapping of DNA around the sides of Fis is primarily mediated by Arg-71 and is further stabilized by Asn-73 (Fig. 25) (314). Fis mutants lacking Arg-71 form complexes that are less bent and dissociate faster, even though equilibrium binding constants measured for different amino acid substitutions at Arg-71 are similar to complexes formed with wild-type Fis (258, 314).

In the absence of an atomic structure of a Fis-DNA complex, a number of related models have been proposed (111, 225, 314, 361, 429, 486). Figure 25 depicts two models that show different extents of overall curvature of the DNA bound to Fis (314). About 50° of curvature is required within the 15-bp core to enable the D helices of Fis to insert into the adjacent major grooves (Fig. 25A). A dynamic association of one or both of the flanking DNAs to enable contact with Arg-71 adds an additional 20° of curvature on each side (Fig. 25B). These models assume that Fis does not undergo a conformational change upon binding that alters the separation between the recognition helices, which is supported by fluorescence resonance energy transfer experiments comparing free and bound Fis dimers (266). Most high-affinity Fis binding sites are not intrinsically bent. However, the ability of a DNA segment to conform to the Fis DNA binding surface is presumably a major determinant specifying a high-affinity binding site and may help explain the ability of Fis to recognize such a wide variety of sequences.

Fis in DNA Supercoiling, Condensation, and Nucleoid Structure

The nonspecific DNA binding and bending properties of Fis imply that it potentially could play a role in DNA condensation. However, incubation of

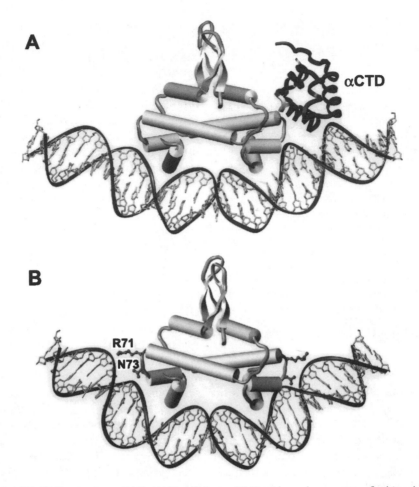

Figure 25. Models of Fis-DNA complexes (314). (A) Fis binding to DNA without the sequences flanking the core binding site contacting the protein. A ribbon representation of residues 249 to 320 of the α subunit of RNA polymerase (178) is also shown docked to the B-C turn region of Fis (2, 256). Residues 271 to 273 specifying the region contacted by Fis on the αCTD are light gray. (B) Fis binding to DNA where the flanking sequences are wrapped along the sides of Fis such that backbone contacts are made by Arg-71 and Asn-73.

relaxed plasmid DNA in vitro with a large amount of Fis in the presence of topoisomerase induces only a few negative supercoils (368; M. Haykinson and R. Johnson, unpublished data). Thus, Fis is rather inefficient at introducing supercoils into circular DNA, particularly when compared with the activity of HU or H-NS in this assay. A study employing atomic force and transmission electron microscopy of Fis-plasmid complexes concluded that the presence of Fis enhances branching of supercoiled plasmids (368). Further evidence for a nonspecific effect on DNA flexibility is obtained from studies on SfiI scission of plasmid DNA. Intermediate concentrations of Fis stimulate SfiI activity, which occurs only when the SfiI tetramer is simultaneously associated with two distinct DNA sites with the intervening DNA looped out (368).

Fis mutants display modest differences in the kinetics of growth-phase-dependent changes in plas-

mid supercoiling. When growth is reinitiated in rich medium from saturated overnight cultures or under long-term exponential-growth conditions, only minor differences in supercoiling densities of plasmids from *fis*+ and *fis* mutant *E. coli* and *S. enterica* serovar Typhimurium cells are observed. Plasmids from *fis* mutant cells display a linking difference of about +1 shortly after growth is reinitiated in rich media from saturated overnight cultures (R. Isaksson, S. McLeod, and R. Johnson, unpublished data). Both isogenic *fis*+ and *fis* mutant cells accumulate plasmid topoisomers of intermediate superhelical densities in prolonged stationary phase, but *fis* mutant cells shift to the more relaxed form earlier (371; S. McLeod, R. Isaksson, and R. Johnson, unpublished data). Since cellular Fis levels are very low at this time, this difference is presumably due to indirect causes. Subculturing from long-term stationary-phase cultures has given mixed results; in some cases the intermediate density form shifts to the highly supercoiled form more rapidly in *fis* mutants (371), while in other cases the highly supercoiled form is obtained more rapidly in *fis*$^+$ cells (255). The variations between *fis*$^+$ and *fis* mutant cells in plasmid supercoiling upon nutrient upshift are largely nullified by a brief rifampin treatment prior to DNA isolation, implying changes in transcription are largely responsible for the differences (242, 255).

Some of the differences in supercoiling observed when *fis*$^+$ and *fis* mutant cells are compared are likely due to and probably moderated by multiple interrelated secondary effects which function to maintain a homeostatic level of supercoiling. As described above, Fis differentially regulates transcription of the two HU subunit genes *hupA* and *hupB* (70) and enhances transcription of *hns* (106, 110). Fis also represses transcription of both *gyrA* and *gyrB* and directly inhibits supercoiling of plasmid DNA by DNA gyrase in vitro (369, 371). Surprisingly, Fis is required for induction of *topA* by hydrogen peroxide (452). Thus, Fis is one of a complex network of factors that control supercoiling levels in the cell (424).

Fis mutant cells tend to be elongated relative to wild type and contain multiple nucleoids. Nucleoids are evenly spaced and have a normal condensed appearance by DAPI staining (Fig. 1D) (112, 318). Fis mutants also have a propensity for minicell formation (402) (T. Paull and R. Johnson, unpublished data). These properties are consistent with a moderate defect in cell division, although the cell cycle asynchrony associated with *fis* mutants may also contribute to this phenotype (43) (see below). Expression of the yeast HMGB protein NHP6A largely reverses the cell morphology phenotype of *fis* mutant cells, suggesting that the nonspecific binding activity of Fis is important at some step leading up to or during the cell division process (318). The cell morphology phenotype of *fis* mutant cells is much less severe than that in *hupAB* cells. *fis hupAB* mutants display an even more severe phenotype than *hupAB* cells, but NHP6A is still able to nearly completely restore the chromosome segregation, nucleoid condensation, and cell size properties to near the wild-type condition.

Immunostaining revealed that Fis is predominantly located in multiple densely staining foci within the nucleoid (411). By contrast, HU, IHF, H-NS, and StpA are uniformly distributed throughout the nucleoid. The preferential localization of Fis suggested from these experiments is surprising. Perhaps it could be related to the clustering of predicted Fis binding sites flanking the terminus (435), or the colocalization of *rrn* or other control regions that contain multiple high-affinity Fis binding sites.

Regulation of Fis Synthesis

The first observation that Fis levels varied with respect to growth phase was made by Thompson et al., who found that Fis binding activity was 70-fold greater in extracts prepared from log-phase cells than in those prepared from stationary-phase cells (417). Subsequent experiments with antibody demonstrated that Fis was one of the most abundant DNA binding proteins in log-phase *E. coli* and *S. enterica* serovar Typhimurium growing under nutrient-rich conditions (21, 309, 412). Peak levels in early- to mid-exponential phase in rich media are 30,000 to 50,000 dimers per cell, which would correspond to over 50 µM Fis or one Fis dimer for every 275 bp, assuming three chromosome equivalents per cell (Table 1). Under batch culture conditions, Fis synthesis ceases well before growth slows, and levels become undetectable as cells are maintained in stationary phase. Peak Fis levels in cells growing in synthetic media batch cultures vary in proportion to the maximal growth rate obtained in the media, and the changes are not as dramatic as those observed in complex media (21, 298). Under steady-state growth conditions, Fis levels vary according to growth rate. For example, cells growing at 2.5 generations/h have about 30,000 dimers/cell, while cells growing at ≤0.5 generations/h have ≤3,000 dimers per cell. There is no evidence that specific environmental factors such as variations in carbon, nitrogen, amino acids, and oxygen directly control Fis levels other than via their indirect effects on growth rate. No changes in the stability of Fis protein with respect to cell physiology have been reported.

The shutoff of Fis synthesis prior to stationary phase is important for survival in stationary phase and/or the ability to recover upon nutrient upshift (309). Cells ectopically expressing 20,000 Fis dimers per cell as they enter stationary phase exhibit low viability with increasing time such that, after 4 days, CFU of *fis* mutants are reduced 10,000-fold relative to cells not expressing *fis*. Cells that do survive resume growth after very long lags. The detrimental effect of Fis expression in stationary phase could be due to repression of genes whose expression is advantageous for long-term survival or the inappropriate expression of genes normally activated in stationary phase. As noted above, *fis*$^+$ cells can respond to a nutrient upshift faster than *fis* mutant cells, particularly when resuming growth after prolonged stationary phase (298, 309).

Fis synthesis is regulated primarily by changes in transcription initiation (21, 299, 309), although there does appear to be a special mechanism for enhancing translation of *fis* over the upstream *dusB* coding regions of the message (J. Xu and R. Johnson, unpublished data). The major promoter responsible for expression of the *dusB-fis* operon initiates transcription 32 bp upstream of the *dusB* ATG. In wild-type cells, transcript levels initiated from this promoter can vary more than 500-fold over the first hour after a nutrient upshift. The *dusB-fis* mRNA displays a constant half-life of 2 min at different stages of growth, indicating that changes in stability are not responsible for the differences in RNA levels (334). A number of *cis*- and *trans*-acting factors collaborate to control the activity of the *fis* promoter. Factors that have been identified thus far include the identity and levels of the initiating nucleotide, autoregulation by Fis, stringent control, IHF, CRP, DNA supercoiling, and the presence of competitive promoters. Growth phase regulation can be achieved with a minimal promoter containing the core sequences from -38 to $+5$ (299, 334). Therefore, the *trans*-acting factors which interact with the *fis* promoter region outside the core sequences are not directly responsible for growth phase regulation, though they modulate the overall activity and, in the case of Fis autoregulation, the kinetics of the shutoff.

The dominant factor responsible for growth-phase-dependent regulation appears related to the nucleotide sensing mechanism that controls ribosomal RNA promoters (124, 447). The *fisP* transcription initiation region is rich in pyrimidines, and the major transcript initiates at a C, which is rare for *E. coli* promoters. When the C at $+1$ is replaced with a purine, growth phase regulation is lost and high levels of *fis* transcription occur into early stationary phase (447). Nucleotide pools vary greatly with respect to the growth phase in batch cultures, with the levels of pyrimidine triphosphates being significantly lower than those of purine triphosphates (292; R. Gourse, personal communication; R. Osuna, personal communication). Moreover, *fis* mRNA levels closely parallel changes in intracellular CTP levels through the growth cycle (Osuna, personal communication). *fisP* is also subject to stringent control and contains a GC-rich discriminator sequence between the -10 promoter region and $+1$ (299, 447). Although amino acid starvation causes a shutoff of *fis* transcription, a *relA spoT* mutant exhibits normal growth phase regulation, indicating that changing ppGpp levels are not the primary determinant controlling *fis* expression as a function of growth phase (21). In addition, a *fis* promoter mutant that has lost stringent control because of multiple A/T substitutions in the discriminator region still displays growth phase regulation, although the kinetics are somewhat altered (447).

Transcription of the *fis* promoter is exquisitely sensitive to the superhelical density of the DNA substrate, being efficient only around a narrow range of physiological densities ($\sigma \sim -0.07$ [370]). This was demonstrated both in vitro, using plasmid substrates with variable superhelical densities, and in vivo, using a combination of topoisomerase mutants and drugs. The close correspondence between the optimal in vivo and in vitro superhelical densities is somewhat surprising, since the level of unconstrained supercoils in vivo is about half the total density, and, at this level, transcription in vitro was poor. Nevertheless, these results imply that changes in supercoiling that occur under different growth conditions would modulate the levels of *fis* transcription (18).

Other factors control the absolute levels of *fis* mRNA but not the temporal transcription patterns exhibited in batch cultures. Transcription of the *fis* operon is subject to autoregulation by Fis protein (21, 299, 309). *dusB-fis* mRNA levels in a *fis* mutant peak at up to 10-fold higher levels and decrease more slowly than in wild-type strains. Fis binds to five sites within or upstream of the promoter and one site immediately downstream of the promoter in *E. coli* (Fig. 26); some of the corresponding sites in the upstream region from *Salmonella* are considerably weaker or absent. Resection experiments indicate that most of the Fis-mediated repression occurs from site II, which overlaps the -35 sigma binding sequence (334). In vitro experiments have shown that Fis effectively competes with RNA polymerase for binding to the primary promoter region (21, 290). IHF binding to a site centered at -114 increases mRNA levels three- to fourfold by an unknown mechanism (334). There are several weak CRP binding sites upstream of the *E. coli fis* promoter, but these are located in regions that are not conserved in the *fis* control region of other enteric bacteria. CRP is reported to have a complex but modest effect on Fis expression in *E. coli* (290). CRP-cAMP augments the repressing activity of Fis, but in the absence of Fis, CRP-cAMP enhances *fisP* activity. Finally, multiple promoters are reported to exist within the intercistronic region between *prmA* and *dusB*, but their transcriptional potential is controversial (289; Osuna, personal communication). A divergently oriented, strong RNA polymerase binding site located immediately upstream of the primary *fis* promoter in *E. coli* (21, 289, 290) is not predicted to be present in the *fis* regulatory regions of other enteric bacteria (25, 309).

Regulation of Transcription by Fis

Fis has been recruited as a transcriptional regulator of a large number of genes, many of which are

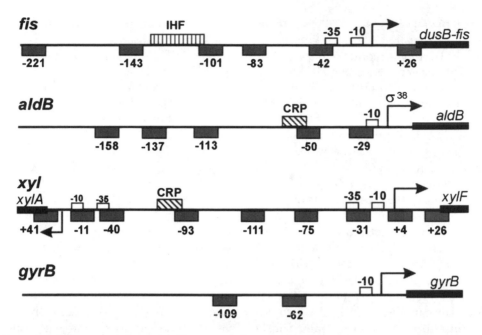

Figure 26. Examples of genes negatively regulated by Fis. Specific Fis binding sites are denoted as gray boxes. The extended *fisP* region from *E. coli* contains six Fis binding sites (21). Also shown is an IHF binding site that negatively regulates transcription (334). Several low-affinity CRP binding sites are not shown (290). In *S. enterica* serovar Typhimurium, Fis binding sites corresponding to −83 and −143 are weaker than those in *E. coli*, and there is no high-affinity site around position −101 (309). *aldB* is expressed preferentially in stationary phase by the RpoS (σ^{38}) form of RNA polymerase. Fis represses whereas CRP-cAMP activates *aldB* transcription (470). The 370-bp intercistronic region between the divergently transcribed *xylA* and *xylF* genes is depicted, and the Fis binding sites are numbered relative to the closest promoter (S. McLeod, J. Xu, and R. Johnson, unpublished data). Fis has been reported to repress transcription at a late step during *gyrB* transcription initiation, since open complex formation is not affected by Fis (369).

controlled in a growth phase or growth rate manner. A two-dimensional gel analysis comparing proteins synthesized from log-phase *fis*⁺ and *fis* mutant cells showed about 5% of the resolved spots increased and 5% decreased (R. Osuna and R. Johnson, unpublished data). An important group of genes whose expression is directly enhanced by *fis* are those that encode stable RNAs used for protein synthesis. These include the *rrn* operons and a number of operons encoding tRNAs, including the *thrU (tufB)* operon encoding both tRNAs and EF-Tu (65, 72, 103, 237, 295, 296, 298, 329, 348, 349, 397, 487). These are genes that must be efficiently transcribed at fast growth rates, when Fis levels are maximal, to maintain the robust protein synthesis potential of the cell. Not all tRNA promoters are activated by Fis, however, as levels of some remain unchanged and some are increased in *fis* mutants (103, 295). A substantial number of genes that are negatively regulated by Fis are among those whose transcription is increased in cells approaching stationary phase or under poor growth conditions (473). Many of these genes encode functions involved in the transport and assimilation of compounds metabolized under suboptimal conditions. Examples include mannitol-1-phosphate dehydrogenase (*mtlD*),

methyl-galactoside permease (*mglA*), D-xylose permease (*xylF*), and the ribose binding periplasmic protein (*rbsB*) (131, 473). Growth-phase regulation of these genes is often maintained in *fis* mutant cells, indicating that Fis often augments but is not solely responsible for their growth-phase regulation. Indeed, transcription of a number of the stationary-phase-expressed genes that are repressed by Fis depends on the RpoS form of RNA polymerase (473).

The *proP* P2 promoter, which transcribes the gene encoding a transporter of proline and glycine betaine, is unusual in that Fis activates a promoter that is transiently expressed in late exponential phase (469, 472). Transcription of the *proP* P2 promoter is nearly completely dependent upon both Fis and RpoS. The decreasing cellular concentrations of Fis in combination with increasing concentrations of the RpoS sigma factor as cells approach stationary phase (231) result in a brief temporal window of transcription.

Examples illustrating the enormous diversity of genes and control systems that have been reported to be directly or indirectly regulated by Fis include lysogeny of phages λ and Mu (positive) (20, 33, 34, 315, 438), nitrogen transport and assimilation (negative) (46, 47, 222, 466), various dehydrogenases (positive and negative) (140, 446, 470), the topoisomerases *gyrA*, *gyrB* (negative), and *topA* (positive), *dnaA* (negative) (123), ribonucleotide reductase (positive) (11, 175, 410), the genes encoding HU (positive at *hupA* and negative at *hupB*) (70) and H-NS (positive) (106, 109), and genes involved in bacterial pathogenesis (positive) (110, 130, 381, 463). Regarding the latter, entry of a pathogen into a host may mimic a nutrient upshift leading to an increase in Fis levels, consistent with the positive regulatory effects of Fis on genes whose products are involved in the early steps of infection. Some of the global effects of Fis on gene expression are indirect and occur via regulation of both specific and broad-spectrum regulators. For example, Fis negatively regulates the transcription of *cytR*, *deoC*, and *glpR* repressor genes (131) and augments transcription of the virulence gene activators *hilA*, *invF*, *ler*, *aggR*, and *virF* (110, 130, 381, 463). Fis also collaborates with CRP-cAMP to autoregulate *crp* transcription, one of the most widely functioning global regulators in the cell (132). Many of the genes that Fis negatively regulates are also activated by CRP-cAMP (131, 473). These two regulators can be thought of as generally having opposing roles in the cell; under optimal growth conditions Fis levels are high and CRP-cAMP levels are (relatively) low, whereas under poor growth conditions the opposite is the case.

Mechanism of Transcriptional Repression by Fis

Repression of transcription by Fis generally reduces RNA levels in the range of 2- to 10-fold. Thus, Fis often functions to negatively modulate gene expression rather than cause a near-complete shutoff of transcription, as is usually the case for specific repressor proteins. A common feature of Fis-repressed promoters is the presence of multiple Fis binding sites within and surrounding the RNA polymerase binding site. Some examples are illustrated in Fig. 26. In most of these cases a high-affinity site overlaps the core promoter sequence and, where tested, appears to mediate most of the repression by directly competing with RNA polymerase binding. The additional sites may amplify repression by forming a higher-order inhibitory nucleoprotein structure, but this has not been experimentally demonstrated. Other, more novel mechanisms of inhibiting transcription by Fis have also been described. In the case of *gyrB*, Fis binds to high-affinity sites centered at −62 and −109 (369). Fis does not appear to interfere with the binding or formation of open complexes at *gyrB* by RNA polymerase, leading Schneider et al. to propose that Fis is preventing promoter escape (369). Fis has been shown

to inhibit transcription from a site as far as −142 from the start of transcription, although in this case both Fis and IHF are required for effective repression (47, 466). There are a number of examples (*bglG*, *crp*, *aldB*, *hupB*, and *lacP*) in which Fis and CRP binding sites overlap, and in some of these cases Fis binding has been shown to interfere with transcriptional activation by CRP-cAMP (55, 70, 132, 315, 470).

Mechanism of Transcriptional Activation by Fis

As is the case with repression, Fis-activated genes often have multiple Fis binding sites within their promoter-regulatory region (Fig. 27). The contributions of individual binding sites on promoter activation can vary greatly, even within related operons. For example, the P1 promoters for all seven *rrn* operons contain from three to five Fis binding sites beginning at −71 (−72 for *rrnE*) and are activated by Fis at levels ranging from 3- to 8-fold in vivo and 3- to 12-fold in vitro (161). Over 80% of activation by Fis is mediated from the proximal site at some of the *rrn* P1 promoters (e.g., *rrnB* and *rrnG*), whereas efficient activation at others (e.g., *rrnA* and *rrnE*) also depends upon the distal sites. Likewise, all three Fis binding sites are required for activation at *tyrT* (284, 285). The relative contributions of the different Fis binding sites and the degree of overall activation depends on the properties of the core and UP element promoter determinants, as well as additional regulatory systems that function at these stable RNA promoters. *proP* P2 is an example of a Fis-activated promoter for which transcription in vivo is essentially dependent upon Fis binding to a site centered at −41 overlapping the RNA polymerase binding site; an upstream binding site at −81 has no detectable role (Fig. 27).

Two mechanisms for Fis-mediated activation of transcription have been described. One is through a direct interaction by Fis with the C-terminal domain of the α subunit (αCTD) of RNA polymerase. The interactions with RNA polymerase have been most thoroughly studied at *rrnB* P1, where most activation occurs via Fis binding centered at −71, and at the class II Fis-activated *proP* P2 promoter, where Fis activates from a binding site centered at −41. In both these cases the binding of Fis and RNA polymerase are mutually cooperative (38, 469). The results from both of these systems indicate that Fis interacts with RNA polymerase at a common site located on one subunit of the Fis dimer and on the αCTD of RNA polymerase. RNA polymerase mutants lacking the αCTD are strongly defective for Fis activation at *rrnB* P1 and abolished at *proP* P2

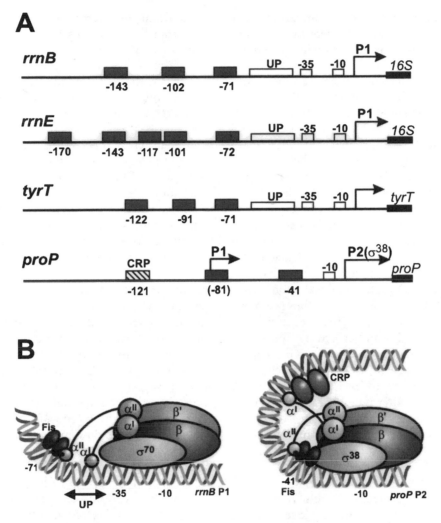

Figure 27. Examples of Fis-activated genes. (A) Fis binds to multiple sites upstream of the promoter regions in *rrnB* (38, 349), *rrnE* (161), and *tyrT* (237). Most of the activation at *rrnB* occurs from the promoter proximal site centered at −71 through a direct interaction with the αCTD of RNA polymerase (2). In addition to the proximal site, the upstream sites are required for efficient activation at *rrnE* and *tyrT* (161, 285). Fis activation at the RpoS-dependent *proP* P2 promoter occurs from the binding site at −41 (469, 472). CRP bound at −121 coactivates P2 transcription in a Fis-dependent manner (259). Transcription from the *proP* P1 promoter is strongly inhibited by CRP by an osmotically dependent mechanism (229, 471) and is weakly inhibited by Fis bound at −81. (B) Schematic models of the Fis-activated transcription initiation complexes at the *rrnB* P1 (left panel) and *proP* P2 (right panel) promoters (2, 256). There is a small preference for Fis activating through contacts with the α^{II}CTD over the α^{I}CTD at *rrnB* but a large preference at *proP*. Coactivation by CRP at *proP* P2 requires both αCTD subunits and the strong preference for Fis activation through the α^{II}CTD.

(37, 258). Alanine scanning of the αCTD defined the contact site on the αCTD as a 240-Å^2 surface-exposed ridge comprising residues Lys-271, Ala-272, and Glu-273 (Fig. 25A), with additional contributions from nearby residues Asp-258 and Leu-289 in the case of the *rrn* promoters (2, 256). A region on Fis located within and immediately surrounding the β-turn between the B and C helices specifies the contact site with RNA polymerase. This transcriptional activation region is defined by Gln-68, Arg-71, Gly-72, and probably Gln-74, which form a 790-Å^2 surface-

exposed ridge (Fig. 23) (37, 258). Crystal structures of mutants defective in transcription, in particular G72A and G72D, support a model in which one side of this ridge, specified by the locations of the mutant side chains, contacts the αCTD (67). Although Arg-71 can contact DNA flanking the Fis core binding site (Fig. 25B), this DNA interaction is believed to be absent in the tripartite complex with αCTD (Fig. 25A). In this regard it is noteworthy that mutants containing aromatic amino acids at residue 71, which do not promote bending of DNA within the flanking sequences, are able to cooperatively interact with RNA polymerase and stimulate transcription, but a lysine substitution, which has wild-type binding characteristics, is defective (258).

Experiments with Fis mutant heterodimers indicate that only one of the Fis subunits in the dimer is contacting αCTD, which has been localized by footprinting to be the downstream subunit in *rrnB* and *rrnE* and is presumably the upstream subunit in *proP* P2 (2, 258). Likewise, mutant αCTD heterodimers indicate that only one αCTD is sufficient for Fis activation (2, 258). In the case of *proP* P2, the α^{II}CTD linked to the β′ subunit is strongly preferred whereas in the class I *rrn* promoters either αCTD is efficient, although a small preference for activation through the α^{II}CTD was observed at *rrnE* P1. CRP coactivates the *proP* P2 promoter by a Fis-dependent mechanism; in this case the α^{II}CTD preferentially interacts with Fis, and the α^{I}CTD interacts with the promoter proximal subunit of CRP (Fig. 27B) (256, 259).

Contact between Fis and the αCTD promotes RNA polymerase binding, leading to open complex formation and initiation (38, 284, 469). An effect on transcription subsequent to recruitment of RNA polymerase to the promoter is implied in experiments at *proP*, where Fis was shown to enhance transcription even under saturating concentrations of RNA polymerase (255, 469). Stimulation of transcription by Fis after initial RNA polymerase binding has also been reported to occur at *tyrT*, *rrnD* P1, and *rrnB* P1 promoters (284, 360; C. Hirvonen and R. Gourse, personal communication). Open complexes formed at *rrnB* P1 are intrinsically very unstable and require high initiating nucleotide concentrations for transcription to proceed. Fis increases the half-life of these open complexes and has been shown to lower the concentration of initiating nucleotide required for transcription using mutant RNA polymerases (23; D. Jin, personal communication; Hirvonen and Gourse, personal communication). By contrast, once formed, open complexes are stable at *proP*, and Fis no longer has any demonstrable role (McLeod and Johnson, unpublished).

A second mechanism proposed for Fis activation of transcription is through the stabilization of a DNA microloop upstream of the promoter (284, 423). As elaborated by Travers and Muskhelishvili, changes in torsional stress introduced into DNA by RNA polymerase may be an integral aspect of the steps leading to productive initiation (423). In the microloop stabilization model, DNA bending by Fis facilitates contacts of DNA upstream of the Fis sites with RNA polymerase to form a topologically constrained microloop. This topological tension, which may be augmented by cooperative binding of Fis to multiple sites within the microloop, could be transmitted to the promoter sequences where melting and elongation occur, thereby enhancing multiple steps in the transcription reaction. Evidence for upstream DNA contacts with RNA polymerase has been obtained by footprinting and laser cross-linking (320, 425). However, a definitive role for Fis in promoting these contacts has not yet been obtained, although Fis-dependent cross-links in the site II region were observed to change during the process of transcription initiation (320). Fis-mediated microloop formation may augment the effects of a direct contact with RNA polymerase through the αCTD to varying extents depending on the specific properties of the promoter. For example, the promoters of stable RNA genes that are stimulated to a large degree by multiple upstream Fis sites may employ microloops or require the action of a DNA bending protein to generate appropriate upstream DNA-RNA polymerase contacts to a greater extent than those promoters that primarily require only the proximal Fis binding site. In addition, a microloop stabilization mechanism may help to compensate for local changes in superhelical density, particularly in those promoters that require high superhelical densities for efficient transcription (345). The relative roles of Fis-induced DNA bending and of Fis contacts with the αCTD and possibly other regions on RNA polymerase by Fis dimers bound to the upstream binding sites at these promoters need to be carefully delineated.

As described above ("Regulation of Transcription by H-NS"), Fis can also function by an indirect mechanism to enhance transcription. Fis binding to high-affinity sites overlapping H-NS binding sites in the *hns*, *rrnB*, and *virF* control region counteracts H-NS–mediated repression, leading to increased transcription (1, 106, 110).

Regulation of DNA Recombination by Fis

Site-specific DNA inversion

Fis is a critical regulatory protein in site-specific DNA inversion reactions catalyzed by the DNA invertase subgroup of serine recombinases (reviewed in reference 182). These reactions occur in bacterial chromosomes as well as in phage and plasmid genomes. The Hin-catalyzed reaction from *Salmonella* sp. controls flagellar phase variation. In this reaction, a 1-kb chromosomal segment containing a promoter, which directs transcription of one of two flagellin genes and a repressor of the alternate flagellin gene, undergoes infrequent inversion (Fig. 8A) (392). This leads to a low rate of switching between two antigenically distinct flagellins, allowing the pathogen to escape an immune response of its host. Site-specific DNA inversion reactions in phage regulate the alternative expression of genes encoding tail fiber proteins which control phage host range. Well-characterized examples in phage include the Gin-catalyzed reaction in the phage Mu genome, where a 3-kb segment inverts (Fig. 8A) (218), and the Cin-catalyzed reaction in the phage P1 genome, where a 4.2-kb segment inverts (154).

The Hin- and Gin-catalyzed DNA inversion reactions are decreased over 1,000-fold in *fis* mutants (183, 220). Likewise, Fis stimulates the rate of inversion over 100-fold in purified in vitro reactions (184, 217). An important feature of these systems is that recombination only occurs between specific sites located on the same DNA molecule and with an overwhelmingly preferred orientation leading to inversion of the intervening DNA (182). Site-specific deletions between recombination sites oriented in a "direct repeat" configuration occur extremely inefficiently. DNA invertase mutants have been isolated that support recombination in the absence of Fis (143, 146, 212). Unlike the Fis-dependent wild-type reaction, reactions performed in the absence of Fis with these mutants no longer display directionality control and do not require DNA supercoiling. These properties help support the view that Fis, combined with the intrinsic features of plectonemically supercoiled DNA, is largely responsible for controlling the directionality preference of the wild-type reaction. The very low expression of the DNA invertase is the primary rate-limiting determinant of the reaction in vivo (52, 190, 328), but the regulation of Fis expression may also impart a growth-phase control on inversion.

Fis mediates its activity through a 63-bp *cis*-acting recombinational enhancer DNA segment, whose location and orientation with respect to the two recombination sites are relatively unimportant (186, 189). The *hin* enhancer contains two Fis binding sites, which are both essential for activity. The spacing of these sites on the DNA helix, 48 bp between their centers, is also critical, consistent with each Fis dimer independently interacting with an invertase dimer bound to a recombination site (51, 185). Cross-linking and immunoelectron microscopy have demonstrated that DNA inversion occurs within a nucleoprotein complex called an invertasome, in which the two recombination sites bound by the DNA invertase are contacting Fis at the enhancer (Fig. 8B) (148). The invertasome is assembled at the base of a branch within plectonemically supercoiled DNA, which stabilizes the three-site complex in a defined configuration. As shown in Fig. 8B, the recombination sites are aligned such that the exchange of DNA strands will result in the inversion of the intervening DNA. The

topological changes in the products of single and multiple recombination reactions provide strong support for this model (149, 193, 194). The requirement for DNA supercoiling to promote and stabilize interactions between the Fis-bound enhancer and the invertase-bound recombination sites prevents reactions between recombination sites on separate DNA molecules (195). The DNA supercoiling requirement also provides a strong energetic barrier to the formation of deletions between directly oriented recombination sites, because the DNA would have to contain an additional loop (193, 275).

Hin and Gin can readily form synaptic complexes with two recombination sites in the absence of Fis or a recombinational enhancer (148). However, these complexes are not catalytically competent, except when a Fis-independent DNA invertase is used (79, 213). These results indicate that the Fis-bound enhancer is regulating the catalytic properties of the DNA invertase, as opposed to facilitating synapsis. Residues within the β-arm region on Fis, probably together with solvent-exposed residues in helix A, are required for activation of the Hin reaction (219, 308). Three residues—Val-16, Asp-20, and Val-22—whose side chains form a contiguous van der Waals surface located near the ends of the β-arms (Fig. 23) are of primary importance and are believed to contact the DNA invertase dimer within the invertasome (357). Surprisingly, experiments employing Fis mutant heterodimers demonstrate that only one of the two β-arms on each Fis dimer bound to the enhancer is sufficient for full stimulation (265).

Contact between Fis and the invertase is postulated to induce a conformational change within the invertase to initiate the chemical steps of the recombination reaction. By analogy to γδ resolvase, a related serine site-specific recombinase (481), the active sites within the invertase dimer are probably not positioned appropriately to enable DNA cleavage in the free protein or when bound to DNA in the absence of Fis. Interactions with Fis within a correctly assembled invertasome are believed to trigger a quaternary change in the structure of the invertase that results in a coordinated attack of the four invertase protomers, leading to double-strand cleavages at the centers of each recombination site (265). The mechanism by which the DNA strands are then exchanged and religated to give the recombinant product is poorly understood, but it is clear that Fis is not required for the ligation reaction (149, 193).

Phage λ site-specific recombination

Fis also functions in the phage λ site-specific recombination reactions, primarily in the excision reaction (13, 230). Phage yields from an induced monolysogen are reduced 100- to 1,000-fold in a *fis* mutant, whereas no effect on lytic growth is evident (19). Southern blot and PCR analyses confirm that the *fis* mutants are defective for *attR* × *attL* recombination. In vitro, the presence of Fis stimulates the excision reaction up to 20-fold, when Xis is limiting, but saturating amounts of Xis alleviate the requirement for Fis (417). In vivo, Fis may be functioning along with IHF as a sensor of growth conditions for λ to ensure that phage development can be completed after excision of its genome from the host chromosome. As noted above, it would be counterproductive for the phage to excise if the metabolic conditions of the lysogen were insufficient to replicate and package progeny viruses, even if the host were subjected to DNA damage. A small stimulation of the formation of lysogens by Fis has been noted in vivo, with part of this stimulation being attributed to a direct effect on the integration reaction (20). However, stimulation of integration by Fis has not been observed in vitro (417).

Fis enhances *attR* × *attL* recombination by binding to a specific site in *attR* called F that is centered 66 bp from the location of strand exchange (Fig. 12). Binding to F both introduces a defined DNA bend into the region and promotes Xis binding to the adjacent X1 site (416, 417). Xis binding to X1, in turn, recruits λ integrase to the adjacent P2 site through specific protein-protein interactions involving residues in the C-terminal end of Xis and the N-terminal DNA binding domain of Int (54, 68, 300, 362, 468). The C-terminal domain of the Int protomer bound at P2 engages in synapsis and DNA exchange with the core sequences at *attL* (209). Thus, Fis and Xis collaborate to recruit Int to its relatively weak binding site at P2 and to mold the architecture of the *attR* DNA such that Int is oriented favorably for synapsis and DNA exchange with *attL*.

The *attR* F site displays not only the highest affinity but also the greatest Fis-induced bending, estimated at 90 to 95°, of native Fis binding sites thus far reported (314, 416). Fis binding to F excludes Xis binding to the overlapping X2 site but strongly stimulates Xis binding to X1 (417). Xis is a monomeric DNA bending protein related to the winged-helix class and binds to an isolated X1 or X2 site on DNA with rather low affinity and specificity (359, 484). In the absence of Fis, Xis cooperatively binds to X1 and X2 with enhanced specificity but with poorer affinity than the F-X1 complex in vitro. This difference is also observed in vivo using challenge phages in which the X1-X2/F region is substituted for the P_{ant} operator in P22 (301). Effective repression of

P_{ant} transcription by Xis is observed only in the presence of Fis. Gel electrophoresis experiments and molecular modeling imply that Xis binding to X1-X2 or Xis + Fis binding to X1-F induces a similar overall direction and amount ($\geq 140°$) of DNA curvature (359, 416). Thus, in vitro reactions containing saturating amounts of Xis mechanistically mimic Xis + Fis reactions (417).

Other specialized recombination reactions

Fis also enhances excision of the lambdoid phage HK022 (97). Lysogens of HK022 often contain tandem prophages, and the DNA within the multiple prophages often rearranges during lysogen formation. Surprisingly, Fis and the phage Xis collaborate to inhibit the formation of these Int-dependent rearrangements.

Fis has also been reported to both positively and negatively influence Tn5/IS50 transposition (451). A Fis binding site is located within the inner end of IS50 that overlaps the tandem GATC methylation sites. Fis binds only to the nonmethylated DNA, and this binding inhibits transposition, presumably by competing with transposase binding. The modest stimulatory effect by Fis on Tn5 transposition appears to be indirect.

Homologous and illegitimate recombination

There have been no reports of Fis directly functioning in homologous recombination reactions and, as noted above, *fis* mutants are not abnormally sensitive to UV irradiation. However, *fis* mutants displayed markedly reduced rates of UV-induced illegitimate recombination in an assay measuring the formation of λ*bio* transducing phage (380). Even though illegitimate recombination rates were reduced in the absence of *fis*, there were no detectable differences from *fis*+ or non-UV-irradiated *fis* mutant cells in the locations of recombination endpoints, which occur at very short (6- to 13-bp) regions of homology.

Control of DNA Replication at *oriC* by Fis

Fis was first shown to be involved in the control of *oriC* replication from the observation that *oriC*-driven plasmids are poorly maintained in *fis* mutants (112, 129). Fis mutations also display synthetic effects with mutations in genes whose products control initiation of chromosome replication such as *dnaA*, *dam*, *gyrB*, and overproduced *seqA* (112, 248). In addition, a deletion of one of the primary DnaA binding sites, R4, in the chromosomal *oriC* locus is not tolerated in the absence of Fis (24). DNA content measurements by flow cytometry after rifampin and cephalexin treatment show that replication initiation in *fis* mutants is asynchronous, as is the case for *ihf* mutants (43).

In vitro footprinting methods by several groups have localized a specific Fis binding site to be between the DnaA binding sites R2 and R3 on the right half of the *oriC* sequence (Fig. 17) (58, 112, 252, 350). Weaker Fis sites, including one to the right of R4, have been noted by some groups (112, 252). In vivo footprinting has shown that the site between R2 and R3 is bound by Fis most of the time in exponentially growing cells but, as expected, is not bound in stationary phase (58, 59). A mutation that prevents Fis binding to this site strongly inhibits *oriC* function on a plasmid. This mutation has only a slight effect within the chromosome, but many *oriC* mutations, including those within primary DnaA binding sites, seem to be remarkably well tolerated in a chromosomal context (8, 450).

A combination of direct and indirect negative effects by Fis may be functioning to control the timing of *oriC* replication. Using synchronized cells prepared from a baby machine, Cassler et al. showed that Fis is bound throughout the cell cycle except at the time of initiation (58). Coincident with the loss of Fis binding, IHF binds to its site near the A/T-rich unwound region, and DnaA binds to R3 and a series of other, weaker sites (Fig. 17). Thus, *oriC* appears to oscillate between at least two different nucleoprotein complexes during the cell cycle, with the Fis-bound complex being incompetent for initiation. Consistent with the temporal binding patterns observed in vivo, experiments employing in vitro *oriC* replication systems have not revealed any positive effects of Fis on initiation (153, 252, 464). Rather, relatively high concentrations of Fis inhibit initiation, with Fis specifically inhibiting DnaA-dependent unwinding at the A/T-rich 13-mers (153). Surprisingly, specific Fis binding to the site between R2 and R3 is not required for this inhibition, since in vitro replication of mutant substrates that do not contain this site are inhibited by Fis to a similar extent as wild-type substrates (464). Thus, the Fis-mediated inhibition of replication observed in vitro appears to be mediated by alternative binding sites or by indirect mechanisms such as general effects on DNA topology.

Acknowledgments. We thank Phoebe Rice for the HU and IHF structure figures along with helpful discussions and Remus Dame, Nora Goosen, and Claire Wyman for Fig. 20.

Work in our laboratories is supported by NIH grants GM38509 to R.C.J. and GM28717 to J.F.G.

ADDENDUM IN PROOF

The X-ray crystal structures of the *Anabaena* HU-DNA complexes discussed in this chapter have now been published (K. K. Swinger, K. M. Lemberg, Y. Zhang, and P. A. Rice, *EMBO J.* **22:**3749–3760, 2003), and K. K. Swinger and P. A. Rice have written a recent review emphasizing structural aspects of HU/IHF proteins (*Curr. Opin. Struct. Biol.* **14:**28–35, 2004). Two groups have performed single-DNA-molecule experiments on HU complexes. These studies show that HU binds in two modes: at low to moderate concentrations, HU forms unstable complexes that compact DNA, but at higher concentrations, HU assembles stable rigid filaments with a helical pitch of 16 nm (J. van Noort, S. Verbrugge, N. Goosen, C. Dekker, and R. T. Dame, *Proc. Natl. Acad. Sci. USA* **101:**6969–6974, 2004; D. Skoko, B. Wong, R. C. Johnson, and J. F. Marko, submitted for publication; see also N. Goosen and R. T. Dame, *FEBS Lett.* **529:**151–156, 2002). The in vivo relevance of the latter mode remains to be determined. The dynamic interplay between Fis, IHF, and DnaA that regulates prereplication assembly at *oriC* (Fig. 17) has now been reproduced in vitro by Ryan et al. (V. T. Ryan, J. E. Grimwade, J. E. Camara, E. Crooke, and A. C. Leonard, *Mol. Microbiol.* **51:**1347–1359, 2004).

Since this chapter was written, atomic structures of the N-terminal domains of *E. coli* H-NS by NMR (V. Bloch, Y. Yang, E. Margeat, A. Chavanieu, M. T. Auge, B. Robert, S. Rimsky, S. Arold, and M. Kochoyan, *Nat. Struct. Biol.* **10:**212–218, 2003) and the *V. cholerae* H-NS homologue VicH by crystallography (R. Cerdan, V. Bloch, Y. Yang, P. Bertin, C. Dumas, S. Rimsky, M. Kochoyan, and S. T. Arold, *J. Mol. Biol.* **334:**179–185, 2003) have been reported. In both structures the two subunits of the dimer are associated in an antiparallel configuration such that the DNA binding domains would be located on opposite ends of the dimer, unlike that depicted in Fig. 19. Dole et al. (S. Dole, V. Nagarajavel, and K. Schnetz, *Mol. Microbiol.* **52:**589–600, 2004) have found that the H-NS silencer within the *bglG* coding region, the first gene of the *bgl* operon (Fig. 21), can block an elongating RNA polymerase in a Rho-dependent manner. H-NS has been found to enhance transposition of IS*903*, Tn*10*, and Tn*552* and to be required for the clustering of IS*903* insertions into hot spots that are normally distributed within the *E. coli* chromosome (B. Swingle, M. O'Carroll, D. Haniford, and K. M. Derbyshire, *Mol. Microbiol.* **52:**1055–1067, 2004). Some reviews emphasizing different aspects of H-NS bave been published (S. Rimsky, *Curr. Opin. Microbiol.* **7:**109–114, 2004; C. Tendeng and P. N. Bertin, *Trends Microbiol.* **11:**511–518, 2003; and C. J. Dorman, *Nat. Rev. Microbiol.* **2:**391–400, 2004).

REFERENCES

1. **Afflerbach, H., O. Schroder, and R. Wagner.** 1999. Conformational changes of the upstream DNA mediated by H-NS and FIS regulate *E. coli rrnB* P1 promoter activity. *J. Mol. Biol.* **286:**339–353.

2. **Aiyar, S. E., S. M. McLeod, W. Ross, C. A. Hirvonen, M. S. Thomas, R. C. Johnson, and R. L. Gourse.** 2002. Architecture of Fis-activated transcription complexes at the *Escherichia coli rrnB* P1 and *rrnE* P1 promoters. *J. Mol. Biol.* **316:**501–516.

3. **Aki, T., and S. Adhya.** 1997. Repressor induced site-specific binding of HU for transcriptional regulation. *EMBO J.* **16:**3666–3674.

4. **Ali, B. M., R. Amit, I. Braslavsky, A. B. Oppenheim, O. Gileadi, and J. Stavans.** 2001. Compaction of single

DNA molecules induced by binding of integration host factor (IHF). *Proc. Natl. Acad. Sci. USA* **98:**10658–10663.

5. **Allingham, J. S., and D. B. Haniford.** 2002. Mechanisms of metal ion action in Tn*10* transposition. *J. Mol. Biol.* **319:** 53–65.

6. **Alonso, J. C., F. Weise, and F. Rojo.** 1995. The *Bacillus subtilis* histone-like protein Hbsu is required for DNA resolution and DNA inversion mediated by the beta recombinase of plasmid pSM19035. *J. Biol. Chem.* **270:** 2938–2945.

7. **Arfin, S. M., A. D. Long, E. T. Ito, L. Tolleri, M. M. Riehle, E. S. Paegle, and G. W. Hatfield.** 2000. Global gene expression profiling in Escherichia coli K12. The effects of integration host factor. *J. Biol. Chem.* **275:**29672–29684.

8. **Asai, T., D. B. Bates, E. Boye, and T. Kogoma.** 1998. Are minichromosomes valid model systems for DNA replication control? Lessons learned from Escherichia coli. *Mol. Microbiol.* **29:**671–675.

9. **Atlung, T., and F. G. Hansen.** 2002. Effect of different concentrations of H-NS protein on chromosome replication and the cell cycle in *Escherichia coli. J. Bacteriol.***184:**1843–1850.

10. **Atlung, T., and H. Ingmer.** 1997. H-NS: a modulator of environmentally regulated gene expression. *Mol. Microbiol.* **24:**7–17.

11. **Augustin, L. B., B. A. Jacobson, and J. A. Fuchs.** 1994. *Escherichia coli* Fis and DnaA proteins bind specifically to the *nrd* promoter region and affect expression of an *nrd-lac* fusion. *J. Bacteriol.* **176:**378–387.

12. **Aviv, M., H. Giladi, G. Schreiber, A. B. Oppenheim, and G. Glaser.** 1994. Expression of the genes coding for the *Escherichia coli* integration host factor are controlled by growth phase, RpoS, ppGpp and by autoregulation. *Mol. Microbiol.* **14:**1021–1031.

13. **Azaro, M. A., and A. Landy.** 2002. λ integrase and the λ Int family, p. 118–148. *In* N. L. Craig, R. Craigie, M. Gellert, and A. M. Lambowitz (ed.), *Mobile DNA II.* ASM Press, Washington, D.C.

14. **Baker, T. A., and A. Kornberg.** 1988. Transcriptional activation of initiation of replication from the *E. coli* chromosomal origin: an RNA-DNA hybrid near *oriC. Cell* **55:** 113–123.

15. **Baker, T. A., and K. Mizuuchi.** 1992. DNA-promoted assembly of the active tetramer of the Mu transposase. *Genes Dev.* **6:**2221–2232.

16. **Balandina, A., L. Claret, R. Hengge-Aronis, and J. Rouviere-Yaniv.** 2001. The *Escherichia coli* histone-like protein HU regulates *rpoS* translation. *Mol. Microbiol.* **39:**1069–1079.

17. **Balandina, A., D. Kamashev, and J. Rouviere-Yaniv.** 2002. The bacterial histone-like protein HU specifically recognizes similar structures in all nucleic acids. DNA, RNA, and their hybrids. *J. Biol. Chem.* **277:**27622–27628.

18. **Balke, V. L., and J. D. Gralla.** 1987. Changes in the linking number of supercoiled DNA accompany growth transitions in *Escherichia coli. J. Bacteriol.* **169:**4499–4506.

19. **Ball, C. A., and R. C. Johnson.** 1991. Efficient excision of phage lambda from the *Escherichia coli* chromosome requires the Fis protein. *J. Bacteriol.* **173:**4027–4031.

20. **Ball, C. A., and R. C. Johnson.** 1991. Multiple effects of Fis on integration and the control of lysogeny in phage lambda. *J. Bacteriol.* **173:**4032–4038.

21. **Ball, C. A., R. Osuna, K. C. Ferguson, and R. C. Johnson.** 1992. Dramatic changes in Fis levels upon nutrient upshift in *Escherichia coli. J. Bacteriol.* **174:**8043–8056.

22. **Barth, M., C. Marschall, A. Muffler, D. Fischer, and R. Hengge-Aronis.** 1995. Role for the histone-like protein

H-NS in growth phase-dependent and osmotic regulation of sigma S and many sigma S-dependent genes in *Escherichia coli. J. Bacteriol.* **177:**3455–3464.

23. **Bartlett, M. S., T. Gaal, W. Ross, and R. L. Gourse.** 2000. Regulation of rRNA transcription is remarkably robust: FIS compensates for altered nucleoside triphosphate sensing by mutant RNA polymerases at *Escherichia coli rrn* P1 promoters. *J. Bacteriol.* **182:**1969–1977.

24. **Bates, D. B., T. Asai, Y. Cao, M. W. Chambers, G. W. Cadwell, E. Boye, and T. Kogoma.** 1995. The DnaA box R4 in the minimal *oriC* is dispensable for initiation of *Escherichia coli* chromosome replication. *Nucleic Acids Res.* **23:**3119–3125.

25. **Beach, M. B., and R. Osuna.** 1998. Identification and characterization of the *fis* operon in enteric bacteria. *J. Bacteriol.* **180:**5932–5946.

26. **Bensaid, A., A. Almeida, K. Drlica, and J. Rouviere-Yaniv.** 1996. Cross-talk between topoisomerase I and HU in *Escherichia coli. J. Mol. Biol.* **256:**292–300.

27. **Bertin, P., N. Benhabiles, E. Krin, C. Laurent-Winter, C. Tendeng, E. Turlin, A. Thomas, A. Danchin, and R. Brasseur.** 1999. The structural and functional organization of H-NS-like proteins is evolutionarily conserved in gram-negative bacteria. *Mol. Microbiol.* **31:**319–329.

28. **Bertin, P., F. Hommais, E. Krin, O. Soutourina, C. Tendeng, S. Derzelle, and A. Danchin.** 2001. H-NS and H-NS-like proteins in Gram-negative bacteria and their multiple role in the regulation of bacterial metabolism. *Biochimie* **83:**235–241.

29. **Bertin, P., P. Lejeune, C. Laurent-Winter, and A. Danchin.** 1990. Mutations in *bglY*, the structural gene for the DNA-binding protein H1, affect expression of several *Escherichia coli* genes. *Biochimie* **72:**889–891.

30. **Bertin, P., E. Terao, E. H. Lee, P. Lejeune, C. Colson, A. Danchin, and E. Collatz.** 1994. The H-NS protein is involved in the biogenesis of flagella in *Escherichia coli. J. Bacteriol.* **176:**5537–5540.

31. **Bertoni, G., N. Fujita, A. Ishihama, and V. de Lorenzo.** 1998. Active recruitment of sigma54-RNA polymerase to the Pu promoter of *Pseudomonas putida*: role of IHF and alphaCTD. *EMBO J.* **17:**5120–5128.

32. **Betermier, M., D. J. Galas, and M. Chandler.** 1994. Interaction of Fis protein with DNA: bending and specificity of binding. *Biochimie* **76:**958–967.

33. **Betermier, M., V. Lefrere, C. Koch, R. Alazard, and M. Chandler.** 1989. The *Escherichia coli* protein, Fis: specific binding to the ends of phage Mu DNA and modulation of phage growth. *Mol. Microbiol.* **3:**459–468.

34. **Betermier, M., I. Poquet, R. Alazard, and M. Chandler.** 1993. Involvement of *Escherichia coli* FIS protein in maintenance of bacteriophage Mu lysogeny by the repressor: control of early transcription and inhibition of transposition. *J. Bacteriol.* **175:**3798–3811.

35. **Bishop, A. C., J. Xu, R. C. Johnson, P. Schimmel, and V. De Crecy-Lagard.** 2002. Identification of the tRNA-dihydrouridine synthase family. *J. Biol. Chem.* **277:**25090–25095.

36. **Bliska, J. B., and N. R. Cozzarelli.** 1987. Use of site-specific recombination as a probe of DNA structure and metabolism in vivo. *J. Mol. Biol.* **194:**205–218.

37. **Bokal, A. J., W. Ross, T. Gaal, R. C. Johnson, and R. L. Gourse.** 1997. Molecular anatomy of a transcription activation patch: FIS-RNA polymerase interactions at the *Escherichia coli rrnB* P1 promoter. *EMBO J.* **16:**154–162.

38. **Bokal, A. J. T., W. Ross, and R. L. Gourse.** 1995. The transcriptional activator protein FIS: DNA interactions and cooperative interactions with RNA polymerase at the

Escherichia coli rrnB P1 promoter. *J. Mol. Biol.* **245:** 197–207.

39. **Bonnefoy, E., and J. Rouviere-Yaniv.** 1991. HU and IHF, two homologous histone-like proteins of *Escherichia coli*, form different protein-DNA complexes with short DNA fragments. *EMBO J.* **10:**687–696.

40. **Bonnefoy, E., and J. Rouviere-Yaniv.** 1992. HU, the major histone-like protein of *E. coli*, modulates the binding of IHF to *oriC. EMBO J.* **11:**4489–4496.

41. **Bonnefoy, E., M. Takahashi, and J. R. Yaniv.** 1994. DNA-binding parameters of the HU protein of *Escherichia coli* to cruciform DNA. *J. Mol. Biol.* **242:**116–129.

42. **Boubrik, F., and J. Rouviere-Yaniv.** 1995. Increased sensitivity to gamma irradiation in bacteria lacking protein HU. *Proc. Natl. Acad. Sci. USA* **92:**3958–3962.

43. **Boye, E., A. Lyngstadaas, A. Löbner-Olesen, K. Skarstad, and S. Wold.** 1993. Regulation of DNA replication in *Escherichia coli*, p. 15–26. *In* E. Fanning, R. Knippers, and E. L. Winnedler (ed.), *DNA Replication and the Cell Cycle*, vol. 43. Springer-Verlag KG, Berlin, Germany.

44. **Bramhill, D., and A. Kornberg.** 1988. A model for initiation at origins of DNA replication. *Cell* **54:**915–918.

45. **Brandi, A., R. Spurio, C. O. Gualerzi, and C. L. Pon.** 1999. Massive presence of the Escherichia coli "major cold-shock protein" CspA under non-stress conditions. *EMBO J.* **18:** 1653–1659.

46. **Browning, D. F., C. M. Beatty, A. J. Wolfe, J. A. Cole, and S. J. Busby.** 2002. Independent regulation of the divergent *Escherichia coli nrfA* and *acs*P1 promoters by a nucleoprotein assembly at a shared regulatory region. *Mol. Microbiol.* **43:**687–701.

47. **Browning, D. F., J. A. Cole, and S. J. Busby.** 2000. Suppression of FNR-dependent transcription activation at the *Escherichia coli nir* promoter by Fis, IHF and H-NS: modulation of transcription initiation by a complex nucleoprotein assembly. *Mol. Microbiol.* **37:**1258–1269.

48. Reference deleted.

49. **Broyles, S. S., and D. E. Pettijohn.** 1986. Interaction of the *Escherichia coli* HU protein with DNA. Evidence for formation of nucleosome-like structures with altered DNA helical pitch. *J. Mol. Biol.* **187:**47–60.

50. **Bruckner, R. C., and M. M. Cox.** 1989. The histone-like H protein of *Escherichia coli* is ribosomal protein S3. *Nucleic Acids Res.* **17:**3145–3161.

51. **Bruist, M. F., A. C. Glasgow, R. C. Johnson, and M. I. Simon.** 1987. Fis binding to the recombinational enhancer of the Hin DNA inversion system. *Genes Dev.* **1:**762–772.

52. **Bruist, M. F., and M. I. Simon.** 1984. Phase variation and the Hin protein: in vivo activity measurements, protein overproduction, and purification. *J. Bacteriol.* **159:**71–79.

53. **Busby, S., A. Kolb, and H. Buc.** 1979. Isolation of plasmid-protein complexes from *Escherichia coli. Eur. J. Biochem.* **99:**105–111.

54. **Bushman, W., S. Yin, L. L. Thio, and A. Landy.** 1984. Determinants of directionality in lambda site-specific recombination. *Cell* **39:**699–706.

55. **Caramel, A., and K. Schnetz.** 2000. Antagonistic control of the *Escherichia coli bgl* promoter by FIS and CAP in vitro. *Mol. Microbiol.* **36:**85–92.

56. **Caramel, A., and K. Schnetz.** 1998. Lac and lambda repressors relieve silencing of the *Escherichia coli bgl* promoter. Activation by alteration of a repressing nucleoprotein complex. *J. Mol. Biol.* **284:**875–883.

57. **Carmona, M., C. Balsalobre, F. Munoa, M. Mourino, Y. Jubete, F. De la Cruz, and A. Juarez.** 1993. *Escherichia coli hha* mutants, DNA supercoiling and expression of the

haemolysin genes from the recombinant plasmid pANN202-312. *Mol. Microbiol.* **9:**1011–1018.

58. **Cassler, M. R., J. E. Grimwade, and A. C. Leonard.** 1995. Cell cycle-specific changes in nucleoprotein complexes at a chromosomal replication origin. *EMBO J.* **14:**5833–5841.

59. **Cassler, M. R., J. E. Grimwade, K. C. McGarry, R. T. Mott, and A. C. Leonard.** 1999. Drunken-cell footprints: nuclease treatment of ethanol-permeabilized bacteria reveals an initiation-like nucleoprotein complex in stationary phase replication origins. *Nucleic Acids Res.* **27:**4570–4576.

60. **Castaing, B., C. Zelwer, J. Laval, and S. Boiteux.** 1995. HU protein of *Escherichia coli* binds specifically to DNA that contains single-strand breaks or gaps. *J. Biol. Chem.* **270:**10291–10296.

61. **Cayley, S., B. A. Lewis, H. J. Guttman, and M. T. Record, Jr.** 1991. Characterization of the cytoplasm of Escherichia coli K-12 as a function of external osmolarity. Implications for protein-DNA interactions in vivo. *J. Mol. Biol.* **222:**281–300.

62. **Ceschini, S., G. Lupidi, M. Coletta, C. L. Pon, E. Fioretti, and M. Angeletti.** 2000. Multimeric self-assembly equilibria involving the histone-like protein H-NS. A thermodynamic study. *J. Biol. Chem.* **275:**729–734.

63. **Chaconas, G., and R. M. Harshey.** 2002. Transposition of phage Mu DNA, p. 384–402. *In* N. L. Craig, R. Craigie, M. Gellert, and A. M. Lambowitz (ed.), *Mobile DNA II.* ASM Press, Washington, D.C.

64. **Chalmers, R., A. Guhathakurta, H. Benjamin, and N. Kleckner.** 1998. IHF modulation of Tn*10* transposition: sensory transduction of supercoiling status via a proposed protein/DNA molecular spring. *Cell* **93:**897–908.

65. **Champagne, N., and J. Lapointe.** 1998. Influence of FIS on the transcription from closely spaced and non-overlapping divergent promoters for an aminoacyl-tRNA synthetase gene (*gltX*) and a tRNA operon (*valU*) in *Escherichia coli. Mol. Microbiol.* **27:**1141–1156.

66. **Cheng, Q., N. Wesslund, N. B. Shoemaker, A. A. Salyers, and J. F. Gardner.** 2002. Development of an in vitro integration assay for the *Bacteroides* conjugative transposon CTnDOT. *J. Bacteriol.* **184:**4829–4837.

67. **Cheng, Y. S., W. Z. Yang, R. C. Johnson, and H. S. Yuan.** 2000. Structural analysis of the transcriptional activation on Fis: crystal structures of six Fis mutants with different activation properties. *J. Mol. Biol.* **302:**1139–1151.

68. **Cho, E. H., R. I. Gumport, and J. F. Gardner.** 2002. Interactions between integrase and excisionase in the phage lambda excisive nucleoprotein complex. *J. Bacteriol.* **184:** 5200–5203.

69. **Churchward, G.** 2002. Conjugative transposons and related mobile elements, p. 177–191. *In* N. L. Craig, R. Craigie, M. Gellert, and A. M. Lambowitz (ed.), *Mobile DNA II.* ASM Press, Washington, D.C.

70. **Claret, L., and J. Rouviere-Yaniv.** 1996. Regulation of HU alpha and HU beta by CRP and FIS in *Escherichia coli. J. Mol. Biol.* **263:**126–139.

71. **Claret, L., and J. Rouviere-Yaniv.** 1997. Variation in HU composition during growth of *Escherichia coli*: the heterodimer is required for long term survival. *J. Mol. Biol.* **273:**93–104.

72. **Condon, C., J. Philips, Z. Y. Fu, C. Squires, and C. L. Squires.** 1992. Comparison of the expression of the seven ribosomal RNA operons in *Escherichia coli. EMBO J.* **11:**4175–4185.

73. **Connolly, K. M., M. Iwahara, and R. T. Clubb.** 2002. Xis protein binding to the left arm stimulates excision of conjugative transposon Tn*916. J. Bacteriol.* **184:**2088–2099.

74. Cornelis, G. R., C. Sluiters, I. Delor, D. Geib, K. Kaniga, C. Lambert de Rouvroit, M. P. Sory, J. C. Vanooteghem, and T. Michiels. 1991. *ymoA*, a *Yersinia enterocolitica* chromosomal gene modulating the expression of virulence functions. *Mol. Microbiol.* 5:1023–1034.

75. Corpet, F. 1988. Multiple sequence alignment with hierarchical clustering. *Nucleic Acids Res.* 16:10881–10890.

76. Craig, N. L., and H. A. Nash. 1984. *E. coli* integration host factor binds to specific sites in DNA. *Cell* 39:707–716.

77. Craigie, R., D. J. Arndt-Jovin, and K. Mizuuchi. 1985. A defined system for the DNA strand-transfer reaction at the initiation of bacteriophage Mu transposition: protein and DNA substrate requirements. *Proc. Natl. Acad. Sci. USA* 82:7570–7574.

78. Crellin, P., and R. Chalmers. 2001. Protein-DNA contacts and conformational changes in the Tn*10* transpososome during assembly and activation for cleavage. *EMBO J.* 20:3882–3891.

79. Crisona, N. J., R. Kanaar, T. N. Gonzalez, E. L. Zechiedrich, A. Klippel, and N. R. Cozzarelli. 1994. Processive recombination by wild-type Gin and an enhancer-independent mutant. Insight into the mechanisms of recombination selectivity and strand exchange. *J. Mol. Biol.* 243:437–457.

80. Cukier, K. R., M. Jacquet, and F. Gros. 1972. Two heat-resistant, low molecular weight proteins from *Escherichia coli* that stimulate DNA-directed RNA synthesis. *Proc. Natl. Acad. Sci. USA* 69:3643–3647.

81. Cunha, S., C. L. Woldringh, and T. Odijk. 2001. Polymer-mediated compaction and internal dynamics of isolated *Escherichia coli* nucleoids. *J. Struct. Biol.* 136:53–66.

82. Cusick, M. E., and M. Belfort. 1998. Domain structure and RNA annealing activity of the *Escherichia coli* regulatory protein StpA. *Mol. Microbiol.* 28:847–857.

83. Dame, R. T., C. Wyman, and N. Goosen. 2000. H-NS mediated compaction of DNA visualised by atomic force microscopy. *Nucleic Acids Res.* 28:3504–3510.

84. Dame, R. T., C. Wyman, R. Wurm, R. Wagner, and N. Goosen. 2002. Structural basis for H-NS-mediated trapping of RNA polymerase in the open initiation complex at the *rrnB* P1. *J. Biol. Chem.* 277:2146–2150.

85. Dattananda, C. S., K. Rajkumari, and J. Gowrishankar. 1991. Multiple mechanisms contribute to osmotic inducibility of *proU* operon expression in *Escherichia coli*: demonstration of two osmoresponsive promoters and of a negative regulatory element within the first structural gene. *J. Bacteriol.* 173:7481–7490.

86. Deighan, P., A. Free, and C. J. Dorman. 2000. A role for the *Escherichia coli* H-NS-like protein StpA in OmpF porin expression through modulation of *micF* RNA stability. *Mol. Microbiol.* 38:126–139.

87. Delgado, J., S. Forst, S. Harlocker, and M. Inouye. 1993. Identification of a phosphorylation site and functional analysis of conserved aspartic acid residues of OmpR, a transcriptional activator for *ompF* and *ompC* in *Escherichia coli*. *Mol. Microbiol.* 10:1037–1047.

88. Delihas, N., and S. Forst. 2001. MicF: an antisense RNA gene involved in response of *Escherichia coli* to global stress factors. *J. Mol. Biol.* 313:1–12.

89. Dersch, P., S. Kneip, and E. Bremer. 1994. The nucleoid-associated DNA-binding protein H-NS is required for the efficient adaptation of *Escherichia coli* K-12 to a cold environment. *Mol. Gen. Genet.* 245:255–259.

90. Dersch, P., K. Schmidt, and E. Bremer. 1993. Synthesis of the *Escherichia coli* K-12 nucleoid-associated DNA-binding protein H-NS is subjected to growth-phase control and autoregulation. *Mol. Microbiol.* 8:875–889.

91. Dhavan, G. M., D. M. Crothers, M. R. Chance, and M. Brenowitz. 2002. Concerted binding and bending of DNA by *Escherichia coli* integration host factor. *J. Mol. Biol.* 315:1027–1037.

92. DiGabriele, A. D., M. R. Sanderson, and T. A. Steitz. 1989. Crystal lattice packing is important in determining the bend of a DNA dodecamer containing an adenine tract. *Proc. Natl. Acad. Sci. USA* 86:1816–1820.

93. Ditto, M. D., D. Roberts, and R. A. Weisberg. 1994. Growth phase variation of integration host factor level in *Escherichia coli*. *J. Bacteriol.* 176:3738–3748.

94. Dixon, N. E., and A. Kornberg. 1984. Protein HU in the enzymatic replication of the chromosomal origin of *Escherichia coli*. *Proc. Natl. Acad. Sci. USA* 81:424–428.

95. Donato, G. M., and T. H. Kawula. 1998. Enhanced binding of altered H-NS protein to flagellar rotor protein FliG causes increased flagellar rotational speed and hypermotility in *Escherichia coli*. *J. Biol. Chem.* 273:24030–24036.

96. Donato, G. M., and T. H. Kawula. 1999. Phenotypic analysis of random *hns* mutations differentiate DNA-binding activity from properties of *fimA* promoter inversion modulation and bacterial motility. *J. Bacteriol.* 181:941–948.

97. Dorgai, L., J. Oberto, and R. A. Weisberg. 1993. Xis and Fis proteins prevent site-specific DNA inversion in lysogens of phage HK022. *J. Bacteriol.* 175:693–700.

98. Dorman, C. J., N. Ni Bhriain, and C. F. Higgins. 1990. DNA supercoiling and environmental regulation of virulence gene expression in *Shigella flexneri*. *Nature* 344:789–792.

99. Dorman, C. J., J. C. Hinton, and A. Free. 1999. Domain organization and oligomerization among H-NS-like nucleoid-associated proteins in bacteria. *Trends Microbiol.* 7:124–128.

100. Dri, A. M., J. Rouviere-Yaniv, and P. L. Moreau. 1991. Inhibition of cell division in *hupA hupB* mutant bacteria lacking HU protein. *J. Bacteriol.* 173:2852–2863.

101. Drlica, K., and J. Rouviere-Yaniv. 1987. Histone-like proteins of bacteria. *Microbiol. Rev.* 51:301–319.

102. Eisenstein, B. I. 1981. Phase variation of type 1 fimbriae in *Escherichia coli* is under transcriptional control. *Science* 214:337–339.

103. Emilsson, V., and L. Nilsson. 1995. Factor for inversion stimulation-dependent growth rate regulation of serine and threonine tRNA species. *J. Biol. Chem.* 270:16610–16614.

104. Engelhorn, M., and J. Geiselmann. 1998. Maximal transcriptional activation by the IHF protein of *Escherichia coli* depends on optimal DNA bending by the activator. *Mol. Microbiol.* 30:431–441.

105. Esposito, D., A. Petrovic, R. Harris, S. Ono, J. F. Eccleston, A. Mbabaali, I. Haq, C. F. Higgins, J. C. Hinton, P. C. Driscoll, and J. E. Ladbury. 2002. H-NS oligomerization domain structure reveals the mechanism for high order self-association of the intact protein. *J. Mol. Biol.* 324:841–850.

106. Falconi, M., A. Brandi, A. La Teana, C. O. Gualerzi, and C. L. Pon. 1996. Antagonistic involvement of FIS and H-NS proteins in the transcriptional control of *hns* expression. *Mol. Microbiol.* 19:965–975.

107. Falconi, M., B. Colonna, G. Prosseda, G. Micheli, and C. O. Gualerzi. 1998. Thermoregulation of *Shigella* and *Escherichia coli* EIEC pathogenicity. A temperature-dependent structural transition of DNA modulates accessibility of *virF* promoter to transcriptional repressor H-NS. *EMBO J.* 17:7033–7043.

108. Falconi, M., M. T. Gualtieri, A. La Teana, M. A. Losso, and C. L. Pon. 1988. Proteins from the prokaryotic nucleoid:

primary and quaternary structure of the 15-kD *Escherichia coli* DNA binding protein H-NS. *Mol. Microbiol.* **2:** 323–329.

109. Falconi, M., N. P. Higgins, R. Spurio, C. L. Pon, and C. O. Gualerzi. 1993. Expression of the gene encoding the major bacterial nucleotide protein H-NS is subject to transcriptional auto-repression. *Mol. Microbiol.* **10:**273–282.

110. Falconi, M., G. Prosseda, M. Giangrossi, E. Beghetto, and B. Colonna. 2001. Involvement of FIS in the H-NS-mediated regulation of *virF* gene of *Shigella* and enteroinvasive *Escherichia coli. Mol. Microbiol.* **42:**439–452.

111. Feng, J.-A., H. S. Yuan, S. E. Finkel, R. C. Johnson, M. Kaczor-Grzeskowiak, and R. E. Dickerson. 1992. The interaction of Fis protein with its DNA binding sequences, p. 1–9. *In* R. H. Sarma and M. H. Sarma (ed.), *Structure and Function, Proceedings of the Seventh Conversation in Biomolecular Stereodynamics*, vol. 2. Adenine Press, New York, N.Y.

112. Filutowicz, M., W. Ross, J. Wild, and R. L. Gourse. 1992. Involvement of Fis protein in replication of the *Escherichia coli* chromosome. *J. Bacteriol.* **174:**398–407.

113. Finkel, S. E., and R. C. Johnson. 1992. The Fis protein: it's not just for DNA inversion anymore. *Mol. Microbiol.* **6:**3257–3265.

114. Flamm, E. L., and R. A. Weisberg. 1985. Primary structure of the *hip* gene of *Escherichia coli* and of its product, the beta subunit of integration host factor. *J. Mol. Biol.* **183:**117–128.

115. Flashner, Y., and J. D. Gralla. 1988. DNA dynamic flexibility and protein recognition: differential stimulation by bacterial histone-like protein HU. *Cell* **54:**713–721.

116. Free, A., and C. J. Dorman. 1995. Coupling of *Escherichia coli hns* mRNA levels to DNA synthesis by autoregulation: implications for growth phase control. *Mol. Microbiol.* **18:**101–113.

117. Free, A., and C. J. Dorman. 1997. The *Escherichia coli stpA* gene is transiently expressed during growth in rich medium and is induced in minimal medium and by stress conditions. *J. Bacteriol.* **179:**909–918.

118. Free, A., M. E. Porter, P. Deighan, and C. J. Dorman. 2001. Requirement for the molecular adapter function of StpA at the *Escherichia coli bgl* promoter depends upon the level of truncated H-NS protein. *Mol. Microbiol.* **42:**903–917.

119. Free, A., R. M. Williams, and C. J. Dorman. 1998. The StpA protein functions as a molecular adapter to mediate repression of the *bgl* operon by truncated H-NS in *Escherichia coli. J. Bacteriol.* **180:**994–997.

120. Frenkiel-Krispin, D., S. Levin-Zaidman, E. Shimoni, S. G. Wolf, E. J. Wachtel, T. Arad, S. E. Finkel, R. Kolter, and A. Minsky. 2001. Regulated phase transitions of bacterial chromatin: a non-enzymatic pathway for generic DNA protection. *EMBO J.* **20:**1184–1191.

121. Freundlich, M., N. Ramani, E. Mathew, A. Sirko, and P. Tsui. 1992. The role of integration host factor in gene expression in *Escherichia coli. Mol. Microbiol.* **6:**2557–2563.

122. Friedrich, K., C. O. Gualerzi, M. Lammi, M. A. Losso, and C. L. Pon. 1988. Proteins from the prokaryotic nucleoid. Interaction of nucleic acids with the 15 kDa *Escherichia coli* histone-like protein H-NS. *FEBS Lett.* **229:**197–202.

123. Froelich, J. M., T. K. Phuong, and J. W. Zyskind. 1996. Fis binding in the *dnaA* operon promoter region. *J. Bacteriol.* **178:**6006–6012.

124. Gaal, T., M. S. Bartlett, W. Ross, C. L. Turnbough, Jr., and R. L. Gourse. 1997. Transcription regulation by initiating NTP concentration: rRNA synthesis in bacteria. *Science* **278:**2092–2097.

125. Gamas, P., M. G. Chandler, P. Prentki, and D. J. Galas. 1987. *Escherichia coli* integration host factor binds specifically to the ends of the insertion sequence IS1 and to its major insertion hot-spot in pBR322. *J. Mol. Biol.* **195:**261–272.

126. Giangrossi, M., A. M. Giuliodori, C. O. Gualerzi, and C. L. Pon. 2002. Selective expression of the beta-subunit of nucleoid-associated protein HU during cold shock in *Escherichia coli. Mol. Microbiol.* **44:**205–216.

127. Giangrossi, M., C. O. Gualerzi, and C. L. Pon. 2001. Mutagenesis of the downstream region of the *Escherichia coli hns* promoter. *Biochimie* **83:**251–259.

128. Giladi, H., S. Koby, G. Prag, M. Engelhorn, J. Geiselmann, and A. B. Oppenheim. 1998. Participation of IHF and a distant UP element in the stimulation of the phage lambda PL promoter. *Mol. Microbiol.* **30:**443–451.

129. Gille, H., J. B. Egan, A. Roth, and W. Messer. 1991. The FIS protein binds and bends the origin of chromosomal DNA replication, *oriC*, of *Escherichia coli. Nucleic Acids Res.* **19:**4167–4172.

130. Goldberg, M. D., M. Johnson, J. C. Hinton, and P. H. Williams. 2001. Role of the nucleoid-associated protein Fis in the regulation of virulence properties of enteropathogenic *Escherichia coli. Mol. Microbiol.* **41:**549–559.

131. Gonzalez-Gil, G., P. Bringmann, and R. Kahmann. 1996. FIS is a regulator of metabolism in *Escherichia coli. Mol. Microbiol.* **22:**21–29.

132. Gonzalez-Gil, G., R. Kahmann, and G. Muskhelishvili. 1998. Regulation of *crp* transcription by oscillation between distinct nucleoprotein complexes. *EMBO J.* **17:**2877–2885.

133. Goodman, S. D., and O. Kay. 1999. Replacement of integration host factor protein-induced DNA bending by flexible regions of DNA. *J. Biol. Chem.* **274:**37004–37011.

134. Goodman, S. D., and H. A. Nash. 1989. Functional replacement of a protein-induced bend in a DNA recombination site. *Nature* **341:**251–254.

135. Goodrich, J. A., M. L. Schwartz, and W. R. McClure. 1990. Searching for and predicting the activity of sites for DNA binding proteins: compilation and analysis of the binding sites for *Escherichia coli* integration host factor (IHF). *Nucleic Acids Res.* **18:**4993–5000.

136. Goransson, M., B. Sonden, P. Nilsson, B. Dagberg, K. Forsman, K. Emanuelsson, and B. E. Uhlin. 1990. Transcriptional silencing and thermoregulation of gene expression in *Escherichia coli. Nature* **344:**682–685.

137. Gowrishankar, J., and D. Manna. 1996. How is osmotic regulation of transcription of the *Escherichia coli proU* operon achieved? A review and a model. *Genetica* **97:**363–378.

138. Granston, A. E., and H. A. Nash. 1993. Characterization of a set of integration host factor mutants deficient for DNA binding. *J. Mol. Biol.* **234:**45–59.

139. Grant, R. A., D. J. Filman, S. E. Finkel, R. Kolter, and J. M. Hogle. 1998. The crystal structure of Dps, a ferritin homolog that binds and protects DNA. *Nat. Struct. Biol.* **5:**294–303.

140. Green, J., M. F. Anjum, and J. R. Guest. 1996. The *ndh*-binding protein (Nbp) regulates the *ndh* gene of *Escherichia coli* in response to growth phase and is identical to Fis. *Mol. Microbiol.* **20:**1043–1055.

141. Griffo, G., A. B. Oppenheim, and M. E. Gottesman. 1989. Repression of the lambda P_{cin} promoter by integrative host factor. *J. Mol. Biol.* **209:**55–64.

142. Grimwade, J. E., V. T. Ryan, and A. C. Leonard. 2000. IHF redistributes bound initiator protein, DnaA, on supercoiled *oriC* of *Escherichia coli*. *Mol. Microbiol.* 35:835–844.

143. Haffter, P., and T. A. Bickle. 1988. Enhancer-independent mutants of the Cin recombinase have a relaxed topological specificity. *EMBO J.* 7:3991–3996.

144. Hales, L. M., R. I. Gumport, and J. F. Gardner. 1994. Determining the DNA sequence elements required for binding integration host factor to two different target sites. *J. Bacteriol.* 176:2999–3006.

145. Hales, L. M., R. I. Gumport, and J. F. Gardner. 1996. Examining the contribution of a dA + dT element to the conformation of *Escherichia coli* integration host factor-DNA complexes. *Nucleic Acids Res.* 24:1780–1786.

146. Haykinson, M. J., L. M. Johnson, J. Soong, and R. C. Johnson. 1996. The Hin dimer interface is critical for Fis-mediated activation of the catalytic steps of site-specific DNA inversion. *Curr. Biol.* 6:163–177.

147. Haykinson, M. J., and R. C. Johnson. 1993. DNA looping and the helical repeat in vitro and in vivo: effect of HU protein and enhancer location on Hin invertasome assembly. *EMBO J.* 12:2503–2512.

148. Heichman, K. A., and R. C. Johnson. 1990. The Hin invertasome: protein-mediated joining of distant recombination sites at the enhancer. *Science* 249:511–517.

149. Heichman, K. A., I. P. Moskowitz, and R. C. Johnson. 1991. Configuration of DNA strands and mechanism of strand exchange in the Hin invertasome as revealed by analysis of recombinant knots. *Genes Dev.* 5:1622–1634.

150. Hengen, P. N., S. L. Bartram, L. E. Stewart, and T. D. Schneider. 1997. Information analysis of Fis binding sites. *Nucleic Acids Res.* 25:4994–5002.

151. Hengge-Aronis, R. 1999. Interplay of global regulators and cell physiology in the general stress response of *Escherichia coli*. *Curr. Opin. Microbiol.* 2:148–152.

152. Hengge-Aronis, R. 2002. Signal transduction and regulatory mechanisms involved in control of the sigma S (RpoS) subunit of RNA polymerase. *Microbiol. Mol. Biol. Rev.* 66:373–395.

153. Hiasa, H., and K. J. Marians. 1994. Fis cannot support *oriC* DNA replication in vitro. *J. Biol. Chem.* 269:24999–25003.

154. Hiestand-Nauer, R., and S. Iida. 1983. Sequence of the site-specific recombinase gene *cin* and of its substrates serving in the inversion of the C segment of bacteriophage P1. *EMBO J.* 2:1733–1740.

155. Higgins, C. F., C. J. Dorman, D. A. Stirling, L. Waddell, I. R. Booth, G. May, and E. Bremer. 1988. A physiological role for DNA supercoiling in the osmotic regulation of gene expression in S. typhimurium and E. coli. *Cell* 52:569–584.

156. Higgins, C. F., J. C. Hinton, C. S. Hulton, T. Owen-Hughes, G. D. Pavitt, and A. Seirafi. 1990. Protein H1: a role for chromatin structure in the regulation of bacterial gene expression and virulence? *Mol. Microbiol.* 4:2007–2012.

157. Higgins, N. P., X. Yang, Q. Fu, and J. R. Roth. 1996. Surveying a supercoil domain by using the gamma delta resolution system in *Salmonella typhimurium*. *J. Bacteriol.* 178:2825–2835.

158. Hillyard, D. R., M. Edlund, K. T. Hughes, M. Marsh, and N. P. Higgins. 1990. Subunit-specific phenotypes of *Salmonella typhimurium* HU mutants. *J. Bacteriol.* 172:5402–5407.

159. Hinton, J. C., D. S. Santos, A. Seirafi, C. S. Hulton, G. D. Pavitt, and C. F. Higgins. 1992. Expression and mutational analysis of the nucleoid-associated protein H-NS of *Salmonella typhimurium*. *Mol. Microbiol.* 6:2327–2337.

160. Hiraga, S., H. Niki, T. Ogura, C. Ichinose, H. Mori, B. Ezaki, and A. Jaffe. 1989. Chromosome partitioning in *Escherichia coli*: novel mutants producing anucleate cells. *J. Bacteriol.* 171:1496–1505.

161. Hirvonen, C. A., W. Ross, C. E. Wozniak, E. Marasco, J. R. Anthony, S. E. Aiyar, V. H. Newburn, and R. L. Gourse. 2001. Contributions of UP elements and the transcription factor FIS to expression from the seven *rrn* P1 promoters in *Escherichia coli*. *J. Bacteriol.* 183:6305–6314.

162. Reference deleted.

163. Hodges-Garcia, Y., P. J. Hagerman, and D. E. Pettijohn. 1989. DNA ring closure mediated by protein HU. *J. Biol. Chem.* 264:14621–14623.

164. Holmes, V. F., and N. R. Cozzarelli. 2000. Closing the ring: links between SMC proteins and chromosome partitioning, condensation, and supercoiling. *Proc. Natl. Acad. Sci. USA* 97:1322–1234.

165. Hommais, F., E. Krin, C. Laurent-Winter, O. Soutourina, A. Malpertuy, J. P. Le Caer, A. Danchin, and P. Bertin. 2001. Large-scale monitoring of pleiotropic regulation of gene expression by the prokaryotic nucleoid-associated protein, H-NS. *Mol. Microbiol.* 40:20–36.

166. Hoover, T. R., E. Santero, S. Porter, and S. Kustu. 1990. The integration host factor stimulates interaction of RNA polymerase with NIFA, the transcriptional activator for nitrogen fixation operons. *Cell* 63:11–22.

167. Hsieh, L. S., J. Rouviere-Yaniv, and K. Drlica. 1991. Bacterial DNA supercoiling and [ATP]/[ADP] ratio: changes associated with salt shock. *J. Bacteriol.* 173:3914–3917.

168. Hu, K. H., E. Liu, K. Dean, M. Gingras, W. DeGraff, and N. J. Trun. 1996. Overproduction of three genes leads to camphor resistance and chromosome condensation in *Escherichia coli*. *Genetics* 143:1521–1532.

169. Huisman, O., P. R. Errada, L. Signon, and N. Kleckner. 1989. Mutational analysis of IS*10*'s outside end. *EMBO J.* 8:2101–2109.

170. Huisman, O., M. Faelen, D. Girard, A. Jaffe, A. Toussaint, and J. Rouviere-Yaniv. 1989. Multiple defects in *Escherichia coli* mutants lacking HU protein. *J. Bacteriol.* 171:3704–3712.

171. Hulton, C. S., A. Seirafi, J. C. Hinton, J. M. Sidebotham, L. Waddell, G. D. Pavitt, T. Owen-Hughes, A. Spassky, H. Buc, and C. F. Higgins. 1990. Histone-like protein H1 (H-NS), DNA supercoiling, and gene expression in bacteria. *Cell* 63:631–642.

172. Hwang, D. S., and A. Kornberg. 1992. Opening of the replication origin of *Escherichia coli* by DnaA protein with protein HU or IHF. *J. Biol. Chem.* 267:23083–23086.

173. Igo, M. M., A. J. Ninfa, and T. J. Silhavy. 1989. A bacterial environmental sensor that functions as a protein kinase and stimulates transcriptional activation. *Genes Dev.* 3:598–605.

174. Ishihama, A. 1999. Modulation of the nucleoid, the transcription apparatus, and the translation machinery in bacteria for stationary phase survival. *Genes Cells* 4:135–143.

175. Jacobson, B. A., and J. A. Fuchs. 1998. Multiple cis-acting sites positively regulate *Escherichia coli nrd* expression. *Mol. Microbiol.* 28:1315–1322.

176. Jaffe, A., D. Vinella, and R. D'Ari. 1997. The *Escherichia coli* histone-like protein HU affects DNA initiation, chromosome partitioning via MukB, and cell division via MinCDE. *J. Bacteriol.* 179:3494–3499.

177. Jaquet, M., K. R. Cukier, J. Pla, and F. Gros. 1971. A thermostable protein factor acting on in vitro DNA transcription. *Biochem. Biophys. Res. Commun.* 45:1597–1607.

178. Jeon, Y. H., T. Negishi, M. Shirakawa, T. Yamazaki, N. Fujita, A. Ishihama, and Y. Kyogoku. 1995. Solution structure of the activator contact domain of the RNA polymerase alpha subunit. *Science* 270:1495–1497.

179. Johansson, J., B. Dagberg, E. Richet, and B. E. Uhlin. 1998. H-NS and StpA proteins stimulate expression of the maltose regulon in *Escherichia coli*. *J. Bacteriol.* 180:6117–6125.

180. Johansson, J., S. Eriksson, B. Sonden, S. N. Wai, and B. E. Uhlin. 2001. Heteromeric interactions among nucleoid-associated bacterial proteins: localization of StpA-stabilizing regions in H-NS of *Escherichia coli*. *J. Bacteriol.* 183:2343–2347.

181. Johansson, J., and B. E. Uhlin. 1999. Differential protease-mediated turnover of H-NS and StpA revealed by a mutation altering protein stability and stationary-phase survival of *Escherichia coli*. *Proc. Natl. Acad. Sci. USA* 96:10776–10781.

182. Johnson, R. C. 2002. Bacterial site-specific DNA inversion systems, p. 230–271. *In* N. L. Craig, R. Craigie, M. Gellert, and A. M. Lambowitz (ed.), *Mobile DNA II*. ASM Press, Washington, D.C.

183. Johnson, R. C., C. A. Ball, D. Pfeffer, and M. I. Simon. 1988. Isolation of the gene encoding the Hin recombinational enhancer binding protein. *Proc. Natl. Acad. Sci. USA* 85:3484–3488.

184. Johnson, R. C., M. F. Bruist, and M. I. Simon. 1986. Host protein requirements for in vitro site-specific DNA inversion. *Cell* 46:531–539.

185. Johnson, R. C., A. C. Glasgow, and M. I. Simon. 1987. Spatial relationship of the Fis binding sites for Hin recombinational enhancer activity. *Nature* 329:462–465.

186. Johnson, R. C., and M. I. Simon. 1985. Hin-mediated site-specific recombination requires two 26 bp recombination sites and a 60 bp recombinational enhancer. *Cell* 41:781–791.

187. Jordi, B. J., A. E. Fielder, C. M. Burns, J. C. Hinton, N. Dover, D. W. Ussery, and C. F. Higgins. 1997. DNA binding is not sufficient for H-NS-mediated repression of *proU* expression. *J. Biol. Chem.* 272:12083–12090.

188. Jordi, B. J., and C. F. Higgins. 2000. The downstream regulatory element of the *proU* operon of *Salmonella typhimurium* inhibits open complex formation by RNA polymerase at a distance. *J. Biol. Chem.* 275:12123–12128.

189. Kahmann, R., F. Rudt, C. Koch, and G. Mertens. 1985. G inversion in bacteriophage Mu DNA is stimulated by a site within the invertase gene and a host factor. *Cell* 41:771–780.

190. Kahmann, R., F. Rudt, and G. Mertens. 1984. Substrate and enzyme requirements for in vitro site-specific recombination in bacteriophage mu. *Cold Spring Harbor Symp. Quant. Biol.* 49:285–294.

191. Kaidow, A., M. Wachi, J. Nakamura, J. Magae, and K. Nagai. 1995. Anucleate cell production by *Escherichia coli* delta *hns* mutant lacking a histone-like protein, H-NS. *J. Bacteriol.* 177:3589–3592.

192. Kamashev, D., and J. Rouviere-Yaniv. 2000. The histone-like protein HU binds specifically to DNA recombination and repair intermediates. *EMBO J.* 19:6527–6535.

193. Kanaar, R., A. Klippel, E. Shekhtman, J. M. Dungan, R. Kahmann, and N. R. Cozzarelli. 1990. Processive recombination by the phage Mu Gin system: implications for the mechanisms of DNA strand exchange, DNA site alignment, and enhancer action. *Cell* 62:353–366.

194. Kanaar, R., P. van de Putte, and N. R. Cozzarelli. 1988. Gin-mediated DNA inversion: product structure and the mechanism of strand exchange. *Proc. Natl. Acad. Sci. USA* 85:752–756.

195. Kanaar, R., P. van de Putte, and N. R. Cozzarelli. 1989. Gin-mediated recombination of catenated and knotted DNA substrates: implications for the mechanism of interaction between cis-acting sites. *Cell* 58:147–159.

196. Kano, Y., and F. Imamoto. 1990. Requirement of integration host factor (IHF) for growth of *Escherichia coli* deficient in HU protein. *Gene* 89:133–137.

197. Reference deleted.

198. Kano, Y., K. Osato, M. Wada, and F. Imamoto. 1987. Cloning and sequencing of the HU-2 gene of *Escherichia coli*. *Mol. Gen. Genet.* 209:408–410.

199. Kano, Y., M. Wada, T. Nagase, and F. Imamoto. 1986. Genetic characterization of the gene *hupB* encoding the HU-1 protein of *Escherichia coli*. *Gene* 45:37–44.

200. Kano, Y., K. Yasuzawa, H. Tanaka, and F. Imamoto. 1993. Propagation of phage Mu in IHF-deficient *Escherichia coli* in the absence of the H-NS histone-like protein. *Gene* 126:93–97.

201. Kano, Y., S. Yoshino, M. Wada, K. Yokoyama, M. Nobuhara, and F. Imamoto. 1985. Molecular cloning and nucleotide sequence of the HU-1 gene of *Escherichia coli*. *Mol. Gen. Genet.* 201:360–362.

202. Kar, S., and S. Adhya. 2001. Recruitment of HU by piggyback: a special role of GalR in repressosome assembly. *Genes Dev.* 15:2273–2281.

203. Katayama, T., M. Takata, and K. Sekimizu. 1996. The nucleoid protein H-NS facilitates chromosome DNA replication in *Escherichia coli dnaA* mutants. *J. Bacteriol.* 178:5790–5792.

204. Kavenoff, R., and B. C. Bowen. 1976. Electron microscopy of membrane-free folded chromosomes from *Escherichia coli*. *Chromosoma* 59:89–101.

205. Kawula, T. H., and P. E. Orndorff. 1991. Rapid site-specific DNA inversion in *Escherichia coli* mutants lacking the histonelike protein H-NS. *J. Bacteriol.* 173:4116–4123.

206. Kellenberger, E. 1990. Intracellular organization of the bacterial genome, p. 173–186. *In* K. Drlica and M. Riley (ed.), *The Bacterial Chromosome*. American Society for Microbiology, Washington, D.C.

207. Kikuchi, A., E. Flamm, and R. A. Weisberg. 1985. An *Escherichia coli* mutant unable to support site-specific recombination of bacteriophage lambda. *J. Mol. Biol.* 183:129–140.

208. Kikuchi, Y., and H. A. Nash. 1978. The bacteriophage lambda int gene product. A filter assay for genetic recombination, purification of Int, and specific binding to DNA. *J. Biol. Chem.* 253:7149–7157.

209. Kim, S., and A. Landy. 1992. Lambda Int protein bridges between higher order complexes at two distant chromosomal loci *attL* and *attR*. *Science* 256:198–203.

210. Reference deleted.

211. Kim, S., L. Moitoso de Vargas, S. E. Nunes-Duby, and A. Landy. 1990. Mapping of a higher order protein-DNA complex: two kinds of long-range interactions in lambda *attL*. *Cell* 63:773–781.

212. Klippel, A., K. Cloppenborg, and R. Kahmann. 1988. Isolation and characterization of unusual Gin mutants. *EMBO J.* 7:3983–3989.

213. Klippel, A., R. Kanaar, R. Kahmann, and N. R. Cozzarelli. 1993. Analysis of strand exchange and DNA binding of enhancer-independent Gin recombinase mutants. *EMBO J.* 12:1047–1057.

214. Klose, K. E., A. K. North, K. M. Stedman, and S. Kustu. 1994. The major dimerization determinants of the nitrogen regulatory protein NTRC from enteric bacteria lie in its carboxy-terminal domain. *J. Mol. Biol.* 241:233–245.

215. Ko, M., and C. Park. 2000. Two novel flagellar components and H-NS are involved in the motor function of *Escherichia coli*. *J. Mol. Biol.* 303:371–382.

216. Kobryn, K., B. D. Lavoie, and G. Chaconas. 1999. Supercoiling-dependent site-specific binding of HU to naked Mu DNA. *J. Mol. Biol.* 289:777–784.

217. Koch, C., and R. Kahmann. 1986. Purification and properties of the *Escherichia coli* host factor required for inversion of the G segment in bacteriophage Mu. *J. Biol. Chem.* 261:15673–15678.

218. Koch, C., G. Mertens, F. Rudt, R. Kahmann, R. Kanaar, R. H. Plasterk, P. van de Putte, R. Sandulache, and D. Kamp. 1987. The invertible G segment, p. 75–91. *In* N. Symonds, A. Toussaint, P. van de Putte, and M. M. Howe (ed.), *Phage Mu*. Cold Spring Harbor Laboratory, Cold Spring Harbor, N.Y.

219. Koch, C., O. Ninnemann, H. Fuss, and R. Kahmann. 1991. The N-terminal part of the *E. coli* DNA binding protein FIS is essential for stimulating site-specific DNA inversion but is not required for specific DNA binding. *Nucleic Acids Res.* 19:5915–5922.

220. Koch, C., J. Vandekerckhove, and R. Kahmann. 1988. *Escherichia coli* host factor for site-specific DNA inversion: cloning and characterization of the *fis* gene. *Proc. Natl. Acad. Sci. USA* 85:4237–4241.

221. Kohno, K., M. Wada, Y. Kano, and F. Imamoto. 1990. Promoters and autogenous control of the *Escherichia coli hupA* and *hupB* genes. *J. Mol. Biol.* 213:27–36.

222. Kolesnikow, T., I. Schroder, and R. P. Gunsalus. 1992. Regulation of *narK* gene expression in *Escherichia coli* in response to anaerobiosis, nitrate, iron, and molybdenum. *J. Bacteriol.* 174:7104–7111.

223. Kornberg, T., A. Lockwood, and A. Worcel. 1974. Replication of the *Escherichia coli* chromosome with a soluble enzyme system. *Proc. Natl. Acad. Sci. USA* 71:3189–3193.

224. Kostrewa, D., J. Granzin, C. Koch, H. W. Choe, S. Raghunathan, W. Wolf, J. Labahn, R. Kahmann, and W. Saenger. 1991. Three-dimensional structure of the *E. coli* DNA-binding protein FIS. *Nature* 349:178–180.

225. Kostrewa, D., J. Granzin, D. Stock, H. W. Choe, J. Labahn, and W. Saenger. 1992. Crystal structure of the factor for inversion stimulation FIS at 2.0 Å resolution. *J. Mol. Biol.* 226:209–226.

226. Krause, H. M., and N. P. Higgins. 1986. Positive and negative regulation of the Mu operator by Mu repressor and *Escherichia coli* integration host factor. *J. Biol. Chem.* 261:3744–3752.

227. Kur, J., N. Hasan, and W. Szybalski. 1989. Physical and biological consequences of interactions between integration host factor (IHF) and coliphage lambda late p'R promoter and its mutants. *Gene* 81:1–15.

228. Lambert de Rouvroit, C., C. Sluiters, and G. R. Cornelis. 1992. Role of the transcriptional activator, VirF, and temperature in the expression of the pYV plasmid genes of *Yersinia enterocolitica*. *Mol. Microbiol.* 6:395–409.

229. Landis, L., J. Xu, and R. C. Johnson. 1999. The cAMP receptor protein CRP can function as an osmoregulator of transcription in *Escherichia coli*. *Genes Dev.* 13:3081–3091.

230. Landy, A. 1989. Dynamic, structural, and regulatory aspects of lambda site-specific recombination. *Annu. Rev. Biochem.* 58:913–949.

231. Lange, R., and R. Hengge-Aronis. 1994. The cellular concentration of the sigma S subunit of RNA polymerase in *Escherichia coli* is controlled at the levels of transcription, translation, and protein stability. *Genes Dev.* 8:1600–1612.

232. La Teana, A., A. Brandi, M. Falconi, R. Spurio, C. L. Pon, and C. O. Gualerzi. 1991. Identification of a cold shock transcriptional enhancer of the *Escherichia coli* gene encoding nucleoid protein H-NS. *Proc. Natl. Acad. Sci. USA* 88:10907–10911.

233. Laurent-Winter, C., P. Lejeune, and A. Danchin. 1995. The *Escherichia coli* DNA-binding protein H-NS is one of the first proteins to be synthesized after a nutritional upshift. *Res. Microbiol.* 146:5–16.

234. Lavoie, B. D., and G. Chaconas. 1990. Immunoelectron microscopic analysis of the A, B, and HU protein content of bacteriophage Mu transpososomes. *J. Biol. Chem.* 265:1623–1627.

235. Lavoie, B. D., and G. Chaconas. 1993. Site-specific HU binding in the Mu transpososome: conversion of a sequence-independent DNA-binding protein into a chemical nuclease. *Genes Dev.* 7:2510–2519.

236. Lavoie, B. D., G. S. Shaw, A. Millner, and G. Chaconas. 1996. Anatomy of a flexer-DNA complex inside a higher-order transposition intermediate. *Cell* 85:761–771.

237. Lazarus, L. R., and A. A. Travers. 1993. The *Escherichia coli* FIS protein is not required for the activation of *tyrT* transcription on entry into exponential growth. *EMBO J.* 12:2483–2494.

238. Lease, R. A., and M. Belfort. 2000. Riboregulation by DsrA RNA: trans-actions for global economy. *Mol. Microbiol.* 38:667–672.

239. Lease, R. A., M. E. Cusick, and M. Belfort. 1998. Riboregulation in *Escherichia coli*: DsrA RNA acts by RNA:RNA interactions at multiple loci. *Proc. Natl. Acad. Sci. USA* 95:12456–12461.

240. Lee, E. C., L. M. Hales, R. I. Gumport, and J. F. Gardner. 1992. The isolation and characterization of mutants of the integration host factor (IHF) of *Escherichia coli* with altered, expanded DNA-binding specificities. *EMBO J.* 11:305–313.

241. Lejeune, P., and A. Danchin. 1990. Mutations in the *bglY* gene increase the frequency of spontaneous deletions in *Escherichia coli* K-12. *Proc. Natl. Acad. Sci. USA* 87:360–363.

242. Leng, F., and R. McMacken. 2002. Potent stimulation of transcription-coupled DNA supercoiling by sequence-specific DNA-binding proteins. *Proc. Natl. Acad. Sci. USA* 99:9139–9144.

243. Lewis, D. E., and S. Adhya. 2002. In vitro repression of the *gal* promoters by GalR and HU depends on the proper helical phasing of the two operators. *J. Biol. Chem.* 277:2498–2504.

244. Lewis, D. E., M. Geanacopoulos, and S. Adhya. 1999. Role of HU and DNA supercoiling in transcription repression: specialized nucleoprotein repression complex at *gal* promoters in *Escherichia coli*. *Mol. Microbiol.* 31:451–461.

245. Li, S., and R. Waters. 1998. *Escherichia coli* strains lacking protein HU are UV sensitive due to a role for HU in homologous recombination. *J. Bacteriol.* 180:3750–3756.

246. Liu, Q., and C. C. Richardson. 1993. Gene 5.5 protein of bacteriophage T7 inhibits the nucleoid protein H-NS of *Escherichia coli*. *Proc. Natl. Acad. Sci. USA* 90:1761–1765.

247. Lorenz, M., A. Hillisch, S. D. Goodman, and S. Diekmann. 1999. Global structure similarities of intact and nicked DNA complexed with IHF measured in solution by fluorescence resonance energy transfer. *Nucleic Acids Res.* 27:4619–4625.

248. Lu, M., J. L. Campbell, E. Boye, and N. Kleckner. 1994. SeqA: a negative modulator of replication initiation in *E. coli*. *Cell* 77:413–426.

249. Lucht, J. M., P. Dersch, B. Kempf, and E. Bremer. 1994. Interactions of the nucleoid-associated DNA-binding protein H-NS with the regulatory region of the osmotically controlled *proU* operon of *Escherichia coli*. *J. Biol. Chem.* **269:**6578–6578.

250. MacWilliams, M. P., R. I. Gumport, and J. F. Gardner. 1996. Genetic analysis of the bacteriophage lambda *attL* nucleoprotein complex. *Genetics* **143:**1069–1079.

251. Malik, M., A. Bensaid, J. Rouviere-Yaniv, and K. Drlica. 1996. Histone-like protein HU and bacterial DNA topology: suppression of an HU deficiency by gyrase mutations. *J. Mol. Biol.* **256:**66–76.

252. Margulies, C., and J. M. Kaguni. 1998. The FIS protein fails to block the binding of DnaA protein to *oriC*, the *Escherichia coli* chromosomal origin. *Nucleic Acids Res.* **26:**5170–5175.

253. Marykwas, D. L., S. A. Schmidt, and H. C. Berg. 1996. Interacting components of the flagellar motor of *Escherichia coli* revealed by the two-hybrid system in yeast. *J. Mol. Biol.* **256:**564–576.

254. Mattison, K., R. Oropeza, N. Byers, and L. J. Kenney. 2002. A phosphorylation site mutant of OmpR reveals different binding conformations at *ompF* and *ompC*. *J. Mol. Biol.* **315:**497–511.

255. McLeod, S. M. 2001. *Architecture of the Fis and CRP Co-activated proP P2 Transcription Initiation Complex.* Ph.D. thesis. University of California, Los Angeles.

256. McLeod, S. M., S. E. Aiyar, R. L. Gourse, and R. C. Johnson. 2002. The C-terminal domains of the RNA polymerase alpha subunits: contact site with Fis and localization during co-activation with CRP at the *Escherichia coli proP* P2 promoter. *J. Mol. Biol.* **316:**517–529.

257. McLeod, S. M., and R. C. Johnson. 2001. Control of transcription by nucleoid proteins. *Curr. Opin. Microbiol.* **4:**152–159.

258. McLeod, S. M., J. Xu, S. E. Cramton, T. Gaal, R. L. Gourse, and R. C. Johnson. 1999. Localization of amino acids required for Fis to function as a class II transcriptional activator at the RpoS-dependent *proP* P2 promoter. *J. Mol. Biol.* **294:**333–346.

259. McLeod, S. M., J. Xu, and R. C. Johnson. 2000. Coactivation of the RpoS-dependent *proP* P2 promoter by Fis and cyclic AMP receptor protein. *J. Bacteriol.* **182:**4180–4187.

260. Mechulam, Y., S. Blanquet, and G. Fayat. 1987. Dual level control of the *Escherichia coli pheST-himA* operon expression. tRNA(Phe)-dependent attenuation and transcriptional operator-repressor control by *himA* and the SOS network. *J. Mol. Biol.* **197:**453–470.

261. Megraw, T. L., and C. B. Chae. 1993. Functional complementarity between the HMG1-like yeast mitochondrial histone HM and the bacterial histone-like protein HU. *J. Biol. Chem.* **268:**12758–12763.

262. Membrillo-Hernandez, J., O. Kwon, P. De Wulf, S. E. Finkel, and E. C. Lin. 1999. Regulation of *adhE* (encoding ethanol oxidoreductase) by the Fis protein in *Escherichia coli*. *J. Bacteriol.* **181:**7390–7393.

263. Mengeritsky, G., D. Goldenberg, I. Mendelson, H. Giladi, and A. B. Oppenheim. 1993. Genetic and biochemical analysis of the integration host factor of *Escherichia coli*. *J. Mol. Biol.* **231:**646–657.

264. Mensa-Wilmot, K., K. Carroll, and R. McMacken. 1989. Transcriptional activation of bacteriophage lambda DNA replication in vitro: regulatory role of histone-like protein HU of *Escherichia coli*. *EMBO J.* **8:**2393–2402.

265. Merickel, S. K., M. J. Haykinson, and R. C. Johnson. 1998. Communication between Hin recombinase and Fis regulatory subunits during coordinate activation of Hin-catalyzed site-specific DNA inversion. *Genes Dev.* **12:**2803–2816.

266. Merickel, S. K., E. R. Sanders, J. L. Vazquez-Ibar, and R. C. Johnson. 2002. Subunit exchange and the role of dimer flexibility in DNA binding by the Fis protein. *Biochemistry* **41:**5788–5798.

267. Miller, H. I. 1984. Primary structure of the *himA* gene of Escherichia coli: homology with DNA-binding protein HU and association with the phenylalanyl-tRNA synthetase operon. *Cold Spring Harbor Symp. Quant. Biol.* **49:**691–698.

268. Miller, H. I., J. Abraham, M. Benedik, A. Campbell, D. Court, H. Echols, R. Fischer, J. M. Galindo, G. Guarneros, T. Hernandez, D. Mascarenhas, C. Montanez, D. Schindler, U. Schmeissner, and L. Sosa. 1981. Regulation of the integration-excision reaction by bacteriophage lambda. *Cold Spring Harbor Symp. Quant. Biol.* **45:**439–445.

269. Miller, H. I., A. Kikuchi, H. A. Nash, R. A. Weisberg, and D. I. Friedman. 1979. Site-specific recombination of bacteriophage lambda: the role of host gene products. *Cold Spring Harbor Symp. Quant. Biol.* **43:**1121–1126.

270. Miller, H. I., and H. A. Nash. 1981. Direct role of the *himA* gene product in phage lambda integration. *Nature* **290:**523–526.

271. Mizushima, T., K. Kataoka, Y. Ogata, R. Inoue, and K. Sekimizu. 1997. Increase in negative supercoiling of plasmid DNA in Escherichia coli exposed to cold shock. *Mol. Microbiol.* **23:**381–386.

272. Moitoso de Vargas, L., S. Kim, and A. Landy. 1989. DNA looping generated by DNA bending protein IHF and the two domains of lambda integrase. *Science* **244:**1457–1461.

273. Mojica, F. J., and C. F. Higgins. 1997. In vivo supercoiling of plasmid and chromosomal DNA in an *Escherichia coli hns* mutant. *J. Bacteriol.* **179:**3528–3533.

274. Morett, E., and P. Bork. 1998. Evolution of new protein function: recombinational enhancer Fis originated by horizontal gene transfer from the transcriptional regulator NtrC. *FEBS Lett.* **433:**108–112.

275. Moskowitz, I. P., K. A. Heichman, and R. C. Johnson. 1991. Alignment of recombination sites in Hin-mediated site-specific DNA recombination. *Genes Dev.* **5:**1635–1645.

276. Mourino, M., C. Madrid, C. Balsalobre, A. Prenafeta, F. Munoa, J. Blanco, M. Blanco, J. E. Blanco, and A. Juarez. 1996. The Hha protein as a modulator of expression of virulence factors in *Escherichia coli*. *Infect. Immun.* **64:**2881–2884.

277. Mukerji, M., and S. Mahadevan. 1997. Characterization of the negative elements involved in silencing the *bgl* operon of *Escherichia coli*: possible roles for DNA gyrase, H-NS, and CRP-cAMP in regulation. *Mol. Microbiol.* **24:**617–627.

278. Murphy, L. D., J. L. Rosner, S. B. Zimmerman, and D. Esposito. 1999. Identification of two new proteins in spermidine nucleoids isolated from *Escherichia coli*. *J. Bacteriol.* **181:**3842–3844.

279. Murphy, L. D., and S. B. Zimmerman. 1997. Isolation and characterization of spermidine nucleoids from *Escherichia coli*. *J. Struct. Biol.* **119:**321–335.

280. Murphy, L. D., and S. B. Zimmerman. 2001. A limited loss of DNA compaction accompanying the release of cytoplasm from cells of *Escherichia coli*. *J. Struct. Biol.* **133:**75–86.

281. Murphy, L. D., and S. B. Zimmerman. 1994. Macromolecular crowding effects on the interaction of DNA with *Escherichia coli* DNA-binding proteins: a model for bacterial nucleoid stabilization. *Biochim. Biophys. Acta* **1219:**277–284.

282. Murphy, L. D., and S. B. Zimmerman. 1997. Stabilization of compact spermidine nucleoids from *Escherichia coli*

under crowded conditions: implications for in vivo nucleoid structure. *J. Struct. Biol.* **119:**336–346.

283. **Murtin, C., M. Engelhorn, J. Geiselmann, and F. Boccard.** 1998. A quantitative UV laser footprinting analysis of the interaction of IHF with specific binding sites: re-evaluation of the effective concentration of IHF in the cell. *J. Mol. Biol.* **284:**949–961.

284. **Muskhelishvili, G., M. Buckle, H. Heumann, R. Kahmann, and A. A. Travers.** 1997. FIS activates sequential steps during transcription initiation at a stable RNA promoter. *EMBO J.* **16:**3655–3665.

285. **Muskhelishvili, G., A. A. Travers, H. Heumann, and R. Kahmann.** 1995. FIS and RNA polymerase holoenzyme form a specific nucleoprotein complex at a stable RNA promoter. *EMBO J.* **14:**1446–1452.

286. **Nash, H. A.** 1996. The E. coli HU and IHF proteins: accessory factors for complex protein-DNA assemblies, p. 149–179. *In* E. E. C. Lin and A. S. Lynch (ed.), *Regulation of Gene Expression in* Escherichia coli. R. G. Landes Co., Austin, Tex.

287. **Nash, H. A.** 1981. Integration and excision of bacteriophage lambda: the mechanism of conservation site specific recombination. *Annu. Rev. Genet.* **15:**143–167.

288. **Nasser, W., and S. Reverchon.** 2002. H-NS-dependent activation of pectate lyases synthesis in the phytopathogenic bacterium *Erwinia chrysanthemi* is mediated by the PecT repressor. *Mol. Microbiol.* **43:**733–748.

289. **Nasser, W., M. Rochman, and G. Muskhelishvili.** 2002. Transcriptional regulation of *fis* operon involves a module of multiple coupled promoters. *EMBO J.* **21:**715–724.

290. **Nasser, W., R. Schneider, A. Travers, and G. Muskhelishvili.** 2001. CRP modulates *fis* transcription by alternate formation of activating and repressing nucleoprotein complexes. *J. Biol. Chem.* **276:**17878–17886.

291. **Nelson, H. C., J. T. Finch, B. F. Luisi, and A. Klug.** 1987. The structure of an oligo(dA).oligo(dT) tract and its biological implications. *Nature* **330:**221–226.

292. **Neuhard, J., and P. Nygaard.** 1987. Purines and pyrimidines, p. 445–473. *In* F. C. Neidhardt, J. L. Ingraham, K. B. Low, B. Magasanik, M. Schaechter, and H. E. Umbarger (ed.), Escherichia coli *and* Salmonella typhimurium: *Cellular and Molecular Biology*, 1st ed., vol. 1. American Society for Microbiology, Washington, D.C.

293. **Nieto, J. M., C. Madrid, E. Miquelay, J. L. Parra, S. Rodriguez, and A. Juarez.** 2002. Evidence for direct protein-protein interaction between members of the enterobacterial Hha/YmoA and H-NS families of proteins. *J. Bacteriol.* **184:**629–635.

294. **Nieto, J. M., M. Mourino, C. Balsalobre, C. Madrid, A. Prenafeta, F. J. Munoa, and A. Juarez.** 1997. Construction of a double *hha hns* mutant of *Escherichia coli*: effect on DNA supercoiling and alpha-haemolysin production. *FEMS Microbiol. Lett.* **155:**39–44.

295. **Nilsson, L., and V. Emilsson.** 1994. Factor for inversion stimulation-dependent growth rate regulation of individual tRNA species in *Escherichia coli. J. Biol. Chem.* **269:**9460–9465.

296. **Nilsson, L., A. Vanet, E. Vijgenboom, and L. Bosch.** 1990. The role of FIS in trans activation of stable RNA operons of *E. coli. EMBO J.* **9:**727–734.

297. **Nilsson, L., H. Verbeek, U. Hoffmann, M. Haupt, and L. Bosch.** 1992. Inactivation of the *fis* gene leads to reduced growth rate. *FEMS. Microbiol. Lett.* **78:**85–88.

298. **Nilsson, L., H. Verbeek, E. Vijgenboom, C. van Drunen, A. Vanet, and L. Bosch.** 1992. FIS-dependent trans activation of stable RNA operons of *Escherichia coli* under various growth conditions. *J. Bacteriol.* **174:**921–929.

299. **Ninnemann, O., C. Koch, and R. Kahmann.** 1992. The E. coli fis promoter is subject to stringent control and autoregulation. *EMBO J.* **11:**1075–1083.

300. **Numrych, T. E., R. I. Gumport, and J. F. Gardner.** 1992. Characterization of the bacteriophage lambda excisionase (Xis) protein: the C-terminus is required for Xis-integrase cooperativity but not for DNA binding. *EMBO J.* **11:**3797–3806.

301. **Numrych, T. E., R. I. Gumport, and J. F. Gardner.** 1991. A genetic analysis of Xis and FIS interactions with their binding sites in bacteriophage lambda. *J. Bacteriol.* **173:**5954–5963.

302. **Nystrom, T.** 1995. Glucose starvation stimulon of *Escherichia coli*: role of integration host factor in starvation survival and growth phase-dependent protein synthesis. *J. Bacteriol.* **177:**5707–5710.

303. **O'Gara J, P., and C. J. Dorman.** 2000. Effects of local transcription and H-NS on inversion of the *fim* switch of *Escherichia coli. Mol. Microbiol.* **36:**457–466.

304. **Ogata, Y., R. Inoue, T. Mizushima, Y. Kano, T. Miki, and K. Sekimizu.** 1997. Heat shock-induced excessive relaxation of DNA in *Escherichia coli* mutants lacking the histone-like protein HU. *Biochim. Biophys. Acta* **1353:**298–306.

305. **Ogawa, T., M. Wada, Y. Kano, F. Imamoto, and T. Okazaki.** 1989. DNA replication in *Escherichia coli* mutants that lack protein HU. *J. Bacteriol.* **171:**5672–5679.

306. **Olsen, P. B., and P. Klemm.** 1994. Localization of promoters in the *fim* gene cluster and the effect of H-NS on the transcription of *fimB* and *fimE. FEMS Microbiol. Lett.* **116:**95–100.

307. **Olsen, P. B., M. A. Schembri, D. L. Gally, and P. Klemm.** 1998. Differential temperature modulation by H-NS of the *fimB* and *fimE* recombinase genes which control the orientation of the type 1 fimbrial phase switch. *FEMS Microbiol. Lett.* **162:**17–23.

308. **Osuna, R., S. E. Finkel, and R. C. Johnson.** 1991. Identification of two functional regions in Fis: the N-terminus is required to promote Hin-mediated DNA inversion but not lambda excision. *EMBO J.* **10:**1593–1603.

309. **Osuna, R., D. Lienau, K. T. Hughes, and R. C. Johnson.** 1995. Sequence, regulation, and functions of Fis in *Salmonella typhimurium. J. Bacteriol.* **177:**2021–2032.

310. **Otwinowski, Z., R. W. Schevitz, R. G. Zhang, C. L. Lawson, A. Joachimiak, R. Q. Marmorstein, B. F. Luisi, and P. B. Sigler.** 1988. Crystal structure of Trp repressor/operator complex at atomic resolution. *Nature* **335:**321–329.

311. **Overdier, D. G., and L. N. Csonka.** 1992. A transcriptional silencer downstream of the promoter in the osmotically controlled *proU* operon of *Salmonella typhimurium. Proc. Natl. Acad. Sci. USA* **89:**3140–3144.

312. **Owen-Hughes, T. A., G. D. Pavitt, D. S. Santos, J. M. Sidebotham, C. S. Hulton, J. C. Hinton, and C. F. Higgins.** 1992. The chromatin-associated protein H-NS interacts with curved DNA to influence DNA topology and gene expression. *Cell* **71:**255–265.

313. **Pan, C. Q., J. A. Feng, S. E. Finkel, R. Landgraf, D. Sigman, and R. C. Johnson.** 1994. Structure of the *Escherichia coli* Fis-DNA complex probed by protein conjugated with 1,10-phenanthroline copper(I) complex. *Proc. Natl. Acad. Sci. USA* **91:**1721–1725.

314. **Pan, C. Q., S. E. Finkel, S. E. Cramton, J. A. Feng, D. S. Sigman, and R. C. Johnson.** 1996. Variable structures of Fis-DNA complexes determined by flanking DNA-protein contacts. *J. Mol. Biol.* **264:**675–695.

315. Pan, C. Q., R. C. Johnson, and D. S. Sigman. 1996. Identification of new Fis binding sites by DNA scission with Fis-1,10-phenanthroline-copper(I) chimeras. *Biochemistry* 35:4326–4333.

316. Paull, T. T., M. J. Haykinson, and R. C. Johnson. 1994. HU and functional analogs in eukaryotes promote Hin invertasome assembly. *Biochimie* 76:992–1004.

317. Paull, T. T., M. J. Haykinson, and R. C. Johnson. 1993. The nonspecific DNA-binding and -bending proteins HMG1 and HMG2 promote the assembly of complex nucleoprotein structures. *Genes Dev.* 7:1521–1534.

318. Paull, T. T., and R. C. Johnson. 1995. DNA looping by *Saccharomyces cerevisiae* high mobility group proteins NHP6A/B. Consequences for nucleoprotein complex assembly and chromatin condensation. *J. Biol. Chem.* 270:8744–8754.

319. Pelton, J. G., S. Kustu, and D. E. Wemmer. 1999. Solution structure of the DNA-binding domain of NtrC with three alanine substitutions. *J. Mol. Biol.* 292:1095–1110.

320. Pemberton, I. K., G. Muskhelishvili, A. A. Travers, and M. Buckle. 2002. FIS modulates the kinetics of successive interactions of RNA polymerase with the core and upstream regions of the *tyrT* promoter. *J. Mol. Biol.* 318:651–663.

321. Perkins-Balding, D., D. P. Dias, and A. C. Glasgow. 1997. Location, degree, and direction of DNA bending associated with the Hin recombinational enhancer sequence and Fis-enhancer complex. *J. Bacteriol.* 179:4747–4753.

322. Petit, M. A., D. Ehrlich, and L. Janniere. 1995. pAM beta 1 resolvase has an atypical recombination site and requires a histone-like protein HU. *Mol. Microbiol.* 18:271–282.

323. Pettijohn, D. E. 1988. Histone-like proteins and bacterial chromosome structure. *J. Biol. Chem.* 263:12793–12796.

324. Pettijohn, D. E. 1996. The nucleoid, p. 158–166. *In* F. C. Neidhardt, R. Curtiss III, J. L. Ingraham, E. C. C. Lin, K. B. Low, B. Magasanik, W. S. Reznikoff, M. Riley, M. Schaechter, and H. E. Umbarger (ed.), Escherichia coli *and* Salmonella: *Cellular and Molecular Biology*, 2nd ed., vol. 1. ASM Press, Washington, D.C.

325. Pettijohn, D. E. 1982. Structure and properties of the bacterial nucleoid. *Cell* 30:667–669.

326. Pettijohn, D. E., and O. Pfenninger. 1980. Supercoils in prokaryotic DNA restrained in vivo. *Proc. Natl. Acad. Sci. USA* 77:1331–1335.

327. Pinson, V., M. Takahashi, and J. Rouviere-Yaniv. 1999. Differential binding of the *Escherichia coli* HU, homodimeric forms and heterodimeric form to linear, gapped and cruciform DNA. *J. Mol. Biol.* 287:485–497.

328. Plasterk, R. H., and P. van de Putte. 1984. Inversion of DNA in vivo and in vitro by Gin and Pin proteins. *Cold Spring Harbor Symp. Quant. Biol.* 49:295–300.

329. Pokholok, D. K., M. Redlak, C. L. Turnbough, Jr., S. Dylla, and W. M. Holmes. 1999. Multiple mechanisms are used for growth rate and stringent control of *leuV* transcriptional initiation in *Escherichia coli*. *J. Bacteriol.* 181:5771–5782.

330. Polaczek, P. 1990. Bending of the origin of replication of *E. coli* by binding of IHF at a specific site. *New Biol.* 2:265–271.

331. Polaczek, P., K. Kwan, and J. L. Campbell. 1998. Unwinding of the *Escherichia coli* origin of replication (oriC) can occur in the absence of initiation proteins but is stabilized by DnaA and histone-like proteins IHF or HU. *Plasmid* 39:77–83.

332. Pon, C. L., R. A. Calogero, and C. O. Gualerzi. 1988. Identification, cloning, nucleotide sequence and chromosomal

map location of *hns*, the structural gene for *Escherichia coli* DNA-binding protein H-NS. *Mol. Gen. Genet.* 212:199–202.

333. Pontiggia, A., A. Negri, M. Beltrame, and M. E. Bianchi. 1993. Protein HU binds specifically to kinked DNA. *Mol. Microbiol.* 7:343–350.

333a. Pratt, L. A., and T. J. Silhavy. 1995. Porin regulon of *Escherichia coli*, p. 105–127. *In* J. A. Hoch and T. J. Silhavy (ed.), *Two-Component Signal Transduction*. American Society for Microbiology, Washington, D.C.

334. Pratt, T. S., T. Steiner, L. S. Feldman, K. A. Walker, and R. Osuna. 1997. Deletion analysis of the *fis* promoter region in *Escherichia coli*: antagonistic effects of integration host factor and Fis. *J. Bacteriol.* 179:6367–6377.

335. Preobrajenskaya, O., A. Boullard, F. Boubrik, M. Schnarr, and J. Rouviere-Yaniv. 1994. The protein HU can displace the LexA repressor from its DNA-binding sites. *Mol. Microbiol.* 13:459–467.

336. Rabin, R. S., L. A. Collins, and V. Stewart. 1992. In vivo requirement of integration host factor for *nar* (nitrate reductase) operon expression in *Escherichia coli* K-12. *Proc. Natl. Acad. Sci. USA* 89:8701–8705.

337. Rajkumari, K., S. Kusano, A. Ishihama, T. Mizuno, and J. Gowrishankar. 1996. Effects of H-NS and potassium glutamate on sigmaS- and sigma70-directed transcription in vitro from osmotically regulated P1 and P2 promoters of *proU* in *Escherichia coli*. *J. Bacteriol.* 178:4176–4181.

338. Read, E. K., R. I. Gumport, and J. F. Gardner. 2000. Specific recognition of DNA by integration host factor. Glutamic acid 44 of the beta-subunit specifies the discrimination of a T:A from an A:T base pair without directly contacting the DNA. *J. Biol. Chem.* 275:33759–33764.

339. Record, M. T., Jr., E. S. Courtenay, D. S. Cayley, and H. J. Guttman. 1998. Responses of *E. coli* to osmotic stress: large changes in amounts of cytoplasmic solutes and water. *Trends Biochem. Sci.* 23:143–148.

340. Renzoni, D., D. Esposito, M. Pfuhl, J. C. Hinton, C. F. Higgins, P. C. Driscoll, and J. E. Ladbury. 2001. Structural characterization of the N-terminal oligomerization domain of the bacterial chromatin-structuring protein, H-NS. *J. Mol. Biol.* 306:1127–1137.

341. Rice, P. A. 1997. Making DNA do a U-turn: IHF and related proteins. *Curr. Opin. Struct. Biol.* 7:86–93.

342. Rice, P. A., S. Yang, K. Mizuuchi, and H. A. Nash. 1996. Crystal structure of an IHF-DNA complex: a protein-induced DNA U-turn. *Cell* 87:1295–1306.

343. Rimsky, S., F. Zuber, M. Buckle, and H. Buc. 2001. A molecular mechanism for the repression of transcription by the H-NS protein. *Mol. Microbiol.* 42:1311–1323.

344. Robertson, C. A., and H. A. Nash. 1988. Bending of the bacteriophage lambda attachment site by *Escherichia coli* integration host factor. *J. Biol. Chem.* 263:3554–3557.

345. Rochman, M., M. Aviv, G. Glaser, and G. Muskhelishvili. 2002. Promoter protection by a transcription factor acting as a local topological homeostat. *EMBO Rep.* 3:355–360.

346. Ross, W., and A. Landy. 1982. Bacteriophage lambda Int protein recognizes two classes of sequence in the phage *att* site: characterization of arm-type sites. *Proc. Natl. Acad. Sci. USA* 79:7724–7728.

347. Ross, W., A. Landy, Y. Kikuchi, and H. Nash. 1979. Interaction of Int protein with specific sites on lambda *att* DNA. *Cell* 18:297–307.

348. Ross, W., J. Salomon, W. M. Holmes, and R. L. Gourse. 1999. Activation of *Escherichia coli leuV* transcription by FIS. *J. Bacteriol.* 181:3864–3868.

349. Ross, W., J. F. Thompson, J. T. Newlands, and R. L. Gourse. 1990. *E. coli* Fis protein activates ribosomal RNA transcription in vitro and in vivo. *EMBO J.* **9:**3733–3742.

350. Roth, A., B. Urmoneit, and W. Messer. 1994. Functions of histone-like proteins in the initiation of DNA replication at *oriC* of Escherichia coli. *Biochimie* **76:**917–923.

351. Rouviere-Yaniv, J. 1978. Localization of the HU protein on the *Escherichia coli* nucleoid. *Cold Spring Harbor Symp. Quant. Biol.* **42:**439–447.

352. Rouviere-Yaniv, J., and F. Gros. 1975. Characterization of a novel, low-molecular-weight DNA-binding protein from *Escherichia coli. Proc. Natl. Acad. Sci. USA* **72:**3428–3432.

353. Rouviere-Yaniv, J., and N. O. Kjeldgaard. 1979. Native *Escherichia coli* HU protein is a heterotypic dimer. *FEBS Lett.* **106:**297–300.

354. Rouviere-Yaniv, J., M. Yaniv, and J. E. Germond. 1979. *E. coli* DNA binding protein HU forms nucleosomelike structure with circular double-stranded DNA. *Cell* **17:**265–274.

355. Rowland, S. J., W. M. Stark, and M. R. Boocock. 2002. Sin recombinase from *Staphylococcus aureus*: synaptic complex architecture and transposon targeting. *Mol. Microbiol.* **44:**607–619.

356. Ryan, V. T., J. E. Grimwade, C. J. Nievera, and A. C. Leonard. 2002. IHF and HU stimulate assembly of pre-replication complexes at *Escherichia coli oriC* by two different mechanisms. *Mol. Microbiol.* **46:**113–124.

357. Safo, M. K., W. Z. Yang, L. Corselli, S. E. Cramton, H. S. Yuan, and R. C. Johnson. 1997. The transactivation region of the Fis protein that controls site-specific DNA inversion contains extended mobile beta-hairpin arms. *EMBO J.* **16:**6860–6873.

358. Sakai, J., R. M. Chalmers, and N. Kleckner. 1995. Identification and characterization of a pre-cleavage synaptic complex that is an early intermediate in Tn*10* transposition. *EMBO J.* **14:**4374–4383.

359. Sam, M. D., C. Papagiannis, K. M. Connolly, L. Corselli, J. Iwahara, J. Lee, M. Phillips, J. M. Wojciak, R. C. Johnson, and R. T. Clubb. 2002. Regulation of directionality in bacteriophage lambda site-specific recombination: structure of the Xis protein. *J. Mol. Biol.* **324:**791–805.

360. Sander, P., W. Langert, and K. Mueller. 1993. Mechanisms of upstream activation of the *rrnD* promoter P1 of *Escherichia coli. J. Biol. Chem.* **268:**16907–16916.

361. Sandmann, C., F. Cordes, and W. Saenger. 1996. Structure model of a complex between the factor for inversion stimulation (FIS) and DNA: modeling protein-DNA complexes with dyad symmetry and known protein structures. *Proteins* **25:**486–500.

362. Sarkar, D., C. Papagiannis, R. C. Johnson, and A. Landy. 2002. Differential affinity and cooperativity functions of the amino-terminal 70 residues of λ integrase. *J. Mol. Biol.* **324:**775–789.

363. Sawitzke, J. A., and S. Austin. 2000. Suppression of chromosome segregation defects of *Escherichia coli muk* mutants by mutations in topoisomerase I. *Proc. Natl. Acad. Sci. USA* **97:**1671–1676.

364. Schembri, M. A., P. B. Olsen, and P. Klemm. 1998. Orientation-dependent enhancement by H-NS of the activity of the type 1 fimbrial phase switch promoter in *Escherichia coli. Mol. Gen. Genet.* **259:**336–344.

365. Schmid, M. B. 1990. More than just "histone-like" proteins. *Cell* **63:**451–453.

366. Schmid, M. B. 1988. Structure and function of the bacterial chromosome. *Trends Biochem. Sci.* **13:**131–135.

367. Schmid, M. B., and R. C. Johnson. 1991. Southern revival—news of bacterial chromatin. Prokaryotic chromosomes: structure and function in genome design. *New Biol.* **3:**945–950.

368. Schneider, R., R. Lurz, G. Luder, C. Tolksdorf, A. Travers, and G. Muskhelishvili. 2001. An architectural role of the *Escherichia coli* chromatin protein FIS in organising DNA. *Nucleic Acids Res.* **29:**5107–5114.

369. Schneider, R., A. Travers, T. Kutateladze, and G. Muskhelishvili. 1999. A DNA architectural protein couples cellular physiology and DNA topology in *Escherichia coli. Mol. Microbiol.* **34:**953–964.

370. Schneider, R., A. Travers, and G. Muskhelishvili. 2000. The expression of the *Escherichia coli fis* gene is strongly dependent on the superhelical density of DNA. *Mol. Microbiol.* **38:**167–175.

371. Schneider, R., A. Travers, and G. Muskhelishvili. 1997. FIS modulates growth phase-dependent topological transitions of DNA in *Escherichia coli. Mol. Microbiol.* **26:**519–530.

372. Schnetz, K. 1995. Silencing of *Escherichia coli bgl* promoter by flanking sequence elements. *EMBO J.* **14:**2545–2550.

373. Schroder, I., S. Darie, and R. P. Gunsalus. 1993. Activation of the *Escherichia coli* nitrate reductase (*narGHJI*) operon by NarL and Fnr requires integration host factor. *J. Biol. Chem.* **268:**771–774.

374. Schroder, O., D. Tippner, and R. Wagner. 2001. Toward the three-dimensional structure of the *Escherichia coli* DNA-binding protein H-NS: a CD and fluorescence study. *Biochem. Biophys. Res. Commun.* **282:**219–227.

375. Schroder, O., and R. Wagner. 2000. The bacterial DNA-binding protein H-NS represses ribosomal RNA transcription by trapping RNA polymerase in the initiation complex. *J. Mol. Biol.* **298:**737–748.

376. Schwan, W. R., J. L. Lee, F. A. Lenard, B. T. Matthews, and M. T. Beck. 2002. Osmolarity and pH growth conditions regulate *fim* gene transcription and type 1 pilus expression in uropathogenic *Escherichia coli. Infect. Immun.* **70:**1391–1402.

377. Segall, A. M., S. D. Goodman, and H. A. Nash. 1994. Architectural elements in nucleoprotein complexes: interchangeability of specific and non-specific DNA binding proteins. *EMBO J.* **13:**4536–4548.

378. Shanado, Y., K. Hanada, and H. Ikeda. 2001. Suppression of gamma ray-induced illegitimate recombination in *Escherichia coli* by the DNA-binding protein H-NS. *Mol. Genet. Genomics* **265:**242–248.

379. Shanado, Y., J. Kato, and H. Ikeda. 1998. *Escherichia coli* HU protein suppresses DNA-gyrase-mediated illegitimate recombination and SOS induction. *Genes Cells* **3:**511–520.

380. Shanado, Y., J. Kato, and H. Ikeda. 1997. Fis is required for illegitimate recombination during formation of lambda *bio* transducing phage. *J. Bacteriol.* **179:**4239–4245.

381. Sheikh, J., S. Hicks, M. Dall'Agnol, A. D. Phillips, and J. P. Nataro. 2001. Roles for Fis and YafK in biofilm formation by enteroaggregative *Escherichia coli. Mol. Microbiol.* **41:**983–997.

382. Shellman, V. L., and D. E. Pettijohn. 1991. Introduction of proteins into living bacterial cells: distribution of labeled HU protein in *Escherichia coli. J. Bacteriol.* **173:**3047–3059.

383. Sheridan, S. D., C. J. Benham, and G. W. Hatfield. 1998. Activation of gene expression by a novel DNA structural

transmission mechanism that requires supercoiling-induced DNA duplex destabilization in an upstream activating sequence. *J. Biol. Chem.* **273:**21298–21308.

384. **Shi, X., and G. N. Bennett.** 1994. Plasmids bearing *hfq* and the *hns*-like gene *stpA* complement *hns* mutants in modulating arginine decarboxylase gene expression in *Escherichia coli. J. Bacteriol.* **176:**6769–6775.

385. **Shiga, Y., Y. Sekine, Y. Kano, and E. Ohtsubo.** 2001. Involvement of H-NS in transpositional recombination mediated by IS*1. J. Bacteriol.* **183:**2476–2484.

386. **Shiga, Y., Y. Sekine, and E. Ohtsubo.** 1999. Transposition of IS*1* circles. *Genes Cells* **4:**551–561.

387. **Shimizu, M., M. Miyake, F. Kanke, U. Matsumoto, and H. Shindo.** 1995. Characterization of the binding of HU and IHF, homologous histone-like proteins of *Escherichia coli*, to curved and uncurved DNA. *Biochim. Biophys. Acta* **1264:**330–336.

388. **Shindo, H., T. Iwaki, R. Ieda, H. Kurumizaka, C. Ueguchi, T. Mizuno, S. Morikawa, H. Nakamura, and H. Kuboniwa.** 1995. Solution structure of the DNA binding domain of a nucleoid-associated protein, H-NS, from *Escherichia coli. FEBS Lett.* **360:**125–131.

389. **Shindo, H., A. Ohnuki, H. Ginba, E. Katoh, C. Ueguchi, T. Mizuno, and T. Yamazaki.** 1999. Identification of the DNA binding surface of H-NS protein from *Escherichia coli* by heteronuclear NMR spectroscopy. *FEBS Lett.* **455:**63–69.

390. **Shoemaker, N. B., R. D. Barber, and A. A. Salyers.** 1989. Cloning and characterization of a *Bacteroides* conjugal tetracycline-erythromycin resistance element by using a shuttle cosmid vector. *J. Bacteriol.* **171:**1294–1302.

391. **Signon, L., and N. Kleckner.** 1995. Negative and positive regulation of Tn10/IS*10*-promoted recombination by IHF: two distinguishable processes inhibit transposition off of multicopy plasmid replicons and activate chromosomal events that favor evolution of new transposons. *Genes Dev.* **9:**1123–1136.

392. **Silverman, M., J. Zieg, G. Mandel, and M. Simon.** 1981. Analysis of the functional components of the phase variation system. *Cold Spring Harbor Symp. Quant. Biol.* **45:**17–26.

393. **Sinden, R. R., J. O. Carlson, and D. E. Pettijohn.** 1980. Torsional tension in the DNA double helix measured with trimethylpsoralen in living *E. coli* cells: analogous measurements in insect and human cells. *Cell* **21:**773–783.

394. **Sinden, R. R., and D. E. Pettijohn.** 1981. Chromosomes in living *Escherichia coli* cells are segregated into domains of supercoiling. *Proc. Natl. Acad. Sci. USA* **78:**224–228.

395. **Skarstad, K., T. A. Baker, and A. Kornberg.** 1990. Strand separation required for initiation of replication at the chromosomal origin of *E. coli* is facilitated by a distant RNA–DNA hybrid. *EMBO J.* **9:**2341–2348.

396. **Skarstad, K., B. Thony, D. S. Hwang, and A. Kornberg.** 1993. A novel binding protein of the origin of the *Escherichia coli* chromosome. *J. Biol. Chem.* **268:**5365–5370.

397. **Slany, R. K., and H. Kersten.** 1992. The promoter of the *tgt/sec* operon in *Escherichia coli* is preceded by an upstream activation sequence that contains a high affinity FIS binding site. *Nucleic Acids Res.* **20:**4193–4198.

398. **Smyth, C. P., T. Lundback, D. Renzoni, G. Siligardi, R. Beavil, M. Layton, J. M. Sidebotham, J. C. Hinton, P. C. Driscoll, C. F. Higgins, and J. E. Ladbury.** 2000. Oligomerization of the chromatin-structuring protein H-NS. *Mol. Microbiol.* **36:**962–972.

399. **Sonden, B., and B. E. Uhlin.** 1996. Coordinated and differential expression of histone-like proteins in *Escherichia coli*: regulation and function of the H-NS analog StpA. *EMBO J.* **15:**4970–4980.

400. **Sonnenfield, J. M., C. M. Burns, C. F. Higgins, and J. C. Hinton.** 2001. The nucleoid-associated protein StpA binds curved DNA, has a greater DNA-binding affinity than H-NS and is present in significant levels in *hns* mutants. *Biochimie* **83:**243–249.

401. **Soutourina, O., A. Kolb, E. Krin, C. Laurent-Winter, S. Rimsky, A. Danchin, and P. Bertin.** 1999. Multiple control of flagellum biosynthesis in *Escherichia coli*: role of H-NS protein and the cyclic AMP-catabolite activator protein complex in transcription of the *flhDC* master operon. *J. Bacteriol.* **181:**7500–7508.

402. **Spaeny-Dekking, L., L. Nilsson, A. von Euler, P. van de Putte, and N. Goosen.** 1995. Effects of N-terminal deletions of the *Escherichia coli* protein Fis on growth rate, tRNA(2Ser) expression and cell morphology. *Mol. Gen. Genet.* **246:**259–265.

403. **Spassky, A., S. Rimsky, H. Garreau, and H. Buc.** 1984. H1a, an *E. coli* DNA-binding protein which accumulates in stationary phase, strongly compacts DNA in vitro. *Nucleic Acids Res.* **12:**5321–5340.

404. **Spears, P. A., D. Schauer, and P. E. Orndorff.** 1986. Metastable regulation of type 1 piliation in *Escherichia coli* and isolation and characterization of a phenotypically stable mutant. *J. Bacteriol.* **168:**179–185.

405. **Speck, C., and W. Messer.** 2001. Mechanism of origin unwinding: sequential binding of DnaA to double- and single-stranded DNA. *EMBO J.* **20:**1469–1476.

406. **Spurio, R., M. Durrenberger, M. Falconi, A. La Teana, C. L. Pon, and C. O. Gualerzi.** 1992. Lethal overproduction of the *Escherichia coli* nucleoid protein H-NS: ultramicroscopic and molecular autopsy. *Mol. Gen. Genet.* **231:**201–211.

407. **Spurio, R., M. Falconi, A. Brandi, C. L. Pon, and C. O. Gualerzi.** 1997. The oligomeric structure of nucleoid protein H-NS is necessary for recognition of intrinsically curved DNA and for DNA bending. *EMBO J.* **16:**1795–1805.

408. **Staczek, P., and N. P. Higgins.** 1998. Gyrase and Topo IV modulate chromosome domain size in vivo. *Mol. Microbiol.* **29:**1435–1448.

409. **Studier, F. W.** 1981. Identification and mapping of five new genes in bacteriophage T7. *J. Mol. Biol.* **153:**493–502.

410. **Sun, L., and J. A. Fuchs.** 1994. Regulation of the *Escherichia coli nrd* operon: role of DNA supercoiling. *J. Bacteriol.* **176:**4617–4626.

411. **Talukder, A. A., S. Hiraga, and A. Ishihama.** 2000. Two types of localization of the DNA-binding proteins within the *Escherichia coli* nucleoid. *Genes Cells* **5:**613–626.

412. **Talukder, A. A., A. Iwata, A. Nishimura, S. Ueda, and A. Ishihama.** 1999. Growth phase-dependent variation in protein composition of the *Escherichia coli* nucleoid. *J. Bacteriol.* **181:**6361–6370.

413. **Tanaka, H., K. Yasuzawa, K. Kohno, N. Goshima, Y. Kano, T. Saiki, and F. Imamoto.** 1995. Role of HU proteins in forming and constraining supercoils of chromosomal DNA in *Escherichia coli. Mol. Gen. Genet.* **248:**518–526.

414. **Tanaka, I., K. Appelt, J. Dijk, S. W. White, and K. S. Wilson.** 1984. 3-Å resolution structure of a protein with histone-like properties in prokaryotes. *Nature* **310:**376–381.

415. **Thompson, J. F., L. M. de Vargas, S. E. Skinner, and A. Landy.** 1987. Protein-protein interactions in a higher-order structure direct lambda site-specific recombination. *J. Mol. Biol.* **195:**481–493.

416. **Thompson, J. F., and A. Landy.** 1988. Empirical estimation of protein-induced DNA bending angles: applications to

lambda site-specific recombination complexes. *Nucleic Acids Res.* **16**:9687–9705.

417. Thompson, J. F., L. Moitoso de Vargas, C. Koch, R. Kahmann, and A. Landy. 1987. Cellular factors couple recombination with growth phase: characterization of a new component in the lambda site-specific recombination pathway. *Cell* **50**:901–908.

418. Reference deleted.

419. Thompson, J. F., U. K. Snyder, and A. Landy. 1988. Helical-repeat dependence of integrative recombination of bacteriophage lambda: role of the P1 and H1 protein binding sites. *Proc. Natl. Acad. Sci. USA* **85**:6323–6327.

420. Thompson, J. F., D. Waechter-Brulla, R. I. Gumport, J. F. Gardner, L. Moitoso de Vargas, and A. Landy. 1986. Mutations in an integration host factor-binding site: effect on lambda site-specific recombination and regulatory implications. *J. Bacteriol.* **168**:1343–1351.

421. Tippner, D., H. Afflerbach, C. Bradaczek, and R. Wagner. 1994. Evidence for a regulatory function of the histone-like *Escherichia coli* protein H-NS in ribosomal RNA synthesis. *Mol. Microbiol.* **11**:589–604.

422. Tippner, D., and R. Wagner. 1995. Fluorescence analysis of the *Escherichia coli* transcription regulator H-NS reveals two distinguishable complexes dependent on binding to specific or nonspecific DNA sites. *J. Biol. Chem.* **270**:22243–22247.

423. Travers, A., and G. Muskhelishvili. 1998. DNA microloops and microdomains: a general mechanism for transcription activation by torsional transmission. *J. Mol. Biol.* **279**:1027–1043.

424. Travers, A., R. Schneider, and G. Muskhelishvili. 2001. DNA supercoiling and transcription in *Escherichia coli*: the FIS connection. *Biochimie* **83**:213–217.

425. Travers, A. A., A. I. Lamond, H. A. Mace, and M. L. Berman. 1983. RNA polymerase interactions with the upstream region of the *E. coli tyrT* promoter. *Cell* **35**:265–273.

426. Trun, N. J., and J. F. Marko. 1998. Architecture of a bacterial chromosome. *ASM News* **64**:276–283.

427. Tsui, P., L. Huang, and M. Freundlich. 1991. Integration host factor binds specifically to multiple sites in the *ompB* promoter of *Escherichia coli* and inhibits transcription. *J. Bacteriol.* **173**:5800–5807.

428. Tupper, A. E., T. A. Owen-Hughes, D. W. Ussery, D. S. Santos, D. J. Ferguson, J. M. Sidebotham, J. C. Hinton, and C. F. Higgins. 1994. The chromatin-associated protein H-NS alters DNA topology in vitro. *EMBO J.* **13**:258–268.

429. Tzou, W. S., and M. J. Hwang. 1999. Modeling helix-turn-helix protein-induced DNA bending with knowledge-based distance restraints. *Biophys. J.* **77**:1191–1205.

430. Ueguchi, C., M. Kakeda, and T. Mizuno. 1993. Autoregulatory expression of the *Escherichia coli hns* gene encoding a nucleoid protein: H-NS functions as a repressor of its own transcription. *Mol. Gen. Genet.* **236**:171–178.

431. Ueguchi, C., M. Kakeda, H. Yamada, and T. Mizuno. 1994. An analogue of the DnaJ molecular chaperone in *Escherichia coli*. *Proc. Natl. Acad. Sci. USA* **91**:1054–1058.

432. Ueguchi, C., and T. Mizuno. 1996. Purification of H-NS protein and its regulatory effect on transcription in vitro. *Methods Enzymol.* **274**:271–276.

433. Ueguchi, C., C. Seto, T. Suzuki, and T. Mizuno. 1997. Clarification of the dimerization domain and its functional significance for the *Escherichia coli* nucleoid protein H-NS. *J. Mol. Biol.* **274**:145–151.

434. Ueguchi, C., T. Suzuki, T. Yoshida, K. Tanaka, and T. Mizuno. 1996. Systematic mutational analysis revealing

the functional domain organization of *Escherichia coli* nucleoid protein H-NS. *J. Mol. Biol.* **263**:149–162.

435. Ussery, D., T. S. Larsen, K. T. Wilkes, C. Friis, P. Worning, A. Krogh, and S. Brunak. 2001. Genome organisation and chromatin structure in *Escherichia coli*. *Biochimie* **83**:201–212.

436. Ussery, D. W., J. C. Hinton, B. J. Jordi, P. E. Granum, A. Seirafi, R. J. Stephen, A. E. Tupper, G. Berridge, J. M. Sidebotham, and C. F. Higgins. 1994. The chromatin-associated protein H-NS. *Biochimie* **76**:968–980.

437. VanBogelen, R. A., E. R. Olson, B. L. Wanner, and F. C. Neidhardt. 1996. Global analysis of proteins synthesized during phosphorus restriction in *Escherichia coli*. *J. Bacteriol.* **178**:4344–4366.

438. van Drunen, C. M., C. van Zuylen, E. J. Mientjes, N. Goosen, and P. van de Putte. 1993. Inhibition of bacteriophage Mu transposition by Mu repressor and Fis. *Mol. Microbiol.* **10**:293–298.

439. van Ulsen, P., M. Hillebrand, L. Zulianello, P. van de Putte, and N. Goosen. 1996. Integration host factor alleviates the H-NS-mediated repression of the early promoter of bacteriophage Mu. *Mol. Microbiol.* **21**:567–578.

440. van Ulsen, P., M. Hillebrand, L. Zulianello, P. van de Putte, and N. Goosen. 1997. The integration host factor-DNA complex upstream of the early promoter of bacteriophage Mu is functionally symmetric. *J. Bacteriol.* **179**:3073–3075.

441. Varshavsky, A. J., V. V. Bakayev, S. A. Nedospasov, and G. P. Georgiev. 1978. On the structure of eukaryotic, prokaryotic, and viral chromatin. *Cold Spring Harbor Symp. Quant. Biol.* **42**:457–473.

442. Varshavsky, A. J., S. A. Nedospasov, V. V. Bakayev, T. G. Bakayeva, and G. P. Georgiev. 1977. Histone-like proteins in the purified *Escherichia coli* deoxyribonucleoprotein. *Nucleic Acids Res.* **4**:2725–2745.

443. Vis, H., M. Mariani, C. E. Vorgias, K. S. Wilson, R. Kaptein, and R. Boelens. 1995. Solution structure of the HU protein from *Bacillus stearothermophilus*. *J. Mol. Biol.* **254**:692–703.

444. Vologodskii, A. V., and N. R. Cozzarelli. 1994. Conformational and thermodynamic properties of supercoiled DNA. *Annu. Rev. Biophys. Biomol. Struct.* **23**:609–643.

445. Von Freiesleben, U., K. V. Rasmussen, T. Atlung, and F. G. Hansen. 2000. Rifampicin-resistant initiation of chromosome replication from *oriC* in *ihf* mutants. *Mol. Microbiol.* **37**:1087–1093.

446. Wackwitz, B., J. Bongaerts, S. D. Goodman, and G. Unden. 1999. Growth phase-dependent regulation of *nuoA-N* expression in *Escherichia coli* K-12 by the Fis protein: upstream binding sites and bioenergetic significance. *Mol. Gen. Genet.* **262**:876–883.

447. Walker, K. A., C. L. Atkins, and R. Osuna. 1999. Functional determinants of the *Escherichia coli fis* promoter: roles of −35, −10, and transcription initiation regions in the response to stringent control and growth phase-dependent regulation. *J. Bacteriol.* **181**:1269–1280.

448. Wang, S., R. Cosstick, J. F. Gardner, and R. I. Gumport. 1995. The specific binding of *Escherichia coli* integration host factor involves both major and minor grooves of DNA. *Biochemistry* **34**:13082–13090.

449. Watson, M. A., and G. Chaconas. 1996. Three-site synapsis during Mu DNA transposition: a critical intermediate preceding engagement of the active site. *Cell* **85**:435–445.

450. Weigel, C., W. Messer, S. Preiss, M. Welzeck, Morigen, and E. Boye. 2001. The sequence requirements for a functional *Escherichia coli* replication origin are different for the

chromosome and a minichromosome. *Mol. Microbiol.* 40:498–507.

451. **Weinreich, M. D., and W. S. Reznikoff.** 1992. Fis plays a role in Tn*5* and IS*50* transposition. *J. Bacteriol.* 174:4530–4537.

452. **Weinstein-Fischer, D., M. Elgrably-Weiss, and S. Altuvia.** 2000. *Escherichia coli* response to hydrogen peroxide: a role for DNA supercoiling, topoisomerase I and Fis. *Mol. Microbiol.* 35:1413–1420.

453. **Weitao, T., K. Nordstrom, and S. Dasgupta.** 2000. *Escherichia coli* cell cycle control genes affect chromosome superhelicity. *EMBO Rep.* 1:494–499.

454. **Werner, M. H., G. M. Clore, A. M. Gronenborn, and H. A. Nash.** 1994. Symmetry and asymmetry in the function of *Escherichia coli* integration host factor: implications for target identification by DNA-binding proteins. *Curr. Biol.* 4:477–487.

455. **Wery, M., C. L. Woldringh, and J. Rouviere-Yaniv.** 2001. HU-GFP and DAPI co-localize on the *Escherichia coli* nucleoid. *Biochimie* 83:193–200.

456. **Westermark, M., J. Oscarsson, Y. Mizunoe, J. Urbonaviciene, and B. E. Uhlin.** 2000. Silencing and activation of ClyA cytotoxin expression in *Escherichia coli*. *J. Bacteriol.* 182:6347–6357.

457. **White, S. W., K. Appelt, K. S. Wilson, and I. Tanaka.** 1989. A protein structural motif that bends DNA. *Proteins* 5:281–288.

458. **White, S. W., K. S. Wilson, K. Appelt, and I. Tanaka.** 1999. The high-resolution structure of DNA-binding protein HU from *Bacillus stearothermophilus*. *Acta Cryst.* 55:801–809.

459. **Whittle, G., B. D. Hund, N. B. Shoemaker, and A. A. Salyers.** 2001. Characterization of the 13-kilobase *ermF* region of the *Bacteroides* conjugative transposon CTnDOT. *Appl. Environ. Microbiol.* 67:3488–3495.

460. **Wiater, L. A., and N. D. Grindley.** 1988. Gamma delta transposase and integration host factor bind cooperatively at both ends of gamma delta. *EMBO J.* 7:1907–1911.

461. **Williams, R. M., and S. Rimsky.** 1997. Molecular aspects of the *E. coli* nucleoid protein, H-NS: a central controller of gene regulatory networks. *FEMS Microbiol. Lett.* 156:175–185.

462. **Williams, R. M., S. Rimsky, and H. Buc.** 1996. Probing the structure, function, and interactions of the *Escherichia coli* H-NS and StpA proteins by using dominant negative derivatives. *J. Bacteriol.* 178:4335–4343.

463. **Wilson, R. L., S. J. Libby, A. M. Freet, J. D. Boddicker, T. F. Fahlen, and B. D. Jones.** 2001. Fis, a DNA nucleoid-associated protein, is involved in *Salmonella typhimurium* SPI-1 invasion gene expression. *Mol. Microbiol.* 39:79–88.

464. **Wold, S., E. Crooke, and K. Skarstad.** 1996. The *Escherichia coli* Fis protein prevents initiation of DNA replication from *oriC* in vitro. *Nucleic Acids Res.* 24:3527–3532.

465. **Worcel, A., and E. Burgi.** 1972. On the structure of the folded chromosome of *Escherichia coli*. *J. Mol. Biol.* 71:127–147.

466. **Wu, H., K. L. Tyson, J. A. Cole, and S. J. Busby.** 1998. Regulation of transcription initiation at the *Escherichia coli nir* operon promoter: a new mechanism to account for co-dependence on two transcription factors. *Mol. Microbiol.* 27:493–505.

467. **Wu, Y. F., and P. Datta.** 1992. Integration host factor is required for positive regulation of the *tdc* operon of *Escherichia coli*. *J. Bacteriol.* 174:233–240.

468. **Wu, Z., R. I. Gumport, and J. F. Gardner.** 1998. Defining the structural and functional roles of the carboxyl region of the bacteriophage lambda excisionase (Xis) protein. *J. Mol. Biol.* 281:651–661.

469. **Xu, J., and R. C. Johnson.** 1997. Activation of RpoS-dependent *proP* P2 transcription by the Fis protein in vitro. *J. Mol. Biol.* 270:346–359.

470. **Xu, J., and R. C. Johnson.** 1995. *aldB*, an RpoS-dependent gene in *Escherichia coli* encoding an aldehyde dehydrogenase that is repressed by Fis and activated by Crp. *J. Bacteriol.* 177:3166–3175.

471. **Xu, J., and R. C. Johnson.** 1997. Cyclic AMP receptor protein functions as a repressor of the osmotically inducible promoter *proP* P1 in *Escherichia coli*. *J. Bacteriol.* 179:2410–2417.

472. **Xu, J., and R. C. Johnson.** 1995. Fis activates the RpoS-dependent stationary-phase expression of *proP* in *Escherichia coli*. *J. Bacteriol.* 177:5222–5231.

473. **Xu, J., and R. C. Johnson.** 1995. Identification of genes negatively regulated by Fis: Fis and RpoS comodulate growth-phase-dependent gene expression in *Escherichia coli*. *J. Bacteriol.* 177:938–947.

474. **Yamada, H., S. Muramatsu, and T. Mizuno.** 1990. An *Escherichia coli* protein that preferentially binds to sharply curved DNA. *J. Biochem. (Tokyo)* 108:420–425.

475. **Yamada, H., T. Yoshida, K. Tanaka, C. Sasakawa, and T. Mizuno.** 1991. Molecular analysis of the *Escherichia coli hns* gene encoding a DNA-binding protein, which preferentially recognizes curved DNA sequences. *Mol. Gen. Genet.* 230:332–336.

476. **Yamashino, T., C. Ueguchi, and T. Mizuno.** 1995. Quantitative control of the stationary phase-specific sigma factor, sigma S, in *Escherichia coli*: involvement of the nucleoid protein H-NS. *EMBO J.* 14:594–602.

477. **Yamazaki, K., A. Nagata, Y. Kano, and F. Imamoto.** 1984. Isolation and characterization of nucleoid proteins from *Escherichia coli*. *Mol. Gen. Genet.* 196:217–224.

478. **Yang, C. C., and H. A. Nash.** 1989. The interaction of *E. coli* IHF protein with its specific binding sites. *Cell* 57:869–880.

479. **Yang, S. W., and H. A. Nash.** 1995. Comparison of protein binding to DNA in vivo and in vitro: defining an effective intracellular target. *EMBO J.* 14:6292–6300.

480. **Yang, S. W., and H. A. Nash.** 1994. Specific photocross-linking of DNA-protein complexes: identification of contacts between integration host factor and its target DNA. *Proc. Natl. Acad. Sci. USA* 91:12183–12187.

481. **Yang, W., and T. A. Steitz.** 1995. Crystal structure of the site-specific recombinase gammadelta resolvase complexed with a 34 bp cleavage site. *Cell* 82:193–207.

482. **Yang, W. Z., T. P. Ko, L. Corselli, R. C. Johnson, and H. S. Yuan.** 1998. Conversion of a beta-strand to an alpha-helix induced by a single-site mutation observed in the crystal structure of Fis mutant Pro26Ala. *Protein Sci.* 7:1875–1883.

483. **Yasuzawa, K., N. Hayashi, N. Goshima, K. Kohno, F. Imamoto, and Y. Kano.** 1992. Histone-like proteins are required for cell growth and constraint of supercoils in DNA. *Gene* 122:9–15.

484. **Yin, S., W. Bushman, and A. Landy.** 1985. Interaction of the lambda site-specific recombination protein Xis with attachment site DNA. *Proc. Natl. Acad. Sci. USA* 82:1040–1044.

485. **Yoshida, T., C. Ueguchi, H. Yamada, and T. Mizuno.** 1993. Function of the *Escherichia coli* nucleoid protein, H-NS: molecular analysis of a subset of proteins whose expression is enhanced in a *hns* deletion mutant. *Mol. Gen. Genet.* 237:113–122.

486. **Yuan, H. S., S. E. Finkel, J. A. Feng, M. Kaczor-Grzeskowiak, R. C. Johnson, and R. E. Dickerson.** 1991. The

molecular structure of wild-type and a mutant Fis protein: relationship between mutational changes and recombinational enhancer function or DNA binding. *Proc. Natl. Acad. Sci. USA* **88:**9558–9562.

487. Zacharias, M., H. U. Goringer, and R. Wagner. 1992. Analysis of the Fis-dependent and Fis-independent transcription activation mechanisms of the *Escherichia coli* ribosomal RNA P1 promoter. *Biochemistry* **31:**2621–2628.

488. Zechiedrich, E. L., A. B. Khodursky, S. Bachellier, R. Schneider, D. Chen, D. M. Lilley, and N. R. Cozzarelli. 2000. Roles of topoisomerases in maintaining steady-state DNA supercoiling in *Escherichia coli. J. Biol. Chem.* **275:** 8103–8113.

489. Zhang, A., and M. Belfort. 1992. Nucleotide sequence of a newly-identified *Escherichia coli* gene, *stpA*, encoding an H-NS-like protein. *Nucleic Acids Res.* **20:**6735.

490. Zhang, A., V. Derbyshire, J. L. Salvo, and M. Belfort. 1995. *Escherichia coli* protein StpA stimulates self-splicing by promoting RNA assembly in vitro. *RNA* **1:**783–793.

491. Zhang, A., S. Rimsky, M. E. Reaban, H. Buc, and M. Belfort. 1996. *Escherichia coli* protein analogs StpA and H-NS: regulatory loops, similar and disparate effects on nucleic acid dynamics. *EMBO J.* **15:**1340–1349.

492. Zimmerman, S. B., and L. D. Murphy. 1996. Macromolecular crowding and the mandatory condensation of DNA in bacteria. *FEBS Lett.* **390:**245–248.

493. Zimmerman, S. B., and S. O. Trach. 1991. Estimation of macromolecule concentrations and excluded volume effects for the cytoplasm of *Escherichia coli. J. Mol. Biol.* **222:**599–620.

494. Zuber, F., D. Kotlarz, S. Rimsky, and H. Buc. 1994. Modulated expression of promoters containing upstream curved DNA sequences by the *Escherichia coli* nucleoid protein H-NS. *Mol. Microbiol.* **12:**231–240.

495. Zulianello, L., E. de la Gorgue de Rosny, P. van Ulsen, P. van de Putte, and N. Goosen. 1994. The HimA and HimD subunits of integration host factor can specifically bind to DNA as homodimers. *EMBO J.* **13:**1534–1540.

496. Zulianello, L., P. van Ulsen, P. van de Putte, and N. Goosen. 1995. Participation of the flank regions of the integration host factor protein in the specificity and stability of DNA binding. *J. Biol. Chem.* **270:**17902–17907.

The Bacterial Chromosome
Edited by N. Patrick Higgins
© 2005 ASM Press, Washington, D.C.

Chapter 6

Domain Behavior and Supercoil Dynamics in Bacterial Chromosomes

N. Patrick Higgins, Shuang Deng, Zhenhua Pang, Richard A. Stein, Keith Champion, and Dipankar Manna

In the 50 years since the double helix was unveiled, giant strides have been made to decipher the primary sequence of complete chromosomes from organisms as diverse as mycoplasma and humans. The availability of many complete sequences changes the way that molecular science is performed. It is now possible to use evolutionary perspective to analyze many genetic and metabolic problems (see the introduction, chapter 1, and chapter 2). The era of microarrays (see chapter 4) makes it possible to understand regulation by accounting for the RNA expression abundance of most genes under various environmental conditions (10, 130, 150, 161). This progress stands in contrast to relatively few experiments that inform about how chromosomes are organized in a small space, what physical limitations govern DNA motion in living cells, and how newly replicated chromosomal copies keep from becoming inextricably tangled together. These physical problems were recognized in the early 1960s, once the implications of DNA and RNA structure were clear and their biochemical synthetic patterns were known (89). Understanding chromosome dynamics will be required to bring our models of transcription, replication, recombination, and chromosome segregation (chapters 2 to 5) up to standards that will be necessary for many new technologies in science and medicine.

A second letter closely followed Watson and Crick's report of a model DNA structure in which the authors explained important implications of their model. One concern was that a plectonemic (interwound) helical structure might require a chromosome to rotate at great speed during strand separation and therefore the products of replication and segregation would likely become tangled (see reference 62). Watson and Crick finessed:

As they (DNA strands) make one complete turn around each other in 34 Å, there will be about 150 turns per million molecular weight, so that whatever the precise structure of the chromosome a considerable amount of uncoiling would be necessary. It is well known from microscopic observation that much coiling and uncoiling occurs during mitosis, and though this is on a much larger scale it probably reflects similar processes on a molecular level. Although it is difficult at the moment to see how these processes occur without everything getting tangled, we do not feel that this objection will be insuperable.

J. D. Watson and F. H. C. Crick (149a)

The aims of this chapter are first to review classical results and incorporate a selected set of facts into models of chromosome dynamics. We have emphasized the topology of DNA during replication and transcription because these two processes exert the most dramatic topological influence on DNA. One focus is on structural properties of bacterial chromosomes that allow rapid adaptability. The final section discusses chromosome structure considered from a stochastic as opposed to a highly ordered perspective. We consider several examples in which DNA movement aids macromolecular assembly much as molecular chaperones aid protein folding.

THE NUCLEOID

The view of a bacterial cell as a bag of enzymes that work through the laws of chemical mass action has been eclipsed by new models in which complex

N. Patrick Higgins, Shuang Deng, Zhenhua Pang, Richard A. Stein, Keith Champion, and Dipankar Manna • Department of Biochemistry and Molecular Genetics, University of Alabama at Birmingham, 720 20th St. South, Birmingham, AL 35294-2170.

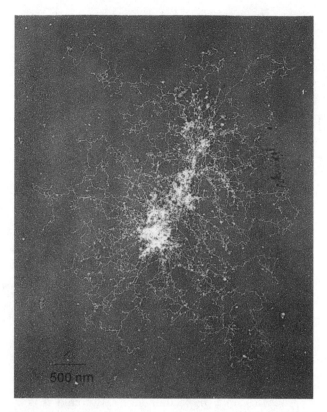

Figure 1. Image of an *E. coli* chromosome. Exponential cells treated with lysozyme and Brij 58 in the presence of 1 M NaCl were layered onto a 10 to 30% sucrose gradient and subjected to centrifugation. An aliquot of the fraction with highest DNA content was applied to a glow-discharged carbon-coated grid, stained with uranyl acetate, rotary shadowed, and photographed using a JEOL 1200EXII electron microscope. Micrograph by Christine Hardy, University of California at Berkeley.

and finely regulated protein machines catalyze the processes of DNA replication, DNA repair, DNA recombination, RNA transcription, and cell division (5, 37). We now think of DNA as thread going through stationary machinery rather than imagine enzymes like RNA and DNA polymerase whirling around the double helix (35, 81, 82, 120). Merging the laws of mass action with the new perspectives on enzyme architecture and cellular protein localization is required to improve our understanding of global chromosome behavior.

In the early 1970s, methods were developed to release bacterial chromosomes, which were called nucleoids, from the confines of the rigid peptidoglycan cell wall and cell membrane. The analysis of these structures employed both physical and visual techniques. When a culture of *Escherichia coli* was grown in defined liquid medium, treated with lysozyme, and then lysed with a mild ionic detergent, the bacterial chromosome could be liberated in a compact form that was not viscous in dilute solution. Fig. 1 shows a

contemporary image captured by electron microscopy. This type of chromosome preparation is called a nucleoid because the bacterial DNA is not contained in a complete encircling membrane like a true eukaryotic nucleus.

The pioneering work of Stonington and Pettijohn in 1971 (135) and of Worcel and Burgi in 1972 (153) was followed by studies in other laboratories that demonstrated several intriguing properties of nucleoids. Sucrose gradient sedimentation was one method for analyzing nucleoids, and three observations linked the sedimentation behavior of nucleoids to important biochemical processes. First, the sedimentation rate of nucleoids increased (measured as an S value) as cells increased in size. When cells were starved for an essential amino acid, nucleoids sedimented at 1,300S. However, once cells were allowed to start growing by addition of limiting nutrients, nucleoid sedimentation values gradually increased to 2,200S, and then a new 1,300S peak emerged. Thus, nucleoids reflected the replication status of the chromosome. Second, a compact nucleoid was altered by incubation with nonspecific single-strand nucleases like pancreatic DNase I. The introduction of a limited number of single-strand nicks caused nucleoid sedimentation values to decrease in gradual stages. The number of nicks needed to unfold the bacterial nucleoid was estimated to be between 5 and 30 (Table 1). This behavior was explained by the work on plasmid DNA structure in the Vinograd laboratory. They showed that plasmid DNA from many sources could be isolated in two distinct forms that were called I and II. Form I had the faster sedimentation value due to supercoiling, and a single chain break caused plasmid DNA to expand and sediment as a relaxed structure (form II) (143, 144).

Nucleoid S values were also sensitive to the addition of ethidium bromide, an agent that intercalates into double-stranded DNA and unwinds the Watson-Crick helix. However, unlike nucleases that only decreased S value, ethidium bromide titration exhibited a critical phase change. At low concentrations, ethidium bromide caused the S value to fall, like nuclease digestion. At higher concentrations, ethidium bromide intercalation caused sedimentation values to increase, reaching values that exceeded the value of untreated samples. These classic experiments proved nucleoids to be highly supercoiled entities, and the ethidium concentration producing the minimum S value reflected the chromosomal superhelix density (33). Nucleoids were understood as multiply looped structures with 20 to 30 domains per genome equivalent (Table 1).

Transcription (or the presence of nascent RNA) was implicated in maintaining a folded nucleoid

Table 1. Thirty years of supercoil domain studies in vitro and in vivo: evidence of short- and long-range interaction in a highly dynamic yet organized bacterial chromosome whose structure is determined by biochemical functionality[a]

Report	Method	Domains per GE	RNA effects
Worcel and Burgi, 1972 (153)	In vitro DNase I and EtBr	4–30	RNase sensitive
Pettijohn and Hecht, 1973 (113)	In vitro DNase I	7–30	RNase sensitive in vitro; rifampin sensitive in vivo
Kavenoff and Bowen, 1976 (71)	In vitro EM	50	RNase sensitive in vitro
Sinden and Pettijohn, 1981 (127)	In vivo psoralen cross-linking	40 uniform[b]	Rifampin resistant in vivo
Higgins, 1999 (52a)	In vivo Tn3 WT Res	150 stochastic	Constitutive promoters block resolution
Deng et al., 2004 (32)	In vivo γδ Res-SsrA	350 stochastic	~30 operons form domains in vivo

[a]Abbreviations: GE, genome equivalent; EtBr, ethidium bromide; EM, electron microscopy; WT, wild type.
[b]The model used by Sinden and Pettijohn assumed that all domains would be of equal size and have equivalent exposure to X-ray-induced breaks.

structure in vitro (111, 113). When nucleoids were incubated with RNase, they lost their high sedimentation character. Similarly, nucleoids became unfolded if cells were treated with rifampin prior to lysis. These experiments raised the hope that some form of RNA might serve as an organizing factor of chromosomal structure. However, failure to discover a specific organizational RNA species left RNA in scientific limbo.

Loop organization of bacterial chromosomes was supported by electron microscopy (31, 48, 71, 72), which confirmed the presence of extended interwound branches of supercoiled DNA. Although statistical analyses were not included, the loop numbers in some published micrographs seemed consistent with estimates of domain numbers measured by limited DNase treatments. Both supercoiled and relaxed loops were evident (72). However, in preparations used for sedimentation analysis, like the one shown in Fig. 1, an important limitation can be inferred from the amount of visible DNA. Less than 15% of cellular DNA becomes extended as interwound looped structures. This image from the center of the DNA mass raises concern that the release of DNA from the dense peptidoglycan core and inner and outer membranes is incomplete (see models below). Are the visible loops a statistically relevant sample of the entire chromosome structure, or do forces that have no in vivo relevance cause them?

CHROMOSOMAL PROTEINS

Work on chromosome behavior stimulated biochemical searches for molecules that condense DNA. In eukaryotes, DNA supercoiling is stabilized by the tight association with nucleosomes, which wrap two turns of DNA as solenoidal loops around the surface of a histone octamer (Fig. 2). High-resolution X-ray crystal structures show that, in addition to looping, DNA in a nucleosome is underwound relative to the B conformation of DNA (119). Removing the histones from covalently closed DNA molecules releases freely diffusible interwound supercoils. In bacteria, DNA supercoiling has both dynamic and static components. Roughly half of the negative supercoils are constrained by association with a group of genetically conserved bacterial proteins (see chapter 5) and half exist as dynamic interwound supercoils (13, 114). The division of negative supercoiling into these two components is under genetic control (55, 114) (Fig. 2).

How are chromosomes organized into a matrix of constrained and unconstrained supercoils? The answer is not completely defined in any bacteria, but in *E. coli* four topoisomerases work in concert with at least 10 abundant proteins to organize chromosome structure in vivo. Topoisomerase I (Topo I) (the ω protein) was the first enzyme discovered with the ability to change DNA linking number (148). The biochemical properties of this enzyme led eventually to our current understanding of four topoisomerases that operate in gram-negative bacteria (25, 55, 115, 145, 147). DNA supercoiling is a form of energy in which the force required to form a new supercoil is an exponential function of superhelix density. Dynamic and diffusible interwound negative supercoils are regulated by three topoisomerase activities (159). Negative supercoils generated by the ATP-dependent catalytic mechanism of DNA gyrase are balanced by the supercoil relaxing activity of Topo I, encoded by *topA*, and Topo IV, which is related to gyrase and is encoded by the *parE* and *parC* genes (95, 159).

One protein that constrains DNA structure is RNA polymerase ($\alpha_2\beta\beta'$) with a molecular mass of 400,000 Da (see chapter 15). During exponential growth about 1,500 molecules of RNA polymerase are engaged in transcribing different genes around the genome (see reference 42). Each enzyme constrains about 1.7 supercoils in the form of a denatured segment of the DNA template (44, 77, 158). A second protein that constrains topology is gyrase. About 1,000 gyrase molecules are bound to chromosomal

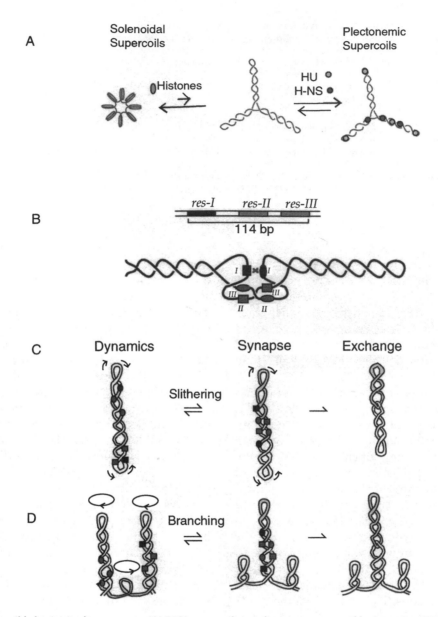

Figure 2. Supercoil behavior in chromosomes. (A) DNA supercoils in eukaryotes are created by wrapping DNA on the surface of the highly conserved histone octamer. Eubacteria introduce supercoils enzymatically with gyrase, which causes DNA to adopt the interwound or plectonemic supercoiled conformation. Proteins like HU and H-NS restrain half of bacterial supercoiling. (B) A 114-bp *res* site includes three subsites, I, II, and III, that each bind a resolvase dimer. Recombination occurs within the *res* I site. A three-node synapse precisely juxtaposes two *res* I sites for catalytic exchange within a protein-DNA recombination complex. Slithering (C) and branching (D) are movements that allow synapse of two *res* sites.

DNA, and gyrase wraps positive supercoils on its surface (74) and enzymatically converts them to negative supercoils (18). The remaining topology is associated with a group of proteins that include HU, IHF, FIS, H-NS, STPA, and SEQA (see chapters 5 and 11). In addition to these small proteins, a three-protein complex of MukB, MukE, and MukF (121, 151), which share their evolutionary history with eukaryotic condensins (50), plays a still undefined role in condensation and segregation.

The diffusible component of negative supercoiling in bacteria represents about half of the total value of sigma (37 to 43% in different *E. coli* strains [67]). A compelling example of the importance of mobile interwound negative supercoiling was discovered in biochemical studies of phage Mu transposition. Craigie and Mizuuchi found that in vitro synapsis of phage Mu ends, which is required for transposition, requires negative supercoiling and a specific orientation of sites (26). Recently a precise structure of this protein-DNA complex, also called a plectosome, was determined (107) (see below). Many site-specific recombination systems rely on mobile interwound supercoiling to form synaptic intermediates (Fig. 2).

DOMAIN BOUNDARY SEARCHES

In vitro nucleoid studies stimulated searches for the controlling elements of domain behavior. In 1980 Sinden et al. reported a scheme to measure torsional effects of both single- and double-chain breaks (126). They combined X-ray treatment, which breaks DNA inside living cells, with UV-induced psoralen cross-linking to study supercoiling in vivo. This assay was based on the fact that psoralen intercalates more readily into torsionally strained, negatively supercoiled DNA than into relaxed DNA. By measuring the number of psoralen-DNA adducts as a function of X-ray dose, they simulated the in vitro nucleoid DNase nicking assay (see above). Calculations indicated that the circular chromosomes in living cells had 40 independent domains (112, 127). The calculation assumed that all domains are equal in size and the probability of an X-ray hit in every domain is equivalent.

Sinden and coworkers' results encouraged other laboratories to search for a mechanical explanation of domains. Because the expression of several genes responds to changes in supercoiling levels (52), two groups sought to identify chromosomal regions that maintain different superhelical densities. Supercoil-sensitive promoters coupled to easily quantified reporter genes were used to search for domains. Supercoiling behavior was remarkably uniform throughout the chromosomes of both *E. coli* (108) and *Salmonella enterica* serovar Typhimurium (97). Aside from expression differences that could be attributed to gene dosage effects of placing modules near the origin or terminus, no statistically significant domain differences were found. These studies proved that topoisomerases (see above) do an excellent job in maintaining an average distribution of σ throughout the genome irrespective of domain boundaries.

In 1996 a new assay permitted analysis of DNA dynamics inside living cells (56). This assay used γδ resolvase and 114-bp resolution sites (*res*) that could be introduced into the chromosome at specified locations. If two *res* sites form a plectonemic synapse by supercoil-dependent slithering or branching (Fig. 2C and D), then a recombination event that deletes the intervening DNA sequence occurs. The assay monitors the disappearance of a drug resistance gene, the change in a colony color phenotype, the

appearance of a new restriction fragment, or a novel PCR product generated in a recombinant chromosome (132). Systematic resolution assays spanning 2% of the genome demonstrated that recombination efficiency follows a first-order decay with respect to DNA distance. In log-phase cells, the resolution efficiency declines 50% for every 20-kb increment separating two sites. This first-order decay rule indicates that domain barriers are random in the genome. For example, a population of log-phase bacteria includes some cells with 100-kb domains and other cells with 10-kb domains for the same interval. These results also raised the estimated number of domains from 40 to 150 (Table 1).

We provide below several models of chromosome structure that attempt to link classic and recent experimental results. (For different perspectives see references 9 and 112.) Because bacteria exist in environments that can demand rapid change, the ability to modulate chromosomal structure on different time scales is emphasized. In the next sections, chromosomal behavior relevant to DNA replication and RNA transcription is discussed.

DNA REPLICATION

DNA synthesis presents the cell with several topological challenges. One problem is that chromosome duplication requires the separation and complete unlinking of the template Watson and Crick strands. For a circular molecule the size of the *E. coli* chromosome (4.64 Mbp) there are 440,000 links between complementary DNA strands and numerous links between sister chromosomes. These links must be completely eliminated to allow replication and segregation. Two forks initiate at *oriC* and then traverse DNA in opposite directions at a rate of 800 nucleotides per second, terminating near the *dif* site (78). Genome halves with opposing replication polarity are called replichores (12). For bacteria growing in rich medium, cells can reinitiate replication prior to chromosome segregation. Under this condition, daughter cells inherit one complete and multiple incomplete chromosomes that can be enumerated using flow cytometry (128). Significant issues in replication involve topological problems in front of the fork, problems behind the fork, and problems at the replication terminus or other places where two forks come close together.

Problems before a Fork

A replisome represents a mobile domain boundary that separates two topological conditions. Positive supercoils in front of the fork are caused by the replication-associated helicase, which causes DNA overwinding (92). Negative supercoils behind the fork must be generated to condense DNA and reestablish the correct balance of constrained and unconstrained negative superhelicity (22, 88, 155, 157). The positive supercoils ahead of the DnaB helicase must dissipate to allow fork movement. Because of the impressive speed of fork progression, all impediments to DNA unwinding, including domain boundaries and double-strand-specific DNA-binding proteins, must be removed at least transiently to allow the helicase and associated replication proteins to pass. Some proteins may survive replication fork progression without becoming displaced. For example, eukaryotic nucleosomes appear to survive in vitro replication with the T4 replisome (16). A more complex situation exists with RNA polymerase. In bacteria, RNA polymerase moves at a rate of about 50 bp/s, which is less than one-tenth the speed of a replication fork. When RNA polymerase is overtaken by the T4 polymerase moving in the same direction, RNA polymerase appears to survive the replisome's passage, remaining bound to both the DNA template and its RNA transcription product (86, 87). New structural data on a transcribing RNA polymerase makes this difficult to understand (see chapter 14). Nonetheless, a head-on collision between a replisome and a transcribing RNA polymerase is a dramatic situation (85). French found that an efficient promoter oriented in the opposite direction of DNA synthesis fork movement caused a delay in DNA replication (41). Most of the sequenced genomes, which include *E. coli* and *S. enterica* serovar Typhimurium (12, 93), have operons that produce abundant RNAs oriented in the direction of replication fork progression.

The sole example of a persistent natural barrier to replication fork progression in *E. coli* is the Tus-Ter complex (57, 59, 60). When complexed with the Tus protein, Ter sites are directional gates that allow replication forks to pass in one direction and cause the DnaB helicase to stop or pause in the opposite direction (6, 73, 90, 129). Numerous Ter sites are organized so that forks move unimpeded from *oriC* to *dif* in each replichore (58); forks stall only when they extend synthesis through the *dif* site and begin to replicate a segment of chromosome in the direction from *dif* toward *oriC*. When a Ter site is placed into a replichore in the unnatural orientation, it extends the time required to complete genome replication (124). Moreover, a Tus-Ter complex in the wrong orientation stimulates the production of chromosomal deletions (11) and generates small amplified circular DNAs that include sequences adjacent to the stall point (76). This type of amplified DNA has been called a "hot segment." Formation of similar circular DNA in the ribosomal genes of yeasts results when replication forks encounter the Fob1 protein, which also acts as a barrier to replication fork movement (69, 75). Therefore, in both eukaryotes and prokaryotes, free-moving replication forks are important for maintaining the genetic stability of the chromosome. Examples of different types of barriers caused by normal cell biochemical reactions are discussed in chapter 24.

Behind the Fork

The topological problem behind the fork is negative supercoiling. The pattern of replication is semidiscontinuous because chain growth on the strand that grows 5′-3′ in the same direction as fork movement can be highly processive. On the opposite strand, replication is discontinuous and requires multiple primers and a DNA repair process to stitch Okazaki fragments together (see chapter 10). On the discontinuous strand (and possibly on the continuous strand as well) all negative supercoiling is lost to rotation at nicks. Supercoil diffusion is fast enough that, without domain barriers, negative supercoiling would not exist in chromosomes undergoing DNA synthesis, and gyrase would cycle futilely.

How many domains insulate chromosomes from negative supercoiling loss behind forks? New technology increases the domain number 10-fold relative to the classic view of 40 domains per genome equivalent (Table 1; see also discussion below). One region of serovar Typhimurium has been studied in detail. This region includes two large operons, *cob* and *pdu*, which are not efficiently transcribed when cells grow aerobically in rich medium (4, 14, 20, 21, 146). Three generalizations of the *his-cob* interval have been extended to 2% of the chromosome (43). First, supercoil domains are more abundant when cells are undergoing DNA replication than when DNA replication is suppressed. As cells enter stationary phase, the resolution efficiency increases continuously for about 2 days (56). Second, the resolution efficiency diminishes as a first-order function of distance along the chromosome. The characteristic that describes a first-order function for resolution is 1/2 D, which is the distance along the chromosome causing a 50% reduction in resolution efficiency. A first-order rule for barriers is a strong indication that barriers occur stochastically throughout the interval (56, 132). More is explained about stochastic chromosome behavior in a separate section below.

Third, the number of domains detected depends on the time period of the assay. The 1/2 D for

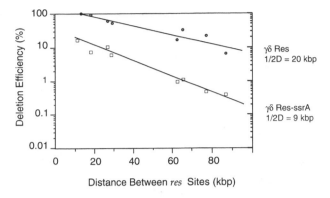

Figure 3. Resolution analysis in a 100-kb interval using two resolvase enzymes, the WT Res (top curve) with a half-life of more than 1 h or the modified Res-SsrA protein (bottom curve), which is a substrate for the ClpXP protease and has a 5-min half-life in exponential cultures of *E. coli* or *Salmonella*.

wild-type (WT) γδ Res protein (Fig. 3, top curve) is 20 kb. Because the γδ Res protein is stable, it has the potential to catalyze resolution for more than 1 h after it is made. Therefore, barriers lasting 10 min are underestimated using the wild-type enzyme to assay chromosome dynamics. An assay with a shorter time interval was made possible by decreasing the Res protein life span (R. Stein, S. Deng, and N. P. Higgins, submitted for publication). By inserting an 11-amino-acid sequence tag of the *ssrA* protease system onto the C terminus of the resolvase (Res-SsrA), an enzyme with a 5-min half-life was obtained (32). Using the Res-SsrA protein, a 7- to 12-min window of chromosome movement can be analyzed (Fig. 3, bottom curve.) Assays with Res-SsrA produce a first-order decay curve, but the 1/2 D is reduced from 20 to 9 kb. Extrapolating from the *his-cob* interval to the rest of the chromosome, the median domain size measured in a 10-min resolution window is 15 kb, which predicts > 350 "domains" per genome equivalent. This estimate is 10-fold higher than the early measurements of Sinden and Pettijohn, who found 40 domains of about 100 kb using psoralen cross-linking (127). Two considerations may partly explain the differences. (i) The initial slope of the X-ray data was used to estimate domain size. In this part of the curve, the largest domains dominate because they yield the biggest change in supercoiling. Because their model assumed all domains would be of equal size, it leads to low accounting. A complex tail region of their data set, which could have many small domains, was not considered. (ii) The time factor is difficult to estimate in these experiments. As shown above for γδ resolvase, barriers are transient, so longer times of assay tend to underestimate barrier frequency and inflate domain size. Evidence for a

chromosome with 400 domains per genome equivalent also comes from two independent experimental estimates, one measuring loop size in nucleoids like the one shown in Fig. 1 (C. Hardy, personal communication) and another measuring the impact of in vivo restriction nuclease expression on supercoiling sensitive promoters (L. Postow, personal communication).

Stalled Forks and Problems at the Terminus

In a circular *E. coli* chromosome, forks from *oriC* traverse DNA divergently at 800 nucleotides per second, generating positive supercoils ahead of the fork that must dissipate to allow fork movement. The chromosome has a supercoil density of σ = −0.06, so 26,000 negative supercoils per genome equivalent lie ahead of the two forks at the start of replication. In addition to a reservoir of negative superhelicity, 1,000 molecules of DNA gyrase are randomly distributed on the chromosome. Two forks generate 9,000 positive supercoils per minute, whereas one gyrase molecule can generate 100 negative supercoils per minute (54). Thus, collectively, gyrase can produce 100,000 negative supercoils per minute. This supercoiling potential will support fork movement at early and middle stages of replication. However, as two replisomes converge, or when one fork becomes stalled, the number of gyrase molecules (and any other topoisomerases) between two forks becomes restricted (Fig. 4). Eventually there will not be space for 90 gyrase molecules to allow maximal fork progression as it converges on the terminus. A negative-supercoil deficit could influence the replication rate whenever two forks (or a fork and a transcribing RNA polymerase) converge.

There are three nonexclusive solutions to the supercoiling dilemma: (i) positive supercoils may diffuse behind the fork to become precatenanes, (ii) an enzyme other than gyrase may dissipate positive supercoils in front of the fork, and (iii) forks may slow down to allow gyrase time to catalyze negative supercoiling.

Evidence for precatenanes in plasmid DNA has been provided by stalling forks with Tus-Ter complexes in vivo (110). This experiment demonstrates the appearance of precatenanes, and serves as proof of principle. However, stalling a fork could promote supercoil diffusion to allow precatenane formation. Khodursky et al. showed that either gyrase or Topo IV allowed fork progression over most of the genome (73a). Only when both gyrase and Topo IV were simultaneously inactivated did replication stop quickly, suggesting that both enzymes can provide a "swivel" for the replisome. Crisona et al. (28) used braids of

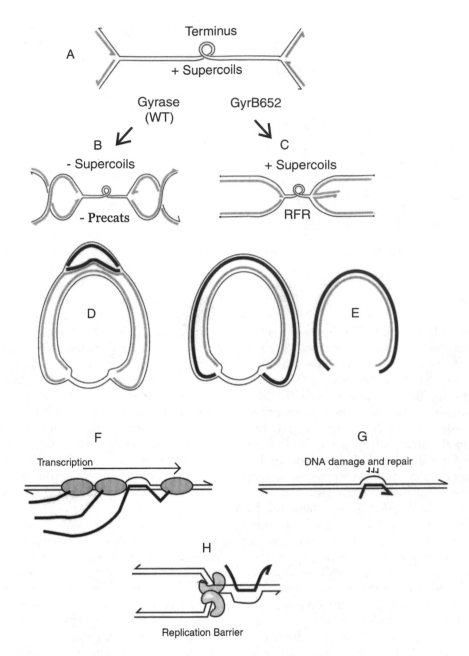

Figure 4. Topological problems at the terminus. (A) As replication forks converge, positive supercoiling builds up to slow fork progression. (B) In WT strains, gyrase can remove the topological barrier, but gyrase mutants may fail to complete DNA synthesis due to fork regression (C) or replication-mediated linearization of one chromosome (D and E). An R loop impedes further transcription (F), exposes the displaced strand of DNA to chemical and enzymatic attack (G), and creates a barrier to replication fork progression (H).

duplex DNA and showed that Topo IV removes positive supercoils very efficiently. However, the cellular location of Topo IV and the timing of how and when this enzyme is at maximal catalytic efficiency are under debate. One model posits that Topo IV is fully active only in terminal stages of replication (39).

Evidence for special characteristics of replication at the terminus comes from numerous observations: (i) Kuempel analyzed DNA synthesis using

CsCl density gradients. Cells grown in heavy medium were shifted to light medium. They used density anomalies generated by recombination at the *dif* site to estimate the frequency of site-specific recombination at *dif*. In the course of these analyses, they found that the *dif* site in dimer chromosomes did not recombine until after most cells had undergone cell division (51). Either replication or recombination at *dif* lags well behind the average time for first-generation cell division. (ii) Li et al. analyzed the distribution of chromosome segments using fluorescent tags that identify both the *oriC* region and the *dif* region (83). They observed many cells with four origin copies, but very few cells (less than 1%) with two *dif* regions. A late replication time for the *dif* segment in dimer chromosomes or replication followed by cohesion are possible explanations for these results (51, 84). (iii) A region of about 50 to 100 kb immediately surrounding the *dif* site has special properties related to XerC/D site-specific recombination (79). This region is susceptible to frequent chromosome breakage, which stimulates homologous recombination (109) (see chapter 13). (iv) Although gyrase is not needed to decatenate the products of replication (30, 160), experimental data suggest an important gyrase role(s) in chromosomal segregation (61, 133). The terminus is susceptible to dramatic loss of diffusible negative supercoiling, especially when a cell carries certain alleles of gyrase (Z. Pang, R. Chen, and N. P. Higgins, submitted for publication). These gyrase mutants are synthetically lethal with recA; they show constitutive induction of the SOS response and exhibit significant RecBCD-dependent degradation of chromosomal DNA (45).

Two explanations of replication anomalies near the terminus can fit the complex phenotypes described above. First, the positive supercoils before converging forks may stimulate replication fork regression, a reaction in which the nascent strands anneal together and branch migration forms a Holliday-like junction or "chicken foot." Fork regression was discovered in 1976 (53), but Postow et al. demonstrated that positive supercoiling can cause replication fork reversal for complexes stalled by the Tus protein at *ter* sites on plasmid DNA (116). Evidence for in vivo replication fork reversal of UV-damaged plasmid DNA has also been reported (24). Fork regression at the terminus may be more frequent when gyrase activity is impaired (Fig. 4C). The consequences of replication fork reversal include degradation of nascent DNA by the RecBCD nuclease and resolution by Holliday junction enzymes requiring replication fork restart (see chapters 12 and 20). Second, because replication initiation frequency is determined by growth rate, replication forks may overtake terminal synthesis (Fig. 4E). This could be especially important under conditions where gyrase activity is suboptimal. Replication overrun would generate nearly complete linear chromosomes (49; Pang et al., submitted). The intermediates in Fig. 4C and E would account for a supercoil deficiency at the terminus, would account for RecBCD degradation, and would require mechanisms to restart replication forks or lose large linear replication products.

TRANSCRIPTION EFFECTS ON DOMAIN BEHAVIOR

Bacteria can respond within seconds to changes in environmental conditions. This requires great flexibility in chromosome structure because transcription shares with replication some similar topological problems. The RNA polymerase of *E. coli* is a large protein (see chapter 15). Twin domains of supercoiling were originally described during transcription, which induces positive supercoils before the enzyme and negative supercoils in the enzyme's wake (22, 23, 44, 88, 155). Assays of torsional strain on DNA confirmed this effect in plasmids (117). However, little is known about the impact of transcription on the topological state of the large bacterial chromosome. Recent research shows that transcription can cause formation of new topological domains (32) and that strong promoters cause transcription-associated recombination and transcription-associated mutations (34).

The most frequent barriers detected by γδ analysis are stochastic and not explained by any distinguishable chromosomal feature; they exist within regions refractory to gene expression (i.e., regions lacking highly transcribed genes). When a 40-kb Mu prophage, which expresses no genes in the lysogenic state, was introduced into the *his-cob* interval, stochastic barriers were just as likely to exist within Mu as in the adjoining *cob* operon region. Scheirer discovered that transcription from the strong early promoter of Mu caused the appearance of a new barrier to Tn3 resolution in the *his-cob* region (122). This effect on resolution was not random; rather, resolution reactions were inhibited between a *res* site near the end of Mu and all *res* sites downstream of the transcription flow. However, resolution involving the *res* site near the Mu end and a *res* site upstream of the induced transcript was unperturbed. This showed that transcription alters plectonemic DNA structure in a bacterial chromosome.

The Mu early promoter is comparable in strength to the near-consensus phage λ promoter p_L. Once the impact of a strong promoter was realized, the question became, how significant is transcription as a general force in channeling chromosome movement in vivo? Deng et al. analyzed transcription by modulating gene expression of a *lacZ* reporter at different chromosomal locations using the TetR repressor system (32). A 14-kb interval in the serovar Typhimurium chromosome is shown in Fig. 5. Here, the TetR repressor modulates expression of *lacZ* followed by a strong transcription terminator from the ribosomal operons (here abbreviated *rrnB*). Expression of the Res protein stimulates site-specific

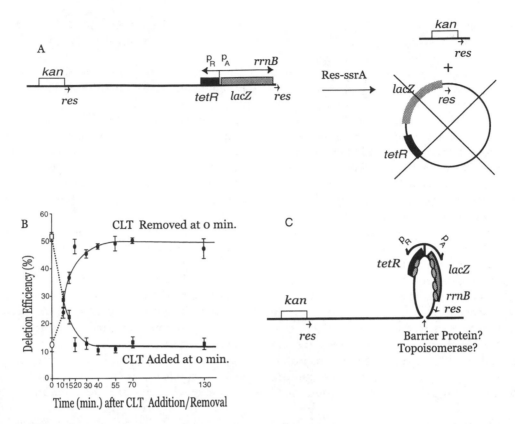

Figure 5. Resolution efficiency in a 14-kb segment of the *S. enterica* serovar Typhimurium chromosome changes dramatically depending on the state of *lacZ* expression. (A) Genetic map of a 14-kb deletion interval, which leads to a deleted circle containing a Tet-regulated copy of *lacZ*. (B) Effect of transcription on site-specific resolution. Cultures of bacteria harboring the genetic interval shown in panel A were grown in log phase with continuous presence (open square) or absence (open circle) of 5 µg/ml CLT. At time 0, CLT was added (solid squares) or washed out (solid circles) of the medium, and cells were exposed to a 10-min expression period of a resolvase with a 5-min half-life. Recombination efficiency is plotted against the time of transcription induction or repression (see reference 32). (C) Model for transcription-induced domains in bacterial chromosomes. Addition of CLT causes RNA polymerase binding at promoters p_R and p_A and transcription of the *tetR* and *lacZ* genes. Persistent transcription induces formation of a domain in which an unknown protein(s) stabilizes a loop that isolates DNA associated with the transcribing RNA polymerases (ovals) from the rest of the genome. Inclusion of the *res* site adjacent to the transcription terminator (*rrnB*) inhibits recombination with the *res* site near the *kan* gene to the left.

recombination, producing a bacterial chromosome with a single *res* site next to the *kan* gene plus a recombinant circle with the *lacZ* reporter that cannot replicate. Addition of chlorotetracycline (CLT) stimulates transcription from both the p_R promoter, which controls *tetR*, and p_A, which controls *lacZ*. After addition of CLT, the deletion efficiency was reduced by half after 10 min and fivefold after 20 min (Fig. 5B). Inhibition of resolution remained strong as long as transcription remained fully induced.

When CLT was washed out of the culture, transcription of *lacZ* stopped and the barrier to resolution disappeared over a 10- to 20-min interval. Thus, a block to resolution caused by transcription responds to but is not contemporaneous with transcription. One model for how this occurs is shown in Fig. 5C. During high transcription, the *res* site adjacent to the *lacZ* gene becomes organized in a DNA

loop that prevents its plectonemic entanglement with the *res* site next to the *kan* gene. What proteins (or other factors) organize the loop is unknown, but one attractive candidate is the condensin-like protein complex MukB, MukE, and MukF, perhaps working together with gyrase (132).

The magnitude of inhibition of resolution depends on the level and persistence of transcription. For WT cells growing in rich medium, 80% of the active RNA polymerase molecules synthesize rRNA and tRNA, which represent <1% of the genome and account for over 85% of total cellular RNA (17). This result was visually illustrated in electron micrographs by French and Miller, who showed that bacterial chromosomes had many transcribing RNA polymerases at the ribosomal and tRNA operons but that most of the chromosome was barren, having one transcribing polymerase per 20 to 40 kb (42).

Microarray assays of transcription output for all genes in the chromosomes of *E. coli* confirm that about 90% of the transcriptional output is focused on less than 3% of all genes (150). A back-of-the-envelope calculation suggests that transcription-driven barriers that inhibit resolution by fourfold or more would be found at only about 20 to 40 chromosomal locations. Together with the stochastic barriers associated with replication discussed above, about 400 domains probably exist in the average *E. coli* or serovar Typhimurium chromosome. However, at any point in time, each organism would have a unique set of chromosome domain coordinates. The involvement of transcription in domain formation is not limited to bacteria. Microarray analyses of *Drosophila melanogaster* have recently shown that 200 clusters of genes are coregulated and probably reside in transcriptionally modulated domains (102, 131).

TRANSCRIPTION-ASSOCIATED RECOMBINATION AND MUTATION

One aspect of transcription is different from replication. During transcription, the single-stranded RNA that is generated carries the potential to make a stable complex with DNA by forming a DNA-RNA heteroduplex, or R loop. For genes that are transcribed at a high rate, this has important consequences for chromosome physical and genetic stability. Transcription was first noted to influence recombination in studies of phage lambda (65). Strong transcription of the early genes of lambda caused a dramatic increase in the frequency of crossovers in the transcribed region. This hyperrecombination phenotype depended on chain elongation rather than initiation, and *rho* mutants, which make longer than normal transcripts, had a strong influence on this phenomenon. A similar effect was reported to occur in the yeast *Saccharomyces cerevisiae* 10 years later (138).

Mellon and Hanawalt discovered another effect of transcription—a selective increase in DNA repair (94). Bacterial plasmids were also found to form transcription-associated deletions (141, 142). A unifying mechanism in both prokaryotes and eukaryotes has recently been proposed to account for these observations (3, 34). During strong transcription, which is particularly relevant during the generation of long transcripts, the RNA sometimes associates with the template DNA. This may or may not involve negative supercoiling caused by the action of RNA polymerase (see above), but once the R loop forms, it prevents other RNA polymerases from completing their transcripts (Fig. 4F). In both prokaryotes and eukaryotes, Topo I is involved in stopping or reversing R-loop formation. Topo I mutants of *E. coli* are very sick, and the sickness can be suppressed by overexpressing RNase H, a nuclease that degrades only RNA that is in complex with DNA (63). Huertas and Aguilera showed in vivo that R loops impeded transcription and stimulated transcription-associated recombination in *S. cerevisiae* (64). Formation of persistent DNA-RNA hybrids is even involved in one case of developmentally programmed recombination. Transcription of the immunoglobulin switch region generates stable R loops in vitro (118) and in vivo (98) during immunoglobulin recombination.

DNA-MEMBRANE ATTACHMENTS

The complex conditions required to generate nucleoids lead to problems in analysis. Cells treated with lysozyme are notoriously difficult to lyse completely, and remnants of the cell wall are difficult to analyze. The image in Fig. 1 was isolated as a "salt-stabilized" nucleoid from the center of the DNA peak in a nonviscous nucleoid preparation. Although numerous interwound loops emanate from a central core, the fraction of cellular DNA extended into loop structures is small. More than 80% of the DNA remains condensed. Is the exposed fraction of cell DNA some specific fraction or a random sample of the population? Systematic statistical studies were not classically done to compare images of nucleoids treated with DNase I, RNase, or various concentrations of ethidium bromide. However, estimates by Hardy (Fig. 1) on the extended loops shows large size variation; assuming the loops reflect in vivo structure, they agree with a 400-domain chromosome (C. Hardy, personal communication).

A landmark manuscript of 1963 proposed a mechanism to coordinate regulation of plasmid and chromosome replication in a bacterial cell (66). Jacob et al. proposed rules for initiating DNA replication in which a diffusible initiator protein (now known as DnaA protein) binds to a unique replicator DNA sequence (now known as *oriC*) to coordinate the timing of DNA synthesis (see chapter 9). This paper also recognized the difficult problem of chromosome segregation and included a hypothesis for membrane attachment to allow efficient chromosome partitioning between two daughter cells. The insights of this paper into mechanics of chromosome initiation were prescient, but assumptions about membrane biogenesis were naive. Understanding chromosomal membrane attachment has been a difficult challenge for the past 40 years.

Early nucleoid experiments in the Worcel lab showed that, if cells were lysed at low temperature, bacterial nucleoids had a visible membranous central core (31, 154). This apparent nucleoid-membrane association required cells to be engaged in both transcription and protein synthesis prior to cell lysis. Other experiments showed that chromosomal DNA could be isolated in a form firmly attached to membranes, and, surprisingly, both inner and outer membranes were present. The Schaechter lab used the M-band technique (lysis with lysozyme and Mg-Sarkosyl) to examine DNA association with the outer membrane (1, 8, 36, 80). Like the nucleoid, DNA-membrane interactions were sensitive to treatment with a single-stranded nuclease, S1, which can degrade both DNA and RNA. At low levels of S1 digestion, 90% of the chromosome could be released from the membrane in average sizes of about 1.2×10^8 Da, or fragments slightly larger than T4 phage DNA. The estimated number of chromosome-membrane attachments was similar to the number of nucleoid domains: 26 ± 11 (1). This similarity begs the question, where are the membrane attachments?

The origin of replication and the terminus both provide examples of specific mechanisms in which chromosomal sequences are brought to the membrane. At oriC, the initiator of DNA replication, the DnaA protein, is membrane associated and it becomes activated to initiate a round of replication through interactions with phospholipids (see chapter 9). At the terminus of replication, the membrane-associated FtsA-Z complex is localized at the septal membrane and division plane (see chapter 28). The FtsK component of this complex is a DNA "translocase" that stimulates both positive and negative DNA super-coiling in the presence of ATP, and it associates with the terminus during the final stages of DNA replication (7). However, these two mechanisms do not explain the large fraction of DNA in membrane preparations; other mechanisms seem likely.

Experiments in the Wang laboratory uncovered a mechanism linking both membranes and RNA synthesis to chromosomal structure. The underlying mechanism turned out much different from the initial hypotheses of specific DNA-binding membrane proteins and suggested instead that chromosomes become tethered to membranes due to a specific pattern of gene expression. Lynch and Wang found one group of genes, which included tetA, lacY, and phoA, to encode integral membrane proteins that become cotranscriptionally inserted via the SecA pathway. A second class included genes like tolC and ampC, which encode proteins exported through the cytoplasmic membrane to the outer membrane. Lynch and Wang detected chromosome-membrane interac-tions as a hypernegative supercoiling phenotype in cells carrying an inactive form of Topo I. Although hypersupercoiling of plasmid DNA can occur in WT cells when a strong promoter like the hybrid tac promoter is derepressed (32, 40), less efficient pro-moters can cause hypersupercoiling in a topA mutant if the encoded gene is an integral membrane protein. The explanation for this effect involves a special con-sequence of molecular handcuffing. Because nascent membrane proteins are transported to the periplasm, they rapidly attach to secretion machinery as nascent peptides while RNA polymerase is still synthesizing mRNA. Drolet showed that TopA mutants produced stable R loops, causing persistent unwinding of DNA (see above). This link between DNA and the mem-brane involves complex interactions between RNA polymerase, mRNA, ribosomes, nascent protein, and the membrane secretion machinery, which is consis-tent with many features of observed nucleoid be-havior (Fig. 6).

Although hypersupercoiling of plasmid DNA is not observed from average-strength promoters in strains carrying a WT topA gene, membrane an-choring of DNA in WT cells is a solid bet. DNA-membrane association has been indicated using other assays, such as high-resolution electron microscopy following rapid cryofixation to prepare cells for mi-croscopy. Bohrmann et al. found a coralline orga-nization of cellular DNA (15). Like a coral reef, chromosomal DNA in vivo had a central dense core (often called the in vivo nucleoid) with multi-ple ribosome-rich DNA extensions leading near the membrane. Spectroscopic observations in which the central masses of bacterial nucleoids are visualized show dramatic changes when either RNA transcrip-tion or protein synthesis is disrupted. For example, inhibition of protein synthesis by addition of either rifampin or chloramphenicol caused lobes of in vivo "nucleoids" to condense into spherical structures (140). And in bacteria grown as long filaments, separated nucleoids rapidly blended together after chloramphenicol treatment (139).

Lynch and Wang's observations suggest an al-ternative mechanism for nucleoid sensitivity to RNase and S1 nuclease. Membrane-DNA linkage is caused by cotranscriptional inner membrane attachment or transport to the outer membrane. Chromosomal connections to this are at the nucleoid periphery in vivo, but they invert and extrude loops during lysis. This hypothesis accounts for sensitivity to single-strand-specific nuclease digestion because RNA links the DNA to translation products.

One version of this hypothesis is shown in Fig. 6. With respect to the number of domains, many factors could contribute to stability of RNA-membrane

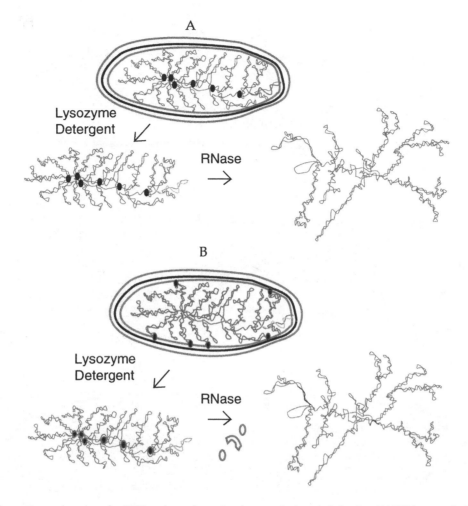

Figure 6. Alternative explanations for RNA and membrane involvement in domain behavior. (A) RNA transcription produces novel RNA species that organize specific domains within the bacterial chromosome. Lysis releases a structure, and digestion with RNase unfolds the nucleoid by digesting "organizational" RNA. (B) Cotranscriptional translation of integral membrane proteins transiently handcuffs DNA to the membrane through a link involving RNA polymerase, mRNA, ribosomes, and nascent membrane protein. Lysis causes the membrane-attached complexes to collapse to the center, forming an inverted structure. Digestion by RNase releases DNA from the membrane, allowing nucleoid expansion.

connections, including the number of genes within multigenic operons that are membrane associated, the level of transcription, and the efficiency of and rate of translation. In vivo, membrane attachment must account for only a fraction of the observed domains (32).

Several investigators have proposed that the observations of Lynch and Wang suggest a role for membrane proteins in chromosome development. According to these hypotheses, cotranslational DNA attachment allows the sister chromosomes to become configured into new domains following DNA replication (100, 152) (Fig. 7). In this scenario, RNA polymerases transcribe many (most?) genes after replication. Thus, the transient expression (35) and membrane association of many nascent proteins could provide a framework for the dynamic process of chromosome reorganization in rapidly dividing cells (35).

STOCHASTIC BEHAVIOR AND REGULATED GENE EXPRESSION

The ability of bacterial cultures to adapt rapidly to changing environmental conditions creates a chromosomal dilemma. Replication and transcription occur at the same time. Replication is regulated by cell cycle components, but transcriptional programs change on a moment's notice. With two potent engines of chromosomal movement regulated by different temporal and spatial rules, how do bacteria avoid the potential for topological chaos? Cell biology experiments show that stochastic behavior,

A

B

C

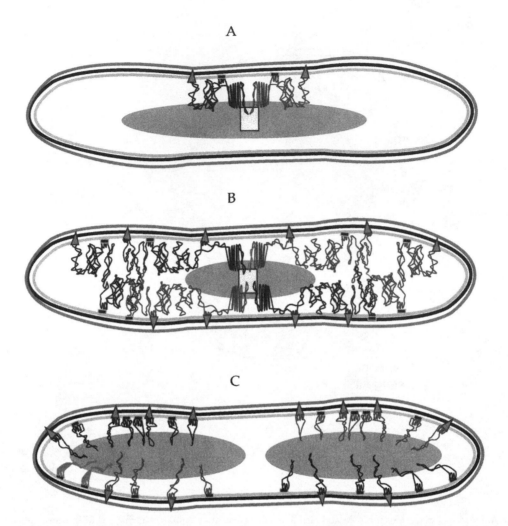

Figure 7. Hypothetical use of transcription and membrane attachment to reform nucleoids after DNA replication. (A) Chromosomal replication of the parental nucleoid (large gray oval) is initiated in a DNA factory (open square) positioned at mid cell. A replication fork initiated at OriC loops out one continuously synthesized strand leftward (black) and a complementary strand rightward that is discontinuously synthesized as Okazaki fragments (gray). After DNA passes through a SeqA zone (aggregate), expression of genes encoding integral membrane proteins (black bars) or proteins exported to the periplasm and outer membrane (gray triangles) handcuff DNA to the membrane-forming domains. (B) A second replisome extrudes the second replichore rightward and leftward. The template chromosome diminishes as the replisomes approach the terminus. (C) Two new nucleoids remain intermittently and dynamically linked to the membrane.

which is suggested by the random-sized loops in Fig. 1 and the first-order 1/2 D of Fig. 3, is an inherent part of gene expression in several well-studied circuits.

The *lac* operon of *E. coli* is possibly the best-studied regulatory system in genetics. Expression output from the *lac* operon can be assayed using a chromogenic substrate to follow the kinetics of β-galactosidase activity with time. The *lacZ* gene has been used in numerous genetic gadgets designed to report levels of transcription and translation, and the *lacI-lacO* genetic module has been used to control expression of foreign genes in *E. coli*. But the *lac* operon provides an important and largely unappreciated example of stochastic behavior. The measur-

able activity of a WT *lac* operon in *E. coli* extends over a 1,000-fold range, going from repressed to strong induction conditions (96). One could wrongly assume that in order to achieve such stunning reproducibility, every cell must respond to inducer in the same way. However, in 1957, Novick and Weiner demonstrated that *lac* induction is largely an all-or-none phenomenon (101). When the system achieves a level of 5% of maximum LacZ expression, roughly 5% of the cells are highly derepressed while 95% of the cells synthesize no enzyme at all. The explanation for this response is that one of the genes in the operon, the *lacY* permease, is a membrane-bound transporter that rapidly concentrates small amounts of inducer present in the medium. Random transcription of the *lac* operon produces rare cells that transiently contain LacY. At low levels of inducer, all cells that have not synthesized LacY are unable to respond to the inducer, while LacY-containing cells concentrate inducer, which results in strong single-cell operon induction.

The all-or-none response is not a unique property of the *lac* operon. In the *araBAD* operon, low arabinose levels trigger a response similar to that found in the *lac* operon. Using single-cell analyses of a fast-folding form of green fluorescent protein (GFP), which was expressed under *araC* control, Siegele and Hu demonstrated that rare bright-fluorescing cells were found in fields of very dark cells (125). The number of regulatory systems studied on the single-cell level is very small, but this response could well be the rule in gene expression.

One problem with stochastic expression output is that widely varying cellular amounts of one protein may be acceptable, but some proteins are produced within a narrow range (e.g., components of the replisome-associated polymerase). The cell-to-cell variability in the *lac* system is subject to genetic control. One recent report used a fast-folding form of GFP controlled by the *lac* operon to quantify operon output in individual cells (105). An operon with a strong promoter and an efficient ribosome binding site had more cell-to-cell variation than the same operon with a poor promoter and an inefficient ribosome binding site. The explanation for this effect is that operons with strong promoters and optimal Shine-Dalgarno sequences are susceptible to large bursts of mRNA and large outputs of translated protein. Weaker promoters with less efficient Shine-Dalgarno sequences caused a lower output but more uniform cell-cell expression levels.

The second example of cell-to-cell variation is more difficult to explain. A strain of *E. coli* was constructed with two copies of the *lac* operon, one expressing GFP and the other expressing cyano-fluorescent proteins (38). Both operons were placed in the chromosome in single copy at points equidistant from the origin of replication but in different replichores so that gene dosage was equivalent. Fluorescence-activated cell sorting analysis showed that low to moderate inducer levels produce cells with a striking range in color as well as fluorescence intensity. Cell-to-cell variation ranged from pure red to pure green, through a broad range of fluorescent red and green mixtures. This is not predicted by the all-or-none mechanism described above. The cellular concentration of inducer should be equivalent for both operons; stochastic barriers may temporarily block one operon, leading one of two nearly identical operons to exclusively express protein.

DNA AS A MOLECULAR CHAPERONE

Aside from a few archaebacterial species that make both histones and DNA gyrase (29), most prokaryotes have gyrase and lack true histones whereas eukaryotes lack gyrase and supercoil DNA on the surface of nucleosomes. One advantage of having gyrase is that diffusible negative supercoils are dynamic and this movement allows DNA to promote rapid and efficient protein-protein interaction. This role for DNA is reminiscent of chaperones, molecules that aid protein folding and that facilitate assembly of multicomponent protein structures like pili (19). Examples of how supercoil movement can promote the assimilation or breakdown of complex protein ensembles are shown in Fig. 8. In the γδ resolution reaction, supercoil movement stimulates assembly of synaptic recombination complexes (Fig. 8A [top] and Fig. 2). In vitro experiments demonstrate the impressive speed of the synapse. After addition of resolvase to a reaction containing a 5-kb plasmid, DNA binding occurs in milliseconds, and most of the plasmids have formed a synapse within 1 s (103, 104, 123). The slow step of the reaction involves conformation changes in the protein-DNA complex, which takes a minute or two to complete the reaction.

A more complex example is the transposition reaction of phage Mu. In this case, supercoiling stimulates a three-site association to generate the strand transfer complex. The three sites are *attL* (the left end of Mu), which binds three protomers of MuA protein plus the Hu protein, *attR* (the right end of Mu), which also binds three MuA proteins, and the IAS (internal activation sequence), which is located 1 kb from the Mu left end (99, 137). Random collisions between ends with six bound MuA proteins do not lead to synapse; the supercoiled substrate must be organized into a plectosome (27). Plectosome assembly is aided by the IAS making contact with two MuA protomers to stimulate tetramerization and strand cleavage of the Mu host junctions (149, 156). Recently, the precise topological structure of this assembly intermediate was defined (107). Five super-helical links define the plectosome, which are shown as nodes in the thick strands of *attL*, *attR*, and the IAS (Fig. 8B). Other examples of multiple-site interactions that are stimulated by negative supercoil dynamics include inversion reactions catalyzed by the Hin and Gin proteins. In these cases a recombination enhancer sequence bound by the Fis protein promotes efficient DNA inversion reactions (68, 70).

In addition to stimulating protein assembly, DNA dynamics can facilitate disassembly of protein-DNA ensembles. An example of a complex reaction leading to the disassembly of a multiprotein complex

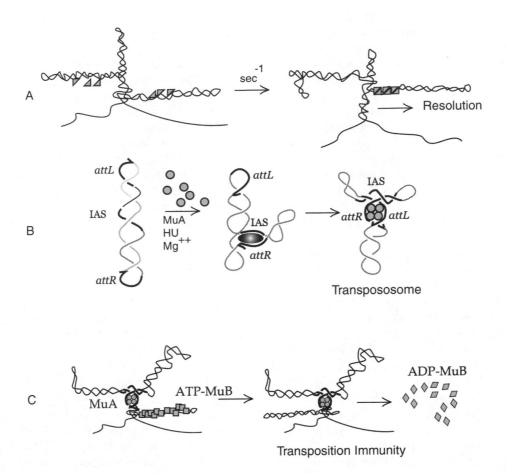

Figure 8. Examples of supercoil-assisted protein assembly and disassembly. (A) Six dimers of the Res protein (triangles) form a synapse that contains three negative supercoils. The time required to form this alignment on a supercoiled 4-kb plasmid is about 1 s. (B) Phage Mu transposition requires interactions of three sites in phage DNA. These sites include the left end of the virus (*attL*), the right end of the virus (*attR*), and the internal activation sequence (IAS). The plectosome or transposition intermediate contains a tetramer of MuA protein and five plectonemic crossings of DNA. (C) Mu transposition immunity involves interactions between a Mu transpososome and a DNA complex of the target selector, MuB protein. MuB binding to DNA is highly cooperative and requires an ATP. Interactions between MuA and MuB stimulate hydrolysis and a conformation change that displaces MuB from DNA.

is provided by the phage MuB protein. During "transpososome" assembly, interactions between MuA and MuB stimulate the overall reaction rate (136), and this interaction generates a zone of immunity that prevents Mu from inserting into itself and distributes the newly replicated copies over the entire genome (2, 91). In vivo, immunity is a strong effect for about 10 kb and is detectable 20 kb from a Mu end (91). In vitro, MuA protein promotes the disassembly of MuB filaments by stimulating hydrolysis of DNA-bound MuB-ATP complexes. The release of phosphate leads to a conformation change in MuB-ADP. Single DNA molecule experiments (46) provide strong evidence that a MuA tetramer, in the form of a transpososome, exploits DNA mobility to cause the processive release of polymeric MuB patches from nearby DNA sequences (47) (Fig. 8C).

SUMMARY

In the bacterial chromosome, two dynamic forces dominate topological behavior: DNA replication and RNA transcription. Topological complications arise whenever two replication forks converge head to head, as occurs at the terminus of replication, or head to tail, when a second fork overtakes a stalled fork during dichotomous growth (Fig. 4). Factors that cause fork stalling include chemical- and radiation-induced changes to DNA nucleotides that modify or eliminate their potential to provide a suitable template for the replisome-associated DNA polymerase. The interactions between transcription machinery and replication machinery change domain structure in several ways. Topological interactions between a replisome and RNA polymerase occur when forks meet transcribing RNA polymerases in head-to-tail or head-to-head orientations. Recent experiments reveal that R loops created by high levels of transcription cause elevated rates of recombination, point mutation, and deletions. Variations in DNA and RNA synthesis patterns require dynamic chromosomal rearrangement on a minute-by-minute time scale. Global structure provided by four topoisomerases and a group of about 10 DNA-binding proteins allows a plasticity that is useful for adaptation to different physiological conditions. These observations indicate that the organizational framework of a bacterial chromosome must be described in statistical terms rather than the highly ordered and predictable models that represent the crystal structures of many folded proteins that do biochemical work on the chromosome.

Acknowledgments. Work in the Higgins laboratory is supported by the National Institutes of Health (GM 33143) and the National Science Foundation (MCB 9122048).

We thank Lisa Postow for critical comments on the manuscript and Darlene Higgins for critical advice and editorial help.

REFERENCES

1. Abe, M., C. Brown, W. G. Hendrickson, D. H. Boyd, P. Clifford, R. H. Cote, and M. Schaechter. 1977. Release of *Escherichia coli* DNA from membrane complexes by single-strand endonucleases. *Proc. Natl. Acad. Sci. USA* **74:**2756–2760.
2. Adzuma, K., and K. Mizuuchi. 1988. Target immunity of Mu transposition reflects a differential distribution of Mu B protein. *Cell* **53:**257–266.
3. Aguilera, A. 2002. The connection between transcription and genomic instability. *EMBO J.* **21:**195–201.
4. Ailion, M., and J. R. Roth. 1997. Repression of the *cob* operon of *Salmonella typhimurium* by adenosylcobalamin is influenced by mutations in the *pdu* operon. *J. Bacteriol.* **179:**6084–6091.
5. Alberts, B. M. 1984. The DNA enzymology of protein machines. *Cold Spring Harbor Symp. Quant. Biol.* **49:**1–112.
6. Andersen, P. A., A. A. Griffiths, I. G. Duggin, and R. G. Wake. 2000. Functional specificity of the replication fork-arrest complexes of *Bacillus subtilis* and *Escherichia coli*: significant specificity for Tus-Ter functioning in *E. coli*. *Mol. Microbiol.* **36:**1327–1335.
7. Aussel, L., F.-X. Barre, M. Aryoy, A. Stasiak, A. Z. Stasiak, and D. J. Sherratt. 2002. FtsK is a DNA motor protein that activates chromosome dimer resolution by switching the catalytic state of the XerC and XerD recombinases. *Cell* **108:**195–205.
8. Ballesta, J. P., E. Cundliffe, M. J. Daniels, J. L. Silverstein, M. M. Susskind, and M. Schaechter. 1972. Some unique properties of the deoxyribonucleic acid-bearing portion of the bacterial membrane. *J. Bacteriol.* **112:**195–199.
9. Bendich, A. J., and K. Drlica. 2000. Prokaryotic and eukaryotic chromosomes: what's the difference? *Bioessays* **22:**481–486.
10. Bernstein, J. A., A. B. Khodursky, P.-H. Lin, S. L. Lin-Chao, and S. N. Cohen. 2002. Global analysis of mRNA decay and abundance in *Escherichia coli* at single-gene resolution using two-color fluorescent DNA microarrays. *Proc. Natl. Acad. Sci. USA* **99:**9697–9702.
11. Bierne, H., S. D. Ehrlich, and B. Michel. 1997. Deletions at stalled replication forks occur by two different pathways. *EMBO J.* **16:**3332–3340.
12. Blattner, F. R., G. Plunkett, C. A. Bloch, N. T. Perna, V. Burland, M. Riley, J. Collado-Vides, J. D. Glasner, C. K. Rode, G. F. Mayhew, J. Gregor, N. W. Davis, H. A. Kirkpatrick, M. A. Goeden, D. J. Rose, B. Mau, and Y. Shao. 1997. The complete genome sequence of *Escherichia coli* K-12. *Science* **277:**1453–1474.
13. Bliska, J. B., and N. R. Cozzarelli. 1987. Use of site-specific recombination as a probe of DNA structure and metabolism in vivo. *J. Mol. Biol.* **194:**205–218.
14. Bobik, T. A., Y. Xu, R. M. Jeter, K. E. Otto, and J. R. Roth. 1997. Propanediol utilization genes (*pdu*) of *Salmonella typhimurium*: three genes for the propanediol dehydratase. *J. Bacteriol.* **179:**6633–6639.
15. Bohrmann, B., W. Villiger, R. Johansen, and E. Kellenberger. 1991. Coralline shape of the bacterial nucleoid after cryofixation. *J. Bacteriol.* **173:**3149–3158.
16. Bonne-Andrea, C., M. L. Wong, and B. M. Alberts. 1990. In vitro replication through nucleosomes without histone displacement. *Nature* **343:**720.
17. Bremer, H., and P. Dennis. 1996. Modulation of chemical composition and other parameters of the cell by growth rate, p. 1553–1569. *In* F. C. Neidhardt, R. Curtiss III, J. L. Ingraham, E. C. C. Lin, K. B. Low, B. Magasanik, W. S. Reznikoff, M. Riley, M. Schaechter, and H. E. Umbarger (ed.), Escherichia coli *and* Salmonella: *Cellular and Molecular Biology.* ASM Press, Washington, D.C.
18. Brown, P. O., and N. R. Cozzarelli. 1979. A sign inversion mechanism for enzymatic supercoiling of DNA. *Science* **206:**1081–1083.
19. Bullitt, E., C. H. Jones, R. Striker, G. Soto, F. Jacob-Dubuisson, J. Pinkner, M. J. Wick, L. Makowski, and S. J. Hultgren. 1996. Development of pilus organelle subassemblies in vitro depends on chaperone uncapping of a beta zipper. *Proc. Natl. Acad. Sci. USA* **93:**12890–12895.
20. Chen, P., M. Ailion, T. Bobik, G. Stormo, and J. Roth. 1995. Five promoters integrate control of the *cob/pdu* regulon in *Salmonella typhimurium*. *J. Bacteriol.* **177:**5401–5410.
21. Chen, P., D. I. Anderson, and J. R. Roth. 1994. The control region of the *pdu/cob* regulon in *Salmonella typhimurium*. *J. Bacteriol.* **176:**5474–5482.
22. Cook, D. N., D. Ma, N. G. Pon, and J. E. Hearst. 1992. Dynamics of DNA supercoiling by transcription in *E. coli*. *Proc. Natl. Acad. Sci. USA* **89:**10603–10607.
23. Cook, P., and F. Gove. 1992. Transcription by an immobilized RNA polymerase from bacteriophage T7 and the topology of transcription. *Nucleic Acids Res.* **20:**3591–3598.
24. Courcelle, J., J. R. Donaldson, K.-H. Chow, and C. T. Courcelle. 2003. Replication fork regression and processing in *Escherichia coli*. *Science* **299:**1064–1067.
25. Cozzarelli, N. R., and J. C. Wang. 1990. *DNA Topology and Its Biological Effects.* Cold Spring Harbor Press, Cold Spring Harbor, N.Y.
26. Craigie, R., and K. Mizuuchi. 1985. Mechanism of transposition of bacteriophage Mu: structure of a transposition intermediate. *Cell* **41:**867–876.
27. Craigie, R., and K. Mizuuchi. 1986. Role of DNA topology in Mu transposition: mechanism of sensing the relative orientation of two DNA segments. *Cell* **45:**793–800.

28. Crisona, N. J., T. R. Strick, D. Bensimon, V. Croquette, and N. R. Cozzarelli. 2000. Preferential relaxation of positively supercoiled DNA by Escherichia coli topoisomerase IV in single-molecule and ensemble measurements. *Genes Dev.* **14:**2881–2892.

29. Decanniere, K., A. M. Babu, K. Sandman, J. N. Reeve, and U. Heinemann. 2000. Crystal structures of recombinant histones HMfA and HMfB from the hyperthermophilic archaeon Methanothermus fervidus. *J. Mol. Biol.* **303:**35–47.

30. Deibler, R. W., S. Rahmati, and E. L. Zechiedrich. 2001. Topoisomerase IV, alone, unknots DNA in E. coli. *Genes Dev.* **15:**748–761.

31. Delius, H., and A. Worcel. 1974. Electron microscopic visualization of the folded chromosome of Escherichia coli. *J. Mol. Biol.* **82:**107–109.

32. Deng, S., R. A. Stein, and N. P. Higgins. 2004. Transcription-induced barriers to supercoil diffusion in the *Salmonella typhimurium* chromosome. *Proc. Natl. Acad. Sci. USA* **101:**3398–3403.

33. Drlica, K., and A. Worcel. 1975. Conformational transitions in the *Escherichia coli* chromosome: analysis by viscometry and sedimentation. *J. Mol. Biol.* **98:**393–411.

34. Drolet, M., S. Broccoli, F. Rallu, C. Hraiky, C. Fortin, E. Masse, and I. Baaklini. 2003. The problem of hypernegative supercoiling and R-loop formation in transcription. *Front. Biosci.* **8:**D210–D221.

35. Dworkin, J., and R. Losick. 2002. Does RNA polymerase help drive chromosome segregation in bacteria? *Proc. Natl. Acad. Sci. USA* **99:**14089–14094.

36. Earhart, C. F., G. Y. Tremblay, M. J. Daniels, and M. Schaechter. 1968. DNA replication studied by a new method for the isolation of cell membrane-DNA complexes. *Cold Spring Harbor Symp. Quant. Biol.* **33:**707–710.

37. Echols, H. 1986. Multiple DNA-protein interactions governing high-precision DNA transactions. *Science* **233:**1050–1056.

38. Elowitz, M. B., A. J. Levine, E. D. Siggia, and P. S. Swain. 2002. Stochastic gene expression in a single cell. *Science* **297:**1183–1186.

39. Espeli, O., C. Levine, H. Hassing, and K. J. Marians. 2003. Temporal regulation of Topoisomerase IV activity in E. coli. *Mol. Cell* **11:**189–201.

40. Figueroa, N., and L. Bossi. 1988. Transcription induces gyration of the DNA template in Escherichia coli. *Proc. Natl. Acad. Sci. USA* **85:**9416–9420.

41. French, S. 1992. Consequences of replication fork movement through transcription units in vivo. *Science* **258:**1362–1365.

42. French, S. L., and O. L. Miller. 1989. Transcription mapping of the *Escherichia coli* chromosome by electron microscopy. *J. Bacteriol.* **171:**4207–4216.

43. Fu, Q. 1998. *Identification and Characterization of Barriers to Supercoil Diffusion in the 17-20 Centisome Region of the* Salmonella typhimurium *Chromosome.* M.S. thesis. University of Alabama at Birmingham, Birmingham.

44. Gamper, H. B., and J. E. Hearst. 1982. A topological model for transcription based on unwinding angle analysis of E. coli RNA polymerase binary, initiation and ternary complexes. *Cell* **29:**81–90.

45. Gari, E., N. Figueroa-Bossi, A.-B. Blanc-Potard, F. Spirito, M. B. Schmid, and L. Bossi. 1996. A class of gyrase mutants of *Salmonella typhimurium* show quinolone-like lethality and require Rec functions for viability. *Mol. Microbiol.* **21:**111–122.

46. Greene, E. C., and K. Mizuuchi. 2002. Direct observation of single MuB polymers: evidence for a DNA-dependent

47. Greene, E. C., and K. Mizuuchi. 2002. Target immunity during Mu DNA transposition. Transpososom assembly and DNA looping enhance MuA-mediated disassembly of the MuB target complex. *Mol. Cell* **6:**1367–1378.

48. Griffith, J. D. 1976. Visualization of prokaryotic DNA in a regularly condensed chromatin like fiber. *Proc. Natl. Acad. Sci. USA* **73:**563–567.

49. Grompone, G., S. D. Ehrlich, and B. Michel. 2003. Replication restart in *gyrB* Escherichia coli mutants. *Mol Microbiol.* **48:**845–854.

50. Heck, M. 1997. Condensins, cohesins, and chromosome architecture: how to make and break a mitotic chromosome. *Cell* **91:**5–8.

51. Hendricks, E. C., H. Szerlong, T. Hill, and P. Kuempel. 2000. Cell division, guillotining of dimer chromosomes and SOS induction in resolution mutants (*dif, xerC* and *xerD*) of *Escherichia coli. Mol. Microbiol.* **36:**973–981.

52. Higgins, C. F., C. J. Dorman, D. A. Stirling, L. Waddell, I. R. Booth, G. May, and E. Bremer. 1988. A physiological role for DNA supercoiling in the osmotic regulation of gene expression in S. typhimurium and E. coli. *Cell* **52:**569–584.

52a. Higgins, N. P. 1999. DNA supercoiling and its consequences for chromosome structure and function, p. 189–202. *In* R. L. Charlebois (ed.), *Organization of the Prokaryotic Genome.* ASM Press, Washington, D.C.

53. Higgins, N. P., K. H. Kato, and B. S. Strauss. 1976. A model for replication repair in mammalian cells. *J. Mol. Biol.* **101:**417–425.

54. Higgins, N. P., C. L. Peebles, A. Sugino, and N. R. Cozzarelli. 1978. Purification of the subunits of Escherichia coli DNA gyrase and reconstitution of enzymatic activity. *Proc. Natl. Acad. Sci. USA* **75:**1773–1777.

55. Higgins, N. P., and A. Vologodskii. 2004. Topological behavior of plasmid DNA, p. 181–201. *In* B. E. Funnell and G. J. Phillips (ed.), *Plasmid Biology.* ASM Press, Washington, D.C.

56. Higgins, N. P., X. Yang, Q. Fu, and J. R. Roth. 1996. Surveying a supercoil domain by using the γδ resolution system in *Salmonella typhimurium. J. Bacteriol.* **178:**2825–2835.

57. Hill, T. M. 1992. Arrest of bacterial DNA replication. *Annu. Rev. Microbiol.* **46:**603–633.

58. Hill, T. M. 1996. Features of the chromosomal terminus region, p. 1602–1614. *In* F. C. Neidhardt, R. Curtiss III, J. L. Ingraham, E. C. C. Lin, K. B. Low, B. Magasanik, W. S. Reznikoff, M. Riley, M. Schaechter, and H. E. Umbarger (ed.), Escherichia coli *and* Salmonella: *Cellular and Molecular Biology,* 2nd ed., vol. 2. ASM Press, Washington, D.C.

59. Hill, T. M., and K. J. Marians. 1990. Escherichia coli Tus protein acts to arrest the progression of DNA replication forks in vitro. *Proc. Natl. Acad. Sci. USA* **87:**2481–2485.

60. Hill, T. M., M. L. Tecklenburg, A. J. Pelletier, and P. L. Kuempel. 1989. *tus,* the trans-acting gene required for termination of DNA replication in *Escherichia coli,* encodes a DNA-binding protein. *Proc. Natl. Acad. Sci. USA* **86:**1593–1597.

61. Hiraga, S., H. Niki, T. Ogura, D. Ichinose, H. Mori, B. Ezaki, and A. Jaffe. 1989. Chromosome partitioning in *Escherichia coli:* novel mutants producing anucleate cells. *J. Bacteriol.* **171:**1496–1505.

62. Holmes, F. L. 1998. The DNA replication problem, 1953–1958. *Trends Biochem. Sci.* **23:**117–120.

63. Hraiky, C., M.-A. Raymond, and M. Drolet. 2000. RNase H overproduction corrects a defect at the level of

transcription elongation during rRNA synthesis in the absence of DNA topoisomerase I in *Escherichia coli*. *J. Biol. Chem.* **275**:11257–11263.

64. **Huertas, P., and A. Aguilera.** 2003. Cotranscriptionally formed DNA:RNA hyprids mediate transcription elongation impairment and transcription-associated recombination. *Mol. Cell* **12**:711–721.

65. **Ikeda, H., and T. Matsumoto.** 1979. Transcription promotes recA-independent recombination mediated by DNA-dependent RNA polymerase in *Escherichia coli*. *Proc. Natl. Acad. Sci. USA* **76**:4571–4575.

66. **Jacob, F., S. Brenner, and F. Cuzin.** 1963. On the regulation of DNA replication in bacteria. *Cold Spring Harbor Symp. Quant. Biol.* **28**:329–348.

67. **Jaworski, A., N. P. Higgins, R. D. Wells, and W. Zacharias.** 1991. Topoisomerase mutants and physiological conditions control supercoiling and Z-DNA formation in vivo. *J. Biol. Chem.* **266**:2576–2581.

68. **Johnson, R. C., and M. F. Bruist.** 1989. Intermediates in hin-mediated DNA inversion: a role for Fis and the recombinational enhancer in the strand exchange reaction. *EMBO J.* **8**:1581–1590.

69. **Johzuka, K., and T. Horiuchi.** 2002. Replication fork block protein, Fob1, acts as an rDNA region specific recombinator in *S. cerevisiae*. *Genes Cells* **7**:99–113.

70. **Kanaar, R., A. Klippel, E. Shekhtman, J. M. Dungan, R. Kahmann, and N. R. Cozzarelli.** 1990. Processive recombination by the phage Mu Gin system: implications for the mechanisms of DNA strand exchange, DNA site alignment, and enhancer action. *Cell* **62**:353–366.

71. **Kavenoff, R., and B. Bowen.** 1976. Electron microscopy of membrane-free folded chromosomes from *Escherichia coli*. *Chromosoma* **59**:89–101.

72. **Kavenoff, R., and O. Ryder.** 1976. Electron microscopy of membrane-associated folded chromosomes of *Escherichia coli*. *Chromosoma* **55**:13–25.

73. **Khatri, G. S., T. MacAllister, P. R. Sista, and D. Bastia.** 1989. The replication terminator protein of E. coli is a DNA sequence-specific contra-helicase. *Cell* **59**:667–674.

73a. **Khodursky, A. B., B. J. Peter, M. B. Schmid, J. DeRisi, D. Botstein, P. O. Brown, and N. R. Cozzarelli.** 2000. Analysis of topoisomerase function in bacterial replication fork movement: use of DNA microarrays. *Proc. Natl. Acad. Sci. USA* **97**:9419–9424.

74. **Kirkegaard, K., and J. C. Wang.** 1981. Mapping the topography of DNA wrapped around gyrase by nucleolytic and chemical probing of complexes of unique DNA sequences. *Cell* **23**:721–729.

75. **Kobayashi, T., J. D. Heck, M. Nomura, and T. Horiuchi.** 1998. Expansion and contraction of ribosomal DNA repeats in *Saccharomyces cerevisiae*. Requirement of replication fork blocking (Fob1) protein and the role of RNA polymerase I. *Genes Dev.* **12**:3821–3830.

76. **Kodama, K., T. Kobayashi, H. Niki, S. Hiraga, T. Oshima, H. Mori, and T. Horiuchi.** 2002. Amplification of hot DNA segments in *Escherichia coli*. *Mol. Microbiol.* **45**:1575–1588.

77. **Korzheva, N., A. Mustaev, M. Kozlov, A. Malhotra, V. Nikiforov, A. Goldfarb, and S. A. Darst.** 2000. A structural model of transcription elongation. *Science* **289**:619–625.

78. **Kuempel, P., J. Henson, L. Dircks, M. Tecklenburg, and D. Lim.** 1991. *dif*, a *recA*-independent recombination site in the terminus region of the chromosome of *Escherichia coli*. *New Biol.* **3**:799–811.

79. **Kuempel, P., A. Hogaard, M. Nielsen, O. Nagappan, and M. Tecklenburg.** 1996. Use of a transposon (Tn*dif*) to obtain suppressing and nonsuppressing insertions of the *dif* resolvase site of *Escherichia coli*. *Genes Dev.* **10**:1162–1171.

80. **Leibowitz, P. J., and M. Schaechter.** 1975. The attachment of the bacterial chromosome to the cell membrane. *Int. Rev. Cytol.* **41**:1–28.

81. **Lemon, K. P., and A. D. Grossman.** 2001. The extrusion-capture model for chromosome partitioning in bacteria. *Genes Dev.* **15**:2031–2041.

82. **Lemon, K. P., and A. D. Grossman.** 1998. Localization of bacterial DNA polymerase: evidence for a factory model of replication. *Science* **282**:1516–1519.

83. **Li, Y., K. Sergueev, and S. Austin.** 2002. The segregation of the *Escherichia coli* origin and terminus of replication. *Mol. Microbiol.* **46**:985–995.

84. **Li, Y., B. Youngren, K. Sergueev, and S. Austin.** 2003. Segregation of the *Escherichia coli* chromosome terminus. *EMBO J.* **50**:825–834.

85. **Liu, B., and B. M. Alberts.** 1995. Head-on collision between a DNA replication apparatus and RNA polymerase transcription complex. *Science* **267**:1131–1137.

86. **Liu, B., M. Wong, R. Tinker, E. Geiduschek, and B. Alberts.** 1993. The DNA replication fork can pass RNA polymerase without displacing the nascent transcript. *Nature* **366**:33–39.

87. **Liu, B., M. L. Wong, and B. Alberts.** 1994. A transcribing RNA polymerase molecule survives DNA replication without aborting its growing RNA chain. *Proc. Natl. Acad. Sci. USA* **91**:10660–10664.

88. **Liu, L. F., and J. C. Wang.** 1987. Supercoiling of the DNA template during transcription. *Proc. Natl. Acad. Sci. USA* **84**:7024–7027.

89. **Maaloe, O., M. Schaechter, and N. O. Kjeldgaard.** 1966. The bacterial nucleus, p. 188–197. *In* O. Maaloe and N. O. Kjeldgaard (ed.), *Control of Macromolecular Synthesis*. W. A. Benjamin, Inc., New York, N.Y.

90. **Manna, A. C., S. P. Karnire, D. E. Bussiere, C. Davies, C. W. White, and D. Bastia.** 1996. Helicase-contrahelicase interaction and the mechanism of termination of DNA replication. *Cell* **87**:881–891.

91. **Manna, D., and N. P. Higgins.** 1999. Phage Mu transposition immunity reflects supercoil domain structure of the chromosome. *Mol. Microbiol.* **32**:595–606.

92. **Marians, K. J.** 1997. Helicase structures: a new twist on DNA unwinding. *Structure* **5**:1129–1134.

93. **McClelland, M., K. E. Sanderson, J. Spieth, S. W. Clifton, P. Latreille, L. Courtney, S. Porwollik, J. Ali, M. Dante, F. Du, S. Hou, D. Layman, S. Leonard, C. Nguyen, K. Scott, A. Holmes, N. Grewal, E. Mulvaney, E. Ryan, H. Sun, L. Florea, W. Miller, T. Stoneking, M. Nhan, R. Waterson, and R. K. Wilson.** 2001. The complete genome sequence of *Salmonella enterica* serovar Typhimurium LT2. *Nature* **413**:852–856.

94. **Mellon, I., and P. C. Hanawalt.** 1989. Induction of the *Escherichia coli* lactose operon selectively increases repair of its transcribed DNA strand. *Nature* **342**:95–98.

95. **Menzel, R., and M. Gellert.** 1983. Regulation of the genes for E. coli DNA gyrase: homeostatic control of DNA supercoiling. *Cell* **34**:105–113.

96. **Miller, J. H.** 1972. *Experiments in Molecular Genetics*. Cold Spring Harbor Laboratory, Cold Spring Harbor, N.Y.

97. **Miller, W. G., and R. W. Simons.** 1993. Chromosomal supercoiling in *Escherichia coli*. *Mol. Microbiol.* **10**:675–684.

98. **Mizuta, R., K. Iwai, M. Shigeno, M. Mizuta, T. Uemura, T. Ushiki, and D. Kitamura.** 2003. Molecular visualization of immunoglobulin switch region RNA/DNA complex by atomic force microscope. *J. Biol. Chem.* **278**:4431–4434.

99. Mizuuchi, M., and K. Mizuuchi. 1989. Efficient Mu transposition requires interaction of transposase with a DNA sequence at the Mu operator: implications for regulation. *Cell* **58**:399–408.

100. Norris, V. 1995. Hypothesis: chromosome separation in *E. coli* involves autocatalytic gene expression, transertion and membrane domain formation. *Mol. Microbiol.* **16**: 1051–1057.

101. Novick, A., and M. Weiner. 1957. Enzyme induction as an all-or-none phenomenon. *Proc. Natl. Acad. Sci. USA* **43**: 553–566.

102. Oliver, B., M. Parisi, and D. Clark. 2002. Gene expression neighborhoods. *J. Biol.* **1**:4.

103. Oram, M., J. F. Marko, and S. E. Halford. 1997. Communications between distant sites on supercoiled DNA from non-exponential kinetics for DNA synapsis by resolvase. *J. Mol. Biol.* **270**:396–412.

104. Oram, M., E. Shipstone, and S. E. Halford. 1994. Synapsis by Tn3 resolvase: speed and dependence on DNA supercoiling. *Biochem. Soc. Trans.* **22**:303.

105. Ozbudak, E. M., M. Thattai, I. Kurtser, A. D. Grossman, and A. van Oudenaarden. 2002. Regulation of noise in the expression of a single gene. *Nat. Genet.* **31**:69–73.

106. Reference deleted.

107. Pathania, S., M. Jayaram, and R. M. Harshey. 2002. Path of DNA within the Mu transpososome: transposase interactions bridging two Mu ends and the enhancer trap five DNA supercoils. *Cell* **109**:425–436.

108. Pavitt, G. D., and C. F. Higgins. 1993. Chromosomal domains of supercoiling in *Salmonella typhimurium*. *Mol. Microbiol.* **10**:685–696.

109. Perals, K., F. Cornet, Y. Merlet, and J.-M. Louarn. 2000. Functional polarization of the *Escherichia coli* chromosome terminus. The *dif* site acts in chromosome dimer resolution only when located between long stretches of opposite polarities. *Mol. Microbiol.* **36**:33–43.

110. Peter, B. J., C. Ullsperger, H. Hiasa, K. J. Marians, and N. R. Cozzarelli. 1998. The structure of supercoiled intermediates in DNA replication. *Cell* **94**:819–827.

111. Pettijohn, D., R. M. Hecht, O. G. Stonington, and T. D. Stamato. 1973. Factors stabilizing DNA folding in bacterial chromosomes, p. 145–162. *In* R. D. Wells and R. B. Innman (ed.), *DNA Synthesis in Vitro*. University Park Press, Baltimore, Md.

112. Pettijohn, D. E. 1996. The nucleoid, p. 158–166. *In* F. C. Neidhardt, R. Curtiss III, J. L. Ingraham, E. C. C. Lin, K. B. Low, B. Magasanik, W. S. Reznikoff, M. Riley, M. Schaechter, and H. E. Umbarger (ed.), Escherichia coli *and* Salmonella: *Cellular and Molecular Biology*, 2nd ed., vol. 1. ASM Press, Washington, D.C.

113. Pettijohn, D. E., and R. Hecht. 1973. RNA molecules bound to the folded bacterial genome stabilize DNA folds and segregate domains of supercoiling. *Cold Spring Harbor Symp. Quant. Biol.* **38**:31–41.

114. Pettijohn, D. E., and O. Pfenninger. 1980. Supercoils in prokaryotic DNA restrained in vivo. *Proc. Natl. Acad. Sci. USA* **77**:1331–1335.

115. Postow, L., N. J. Crisona, B. J. Peter, C. D. Hardy, and N. R. Cozzarelli. 2001. Topological challenges to DNA replication: conformations at the fork. *Proc. Natl. Acad. Sci. USA* **98**:8219–8226.

116. Postow, L., C. Ullsperger, R. Keller, C. Bustamante, A. F. Vologodskii, and N. R. Cozzarelli. 2001. Positive torsional strain causes the formation of a four-way junction at replication forks. *J. Biol. Chem.* **276**:2790–2796.

117. Rahmouni, A. R., and R. D. Wells. 1992. Direct evidence for the effect of transcription on local DNA supercoiling *in vivo*. *J. Mol. Biol.* **223**:131–144.

118. Reaban, M. E., J. Lebowitz, and J. A. Griffin. 1994. Transcription induces the formation of a stable RNA-DNA hybrid in the immunoglobulin alpha switch region. *J. Biol. Chem.* **269**:21850–21857.

119. Richmond, T. J., and C. A. Davey. 2003. The structure of DNA in the nucleosome core. *Nature* **423**:145–150.

120. Sawitzke, J., and S. Austin. 2001. An analysis of the factory model for chromosome replication and segregation in bacteria. *Mol. Microbiol.* **40**:786–794.

121. Sawitzke, J. A., and S. Austin. 2000. Suppression of chromosome segregation defects of Escherichia coli muk mutants by mutations in topoisomerase I. *Proc. Natl. Acad. Sci. USA* **97**:1671–1676.

122. Scheirer, K., and N. P. Higgins. 2001. Transcription induces a supercoil domain barrier in bacteriophage Mu. *Biochimie* **83**:155–159.

123. Sessions, R. B., M. Oram, M. D. Szczelkun, and S. E. Halford. 1997. Random walk models for DNA synapsis by resolvase. *J. Mol. Biol.* **270**:413–425.

124. Sharma, B., and T. M. Hill. 1995. Insertion of inverted *Ter* sites into the terminus region of the *Escherichia coli* chromosome delays completion of DNA replication and disrupts the cell cycle. *Mol. Microbiol.* **18**:45–61.

125. Siegele, D. A., and J. C. Hu. 1997. Gene expression from plasmids containing the araBAD promoter at subsaturating inducer concentrations represents mixed populations. *Proc. Natl. Acad. Sci. USA* **94**:8168–8172.

126. Sinden, R. R., J. O. Carlson, and D. E. Pettijohn. 1980. Torsional tension in the DNA double helix measured with trimethylpsoralen in living *E. coli* cells: analogous measurements in insect and human cells. *Cell* **21**:773–783.

127. Sinden, R. R., and D. E. Pettijohn. 1981. Chromosomes in living *Escherichia coli* cells are segregated into domains of supercoiling. *Proc. Natl. Acad. Sci. USA* **78**:224–228.

128. Skarstad, K., H. B. Steen, and E. Boye. 1985. *Escherichia coli* DNA distributions measured by flow cytometry and compared with theoretical computer simulations. *J. Bacteriol.* **163**:661–668.

129. Skokotas, A., H. Hiasa, K. J. Marians, L. O'Donnell, and T. M. Hill. 1995. Mutations in the Escherichia coli Tus protein define a domain positioned close to the DNA in the Tus-Ter complex. *J. Biol. Chem.* **270**:30941–30948.

130. Snyder, M., and M. Gerstein. 2003. Defining genes in the genomics era. *Science* **300**:258–260.

131. Spellman, P. T., and G. M. Rubin. 2002. Evidence for large domains of similarly expressed genes in the *Drosophila* genome. *J. Biol.* **1**:8.

132. Staczek, P., and N. P. Higgins. 1998. DNA gyrase and Topoisomerase IV modulate chromosome domain size in vivo. *Mol. Microbiol.* **29**:1435–1448.

133. Steck, T. R., and K. Drlica. 1984. Bacterial chromosome segregation: evidence for DNA gyrase involvement in decatenation. *Cell* **36**:1081–1088.

134. Reference deleted.

135. Stonington, G. O., and D. Pettijohn. 1971. The folded genome of *Escherichia coli* isolated in a protein-DNA-RNA complex. *Proc. Natl. Acad. Sci. USA* **68**:6–9.

136. Surette, M. G., and G. Chaconas. 1991. Stimulation of the Mu DNA strand cleavage and intramolecular strand transfer reactions by the Mu B protein is independent of stable

binding of the Mu B protein to DNA. *J. Biol. Chem.* **266:**17306–17313.

137. **Surette, M. G., B. D. Lavoie, and G. Chaconas.** 1989. Action at a distance in Mu DNA transposition: an enhancer-like element is the site of action of supercoiling relief activity by IHF. *EMBO J.* **8:**3483–3489.

138. **Thomas, B. J., and R. Rothstein.** 1989. Elevated recombination rates in transcriptionally active DNA. *Cell* **56:** 619–630.

139. **Van Helvoort, J. M., J. Kool, and C. L. Woldringh.** 1996. Chloramphenicol causes fusion of separated nucleoids in *Escherichia coli* K-12 cells and filaments. *J. Bacteriol.* **178:** 4289–4293.

140. **Van Helvoort, J. M., and C. L. Woldringh.** 1994. Nucleoid partitioning in *Escherichia coli* during steady state growth and upon recovery from chloroamphenicol treatment. *Mol. Microbiol.* **13:**577–583.

141. **Vilette, D., S. D. Ehrlich, and B. Michel.** 1995. Transcription-induced deletions in *Escherichia coli* plasmids. *Mol. Microbiol.* **17:**493–504.

142. **Vilette, D., M. Uzest, S. D. Ehrlich, and B. Michel.** 1992. DNA transcription and repressor binding affect deletion formation in *Escherichia coli* plasmids. *EMBO J.* **11:** 3629–3634.

143. **Vinograd, J., and J. Lebowitz.** 1966. Physical and topological properties of circular DNA. *J. Gen. Physiol.* **49:**103–125.

144. **Vinograd, J., J. Lebowitz, and R. Watson.** 1968. Early and late helix-coil transitions in closed circular DNA. The number of superhelical turns in polyoma DNA. *J. Mol. Biol.* **33:**173–197.

145. **Vologodskii, A. V., W. Zhang, V. V. Rybenkov, A. A. Podtelezhnikov, D. Subramanian, J. D. Griffith, and N. R. Cozzarelli.** 2001. Mechanism of topology simplification by type II DNA topoisomerases. *Proc. Natl. Acad. Sci. USA* **98:**3045–3049.

146. **Walter, D., M. Ailion, and J. Roth.** 1997. Genetic characterization of the *pdu* operon: use of 1,2-propanediol in *Salmonella typhimurium.* *J. Bacteriol.* **179:**1013–1022.

147. **Wang, J. C.** 1991. DNA toperisomerases: why so many? *J. Biol. Chem.* **266:**6659–6662.

148. **Wang, J. C.** 1971. Interaction between DNA and an *Escherichia coli* protein omega. *J. Mol. Biol.* **55:**523–533.

149. **Wang, Z., S.-Y. Namgoong, X. Zhang, and R. M. Harshey.** 1996. Kinetic and structural probing of the precleavage synaptic complex (type 0) formed during phage Mu transposition. *J. Biol. Chem.* **271:**9619–9626.

149a. **Watson, J. D., and F. H. C. Crick.** 1953. Genetical implications of the structure of deoxyribonucleic acid. *Nature* **171:**964–967.

150. **Wei, Y., J.-M. Lee, C. Richmond, F. R. Blattner, J. A. Rafalski, and R. A. LaRossa.** 2001. High density microarray-mediated gene expression profiling of *Escherichia coli.* *J. Bacteriol.* **183:**545–556.

151. **Weitao, T., K. Nordstrom, and S. Dasgupta.** 2000. Escherichia coli cell cycle control genes affect chromosome superhelicity. *EMBO Rep.* **1:**494–499.

152. **Woldringh, C. L., P. R. Jensen, and H. V. Westerhoff.** 1995. Structure and partitioning of bacterial DNA: determined by a balance of compaction and expansion forces? *FEMS Microbiol. Lett.* **131:**235–242.

153. **Worcel, A., and E. Burgi.** 1972. On the structure of the folded chromosome of *Escherichia coli.* *J. Mol. Biol.* **71:**127–147.

154. **Worcel, A., and E. Burgi.** 1974. Properties of a membrane-attached form of the folded chromosome of *Escherichia coli.* *J. Mol. Biol.* **82:**91–105.

155. **Wu, H.-Y., S. Shyy, J. C. Wang, and L. F. Liu.** 1988. Transcription generates positively and negatively supercoiled domains in the template. *Cell* **53:**433–440.

156. **Yang, J.-Y., M. Jayaram, and R. M. Harshey.** 1996. Positional information within the Mu transpose tetramer: catalytic contributions of individual monomers. *Cell* **85:**447–455.

157. **Yang, L., C. B. Jessee, K. Lau, H. Zhang, and L. F. Liu.** 1989. Template supercoiling during ATP-dependent DNA helix tracking: studies with simian virus 40 large tumor antigen. *Proc. Natl. Acad. Sci. USA* **86:**6121–6125.

158. **Yin, Y. W., and T. A. Steitz.** 2002. Structural basis for the transition from initiation to elongation transcription in T7 RNA polymerase. *Science* **298:**1387–1395.

159. **Zechiedrich, E. L., B. K. Arkady, S. Bachellier, D. Chen, D. M. Lilley, and N. R. Cozzarelli.** 2000. Roles of topoisomerases in maintaining steady-state DNA supercoiling in *Escherichia coli.* *J. Biol. Chem.* **275:**8103–8113.

160. **Zechiedrich, E. L., A. B. Khodursky, and N. R. Cozzarelli.** 1997. Topoisomerase IV, not gyrase, decatenates products of site-specific recombination in *Escherichia coli.* *Genes Dev.* **11:**2580–2592.

161. **Zhu, H., M. Bilgin, R. Bangham, D. Hall, A. Casamayor, P. Bertone, N. Lan, R. Jansen, S. Bidlingmaier, T. Houfek, T. Mitchell, P. Miller, R. A. Dean, M. Gerstein, and M. Snyder.** 2001. Global analysis of protein activities using proteome chips. *Science* **293:**2101–2105.

Chapter 7

Stationary-Phase Chromosomes

Abraham Minsky and Roberto Kolter

One of the most remarkable features of prokaryotes is their ability to remain viable for very long periods under conditions that are not propitious for growth. Perhaps even more remarkable is the finding that under such conditions, bacteria become resistant to multiple and diverse environmental assaults in doses that would have been lethal during rapid growth. In response to starvation, some bacteria form extremely resistant dormant spores (43), while others assemble into multicellular aggregates such as the myxobacterial fruiting bodies (28). But even without such extensive changes, many bacterial species, including *Escherichia*, *Salmonella*, and *Vibrio* spp., differentiate into highly resistant states as a result of starvation (21, 22, 25, 32). These phenotypic changes have far-reaching consequences, in particular when microbial pathogenesis is considered.

STATIONARY STATE, STRESS RESISTANCE, AND MICROBIAL PATHOGENESIS

During their life cycle, most microbial pathogens alternate between milieus within their host and aquatic or soil niches, where they persist as free-living organisms. Following initial infection, many intracellular pathogens become restricted to host cell phagosomes, within which they are confronted with a wide assortment of harsh conditions and detrimental agents, including low oxygen levels, highly reactive oxygen and nitrogen species, acidic pH, and antimicrobial peptides (42). Endurance in such an adverse environment is mediated by alternative transcription factors, some of which are specifically induced by the nutritional deprivation encountered upon prolonged residence in the phagosomes. Thus, the "stationary-phase-like" state assumed by *Mycobacterium tuberculosis* within phagosomes results in

extreme resistance to diverse assaults, including practically all available antibiotics (10). In virulent *Salmonella*, as well as in *Legionella pneumophila*, *Staphylococcus aureus*, and *Brucella abortus*, nutrient depletion induces the expression of genes that promote virulence and dramatically enhance stress resistance (13, 20, 58). Extracellular pathogens such as *Escherichia coli* and *Yersinia enterocolitica* can encounter extreme pH values, reactive electrophilic agents, osmotic stress, and deficiency of specific nutrients. Resistance against these assaults, as well as the expression of virulence determinants, reaches its highest levels when nutrients become limiting and cells cease growth and enter a stationary-phase-like state (14). Furthermore, the ability of bacteria to survive and disseminate in habitats outside the host, where they are regularly exposed to oxidative and osmotic stresses, UV radiation, and sharp temperature alterations, was found to be causally related to the dearth of nutrients that characterizes such soil or aquatic environments (8, 77).

Starvation, and the resulting stationary state, thus represents a critical facet of virulence, as it is regularly encountered by bacteria in their various habitats and significantly promotes their ability to endure continuous exposure to diverse assaults. How is such a broad and effective stationary state-dependent resistance achieved?

PROTECTION STRATEGIES IN STATIONARY-STATE BACTERIA: A CONCEPTUAL ENIGMA

All currently known prokaryotic defense pathways rely on inducible mechanisms that are activated by specific signals. Since inducible responses need to cope with assaults that are already present, they must

Abraham Minsky • Department of Organic Chemistry, The Weizmann Institute of Science, Rehovot 76100, Israel. **Roberto Kolter** • Department of Microbiology and Molecular Genetics, Harvard Medical School, 200 Longwood Ave., Boston, MA 02115.

evolve and reach full capacity as fast and as efficiently as possible. To meet this requisite, bacterial stress strategies are based on elaborate integrated circuits and branched regulatory cascades that enable a burstlike induction of one or several regulons as a response to a particular assault. Activation of inducible stress responses is thus associated with massive and rapid metabolic readjustments that require de novo synthesis of numerous proteins, some in very large quantities. Such metabolic alterations are fundamentally incompatible with a state of prolonged starvation, during which the supply of exogenous substrates is limited or nonexistent. The ability of starved bacteria to endure environmental assaults is therefore generally assigned to a maintenance mode that is mediated by the activities of a group of enzymes whose synthesis is triggered at the onset of starvation and proceeds for several hours (12, 21, 22, 25). For this maintenance strategy to be effective, two fundamental requirements must, however, be met. Enzymes must remain functional throughout the period of no growth, and the energy required for the chemical processes they catalyze must be available.

Neither of these requirements can be effectively fulfilled during prolonged periods of starvation. A growing body of evidence points to an accelerated oxidation and degradation of proteins during stationary phase. Significantly, proteins that are directly involved in bacterial stress responses, including the heat shock protein DnaK and the DNA-binding protein H-NS, were found to be particularly susceptible to oxidative damage (11, 12). Thus, proteins are continuously depleted as starvation proceeds, leading to a severely and progressively impaired ability of the cells to mount stress responses and maintain stress resistance.

Even more imposing are considerations that pertain to energy supplies. Environmental assaults are generally met through enzymatically catalyzed chemical reactions that act to neutralize detrimental agents or to repair damaged cellular components. For these reactions to be effective, their intrinsic thermodynamic equilibrium must, in most cases, be perturbed toward desired products. This bias is commonly achieved by coupling the reaction to the hydrolysis of high-energy species such as ATP. Yet, in nongrowing bacteria the main source of high-energy compounds is the degradation of endogenous components, predominantly ribosomes and membranes (49). Thus, unless tightly regulated, enzymatic repair activities may, by themselves, lead to an irreversible loss of cell integrity or of essential cellular functions (49). It follows that inducible enzymatic defense pathways that are highly effective during active growth or short periods of nutrient depletion are fundamentally incompatible with prolonged starvation. This assertion seems to be particularly relevant for DNA repair and protection.

DNA PROTECTION IN STARVED BACTERIA

The intrinsic chemical and physical vulnerability of DNA molecules (37) and the lethal effects caused by unrepaired DNA lesions, even if they occur at low frequency, highlight the need for particularly efficient DNA protection mechanisms. Cells can maintain the integrity of their DNA through two distinct pathways. Detrimental chemicals are neutralized and damage is repaired through enzymatically catalyzed chemical processes. In eukaryotic cells these biochemical pathways are amplified by the tight nucleosomal assembly that has been shown to provide a highly effective structural protection against DNA-modifying agents (41). Such a lasting mode of DNA protection is, however, absent in prokaryotes which lack a nucleosomal organization and whose chromatin is characterized by a significantly lower ratio of structural DNA-binding proteins to DNA than that found in eukaryotes (30).

In natural habitats, bacterial exposure to DNA-damaging factors such as oxidating and alkylating agents, radicals, or UV irradiation is often accompanied by nutrient depletion. Recombination and excision DNA repair pathways require rapid synthesis of numerous enzymes, whose exquisitely regulated activities are heavily ATP dependent (33, 36). The extravagance of DNA repair pathways is underscored by the adaptive response to DNA alkylation by which an entire protein molecule is expended in order to repair a single damaged base (60). Although such imposing requirements can be effectively met as long as nutrients are plentiful, they can hardly be answered in starved cells, where enzymes are rapidly and progressively degraded (11, 67), and de novo protein synthesis as well as energy-generating processes are seriously impaired (25). Thus, basic kinetic and thermodynamic considerations point toward a fundamental enigma: how is the broad and constitutive DNA protection that characterizes nongrowing bacteria achieved? Extensive studies conducted on dormant bacterial spores indicated that sporal DNA protection is almost exclusively based on prevention of DNA damage through structural sequestration, rather than on active repair mechanisms, which are precluded by the virtual inactivity of all enzymes present in the spore core (62). Recent observations imply that such a structure-dependent DNA protection strategy is also deployed by nonsporulating bacteria in the stationary state (15).

CHROMATIN ORGANIZATION IN RAPIDLY GROWING BACTERIA

In the next sections we present a concise survey of data derived from light and electron microscopy techniques on the architecture of the chromosome in actively growing bacteria, and proceed to briefly describe phase transitions that characterize DNA molecules. We consider these two issues a prerequisite for a deeper understanding of the factors that determine the structure of chromatin in stationary-state bacteria.

The organization of the chromosome within bacterial cells and the factors that dictate and modulate this organization remain, rather surprisingly, less thoroughly understood than the structure of eukaryotic chromosomes. Several factors conspired to bring about this situation. These include the small size of the prokaryotic cell that restricts the effectiveness of light microscopy, the relatively low content of DNA-binding proteins associated to the bacterial chromosome which resulted, as will be discussed below, in severe artifacts in conventional electron microscopic studies, as well as the apparent absence of structural order and architectural hierarchy within the nucleoid. This last feature represents a fundamental difference between bacterial DNA organization and that of eukaryotic DNA, which is orderly packed into a nucleosomal assembly that is further folded into higher-order structures.

Observations using various light microscopy techniques, including phase-contrast, confocal, and fluorescence microscopy, indicate that the chromatin in rapidly growing bacteria is confined to several lobes with distinct boundaries (31, 59, 70, 75). The recent application of cytological methods in conjunction with time-lapse fluorescence microscopy studies, including fluorescence in situ hybridization and the use of green fluorescent protein, has shown that in actively growing bacteria the chromosome is folded in a regular manner, with the origin of replication near one pole of the cell and the terminus near the other (69). Such an orderly packaging notwithstanding, the nucleoid is demonstrated by these techniques to be highly dynamic, with rapid movements of individual sites that can be detected at a variety of cellular locations (17, 47, 63, 64, 69).

In their comprehensive review on the bacterial nucleoid, Robinow and Kellenberger (59) emphasized the high sensitivity of bacterial chromatin to dehydration procedures necessary for preparation of electron microscopy specimens. Specifically, it has been demonstrated that due to the relatively low protein content in the nucleoid, DNA molecules undergo extensive dehydration-induced aggregation.

The authors argued that the confined globular shape of the nucleoid detected by electron microscopy following conventional OsO_4 fixation, which is reminiscent of the shape observed in light microscopes, reflects a preparation-dependent artifact. A rather different image of the bacterial chromatin was obtained by cryofixation of the cells, followed by freeze-substitution (59). In this method, living cells are very rapidly frozen in liquid ethane, and the amorphous ice in the frozen specimen is then slowly replaced with acetone containing low concentrations of OsO_4 at $-90°C$. It has been argued that such a procedure rapidly and effectively immobilizes the cellular components and prevents aggregation (29). Following the cryofixation-substitution procedure, bacterial chromatin is visualized as amorphous ribosome-free spaces irregularly dispersed over large parts of the cytoplasm (Fig. 1A) (23, 29, 59). The irregular "coralline-like" DNA organization in actively growing bacteria has been further demonstrated by DNA immunolabeling using DNA-specific antibodies (6, 24), as well as by specific DNA staining agents applied on cryofixed and cryosubstituted specimens (Fig. 1B) (15).

On the basis of the images obtained by the cryofixation-substitution technique, the architecture of the nucleoid in rapidly growing bacteria has been suggested to comprise a dense confined interior from which numerous long projections extend deep into the cytoplasm and may reach the inner face of the cytoplasmic membrane. The irregular shape of the chromatin was proposed to reflect its highly dynamic nature, associated with continuous replication and

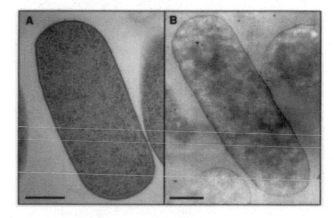

Figure 1. Electron microscopy of cryofixed *E. coli* cells. (A) Actively growing wild-type *E. coli* cell at mid-logarithmic phase. The dark particles are ribosomes. The ribosome-free spaces contain chromatin. (B) *E. coli* at mid-logarithmic phase, stained solely with the DNA-specific reagent osmium-amine-SO_2. The irregular spreading of the chromatin over large parts of the cytoplasm is indicated. Scale bars, 400 nm.

segregation, and the fine projections were suggested to represent stretches of the genome that are engaged in transcription (23, 59). Coupled transcription-translation and cotranslational insertion or translocation of membrane proteins (75) are deemed to be responsible for DNA-membrane contacts indicated by DNA immunostaining. Notably, the results obtained following the cryofixation-substitution technique are consistent with the highly active image of the chromatin derived from the recent time-lapse fluorescence studies using fluorescence in situ hybridization and green fluorescent protein. Moreover, it should be emphasized that, within the limits of resolution, there are no significant inconsistencies between the shape of the nucleoid in images obtained by fluorescence and electron microscopy, as the fine excrescences projected from the core would not be detected in the light microscope.

FORCES INVOLVED IN BACTERIAL DNA ARCHITECTURE

Phase Separation and Liquid Crystalline Phases of DNA Molecules

DNA molecules are characterized by a large intrinsic capacity to undergo monomolecular collapse and intermolecular aggregation from dispersed structures into ordered, highly condensed states (5, 40, 71, 78). The common denominator of these phase transitions is associated with large DNA densities that are obtained either due to modulations of the solvent properties, reflected by the Flory-Huggins parameter χ, or upon DNA confinement in a restricted space.

The χ parameter is proportional to the difference in free energy between the interactions of DNA molecules with the solvent (or with other solutes) and DNA-DNA interactions. At low χ values, DNA interacts preferentially with the solvent, whereas at higher χ values, DNA-DNA interactions prevail, leading to DNA phase separation (45, 53). χ is affected by multivalent cations such as polyamines which effectively screen electrostatic repulsions, enable cross-links, and produce DNA-DNA attraction through correlated fluctuation of the ion atmosphere (4, 5, 51, 65). χ is further affected by the presence of cosolutes that modulate the properties of the solvent. In addition, spontaneous DNA aggregation and phase separation have been shown to result from the mere confinement of large amounts of DNA, due to excluded volume effects (38–40, 57, 68). Specifically, in confined and highly crowded solutions of species characterized by a large axial ratio such as DNA molecules, an isotropic dispersion no longer represents the state of maximum entropy (50). For such elongated and rigid species, the crowding-induced decrease in orientational freedom is more than compensated for by the gain in translational entropy, resulting in ordered states and liquid crystallinity (34). In solutions containing relatively low DNA concentrations, a chiral cholesteric phase is particularly stable since, due to the DNA helical conformation, a twisted helical supramolecular assembly is associated with lower excluded volume and hence a higher translational entropy (19), as has indeed been demonstrated (35). Notably, the tight DNA packaging within a cholesteric organization has been shown to substantially reduce the accessibility of DNA molecules to a variety of damaging factors, including irradiation, radicals, and nucleases (46, 66).

DNA Phase Separation and Condensation in Bacteria

Although bacterial DNA is not confined by a nuclear membrane, the extremely crowded cellular milieu containing large concentrations of proteins is likely to favor phase separation and DNA condensation (45, 73, 79). These processes are effected by the presence of ions and macromolecules that modulate the χ value of the cytoplasm and cause large excluded volume effects. DNA phase separation and condensation are further promoted by DNA supercoiling, by the presence of specific DNA condensing ligands such as histone-like proteins, as well as by the effects of crowding on DNA interactions with these condensing agents. Local DNA densities within bacteria were evaluated by means of ratio-contrast microscopic imaging and were found to correspond to concentrations of approximately 80 to 100 mg/ml (7). The structural considerations presented above predict that such densities would result in the formation of a highly ordered liquid crystalline DNA phase (75, 79). Indeed, a tightly packed, presumably cholesteric assembly of plasmid DNA molecules was detected within *E. coli* bacteria harboring high-copy plasmids, by means of X-ray scattering measurements conducted on intact cells (55, 56).

On the basis of the observations obtained from the various microscopy techniques, and the considerations pertaining to DNA phase separation and condensation in crowded environments, an inclusive theory concerning the factors that determine the structure of bacterial DNA has been proposed (74, 75). According to this theory, the nucleoid architecture is dictated by a balance between two opposing forces: compaction and expansion. Compaction is imposed by DNA supercoiling, structural DNA-binding proteins such as HU and H-NS, as well as crowding. These factors, in particular the crowding effects, were suggested to be potent enough to result in persistent phase separation and mandatory compaction of the nucleoid (75, 79). Such a tight organization is indeed reflected by the dense core of the chromatin that is detected in both light and electron microscopes. Expansion forces derive, on the other hand, from the numerous ongoing transactions in which DNA molecules are engaged in actively growing cells: replication and segregation that occur simultaneously in rapidly growing bacteria, coupled transcription-translation, and cotranslational translocation of membrane proteins. Such transactions and the ensuing DNA expansion are demonstrated by the irregular and partially dispersed shape of the chromatin revealed in growing cells, the branched coralline DNA projections extending deep into the cytoplasm, as well as by the rapid delocalization of bacterial DNA sites detected in time-lapse fluorescence microscopy.

CHROMATIN ORGANIZATION IN NONGROWING BACTERIA

DNA-Dps Cocrystals

Under conditions of nutritional stress, *E. coli* cells produce a nonspecific DNA-binding protein termed Dps, whose expression is regulated by the alternative sigma factor σ^s (1, 2). Dps accumulates to a very large amount and constitutes the major protein component of the chromatin in late stationary bacteria (3, 27). Within the DNA-Dps complexes, the stability of Dps is substantially enhanced relative to the stability of the free protein (1), and the DNA is effectively protected against oxidizing agents and nucleases (44). Closely related Dps sequence homologs were identified in many distantly related bacteria (1, 9, 52), implying that this protein maintains a general and crucial function. The avid binding between the Dps and DNA, as well as the resulting enhanced protection of both components, is notable, since members of the Dps family do not share salient sequence similarities to any other currently known DNA-binding proteins and do not reveal any recognizable DNA-binding motifs (1, 18). The finding that the surface of the Dps dodecameric assembly, which is the plausible DNA-binding species, is dominated by negative charges (18), is intriguing in light of the negatively charged backbone of nucleic acids. The interaction between purified Dps and DNA molecules was found to result in an extremely rapid formation of DNA-Dps cocrystals (Fig. 2), within which the DNA is effectively protected against oxidative agents and various nucleases (76).

Cryofixed and cryosubstituted specimens of starved *E. coli* cells reveal a strikingly different morphology than that exhibited by rapidly growing cells prepared for electron microscopy studies through an

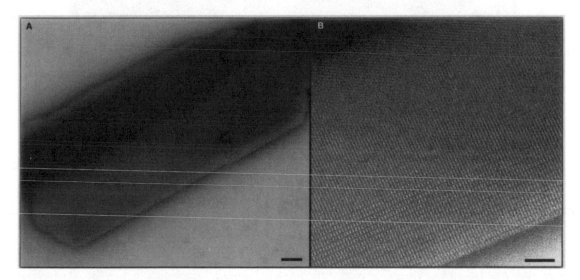

Figure 2. Electron microscopy of Dps-DNA cocrystals. (A) Purified Dps (50 µg/ml), incubated with closed circular supercoiled DNA molecules (Dps/DNA ratio, 1:5 [wt/wt]). Multiple crystals of dimensions and morphology similar to those revealed by the crystal depicted in this panel are detected following several seconds of incubation of the protein with linear, nicked-circular, and supercoiled DNA, as well as with single-stranded RNA molecules. Fluorescence studies using labeled DNA molecules indicate that DNA is incorporated within the crystals. (B) Higher magnification of the crystal shown in panel A. Scale bars, 100 nm (A) and 40 nm (B).

identical method. Two days following the onset of stationary phase, bacteria that overexpress Dps, as well as wild-type *E. coli* cells, exhibit a prominent demixing of the chromatin and the ribosomes, with the ribosomes localized in the periphery of the cytoplasm (15). Large Dps-DNA cocrystals, revealing a morphology similar to that observed in the in vitro generated DNA-Dps cocrystals, are formed in starved bacteria that overexpress Dps (Fig. 3A and B). Such prominent structures are only seldom encountered in starved wild-type cells. In the majority of these cells, a dense mass from which the ribosomes are

completely excluded is observed in the center of the cytoplasm (Fig. 3C). To assess the composition of this condensed mass, the DNA-specific stain osmium-amine has been applied on thin sections of starved bacteria (Fig. 3D). In sharp contrast to the DNA spreading observed using this method in actively growing cells (Fig. 1B), starved bacteria exhibit heavy staining that exclusively colocalizes with the central dense mass, indicating massive DNA segregation within this region. Notably, when DNA is isolated from starved bacteria, Dps is found to be the predominant protein bound to DNA, confirming previous

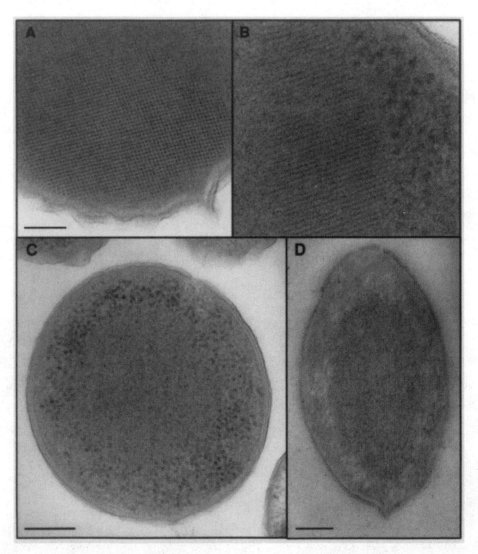

Figure 3. Electron microscopy of starved cryofixed *E. coli* cells. (A and B) Sections of Dps-overproducing *E. coli* cells induced to express Dps at mid-logarithmic phase and then incubated for 48 h. The two different morphologies of the intracellular Dps-DNA crystals correspond to sections that are parallel (A) or perpendicular (B) to the layered structure of the cocrystal (see Fig. 4B). (C) Wild-type *E. coli* cells incubated for 48 h after the onset of the stationary phase. A clear demixing of the ribosomes, located in the periphery of the cytoplasm, and the DNA at the center, is detected. (D) Wild-type *E. coli* incubated for 48 h after the onset of the stationary phase, and stained solely with the DNA-specific reagent osmium-amine-SO_2. The localization of DNA in the center, as opposed to DNA spreading during active growth (Fig. 1B), is clearly indicated. Scale bars, 50 nm (A) and 150 nm (C and D).

observations that Dps is the most abundant protein component of the nucleoid in stationary-phase *E. coli* cells (3, 27).

Ordered intracellular structures can be detected and characterized by X-ray scattering measurements conducted on intact cells (55). To gain insight into the structural features of the dense central region detected in stationary-state bacteria, this noninvasive technique was applied on actively growing and starved *E. coli* cells (15). Starved wild-type bacteria exhibit two X-ray bands: a relatively narrow band that corresponds to a spacing of 94 Å, and a broad band in which two maxima, at 49 and 47 Å, could be discerned (Fig. 4A). Similar diffraction patterns are revealed by stationary-state bacteria that slightly overproduce the Dps protein, whereas actively growing cells or starved bacteria incubated for 2 h in fresh media do not exhibit any X-ray diffraction maxima. The diffraction patterns are consistent with a DNA-Dps structure in which Dps and DNA form stacked alternating layers (Fig. 4B). According to this model, the diffraction at 94 Å corresponds to the intralayer spacing between Dps dodecamers whose diameter is about 90 Å (18). The broad band centered at 47 to 49 Å may represent a superposition of a second-order Dps-Dps diffraction and DNA-DNA spacing. Notably, similar DNA-DNA spacings were detected in an intracellular cholesteric organization of plasmid DNA molecules (55, 56). On the basis of the X-ray scattering patterns, the intracellular DNA staining, and the finding that Dps is the most abundant DNA-binding protein in starved bacteria (3), it has been proposed that the central region in stationary-phase wild-type cells consists of tightly packed DNA-Dps microcrystals within which the DNA is sequestered and physically protected (15).

DNA Cholesteric Liquid Crystalline Phase in Starved *dps⁻* Bacteria

Wild-type *E. coli* cells starved for 3 days have been shown to be essentially unaffected by relatively high doses of oxidating agents, whereas identical treatment led to a rapid loss of viability of more than 7 orders of magnitude in similarly starved mutants lacking *dps* (1). Six-day-starved *dps⁻* mutant cells exposed to H_2O_2 were found, however, to exhibit a viability loss of less than 2 orders of magnitude, indicating the acquisition of an effective Dps-independent mode of protection following prolonged starvation (15). The morphological features of actively growing *dps⁻* bacteria are indistinguishable from those of rapidly growing wild-type *E. coli*. While 5-day-old *dps⁻* bacteria still exhibit the same morphology, a dramatically different structure is revealed after an additional day. As observed for stationary wild-type *E. coli*, 6-day-old *dps⁻* cells exhibit a total separation of the ribosomes, localized at the periphery of the cell, from the DNA which occupies the central part of the cytoplasm (Fig. 5). Yet, the morphology of the DNA in the starved *dps⁻* mutant conspicuously differs from that revealed by starved wild-type cells, assuming a configuration of parallel rows of nested arcs. Very similar DNA patterns were previously detected in slowly growing bacteria and in primitive algae (39) and were found to represent

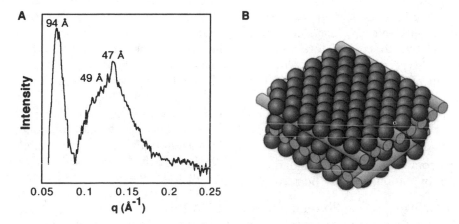

Figure 4. Dps-DNA cocrystal. (A) X-ray scattering patterns from intact wild-type *E. coli* cells, presented as a difference profile that is obtained by subtracting the scattering curve of mid-logarithmic-phase bacteria, in which no bands are discerned, from the scattering profile of *E. coli* cells incubated for 48 h following the onset of stationary phase. (B) A proposed schematic model of the Dps-DNA cocrystal. Dps dodecamers are depicted as spheres, and DNA molecules are represented as rods. The layered structure is consistent with the cocrystal morphology detected within Dps-overproducing *E. coli* cells (Fig. 3A and B), as well as with the X-ray scattering patterns exhibited by starved wild-type *E. coli* cells.

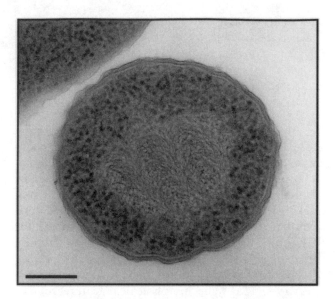

Figure 5. Electron microscopy of a *dps⁻ E. coli* cell. *dps⁻* cells from cultures incubated for 6 days following the onset of a stationary phase are shown. A salient demixing of the DNA in the center and ribosomes at the periphery is detected. The nested arcs revealed by the DNA are characteristic of a cholesteric liquid crystalline organization. Scale bar, 150 nm.

a cholesteric liquid crystalline DNA organization (35). In this arrangement, DNA segments are ordered in pseudoplanes which rotate with respect to each other to form a helicoidal structure. A projection into a plane of an oblique section of such an organization yields the observed parallel rows of nested arcs (Fig. 5).

As is the case for wild-type bacteria, the morphological traits detected in starved *dps⁻* cells are completely reversible: the morphology of 6-day-old cells incubated in fresh medium for 6 h is identical to that of actively growing bacteria. Yet, appearance of visible colonies from 6-day-old *dps⁻* cells is delayed by several hours relative to colony formation by similarly starved wild-type cells (15). Since the exponential growth rate of the two strains in either rich or poor medium is identical, this observation suggests that wild-type cells can resume growth after long-term starvation faster than *dps⁻* bacteria.

Biochemical Factors Determining Chromatin Reorganization in Stationary-State Bacteria

The in vitro DNA-Dps cocrystallization process was found to be very sensitive to the concentration of magnesium ions (15). In a solution containing 20 mM NaCl and 1.0 mM of the divalent ion chelator EDTA, no crystalline structures could be detected, whereas large and highly ordered crystals were rapidly formed when EDTA concentration was reduced

to 0.4 mM. Only small crystals were observed when Mg^{2+} was added to a final concentration of 1 mM, and no assemblies were obtained when the concentration of the doubly charged ion was raised to 3.0 mM. Notably, elevated concentrations of either EDTA (presumed to chelate Mg^{2+} ions that remain bound to the Dps during its purification) or Mg^{2+} not only inhibit the formation of the cocrystals but also act to destroy preexisting ordered complexes, thus demonstrating the reversibility of the cocrystallization process. The salient effects exerted by magnesium ions upon DNA-Dps interactions were further demonstrated by DNA protection studies. When DNA is exposed to nucleases following incubation with Dps, progressive DNA protection is observed as the Dps-to-DNA ratio is increased. Yet, when Mg^{2+} ions are included in the reaction mixture in increasing amounts, DNA protection is progressively eliminated and is completely abolished at 5.0 mM Mg^{2+}. Thus, both DNA-Dps cocrystallization and Dps-related DNA protection depend upon the specific concentration of Mg^{2+} ions. On the basis of these observations, it has been proposed that DNA-Dps complex formation is mediated through multiple ion bridges that are maintained by doubly charged cations. Significantly, such ion bridges can be formed only in a particular range of Mg^{2+} concentration. Above or below this range, both DNA and Dps dodecamers exhibit surfaces that are predominantly positively or negatively charged, respectively, and hence electrostatically repel each other.

The effects of Mg^{2+} and Fe^{2+} ions upon the morphology of starved *E. coli* cells was examined (15). Whereas stationary-state bacteria assemble their DNA in DNA-Dps micrococrystals that segregate in the central part of the cytoplasm, no DNA-Dps dense assembly or DNA-ribosome separation could be detected in cells that were starved for 2 days in media supplemented with either 5 mM Mg^{2+} or 10 μM Fe^{2+}. Similarly, no crystalline structures were observed under these conditions in starved bacteria that overproduce Dps. In both cases, the morphology is indistinguishable from that revealed by actively growing bacteria. In addition, starved wild-type or Dps-overproducing bacteria incubated in magnesium or iron-enriched media did not reveal any discrete X-ray scattering peaks, indicating that such conditions completely prevent the formation of ordered DNA-Dps complexes. On the basis of these results, and the observation that the concentration of free magnesium ions within *E. coli* cells in depleted media is essentially identical to the concentration of these ions in the environment (26), it was suggested that extracellular divalent cations provide an on-off signal for intracellular DNA-Dps cocrystallization (15). As long as the medium contains nutrients that sustain fast and active growth, including divalent ions in relatively high concentrations, DNA-Dps complex formation is prevented due to electrostatic repulsion. Yet, bacterial proliferation in defined niches within the host or in ex vivo environments leads to a progressive depletion of divalent ions due to rapid consumption or host defense activities. As the concentration of the ions decreases below a threshold value, ion bridges between DNA and Dps dodecamers can be formed, resulting in binding. When fresh nutrients—which include divalent ions—are available, the reverse process occurs, leading to the release of DNA from the DNA-Dps cocrystals.

The proposed role of divalent ions as an on-off switch is of fundamental physiological significance: upon bacterial infection, the host actively reduces the availability of iron ions in the phagosomes as part of its defense strategy against pathogens (42). Moreover, it has been suggested that phagosome maturation is accompanied by a decrease of Mg^{2+} ions within the phagosomal milieu to very low levels, and that this decrease acts as a primary environmental signal for the activation of the PhoPQ virulence system in *Salmonella* (72). As such, DNA protection by means of structural sequestration within DNA-Dps cocrystals provides a new aspect to the notion that pathogenic bacteria have evolved effective means to exploit the host defense mechanisms.

Chromatin reorganization into a cholesteric phase in starved *dps⁻* bacteria is unaffected by the extracellular concentration of doubly charged cations, but is substantially accelerated by the presence of polyamines in the growth media. Whereas a cholesteric DNA phase is detected in *dps⁻* cells following 5 to 6 days of "regular" starvation, bacteria supplemented with spermidine at the onset of the stationary phase reveal a cholesteric organization after only 4 days. This finding is consistent with the fact that polyamines promote an isotropic-to-cholesteric DNA transition in vitro (51, 65). Notably, phosphate starvation has been shown to effect enhanced conversion of threonine and arginine to spermidine, resulting in the accumulation of this polyamine in bacteria during stationary phase (16).

PHYSICAL ASPECTS OF CHROMATIN REORGANIZATION IN STATIONARY-STATE BACTERIA

In the first sections of this chapter, an enigma associated with DNA repair and protection in stationary-state bacteria was pointed out. The chemistry and structure of living systems, characterized by directed transitions, nonrandom spatial organization, and ever-evolving asymmetric assemblies, are all heavily energy dependent. Yet, bacteria are capable of outlasting prolonged periods of starvation during which energy-generating processes are rendered highly inefficient and the cellular energy balance is severely and progressively disturbed (25).

The architecture of bacterial stationary-phase chromatin appears to reflect the ability of bacteria to protect vital components by means of an energy-independent formation of ordered intracellular assemblies. This process can be considered in terms of the interplay between DNA compaction and expansion forces that are operative in bacterial systems. As detailed in the previous sections, mandatory DNA condensation that derives from DNA supercoiling, packaging proteins, and crowding is balanced in rapidly growing bacteria by the numerous transactions in which DNA molecules are continuously engaged (74, 75). This balance is, however, perturbed in stationary-state cells. The progressive degradation of DNA-binding proteins, as well as the gradual depletion of energy sources that accompanies prolonged starvation, results in a progressively attenuated frequency of DNA transactions. This massive attenuation in cellular activities tips the balance of forces toward DNA condensation. In *dps⁻* cells where the stationary-phase chromatin is highly depleted of proteins, such a change leads to a spontaneous, entropically driven DNA collapse into a tightly packed cholesteric liquid crystalline structure. In wild-type starved bacteria containing large amounts of the DNA-binding protein Dps, a phase transition that culminates in the formation of DNA-Dps cocrystals occurs. This transition can be considered in terms of DNA interactions with its environment. DNA-Dps complex formation dramatically attenuates interactions between DNA molecules and other cytoplasmic factors, resulting in phase separation, segregation, and cocrystallization.

DNA reorganization into crystalline or liquid crystalline morphologies of bacterial chromatin in starved bacteria promotes DNA protection through effective physical sequestration and shielding. Since Dps is produced at the very onset of starvation, yet does not bind DNA until a starvation-related signal (e.g., attenuation of cytoplasmic dication concentration) is produced, these transitions do not require de novo protein synthesis or enzymatically catalyzed chemical transformations, and hence no input of energy. Moreover, the crystalline and liquid crystalline DNA states can be maintained for indefinite time periods without energy consumption and dissipation, providing a uniquely adequate protection strategy in quiescent bacteria.

Although this survey is concerned with stationary-phase chromosomes, a sideways glance at sporulating bacteria seems to be in place. During sporulation, bacteria are divided into two unequally sized compartments consisting of the mother cell and the smaller forespore. The forespore chromatin assumes an extremely compact structure whereas the nucleoid in the mother cell retains its diffuse lobular morphology (54, 61). The factors that induce sporal DNA condensation, as well as the details of the resulting morphology, remain mostly unknown. It has, however, been shown that while the final ringlike structure of the spore chromatin is dictated by the major spore DNA-binding proteins, i.e., the small acid-soluble spore proteins (SASPs), initial DNA compaction occurs very early in the sporulation process and is independent of SASPs (61). According to the model of autocatalytic gene expression (48), when bacteria contain two chromosomes, one copy will be highly expressed in comparison with the other copy, which will sustain a progressive silencing. Following the considerations presented above, a partial silencing of the spore chromatin may lead to an entropically driven phase separation and phase transition of the spore DNA into a highly condensed liquid crystal structure, akin to that encountered in dps^- cells. Subsequent binding of the SASP proteins results in DNA reorganization into a different, yet still tightly packed and physically sequestered morphology.

REFERENCES

1. Almiron, M., A. J. Link, D. Furlong, and R. Kolter. 1992. A novel DNA-binding protein with regulatory and protective roles in starved *Escherichia coli*. *Genes Dev.* 6:2646–2654.
2. Altuvia, S., M. Almiron, G. Huisman, R. Kolter, and G. Storz. 1994. The *dps* promoter is activated by OxyR during growth and by IHF and sigma S in stationary phase. *Mol. Microbiol.* 13:265–272.
3. Azam, A. T., A. Iwata, A. Nishimura, S. Ueda, and A. Ishihama. 1999. Growth phase-dependent variation in protein composition of the *Escherichia coli* nucleoid. *J. Bacteriol.* 181:6361–6370.
4. Bloomfield, V. A. 1991. Condensation of DNA by multivalent cations: considerations on mechanism. *Biopolymers* 31:1471–1481.
5. Bloomfield, V. A. 1996. DNA condensation. *Curr. Opin. Struct. Biol.* 6:334–341.
6. Bohrmann, B., W. V. R. Johansen, and E. Kellenberger. 1991. Corraline shape of the bacterial nucleoid after cryofixation. *J. Bacteriol.* 173:3149–3158.
7. Bohrmann, B., M. Haider, and E. Kellenberger. 1993. Concentration evaluation of chromatin in unstained resin-embedded sections by means of low-dose ratio-contrast imaging in STEM. *Ultramicroscopy* 49:235–251.
8. Byrne, B., and M. S. Swanson. 1998. Expression of *Legionella pneumophila* virulence traits in response to growth conditions. *Infect. Immun.* 66:3029–3034.
9. Chen, L., and J. D. Helmann. 1995. *Bacillus subtilis* MrgA is a Dps(PexB) homologue: evidence for metalloregulation of an oxidative-stress gene. *Mol. Microbiol.* 18:295–300.
10. DeMaio, J., Y. Zhang, C. Ko, D. B. Young, and W. R. Bishai. 1996. A stationary-phase stress-response sigma factor from *Mycobacterium tuberculosis*. *Proc. Natl. Acad. Sci. USA* 93:2790–2794.
11. Dukan, S., and T. Nyström. 1998. Bacterial senescence: stasis results in increased and differential oxidation of cytoplasmic proteins leading to developmental induction of the heat shock regulon. *Genes Dev.* 12:3431–3441.
12. Dukan, S., and T. Nyström. 1999. Oxidative stress defense and deterioration of growth-arrested Escherichia coli cells. *J. Biol. Chem.* 274:26027–26032.
13. Fang, F. C., S. J. Libby, N. A. Buchmeier, P. C. Loewen, J. Switala, J. Harwood, and D. G. Guiney. 1992. The alternative sigma factor katF (rpoS) regulates *Salmonella* virulence. *Proc. Natl. Acad. Sci. USA* 89:11978–11982.
14. Ferguson, G. P., R. I. Creighton, Y. Nikolaev, and I. R. Booth. 1998. Importance of RpoS and Dps in survival of exposure of both exponential- and stationary-phase *Escherichia coli* cells to the electrophile N-ethylmaleimide. *J. Bacteriol.* 180:1030–1036.
15. Frenkiel-Krispin, D., S. Levin-Zaidman, E. Shimoni, S. G. Wolf, E. J. Wachtel, T. Arad, S. E. Finkel, R. Kolter, and A. Minsky. 2001. Regulated phase transitions of bacterial chromatin: a non-enzymatic pathway for generic DNA protection. *EMBO J.* 20:1184–1191.
16. Gérard, F., A. M. Dri, and P. L. Moreau. 1999. Role of Escherichia coli RpoS, LexA and H-NS global regulators in metabolism and survival under aerobic, phosphate-starvation conditions. *Microbiology* 145:1547–1562.
17. Gordon, G. S., D. Sitnikov, C. D. Webb, A. Teleman, A. Straight, R. Losick, A. W. Murray, and A. Wright. 1997. Chromosome and low copy plasmid segregation in *E. coli*: visual evidence for distinct mechanisms. *Cell* 90:1113–1121.
18. Grant, R. A., D. J. Filman, S. E. Finkel, R. Kolter, and J. M. Hogle. 1998. The crystal structure of Dps, a ferritin homolog that binds and protects DNA. *Nat. Struct. Biol.* 5:294–303.
19. Green, M. M., N. C. Peterson, T. Sato, A. Teramoto, R. Cook, and S. Lifson. 1995. A helical polymer with a cooperative response to chiral information. *Science* 268:1860–1865.
20. Hammer, B. K., and M. S. Swanson. 1999. Co-ordination of *Legionella pneumophila* virulence with entry into stationary phase by ppGpp. *Mol. Microbiol.* 33:721–731.
21. Hengge-Aronis, R. 1993. Survival of hunger and stress: the role of rpoS in early stationary phase gene regulation in *E. coli*. *Cell* 72:165–168.
22. Hengge-Aronis, R. 1996. Regulation of gene expression during entry into stationary phase, p. 1497–1512. *In* F. C. Neidhardt, R. Curtiss III, J. L. Ingraham, E. C. C. Lin, K. B. Low, B. Magasanik, W. S. Reznikoff, M. Riley, M. Schaechter, and H. E. Umbarger (ed.), Escherichia coli *and* Salmonella: *Cellular and Molecular Biology*, 2nd ed., vol. 1. ASM Press, Washington, D.C.
23. Hobot, J. A., W. Villiger, J. Escaig, M. Maeder, A. Ryter, and E. Kellenberger. 1985. Shape and fine structure of nucleoids observed on sections of ultrarapidly frozen and cryosubstituted bacteria. *J. Bacteriol.* 162:960–971.
24. Hobot, J. A., M. A. Bjornsti, and E. Kellenberger. 1987. Use of on-section immunolabeling and cryosubstitution for studies of bacterial DNA distribution. *J. Bacteriol.* 169:2055–2062.
25. Huisman, G. W., D. A. Siegele, M. M. Zambrano, and R. Kolter. 1996. Morphological and physiological changes during stationary phase, p. 1672–1682. *In* F. C. Neidhardt,

R. Curtiss III, J. L. Ingraham, E. C. C. Lin, K. B. Low, B. Magasanik, W. S. Reznikoff, M. Riley, M. Schaechter, and H. E. Umbarger (ed.), Escherichia coli and Salmonella: Cellular and Molecular Biology, 2nd ed., vol. 2. ASM Press, Washington, D.C.

26. Hurwitz, C., and C. L. Rosano. 1967. The intracellular concentration of bound and unbound magnesium ions in Escherichia coli. J. Biol. Chem. 242:3719–3722.

27. Ishihama, A. 1999. Modulation of the nucleoid, the transcription apparatus, and the translation machinery in bacteria for stationary phase survival. Genes Cells 4:135–143.

28. Kaiser, D. 1998. How and why myxobacteria talk to each other. Curr. Opin. Microbiol. 1:663–668.

29. Kellenberger, E. 1991. The potential of cryofixation and freeze substitution: observations and theoretical considerations. J. Microsc. 161:183–203.

30. Kellenberger, E., and S. G. B. Arnold. 1992. Chromatins of low-protein content: special features of their compaction and condensation. FEMS Microbiol. Lett. 100:361–370.

31. Kellenberger, E., and C. Kellenberger-Van der Kamp. 1994. Unstained and in-vivo fluorescently stained bacterial nucleoids and plasmolysis observed by a new specimen preparation method for high-power light microscopy of metabolically active cells. J. Microsc. 176:132–142.

32. Kolter, R., D. A. Siegele, and A. Tormo. 1993. The stationary phase of the bacteria life cycle. Annu. Rev. Microbiol. 47:855–874.

33. Kowalczykowski, S. C. 1991. Biochemistry of genetic recombination: energetics and mechanism of DNA strand exchange. Annu. Rev. Biophys. Biophys. Chem. 20:539–575.

34. Kulp, D. T., and J. Herzfeld. 1995. Crowding-induced organization of cytoskeletal elements. III. Spontaneous bundling and sorting of self-assembled filaments with different flexibilities. Biophys. Chem. 57:93–102.

35. Leforestier, A., and F. Livolant. 1993. Supramolecular ordering of DNA in the cholesteric liquid crystalline phase: an ultrastructural study. Biophys. J. 65:56–72.

36. Lin, J. J., and A. Sancar. 1992. (A)BC excinuclease: the Escherichia coli nucleotide excision repair enzyme. Mol. Microbiol. 6:2219–2224.

37. Lindahl, T. 1993. Instability and decay of the primary structure of DNA. Nature 362:709–715.

38. Livolant, F., A. M. Levelut, J. Doucet, and J. P. Benoit. 1989. The highly concentrated liquid-crystalline phase of DNA is columnar hexagonal. Nature 339:724–726.

39. Livolant, F. 1991. Supramolecular organization of double-stranded DNA molecules in the columnar hexagonal liquid crystalline phase. An electron microscopic analysis using freeze-fracture methods. J. Mol. Biol. 218:165–181.

40. Livolant, F., and A. Leforestier. 1996. Condensed phases of DNA: structure and phase transitions. Prog. Polymer Sci. 21:1115–1164.

41. Ljungman, M., and P. C. Hanawalt. 1992. Efficient protection against oxidative DNA damage in chromatin. Mol. Carcinog. 5:264–269.

42. Mahan, M. J., J. M. Slauch, and J. J. Mekalanos. 1996. Environmental regulation of virulence gene expression in Escherichia, Salmonella, and Shigella spp., p. 2803–2815. In F. C. Neidhardt, R. Curtiss III, J. L. Ingraham, E. C. C. Lin, K. B. Low, B. Magasanik, W. S. Reznikoff, M. Riley, M. Schaechter, and H. E. Umbarger (ed.), Escherichia coli and Salmonella: Cellular and Molecular Biology. ASM Press, Washington, D.C.

43. Margolis, P., A. Driks, and R. Losick. 1991. Differentiation and the establishment of cell type during sporulation in Bacillus subtilis. Curr. Opin. Genet. Dev. 1:330–335.

44. Martinez, A., and R. Kolter. 1997. Protection of DNA during oxidative stress by the nonspecific DNA-binding protein Dps. J. Bacteriol. 179:5188–5194.

45. Murphy, L. D., and S. B. Zimmerman. 1995. Condensation and cohesion of lambda DNA in cell extracts and other media: implications for the structure and function of DNA in prokaryotes. Biophys. Chem. 57:71–92.

46. Newton, G. L., J. A. Aguilera, J. F. Ward, and R. C. Fahey. 1996. Polyamine-induced compaction and aggregation of DNA: a major factor in radioprotection of chromatin under physiological conditions. Radiat. Res. 145:776–780.

47. Niki, H., Y. Yamaichi, and S. Hiraga. 2000. Dynamic organization of chromosomal DNA in E. coli. Genes Dev. 14:212–223.

48. Norris, V., and M. S. Madsen. 1995. Autocatalytic gene expression occurs via transertion and membrane domain formation and underlies differentiation in bacteria: a model. J. Mol. Biol. 253:739–748.

49. Nyström, T., C. Larsson, and L. Gustafsson. 1996. Bacterial defense against aging: role of the Escherichia coli ArcA regulator in gene expression, readjusted energy flux and survival during stasis. EMBO J. 15:3219–3228.

50. Onsager, L. 1949. The effects of shape on the interaction of colloidal particles. Ann. N. Y. Acad. Sci. 51:627–659.

51. Pelta, J., D. Durand, J. Doucet, and F. Livolant. 1996. DNA mesophases induced by spermidine: structural properties and biological implications. Biophys. J. 71:48–63.

52. Pena, M. M., W. Burkhart, and G. S. Bullerjahn. 1995. Purification and characterization of a Synechococcus sp. strain PCC 7942 polypeptide structurally similar to the stress-induced Dps/PexB protein of Escherichia coli. Arch. Microbiol. 163:337–344.

53. Post, C. B., and B. H. Zimm. 1982. Theory of DNA condensation: collapse versus aggregation. Biopolymers 21:2123–2137.

54. Ragkousi, K., A. E. Cowan, M. A. Ross, and P. Setlow. 2000. Analysis of nucleoid morphology during germination and outgrowth of spores of Bacillus species. J. Bacteriol. 182:5556–5562.

55. Reich, Z., E. J. Wachtel, and A. Minsky. 1994. Liquid-crystalline mesophases of plasmid DNA in bacteria. Science 264:1460–1463.

56. Reich, Z., E. J. Wachtel, and A. Minsky. 1995. In vivo quantitative characterization of intermolecular interactions. J. Biol. Chem. 270:7045–7046.

57. Rill, R. L. 1986. Liquid crystalline phases in concentrated aqueous solutions of Na⁺ DNA. Proc. Natl. Acad. Sci. USA 83:342–346.

58. Robertson, G. T., and R. Roop. 1999. The Brucella abortus host factor I (HF-I) protein contributes to stress resistance during stationary phase and is a major determinant of virulence in mice. Mol. Microbiol. 34:690–700.

59. Robinow, C., and E. Kellenberger. 1994. The bacterial nucleoid revisited. Microbiol. Rev. 58:211–232.

60. Samson, L. 1992. The suicidal DNA repair methyltransferases of microbes. Mol. Microbiol. 6:825–831.

61. Setlow, B., N. Magill, P. Febbroriello, L. Nakhimovsky, D. E. Koppel, and P. Setlow. 1991. Condensation of the forespore nucleoid early in sporulation of Bacillus species. J. Bacteriol. 173:6270–6278.

62. Setlow, P. 1995. Mechanisms for the prevention of damage to DNA in spores of Bacillus species. Annu. Rev. Microbiol. 49:29–54.

63. Shapiro, L., and R. Losick. 2000. Dynamic spatial regulation in the bacterial cell. Cell 100:89–98.

64. Sharpe, M. E., and J. Errington. 1999. Upheaval in the bacterial nucleoid. *Trends Genet.* **15:**70–74.

65. Sikorav, J. L., J. Pelta, and F. Livolant. 1994. A liquid crystalline phase in spermidine-condensed DNA. *Biophys. J.* **67:**1387–1392.

66. Spotheim-Maurizot, M., F. Garnier, R. Sabattier, and M. Charlier. 1992. Metal ions protect DNA against strand breakage induced by fast neutrons. *Int. J. Radiat. Biol.* **62:**659–666.

67. Stadtman, E. R. 1992. Protein oxidation and aging. *Science* **257:**1220–1224.

68. Strzelecka, T. E., and R. L. Rill. 1990. Phase transitions of concentrated DNA solutions in low concentrations of 1:1 supporting electrolyte. *Biopolymers* **30:**57–71.

69. Teleman, A. A., P. L. Graumann, D. C.-H. Lin, A. D. Grossman, and R. Losick. 1998. Chromosome arrangement within a bacterium. *Curr. Biol.* **8:**1102–1109.

70. Valkenburg, J. A. C., C. L. Woldringh, G. J. Brakenhoff, H. T. M. van der Voort, and N. Nanninga. 1985. Confocal scanning light microscopy of the *E. coli* nucleoid: comparison with phase-contrast and electron microscope images. *J. Bacteriol.* **161:**478–483.

71. Vasilevskaya, V. V., A. R. Khokhlov, Y. Matsuzawa, and K. Yoshikawa. 1995. Collapse of single DNA molecule in poly(ethylene glycol) solutions. *J. Chem. Phys.* **102:**6595–6602.

72. Vescovi, E. G., F. C. Soncini, and E. A. Groisman. 1996. Mg^{2+} as an extracellular signal: environmental regulation of *Salmonella* virulence. *Cell* **84:**165–174.

73. Walter, H., and D. E. Brooks. 1995. Phase separation in cytoplasm, due to macromolecular crowding, is the basis for microcompartmentation. *FEBS Lett.* **361:**135–139.

74. Woldringh, C. L., P. R. Jensen, and H. S. Westerhoff. 1995. Structure and partitioning of bacterial DNA: determined by a balance of compaction and expansion forces? *FEMS Microbiol. Lett.* **131:**235–242.

75. Woldringh, C. L., and T. Odijk. 1999. Structure of DNA within the bacterial cell: physics and physiology, p. 171–187. *In* R. L. Charlebois (ed.), *Organization of the Prokaryotic Genome.* ASM Press, Washington, D.C.

76. Wolf, S. G., D. Frenkiel, T. Arad, S. E. Finkel, R. Kolter, and A. Minsky. 1999. DNA protection by stress-induced biocrystallization. *Nature* **400:**83–85.

77. Yildiz, F. H., and G. K. Schoolnik. 1998. Role of rpoS in stress survival and virulence of *Vibrio cholerae. J. Bacteriol.* **180:**773–784.

78. Yoshikawa, K., and Y. Matsuzawa. 1995. Discrete phase transition of giant DNA dynamics of globule formation from a single molecular chain. *Physica D* **84:**220–227.

79. Zimmerman, S. B., and L. D. Murphy. 1996. Macromolecular crowding and the mandatory condensation of DNA in bacteria. *FEBS Lett.* **390:**245–248.

II. REPLICATION MACHINES

The Bacterial Chromosome
Edited by N. Patrick Higgins
© 2005 ASM Press, Washington, D.C.

Chapter 8

Replication Hits 50

KENNETH J. MARIANS

DNA replication has been studied intensively now for over 50 years. Indeed, the spring of 2003 brought several celebrations of the 50th anniversary of the publication of Watson and Crick's article (60) describing the structure of the double helix. Progress has been continuous and impressive: essentially all of the enzymes required for DNA replication in *Escherichia coli* have been identified and purified, their activities have been characterized biochemically, the effects of defects in their activities have been characterized genetically, and their role in the replisome has been defined. The chapters in this section focus on providing a current description of the state of the art and, while extensively referenced, should not be viewed as comprehensive treatises. In general, the chapters tend to highlight the last decade of research in the field. In this overview I will try to illustrate some conceptual visions that have been realized, as well as the dominant strains in research over the last 10 years. While it is easy to debate what the latter are, I consider them to be the role of protein-protein interactions in directing replication operations, the contribution of structural biology to our understanding of mechanism, the application of cell biological techniques to probe chromosome dynamics and its relationship to DNA metabolism in the cell, and the realization that the classical description of how chromosomal replication was executed was an oversimplification.

CONCEPTUAL VISIONS FULFILLED

Subsequent to Okazaki's proposal for discontinuous lagging strand synthesis (42), perhaps the most significant conceptual vision in research in DNA replication was that of Alberts and colleagues, who proposed that the machinery that synthesized the leading and lagging strands of DNA at the replica-

tion fork were likely to be coupled physically (54). In studies with replication forks reconstituted with purified bacteriophage T4 proteins, Alberts and his colleagues noted that even when reaction conditions restricted the fraction of active templates severely, both the nascent leading and lagging strands were still made. They suggested that this implied the existence of a mechanism to ensure the rapid retargeting of the lagging strand polymerase to the replication fork after the synthesis of an Okazaki fragment was completed and the polymerase was dissociated from the template.

In the original formulation of the trombone model of a replication fork (1), the solution to this problem was the proposal that the replication fork polymerase was dimeric, with two active catalytic centers, one synthesizing the leading strand and the other synthesizing the lagging strand. The lagging strand polymerase was proposed to be physically tethered to the leading strand polymerase, creating a loop in the lagging strand template. This resulted in the juxtaposition in space of the site of termination of synthesis of the current Okazaki fragment and synthesis of the primer for the next Okazaki fragment. Because the lagging strand polymerase did not leave the replication fork after it completed synthesis of an Okazaki fragment and dissociated from the lagging strand template, the time required to find the next primer to initiate Okazaki fragment synthesis decreased dramatically. The loop in the lagging strand template alternated between growth (during synthesis of the Okazaki fragment) and complete release (when the lagging strand polymerase moved to the new primer). The model derived its name from the trombone slide-like action of this loop.

One key aspect of this model has been proved correct for replication forks formed in vitro using either *E. coli*, bacteriophage T4, or bacteriophage T7 replication proteins. That is, there are two polymerases

Kenneth J. Marians • Molecular Biology Program, Memorial Sloan-Kettering Cancer Center, 1275 York Ave., New York, NY 10021.

at the replication fork, and the lagging strand polymerase is physically anchored to the fork. The mechanism of coupling, however, is only clear in the *E. coli* system, where the polymerases are coupled via interactions between the τ and α subunits of the DNA polymerase III holoenzyme (Pol III HE) (22, 61). It has been suggested that coupling in the bacteriophage T7 (7, 28) and bacteriophage T4 (50) systems may be mediated via interactions between the polymerases and the helicases. Very recent evidence does indicate, though, that the two bacteriophage T4 gene 43 polymerase subunits are in physical contact at the fork (16).

Interestingly, it is the same polymerase that synthesizes both the nascent leading and lagging strands of DNA—a situation that appears not to be preserved in *Bacillus subtilis* (8). In the case of *E. coli*, the polymerase is the combination of the core polymerase (αεθ) and the processivity factor, β. Thus, to account for the drastic differences in processivity called for on the leading and lagging strands, both McHenry and Johanson (39) and Maki et al. (36) proposed that asymmetric assortment of the subunits of the Pol III HE might generate two coupled polymerases at the replication fork with differential processivities. Recent observations, some of which are summarized in chapter 10, demonstrate that the Pol III HE indeed has a nonsymmetrical assortment of subunits, with a probable stoichiometry of $(\alpha\varepsilon\theta)_2\beta_4\tau_2\gamma\delta\delta'\chi\psi$ (44), with δ' and ψ being bound to γ, not to τ (12). Thus, even though both γ and τ are products of *dnaX*, the asymmetry arises because δ is bound to δ' and χ is bound to ψ. The arrangement of subunits in the holoenzyme, however, is likely to have more to do with the architecture of the replisome than with differential processivity (more about this later). It is quite clear that processivity on the lagging strand is limited by one of two events (32): either the collision of the lagging strand polymerase with the 5' end of the penultimate Okazaki fragment, as posited in the trombone model (1), or the interaction of primase with DnaB at the fork, as proposed by Wu et al. (62).

The replicon model (17) proposed that replication is controlled by an initiator protein that binds to a specific initiator region and that chromosome segregation is accomplished by binding of the newly replicated origin regions to the cell membrane, which then pulls the daughter chromosomes apart as it grows. The discovery that DnaA bound to *oriC* (11) confirmed the first part of this model, and, although the conception of how membranes grew that was embraced in the replicon model proved incorrect, investigators have long searched for a definitive connection between replication and the cell membrane. As summarized in chapter 9, it is clear that the cell

membrane plays an important role in the regulation of the initiation activity of DnaA and thus in the regulation of initiation at *oriC*. The protein is found at the membrane when immunolocalized in vivo, a region on the protein has been demonstrated to be required for this interaction, and phospholipid stimulates hydrolysis of bound ATP, thereby inactivating the initiation potential of the protein. The involvement of the cell membrane in chromosome segregation is still problematic; however, an emerging appreciation of cellular organization in *E. coli*, discussed below, may yet validate this aspect of the replicon hypothesis as well.

PROTEIN-PROTEIN INTERACTIONS ORDER REPLICATION FORK FORMATION AND FUNCTION

It seems reasonable that replication would be controlled in the cell. A complete divorce between the cycle of replication and the cycle of cell division would be detrimental to the population, leading to cells whose function was compromised by either the lack of a chromosome or the presence of too many of them. Although the direct links between replication and cell division still prove elusive, it has become clear that replication fork formation and function are driven to a large extent by an ordered series of protein-protein interactions.

One way that replication can be limited is to control where (and when) a replication fork can form. In *E. coli*, forming a replication fork is all about getting DnaB, the replication fork helicase, onto the DNA. This event is dependent on protein-protein interactions, as is subsequent replisome assembly. As discussed in chapter 9, a specific interaction between DnaB and DnaA (5, 37) delivers DnaB to the latter protein bound at the origin, leading to the displacement of DnaB from its resting interaction with DnaC in the cytoplasm. This latter interaction serves to limit promiscuous access of DnaB to the DNA, allowing fork loading only at the origin or by the replication restart pathway (discussed below).

Primase is then recruited via an interaction with DnaB (58). The synthesis of primers on the leading and lagging strand templates recruits the Pol III HE. Replisome assembly then requires the establishment of the τ-DnaB interaction (22, 65). The τ subunit of the Pol III HE forms the central core of the replisome. Interactions between τ and the α subunit of the polymerase core hold the leading and lagging strand polymerases together at the replication fork (22). This interaction forms the basis of the coupling of the leading and lagging strand polymerases discussed above. The τ-DnaB interaction also mediates rapid replication fork movement (22) and, by preventing cycling of β from the core-β complex (23), contributes to assigning which of the two polymerase assemblies becomes the leading strand polymerase (23, 65). τ also forms the core of the DnaX complex ($\tau_2\gamma\delta\delta'\chi\psi$), the actual form of the β clamp-loading machine in the replisome.

Once formed, function of the replication fork is directed by protein-protein interactions as well. The affinity of the primase-DnaB interaction governs Okazaki fragment size (57) and, as discussed above, can also direct cycling of the lagging strand polymerase. An interaction between primase and the Pol III HE limits the size of primers synthesized at the fork to 12 nucleotides (67). And, as discussed in chapter 10, an interaction between primase and single-stranded DNA-binding protein (SSB) stabilizes the newly synthesized primer on the DNA. This interaction is then displaced by a χ-SSB interaction that acts to deliver the primer to the lagging strand polymerase (64).

STRUCTURAL INSIGHTS INTO THE ACTIVITIES AND FUNCTIONS OF REPLICATION PROTEINS

An impressive number of structures of replication proteins have been solved, primarily by X-ray crystallography, leading not only to molecular portraits of the actors involved, but also to fundamental insights into the mechanisms of catalytic activity. The chemical mechanism of the granddaddy of all replication enzymes, DNA polymerase I, described in chapter 10, is clear, as is the explanation for nucleotide selectivity: induced fit could unmistakably be observed when comparing several of the structures (43). And it is indeed tight in that catalytic site, leaving little room for other than the correct incoming nucleotide-template base pair. Remarkably, relaxation of induced fit—the evolution of a roomier active site—turns out to be the means by which the lesion-bypass polymerases enable incorporation of a nucleotide opposite a bulky lesion (33).

The picture of how DNA strands are managed at the replication fork has begun to emerge as a result of the crystal structures of the SSB bound to single-stranded DNA (ssDNA) (45), the bacteriophage T7 gene 4 hexameric replication fork DNA helicase (47, 53), and the RNA polymerase domain of DnaG (21), the replication fork primase. The SSB structure shows us the path of ssDNA around the protomers in the tetramer in what is likely to be both forms of SSB binding to DNA (6, 34). Model building allows one

to template the crystal structure onto the electron micrographs of these different binding modes in a very satisfactory manner.

The asymmetrical hexameric structure of the helicase domain of the T7 gene 4 protein bound to ADPNP is likely to be the industry standard for replication fork helicases. A (not yet proven) model for translocation of the hexameric helicase on ssDNA links a binding change ATPase activity to movement of paddlelike DNA-binding loops in the center of the toroid, effecting a corkscrewlike motion of the lagging strand template (the screw) through the circular washer (the helicase) (53). Anchoring of the displaced leading strand template on the outside of the toroid is all that is required to effect unwinding. This model also addresses a potential topological problem at the replication fork. Because the helicase is anchored to the polymerase by protein-protein interactions, rotation of the helicase to effect unwinding would have specified a requirement that the interaction effectively be like a slipped differential, alternately breaking and reforming contacts. If the above model holds, no rotation of the helicase relative to the lagging strand template is required.

All polymerases appear to conform roughly to the "classic" right hand-like structure of DNA polymerase I, except DnaG. This protein has a different fold around the active site more characteristic of the structure of the TOPRIM (topoisomerase-primase [2]) motif as found in topoisomerases. This region is known to be involved in metal binding during catalysis. Modeling of the RNA polymerase domain of primase together with the helicase domain of the T7 gene 4 protein argues that the primase catalytic site is on the outside of the toroid composed of the primase and helicase domains of T7 gene 4 and, by inference, the associated DnaG and DnaB proteins at the replication fork (59). This suggestion is supported by biochemical data from Richardson's laboratory (26). This is an important model, because it solves the long-standing problem of how primase could go "backward" on the DNA template. The lagging strand template is exposed as ssDNA in the $5' \rightarrow 3'$ direction; thus, if the primase active site were in the center of the toroid, the enzyme would see its initiation signal after the downstream template had left. With the catalytic site being on the outside of the toroid, looping of the lagging strand template back on itself while it was bound would provide a normal disposition of template for the RNA polymerase activity of primase.

Particularly revealing have been the series of structures of the β clamp and clamp loader subunits from the Kuriyan lab (13, 18, 19, 25). The landmark structure of β revealed in an elegant picture the molecular basis for the sliding-clamp concept as proposed originally by Alberts and colleagues (15) and the prediction that the sliding clamp was tethered topologically to the DNA from the experiments of Stukenberg et al. (56). Of course, the circular nature of β raised another problem: how was the circle cracked to load β onto the DNA template? This question seems to have been answered by the recent structures of a portion of the clamp loader, taking the form of a pentameric complex in the order $\delta\gamma_3\delta'$, with all of the subunits associated by their N-terminal domains, and of a monomer of β bound to δ. γ, δ, and δ' all take the shape of AAA proteins, C-shaped proteins that appear to be able to hinge in the middle in a scissorslike action in response to binding and hydrolysis of ATP. Sequential ATP hydrolysis around the pentameric ring, starting at δ', frees the β-binding element of δ from δ', causing the pentamer to swing apart. δ now can interact with β, wedging itself into the protein and distorting an α-helix near the β dimer interface. Preexisting tension in the β ring causes it to spring open. It is still not clear if the primer-template is positioned specifically into the opened ring or simply diffuses in. Once in place, completion of ATP hydrolysis causes the ring to close around the DNA and causes the clamp loader to dissociate. Such a process may be common to many situations where rings have to be opened. It is therefore of considerable interest that the structure of the DnaA protein (9), another AAA protein, also falls into this C-shaped category. Thus, a similar process may be engaged when DnaA loads the DnaB ring to DNA at *oriC*.

CHROMOSOME DYNAMICS AND DNA REPLICATION

In eukaryotes, DNA replication appears to occur in stationary factories that can be visualized as foci in the nucleus via either immunofluorescence microscopy of replication proteins or fluorescence microscopy of live cells carrying green fluorescent protein-tagged replication proteins. Dedicated cellular organelles are also responsible for ensuring the proper segregation of chromosomes into daughter cells during mitosis. The underlying basis for chromosome segregation and the manner in which replication is executed in bacteria are less clear, with no obvious mitotic apparatus evident by various forms of microscopy. Absent the replicon hypothesis, with its proposal of attachment of the newly replicated origins to the cell membrane, most thinking was that replication was accomplished by the replisome traveling along the chromosome and that chromosome segregation was a passive process. Recent cell biological observations have revealed that bacterial cells are much more like eukaryotic ones in these aspects than was previously appreciated and that they exhibit a significant level of structural organization and compartmentalization.

In bacteria, the division plane at midcell is a crucial locus where most of the events leading to cytokinesis take place. Using green fluorescent protein-tagged subunits of the Pol III HE, Grossman and colleagues have shown that the DNA polymerases assemble (30) and replicate DNA at midcell (31) in *B. subtilis*. This replication factory remains stationary, suggesting that the DNA is drawn through the factory during replication. These observations have led to the development of the "extrusion-capture" model of chromosome partitioning (29), whereby replication itself drives DNA segregation by pushing the newly replicated DNA outward from the anchored factory toward opposite cell poles. Using immunofluorescence microscopy of various subunits of the Pol III HE, my laboratory has confirmed that replication in *E. coli* also occurs in factories (10).

In *E. coli*, recent evidence, summarized in chapter 11, suggests that as the newly replicated DNA is extruded from the factory, it becomes coated with SeqA, a protein that was found originally because of its effect on initiation of chromosomal replication at *oriC* (35). SeqA binds hemimethylated DNA (3, 55) and appears to be membrane associated (52). For much of the cell cycle, SeqA can be localized in the cell by immunofluorescence microscopy as discrete pairs of foci positioned such that they are flanking the replication factory (4, 14). It has been suggested that the combination of SeqA binding to the membrane and to the nascent DNA may help direct the daughter chromosomes away from the cell center (48). SeqA foci themselves appear to be dependent on the presence of MukB (41), an SMC-like protein found in bacteria that do not have bona fide SMC family members, which is likely involved in recondensation of the newly replicated chromosomes as they move away from the replication factory. This recondensation process also requires that the daughter chromosomes be supercoiled (49).

The role of SeqA in chromosome dynamics outlined above and the fact that its binding to hemimethylated DNA is clearly not restricted to the origin region must force a reevaluation of its proposed role in initiation of replication, as espoused in chapter 9. It seems more likely that all of the proposed effects of SeqA on initiation of replication at *oriC* are simply a consequence of its role in chromosome dynamics.

Not only does the chromosome appear to be managed in an active way during replication, but, as described in chapter 13, the structure of the chromosome itself appears to impart to it specific properties that are sensed during chromosome segregation and the resolution of dimeric chromosomes at *dif* by the XerCD system (described in chapter 28). Studies on the effect of moving *dif* on chromosomal dimer resolution and on the tolerance of deletions and inversions in the terminus region have led to the suggestion that the chromosome is polarized, meaning that the directionality of the path taken along the chromosome away from *oriC* has some physical manifestation. Such polarization implies that there are sequence elements (called *dif* activity zone-organizing polar elements) embedded in the chromosome that impart the first level of polarization and that there are likely to be proteins that somehow recognize these elements and translate the sequence polarization into a physical presence in the cell. A possible candidate for these sequence elements has been identified: the RAG sequence motif. It has been suggested that polarization might manifest itself physically through interaction with the C-terminal motor domain of the cell division protein FtsK.

Thus, one can envision an interesting picture of the chromosome being actively moved through the cell in a directional fashion by virtue of association with FtsK that is fixed at the septum. Indeed, this domain of FtsK has already been shown to affect the direction of resolution at *dif* by XerCD (see chapter 28). Furthermore, we have shown that the activity of topoisomerase IV, the enzyme that decatenates the daughter chromosomes, is temporally regulated, being manifested only late in the cell cycle and primarily at the septal-proximal regions of the nucleoids (10). We have also observed that the C-terminal domain of FtsK stimulates topoisomerase IV-catalyzed, but not gyrase-catalyzed, decatenation of multiply linked DNA dimers in vitro (unpublished data), the very activity required to decatenate the daughter chromosomes. It will be most interesting to see if these observations represent the first peek at a bacterial mitotic-like apparatus and terminosome.

REPLICATION IS A SALTATORY PROCESS

Most textbooks describe replication of the *E. coli* chromosome as a process that begins with the formation of two replication forks at *oriC* and proceeds with the progression of those two forks around the circular chromosome to meet somewhere in the terminus region. Implicit in this picture is the assumption that the forks that start replication are the same ones that complete it. Intensive biochemical and genetic research over the past decade, summarized in chapter 12, has demonstrated that this picture is an oversimplification of the process and that, in fact, there is a high probability that a fork formed at *oriC* will not complete replication of the chromosome. Instead, forks either become arrested by virtue of a collision with endogenous template damage or may collapse completely, either because of an encounter with a nick in one of the template strands or because of breakage at a stalled fork. It is likely that replication fork arrest also causes the replisome to dissociate. Thus, in these instances, not only must the template damage be repaired, but replication has to be restarted in an origin-independent manner. In addition, with certainty in the case of replication fork collapse, and possibly in some instances of replication fork arrest, replication restart is dependent on the action of the homologous recombination (HR) machinery to generate the substrate for restart. This recombination-directed replication is essential for survival, represents a previously unanticipated marriage of these two major pathways of DNA metabolism, and is an echo of the manner in which bacteriophage T4 is replicated.

All forms of replication restart illuminated thus far employ assembly of variations of the replication restart primosome (46) as the means of reactivating replication. The proteins required for restart primosome assembly—PriA, PriB, PriC, DnaT, DnaB, DnaC, and DnaG—were discovered originally because they acted to assemble a primosome on φX174 viral DNA during its conversion to the replicative form. A study by Kaguni and Kornberg (20) made it clear that whereas the last three proteins are required for initiation at *oriC*, the first four are not. A role for these orphaned proteins finally emerged as a result of studies on the phenotype of *priA* mutations (27, 40) and on alternate DnaA- and *oriC*-independent pathways of initiation of chromosomal replication (38). The convergence of these observations led to Kogoma's proposal that PriA was involved in assembling a replication fork on D loops formed by the HR enzymes (24). This proposal was itself an extension of previous suggestions that PriA was involved in restart (40, 51, 66).

While the Kogoma proposal related to replication fork collapse and the repair of double-strand breaks, a reaction that we have reconstituted in vitro (63), as described in chapter 12, subsequent studies demonstrated that the PriA-dependent pathway was involved in restart at arrested forks as well. In fact, it appears that there are multiple and redundant pathways of restart involving various assortments of the primosomal proteins. It is likely, therefore, that φX phage combined multiple pathways of fork loading to compete more efficiently with the cellular apparatus for DnaB.

Actual replication fork restart probably occurs at one of two types of substrates: joint molecules formed between a broken and an intact daughter chromosome arm, and a protein-free replication fork itself (i.e., a Y structure). Processing of stalled forks appears complex and is likely to involve a reaction that has been termed replication fork reversal (or regression). This interesting reaction occurs spontaneously in positively supercoiled DNA and involves the rewinding of the parental stands and the pairing of the nascent leading and lagging strands so that a Holliday junction is formed and a double-stranded end (formed by the pairing of the nascent strands) is generated. Many different enzymes can operate on this structure, and evidence exists for the potential involvement of RecBCD, RuvAB, RecG, and RuvC under various conditions. Metabolism of stalled forks must, however, lead to correction of the initial impediment and regeneration of a substrate for PriA-directed restart. These are intriguing reactions that are only now being unraveled biochemically.

The evolutionary implications for the coupling of HR and replication are interesting as well. One could argue that HR arose because of the importance of repairing stalled and broken replication forks and was then coopted into other types of DNA metabolism.

Another huge step forward has been the discovery of the lesion bypass polymerases. This group of proteins—the Y family of polymerases, described in chapter 10—has risen to prominence in a few short years of research. They appear to play a role in bypass of all forms of template damage and in hypervariation of immunoglobulin and T-cell-receptor genes in eukaryotic cells.

The picture as it emerges now is that replication forks are beset by all forms of hardship and succumb with a surprisingly high frequency. In *E. coli*, estimates of the probability that a replication fork formed at *oriC* will be arrested before completing replication of one arm of the chromosome vary from 0.15 to almost certainty. Clearly, this is a major sea change in our concept of how replication proceeds.

FORECAST

The past 50 years of research appears to have increased the apparent complexity of a seemingly simple reaction, DNA replication. In large part the reason for this contention is that our ability to appreciate and probe the complexity of the reaction is what has actually improved. It is surely safe to predict that this trend will continue. Biochemical reconstitution experiments, complemented by genetic analysis, should shed light on the currently byzantine

pathways of replication fork restart and how this process is connected to the cell cycle. For that matter, the connection between the cell cycle, cell division, and DNA replication should begin to emerge as well. Given the success in solving the crystal structure of the ribosome, it is exciting to anticipate the solution of structures of functional units of proteins at the replication fork. It does also seem that focus is starting to shift away from the strictly reductionist approach of understanding how one protein works or one reaction pathway operates to appreciating more global issues. How is the chromosome organized? How does that organization affect DNA metabolism? How is chromosome dynamics influenced by chromosome organization? And, how does chromosome dynamics affect DNA metabolism? Equally intriguing and becoming more approachable are questions such as, what is the nature and composition of organelles such as the replication factory? And, how is chromosome segregation actually accomplished? It should be an interesting next 50 years!

REFERENCES

1. **Alberts, B. M., J. Barry, P. Bedinger, T. Formosa, C. V. Jongeneel, and K. N. Kreuzer.** 1983. Studies on DNA replication in the bacteriophage T4 in vitro system. *Cold Spring Harbor Symp. Quant. Biol.* **47:**655–668.

2. **Aravind, L., D. D. Leipe, and E. V. Koonin.** 1998. Toprim—a conserved catalytic domain in type IA and II topoisomerases, DnaG-type primases, OLD family nucleases and RecR proteins. *Nucleic Acids Res.* **26:**4205–4213.

3. **Brendler, T., A. Abeles, and S. Austin.** 1995. A protein that binds to the P1 origin core and the oriC 13mer region in a methylation-specific fashion is the product of the host seqA gene. *EMBO J.* **14:**4083–4089.

4. **Brendler, T., J. Sawitzke, K. Sergueev, and S. Austin.** 2000. A case for sliding SeqA tracts at anchored replication forks during *Escherichia coli* chromosome replication and segregation. *EMBO J.* **19:**6249–6258.

5. **Carr, K. M., and J. M. Kaguni.** 2002. *Escherichia coli* DnaA protein loads a single DnaB helicase at a DnaA box hairpin. *J. Biol. Chem.* **277:**39815–39822.

6. **Chrysogelos, S., and J. Griffith.** 1982. *Escherichia coli* single-strand binding protein organizes single-stranded DNA in nucleosome-like units. *Proc. Natl. Acad. Sci. USA* **79:**5803–5807.

7. **Debyser, Z., S. Tabor, and C. C. Richardson.** 1994. Coordination of leading and lagging strand DNA synthesis at the replication fork of bacteriophage T7. *Cell* **77:**157–166.

8. **Dervyn, E., C. Suski, R. Daniel, C. Bruand, J. Chapuis, J. Errington, L. Janniere, and S. D. Ehrlich.** 2001. Two essential DNA polymerases at the bacterial replication fork. *Science* **294:**1716–1719.

9. **Erzberger, J. P., M. M. Pirruccello, and J. M. Berger.** 2002. The structure of bacterial DnaA: implications for general mechanisms underlying DNA replication initiation. *EMBO J.* **21:**4763–4773.

10. **Espeli, O., C. Levine, H. Hassing, and K. J. Marians.** 2003. Temporal regulation of topoisomerase IV activity in *E. coli*. *Mol. Cell* **11:**189–201.

11. Fuller, R. S., and A. Kornberg. 1983. Purified *dnaA* protein in initiation of replication at the *Escherichia coli* chromosomal origin of replication. *Proc. Natl. Acad. Sci. USA* 80:5817–5821.

12. Glover, B. P., and C. S. McHenry. 2000. The DnaX-binding subunits δ′ and ψ are bound to γ and not τ in the DNA polymerase III holoenzyme. *J. Biol. Chem.* 275:3017–3020.

13. Guenther, B., R. Onrust, A. Sali, M. O'Donnell, and J. Kuriyan. 1997. Crystal structure of the δ′ subunit of the clamp-loader complex of *E. coli* DNA polymerase III. *Cell* 91:335–345.

14. Hiraga, S., C. Ichinose, H. Niki, and M. Yamazoe. 1998. Cell cycle-dependent duplication and bidirectional migration of SeqA-associated DNA-protein complexes in *E. coli. Mol. Cell* 1:381–387.

15. Huang, C. C., J. E. Hearst, and B. M. Alberts. 1981. Two types of replication proteins increase the rate at which T4 DNA polymerase traverses the helical regions in a single-stranded DNA template. *J. Biol. Chem.* 256:4087–4094.

16. Ishmael, F. T., M. A. Trakselis, and S. J. Benkovic. 2003. Protein-protein interactions in the bacteriophage T4 replisome. The leading strand holoenzyme is physically linked to the lagging strand holoenzyme and the primosome. *J. Biol. Chem.* 278:3145–3152.

17. Jacob, F. S., S. Brenner, and F. Cuzin. 1963. On the regulation of DNA replication in bacteria. *Cold Spring Harbor Symp. Quant. Biol.* 28:329–348.

18. Jeruzalmi, D., M. O'Donnell, and J. Kuriyan. 2001. Crystal structure of the processivity clamp loader γ complex of *E. coli* DNA polymerase III. *Cell* 106:429–441.

19. Jeruzalmi, D., O. Yurieva, Y. Zhao, M. Young, J. Stewart, M. Hingorani, M. O'Donnell, and J. Kuriyan. 2001. Mechanism of processivity clamp opening by the δ subunit wrench of the clamp loader complex of *E. coli* DNA polymerase III. *Cell* 106:417–428.

20. Kaguni, J. M., and A. Kornberg. 1984. Replication initiated at the origin (*oriC*) of the *E. coli* chromosome reconstituted with purified enzymes. *Cell* 38:183–190.

21. Keck, J. L., D. D. Roche, A. S. Lynch, and J. M. Berger. 2000. Structure of the RNA polymerase domain of *E. coli* primase. *Science* 287:2482–2486.

22. Kim, S., H. G. Dallmann, C. S. McHenry, and K. J. Marians. 1996. τ couples the leading- and lagging-strand polymerases at the *Escherichia coli* DNA replication fork. *J. Biol. Chem.* 271:21406–21412.

23. Kim, S., H. G. Dallmann, C. S. McHenry, and K. J. Marians. 1996. τ protects beta in the leading-strand polymerase complex at the replication fork. *J. Biol. Chem.* 271:4315–4318.

24. Kogoma, T. 1996. Recombination by replication. *Cell* 85:625–627.

25. Kong, X. P., R. Onrust, M. O'Donnell, and J. Kuriyan. 1992. Three-dimensional structure of the β subunit of *E. coli* DNA polymerase III holoenzyme: a sliding DNA clamp. *Cell* 69:425–437.

26. Kusakabe, T., K. Baradaran, J. Lee, and C. C. Richardson. 1998. Roles of the helicase and primase domain of the gene 4 protein of bacteriophage T7 in accessing the primase recognition site. *EMBO J.* 17:1542–1552.

27. Lee, E. H., and A. Kornberg. 1991. Replication deficiencies in *priA* mutants of *Escherichia coli* lacking the primosomal replication n′ protein. *Proc. Natl. Acad. Sci. USA* 88:3029–3032.

28. Lee, J., P. D. Chastain II, T. Kusakabe, J. D. Griffith, and C. C. Richardson. 1998. Coordinated leading and lagging strand DNA synthesis on a minicircular template. *Mol. Cell* 1:1001–1010.

29. Lemon, K. P., and A. D. Grossman. 2001. The extrusion-capture model for chromosome partitioning in bacteria. *Genes Dev.* 15:2031–2041.

30. Lemon, K. P., and A. D. Grossman. 1998. Localization of bacterial DNA polymerase: evidence for a factory model of replication. *Science* 282:1516–1519.

31. Lemon, K. P., and A. D. Grossman. 2000. Movement of replicating DNA through a stationary replisome. *Mol. Cell* 6:1321–1330.

32. Li, X., and K. J. Marians. 2000. Two distinct triggers for cycling of the lagging strand polymerase at the replication fork. *J. Biol. Chem.* 275:34757–34765.

33. Ling, H., F. Boudsocq, R. Woodgate, and W. Yang. 2001. Crystal structure of a Y-family DNA polymerase in action: a mechanism for error-prone and lesion-bypass replication. *Cell* 107:91–102.

34. Lohman, T. M., and L. B. Overman. 1985. Two binding modes in *Escherichia coli* single strand binding protein-single stranded DNA complexes. Modulation by NaCl concentration. *J. Biol. Chem.* 260:3594–3603.

35. Lu, M., J. L. Campbell, E. Boye, and N. Kleckner. 1994. SeqA: a negative modulator of replication initiation in *E. coli. Cell* 77:413–426.

36. Maki, H., S. Maki, and A. Kornberg. 1988. DNA polymerase III holoenzyme of *Escherichia coli*. IV. The holoenzyme is an asymmetric dimer with twin active sites. *J. Biol. Chem.* 263:6570–6578.

37. Marszalek, J., W. Zhang, T. R. Hupp, C. Margulies, K. M. Carr, S. Cherry, and J. M. Kaguni. 1996. Domains of DnaA protein involved in interaction with DnaB protein, and in unwinding the *Escherichia coli* chromosomal origin. *J. Biol. Chem.* 271:18535–18542.

38. Masai, H., T. Asai, Y. Kubota, K. Arai, and T. Kogoma. 1994. *Escherichia coli* PriA protein is essential for inducible and constitutive stable DNA replication. *EMBO J.* 13:5338–5345.

39. McHenry, C. S., and K. O. Johanson. 1984. DNA polymerase III holoenzyme of *Escherichia coli*: an asymmetric dimeric replicative complex containing distinguishable leading and lagging strand polymerases. *Adv. Exp. Med. Biol.* 179:315–319.

40. Nurse, P., K. H. Zavitz, and K. J. Marians. 1991. Inactivation of the *Escherichia coli priA* DNA replication protein induces the SOS response. *J. Bacteriol.* 173:6686–6693.

41. Ohsumi, K., M. Yamazoe, and S. Hiraga. 2001. Different localization of SeqA-bound nascent DNA clusters and MukF-MukE-MukB complex in *Escherichia coli* cells. *Mol. Microbiol.* 40:835–845.

42. Okazaki, R., T. Okazaki, K. Sakabe, K. Sugimoto, and A. Sugino. 1968. Mechanism of DNA chain growth. I. Possible discontinuity and unusual secondary structure of newly synthesized chains. *Proc. Natl. Acad. Sci. USA* 59:598–602.

43. Patel, P. H., M. Suzuki, E. Adman, A. Shinkai, and L. A. Loeb. 2001. Prokaryotic DNA polymerase I: evolution, structure, and "base flipping" mechanism for nucleotide selection. *J. Mol. Biol.* 308:823–837.

44. Pritchard, A. E., H. G. Dallmann, B. P. Glover, and C. S. McHenry. 2000. A novel assembly mechanism for the DNA polymerase III holoenzyme DnaX complex: association of δδ′ with DnaX₄ forms DnaX₃δδ′. *EMBO J.* 19:6536–6545.

45. Raghunathan, S., A. G. Kozlov, T. M. Lohman, and G. Waksman. 2000. Structure of the DNA binding domain of *E. coli* SSB bound to ssDNA. *Nat. Struct. Biol.* 7:648–652.

46. Sandler, S. J., and K. J. Marians. 2000. Role of PriA in replication fork reactivation in *Escherichia coli. J. Bacteriol.* 182:9–13.

47. Sawaya, M. R., S. Guo, S. Tabor, C. C. Richardson, and T. Ellenberger. 1999. Crystal structure of the helicase domain from the replicative helicase-primase of bacteriophage T7. *Cell* 99:167–177.

48. Sawitzke, J., and S. Austin. 2001. An analysis of the factory model for chromosome replication and segregation in bacteria. *Mol. Microbiol.* **40:**786–794.

49. Sawitzke, J. A., and S. Austin. 2000. Suppression of chromosome segregation defects of *Escherichia coli muk* mutants by mutations in topoisomerase I. *Proc. Natl. Acad. Sci. USA* **97:**1671–1676.

50. Selick, H. E., J. Barry, T. A. Cha, M. Munn, M. Nakanishi, M. L. Wong, and B. M. Alberts. 1987. Studies on the T4 bacteriophage DNA replication system. *UCLA Symp. Mol. Cell Biol.* **47:**183–214.

51. Seufert, W., and W. Messer. 1986. Initiation of *Escherichia coli* minichromosome replication at *oriC* and at protein n' recognition sites. Two modes for initiating DNA synthesis in vitro. *EMBO J.* **5:**3401–3406.

52. Shakibai, N., K. Ishidate, E. Reshetnyak, S. Gunji, M. Kohiyama, and L. Rothfield. 1998. High-affinity binding of hemimethylated *oriC* by *Escherichia coli* membranes is mediated by a multiprotein system that includes SeqA and a newly identified factor, SeqB. *Proc. Natl. Acad. Sci. USA* **95:**11117–11121.

53. Singleton, M. R., M. R. Sawaya, T. Ellenberger, and D. B. Wigley. 2000. Crystal structure of T7 gene 4 ring helicase indicates a mechanism for sequential hydrolysis of nucleotides. *Cell* **101:**589–600.

54. Sinha, N. K., C. F. Morris, and B. M. Alberts. 1980. Efficient in vitro replication of double-stranded DNA templates by a purified T4 bacteriophage replication system. *J. Biol. Chem.* **255:**4290–4293.

55. Slater, S., S. Wold, M. Lu, E. Boye, K. Skarstad, and N. Kleckner. 1995. *E. coli* SeqA protein binds *oriC* in two different methyl-modulated reactions appropriate to its roles in DNA replication initiation and origin sequestration. *Cell* **82:**927–936.

56. Stukenberg, P. T., P. S. Studwell-Vaughan, and M. O'Donnell. 1991. Mechanism of the sliding β-clamp of DNA polymerase III holoenzyme. *J. Biol. Chem.* **266:**11328–11334.

57. Tougu, K., and K. J. Marians. 1996. The interaction between helicase and primase sets the replication fork clock. *J. Biol. Chem.* **271:**21398–21405.

58. Tougu, K., H. Peng, and K. J. Marians. 1994. Identification of a domain of *Escherichia coli* primase required for functional interaction with the DnaB helicase at the replication fork. *J. Biol. Chem.* **269:**4675–4682.

59. VanLoock, M. S., Y.-J. Chen, X. Yu, S. S. Patel, and E. H. Egelman. 2001. The primase active site is on the outside of the hexameric bacteriophage T7 gene 4 helicase-primase ring. *J. Mol. Biol.* **311:**951–956.

60. Watson, J. D., and F. H. C. Crick. 1953. Genetic implications of the structure of deoxyribonucleic acid. *Nature* **171:**964–967.

61. Wu, C. A., E. L. Zechner, A. J. Hughes, Jr., M. A. Franden, C. S. McHenry, and K. J. Marians. 1992. Coordinated leading- and lagging-strand synthesis at the *Escherichia coli* DNA replication fork. IV. Reconstitution of an asymmetric, dimeric DNA polymerase III holoenzyme. *J. Biol. Chem.* **267:**4064–4073.

62. Wu, C. A., E. L. Zechner, J. A. Reems, C. S. McHenry, and K. J. Marians. 1992. Coordinated leading- and lagging-strand synthesis at the *Escherichia coli* DNA replication fork. V. Primase action regulates the cycle of Okazaki fragment synthesis. *J. Biol. Chem.* **267:**4074–4083.

63. Xu, L., and K. J. Marians. 2003. PriA mediates DNA replication pathway choice at recombination intermediates. *Mol. Cell* **11:**817–826.

64. Yuzhakov, A., Z. Kelman, and M. O'Donnell. 1999. Trading places on DNA—a three-point switch underlies primer handoff from primase to the replicative DNA polymerase. *Cell* **96:**153–163.

65. Yuzhakov, A., J. Turner, and M. O'Donnell. 1996. Replisome assembly reveals the basis for asymmetric function in leading and lagging strand replication. *Cell* **86:**877–886.

66. Zavitz, K. H., and K. J. Marians. 1991. Dissecting the functional role of PriA protein-catalysed primosome assembly in *Escherichia coli* DNA replication. *Mol. Microbiol.* **5:**2869–2873.

67. Zechner, E. L., C. A. Wu, and K. J. Marians. 1992. Coordinated leading- and lagging-strand synthesis at the *Escherichia coli* DNA replication fork. III. A polymerase-primase interaction governs primer size. *J. Biol. Chem.* **267:**4054–4063.

The Bacterial Chromosome
Edited by N. Patrick Higgins
© 2005 ASM Press, Washington, D.C.

Chapter 9

Initiation of Chromosomal Replication

JOHANNA ELTZ CAMARA AND ELLIOTT CROOKE

A bacterium's genome must be duplicated faithfully once, and only once, during the cell cycle. The complex process of bacterial chromosomal replication can be divided into several stages: initiation, priming of chain starts, chain elongation, and termination. Since much of what is known about the initiation of bacterial chromosomal replication comes from studies of *Escherichia coli*, this chapter concentrates on initiation in that organism. Initiation, which occurs at a unique replication origin sequence (*oriC*) on the *E. coli* chromosome, is a highly controlled point in the regulation of the bacterial cell cycle. Genetic studies and the reconstitution of in vitro replication systems for *oriC*-containing plasmids helped define required components and molecular mechanisms of chromosomal initiation. Included have been the identification of key sequence elements contained within *oriC* and proteins that act upon the origin and how, working in concert, they promote the initiation of DNA synthesis. Many aspects of these topics have been covered in comprehensive reviews (40, 52, 63, 68, 97) that make extensive reference to original sources. Here, an emphasis is placed on some recent advances regarding the major participating factors, *oriC* and DnaA protein, the temporal order of initiation events, and mechanisms that regulate the initiation of chromosomal replication.

THE ORIGIN OF REPLICATION, *oriC*

The circular chromosome of *E. coli* has a single replication origin, *oriC*, located at 84.3 min. Definitive identification of *oriC* came with the construction of autonomously replicating minichromosomes, plasmids that contain *oriC* as the sole origin of replication (29, 120). Deletion analyses done independently from the left and right sides of *oriC* on minichromosomes defined the minimal origin as 245 bp (84). However, subsequent studies uncovered the necessity of a sequence-independent 12-bp AT-rich region flanking the left side of the 245-bp segment, thus redefining the minimal origin as being 258 bp in length (1).

Biochemical studies and comparisons of the sequence of the *E. coli* origin with those of other bacterial origins revealed crucial sequence elements within *oriC*. Included are the five DnaA boxes recognized by the DnaA initiator protein, three AT-rich 13-mers located on the left side of the origin, a binding site for integration host factor (IHF) to the right of DnaA box R1, a binding site for factor for inversion stimulation (FIS) between DnaA boxes R2 and R3, and 11 GATC Dam methyltransferase recognition sites (Fig. 1). The roles of these elements in the initiation and regulation of chromosomal replication are discussed below. Other factors found to bind sequence-specific elements in *oriC* are IciA (34, 37, 109) and Rob (96). However, the significance of these factors in initiating or controlling initiation of chromosomal replication in vivo is unknown.

Comparative alignment of bacterial origins further revealed that the conserved DnaA boxes and AT-rich 13-mers in *oriC* are separated by segments of DNA that are variable in sequence but not in length, suggesting that helical phasing between the elements is critical. In agreement, replication from origins constructed with altered spacing is perturbed. Insertions and deletions between DnaA boxes R3 and R4, or insertions between DnaA boxes R2 and R3, are tolerated only if they constitute a complete helical turn of the DNA (67). Spacing between the AT-rich 13-mers and the DnaA boxes is also critical: any changes in the distance between the clusters of these two elements cause a severe decrease in replication (33).

The context and composition of common origin elements in several different bacteria have been investigated and compared. In *Bacillus subtilis*, a 10-kb fragment from the origin region was found to resemble

Johanna Eltz Camara and Elliott Crooke • Department of Biochemistry and Molecular Biology, Georgetown University Medical Center, 3900 Reservoir Rd. NW, Washington, DC 20007.

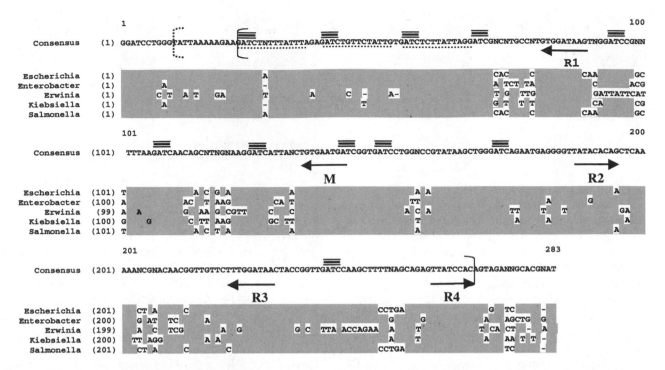

Figure 1. Alignment of bacterial *oriC* DNA sequences. The sequences were acquired by performing a standard nucleotide-nucleotide BLAST search with the *E. coli oriC* sequence (gi | 42154) on the National Center for Biotechnology Information website (http://www.ncbi.nlm.nih.gov/BLAST/). Matches were found for *Enterobacter* (gi | 148371), *Erwinia* (gi | 40915), *Klebsiella* (gi | 149827), and *Salmonella* (gi | 154217). The alignment was performed using AlignX from Vector NTI software (InforMax, Inc.) The solid brackets highlight the traditional 245-bp *oriC* region. The dotted bracket encompasses an AT-rich region also found important for *oriC* function. AT-rich 13-mers are indicated with a dotted underline, GATC methylation sites are indicated with a triple bar, and 9-mer DnaA-binding sites are indicated with underlining arrows and are labeled (R1 through R4, and M). Shaded regions represent identity to the consensus sequence. Variations from the consensus sequence are indicated by a letter, or by a hyphen (-) for a gap in the sequence. Regions of *oriC* that lack consensus are represented by N. (Fewer than three sequences contain the same nucleotide at that position.)

the *rnpA-dnaA-gyrB* gene organization of *E. coli* and to contain open reading frames for *rnpA*, *dnaA*, *dnaN*, *recF*, and *gyrB*. Two regulatory regions flanking *B. subtilis dnaA*, one or both of which are proposed to be replication origins, contain varying numbers of DnaA boxes. The conserved *dnaA-gyrB* region of this fragment is believed to be part of the replication origin of a primordial replicon (82).

Initial attempts to locate autonomously replicating fragments from the regulatory regions of *B. subtilis dnaA* failed due to the presence of DnaA boxes, which inhibit plasmid transformation (74). When these difficulties were overcome, identification of an autonomously replicating sequence from the *B. subtilis* chromosome was possible (75), facilitating the generation of an in vitro replication system for *B. subtilis* DNA synthesis (76). Use of this system has aided in showing that replication in *B. subtilis* is semiconservative and bidirectional (76, 77).

In *Pseudomonas putida*, the genes *dnaA*, *dnaN*, *recF*, and *gyrB* were found to be highly homologous to genes in the *dnaA* regions of both *E. coli* and

B. subtilis. *P. putida*, *E. coli*, and *B. subtilis* contain a 600-bp region upstream of *dnaA* containing varying numbers of DnaA boxes. It appears that the *dnaA*-DnaA box combination is conserved in most eubacteria and represents the ancestral replication origin (20). *E. coli oriC* is 44 kb away from this primordial origin as the result of either translocation or independent evolution.

The *oriC* of *Mycobacterium smegmatis* has been hypothesized to be in a 3.5-kb region surrounding *dnaA*. Deletion analysis of this region revealed a 531-bp fragment that exhibited *oriC* activity. However, the activity was lower than that obtained using the entire *dnaA* region, which contains flanking regions that most likely stimulate activity. A 495-bp fragment from the same region failed to produce *oriC* activity. This active 531-bp fragment contains putative DnaA-recognition sites as well as an AT-rich cluster, both of which appear to be essential. Mutations in the DnaA boxes resulted in decreased *oriC* activity, and deletion of the AT-rich region completely abolished activity (86).

The *oriC-dnaA* region of *Thermus thermophilus* has also been identified. It consists of 13 characteristically arranged DnaA boxes and an AT-rich stretch, followed by *dnaN*. A typical four-domain DnaA protein is expressed from *dnaA* upstream of the origin. Unlike *E. coli*, *Tth*-DnaA specifically binds three tandemly repeated DnaA boxes. This high-affinity binding requires the presence of ATP (91).

The chromosomes of streptomycetes differ from those of other prokaryotes in that they are large, linear, and GC rich. A comparison of three *Streptomyces* species revealed that their *oriC* regions contain 19 DnaA boxes with varying levels of specific binding to DnaA, all with conserved location, orientation, and spacing. *Streptomyces oriC* contains short, AT-rich sequences, unlike the AT-rich 13-mers found in *E. coli oriC*. Attempts to unwind *Streptomyces* have been unsuccessful, possibly due to the lack of accessory proteins that may interact with DnaA in the unwinding process (39).

INITIATOR PROTEIN, DnaA

DnaA protein, a sequence-specific DNA-binding protein, is responsible for setting in motion the cascade of events for initiating chromosomal replication, including origin recognition, strand opening, and loading of the replicative helicase at the sites of the future bidirectional replication forks (Fig. 2). How DnaA, with its multiple functions, participates in this initiation process is discussed below in the outline of sequential initiation events.

Structural considerations of *E. coli* DnaA have been derived largely from genetic analyses of *dnaA* mutants, biochemical and biophysical examinations of DnaA protein, and work with the primary sequence of DnaA through sequence alignments and structurally predictive algorithms (69, 90, 104). In general, DnaA has been predicted to have four distinct domains. The amino-terminal domain I (amino acids 1 to 86) is involved in the protein-protein interactions of DnaA oligomerization and binding of DnaB helicase. Domain II (amino acids 87 to 134) seems to comprise a flexible loop whose length varies between different bacterial DnaA proteins. Domain III (amino acids 135 to 373) adopts an open twisted α/β structure that serves as the core of DnaA. Within domain III are the high-affinity nucleotide-binding site and a second site for interaction with DnaB. The carboxy-terminal remainder of DnaA, domain IV, possesses the site for sequence-specific DNA binding.

A milestone in the understanding of DnaA structure and function was reached with the deter-

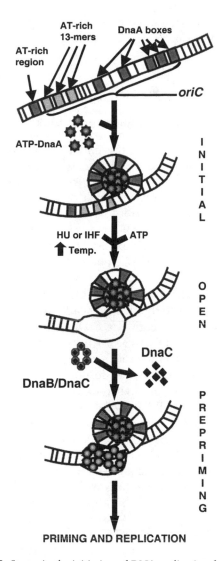

Figure 2. Stages in the initiation of DNA replication from *oriC*.

mination of the three-dimensional structure of the conserved core region of a bacterial DnaA protein to 2.7-Å resolution (19). A truncated version of *Aquifex aeolicus* DnaA, one having domains III and IV (amino acids 76 to 399 of *A. aeolicus* DnaA), was crystallized in its ADP-bound form. Within domain III the protein exhibits the nucleotide-binding fold motif typical of the AAA+ family of ATPases (78). The DNA-binding domain IV was seen to be a long helix followed by a helix-turn-helix motif and two additional helices. Modeling showed that domain IV can associate with DNA, and only slight swiveling about the helix that connects domains III and IV is required. Moreover, modeling guided by structural homologues of oligomeric AAA+ proteins predicts that protomer-protomer interactions involving the nucleotide-binding cleft may occur, providing an explanation of how domains III and IV of DnaA can support homo-oligomerization.

INITIATION EVENTS

Initiator Binding to the Origin

Multiple copies of the initiator protein, DnaA, recognize and bind to 9-mer sequences in *oriC*, known as DnaA boxes (Fig. 2). As mentioned above, DNA footprinting revealed that there are four DnaA boxes, known as R1 to R4, which have the common sequence 5'-TTAT(C/A)CA(C/A)A, within *oriC* (21, 32, 35, 116). A fifth box, M box, with the related sequence 5'-TCATTCACA, binds DnaA in vitro with less efficiency (66). It has also been shown by DNA footprinting that DnaA must be bound to ADP or ATP in order to bind specifically to these 9-mer DnaA boxes; nucleotide-free DnaA binds less specifically to *oriC* DNA (35). All five DnaA boxes are essential for *oriC* function, although point mutations in individual DnaA boxes do not impair replication from *oriC* in vivo or in vitro (32, 57). A combination of mutations in R1 and R4 results in the slowed growth of *oriC* plasmid-bearing host cells. However, mutations in these DnaA boxes do not affect the ability of DnaA to bind to other DnaA boxes (32).

When mutational analysis of chromosomal *oriC* was performed, no mutants with an altered R1 were obtained, pointing to the importance of this DnaA box for *oriC* function. Mutations in DnaA boxes R2, R3, and M were obtained, but flow cytometry revealed that they were accompanied by asynchronous initiation. Also, mutant cells in which the sequence of R4 was scrambled or the spacing between R3 and R4 was altered were severely asynchronous. Many of these chromosomal *oriC* mutations were not able to support minichromosome replication, bringing the definition of the minimal *oriC* into question as it is defined by minichromosomal *oriC* function (115). Other work revealed that inverting or scrambling R1 or M DnaA boxes inactivated DNA replication, and replacing R3 with R1, which possesses a higher DnaA binding affinity, upset the regulation of initiation. These results suggest strict binding requirements for optimal formation of the replication initiation complex (57).

DnaA binds in an ordered manner to DnaA boxes in reactions containing ATP or ADP, the nucleotides bound to DnaA in vivo. DnaA box R4 is the first site bound and has a greater affinity than R1, which is the next DnaA box bound. DnaA subsequently occupies R2 and R3. Little or no binding was observed for DnaA box M. The binding to R3 may be critical in promoting replication since it is the last DnaA box bound and it occurs at the time replication is initiated, while DnaA boxes R1, R2, and R4 are bound throughout the cell cycle (13).

The 94-amino-acid carboxy-terminal region of DnaA is responsible for the specific binding to DnaA boxes. Mutational studies have provided insight into this region of DnaA protein. Mutations found in the temperature-sensitive alleles dnaA204, dnaA205, and dnaA211 are able to completely abolish DNA binding (88). Further studies demonstrated that mutations causing defective DNA binding could be found all over the carboxy-terminal region and revealed, through surface plasmon resonance, that the mutant proteins have varied characteristics, including those with no or severely reduced binding and those with altered binding specificities (5). DnaA proteins with missense mutations have been isolated based on their inability to support replication of the DnaA-dependent plasmid pSC101. Analysis of these mutant DnaA proteins pointed to four functional domains in DnaA, including a carboxy-terminal DNA-binding domain (104).

When investigating which residues in the C-terminal region were responsible for specific binding, 11 missense mutations displayed defective DNA binding. One particular mutation, T435M, retained only nonspecific DNA binding capability, suggesting that this residue is responsible for specific DnaA-DNA binding. When this mutant form of DnaA was used as the sole source of DnaA in an in vitro DNA replication system, it was not able to support replication. However, T435M DnaA was able to augment limiting amounts of wild-type DnaA in vitro, supporting the hypothesis that not all DnaA monomers are specifically bound to the origin and that there is direct interaction among monomers at the origin (105).

Helicobacter pylori DnaA, whose expression has been verified through examination by Western blot and reverse transcription-PCR, also possesses a DNA-binding domain in its C terminus. This domain binds specifically to five DnaA boxes present in the *H. pylori* putative *oriC* region, which resides upstream of *dnaA*, as analyzed by DNase I and gel retardation studies (122).

Strand Opening

In *E. coli*, once DnaA has bound to the DnaA boxes, certain regions of *oriC* undergo unwinding (Fig. 2). Deletion studies of *oriC* and DNA nuclease probing identified three tandem 13-mer AT-rich regions involved in initial duplex unwinding for the formation of the open complex (8, 9, 14). Mutational analysis of these 13-mers in vitro and in vivo has shown that oriC function requires AT richness in the leftmost 13-mer and sequence specificity in the middle and right 13-mers. Moreover, interaction of the DNA with DnaA was crucial for opening in the middle and right 13-mers (36).

ATP-DnaA binds to these AT-rich 13-mers in 6-mer segments known as ATP-DnaA boxes (5'-AGatct-3'). This binding requires binding of DnaA to the R1 9-mer DnaA box, as seen through DNase I footprinting and surface plasmon resonance. ATP-DnaA binds with an affinity of 1 nM to the R1 DnaA box, while it binds with affinities of 400 and 40 nM to double-stranded and single-stranded ATP-DnaA boxes, respectively. Cooperative binding to these regions of varying binding affinities may be a general property of replication initiator proteins (102).

Other independent methods have also pointed to a role for DnaA in the unwinding of oriC. Monoclonal antibodies to DnaA can inhibit oriC unwinding by blocking the interaction between the protein and DNA. Two particular DnaA mutations, R334H and R342H, have been shown to make the DnaA activity for unwinding oriC more sensitive at low temperatures due to a defect in the potential to unwind (65, 108). Potassium permanganate probing of the oriC region provided the first evidence that ATP-DnaA induced duplex melting and a conformational change of AT-rich regions during initiation, severely distorting the DNA between 13-mers (24).

The construction of a hybrid origin that contains the DnaA-box region of E. coli oriC and the AT-rich region of B. subtilis oriC was unwound by E. coli DnaA, but not by B. subtilis DnaA. The AT-rich region of B. subtilis contains a 27-mer exclusively made up of A and T with AT-rich sequences close by, differing from the AT-rich 13-mers of E. coli. Although the AT-rich regions of these bacteria are very different in sequence, they nonetheless are very similar in their spatial pattern of unwinding. It appears that species specificity for unwinding in the AT-rich regions of oriC lies in the DnaA boxes (92).

For DnaA to trigger duplex opening on oriC minichromosomes in vitro, DnaA also requires the presence of accessory protein HU or IHF, physiological concentrations of ATP, and negative superhelicity in the plasmid DNA. The extent to which these minichromosomes open correlates to the subsequent amounts of plasmid replication (35). HU protein is a heterotypic dimer made up of alpha and beta subunits and may play a role in regulating the initiation of DNA replication. Synchrony studies performed on cells in the absence of one or both subunits revealed asynchronous initiation in the absence of both subunits, while the absence of the alpha subunit resulted in reduced cell synchrony and the absence of the beta subunit resulted in a normal phenotype. The ratio of alpha to beta subunits in the dnaA46ts mutant is greatly perturbed, with little or no accumulation of beta. Therefore, it appears that HU composition adjusts to different states of DnaA (4).

IHF binds to a specific site on oriC at the time of DNA replication initiation and stimulates open complex formation of supercoiled oriC in vitro. Footprinting studies reveal that IHF is able to redistribute DnaA prebound to supercoiled oriC plasmids. IHF stimulated DnaA binding to the DnaA boxes R2, R3, and R5 (M box), as well as three other previously unidentified non-R sites, when bound to its cognate site and accompanied by a functional R4 DnaA box. IHF functionally reduces the levels of DnaA required to open DNA strands and trigger initiation, which eliminates competition between sites with stronger and weaker affinities for free DnaA (26).

There appears to be interplay between IHF and another protein, FIS, at the origin throughout the cell cycle as well as in different stages of growth. Mutations in FIS or IHF binding sites result in asynchronous initiation as measured by flow cytometry (115). Initially, purified FIS protein appeared to have an inhibitory effect on in vitro oriC DNA replication, which could be modulated by varying concentrations of DnaA and RNA polymerase. This inhibition by FIS did not depend on the binding of FIS to its primary binding site between DnaA boxes R2 and R3. It was proposed that FIS was involved in an alternate, initiation-preventing complex (28, 117).

In situ footprinting of complexes containing DnaA and FIS showed that when FIS was bound to its primary binding site, it did not block the binding of DnaA to DnaA boxes R2 and R3. Furthermore, when FIS was stably bound to its binding site, it did not inhibit in vitro oriC DNA replication, contrary to previous observations. Inhibition of oriC DNA replication was observed only at high FIS levels, which results in the absorption of the negative superhelicity of oriC plasmid essential for replication (62).

FIS and IHF appear to alternate in their binding at the origin during different stages of the cell cycle and cell growth. At least two different nucleoprotein complexes have been identified in cycling cells. The first complex is found during most of the cell cycle and contains FIS. The nucleoprotein complex then switches to an IHF-bound form at the time that DNA replication is initiated. IHF binding coincides with DnaA binding to its previously unbound R3 DnaA box. In stationary-phase cells, a third complex, in which FIS is absent and IHF is bound, forms at inactive origins; this complex is not observed in proliferating cells (13). This observation agrees with results of an independent method in which cells were permeabilized to allow nuclease access to the nucleoprotein complex. These results showed that FIS binds *oriC* during exponential cell growth, while IHF binds *oriC* when cells are in stationary phase, accompanied by increased DnaA binding to *oriC*. Enhanced single-strand nuclease cleavage of DNA in wild-type origins during stationary phase was also observed, suggesting that most *oriC* copies are unwound as part of an initiation-like complex at this stage (14).

Helicase Loading

In *E. coli*, once the open complex is formed, the next step in the initiation of DNA replication involves the loading of the helicase DnaB, which has been shown to stabilize the open structure in vitro (9). The DnaB protein interacts with DnaA and DnaC in assembling a complex at *oriC* on plasmids (Fig. 2). DnaA and DnaB are found at the *oriC* complex by antibody binding studies and electron microscopy, while DnaC is not. Entry of DnaB into such complexes requires DnaA and DnaC as well as a supercoiled template (22).

The use of a monoclonal antibody that appears to interfere with the physical interaction between DnaA and DnaB prevents the formation of a prepriming complex when DnaB is complexed with DnaC. Cross-linking studies have verified a direct interaction between DnaA and DnaB. This was the first direct evidence that DnaA played a role as a site for binding of DnaB helicase to DNA and that it may also orient DnaB, accounting for the directionality of replication fork movement (64).

It was shown through surface plasmon resonance that DnaA actively mediates DnaB entry at *oriC* through the transient binding of DnaB to a region of DnaA between amino acids 111 and 148. A second functional domain of DnaA that stabilizes the binding of DnaB during initiation and may be involved in a unique nucleoprotein complex was identified by using a form of DnaA lacking 62 amino acids from the N-terminal end, which is inactive in forming the prepriming complex at *oriC* (106). The stoichiometry of the DnaA initiator protein and DnaB helicase has been investigated with quantitative immunoblot. In these studies, DnaA alone appears to bind in a ratio of 4 to 5 monomers per *oriC* plasmid. However, when both DnaA and DnaB are present, they bind in ratios of 10 DnaA monomers per *oriC* plasmid and 2 DnaB hexamers per *oriC* plasmid. It has been suggested that the additional DnaA present probably restricts any additional binding of DnaB hexamers to *oriC* (12).

In *B. subtilis*, DnaC, the counterpart to *E. coli* DnaB helicase, has been genetically proven to be necessary for initiation of DNA replication from the origin. DnaD appears to play an important role in the loading of DnaC helicase onto the *B. subtilis oriC* due to its interaction with DnaA, as revealed through yeast two-hybrid analysis. This analysis also revealed that DnaD interacts with itself. A mutant form of the protein, DnaD23, which affected replication initiation at restrictive temperatures, was able to interact with wild-type DnaD, but not with DnaA. Also, a tagged version of DnaD appears to form stable dimers (38).

MECHANISMS REGULATING THE TIMING OF INITIATION

Origin Sequestration

E. coli possesses several mechanisms that serve to regulate the timing of initiation and, thus, ensure that the bacterial chromosome replicates once, and only once, during a given cell cycle (reviewed in reference 46) (Fig. 3).

One of these mechanisms involves the sequestration of *oriC* immediately following its replication. Origin DNA is remarkably rich in GATC sequences, the target of Dam methylase. Hemimethylated origins, which arise with synthesis of the newly replicated strands, are tightly bound by SeqA protein, a negative modulator of initiation, and therefore cannot be acted upon by replicatively active DnaA protein for a period of time equal to approximately one-third of the cell cycle (Fig. 3) (60, 100, 112; reviewed in reference 17). SeqA is absolutely required during this period of hemimethylation, also known as the eclipse period, to prevent immediate reinitiation (113).

In vitro, SeqA protein exists as a homotetramer and aggregates or multimerizes reversibly in a concentration-dependent manner. SeqA tetramers must interact properly and form active aggregates for binding to hemimethylated DNA to occur. This SeqA aggregation may be important not only for regulating chromosomal replication, but also for chromosomal segregation (59).

E. coli seqA mutants initiate replication at a cell mass 10 to 20% lower than that of wild-type cells, and origins within *seqA* mutants often go through more than one initiation event per cell cycle; 30% of mutant cells contain two or more genome equivalents. With *E. coli*, the time required to replicate the chromosome can be significantly longer than the

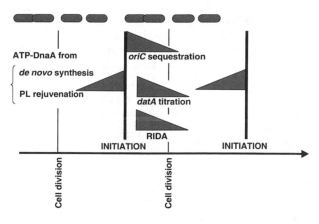

Figure 3. Mechanisms controlling the initiation of chromosomal replication. Prior to the initiation of a round of chromosomal replication, the amount of ATP-DnaA increases through de novo protein synthesis and possibly through the membrane rejuvenation of ADP-DnaA into ATP-DnaA (indicated by the left-to-right increasing wedge shapes). Once a critical level of DnaA has been reached, replication is initiated at *oriC*. Three mechanisms then come into play to prevent untimely reinitiation (indicated by the left-to-right decreasing wedge shapes). Immediately following its replication, *oriC* is sequestered away from the activity of DnaA, with the sequestration of *oriC* lasting for approximately one-third of the cell cycle. Before sequestration ends, the availability of active DnaA must be decreased to a level too low to promote reinitiation of the replicated origin. This occurs through the binding of DnaA protein to the *datA* locus and RIDA conversion of ATP-DnaA protein to inactive ADP-DnaA. The onset of *datA* titration of DnaA and the onset of DnaA inactivation by RIDA happen relatively soon after initiation. When in a cell cycle, whether they cease to contribute significantly to the decrease of DnaA potential is unclear. PL, phospholipid.

doubling time for fast-growing cells. Therefore, such rapidly growing cells will have several origins of chromosomal replication, and in wild-type cells all the origins will initiate synchronously (95). However, initiation of replication at the multiple origins in *seqA* mutants does not occur synchronously. Thus, while SeqA protein appears to be a timing factor, it is not clear if the resulting asynchrony in *seqA* mutants is due to a disruption in a coordinated response of all origins to an internal cue or a disruption of a direct communication between the several copies of origins present in rapidly growing cells (6).

In an in vitro *oriC* DNA replication system, purified SeqA protein blocks the formation of prepriming complexes. Similarly, SeqA inhibits replication from established prepriming complexes, but without their disruption. Interestingly, it appears that SeqA alters the dependence of this replication system on the concentration of DnaA, allowing replication at lower DnaA levels. Thus, it appears that SeqA is involved in the assembly of initiation complexes, but then has a negative regulatory affect on their fate (118).

Probing with potassium permanganate has shown that SeqA inhibits open complex formation on *oriC* plasmids. It was proposed that SeqA may prevent strand opening, either through a direct interaction with AT-rich regions or, indirectly, by changing the plasmids' topology. Experimental results suggest that the inhibition is not the result of DnaA's being unable to bind, but is most likely due to the effect SeqA has on plasmid topology; SeqA restrains negative supercoils in a population of the *oriC* plasmids. This change in topology occurs regardless of plasmid methylation, indicating that the effect is not due to SeqA's specifically binding to *oriC* (99, 110). Electron microscopy has revealed that SeqA binds in a highly cooperative manner specifically to fully and hemimethylated *oriC* at two locations on either side of the R1 DnaA box. It was suggested from these results that SeqA binds two nucleation sites in *oriC* through protein-protein interactions, spreading to adjacent regions and recruiting additional *oriC* into larger aggregates (98, 99). This cooperative binding of SeqA is in contrast to that of HU protein, which also is able to restrain negative supercoils (99, 110). HU has been shown to be a negative modulator of SeqA expression. In wild-type cells, levels of SeqA are maximal at mid-log phase and then decrease as the cells enter late log phase. In *hup* mutant cells lacking HU protein, while the maximal levels of SeqA are still observed at mid-log phase, the levels are increased up to fourfold over those in wild-type cells (58).

Purified SeqA binds to the AT-rich and 13-mer regions of hemimethylated *oriC* as well as the DnaA boxes R1 and M. SeqA is also more efficient at inhibiting replication of hemimethylated *oriC* than of fully methylated *oriC* in vitro. The mechanism of SeqA inhibition has been described as the competitive binding of SeqA protein to a long region on hemimethylated *oriC*, from the AT-rich region to the M DnaA box, thus releasing DnaA molecules; this mechanism has also been observed in titration studies (107, 113). In mutant cells lacking SeqA, the *dnaA204* mutant protein, which is usually rapidly degraded, is stabilized. This results in increased concentrations of this mutant DnaA in cells and the normalization of cell mass at the time of initiation. This suggests that this mutant form of DnaA can form a stable protein-DNA interaction, which is usually disrupted by the presence of SeqA in wild-type cells (111).

Gram-positive bacteria, such as *B. subtilis*, do not possess a methylation system or a SeqA homologue, although their DNA replication is under a control equal to that of *E. coli*. Since these cells lack SeqA, the origin may be able to bind more DnaA. Newly replicated origins would have a high capacity to bind

DnaA, acting as a sink to control free DnaA levels. This may be the main form of replication control in gram-positive bacteria (reviewed in reference 7).

The Nucleotide State of DnaA

DnaA protein has the capacity to tightly bind ADP or ATP (K_ds of 100 and 30 nM, respectively) in vitro, and likely exists in these nucleotide-bound forms in vivo. While both forms can bind the origin, only the ATP-bound form can activate replication. Moreover, DnaA has a slow intrinsic DNA-dependent ATPase activity. The resulting ADP remains tightly bound, thus generating inactive ADP-DnaA (see more below). The exchange of ATP for ADP in order to reactivate the protein is a slow process; after 40 min only 50% of the bound ADP has exchanged with excess ATP in vitro (93). Through site-directed mutagenesis, a DnaA mutant was created with a lower ATPase activity. When examined in vivo and in vitro, the Glu204Gln mutant had ATPase activity only one-third that of wild-type, although DnaA-nucleotide–binding affinity was not affected. This mutant *dnaA* gene is lethal to cells, unless they replicate DNA independently of *oriC*. Also, induction of this mutant results in overinitiation of DNA replication, showing that the intrinsic ATPase activity of DnaA negatively regulates DNA replication (71).

Adenine nucleotides are allosteric effectors that profoundly affect the conformation of DnaA. Limited proteolytic digestion mapping reveals that while the ATP and ADP forms of DnaA are structurally very similar, they are quite distinct from nucleotide-free DnaA (23). Circular dichronism and intrinsic fluorescence studies provided evidence that ATP binding decreased the content of α-helices in the DnaA molecule (53). Physiologically, fluctuations in the nucleotide-bound form of DnaA change in coordination with the cellular replication cycle. ATP-DnaA is usually repressed to ~20% of total cellular DnaA, but a temporal increase is observed around the time of initiation, pointing to a tight link between DNA replication and the cell cycle (56).

Cells harboring the *dnaAcos* mutation overinitiate replication at the nonpermissive temperature of 30°C. DnaAcos protein binds DNA with similar affinity to that of wild type but appears to be defective in nucleotide binding. This mutant form of DnaA is able to support in vitro replication of single-stranded DNA that contains a DnaA-binding hairpin, but is inactive for initiating replication from *oriC*, suggesting the importance of DnaA nucleotide binding in replication from the chromosomal origin (42).

Two different mutant forms of DnaA altered in the ATP-binding domain were created by site-directed mutagenesis. The change K178I or D235N causes the DnaA proteins to lose affinity for ADP and ATP but maintain their capacity to bind *oriC*. These mutant DnaA proteins have only 10% of the specific activity of wild-type DnaA in initiating replication from *oriC* in vitro and appear to be defective in duplex opening at *oriC*, suggesting that these particular amino acids are essential and that ATP binding is also required in DnaA for functional DNA replication (72).

Membrane Rejuvenation

Evidence has accumulated in vitro and in vivo supporting the importance of membrane phospholipids in modulating DnaA initiation activity via a rejuvenation process (reviewed in reference 18) (Fig. 3 and 4). Phospholipids with acidic headgroups are able to catalyze the rapid release of nucleotide from DnaA protein (94). These phospholipids must be above their transition temperature and maintain a certain degree of fluidity in order to catalyze the release of nucleotide (15, 121). Along with the DnaA-nucleotide release process, DnaA rejuvenation also requires the subsequent binding of ATP. This is accomplished in vitro in the presence of *oriC* DNA, which stabilizes DnaA in the presence of phospholipids and facilitates high-affinity nucleotide binding (16).

Genetic manipulations that alter the in vivo phospholipid composition of *E. coli* demonstrate the importance of acidic membrane phospholipids for normal cell growth, likely through a direct effect on chromosomal replication. Production of acidic phospholipids can be controlled by placing the expression of the gene encoding phosphatidylglycerol synthase,

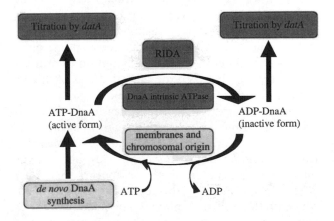

Figure 4. Positive and negative effects on the initiation potential of DnaA protein. Processes that decrease the initiation potential of DnaA protein are shaded in dark gray. Processes that contribute to the initiation potential of DnaA are shaded in light gray.

pgsA, under inducible control. While such cells grow normally in the presence of the inducer, removal of the inducer results in a marked decrease in acidic phospholipid levels as the cells pass through several generations; after approximately 10 generations the acidic phospholipid content in these genetically altered *E. coli* cells constitutes only 8 to 10% of the total cellular phospholipid. At this point, although the cells remain viable, they undergo a growth arrest (27). Bypassing the normal, DnaA-dependent initiation of chromosomal replication from *oriC* via constitutive stable DNA replication (51) relieves the growth arrest of cells lacking sufficient acidic phospholipids (119). The underlying cause of the arrested growth of acidic phospholipid-deficient cells may involve DnaA protein directly. Certain amino acid substitutions and small deletions, within a region of DnaA previously shown to be essential for productive membrane binding (23), are able to alleviate the growth arrest in cells with insufficient levels of acidic phospholipids even though the cells cannot carry out constitutive stable DNA replication and must initiate replication from *oriC* (123). Moreover, expression of two of these mutant forms of DnaA protein, DnaA(L366K) and DnaA(Δ363-367), in cells with normal membrane lipid composition perturbs chromosomal replication by reducing the frequency of initiation and disrupting initiation synchrony (123).

Photolabeling of DnaA protein with phospholipid analogues has shown that membrane-mediated release of nucleotide from DnaA accompanies the insertion of a distinct region of DnaA into the hydrophobic region of the lipid bilayer (23). The coupled membrane insertion and dissociation of bound adenine nucleotide requires that the membrane surface be anionic in nature and that the bilayer have adequate fluidity. The region of DnaA that inserts into the lipid bilayer encompasses a putative amphipathic helix (23), a motif which can serve as a membrane-seeking structure. The binding and insertion of DnaA, as well as the release of bound nucleotide, are inhibited by increasing ionic strength. Thus, it appears that at least one role of the acidic phospholipid headgroups is to recruit DnaA, a protein rich in basic residues, to the membrane through electrostatic interactions (50). In support of this model is the observation that membranes composed of sialic acid-containing eukaryotic ganglioside GM_1 are as active as acidic glycerophospholipids at promoting release of bound nucleotide, while membranes of asialo-GM_1 are totally inactive (23). This ability of anionic sphingolipids to trigger the release of bound nucleotide suggests that an acidic surface on a fluid bilayer is the critical feature, rather than a specific structural recognition between DnaA protein

and certain phospholipids. More recently, a mutational analysis indicated that DnaA residues Arg-328 and Lys-372 are involved in the binding to acidic phospholipids, leading to the proposal that the binding causes a change in the higher-order structure of the ATP-binding pocket, resulting in the release of ADP from DnaA (61). This is a variation of an earlier model wherein electrostatic forces serve to recruit or properly orient DnaA at the membrane, and the hydrophobic force-driven insertion of a domain of DnaA into the nonpolar environment of the lipid bilayer causes a conformational change in the protein, weakening the binding of ADP (50).

RIDA-Mediated Decrease in DnaA Activity

Another mechanism that regulates the initiation activity of DnaA protein has been termed RIDA (regulatory inactivation of DnaA) (reviewed in reference 44) (Fig. 3 and 4). This activity was found in a soluble extract that specifically inhibits an in vitro *oriC* replication system (41, 55). Components in the extract responsible for this inhibitory RIDA activity include the β subunit of DNA polymerase III loaded onto DNA as a sliding clamp and a partially purified factor IdaB (43). RIDA activity stimulates the hydrolysis of ATP bound to the DnaA protein, which results in the formation of the inactive ADP-DnaA, and the RIDA activity is further stimulated by DNA synthesis. In vivo, the majority of DnaA in wild-type cells has been seen to exist in the ADP-bound form. However, in the absence of functional β subunit, there is a dramatic increase in the cellular ratio of ATP-DnaA to ADP-DnaA (43, 55, 103).

The novel DnaA-related protein Hda (homologous to DnaA) is a multicopy suppressor of a β-subunit mutant and appears to contain IdaB activity (47). Hda belongs to same protein family as DnaA, the AAA+ proteins (78), with the highest homology in the domain III ATP-binding region. One study reported that the disruption of *hda* results in an increase in cellular levels of ATP-DnaA from 20 to 70% due to the disruption in the hydrolyzing RIDA activity and that the *hda* gene is essential for cell growth as observed through P1-mediated transduction of a Δ*hda* construct (47). However, Camara et al. did not observe the *hda* gene to be essential for cell growth or viability after performing similar experiments that utilized a different Δ*hda* construct (11). Although there are conflicting data on the role Hda plays as an essential factor in cells (11, 47), both studies reveal that disruption of *hda* causes the loss of controlled replication through hyperinitiation.

The previously mentioned DnaAcos mutant, which overinitiates in vivo and is defective in nucleotide binding, is insensitive to RIDA activity (41). Also, the R334H DnaA mutant severely inhibits in vitro ATP hydrolysis and may be located at or near the DnaA ATP-binding site (103). This residue is highly conserved in other proteins homologous to DnaA and lies in Box VIII of the AAA+ ATPase protein family.

Initiator Titration, *datA* Locus, and Transcription Regulation

Initiation activity of DnaA can also be regulated through the availability of free DnaA in the cell (Fig. 3 and 4), which may be controlled through several different mechanisms. DnaA has the ability to bind *oriC* DNA on the *E. coli* chromosome, as discussed previously. Other regions on the chromosome also have an affinity for this protein and can titrate free DnaA, affecting its availability to initiate replication at *oriC*. There is a particular 1-kb sequence located at 94.7 min on the genetic map with an unusually high affinity for DnaA (48, 49). This region, known as the *datA* locus, is able to bind approximately 370 DnaA monomers when present on a plasmid as revealed by immunoassays. It has been estimated that 60% of a cell's total DnaA is bound to the chromosomal *datA* locus (48).

Normal cell growth can occur in an *E. coli* strain lacking *datA*, but it is accompanied by asynchronous initiation and extra initiation events. This phenotype could be suppressed by the addition of DnaA-titrating plasmids (49). It has been proposed that as the *datA* locus in wild-type cells is replicated, it becomes capable of titrating twice as much DnaA. This would lower the intracellular concentrations of DnaA below a threshold needed for initiation, and thus contribute to the prevention of reinitiations at *oriC* after sequestration (48, 49). Other regions on the chromosome, in addition to *oriC* and *datA*, have been reported to have a higher than usual affinity for DnaA. These regions contain 1 to 2 DnaA boxes and appear to be arranged randomly around the chromosome (89).

DnaA titration by the *datA* locus works to complement the other mechanisms previously discussed in the regulation of initiation. Although SeqA appears to compete for binding with DnaA at *oriC*, SeqA is not expected to compete for binding of the *datA* locus due to the lack of GATC sites in this region (113). Also, RIDA activity appears to proceed independently of the *datA* mechanism since ATP-DnaA levels in both *datA*$^+$ and *datA*$^-$ strains could be maintained within a range of 20 to 30% of total bound DnaA (45).

The level of free DnaA in a cell can also be controlled through autoregulation of *dnaA* operon transcription. DnaA protein binds to the DnaA box between the two *dnaA* promoters, *dnaA*1p and *dnaA*2p, and is responsible for transcriptional repression (2, 3, 10, 54, 85, 114). It has been shown that the ATP-bound form of DnaA represses *dnaA* transcription more efficiently than the ADP-bound form. Therefore, it appears that the alternating nucleotide-bound forms of DnaA can fulfill different functions by binding different sequences (101).

Transcription of the *dnaA-dnaN* operon in *B. subtilis* is also downregulated in the presence of increased DnaA. These cells were able to grow at high DnaA levels when a copy of *dnaN* (β subunit) was placed downstream of an extra copy of *dnaA*. Repression of the native operon by high levels of DnaA appears to deplete β-subunit supplies in the cell. This autoregulation of *dnaA-dnaN* transcription could act as a rate-limiting factor in DNA replication if DnaA levels increase during cell cycle progression (83).

In *E. coli*, autoregulation of *dnaA* transcription appears to be less important for the regulation of initiation at *oriC* than the *datA* locus, as shown by studies which employed cells with extra *datA* copy-plasmids and *oriC* minichromosomes (73).

CELLULAR SITE OF INITIATION

Regulated initiation of chromosomal replication likely involves not only its timing during the cell cycle, but also where it happens within the cell. As such, much effort has been given to establish the intracellular location of key components involved in the initiation of DNA replication.

To facilitate visualization of the origin region of the chromosome, tandem repeats of the *lac* operon were placed near *oriC*. An expressed GFP-LacI fusion protein that binds to the repeat elements was found to localize at or near the poles in living cells examined with fluorescence microscopy (25).

Other studies have employed fluorescence in situ hybridization to localize *oriC* in fixed cells. In relatively rapidly growing newborn cells, *oriC* was seen at the cell poles. However, *oriC* moved toward midcell before initiation in slow-growing cells that contain only one origin (80, 81). Similarly, experiments utilizing fluorescence in situ hybridization-labeled *oriC* in cells containing two origin foci at birth revealed that the origins appeared to move near the center of prospective daughter cells at the time of DNA replication initiation (87).

In addition to the origin of replication, the initiator protein DnaA has also been visualized in cells. DnaA appears to be localized at the membrane as revealed through immunogold cryothin-section electron microscopy and immunofluorescence. These results suggest that the initiation of chromosomal replication is affiliated with the membrane in *E. coli* (79). As mentioned previously, acidic phospholipids rejuvenate DnaA protein in vitro, and the synthesis of acidic phospholipids is necessary for proper cell growth in vivo. Acidic phospholipid domains have been identified with the cardiolipin-specific stain 10-*N*-nonyl acridine orange in the septal and polar regions of the cell (70). Although DnaA and acidic phospholipid domains have not yet been colocalized, these two cellular components may interact as part of the overall localization and regulation of replication from *oriC*.

The cellular location of the negative regulator of initiation, SeqA, has also been determined by immunofluorescence microscopy. Surprisingly, SeqA is localized as discrete foci in both *oriC*⁺ and *oriC*⁻ cells. However, this localization is dependent upon DNA methylation. Unlike *oriC*, SeqA appears to be localized at midcell in newborn cells and proceeds to migrate quite differently from *oriC* (30, 31; reviewed in reference 27).

SUMMARY AND FUTURE PERSPECTIVES

Recently, significant advances have been made in our knowledge of the initiation of chromosomal replication. A detailed understanding of the assembly and remodeling of the nucleoprotein complexes at *oriC* associated with the onset of DNA synthesis is emerging. Similarly, it is becoming clear that three independent mechanisms exist to prevent replication from reinitiating at *oriC*: (i) sequestration of *oriC*, (ii) titration of DnaA by the *datA* locus, and (iii) conversion of active ATP-DnaA to its inactive ADP-form.

However, major questions still remain. What are the structural differences between ATP-DnaA and ADP-DnaA that result in one form's being active and the other inert? Generation and comparison of three-dimensional crystallographic structures of DnaA with different bound ligands will hopefully provide answers to this question.

What is the physiological relevance of phospholipid rejuvenation of ADP-DnaA? Although it is now clear that there is a close link between DnaA function and cells having adequate levels of acidic phospholipids, the underlying reason is still obscure.

Is there spatial control of initiation (i.e., is "where" as important as "when"?) and does where initiation occurs play an important part in the later processes of chromosomal replication and segregation? Continued studies to localize components in-

volved in the initiation, replication, and segregation of the chromosome should prove fruitful.

While we now have a better understanding of mechanisms that negatively regulate initiation, our knowledge of what promotes initiation is likely incomplete. Are there other physiological signals besides adequate levels of DnaA necessary to trigger a round of replication? What aspects of the cell's physiology are being sensed to link the chromosome replication cycle to cell growth?

Answers to these and other important questions, including what similarities and differences exist between *E. coli* and other bacteria and eukaryotic cells, should be of great interest.

REFERENCES

1. Asai, T., M. Takanami, and M. Imai. 1990. The AT richness and gid transcription determine the left border of the replication origin of the E. coli chromosome. *EMBO J.* 9:4065–4072.

2. Atlung, T., E. Clausen, and F. G. Hansen. 1984. Autorepression of the dnaA gene of Escherichia coli, p. 199–207. *In* U. Huebscher and S. Spardi (ed.), *Proteins Involved in DNA Replication.* Plenum Press, New York, N.Y.

3. Atlung, T., E. Clausen, and F. G. Hansen. 1985. Autoregulation of the dnaA gene of Escherichia coli. *Mol. Gen. Genet.* 200:442–450.

4. Bahloul, A., F. Boubrik, and J. Rouviere-Yaniv. 2001. Roles of Escherichia coli histone-like protein HU in DNA replication: HU-beta suppresses the thermosensitivity of dnaA46ts. *Biochimie* 83:219–229.

5. Blaesing, F., C. Weigel, M. Welzeck, and W. Messer. 2000. Analysis of the DNA-binding domain of Escherichia coli DnaA protein. *Mol. Microbiol.* 36:557–569.

6. Boye, E., T. Stokke, N. Kleckner, and K. Skarstad. 1996. Coordinating DNA replication initiation with cell growth: differential roles for DnaA and SeqA proteins. *Proc. Natl. Acad. Sci. USA* 93:12206–12211.

7. Boye, E., A. Lobner-Olesen, and K. Skarstad. 2000. Limiting DNA replication to once and only once. *EMBO Rep.* 1:479–483.

8. Bramhill, D., and A. Kornberg. 1988. Duplex opening by dnaA protein at novel sequences in initiation of replication at the origin of the E. coli chromosome. *Cell* 52:743–755.

9. Bramhill, D., and A. Kornberg. 1988. A model for initiation at origins of DNA replication. *Cell* 54:915–918.

10. Braun, R. E., K O'Day, and A. Wright. 1985. Autoregulation of the DNA replication gene dnaA in E. coli. *Cell* 40:159–169.

11. Camara, J. E., K. Skarstad, and E. Crooke. 2003. Controlled initiation of chromosomal replication in *Escherichia coli* requires functional Hda protein. *J. Bacteriol.* 185:3244–3248.

12. Carr, K. M., and J. M. Kaguni. 2001. Stoichiometry of DnaA and DnaB proteins in initiation of the E. coli chromosomal origin. *J. Biol. Chem.* 276:44919–44925.

13. Cassler, M. R., J. E. Grimwade, and A. C. Leonard. 1995. Cell cycle-specific changes in nucleoprotein complexes at a chromosomal replication origin. *EMBO J.* 14:5833–5841.

14. Cassler, M. R., J. E. Grimwade, K. C. McGarry, R. T. Mott, and A. C. Leonard. 1999. Drunken-cell footprints: nuclease treatment of ethanol-permeabilized bacteria reveals an initiation-like

nucleoprotein complex in stationary phase replication origins. *Nucleic Acids Res.* **27:**4570–4576.

15. **Castuma, C. E., E. Crooke, and A. Kornberg.** 1993. Fluid membranes with acidic domains activate DnaA, the initiator protein of replication in Escherichia coli. *J. Biol. Chem.* **268:**24665–24668.

16. **Crooke, E., C. E. Castuma, and A. Kornberg.** 1992. The chromosome origin of Escherichia coli stabilizes DnaA protein during rejuvenation by phospholipids. *J. Biol. Chem.* **267:**16779–16782.

17. **Crooke, E.** 1995. Regulation of chromosomal replication in E. coli: sequestration and beyond. *Cell* **82:**877–880.

18. **Crooke, E.** 2001. Escherichia coli DnaA protein-phospholipid interactions: in vitro and in vivo. *Biochimie* **83:**19–23.

19. **Erzberger, J. P., M. M. Pirruccello, and J. M. Berger.** 2002. The structure of bacterial DnaA: implications for general mechanisms underlying DNA replication initiation. *EMBO J.* **21:**4763–4773.

20. **Fujita, M. Q., H. Yoshikawa, and N. Ogasawara.** 1989. Structure of the dnaA region of Pseudomonas putida: conservation among three bacteria, Bacillus subtilis, Escherichia coli and P. putida. *Mol. Gen. Genet.* **215:**381–387.

21. **Fuller, R. S., B. E. Funnell, and A. Kornberg.** 1984. The dnaA protein complex with the E. coli chromosomal replication origin (oriC) and other DNA sites. *Cell* **38:**889–900.

22. **Funnell, B. E., T. A. Baker, and A. Kornberg.** 1987. In vitro assembly of a prepriming complex at the origin of the Escherichia coli chromosome. *J. Biol. Chem.* **262:**10327–10334.

23. **Garner, J., P. Durrer, J. Kitchen, J. Brunner, and E. Crooke.** 1998. Membrane-mediated release of nucleotide from an initiator of chromosomal replication, Escherichia coli DnaA, occurs with insertion of a distinct region of the protein into the lipid bilayer. *J. Biol. Chem.* **273:**5167–5173.

24. **Gille, H., and W. Messer.** 1991. Localized DNA melting and structural perturbations in the origin of replication, oriC, of Escherichia coli in vitro and in vivo. *EMBO J.* **10:**1579–1584.

25. **Gordon, G. S., D. Sitnikov, C. D. Webb, A. Teleman, A. Straight, R. Losick, A. W. Murray, and A. Wright.** 1997. Chromosome and low copy plasmid segregation in E. coli: visual evidence for distinct mechanisms. *Cell* **90:**1113–1121.

26. **Grimwade, J. E., V. T. Ryan, and A. C. Leonard.** 2000. IHF redistributes bound initiator protein, DnaA, on supercoiled oriC of Escherichia coli. *Mol. Microbiol.* **35:**835–844.

27. **Heacock, P. N., and W. Dowhan.** 1989. Alteration of the phospholipid composition of Escherichia coli through genetic manipulation. *J. Biol. Chem.* **264:**14672–14677.

28. **Hiasa, H., and K. J. Marians.** 1994. Fis cannot support oriC DNA replication in vitro. *J. Biol. Chem.* **269:**24999–25003.

29. **Hiraga, S.** 1976. Novel F prime factors able to replicate in Escherichia coli Hfr strains. *Proc. Natl. Acad. Sci. USA* **73:**198–202.

30. **Hiraga, S., C. Ichinose, H. Niki, and M. Yamazoe.** 1998. Cell cycle-dependent duplication and bidirectional migration of SeqA-associated DNA-protein complexes in E. coli. *Mol. Cell* **1:**381–387.

31. **Hiraga, S., C. Ichinose, T. Onogi, H. Niki, and M. Yamazoe.** 2000. Bidirectional migration of SeqA-bound hemimethylated DNA clusters and pairing of oriC copies in Escherichia coli. *Genes Cells* **5:**327–341.

32. **Holz, A., C. Shaefer, H. Gille, W. R. Jueterbock, and W. Messer.** 1992. Mutations in the DnaA binding sites of the replication origin of Escherichia coli. *Mol. Gen. Genet.* **233:**81–88.

33. **Hsu, J., D. Bramhill, and C. M. Thompson.** 1994. Open complex formation by DnaA initiation protein at the Escherichia coli chromosomal origin requires the 13-mers precisely spaced relative to the 9-mers. *Mol. Microbiol.* **11:**903–911.

34. **Hwang, D. S., and A. Kornberg.** 1990. A novel protein binds a key origin sequence to block replication of an E. coli minichromosome. *Cell* **63:**325–331.

35. **Hwang, D. S., and A. Kornberg.** 1992. Opening of the replication origin of Escherichia coli by DnaA protein with protein HU or IHF. *J. Biol. Chem.* **267:**23083–23086.

36. **Hwang, D. S., and A. Kornberg.** 1992. Opposed actions of regulatory proteins, DnaA and IciA, in opening the replication origin of Escherichia coli. *J. Biol. Chem.* **267:**23087–23091.

37. **Hwang, D. S., B. Thony, and A. Kornberg.** 1992. IciA protein, a specific inhibitor of initiation of Escherichia coli chromosomal replication. *J. Biol. Chem.* **267:**2209–2213.

38. **Ishigo-oka, D., N. Ogasawara, and S. Moriya.** 2001. DnaD protein of *Bacillus subtilis* interacts with DnaA, the initiator protein of replication. *J. Bacteriol.* **183:**2148–2150.

39. **Jakimowicz, D., J. Majka, W. Messer, C. Speck, M. Fernandez, M. C. Martin, J. Sanchez, F. Schauwecker, U. Keller, H. Schrempf, and J. Zakrzewska-Czerwinska.** 1998. Structural elements of the Streptomyces oriC region and their interactions with the DnaA protein. *Microbiology* **144:**1281–1290.

40. **Kaguni, J. M.** 1997. Escherichia coli DnaA protein: the replication initiator. *Mol. Cell* **7:**145–157.

41. **Katayama, T., and E. Crooke.** 1995. DnaA protein is sensitive to a soluble factor and is specifically inactivated for initiation of in vitro replication of the Escherichia coli minichromosome. *J. Biol. Chem.* **270:**9265–9271.

42. **Katayama, T., E. Crooke, and K. Sekimizu.** 1995. Characterization of Escherichia coli DnaAcos protein in replication systems reconstituted with highly purified proteins. *Mol. Microbiol.* **18:**813–820.

43. **Katayama, T., T. Kubota, K. Kurokawa, E. Crooke, and K. Sekimizu.** 1998. The initiator function of DnaA protein is negatively regulated by the sliding clamp of the E. coli chromosomal replicase. *Cell* **94:**61–71.

44. **Katayama, T., and K. Sekimizu.** 1999. Inactivation of Escherichia coli DnaA protein by DNA polymerase III and negative regulations for initiation of chromosomal replication. *Biochimie* **81:**835–840.

45. **Katayama, T., K. Fujimitsu, and T. Ogawa.** 2001. Multiple pathways regulating DnaA function in Escherichia coli: distinct role for DnaA titration by the datA locus and the regulatory inactivation of DnaA. *Biochimie* **83:**13–17.

46. **Katayama, T.** 2001. Feedback controls restrain the initiation of Escherichia coli chromosomal replication. *Mol. Microbiol.* **41:**9–17.

47. **Kato, J., and T. Katayama.** 2001. Hda, a novel DnaA-related protein, regulates the replication cycle in Escherichia coli. *EMBO J.* **20:**4253–4262.

48. **Kitagawa, R., H. Mitsuki, T. Okazaki, and T. Ogawa.** 1996. A novel DnaA protein-binding site at 94.7 min on the Escherichia coli chromosome. *Mol. Microbiol.* **19:**1137–1147.

49. **Kitagawa, R., T. Ozaki, S. Moriya, and T. Ogawa.** 1998. Negative control of replication initiation by a novel chromosomal locus exhibiting exceptional affinity for Escherichia coli DnaA protein. *Genes Dev.* **12:**3032–3043.

50. **Kitchen, J. L., Z. Li, and E. Crooke.** 1999. Electrostatic interactions during acidic phospholipid reactivation of DnaA protein, the Escherichia coli initiator of chromosomal replication. *Biochemistry* **38:**6213–6221.

51. Kogoma, T., and K. von Meyenburg. 1983. The origin of replication, oriC, and the dnaA protein are dispensable in stable DNA replication (sdrA) mutants of Escherichia coli K-12. *EMBO J.* **2**:463–468.

52. Kornberg, A., and T. A. Baker. 1992. *DNA Replication*, 2nd ed. W. H. Freeman & Co., New York, N.Y.

53. Kubota, T., T. Katayama, Y. Ito, T. Mizushima, and K. Sekimizu. 1997. Conformational transition of DnaA protein by ATP: structural analysis of DnaA protein, the initiator of Escherichia coli chromosome replication. *Biochem. Biophys. Res. Commun.* **232**:130–135.

54. Kucherer, C., H. Lother, R. Kolling, M. A. Schauzu, and W. Messer. 1986. Regulation of transcription of the chromosomal dnaA gene of Escherichia coli. *Mol. Gen. Genet.* **205**:115–121.

55. Kurokawa, K., T. Mizushima, T. Kubota, T. Tsuchiya, T. Katayama, and K. Sekimizu. 1998. A stimulation factor for hydrolysis of ATP bound to DnaA protein, the initiator of chromosomal DNA replication in Escherichia coli. *Biochem. Biophys. Res. Commun.* **243**:90–95.

56. Kurokawa, K., S. Nishida, A. Emoto, K. Sekimizu, and T. Katayama. 1999. Replication cycle-coordinated change of the adenine nucleotide-bound forms of DnaA protein in Escherichia coli. *EMBO J.* **18**:6642–6652.

57. Langer, U., S. Richter, A. Roth, C. Weigel, and W. Messer. 1996. A comprehensive set of DnaA-box mutations in the replication origin, oriC, of Escherichia coli. *Mol. Microbiol.* **21**:301–311.

58. Lee, H., H. K. Kim, S. Kang, C. B. Hong, J. Yim, and D. S. Hwang. 2001. Expression of the seqA gene is negatively modulated by the HU protein in Escherichia coli. *Mol Gen. Genet.* **264**:931–935.

59. Lee, H., S. Kang, S. H. Bae, B. S. Choi, and D. S. Hwang. 2001. SeqA protein aggregation is necessary for SeqA function. *J. Biol. Chem.* **276**:34600–34606.

60. Lu, M., J. L. Campbell, E. Boye, and N. Kleckner. 1994. SeqA: a negative modulator of replication initiation in E. coli. *Cell* **77**:413–426.

61. Makise, M., S. Mima, T. Tsuchiya, and T. Mizushima. 2001. Molecular mechanism for functional interaction between DnaA protein and acidic phospholipids. *J. Biol. Chem.* **276**:7450–7456.

62. Margulies, C., and J. M. Kaguni. 1998. The FIS protein fails to block the binding of DnaA protein to oriC, the Escherichia coli chromosomal origin. *Nucleic Acids Res.* **26**:5170–5175.

63. Marians, K. J. 1992. Prokaryotic DNA replication. *Annu. Rev. Biochem.* **61**:673–719.

64. Marszalek, J., and J. M. Kaguni. 1994. DnaA protein directs the binding of DnaB protein in initiation of DNA replication in Escherichia coli. *J. Biol. Chem.* **269**:4883–4890.

65. Marszalek, J., W. Zhang, T. R. Hupp, C. Marguiles, K. M. Carr, S. Cherry, and J. M. Kaguni. 1996. Domains of DnaA protein involved in interaction with DnaB protein, and in unwinding the Escherichia coli chromosomal origin. *J. Biol. Chem.* **271**:18535–18542.

66. Matsui, M., A. Oka, M. Takanami, S. Yasuda, and Y. Hirota. 1985. Sites of dnaA protein-binding in the replication origin of Escherichia coli K-12 chromosome. *J. Mol. Biol.* **184**:529–533.

67. Messer, W., H. Hartmann-Kuhlein, U. Langer, E. Mahlow, A. Roth, S. Schaper, B. Urmoneit, and B. Woelker. 1992. The complex for replication initiation of Escherichia coli. *Chromosoma* **102**:S1–S6.

68. Messer, W., and C. Weigel. 1996. Initiation of chromosome replication, p. 1579–1601. *In* F. C. Neidhardt, R. Curtiss III, J. L. Ingraham, E. C. C. Lin, K. B. Low, B. Magasanik, W. S.

Reznikoff, M. Riley, M. Schaechter, and H. E. Umbarger (ed.), Escherichia coli *and* Salmonella: *Cellular and Molecular Biology*, 2nd ed., vol. 2. ASM Press, Washington, D.C.

69. Messer, W., F. Blaesing, J. Majka, J. Nardmann, S. Schaper, A. Schmidt, H. Seitz, C. Speck, D. Tungler, G. Wegrzyn, C. Weigel, M. Welzeck, and J. Zakrzewska-Czerwinska. 1999. Functional domains of DnaA proteins. *Biochimie* **81**:819–825.

70. Mileykovskaya, E. and W. Dowhan. 2000. Visualization of phospholipid domains in *Escherichia coli* by using the cardiolipin-specific dye 10-N-nonyl acridine orange. *J. Bacteriol.* **182**:1172–1175.

71. Mizushima, T., S. Nishida, K. Kurokawa, T. Katayama, T. Miki, and K. Sekimizu. 1997. Negative control of DNA replication by hydrolysis of ATP bound to DnaA protein, the initiator of chromosomal DNA replication in Escherichia coli. *EMBO J.* **16**:3724–2730.

72. Miszushima, T., T. Takaki, T. Kubota, T. Tsuchiya, T. Miki, T. Katayama, and K. Sekimizu. 1998. Site-directed mutational analysis for the ATP binding of DnaA protein. Functions of two conserved amino acids (Lys-178 and Asp-235) located in the ATP-binding domain of DnaA protein in vitro and in vivo. *J. Biol. Chem.* **273**:20847–20851.

73. Morigen, E. Boye, K. Skarstad, and A. Lobner-Olesen. 2001. Regulation of chromosomal replication by DnaA protein availability in Escherichia coli: effects of the datA region. *Biochim. Biophys. Acta* **1521**:73–80.

74. Moriya, S., T. Fukuoka, N. Ogasawara, and H. Yoshikawa. 1988. Regulation of initiation of the chromosomal replication by DnaA-boxes in the origin region of the Bacillus subtilis chromosome. *EMBO J.* **7**:2911–2917.

75. Moriya, S., T. Atlung, F. G. Hansen, H. Yoshikawa, and N. Ogasawara. 1992. Cloning of an autonomously replicating sequence (ars) from the Bacillus subtilis chromosome. *Mol. Microbiol.* **6**:309–315.

76. Moriya, S., W. Firshein, H. Yoshikawa, and N. Ogasawara. 1994. Replication of a Bacillus subtilis oriC plasmid in vitro. *Mol. Microbiol.* **12**:469–478.

77. Moriya, S., and N. Ogasawara. 1996. Mapping of the replication origin of the Bacillus subtilis chromosome by the two-dimensional gel method. *Gene* **176**:81–84.

78. Neuwald, A. F., L. Aravind, J. L. Spouge, and E. V. Koonin. 1999. AAA+: a class of chaperone-like ATPases associated with the assembly, operation, and disassembly of protein complexes. *Genome Res.* **9**:27–43.

79. Newman, G., and E. Crooke. 2000. DnaA, the initiator of *Escherichia coli* chromosomal replication, is located at the cell membrane. *J. Bacteriol.* **182**:2604–2610.

80. Niki, H., and S. Hiraga. 1998. Polar localization of the replication origin and terminus in Escherichia coli nucleoids during chromosome partitioning. *Genes Dev.* **12**:1036–1045.

81. Niki, H., Y. Yamaichi, and S. Hiraga. 2001. Dynamic organization of chromosomal DNA in Escherichia coli. *Genes Dev.* **14**:212–223.

82. Ogasawara, N., S. Moriya, K. von Meyenburg, F. G. Hansen, and H. Yoshikawa. 1985. Conservation of genes and their organization in the chromosomal replication origin region of Bacillus subtilis and Escherichia coli. *EMBO J.* **4**:3345–3350.

83. Ogura, Y., Y. Imai, N. Ogasawara, and S. Moriya. 2001. Autoregulation of the dnaA-dnaN operon and effects of DnaA protein levels on replication initiation in *Bacillus subtilis*. *J. Bacteriol.* **183**:3833–3841.

84. Oka, A., K. Sugimoto, M. Takanami, and Y. Hirota. 1980. Replication origin of the Escherichia coli K-12 chromosome:

the size and structure of the minimum DNA segment carrying the information for autonomous replication. *Mol. Gen. Genet.* **178:**9–20.

85. **Polaczek, P., and A. Wright.** 1990. Regulation of expression of the dnaA gene in Escherichia coli: role of the two promoters and the DnaA box. *New Biol.* **2:**574–582.

86. **Qin, M.-H., M. V. V. S. Madiraju, S. Zachariah, and M. Rajagopalan.** 1997. Characterization of the *oriC* region of *Mycobacterium smegmatis. J. Bacteriol.* **179:**6311–6317.

87. **Roos, M., A. B. van Geel, M. E. Aarsman, J. T. Veuskens, C. L. Woldringh, and N. Nanninga.** 1999. Cellular localization of oriC during the cell cycle of Escherichia coli as analyzed by fluorescent in situ hybridization. *Biochimie* **81:**797–802.

88. **Roth, A., and W. Messer.** 1995. The DNA binding domain of the initiator protein DnaA. *EMBO J.* **14:**2106–2111.

89. **Roth, A., and W. Messer.** 1998. High-affinity binding sites for the initiator protein DnaA on the chromosome of Escherichia coli. *Mol. Microbiol.* **28:**395–401.

90. **Schaper, S., and W. Messer.** 1997. Prediction of the structure of the replication initiator protein DnaA. *Proteins* **28:**1–9.

91. **Schaper, S., J. Nardmann, G. Luder, R. Lurz, C. Speck, and W. Messer.** 2000. Identification of the chromosomal replication origin from Thermus thermophilus and its interaction with the replication initiator DnaA. *J. Mol. Biol.* **299:**655–665.

92. **Seitz, H., M. Welzeck, and W. Messer.** 2001. A hybrid bacterial replication origin. *EMBO Rep.* **2:**1003–1006.

93. **Sekimizu, K., D. Bramhill, and A. Kornberg.** 1987. ATP activates dnaA protein in initiating replication of plasmids bearing the origin of the E. coli chromosome. *Cell* **50:**259–265.

94. **Sekimizu, K., and A. Kornberg.** 1988. Cardiolipin activation of DnaA protein, the initiation protein in E. coli. *J. Biol. Chem.* **263:**7131–7135.

95. **Skarstad, K., E. Boye, and H. B. Steen.** 1986. Timing of initiation of chromosome replication in individual Escherichia coli cells. *EMBO J.* **5:**1711–1717.

96. **Skarstad, K., B. Thony, D. S. Hwang, and A. Kornberg.** 1993. A novel binding protein of the origin of the Escherichia coli chromosome. *J. Biol. Chem.* **268:**5365–5370.

97. **Skarstad, K., and E. Boye.** 1994. The initiator protein DnaA: evolution, properties and function. *Biochim. Biophys. Acta* **1217:**111–130.

98. **Skarstad, K., G. Lueder, R. Lurz, C. Speck, and W. Messer.** 2000. The Escherichia coli SeqA protein binds specifically and co-operatively to two sites in hemimethylated and fully methylated oriC. *Mol. Microbiol.* **36:**1319–1326.

99. **Skarstad, K., N. Torheim, S. Wold, R. Lurz, W. Messer, S. Fossum, and T. Bach.** 2001. The Escherichia coli SeqA protein binds to two sites in fully and hemimethylated oriC and has the capacity to inhibit DNA replication and effect chromosome topology. *Biochimie* **83:**49–51.

100. **Slater, S., S. Wold, M. Lu, E. Boye, K. Skarstad, and N. Kleckner.** 1995. E. coli SeqA protein binds oriC in two different methyl-modulated reactions appropriate to its roles in DNA replication initiation and origin sequestration. *Cell* **82:**927–936.

101. **Speck, C., C. Weigel, and W. Messer.** 1999. ATP- and ADP-DnaA protein, a molecular switch in gene regulation. *EMBO J.* **18:**6169–6176.

102. **Speck, C., and W. Messer.** 2001. Mechanism of origin unwinding: sequential binding of DnaA to double- and single-stranded DNA. *EMBO J.* **20:**1469–1476.

103. **Su'etsugu, M., H. Kawakami, K. Kurokawa, T. Kubota, M. Takata, and T. Katayama.** 2001. DNA replication-coupled inactivation of DnaA protein in vitro: a role for DnaA ar-

ginine-334 of the AAA+ Box VIII motif in ATP hydrolysis. *Mol. Microbiol.* **40:**376–386.

104. **Sutton, M. D., and J. M. Kaguni.** 1997. The Escherichia coli dnaA gene: four functional domains. *J. Mol. Biol.* **274:**546–561.

105. **Sutton, M. D., and J. M. Kaguni.** 1997. Threonine 435 of Escherichia coli DnaA protein confers sequence-specific DNA binding activity. *J. Biol. Chem.* **272:**48824–51319.

106. **Sutton, M. D., K. M. Carr, M. Vicente, and J. M. Kaguni.** 1998. Escherichia coli DnaA protein: the N-terminal domain and loading of DnaB helicase at the E. coli chromosomal origin. *J. Biol. Chem.* **273:**34255–34262.

107. **Taghbalout, A., A. Landoulsi, R. Kern, M. Yamazoe, S. Hiraga, B. Holland, M. Kohiyama, and A. Malki.** 2000. Competition between the replication initiator DnaA and the sequestration factor SeqA for binding to the hemimethylated chromosomal origin of E. coli in vitro. *Genes Cells* **5:**873–884.

108. **Takata, M., L. Guo, T. Katayama, M. Hase, Y. Seyama, T. Miki, and K. Sekimizu.** 2000. Mutant DnaA proteins defective in duplex opening of oriC, the origin of chromosomal DNA replication in Escherichia coli. *Mol. Microbiol.* **35:**454–462.

109. **Thony, B., D. S. Hwang, L. Fradkin, and A. Kornberg.** 1991. IciA, an Escherichia coli gene encoding a specific inhibitor of chromosomal initiation of replication in vitro. *Proc. Natl. Acad. Sci. USA* **88:**4066–4070.

110. **Torheim, N. K., and K. Skarstad.** 1999. Escherichia coli SeqA protein affects DNA topology and inhibits open complex formation at oriC. *EMBO J.* **18:**4882–4888.

111. **Torheim, N. K., E. Boye, A. Lobner-Olesen, T. Stokke, and K. Skarstad.** 2000. The Escherichia coli SeqA protein destabilizes mutant DnaA204 protein. *Mol. Microbiol.* **37:**629–638.

112. **von Freiesleben, U., K. V. Rasmussen, and M. Schaechter.** 1994. SeqA limits DnaA activity in replication from oriC in Escherichia coli. *Mol. Microbiol.* **14:**763–772.

113. **von Freiesleben, U., M. A. Krekling, F. G. Hansen, and A. Lobner-Olesen.** 2000. The eclipse period of Escherichia coli. *EMBO J.* **19:**6240–6248.

114. **Wang, Q., and J. M. Kaguni.** 1987. Transcriptional repression of the dnaA gene of Escherichia coli by dnaA protein. *Mol. Gen. Genet.* **209:**518–525.

115. **Weigel, C., W. Messer, S. Preiss, M. Welzeck, Morigen, and E. Boye.** 2001. The sequence requirements for a functional Escherichia coli replication origin are different for the chromosome and a minichromosome. *Mol. Microbiol.* **40:**498–507.

116. **Woelker, B., and W. Messer.** 1993. The structure of the initiation complex at the replication origin, oriC, of Escherichia coli. *Nucleic Acids Res.* **21:**23087–23091.

117. **Wold, S., E. Crooke, and K. Skarstad.** 1996. The Escherichia coli Fis protein prevents initiation of DNA replication from oriC in vitro. *Nucleic Acids Res.* **24:**3527–3532.

118. **Wold, S., E. Boye, S. Slater, N. Kleckner, and K. Skarstad.** 1998. Effects of purified SeqA protein on oriC-dependent DNA replication in vitro. *EMBO J.* **17:**4158–4165.

119. **Xia, W., and W. Dowhan.** 1995. In vivo evidence for the involvement of anionic phospholipids in initiation of DNA replication in Escherichia coli. *Proc. Natl. Acad. Sci. USA* **92:**783–787.

120. **Yasuda, S., and Y. Hirota.** 1977. Cloning and mapping of the replication origin of Escherichia coli. *Proc. Natl. Acad. Sci. USA* **74:**5458–5462.

121. **Yung, B. Y., and A. Kornberg.** 1988. Membrane attachment activates dnaA protein, the initiation protein of chromosome

replication in Escherichia coli. *Proc. Natl. Acad. Sci. USA* 85:7202–7205.

122. **Zawilak, A., S. Cebrat, P. Mackiewicz, A. Krol-Hulewicz, D. Jakimowicz, W. Messer, G. Gosciniak, and J. Zakrzewska-Czerwinska.** 2001. Identification of a putative chromosomal replication origin from Helicobacter pylori and its inter-

action with the initiator protein DnaA. *Nucleic Acids Res.* 29:2251–2259.

123. **Zheng, W., Z. Li, K. Skarstad, and E. Crooke.** 2001. Mutations in DnaA protein suppress the growth arrest of acidic phospholipid-deficient Escherichia coli cells. *EMBO J.* **20:** 1164–1172.

The Bacterial Chromosome
Edited by N. Patrick Higgins
© 2005 ASM Press, Washington, D.C.

Chapter 10

DNA Elongation

Manju M. Hingorani and Mike O'Donnell

GENERAL FEATURES OF DNA REPLICATION

During initiation of replication at the origin, an array of proteins is assembled into two replication forks that synthesize DNA bidirectionally and produce a faithful copy of the chromosome. Each protein provides a key activity required for rapid and accurate DNA synthesis (Table 1). Foremost at the replication fork junction, the DNA helicase unwinds double-stranded DNA (dsDNA) to create transient single-stranded DNA (ssDNA) templates for DNA synthesis (Fig. 1). ssDNA-binding protein (SSB) stabilizes the ssDNA, melting out secondary structures in the template ahead of DNA polymerase. Primase synthesizes short RNA primers that serve as starting points for DNA synthesis. Due to the antiparallel nature of DNA and the 5'-3' directionality of the polymerase, a primer is extended continuously on only one template (leading strand synthesis). On the opposite template, the primase repeatedly synthesizes RNA primers that are extended into 1- to 2-kb Okazaki fragments by a second DNA polymerase (lagging strand synthesis). Eventually, the RNA primers are replaced with DNA, and the discontinuous fragments are joined by DNA ligase into a single DNA strand. The two polymerases also edit errors in the newly synthesized DNAs with an associated (or inherent) 3'-5' exonuclease activity. Two additional proteins are critical for polymerase function: a ring-shaped clamp that serves as a sliding tether for the polymerase on DNA (thus increasing the overall speed of DNA replication), as well as a clamp loader machine that is required to assemble the clamp onto DNA. Together with these proteins, the polymerases are capable of simultaneously replicating leading and lagging DNA strands at high speeds approaching a thousand nucleotides per second and an error frequency as low as 1 in 10^6 to 1 in 10^7 bp (84).

Since the first discovery of a DNA polymerase—DNA polymerase I (Pol I)—from *Escherichia coli* about 50 years ago (83), the DNA replication machinery of several organisms ranging from viruses to complex eukaryotes has been investigated. One striking finding of this research is the remarkable conservation in structure and function of DNA replication proteins from diverse organisms. As a result, knowledge gained from the investigation of prokaryotic DNA replication mechanisms has led to as many significant advances in our understanding of the process in mammals as it has in microbes. The following sections describe the structure and mechanisms of action of bacterial DNA replication proteins, and how their activities are coordinated for efficient duplication of chromosomal DNA.

THE PROTEIN MACHINES

Holoenzyme/Replicase

The proteins listed above function together as catalytic or structural components of the chromosomal DNA replication machinery known as a replisome (Fig. 1). At the heart of the replisome is the replicase, or DNA polymerase holoenzyme—itself a complex protein machine comprising DNA polymerase and the accessory clamp and clamp loader proteins. Thus far, detailed studies of such prokaryotic DNA polymerase holoenzymes have been performed mainly in the gram-negative *E. coli* model system (75) and the *E. coli* bacteriophage T4 (8). However, homologues of the three holoenzyme components have been found in other eubacteria, i.e., gram-positive bacteria (e.g., *Bacillus subtilis* [106], *Streptococcus pyogenes* [18], and *Staphylococcus aureus* [80]) and thermophiles (e.g., *Thermus thermophilus*

Manju M. Hingorani • Wesleyan University, 205 Hall-Atwater Laboratories, Middletown, CT 06459. **Mike O'Donnell** • Rockefeller University and Howard Hughes Medical Institute, 1230 York Ave., New York, NY 10021.

Table 1. DNA replication proteins in *E. coli*

Protein	Gene(s)	Monomer mass (kDa)	Function
DNA elongation			
DnaB	*dnaB*	52 (hexamer)	Replicative helicase
DnaC	*dnaC*	28	Helicase loader
SSB	*ssb*	19 (tetramer)	ssDNA binding protein
DnaG	*dnaG*	66	Primase
Pol III core			Replicative polymerase
(αεθ)	α, *dnaE*; ε, *dnaQ*	α: 130; ε: 28	α: polymerase; ε: 3′-5′ exonuclease
	θ, *holE*	θ: 9	θ: unknown
β	*dnaN*	41	Circular clamp
γ/τ complex (γδ,δ′,χ,ψ)	γ/τ, *dnaX*	γ: 48; τ: 71	γ/τ: motor, τ: replisome organizer
	δ, *holA*; δ′, *holB*	δ: 39; δ′: 37	δ: clamp opener; δ′: stator
	χ, *holC*; ψ, *holD*	χ: 17; ψ: 15	χ: switching protein; ψ: stabilizer
Pol I	*pol I/polA*	103	Primer removal
DNA ligase I	*lig*	74	Okazaki fragment linkage
Topo II (gyrase)	*gyrA*, *gyrB*	GyrA: 97; GyrB: 90	Negative supercoiling, decatenation
Translesion DNA synthesis			
Pol II/DinA	*dinA/polB*	90	Translesion synthesis, replisome reactivation
Pol IV/DinB	*dinB*	42	Translesion synthesis, adaptive mutagenesis
Pol V/UmuD′₂C	*umuC*	48	Translesion synthesis (SOS mutagenesis)
	umuD	15	
Recombinative repair			
RecA	*recA*	37	Recombinase
RecF	*recF*	41	ATPase, binds ss+dsDNA, RecR (RecO)
RecO	*recO*	28	Binds ssDNA, dsDNA, RecR, SSB; mediates RecA-SSB/ssDNA binding
RecR	*recR*	22	Binds RecF
RecJ	*recJ*	60	5′-3′ ssDNA exonuclease
RecQ	*recQ*	74	3′-5′ DNA helicase
RecBCD	*recB*	RecB: 134	Helicase, 3′-5′ exonuclease
	recC	RecC: 22	5′-3′ exonuclease, ssDNA endonuclease
	recD	RecD: 67	dsDNA exonuclease
Replication restart			
PriA	*priA*	81.6	D-loop recognition, 3′-5′ helicase
PriB	*priB*	11.5	Mediates PriA-DnaT complex formation
PriC	*priC*	21	Primosome component
DnaT	*dnaT*	20	Primosome component

[110, 175] and *Aquifex aeolicus* [18a]), and in archaebacteria (e.g., *Methanobacterium thermautotrophicum* and *Sulfolobus solfataricus* [72]) as well. Thus, the strategy of utilizing a combination of polymerase, clamp, and clamp loader proteins for efficient genomic DNA replication is common in the diverse bacterial spectrum (not surprisingly, these proteins enjoy widespread use in eukaryotes as well [165]).

The *E. coli* DNA polymerase III holoenzyme comprises 10 proteins, including the core polymerase (α, DNA polymerase; ε, proofreading 3′-5′ exonuclease; and θ, unknown function), the sliding clamp (β), and the five-subunit clamp loader γ complex (γ, δ, δ′, χ, ψ) (75). The 10th protein, τ, is a larger version of the γ subunit encoded by the same *dnaX* gene (a translational frameshift produces the truncated γ protein). Unlike γ, τ can bind the core polymerase and DnaB helicase via its additional C-terminal domains

(42, 43). τ can also replace γ in γ complex (123, 132); thus the clamp loader within the holoenzyme contains both τ and γ subunits, with τ playing the additional role of organizing the holoenzyme components as well as other replisomal proteins.

Figure 1 depicts the architecture of the *E. coli* holoenzyme and its function during DNA replication. Two τ subunits within the clamp loader hold together two copies of the core polymerase αεθ (likely via a flexible connection), each secured to the primer-template by a β clamp (151). This arrangement allows simultaneous and coordinated synthesis of the leading and lagging DNA strands. The contact between τ and DnaB helicase has dual functions: it stimulates helicase activity and helps tether the leading strand polymerase to the template for continuous DNA synthesis (the lagging strand polymerase is free to initiate and complete the numerous Okazaki fragments)

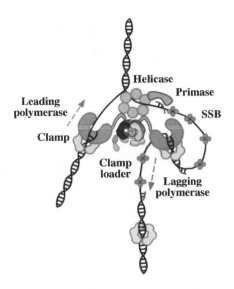

Figure 1. Replisome. At the DNA replication fork the helicase unwinds dsDNA, making ssDNA templates that are stabilized by SSB. Primase synthesizes RNA primers around which a clamp loader assembles circular clamps for use by the DNA polymerase. The holoenzyme, comprising two core DNA polymerases (αεθ), a connector protein (τ) and clamp loader (γ/τ complex), and two clamps (β), replicates leading and lagging DNA strands simultaneously. A connection between the holoenzyme (via τ) and DnaB helicase stimulates DNA unwinding up to the speed of the DNA polymerase. Additional connections between the helicase and primase, the clamp loader, and SSB help coordinate the activities of the various protein components and help stabilize the replisome at the fork junction.

(79, 178). Upon completion of a 1- to 2-kb Okazaki fragment, the lagging strand polymerase releases both the β clamp and DNA and starts anew at an upstream RNA primer on which β has been assembled by the clamp loader (98, 153). Thus, β-clamp assembly is required frequently (every 1 to 2 s for a polymerase speed of ∼800 bases/s) during lagging strand synthesis (150). The τ-mediated connection between the clamp loader and the polymerase organizes clamp loading and DNA synthesis activities within one holoenzyme particle, likely facilitating timely assembly of β clamps on DNA for use by the polymerase. As discussed in the following sections, this organizational model of a DNA replicase is more or less adhered to by a variety of polymerases that duplicate chromosomal DNA in other organisms.

DNA polymerase

As it catalyzes DNA synthesis, DNA polymerase goes through the following reaction cycle: (i) it binds primer-template DNA and (ii) a deoxynucleoside triphosphate (dNTP) in the active site, (iii) catalyzes the nucleotidyl transfer reaction, in which a covalent link is formed between the 3′-hydroxyl group on the pri-

mer and the dNTP, and finally (iv) moves on DNA to bring the new 3′-hydroxyl group into the active site for the next reaction cycle. This reaction presents an interesting challenge for the polymerase, as it must bind all four dNTPs with high affinity, yet in each catalytic cycle it must select only one dNTP that forms a correct base pair with the template. Moreover, the polymerase must also bind the DNA with high affinity, yet release and reposition it within the active site at the end of each catalytic cycle without complete dissociation. Structural and kinetic studies of DNA polymerases reveal how these enzymes perform such complex tasks with high speed and fidelity.

Based on similarities in sequence, DNA polymerases from various organisms have been classified into different families: the polymerase I/Pol A family (e.g., *E. coli* Pol I, bacteriophage T7 DNA polymerase, and eukaryotic Pol γ), the polymerase α/Pol B family (e.g., *E. coli* Pol II, bacteriophage T4 and RB69 polymerases, eukaryotic Pol α, δ, and ε, and archaeal type B polymerases), the Pol C family (e.g., *E. coli* Pol III α subunit and *B. subtilis* Pol C), the Pol X family (e.g., eukaryotic Pol β), the RT family (e.g., human immunodeficiency virus [HIV] reverse transcriptase and telomerase), the Pol D family (e.g., archaeal DP2-type polymerases), and the Pol Y family comprising recently discovered lesion bypass polymerases (e.g., *E. coli* UmuC, DinB, and eukaryotic Pol η) (21, 29, 67). The first crystal structure of a DNA polymerase—Klenow fragment of *E. coli* DNA Pol I—revealed key features that were subsequently found in all other DNA polymerase structures, irrespective of their classification (except for slight differences in Pol β) (17, 122). Thus, although structural information on the α subunit of the replicative *E. coli* Pol III holoenzyme is not yet available, its characteristics can be inferred from the other polymerase structures.

E. coli Pol I resembles a half-open right hand with a "palm" domain that forms a cleft flanked by "fingers" and "thumb" domains (Fig. 2A). The three domains hold the primer-template within the cleft and position the incoming dNTP for reaction with the primer 3′-hydroxyl group. Specifically, the fingers interact with incoming dNTP and the ssDNA template, the thumb interacts with dsDNA, and the palm domain contains the catalytic site where the new base pair and phosphodiester bond is formed (Fig. 2B and C). The site is characterized by two metal cations (Mg^{2+}) ligated by carboxylate oxygens from two absolutely conserved aspartate residues (67). Ion A lowers the pK_a of the 3′-hydroxyl group, facilitating formation of the hydroxide nucleophile that attacks α-phosphate of dNTP (Fig. 2D). Ion A also stabilizes the pentacovalent transition state formed during the reaction. Ion B ligates the phosphate group oxygens

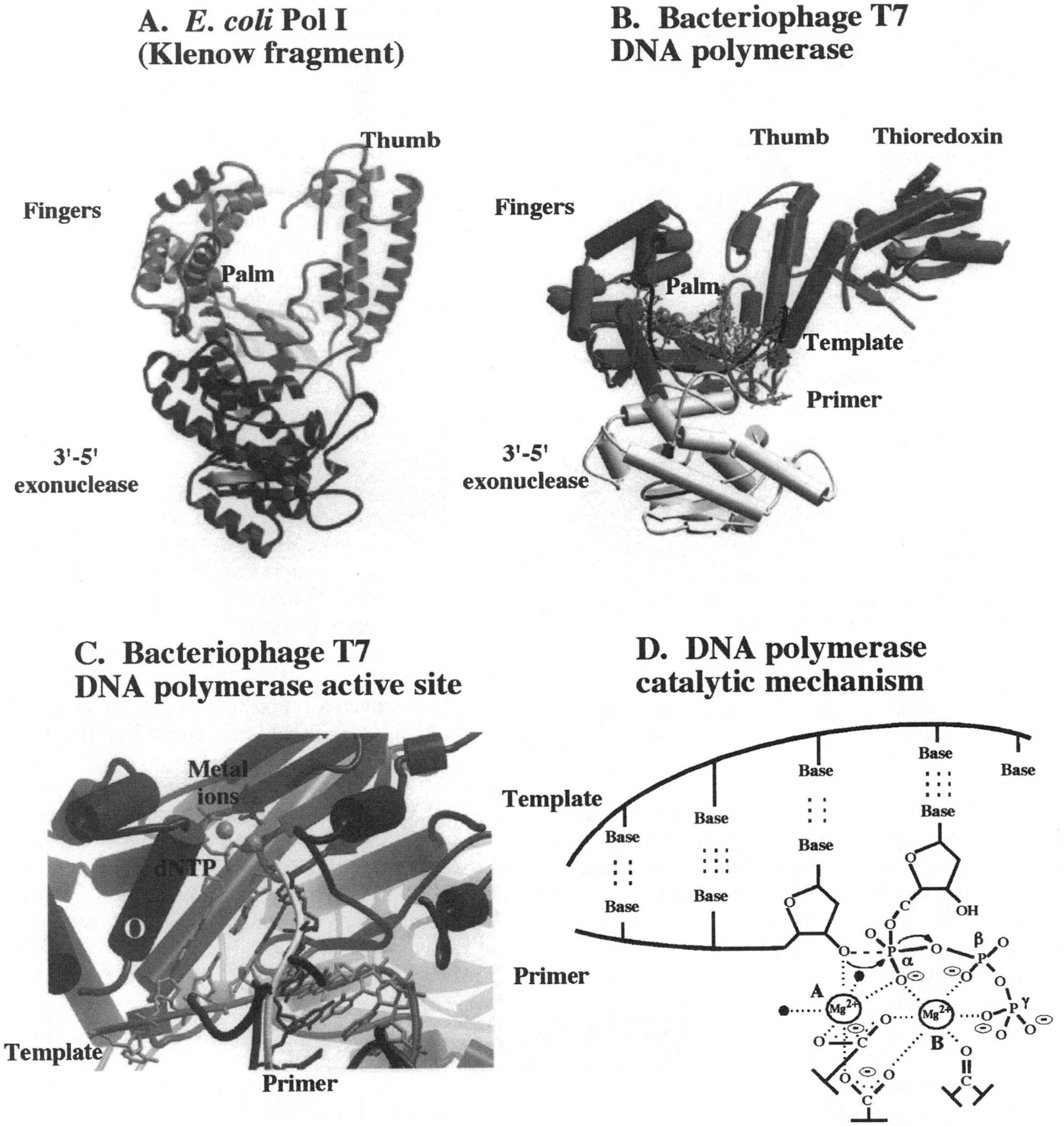

A. *E. coli* Pol I (Klenow fragment)

B. Bacteriophage T7 DNA polymerase

C. Bacteriophage T7 DNA polymerase active site

D. DNA polymerase catalytic mechanism

Figure 2. DNA polymerase structure and catalytic site geometry. (A and B) The polymerase structure comprises the "fingers," "palm," and "thumb" domains arranged to form a DNA binding cleft with the active site at the base (in the palm domain). (B and C) The template strand is bent to expose a nucleotide for base pairing with the incoming dNTP (shown for T7 DNA polymerase), while the primer 3′-OH end and the new base pair are held snugly within the active site to facilitate correct base pairing and nucleotidyl transfer. (D) At the active site, two metal ions (A and B) are coordinated by conserved acidic residues and water molecules, and in turn coordinate the phosphate group oxygens of the incoming dNTP. The primer 3′-OH initiates nucleophilic attack on the α phosphate of dNTP, resulting in phosphodiester bond formation and pyrophosphate release.

on the dNTP, likely aligning the molecule for nucleophilic attack as well as stabilizing the charge on the transition state. Other polar groups and ion B may also help stabilize the pyrophosphate product of nu-

cleotidyl transfer as it leaves the active site at the end of the reaction. The two-metal-ion-catalyzed chemistry of nucleotidyl transfer was proposed by Steitz and coworkers, based on a similar mechanism proposed for the 3'-5' exonuclease of Pol I (7). The *E. coli* Pol I active-site geometry and corresponding catalytic mechanism have been found in all DNA polymerase structures solved to date whether they share sequence homology with Pol I or not—reflecting strong conservation of DNA polymerase mechanism of action throughout evolution (Fig. 2B) (17, 67).

The Klenow fragment also served as an initial model system for understanding the kinetic pathway followed by polymerases when catalyzing DNA synthesis (19, 63). These studies, together with detailed analyses of the bacteriophage T7 and T4 DNA polymerases, as well as HIV reverse transcriptase, revealed a reaction pathway that appears common to DNA polymerases (shown in Fig. 3A for the T7 DNA polymerase) (30, 41, 63, 69, 126, 170). The polymerase binds a primer-template with high affinity in the active site (structural data indicate that during this step the thumb rotates in toward the palm, allowing the thumb tip to "grip" DNA). Next, the dNTP is bound, resulting in a ternary Pol•DNA•dNTP complex. The polymerase binds the correct nucleotide with about 200- to 400-fold-higher affinity than the incorrect nucleotide, apparently able to distinguish a correct base pair from a mismatch during this initial step in the reaction (126). The structures of bacteriophage T7 (33), *Geobacillus stearothermo-*

philus (78), and RB69 polymerases (40), among others, reveal how this selectivity may come about. At the active site, the single-stranded template DNA is kinked such that the next base is exposed for pairing with the incoming dNTP (Fig. 2B). The new base pair fits in a tight binding pocket formed between residues of the fingers domain and the 3' end of the primer (Fig. 2C). Most DNA polymerases do not appear to have specific hydrogen bonding or electrostatic interactions with the new base pair; however, they exhibit exquisite sensitivity to the geometry of the base pair (32, 82, 100). This is likely due to the sterically constrained nature of the binding pocket and van der Waals contacts between the polymerase and the base pair. Accordingly, the polymerase selects for a dNTP that forms a base pair with correct Watson-Crick geometry and against dNTPs that form a distorted, mismatched base pair.

The kinetic studies predict that a slow, rate-limiting conformational change in the polymerase occurs following dNTP binding, in which the polymerase "closes in" around the DNA and the nascent base pair in preparation for the nucleotidyl transfer reaction (126) (Pol*•DNA•dNTP complex in Fig. 3A). Comparisons of DNA polymerase structures in apo form, in binary complex (with DNA), or in ternary complex (with DNA and dNTP) reveal that on

$$\text{Pol} \rightleftharpoons \text{Pol•DNA} \rightleftharpoons \text{Pol•DNA•dNTP} \rightleftharpoons \text{Pol'•DNA•dNTP} \rightleftharpoons \text{Pol'•DNA}_{+1}\text{•P} \rightleftharpoons \text{Pol•DNA}_{+1}\text{•P} \rightleftharpoons \text{Pol•DNA}_{+1}$$

A. Klentaq1•DNA complex

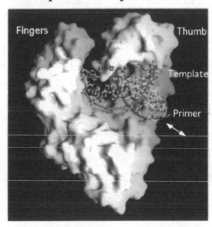

B. Klentaq1*•DNA•dNTP complex

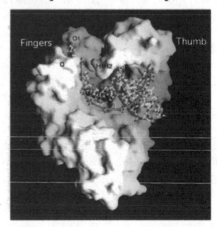

Figure 3. DNA polymerase mechanism of action. The DNA synthesis pathway (shown here for T7 DNA polymerase) initiates with rapid and high-affinity binding of primer-template DNA and dNTP to the polymerase. Next, the polymerase undergoes a slow, rate-limiting conformation in preparation for nucleotidyl transfer. The fast chemistry step is followed by another conformational change that allows product release and translocation of the new 3'-OH terminus into the active site. (A) An image of the Klentaq1•DNA "open" complex with primed DNA in the binding cleft. The arrow predicts translocation of the primer-template following nucleotidyl transfer. (B) A Klentaq1*•DNA•dNTP "closed" ternary complex, with DNA and dNTP trapped securely within the active site due to the inward movement of the fingers domain.

binding dNTP the polymerase switches to a "closed" form in which helices in the fingers domain rotate in toward the active site (99, 127, 141, 142). Figure 3B and C show two of the conformations adopted by the Klenow fragment of *Thermus aquaticus* DNA polymerase (Klentaq1) (99). The Klentaq1•DNA form (absent dNTP) is fully open, with the tip of the fingers domain (O helix) turned outward such that a crevice is visible between the fingers and thumb domains. In the Klentaq1*•DNA•dNTP form the crevice is sealed by movement of the O helix closer to the active site, where it helps orient the DNA and dNTP for base-pair formation. The "closed" conformation also buries the dNTP in a solvent-inaccessible environment that is favorable for chemistry. This rate-limiting transition of the polymerase from open to closed conformation is severely inhibited by the presence of a mismatched base pair in the active site, which in turn slows down incorporation of the incorrect nucleotide into the DNA (170). Thus, the polymerase uses an "induced-fit" mechanism to select the correct nucleotide and increase the fidelity of DNA replication to about 10^5 bases.

The DNA polymerase structure is also designed to effectively distinguish dNTPs from rNTPs (ribonucleoside triphosphates), facilitating DNA—not RNA—synthesis. In the Pol I family polymerases (e.g., Klenow fragment), a glutamate residue forms a steric block against the 2'-OH group of an incoming rNTP while a nearby phenylalanine residue constrains the nucleotide to enhance the blocking effect (3). A similar steric clash between the 2'-OH and phenylalanine in Moloney murine leukemia virus reverse transcriptase (44) and the 2'-OH and tyrosine in Pol α family polymerases (16, 40, 45) as well as HIV-1 RT (22) increases selectivity for dNTPs over rNTPs. In T7 DNA polymerase, the C-2 carbon of the dNTP ribose occupies a compact space, leaving no room for a 2'-OH group in the nucleotide-binding site (33). Thus, DNA polymerases synthesize DNA instead of RNA despite 10-fold-higher concentrations of rNTPs over dNTPs in vivo.

Following the decisive conformational change in the polymerase after correct dNTP binding, nucleotidyl transfer occurs rapidly, extending the primer by one nucleotide. The polymerase then undergoes another conformational change that likely returns it to a more open state, facilitating pyrophosphate release from the active site (at least in Pol I family polymerases) and repositioning the primer-terminus in the polymerase active site to continue DNA synthesis (126). Importantly, the DNA polymerase structure appears designed to minimize its complete dissociation from the primer-template during the repositioning step. The Klenow fragment (6) and T7 DNA

polymerase (33) structures, among others, reveal that the thumb domain forms a caplike structure that partially covers the DNA binding cleft. Removal of a 24-amino-acid domain from the tip of the Klenow fragment thumb substantially reduces its affinity for DNA, implying that contacts between the cap structure and the primer-template stabilize the polymerase on DNA (112). Moreover, the Klentaq1 structure indicates that the DNA binding cleft is like a cylinder in which the primer-template can slide freely without complete dissociation from the polymerase (99). The result is that the polymerase can extend the primer by several nucleotides per template binding event—a property defined as processivity—that significantly increases the efficiency of DNA replication.

DNA polymerases possess another property important for efficient DNA replication: a 3'-5' exonuclease that proofreads the newly synthesized DNA, thus minimizing errors and reducing the load on postreplication correction mechanisms (67). Most polymerases contain or are tightly associated with a proofreading exonuclease activity. For example, *E. coli* Pol I contains both polymerase and exonuclease activities within a single polypeptide (122), and the Pol III α subunit associates with ε, a 3'-5' exonuclease, within the polymerase core (108, 143); in conjunction with ε, DNA Pol III functions with very high fidelity, on the order of 1 error in 10^7 nucleotides incorporated into DNA (13). Structural data indicate that both Pol I and Pol α family polymerases can probe the minor groove of the two newest base pairs on the primer-template and sense distortions caused by a mismatch (33, 40, 99). A mismatch in these positions severely inhibits further DNA synthesis (30), and during this pause the primer can partially unwind from the template and partition into the exonuclease active site for editing (perhaps assisted by movement of the thumb domain) (135, 144). Similar to the polymerase catalytic mechanism, the exonuclease utilizes two metal ions to help generate a nucleophile for attack on the phosphodiester bond (a hydroxide ion from water in this case); the ions also stabilize the geometry and charge of the transition state, which facilitates removal of nucleotides from DNA (7). However, unlike DNA synthesis, the exonuclease reaction is not processive, primarily because the primer can rapidly move back and forth between the exonuclease and polymerase active sites, and a correctly base-paired primer-template binds the polymerase with higher affinity than the exonuclease (30, 63). Therefore, once the incorrect nucleotides are excised, the DNA preferentially occupies the polymerase site where DNA elongation can continue.

Circular clamp

As described above, DNA polymerases have an inherent ability to elongate DNA by more than one nucleotide without dissociating from the template (processive DNA synthesis). However, most polymerases, including Pol III core, can extend DNA by only 10 or so nucleotides per template binding event. Therefore, these polymerases require accessory proteins to achieve the processivity of several thousand nucleotides necessary for chromosomal DNA replication within a short time period (37, 59, 161). The most common processivity factor is a "sliding clamp" that tethers the polymerase and thus minimizes its dissociation from DNA during synthesis (56). Early biochemical studies of sliding clamps revealed that these proteins move freely on duplex DNA and fall off the ends of linear DNA molecules (152). Therefore, sliding clamps were predicted to form a sequence-independent topological link with duplex DNA. This hypothesis was confirmed when the crystal structure of the E. coli β clamp revealed a ring-shaped protein with a central channel large enough to accommodate duplex DNA without steric hindrance (81)

(35-Å diameter [Fig. 4A]). There are no specific DNA-binding elements inside the ring; however, the inner surface has a positive electrostatic potential that complements the phosphate backbone of DNA. Therefore, when the ring-shaped clamp encircles DNA, the nonspecific yet stable topological interaction allows it to slide freely on the duplex and serve as a mobile tether to secure the polymerase on DNA during replication (88).

The circular structure of E. coli β was soon found to be common among sliding clamps from a variety of organisms, including Saccharomyces cerevisiae (yPCNA [87]), humans (hPCNA [52]), and bacteriophages T4 (gp45 [113]) and RB69 (144) (Fig. 4A). Despite minimal homology in their amino acid sequences, these proteins form very similar tertiary and quaternary structures. The β clamp is a dimer formed by head-to-tail interaction between two semicircular monomers. Each monomer consists of three domains that contain two β-α-β-β-β motifs, thus repeating the motif 12 times around the ring. The β strands form six β sheets that serve as a scaffold for the α helices that line the inner surface of the ring. The domains are linked by long connector loops that enhance the

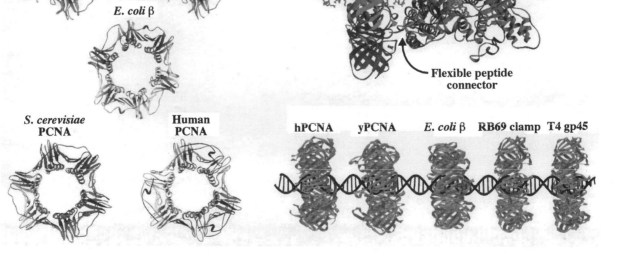

A. Circular sliding clamps

T4 gp45   RB69 clamp

E. coli β

S. cerevisiae PCNA   Human PCNA

B. Model structure of RB69 DNA polymerase and clamp on DNA

Clamp   DNA polymerase

Flexible peptide connector

hPCNA   yPCNA   E. coli β   RB69 clamp   T4 gp45

Figure 4. Circular clamps. (A) Circular sliding clamps from E. coli (β dimer), bacteriophages T4 (gp45 trimer) and RB69, and S. cerevisiae and humans (PCNA trimer). The central cavity averages 35 Å in diameter, sufficient to encircle dsDNA without any steric hindrance. (B) A model structure of the bacteriophage RB69 DNA polymerase tethered by its circular clamp (onto DNA) via a short C-terminal polymerase peptide connector.

stability of the ring and are involved in interactions with other proteins (sliding clamps bind a variety of proteins from DNA replication and other metabolic processes [105, 168]). Unlike the dimeric β clamp, PCNA and bacteriophage T4 and RB69 clamps are formed by three monomers; however, these monomers together also present a total of 12 β-α-β-β-β motifs in a ring that appears remarkably similar to β (Fig. 4A). Moreover, these clamps have an ~35-Å-wide central channel and a positive electrostatic potential on the inner surface, like β, that allows them to encircle duplex DNA and slide freely on it.

A structural model of the RB69 clamp and DNA polymerase complex provides a view of how the clamp secures the polymerase onto primed DNA (144) (Fig. 4B). The RB69 clamp structure was solved in complex with an 11-amino-acid carboxyl-terminal peptide from the RB69 DNA polymerase, which is an essential point of contact between the two proteins (1, 50). This C terminus extends out behind the DNA polymerase and anchors it to the clamp encircled around DNA. Hydrophobic residues (leucine and phenylalanine) near the end of the peptide form tight contacts with a hydrophobic pocket in the clamp, while the remaining residues form a flexible tether between the polymerase and clamp. It is hypothesized that a flexible connection may be advantageous during replication, to allow the polymerase some freedom of movement on DNA without necessitating complete dissociation from the clamp and DNA (e.g., when switching between the polymerase and exonuclease active sites) (144).

One defining structural feature of the sliding clamp is the continuous β sheet formed across the dimer interface by hydrogen bonding between β strands from each monomer. These interactions, along with a small hydrophobic core and six ion pairs at the dimer interface, confer high stability to the β dimer. Thus, the closed form of the clamp is energetically favorable (β dimer K_d <50 nM), and work must be done to open it and insert DNA into the central channel (148). This work is performed by a "molecular matchmaker"—a clamp loader protein that catalyzes β assembly onto primed DNA in a reaction fueled by ATP (139).

Clamp loader

To forge a topological link between the β clamp and DNA, the clamp loader has to open the β clamp, direct DNA into the central channel, facilitate closure of the clamp around DNA, and release the clamp and DNA, coupled to ATP binding and hydrolysis (162). All known clamp loaders are composed of multiple proteins that presumably work together to accomplish these tasks. The *E. coli* clamp loader complex contains three copies of the γ protein (or two τ and one γ in the holoenzyme) and one each of the δ, δ', χ, and ψ proteins (60, 123, 132). δ is the key to clamp assembly as the only subunit of γ complex that binds β with high affinity and opens the β ring (61, 162) (γ also binds β, but the interaction is weak compared with the δ-β interaction [95]). γ is the motor subunit that harnesses energy from ATP to fuel the clamp assembly reaction (92, 160). The γ/τ, δ', and δ proteins belong to the AAA+ ATPase family of proteins that generally function as oligomers and utilize energy from ATP to perform work (120) (γ and δ' share significant sequence homology, but δ' cannot bind or hydrolyze ATP and neither can δ). The δ' subunit appears to be a stator that alternately blocks or allows interaction between δ and β (and between γ and β). The χ and ψ subunits are not directly involved in clamp loading; χ assists assembly of the polymerase onto the primer-template (177) (described in a following section), and ψ confers additional stability to the γ complex structure (121) (as evidenced by resistance to high salt concentrations). γ complex has very low affinity for β in the absence of ATP, implying that the clamp-binding site on the δ subunit is buried within γ complex (occluded by δ') (117). Clamp assembly begins when the clamp loader binds ATP and undergoes a conformational change that allows δ access to β (55, 117) (under these conditions, γ complex binds β with affinity similar to that of the free δ subunit; K_d, ~0.4 μM [94]). This ATP binding-driven γ complex-β interaction is sufficient to open the clamp for assembly around DNA (162). The ATP-bound γ complex•β binds primed DNA with high affinity, which stimulates its ATPase activity (54, 55, 162). Following ATP hydrolysis, γ complex releases β, now linked topologically to DNA, and upon releasing ADP, γ complex reverts to its original conformation (11, 54) (Fig. 5A).

Intensive biochemical studies and recent structural data reveal the detailed mechanics of γ-complex-mediated β-clamp opening (61, 162). In γ complex, one δ', one δ, and three γ/τ subunits are arranged in a pentameric ring (60) (Fig. 5B). Each subunit has an overall "C"-shape, with a RecA-like N-terminal domain I (containing an ATP-binding P-loop and a sensor I motif that likely transduces ATP binding energy into movement) connected to a helical domain II (containing a sensor II motif also sensitive to ATP binding and hydrolysis), which is in turn connected to a helical C-terminal domain III (the five C-terminal domains interact to form a ring). Although the conformity in γ and δ' protein structures was expected, given their sequence homology (51), the similarity between δ and γ/δ' was a surprise since

A. Pathway of clamp assembly on DNA

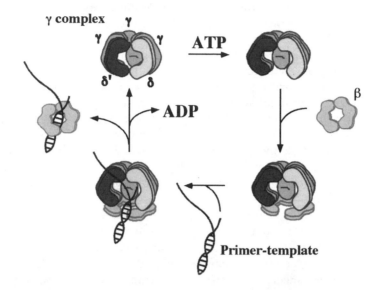

B. *E. coli* γ complex clamp loader structure

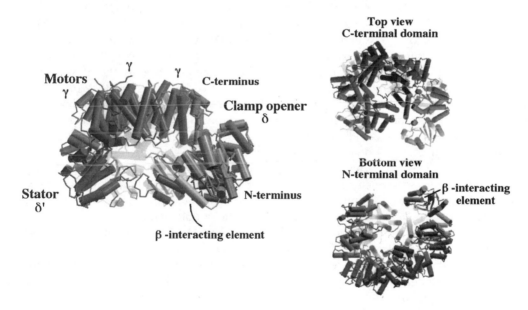

Figure 5. Clamp loader. (A) A model pathway of clamp assembly by the clamp loader begins with ATP binding to the γ/τ subunits, which triggers a change in the complex conformation from "closed" (in which the β-binding element on δ is buried) to "open," allowing interaction between γ/τ complex and β. The clamp loader•clamp complex binds primed DNA with high affinity, which triggers rapid ATP hydrolysis and clamp assembly around DNA. The ADP-bound γ/τ complex releases both β and DNA linked topologically to each other. ADP dissociation recycles the clamp loader for the next round of clamp assembly on DNA. (B) A crystal structure of the *E. coli* γδδ′ clamp loader (χ and ψ are not essential for clamp assembly) reveals three γ subunits, one δ, and one δ′ subunit arranged in a pentamer. The C-terminal domains of all five proteins are arranged in a closed ring, but a large separation between the N-terminal domains of δ and δ′ in an open conformation exposes the β-binding element on δ.

these proteins share only 6 to 8% sequence identity. There are, however, structural features in δ that explain its special ability to bind and open the β clamp for assembly onto DNA. Domain I of the δ subunit contains a structural element in which two amino acids (leucine and phenylalanine) form a small but critical contact with a hydrophobic pocket on β (similar to the hydrophobic contact between RB69 DNA polymerase and clamp). Furthermore, the connection between δ domain II and domain III appears to be flexible; thus the N-terminal domains I and II can swing out from the γ complex without complete disruption of the pentamer (domains III of the γ, δ, and δ′ subunits maintain the ringlike structure). This structure supports the model that when the γ complex is in an inactive conformation, in the absence of ATP, the β-binding element of δ is occluded by δ′. During clamp assembly, ATP binding-driven changes in the γ subunits could cause the N-terminal domains of the pentamer to move and expose the β-binding element on δ (60) (Fig. 5A and B).

The clamp loader does not break the β dimer apart completely into monomers for assembly on DNA. Instead, the clamp is cracked open at only one of its two dimer interfaces, and this opening is sufficient for clamp assembly, as γ complex can load a β clamp covalently closed at one interface onto DNA (162). δ binds a monomeric form of β (equivalent to an open clamp) with ~50-fold-higher affinity than the β dimer (a closed clamp), which suggests that the δ-β dimer binding energy is harnessed for the work of clamp opening (148). The crystal structure of δ·β reveals how this binding energy might be utilized (61). When δ binds β, the β-binding element wedges its hydrophobic residues into one β subunit. The interaction bends a loop in β, which in turn distorts an α helix near the dimer interface, disrupting the contacts between the β subunits at that interface. Preexisting tension in the β ring causes it to relax easily into an open conformation when not constrained by the tight interface contacts. This "spring-loaded" feature of the β dimer structure coupled with δ action triggers clamp opening. δ (and likely the other γ complex components as well) stabilizes the open conformation until DNA is positioned within the clamp.

Binding of primer-template DNA to γ complex stimulates rapid ATP hydrolysis, followed by much slower release of the ADP product (11, 55). During the pause between ATP hydrolysis and ADP dissociation, γ complex may properly position β and DNA for topological linkage, and/or revert to a conformation that has low affinity for both β and DNA. As a result, the β·DNA complex is released—ready for interaction with DNA polymerase—and γ complex can start another cycle of clamp assembly at a newly primed template.

Other Replisomal Proteins

As depicted in Fig. 1, the holoenzyme works in coordination with several other proteins at the replication fork that have functions ranging from initiation (e.g., DNA helicase-catalyzed creation of ssDNA templates) to completion (e.g., DNA ligase-catalyzed joining of Okazaki fragments) of the DNA elongation process. As in the case of Pol III holoenzyme, bacteria (primarily E. coli) have served as a useful model system for understanding the structure and mechanisms of action of many of these DNA replication proteins.

DNA helicase

DNA helicases catalyze unwinding of the double helix, using energy from nucleoside 5′-triphosphate (NTP) binding and hydrolysis to destabilize hydrogen bonding between base pairs and separate the two DNA strands (125). E. coli DnaB was among the first helicases discovered to function as a hexamer (138), and later structural studies of hexameric helicases such as SV40 large T antigen and bacteriophage T7 gp4 revealed that the six subunits are arranged in the form of a ring (34, 125, 140) (Fig. 6A). Although helicases can function as monomers and dimers as well, replicative DNA helicases are predominantly ring-shaped hexamers with a central channel that is large enough to accommodate DNA (20 to 40 Å in diameter). DNA binding studies of T7 gp4 and DnaB indicate that replicative DNA helicase rings bind one strand of ssDNA inside the central channel (34, 62). At a DNA replication fork, the unwinding activity of both T7 gp4 and T4 gp41 helicases is severely inhibited by a block in the 5′ ssDNA strand, but is unaffected by a similar block in the 3′ ssDNA strand (53, 134, 173). These data predict that the helicase ring unwinds duplex DNA by translocating on the 5′ strand (in the 5′-3′ direction) and excluding the 3′ strand from the central channel (Fig. 6A). It is not clear yet whether there are additional interactions between hexameric helicases and the excluded ssDNA strand and dsDNA at the fork that facilitate melting of the double helix.

The DNA unwinding activity of a helicase is intimately coupled to its NTPase activity. Helicases bind DNA with higher affinity in the presence of NTP than in the presence of its hydrolysis product, NDP. For example, the dTTP-bound form of T7 gp4 binds ssDNA with a K_d of 0.2 μM, but there is almost no detectable interaction when the helicase is bound to dTDP (125). These data present a mechanism for DNA unwinding in which NTP binding and hydrolysis drive DNA binding and release, respectively, which in turn drives movement of the helicase on DNA. However, complete dissociation of DNA from the helicase at the end of each NTPase cycle would present a serious problem, as helicases must translocate rapidly over thousands of bases during DNA replication. Entrapment of DNA within the helicase ring presumably lowers the probability of complete dissociation and facilitates processive DNA unwinding. Additionally, it appears that ATP hydrolysis

A. Bacteriophage T7 DNA helicase

Top view

DNA modelled
in central channel

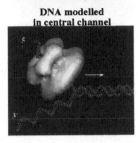

B. Catalytic domain of
E. coli DnaG primase

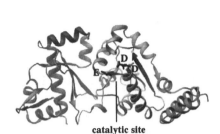

catalytic site

C. E. coli SSB tetramer

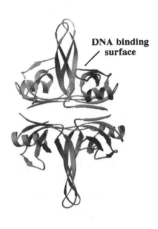

DNA binding
surface

D. NAD⁺-dependent DNA ligase

DNA modelled
in central channel

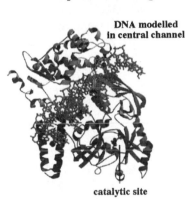

catalytic site

E. E. coli DNA topoisomerase III

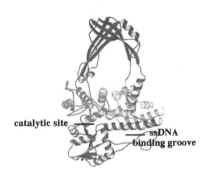

catalytic site

ssDNA
binding groove

F. S. cerevisiae topoisomerase II
DNA-binding/cleavage fragment

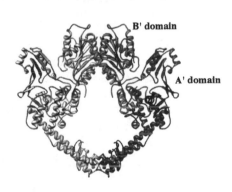

B' domain

A' domain

Figure 6. Additional DNA replication proteins. (A) Crystal structure of the phage T7 gp4 helicase domain, looking down the 6_1-symmetry axis, and a model of the gp4 helicase/primase at the DNA replication fork junction (based on electron microscopy data). (B) Crystal structure of the *E. coli* primase catalytic domain indicating the active site. (C) Crystal structure of the *E. coli* SSB tetramer showing the oligonucleotide-binding domain (five-stranded antiparallel β barrel) per monomer. (D) Crystal structure of the *T. filiformis* NAD⁺-dependent DNA ligase with DNA modeled within the central channel. (E) Crystal structure of type IA *E. coli* topoisomerase III. (F) The DNA binding/cleavage fragment of type II *S. cerevisiae* topoisomerase II (analogous to *E. coli* DNA gyrase).

occurs at only one or a subset of the helicase subunits at a time, implying that only one or a few of the subunits release DNA at any given time. Interestingly, these helicases share structural similarities with the F_1-ATPase hexamer, and the alternating site mechanism of ATP hydrolysis among the F_1-ATPase subunits has also been detected for the T7 gp4 helicase in the absence of DNA (57). These observations suggest that hexameric helicase subunits may sequentially bind and release DNA—concomitant with NTP binding, hydrolysis, and NDP release—to propel processive helicase movement on DNA and duplex DNA unwinding.

Recent discoveries of eukaryotic and archaeal helicases that are essential for replication initiation reveal that these organisms also utilize ring-shaped oligomeric enzymes for DNA unwinding. For example, the *Schizosaccharomyces pombe* (90) and mammalian (58, 174) minichromosomal maintenance proteins (human/mouse MCM 4/6/7 complex) and *M. thermautotrophicum* MCM (25, 74, 145) (an archaeal homologue of eukaryotic MCM4) have demonstrable helicase activity and form ring-shaped hexamers and double hexamers. Studies of the *S. pombe* system have shown that a forked DNA substrate stimulates processive DNA unwinding by the MCM4-MCM6-MCM7 complex, suggesting that MCM helicases are also responsible for unwinding DNA during the elongation phase of DNA replication (90).

Primase

Since DNA polymerases are incapable of linking together two dNTPs to initiate DNA synthesis de novo, primase activity is an essential feature of DNA replication. Primases are specialized RNA polymerases that synthesize short RNA or RNA/DNA segments for elongation by DNA polymerases (70, 84). Prokaryotic primases are generally known to associate with other replication proteins, particularly DNA helicases; for example, E. coli DnaG primase forms a complex with DnaB helicase (159), and in bacteriophage T7 a single gp4A polypeptide has both primase and helicase activities (155). DnaG primase preferentially binds 5'-CTG-3'/5'-CAG-3' sequences on ssDNA templates and synthesizes 8- to 12-nucleotide-long RNA primers (12, 172). Interaction between DnaG and DnaB results in $\sim10^3$-fold stimulation of primase activity (66), presumably because the interaction positions the primase within the replisome where it can accept ssDNA directly from the helicase at the replication fork junction. DnaG is a monomer composed of three distinct domains: an N-terminal zinc-binding domain that is required for primase activity and may be involved in template sequence recognition (124), a central RNA polymerase catalytic domain, and a C-terminal DnaB helicase binding domain (159). Crystal structures of the DnaG catalytic core reveal a cashew-shaped protein with a large central groove that is predicted to bind the DNA template and RNA/DNA duplex, and contains the active site (71, 131) (Fig. 6B). The architecture around the active site is not like the palm domains of DNA polymerases (and RNA polymerases) and is related instead to the "TOPRIM" fold found in topoisomerases (2). A cluster of three conserved acidic residues in the TOPRIM fold mark the likely site of catalysis, suggesting that metal ions are involved in the nucleotidyl transfer reaction, as observed for the DNA polymerase mechanism.

Eukaryotic and archaeal primases differ from bacterial and bacteriophage primases in both sequence and structure. Eukaryotic primases are generally multiprotein complexes (e.g., the four-subunit Pol α-primase), and they synthesize hybrid RNA-DNA primers containing 5' RNA and 3' DNA segments (165). No eukaryotic primase structure has been solved to date; however, the RNA polymerase subunit of archaeal Pyrococcus furiosus primase reveals a unique structure lacking generic polymerase features such as the fingers and thumb domains (4).

Nonetheless, the active site contains three conserved acidic residues, indicating that eukaryotic/archaeal primases also utilize metal ions in a catalytic mechanism common to bacterial primases and all known DNA polymerases.

SSBs

SSBs bind ssDNA with high affinity and stabilize template DNAs formed by the unwinding action of DNA helicase (104). E. coli SSB is a stable heterotetramer that forms long protein clusters on DNA under low-salt conditions (unlimited cooperativity), but not under high-salt conditions. These differing interactions reflect the multiple DNA binding modes of SSB; the unlimited cooperativity/clustering type of interaction involves two SSB subunits of the tetramer binding 35 ssDNA bases, while a more limited cooperativity is associated with all four subunits binding 65 ssDNA bases. Presumably these DNA binding modes serve different functions during DNA metabolism. For example, during DNA replication SSB likely saturates ssDNA templates using the clustering mode, whereas during repair of short stretches of DNA, limited binding may be more suitable (104).

The SSB tetramer crystal structure reveals oligonucleotide/oligosaccharide/oligopeptide binding folds that are responsible for major interactions with ssDNA (one O binding fold per monomer) (133). Recent crystal structures of eukaryotic ssDNA-binding proteins (replication protein A trimer [RPA]) also show multiple O binding folds that bind ssDNA and are implicated in protein-protein interactions (14, 15). Prokaryotic SSB and eukaryotic RPA are known to bind several proteins involved in DNA metabolism. The interactions between E. coli SSB and primase and between SSB and χ protein of γ/τ complex are critical for assembly of DNA polymerase onto primed DNA (77, 177) (described in detail in a following section). E. coli SSB is also known to bind Pol II (114), exonuclease I (46), and the RecO (163)—proteins that are important for DNA repair and recombination. RPA binds Pol α/primase (31), DNA repair proteins such as XPA and UNG (111), DNA recombination proteins such as Rad51 and Rad52 (48, 111), as well as the cell cycle control protein p53 (103). These interactions likely facilitate coordination of various protein activities in DNA metabolism and other cellular processes.

DNA ligase

DNA ligases catalyze phosphodiester bond formation at a nick (or dsDNA break), between the 3'-hydroxyl end of one DNA and the 5'-phosphate end of an adjacent DNA molecule (93, 157). Thus, during DNA replication, ligase activity is essential for joining the discontinuous Okazaki fragments into a single lagging strand. DNA ligases utilize either ATP or NAD^+ as cofactors during catalysis. ATP-dependent ligases have been found in bacteriophages, eukaryotes, and archaea. In contrast, NAD^+-dependent ligases have been found solely in bacteria (except for the recent discovery of NAD^+-dependent ligase in a eukaryotic virus) (147, 157). There is no significant sequence similarity among the ATP- and NAD^+-dependent DNA ligases, yet their reaction mechanisms are essentially identical. First, a conserved lysine residue in the active site attacks the α-phosphate of ATP (or adenylyl-phosphate of NAD^+), forming a covalent enzyme-AMP intermediate and pyrophosphate (or nicotinamide mononucleotide) product. Next, the AMP is transferred from the ligase to the 5′-phosphate group of nicked DNA. Finally, the 3′-OH terminus of an adjacent DNA attacks the activated 5′-phosphate, resulting in phosphodiester bond formation and AMP release. This final step is analogous to DNA polymerase-catalyzed phosphodiester bond formation, and like polymerases, DNA ligases appear to utilize two metal ions (ligated by conserved acidic residues) in the reaction mechanism (93, 157).

Bacterial NAD^+-dependent DNA ligases exhibit high sequence similarity, and according to the *Thermus filiformis* DNA ligase structure these enzymes contain four domains arranged in the form of a toroid with a central cavity large enough to accommodate duplex DNA (Fig. 6D) (91). An N-terminal domain contains the NAD^+ binding site and the adenylation site with the conserved lysine residue. The C-terminal domains contain a zinc finger motif and are responsible for binding DNA (far from the N-terminal catalytic site in the *T. filiformis* ligase structure). A proposed mechanism for ligase action suggests that the enzyme encircles dsDNA and tracks along the duplex until it encounters a nick. When the ligase recognizes a nick, it might undergo large conformational changes that kink the DNA and insert the nick into the catalytic site where ligation occurs. The modular structure of DNA ligase may favor such conformational changes, although there is no direct evidence for these changes or for ligase tracking on DNA as yet.

Eukaryotes utilize multiple ATP-dependent DNA ligases that appear to be specialized for different pathways of DNA replication, repair, and recombination (157, 158). These enzymes share conserved ligase domains—indicating a common catalytic core—and contain a variety of additional N-terminal or C-terminal domains that might confer specialization

(e.g., through interactions with different proteins). DNA ligase I appears to be primarily responsible for ligation of Okazaki fragments in eukaryotic DNA replication (164). Ligase I interacts with the circular clamp PCNA and inhibits Pol δ activity, leading to the hypothesis that ligase displaces Pol δ from PCNA at the nick between two fragments and ligates them to complete lagging strand DNA synthesis (96). The physiological functions of DNA ligases II, III, IV, and V are not exactly clear yet, although their activity has been implicated in various pathways of DNA repair and recombination.

Topoisomerase

Topoisomerases catalyze changes in the topological state of DNA by breaking one or both strands of duplex DNA, passing one strand or a duplex (respectively) through the break, and resealing the break. This property is important for resolving DNA tangles formed during the processes of DNA replication, recombination, repair, and transcription (23, 167). As the helicase unwinds the double helix during DNA replication, positively supercoiled DNA accumulates ahead of the replication fork (in the absence of free DNA ends) and resists further unwinding. Therefore, topoisomerase activity is essential to relax the DNA and allow replication fork progression during the elongation phase of DNA replication. Initiation of DNA replication also depends on topoisomerase action, as these enzymes maintain DNA in a negatively supercoiled state that facilitates initial unwinding of the origin and assembly of the replication complex.

Topoisomerases are broadly classified into two families: type I enzymes (subtypes IA and IB), which break ssDNA, and type II enzymes, which break dsDNA during the catalytic cycle. *E. coli* topoisomerases I and III are type IA enzymes, while DNA gyrase (topoisomerase II) is a type II enzyme. Both type I and type II topoisomerases have a toroidal shape with a central cavity large enough to accommodate DNA—an important feature in their mechanisms of action (Fig. 6E and F). In *E. coli* topoisomerases I and III, four domains of a single polypeptide form a ring ~25 Å in diameter (24, 102). The proposed reaction mechanism initiates with binding of ssDNA (one strand of a duplex) to a groove on the outer surface of the toroid. This event triggers conformational changes that expose the active site and open a gate into the central cavity. Next the enzyme nicks the ssDNA, passes the other strand through the break, and holds it in the central cavity until the nicked strand is religated. The topoisomerase then releases the relaxed DNA molecule and can restart another reaction cycle (38, 101). Type IB enzymes (e.g., human topoisomerase I) function differently in that the toroid first encircles duplex DNA, and then single-strand cleavage, strand passage, and re-ligation all occur within the central cavity (136, 137, 149).

Type II topoisomerases utilize energy from ATP binding and hydrolysis to catalyze transport of one duplex DNA through another and change the topology of DNA (23, 167) (Fig. 6F). The E. coli DNA gyrase is composed of two subunits (GyrA and GyrB) that associate to form an A_2B_2 tetramer [in eukaryotic type II topoisomerases, A and B are part of a single polypeptide and thus form an $(AB)_2$ dimer]. The B subunit/domain binds and hydrolyzes ATP while the A subunit/domain binds and cleaves DNA. The E. coli GyrA and S. cerevisiae topoisomerase II structures reveal that the DNA binding/cleavage domains form a large cavity ($\sim$50 Å in diameter), which is used to trap DNA molecules during the reaction (10, 36, 115). In the proposed mechanism of action, the topoisomerase binds a duplex DNA (gate or G segment) and cleaves both strands, resulting in covalent attachment of a 5'-phosphate end to a tyrosine residue on each monomer. ATP binding allows entry of a second dsDNA molecule (transport or T segment) into the enzyme. The ends of the G segment are parted to allow ATPase-coupled transport of the T segment into a large central cavity formed by DNA binding/cleavage domains. The G segment is then closed, and the T segment is released from the cavity to complete the reaction cycle (5, 23).

In E. coli all three topoisomerases—I, II, and III—have been implicated in DNA replication, but gyrase/topoisomerase II activity appears essential throughout the process: to negatively supercoil the parent molecule for replication initiation, to untwist DNA ahead of the replication fork, and to decatenate and supercoil the daughter DNA molecules at the end of replication.

COORDINATED ACTION OF REPLISOMAL PROTEINS

Clamp assembly on DNA is only one of an ordered series of reactions that occur during the elongation phase of DNA replication. Each replisomal protein provides a specific function or interaction at a specific point in the DNA replication pathway (akin to an assembly line), enabling highly efficient DNA synthesis. The following are examples of coordinated action by the E. coli helicase, primase, SSB, clamp loader, clamp, and DNA polymerase, among other proteins.

Primase-to-Polymerase Switch

The leading strand polymerase in the holoenzyme is linked to DnaB helicase (via τ), coupling continuous DNA synthesis to rapid helicase-mediated movement of the replication fork. The other polymerase initiates and completes an Okazaki fragment every 1 to 2 s for lagging strand synthesis (Fig. 7). As DnaB unwinds duplex DNA, DnaG primase creates 8- to 12-nucleotide-long RNA primers on the ssDNA templates. The primase retains a tight grip on the primed site aided by interaction with SSB, possibly to stabilize the primer on DNA and likely protect it from the action of nucleases as well (177) (Fig. 7A). Next, the γ complex clamp loader binds the primer-template and displaces the primase, mainly because interaction between the χ subunit of γ complex and SSB competes effectively with the primase-SSB interaction (77, 177) (Fig. 7B). This protein switch ensures that the clamp loader is in place to assemble β on DNA and the primase is freed to synthesize multiple primers for lagging DNA synthesis. Clamp assembly is followed by a second protein switch, this time between the clamp loader and DNA polymerase (118) (Fig. 7C). Both the α subunit of core polymerase and the δ subunit of γ complex compete for the same binding site on the β clamp. In solution, the γ complex has higher affinity for β compared with the DNA polymerase, but once β is loaded on DNA the interaction is weakened, allowing the polymerase to compete successfully for the clamp. The polymerase, thus tethered onto the primer-template by β, is ready to initiate processive DNA replication. This coordinated three-protein switch facilitates an orderly handoff of RNA primer from the primase to polymerase during the multiple rounds of replication initiation necessary for lagging strand DNA synthesis.

Completion of Lagging Strand Synthesis

When the lagging strand polymerase arrives at a nick (against a downstream fragment), its connection to the β clamp and DNA is weakened and it dissociates, leaving β behind on the completed fragment (interaction of primase with the replisome and primer synthesis at an upstream template site may also trigger this polymerase cycling) (98, 153). Since β is a stably closed clamp, it remains associated with the fragment over a prolonged period ($t_{1/2}$ on DNA > 100 min), unless removed actively by a clamp unloading mechanism (94, 171). There are only about 300 β clamps in the cell—an insufficient abundance for stoichiometric use on the 2,000 to 4,000 Okazaki fragments synthesized. Therefore, they must be recycled during lagging strand synthesis (94). Both γ

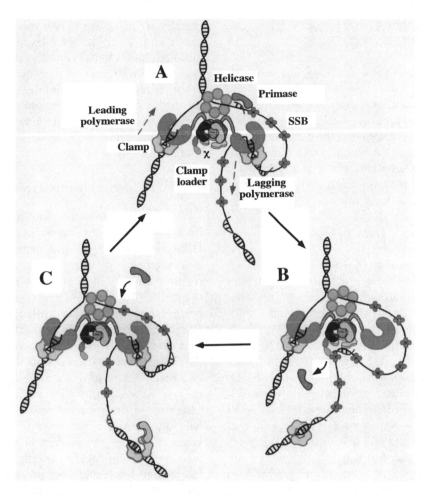

Figure 7. Protein switching during lagging DNA synthesis. (A) After synthesizing a primer, the primase maintains its grip on the site aided by interaction with SSB. (B) The χ subunit disrupts the primase-SSB contact, which triggers primase recycling to an upstream site and allows the clamp loader to assemble a circular clamp around the primed DNA. Meanwhile, the lagging strand polymerase finishes an Okazaki fragment and releases its clamp and DNA. The clamp on DNA is likely bound next by Pol I and DNA ligase for processing of Okazaki fragments into a continuous lagging strand. (C) The lagging strand polymerase replaces the clamp loader at the upstream primed DNA (with β on it) and initiates a new fragment. Meanwhile, the leftover β clamp on DNA is free to be recycled by the δ clamp unloader.

complex and the δ subunit (alone) can open β and unload it from DNA; however, given the 5- to 10-fold excess of δ in the cell, it is likely that δ serves as the clamp recycler in vivo (94). The core polymerase and δ bind the same site on β; therefore δ can recycle β clamps only after they are abandoned on DNA by the polymerase (118). Before a β clamp is recycled, however, it is likely utilized during Okazaki fragment maturation by Pol I and DNA ligase enzymes, which are known to interact with β (96, 105). Pol I 5′-3′ exonuclease activity digests the RNA primer at the 5′ end of each fragment and its polymerase activity fills the gap between adjacent fragments by nick translation, and finally DNA ligase seals the nick (84). Presumably these enzymes target to their sites of action on the lagging strand assisted by interaction with β clamps left behind by Pol III. The ordered nature of

protein-protein switches occurring from Okazaki fragment initiation to maturation likely contributes to the overall efficiency of chromosomal DNA replication.

The protein switching mechanisms described above, in which completion of one action lays up another, apparently occur in an ordered series because the upcoming protein component has high affinity for the product of a preceding action. Similar mechanisms have been observed in other prokaryotic and eukaryotic DNA replication systems (discussed in a following section). Furthermore, such protein switching mechanisms likely occur in other DNA metabolic pathways that employ multiple sequentially acting protein machines as well, such as DNA repair, recombination, and replication restart (described below).

ELONGATION PAST A LESION OR BREAK IN DNA

Under normal cell growth conditions, an estimated 10 to 50% of replication forks initiated at the replication origin encounter lesions or breaks in DNA that halt their forward progress. Cell survival depends on the ability of the replisomal protein machinery to overcome such blocks in the path of DNA elongation. An elaborate array of yet other protein machines work in conjunction with replisomal proteins to perform this critical task.

Translesion DNA Synthesis

One mechanism for bypassing lesions in DNA involves specialized DNA polymerases that can replicate flawed templates that contain pyrimidine-pyrimidine photoadducts, abasic sites, and cyclobutane dimers, among other defects (154). These enzymes belong to the UmuC-DinB-Rad30-Rev1 superfamily of DNA polymerases that can insert dNTPs into the primer across a lesion (translesion DNA synthesis), unlike replicative polymerases. In *E. coli* the DinB (Pol IV) and UmuD$'_2$C (Pol V) and DinA (Pol II) polymerases are responsible for translesion synthesis (119, 128), and in eukaryotes (e.g., humans) at least four such polymerases are involved in the process (65) (Pol η/Rad30Ap/XP-V, Pol ι/Rad30Bp, Pol κ/DinB1p/Pol Q/Pol θ, and Rev1p). A minimal pathway for translesion synthesis envisions that when the replisome stalls at a lesion, the replicative DNA polymerase (and perhaps other replisomal proteins) dissociates from DNA, allowing lesion bypass polymerase access to the primer-template. Bypass polymerase activity appears to be distributive (e.g., Pol V [128]), suggesting that once the polymerase traverses the lesion it "falls off," allowing reassembly of the replicative DNA polymerase and continued DNA elongation. Interestingly, Pol IV (166) and Pol V (156) both interact with the β sliding clamp, which perhaps helps target them to the site of action on DNA. These interactions may be indicative of a protein-protein switching mechanism in which the damaged DNA is handed off from the replicative DNA polymerase to lesion bypass polymerase and back, analogous to the primase-to-polymerase switch observed during lagging strand synthesis.

It is important to note that translesion DNA synthesis can be highly error prone, given that many bypass polymerases are low-fidelity enzymes that incorporate non-base-paired dNTPs across the lesion (49); however, bypass polymerases such as the xeroderma pigmentosum variant gene product XP-V/Pol η can perform error-free translesion synthesis by preferentially inserting dATPs across a thymine-thymine dimer and dCTP across 8-oxoguanine (169). Thus, although lesion bypass enables replication fork progress, the resulting errors have to be repaired by other DNA repair mechanisms or result in potentially disastrous mutations in genomic DNA. It is not surprising, therefore, that there exist additional mechanisms that allow replication fork progression in a more error-free fashion.

Recombinative DNA Repair

Recombinative error-free bypass of lesions and breaks in DNA is dependent on proteins that catalyze DNA recombination, and genetic and biochemical evidence points to at least two recombinational DNA repair pathways in bacteria. One pathway involves the RecBCD (and RecA) proteins that facilitate repair of double-strand breaks, while the other involves the RecFOR (and RecA) proteins that facilitate gap repair across DNA lesions (27). As before, a lesion in the DNA template (that results in a gap in new DNA) or a nick (that results in a double-strand break) stalls the replisome, prompting partial or complete disassembly of the proteins from DNA. The RecBCD helicase/nuclease initiates double-strand-break repair when it invades the broken DNA end, melting and incising it (at chi sites) to create a 3' terminus. This is followed by RecA-mediated invasion of the 3' terminus into a complementary DNA, migration, and resolution of the crossover by a Holliday junction resolvase (e.g., RecG and/or RuvABC resolvase) (35, 85). In the case of gaps formed in DNA following polymerase dissociation at a lesion, the RecFOR proteins facilitate RecA-mediated recombination repair (as described above) or backward movement of the fork over the lesion (which would allow excision repair enzymes to repair the lesion) (27, 85). In both types of recombinative repair, when the DNA crossovers are resolved, the remaining gaps in DNA are filled in by origin-independent restart of DNA replication. The PriA protein directs reassembly of replication forks (in a process involving PriB, PriC, DnaB, DnaC, DnaG, and DnaT proteins), and Pol III holoenzyme elongates DNA to complete the repair process (109).

The processes described above clearly indicate that the pathways of DNA replication, recombination, and repair are intimately intertwined. The strong link between DNA replication and recombination was clearly recognized first in the bacteriophage T4 system (86), and it is likely that in T4, as in *E. coli*, recombinative DNA repair is critical for rescuing stalled replication forks. Detailed understanding of how the numerous DNA replication/recombination/

repair proteins coordinate their action at the replication fork in *E. coli* and other organisms awaits further investigation.

GENERAL APPLICATION OF THE *E. COLI* REPLISOME MODEL

Decades of research on prokaryotic organisms such as *E. coli* and the T7 and T4 bacteriophages have revealed the proteins responsible for chromosomal DNA replication as well as some of their mechanisms of action. The studies also show that, despite their diversity, these organisms utilize the same basic protein machinery to initiate and coordinate rapid synthesis of leading and lagging DNA strands (although individual proteins differ in sequence and structure to varying extents). Moreover, the DNA replication mechanisms appear generally conserved even in significantly more complex creatures, such as humans. Consequently, these prokaryotes serve as useful model systems to understand chromosomal replication across the entire spectrum of bacteria, archaea, and eukarya, which is especially useful at the present time for filling in molecular details in the genome-scale analysis of hundreds of living organisms.

Bacteria

A comparison of DNA replication proteins from gram-positive and gram-negative bacteria highlights the evolutionary success of key components of the DNA replication machinery. The evolutionary split between these bacteria occurred more than a billion years ago, and over time individual proteins such as DNA polymerase have diverged significantly in some aspects. For example, the gram-positive replicative DNA polymerase (Pol C) contains both polymerase and proofreading exonuclease activities within a single polypeptide (unlike the gram-negative *E. coli* α polymerase and ε exonuclease). However, the overall geometry and catalytic mechanism of the polymerase remain conserved (as observed from the gram-positive *G. stearothermophilus* polymerase structure [78]), as is the strategy of using a sliding clamp and clamp loader for processive DNA synthesis (18, 80). Moreover, all bacteria also utilize homologues of the DnaB helicase, primase, SSB, DNA ligase, and topoisomerase proteins in DNA replication.

The gram-positive organism *S. pyogenes*, for example, utilizes Pol C DNA polymerase and τ, δ, δ', and β proteins for chromosomal DNA replication (18). Unlike *E. coli*, the *S. pyogenes dnaX* gene encodes only full-length τ and no γ subunit, and there are no recognizable χ and ψ homologues in gram-

positive or most other bacteria (although functional homologues of χ and ψ may exist in all bacteria). The τ, δ, and δ' proteins form a clamp loader (analogous to *E. coli* γ complex) that assembles β clamps around primed DNA. τ (a homologue of *E. coli* γ and τ proteins) also binds Pol C, forming a Pol C holoenzyme together with the β clamp. A major difference in the *S. pyogenes* system compared with *E. coli* is weak interaction between τ and Pol C, and between τ and $\delta\delta'$; however, at high concentrations these proteins assemble into a holoenzyme, and even when dilute, they function much like the *E. coli* Pol III holoenzyme. All the genes encoding *S. pyogenes* replication proteins share sequence homology with their *E. coli* counterparts, except for the *holA* gene that encodes the δ subunit. In fact, *holA* is not detectable in any bacterial genome search (except in *Haemophilus influenzae* [39] and other relatives of *E. coli*), although homologues of *dnaX* (τ/γ) and *holB* (δ') can be readily found. In some organisms, a complex of proteins homologous to τ/γ and δ' is sufficient for a functional clamp loader (e.g., archaeal [72] and eukaryotic replication factor C (RFC) clamp loaders [116]). Nonetheless, since δ serves an essential function as the clamp opener in *E. coli*, a δ-like protein (with a clamp-opening element) is predicted to be a necessary component of DNA polymerase holoenzymes. The *S. pyogenes* study substantiates this hypothesis by demonstrating that the closely related τ and δ' proteins are not sufficient, and the δ protein is required for clamp assembly on DNA (18).

The *S. pyogenes* study also confirms that a multicomponent holoenzyme replicates chromosomal DNA in bacteria other than the *E. coli* model system. The *E. coli* replicative polymerase purifies from cells as a holoenzyme, complete with its 10 component proteins. In contrast, replicative polymerases from several other bacteria (particularly gram-positive Pol C polymerases) purify as single proteins. Although these polymerases were known to utilize accessory proteins, such as the clamp and clamp loader, the existence of a single holoenzyme particle that coordinates leading and lagging DNA synthesis had not been demonstrated directly (106). As mentioned above, *S. pyogenes* Pol C can be reconstituted in vitro into a Pol C-$\tau\delta\delta'$ complex containing two copies of the polymerase. The interaction between Pol C and τ is weak compared with the *E. coli* $\alpha\varepsilon\theta$ and τ interaction, explaining the difficulty in purifying *S. pyogenes* holoenzyme as a single particle. Nonetheless, like *E. coli* Pol III holoenzyme, the *S. pyogenes* holoenzyme replicates DNA processively, at a speed of $\sim$800 nucleotides/s, consistent with the similarity in their chromosome size and cell division time (18).

Bacteriophages

Studies of the bacteriophage T4 and T7 systems have contributed substantially to the understanding of replisomal protein function in DNA replication. The bacteriophage T4 DNA replication system contains the following protein components: gp41 DNA helicase, gp61 primase, gp32 ssDNA binding protein, gp43 DNA polymerase, gp45 circular clamp, gp44/62 clamp loader; in addition, gp59 aids helicase assembly on DNA (analogous to *E. coli* DnaC, the DnaB helicase loader), and RNase H and ligase are required to process Okazaki fragments. The T7 system is comparatively simple, comprising gp5, the DNA polymerase that forms a 1:1 complex with *E. coli* thioredoxin (processivity factor), gp4 helicase/primase protein, and gp2.5 single-strand DNA binding protein (8).

As in *E. coli*, replication of the leading and lagging DNA strands appears coordinated in bacteriophage (9, 89), and synthesis at both strands is resistant to dilution and template challenge experiments, indicating that both polymerases (and likely other replication proteins) remain stably associated with the replication fork (68). There is no evidence for a central organizer protein like *E. coli* τ, although phage ssDNA-binding proteins appear to bind several proteins at the replication fork, including the polymerases and primase/helicase, and may help organize the holoenzyme and coordinate the various protein functions. Moreover, a DNA polymerase-polymerase interaction has been reported in phage T4, suggesting that the two polymerases may associate directly with each other within a holoenzyme (8).

Instead of a circular clamp, the phage T7 DNA polymerase uses thioredoxin to increase its processivity on DNA (thioredoxin may function by capping the DNA binding cleft and trapping primer-template within the polymerase, or serve as an electrostatic tether between the polymerase and DNA [33]) (Fig. 2B). In contrast, bacteriophage T4 (and the closely related RB69 phage) possesses a clamp and clamp loader mechanism for polymerase processivity much like that of *E. coli*. The T4 polymerase has a C-terminal peptide extension that hooks into a small hydrophobic pocket on the gp45 clamp to form a flexible connection between the two proteins (Fig. 4). This flexible connection likely allows small movements of the polymerase on DNA without completely disrupting the link between the polymerase and clamp, such as at pause sites on DNA, or when the primer is switched between the polymerase and exonuclease active sites for proofreading. The clamp loader is a pentamer of four gp44 subunits and one gp62 subunit. Given that gp44 shares sequence

homology with the *E. coli* γ/τ and δ' proteins (AAA+ ATPase family [120]), it is likely that the gp44/62 pentamer has a quaternary structure similar to that of γ complex (Fig. 5B). The T4 clamp loader mechanism is also ATP driven, with ATP binding and hydrolysis at the gp44 subunits coupled to clamp opening and assembly onto a primed DNA template (8, 129). Once the clamp is assembled, the clamp loader apparently chaperones the polymerase onto DNA, forming a processive holoenzyme competent for DNA elongation.

Eukarya

The DNA replication fork in eukaryotic cells has much the same protein machinery associated with it as in prokaryotic organisms (165). In model systems such as *S. cerevisiae*, as well as in humans, Pol α, a four-subunit primase-DNA polymerase hybrid, synthesizes an RNA-DNA primer of about 35 to 50 nucleotides on ssDNA. Next, the five-subunit RFC clamp loader assembles a PCNA clamp on the primed template (116). The RFC clamp loading mechanism is likely similar to that of *E. coli* γ complex, given that RFC subunits share sequence homology with the *E. coli* γ/τ and δ' proteins (AAA+ ATPase family [120]), and electron microscopy data indicate that RFC resembles the ring/U-shaped pentameric γ complex (60, 146). The replicative polymerase, Pol δ, consists of two to five different subunits (depending on the source organism) that provide PCNA-dependent processive polymerase activity and 3'-5' proofreading exonuclease activity (20). Pol δ, PCNA, and RFC appear to form a holoenzyme; however, it is not known yet whether this complex contains two polymerases for simultaneous synthesis of the leading and lagging DNA (RFC clamp loader subunits could serve as a bridge for two Pol δ molecules as observed with τ in *E. coli* [151]). Pol ε, another polymerase with PCNA-stimulated activity, may also be responsible for lagging strand DNA replication (20).

Assembly of the DNA replicase on a primer template requires coordinated competitive interactions between Pol α, RFC, RPA (ssDNA binding protein), and Pol δ (176). Following primer synthesis Pol α remains associated with DNA (presumably to protect the primer from nucleases), aided by interaction with RPA proteins coating the ssDNA template. In the first protein switch, RFC competes with Pol α for RPA and disrupts their interaction, freeing the primed site for PCNA clamp assembly (107, 176). After loading PCNA, RFC remains on the primer-template, in contact with PCNA and RPA (and DNA) until the second protein switch.

Subsequently, Pol δ competes with RFC for both PCNA and RPA and assembles at the site to initiate DNA replication (176). As in *E. coli*, these competition reactions likely occur in an ordered fashion in the cell because each successive protein has higher affinity for the product of the previous reaction. Thus, RFC likely binds RPA with higher affinity at a primer-template junction compared with RPA on ssDNA, resulting in successful competition with Pol α at the primer-template; Pol δ likely does not bind the primer-template with high affinity until RFC has loaded PCNA onto it. These ordered competition reactions are analogous to those observed in the bacterial DNA replication system, emphasizing the conservation through evolution of these DNA replication mechanisms.

Also similar is the participation of key DNA replication proteins in multiple DNA metabolic processes. For example, PCNA is bound by DNA ligase I (which seals nicks in DNA) (96), Fen1 (which processes Okazaki fragments) (97), and DNA cytosine-5 methyltransferase (which methylates new DNA) (26), among other proteins, that perhaps utilize the clamp to efficiently target their site of action on DNA and/or coordinate their activity with DNA replication (168). The interaction observed between PCNA and the Msh2-6 mismatch repair proteins may also be important for coupling DNA mismatch repair to DNA synthesis (64) (like the analogous connection between *E. coli* β and MutS [105]). Another important interaction occurs between PCNA and the human cell cycle regulatory protein P21[CIP1], which inhibits DNA replication by competing effectively with Pol δ for PCNA (47). RFC may also participate in cell cycle regulatory pathways, as indicated by its cell cycle-dependent phosphorylation by cyclin-dependent and DNA-dependent kinases (116). The central role of RFC and PCNA in chromosomal DNA replication makes these proteins useful contact points for coordination of cell cycle regulation with DNA replication. The eukaryotic ssDNA binding protein RPA (a three-subunit complex) is also a multifunctional component in DNA metabolism. As described above, RPA is one of the key players in the pathway of holoenzyme assembly on primed DNA. In addition, an RPA subunit (RPA32) interacts with DNA repair proteins uracil DNA glycosylase, XPA (111), as well as Rad51 and Rad52 (48, 111), implicating its function in nucleotide excision, base excision, and recombination pathways of fixing damaged DNA. RPA may coordinate protein function in these three distinct repair pathways via ordered competition reactions, as observed during holoenzyme assembly on DNA.

Archaea

Archaeal proteins responsible for DNA replication appear more closely related in amino acid sequence (and organization) to eukarya than to bacteria, as revealed by the genomic sequences of several archaea, including *M. thermautotrophicum*, *Methanococcus jannaschii*, and *P. furiosus* (72). Replicative archaeal DNA polymerases are classified into the Pol B family or the DP2 family, and both polymerases have been identified in all archaea whose genomes have been sequenced. Both enzymes replicate DNA processively and possess proofreading activity, and could therefore catalyze chromosomal DNA replication. Thus far, however, only B-type polymerase activity is found to be stimulated by archaeal homologues of PCNA and RFC proteins (76) (which is a hallmark of chromosomal DNA replication). The PCNAs are very similar in sequence and structure to eukaryotic PCNA; unlike eukaryotic cells, however, some archaea possess more than one PCNA homologue (e.g., *S. solfataricus* [28]). Archaeal RFC clamp loaders are also highly homologous to their eukaryotic counterparts, although they appear composed of only two RFC proteins (as in *S. solfataricus* [130] and *M. thermautotrophicum* [73]) instead of five, as in *S. cerevisiae* and humans. It is likely, however, that these two proteins assemble into circular pentamers similar to the *E. coli* γ complex. Like the *E. coli* γ complex and eukaryotic RFC, archaeal RFC proteins use energy from ATP binding and hydrolysis to catalyze PCNA assembly onto primed DNA. It is conceivable that all these accessory proteins form a single holoenzyme particle along with the DNA polymerase, but there is no information as yet on such an entity in archaea. Other replisomal protein homologues, including DNA helicase, ssDNA binding protein RPA, Okazaki fragment processing protein Fen1, DNA ligase I, and topoisomerases have also been identified in archaea (72), once again highlighting the evolutionary success of the general mechanism of chromosomal DNA replication elucidated first in bacteria.

REFERENCES

1. **Alley, S. C., A. D. Jones, P. Soumillion, and S. J. Benkovic.** 1999. The carboxyl terminus of the bacteriophage T4 DNA polymerase contacts its sliding clamp at the subunit interface. *J. Biol. Chem.* 274:24485–24489.

2. **Aravind, L., D. D. Leipe, and E. V. Koonin.** 1998. Toprim—a conserved catalytic domain in type IA and II topoisomerases, DnaG-type primases, OLD family nucleases and RecR proteins. *Nucleic Acids Res.* 26:4205–4213.

3. **Astatke, M., K. Ng, N. D. Grindley, and C. M. Joyce.** 1998. A single side chain prevents Escherichia coli DNA polymerase

I (Klenow fragment) from incorporating ribonucleotides. *Proc. Natl. Acad. Sci. USA* **95**:3402–3407.

4. **Augustin, M. A., R. Huber, and J. T. Kaiser.** 2001. Crystal structure of a DNA-dependent RNA polymerase (DNA primase). *Nat. Struct. Biol.* **8**:57–61.

5. **Baird, C. L., T. T. Harkins, S. K. Morris, and J. E. Lindsley.** 1999. Topoisomerase II drives DNA transport by hydrolyzing one ATP. *Proc. Natl. Acad. Sci. USA* **96**:13685–13690.

6. **Beese, L. S., V. Derbyshire, and T. A. Steitz.** 1993. Structure of DNA polymerase I Klenow fragment bound to duplex DNA. *Science* **260**:352–355.

7. **Beese, L. S., and T. A. Steitz.** 1991. Structural basis for the 3′-5′ exonuclease activity of Escherichia coli DNA polymerase I: a two metal ion mechanism. *EMBO J.* **10**:25–33.

8. **Benkovic, S. J., A. M. Valentine, and F. Salinas.** 2001. Replisome-mediated DNA replication. *Annu. Rev. Biochem.* **70**:181–208.

9. **Berdis, A. J., and S. J. Benkovic.** 1998. Simultaneous formation of functional leading and lagging strand holoenzyme complexes on a small, defined DNA substrate. *Proc. Natl. Acad. Sci. USA* **95**:11128–11133.

10. **Berger, J. M., S. J. Gamblin, S. C. Harrison, and J. C. Wang.** 1996. Structure and mechanism of DNA topoisomerase II. *Nature* **379**:225–232.

11. **Bertram, J. G., L. B. Bloom, M. M. Hingorani, J. M. Beechem, M. O'Donnell, and M. F. Goodman.** 2000. Molecular mechanism and energetics of clamp assembly in Escherichia coli. The role of ATP hydrolysis when gamma complex loads beta on DNA. *J. Biol. Chem.* **275**:28413–28420.

12. **Bhattacharyya, S., and M. A. Griep.** 2000. DnaB helicase affects the initiation specificity of Escherichia coli primase on single-stranded DNA templates. *Biochemistry* **39**:745–752.

13. **Bloom, L. B., X. Chen, D. K. Fygenson, J. Turner, M. O'Donnell, and M. F. Goodman.** 1997. Fidelity of Escherichia coli DNA polymerase III holoenzyme. The effects of beta, gamma complex processivity proteins and epsilon proofreading exonuclease on nucleotide misincorporation efficiencies. *J. Biol. Chem.* **272**:27919–27930.

14. **Bochkarev, A., E. Bochkareva, L. Frappier, and A. M. Edwards.** 1999. The crystal structure of the complex of replication protein A subunits RPA32 and RPA14 reveals a mechanism for single-stranded DNA binding. *EMBO J.* **18**:4498–4504.

15. **Bochkarev, A., R. A. Pfuetzner, A. M. Edwards, and L. Frappier.** 1997. Structure of the single-stranded-DNA-binding domain of replication protein A bound to DNA. *Nature* **385**:176–181.

16. **Bonnin, A., J. M. Lazaro, L. Blanco, and M. Salas.** 1999. A single tyrosine prevents insertion of ribonucleotides in the eukaryotic-type phi29 DNA polymerase. *J. Mol. Biol.* **290**:241–251.

17. **Brautigam, C. A., and T. A. Steitz.** 1998. Structural and functional insights provided by crystal structures of DNA polymerases and their substrate complexes. *Curr. Opin. Struct. Biol.* **8**:54–63.

18. **Bruck, I., and M. O'Donnell.** 2000. The DNA replication machine of a gram-positive organism. *J. Biol. Chem.* **275**:28971–28983.

18a. **Bruck, I., A. Yuzhakov, O. Yurieva, D. Jeruzalmi, M. Skangalis, J. Kuriyan, and M. O'Donnell.** 2002. Analysis of multicomponent thermostable DNA polymerase III replicase from an extreme thermophile. *J. Biol. Chem.* **277**:17334–17348.

19. **Bryant, F. R., K. A. Johnson, and S. J. Benkovic.** 1983. Elementary steps in the DNA polymerase I reaction pathway. *Biochemistry* **22**:3537–3546.

20. **Burgers, P. M.** 1998. Eukaryotic DNA polymerases in DNA replication and DNA repair. *Chromosoma* **107**:218–227.

21. **Burgers, P. M. J., E. V. Koonin, E. Bruford, L. Blanco, K. C. Burtis, M. F. Christman, W. C. Copeland, E. C. Friedberg, F. Hanaoka, D. C. Hinkle, C. W. Lawrence, M. Nakanishi, H. Ohmori, L. Prakash, S. Prakash, C. A. Reynaud, A. Sugino, T. Toda, Z. Wang, J. C. Weill, and R. Woodgate.** 2001. Eukaryotic DNA polymerases: proposal for a revised nomenclature. *J. Biol. Chem.* **276**:43487–43490.

22. **Cases-Gonzalez, C. E., M. Gutierrez-Rivas, and L. Menendez-Arias.** 2000. Coupling ribose selection to fidelity of DNA synthesis. The role of Tyr-115 of human immunodeficiency virus type 1 reverse transcriptase. *J. Biol. Chem.* **275**:19759–19767.

23. **Champoux, J. J.** 2001. DNA topoisomerases: structure, function, and mechanism. *Annu. Rev. Biochem.* **70**:369–413.

24. **Changela, A., R. J. DiGate, and A. Mondragon.** 2001. Crystal structure of a complex of a type IA DNA topoisomerase with a single-stranded DNA molecule. *Nature* **411**:1077–1081.

25. **Chong, J. P., M. K. Hayashi, M. N. Simon, R. M. Xu, and B. Stillman.** 2000. A double-hexamer archaeal minichromosome maintenance protein is an ATP-dependent DNA helicase. *Proc. Natl. Acad. Sci. USA* **97**:1530–1535.

26. **Chuang, L. S., H. I. Ian, T. W. Koh, H. H. Ng, G. Xu, and B. F. Li.** 1997. Human DNA-(cytosine-5) methyltransferase-PCNA complex as a target for p21WAF1. *Science* **277**:1996–2000.

27. **Cox, M. M.** 2001. Historical overview: searching for replication help in all of the rec places. *Proc. Natl. Acad. Sci. USA* **98**:8173–8180.

28. **De Felice, M., C. W. Sensen, R. L. Charlebois, M. Rossi, and F. M. Pisani.** 1999. Two DNA polymerase sliding clamps from the thermophilic archaeon Sulfolobus solfataricus. *J. Mol. Biol.* **291**:47–57.

29. **Delarue, M., O. Poch, N. Tordo, D. Moras, and P. Argos.** 1990. An attempt to unify the structure of polymerases. *Protein Eng.* **3**:461–467.

30. **Donlin, M. J., S. S. Patel, and K. A. Johnson.** 1991. Kinetic partitioning between the exonuclease and polymerase sites in DNA error correction. *Biochemistry* **30**:538–546.

31. **Dornreiter, I., L. F. Erdile, I. U. Gilbert, D. von Winkler, T. J. Kelly, and E. Fanning.** 1992. Interaction of DNA polymerase alpha-primase with cellular replication protein A and SV40 T antigen. *EMBO J.* **11**:769–776.

32. **Doublie, S., and T. Ellenberger.** 1998. The mechanism of action of T7 DNA polymerase. *Curr. Opin. Struct. Biol.* **8**:704–712.

33. **Doublie, S., S. Tabor, A. M. Long, C. C. Richardson, and T. Ellenberger.** 1998. Crystal structure of a bacteriophage T7 DNA replication complex at 2.2 A resolution. *Nature* **391**:251–258.

34. **Egelman, H. H., X. Yu, R. Wild, M. M. Hingorani, and S. S. Patel.** 1995. Bacteriophage T7 helicase/primase proteins form rings around single-stranded DNA that suggest a general structure for hexameric helicases. *Proc. Natl. Acad. Sci. USA* **92**:3869–3873.

35. **Eggleston, A. K., and S. C. West.** 1997. Recombination initiation: easy as A, B, C, D . . . chi? *Curr. Biol.* **7**:R745–R749.

36. **Fass, D., C. E. Bogden, and J. M. Berger.** 1999. Quaternary changes in topoisomerase II may direct orthogonal movement of two DNA strands. *Nat. Struct. Biol.* **6**:322–326.

37. Fay, P. J., K. O. Johanson, C. S. McHenry, and R. A. Bambara. 1981. Size classes of products synthesized processively by DNA polymerase III and DNA polymerase III holoenzyme of Escherichia coli. *J. Biol. Chem.* 256:976–983.

38. Feinberg, H., A. Changela, and A. Mondragon. 1999. Protein-nucleotide interactions in E. coli DNA topoisomerase I. *Nat. Struct. Biol.* 6:961–968.

39. Fleischmann, R. D., M. D. Adams, O. White, R. A. Clayton, E. F. Kirkness, A. R. Kerlavage, C. J. Bult, J. F. Tomb, B. A. Dougherty, J. M. Merrick, et al. 1995. Whole-genome random sequencing and assembly of Haemophilus influenzae Rd. *Science* 269:496–512.

40. Franklin, M. C., J. Wang, and T. A. Steitz. 2001. Structure of the replicating complex of a pol alpha family DNA polymerase. *Cell* 105:657–667.

41. Frey, M. W., L. C. Sowers, D. P. Millar, and S. J. Benkovic. 1995. The nucleotide analog 2-aminopurine as a spectroscopic probe of nucleotide incorporation by the Klenow fragment of Escherichia coli polymerase I and bacteriophage T4 DNA polymerase. *Biochemistry* 34:9185–9192.

42. Gao, D., and C. S. McHenry. 2001. tau binds and organizes Escherichia coli replication proteins through distinct domains. Domain IV, located within the unique C terminus of tau, binds the replication fork helicase, DnaB. *J. Biol. Chem.* 276:4441–4446.

43. Gao, D., and C. S. McHenry. 2001. tau binds and organizes Escherichia coli replication through distinct domains. Partial proteolysis of terminally tagged tau to determine candidate domains and to assign domain V as the alpha binding domain. *J. Biol. Chem.* 276:4433–4440.

44. Gao, G., M. Orlova, M. M. Georgiadis, W. A. Hendrickson, and S. P. Goff. 1997. Conferring RNA polymerase activity to a DNA polymerase: a single residue in reverse transcriptase controls substrate selection. *Proc. Natl. Acad. Sci. USA* 94:407–411.

45. Gardner, A. F., and W. E. Jack. 1999. Determinants of nucleotide sugar recognition in an archaeon DNA polymerase. *Nucleic Acids Res.* 27:2545–2553.

46. Genschel, J., U. Curth, and C. Urbanke. 2000. Interaction of E. coli single-stranded DNA binding protein (SSB) with exonuclease I. The carboxy-terminus of SSB is the recognition site for the nuclease. *Biol. Chem.* 381:183–192.

47. Gibbs, E., Z. Kelman, J. M. Gulbis, M. O'Donnell, J. Kuriyan, P. M. Burgers, and J. Hurwitz. 1997. The influence of the proliferating cell nuclear antigen-interacting domain of p21(CIP1) on DNA synthesis catalyzed by the human and Saccharomyces cerevisiae polymerase delta holoenzymes. *J. Biol. Chem.* 272:2373–2381.

48. Golub, E. I., R. C. Gupta, T. Haaf, M. S. Wold, and C. M. Radding. 1998. Interaction of human rad51 recombination protein with single-stranded DNA binding protein, RPA. *Nucleic Acids Res.* 26:5388–5393.

49. Goodman, M. F., and B. Tippin. 2000. Sloppier copier DNA polymerases involved in genome repair. *Curr. Opin. Genet. Dev.* 10:162–168.

50. Goodrich, L. D., T. C. Lin, E. K. Spicer, C. Jones, and W. H. Konigsberg. 1997. Residues at the carboxy terminus of T4 DNA polymerase are important determinants for interaction with the polymerase accessory proteins. *Biochemistry* 36:10474–10481.

51. Guenther, B., R. Onrust, A. Sali, M. O'Donnell, and J. Kuriyan. 1997. Crystal structure of the delta' subunit of the clamp-loader complex of E. coli DNA polymerase III. *Cell* 91:335–345.

52. Gulbis, J. M., Z. Kelman, J. Hurwitz, M. O'Donnell, and J. Kuriyan. 1996. Structure of the C-terminal region of p21(WAF1/CIP1) complexed with human PCNA. *Cell* 87:297–306.

53. Hacker, K. J., and K. A. Johnson. 1997. A hexameric helicase encircles one DNA strand and excludes the other during DNA unwinding. *Biochemistry* 36:14080–14087.

54. Hingorani, M. M., L. B. Bloom, M. F. Goodman, and M. O'Donnell. 1999. Division of labor—sequential ATP hydrolysis drives assembly of a DNA polymerase sliding clamp around DNA. *EMBO J.* 18:5131–5144.

55. Hingorani, M. M., and M. O'Donnell. 1998. ATP binding to the Escherichia coli clamp loader powers opening of the ring-shaped clamp of DNA polymerase III holoenzyme. *J. Biol. Chem.* 273:24550–24563.

56. Hingorani, M. M., and M. O'Donnell. 2000. Sliding clamps: a (tail)ored fit. *Curr. Biol.* 10:R25–R29.

57. Hingorani, M. M., M. T. Washington, K. C. Moore, and S. S. Patel. 1997. The dTTPase mechanism of T7 DNA helicase resembles the binding change mechanism of the F1-ATPase. *Proc. Natl. Acad. Sci. USA* 94:5012–5017.

58. Ishimi, Y. 1997. A DNA helicase activity is associated with an MCM4, -6, and -7 protein complex. *J. Biol. Chem.* 272:24508–24513.

59. Jarvis, T. C., J. W. Newport, and P. H. von Hippel. 1991. Stimulation of the processivity of the DNA polymerase of bacteriophage T4 by the polymerase accessory proteins. The role of ATP hydrolysis. *J. Biol. Chem.* 266:1830–1840.

60. Jeruzalmi, D., M. O'Donnell, and J. Kuriyan. 2001. Crystal structure of the processivity clamp loader gamma (gamma) complex of E. coli DNA polymerase III. *Cell* 106:429–441.

61. Jeruzalmi, D., O. Yurieva, Y. Zhao, M. Young, J. Stewart, M. Hingorani, M. O'Donnell, and J. Kuriyan. 2001. Mechanism of processivity clamp opening by the delta subunit wrench of the clamp loader complex of E. coli DNA polymerase III. *Cell* 106:417–428.

62. Jezewska, M. J., S. Rajendran, D. Bujalowska, and W. Bujalowski. 1998. Does single-stranded DNA pass through the inner channel of the protein hexamer in the complex with the Escherichia coli DnaB Helicase? Fluorescence energy transfer studies. *J. Biol. Chem.* 273:10515–10529.

63. Johnson, K. A. 1993. Conformational coupling in DNA polymerase fidelity. *Annu. Rev. Biochem.* 62:685–713.

64. Johnson, R. E., G. K. Kovvali, S. N. Guzder, N. S. Amin, C. Holm, Y. Habraken, P. Sung, L. Prakash, and S. Prakash. 1996. Evidence for involvement of yeast proliferating cell nuclear antigen in DNA mismatch repair. *J. Biol. Chem.* 271:27987–27990.

65. Johnson, R. E., M. T. Washington, S. Prakash, and L. Prakash. 1999. Bridging the gap: a family of novel DNA polymerases that replicate faulty DNA. *Proc. Natl. Acad. Sci. USA* 96:12224–12226.

66. Johnson, S. K., S. Bhattacharyya, and M. A. Griep. 2000. DnaB helicase stimulates primer synthesis activity on short oligonucleotide templates. *Biochemistry* 39:736–744.

67. Joyce, C. M., and T. A. Steitz. 1994. Function and structure relationships in DNA polymerases. *Annu. Rev. Biochem.* 63:777–822.

68. Kadyrov, F. A., and J. W. Drake. 2001. Conditional coupling of leading-strand and lagging-strand DNA synthesis at bacteriophage T4 replication forks. *J. Biol. Chem.* 276:29559–29566.

69. Kati, W. M., K. A. Johnson, L. F. Jerva, and K. S. Anderson. 1992. Mechanism and fidelity of HIV reverse transcriptase. *J. Biol. Chem.* 267:25988–25997.

70. Keck, J. L., and J. M. Berger. 2001. Primus inter pares (first among equals). *Nat. Struct. Biol.* 8:2–4.

71. Keck, J. L., D. D. Roche, A. S. Lynch, and J. M. Berger. 2000. Structure of the RNA polymerase domain of E. coli primase. *Science* **287:**2482–2486.

72. Kelman, Z. 2000. DNA replication in the third domain (of life). *Curr. Protein Pept. Sci.* **1:**1–25.

73. Kelman, Z., and J. Hurwitz. 2000. A unique organization of the protein subunits of the DNA polymerase clamp loader in the archaeon Methanobacterium thermoautotrophicum deltaH. *J. Biol. Chem.* **275:**7327–7336.

74. Kelman, Z., J. K. Lee, and J. Hurwitz. 1999. The single minichromosome maintenance protein of Methanobacterium thermoautotrophicum DeltaH contains DNA helicase activity. *Proc. Natl. Acad. Sci. USA* **96:**14783–14788.

75. Kelman, Z., and M. O'Donnell. 1995. DNA polymerase III holoenzyme: structure and function of a chromosomal replicating machine. *Annu. Rev. Biochem.* **64:**171–200.

76. Kelman, Z., S. Pietrokovski, and J. Hurwitz. 1999. Isolation and characterization of a split B-type DNA polymerase from the archaeon Methanobacterium thermoautotrophicum deltaH. *J. Biol. Chem.* **274:**28751–28761.

77. Kelman, Z., A. Yuzhakov, J. Andjelkovic, and M. O'Donnell. 1998. Devoted to the lagging strand—the subunit of DNA polymerase III holoenzyme contacts SSB to promote processive elongation and sliding clamp assembly. *EMBO J.* **17:**2436–2449.

78. Kiefer, J. R., C. Mao, J. C. Braman, and L. S. Beese. 1998. Visualizing DNA replication in a catalytically active Bacillus DNA polymerase crystal. *Nature* **391:**304–307.

79. Kim, S., H. G. Dallmann, C. S. McHenry, and K. J. Marians. 1996. Coupling of a replicative polymerase and helicase: a tau-DnaB interaction mediates rapid replication fork movement. *Cell* **84:**643–650.

80. Klemperer, N., D. Zhang, M. Skangalis, and M. O'Donnell. 2000. Cross-utilization of the beta sliding clamp by replicative polymerases of evolutionary divergent organisms. *J. Biol. Chem.* **275:**26136–26143.

81. Kong, X. P., R. Onrust, M. O'Donnell, and J. Kuriyan. 1992. Three-dimensional structure of the beta subunit of E. coli DNA polymerase III holoenzyme: a sliding DNA clamp. *Cell* **69:**425–437.

82. Kool, E. T. 1998. Replication of non-hydrogen bonded bases by DNA polymerases: a mechanism for steric matching. *Biopolymers* **48:**3–17.

83. Kornberg, A. 1960. Biologic synthesis of deoxyribonucleic acid. *Science* **131:**1503–1508.

84. Kornberg, A., and T. A. Baker. 1992. *DNA Replication*, 2nd ed. W. H. Freeman & Co., New York, N.Y.

85. Kowalczykowski, S. C. 2000. Initiation of genetic recombination and recombination-dependent replication. *Trends Biochem. Sci.* **25:**156–165.

86. Kreuzer, K. N. 2000. Recombination-dependent DNA replication in phage T4. *Trends Biochem. Sci.* **25:**165–173.

87. Krishna, T. S., X. P. Kong, S. Gary, P. M. Burgers, and J. Kuriyan. 1994. Crystal structure of the eukaryotic DNA polymerase processivity factor PCNA. *Cell* **79:**1233–1243.

88. Kuriyan, J., and M. O'Donnell. 1993. Sliding clamps of DNA polymerases. *J. Mol. Biol.* **234:**915–925.

89. Lee, J., P. D. Chastain II, T. Kusakabe, J. D. Griffith, and C. C. Richardson. 1998. Coordinated leading and lagging strand DNA synthesis on a minicircular template. *Mol. Cell* **1:**1001–1010.

90. Lee, J. K., and J. Hurwitz. 2001. Processive DNA helicase activity of the minichromosome maintenance proteins 4, 6, and 7 complex requires forked DNA structures. *Proc. Natl. Acad. Sci. USA* **98:**54–59.

91. Lee, J. Y., C. Chang, H. K. Song, J. Moon, J. K. Yang, H. K. Kim, S. T. Kwon, and S. W. Suh. 2000. Crystal structure of NAD(+)-dependent DNA ligase: modular architecture and functional implications. *EMBO J.* **19:**1119–1129.

92. Lee, S. H., and J. R. Walker. 1987. Escherichia coli DnaX product, the tau subunit of DNA polymerase III, is a multifunctional protein with single-stranded DNA-dependent ATPase activity. *Proc. Natl. Acad. Sci. USA* **84:**2713–2717.

93. Lehman, I. R. 1974. DNA ligase: structure, mechanism, and function. *Science* **186:**790–797.

94. Leu, F. P., M. M. Hingorani, J. Turner, and M. O'Donnell. 2000. The delta subunit of DNA polymerase III holoenzyme serves as a sliding clamp unloader in Escherichia coli. *J. Biol. Chem.* **275:**34609–34618.

95. Leu, F. P., and M. O'Donnell. 2001. Interplay of a clamp loader subunits in opening the beta sliding clamp of Escherichia coli DNA polymerase III holoenzyme. *J. Biol. Chem.* **276:**47185–47194.

96. Levin, D. S., W. Bai, N. Yao, M. O'Donnell, and A. E. Tomkinson. 1997. An interaction between DNA ligase I and proliferating cell nuclear antigen: implications for Okazaki fragment synthesis and joining. *Proc. Natl. Acad. Sci. USA* **94:**12863–12868.

97. Li, X., J. Li, J. Harrington, M. R. Lieber, and P. M. Burgers. 1995. Lagging strand DNA synthesis at the eukaryotic replication fork involves binding and stimulation of FEN-1 by proliferating cell nuclear antigen. *J. Biol. Chem.* **270:**22109–22112.

98. Li, X., and K. J. Marians. 2000. Two distinct triggers for cycling of the lagging strand polymerase at the replication fork. *J. Biol. Chem.* **275:**34757–34765.

99. Li, Y., S. Korolev, and G. Waksman. 1998. Crystal structures of open and closed forms of binary and ternary complexes of the large fragment of Thermus aquaticus DNA polymerase I: structural basis for nucleotide incorporation. *EMBO J.* **17:**7514–7525.

100. Li, Y., and G. Waksman. 2001. Crystal structures of a ddATP-, ddTTP-, ddCTP-, and ddGTP-trapped ternary complex of Klentaq1: insights into nucleotide incorporation and selectivity. *Protein Sci.* **10:**1225–1233.

101. Li, Z., A. Mondragon, and R. J. DiGate. 2001. The mechanism of type IA topoisomerase-mediated DNA topological transformations. *Mol. Cell* **7:**301–307.

102. Lima, C. D., J. C. Wang, and A. Mondragon. 1994. Three-dimensional structure of the 67K N-terminal fragment of E. coli DNA topoisomerase I. *Nature* **367:**138–146.

103. Lin, Y. L., C. Chen, K. F. Keshav, E. Winchester, and A. Dutta. 1996. Dissection of functional domains of the human DNA replication protein complex replication protein A. *J. Biol. Chem.* **271:**17190–17198.

104. Lohman, T. M., and M. E. Ferrari. 1994. Escherichia coli single-stranded DNA-binding protein: multiple DNA-binding modes and cooperativities. *Annu. Rev. Biochem.* **63:**527–570.

105. Lopez de Saro, F. J., and M. O'Donnell. 2001. Interaction of the beta sliding clamp with MutS, ligase, and DNA polymerase I. *Proc. Natl. Acad. Sci. USA* **98:**8376–8380.

106. Low, R. L., S. A. Rashbaum, and N. R. Cozzarelli. 1976. Purification and characterization of DNA polymerase III from Bacillus subtilis. *J. Biol. Chem.* **251:**1311–1325.

107. Maga, G., M. Stucki, S. Spadari, and U. Hubscher. 2000. DNA polymerase switching: I. Replication factor C displaces DNA polymerase alpha prior to PCNA loading. *J. Mol. Biol.* **295:**791–801.

108. Maki, H., and A. Kornberg. 1987. Proofreading by DNA polymerase III of Escherichia coli depends on cooperative

interaction of the polymerase and exonuclease subunits. *Proc. Natl. Acad. Sci. USA* 84:4389–4392.

109. Marians, K. J. 2000. PriA-directed replication fork restart in Escherichia coli. *Trends Biochem. Sci.* 25:185–189.

110. McHenry, C. S., M. Seville, and M. G. Cull. 1997. A DNA polymerase III holoenzyme-like subassembly from an extreme thermophilic eubacterium. *J. Mol. Biol.* 272:178–189.

111. Mer, G., A. Bochkarev, R. Gupta, E. Bochkareva, L. Frappier, C. J. Ingles, A. M. Edwards, and W. J. Chazin. 2000. Structural basis for the recognition of DNA repair proteins UNG2, XPA, and RAD52 by replication factor RPA. *Cell* 103:449–456.

112. Minnick, D. T., M. Astatke, C. M. Joyce, and T. A. Kunkel. 1996. A thumb subdomain mutant of the large fragment of Escherichia coli DNA polymerase I with reduced DNA binding affinity, processivity, and frameshift fidelity. *J. Biol. Chem.* 271:24954–24961.

113. Moarefi, I., D. Jeruzalmi, J. Turner, M. O'Donnell, and J. Kuriyan. 2000. Crystal structure of the DNA polymerase processivity factor of T4 bacteriophage. *J. Mol. Biol.* 296: 1215–1223.

114. Molineux, I. J., and M. L. Gefter. 1974. Properties of the Escherichia coli in DNA binding (unwinding) protein: interaction with DNA polymerase and DNA. *Proc. Natl. Acad. Sci. USA* 71:3858–3862.

115. Morais Cabral, J. H., A. P. Jackson, C. V. Smith, N. Shikotra, A. Maxwell, and R. C. Liddington. 1997. Crystal structure of the breakage-reunion domain of DNA gyrase. *Nature* 388:903–906.

116. Mossi, R., and U. Hubscher. 1998. Clamping down on clamps and clamp loaders—the eukaryotic replication factor C. *Eur. J. Biochem.* 254:209–216.

117. Naktinis, V., R. Onrust, L. Fang, and M. O'Donnell. 1995. Assembly of a chromosomal replication machine: two DNA polymerases, a clamp loader, and sliding clamps in one holoenzyme particle. II. Intermediate complex between the clamp loader and its clamp. *J. Biol. Chem.* 270:13358–13365.

118. Naktinis, V., J. Turner, and M. O'Donnell. 1996. A molecular switch in a replication machine defined by an internal competition for protein rings. *Cell* 84:137–145.

119. Napolitano, R., R. Janel-Bintz, J. Wagner, and R. P. Fuchs. 2000. All three SOS-inducible DNA polymerases (Pol II, Pol IV and Pol V) are involved in induced mutagenesis. *EMBO J.* 19:6259–6265.

120. Neuwald, A. F., L. Aravind, J. L. Spouge, and E. V. Koonin. 1999. AAA+: a class of chaperone-like ATPases associated with the assembly, operation, and disassembly of protein complexes. *Genome Res.* 9:27–43.

121. O'Donnell, M., and P. S. Studwell. 1990. Total reconstitution of DNA polymerase III holoenzyme reveals dual accessory protein clamps. *J. Biol. Chem.* 265:1179–1187.

122. Ollis, D. L., P. Brick, R. Hamlin, N. G. Xuong, and T. A. Steitz. 1985. Structure of large fragment of Escherichia coli DNA polymerase I complexed with dTMP. *Nature* 313:762–766.

123. Onrust, R., J. Finkelstein, V. Naktinis, J. Turner, L. Fang, and M. O'Donnell. 1995. Assembly of a chromosomal replication machine: two DNA polymerases, a clamp loader, and sliding clamps in one holoenzyme particle. I. Organization of the clamp loader. *J. Biol. Chem.* 270:13348–13357.

124. Pan, H., and D. B. Wigley. 2000. Structure of the zinc-binding domain of Bacillus stearothermophilus DNA primase. *Struct. Fold Des.* 8:231–239.

125. Patel, S. S., and K. M. Picha. 2000. Structure and function of hexameric helicases. *Annu. Rev. Biochem.* 69:651–697.

126. Patel, S. S., I. Wong, and K. A. Johnson. 1991. Pre-steady-state kinetic analysis of processive DNA replication including complete characterization of an exonuclease-deficient mutant. *Biochemistry* 30:511–525.

127. Pelletier, H., M. R. Sawaya, A. Kumar, S. H. Wilson, and J. Kraut. 1994. Structures of ternary complexes of rat DNA polymerase beta, a DNA template-primer, and ddCTP. *Science* 264:1891–1903.

128. Pham, P., S. Rangarajan, R. Woodgate, and M. F. Goodman. 2001. Roles of DNA polymerases V and II in SOS-induced error-prone and error-free repair in Escherichia coli. *Proc. Natl. Acad. Sci. USA* 98:8350–8354.

129. Pietroni, P., M. C. Young, G. J. Latham, and P. H. von Hippel. 2001. Dissection of the ATP-driven reaction cycle of the bacteriophage T4 DNA replication processivity clamp loading system. *J. Mol. Biol.* 309:869–891.

130. Pisani, F. M., M. De Felice, F. Carpentieri, and M. Rossi. 2000. Biochemical characterization of a clamp-loader complex homologous to eukaryotic replication factor C from the hyperthermophilic archaeon Sulfolobus solfataricus. *J. Mol. Biol.* 301:61–73.

131. Podobnik, M., P. McInerney, M. O'Donnell, and J. Kuriyan. 2000. A TOPRIM domain in the crystal structure of the catalytic core of Escherichia coli primase confirms a structural link to DNA topoisomerases. *J. Mol. Biol.* 300:353–362.

132. Pritchard, A. E., H. G. Dallmann, B. P. Glover, and C. S. McHenry. 2000. A novel assembly mechanism for the DNA polymerase III holoenzyme DnaX complex: association of deltadelta' with DnaX(4) forms DnaX(3)deltadelta'. *EMBO J.* 19:6536–6545.

133. Raghunathan, S., A. G. Kozlov, T. M. Lohman, and G. Waksman. 2000. Structure of the DNA binding domain of E. coli SSB bound to ssDNA. *Nat. Struct. Biol.* 7:648–652.

134. Raney, K. D., T. E. Carver, and S. J. Benkovic. 1996. Stoichiometry and DNA unwinding by the bacteriophage T4 41:59 helicase. *J. Biol. Chem.* 271:14074–14081.

135. Reddy, M. K., S. E. Weitzel, and P. H. von Hippel. 1992. Processive proofreading is intrinsic to T4 DNA polymerase. *J. Biol. Chem.* 267:14157–14166.

136. Redinbo, M. R., J. J. Champoux, and W. G. Hol. 1999. Structural insights into the function of type IB topoisomerases. *Curr. Opin. Struct. Biol.* 9:29–36.

137. Redinbo, M. R., L. Stewart, P. Kuhn, J. J. Champoux, and W. G. Hol. 1998. Crystal structures of human topoisomerase I in covalent and noncovalent complexes with DNA. *Science* 279:1504–1513.

138. Reha-Krantz, L. J., and J. Hurwitz. 1978. The dnaB gene product of Escherichia coli. I. Purification, homogeneity, and physical properties. *J. Biol. Chem.* 253:4043–4050.

139. Sancar, A., and J. E. Hearst. 1993. Molecular matchmakers. *Science* 259:1415–1420.

140. San Martin, M. C., C. Gruss, and J. M. Carazo. 1997. Six molecules of SV40 large T antigen assemble in a propeller-shaped particle around a channel. *J. Mol. Biol.* 268:15–20.

141. Sawaya, M. R., H. Pelletier, A. Kumar, S. H. Wilson, and J. Kraut. 1994. Crystal structure of rat DNA polymerase beta: evidence for a common polymerase mechanism. *Science* 264:1930–1935.

142. Sawaya, M. R., R. Prasad, S. H. Wilson, J. Kraut, and H. Pelletier. 1997. Crystal structures of human DNA polymerase beta complexed with gapped and nicked DNA: evidence for an induced fit mechanism. *Biochemistry* 36: 11205–11215.

143. Scheuermann, R., S. Tam, P. M. Burgers, C. Lu, and H. Echols. 1983. Identification of the epsilon-subunit of

Escherichia coli DNA polymerase III holoenzyme as the dnaQ gene product: a fidelity subunit for DNA replication. *Proc. Natl. Acad. Sci. USA* 80:7085–7089.

144. Shamoo, Y., and T. A. Steitz. 1999. Building a replisome from interacting pieces: sliding clamp complexed to a peptide from DNA polymerase and a polymerase editing complex. *Cell* 99:155–166.

145. Shechter, D. F., C. Y. Ying, and J. Gautier. 2000. The intrinsic DNA helicase activity of Methanobacterium thermoautotrophicum delta H minichromosome maintenance protein. *J. Biol. Chem.* 275:15049–15059.

146. Shiomi, Y., J. Usukura, Y. Masamura, K. Takeyasu, Y. Nakayama, C. Obuse, H. Yoshikawa, and T. Tsurimoto. 2000. ATP-dependent structural change of the eukaryotic clamp-loader protein, replication factor C. *Proc. Natl. Acad. Sci. USA* 97:14127–14132.

147. Sriskanda, V., R. W. Moyer, and S. Shuman. 2001. Nad+-dependent DNA ligase encoded by a eukaryotic virus. *J. Biol. Chem.* 276:36100–36109.

148. Stewart, J., M. M. Hingorani, Z. Kelman, and M. O'Donnell. 2001. Mechanism of beta clamp opening by the delta subunit of Escherichia coli DNA polymerase III holoenzyme. *J. Biol. Chem.* 276:19182–19189.

149. Stewart, L., M. R. Redinbo, X. Qiu, W. G. Hol, and J. J. Champoux. 1998. A model for the mechanism of human topoisomerase I. *Science* 279:1534–1541.

150. Studwell, P. S., and M. O'Donnell. 1990. Processive replication is contingent on the exonuclease subunit of DNA polymerase III holoenzyme. *J. Biol. Chem.* 265:1171–1178.

151. Stukenberg, P. T., and M. O'Donnell. 1995. Assembly of a chromosomal replication machine: two DNA polymerases, a clamp loader, and sliding clamps in one holoenzyme particle. V. Four different polymerase-clamp complexes on DNA. *J. Biol. Chem.* 270:13384–13391.

152. Stukenberg, P. T., P. S. Studwell-Vaughan, and M. O'Donnell. 1991. Mechanism of the sliding beta-clamp of DNA polymerase III holoenzyme. *J. Biol. Chem.* 266:11328–11334.

153. Stukenberg, P. T., J. Turner, and M. O'Donnell. 1994. An explanation for lagging strand replication: polymerase hopping among DNA sliding clamps. *Cell* 78:877–887.

154. Sutton, M. D., and G. C. Walker. 2001. Managing DNA polymerases: coordinating DNA replication, DNA repair, and DNA recombination. *Proc. Natl. Acad. Sci. USA* 98:8342–8349.

155. Tabor, S., and C. C. Richardson. 1981. Template recognition sequence for RNA primer synthesis by gene 4 protein of bacteriophage T7. *Proc. Natl. Acad. Sci. USA* 78:205–209.

156. Tang, M., P. Pham, X. Shen, J. S. Taylor, M. O'Donnell, R. Woodgate, and M. F. Goodman. 2000. Roles of E. coli DNA polymerases IV and V in lesion-targeted and untargeted SOS mutagenesis. *Nature* 404:1014–1018.

157. Timson, D. J., M. R. Singleton, and D. B. Wigley. 2000. DNA ligases in the repair and replication of DNA. *Mutat. Res.* 460:301–318.

158. Tomkinson, A. E., and D. S. Levin. 1997. Mammalian DNA ligases. *Bioessays* 19:893–901.

159. Tougu, K., H. Peng, and K. J. Marians. 1994. Identification of a domain of Escherichia coli primase required for functional interaction with the DnaB helicase at the replication fork. *J. Biol. Chem.* 269:4675–4682.

160. Tsuchihashi, Z., and A. Kornberg. 1989. ATP interactions of the tau and gamma subunits of DNA polymerase III holo-

enzyme of Escherichia coli. *J. Biol. Chem.* 264:17790–17795.

161. Tsurimoto, T., and B. Stillman. 1990. Functions of replication factor C and proliferating-cell nuclear antigen: functional similarity of DNA polymerase accessory proteins from human cells and bacteriophage T4. *Proc. Natl. Acad. Sci. USA* 87:1023–1027.

162. Turner, J., M. M. Hingorani, Z. Kelman, and M. O'Donnell. 1999. The internal workings of a DNA polymerase clamp-loading machine. *EMBO J.* 18:771–783.

163. Umezu, K., N. W. Chi, and R. D. Kolodner. 1993. Biochemical interaction of the Escherichia coli RecF, RecO, and RecR proteins with RecA protein and single-stranded DNA binding protein. *Proc. Natl. Acad. Sci. USA* 90:3875–3879.

164. Waga, S., G. Bauer, and B. Stillman. 1994. Reconstitution of complete SV40 DNA replication with purified replication factors. *J. Biol. Chem.* 269:10923–10934.

165. Waga, S., and B. Stillman. 1998. The DNA replication fork in eukaryotic cells. *Annu. Rev. Biochem.* 67:721–751.

166. Wagner, J., S. Fujii, P. Gruz, T. Nohmi, and R. P. Fuchs. 2000. The beta clamp targets DNA polymerase IV to DNA and strongly increases its processivity. *EMBO Rep.* 1:484–488.

167. Wang, J. C. 1996. DNA topoisomerases. *Annu. Rev. Biochem.* 65:635–692.

168. Warbrick, E. 2000. The puzzle of PCNA's many partners. *Bioessays* 22:997–1006.

169. Washington, M. T., R. E. Johnson, S. Prakash, and L. Prakash. 2001. Mismatch extension ability of yeast and human DNA polymerase eta. *J. Biol. Chem.* 276:2263–2266.

170. Wong, I., S. S. Patel, and K. A. Johnson. 1991. An induced-fit kinetic mechanism for DNA replication fidelity: direct measurement by single-turnover kinetics. *Biochemistry* 30:526–537.

171. Yao, N., J. Turner, Z. Kelman, P. T. Stukenberg, F. Dean, D. Shechter, Z. Q. Pan, J. Hurwitz, and M. O'Donnell. 1996. Clamp loading, unloading and intrinsic stability of the PCNA, beta and gp45 sliding clamps of human, E. coli and T4 replicases. *Genes Cells* 1:101–113.

172. Yoda, K., H. Yasuda, X. W. Jiang, and T. Okazaki. 1988. RNA-primed initiation sites of DNA replication in the origin region of bacteriophage lambda genome. *Nucleic Acids Res.* 16:6531–6546.

173. Yong, Y., and L. J. Romano. 1996. Benzo[a]pyrene-DNA adducts inhibit the DNA helicase activity of the bacteriophage T7 gene 4 protein. *Chem. Res. Toxicol.* 9:179–187.

174. You, Z., Y. Komamura, and Y. Ishimi. 1999. Biochemical analysis of the intrinsic Mcm4-Mcm6-mcm7 DNA helicase activity. *Mol. Cell. Biol.* 19:8003–8015.

175. Yurieva, O., M. Skangalis, J. Kuriyan, and M. O'Donnell. 1997. Thermus thermophilus dnaX homolog encoding gamma- and tau-like proteins of the chromosomal replicase. *J. Biol. Chem.* 272:27131–27139.

176. Yuzhakov, A., Z. Kelman, J. Hurwitz, and M. O'Donnell. 1999. Multiple competition reactions for RPA order the assembly of the DNA polymerase delta holoenzyme. *EMBO J.* 18:6189–6199.

177. Yuzhakov, A., Z. Kelman, and M. O'Donnell. 1999. Trading places on DNA—a three-point switch underlies primer handoff from primase to the replicative DNA polymerase. *Cell* 96:153–163.

178. Yuzhakov, A., J. Turner, and M. O'Donnell. 1996. Replisome assembly reveals the basis for asymmetric function in leading and lagging strand replication. *Cell* 86:877–886.

Chapter 11

SeqA Protein Binding and the *Escherichia coli* Replication Fork

THERESE BRENDLER AND STUART AUSTIN

The chromosome of *Escherichia coli* is a circular DNA molecule of approximately 4.7 million bp that has a contour length of about 1.5 mm. With a cell length of 1 to 2 μm and the DNA mass (the nucleoid) occupying about half this length, it is evident that the chromosome must be very extensively folded. In every cell generation, the entire chromosome is replicated, and the two duplicate products must be separated and segregated away from each other. Cell division then results in daughter cells with complete chromosomes. As every aspiring angler will recognize, long, thin, highly folded structures are prone to knots and tangles that appear to defy all efforts to resolve them. Involvement of two lines can complicate the situation considerably. The bacterial case is far more complicated than that normally encountered by anglers. Due to the helical form of DNA and the semiconservative nature of replication, the daughter chromosomes start out as single strands that are wound around each other every 11 bp or so. In addition, the daughters frequently recombine with each other, generating knots and linking them into dimers. Yet simple bacterial cells efficiently resolve their replicated chromosomes once in every generation. The likely mechanisms used to resolve topological intertwining and recombination between daughter chromosomes are dealt with elsewhere in this book. However, it is still poorly understood how two separate DNA masses are produced following replication and how as yet unrecognized cellular components and activities are likely to be involved. One such potential player is the *E. coli* SeqA protein.

THE SeqA PROTEIN

The gene for the SeqA protein was originally identified as a locus involved in a process known as origin sequestration (10, 24). Sequestration regulates the timing of initiation by recognizing newly replicated origins of replication. It imposes a block to further replication until the system is reset for the next cell cycle (3, 4, 10, 24, 27). This process has some similarity to the origin "licensing" which ensures that any given segment of eukaryotic DNA is not replicated more than once in each cell cycle. The *E. coli* chromosome is replicated bidirectionally from a single origin, *oriC*. The timing of initiation from *oriC* in the *E. coli* cell cycle is disrupted by mutations in the gene termed *seqA* (22, 32). The gene encodes a negative regulator of replication, because the mutant cells have excess DNA content. The timing of replication is disrupted such that initiation often occurs more than once per cell cycle (5, 22). It was proposed that the *seqA* product is involved in sequestering *oriC* following one round of replication, so that a second round cannot immediately follow.

The newly replicated DNA is recognized by monitoring the methylation state of the adenine bases in the sequence GATC (27). The two adenine bases on opposite strands of the GATC sequences in *E. coli* are normally methylated at the N^6 position (Fig. 1). This occurs by the action of an adenine methyltransferase, Dam methylase, on the completed DNA. It is the product of the *E. coli dam* gene.

There is a cluster of GATC sites within the boundaries of the *oriC* sequence. When the origin is replicated, the adenine bases incorporated into the new DNA strands are not methylated. Thus, immediately following initiation, the newly replicated origin sequences are methylated on one strand only (hemimethylated) (Fig. 1). These hemimethylated GATC sequences in the origin are recognized and bound by the SeqA protein (7, 31). SeqA appears to cause sequestration by binding specifically to the newly replicated DNA at *oriC* and preventing reinitiation.

Therese Brendler and Stuart Austin • Gene Regulation and Chromosome Biology Laboratory, Division of Basic Sciences, NCI-Frederick, Frederick, MD 21702-1201.

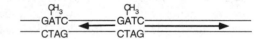

Figure 1. Replication creates hemimethylated GATC sequences. In *E. coli*, the sequence GATC is methylated on the adenine bases of both strands by the action of the Dam methylase. As the sequence is replicated, the newly synthesized strands are unmethylated, so that the product duplexes are hemimethylated for a period of time, until Dam methylase can act to restore full methylation.

It also delays the ability of the Dam methylase to restore full methylation to the bound GATC sequences. The latter activity is presumably due to the bound SeqA protein blocking Dam binding to the same sequence. The mechanism by which bound SeqA blocks *oriC* function involves blocking the binding or activity of the essential initiation factor, DnaA (5, 30).

SeqA BINDS TO REPLICATION ORIGINS

Consistent with its role in modulating origin function, the SeqA protein was found to bind to the chromosomal origin, *oriC*, in vitro. The minimal origin contains a cluster of 12 GATC sites. Binding was preferential to the hemimethylated form of the sequence (7, 31). Binding also occurred to the hemimethylated form of the plasmid replication origin of bacteriophage P1 (7). The P1 origin also contains a cluster of GATC sites, although its organization and most of its sequence differ radically from those of *oriC*. As we will describe below, further in vitro studies showed that binding is not exclusive to replication origins.

Sequestration of *oriC* is accompanied by binding of the hemimethylated origin sequences to the cell membrane (24). Membrane fractions can specifically inhibit DNA replication in vitro when the *oriC* template is hemimethylated (19). The hemimethylated *oriC* binding activity of the membrane can be resolved into two components, SeqA protein and a second protein component termed SeqB (29). SeqB does not bind to *oriC* independently, but enhances the ability of SeqA to do so in a hemimethylation-specific manner. Thus, the block to reinitiation from

oriC imposed on newly replicated DNA by the SeqA protein appears to involve formation of a complex at the cell membrane. The membrane affinity of the SeqA protein tends to make it insoluble in whole-cell extracts unless the salt concentration is elevated (7).

BINDING RULES FOR THE SeqA PROTEIN

Further in vitro binding studies using purified SeqA protein have shown that SeqA binding is not exclusively directed to origins of replication (7, 8). A double-stranded oligonucleotide was constructed that has three GATC sequences spaced one turn of the helix apart. All other bases in this sequence were randomly chosen. This sequence bound SeqA efficiently when the GATC sites were hemimethylated. Fully methylated and unmethylated oligonucleotides failed to bind to the protein. By modifying the sequence of the oligonucleotide, the number and spacing of the GATC sites could be varied (Fig. 2). The following rules for binding were established.

Oligonucleotides with a single hemimethylated GATC site do not bind SeqA at all. Two sites are sufficient, but the spacing between the sites is important.

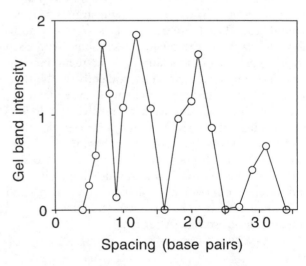

Figure 2. The effect of the spacing of GATC sites on SeqA binding. (Top) A series of oligonucleotide binding substrates have two hemimethylated GATC sequences in an otherwise randomly chosen sequence. The position of the first GATC sequence is fixed. The position of the second GATC sequence is varied to create a series of spacings between the adenine bases of 4 bp (when the GATCs are immediately adjacent to each other) to 34 bp. (Bottom) The results of electrophoretic mobility shift assays with purified SeqA protein. Binding is optimal when the two sites are very close (7 bp) or when they are on the same face of the helix (11, 21, and 31 bp).

The spacing is optimal when two GATC sites are an integral number of turns of the helix apart (Fig. 2). Spacing shows considerable flexibility, but binding does not occur when sites are on opposite helical faces or when the spacing is very large (seven turns of the helix apart). The methyl groups of two hemimethylated sites can be either in *cis* or in *trans* to each other, and when three sites are present on a fragment, all three positions appear to be occupied at the same time (8).

From these binding experiments, and from footprinting experiments that probe the configurations of the bound molecules, it was concluded that SeqA can recognize individual GATC sites but must make contact with adjacent bound molecules in order to achieve a stably bound configuration (Fig. 3). The studies imply that when SeqA is bound to clustered GATC sites, it forms a protein filament, with the DNA intervening between the GATC sites looped out (8) (Fig. 3).

SeqA BINDS MORE EFFICIENTLY TO HEMIMETHYLATED SEQUENCES WHEN THEY ARE PRESENT IN RELAXED DNA

A mutant host that is temperature sensitive for Dam methylase can be used to prepare circular, supercoiled plasmid DNA that is enriched for hemimethylated GATC sites. Using a nitrocellulose filter binding assay, it was found that this type of substrate binds SeqA, but does so with relatively low affinity. However, when the DNA was cut with a restriction enzyme to make it linear, or nicked on one strand by topoisomerase I to give a relaxed circular molecule, the efficiency of binding to the hemimethylated DNA increased markedly (9). With each of these substrates, SeqA retained its ability to discriminate between hemimethylation and the other methylation states. However, the protein binds more efficiently to relaxed or linear DNA and shows optimum discrimination

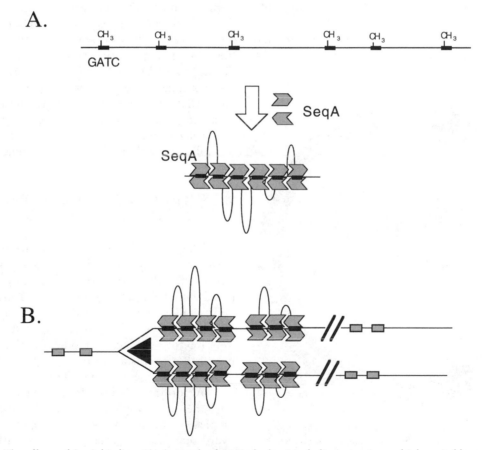

Figure 3. The effects of SeqA binding. (A) A stretch of DNA duplex (single line) contains multiple, suitably spaced, hemimethylated GATC sequences (black boxes). It will bind SeqA efficiently. The bound form of the protein is thought, on theoretical grounds, to be a symmetrical dimer. Binding studies suggest that the product will be a SeqA filament with the intervening sequences looped out as shown below. (B) As the replication fork passes around the chromosome, fully methylated GATC sites (shaded boxes at left) become transiently hemimethylated (black boxes) until Dam methylase restores full methylation (shaded boxes at right). This would result in a tract of SeqA bound to the newly replicated DNA following the fork.

for the hemimethylated form of these relaxed substrates. The finding that SeqA tends to reduce negative supercoiling of the chromosome in vivo (33) is consistent with its preference for relaxed substrates in vitro.

A WAVE OF BOUND SeqA PROTEIN SHOULD FOLLOW THE REPLICATION FORK

From the oligonucleotide binding experiments, it can be concluded that any sequence containing two or more hemimethylated GATC sites will bind SeqA protein unless the spacing between the sites specifically precludes it. Analysis of the complete sequence of the *E. coli* chromosome shows that there are at least 1,750 pairs of GATC sites on the chromosome that would be suitable for binding if hemimethylated (Fig. 4). These potential binding sites are fairly evenly distributed around the chromosome.

The newly replicated DNA produced when the replication forks pass around the chromosome is transiently hemimethylated (24). Few data are available for the average half-life of hemimethylated sites, but it is probably of the order of a few minutes after the passage of the fork. The Dam methylase then re-

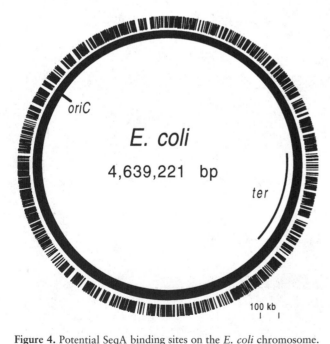

Figure 4. Potential SeqA binding sites on the *E. coli* chromosome. The positions of 1,750 potential SeqA binding sites are shown that would bind SeqA if hemimethylated. The bars represent the positions of pairs of GATC sites which are less than 34 bp apart and have a spacing which is known to promote SeqA binding when hemimethylated (see Fig. 2). As somewhat larger spacings are likely to bind also, the actual number of potential binding sites is probably greater.

stores full methylation (10). Assuming a 2-min lifetime for the average hemimethylated site and a 40-min replication time for the chromosome, a tract containing approximately 100 suitably spaced pairs of hemimethylated GATC sites would be expected to follow the progression of each fork in a dynamic wave. Tracts of bound SeqA should therefore continuously decorate the DNA behind the replication forks, as illustrated in Fig. 3.

If SeqA protein binding forms a coherent filament of protein on the DNA with the sequences intervening between the GATC sites looped out, the newly replicated DNA would be organized and compacted as it emerges from the replication fork (Fig. 3). The bound protein would dissociate as it regresses from the fork due to completion of methylation as the DNA matures. New binding of SeqA would occur adjacent to the fork as new hemimethylated sites are formed by replication. This would result in a net motion of the focus along the DNA in concert with the fork. Note that the newly emerging DNA is in a relaxed state due to unwinding of the helix during replication and the discontinuous nature of the replication process. As SeqA binds preferentially to relaxed DNA, it should be targeted to this region not only by the methylation state, but also by the topological state of the DNA. The fact that *seqA* mutants have increased chromosomal supercoiling (33) suggests that SeqA not only binds to the relaxed DNA emerging from the fork, but also constrains this DNA in a relaxed form for a sufficient time to influence the overall supercoiling of the chromosome.

WHERE ARE REPLICATION FORKS LOCATED IN THE CELL?

The *E. coli* chromosome replication forks are not thought to move about the nucleoid as they progress around the chromosome. Rather, they appear to be fixed within the cell, with the replicating DNA being fed through them. This idea was first suggested in the early 1970s when Dingman proposed a model (now referred to as the factory model for DNA synthesis) in which both replication forks were anchored to the membrane at the cell center (12).

According to this model, the old DNA feeds into the fork complex, and the two newly replicated duplexes emerging from it are directed outward toward opposite cell poles (Fig. 5). Multifork replication, which occurs at higher growth rates, is also accommodated in this model. In this case, new fork anchoring sites are formed at the cell 1/4 and 3/4 positions in preparation for the next generation. This would allow the segregation of chromosomes that

A.

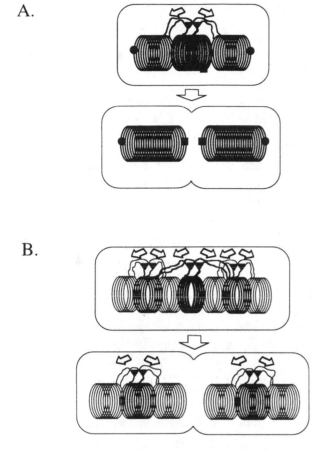

B.

Figure 5. A representation of the factory model for chromosome replication and segregation. (A) In slow-growing cells, the replication forks (black triangles) are anchored at the cell center in a replication "factory" which contains all the proteins necessary for ongoing replication. The old DNA (darkest coils) is fed into the factory, and the newly replicated DNA (lighter coils) emerges from it and is directed outward toward the cell poles, driven by the replication process itself. Condensation of the newly replicated DNA forms two new nucleoid structures (lighter coils). The newly replicated DNA is deposited on the inner faces of the nascent nucleoids, such that the origins (black disks) end up on the outer faces and the termini (black squares) on the inner faces of the completed nucleoids. (B) At fast growth rates, new rounds of replication initiate before the previous round is completed. New factory sites form in the cell at 1/4 and 3/4 positions, and the nucleoids segregate to daughter cells while still undergoing replication.

have already initiated new rounds of replication (Fig. 5).

A very similar model has been outlined, based on observations made in *Bacillus subtilis* (20). Chromosome segregation is a direct consequence of replication and occurs concomitantly with it. The "factory" consists of a protein complex containing the enzymes necessary for ongoing replication. During replication, the factory is fixed at the plane of cell division. The DNA replication process itself drives DNA segregation, pushing the newly replicated DNA outward

from the anchored replication forks toward opposite cell poles. This results in two substantially separated nucleoid masses on termination of replication (15, 20). Evidence for this model consists of the finding that a distinct mass of replicative DNA polymerase is present at the cell center in *B. subtilis* (20). In addition, the DNA at the replication forks is located at the cell center. This result was obtained using *B. subtilis* cells in which the replication forks were stalled at specific positions on the chromosome (21). In *E. coli*, the forks also appear to be tethered at the cell center, because the most recently synthesized DNA is found there (18). If SeqA forms tracts of bound protein on the newly replicated DNA as it emerges from the replication forks, these SeqA tracts would also be associated with the cell center and would appear to remain stationary during replication as the DNA passes through the replication factory.

WHERE IS SeqA PROTEIN LOCATED IN THE CELL?

Proteins that are localized to discrete positions within the cell can often be seen by fluorescence microscopy. SeqA localization has been studied in two different ways. Immunofluorescence labeling involves the use of a fluorescence-labeled anti-SeqA antibody which is introduced into fixed cells that have been made permeable to proteins. The other method makes use of a fusion of the SeqA protein to green fluorescent protein (GFP). This fusion is produced within the cell and allows the study of SeqA location in living cells. Neither technique is sensitive enough to detect individual protein molecules. Rather, they detect clusters (foci) containing many molecules that are grouped in a small region of the cell.

If SeqA acts primarily at the origin of replication, and sufficient SeqA molecules are bound there, visible foci might be expected at the origin. Fluorescence hybridization techniques have shown that the origin of replication localizes to the outside border of the nucleoid in most cells, near one or both of the cell poles (14, 23, 26). On the other hand, we have argued that tracts of 100 or so bound SeqA molecules may follow the replication forks around the chromosome. Thus, discrete foci of SeqA molecules corresponding to these tracts would be expected at or near the forks rather than at some fixed chromosomal position. If replication forks are anchored in the cell in the way envisioned in the factory model, the foci should be found at the centers of cells, or at the 1/4 and 3/4 positions in faster-growing cells (Fig. 5 and 6).

SeqA protein forms discrete foci within the cell as detected by immunofluorescence microscopy in

Growth rate

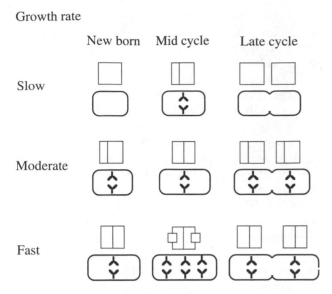

Figure 6. The spatial distribution of replication forks in the cell. Cells growing at three different growth rates are illustrated. They correspond approximately to 80-, 40-, and 20-min generation times at 37°C. Above each cell, the configuration of the chromosome is diagrammed. The expected numbers and positions of the replication forks (forked lines) are shown for each cell. Newborn cells, late-cycle cells near division, and typical mid-cycle cells are illustrated. Note that at faster growth rates, some cells may have three pairs of forks, placed at the 1/4, center, and 3/4 cell positions as shown. However, if termination of replication occurs just after cell division, such cells will be rare, and most mid-cycle cells will have two pairs of forks at the 1/4 and 3/4 cell positions. At very fast growth rates, a third round of replication may start before cell division. In this case, most cells will have six pairs of forks, placed at the 1/8, 1/4, 3/8, 5/8, 3/4, and 7/8 positions.

fixed cells (16) or by visualization of a SeqA-GFP in living cells (9, 25). Formation of foci is dependent on the function of Dam methylase. As these foci persist in *oriC* deletion strains, they do not colocalize with *oriC* (25). Also, separate locations for SeqA and *oriC* were seen in many cells in double-labeling experiments (1). Thus, SeqA foci do not mark the positions of the origins, at least in most cells. The currently available data are supportive of the idea that SeqA foci mark the positions of replication forks.

THE RELATIONSHIP BETWEEN THE NUMBER OF FLUORESCENT SeqA FOCI IN LIVING CELLS AND THE NUMBER OF REPLICATION FORKS

Replication of the *E. coli* chromosome takes about 40 min at 37°C. Nevertheless, the doubling time of the cells can be as low as 20 min when nutritional conditions are good. In these circumstances, a second round of replication is initiated before the

first is completed (11). Such cells contain six replication forks during part of the cell cycle, and four prior to division (Fig. 6). At very high growth rates, some cells may undergo yet another initiation cycle and contain 12 or 14 forks (6). If SeqA foci mark the forks, and if the bound tracts are spatially separated in the cells, the number of replication forks and the number of SeqA foci should increase in parallel with increasing growth rate. In fixed cells using immunofluorescence, the number of foci increased sharply with growth rate (16), but the distribution of numbers of foci did not appear to fit well with the expected number of replication forks. In living cells, the distribution of foci using a GFP-SeqA fusion was measured (Fig. 7). SeqA foci were counted in cells growing at different growth rates at 30°C, a temperature at which GFP folds into its active form. The average number of foci per cell increased sharply with increasing growth rate. The distributions of numbers of foci were fairly smooth (Fig. 7). The average number of foci per cell at each growth rate corresponded to estimates for the average number of forks per cell reasonably well (9). In very slow growing cells (0.4 doublings per hour at 30°C), a proportion of the cells should have no replication forks, as there is a significant gap between the termination of replication and the initiation of the next round. Under these conditions, approximately half of the cells were found to have no focus at all (9) (Fig. 6 and 7).

THE CORRELATION OF THE LOCATIONS OF SeqA FOCI AND REPLICATION FORKS

Using the predictions of the factory model and the known parameters of the cell cycle, the positions of the replication forks in the cell can be predicted (Fig. 6). If the SeqA foci seen in the cells mark regions close to the forks, then some correspondence between the positions of foci and the theoretical locations of the forks should be evident. Note that the factory model implies that both forks formed by the initiation event are in the same general location. It is not clear whether the corresponding SeqA tracts would form two closely spaced foci or would coalesce into a single focus at these positions.

Slow Growth Rates

At very slow growth rates (less than one doubling per hour at 37°C), some cells in the population should not be replicating at all and should have no forks. In the replicating portion of the population, each cell should have a pair of forks tethered to the cell center. We examined cells growing at 0.4 doublings per hour

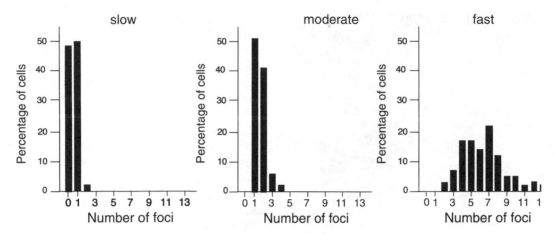

Figure 7. The numbers of SeqA foci per cell at three different growth rates. The numbers of SeqA foci in ~100 cells were counted from cultures growing at three different growth rates at 30°C. Slow: cells growing in M63 glycerol medium; generation time 160 min. Moderate: cells growing in M63 glucose medium; generation time 50 min. Fast: cells growing in L broth; generation time 30 min. These generation times give cell cycle parameters roughly equivalent to those illustrated in Fig. 6, although the actual times are greater here due to the low temperature which is required for GFP visualization.

at 30°C (Fig. 8). Approximately half of the cells had no visible focus. Almost all of the rest of the population consisted of cells with a single focus at or near the cell center (Fig. 8a).

Note that when the forks meet at termination, the two SeqA tracts that are following the forks should coalesce. The terminal regions of both daughter chromosomes should have tracts of bound SeqA at this time. As the completed nucleoids segregate away from each other, the bound protein should persist for at least a short time so that the focus would appear to split before fading out, with the two halves regressing from each other as the inner borders of the nucleoid move apart. This can account for the occasional cells that have two foci symmetrically placed across from and near the cell center (not shown). It could also explain the observed splitting of central SeqA foci and the rapid motion of the halves away from the center line as seen by time-lapse microscopy (25).

Moderate Growth Rates

In cells growing at moderate growth rates, chromosome initiation cycles overlap somewhat, and the cells are constantly replicating their DNA. All cells should have replication forks. Newborn cells should contain a partially replicated chromosome. Through most of the cell cycle, the chromosomes have one pair of forks. For a short period after initiation at mid cell cycle, three pairs may be present, but the first round of replication completes soon afterward, and most of the larger cells all contain two separate, partly replicated chromosomes (Fig. 6).

Cells in this growth rate range almost invariably contain SeqA foci. They generally have a single focus in the cell center, or two foci as pairs symmetrically displaced from the center and often at the 1/4 and 3/4 positions (16, 25). In living cells, a prominent class had pairs of foci at the cell center, either aligned with the short axis of the cell or at an oblique angle to it, or had similar pairs at one or both of the 1/4 and 3/4 positions (Fig. 8b). More than 60% of cells with four foci had two pairs where each pair was aligned and spaced similarly at the 1/4 and 3/4 positions (Fig. 8). Closely spaced pairs of foci at the center or quarter cell positions might indicate cells in which the two tracts of bound SeqA molecules that emerge from the forks are sufficiently separated in the line of view to appear as separate entities. In rare cases, two foci are symmetrically disposed about the cell center, but are quite well separated without being placed at the cell quarters. These appear to be larger cells and probably represent cases where the SeqA tracts have reached the terminus and persist at the inside borders of the two nucleoids as they are segregating away from each other. Thus, most of the configurations of SeqA foci seen in cells growing at moderate growth rates appear to be consistent with the factory model and with the localization of SeqA to the replication forks.

Higher Growth Rates

At higher growth rates the chromosome cycle becomes quite complex. This is because second rounds of replication begin well before the first has terminated (Fig. 6). At extreme growth rates, a third round

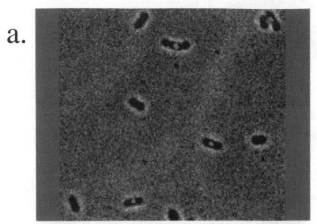

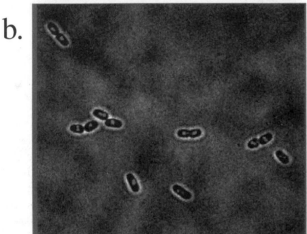

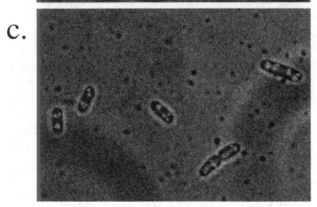

Figure 8. SeqA-GFP expressed in cells growing at three different growth rates at 30°C. (a) Cells growing in minimal glycerol medium (160-min doubling time). (b) Cells growing in minimal glucose medium (50-min doubling time). (c) Cells growing in LB broth (30-min doubling time).

may initiate on the chromosomes, giving some cells with seven pairs of forks. The extension of the factory model to high growth rates (Fig. 5 and 6) predicts that the first-generation forks should be at the cell center, and the second-generation forks should be at the cell 1/4 and 3/4 positions. Third-generation forks, if present, should occupy the cell eighth positions.

Virtually all fast-growing cells have multiple foci (9, 16). Their numbers and positions are often hard to determine as they are crowded together. Some cells have 10 or more foci under these conditions (Fig. 8). The number of foci tends to increase as a function of cell size. It is difficult to say whether these multiple foci are placed at specific locations in the cell. However, at least some cells appear to have regular patterns, with paired foci at cell quarters and so on (Fig. 8). Also, more than 90% of the cells have an equitable distribution of foci in each half of the cell. The cytoplasmic space between the separating nucleoids is free of foci in the large majority of cells. This probably indicates that termination of replication is occurring soon after cell division, so that most of the cells have two separate nucleoids and have no forks at the cell center. This may explain why relatively few cells were seen with foci both at the center and at the 1/4 and 3/4 positions.

The complexity of the patterns of SeqA foci at higher growth rates makes analysis difficult due to overlapping foci, for example. Also, further complications will arise if, as seems likely, there are minor variations in the timing of chromosome cycle events and cell division in individual cells. Added to this is the likelihood that some foci will be at the terminus and will segregate away from each other as the nucleoids segregate. Thus, both the patterns themselves and the theoretical predictions for them are complex, and the analysis of cells at higher growth rates is unlikely to provide strong enough evidence to confirm theoretical predictions. However, the observed patterns are not inconsistent with those expected for the extended factory model with SeqA marking the replication forks.

Interpretation of localization of SeqA foci is not straightforward. The numbers of foci are likely to be underestimated due to the inability of the microscope to resolve foci that are close together or are placed one above the other in the line of view. It is also possible that the positions of foci seen by immunofluorescence microscopy are altered by the fixation of the cells. Nevertheless, it can be reasonably concluded from most of the extant studies that SeqA foci mark the positions of the replication forks, and that the forks are placed as predicted by the factory model (Fig. 5).

NUCLEOIDS IN SeqA MUTANTS

Null mutations in *seqA* have a profound effect on the nucleoid. Many cells have nucleoids that are misshapen or misplaced (2, 22). In addition, *seqA* mutant cultures contain an elevated number of anucleate cells (25). However, it is not clear whether these problems are due to a fundamental defect in nucleoid formation or to segregation. The mutants are deficient in controlling the timing of replication initiation and often have excess DNA content. It is possible that this has an indirect effect on nucleoid form, position, and segregation.

DOES SeqA PLAY A MECHANICAL ROLE IN CHROMOSOME SEGREGATION?

We have presented the argument that SeqA protein foci mark the positions of the replication forks in *E. coli* cells and that the positions so revealed are in reasonable agreement with the factory model for DNA replication and segregation. The model requires the newly replicated DNA emerging from the replication factory to be directed away from the cell center in such a fashion that two separate daughter nucleoids are formed prior to cell division (Fig. 5). To achieve this, it is important that the newly replicated DNA exits the factory in an orderly fashion (28). The following considerations are relevant. The segregation mechanism should differentiate between the newly replicated DNA and the old DNA and keep them separate. In addition, the newly replicated sequences should be organized in such a way as to prevent them from becoming entangled with each other. Also, it may be necessary to channel the sequences destined to form the two new nucleoids so that they move out from the factory toward the cell poles in the appropriate directions. Last, it may be important to convert the newly replicated DNA into a form that has sufficient mechanical rigidity to be "pushed" by replication for a considerable distance within the cell. The properties of the SeqA protein and its selective binding to the newly replicated DNA at the replication forks suggest that it might be directly involved in some or all of these processes.

Replication of the chromosome is bidirectional, and both forks contribute to the synthesis of each daughter nucleoid (Fig. 4). It is likely, then, that both forks are spatially coordinated within the replication factory so that the products can be led away in the appropriate directions (Fig. 9). This may occur by anchoring the forks closely to each other, as illustrated in Fig. 9A. However, another way of coordinating the two forks is possible (13, 28). The leading and lagging strand DNA polymerases of *E. coli* are physically associated via a dimer of the tau subunit in a simple replication fork (17). Coordinated forks should contain a tetrameric assembly of polymerases. In this assembly, the two polymerase dimers carrying out synthesis of the leading and lagging strand of each fork might isomerize by exchange of subunits.

In this form, each dimer is synthesizing the leading strand of one fork and the lagging strand of the other. In this way, the two forks would be held together, and the products for each nucleoid could be directed outward toward the cell poles in a coordinated fashion, as illustrated in Fig. 9B.

In either of the cases illustrated in Fig. 9, a role for SeqA in organizing the DNA as it exits the factory seems probable. SeqA is highly self-associating in vitro and, once bound specifically to DNA fragments, can cause them to associate into large aggregates (30). Thus, stretches of newly synthesized DNA with SeqA bound to them are likely to associate with each other. This may have the effect of sequestering the newly replicated DNA destined for one daughter nucleoid into a coherent filament, as illustrated in Fig. 9B. Note that the DNA of the old nucleoid would be excluded from this complex, because it cannot bind SeqA. Thus, old and new DNA would be effectively separated. The directionality imposed by the configuration of the factory could also keep the two nascent nucleoids from being entangled. It is also possible that the form and rigidity of the putative SeqA-associated filaments ensure that the newly replicated DNA is pushed out in the appropriate directions and that the emerging filament does not buckle and cross back over the plane of cell division to cause entanglement of the nascent nucleoids. Last, SeqA appears to have affinity for the cell membrane. The directionality and mechanical effectiveness of the process may be further enhanced by membrane association of the SeqA tracts. This would constrain the movement of the emerging DNA to sliding in two dimensions, thus greatly simplifying the problem of keeping the filaments and nascent nucleoids spatially separated.

CONSERVATION OF THE SeqA SYSTEM IN OTHER ORGANISMS

The presence of a SeqA gene is not universal among the eubacteria. None of the gram-positive species have the gene. They also lack a Dam methylase and have no adenine methylation of their GATC sequences. This is also true of at least some gram-negative species. How, then, is DNA replication and segregation achieved in these organisms? As we have outlined, there are good reasons to believe that the factory model is applicable to gram-positive organisms such as *B. subtilis*. Perhaps some methylation-independent system for marking the newly replicated DNA exists in these organisms. One possibility that comes to mind is the presence of a dimeric DNA-binding protein that binds to double-stranded DNA

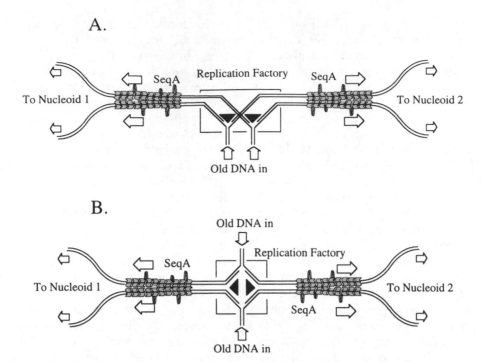

Figure 9. The replication factory and SeqA. The factory model assumes that the two replication forks for bidirectional replication are anchored at the cell center. The forks could be arranged in two different ways. (A) The forks are coordinated as shown. The old DNA feeds into the replication factory, and the newly synthesized DNA is channeled away in opposite directions along the long axis of the cell. As the newly synthesized DNA is hemimethylated, SeqA binds to it. Self-association of the bound SeqA keeps the two duplexes destined to form each nucleoid together as a coherent filament. This facilitates the movement of the products outward toward the cell poles. Further out toward the poles, full methylation of the DNA is restored, SeqA protein dissociates, and the DNA is condensed to form a new nucleoid structure. (B) An alternative configuration for the forks is shown. Here, it is assumed that the two polymerase dimers that are acting at the two forks isomerize so that each dimer synthesizes the leading strand of one fork and the lagging strand of the other (13). This idea is attractive, because this configuration naturally holds the forks together, spatially coordinates the two duplexes that will form each nucleoid, and ensures that they emerge from the factory in opposite directions.

such that each subunit recognizes one strand. This binding would mimic the binding configuration suggested for SeqA. If the single-strand contact persisted through replication, newly replicated DNA would be uniquely marked by having protein subunits on one strand only. If this "hemidecorated" DNA were self-associating, a mechanism similar to that proposed for SeqA might operate in the absence of any methylation signals.

REFERENCES

1. **Azam, T. A., S. Hiraga, and A. Ishihama.** 2000. Two types of localization of the DNA-binding proteins within the *Escherichia coli* nucleoid. *Genes Cells* 5:613–626.
2. **Bahloul, A., J. Meury, R. Kern, J. Garwood, S. Guha, and M. Kohiyama.** 1996. Co-ordination between membrane oriC sequestration factors and a chromosome partitioning protein, TolC (MukA). *Mol. Microbiol.* 22:275–282.
3. **Bakker, A., and D. Smith.** 1989. Methylation of GATC sites is required for precise timing between rounds of replication in *Escherichia coli. J. Bacteriol.* 171:5738–5742.
4. **Boye, E., and A. Lobner-Olesen.** 1990. The role of *dam* methyl transferase in the control of DNA replication in *E. coli. Cell* 62:981–989.
5. **Boye, E., A. Lobner-Olesen, and K. Skarstad.** 2000. Limiting DNA replication to once and only once. *EMBO Rep.* 1:479–483.
6. **Bremer, H., and P. B. Dennis.** 1987. Modulation of chemical composition and other parameters of the cell by growth rate, p. 1527–1542. *In* F. C. Neidhardt, J. L. Ingraham, K. B. Low, B. Magasanik, M. Schaechter, and H. E. Umbarger (ed.), *Escherichia coli and Salmonella typhimurium: Cellular and Molecular Biology*, 1st ed. American Society for Microbiology, Washington, D.C.
7. **Brendler, T., A. Abeles, and S. Austin.** 1995. A protein that binds to the P1 origin core and the *oriC* 13mer region in a methylation-specific fashion is the product of the host *seqA* gene. *EMBO J.* 14:4083–4089.
8. **Brendler, T., and S. Austin.** 1999. Binding of SeqA protein to DNA requires interaction between two or more complexes bound to separate hemimethylated GATC sequences. *EMBO J.* 18:2304–2310.
9. **Brendler, T., J. Sawitzke, K. Sergueev, and S. Austin.** 2000. A case for sliding SeqA tracts at anchored replication forks during *E. coli* chromosome replication and segregation. *EMBO J.* 19:6249–6258.

10. Campbell, J. L., and N. Kleckner. 1990. *E. coli oriC* and the *dnaA* gene promoter are sequestered from *dam* methyltransferase following the passage of the chromosomal replication fork. *Cell* **62:**967–979.

11. Cooper, S., and C. E. Helmstetter. 1968. Chromosome replication and the division cycle of *Escherichia coli B/r. J. Mol. Biol.* **31:**519–540.

12. Dingman, C. W. 1974. Bidirectional chromosome replication: some topological considerations. *J. Theor. Biol.* **43:**187–195.

13. Fang, L., M. J. Davey, and M. O'Donnell. 1999. Replisome assembly at *oriC*, the replication origin of *E. coli*, reveals an explanation for initiation sites outside an origin. *Mol. Cell* **4:**541–553.

14. Gordon, G. S., D. Sitnikov, C. D. Webb, A. Teleman, A. Straight, R. Losick, A. W. Murray, and A. Wright. 1997. Chromosome and low copy plasmid segregation in *E. coli*: visual evidence for distinct mechanisms. *Cell* **90:**1113–1121.

15. Gordon, G. S., and A. Wright. 1998. DNA segregation: putting chromosomes in their place. *Curr. Biol.* **8:**R925–R927.

16. Hiraga, S., C. Ichinose, H. Niki, and M. Yamazoe. 1998. Cell cycle-dependent duplication and bidirectional migration of SeqA-associated DNA-protein complexes in *E. coli. Mol. Cell* **1:**381–387.

17. Kim, S., H. G. Dallmann, C. S. McHenry, and K. J. Marians. 1996. Tau couples the leading- and lagging-strand polymerases at the *Escherichia coli* DNA replication fork. *J. Biol. Chem.* **271:**21406–21412.

18. Koppes, L. J., C. L. Woldringh, and N. Nanninga. 1999. *Escherichia coli* contains a DNA replication compartment in the cell center. *Biochimie* **81:**803–810.

19. Landouisi, A., A. Malke, R. Kern, M. Kohiyama, and P. Hughes. 1990. The *E. coli* cell surface specifically prevents the initiation of DNA replication at *oriC* on hemimethylated DNA templates. *Cell* **63:**1053–1060.

20. Lemon, K. P., and A. D. Grossman. 1998. Localization of bacterial DNA polymerase: evidence for a factory model of replication. *Science* **282:**1516–1519.

21. Lemon, K. P., and A. D. Grossman. 2000. Movement of replicating DNA through a stationary replisome. *Mol. Cell* **6:**1321–1330.

22. Lu, M., J. L. Campbell, E. Boye, and N. Kleckner. 1994. SeqA: a negative modulator of initiation in *E. coli. Cell* **77:**413–426.

23. Niki, H., and S. Hiraga. 1998. Polar localization of the replication origin and terminus in *Escherichia coli* nucleoids during chromosome partitioning. *Genes Dev.* **12:**1036–1045.

24. Ogden, G. B., M. J. Pratt, and M. Schaechter. 1988. The replicative origin of the *E. coli* chromosome binds to cell membranes only when hemimethylated. *Cell* **54:**127–135.

25. Onogi, T., H. Niki, M. Yamazoe, and S. Hiraga. 1999. The assembly and migration of SeqA-Gfp fusion in living cells of *Escherichia coli. Mol. Microbiol.* **31:**1775–1782.

26. Roos, M., A. B. van Geel, M. E. Aarsman, J. T. Veuskens, C. L. Woldringh, and N. Nanninga. 1999. Cellular localization of oriC during the cell cycle of *Escherichia coli* as analyzed by fluorescent in situ hybridization. *Biochimie* **81:**797–802.

27. Russell, D. W., and N. D. Zinder. 1987. Hemimethylation prevents DNA replication in *E. coli. Cell* **50:**1071–1079.

28. Sawitzke, J., and S. Austin. 2001. An analysis of the factory model for chromosome replication and segregation in bacteria. *Mol. Microbiol.* **40:**786–794.

29. Shakibai, N., K. Ishidate, E. Reshetnyak, S. Gunji, M. Kohiyama, and L. Rothfield. 1998. High-affinity binding of hemimethylated *oriC* by *Escherichia coli* membranes is mediated by a multiprotein system that includes SeqA and a newly identified factor, SeqB. *Proc. Natl. Acad. Sci. USA* **95:**11117–11121.

30. Skarstad, K., G. Lueder, R. Lurz, C. Speck, and W. Messer. 2000. The *Escherichia coli* SeqA protein binds specifically and co-operatively to two sites in hemimethylated and fully methylated *oriC. Mol. Microbiol.* **36:**1319–1326.

31. Slater, S., S. Wold, M. Lu, E. Boye, K. Skarstad, and N. Kleckner. 1995. *E. coli* SeqA protein binds *oriC* in two different methyl-modulated reactions appropriate to its roles in DNA replication initiation and origin sequestration. *Cell* **82:**927–936.

32. von Freiesleben, U., K. V. Rasmussen, and M. Schaechter. 1994. SeqA limits DnaA activity in replication from *oriC* in *Escherichia coli. Mol. Microbiol.* **14:**763–772.

33. Weitao, T., K. Nordstrom, and S. Dasgupta. 2000. *Escherichia coli* cell cycle control genes affect chromosome super-helicity. *EMBO Rep.* **1:**494–499.

Chapter 12

Reinitiation of DNA Replication

KENNETH N. KREUZER AND BÉNÉDICTE MICHEL

Contrary to what most current textbooks state, replication forks initiated at bacterial origins often fail to replicate all the way to the terminus on the opposite side of the chromosome. During its traverse from the origin toward the terminus, the replication complex may encounter tightly bound proteins, nicks in the template, and various forms of DNA damage (even in cells growing without exogenous DNA-damaging agents). Furthermore, the replication complex itself may sometimes fail to function even when the template is unaffected, for example, when the replicative helicase or DNA polymerase becomes damaged. Our understanding of how the replication fork responds to each of the problems is very rudimentary. However, it is clear that in each case, the replication complex becomes inactivated at some frequency. It is also clear that bacterial cells, viruses, and eukaryotic cells have evolved sophisticated pathways to reconstitute replication forks that have become inactivated. As will be described below, bacterial cells have multiple pathways for replication restart, some of which involve homologous recombination reactions and some of which do not.

The main importance of replication restart is obvious, namely, the completion of genomic replication for cell division. However, the phenotypes of various mutants that are deficient in one or more restart pathways reveal additional important roles. As will be described in more detail below, a common phenotype is hypersensitivity to DNA-damaging agents, implying that restart pathways are particularly important when the template is heavily damaged. A second common phenotype is genome instability. Apparently, when the normal pathways for restarting forks are unavailable, some of the damaged forks enter nonhomologous recombination pathways and/or homologous recombination pathways with repeated sequences elsewhere in the genome. Indeed, recent studies with mammalian cells strongly suggest that defects in replication restart can contribute to cancer formation via genomic instability (9, 17).

Particular fork restart pathways can also be important for homologous recombination reactions. In this case, there is no "old fork" that became inactivated, but rather a new replication fork is assembled onto a recombination intermediate. The strongest evidence for this view is that *Escherichia coli priA* mutants, which are deficient in replication restart, are also deficient in conjugation and transduction by phage P1 (see below). The implication is that in both conjugation and transduction, the 3' ends of the entering DNA fragment initiate new replication forks within the chromosome of the recipient after the appropriate strand-invasion reactions. It is possible, though not proven, that the forks initiated in this way replicate the entire bacterial chromosome to complete the genetic recombination event.

One or more pathways of replication restart also emerged from studies of unusual chromosomal replication in *E. coli*. Kogoma and colleagues analyzed replication that could occur in *E. coli* in the absence of the functional initiator protein DnaA (or other treatments that block normal initiation from *oriC*) (59). This unusual form of replication was called stable DNA replication (SDR). One kind of SDR is inducible by DNA damage and is therefore called iSDR. The process of iSDR is believed to involve damage-induced breakage of the chromosome, followed by recombination of the broken ends and initiation of new replication forks upon the recombination intermediates. Recent evidence suggests that iSDR, or some closely related process, is also important in the process of stationary-phase mutation in the absence of DNA damage (14). A second form of SDR, cSDR, occurs constitutively when either the major RNase H or the RecG protein is inactivated.

Kenneth N. Kreuzer • Department of Biochemistry, Duke University Medical Center, Box 3711, Durham, NC 27710. **Bénédicte Michel** • Génétique Microbienne, Institut National de la Recherche Agronomique, 78352 Jouy-en-Josas, France.

Table 1. Proteins implicated in replication restart in phage T4 and *E. coli*

Protein function	Phage T4 protein(s)	*E. coli* protein(s)
Replicative helicase	gp41	DnaB
Loader of helicase	gp59	PriA, PriB, PriC, DnaC, DnaT
Primase	gp61	DnaG
Single-strand DNA binding	gp32	SSB
Strand invasion	UvsX	RecA
Loader of strand-invasion protein	UvsY	RecO, RecR
Branch migration	gp41, UvsW	RuvAB, RecG
Holliday junction cleaving	Endo VII	RuvC

These and other results suggest that cSDR involves the assembly of replication forks on persistent R-loops in the chromosome. Although cSDR does not appear to involve initiation of replication at recombination intermediates, the similarity between R-loops and D-loops suggests a fairly close relationship between cSDR and pathways of replication restart.

In this chapter, we highlight our current knowledge of replication restart in *E. coli* and T4-infected *E. coli* (see Table 1 for a list of relevant proteins and Table 2 for a summary of key terms and abbreviations). We begin with the phage T4 system. T4 uses homologous recombination to initiate most of its DNA replication. Thus, the process of recombination-dependent DNA replication (RDR) in T4 has been studied for several decades, and is relatively easy to

study since it occurs so frequently. For these reasons, T4 currently provides the clearest view of one particular pathway of replication restart. Most studies of bacterial replication restart have been conducted in the *E. coli* system, which provides our best current understanding of the multiple pathways of fork restart and their integration into the cell cycle and other aspects of bacterial physiology (e.g., homologous genetic recombination). However, the reductionist approach of detailed study in the *E. coli* model system always raises a question: does what we learn in *E. coli* apply to other bacteria (much less elephants)? Recent studies with the gram-positive bacterium *Bacillus subtilis* therefore provide a broader perspective on replication restart, and these are summarized briefly below.

REPLICATION RESTART VIA RECOMBINATION IN BACTERIOPHAGE T4

Replication Restart as a Major Mode of T4 DNA Replication

For several decades, it has been clear that recombination and DNA replication are tightly coupled during the infection cycle of bacteriophage T4 (97). Indeed, studies with phage T4 have led the way in our understanding of the mechanism of RDR, and we summarize some of the mechanistic lessons learned from this system below. We begin, however, by considering DNA metabolism in the life cycle of T4. With our current understanding of the role of

Table 2. Terminology used in this chapter

Term	Definition
Replication fork arrest	An event in which the replisome halts its normal progress through the parental template; it is possible that the replisome is inactivated during only a subset of replication fork arrest events.
Replication fork collapse	An event in which the replication fork breaks due to an encounter with a nick or gap in the parental template; collapse results in generation of a double-strand end (not a double-strand break) involving one of the two daughter duplexes behind the fork.
Replication fork reversal (RFR)	The extrusion of the two product strands by a branch-migration reaction to form a Holliday junction; also called replication fork regression by some authors
Replication restart	The multiple pathways for restarting replication forks that have become inactivated
Recombination-dependent replication (RDR)	Any process that initiates DNA replication by assembling a new replisome on an intermediate in homologous recombination
Stable DNA replication (SDR)	Two pathways of *E. coli* chromosomal replication that are independent of DnaA and the normal initiation pathway at *oriC*
Double-strand end	The end of a duplex DNA molecule that is either blunt or has a single-strand extension
Double-strand break (DSB)	A pair of DNA ends generated by a breakage event

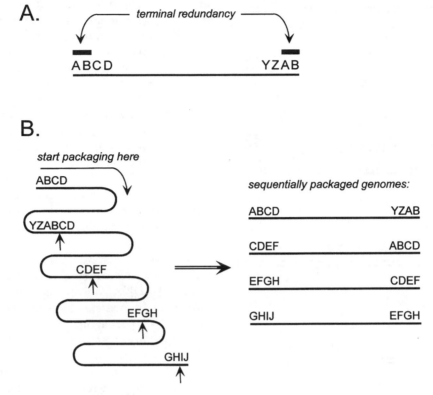

Figure 1. The T4 genome is terminally redundant and circularly permuted. (A) Every packaged genomic DNA molecule of phage T4 has the same sequence at the two ends of the duplex. (B) DNA packaging in T4 occurs by sequential headful packaging. Since the phage head holds slightly more than 100% of the unique genomic sequence, each packaged DNA is terminally redundant. In addition, since packaging is sequential from concatemeric DNA, different packaged molecules have different end sequences (circular permutation).

recombination in restarting replication forks, we can now reinterpret the T4 life cycle in simple terms.

Let us begin with an infection by a single phage particle. The infecting DNA from a T4 particle is a linear duplex of just over 170,000 bp, with terminal repeats of about 3,000 bp (Fig. 1A; genome referred to as terminally redundant). Soon after phage gene expression commences, DNA replication is initiated from internal replication origins, and the forks so generated travel toward the ends of the infecting genome (Fig. 2A). Soon, each origin-initiated fork will reach a double-strand end and thereby become inactivated (Fig. 2B). The simple reinterpretation of DNA metabolism in the T4 life cycle comes from the realization that the next phase of T4 DNA replication, RDR, is exactly equivalent to recombinational restart of these inactivated replication forks.

Mosig (97) proposed a molecular model for this RDR. According to the model, the 3' ends of the parental strand in each of the two daughter molecules cannot be completely replicated because of the polarity of DNA polymerase and the need for a primer. The phage-encoded strand-exchange protein can bind

to this single-stranded 3' end (see below), directing the search for homologous duplex DNA. Thanks to the terminal redundancy, homologous DNA can be found at the other end of the genome in either of the two daughter molecules (Fig. 2C). In the Mosig model, the D-loop so generated becomes the site of assembly of a new replication complex, with the invading 3' end serving as the primer for the new leading strand (Fig. 2D). Note that the new fork is traveling inward and thereby has the potential to replicate the entire genome. Also note that the product of this round of replication will be roughly twice the original genome length, i.e., a linear concatemer of the phage chromosome.

When the new fork traverses the entire genome and reaches another genome end, the same process can happen again. Thus, the process of RDR in T4 is self-regenerating: forks are continually restarted whenever they reach a genome end. This is half the explanation for the long-standing observation that continued DNA replication at late times is dependent on phage-encoded recombination proteins (21, 26, 92). The second half of the explanation is that phage

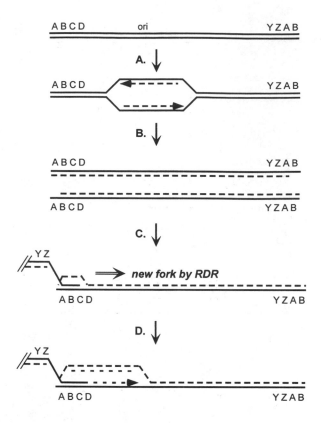

ABCD   ori   YZAB

A. ↓

ABCD   YZAB

B. ↓

ABCD   YZAB

ABCD   YZAB

C. ↓

YZ
                    → new fork by RDR
ABCD   YZAB

D. ↓

YZ

ABCD   YZAB

E. Multiple infection:

YZ
                              ⟹ new fork by RDR
MNOP   ABCD   KLMN

Figure 2. DNA replication in phage T4 infections. As described in the text, origin replication leads to unreplicated 3' ends in the daughter molecules (steps A and B). Each single-stranded 3' end can undergo a strand-invasion reaction with the opposite end of one of the daughter molecules (C), leading to the assembly of a new replication fork (D). When two or more phage infect the same cell, D-loop formation can occur with homologous DNA in more central regions of the genome (E), but the consequences are very similar.

replication origins become repressed at late times and therefore cannot contribute to late DNA replication. Repression of the origins depends on the UvsW protein, which is expressed from a late promoter (22). UvsW is an RNA-DNA helicase that disrupts the origin R-loop, an essential intermediate in origin replication (25). Interestingly, UvsW also plays a positive role in RDR (see below).

Many rounds of RDR convert the intracellular form of T4 DNA into a long concatemer. This concatemer is the competent substrate for DNA pack-

aging into the preformed head particles (Fig. 1B) (for a review, see reference 10). DNA packaging occurs by a headful mechanism, and the T4 head can hold several percent more DNA than the unit length of genetic material. This explains the terminal redundancy of the viral DNA molecule. In addition, another feature of T4 genome structure, circular permutation, is explained by the packaging mechanism. During packaging, multiple genomes are sequentially cleaved from a single concatemer, and each packaging event cuts off more than one genome length (Fig. 1B). Therefore, individual molecules of packaged T4 DNA have different (random) end sequences. The term circular permutation is used because permuted linear molecules with random ends can be generated by randomly cleaving a circle (but note that cleaving a circle will generate molecules without terminal redundancy). An added complexity is that the intracellular T4 DNA concatemer contains many recombinational branches, which are resolved before or during DNA packaging.

With this more complete understanding of the T4 genome structure, we can now consider DNA metabolism in an infection by multiple phage particles. The only difference from the solo infection is that there are more locations of homologous DNA for the unreplicated 3' ends after origin replication. The end of one infecting DNA molecule will likely be homologous to a more central region of another infecting DNA molecule (Fig. 2E). This does not change the mechanism or outcome of RDR: the newly initiated fork will simply reach another genome end more quickly if it does not need to traverse the entire genome.

In the mechanisms discussed so far, we have focused on the newly initiated replication fork that was primed by the invading 3' end. In this view, the process of RDR is a unidirectional replication event that leaves a Y-branch behind at the point of the original strand invasion. These Y-branches likely contribute to the branched nature of the intracellular T4 concatemer, eventually being resolved by T4 endonuclease VII (the first Holliday junction-resolving enzyme discovered [96]). It is also possible to imagine that the D-loop assembles a bidirectional or even a tridirectional replication fork (see references 66 and 98 for detailed mechanism). In these cases, the rate of DNA replication would accelerate as more active forks are brought on line. It is not yet clear whether in vivo T4 RDR is uni-, bi-, or tridirectional, or some combination thereof.

Mosig et al. (99) have also discussed multiple pathways of T4 RDR, some of which may be restricted to certain mutant infections. They also addressed the possible significance of RDR pathways in the evolution of phage.

To summarize, the extensive use of RDR in T4 DNA metabolism can be viewed simply as replication restart in a system with a linear genome. Indeed, phage T4 has been telling us for decades that inactivated replication forks can be restarted by recombination.

Replication Restart in T4 at Internal Locations, Including Sites of DNA Damage

Since T4 has such an active and proficient pathway for restarting replication forks after they encounter genome ends, it would not be surprising to find that the pathway is used for other purposes. We briefly summarize evidence that T4 RDR is indeed used to reinitiate replication forks that have run into difficulty at sites of DNA damage from exogenous agents, and also from spontaneous processes that disrupt DNA replication.

The general process of recombinational repair was first uncovered using the phage T4 model system, when Luria (79) found that UV-damaged genomes could give rise to a productive burst only if more than one damaged genome infected a given cell. Decades of investigation have revealed that T4 mutants deficient in RDR are also deficient in repair of damage from UV, X rays, mitomycin C, and DNA-alkylating agents (for reviews, see references 8 and 65). A reasonable but unproven hypothesis is that a large component of this recombinational repair involves the restart of replication forks by RDR.

As in the E. coli system (see below), there may be multiple pathways that can be triggered when a T4 replication fork encounters damage, depending on the exact nature of the damage and the strand on which it is located. One interesting pathway that was recently proposed is based on experiments with inhibitors of type II topoisomerase that stabilize a covalent enzyme-DNA intermediate (the cleavage complex). This hypothetical pathway is based on the observation that a blocked replication fork can be cleaved in vitro, by the phage-encoded endonuclease VII, into either three fragments or an intact linear plus one broken arm (44; G. Hong and K. N. Kreuzer, unpublished data [cited in reference 66]; also see reference 40). In the proposed pathway, the resulting double-strand ends trigger new rounds of RDR by invading homologous duplex DNA molecules. Because the T4 infection has multiple copies of the genome and an avid system that initiates replication from double-strand breaks (DSBs), this seemingly sloppy system may be very well suited to the viral life cycle. Perhaps the site of DNA damage that blocked the replication fork is destroyed by exo- or endonuclease action prior to the new round of RDR, making this a true repair pathway.

Investigations of phage T4 hotspots for marker rescue recombination provided a dramatic demonstration of the coupling between replication, DNA damage, and recombination. The hotspots were first detected as regions of the genome where genetic markers could be rescued by homologous recombination from UV-irradiated phage at an inflated frequency (141). The same hotspots can be detected with other forms of DNA damage, including X rays and decay of incorporated radioactive ^{32}P (15, 75). More recently, the phage replication origins were shown to be both necessary and sufficient for the activity of these hotspots (23, 143). Furthermore, hotspot activity requires that the origin be located in the damaged DNA—the presence or absence of the origin in the nondamaged DNA (the DNA that is recipient of the genetic marker) is irrelevant. These results strongly suggest that replication forks initiated at the origin trigger recombination when they encounter DNA damage. Because the extent of DNA damage is very high, the forks are likely to encounter damage a short distance from the origin, leading to the localized hotspot region. The mechanistic details of the inflated recombination remain to be established, but they likely involve the recombinational restart of replication. One speculative model follows the pathway mentioned above, namely, cleavage of the blocked fork by endonuclease VII, followed by DSB-directed replication.

Recent evidence indicates that T4 replication forks sometimes need to be restarted even when they do not encounter a DNA end or exogenous (i.e., experimentally introduced) DNA lesion. The experimental system involves T4 infection of cells containing two plasmids that have homology to each other. The key finding was that the presence of a T4 replication origin on one plasmid stimulated both recombination between the two plasmids and replication of the second plasmid (which does not contain an origin) (34). These results indicate that replication forks initiated from the origin on the first plasmid sometimes run into difficulty and reinitiate replication via RDR, with strand invasion into the second plasmid. Presumably, similar events happen during replication of the phage genome, and contribute to the high basal level of homologous recombination in this system.

In summary, DNA metabolism in phage T4 relies very heavily on the recombinational restart of DNA replication. Replication forks initiated from origins quickly reach the genome ends and are then restarted on homologous DNA by the process of RDR. RDR itself is self-regenerating, because replication forks initiated by RDR will eventually reach another genome end and trigger the process once again. Finally, one or more pathways of recombinational repair involve the restart of replication by RDR. This provides the phage with a strong resistance to DNA-damaging agents and is also probably used to solve routine problems that replication forks may encounter.

Mechanistic Lessons from Phage T4

Because of the prominence of RDR in the T4 life cycle, the T4 system has led the way in mechanistic studies of the process. The first in vitro RDR reaction was reconstituted with T4 proteins (31), and numerous studies of purified replication and recombination proteins have provided important insights about the interconnections between replication and recombination. As we describe the overall mechanism of T4 RDR, we attempt to illustrate some of these insights and also point out some of the gaps in our understanding.

What is the nature of the invasive end?

Strand-invasion models generally involve a single-stranded DNA (ssDNA) end, because ssDNA is the preferred substrate for loading strand-invasion proteins (see below). The original model of Mosig (97) proposed that these ssDNA ends are generated simply by the act of DNA replication: the 3′ ends of the two parental strands should remain single stranded after replication because of the polarity of DNA polymerase and the need for a primer (see above). Although this seems a very reasonable model, one experimental fact suggests that it may not be complete: mutations in T4 genes *46* or *47* abolish T4 RDR (26, 46).

The products of genes *46* and *47* are involved in host DNA breakdown and have long been suspected of being a nuclease (95). The gp46/47 complex has been very difficult to analyze biochemically, but very recent studies with purified protein did reveal a nuclease activity (11). In addition, gp47 has a conserved nuclease motif (126), and even conservative mutations in the highly conserved histidine residue of that motif (His-10) abolish biological function (A. Panigrahi and K. N. Kreuzer, unpublished data). These results strongly suggest that the putative nuclease activity is important for T4 RDR. One possibility is that gp46/47 plays a critical role in digesting 5′ ends to create the invasive 3′ ends for RDR (Fig. 3A). Another interesting possibility is that gp46/47 plays a critical role in removing the gp2 protein, which is bound to the T4 genome ends and which protects

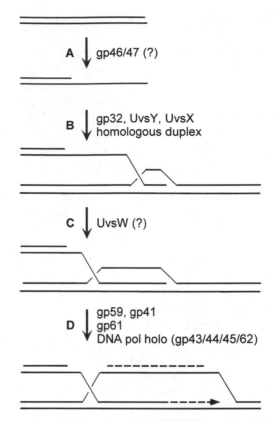

Figure 3. Proposed pathway for T4 RDR. The proposed steps in T4 RDR are depicted, including 5′-end resection (A), D-loop formation (B), D-loop stabilization (C), and replication fork assembly and function (D). See text for further description.

against the action of the host RecBCD enzyme (2). Further work is clearly needed to elucidate the role of prior DNA replication and/or nucleases such as gp46/47 in preparation of the invasive 3′ ends.

Importance of RMP protein UvsY in the strand-invasion step

ssDNA within a T4-infected cell is quickly covered with the ssDNA-binding protein gp32 (for a review, see reference 139). A major conundrum in considering the mechanism of T4 recombination is that the strand-exchange protein UvsX (RecA homologue) has a much lower affinity for ssDNA than does gp32, and therefore UvsX cannot bind to gp32-coated ssDNA under physiological conditions (30, 144). The solution to this problem came from studies of the UvsY protein, which binds gp32-coated ssDNA and thereby allows assembly of the UvsX protein presynaptic filament onto the DNA (42, 57, 145). UvsY was thereby classified as a replication/recombination mediator protein, or RMP (7). UvsY appears to modify the structure of the ssDNA within the gp32-DNA complex, thus favoring displacement of the gp32 by UvsX (130). Once the presynaptic filament is formed, UvsX promotes the search for homology within duplex DNA, leading to D-loop formation (Fig. 3B). UvsY protein has served as a prototype for recombination mediator proteins. Since the role of UvsY in the T4 strand-invasion reaction was elucidated, similar roles for *E. coli* RecO and RecR proteins and the eukaryotic Rad52 protein were also uncovered (129, 135).

Interestingly, UvsY appears to play additional roles in the early stages of T4 recombination. First, UvsY stabilizes the UvsX-ssDNA presynaptic filament, with maximum stability at a 1:1 protein ratio (57). Second, UvsY interacts with gp46/47, which has led to the suggestion that UvsY may directly couple the strand resection step to the strand-invasion step (11). With all of these important roles in the process of recombination, it is now clear why *uvsY* mutants are just as defective in vivo as *uvsX* mutants.

UvsW is the key switch between origin-dependent and recombination-dependent replication

As mentioned above, the phage-encoded UvsW protein is a helicase that can unwind the RNA component from an origin R-loop (25). The protein thereby represses origin replication at late times, when UvsW is synthesized from its late promoter. However, as its name implies, the *uvsW* gene was first discovered based on a very different function. Mutations in *uvsW* cause UV sensitivity because they abrogate the recombinational repair of UV damage (41). Furthermore, *uvsW* mutations act in the same UV repair pathway as mutations in *uvsX* and *uvsY*, arguing that the UvsW protein plays a positive role in the strand-invasion reaction.

More direct evidence for a positive role of UvsW in T4 RDR has also been obtained. Mutations that inactivate the UvsW protein were found to greatly decrease DSB-stimulated replication in plasmid model systems (16, 34). Biochemical studies demonstrate that UvsW is a helicase that requires branched DNA substrates (16), suggesting that the protein might play a role in some branch-migration step. One simple model is that UvsW drives three-stranded branch migration during the strand-invasion step, stabilizing the resulting D-loop (Fig. 3C). Regardless of the mechanistic details, UvsW clearly plays a critical role in switching the initiation of T4 DNA replication from R-loop substrates to D-loop substrates. Interestingly, the T4 UvsW protein can substitute for at least some function(s) of *E. coli* RecG (16), which is also proposed to inhibit replication

from R-loops and to promote homologous recombination (59, 138).

RMP protein gp59 couples recombination to replication

Once the D-loop is formed, the next step in RDR involves assembly of the replication complex onto the D-loop (Fig. 3D). A second RMP protein, gp59, plays a critical role in this step, coupling DNA replication to recombination. Like UvsX protein, the T4 replicative helicase gp41 binds very poorly to gp32-coated ssDNA. When the gp59 protein was purified, it was found to overcome this inhibition and greatly stimulate the loading of gp41 (6). Once loaded, the replicative helicase increases the rate of leading strand synthesis because it catalyzes unwinding of the parental duplex and also permits lagging strand synthesis by assisting the T4 primase (gp61) in Okazaki fragment primer synthesis (108).

gp59 plays a critical role in T4 RDR in vivo, because gene 59 mutations abolish RDR (21). gp59 binds to branched-DNA structures, which has led to a model in which the replicative helicase is targeted to the displaced strand of the D-loop by gp59 (49). The interaction of gp59 with gp32 on the displaced strand is also likely to be critical in this targeting reaction (11). The crystal structure of gp59 was recently solved, and a speculative model for binding to branched DNA was presented (102).

Assembly of the replicative helicase allows semiconservative replication and thereby prevents bubble-migration synthesis

The original in vitro T4 RDR reaction occurred by an unusual mechanism in which the product strand was extruded from the back edge of a D-loop as it was extended at the front edge of the D-loop (31). This "bubble-migration synthesis" therefore resembles the mechanism of transcription, and it is formally conservative since the parental DNA is recovered intact (Fig. 4A).

Various models for RDR and DSB repair invoke this sort of bubble-migration reaction (27, 101, 104). It is important to note, however, that the bubble-migration reaction is completely suppressed when the replicative helicase is loaded by gp59 (Fig. 4B). As mentioned above, the replicative helicase is loaded onto the displaced parental strand of the D-loop, not onto the product strand that is being extruded at the back of the D-loop. With this more complete reaction, the gp41/61 complex synthesizes the RNA primers for Okazaki fragment synthesis upon the

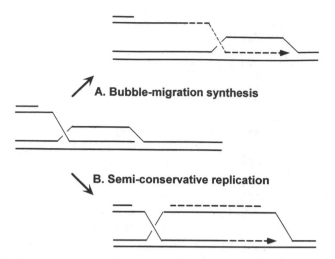

Figure 4. Two modes of DNA synthesis from a D-loop. (A) In the absence of a replicative helicase, T4 replication proteins can catalyze bubble-migration synthesis, in which only one product strand (equivalent to leading strand) is generated. (B) When the replicative helicase is successfully loaded, bubble-migration synthesis is suppressed in favor of normal semiconservative replication.

displaced strand of the D-loop, and normal semiconservative replication ensues. As expected for normal semiconservative replication, a type II topoisomerase is required for extensive elongation on a circular template (J. Barry and B. Alberts, unpublished data [cited in reference 64]).

gp59 also serves as a molecular gatekeeper of the replication complex

Studies of in vitro origin-dependent replication have suggested a second role for the gp59 protein (48, 109). Surprisingly, replication from an origin R-loop was strongly inhibited by the presence of gp59 when gp41 was missing. That is, gp59 inhibited the progress of DNA polymerase when the replicative helicase was not available. These results suggest that gp59 acts as a molecular gatekeeper, holding DNA polymerase at the starting gate until the replicative helicase/primase complex is loaded and ready to function. The function of such a gatekeeper might be important in preventing an excess accumulation of ssDNA products, or perhaps because it is more difficult to load helicase/primase once the replication complex has begun its travels. Considering both the RMP and gatekeeper functions of gp59, this protein is clearly a central player in allowing a coordinated RDR reaction in which a D-loop is efficiently converted into a replication fork with the proper assembly of proteins for efficient leading and lagging strand synthesis.

Is DSB repair in T4 just RDR in a slight disguise?

Phage T4 is very efficient at repairing frank DSBs, and the natural process of intron mobility takes advantage of this efficient repair reaction (18). Studies of DSB repair in T4 have supported three different models: the Szostak et al. (131) model, synthesis-dependent strand annealing (101), and extensive chromosomal replication (ECR) (33).

Without going into all the mechanistic details, the first two models propose localized DNA synthesis to restore the genetic information missing at the break (see reference 34 for a more complete discussion of all three models). The ECR model, on the other hand, proposes that each of the two broken ends initiates a new replication fork after strand invasion into an intact homologous duplex. Each of the new replication forks has a fully functional replication complex that promotes normal semiconservative replication, potentially until a genome end is reached. In the ECR model, the two broken ends are never reconnected, but ECR is nonetheless a legitimate repair reaction since the breaks disappear and intact DNA is restored. As discussed in more detail by George et al. (34), analysis of plasmid model systems for DSB repair provide some support for the ECR model, but definitive evidence distinguishing between the three major models is still lacking. The ECR model proposes that the Mosig model for RDR can be used to repair a frank DSB, and thereby provides a unified view of T4 replication, recombination, and DSB repair.

REPLICATION RESTART IN *E. COLI* WITH AND WITHOUT RECOMBINATION

The *E. coli* Replication Fork

As a prelude to discussing replication restart in bacteria, we briefly review the protein machinery at the *E. coli* replication fork. Progression of replication forks requires the coordinated action of three activities: DNA polymerase, DNA helicase, and primase. In *E. coli*, the chromosome is duplicated by the DNA polymerase III holoenzyme (reviewed in reference 55 and chapter 10). This multipolypeptide complex is composed of 10 different subunits. The two core polymerases synthesize leading and lagging strands in a coordinated fashion. Each of the polymerases is tethered to the DNA by a β clamp. The γ complex loads and unloads the β clamp on the leading and lagging strands. Finally, the τ polypeptide is a scaffold subunit to which both core and γ complexes are attached (reviewed in reference 55 and chapter 10). The DnaB helicase migrates in front of the polymerase on the lagging strand, in the 5′ to 3′ direction. It interacts with Pol III holoenzyme via the τ subunit (56). The DnaG primase catalyzes the synthesis of Okazaki fragment RNA primers; it interacts directly with DnaB (133). Although the polypeptides that compose the polymerase holoenzyme may vary in different bacteria, the basic composition of the replisome is highly conserved.

The replisome is anchored to the DNA template chromosome via three polypeptide rings: the DnaB helicase, which forms a hexameric ring that encircles the lagging strand template in front of the replisome, and the two β-clamp dimers that encircle the sister chromatids. Loading of a ring structure requires a specific loading apparatus and uses the energy of ATP hydrolysis (72, 134). The loading apparatus for the β clamp is the γ complex (reviewed in reference 55 and chapter 10), which is an intrinsic part of the polymerase III holoenzyme. The loading apparatus for the DnaB helicase is a protein called DnaC in E. coli; DnaC does not remain associated with the replisome during replication progression (reviewed in reference 62).

Loading of DnaB by DnaC triggers the assembly of a replisome, as opening of DNA strands allows the binding of polymerase III and primase. For several years, loading of DnaB by DnaC was thought to be permitted on the chromosome only at the origin oriC, after formation of a preinitiation complex by binding of several DnaA molecules to specific sites (reviewed in reference 62 and chapter 9). The finding that replication could be initiated independently of DnaA, at sites that are not a specific origin sequence, opened a new field of investigation (reviewed in reference 59). Studies by several laboratories have led to the idea that DnaA-independent replication initiation plays a critical role in bacteria, restarting arrested replication forks and repairing chromosomal DSBs that occur during replication (reviewed in references 20 and 63). Restart of arrested replication fork is a function presumably needed in all organisms (reviewed in references 47 and 116).

The coordinated action of several proteins—PriA, PriB, PriC, DnaT, and DnaC—is required for replication reinitiation at nonorigin sequences in E. coli (reviewed in reference 82). The preprimosome proteins PriA, PriB, PriC, and DnaT assemble to form a molecular scaffold that allows DnaC to load the DnaB helicase onto the DNA, which, in turn, leads to the assembly of a complete replisome by the loading of DnaG and Pol III. The molecular mechanism of primosome assembly in vitro, the targets for the assembly in vitro and in vivo, and the physiological roles of primosome-dependent replication restart have been the subject of intense studies in

E. coli in the past 10 years (reviewed in references 3, 52, 58, 59, 82, and 119).

Primosome Proteins and Primosome Assembly Sites

The PriA, PriB, and PriC proteins were originally identified as essential for the conversion of the single-stranded form of bacteriophage φX174 DNA to the double-stranded replicative form (reviewed in reference 81). They were also shown to be required for the initiation of the lagging strand synthesis of the vector plasmid pBR322. On these two replicons, they recognize a specific site called primosome assembly site (PAS). As discussed below, the precise role of these proteins in the replication of the E. coli chromosome turned out to be independent of PAS recognition and was not understood until more recently.

The gene encoding PriA was cloned by reverse genetics (74, 111). The gene maps to 88.7 min on the chromosome and encodes a 732-amino-acid protein with a molecular mass of 81.7 kDa. PriA is the key protein in assembly of the primosome complex, the base of the scaffold that leads to the loading of DnaB. PriA binding specificity therefore determines the primosome assembly targets. In addition to its primosome assembly activity, PriA has an ssDNA-dependent ATPase activity, which is also dependent on PAS sites in the presence of single-strand binding protein (SSB). PriA has a 3′ to 5′ helicase activity, translocating along ssDNA in the direction opposite to the replication fork helicase DnaB. The PriA protein has Walker A and B motifs, along with other highly conserved helicase residues that place PriA in the SF2 superfamily of helicases (37). Primosome assembly and helicase activity can be uncoupled, because point mutations (e.g., priA300) that specifically inactivate the 3′ to 5′ helicase activity of PriA do not affect the primosome assembly activity (148). PriA is characterized by the presence of a Cys-metal binding motif, which is important for helicase activity and essential for the assembly of the primosomal proteins (77, 149). The intracellular amount of PriA is estimated to be 50 to 100 molecules per cell (122).

The genes encoding PriB and PriC, which map at 95.5 min and 10.6 min on the chromosome, respectively, were also identified by reverse genetics (1, 146). The priB gene is part of a ribosomal protein operon. However, presumably due to poor codon usage, the PriB protein is estimated at only 80 copies per cell. It is a small protein of 104 amino acids (12.5 kDa) that sediments as a dimer. PriC is a protein of 175 amino acids (20 kDa).

In contrast to the priA, priB, and priC genes, the dnaT and dnaC genes were identified in the search for genes essential for DNA replication (71, 86). Most *dnaC^{ts}* thermosensitive mutants are slow-stop mutants, indicating a role for DnaC in replication initiation (136, 140). The *dnaT* and *dnaC* genes form an operon, with two other genes encoding proteins of unknown function, at 98 min on the *E. coli* map (84). The role of DnaC in the initiation of replication at *oriC* was characterized biochemically. Six molecules of DnaC interact with a DnaB hexamer to load DnaB on an SSB-coated DNA (reviewed in references 5 and 62). DnaC requires the assistance of either DnaA or PriA/PriB/PriC/DnaT to overcome the inhibition of DnaB loading by SSB (reviewed in reference 119).

The molecular assembly of the primosome was first reconstituted in vitro on the bacteriophage φX174 PAS sequence (105, 106). The ordered assembly of the primosome on PAS involves PriA recognizing and binding to the PAS sequence, followed by PriB stabilizing the PriA-PAS complex. The cysteine-rich region of PriA was shown to be essential for the stabilization of the PriA-PAS complex by PriB (77). Addition of DnaT then leads to the formation of a PriA-PriB-DnaT-PAS complex. Loading of DnaB from a $DnaC_6$-$DnaB_6$ complex and subsequent loading of DnaG generates the complete primosome complex. PriC is the only protein that is dispensable for the primosome assembly in vitro: the presence of PriC improves the efficiency of priming and DNA synthesis by twofold or less. In vitro, the primosome remains associated with the replisome during DNA synthesis initiated at a PAS site. Whether the primosome is directly associated with the replication machinery on chromosomes in vivo is unknown.

Primosome Assembly Is Required for SDR and Homologous Recombination In Vivo

priA mutants exhibit a severe growth defect and are constitutively induced for the SOS response, indicating that primosome assembly is an important component of bacterial life (73, 112, 147). The characterization of primosome assembly on PAS suggested that this process might occur in the *E. coli* chromosome for the initiation of Okazaki fragments during lagging strand synthesis. However, no PAS sites were found in the chromosome. In addition, although PriA was essential for the replication of pBR322, deletion of the PAS site in this plasmid had little effect, suggesting that some other DNA structure or sequence could substitute for PAS and act as a nucleation site for PriA-dependent primosome assembly. It was therefore hypothesized that completion of chromosome replication may become dependent on PriA in the event that the replication complex stalls or dissociates (112, 147).

The first direct evidence that PriA-dependent primosome assembly could promote chromosome replication came from the study of SDR. cSDR and iSDR are processes that allow the replication of the entire *E. coli* chromosome in a DnaA- and *oriC*-independent manner (see above) (reviewed in references 3 and 59). Because iSDR requires the RecBC and RecA proteins, the enzymes that catalyze recombinational repair of DSBs (4; reviewed in reference 68), iSDR was proposed to result from the initiation of chromosome replication at recombination intermediates. Although SDR is independent of DnaA, it requires DnaC, indicating that loading of DnaB is essential for this type of chromosome replication (84). Both cSDR and iSDR were shown to require PriA, suggesting that SDR involves a primosome assembly process similar to that occurring at PAS sites (85). Genetic studies of SDR led to a model in which PriA would promote primosome assembly, and in turn the assembly of a functional replisome, at R-loop and D-loop structures in vivo (reviewed in references 58 and 59).

Further insight into the nature of the substrate for PriA-dependent primosome assembly in vivo came from the observation that *priA* mutations are defective for homologous recombination (58, 121). *priA* mutations were found to increase the sensitivity of wild-type cells to UV irradiation, γ irradiation, and mitomycin C and to render cells defective for P1 transduction and to a lesser extent for Hfr conjugation, all processes that require RecBCD-promoted homologous recombination. The *priA300* mutant, which lacks only the helicase activity, could promote most of the homologous recombination reactions, indicating that the primosome assembly activity of PriA is required for recombination (58, 121). The essential role of PriA in RecBCD-mediated recombination in vivo indicates that homologous recombination mediated by RecBCD could lead to a viable product only if accompanied by primosome assembly, hence by initiation of DNA replication. Conversely, this led to the notion that an important role for PriA in vivo might be to initiate replication from recombination intermediates.

Primosome Assembly at the D-Loop In Vitro

PriA binds to synthetic D-loop structures in vitro and unwinds the D-loops whether they were formed by a 3′ invasion or a 5′ invasion (90). The minimal requirement for PriA binding in vitro appeared to be a duplex with an unpaired single-stranded tail at one end. The PriA helicase activity is not required for PriA binding to oligonucleotide substrates but is essential for unwinding. Upon binding to a forked structure, PriA unwinds either the duplex ahead of the fork or the lagging strand duplex (depending on whether it translocates on the leading or lagging strand template). In vitro, PriA action depends on its binding mode and on the presence of SSB. Unwinding of the duplex ahead of the fork involves binding of only the helicase domain of PriA and is inhibited by SSB. In contrast, unwinding of the lagging strand involves binding of two PriA domains, the helicase and fork binding domains, and is not inhibited by SSB (53, 110). PriA would presumably favor unwinding of the lagging strand in vivo, because ssDNA at arrested replication forks should be covered with SSB.

The Marians laboratory reconstituted in vitro the complete assembly of primosomes at D-loops (76, 78) (Fig. 5). Proteins essential for the reaction were PriA, PriB, DnaT, DnaC, DnaB, and DnaG. Importantly, primosomes assembled at such D-loops promoted the initiation of concerted DNA replication on both strands. Helicase-deficient PriA protein sup-

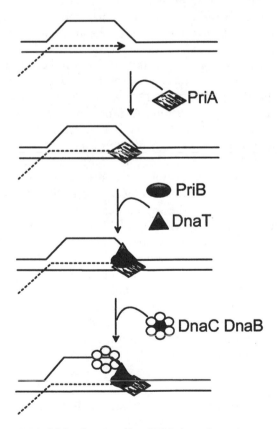

Figure 5. Model for the assembly of PriA-dependent primosome at a D-loop structure. The PriA protein (diamond) first recognizes the D-loop. Binding of PriB (oval) and DnaT (triangle) stabilizes the complex. In the last step, DnaC (closed circle) loads DnaB (open circles). Full lines represent template DNA strands, and dashed lines represent the invading strand in the D-loop. Adapted from reference 82.

ported replication on a D-loop template, even to a greater extent than the wild-type protein, indicating that only the primosome assembly function of PriA was required for the reaction (78). Similar to its minor role in primosome assembly at a PAS site, PriC increased the efficiency of DNA synthesis threefold but was not essential for the reaction. The in vitro reconstitution of replication initiation from a D-loop structure by PriA-dependent primosome assembly strongly supported the models drawn from in vivo studies.

Primosome Assembly Is Required for Bacterial Growth

Although PriA is required for efficient RecBCD-mediated homologous recombination in *E. coli*, it was immediately apparent that PriA was not simply a recombination protein, because the phenotype of *priA* mutants is very different from that of other recombination-deficient mutants. Unlike *recA*- or *recBC*-deficient cells, *priA* mutants are poor growers, sensitive to rich medium, chronically induced for the SOS response, and filamentous (reviewed in reference 147). Moreover, inactivation of *recA* or *recBC* in *priA* mutants does not suppress the growth defect associated with the *priA* mutation (87). Therefore, the growth defect is not caused simply by a lack of processing of recombination intermediates. PriA protein might also be required in vivo for replication restart from D-loops formed by means other than homologous recombination, or from DNA structures other than D-loops, perhaps stalled replication forks.

The viability of *priA* mutants was improved by inactivation of the *sfiA* gene. SfiA acts as a checkpoint protein: it is an SOS-induced protein that prevents the assembly of the septum protein FtsZ, allowing DNA repair to take place prior to cell division in SOS-induced cells (103). Although *sfiA* inactivation does not suppress the recombination defects of *priA* mutants and does not allow growth on rich medium, the improved viability of *sfiA priA* double mutants compared with *priA* single mutants suggests that part of the poor growth of *priA* mutants results from the division block subsequent to SOS induction (112). Most of the studies of *priA* mutants were performed in *sfiA* null cells, ensuring that the observed phenotype resulted from the primary replication defect and not from the blockage of cell division by induced levels of SfiA.

The role of PriB and PriC proteins in vivo was studied with null mutants. Inactivation of either *priB* or *priC* alone caused no detectable phenotype, whereas inactivation of both was lethal. Because PriB and PriC were originally discovered as partners in the same biological process, this observation led to the conclusion that PriB and PriC proteins have redundant functions and can substitute for each other in vivo (120). However, the PriA PriB and PriA PriC pathways are not entirely equivalent. The *priA300* (K230R) mutation interferes with the PriA PriC pathway but not with the PriA PriB pathway, indicating that the helicase activity of PriA is important only in the PriA PriC pathway (117). Furthermore, only the PriA PriB pathway allows efficient replication restart in a strain in which the frequency of replication arrest is increased due to a mutation in the replicative polymerase holoenzyme (28).

Additional pathways of primosome assembly can also function, at least in certain mutant cells. In a *priA* mutant, *priB* inactivation does not affect viability, whereas *priC* inactivation is lethal, leading to the conclusion that some residual primosome assembly occurs by a PriC-dependent pathway that does not require PriA (118). Although this PriC-dependent pathway of primosome assembly allows residual growth of *priA* mutants in synthetic medium, this pathway leads to SOS induction and does not allow growth in rich medium, cSDR, iSDR, or homologous recombination. Therefore, the PriC-dependent pathway does not efficiently fulfill most of the *priA* functions.

Secondary mutations in the *dnaC* gene can suppress the poor viability of *priA* null single mutants, indicating that a modification of DnaC helps this protein to load DnaB in the absence of PriA (121). One such mutant protein, DnaC810, was studied in vitro and shown to load DnaB onto a D-loop structure in the absence of all PriA, PriB, PriC, and DnaT proteins (78). DnaC810 can also bypass the requirement for preprimosome proteins in loading DnaB onto SSB-coated DNA (142). The biochemical properties of the DnaC810 protein thus confirmed the suggestion drawn from genetic data that this mutation confers a gain of function to the DnaC protein.

In contrast to the complete suppression of the poor viability of *priA* null single mutants, the *dnaC809* mutation (same amino acid substitution as *dnaC810*) only partly suppresses the lethality of *priB priC* double mutants or that of the *priA priB priC* triple mutant. It is interesting to note that DnaC810 can bypass the need for PriA, PriB, and PriC proteins to load DnaB on D-loops in vitro, but the equivalent DnaC809 cannot bypass the need for these three proteins in vivo. These results suggest that some substrate other than a D-loop is important for primosome-initiated replication in bacteria, or alternatively, that the presence of recombination proteins in vivo modulates the accessibility of primosomal proteins to

recombination intermediates. An additional mutation in the *dnaC* gene of the *priB priC dnaC809* triple mutant can restore all phenotypes to the wild-type levels (*dnaC809 820* [120]). Therefore, multiple mutations in *dnaC* can lead to additive gains of function. The DnaC809 820 mutant protein restores the viability of *priA priC* double mutants and *priA priB priC* triple mutants and can therefore readily bypass the need for all PriA, PriB, and PriC proteins in vivo (118, 120).

Primosome Assembly at Replication Forks during Bacteriophage Mu Replicative Transposition

Further insight into the chromosomal sites targeted by PriA, and, in turn, into the physiological role of primosome assembly, came from studies of the function of primosome assembly in the physiology of *E. coli* bacteriophage Mu. Phage Mu replicates its DNA by a transposition process that creates a replication forklike structure onto which a replication complex is assembled (reviewed in reference 52). A specific set of host proteins is required to replicate Mu DNA from the transposition intermediate in vitro. These host factors include the preprimosome proteins PriA, PriB, DnaT, and DnaC along with DnaB, DnaG, and DNA Pol III holoenzyme, indicating that primosome assembly leads to the formation of a replication fork (50). Mu replication by transposition was also shown to require functional PriA protein in vivo (50). Interestingly, the replication-forklike structure made by Mu transposase lacks a single-stranded region on the lagging strand template, which is normally required for the loading of DnaB. Jones and Nakai (51) showed that the helicase activity of PriA can promote the unwinding of the lagging strand arm, allowing the loading of DnaB on the exposed single-stranded region. Loss of PriA helicase activity impaired Mu DNA replication in vivo and in vitro. The demonstration that PriA helicase activity could open duplex DNA to allow the entry of DnaB was the first indication of a potentially important function for the PriA helicase activity. It was proposed that concerted helicase activity and primosome assembly could allow replication restart from fully double-stranded forked structures (51; reviewed in reference 52).

Replication Fork Collapse and Replication Restart

Because *priA* mutants are defective for RecABC-mediated homologous recombination (i.e., the recombinational repair of DSBs), the loading of a replisome by PriA-dependent primosome assembly is likely essential for the conversion of recombination intermediates into viable products (60, 121). Results with lambda phage model systems confirmed that recombination intermediates are indeed transformed into replication forks in vivo (70, 100). Although the evolutionary advantage of coupling DSB repair and replication initiation is not obvious when the double-stranded ends are from exogenous DNA (as in P1 transduction or Hfr recombination; but see reference 128), the need for such a direct link is immediately apparent if a substantial proportion of the double-stranded ends formed spontaneously during bacterial growth are at replication forks.

Conversion of a growing replication fork into a double-stranded end has been proposed to occur when the replication fork runs into a single-strand interruption, a process termed collapse (67) (Fig. 6). Direct evidence for replication fork collapse in vivo was recently obtained: Kuzminov (69) showed that a site-specific nick could be converted into a double-stranded end in a process that required DNA replication. In this model system, the site-specific nicks were generated with the M13 gene II nicking protein, which binds tightly to the nick. Further experiments are needed to test whether protein-free and protein-bound nicks are converted into double-stranded ends by the same pathway.

Although homologous recombination is not essential for viability in wild-type E. coli, recombinational repair of DSBs is essential for viability of certain mutants impaired for DNA replication or DNA repair (see references 45 and 67). For example, mutants defective for polymerase I (polA) or ligase (lig) have defects in the closing of Okazaki fragments and require DSB repair proteins (RecA and RecBC) for viability. Such mutants were proposed to suffer an increased level of DSBs resulting from the collapse of replication forks upon encounter with single-strand interruptions left by previous replication forks. polA priA double mutants were found to be lethal (73), in support of the idea that DSB repair is coupled with replication restart in this strain. It was proposed that reincorporation of a "collapsed" chromosome into the homologous chromatid by recombination is followed by PriA-dependent primosome assembly at the recombination intermediate, allowing replication to restart from the region where the fork has been originally disrupted (67, 68) (Fig. 6).

Replication Fork Arrest and Replication Restart

The cause and frequency of spontaneous replication arrests in wild-type cells grown under various conditions are not known. Therefore, the consequences of arresting replication have been studied with mutants defective for a known replication protein. Inactivation of a replicative helicase leads to a set of events that are best explained by a model involving replication fork reversal (RFR) (Fig. 7) (123; reviewed in reference 93). According to the RFR model, replication fork arrest is followed by reannealing of the nascent leading and lagging strands, allowing the template strands to pair, and leading to the formation of a Holliday junction (Fig. 7) (also referred to as replication fork regression by some authors). RFR was proposed to occur upon replication arrest (i) in rep mutants, defective for an accessory replicative helicase thought to remove protein roadblocks from the path of replication forks (19, 123), (ii) in a holD^{G10} mutant, impaired for the Ψ subunit of the holoenzyme polymerase III γ complex (29), (iii) upon a shift to restrictive temperature of dnaB^{ts} or dnaE^{ts} mutants, which carry a conditional mutation in genes encoding the replisome helicase and Pol III catalytic subunit, respectively (123, 124; G. Grompone, M. Seigneur, and B. Michel, unpublished results).

The key features of a reversed fork are the formation of a double-stranded end, a substrate for RecBCD, and the formation of a Holliday junction,

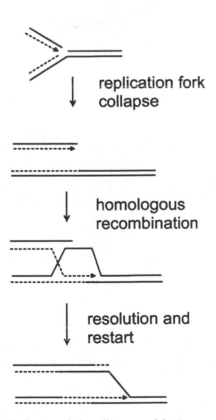

Figure 6. Replication fork collapse model. Encounter of the replication fork with a gap or nick leads to the generation of a double-stranded end. Reincorporation of the broken chromosome end into the homologue sister is mediated by RecBCD-RecA homologous recombination. Resolution of the recombination intermediate and assembly of replication proteins leads to the reconstitution of a replication fork. Adapted from reference 67.

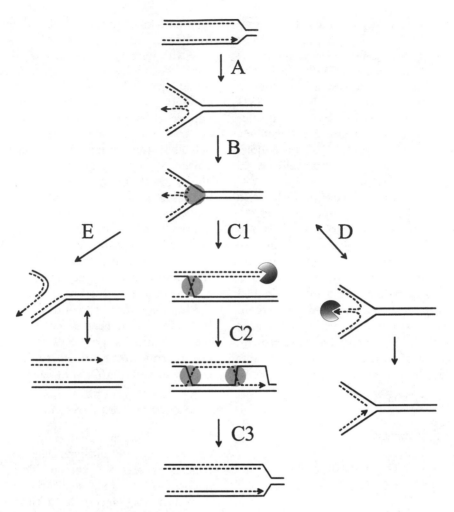

Figure 7. RFR model. In the first step (A), the replication fork is arrested and the two newly synthesized strands anneal, forming a Holliday junction. In the second step (B), the junction is stabilized by RuvAB binding. In recombination-proficient strains, RecBCD binds to the double-stranded end (C1); degradation takes place until the first recognized χ site (χ is an octameric sequence that switches RecBCD from an exonuclease to a recombinase enzyme), which facilitates a genetic exchange mediated by RecA (C2); finally, RuvC resolves Holliday junctions bound by RuvAB (C3). In this manner, recombination enzymes reconstitute the replication fork. Alternatively, RecBCD-mediated degradation of the tail progresses up to the RuvAB-bound Holliday junction (D). Replication can restart when RecBCD has displaced the RuvAB complex. This pathway can take place in recombination-proficient strains if RecBCD reaches RuvAB before encountering a χ site; it is the only pathway that leads to a viable chromosome in *recA* and *ruvC* mutants. In the absence of RecBCD, RuvC resolves the RuvAB-bound Holliday junction, which leads to the RuvABC-dependent DSBs observed in *recBC* mutants (E). The gray circles represent RuvAB, and the partial circles represent RecBCD. Continuous and discontinuous lines represent the template and the newly synthesized strand of the chromosome, respectively; the arrowheads indicate the 3' end of the growing strands. Adapted from references 93 and 123.

a substrate for the RuvABC complex (reviewed in references 68 and 137). In cells deficient for RecBCD, the extruded double-stranded end is not acted upon. Instead, the Holliday junction is resolved by RuvABC, resulting in the formation of a broken chromosome arm, identical to the product of a direct breakage of the replication fork (Fig. 7E). Indeed, RuvABC-dependent breakage of chromosomes could be directly detected by pulse field gel electrophoresis in *rep*, *holD^{G10}*, *dnaB^{ts}*, and *dnaE^{ts}* mutant strains lacking RecBC (reviewed in references 93 and 94).

When RFR occurs in cells proficient for homologous recombination, reincorporation of the double-stranded end into the chromosome leads to the formation of a recombination intermediate, substrate for PriA-dependent primosome assembly and replication restart (Fig. 7C). Formation of such a recombination intermediate may be the major pathway for the reconstitution of a replication fork after reversal. However, both RecA and RecBC are required for this reaction, while only RecBC is essential for viability in *holD^{G10}* and *rep* mutants. Because *rep recA* and *holD*^{G10} *recA* double mutants are viable, reversed forks can apparently be processed to replicating chromosomes by an alternative, nonrecombinogenic pathway. The exonuclease V activity of RecBCD is essential in *rep recA* and *holD*^{G10} *recA* double mutants, suggesting that degradation of the double-stranded tail formed by RFR creates a chromosome on which replication can restart (Fig. 7D) (29, 123). Formation of the Holliday junction at blocked forks requires RecA in the *dnaB*^{ts} mutant, whereas it occurs in the absence of RecA in *rep* and DNA polymerase III mutants (29, 124; G. Grompone, M. Seigneur, and B. Michel, unpublished results). The mechanism of RFR upon replication arrest in *rep* and *pol III* mutants remains to be elucidated.

In conclusion, the RFR reaction can promote the reactivation of a disabled replication fork in at least two ways: reincorporation of the reversed fork by homologous recombination or degradation of the double-stranded tail. Both of these presumably allow replication to resume without actual breakage of the DNA. As expected, PriA-dependent primosome assembly is involved in replication restart after fork reversal since PriA is required for the viability of *rep* and *holD*^{G10} mutants (118, 123; M. J. Flores, M. Petranovic, and B. Michel, unpublished results).

Replication Fork Reversal In Vitro

Possible mechanisms for the branch migration reaction in RFR have been revealed by in vitro experiments. More than 2 decades ago, RFR was shown to occur spontaneously when DNA was extracted from cells under certain conditions (107, 132). More recently, positive supercoiling was shown to promote in vitro RFR, which is sensible because RFR allows rewinding of the parental strands and thereby relieves positive supercoiling (114). However, actively replicating DNA within the cell should generally be negatively supercoiled, and negative superhelicity should strongly inhibit RFR without a protein catalyst.

Purified proteins, notably RecA and RecG, can promote RFR with certain substrates. In substrates that were not topologically constrained and that contained a single-strand gap on the leading strand, RecA protein catalyzed extrusion of the lagging strand product as a single strand and then full-fork reversal through a Holliday junction intermediate (115). The RecG helicase was likewise reported to promote RFR, even when the template DNA was negatively supercoiled (91; also see reference 89). RecG strongly prefers forks with a single-strand gap on the leading strand; this and other results argue for a special role of RecG in the replication of damaged DNA (89, 127) (see below).

DNA Lesions and Replication Restart

In addition to single-strand interruptions and protein roadblocks that may impair the progression of replication forks, one of the main causes of replication arrest or collapse is likely to be the encounter of replication forks with DNA lesions. The arrest and restart of replication forks at DNA damage may promote survival by providing a window of time in which DNA repair mechanisms (e.g., excision repair) can correct the damaged site before any DNA polymerase replicates through the damage.

DNA damage on the lagging strand template presumably can block the lagging strand DNA polymerase and thereby leave a gap. However, because lagging strand replication can be continually reprimed with the replisome-associated primase, the fork may continue unabated and simply leave a gap in the daughter molecule. Damage on the leading strand template is a more serious issue a priori, because there is no obvious mechanism for repriming the leading strand after a blocking DNA lesion. The lagging strand 5′ end was proposed to progress past the leading strand 3′ end upon leading strand blockage, creating a fork with a single-strand gap on the leading strand arm (43) (Fig. 8). As mentioned above, this is a preferred substrate for RecG. Furthermore, a role of RecG in reversal of such forks might play an important role in the UV sensitivity of *recG* mutants (38, 88).

RFR provides at least two possible mechanisms that might overcome leading strand template damage. One possibility, originally proposed by Higgins et al. (43), involves template switching, with the leading strand product being extended by DNA polymerase using the (extruded) lagging strand product as template (Fig. 8, left). Readvance of the reversed fork results in the leading strand product now extending past the site of damage. The extension reaction has been correctly templated (and thereby generally mutation free), and the cell has gained a long period of time to effect repair of the original DNA damage. Since *E. coli* has potent exonucleases that act upon single-stranded and double-stranded DNA ends, this pathway would require careful coordination between the RFR reaction and the polymerase extension reaction. The second possibility is that RFR moves the lesion back into double-stranded parental DNA, allowing efficient repair by pathways that require a duplex template (e.g., excision repair pathways [Fig. 8, right]) (24, 38, 94).

In aerobically grown bacteria without exogenous DNA-damaging agents, the main source of DNA damage is oxidative stress. Bacteria possess a battery of repair pathways that rapidly remove oxidative

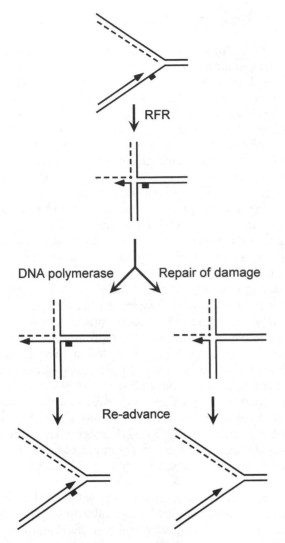

Figure 8. Two models for the role of RFR in response to DNA damage. When the fork is arrested by damage on the leading strand, the lagging strand polymerase may replicate somewhat beyond the region of the damage. RFR can thereby extrude the two product strands, moving the site of DNA damage back into parental duplex DNA. In the model on the left, DNA polymerase extends the nascent leading strand product using the nascent lagging strand as template (43). Upon readvance of the fork by branch migration, the leading strand product has been extended past the damage, allowing the replication fork to be reestablished. In the model on the right, excision repair (or other conventional repair pathways) corrects the leading strand template lesion (see references 24 and 94). This repair is possible only after the lesion has branch migrated back into duplex parental DNA via the RFR reaction. The replication fork can be reestablished after the fork is readvanced by branch migration.

lesions, as well as lesions caused by exogenous DNA-damaging agents (32). Nevertheless, in conditions of increased levels of DNA damage, replication forks do encounter unrepaired damages, and such encounters are thought to block fork progression. The importance of replication restart in response to DNA damage is clearly indicated by the fact that *priA* mutants are sensitive to UV irradiation, γ irradiation, and mitomycin C (60, 121). The UV sensitivity of *priA* mutants is suppressed by expression of the helicase-deficient PriA300 protein, indicating that only primosome assembly is required for the formation of viable chromosomes after UV irradiation (121). Primosome assembly is expected to be required for replication restart when lesions cause the formation of DSBs, which are repaired by RecBCD and RecA. Indeed, DSB recombinational repair plays an important role in the repair of both oxidative and UV lesions (reviewed in reference 68), which may be the reason for the sensitivity of *priA* mutants to DNA-damaging agents.

Whether a single UV lesion leads to the complete disassembly of the replisome and therefore to a need for PriA-dependent reloading of the replicative helicase is unknown. Translesion DNA synthesis in UV-irradiated cells is catalyzed by specific DNA polymerases (reviewed in references 35 and 36). PriA has been shown to be required for rapid resumption of replication following UV irradiation (M. Goodman, unpublished results [cited in reference 35]), suggesting that loss of DnaB may occur. Primosome assembly may also be important after a protein bound to a DNA lesion blocks the fork and thereby causes the disassembly of the replisome and its constituent helicase (88). At present, we know very little about the precise and possibly multiple roles of PriA-dependent replication restart when replication forks are arrested by DNA damage. The DNA repair machinery of the cell is likely to be closely coupled to the process of RFR in particular and replication restart in general.

THE PRIMOSOME PROTEINS OF *B. SUBTILIS*

PriA homologues are found in several unrelated bacteria, suggesting a high conservation of PriA structure and function (52). In vivo characterization of *priA* mutants in the gram-positive organism *B. subtilis* and in vitro studies of the PriA$_{Bs}$ protein revealed that the *E. coli* and *B. subtilis* enzymes share several properties, in agreement with the observation that the two proteins are 32% identical (113). In contrast, other primosome proteins are apparently not conserved; detailed studies of primosome assembly in *B. subtilis* revealed that different bacteria may indeed have evolved different ways of promoting replication restart (13, 83).

Purified *B. subtilis* PriA$_{Bs}$ protein recognizes two specific DNA substrates (113). First, PriA$_{Bs}$ binds to a specific site named *ssiA* (single-strand initiation site) identified on plasmids as required for the initiation of lagging strand synthesis on theta and rolling-circle replicating plasmids (12). Second, PriA$_{Bs}$ binds specifically to D-loop structures made with oligonucleotides (113). The similarity of target structures for the PriA proteins of *B. subtilis* and *E. coli* led to the proposal of a conserved function for these enzymes. *B. subtilis priA* mutants are poorly viable and sensitive to rich medium, indicating that replication restart is an important function in *B. subtilis*, as it is in *E. coli* (113). In *B. subtilis*, *priA* mutations also increase sensitivity to UV irradiation, suggesting that PriA$_{Bs}$ is required for homologous recombination. However, despite the similarity of function, expressing the *B. subtilis* PriA$_{Bs}$ protein did not complement the defects of a *priA E. coli* mutant (113), and the subsequent steps of primosome assembly turned out to be promoted in a species-specific way.

No clear homologues of the *E. coli* PriB, PriC, DnaT, and DnaC proteins are found in *B. subtilis*, although weak similarities between *E. coli* DnaC and *B. subtilis* DnaI$_{Bs}$ proteins have been reported (61). The different steps of primosome assembly, from PriA recognition of its target to loading of the replisome helicase, are therefore species specific. (The homologue of the *E. coli* DnaB helicase is named DnaC in *B. subtilis*; to avoid confusion, it is referred to here as the replisome helicase). The identification of *B. subtilis* primosome proteins originated from studies of extrachromosomal elements. Three *B. subtilis* replication proteins—DnaB$_{Bs}$, DnaD$_{Bs}$, and DnaI$_{Bs}$—were proposed to be required for primosome assembly based on their requirement for efficient synthesis of the complementary strand on *ssiA*-carrying plasmids (12). Interestingly, all three genes are essential for viability. Mutants thermosensitive for growth were shown to be impaired in the initiation of replication at *B. subtilis oriC* (39, 47a, 54), indicating an essential role for these proteins in the loading of the replisome in DnaA$_{Bs}$-dependent replication initiation.

Detailed biochemical studies demonstrated the sequential binding of PriA$_{Bs}$, DnaD$_{Bs}$, and DnaB$_{Bs}$ to a forked DNA molecule (83). Furthermore, the same sequential binding reaction could be reconstituted on D-loop structures made with oligonucleotides (83). The preferred DNA substrate for this assembly in vitro, a D-loop, mimics an arrested replication fork with an unreplicated lagging strand region, or a recombination intermediate. It was proposed that the assembly of PriA$_{Bs}$, DnaD$_{Bs}$, and DnaB$_{Bs}$ promotes the loading of the replicative helicase in vivo, with the help of DnaI$_{Bs}$. This model was further supported by the observation that mutations that suppress the poor viability of *priA B. subtilis* null mutants lie in the

dnaB$_{Bs}$ gene. Several mutations suppressing the poor growth of *priA$_{Bs}$* mutants were isolated, and all were located in the 3′ end of the *dnaB$_{Bs}$* gene (13). One such mutant, *dnaB75*, was studied more thoroughly. In addition to restoring wild-type levels of viability and UV resistance to *priA$_{Bs}$* mutants, the *dnaB75* mutation allowed replication initiation at *ssiA* plasmid sites independently from *priA$_{Bs}$*. Altogether, these results led to the conclusion that, in addition to their role in *oriC* replication initiation, DnaD$_{Bs}$, DnaB$_{Bs}$, and DnaI$_{Bs}$ proteins are essential for PriA-dependent primosome assembly at *ssiA* and D-loops in vivo (13, 83, 113). In addition, the requirement for these three proteins is increased by *priA* inactivation, suggesting that they may allow replication restart in the absence of PriA, a model supported by in vitro observations (13, 83). Therefore, in *B. subtilis* as in *E. coli*, the viability of *priA* null mutants may rely on the functionality of an alternative, poorly efficient, replication restart pathway.

In summary, replication restart in *B. subtilis* appears to use the same preprimosome proteins that load the replicative helicase in DnaA-dependent replication, which is quite different from the situation in *E. coli*, where restart-specific proteins are used. At present, we do not know whether the *E. coli* paradigm or the *B. subtilis* paradigm for restart proteins is more common throughout the bacterial world.

DISCUSSION

Comparing processing of arrested replication forks in T4 and *E. coli* reveals striking similarities and some interesting possible differences. Both phage and bacteria clearly restart replication forks from recombination intermediates. This reaction is an intrinsic constituent of phage propagation, and, consequently, T4 recombination mutants are quite profoundly deficient in DNA replication. In contrast, homologous recombination per se is not essential in *E. coli*. Nonetheless, growth of *E. coli* does require fork restart, which occurs by both recombination-dependent and -independent pathways.

The differing specificities of the Holliday junction-resolving enzymes of phage T4 and *E. coli* might provide a hint to explain the fact that phage T4 relies more heavily on recombination to restart replication forks. Like most phage enzymes that cleave Holliday junctions, T4 endonuclease VII is able to cleave simple forked structures, including arrested replication forks (see above). In contrast, Holliday junction resolvases in bacteria do not cleave forklike structures (125). As described above, T4 replication forks blocked at sites of DNA damage are apparently cleaved by endonuclease VII, generating double-stranded ends. Since T4 is very proficient at RDR initiated from ends, and does not need to carefully regulate chromosome copy number, cleavage and recombinational restart may be perfectly adequate for processing blocked forks. In contrast, in wild-type bacteria, the absence of a fork-debranching activity ensures that blocked forks escape direct breakage. Instead, blocked forks can be rescued either by direct action of primosome loading proteins, or by RFR coupled to a silent recombination event of the extruded strand with the parental template ahead of the fork (Fig. 7, step C2). The major advantage of RFR is that the double-stranded end remains covalently linked to the homologous chromosome. The intramolecular recombination reaction promoted by RFR may thus provide an important advantage in all organisms that carry repeated sequences, in which a free broken chromatid is a potential source of gross rearrangements.

In addition to replication restart from recombination intermediates, replication in bacteria can restart from forked structures that have not recombined. The existence of recombination-independent replication restart is based on the observation that *priA* inactivation affects the viability of *recA* single mutants and is lethal in *recA* derivatives of cells that undergo RFR at high frequency. Recombination-independent restart in bacteria correlates with the property of PriA to unwind the lagging strand, hence to expose the single-stranded DNA on which the replicative helicase is reloaded. The phage-encoded helicase loading protein gp59 does not carry this activity. This major difference between PriA and gp59 may again be related to the propensity of T4 to restart replication from recombination intermediates, i.e., D-loops created when a broken end invades homologous DNA. It is not clear whether T4 is even capable of restarting replication forks without using this recombinational pathway. Since in most organisms homologous recombination is not essential for viability, it can be speculated that replication restart from a nonrecombined structure is a general phenomenon. This in turn predicts that the enzymatic machinery that promotes restart of arrested forks in these organisms will generally have helicase activity to remove the 5' end of the nascent lagging strand.

Several pathways for restarting forks have been characterized in bacteria, and inactivating all of them simultaneously is lethal, i.e., prevents colony formation. This observation suggests that spontaneous replication arrest with replisome disassembly occurs quite frequently, on the order of once per DNA replication cycle (in *E. coli*, once per 4 Mb). On the other hand, recent experiments analyzing DNA content in synchronized cells using flow cytometry suggests that DnaC-dependent replication restart is needed to complete about 20% of *E. coli* chromosome replication cycles (80). The direct analysis of replication fork arrest and replication restart within the chromosomes of wild-type cells has just begun, and we can anticipate that the frequency of arrest may vary during the cell cycle, in different growth conditions, and between different bacteria. Nevertheless, the ubiquitous nature of PriA in the bacterial world suggests that replication restart is an essential process in all prokaryotic organisms.

REFERENCES

1. **Allen, G. C., and A. Kornberg.** 1991. The *priB* gene encoding the primosomal replication N-protein of *Escherichia coli*. *J. Biol. Chem.* **266:**11610–11613.
2. **Appasani, K., D. S. Thaler, and E. B. Goldberg.** 1999. Bacteriophage T4 gp2 interferes with cell viability and with bacteriophage lambda red recombination. *J. Bacteriol.* **181:**1352–1355.
3. **Asai, T., and T. Kogoma.** 1994. D-loops and R-loops: alternative mechanisms for the initiation of chromosome replication in *Escherichia coli*. *J. Bacteriol.* **176:**1807–1812.
4. **Asai, T., S. Sommer, A. Bailone, and T. Kogoma.** 1993. Homologous recombination-dependent initiation of DNA replication from DNA damage-inducible origins in *Escherichia coli*. *EMBO J.* **12:**3287–3295.
5. **Barcena, M., T. Ruiz, L. E. Donate, S. E. Brown, N. E. Dixon, W. Radermacher, and J. M. Carazo.** 2001. The DnaB-DnaC complex: a structure based on dimers assembled around an occluded channel. *EMBO J.* **20:**1462–1468.
6. **Barry, J., and B. Alberts.** 1994. Purification and characterization of bacteriophage T4 gene 59 protein. A DNA helicase assembly protein involved in DNA replication. *J. Biol. Chem.* **269:**33049–33062.
7. **Beernink, H. T., and S. W. Morrical.** 1999. RMPs: recombination/replication mediator proteins. *Trends Biochem. Sci.* **24:**385–389.
8. **Bernstein, C.** 1981. Deoxyribonucleic acid repair in bacteriophage. *Microbiol. Rev.* **45:**72–98.
9. **Bishop, A. J. R., and R. H. Schiestl.** 2000. Homologous recombination as a mechanism for genome rearrangements: environmental and genetic effects. *Hum. Mol. Genet.* **9:**2427–2434.
10. **Black, L. W., and M. K. Showe.** 1983. Morphogenesis of the T4 head, p. 219–245. *In* C. K. Mathews, E. M. Kutter, G. Mosig, and P. B. Berget (ed.), *Bacteriophage T4.* American Society for Microbiology, Washington, D.C.
11. **Bleuit, J. S., H. Xu, Y. Ma, T. Wang, J. Liu, and S. W. Morrical.** 2001. Mediator proteins orchestrate enzyme-ssDNA assembly during T4 recombination-dependent DNA replication and repair. *Proc. Natl. Acad. Sci. USA* **98:**8298–8305.
12. **Bruand, C., S. D. Ehrlich, and L. Jannière.** 1995. Primosome assembly site in *Bacillus subtilis. EMBO J.* **14:**2642–2650.
13. **Bruand, C., M. Farache, S. McGovern, S. D. Ehrlich, and P. Polard.** 2001 DnaB, DnaD and DnaI proteins are components of the *Bacillus subtilis* replication restart primosome. *Mol. Microbiol.* **42:**245–255.
14. **Bull, H. J., M. J. Lombardo, and S. M. Rosenberg.** 2001. Stationary-phase mutation in the bacterial chromosome:

recombination protein and DNA polymerase IV dependence. *Proc. Natl. Acad. Sci. USA* **98:**8334–8341.

15. **Campbell, D. A.** 1969. *On the Mechanism of the Recombinant Increase in X-Irradiated Bacteriophage T4D.* Ph.D. thesis. University of Washington, Seattle.

16. **Carles-Kinch, K., J. W. George, and K. N. Kreuzer.** 1997. Bacteriophage T4 UvsW protein is a helicase involved in recombination, repair, and the regulation of DNA replication origins. *EMBO J.* **16:**4142–4151.

17. **Chakraverty, R. K., and I. D. Hickson.** 1999. Defending genome integrity during DNA replication: a proposed role for RecQ family helicases. *Bioessays* **21:**286–294.

18. **Clyman, J., S. Quirk, and M. Belfort.** 1994. Mobile introns in the T-even phages, p. 83–88. *In* J. D. Karam (ed. in chief), *Molecular Biology of Bacteriophage T4.* ASM Press, Washington, D.C.

19. **Colasanti, J., and D. T. Denhardt.** 1987. The *Escherichia coli rep* mutation. Consequences of increased and decreased Rep protein levels. *Mol. Gen. Genet.* **209:**382–390.

20. **Cox, M. M.** 2001. Recombinational repair of damaged replication forks in *Escherichia coli. Annu. Rev. Genet.* **35:**53–82.

21. **Cunningham, R. P., and H. Berger.** 1977. Mutations affecting genetic recombination in bacteriophage T4D. I. Pathway analysis. *Virology* **80:**67–82.

22. **Derr, L. K., and K. N. Kreuzer.** 1990. Expression and function of the *uvsW* gene of bacteriophage T4. *J. Mol. Biol.* **214:**643–656.

23. **Doan, P. L., K. G. Belanger, and K. N. Kreuzer.** 2001. Two types of recombination hotspots in bacteriophage T4: one requires DNA damage and a replication origin and the other does not. *Genetics* **157:**1077–1087.

24. **Doe, C. L., J. Dixon, F. Osman, and M. C. Whitby.** 2000. Partial suppression of the fission yeast rqh1(-) phenotype by expression of the bacterial Holliday junction resolvase. *EMBO J.* **19:**2751–2762.

25. **Dudas, K. C., and K. N. Kreuzer.** 2001. UvsW protein regulates bacteriophage T4 origin-dependent replication by unwinding R-loops. *Mol. Cell. Biol.* **21:**2706–2715.

26. **Edgar, R. S., G. H. Denhardt, and R. H. Epstein.** 1964. A comparative study of conditional lethal mutations of bacteriophage T4D. *Genetics* **49:**635–648.

27. **Ferguson, D. O., and W. K. Holloman.** 1996. Recombinational repair of gaps in DNA is asymmetric in *Ustilago maydis* and can be explained by a migrating D-loop model. *Proc. Natl. Acad. Sci. USA* **93:**5419–5424.

28. **Flores, M. J., S. D. Ehrlich, and B. Michel.** 2002. Primosome assembly requirement for replication restart in the *Escherichia coli holD*^{G10} replication mutant. *Mol. Microbiol.* **44:**783–792.

29. **Flores, M. J., H. Bierne, S. D. Ehrlich, and B. Michel.** 2001. Impairment of lagging strand synthesis triggers the formation of a RuvABC substrate at replication forks. *EMBO J.* **20:**619–629.

30. **Formosa, T., and B. M. Alberts.** 1986. Purification and characterization of the T4 bacteriophage uvsX protein. *J. Biol. Chem.* **261:**6107–6118.

31. **Formosa, T., and B. M. Alberts.** 1986. DNA synthesis dependent on genetic recombination: characterization of a reaction catalyzed by purified T4 proteins. *Cell* **47:**793–806.

32. **Friedberg, E. C., G. C. Walker, and W. Siede.** 1995. *DNA Repair and Mutagenesis.* ASM Press, Washington, D.C.

33. **George, J. W., and K. N. Kreuzer.** 1996. Repair of double-strand breaks in bacteriophage T4 by a mechanism that involves extensive DNA replication. *Genetics* **143:**1507–1520.

34. **George, J. W., B. A. Stohr, D. J. Tomso, and K. N. Kreuzer.** 2001. The tight linkage between DNA replication and double-strand break repair in bacteriophage T4. *Proc. Natl. Acad. Sci. USA* **98:**8290–8297.

35. **Goodman, M. F.** 2000. Coping with replication "train wrecks" in *Escherichia coli* using Pol V, Pol II and RecA proteins. *Trends Biochem. Sci.* **25:**189–195.

36. **Goodman, M. F., and B. Tippin.** 2000. Sloppier copier DNA polymerases involved in genome repair. *Curr. Opin. Genet. Dev.* **10:**162–168.

37. **Gorbalenya, A. E., and E. V. Koonin.** 1993. Helicases: amino acid sequence comparisons and structure-function relationships. *Curr. Opin. Struct. Biol.* **3:**419–429.

38. **Gregg, A. V., P. McGlynn, R. P. Jaktaji, and R. G. Lloyd.** 2002. Direct rescue of stalled replication forks via the combined action of PriA and RecG helicase activities. *Mol. Cell* **9:**241–251.

39. **Gross, J. D., D. Karamata, and P. G. Hempstead.** 1968. Temperature sensitive mutants of *Bacillus subtilis* defective in DNA synthesis. *Cold Spring Harbor Symp. Quant. Biol.* **33:**307–312.

40. **Gruber, M., R. E. Wellinger, and J. M. Sogo.** 2000. Architecture of the replication fork stalled at the 3′ end of yeast ribosomal genes. *Mol. Cell. Biol.* **20:**5777–5787.

41. **Hamlett, N. V., and H. Berger.** 1975. Mutations altering genetic recombination and repair of DNA in bacteriophage T4. *Virology* **63:**539–567.

42. **Harris, L. D., and J. D. Griffith.** 1989. UvsY protein of bacteriophage T4 is an accessory protein for in vitro catalysis of strand exchange. *J. Mol. Biol.* **206:**19–27.

43. **Higgins, N. P., K. Kato, and B. Strauss.** 1976. A model for replication repair in mammalian cells. *J. Mol. Biol.* **101:**417–425.

44. **Hong, G., and K. N. Kreuzer.** 2000. An antitumor drug-induced topoisomerase cleavage complex blocks a bacteriophage T4 replication fork in vivo. *Mol. Cell. Biol.* **20:**594–603.

45. **Horiuchi, T., and Y. Fujimura.** 1995. Recombinational rescue of the stalled DNA replication fork: a model based on analysis of an *Escherichia coli* strain with a chromosome region difficult to replicate. *J. Bacteriol.* **177:**783–791.

46. **Hosoda, J., E. Mathews, and B. Jansen.** 1971. Role of genes 46 and 47 in bacteriophage T4 reproduction. *J. Virol.* **8:**372–387.

47. **Hyrien, O.** 2000. Mechanisms and consequences of replication fork arrest. *Biochimie* **82:**5–17.

47a. **Imai, Y., N. Ogasawara, D. Ishigo-oka, R. Kadoya, T. Daito, and S. Moriya.** 2000. Subcellular localization of DNA-initiation proteins in *Bacillus subtilis*: evidence that chromosome replication begins at either edge of the nucleoids. *Mol. Microbiol.* **36:**1037–1048.

48. **Jones, C. E., T. C. Mueser, K. C. Dudas, K. N. Kreuzer, and N. G. Nossal.** 2001. Bacteriophage T4 gene 41 helicase and gene 59 helicase-loading protein: a versatile couple with roles in replication and recombination. *Proc. Natl. Acad. Sci. USA* **98:**8312–8318.

49. **Jones, C. E., T. C. Mueser, and N. G. Nossal.** 2000. Interaction of the bacteriophage T4 gene 59 helicase loading protein and gene 41 helicase with each other, and with fork, flap, and cruciform DNA. *J. Biol. Chem.* **275:**27145–27154.

50. **Jones, J. M., and H. Nakai.** 1997. The φX174-type primosome promotes replisome assembly at the site of recombination in bacteriophage Mu transposition. *EMBO J.* **16:**6886–6895.

51. **Jones, J. M., and H. Nakai.** 1999. Duplex opening by primosome protein PriA for replisome assembly on a recombination intermediate. *J. Mol. Biol.* **289:**503–516.

52. Jones, J. M., and H. Nakai. 2000. PriA and phage T4 gp59: factors that promote DNA replication on forked substrates. *Mol. Microbiol.* **36:**519–527.

53. Jones, J. M., and H. Nakai. 2001. *Escherichia coli* PriA helicase: fork binding orients the helicase to unwind the lagging strand side of arrested replication forks. *J. Mol. Biol.* **312:** 935–947.

54. Karamata, D., and J. D. Gross. 1970. Isolation and genetic analysis of temperature sensitive mutants of *B. subtilis* defective in DNA synthesis. *Mol. Gen. Genet.* **108:**277–287.

55. Kelman, Z., and M. O' Donnell. 1995. DNA polymerase III holoenzyme: structure and function of a chromosomal replicating machine. *Annu. Rev. Biochem.* **64:**171–200.

56. Kim, S., H. G. Dallmann, C. S. McHenry, and K. J. Marians. 1996. Coupling of a replicative polymerase and helicase: a tau-DnaB interaction mediates rapid replication fork movement. *Cell* **84:**643–650.

57. Kodadek, T., D. C. Gan, and K. Stemke-Hale. 1989. The phage T4 uvsY recombination protein stabilizes presynaptic filaments. *J. Biol. Chem.* **264:**16451–16457.

58. Kogoma, T. 1996. Recombination by replication. *Cell* **85:**625–627.

59. Kogoma, T. 1997. Stable DNA replication: interplay between DNA replication, homologous recombination, and transcription. *Microbiol. Mol. Biol. Rev.* **61:**212–238.

60. Kogoma, T., G. W. Cadwell, K. G. Barnard, and T. Asai. 1996. The DNA replication priming protein, PriA, is required for homologous recombination and double-strand break repair. *J. Bacteriol.* **178:**1258–1264.

61. Koonin, E. V. 1992. A new group of putative RNA helicases. *Trends Biochem. Sci.* **17:**495–497.

62. Kornberg, R. D., and T. Baker. 1992. *DNA Replication.* W. H. Freeman & Co., New York, N.Y.

63. Kowalczykowski, S. C. 2000. Initiation of genetic recombination and recombination-dependent replication. *Trends Biochem. Sci.* **25:**156–165.

64. Kreuzer, K. N., and S. W. Morrical. 1994. Initiation of DNA replication, p. 28–42. *In* J. D. Karam (ed. in chief), *Molecular Biology of Bacteriophage T4.* ASM Press, Washington, D.C.

65. Kreuzer, K. N., and J. W. Drake. 1994. Repair of lethal DNA damage, p. 89–97. *In* J. D. Karam (ed. in chief), *Molecular Biology of Bacteriophage T4.* ASM Press, Washington, D.C.

66. Kreuzer, K. N. 2000. Recombination-dependent DNA replication in phage T4. *Trends Biochem. Sci.* **25:**165–173.

67. Kuzminov, A. 1995. Collapse and repair of replication forks in *Escherichia coli. Mol. Microbiol.* **16:**373–384.

68. Kuzminov, A. 1999. Recombinational repair of DNA damage in *Escherichia coli* and bacteriophage lambda. *Microbiol. Mol. Biol. Rev.* **63:**751–813.

69. Kuzminov, A. 2001. Single-strand interruptions in replicating chromosomes cause double-strand breaks. *Proc. Natl. Acad. Sci. USA* **98:**8241–8246.

70. Kuzminov, A., and F. W. Stahl. 1999. Double-strand end repair via the RecBC pathway in *Escherichia coli* primes DNA replication. *Genes Dev.* **13:**345–356.

71. Lark, C. A., J. Riazi, and K. G. Lark. 1978. *dnaT,* dominant conditional-lethal mutation affecting DNA replication in *Escherichia coli. J. Bacteriol.* **136:**1008–1017.

72. Lee, D. G., and S. P. Bell. 2000. ATPase switches controlling DNA replication initiation. *Curr. Opin. Cell Biol.* **12:**280–285.

73. Lee, E. H., and A. Kornberg. 1991. Replication deficiencies in *priA* mutants of *Escherichia coli* lacking the primosomal replication n'-protein. *Proc. Natl. Acad. Sci. USA* **88:**3029–3032.

74. Lee, E. H., H. Masai, G. C. Allen, and A. Kornberg. 1990. The *priA* gene encoding the primosomal replicative n' protein

75. of *Escherichia coli. Proc. Natl. Acad. Sci. USA* **87:**4620–4624.

75. Levy, J. N. 1975. Effects of radiophosphorus decay in bacteriophage T4D. II. The mechanism of marker rescue. *Virology* **68:**14–26.

76. Liu, J., and K. J. Marians. 1999. PriA-directed assembly of a primosome on D loop DNA. *J. Biol. Chem.* **274:**25033–25041.

77. Liu, J., P. Nurse, and K. J. Marians. 1996. The ordered assembly of the φX174-type primosome. 3. PriB facilitates complex formation between PriA and DnaT. *J. Biol. Chem.* **271:**15656–15661.

78. Liu, J. I., L. W. Xu, S. J. Sandler, and K. J. Marians. 1999. Replication fork assembly at recombination intermediates is required for bacterial growth. *Proc. Natl. Acad. Sci. USA* **96:**3552–3555.

79. Luria, S. 1947. Reactivation of irradiated bacteriophage by transfer of self-reproducing units. *Proc. Natl. Acad. Sci. USA* **33:**253–264.

80. Maisnier-Patin, S., K. Nordstrom, and S. Dasgupta. 2001. Replication arrests during a single round of replication of the Escherichia coli chromosome in the absence of DnaC activity. *Mol. Microbiol.* **42:**1371–1382.

81. Marians, K. J. 1992. Prokaryotic DNA replication. *Annu. Rev. Biochem.* **61:**673–719.

82. Marians, K. J. 2000. PriA-directed replication fork restart in *Escherichia coli. Trends Biochem. Sci.* **25:**185–189.

83. Marsin, S., S. McGovern, S. D. Ehrlich, C. Bruand, and P. Polard. 2001. Early steps of *Bacillus subtilis* primosome assembly. *J. Biol. Chem.* **276:**45818–45825.

84. Masai, H., and K. Arai. 1988. Operon structure of *dnaT* and *dnaC* genes essential for normal and stable DNA replication of *Escherichia coli* chromosome. *J. Biol. Chem.* **263:**15083–15093.

85. Masai, H., T. Asai, Y. Kubota, K. Arai, and T. Kogoma. 1994. *Escherichia coli* PriA protein is essential for inducible and constitutive stable DNA replication. *EMBO J.* **13:**5338–5345.

86. Masai, H., M. W. Bond, and K. Arai. 1986. Cloning of the *Escherichia coli* gene for primosomal protein i: the relationship to *dnaT,* essential for chromosomal DNA replication. *Proc. Natl. Acad. Sci. USA* **83:**1256–1260.

87. McCool, J. D., and S. J. Sandler. 2001. Effects of mutations involving cell division, recombination, and chromosome dimer resolution on a *priA2:kan* mutant. *Proc. Natl. Acad. Sci. USA* **98:**8203–8210.

88. McGlynn, P., and R. G. Lloyd. 2000. Modulation of RNA polymerase by (P)ppGpp reveals a RecG-dependent mechanism for replication fork progression. *Cell* **101:**35–45.

89. McGlynn, P., and R. G. Lloyd. 2001. Rescue of stalled replication forks by RecG: simultaneous translocation on the leading and lagging strand templates supports an active DNA unwinding model of fork reversal and Holliday junction formation. *Proc. Natl. Acad. Sci. USA* **98:**8227–8234.

90. McGlynn, P., A. A. Al-Deib, J. Liu, K. J. Marians, and R. G. Lloyd. 1997. The DNA replication protein PriA and the recombination protein RecG bind D-loops. *J. Mol. Biol.* **270:**212–221.

91. McGlynn, P., R. G. Lloyd, and K. J. Marians. 2001. Formation of Holliday junctions by regression of nascent DNA in intermediates containing stalled replication forks: RecG stimulates regression even when the DNA is negatively supercoiled. *Proc. Natl. Acad. Sci. USA* **98:**8235–8240.

92. Melamede, R. J., and S. S. Wallace. 1977. Properties of the nonlethal recombinational repair *x* and *y* mutants of bacteriophage T4. II. DNA synthesis. *J. Virol.* **24:**28–40.

93. Michel, B. 2000. Replication fork arrest and DNA recombination. *Trends Biochem. Sci.* **25**:173–178.

94. Michel, B., M. J. Flores, E. Viguera, G. Grompone, M. Seigneur, and V. Bidnenko. 2001. Rescue of arrested replication forks by homologous recombination. *Proc. Natl. Acad. Sci. USA* **98**:8181–8188.

95. Mickelson, C., and J. S. Wiberg. 1981. Membrane-associated DNase activity controlled by genes 46 and 47 of bacteriophage T4D and elevated DNase activity associated with the T4 *das* mutation. *J. Virol.* **40**:65–77.

96. Mizuuchi, K., B. Kemper, J. Hays, and R. A. Weisberg. 1982. T4 endonuclease VII cleaves Holliday structures. *Cell* **29**:357–365.

97. Mosig, G. 1983. Relationship of T4 DNA replication and recombination, p. 120–130. *In* C. K. Mathews, E. M. Kutter, G. Mosig, and P. B. Berget (ed.), *Bacteriophage T4*. American Society for Microbiology, Washington, D.C.

98. Mosig, G. 1994. Homologous recombination, p. 54–82. *In* J. D. Karam (ed. in chief), *Molecular Biology of Bacteriophage T4*. ASM Press, Washington, D.C.

99. Mosig, G., J. Gewin, A. Luder, N. Colowick, and D. Vo. 2001. Two recombination-dependent DNA replication pathways of bacteriophage T4, and their roles in mutagenesis and horizontal gene transfer. *Proc. Natl. Acad. Sci. USA* **98**:8306–8311.

100. Motamedi, M. R., S. K. Szigety, and S. M. Rosenberg. 1999. Double-strand-break repair recombination in *Escherichia coli*: physical evidence for a DNA replication mechanism in vivo. *Genes Dev.* **13**:2889–2903.

101. Mueller, J. E., T. Clyman, Y. J. Huang, M. M. Parker, and M. Belfort. 1996. Intron mobility in phage T4 occurs in the context of recombination-dependent DNA replication by way of multiple pathways. *Genes Dev.* **10**:351–364.

102. Mueser, T. C., C. E. Jones, N. G. Nossal, and C. C. Hyde. 2000. Bacteriophage T4 gene 59 helicase assembly protein binds replication fork DNA. The 1.45 Å resolution crystal structure reveals a novel α-helical two-domain fold. *J. Mol. Biol.* **296**:597–612.

103. Mukherjee, A., C. Cao, and J. Lutkenhaus. 1998. Inhibition of FtsZ polymerization by SulA, an inhibitor of septation in *Escherichia coli*. *Proc. Natl. Acad. Sci. USA* **95**:2885–2890.

104. Nassif, N., J. Penney, S. Pal, W. R. Engels, and G. B. Gloor. 1994. Efficient copying of nonhomologous sequences from ectopic sites via P-element-induced gap repair. *Mol. Cell. Biol.* **14**:1613–1625.

105. Ng, J. Y., and K. J. Marians. 1996. The ordered assembly of the φX174-type primosome. 1. Isolation and identification of intermediate protein-DNA complexes. *J. Biol. Chem.* **271**:15642–15648.

106. Ng, J. Y., and K. J. Marians. 1996. The ordered assembly of the φX174-type primosome. 2. Preservation of primosome composition from assembly through replication. *J. Biol. Chem.* **271**:15649–15655.

107. Nilsen, T., and C. Baglioni. 1979. Unusual base-pairing of newly synthesized DNA in HeLa cells. *J. Mol. Biol.* **133**:319–338.

108. Nossal, N. G. 1994. The bacteriophage T4 DNA replication fork, p. 43–53. *In* J. D. Karam (ed. in chief), *Molecular Biology of Bacteriophage T4*. ASM Press, Washington, D.C.

109. Nossal, N. G., K. C. Dudas, and K. N. Kreuzer. 2001. Bacteriophage T4 proteins replicate plasmids with a preformed R loop at the T4 *ori(uvsY)* replication origin in vitro. *Mol. Cell* **7**:31–41.

110. Nurse, P., J. Liu, and K. J. Marians. 1999. Two modes of PriA binding to DNA. *J. Biol. Chem.* **274**:25026–25032.

111. Nurse, P., R. J. DiGate, K. H. Zavitz, and K. J. Marians. 1990. Molecular cloning and DNA sequence analysis of *Escherichia coli priA*, the gene encoding the primosomal protein replication factor Y. *Proc. Natl. Acad. Sci. USA* **87**:4615–4619.

112. Nurse, P., K. H. Zavitz, and K. J. Marians. 1991. Inactivation of the *Escherichia coli* PriA DNA replication protein induces the SOS response. *J. Bacteriol.* **173**:6686–6693.

113. Polard, P., S. Marsin, S. McGovern, M. Velten, D. Wigley, S. D. Ehrlich, and C. Bruand. 2002. Restart of DNA replication in Gram-positive bacteria: functional characterisation of the *Bacillus subtilis* PriA initiator. *Nucleic Acids Res.* **30**:1593–1605.

114. Postow, L., N. J. Crisona, B. J. Peter, C. D. Hardy, and N. R. Cozzarelli. 2001. Topological challenges to DNA replication: conformations at the fork. *Proc. Natl. Acad. Sci. USA* **98**:8219–8226.

115. Robu, M. E., R. B. Inman, and M. M. Cox. 2001. RecA protein promotes the regression of stalled replication forks in vitro. *Proc. Natl. Acad. Sci. USA* **98**:8211–8218.

116. Rothstein, R., B. Michel, and S. Gangloff. 2000. Replication fork pausing and recombination or "gimme a break." *Genes Dev.* **14**:1–10.

117. Sandler, S. J., J. D. McCool, and R. Johansen. 2001. PriA mutations that affect PriA-PriC function during replication restart. *Mol. Microbiol.* **41**:697–704.

118. Sandler, S. J. 2000. Multiple genetic pathways for restarting DNA replication forks in *Escherichia coli* K-12. *Genetics* **155**:487–497.

119. Sandler, S. J., and K. J. Marians. 2000. Role of PriA in replication fork reactivation in *Escherichia coli*. *J. Bacteriol.* **182**:9–13.

120. Sandler, S. J., K. J. Marians, K. H. Zavitz, J. Coutu, M. A. Parent, and A. J. Clark. 1999. *dnaC* mutations suppress defects in DNA replication- and recombination-associated functions in *priB* and *priC* double mutants in *Escherichia coli* K-12. *Mol. Microbiol.* **34**:91–101.

121. Sandler, S. J., H. S. Samra, and A. J. Clark. 1996. Differential suppression of *priA2::kan* phenotypes in *Escherichia coli* K-12 by mutations in *priA*, *lexA*, and *dnaC*. *Genetics* **143**:5–13.

122. Schlomai, J., and A. Kornberg. 1980. An *Escherichia coli* replication protein that recognizes a unique sequence within a hairpin region in φX174 DNA. *Proc. Natl. Acad. Sci. USA* **77**:799–803.

123. Seigneur, M., V. Bidnenko, S. D. Ehrlich, and B. Michel. 1998. RuvAB acts at arrested replication forks. *Cell* **95**:419–430.

124. Seigneur, M., S. D. Ehrlich, and B. Michel. 2000. RuvABC-dependent double-strand breaks in *dnaBts* mutants require RecA. *Mol. Microbiol.* **38**:565–574.

125. Sharples, G. J. 2001. The X philes: structure-specific endonucleases that resolve Holliday junctions. *Mol. Microbiol.* **39**:823–834.

126. Sharples, G. J., and D. R. F. Leach. 1995. Structural and functional similarities between the SbcCD proteins of *Escherichia coli* and the RAD50 and MRE11 (RAD32) recombination and repair proteins of yeast. *Mol. Microbiol.* **17**:1215–1220.

127. Singleton, M. R., S. Scaife, and D. B. Wigley. 2001. Structural analysis of DNA replication fork reversal by RecG. *Cell* **107**:79–89.

128. Smith, G. R. 1991 Conjugational recombination in *E. coli*— myths and mechanisms. *Cell* **64**:19–27.

129. Sung, P. 1997. Function of yeast Rad52 protein as a mediator between replication protein A and the Rad51 recombinase. *J. Biol. Chem.* **272**:28194–28197.

130. Sweezy, M. A., and S. W. Morrical. 1999. Biochemical interactions within a ternary complex of the bacteriophage T4 recombination proteins uvsY and gp32 bound to single-stranded DNA. *Biochemistry* **38**:936–944.

131. Szostak, J. W., T. L. Orr-Weaver, R. J. Rothstein, and F. W. Stahl. 1983. The double-strand-break repair model for recombination. *Cell* **33**:25–35.

132. Tatsumi, K., and B. Strauss. 1978. Production of DNA bifilarly substituted with bromodeoxyuridine in the first round of synthesis: branch migration during isolation of cellular DNA. *Nucleic Acids Res.* **5**:331–347.

133. Tougu, K., H. Peng, and K. J. Marians. 1994. Identification of a domain of *Escherichia coli* primase required for functional interaction with the DnaB helicase at the replication fork. *J. Biol. Chem.* **269**:4675–4682.

134. Turner, J., M. M. Hingorani, Z. Kelman, and M. O'Donnell. 1999. The internal workings of a DNA polymerase clamp-loading machine. *EMBO J.* **18**:771–783.

135. Umezu, K., and R. D. Kolodner. 1994. Protein interactions in genetic recombination in *Escherichia coli*. Interactions involving RecO and RecR overcome the inhibition of RecA by single-stranded DNA-binding protein. *J. Biol. Chem.* **269**:30005–30013.

136. Wechsler, J. A. 1975. Genetic and phenotypic characterization of *dnaC* mutations. *J. Bacteriol.* **121**:594–599.

137. West, S. C. 1997. Processing of recombination intermediates by the RuvABC proteins. *Annu. Rev. Genet.* **31**:213–244.

138. Whitby, M. C., G. J. Sharples, and R. G. Lloyd. 1995. The RuvAB and RecG proteins of *Escherichia coli*. *Nucleic Acids Mol. Biol.* **9**:66–83.

139. Williams, K. R., and W. H. Konigsberg. 1983. Structure-function relationships in the T4 single-stranded DNA binding protein, p. 82–89. *In* C. K. Mathews, E. M. Kutter, G. Mosig, and P. B. Berget (ed.), *Bacteriophage T4*. American Society for Microbiology, Washington, D.C.

140. Withers, H. L., and R. Bernander. 1998. Characterization of *dnaC2* and *dnaC28* mutants by flow cytometry. *J. Bacteriol.* **180**:1624–1631.

141. Womack, F. C. 1965. Cross reactivation differences in bacteriophage T4D. *Virology* **26**:758–761.

142. Xu, L. W., and K. J. Marians. 2000. Purification and characterization of DnaC810, a primosomal protein capable of bypassing PriA function. *J. Biol. Chem.* **275**:8196–8205.

143. Yap, W. Y., and K. N. Kreuzer. 1991. Recombination hotspots in bacteriophage T4 are dependent on replication origins. *Proc. Natl. Acad. Sci. USA* **88**:6043–6047.

144. Yonesaki, T., and T. Minagawa. 1985. T4 phage gene uvsX product catalyzes homologous DNA pairing. *EMBO J.* **4**:3321–3327.

145. Yonesaki, T., and T. Minagawa. 1989. Synergistic action of three recombination gene products of bacteriophage T4, uvsX, uvsY, and gene 32 proteins. *J. Biol. Chem.* **264**:7814–7820.

146. Zavitz, K. H., R. J. Digate, and K. J. Marians. 1991. The PriB and PriC replication proteins of *Escherichia coli* genes, DNA sequence, overexpression, and purification. *J. Biol. Chem.* **266**:13988–13995.

147. Zavitz, K. H., and K. J. Marians. 1991. Dissecting the functional role of PriA protein-catalysed primosome assembly in *Escherichia coli* DNA replication. *Mol. Microbiol.* **5**:2869–2873.

148. Zavitz, K. H., and K. J. Marians. 1992. ATPase-deficient mutants of the *Escherichia coli* DNA replication protein PriA are capable of catalyzing the assembly of active primosomes. *J. Biol. Chem.* **267**:6933–6940.

149. Zavitz, K. H., and K. J. Marians. 1993. Helicase-deficient cysteine to glycine substitution mutants of *Escherichia coli* replication protein PriA retain single-stranded DNA-dependent ATPase activity. Zn2+ stimulation of mutant PriA helicase and primosome assembly activities. *J. Biol. Chem.* **268**:4337–4346.

Chapter 13

The Terminus Region of the *Escherichia coli* Chromosome, or, All's Well That Ends Well

JEAN-MICHEL LOUARN, PETER KUEMPEL, AND FRANÇOIS CORNET

Research on termination of the replication cycle and the terminus region was prompted by the discovery of bidirectional replication of the *Escherichia coli* chromosome (7, 70, 88). Early studies demonstrated that termination involves the collision of two replication forks moving in opposite directions (53), and it was proposed that termination might control expression of genes involved in cell division (48). Genes and sites involved in termination were sought, and a region which blocked passage of replication forks was identified by two groups (54, 64). The mechanism inhibiting fork movement, which has been co-opted by some plasmids (25), was thoroughly worked on during the next 15 years. Inhibition requires the 23-bp *Ter* sites, as well as binding by the Tus protein. The *Ter* sites have a polarity, and forks are blocked only at sites that are in the nonpermissive orientation. Despite the elegance of the Tus-*Ter* system (for reviews, see references 13 and 41), it is not important for the normal growth and division of *E. coli*. The relationship between completion of the replication cycle and division still remains unclear.

Studies of the Tus-*Ter* system and the terminus region had other fruitful consequences, since they led to the recognition of phenomena important in the cell cycle, which are restricted to the terminus region by features of the global organization of the bacterial nucleoid. These are the main topics of this chapter, which is devoted exclusively to *E. coli*. The features of the terminus of *Bacillus subtilis*, the only other organism in which they have been extensively studied, have been reviewed recently (29).

THE TERMINUS AND EVENTS THAT HAPPEN THERE

The chromosome terminus is the region diametrically opposed to *oriC* (Fig. 1). This simple definition covers a number of features. With respect to replication, the "true" terminus is the region between polar pause sites *TerA* and *TerC*, which delimit a replication fork trap from which replication forks do not normally exit (27, 40). The recent cytological studies of the *E. coli* chromosome provide a different definition (78): the terminus is the rather large region (about 15 to 20% of the chromosome, including the replication fork trap) that behaves as a unit during nucleoid movements associated with the cell cycle. Large parts of this region are not essential for bacterial growth, as it contains several cryptic prophages (12), and cells can survive a 350-kb deletion between *TerA* and *TerB* (39). Examination of the sequence of the terminus region reveals several peculiarities, in particular a polarization in which short repeated sequences have opposing orientation on the two sides of the *dif* site, which maps diametrically opposed to *oriC* (14).

The function which is the most specific to the terminus is resolution of chromosome dimers to monomers (9, 17, 55). This involves site-specific recombination occurring at the *dif* locus under the control of the XerC/XerD recombinases and of the FtsK cell division protein. Resolution of chromosome dimers exhibits an unusual regional control, as the *dif* site is active only when it is located within a small *dif* activity zone (DAZ) (20, 56). In addition, the *dif* locus is potentially implicated in decatenation of interlinked chromosomes, since topoisomerase IV binds to the site (43). Another remarkable property of the

Jean-Michel Louarn and François Cornet • Laboratoire de Microbiologie et de Génétique moléculaire du CNRS, 118 route de Narbonne, 31062 Toulouse Cedex, France.   **Peter Kuempel** • Department of Molecular, Cellular and Developmental Biology, University of Colorado, Boulder, CO 80309.

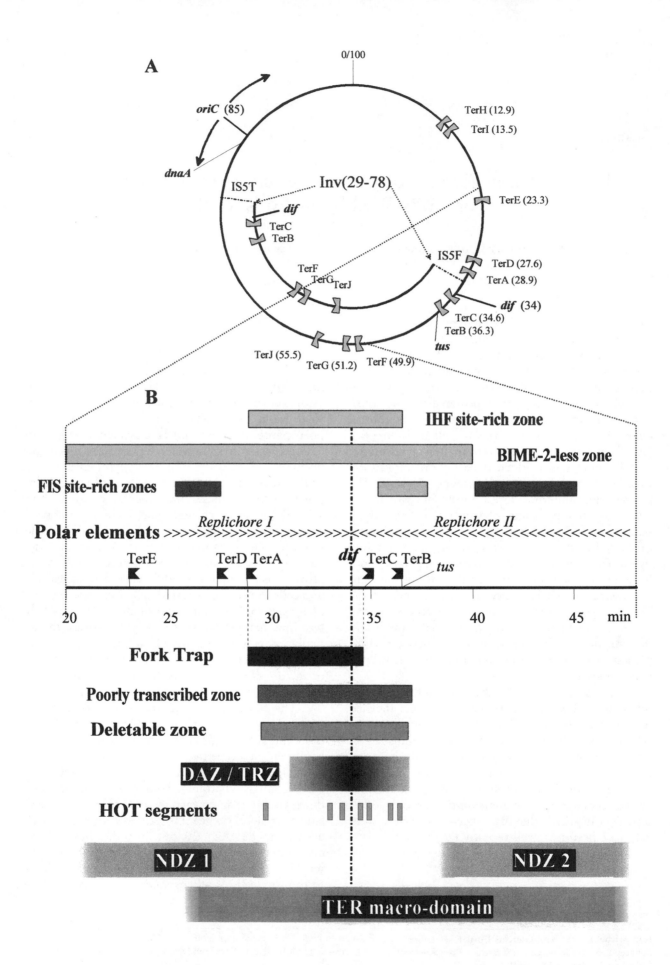

terminus region is an elevated frequency of homologous recombination which is observed when dimer resolution is inhibited (22, 67), as well as in *rnh* and *recD* mutants (22, 80). Finally, the regions peripheral to the DAZ display an organization into nondivisible zones (NDZ) in which inversions are poorly tolerated (90).

This chapter describes our current understanding of the features of the terminus region. In contrast with the preceding review on the same subject (41), the Tus-*Ter* termination system is only marginally treated. The core of this chapter deals with resolution of chromosome dimers by the *dif*-XerCD/FtsK system and its regional control. Other aspects, such as the organization in NDZ, local hyperrecombination, and evolution of terminus sequences, are also presented. We stress the role of the terminus in the ultimate operations of the replication cycle and the postreplicative processing of sister chromosomes.

THE TER MACRODOMAIN

The bacterial chromosome is no longer considered a mass of unstructured DNA stuffed inside the cell. Using either fluorescent proteins binding to specific chromosome positions or fluorescent in situ hybridization, it has been demonstrated that regions of the chromosome move in specific patterns during the cell cycle (35, 42). Using fluorescent in situ hybridization, Niki et al. studied the positioning in the cell of DNA fragments representative of every 5% of the chromosome (78). Their results permitted them to identify four large domains: the ORI macrodomain (about 20% of the chromosome centered on *oriC*), the TER macrodomain (from about 25 min to 50 min on the map [Fig. 1]), and two intermediate junction regions. The different patterns of behavior of the ORI and TER domains are most simply described by an analysis of the cells which display two separate foci for each of the probed regions. In most cells, both ORI foci migrate toward opposite poles during the cell cycle. In contrast, both TER foci keep a midcell position, and they are consequently located where the

septum will be formed at division. The boundaries between the large chromosome domains are presumably indistinct, and markers of the large junction regions display intermediate subcellular locations depending on their relative distance to *oriC* and *dif*.

The distinct patterns of movement of the ORI and TER macrodomains indicate that each must have its own system for positioning itself in the cell. This is best demonstrated by results obtained with a strain containing a large inversion, Inv(29-78), which places *dif* in the vicinity of *oriC* (Fig. 1). In most cells, the cellular locations of these macrodomains were not grossly altered. Some cells were anucleate or displayed mispositioned *oriC* or *dif* regions, however, indicating that chromosome movement and placement were impaired in these cells.

The rules dictating the remodeling and repositioning of nucleoid regions during replication and postreplication processes are still being disputed (92). Regardless, it is obvious that the terminus behaves differently from the origin, and, once replicated, its copies remain central in the cell and in the vicinity of the future septum. Observations obtained with other procedures support this conclusion. The fragility of the *dif* region when XerC/D resolution is abolished indicates that chromosome structure places the terminus region at the septum, and DNA-DNA hybridization also demonstrates that in this situation the septum shears the terminus region and not elsewhere in the dimer chromosome (see "Fragility of the DAZ When Resolution Is Inhibited"). The consequences of inversions in the regions peripheral to the terminus also suggest that these regions are committed to a specific organization (see "NDZ, a Further Indication of Structure").

BIPOLARIZATION OF THE TERMINUS REGION AS INDICATED BY DNA SEQUENCE AND OTHER STRUCTURAL FEATURES

The first evidence for bipolarization of the terminus region was provided by the distribution of the

---

Figure 1. Main features of the terminus region. (A) Location of *dif* and Ter sites on the *E. coli* chromosome. The inner circle shows chromosome organization in the strain carrying the large inversion from 29 to 78 min. (B) Enlargement of the terminus region. Above the linear map are indicated some striking features of the sequence. From bottom to top: presence of polar elements, with change of orientation at *dif* (14); the regions rich in potential target sites for nucleoid-associated FIS protein (104); the region deprived of BIME-2 elements (2); the regions rich in potential target sites for nucleoid-associated IHF protein (104). Below the linear map are shown zones of occurrence of phenomena characteristic of the terminus. From bottom to top: the TER macrodomain (78); the regions refractory to inversion (NDZ) (36); the HOT segments (80); the region where *dif* maps when active in dimer resolution (DAZ), or which becomes fragile when *dif* is inactive (TRZ) (20, 21, 56); the deletable zone, which carries no essential gene (39); the poorly transcribed zone (104); the replication fork trap (27, 40). Graded ends indicate that domain limits are not precisely known (for TER and NDZ) or that phenomenon intensity decreases gradually on either side (for DAZ and TRZ).

Ter pause sites. These 23-bp polar elements are all identically oriented on their respective half of the chromosome (replichore) (Fig. 1). To date, 10 functional pause sites have been mapped, all in the terminus half of the chromosome (24, 41).

Examination of the *E. coli* genome sequence has demonstrated other types of polarity in the terminus region. The strands running 5′ to 3′ from *oriC* to *dif* (the templates for lagging strand replication) display a 3% overrepresentation of G with respect to C (for review, see reference 71). This G/C asymmetry between strands is referred to as a chirochore or replichore structure (10, 62), and it generates a general tendency of G-rich oligomers to present an orientation bias (103). The chi sequence (5′GCTGGTGG3′) provides a well-known example of this bias (10). The tetramer AGGG provides another example. If one considers octamers in which the replichore-associated skew is greater than 70% (http://www.tigr.org/~Salzberg), 60% of them contain this tetramer.

A major family of strongly skewed sequences is the motif 5′RRNAGGGS3′ (R = any purine, N = any base, S = G or C) (14, 93). No role has yet been definitely assigned to this motif, which we call RAG, but it displays a skew which increases in intensity from *oriC* to *dif* (Fig. 2A) and levels off in the 400 kb surrounding *dif*. The skew shifts sign abruptly at *dif* (Fig. 2B), and this shift is much more abrupt than that which occurs in the *oriC* region. The motif RGNAGGGS displays the highest skew, which reaches 97% of the region from 26.5 min to 40.5 min, a region coincident with the TER domain (23). The chirochore structure of the chromosome has been attributed to different mutational pressures on leading and lagging strands (for reviews, see references 34 and 63). This model hardly explains why the skew of a motif as complex as RAG increases regularly with the distance from *oriC*, or why it shifts sign so abruptly at *dif*, in the sole region replicated in either direction. Some physiological role is expected. In "Resolution of Chromosome Dimers: the Main Terminus Function," we propose that polarization of both sides of *dif* is required for *dif* activity. RAG or some similar motif may generate this polarization.

Analysis of the genome sequence has revealed other types of sites which display nonrandom distribution with respect to the terminus region (Fig. 1). Bacterial interspersed mosaic elements (BIMEs) are highly repetitive sequences (more than 300 copies), which are located at transcribed but extragenic locations (2). Composed of various motifs with imperfect palindromic organization, they play structural roles. BIME-1 sites carry a binding site for the nucleoid-associated IHF protein (integration host factor) (11), whereas BIME-2 sites are high-affinity sites

for DNA gyrase (32, 107). Both types of sites can provoke attenuation of transcription (32). Strikingly, BIME-2 sites are absent from the region between 20 and 40 min (largely coincident with the TER macrodomain), and BIME-1 sites and atypical BIME sites are underrepresented in that region (2). The absence of BIME-2 (gyrase) sites might explain the observation that the in vivo density of chromosome-bound gyrase molecules is lower in the terminus region than in the other tested regions of the chromosome (6).

A computational search for sequences displaying extreme parameters for DNA curvature, stability, and flexibility revealed that the terminus harbors several AT-rich regions, which display an abnormally high level of curvature and a low degree of helix stability. The role of these regions is not known (81). Despite the low density of BIME-1 sites, the TER region harbors the highest density of potential (defined by sequence similarities) IHF sites on the chromosome (104). A very high density of potential binding sites for the FIS protein (factor for inversion stimulation) has also been observed in three regions internal or peripheral to the TER domain (104).

UTILIZATION OF THE Tus-*Ter* SYSTEM

As mentioned above, *Ter* sites provided some of the first examples of polarity in the bacterial chromosome. The reader is referred to recent reviews (13, 41) to learn more about the biochemistry of these sites, and how interactions between *Ter* sites and the Tus protein inhibit replication fork movement in a polar manner that involves physical interaction between Tus and DnaB (75). This section deals only with physiological aspects of the system.

Ter sites behave in vivo as temporary pause sites rather than as absolute blocks to replication. The first examples of forks moving through *Ter* sites in the nonpermissive orientation were provided by strains in which integrative suppression of a *dnaA* mutation has altered the normal pattern of replication (79). In extreme cases, the plasmid conducting chromosome replication was inserted in the terminus region and termination occurred in the *oriC* region, after the forks had passed through most terminators in the nonpermissive direction (27, 65). Bacteria of this type had very long replication times, grew poorly, were rich medium sensitive, and formed filaments. The sick phenotype was in part due to replication forks pausing at *Ter* sites, since selection for rich-medium-resistant derivatives from one of these strains resulted in the isolation of the large 29-78 inversion strain shown in Fig. 1 (66). In this inversion strain, replication

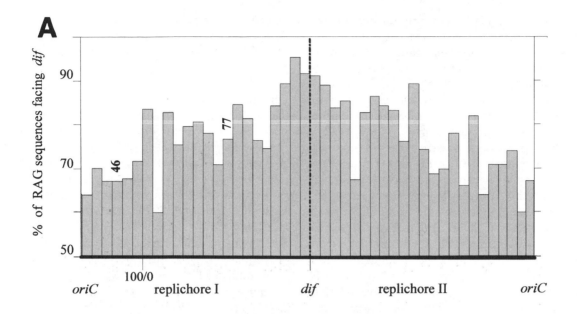

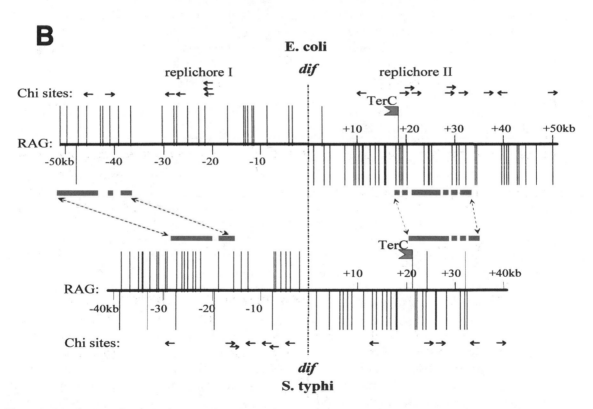

Figure 2. Distribution of polarized sequences. (A) Distribution of RAG (RRNAGGGS) sequences along the *E. coli* K-12 chromosome (N = any base, R = any purine, S = G or C). This map shows the percentage of RAG elements facing *dif* in every 100 kb of each replichore, starting from *dif*. Note the decrease in polarity with distance from *dif*. The number of RAG sequences per 100 kb is randomly distributed, ranging from 46 to 77, with an average of 1 element per 1.6 kb. (B) Comparison between *dif* regions of *E. coli* K-12 (50 kb on each side of *dif*) and serovar Typhi (40 kb on each side of *dif*). The *Salmonella enterica* serovar Typhi sequence was obtained from the Sanger Institute (http://www.sanger.ac.uk). Maps are aligned at *dif*; the flag indicates the position of TerC. Horizontal heavy lines joined by dotted arrows indicate regions of detected homology between the two genomes. Vertical lines indicate RAG sequences; the orientation is indicated by position above or below the maps. Note the dramatic inversion of RAG polarization at *dif* in both genomes. Horizontal arrows indicate chi sites (5′GCTGGTGG3′); the arrowhead indicates the 5′ end. Consequently, the arrows indicate the direction that RecBCD must go through a given chi site in order to stimulate recombination. Note the relative scarcity of chi sites near *dif* in *E. coli*, and that the inversion of chi site polarization seems to occur at TerC.

forks initiated at the plasmid origin encountered fewer *Ter* sites in nonpermissive orientation.

When a replication fork encounters a *Ter* site in the nonpermissive orientation, for how long is it arrested? In the large Inv(29-78) inversion, *TerC* is located near *oriC* (Fig. 1) and the whole series of pause sites of replichore II is in nonpermissive orientation when replication is initiated at *oriC*. The data (27) indicate that, in vivo, fork arrest should not last longer than 5 min at 30°C at *TerC* or *TerB*. These values, measured in bacteria with forks initiated at *oriC*, with a nearly normal balance between pause sites and *tus* gene, and which exhibit no SOS induction, provide the best estimate available for the in vivo effects of active pause sites. The length of fork arrest at a pause site in vivo is quite short, compared with the half-life of a Tus-*Ter* complex in vitro, which is several hours (24). Since at least some replication forks blocked at *Ter* sites in the nonpermissive orientation can resume replication, what is required for this? Transcription of a *Ter* site in permissive orientation has been shown to disrupt Tus-*Ter* interaction, and this could be one event allowing fork restart (73). A 27-kDa antitermination protein, which destabilizes the Tus-*Ter* interaction in vitro, may also be involved, but its role in vivo is not documented (77).

Poor growth is often observed in strains with a combination of pause sites that generates a region displaying delayed replication; this may be alleviated by functions related to homologous recombination and SOS repair. RecA⁻ bacteria are clearly disadvantaged when replication forks have to pass through terminators in nonpermissive orientation. This is the case for integratively suppressed strains replicating from an origin located in the terminus (unpublished observations), for strains in which the suppressor plasmid, inserted in place of *oriC*, allows predominant unidirectional replication (69), for a strain in which a *Ter* site has been inserted in nonpermissive orientation in the *lac* genes (45), and for a strain in which a pair of *Ter* sites in opposite orientation have been inserted in the normal fork trap (95). In the last three examples, growth of the RecA⁻ derivatives was permitted in the absence of Tus. The requirement for RecA for survival suggests that, at least in some conditions, SOS induction is required to solve the problems caused by arrested forks. Arrest of replication can induce SOS, and it has been observed that a *Ter* site in nonpermissive orientation on a pUC plasmid in a Tus⁺ strain may sometimes result in SOS induction (101). It has also been established that replication fork arrest, whatever the reason, may cause double-strand breaks by the action of the Holliday junction resolvase RuvABC (94). This may result in a need to repair the frequent collapse of

stalled replication forks (57); it also explains how the SOS system may be turned on after fork arrest, and the advantage, described above, for recombination-proficient cells.

The rescue of a failed replication fork by recombination does not explain the release of the Tus-*Ter* barrier. A fork rebuilt after collapse would simply encounter the *Ter* site once again, a view which has recently received direct support (46). The SOS system plays probably no role in restarting forks stopped at *Ter* sites, as *Ter* site transgression can occur without SOS induction. This happens, for example, in strains with the large Inv(29-78) inversion. In this strain, *TerC* and *TerB* are transgressed (27), while the low RecA content demonstrates that SOS is not induced (P. Gamas and J.-M. Louarn, unpublished data). This inversion is well tolerated in RecA⁺ strains, but cells cannot grow in rich medium if strains are RecA⁻ or *dnaA46* (66). The benefit for this strain to be RecA⁺ is certain, but this is not due to SOS induction, and there is no proof that the benefit comes from the release of Tus-*Ter* interaction.

The teleonomy of the Tus-*Ter* system in *E. coli* also remains unclear. No disadvantage seems attached to inactivation of the *tus* gene, even when treatments generating DNA lesions make cell viability dependent on extensive DNA repair (19). Conversely, Tus⁻ bacteria are favored in some circumstances. When chromosome replication is placed under the control of an integrated plasmid with predominant unidirectional replication, the requirement for RecA is alleviated by *tus* inactivation (69). Similarly, deleterious inverted pause sites became harmless in Tus⁻ conditions (95). As seen in "NDZ, a Further Indication of Structure," Tus might also act as a warden of chromosome organization. In our present opinion, the arrangement of polar pause sites on the chromosome, which forces replication to terminate in the center of the terminus macrodomain near *dif*, facilitates temporal harmony between end of replication and triggering of events which, like movements of the TER domains away from the septal plane, involve the chromosome polarization centered on *dif* (see "Bipolarization of the Terminus Region as Indicated by DNA Sequence and Other Structural Features" and "Resolution of Chromosome Dimers: the Main Terminus Function"). Even if a positive role of the Tus-pause site system has not yet been detected in lab conditions, in the early times of chromosome evolution the system might have facilitated the achievement of the present configuration, with *dif* located diametrically opposed to *oriC* and fork meeting near *dif*. Despite the current lack of evidence, it seems that there must be some advantage to the Tus-*Ter* system, which is well conserved between *E. coli* and *Salmonella* even though the regions concerned share no homology (see "Evolutionary Pressures at Work in TER Regions").

RESOLUTION OF CHROMOSOME DIMERS: THE MAIN TERMINUS FUNCTION

The function of the terminus region about which the most is known is the resolution of circular dimer chromosomes. Resolution occurs by means of the *dif* site and the XerC and XerD proteins, but it also requires cell division and the FtsK protein. Studies of this function have played an important role in the development of the hypothesis that the chromosome contains a polarity centered about the terminus, and that the terminus is a specialized region with specific properties.

Role of *dif* in Chromosome Dimer Resolution

Circular dimer chromosomes arise by recombination between daughter chromosomes, which is a RecA-dependent process that is sometimes called sister chromatid exchange. Homologous recombination between two circular monomers, the simplest way to form a circular dimer, probably rarely occurs due to the segregation of sister nucleoid masses. The model generally accepted, due largely to the pioneering studies of Kogoma (for a review, see reference 57), is that dimers are generated during the repair by recombination of broken replication forks. Figure 3A shows how a chromosome dimer can form by this chain of events. As the result of a reciprocal exchange that can occur when the Holliday junction is resolved, the newly synthesized DNA at each fork is added onto a strand which is continuous with a template strand (Fig. 3A, part 3). It should be noted that the replicating chromosome now contains only two linear strands, in contrast to the four strands (two template strands, two newly synthesized strands) that were present before recombination (Fig. 3A, part 1). At completion of replication, a circular dimer is present which contains two strands, each of which is twice the length of a chromosome (Fig. 3A, part 4). Once formed, dimers can be resolved to monomers by another recombination event. This could occur by a second breakage and reformation of a replication fork, with the appropriate cleavage of the Holliday junction. The frequency of these double exchanges is not known. We only know that about 15% of the chromosomes are in the dimer form at division (see below). The solution for resolution of these dimers is to use a RecA-independent, site-specific, high-efficiency recombination system.

Cell Cycle Control of Dimer Resolution

The *dif* site provides efficient site-specific recombination. Its location in the middle of the terminus region of the chromosome is ideal for a resolvase site, since it forces resolution, which requires the site to have doubled, to occur after the end of the replication cycle. Actually, chromosome structure has evolved such that interactions between ectopic sites are limited, and they are ineffective for dimer resolution unless they are in the DAZ. The topographical control of *dif* activity is discussed in the next and subsequent sections. This subsection considers cell cycle control of dimer resolution. The more mechanistic aspects of Xer recombination are considered in chapter 28.

The major aspects of the cell cycle control of *dif* activity have been demonstrated by an assay that uses heavy isotopes (^{15}N and ^{13}C) to directly monitor recombination at the *dif* site (98, 99). The concepts are similar to those initially used by Meselson and Stahl (72). One generation after a shift from heavy to light medium, the DNA has hybrid density. After a second generation, one-half has hybrid density and one-half is light (Fig. 3B). If recombination at *dif* occurs between two hybrid density sites, the product still has hybrid density (Fig. 3B, part 1). A second cycle of replication, however, will produce *dif* DNA with approximately semihybrid density, which can be readily detected. Semihybrid density DNA is specific for recombining molecules, and none is detected in bulk DNA. But probing for *dif* demonstrated that such DNA is present, and 15% of the *dif* DNA has semihybrid density (Fig. 3B, part 2a). This unusual density was not observed for other loci or in *xerC* or *xerD* mutants. It also was absent in *recA* mutants, in which dimers should not form, and compensating recombination would not be expected to occur at *dif*. The proportion of semihybrid *dif* DNA indicates that resolution occurs in 15% of the chromosomes. Further experiments demonstrated a direct correlation between Xer recombination at *dif* and homologous recombination elsewhere in the chromosome, so that *dif* recombination can be used to identify mutations that increase or decrease sister chromatid exchange (83, 99).

An unexpected result of the above experiments was the demonstration that cell division was required for dimer resolution. If cell division was blocked by cephalexin, or by use of a *ftsZ*(Ts) mutation, recombination did not occur at *dif* (Fig. 3B, part 2b). The basis of the requirement for cell division was clarified further when the density-label assay was used to demonstrate that FtsK was required for dimer resolution (97). This requirement can best be understood

A: Formation of a chromosome dimer

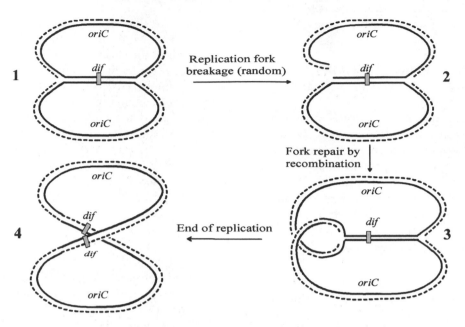

Replication fork
breakage (random)

Fork repair by
recombination

End of replication

B: Measure of recombination at *dif* by density shift experiments

1:

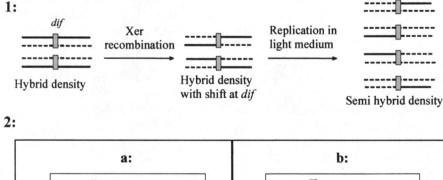

Hybrid density

Xer
recombination

Hybrid density
with shift at *dif*

Replication in
light medium

Semi hybrid density

2:

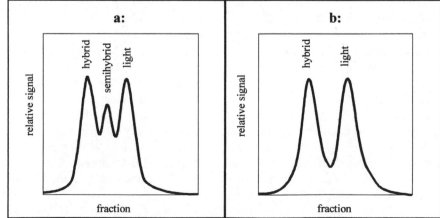

Figure 3. Generation of chromosome dimers and resolution at *dif*. (A) Formation of a circular dimer by homologous recombination. A replicating chromosome (1) suffers breakage of a replication fork (2). Repair of the fork by recombination, which can include resolution of the Holliday junction such that sister chromatid exchange occurs, followed by some further replication, gives structure 3. Note that only two strands of DNA are now present, and further synthesis adds onto the ends of these two strands. Completion of the replication cycle gives a circular dimer (4). (B) Dimer resolution at *dif* can be detected by formation of physical recombinants (98). Cells growing exponentially in heavy (^{13}C ^{15}N) rich medium were transferred to light

in the context of the properties of FtsK. The first 200 amino acids are essential for cell survival, and they contain transmembrane sequences that anchor the protein in the membrane (5). Several studies have shown that FtsK is located in the septal ring (106, 108). It requires ZipA and FtsA for its recruitment (37, 87), and is required for colocalization of FtsQ, FtsL, and FtsI with FtsZ (16). The middle 600 amino acids of the protein, which compose what appears to be a linker region, and the terminal 500 amino acids are not essential for cell survival (5). However, this C-terminal region is the part required for resolution of dimer chromosomes. These 500 amino acids contain a nucleotide binding site, as well as a motif that is found in proteins that transfer DNA, such as the SpoIIIE protein of *B. subtilis*. This is the protein that moves DNA from the mother cell into the prespore during sporogenesis (96).

Since FtsK is located in the septum, it is straightforward to construct models in which the ingrowing septum permits FtsK to interact with the *dif*/XerC/XerD complexes on dimer chromosomes. The colocation model is presented in "The Colocation Model for Dimer Resolution" below. An important aspect of this model is the functional polarization of the terminus region, presented in the next two sections.

Experiments by Barre et al. (4) provided further insights into the crucial role that FtsK plays in dimer resolution. Using two-dimensional gels, they analyzed the formation in vivo of Holliday junctions between *dif* sites. A first-strand exchange catalyzed by XerC may occur in the absence of FtsK. This requires both XerC and XerD, and results in an unstable Holliday junction, since separation may occur without exchange. Also, this initial interaction is not limited to *dif* sites located in the terminus. Since the Holliday junction also forms in *recA* mutants, it appears that a chromosome dimer is not required to bring the sites together so that the initial interaction can occur. The initial interaction with formation of a Holliday junction may occur even on monomer chromosomes, for which no resolution is necessary. To get complete Xer recombination, FtsK is needed

in addition (see chapter 28 for the most recent insights into the reaction). Since in normal cells FtsK is localized at the septum, it might be that the concentration of the required C-term becomes high enough to allow complete Xer recombination between *dif* sites only when the septum forms. This explains in part why dimer resolution requires cell division. Further aspects of this interaction are considered in the next subsection, which considers positional constraints on *dif* activity.

Topographical Control of Dimer Resolution

In addition to being controlled by events of the cell cycle, activity for dimer resolution by the XerCD/FtsK/*dif* system is highly dependent on the location of the site in the chromosome (20, 56, 60, 102). Strains harboring a translocated *dif* site are unable to resolve chromosome dimers unless *dif* is inserted within a narrow zone around its normal position (20, 56). The chromosome region where *dif* sites reach their full activity (DAZ) is no larger than 10 to 20 kb, and *dif* activity decreases progressively on either side of the DAZ along regions of about 20 kb. The activity of transplaced *dif* sites can be determined by a growth competition assay, which measures the efficiency of chromosome dimer resolution, as well as by an assay that measures the rate of loss of kanamycin resistance from a *dif*-Kn-*dif* cassette. Comparable results were obtained with the two assays (Fig. 4) (82, 83).

The inability of transplaced *dif* sites to resolve dimers is not solely due to inefficiency of Xer recombination. Overproduction of FtsK full protein or of its C-terminal domain increased the rate of recombination at, and loss of kanamycin resistance from, *dif*-Kn-*dif* cassettes, but it did not increase the ability of cassettes located outside the DAZ to carry out dimer resolution (4, 83). Complete recombination between sites is not sufficient for resolution; other factors are important, probably features of chromosome structure which facilitate segregation of the monomers resulting from recombination between the *dif* sites of a dimer.

---

medium and grown for two generations. At this time, total chromosomal DNA must be found as hybrid or light material in equal amount. (1) Predictions. Recombination between *dif* sites after the first replication cycle produces single strands that shift from heavy to light density at the site. These can be detected after the next replication cycle (in light medium), which yields semihybrid-density double-stranded DNA. (2) Experimental data. Chromosomal DNA harvested after two rounds of replication in light medium was digested to produce restriction fragments in which *dif* was centered in its fragment. Hybrid, semihybrid, and light material were separated by centrifugation in cesium chloride density gradients, and the *dif*-carrying fragments were identified by hybridization with a radioactive probe. (2a) Typical profile obtained with wild-type bacteria: about 15% of the *dif* DNA was present as semihybrid material. This was unique to *dif*. (2b) Typical profile obtained in *xer* mutants, or when cell division was blocked by cephalexin treatment or temperature shift of an *ftsZ* mutant. No semihybrid material was detected at *dif*.

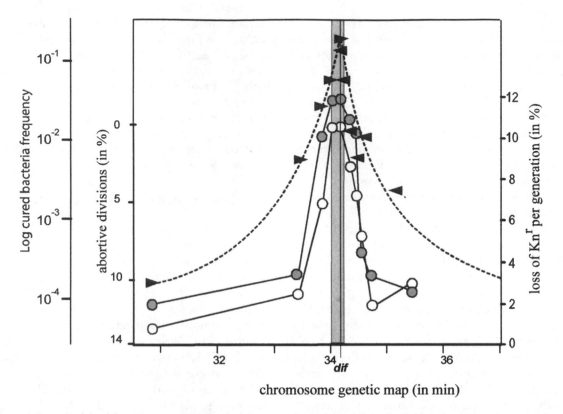

Figure 4. Extents of the *dif* activity zone (DAZ) and of the zone where terminal recombination is detected in absence of dimer resolution (TRZ). The DAZ extent has been determined in three ways. (i) Recovery of a Dif⁺ phenotype after transposition of transposon (Tn*dif*) into the chromosome of Δ*dif* (1 kb) mutant (56). The shaded vertical area spanning the normal location of the *dif* site shows the region of the seven Dif⁺ insertions analyzed in this study. (ii) Measure of increased viability due to an ectopic *dif* site (open circles), with reference to the isogenic *dif*-deleted strain (82). The percentage of abortive divisions increased with the distance between the ectopic site and the *dif* wild-type position, to reach a maximum value of about 15% when this distance is higher than 20 kb and/or *dif* is inactive. (iii) Frequency of loss per generation (closed circles) of a kanamycin resistance determinant located between two *dif* sites inserted at ectopic locations (82). The percentage of excision reached a value of about 12% per generation when the cassette was inserted at the *dif* normal location, and rapidly decreased when the cassette was moved in either direction. The extent of the region involved (the TRZ) was estimated from the frequency of bacteria cured of a prophage inserted by homologous recombination at the indicated positions (closed arrowheads). Data are shown only for prophage insertions that maintained the RAG polarity in the two sides of the terminus region, centered at *dif*.

Since *dif* is in the terminus region, this suggests that nearby meeting of replication forks might be an additional requirement for proper *dif* activity. Interestingly, this is not important for *dif* activity: *dif* still functions when removed from the region where replication terminates (20, 56, 68). The most extreme example of this is the Inv(29-78) strain, in which *dif* is adjacent to *oriC*, and termination occurs far away from this region, at least in Tus⁻ conditions. Note that in the rearranged chromosome, *dif* is still surrounded by terminus region sequences (Fig. 1). Whatever structures are important for resolution, they apparently can still form, even though replication forks are meeting at a considerable distance from *dif*. Although *dif* can still function in this rearranged chromosome, the coincidence of *dif* and the region

in which termination occurs indicates that this has been selected during chromosome evolution. Indeed, secondary rearrangements restoring near-normal genome organization and improved fitness are easily observed in the Inv(29-78) strain (66).

Although the inversion described above does not alter *dif* activity, other rearrangements have provided important clues about the polarity of the terminus and surrounding regions, and how this affects *dif* activity. Although the DAZ occupies less than 20 kb, studies with deletions demonstrated that the ability to form a DAZ extends over a much larger region. For instance, the DAZ and its surrounding region can be deleted by a 185-kb deletion (Fig. 5, deletion type 2), and a 33-bp *dif* sequence inserted at the point of the deletion functions normally. Similar effects have been

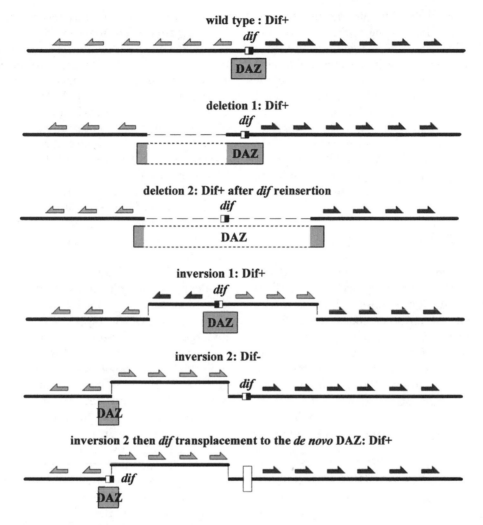

Figure 5. Activity of *dif* is controlled by polarization of flanking regions. Determinants of the DAZ (see Fig. 4) were characterized by analysis of deletions and inversions near *dif* (20, 56, 82), and the location of DAZ in wild-type and rearranged chromosomes is indicated by the boxed region. On the drawings, DOPEs are symbolized by arrows—gray or black, depending on the replichore. The orientation of the arrows indicates the direction in which DNA is proposed to be pulled (or pushed) away from the ingrowing septum (see "The Colocation Model for Dimer Resolution"). Deletions and inversions are described from top to bottom. Deletion type 1: deletions that had both endpoints on the same side of *dif* were wild type for dimer resolution (Dif⁺). Deletion type 2: deletions that had endpoints on opposite sides of *dif* were mutant for dimer resolution (Dif⁻). This phenotype was suppressed when *dif* was reinserted at the deletion junction. These deletion data indicated that no unique site (besides *dif*) is needed for *dif* activity, and that a DAZ is always found at the junction between the severed replichores. Inversion type 1: inversions that had endpoints on opposite sides of *dif* were always wild type for dimer resolution (Dif⁺). Inversion type 2: inversions that had endpoints on the same side of *dif* were mutant for dimer resolution (Dif⁻), provided that the inverted region was large enough and close enough to *dif*. When the proximal endpoint mapped at more than 30 kb away from *dif*, these inversions had no effect on *dif* activity. Suppression of inversion type 2: the inversion generates a new DAZ at the endpoint distal from the normal location of *dif*. When *dif* is transplaced to this location, dimer resolution activity is regained (Dif⁺).

observed with other deletions. This indicates that a new DAZ is formed at the junction between the two severed replichores (Fig. 5) (20, 56, 82, 102).

Inversions have provided the demonstration that local polarity controls *dif* activity. Inversions which have the endpoints on opposite sides of *dif* do not affect *dif* activity. This presumably occurs since the

polarity status at *dif* is unaffected by this type of inversion (Fig. 5, inversion type 1 [Dif⁺]). Other inversions have dramatic effects. Inversion of at least four independent segments, all of them in a 60-kb region around *dif* with two on each side of the DAZ, resulted in inhibition of *dif* activity. These segments may contain polar sequence elements; inversion altered the position of DAZ by altering the location at which opposite polarities meet (Fig. 5, inversion type 2 [Dif$^-$]). Consistent with this, inversion of a 150-kb region adjacent to *dif* inactivates the *dif* site, and activity is regained if *dif* is transferred to the other end of the inversion (Fig. 5, inversion type 2 [Dif$^+$]). In a similar fashion, it has been observed that insertion of λ (48 kb) adjacent to *dif* inactivates *dif*. Since this effect depends on the orientation of the prophage, this indicates that λ also contains polarity elements, which may behave in a way similar to that of the terminus. These results are presented in the next section.

Fragility of the DAZ When Resolution Is Inhibited

Terminal recombination (previously called terminal hyperrecombination [67]) was discovered at the same time as *dif*. It was observed that a λ prophage inserted into the chromosome between directly repeated sequences was excised at a higher frequency (up to 1,000-fold) when the prophage was located in the terminus, compared with the rest of the chromosome. Analysis of terminal recombination provided the first example of a region-specific activity (68), and has been instrumental in developing concepts related to the terminus region. It is known now that terminal recombination is induced by inactivation of dimer resolution at *dif* (22). However, the relationship with dimer resolution was not apparent in the earlier studies, since the cells used were genetically wild type for the *dif*/Xer resolution system. The involvement of dimer resolution turned out to be a consequence of the assay, which was based on λ prophage excision. Eventually it was observed (21, 22) that (i) chromosomal segments of the region, when flanked by repeats identical to those used in the prophage assay, did not excise at high frequency unless Xer recombination was abolished; (ii) prophage-induced terminal recombination depended on the orientation of the λ inserted near *dif*; (iii) intensity of terminal recombination, measured by the prophage excision assay, was largely independent of the prophage orientation in *xer* mutants, at least near *dif*; and (iv) cells exhibiting prophage-induced terminal recombination exhibited a Dif phenotype. Consistent with its connection with dimer resolution, terminal recombination was shown to be independent from the Tus-pause site system and termination (21, 68).

The current model for prophage-induced terminal recombination combines concepts of polarity, dimer resolution, and cell division. It is postulated that the λ prophage is itself polarized by elements similar to those found in the terminus and that its polarization interferes with the correct polarization of the *dif* region. The inactivation of dimer resolution at *dif* due to a prophage inserted in the "wrong" orientation is similar to that caused by inversions adjacent to *dif* (Fig. 5). Inactivation of dimer resolution in turn leads to DNA damage of unresolved dimers at cell division, which is repaired by homologous recombination. The prophage orientation which hinders *dif* activity changes abruptly at *dif*; this polarization may be generated by RAG motifs (see "Bipolarization of the Terminus Region as Indicated by DNA Sequence and Other Structural Features"), since 32 of the 42 present on λ DNA share the same orientation. The polarization is certainly not generated by the well-known skewed element chi, which is not found on λ.

Since terminal recombination is RecABC dependent (21), double-strand ends must be involved. Two models can be proposed. In the guillotine model, chromosomal DNA trapped by the closing septum undergoes direct scission, and the double-strand ends are ports of entry for RecBCD (38, 89); in the garrote model, DNA trapping by the septum inhibits progression of the next coming replication forks. The halted forks are subsequently transformed into recombination forks, with extrusion of the recent strands. Their pairing to yield free-ended double-stranded DNA provides sites of entry for RecBCD (67, 94). It is possible that both processes are present.

Here we use the term guillotining, since this must occur in physically separated daughter cells that contain a dimer at the time of division (38). DNA degradation in the *dif* region was directly detected when the *dif*/Xer system was inhibited (89). Terminal recombination is obviously the successful outcome of attempts to repair lesions associated with this degradation. Terminal recombination, as well as degradation of the terminus region, provides strong support to the model that the *dif* region has a particular cellular location, at least when dimer chromosomes persist due to lack of resolution. The guillotining which occurs in resolvase mutants usually kills both daughter cells. Hendricks et al. (38) used a microcolony analysis to follow cell growth after guillotining, and although both daughters usually formed long filaments after the crucial cell division, none were observed to resume normal growth and division. An important component of this cell death is the rapid loss of DNA at double-strand ends caused by cell division, and cell death was suppressed in RecBC$^+$D$^-$ mutants. In contrast to the absence of survivors observed in a microcolony analysis, the λ lysogens used in tests of terminal recombination give a higher rate of recovery, since up to 20% of the progeny of a guillotining cell division may survive in this case. It should be noted that the insertion of 48 kb of λ DNA at some distance from *dif* generates in the terminus, when prophage orientation perturbs local polarity, two regions where opposing polarities meet. Cells with sheared chromosomes will now sometimes contain an overlapping region, and the chromosome can now be repaired. This explains that in *xer* mutants the region affected by terminal recombination is narrow when the prophage does not perturb polarity (Fig. 4) but is much larger when it interferes with ambient polarity (22, 23).

The Colocation Model for Dimer Resolution

The preceding sections have described the roles of cell division, the FtsK protein, and chromosome polarization in dimer resolution, as well as the outcome of failed resolution. The colocation model (Fig. 6) has been proposed as a unifying explanation for these phenomena and as a guide for future research in these areas. The four statements of this model are as follows. (i) Polarization of *dif*-flanking regions is needed to locate *dif* sites belonging to a chromosome dimer in the immediate vicinity of the septum. The polar elements might be used by the chromosome partition machinery to drag the TER domain away from the ingrowing septum, whether the sister chromosomes are monomers or dimers. Only in the case of a dimer would the dragging result in a tug-of-war, which places the *dif* sites under the septum. (ii) *dif* sites of dimers are consequently placed in the vicinity of the C-terminal domain of the septum-bound FtsK protein. This positioning allows an effective concentration of sites and protein factors, which maximizes exchanges between the *dif* sites. This positioning could also be important for obtaining proper spatial configuration. (iii) When a full exchange between *dif* sites resolves the dimer, the partition process resumes immediately and drags the TER DNA away from the septum (and FtsK) so that no subsequent Xer recombination can be catalyzed. (iv) If resolution does not occur for some reason, the dimer-borne *dif* regions are trapped by the septum and undergo degradation and recombinogenic alterations.

This model, which integrates spatial and temporal features of nucleoid processing and septum formation, conveniently explains the peculiarities of chromosome dimer resolution. It explains, for example, why exchanges at ectopic *dif* sites or between *dif*-kan-*dif* cassette insertions are scarce outside the DAZ (Fig. 4). It also explains why Xer recombination at *dif* has not evolved to be directional, in contrast to what

happens at plasmid-borne Xer sites such as *psi* or *cer* (see chapter 28). The model also provides new clues for thinking about chromosome partitioning and cell division. Though it rests on many predictions, it is supported by recent progress toward an understanding of the role of chromosome polarization: identification of the process making use of polarity and of FtsK as the partner recognizing polar elements, but not yet of the polar elements themselves (see below).

The Nature of DAZ-Organizing Polar Elements (DOPEs)

Attempts to characterize the polar elements have been so far unsuccessful, perhaps because they are heterogeneous or composite, or because their inversion only affects *dif* activity when a sufficient number of them are altered. However, candidates exist among sequence motifs described above. BIME elements, which are not polarized and are absent from the terminus, are ipso facto eliminated. AT-rich regions are not yet excluded, although they are probably not present in sufficient number. The skewed oligomers displaying a strong orientation bias which shifts at *dif* are better candidates. Among these oligomers, chi sequences are rejected because they are absent from segments (including λ) whose inversion inhibits *dif* activity. The RAG motifs are the best candidates, considering their frequency and change of orientation preference at *dif*, and also because they polarize the terminus with the highest skew (Fig. 2). This contention is supported by several indirect arguments: (i) all four *E. coli* sequenced genomes display a shift of polarity of RAG elements at *dif* and a polarization by these elements as strong as that in K-12, although the location of these elements is variable (see "Evolutionary Pressures at Work in TER Regions"); (ii) λ DNA, whose presence near *dif* may disturb the normal polarization, is polarized by RAG motifs to the same extent as the average *E. coli* chromosome (22), and the λ late operon—about 20 kb with a 10:1 skew of RAG elements—is sufficient to disturb the DAZ (J. Corre, unpublished data); (iii) in *Salmonella* the sequences flanking *dif* are as strongly polarized as they are in *E. coli* by RAG elements, despite a nearly total absence of homology between the two species in this region (see "Evolutionary Pressures at Work in TER Regions"); and (iv) motifs belonging to the RAG family also polarize in opposite directions the regions flanking the *dif* sites of many bacterial chromosomes (see "Evolutionary Pressures at Work in TER Regions"). Nevertheless, the direct evidence for the involvement of RAG elements in the control of the DAZ has still to be obtained.

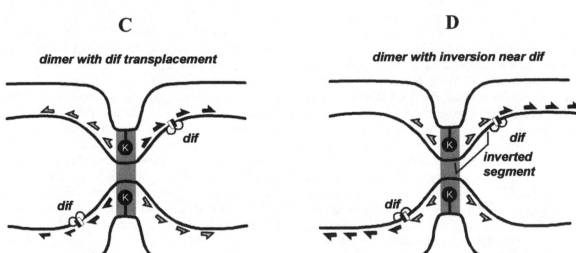

Figure 6. The colocation model for control of dimer resolution. The gray area represents the part of the cell within which septum-anchored FtsK protein (filled tailed circles) can access the other partners of resolution: chromosomal *dif* sites (black and white squares) and XerC and XerD recombinases (open circles). Since Xer recombinases are most probably free in the cytoplasm, the other important topological factor is the mechanism which positions a pair of *dif* sites within the FtsK activity space. It is proposed that this involves the polarization of *dif*-flanking regions. DOPEs (represented as in Fig. 5) are supposed to be sequences recognized by the partitioning process which tends to pull (or push) all chromosomal DNA away from the septum plane. In the case of dimer chromosomes (A), forces acting in opposite directions will place the DAZ at the ingrowing septum. FtsK plays a crucial role in the positioning step. The next step is formation of the synaptic complex *dif*/XerC/XerD/FtsK, and recombination between *dif* sites can then proceed. After exchange, the forces involved in partitioning separate the recombinant sites from each other, thus precluding regeneration of a chromosome dimer. When chromosomes are monomers (B), *dif* is transplaced outside the DAZ (C), or an inversion moved the DAZ away from *dif* (D), *dif* sites do not enter within the FtsK activity space, and Xer recombination is rare. Adapted from reference 83.

Role of FtsK in Terminus Mobilization

Recent data suggest strongly that FtsK plays two roles in dimer resolution. In addition to controlling the structure of the recombination complex between *dif* sites and recombinases, it participates in the last step of the positioning of *dif* sites on a chromosome dimer, in a process which implicates terminus polarization. First, Aussel et al. (1) have succeeded in setting up an in vitro assay for complete Xer recombination at *dif*. The key to this success was the use of a truncated FtsK protein which links the C-terminal domain to 50 amino acids from the junction between the N-terminal domain and the intermediate domain. They obtained evidence that this truncated protein forms multimeric rings around DNA and displays a DNA mobilization activity. Second, it is known that the *loxP*/Cre recombination system can substitute for the *dif*/XerCD system to bring about chromosome dimer resolution (60). Capiaux et al. (15) have shown that chromosome dimer resolution by *loxP*/Cre requires that *loxP* be inserted within the DAZ and that the strain be FtsK$^+$. This strongly suggests that FtsK is responsible for the regional control of chromosome dimer resolution. Third, Corre et al. (22) have shown that terminal recombination is still observed when resolution at *dif* is inhibited by an *ftsK* mutation that removes the C-terminal domain from FtsK. They have more recently shown that in such a mutant the gradient of terminal recombination is broader than in a Xer$^-$ mutant, and, strikingly, that polarity effects are abolished (23). Taken together, these results indicate a role for FtsK as a DNA motor protein displaying a DOPE-oriented mobilization activity. The septal FtsK ring(s) might mobilize each DNA thread passing through the septum in the direction dictated by DOPE polarity. Hence, DNA movement would stop when the region in which polarity shifts sign reaches the ring(s). Since *dif* maps in this region, the final result would be a positioning of both *dif* sites of a dimer close to each other, in a configuration facilitating their synapse. The additional fact that the terminus region remains fragile in the *ftsK* mutants reveals that there must be more than one mechanism involved in terminus positioning at division. The FtsK-independent positioning mechanism has not been identified so far.

DECATENATION

Decatenation of daughter chromosomes is another event that occurs at the end of the replication cycle, and there is evidence that *dif* can be involved. Decatenation is considered in detail in chapter 6, but we want to mention briefly certain aspects of it here. At present, there is strong evidence that topoisomerase IV (Topo IV) is the major decatenating enzyme (109, 122). Although decatenation and resolution both involve breakage and rejoining of DNA, there is no evidence indicating that Topo IV can carry out the type of recombination event required for resolution, or that *dif*/XerC/XerD/FtsK recombination can also be used for decatenation. However, based on the positioning of the almost completely separated daughter

chromosomes and their *dif* sites at the end of the cell cycle (see "The Colocation Model for Dimer Resolution" and Fig. 6), it seems that *dif* would be an ideal location for a decatenation site. Indeed, *dif* is a preferred site for interaction with Topo IV (43). Using an assay based on cutting of DNA at sites at which Topo IV is bound, it was observed that DNA was cut at *dif*, and furthermore, this was the only cut site of this type in an 88-kb fragment that flanked *dif*. XerC and XerD, but not FtsK, were also required for this cleavage.

Interaction of Topo IV and decatenation at *dif* are not essential, however. This is indicated by the properties of *xerC* or *xerD* mutants, or cells in which *dif* has been deleted. Cells containing dimer chromosomes (ca. 15%) guillotine their chromosomes at cell division, form filaments, and die (38). However, the remaining 85% of the cells appear to grow and divide normally, and their cell cycle is unaffected. Since these cells should also contain catenated monomer chromosomes, decatenation does not require a functioning *dif* site.

What is the role of the interaction between *dif* and Topo IV? Since resolution occurs just before cell division, cells with dimer chromosomes might have special requirements for decatenation. Catenating turns might be chased into the terminus region as nucleoid processing occurs, but since the daughter chromosomes are still attached as a circular dimer and cannot be separated, it would be difficult to complete decatenation. Once resolution to monomers has occurred, decatenation and complete separation of the chromosomes could occur, but there would only be a limited time for this before cell division. The presence of a preferred decatenation site at *dif* would ensure that decatenation was completed. In mutants that are unable to resolve dimers, the absence of this decatenation would be difficult to detect, due to the guillotining that occurs in those cells. Absence of decatenation could even provide an additional source of DNA damage in resolution mutants.

NDZ, A FURTHER INDICATION OF STRUCTURE

As described above, inversions have provided important information about the structure of the region that surrounds *dif* and the role of polarity in the function of the DAZ. Analysis of other inversions that involve the periphery of the TER domain provides further evidence of chromosome structure. The elements responsible for the observed effects are still unknown, but in at least some instances the Tus-*Ter* system seems to be involved.

Konrad (50) was the first to discover that a chromosome segment may be refractory to inversion. The analysis of inversion behavior has been extended to many segments covering the entire *E. coli* chromosome (90), and three different behaviors were observed. Segments could be type T, in which the inversion is tolerated and stable over generations; type R, in which inversions can be isolated but are refractory and detrimental; or type N, in which inversions cannot be isolated because the inversion is not feasible or so strongly disabling that it is never detected. The striking observation was that deleterious inversions have one endpoint located in the 20 to 30% of the *E. coli* chromosome flanking the replication terminus. These regions were termed "nondivisible zones," and they coincide with the periphery of the TER domain (Fig. 1).

The regions involved contain a number of replication pause sites, and these obviously could play an important role in defining the NDZ region. In the inversions characterized in the earlier experiments, however, the Tus-*Ter* system was not involved in the NDZ phenomenon. If the *tus* gene was deleted, it was still not possible to isolate the inversions, or they remained deleterious (33). It was proposed that the detriment caused by inverted segments of the NDZs could be an alteration of a higher-order nucleoid organization involving interactions between polar sequences (90).

Guijo et al. (36) recently refined the analysis of NDZ1 (between min 23 and min 30) by subdividing the region into smaller segments. They observed, in addition to a majority of R and N segments, a few T segments. The patchwork of T and R segments observed was interpreted as indicating the existence in NDZ1 of several smaller NDZs, each harboring several elements whose relative orientation is physiologically important. Microscopic observation of bacteria carrying deleterious inversions revealed that many nucleoids had an aberrant morphology, similar to that of Par⁻ mutants. This phenotype was observed regardless of the R segment that was analyzed. NDZ organization might be important for general events such as sister chromosome separation or for more specific events, such as decatenation. It should also be noted that the regions where NDZs are found coincide rather well with the two regions peripheral to the TER domain which harbor a very high density of potential FIS sites (Fig. 1) (104). Whether the presence of these sites is involved in the NDZ phenomenon will require further exploration.

The role of the Tus-*Ter* system in NDZ1 has been revisited by Guijo et al. (36). Attention was focused on segments harboring one pause site only, namely *TerD*, as their inversion would be expected to generate a zone difficult to replicate between inverted *TerD* and *TerA*. In Tus⁺ conditions, one of these *TerD*-containing segments (the smallest) tolerated inversion, but four others were classified N or R. In Tus⁻ conditions, only one of these four segments (the longest) shifted from the N to the T class. At present there is no simple explanation for these puzzling observations, and more than one factor is apparently involved. It should be noted that the *his-trp* segment (between min 38 and min 45) of the *Salmonella enterica* serovar Typhimurium chromosome, which is nonpermissive to inversion in Tus⁺ bacteria, becomes tolerant to inversion in Tus⁻ conditions. In this situation it was demonstrated that the recombination event generating the inversion was prevented by the presence of Tus (R. Camacho and J. R. Roth, unpublished data). Whether this prohibition is mediated through the interaction of Tus with pause sites is an open question.

Another aspect of chromosome structure that might play a role in the NDZ phenomenon is the domain structure of the chromosome (86). The present consensus about the domains is that they are due to supercoiling. Their limits fluctuate, and transient domains are created by intracellular viscosity or by stochastic obstacles associated, for example, with the strain of replication or transcription (see chapter 6). Exceptions are found in the terminus. Using an assay based on the topological requirements of the recombination between Tn3 *res* sites, Higgins and coworkers have observed the presence, near the *dif* site of *S. enterica* serovar Typhimurium, of two loci acting as barriers to free diffusion of supercoiling (N. P. Higgins, personal communication). These barriers could be due to the inherent structure of the chromosome itself, or they could be sites for anchoring the TER domain to some fixed cell structure, such as the envelopes. Barriers of this type may preclude interaction between distant polar elements and behave as limits of NDZ. It is not known whether similar sites are present in *E. coli*, as the sites identified by Higgins and coworkers are in a region which is not found in *E. coli*.

TERMINUS PRONENESS TO RECOMBINATION AND TRANSPOSITION

In "Fragility of the DAZ When Resolution Is Inhibited" we described the terminal recombination that occurs when resolution at *dif* is inhibited. Additional homologous recombination activities have been observed in the terminus region, and they are described in this section. These activities have been observed in *rnh* (RNase H) mutants and in mutants in which the RecD subunit of the RecBCD complex is inactivated. Transposition targeting in the terminus region by Tn7 is also described.

In mutants lacking RNase H, surrogate initiation of replication occurs at a number of loci. These origins are sometimes called *oriK*, and they are able to maintain viable chromosome replication when the normal *oriC* origin is deleted (49). There are indications that several *oriK*s map in the terminus region, that they function frequently in *rnh* Δ*oriC* strains, and that they also function in *rnh oriC*⁺ strains (26, 105). In a search for *oriK* by shotgun cloning, Horiuchi and coworkers selected by transformation circularized DNA fragments that could be maintained as free circular elements by an *rnh* mutant, and they isolated eight such fragments (Hot fragments [80]). Seven of the Hot fragments map in the terminus. It was later shown that their activity was not due to the presence of an intrinsic *oriK* in the Hot fragment, but that it depended on homologous recombination. It was proposed that when circular molecules containing Hot fragments have recombined into the chromosome, they are present as tandem repeats; these repeats are replicated by an adjacent *oriK*, and the eventual processing of the extra DNA by recombination enzymes allows circle formation. Hot activity of some Hot fragments (those mapping near *TerA*, *TerB*, and *TerC*) was abolished by a *tus* mutation. It was suggested that, at least in these cases, blocked replication forks provided the entry point for the RecBCD enzyme, and that recombination in the Hot fragment was increased due to the presence of a chi site. Consistent with this, it was observed that HotA activity required the presence of a chi site (44).

More recently, a Hot activity independent of the *rnh* mutation was constructed (45). A 22-bp *TerA* sequence was inserted in the nonpermissive orientation into *lacZ* (called *TerL*), adjacent to a chi site in *lacY* (called HotL). HotL function was similar to that observed for sites in the terminus region. If cells were transformed with circular molecules harboring HotL, abundant circular HotL-containing molecules could be isolated, provided that replication forks were blocked at *TerL*. The *rnh* mutation and *oriK* activation were not necessary for HotL activity, since pausing replication forks were now provided by *oriC*. These experiments provide strong support for the proposal regarding Hot activity and *oriK* origins in the terminus. In the model of Horiuchi and coworkers, it was proposed that the RecBCD complex entered that DNA at breaks generated by stalled forks. An alternative possibility is that the RuvAB-dependent recombination process described by Seigneur et al. (94) may ensue at the stalled forks. Many questions remain, such as, what exactly is an *oriK*, why are

they found in the terminus region, and what activates Hot sites that do not require a *Ter* site for function?

Another situation in which increased recombination has been observed in the terminus is *recD* mutants: in such strains, deletions by homologous recombination between direct repeats are at least 10-fold more frequent in the terminus than in other regions (22). The increased recombination caused by *recD* is as intense as that due to Xer inactivation. It should be noted that this deletion assay does not use an inserted lambda, and it does not interfere with normal dimer resolution in resolvase-positive strains. How inactivation of exonuclease V triggers this localized hyperrecombination is not yet understood. One possibility is that the normal blockage of forks at *Ter* sites (the most frequently operative should be *TerC* [68]) leads to DNA structures providing opportunities for recombination, such as free-ended DNA due to snapback of replication forks. Elimination of such material might be performed by RecBCD, a phenomenon which is not necessarily recombinogenic. But in a *recD* mutant, such DNA formations may subsist, and their processing by RecBC in the absence of exonuclease V activity can now increase local recombination. Most probably due to differing causes, the localized increases in recombination due to *recD* and *xerC* mutations can be cumulated and, interestingly, the accumulation effect is far more than additive (22). This may be related to the correction by *recD* of some of the deleterious effects caused by *dif* inactivation (89). A RecD⁻ phenotype certainly minimizes the DNA degradation that otherwise occurs in the vicinity of the *dif* locus in XerC⁻ mutants, due to aborted dimer resolution. The continued presence of free-ended DNA may lead to increased opportunities for homologous recombination to restore chromosome circularity and viability, with recombination products becoming consequently detectable.

Another phenomenon that is presumably related to the properties of forks in the terminus region is Tn7 transposition. This transposon may hop to a preferential locus (*attTn7*) or to multiple locations, depending on a factor modulating transposase activity. The general transposition mechanism requires a transposon-encoded DNA binding protein, TnsE. A wild-type version of TsnE directs transposition of Tn7 preferentially into the terminus and, more precisely, near *Ter* sites (84). A mutated and more efficient version of TsnE causes much less regional specificity but generates an absolute orientation specificity related to direction of replication. Remarkably, the only region which does not display orientation specificity is the *TerA-TerB* interval (85). The authors proposed that TsnE recognizes features frequently found at replication forks (they suggested 3′ ends), and that targeting to the terminus by wild-type TsnE reflects the lesser affinity of this protein to DNA, which may favor recognition of stalled replication forks. Though replication forks stalled at terminators for a significant lapse of time might not be frequent, they may constitute the bulk of stalled forks during unperturbed growth and be numerous enough to target Tn7 transposition, which is also a low-frequency event.

EVOLUTIONARY PRESSURES AT WORK IN TER REGIONS

Perhaps a large proportion of the DNA in this region is utilized for maintaining the structural organization of the dividing nucleoid-membrane complex.

B. J. Bachmann et al. (3)

When genomes of bacterial species closely related to each other are compared, maximal divergence appears localized either in the origin or in the terminus regions (100). Comparisons of this type also indicate that chromosomal inversions have occurred in the course of bacterial evolution, and the most frequent have been symmetric inversions around the replication origin-to-terminus axis (30). These global observations support the proposal that TER regions are especially sensitive to variation, and that maintenance of replichore polarization is positively selected.

The information described above suggests that evolutionary pressures having potentially opposite effects have shaped the current structure of *E. coli* termini. On one hand, the localized hyperrecombination and facilitated transposition characters of the TER region may favor divergence due to integration of horizontally transferred material or internal rearrangements. On the other hand, the need for bipolarization near *dif* plus the existence of the orientation constraints revealed by NDZs should be factors of stability. Some of the effects of these antagonistic pressures can be seen in genome sequences. Due to the steadily increasing availability of sequence information, it is now possible to compare the 50 kb surrounding *dif* in four different *E. coli* strains: K-12, two enterohemorrhagic O157:H7 strains, and one uropathogenic CFT073 strain (G. Plunkett III [University of Wisconsin], personal communication). Both O157:H7 strains are quite similar; they differ from each other by a large inversion including *dif* (J. Lobry, personal communication), and when this is taken into account they display a very similar distribution of RAG elements. All four strains display ex-

cellent conservation of the strong polarization of RAG sequences and of their polarity shift at *dif* (as in K-12 [Fig. 2]). Their number and position are less conserved. For instance, the immediate neighborhoods of *dif* in O157:H7 and K-12 are well conserved, except that O157:H7 carries a 3.8-kb insertion starting at about 100 bp to the right of *dif* in K-12 and substituting for about 1 kb of K-12 material. This insertion carries four RAG elements all appropriately oriented. Most of the RAG motifs present in homologous regions are conserved, and the few strain-specific ones result from point mutations. Evolution of the terminus occurred in a context of conservation of at least some of the sequences of the region. It has been pointed out that the genes of clear foreign origin present in the region are among the oldest ones of the K-12 genome (58). However, none of the genes from *TerA* to *TerB* are essential. Some are certainly functional, e.g., the *hipAB* and *marAR* loci (18, 38, 74), but on the average the genes of the region are poorly expressed (104). Interestingly, the *hipAB* locus functions like a poison-antidote system (8) and may consequently act as a killer of bacteria undergoing its loss, thus protecting the region from deletion.

One striking example of terminus evolution is provided by a comparison of *dif* regions of the related enteric bacteria *E. coli* K-12 and *S. enterica* serovars Typhi and Typhimurium. Figure 2 shows that the region surrounding *dif* in *E. coli* shares no sequence similarity (along 50 kb!) with the regions flanking the *dif* sites of *S. enterica* serovars Typhi and Typhimurium (which share extensive homology in the region). However, the organization of these regions with respect to termination and, more important, dimer resolution is well conserved. For instance, in *S. enterica* serovar Typhi, the *TerC* analogue maps at about the same distance from *dif* as it does in *E. coli*, at the beginning of a 15-kb segment conserved between *S. enterica* serovar Typhi and *E. coli*. Remarkably, sequences of *dif* sites are very well conserved, and bipolarization is observed in both species: despite the general lack of homology, the distribution of RAG elements polarizes the *dif*-flanking regions of *S. enterica* serovars Typhi (Fig. 2) and Typhimurium (not shown), as clearly as it does in *E. coli*. One may suspect that the RAG motif and its skew share the same function in these related bacteria: placing *dif* in the right environment for its resolution activity. The genetic structure of TER regions is thus paradoxical: there is excellent conservation of the only important site of the region, *dif*, and of the global distribution of short elements that are supposed to make it active, but besides these common characteristics, the sequences of the regions are highly divergent from species to species.

What may have happened in the sequences of TER regions of these enteric bacteria, and how has the peculiar distribution of the RAG motifs near *dif* been reached in totally different environments? Several factors could favor integration of horizontally transmitted material and be responsible for the strong divergence of TER regions between *E. coli* and *Salmonella*. These include increased recombination due to failed dimer resolution, replication forks aborted at *Ter* sites, or accidental usage of an *oriK*. These may preferentially cause internal rearrangements, but they would also facilitate insertion of foreign DNA. It is worth mentioning another possibility: entrance into a cell of foreign DNA that contains chi sites could convert this cell in a RecD⁻ phenocopy (51, 76). Increased recombination activity in the terminus should be expected from this transient phenotype. In this context of increased recombination in the terminus and decreased DNA degradation, the *dif* region would become a preferential target for integration of the exotic material.

The retrieval of a convenient bipolarization of the *dif*-flanking DNA after horizontal transfer is another question. First, there must have been selection among inserted foreign DNA fragments for those possessing enough DOPEs with sufficient skew to allow near-correct *dif* activity. Eventual improvement of polarization may have followed, by a slower process of mutation selection generating new, conveniently oriented RAG motifs and eliminating preexisting, badly oriented ones. Essential genes present in the region may have been mutated in this process, and this may have provided a pressure toward conservation of individuals having fewer and fewer such genes in the region. Further comparison of sequences of terminus regions in related species will eventually provide better prospects on the intriguing aspect of terminus evolution, and test the hypothesis that the terminus is primarily devoted to the structural maintenance of the dividing chromosome.

Another mechanism for evolution of the *dif* region is simply site-specific recombination involving low efficiency sites. This is exemplified by the mechanism of integration of the φLf phage DNA into the chromosome of *Xanthomonas campestris* (61). The phage harbors a degenerate *dif* site containing only one Xer arm and the central region. It may integrate within the chromosomal *dif* site by a RecA-independent mechanism. Owing to the degenerate structure of the phage site, Xer recombination, both integrative and excisive, probably occurs at low frequency, so that the integrated state may appear relatively stable.

The importance in evolution of this strategy is highlighted by the recent discovery that the phage responsible for pathogenicity of *Vibrio cholerae* integrates into the larger chromosome of its host by Xer recombination at a *dif*-like site (47).

The above discussion has emphasized *dif* and the RAG elements, and how they might have coevolved. But it could be that polarity did not evolve solely for the proper functioning of *dif*. Adaptive recycling of the GC skew, itself probably caused by an asymmetric mutation pressure on leading and lagging strands, might have occurred to satisfy other needs. One is the repair of aborted replication forks (by RecBCD action controlled by polar chi sites [57]). Another is to prevent DNA being trapped by the septum (by FtsK using RAG or similar polarized motifs [reference 23 and unpublished data]). Other processes, such as decatenation, maintenance of proper supercoiling levels, domain formation, ongoing separation of nascent daughter chromosomes during replication, and perhaps replication itself, might also use chi or RAG or another form of polarity. It is possible that recombination at *dif* evolved to take advantage of a preexisting polarity already in use. The existing status of RAG, *dif*, and associated factors would, in this view, be the product of an extended period of coevolution of the multiple transactions involved in chromosome processing. It appears to be a general rule that *dif* sites, recognized or putative, map at the junction between replichores polarized in opposite directions by motifs belonging to the RAG family. This is the case for the chromosomes of *Agrobacterium tumefaciens*, *E. coli*, *Haemophilus influenzae*, *Klebsiella pneumoniae*, *S. enterica* serovars Typhi and Typhimurium, *V. cholerae*, *X. campestris*, and *Yersinia pestis*. Since polarization generates sequence constraints generalized over the whole genome, polarity-based systems would contribute to the general stability of genomes. A robust bacterial taxonomy may emerge as analysis of the growing number of bacterial genome sequences allows us to trace the conservation and evolution of these systems.

FROM TERMINATION TO CELL DIVISION: THE MISSING LINK

Termination of replication is normally followed by cell division. As originally described in the model of Cooper and Helmstetter (19), the period between termination and cell division is called the D period. Not so constant as it was initially thought, it is long (20 min when the generation time is 30 min) and is devoted to a series of operations such as FtsZ ring formation, resolution and decatenation of daughter chromosomes, and structural changes as the daughter chromosomes form, all of which culminate with septation and completion of cell division.

Many operations of the cell cycle proceed in an orderly way, and they depend upon the preceding ones. These operations can require a fair amount of time, and to get them to fit into a generation time, many have to overlap at least at fast growth rates. This makes it difficult to determine when and how they are committed. One of the major unanswered questions concerning the cell cycle is what couples cell division to chromosome replication. One ancient and still pertinent proposal is that termination commits cell division. If such a control exists, it does not involve any unique site of the terminus, and in particular it does not involve the Tus-Ter system or dif. It could be the meeting of replication forks and subsequent dissociation of replisomes. Since this question is more appropriate for a review of bacterial cell cycle control, we close the present discussion with a suggestion. It is inspired by the probable existence of a replication fork factory located in the center of the replicating cell (52, 59) and by the fact that initiation of the FtsZ ring coincides with termination (28). It may be that the main control of the cell cycle (adorned by several backup pathways) is exerted by the event that positions a new replication machinery in the cell center. Subsequently, when the replication forks meet at the end of the replication cycle and the replisomes dissociate, that same topological determinant is used in cascade to position the FtsZ ring.

Acknowledgments. We gratefully thank Richard D'Ari, Tom Hill, and David Lane for careful reading and useful comments. Richard D'Ari suggested the Shakespearian part of the title. J.-M.L. and P.K. also thank the many coworkers (technicians, undergraduate and graduate students, postdocs, and permanent collaborators) who have contributed so much to these studies for so many years. In addition to solid results and dissenting models, they have provided much that made this research so enjoyable and stimulating.

REFERENCES

1. Aussel, L., F. X. Barre, M. Aroyo, A. Stasiak, A. Z. Stasiak, and D. Sherratt. 2002. FtsK is a DNA motor protein that activates chromosome dimer resolution by switching the catalytic state of the XerC and XerD recombinases. *Cell* 108:195–205.
2. Bachellier, S., J. M. Clement, and M. Hofnung. 1999. Short palindromic repetitive DNA elements in enterobacteria: a survey. *Res. Microbiol.* 150:627–639.
3. Bachmann, B. J., K. B. Low, and A. L. Taylor. 1976. Recalibrated linkage map of *Escherichia coli* K-12. *Bacteriol. Rev.* 40:116–167.
4. Barre, F. X., M. Aroyo, S. D. Colloms, A. Helfrich, F. Cornet, and D. J. Sherratt. 2000. FtsK functions in the processing of a Holliday junction intermediate during bacterial chromosome segregation. *Genes Dev.* 14:2976–2988.
5. Begg, K. J., S. J. Dewar, and W. D. Donachie. 1995. A new *Escherichia coli* cell division gene, *ftsK*. *J. Bacteriol.* 177:6211–6222.
6. Béjar, S., and J. P. Bouché. 1984. The spacing of *Escherichia coli* DNA gyrase sites cleaved in vivo by treatment with oxolinic acid and sodium dodecyl sulfate. *Biochimie* 66:693–700.
7. Bird, R. E., J. Louarn, J. Martuscelli, and L. Caro. 1972. Origin and sequence of chromosome replication in *Escherichia coli*. *J. Mol. Biol.* 70:549–566.
8. Black, D. S., B. Irwin, and H. S. Moyed. 1994. Autoregulation of *hip*, an operon that affects lethality due to inhibition of peptidoglycan or DNA synthesis. *J. Bacteriol.* 176:4081–4091.
9. Blakely, G., S. Colloms, G. May, M. Burke, and D. Sherratt. 1991. *Escherichia coli* XerC recombinase is required for chromosomal segregation at cell division. *New Biol.* 3:789–798.
10. Blattner, F. R., G. Plunkett III, C. A. Bloch, N. T. Perna, V. Burland, M. Riley, J. Collado-Vides, J. D. Glasner, C. K. Rode, G. F. Mayhew, J. Gregor, N. W. Davis, H. A. Kirkpatrick, M. A. Goeden, D. J. Rose, B. Mau, and Y. Shao. 1997. The complete genome sequence of *Escherichia coli* K-12. *Science* 277:1453–1474.
11. Boccard, F., and P. Prentki. 1993. Specific interaction of IHF with RIBs, a class of bacterial repetitive DNA elements located at the 3′ end of transcription units. *EMBO J.* 12:5019–5027.
12. Bouché, J. P., J. P. Gelugne, J. Louarn, J. M. Louarn, and K. Kaiser. 1982. Relationships between the physical and genetic maps of a 470 x 10(3) base-pair region around the terminus of *Escherichia coli* K12 DNA replication. *J. Mol. Biol.* 154:21–32.
13. Bussiere, D. E., and D. Bastia. 1999. Termination of DNA replication of bacterial and plasmid chromosomes. *Mol. Microbiol.* 31:1611–1618.
14. Capiaux, H., F. Cornet, J. Corre, M. Guijo, K. Perals, J. E. Rebollo, and J. M. Louarn. 2001. Polarization of the *Escherichia coli* chromosome. A view from the terminus. *Biochimie* 83:161–170.
15. Capiaux, H., C. Lesterlin, K. Perals, J. M. Louarn, and F. Cornet. 2002. Functional replacement of the *E. coli dif* recombination site reveals a dual role for the FtsK protein in chromosome segregation. *EMBO Rep.* 3:523–526.
16. Chen, J. C., and J. Beckwith. 2001. FtsQ, FtsL and FtsI require FtsK, but not FtsN, for co-localization with FtsZ during *Escherichia coli* cell division. *Mol. Microbiol.* 42:395–413.
17. Clerget, M. 1991. Site-specific recombination promoted by a short DNA segment of plasmid R1 and by a homologous segment in the terminus region of the *Escherichia coli* chromosome. *New Biol.* 3:780–788.
18. Cohen, S. P., S. B. Levy, J. Foulds, and J. L. Rosner. 1993. Salicylate induction of antibiotic resistance in *Escherichia coli*: activation of the *mar* operon and a *mar*-independent pathway. *J. Bacteriol.* 175:7856–7862.
19. Cooper, S., and C. E. Helmstetter. 1968. Chromosome replication and the division cycle of *Escherichia coli* B/r. *J. Mol. Biol.* 31:519–540.
20. Cornet, F., J. Louarn, J. Patte, and J. M. Louarn. 1996. Restriction of the activity of the recombination site *dif* to a small zone of the *Escherichia coli* chromosome. *Genes Dev.* 10:1152–1161.
21. Corre, J., F. Cornet, J. Patte, and J. M. Louarn. 1997. Unraveling a region-specific hyper-recombination phenomenon: genetic control and modalities of terminal recombination in *Escherichia coli*. *Genetics* 147:979–989.

22. Corre, J., J. Patte, and J. M. Louarn. 2000. Prophage lambda induces terminal recombination in *Escherichia coli* by inhibiting chromosome dimer resolution. An orientation-dependent cis-effect lending support to bipolarization of the terminus. *Genetics* **154:**39–48.

23. Corre, J., and J. M. Louarn. 2002. Evidence from terminal recombination gradients that FtsK uses replichore polarity to control terminus positioning at division. *J. Bacteriol.* **184:** 3801–3807.

24. Coskun-Ari, F. F., and T. M. Hill. 1997. Sequence-specific interactions in the Tus-Ter complex and the effect of base pair substitutions on arrest of DNA replication in *Escherichia coli. J. Biol. Chem.* **272:**26448–26456.

25. Crosa, J. H., L. K. Luttropp, and S. Falkow. 1976. Mode of replication of the conjugative R-plasmid RSF1040 in *Escherichia coli. J. Bacteriol.* **126:**454–466.

26. de Massy, B., O. Fayet, and T. Kogoma. 1984. Multiple origin usage for DNA replication in sdrA(rnh) mutants of *Escherichia coli* K-12. Initiation in the absence of oriC. *J. Mol. Biol.* **178:**227–236.

27. de Massy, B., S. Bejar, J. Louarn, J. M. Louarn, and J. P. Bouche. 1987. Inhibition of replication forks exiting the terminus region of the *Escherichia coli* chromosome occurs at two loci separated by 5 min. *Proc. Natl. Acad. Sci. USA* **84:**1759–1763.

28. Den Blaauwen, T., N. Buddelmeijer, M. E. Aarsman, C. M. Hameete, and N. Nanninga. 1999. Timing of FtsZ assembly in *Escherichia coli. J. Bacteriol.* **181:**5167–5175.

29. Duggin, I. G., and R. G. Wake. 2001. Termination of chromosome replication, p. 87–95. *In* A. L. Sonenshein, J. A. Hoch, and R. M. Losick (ed.), Bacillus subtilis *and Its Closest Relatives: from Genes to Cells.* American Society for Microbiology, Washington D.C.

30. Eisen, J. A., J. F. Heidelberg, O. White, and S. L. Salzberg. 2000. Evidence for symmetric chromosomal inversions around the replication origins in bacteria. *Genome Biol.* **1:**1–9.

31. Espeli, O., and F. Boccard. 1997. In vivo cleavage of *Escherichia coli* BIME-2 repeats by DNA gyrase: genetic characterization of the target and identification of the cut site. *Mol. Microbiol.* **26:**767–777.

32. Espeli, O., L. Moulin, and F. Boccard. 2001. Transcription attenuation associated with bacterial repetitive extragenic BIME elements. *J. Mol. Biol.* **314:**375–386.

33. François, V., J. Louarn, J. Patte, J. E. Rebollo, and J. M. Louarn. 1990. Constraints in chromosomal inversions in *Escherichia coli* are not explained by replication pausing at inverted terminator-like sequences. *Mol. Microbiol.* **4:**537–542.

34. Frank, A. C., and J. R. Lobry. 1999. Asymmetric substitution patterns: a review of possible underlying mutational or selective mechanisms. *Gene* **238:**65–77.

35. Gordon, G. S., D. Sitnikov, C. D. Webb, A. Teleman, A. Straight, R. Losick, A. W. Murray, and A. Wright. 1997. Chromosome and low copy plasmid segregation in *E. coli*: visual evidence for distinct mechanisms. *Cell* **90:**1113–1121.

36. Guijo, M. I., J. Patte, M. del Mar Campos, J. M. Louarn, and J. E. Rebollo. 2001. Localized remodeling of the *Escherichia coli* chromosome. The patchwork of segments refractory and tolerant to inversion near the replication terminus. *Genetics* **157:**1413–1423.

37. Hale, C. A, and P. A. de Boer. 2002. ZipA is required for recruitment of FtsK, FtsQ, FtsL, and FtsN to the septal ring in *Escherichia coli. J. Bacteriol.* **184:**2552–2556.

38. Hendricks, E. C., H. Szerlong, T. Hill, and P. Kuempel. 2000. Cell division, guillotining of dimer chromosomes and SOS induction in resolution mutants (*dif, xerC* and *xerD*) of *Escherichia coli. Mol. Microbiol.* **36:**973–981.

39. Henson, J. M., and P. L. Kuempel. 1985. Deletion of the terminus region (340 kilobase pairs of DNA) from the chromosome of *Escherichia coli. Proc. Natl. Acad. Sci. USA* **82:**3766–3770.

40. Hill, T. M., J. M. Henson, and P. L. Kuempel. 1987. The terminus region of the *Escherichia coli* chromosome contains two separate loci that exhibit polar inhibition of replication. *Proc. Natl. Acad. Sci. USA* **84:**1754–1758.

41. Hill, T. M. 1996. Features of the chromosomal terminus region, p. 1602–1614. *In* F. C. Neidhardt, R. Curtiss III, J. L. Ingraham, E. C. C. Lin, K. B. Low, B. Magasanik, W. S. Reznikoff, M. Riley, M. Schaechter, and H. E. Umbarger (ed.), Escherichia coli *and* Salmonella: *Cellular and Molecular Biology*, 2nd ed., vol. 2. American Society for Microbiology, Washington, D.C.

42. Hiraga, S., C. Ichinose, H. Niki, and M. Yamazoe. 1998. Cell cycle-dependent duplication and bidirectional migration of SeqA-associated DNA-protein complexes in *E. coli. Mol. Cell* **1:**381–387.

43. Hojgaard, A., H. Szerlong, C. Tabor, and P. Kuempel. 1999. Norfloxacin-induced DNA cleavage occurs at the *dif* resolvase locus in *Escherichia coli* and is the result of interaction with topoisomerase IV. *Mol. Microbiol.* **33:**1027–1036.

44. Horiuchi, T., Y. Fujimura, H. Nishitani, T. Kobayashi, and M. Hidaka. 1994. The DNA replication fork blocked at the Ter site may be an entrance for the RecBCD enzyme into duplex DNA. *J. Bacteriol.* **176:**4656–4663.

45. Horiuchi, T., H. Nishitani, and T. Kobayashi. 1995. A new type of *E. coli* recombinational hotspot which requires for activity both DNA replication termination events and the Chi sequence. *Adv. Biophys.* **31:**133–147.

46. Hou, R., and T. M. Hill. 2002. Loss of RecA function affects the ability of *Escherichia coli* to maintain recombinant plasmids containing a Ter site. *Plasmid* **47:**36–50.

47. Huber, K. E., and M. K. Waldor. 2002. Filamentous phage integration requires the host recombinases XerC and XerD. *Nature* **417:**656–659.

48. Jones, N. C., and W. D. Donachie. 1973. Chromosome replication, transcription and control of cell division in *Escherichia coli. Nat. New Biol.* **243:**100–103.

49. Kogoma, T. 1978. A novel *Escherichia coli* mutant capable of DNA replication in the absence of protein synthesis. *J. Mol. Biol.* **121:**55–69.

50. Konrad, E. B. 1977. Method for the isolation of *Escherichia coli* mutants with enhanced recombination between chromosomal duplications. *J. Bacteriol.* **130:**167–172.

51. Koppen, A., S. Krobitsch, B. Thoms, and W. Wackernagel. 1995. Interaction with the recombination hot spot chi in vivo converts the RecBCD enzyme of *Escherichia coli* into a chi-independent recombinase by inactivation of the RecD subunit. *Proc. Natl. Acad. Sci. USA* **92:**6249–6253.

52. Koppes, L. J., C. L. Woldringh, and N. Nanninga. 1999. *Escherichia coli* contains a DNA replication compartment in the cell center. *Biochimie* **81:**803–810.

53. Kuempel, P. L., P. Maglothin, and D. M. Prescott. 1973. Bidirectional termination of chromosome replication in *Escherichia coli. Mol. Gen. Genet.* **125:**1–8.

54. Kuempel, P. L., S. A. Duerr, and N. R. Seeley. 1977. Terminus region of the chromosome in *Escherichia coli* inhibits replication forks. *Proc. Natl. Acad. Sci. USA* **74:**3927–3931.

55. Kuempel, P. L., J. M. Henson, L. Dircks, M. Tecklenburg, and D. F. Lim. 1991. *dif*, a recA-independent recombination site in the terminus region of the chromosome of *Escherichia coli. New Biol.* **3:**799–811.

56. Kuempel, P., A. Hogaard, M. Nielsen, O. Nagappan, and M. Tecklenburg. 1996. Use of a transposon (Tndif) to obtain

suppressing and nonsuppressing insertions of the *dif* resolvase site of *Escherichia coli*. *Genes Dev.* **10:**1162–1171.

57. **Kuzminov, A.** 1999. Recombinational repair of DNA damage in *Escherichia coli* and bacteriophage lambda. *Microbiol. Mol. Biol. Rev.* **63:**751–813.

58. **Lawrence, J. G., and H. Ochman.** 1998. Molecular archaeology of the *Escherichia coli* genome. *Proc. Natl. Acad. Sci. USA* **95:**9413–9417.

59. **Lemon, K. P., and A. D. Grossman.** 1998. Localization of bacterial DNA polymerase: evidence for a factory model of replication. *Science* **282:**1516–1519.

60. **Leslie, N. R., and D. J. Sherratt.** 1995. Site-specific recombination in the replication terminus region of *Escherichia coli*: functional replacement of *dif*. *EMBO J.* **14:**1561–1570.

61. **Lin, N. T., R. Y. Chang, S. J. Lee, and Y. H. Tseng.** 2001. Plasmids carrying cloned fragments of RF DNA from the filamentous phage φLf can be integrated into the host chromosome via site-specific integration and homologous recombination. *Mol. Genet. Genomics* **266:**425–435.

62. **Lobry, J. R.** 1995. Properties of a general model of DNA evolution under no-strand-bias conditions. *J. Mol. Evol.* **40:**326–330.

63. **Lobry, J. R., and J. M. Louarn.** 2003. Polarisation of prokaryotic chromosomes. *Curr. Opin. Microbiol.* **6:**101–107.

64. **Louarn, J., J. Patte, and J. M. Louarn.** 1977. Evidence for a fixed termination site of chromosome replication in *Escherichia coli* K12. *J. Mol. Biol.* **115:**295–314.

65. **Louarn, J., J. Patte, and J. M. Louarn.** 1982. Suppression of *Escherichia coli dnaA46* mutations by integration of plasmid R100.1 derivatives: constraints imposed by the replication terminus. *J. Bacteriol.* **151:**657–667.

66. **Louarn, J. M., J. P. Bouche, F. Legendre, J. Louarn, and J. Patte.** 1985. Characterization and properties of very large inversions of the *E. coli* chromosome along the origin-to-terminus axis. *Mol. Gen. Genet.* **201:**467–476.

67. **Louarn, J. M., J. Louarn, V. Francois, and J. Patte.** 1991. Analysis and possible role of hyperrecombination in the termination region of the *Escherichia coli* chromosome. *J. Bacteriol.* **173:**5097–5104.

68. **Louarn, J., F. Cornet, V. Francois, J. Patte, and J. M. Louarn.** 1994. Hyperrecombination in the terminus region of the *Escherichia coli* chromosome: possible relation to nucleoid organization. *J. Bacteriol.* **176:**7524–7531.

69. **Maisnier-Patin, S., K. Nordstrom, and S. Dasgupta.** 2001. RecA-mediated rescue of *Escherichia coli* strains with replication forks arrested at the terminus. *J. Bacteriol.* **183:**6065–6073.

70. **Masters, M., and P. Broda.** 1971. Evidence for the bidirectional replications of the *Escherichia coli* chromosome. *Nat. New Biol.* **232:**137–140.

71. **McLean, M. J., K. H. Wolfe, and K. M. Devine.** 1998. Base composition skews, replication orientation, and gene orientation in 12 prokaryote genomes. *J. Mol. Evol.* **47:**691–696.

72. **Meselson, M., and F. W. Stahl.** 1958. The replication of DNA in *E. coli*. *Proc. Natl. Acad. Sci. USA* **44:**671–675.

73. **Mohanty, B. K., T. Sahoo, and D. Bastia.** 1998. Mechanistic studies on the impact of transcription on sequence-specific termination of DNA replication and vice versa. *J. Biol. Chem.* **273:**3051–3059.

74. **Moyed, H. S., and K. P. Bertrand.** 1983. *hipA*, a newly recognized gene of *Escherichia coli* K-12 that affects frequency of persistence after inhibition of murein synthesis. *J. Bacteriol.* **155:**768–775.

75. **Mulugu, S., A. Potnis, Shamsuzzaman, J. Taylor, K. Alexander, and D. Bastia.** 2001. Mechanism of termination of DNA replication of *Escherichia coli* involves helicase-

contrahelicase interaction. *Proc. Natl. Acad. Sci. USA* **98:**9569–9574.

76. **Myers, R. S., A. Kuzminov, and F. W. Stahl.** 1995. The recombination hot spot chi activates RecBCD recombination by converting Escherichia coli to a recD mutant phenocopy. *Proc. Natl. Acad. Sci. USA* **92:**6244–6248.

77. **Natarajan, S., S. Kaul, A. Miron, and D. Bastia.** 1993. A 27 kd protein of *E. coli* promotes antitermination of replication in vitro at a sequence-specific replication terminus. *Cell* **72:**113–120.

78. **Niki, H., Y. Yamaichi, and S. Hiraga.** 2000. Dynamic organization of chromosomal DNA in *Escherichia coli*. *Genes Dev.* **14:**212–223.

79. **Nishimura, Y., L. Caro, C. M. Berg, and Y. Hirota.** 1971. Chromosome replication in *Escherichia coli*. IV. Control of chromosome replication and cell division by an integrated episome. *J. Mol. Biol.* **55:**441–456.

80. **Nishitani, H., M. Hidaka, and T. Horiuchi.** 1993. Specific chromosomal sites enhancing homologous recombination in *Escherichia coli* mutants defective in RNase H. *Mol. Gen. Genet.* **240:**307–314.

81. **Pedersen, A. G., L. J. Jensen, S. Brunak, H. H. Staerfeldt, and D. W. Ussery.** 2000. A DNA structural atlas for *Escherichia coli*. *J. Mol. Biol.* **299:**907–930.

82. **Perals, K., F. Cornet, Y. Merlet, I. Delon, and J. M. Louarn.** 2000. Functional polarization of the *Escherichia coli* chromosome terminus: the dif site acts in chromosome dimer resolution only when located between long stretches of opposite polarity. *Mol. Microbiol.* **36:**33–43.

83. **Perals, K., H. Capiaux, J. B. Vincourt, J. M. Louarn, D. J. Sherratt, and F. Cornet.** 2001. Interplay between recombination, cell division and chromosome structure during chromosome dimer resolution in *Escherichia coli*. *Mol. Microbiol.* **39:**904–913.

84. **Peters, J. E., and N. L. Craig.** 2000. Tn7 transposes proximal to DNA double-strand breaks and into regions where chromosomal DNA replication terminates. *Mol. Cell* **6:**573–582.

85. **Peters, J. E., and N. L. Craig.** 2001. Tn7 recognizes transposition target structures associated with DNA replication using the DNA-binding protein TnsE. *Genes Dev.* **15:**737–747.

86. **Pettijohn, D. E.** 1996. The nucleoid, p. 158–166. *In* F. C. Neidhardt, R. Curtiss III, J. L. Ingraham, E. C. C. Lin, K. B. Low, B. Magasanik, W. S. Reznikoff, M. Riley, M. Schaechter, and H. E. Umbarger (ed.), Escherichia coli *and* Salmonella: *Cellular and Molecular Biology*, 2nd ed., vol. 1. American Society for Microbiology, Washington, D.C.

87. **Pichoff, S., and J. Lutkenhaus.** 2002. Unique and overlapping roles for ZipA and FtsA in septal ring assembly in *Escherichia coli*. *EMBO J.* **21:**685–693.

88. **Prescott, D. M., and P. L. Kuempel.** 1972. Bidirectional replication of the chromosome in *Escherichia coli*. *Proc. Natl. Acad. Sci. USA* **69:**2842–2845.

89. **Prikryl, J., E. C. Hendricks, and P. L. Kuempel.** 2001. DNA degradation in the terminus region of resolvase mutants of *Escherichia coli*, and suppression of this degradation and the Dif phenotype by *recD*. *Biochimie* **83:**171–176.

90. **Rebollo, J. E., V. Francois, and J. M. Louarn.** 1988. Detection and possible role of two large nondivisible zones on the *Escherichia coli* chromosome. *Proc. Natl. Acad. Sci. USA* **85:**9391–9395.

91. **Roecklein, B. A., and P. L. Kuempel.** 1992. In vivo characterization of tus gene expression in *Escherichia coli*. *Mol. Microbiol.* **6:**1655–1661.

92. **Roos, M., R. Lingeman, C. L. Woldringh, and N. Nanninga.** 2001. Experiments on movement of DNA regions in *Escherichia coli* evaluated by computer simulation. *Biochimie* **83:**67–74.

93. Salzberg, S. L., A. J. Salzberg, A. R. Kerlavage, and J. F. Tomb. 1998. Skewed oligomers and origins of replication. *Gene* **217**:57–67.

94. Seigneur, M., V. Bidnenko, S. D. Ehrlich, and B. Michel. 1998. RuvAB acts at arrested replication forks. *Cell* **95**:419–430.

95. Sharma, B., and T. M. Hill. 1995. Insertion of inverted Ter sites into the terminus region of the *Escherichia coli* chromosome delays completion of DNA replication and disrupts the cell cycle. *Mol. Microbiol.* **18**:45–61.

96. Sharpe, M. E., and J. Errington. 1996. The *Bacillus subtilis soj-spo0J* locus is required for a centromere-like function involved in prespore chromosome partitioning. *Mol. Microbiol.* **21**:501–509.

97. Steiner, W., G. Liu, W. D. Donachie, and P. Kuempel. 1999. The cytoplasmic domain of FtsK protein is required for resolution of chromosome dimers. *Mol. Microbiol.* **31**:579–583.

98. Steiner, W. W., and P. L. Kuempel. 1998. Cell division is required for resolution of dimer chromosomes at the dif locus of *Escherichia coli*. *Mol. Microbiol.* **27**:257–268.

99. Steiner, W. W., and P. L. Kuempel. 1998. Sister chromatid exchange frequencies in *Escherichia coli* analyzed by recombination at the *dif* resolvase site. *J. Bacteriol.* **180**:6269–6275.

100. Suyama, M., and P. Bork. 2001. Evolution of prokaryotic gene order: genome rearrangements in closely related species. *Trends Genet.* **17**:10–13.

101. Taki, K., and T. Horiuchi. 1999. The SOS response is induced by replication fork blockage at a Ter site located on a pUC-derived plasmid: dependence on the distance between ori and Ter sites. *Mol. Gen. Genet.* **262**:302–309.

102. Tecklenburg, M., A. Naumer, O. Nagappan, and P. Kuempel. 1995. The *dif* resolvase locus of the *Escherichia coli* chromosome can be replaced by a 33-bp sequence, but function depends on location. *Proc. Natl. Acad. Sci. USA* **92**:1352–1356.

103. Uno, R., Y. Nakayama, K. Arakawa, and M. Tomita. 2000. The orientation bias of Chi sequences is a general tendency of G-rich oligomers. *Gene* **259**:207–215.

104. Ussery, D., T. S. Larsen, K. T. Wilkes, C. Friis, P. Worning, A. Krogh, and S. Brunak. 2001. Genome organisation and chromatin structure in *Escherichia coli*. *Biochimie* **83**:201–212.

105. von Meyenburg, K., E. Boye, K. Skarstad, L. Koppes, and T. Kogoma. 1987. Mode of initiation of constitutive stable DNA replication in RNase H-defective mutants of *Escherichia coli* K-12. *J. Bacteriol.* **169**:2650–2658.

106. Wang, L., and J. Lutkenhaus. 1998. FtsK is an essential cell division protein that is localized to the septum and induced as part of the SOS response. *Mol. Microbiol.* **29**:731–740.

107. Yang, Y., and G. F. Ames. 1988. DNA gyrase binds to the family of prokaryotic repetitive extragenic palindromic sequences. *Proc. Natl. Acad. Sci. USA* **85**:8850–8854.

108. Yu, X. C., E. K. Weihe, and W. Margolin. 1998. Role of the C terminus of FtsK in *Escherichia coli* chromosome segregation. *J. Bacteriol.* **180**:6424–6428.

109. Zechiedrich, E. L., and N. R. Cozzarelli. 1995. Roles of topoisomerase IV and DNA gyrase in DNA unlinking during replication in *Escherichia coli*. *Genes Dev.* **9**:2859–2869.

110. Zechiedrich, E. L., A. B. Khodursky, and N. R. Cozzarelli. 1997. Topoisomerase IV, not gyrase, decatenates products of site-specific recombination in *Escherichia coli*. *Genes Dev.* **11**:2580–2592.

III. TRANSCRIPTION MACHINES

The Bacterial Chromosome
Edited by N. Patrick Higgins
© 2005 ASM Press, Washington, D.C.

Chapter 14

Overview of Transcription

Jeffrey Roberts

It will be clear from the chapters in this section that the study of bacterial transcription has been transformed by crystallographic structures of RNA polymerase: the core and holoenzyme structures of the bacterial *Thermus aquaticus* and *Thermus thermophilus* enzymes from the laboratories of Seth Darst and Dmitry Vassylyev, as well as a series of yeast polymerase II structures from Roger Kornberg's laboratory. As was predicted from their genetic similarity, the basic enzymatic machineries of the prokaryotic and eukaryotic RNA polymerase are essentially identical (chapter 15). The structure itself is gratifying and interesting, giving shape to years of biochemical studies that characterized the interaction of RNA polymerase with nucleic acids—the central step of gene expression. Models of the open promoter and elongation complexes derived from these structures reveal the active center of the enzyme buried deep in a cleft that also encloses the stabilizing RNA-DNA hybrid; the catalytic metals are apparent, and protein elements that may cycle during steps of translocation have been identified. The initiation process is revealed by the way the sigma initiation factor contacts the −35 and −10 promoter elements in the open promoter complex: one domain of sigma binds the duplex −35 element while another binds the nontemplate single strand of the −10 element to hold the strands apart at the boundary of the transcription bubble and expose the template to the active site. As is true for all dynamic processes for which there exist static views, the next challenge will be to understand the transition from the −10 element recognized as duplex DNA on the surface in the closed complex, to the unwound open complex with the template strand deeply buried in the enzyme.

The structures of RNA polymerase core and, separately, of the loosely connected C-terminal domains of the alpha subunit correlate decades of biochemical and genetic studies on the mechanisms and pathways of transcription activation (chapter 16). Activators generally work by contacting RNA polymerase subunits, usually binding a nearby but not necessarily adjacent DNA site. Such contact may activate initiation either by stabilizing the adjacent binding of RNA polymerase (a process sometimes called recruitment) or by favoring transition from the initial closed complex to the open promoter complex (sometimes called remodeling). Following the important discovery that deletion of the alpha C-terminal domain disables certain activators, the specific contacts between this domain and the preeminent model activator catabolite activator protein were found. As it turns out, catabolite activator protein and other activators can work by either recruitment or remodeling, depending upon their precise situation with respect to RNA polymerase; furthermore, a variety of different subunit contacts, including contacts to sigma, can be made. Other distinct modes of activation exist, including the striking ability of the activators MerR and BmrR to unwind DNA between the promoter elements and shorten the distance between −10 and −35 elements so that it matches the spacing native to holoenzyme. And the variant σ54 holoenzyme uses an ATP-dependent activator to remodel a stable closed complex into an active open promoter complex. With these various examples, there now likely exists a molecular paradigm for any newly discovered mode or agent of transcription activation in bacteria.

There are global modes of RNA polymerase initiation regulation in addition to the specific activation that Dove and Hochschild discuss in chapter 16. First, the variety of sigma factors (Fig. 1) provides switches between whole categories of genes. In particular, sigma factors whose activity or concentration is regulated by cellular conditions act to turn

Jeffrey Roberts • Department of Molecular Biology and Genetics, 349 Biotechnology Building, Cornell University, Ithaca, NY 14853-2703.

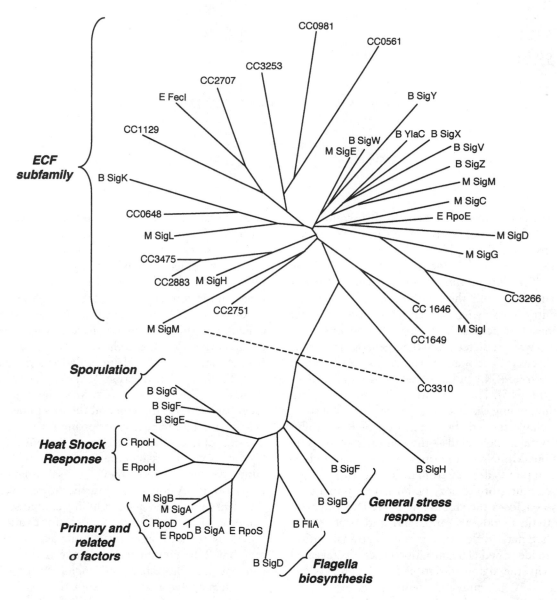

Figure 1. σ70 family members from four diverse bacteria: *E. coli* (E), *Caulobacter crescentus* (CC or C), *Bacillus subtilis* (B), and *Mycobacterium tuberculosis* (M). The seventh sigma factor of *E. coli*, σ54, is unrelated to the main sigma family and is not shown. Reprinted from reference 16 with permission.

on particular gene sets, such as those required for cell maintenance after heat shock, or those involved in flagellar biosynthesis, or those responding to various extracellular contingencies—the extracytoplasmic function (ECF) sigma factors. Although *Mycoplasma genitalium* survives with a single sigma factor and *Escherichia coli* musters up only a modest 7, *Streptomyces coelicolor* boasts 65, mostly ECF sigma factors that presumably deal with the complexities of its extracellular environment (2).

Another important control of transcription initiation involves the intracellular concentration of nucleotides. Magic Spot, a name worth recalling, which

comprises the nucleotides guanosine tetraphosphate (ppGpp) and pentaphosphate (pppGpp), is a signal of starved ribosomes, synthesized by the ribosomal proteins RelA and SpoT (4). Magic Spot inhibits initiation at promoters for stable RNA (ribosomal and transfer RNA) synthesis, likely by binding the RNAP core subunits and further weakening already short-lived open complexes at these promoters; Magic Spot is believed to activate other operons indirectly by increasing the concentration of available RNA polymerase (1). In a second function of small molecules, variations in the concentration of the RNA precursors ATP and GTP contribute to control of initiation at the rRNA gene promoters (18), and, in various subtle variations, determine the efficiency of promoter escape in operons involved in pyrimidine metabolism (12).

DNA structure beyond the immediate promoter influences RNA polymerase function. For example, some structure imposed by the DNA binding protein Hns on surrounding regions silences promoters that otherwise should be completely licensed to work (19). Supercoiling density can be an important determinant of promoter function; it is noteworthy that the first mutation isolated in a gene for a topoisomerase was selected to suppress a promoter mutation—*supX* of *Salmonella*, which turned out to encode topoisomerase I. Variations in supercoiling might act globally, as has been suggested for the response to variations in osmotic pressure (9), or locally in space and time, as may occur when the twin opposing supercoils spewed out of a transcribing RNAP (23) act to change the activity of an adjacent promoter (10).

Bacterial RNA polymerase provides the central model for the transcription elongation complex and its various interesting fates—backtracking and correction by Gre protein-mediated transcript cleavage, transcription termination, and the antitermination controls that were discovered in bacteria (chapter 17). A distinctive property of the incipient elongation complex is strongly illuminated by the structure: abortive initiation, the release of a substantial fraction of failed initial transcripts, which now can be ascribed to the necessity of the emerging transcript to expel part of the sigma initiation factor from the RNA exit channel. A collection of antitermination mechanisms, including the recently discovered riboswitch attenuation control (7, 13, 14), have been worked out in detail. Despite a variety of models, there is no certain way to relate termination and antitermination to RNA polymerase structure, mostly because, in contrast to initiation, there is no structural image of any stages of the processes. Nonetheless, as Artsimovitch describes in chapter 17, the structure gives significant focus to mechanistic studies of intrinsic (hairpin-induced) as well as enzymatic termination. Thus, the Rho termination factor is a "helicase" that uses ATP to attach (and probably traverse) RNA, and likely acts by extracting RNA from the elongation complex—not, as is abundantly clear from the structures, by attacking the RNA-DNA hybrid directly. Mfd is a DNA "helicase" (really a translocase) that interacts with upstream DNA in the complex and is proposed to act by collapsing the transcription bubble as it forces the enzyme forward.

Finally, Kushner (chapter 18) considers the enzymology of mRNA degradation, a process essential to operon function and a topic out of the mists of the origins of molecular biology but one that has been reluctant to yield the unifying mechanistic insights provided by studies of RNA biosynthesis. However, recent understanding of the roles of nucleases discovered decades ago, such as polynucleotide phosphorylase and RNase II, and the discovery of particles devoted to messenger degradation promise a more satisfactory view of the demise of mRNA.

RNA polymerase and its transcription factors have functions beyond their obvious activity to provide RNA molecules to the cell, reflecting the fact that RNA polymerase and the process of transcription must have evolved as DNA arose from the primal RNA world—neither is worth much without the other. There is evidence or informed speculation implicating RNA polymerase and transcription proteins in processes of replication, DNA repair, and cell division. Thus, transcription by RNA polymerase activates the origins of replication of *E. coli* and phage λ in some structural way independent of the RNA product. Classic studies of plasmid ColE1 replication showed the role of transcripts in priming DNA replication and in regulating the process of origin priming (22). That transcription may direct DNA repair was shown first in eukaryotes (3), but is also true in bacteria: RNA polymerase stalls at noncoding DNA lesions and becomes a target of the ATPase and "helicase" activities of Mfd (mutation frequency decline), which both removes RNA polymerase and recruits the machinery of DNA excision repair (20). Thus, RNA polymerase directly detects DNA damage and accelerates its removal. Furthermore, some cellular mutations that impair DNA repair can be suppressed by mutations in RNA polymerase genes, suggesting again an intimate relation of the two processes (15).

Rho and Mfd could have broader roles than currently understood. Rho is essential, and missense mutants have a variety of phenotypes, including sensitivity to DNA damage, recombination deficiency, and protein hyperdegradation (5, 21). These phenotypes could result from termination deficiency affecting expression of a variety of genes, including at least one essential one. However, in addition to the possibility of some uncharacterized activity, Rho (and Mfd as well) might be required to remove transcription complexes that obstruct processes of replication and repair.

One situation requiring RNAP removal is likely to be the encounter of transcription and replication. Although the phage T4 replication fork may pass through a transcription complex (11), our current understanding of RNA polymerase structure makes this seem an unlikely prospect for the bulk of encounters, most seriously in the case of ribosomal gene operons that may be closely packed with transcription complexes. Remarkably, the replication fork overtaking a field of transcription complexes from behind appears to remove them without delay, although in the converging situation the replication fork is considerably obstructed (8). Possibly the replication fork has an inborn mechanism to remove transcription complexes—namely, by the activity of the powerful replication helicases that might force collapse of the transcription bubble through positive supercoiling, similar to the putative winding activity of the Mfd factor in disrupting stalled transcription complexes (e.g., at a DNA lesion) (17). This might work well only for an approach from behind, accounting for the highly oriented structure of ribosomal operons, which point away from the replication origin. Possibly in converging transcription and replication there are other requirements for resolving the conflict, such as the activities of proteins like Mfd and Rho.

Finally, just as transcription and replication coevolved, so did the coordination of chromosome segregation and cell division arise in the context of both. DNA is transcribed as it moves about the cell in an organized fashion during replication (prokaryotes choosing not to shut down transcription while the cell is dividing). And RNA is translated at the same time, causing an added complication when emerging membrane proteins are inserted into the membrane and provide points of fixation for the complex. The cell has figured all this out, and it seems plausible that concurrent transcription and translation have positive as well as negative functions. Thus, Dworkin and Losick (6) propose that RNA polymerase provides an actual engine for chromosome segregation, and that the directionality of the major transcription units orients the process. What interactions might provide the attachments necessary to anchor this force are undetermined, but could involve the chain of connection that leads from RNA polymerase to points of protein insertion into the membrane.

REFERENCES

1. Barker, M. M., T. Gaal, and R. L. Gourse. 2001. Mechanism of regulation of transcription initiation by ppGpp. II. Models for positive control based on properties of RNAP mutants and competition for RNAP. *J. Mol. Biol.* **305:**689–702.
2. Bentley, S. D., K. F. Chater, A.-M. Cerdeño-Tárraga, G. L. Challis, N. R. Thomson, K. D. James, D. E. Harris, M. A. Quail, H. Kieser, D. Harper, A. Bateman, S. Brown, G. Chandra, C. W. Chen, M. Collins, A. Cronin, A. Fraser, A. Goble, J. Hidalgo, T. Hornsby, S. Howarth, C.-H. Huang, T. Kieser, L. Larke, L. Murphy, K. Oliver, S. O'Neil, E. Rabbinowitsch, M.-A. Rajandream, K. Rutherford, S. Rutter, K. Seeger, D. Saunders, S. Sharp, R. Squares, S. Squares, K. Taylor, T. Warren, A. Wietzorrek, J. Woodward, B. G. Barrell, J. Parkhill, and D. A. Hopwood. 2002. Complete genome sequence of the model actinomycete Streptomyces coelicolor A3(2). *Nature* **417:**141–147.
3. Bohr, V. A., C. A. Smith, D. S. Okumoto, and P. C. Hanawalt. 1985. DNA repair in an active gene: removal of pyrimidine dimers from the DHFR gene of CHO cells is much more efficient than in the genome overall. *Cell* **40:**359–369.
4. Cashel, M., D. R. Gentry, V. J. Hernandez, and D. Vinella. 1996. The stringent response, p. 1458–1496. *In* F. C. Neidhardt, R. Curtiss III, J. L. Ingraham, E. C. C. Lin, K. B. Low, B. Magasanik, W. S. Reznikoff, M. Riley, M. Schaechter, and H. E. Umbarger (ed.), Escherichia coli *and* Salmonella: *Cellular and Molecular Biology*, 2nd ed. ASM Press, Washington, D.C.
5. Das, A., D. Court, and S. Adhya. 1976. Isolation and characterization of conditional lethal mutants of *Escherichia coli* defective in transcription termination factor *rho. Proc. Natl. Acad. Sci. USA* **73:**1959–1963.
6. Dworkin, J., and R. Losick. 2002. Does RNA polymerase help drive chromosome segregation in bacteria? *Proc. Natl. Acad. Sci. USA* **99:**14089–14094.
7. Epshtein, V., A. S. Mironov, and E. Nudler. 2003. The riboswitch-mediated control of sulfur metabolism in bacteria. *Proc. Natl. Acad. Sci. USA* **100:**5052–5056.
8. French, S. 1992. Consequences of replication fork movement through transcription units in vivo. *Science* **258:**1362–1365.
9. Jordi, B. J., T. A. Owen-Hughes, C. S. Hulton, and C. F. Higgins. 1995. DNA twist, flexibility and transcription of the osmoregulated proU promoter of *Salmonella typhimurium. EMBO J.* **14:**5690–5700.
10. Lilley, D. M., and C. F. Higgins. 1991. Local DNA topology and gene expression: the case of the leu-500 promoter. *Mol. Microbiol.* **5:**779–783.
11. Liu, B., M. L. Wong, and B. Alberts. 1994. A transcribing RNA polymerase molecule survives DNA replication without aborting its growing RNA chain. *Proc. Natl. Acad. Sci. USA* **91:**10660–10664.
12. Liu, C., L. S. Heath, and C. L. Turnbough. 1994. Regulation of *pyrBI* operon expression in *Escherichia coli* by UTP-sensitive reiterative RNA synthesis during transcriptional initiation. *Genes Dev.* **8:**2904–2912.
13. Mandal, M., B. Boese, J. E. Barrick, W. C. Winkler, and R. R. Breaker. 2003. Riboswitches control fundamental biochemical pathways in *Bacillus subtilis* and other bacteria. *Cell* **113:**577–586.
14. McDaniel, B. A. M., F. J. Grundy, I. Artsimovitch, and T. M. Henkin. 2003. Transcription termination control of the S box system: direct measurement of S-adenosylmethionine by the leader RNA. *Proc. Natl. Acad. Sci. USA* **100:**3083–3088.
15. McGlynn, P., and R. G. Lloyd. 2000. Modulation of RNA polymerase by (p)ppGpp reveals a RecG-dependent mechanism for replication fork progression. *Cell* **101:**35–41.
16. Paget, M. S. B., and J. D. Helmann. 2003. The σ70 family of sigma factors. *Genome Biol.* **4:**203.
17. Park, J. S., M. T. Marr, and J. W. Roberts. 2002. E. coli transcription repair coupling factor (Mfd protein) rescues arrested complexes by promoting forward translocation. *Cell* **109:**757–767.
18. Schneider, D. A., W. Ross, and R. L. Gourse. 2003. Control of rRNA expression in *Escherichia coli. Curr. Opin. Microbiol.* **6:**151–156.
19. Schnetz, K. 1995. Silencing of *Escherichia coli bgl* promoter by flanking sequence elements. *EMBO J.* **14:**2545–2550.

20. **Selby, C. P., and A. Sancar.** 1993. Transcription-repair coupling and mutation frequency decline. *J. Bacteriol.* **175:** 7509–7514.

21. **Simon, L. D., M. Gottesman, K. Tomczak, and S. Gottesman.** 1979. Hyperdegradation of proteins in *Escherichia coli rho* mutants. *Proc. Natl. Acad. Sci. USA* **76:**1623–1627.

22. **Tomizawa, J., T. Itoh, G. Selzer, and T. Som.** 1981. Inhibition of ColE1 RNA primer formation by a plasmid-specified small RNA. *Proc. Natl. Acad. Sci. USA* **78:**1421–1425.

23. **Wu, H.-Y., S. Shyy, J. C. Wang, and L. F. Liu.** 1988. Transcription generates positively and negatively supercoiled domains in the template. *Cell* **53:**433–440.

The Bacterial Chromosome
Edited by N. Patrick Higgins
© 2005 ASM Press, Washington, D.C.

Chapter 15

The Structure of Bacterial RNA Polymerase

KATI GESZVAIN AND ROBERT LANDICK

Recently reported crystal structures of RNA polymerase (RNAP) have begun to shed light on the central enzyme of gene expression. In this chapter, we describe these structures and their implications for understanding the mechanism of transcription and the regulation of key steps in the transcription cycle. To lay a foundation for understanding the structures, we begin with a summary of the main features of the transcription cycle, RNAP's mechanism, and RNAP's subunit composition and primary structure.

THE TRANSCRIPTION CYCLE

There are three main steps in the transcription cycle: initiation, elongation, and termination (see chapters 16 and 17). During initiation, the core RNAP enzyme (subunit composition α_2, β', β, and ω in bacteria [13]) binds to one of the family of σ initiation factors. The resulting holoenzyme is able to bind specifically to the promoter DNA, forming the closed complex (RP_c [14, 47]) in a process called promoter recognition (Color Plate 3A [see color insert]). RP_c isomerizes in two or more steps to the open complex (RP_o), in which the strands of the DNA have melted to allow active-site access to the template strand (78). RP_o is capable of initiating transcription but, in most cases, remains at the promoter in an initial transcription complex that undergoes reiterative rounds of short transcript formation and release, called abortive transcription (Color Plate 3A), before releasing contacts with the DNA and escaping from the promoter (48, 98). After RNAP leaves the promoter, it forms a transcription elongation complex (TEC), the σ subunit is bound less avidly, and σ eventually dissociates. The TEC is processive and extremely stable (49), transcribing at an average rate of 30 to 100 nucleotides (nt)/s for tens of kilobases down the DNA template (54, 99). Transcription ends when RNAP reaches an intrinsic termination signal, char-

acterized by an RNA hairpin in the nascent transcript, or is acted upon by the termination factor ρ, causing the RNA transcript and the DNA to be released and freeing the core RNAP to begin another round of transcription. All steps in this enzymatic cycle of RNA synthesis can be modulated by regulatory molecules; understanding this regulation requires knowledge of the structure of intermediates in the cycle.

THE STRUCTURE OF THE TEC

Much is known about the general architecture of RNAP and the nucleic acid scaffold in the TEC from biochemical experiments (Color Plate 3B). By convention, the DNA that RNAP has yet to transcribe is called the downstream DNA and is designated with positive numbers; the upstream DNA is designated with negative numbers. Approximately 35 bp of DNA are protected within the TEC from DNase I or hydroxyl radical cleavage, with 15 to 20 bp held in the downstream jaws (61, 107). An ~17-bp region of the DNA called the transcription bubble is melted to expose the template strand (52, 108). The nascent RNA transcript forms an 8- or 9-bp RNA-DNA hybrid that is held in the active-site cleft (46, 52, 69, 88). Once the transcript is longer than 9 nt, it is peeled off the template strand and exits from the TEC through the RNA-exit channel, with an additional 5 nt of the RNA protected in the exit channel before the RNA emerges from the TEC (44).

THE NUCLEOTIDE ADDITION CYCLE

The addition of nucleotides to the 3' end of the growing RNA transcript is a multistep process involving nucleoside triphosphate (NTP) binding, phosphodiester bond formation, pyrophosphate release, and

---
Kati Geszvain and Robert Landick • Department of Bacteriology, University of Wisconsin—Madison, Madison, WI 53706.

enzyme translocation (Color Plate 3C). These steps occur in an active center composed of two sites, by convention called the i and $i+1$ sites. At the start of one round of NTP addition, the 3' end of the RNA lies in the i site, and the incoming NTP binds in the $i+1$ site (Color Plate 3C, "posttranslocated"). Two Mg^{2+} ions chelated in the active center are thought to catalyze an S_N2 nucleophilic attack of the 3'-OH group on the end of the RNA transcript on the α phosphate of the incoming NTP by stabilizing a trigonal bipyramidal transition state (Color Plate 3C, inset) (90, 91). Only one Mg^{2+} is stably bound in the active center; the other likely arrives coordinated to the nucleotide triphosphate. After phosphodiester bond formation occurs, the newly formed 3' end is located in the $i+1$ site and a molecule of pyrophosphate (PP_i) has been produced. Before a new round of NTP addition can occur, the PP_i is released from the active center and the TEC translocates one base pair down the DNA template to position the new 3' end of the transcript in the i site (32, 90). The catalytic cycle can be disrupted by regulatory signals intrinsic to the nucleic acid scaffold, such as pause and arrest sites, and by regulatory proteins such as ρ and λQ (described in chapter 17).

CONSERVATION OF RNAP PRIMARY SEQUENCES

The basic architecture of multisubunit RNAPs is conserved throughout the living world, with two large subunits forming the bulk of the enzyme (β' and β in bacteria), a homo- or heterodimer of smaller subunits on the periphery of the enzyme involved in assembly (the α dimer in bacteria), and at least one accessory subunit (ω in bacteria). β' and β are split into two polypeptides in some organisms (84) and can be fused into one polypeptide in others (106). Together, β' and β form the catalytic core of the enzyme and maintain the nucleic acid scaffold of the TEC (Color Plates 3B and 4A [see color insert]). β' and β are homologous to the two largest subunits of eukaryotic RNAPs (RPB1 and RPB2, respectively, in yeast RNAP II). Elements of sequence similarity are present in a conserved order in the primary structure of these subunits: A through H in β' and A through I in β (1, 93). In the three-dimensional structure of core RNAP, these conserved elements cluster around the active center, with the more divergent regions of the subunits located on the periphery of the enzyme (see Fig. 6 in reference 109).

The two α subunits play nonequivalent roles in the core structure (109). One α monomer, referred to as αI, contacts the β subunit (green in Color Plate 4A)

exclusively, whereas the other contacts β' (αII; yellow in Color Plate 4A). This is consistent with the order of assembly of the subunits into RNAP: the α dimer forms first, then β and β' bind consecutively (40). The α subunits are divided into two functional domains connected by a flexible linker (Color Plate 3B); the amino-terminal domain (NTD) is involved in RNAP assembly, and the carboxy-terminal domain (CTD) is involved in binding to the promoter UP element and interaction with transcriptional activators (7). A heterodimer of RPB3 and RPB11 in yeast RNAP II is similar in sequence to the NTD of the α homodimer, and is homologous in both its structure and its function in RNAP assembly (72, 92). The αCTD is related to the helix-hairpin-helix (HhH) family of DNA binding proteins (76). The ω subunit, which may also function in RNAP assembly, is homologous to RPB6 in eukaryotic RNAP II (62).

Currently, crystal structures are available for core RNAP (22, 109), holoenzyme (66, 97), holoenzyme with a "fork junction" DNA that mimics elements of the open complex (65), and the TEC (34). From the crystal structures and biochemical and genetic evidence, models of some intermediates in the transcription cycle for which there are no structures available have also been proposed (45, 65). Both the crystal structures themselves and the models derived from them are necessarily static approximations of the conformations assumed by RNAP during the transcription cycle. When viewed with appropriate caution, however, they afford powerful insight into the behavior of this intriguing enzyme. Some of these insights have been discussed in detail in recent reviews (8, 21, 29, 39, 81, 105). In this chapter, we provide a comprehensive overview of what has been learned from the RNAP structures and models.

CORE RNAP

Structural Conservation

The elucidation of the structure of RNAP from multiple organisms has revealed that the structure of RNAP, as well as its sequence, is conserved among prokaryotes and between prokaryotes and eukaryotes. The X-ray crystal structure of *Thermus aquaticus* RNAP aligns well with a 15-Å cryo-electron microscopy structure of *Escherichia coli* RNAP (23), supporting the idea that the thermophilic and mesophilic RNAPs have similar structures. Therefore, biochemical data derived from work with *E. coli* can be used in conjunction with structural data from *T. aquaticus*.

The overall shape is the same for both yeast and bacterial RNAP, as are discrete elements of the structure. These structural motifs, however, do not correspond to the elements of sequence conservation (β A to I, β′ A to H). Although the names of the conserved sequence elements remain in use, descriptive names such as the rudder or bridge helix (defined below) are needed to identify structural motifs (22, 109). The structural conservation among RNAPs is significantly greater than the sequence conservation (22, 109).

Overall Structure

The structure of *T. aquaticus* core RNAP, solved in 1999, first revealed the general shape of the enzyme (109). RNAP is 150 Å by 115 Å by 110 Å, with a deep cleft 27 Å wide that creates an overall "crab-claw" shape (Color Plate 4A, inset). The β and β′ subunits interact extensively, with part of the β subunit forming one pincer of the crab-claw and part of β′ forming the other (Color Plate 4A). The cleft is lined with positive residues, whereas the outside surface of RNAP is predominantly negative in charge (see Fig. 4 in reference 66). The αNTD dimer is located on the surface of the enzyme, opposite from the deep cleft. The ω subunit wraps around the carboxy-terminal tail of the β′ subunit and contacts β′ conserved regions D and G, conformationally constraining the β′ subunit and aiding its assembly into RNAP (62).

The αCTDs and linkers are disordered in the RNAP structures (65, 66, 109). The linker is unstructured (42). The structure of the αCTD has been determined both by nuclear magnetic resonance (NMR) and, in complex with DNA and catabolite activator protein, by X-ray crystallography (6, 41). A 14-amino-acid unstructured linker, corresponding to the length of the *E. coli* linker that is not seen in either the CTD or core structures, is predicted to span a root mean square distance of 43 Å if unconstrained by protein interactions. Thus, for perspective, the CTDs are depicted in arbitrary positions 43 Å from the NTDs in Color Plate 4A.

There are two Zn^{2+}-binding elements in prokaryotic RNAPs, which are not conserved in eukaryotes. One Zn^{2+}-binding element is a discrete domain (ZBD), previously identified by sequence analysis in region A of β′, located across from the β flap domain (described below). The crystal structure also reveals a novel Zn^{2+}-binding element in β′ (Zn^{2+} II) composed of four cysteine residues from regions F and G (Color Plate 4A). This domain lies on the outside of the cleft, suggesting it plays a role in folding β′ (58, 109).

Mobile Domains

Comparison of the various RNAP crystal structures now available has allowed the identification of several mobile domains in RNAP (23). The bulk of the enzyme, made up of the αNTDs, ω, and portions of β and β′ near the active site, forms an immobile core with four other modules able to move with respect to it. The β′ clamp domain forms the β′ pincer and can close down around the main channel to hold the DNA and the RNA-DNA hybrid more tightly in the active center. The β pincer is made up of two mobile domains called the lobe and protrusion (Color Plate 4B). Movement of these two modules can also open and close the active-site channel. The fourth mobile domain is the β flap, which covers the RNA-exit channel and in the core crystal structure appears to be held away from the body of the enzyme by crystal packing forces (66, 109).

The Active-Site Channel

The active-site (or main) channel, formed by the cleft between β and β′, is highly conserved among RNAPs and is lined with structural elements essential for catalysis and maintaining the nucleic acid scaffold. The active center is marked by a Mg^{2+} ion chelated at the base of the channel by three aspartate residues from the universally conserved NADFDGD motif of β′ region D (Color Plate 4B, Mg^{2+} I). Closing over the $i + 1$ site is a loop called the βD loop II, centered on *E. coli* residue 568 (most easily seen in Color Plate 7B). This loop is immediately adjacent to the β′ bridge helix, which lies just downstream of the $i + 1$ site (Color Plate 4A and B). The binding pocket for the antibiotic rifampin (defined by Rif^R substitutions and an RNAP-rifampin cocrystal [18]) is centered approximately 20 Å upstream from the active center on the wall of the active-site channel (Color Plate 4B). Rifampin positioned in this pocket would block growth of the RNA chain past two or three nucleotides, explaining the bactericidal effect of the antibiotic (18). Further upstream along the active-site channel is a "figure 8"-shaped loop called the β′ rudder. The upstream edge of the active-site channel is formed by the β subunit's flap domain, as well as the β′ lid and zipper domains (109).

The Secondary Channel

Immediately downstream of the active site, the bridge helix separates the main channel into a downstream DNA entry channel and a 10- to 12-Å-wide secondary channel (Color Plate 4A and B). Just inside the secondary channel lies the β′ trigger loop, which is partially disordered in *T. aquaticus* core and RNAP II but ordered in the *Thermus thermophilus* holoenzyme. This channel is too narrow for double-stranded nucleic acids to pass through, but is optimally positioned to allow NTP's access to the active center. Therefore, it has been proposed that the secondary channel serves as the entry site for NTPs (109). In a backtracked elongation complex, in which the RNAP has moved backward along the DNA and RNA, placing the active site over an internal phosphodiester bond, the 3′ end of the RNA transcript inserts into the secondary channel (30).

The RNA-Exit Channel

After the nascent transcript separates from the RNA-DNA hybrid, it is extruded from the TEC through the RNA-exit channel (Color Plate 4B; see Color Plate 7A). Cross-linking data suggest that the flap covers this channel (45), with the RNA passing between the base of the flap and the β′ lid (96). It has been suggested that during elongation the flap is closed down around the RNA in the exit channel, possibly contacting the β′ ZBD, but at hairpin-dependent pause signals the formation of an RNA hairpin underneath the flap opens the RNA-exit channel by clamp or flap movement, causing an allosteric change in the active site that alters the elongation behavior of the enzyme (96).

The Downstream DNA Channel

The downstream DNA is held in a channel formed by the β lobe and the β′ jaw, with 15 to 20 bp in the TEC protected from nuclease cleavage (Color Plate 4B; see Color Plate 7A). At least 9 bp of duplex DNA downstream of the active site is required for TEC stability (69), and the sequence of the downstream DNA can modulate the response to pause (53) and termination signals (74, 94), as well as modulate the rate of elongation (38). This suggests that the interaction between the downstream DNA and RNAP is important for TEC function.

The αCTD

Although the αCTD is not resolved in any of the RNAP crystal structures, structures of the isolated domain are available. An NMR structure for the isolated domain from *E. coli* is available (41), and a 3.1-Å X-ray cocrystal of *E. coli* αCTD with the catabolite activator protein (CAP) and an UP element is available (6). These structures reveal that the αCTD is a compactly folded domain with four α helices and

one nonstandard helix. Four of these five helices are involved in forming two HhH motifs, identifying the αCTD as a member of the HhH family of DNA-binding proteins (Color Plate 4C). This family is characterized by the presence of two antiparallel α-helices connected by a hairpin loop (76).

During transcription initiation, the αCTD makes multiple functional interactions, with the DNA in an UP element, activators such as CAP, and, potentially, σ region 4. In a cocrystal of αCTD, DNA, and CAP, Arg-265 and Asn-294, located at the helix-hairpin junctions of the two HhH motifs in the αCTD, position each other in the narrowed minor groove of three adjacent A/T base pairs in the UP element, with Arg-265 making a base-specific contact to N3 of an adenine in the sequence (Color Plate 4C). This interaction is aided by contacts of several other αCTD side chains to phosphates along the narrowed minor groove. In the cocrystal (6), αCTD and CAP interact through activating region 1 (AR1) of CAP and one of the surfaces of the αCTD available for interaction with transcriptional activators, the 287 determinant (Val-287), as was predicted from genetic experiments (80). When αCTD is bound to a promoter-proximal UP element or interacts with an activator that positions it next to the promoter (i.e., a class I CAP site), it is immediately adjacent to the binding site of σ region 4.2. A model of a DNA-αCTD-σ region 4.2 ternary complex assembled from the *T. aquaticus* σ region 4-DNA complex (19) and the *E. coli* αCTD-DNA complex identifies surfaces of α and σ that lie in close proximity (Color Plate 4C). This suggests that an interaction between the CTD and σ is involved in transcription activation, possibly by stabilizing σ region 4's contact to the −35 promoter element. In agreement with this, substitutions generated at these surfaces in the αCTD at Asp-259 and Glu-261 and σ region 4.2 at Arg-603 can decrease UP element function (20, 77).

HOLOENZYME

The σ^70 Family of Initiation Factors

The multiple members of the σ initiation factor family are divided into two classes, with little sequence conservation between the two. One class is similar to *E. coli*'s "housekeeping" σ, σ^70. The other is similar to σ^54 or σ^N and is responsible for transcribing genes required for nitrogen fixation as well as the stress response (12). The σ^70 class of initiation factors is the better characterized of the two. It is composed of the primary σ factors, which are responsible for transcribing most genes involved in basic cellular metabolism, and the alternative σ factors, which transcribe subsets of genes required under specific growth conditions, such as heat shock, or specific cellular processes, such as flagella production (57, 79). The primary σ's have four regions of sequence conservation (1.1 to 1.2, 2.1 to 2.4, 3.0 to 3.2, and 4.1 to 4.2 [Color Plate 5A (see color insert)]) that are responsible for core binding, promoter recognition, and DNA melting as well as a nonconserved region inserted between regions 1.2 and 2.1 that is found in only some σ's (57, 85). Region 1.1 also is not present in the alternative σ factors (57).

No complete structure for free σ is available, but biochemical data and X-ray crystal structures of proteolytic fragments provide some insight into the structure of the subunit. Limited proteolysis of free σ indicates that the subunit is made up of compactly folded domains joined by flexible linkers (85). This has been confirmed by the publication of the crystal structure of portions of the housekeeping σ, σ^A, from *T. aquaticus* (19). Fortuitous contamination of the crystallization solution with a protease produced fragments of σ containing regions 1.2 through 3.1 and regions 4.1 to 4.2. In this structure, region 1.2 through 2.4 folds into a compact structure, with a flexible linker joining it to region 3. What had been referred to as conserved region 2.5 (reference 5) is actually part of the region 3 domain; therefore it has been renamed region 3.0. Regions 4.1 and 4.2 also fold into a compact structure (Color Plate 5A). Region 1.1 and the linker between regions 3 and 4 were completely proteolyzed and therefore are not visible in this structure. Region 1.1 blocks the ability of region 4.2 to bind to the −35 element of the promoter (25, 26), suggesting that these two regions interact in free σ. However, NMR studies with region 4.2 detect no interaction between it and region 1.1 (17). Therefore, the structure and location of region 1.1 in free σ remain unknown.

σ-Core Interactions

The crystal structures for holoenzyme from *T. aquaticus* and *T. thermophilus* reveal the interactions between σ and core that confer on RNAP the ability to recognize the promoter (66, 97). These interactions are quite extensive, as had been predicted by biochemical studies (8, 35, 86). In holoenzyme, σ is folded into three flexibly linked domains, σ$_2$, σ$_3$, and σ$_4$ (Color Plate 5B), containing conserved regions 1.2 to 2.4, 3.0 to 3.1, and 4.1 to 4.2, respectively (66). σ$_2$ is bound to the β' clamp, with the major contact between σ region 2.2 and the coiled-coil domain of β' (Color Plate 5A), in agreement with biochemical and genetic evidence that these two re-

gions are the primary interface between core and σ (reference 3). σ$_3$ is located within the active-site channel, contacting primarily the β subunit near the active site. σ$_4$ wraps around the flap-tip helix of the β flap domain; a hydrophobic patch on the flap-tip helix thus becomes buried in the hydrophobic core of σ$_4$. Residues in regions 2.4, 3.0, and 4.2 that have been identified as making contacts with the promoter DNA are all surface exposed on the holoenzyme structure (Color Plate 5B).

The compact domains of σ are connected by flexible linkers that allow the subunit to stretch across the upstream face of the enzyme. The short linker between σ$_2$ and σ$_3$ is highly conserved and interacts with the β' zipper (Color Plate 5B), a conserved structural feature also present in eukaryotes. The 45-Å distance between regions 3 and 4 is spanned by conserved region 3.2. The 3.2 linker is almost completely buried in the active-site channel and the RNA-exit channel (Color Plate 5C), first interacting with the β' rudder, zipper, and lid before turning toward the active center (a 9-residue segment of the linker approaches within ∼25 Å of the active-center Mg^{2+}I), then turns back to enter the RNA-exit channel. In the relatively closed holoenzyme structure, the β' lid contacts the inner surface of the β flap, completely surrounding the 3.2 linker (66). The extended conformation of σ across core positions regions 2.4 and 4.2 optimally to bind the −10 and −35 promoter elements spaced 17 bp apart (51).

The position of region 1.1 in the holoenzyme is uncertain, but the available data suggest that it is located in the downstream side of the active-site channel (60). Neither holoenzyme crystal structure includes region 1.1: it has been proteolyzed in the *T. aquaticus* structure (66) and is not resolved in the *T. thermophilus* structure (97). Some evidence has suggested that region 1.1 is bound at the upstream face of RNAP, interacting with the flap and σ region 4 (10, 35). However, if this interaction occurs, it must be a transient intermediate in holoenzyme formation since fluorescence resonance energy transfer experiments that mapped the contacts between core and σ showed that region 1.1 is located within the downstream DNA channel in holoenzyme (60). Also, in the holoenzyme crystal structure (66), the amino-terminal fragment of σ is pointed into the active-site channel, not toward the flap (Color Plate 5B).

Conformational Changes in Holoenzyme Formation

Upon holoenzyme formation, both the core subunits and σ undergo conformational changes. In the core subunits, some regions move, whereas others that were disordered in the core RNAP structure become ordered (66). In the *T. aquaticus* holoenzyme structure, the β′ clamp and the β lobe domains rotate in toward the active-site channel, narrowing the width of the channel by 10 Å relative to core. The positioning of σ domains 2 and 3 on opposite sides of the active-site channel, with σ_2 on the mobile clamp domain, suggests that σ could play a role in opening and closing the active-site channel during promoter binding. The interaction with σ region 4 rotates the flap-tip helix about 15° toward the active site relative to core. The β protrusion domain rotates about 10 Å out from the active-site channel, most likely in response to the changes in the other domains (Color Plate 5B). The β′ ZBD, lid, and zipper domains that were disordered in the core structure all become ordered and visible in the holoenzyme (66, 97).

The DNA-binding ability of σ is unmasked upon binding to core by moving region 1.1 and changing the conformation of the DNA-binding domains themselves. The distances between region 1 and region 2 and between region 4 and region 2 increase upon core binding, suggesting that the conformation of σ in holoenzyme is "stretched out" relative to free σ and that region 1.1 has been moved away from the DNA-binding surfaces of region 2.4 and 4.2 (15, 16). A fragment of σ containing part of conserved region 1.2 through to region 2.4 binds specifically to the nontemplate strand of the promoter DNA only after binding to β′ or a fragment of β′ containing the coiled-coil domain. This suggests that the interaction between σ and core results in a conformational change within region 2 that allows DNA binding (104). In the holoenzyme structure, two rearrangements in σ region 2 are evident. First, a loop that covers the core binding surface in region 2.2 moves out of the way. Second, the bundle of helices made up of regions 1.2 and 2.1 to 2.4 rotates about 12° relative to the nonconserved region. However, it is not clear how these changes facilitate DNA binding. Possibly, further conformational changes occur in region 2 during the process of RP_o formation that enable nontemplate DNA binding, or some other interaction inhibits DNA binding by free σ or the σ fragment (2).

PROMOTER RECOGNITION

Closed Complex Formation

Transcription initiation is a multistep process in which holoenzyme (R) binds to the promoter (P) to form the closed complex (RP_c); then this complex undergoes isomerization through at least one intermediate (RP_i) to the open complex (RP_o) that is capable of binding NTPs and initiating transcription (ITC). The steps in this reaction can be depicted as (78):

$$R + P \leftrightarrow RP_c \leftrightarrow RP_i \leftrightarrow RP_o \rightarrow ITC$$

In the RP_c, σ^{70}-containing holoenzyme engages the −10 and −35 conserved hexamers of the promoter, with the DNA remaining double stranded and protected from both nuclease and hydroxyl radical cleavage from −54 to −6 (47). The spacing between the −10 and −35 elements can vary but is almost always between 16 and 18 bp (37). A separate set of σ^{70}-dependent promoters lacks the −35 element and instead requires recognition of the −16 TG element (a TRTGn motif located immediately upstream of the −10, also referred to as the extended −10 [11]). The structure of the RP_c must accommodate these variable DNA contacts as well as facilitate the transition to the RP_o.

Although no crystal structure of RP_c exists, its structure can be modeled from the holoenzyme structure, as well as from a σ_4-DNA cocrystal (19) and the holoenzyme-fork junction structure (described below) (65). In RP_c, the DNA-binding elements of σ lie across one face of the holoenzyme, defining this as the upstream face (Color Plate 6A [see color insert]). The −10 recognition helix of region 2.4 is spaced about 16 Å away from the −16 TG element recognition helix in region 3.0, which would easily accommodate the 5 bp of DNA separating the two elements (17-Å spacing, assuming straight, B-form DNA). Two residues in σ region 3.0, previously identified genetically to be involved in the recognition of the −16 TG sequence, face into the major groove of the −16 TG element (5). Region 4.2 is located on the β flap domain 76 Å away from region 2.4. This position could accommodate the canonical 17-bp spacing between the −10 and −35 elements with an 8° bend centered at about bp −25. However, the flap-σ region 4 complex can move relative to the DNA by at least 6 Å, allowing holoenzyme to bind promoters with noncanonical spacing (66, 97). The σ_4-DNA cocrystal shows that region 4.2 interacts with the major groove throughout the −35 element, contacting the phosphate backbone as well as making specific interactions with the DNA bases. The insertion of region 4.2's helix-turn-helix (HTH) recognition helix into the major groove induces a 36° bend in the DNA; this distortion may be important for the interaction of transcriptional regulators that bind upstream of the −35 element (19). The DNA in RP_c does not enter the active-site channel; this explains the lack of nuclease protection downstream of −5 in this complex (Color Plate 6A).

Open Complex Structure

Before RNAP can initiate transcription, the downstream DNA must insert into the downstream DNA channel, the DNA around the start site must be melted, and the template strand must insert into the active site to form the open complex (RP$_o$). A crystal structure of holoenzyme with a fork junction DNA template has been used to model this complex (65). The fork junction DNA contains double-stranded DNA from −41 through the −35 element up to the first base pair of the −10 element (the −12 bp), and the single-stranded nontemplate strand of the −10 element from −11 through to base −7. Holoenzyme bound to the fork junction DNA (RF) mimics many properties of the RP$_o$: (i) RF, like RP$_o$, is resistant to DNA binding competitors; (ii) substitutions in the promoter or RNAP that inhibit RP$_o$ formation also inhibit binding of the fork junction DNA; (iii) formation of RF is a multistep process, and some of the intermediates are similar to those in RP$_o$ formation (33, 36). Therefore, the RF structure can be used to model the structure of the RP$_o$ (65).

The RF structure suggests many details of the RP$_o$ structure. The DNA lies along one face of the RNAP, with the 8° bend at about nt −25 and a sharp, 37° turn at about −16 pointing the DNA toward the active center. In the RF structure, all of the RNAP-DNA contacts with the consensus promoter elements are mediated by σ. The clamp domain, along with σ region 2 bound to it, rotates in toward the main channel, causing it to close by 3 Å relative to the holoenzyme. The flap domain also moves about 6 Å downstream relative to the DNA, illustrating the flexibility of this domain. However, the σ region 4-DNA interaction appears to be distorted by crystal packing forces, so the positions of the flap, region 4, and −35 DNA in this crystal likely do not represent their normal positions. Residues in region 2.4 of σ identified as required to recognize the −12 position of the −10 element (89) are surface exposed and contact the −12 base (Color Plate 5A). In σ region 2.3, highly conserved aromatic residues that play a role in promoter melting (71) are surface exposed and positioned to interact with the single-stranded nontemplate DNA. One of these residues, Trp-256, appears to be stacked on the exposed face of the base pair at position −12. This interaction likely forms the upstream edge of the transcription bubble, and its formation may be the defining step in DNA melting (56). Universally conserved positive residues in regions 2.2 and 2.3 that appear to be involved in DNA binding and open complex formation (95) are positioned to interact with the negatively charged phosphate backbone of the nontemplate strand immediately upstream of the −10 element (Color Plate 5A).

Murakami et al. (65) propose a model of the RP$_o$ based on the RF structure, as well as the known structures of B-form DNA, the σ$_4$-DNA complex, and the model of the bacterial ternary elongation complex. The nontemplate DNA is proposed to be held in a groove between the lobe and protrusion based on cross-linking data (67). In the model, nt −6 through −3 of the nontemplate strand are exposed (depicted in the TEC structure, Color Plate 7A), consistent with previous nuclease and hydroxyl radical digestion studies that suggested that this part of the nontemplate strand is accessible in the TEC (101). The template strand of the DNA is inserted into the active site in the RP$_o$ in order to base pair with initiating NTPs. To reach the active site, the template strand passes through a tunnel that is completely enclosed by σ regions 2 and 3, the β' lid and rudder, and the β protrusion. The entrance to this tunnel is lined with highly conserved basic residues from σ regions 2.4 and 3.0 (Color Plate 6D). This positive charge may play a role in directing the negatively charged DNA into the tunnel. Downstream of the active site, the two DNA strands reanneal and are enclosed in the RNAP between the β' jaw and β lobe domain in the downstream DNA channel.

Open Complex Formation

The isomerization of the closed complex to the open complex is a multistep process, with at least one kinetically significant intermediate (78). Analysis of the effect of temperature on the kinetics of RP$_o$ formation suggested that both RNAP and the DNA undergo dramatic conformational changes during this process and led to the proposal that a kink in the DNA at the −10 hexamer forms in an intermediate to RP$_o$ formation (78). Substitution of the −11 template strand base with 2-amino purine results in the inability of RNAP to melt the DNA (56), suggesting that melting starts at this base in the −10 element. Aromatic residues in σ region 2.3 can be seen interacting with nontemplate strand bases in the RF crystal structure and may help pull the strands apart. DNA melting is blocked from proceeding upstream, possibly by the interaction of Trp-256 with the −12 base pair (Color Plate 6B).

The kink formed in the DNA by the initial unwinding at the −10 directs the downstream DNA into the entrance of the downstream DNA channel (Color Plate 6C). The interaction between the downstream DNA and the RNAP triggers further closing of the channel, bringing the DNA further down into the cleft and leading to the subsequent unwinding of the DNA (78). In support of the hypothesis that downstream DNA contacts regulate the DNA-melting step of RP$_o$ formation, in the *T. thermophilus* holoenzyme structure the β gate loop (*E. coli* residues 370 to 381 [Color Plate 6A]) and σ region 1.2 narrow the downstream DNA channel and prevent entry of double-stranded, but not single-stranded, DNA (97). In the model of the RP$_o$, the gate loop would interact with the major groove of the DNA at position +1 to +3, the endpoint of DNA melting (97). A large deletion in β that removes this gate loop as well as a substantial part of the lobe results in an RNAP that cannot melt the DNA downstream of −7, but, unlike wild-type RNAP, can initiate melting upstream of −7 at low temperature (82). This suggests that the gate loop, lobe, or both hinder the entry of the DNA into the downstream DNA channel, limiting the ability of holoenzyme to melt the DNA at low temperatures. Once the DNA enters the channel (Color Plate 6D), however, interaction between it and the gate loop and lobe may prevent rewinding and drive completion of DNA unwinding and RP$_o$ formation by trapping unwinding intermediates generated by thermal fluctuation (a type of thermal ratchet mechanism).

σ region 1.1 also plays a role in RP$_o$ formation. This was originally suggested by the fact that substitutions and deletions in region 1.1 have been shown to impair RP$_o$ formation at some promoters (100, 102). Region 1.1 is proposed to be in the downstream DNA channel in the holoenzyme (Color Plate 6B) and just outside the downstream channel (Color Plate 6D) in the RP$_o$ (60). Possibly, the role of region 1.1 is to hold the downstream DNA channel open in the holoenzyme and RP$_c$ to allow the DNA access to the channel (66, 97). During the transition to the RP$_o$, region 1.1 would exchange places with the DNA to sit outside the channel (Color Plate 6C). This would allow the two sides of the active-site channel to close down around the DNA, with the β' rudder and part of the protrusion interacting across the channel through the middle of the transcription bubble sealing the DNA strands apart (65).

THE ELONGATION COMPLEX

The Model of the Elongation Complex

The structure of yeast RNAP II TEC has been solved to 3.3 Å, but there is as yet no structure for a bacterial TEC (34). However, the high level of sequence and structural conservation between bacterial and eukaryotic RNAP makes it possible to infer the structure of bacterial TEC from the RNAP II TEC structure. Extensive cross-linking studies between the nucleic acids and the TEC have also allowed modeling of the prokaryotic TEC (45). The structures of free and elongating RNAP II differ mainly in the position of the clamp domain, which is closed down around the RNA-DNA hybrid. The downstream DNA enters the complex through a cleft between the β' jaw and β lobe. The 90° turn in the template strand induced during the formation of the RP$_o$ positions the +1 DNA base in the active site for base pairing with the incoming RNA nucleotide. The 8- or 9-bp RNA-DNA hybrid extends through the active-site channel past the β' rudder toward the flap domain (Color Plate 7A [see color insert]). The growing RNA transcript then passes between the β' lid and the base of the β flap domain into the RNA-exit channel underneath the flap before finally exiting RNAP (45). The upstream edge of the transcription bubble may be maintained by the lid and zipper domains of β' and possibly the β flap, leaving ∼17 bases melted in the bubble. The exiting upstream DNA duplex and the entering downstream DNA form a bend angle of about 90° in the TEC, due to the turn between the downstream DNA and the RNA-DNA hybrid (34). Atomic force microscopy measurements have indicated that ∼60 bp of DNA are compacted within the TEC (75); however, this has not been reconciled with the 35-bp nuclease footprint consistently seen with the TEC (61, 107).

Stability of the Elongation Complex

The structure of RNAP provides clues as to how the TEC can be both extremely stable and processive. Biochemical observations suggest that the RNA-DNA hybrid is required for the stability of the elongation complex (43, 88). Deletion of the rudder, which lies on the upstream end of the RNA-DNA hybrid, results in a TEC that is much less stable (50), suggesting that this domain holds the hybrid in the active-site channel. The active-site channel forms a complementary pocket for binding the hybrid, but the specificity is for the phosphate backbone, not specific bases. Several of the residues that contact the backbone interact with two phosphate groups simultaneously, possibly reducing the activation energy required for translocation. Also, the binding pocket is lined with positive residues that may attract the hybrid sequence nonspecifically (34). The single-stranded RNA transcript in the RNA-exit channel has also been shown to be important for the stability of the elongation complex (46, 103), possibly due to the flap domain closing down around the RNA. Together, these contacts allow for tight but sequence-nonspecific binding to the DNA template.

The Mechanism of Catalysis

The RNAP crystal structures reveal the architecture of the active center (Color Plate 7B). The universally conserved NADFDGD motif of the β' subunit chelates a Mg^{2+} ion deep in the active-site channel (Color Plate 4B), positioned between the i and $i+1$ sites (Mg^{2+} I), but the position of the second Mg^{2+} ion (Mg^{2+} II) is not as clear. Mg^{2+} II is not present in the *T. aquaticus* core crystal structure and is in two different positions in the *T. thermophilus* holoenzyme and RNAP II structures (22, 97, 109). This more weakly bound Mg^{2+} ion is thought to be brought into the active site by the incoming NTP. Sosunov et al. (90) have recently proposed an alternative Mg^{2+} binding site based on modeling and substitutions made in the active center (Color Plate 7B). In their model, Mg^{2+} II is coordinated by two of the three aspartate residues of the NADFDGD motif and stabilized by the phosphate of the incoming NTP. Unlike the positions assigned to Mg^{2+} II in the crystal structures, this position fits the requirements for the S_N2 geometry of the phosphodiester bond formation chemistry, as well as the nuclease and pyrophosphorylase activities of the active center (90). NTP gains access to the active center through the secondary channel. A binding site for the incoming NTP at the entrance to the secondary channel ("E site") (Color Plate 7B and C) has been proposed based on modeling and mutagenesis of the active site (90); NTP bound at this site has been detected in crystal structures (70).

The presence of alternative conformations of the bridge helix and trigger loop in the active site in the different crystal structures has led to conjecture about the possible mechanism of catalysis (Color Plate 7B and C). In the RNAP II TEC structure, the bridge helix is straight and the 3' end of the nascent RNA lies next to the helix in the $i+1$ site (34). However, in the bacterial RNAP structures, the bridge helix is bent or unfolded near the active site, with side chains in the unfolded portion of the helix sterically clashing with the presence of NTP or the hybrid in the $i+1$ site (e.g., β' T890 and A791 from the bridge helix clash with the template base at the 3' end of the hybrid [30, 109]). This conformational change in the bridge helix may be coupled to movement of the trigger loop, leading to the suggestion that their movements are interdependent and constitute a "swing-gate" structural element (30). The conformational changes in the swing gate are proposed to be required for translocation of RNAP down the template: bending of the bridge helix after NTP incorporation may drive the 3' end of the transcript from the $i+1$ to the i site. Subsequent straightening of the helix would allow NTP binding in the $i+1$ site and another round of NTP addition (30). Alternatively, the movements of the swing gate may be analogous to that of the O-helix in DNA polymerase, the movement of which facilitates proper alignment of the 3' nucleotide and dNTP substrate required for catalysis (28). In DNA polymerases, accumulation of negative charge on the newly formed pyrophosphate is proposed to drive an active-site rearrangement that translocates the 3' nt from the $i+1$ to the i site and allows PP_i release (27); a similar scenario may occur in RNAP.

Other Functions of the Active Site

The active site of RNAP is responsible not only for nucleotide addition, but also for discriminating ribo- from deoxyribonucleotides, maintaining the fidelity of transcription and in some cases cleaving the RNA transcript, either as a means of proofreading or to escape from a backtracked state. The conformation of the $i+1$ site may be involved in maintaining transcriptional fidelity. A Rif^R substitution in the β subunit that increased misincorporation (*ack-1*) maps between residues 565 and 576, which includes part of the βD loop II (Color Plate 7B) that lies over the $i+1$ site (55). Possibly, altering the structure of this loop allows non-Watson-Crick base pairs to fit better into the active site, increasing the rate of addition of mismatched NTPs. Cleavage of the nascent transcript is also involved in maintaining fidelity (31). The active-site channel is complementary to the conformation of the RNA-DNA hybrid, but not to the conformation of double-stranded DNA. Misincorporation of dNTPs, as well as mismatched rNTPs, into the nascent RNA would distort the hybrid and possibly decrease the stability of the complex, causing the RNAP to backtrack along the RNA and DNA chains to a point where correctly synthesized RNA would be in the hybrid and the misincorporated nucleotides would be in the secondary channel (34). This complex would then be subject to the action of the GreA/B cleavage factors, which bind in the outer entrance of the secondary channel (73) with a coiled-coil domain in their amino termini extending through the channel up to the active site (70), and stimulate the intrinsic nuclease activity of the active site (9). Highly conserved acidic residues in the Gre factors' coiled-coil domain may modify the active site to catalyze the cleavage reaction, either by directly stabilizing a Mg^{2+} II ion in the position proposed by Sosunov et al. or disrupting a salt bridge between an aspartate residue at β 814 and β R1106, allowing β D814 to chelate the second Mg^{2+} ion (70, 90).

INITIATION OF TRANSCRIPTION AND PROMOTER ESCAPE

Formation of the First Phosphodiester Bond

RNAP initiates transcription de novo from two NTPs, not from a primer as the DNA polymerases do. This means that the first phosphodiester bond forms between two nucleotide triphosphates in the i and $i + 1$ sites; each NTP would bind an Mg^{2+} ion. There-Therefore, during transcription initiation, the active center must accommodate three Mg^{2+} ions. The γ phosphate of the initiating NTP cross-links to a section of the σ loop 3.2 (83) that protrudes into the active site (66, 97). Possibly, this loop helps to chelate the Mg^{2+} ion associated with the NTP in the i site through highly conserved acidic residues, one of which is located ~15 Å from the γ phosphate of the initiating NTP. A truncated σ lacking region 3.2 forms a holoenzyme with lower affinity for the initiating NTP (19). Once this first phosphodiester bond is formed, translocation occurs and NTP binding is no longer required in the i site.

Abortive Initiation

The initiation of transcription is characterized by competition between transcript elongation and release. Successful elongation requires that RNAP disengage from contacts with the promoter sequences and begin translocating down the template. Before this happens, RNAP goes through several rounds of abortive transcription in which a short RNA product is synthesized and released, while RNAP remains at the promoter (98). In the holoenzyme structure, the σ region 3.2 loop occupies the active-site channel and the RNA-exit channel (Color Plate 6D), blocking extension of the RNA product past a few nucleotides (66). This suggests that the competition between transcript elongation and release reflects the competition between the 3.2 loop and the RNA to fill the active-site channel and the RNA-exit channel. A holoenzyme formed with a σ^{70} subunit truncated before region 3.2 exhibits decreased abortive initiation, supporting the competition model (66). Once the transcript reaches approximately 12 nt, it will have displaced the 3.2 loop (Color Plate 8A [see color insert]). This may disrupt the interaction between σ region 4 and the flap and result in destabilizing the interaction between σ and the -35 element of the promoter, initiating the process of promoter escape (Color Plate 8B) and the transition into the elongation complex (Color Plate 8C). The rate at which the initiating RNAP escapes from the promoter and the species of abortive products generated are both influenced by the sequences in the promoter (98).

σ Release

The release of σ from the elongating complex is a multistep process. In the open complex, σ is tightly bound to core, with contacts between σ region 2.2 and the β' coiled coil, σ region 3.2 and the RNA-exit channel, and σ region 4 and the flap domain (66, 97). During the process of abortive initiation, region 3.2 is displaced from the RNA-exit channel. This, in turn, may disrupt the interaction of σ region 4 with the flap. In fact, at the σ-dependent promoter-proximal pause site in λ P_R', σ region 4 has been repositioned in the complex by the antiterminator Q such that it has left the flap and is bound to a -35-like sequence located immediately upstream from the pause site. Thus it can be argued that as the polymerase leaves the promoter, the interaction between core and at least parts of σ is weakened (59, 68). Once the contacts with σ region 3 and 4 are lost, the residual region 2 interaction is lost slowly and stochastically (87). Stochastic release of σ weakly bound to the TEC may explain why some studies have detected σ persisting in the elongation complex (4, 63), even though experiments with reconstituted elongation complexes establish that RNA and σ compete for binding to RNAP (24). Furthermore, estimates of the in vivo activity of free σ suggest that it may be capable of rebinding the TEC during the course of transcript elongation (R. A. Mooney and R. Landick, unpublished data).

CONCLUSIONS

The publication of crystal structures for multiple forms of RNAP has made possible a much more detailed examination of the function of the enzyme and the mechanisms of catalysis, promoter recognition, and transcriptional activation. However, many questions remain to be answered. What is the structure of σ region 1.1, and where is it located in free σ, the holoenzyme, and RP$_o$? What is the mechanism of strand separation during open complex formation? How is the first phosphodiester bond formed? What are the conformational changes in the active site, and what are their roles in catalysis and translocation? What features of the active site are required to maintain transcriptional fidelity? How do pause sites dramatically slow the rate of nucleotide addition, and how do termination signals dissociate the TEC? Many of these questions can be addressed with additional crystal structures. For example, solving the structure of the TEC with a nonhydrolyzable nucleotide analogue in the $i+1$ site would be informative by showing the conformation of the active site immediately before nucleotide addition. Solving the first RNAP crystal structures is only the beginning of the road to understanding the function of the enzyme.

REFERENCES

1. Allison, L. A., M. Moyle, M. Shales, and C. J. Ingles. 1985. Extensive homology among the largest subunits of eukaryotic and prokaryotic RNA polymerases. *Cell* **42**:599–610.

2. Anthony, L. C., and R. R. Burgess. 2002. Conformational flexibility in sigma70 region 2 during transcription initiation. *J. Biol. Chem.* **277**:46433–46441.

3. Arthur, T. M., and R. R. Burgess. 1998. Localization of a sigma70 binding site on the N terminus of the Escherichia coli RNA polymerase beta' subunit. *J. Biol. Chem.* **273**:31381–31387.

4. Bar-Nahum, G., and E. Nudler. 2001. Isolation and characterization of sigma(70)-retaining transcription elongation complexes from Escherichia coli. *Cell* **106**:443–451.

5. Barne, K. A., J. A. Bown, S. J. Busby, and S. D. Minchin. 1997. Region 2.5 of the Escherichia coli RNA polymerase sigma70 subunit is responsible for the recognition of the "extended −10" motif at promoters. *EMBO J.* **16**:4034–4040.

6. Benoff, B., H. Yang, C. L. Lawson, G. Parkinson, J. Liu, E. Blatter, Y. W. Ebright, H. M. Berman, and R. H. Ebright. 2002. Structural basis of transcription activation: the CAP-alpha CTD-DNA complex. *Science* **297**:1562–1566.

7. Blatter, E. E., W. Ross, H. Tang, R. L. Gourse, and R. H. Ebright. 1994. Domain organization of RNA polymerase alpha subunit: C-terminal 85 amino acids constitute a domain capable of dimerization and DNA binding. *Cell* **78**:889–896.

8. Borukhov, S., and E. Nudler. 2003. RNA polymerase holoenzyme: structure, function and biological implications. *Curr. Opin. Microbiol.* **6**:93–100.

9. Borukhov, S., V. Sagitov, and A. Goldfarb. 1993. Transcript cleavage factors from E. coli. *Cell* **72**:459–466.

10. Bowers, C. W., and A. J. Dombroski. 1999. A mutation in region 1.1 of sigma70 affects promoter DNA binding by Escherichia coli RNA polymerase holoenzyme. *EMBO J.* **18**:709–716.

11. Bown, J. A., K. A. Barne, S. Minchin, and S. Busby. 1997. Extended −10 promoters. *Nucleic Acids Mol. Biol.* **11**:41–52.

12. Buck, M., M. T. Gallegos, D. J. Studholme, Y. Guo, and J. D. Gralla. 2000. The bacterial enhancer-dependent sigma(54) (sigma(N)) transcription factor. *J. Bacteriol.* **182**:4129–4136.

13. Burgess, R. R., and A. A. Travers. 1970. Escherichia coli RNA polymerase: purification, subunit structure, and factor requirements. *Fed. Proc.* **29**:1164–1169.

14. Burgess, R. R., A. A. Travers, J. J. Dunn, and E. K. Bautz. 1969. Factor stimulating transcription by RNA polymerase. *Nature* **221**:43–46.

15. Callaci, S., E. Heyduk, and T. Heyduk. 1998. Conformational changes of Escherichia coli RNA polymerase sigma 70 factor induced by binding to the core enzyme. *J. Biol. Chem.* **273**:32995–33001.

16. Callaci, S., E. Heyduk, and T. Heyduk. 1999. Core RNA polymerase from E. coli induces a major change in the domain arrangement of the sigma 70 subunit. *Mol. Cell* **3**:229–238.

17. Camarero, J. A., A. Shekhtman, E. A. Campbell, M. Chlenov, T. M. Gruber, D. A. Bryant, S. A. Darst, D. Cowburn, and T. W. Muir. 2002. Autoregulation of a bacterial sigma factor explored by using segmental isotopic labeling and NMR. *Proc. Natl. Acad. Sci. USA* **99**:8536–8541.

18. Campbell, E. A., N. Korzheva, A. Mustaev, K. Murakami, S. Nair, A. Goldfarb, and S. A. Darst. 2001. Structural mechanism for rifampicin inhibition of bacterial RNA polymerase. *Cell* **104**:901–912.

19. Campbell, E. A., O. Muzzin, M. Chlenov, J. L. Sun, C. A. Olson, O. Weinman, M. L. Trester-Zedlitz, and S. A. Darst. 2002. Structure of the bacterial RNA polymerase promoter specificity sigma subunit. *Mol. Cell* **9**:527–539.

20. Chen, H., H. Tang, and R. H. Ebright. 2003. Functional interaction between RNA polymerase alpha subunit C-terminal domain and sigma70 in UP-element- and activator-dependent transcription. *Mol. Cell* **11**:1621–1633.

21. Conaway, J. W., A. Shilatifard, A. Dvir, and R. C. Conaway. 2000. Control of elongation by RNA polymerase II. *Trends Biochem Sci.* **25**:375–380.

22. Cramer, P., D. A. Bushnell, and R. D. Kornberg. 2001. Structural basis of transcription: RNA polymerase II at 2.8 angstrom resolution. *Science* **292**:1863–1876.

23. Darst, S. A., N. Opalka, P. Chacon, A. Polyakov, C. Richter, G. Zhang, and W. Wriggers. 2002. Conformational flexibility of bacterial RNA polymerase. *Proc. Natl. Acad. Sci. USA* **99**:4296–4301.

24. Daube, S. S., and P. H. von Hippel. 1999. Interactions of Escherichia coli sigma(70) within the transcription elongation complex. *Proc. Natl. Acad. Sci. USA* **96**:8390–8395.

25. Dombroski, A. J., W. A. Walter, and C. A. Gross. 1993. Amino-terminal amino acids modulate sigma-factor DNA-binding activity. *Genes Dev.* **7**:2446–2455.

26. Dombroski, A. J., W. A. Walter, M. T. Record, Jr., D. A. Siegele, and C. A. Gross. 1992. Polypeptides containing highly conserved regions of transcription initiation factor sigma 70 exhibit specificity of binding to promoter DNA. *Cell* **70**:501–512.

27. Doublie, S., and T. Ellenberger. 1998. The mechanism of action of T7 DNA polymerase. *Curr. Opin. Struct. Biol.* **8**:704–712.

28. Doublie, S., S. Tabor, A. M. Long, C. C. Richardson, and T. Ellenberger. 1998. Crystal structure of a bacteriophage T7 DNA replication complex at 2.2 Å resolution. *Nature* **391**:251–258.

29. Ebright, R. H. 2000. RNA polymerase: structural similarities between bacterial RNA polymerase and eukaryotic RNA polymerase II. *J. Mol. Biol.* **304**:687–698.

30. Epshtein, V., A. Mustaev, V. Markovtsov, O. Bereshchenko, V. Nikiforov, and A. Goldfarb. 2002. Swing-gate model of nucleotide entry into the RNA polymerase active center. *Mol. Cell* **10**:623–634.

31. Erie, D. A., O. Hajiseyedjavadi, M. C. Young, and P. H. von Hippel. 1993. Multiple RNA polymerase conformations and GreA: control of the fidelity of transcription. *Science* **262**:867–873.

32. Erie, D. A., T. D. Yager, and P. H. von Hippel. 1992. The single-nucleotide addition cycle in transcription: a biophysical and biochemical perspective. *Annu. Rev. Biophys. Biomol. Struct.* **21**:379–415.

33. Fenton, M. S., S. J. Lee, and J. D. Gralla. 2000. Escherichia coli promoter opening and −10 recognition: mutational analysis of sigma70. *EMBO J.* **19**:1130–1137.

34. Gnatt, A. L., P. Cramer, J. Fu, D. A. Bushnell, and R. D. Kornberg. 2001. Structural basis of transcription: an

RNA polymerase II elongation complex at 3.3 A resolution. *Science* **292**:1876–1882.

35. Gruber, T. M., D. Markov, M. M. Sharp, B. A. Young, C. Z. Lu, H. J. Zhong, I. Artsimovitch, K. M. Geszvain, T. M. Arthur, R. R. Burgess, R. Landick, K. Severinov, and C. A. Gross. 2001. Binding of the initiation factor sigma(70) to core RNA polymerase is a multistep process. *Mol. Cell* **8**:21–31.

36. Guo, Y., and J. D. Gralla. 1998. Promoter opening via a DNA fork junction binding activity. *Proc. Natl. Acad. Sci. USA* **95**:11655–11660.

37. Harley, C. B., and R. P. Reynolds. 1987. Analysis of E. coli promoter sequences. *Nucleic Acids Res.* **15**:2343–2361.

38. Holmes, S. F., and D. A. Erie. 2003. Downstream DNA sequence effects on transcription elongation: allosteric binding of nucleoside triphosphates facilitates translocation via a ratchet motion. *J. Biol. Chem.* **278**:35597–35608.

39. Hsu, L. M. 2002. Open season on RNA polymerase. *Nat. Struct. Biol.* **9**:502–504.

40. Ishihama, A., N. Fujita, and R. E. Glass. 1987. Subunit assembly and metabolic stability of E. coli RNA polymerase. *Proteins* **2**:42–53.

41. Jeon, Y. H., T. Negishi, M. Shirakawa, T. Yamazaki, N. Fujita, A. Ishihama, and Y. Kyogoku. 1995. Solution structure of the activator contact domain of the RNA polymerase alpha subunit. *Science* **270**:1495–1497.

42. Jeon, Y. H., T. Yamazaki, T. Otomo, A. Ishihama, and Y. Kyogoku. 1997. Flexible linker in the RNA polymerase alpha subunit facilitates the independent motion of the C-terminal activator contact domain. *J. Mol. Biol.* **267**:953–962.

43. Kireeva, M. L., N. Komissarova, D. S. Waugh, and M. Kashlev. 2000. The 8-nucleotide-long RNA:DNA hybrid is a primary stability determinant of the RNA polymerase II elongation complex. *J. Biol. Chem.* **275**:6530–6536.

44. Komissarova, N., and M. Kashlev. 1998. Functional topography of nascent RNA in elongation intermediates of RNA polymerase. *Proc. Natl. Acad. Sci. USA* **95**:14699–14704.

45. Korzheva, N., A. Mustaev, M. Kozlov, A. Malhotra, V. Nikiforov, A. Goldfarb, and S. A. Darst. 2000. A structural model of transcription elongation. *Science* **289**:619–625.

46. Korzheva, N., A. Mustaev, E. Nudler, V. Nikiforov, and A. Goldfarb. 1998. Mechanistic model of the elongation complex of Escherichia coli RNA polymerase. *Cold Spring Harbor Symp. Quant. Biol.* **63**:337–345.

47. Kovacic, R. T. 1987. The 0 degree C closed complexes between Escherichia coli RNA polymerase and two promoters, T7-A3 and lacUV5. *J. Biol. Chem.* **262**:13654–13661.

48. Krummel, B., and M. J. Chamberlin. 1989. RNA chain initiation by Escherichia coli RNA polymerase. Structural transitions of the enzyme in early ternary complexes. *Biochemistry* **28**:7829–7842.

49. Krummel, B., and M. J. Chamberlin. 1992. Structural analysis of ternary complexes of Escherichia coli RNA polymerase. Deoxyribonuclease I footprinting of defined complexes. *J. Mol. Biol.* **225**:239–250.

50. Kuznedelov, K., N. Korzheva, A. Mustaev, and K. Severinov. 2002. Structure-based analysis of RNA polymerase function: the largest subunit's rudder contributes critically to elongation complex stability and is not involved in the maintenance of RNA-DNA hybrid length. *EMBO J.* **21**:1369–1378.

51. Kuznedelov, K., L. Minakhin, A. Niedziela-Majka, S. L. Dove, D. Rogulja, B. E. Nickels, A. Hochschild, T. Heyduk, and K. Severinov. 2002. A role for interaction of the RNA polymerase flap domain with the sigma subunit in promoter recognition. *Science* **295**:855–857.

52. Lee, D. N., and R. Landick. 1992. Structure of RNA and DNA chains in paused transcription complexes containing Escherichia coli RNA polymerase. *J. Mol. Biol.* **228**:759–777.

53. Lee, D. N., L. Phung, J. Stewart, and R. Landick. 1990. Transcription pausing by Escherichia coli RNA polymerase is modulated by downstream DNA sequences. *J. Biol. Chem.* **265**:15145–15153.

54. Levin, J. R., B. Krummel, and M. J. Chamberlin. 1987. Isolation and properties of transcribing ternary complexes of Escherichia coli RNA polymerase positioned at a single template base. *J. Mol. Biol.* **196**:85–100.

55. Libby, R. T., J. L. Nelson, J. M. Calvo, and J. A. Gallant. 1989. Transcriptional proofreading in Escherichia coli. *EMBO J.* **8**:3153–3158.

56. Lim, H. M., H. J. Lee, S. Roy, and S. Adhya. 2001. A "master" in base unpairing during isomerization of a promoter upon RNA polymerase binding. *Proc. Natl. Acad. Sci. USA* **98**:14849–14852.

57. Lonetto, M., M. Gribskov, and C. A. Gross. 1992. The sigma 70 family: sequence conservation and evolutionary relationships. *J. Bacteriol.* **174**:3843–3849.

58. Markov, D., T. Naryshkina, A. Mustaev, and K. Severinov. 1999. A zinc-binding site in the largest subunit of DNA-dependent RNA polymerase is involved in enzyme assembly. *Genes Dev.* **13**:2439–2448.

59. Marr, M. T., S. A. Datwyler, C. F. Meares, and J. W. Roberts. 2001. Restructuring of an RNA polymerase holoenzyme elongation complex by lambdoid phage Q proteins. *Proc. Natl. Acad. Sci. USA* **98**:8972–8978.

60. Mekler, V., E. Kortkhinjia, J. Mukhopadhyay, J. Knight, A. Revyakin, A. N. Kapanidis, W. Niu, Y. W. Ebright, R. Levy, and R. H. Ebright. 2002. Structural organization of bacterial RNA polymerase holoenzyme and the RNA polymerase-promoter open complex. *Cell* **108**:599–614.

61. Metzger, W., P. Schickor, and H. Heumann. 1989. A cinematographic view of Escherichia coli RNA polymerase translocation. *EMBO J.* **8**:2745–2754.

62. Minakhin, L., S. Bhagat, A. Brunning, E. A. Campbell, S. A. Darst, R. H. Ebright, and K. Severinov. 2001. Bacterial RNA polymerase subunit omega and eukaryotic RNA polymerase subunit RPB6 are sequence, structural, and functional homologs and promote RNA polymerase assembly. *Proc. Natl. Acad. Sci. USA* **98**:892–897.

63. Mukhopadhyay, J., A. N. Kapanidis, V. Mekler, E. Kortkhonjia, Y. W. Ebright, and R. H. Ebright. 2001. Translocation of sigma(70) with RNA polymerase during transcription: fluorescence resonance energy transfer assay for movement relative to DNA. *Cell* **106**:453–463.

64. Murakami, K. S., and S. A. Darst. 2003. Bacterial RNA polymerases: the wholo story. *Curr. Opin. Struct. Biol.* **13**:31–39.

65. Murakami, K. S., S. Masuda, E. A. Campbell, O. Muzzin, and S. A. Darst. 2002. Structural basis of transcription initiation: an RNA polymerase holoenzyme-DNA complex. *Science* **296**:1285–1290.

66. Murakami, K. S., S. Masuda, and S. A. Darst. 2002. Structural basis of transcription initiation: RNA polymerase holoenzyme at 4 A resolution. *Science* **296**:1280–1284.

67. Naryshkin, N., A. Revyakin, Y. Kim, V. Mekler, and R. H. Ebright. 2000. Structural organization of the RNA polymerase-promoter open complex. *Cell* **101**:601–611.

68. Nickels, B. E., C. W. Roberts, H. Sun, J. W. Roberts, and A. Hochschild. 2002. The sigma(70) subunit of RNA polymerase is contacted by the (lambda)Q antiterminator during early elongation. *Mol. Cell* **10**:611–622.

69. Nudler, E., E. Avetissova, V. Markovtsov, and A. Goldfarb. 1996. Transcription processivity: protein-DNA interactions holding together the elongation complex. *Science* 273:211–217.

70. Opalka, N., M. Chlenov, P. Chacon, W. J. Rice, W. Wriggers, and S. A. Darst. 2003. Structure and function of the transcription elongation factor GreB bound to bacterial RNA polymerase. *Cell* 114:272–274.

71. Panaghie, G., S. E. Aiyar, K. L. Bobb, R. S. Hayward, and P. L. de Haseth. 2000. Aromatic amino acids in region 2.3 of Escherichia coli sigma 70 participate collectively in the formation of an RNA polymerase-promoter open complex. *J. Mol. Biol.* 299:1217–1230.

72. Pati, U. K. 1994. Human RNA polymerase II subunit hRPB14 is homologous to yeast RNA polymerase I, II, and III subunits (AC19 and RPB11) and is similar to a portion of the bacterial RNA polymerase alpha subunit. *Gene* 145:289–292.

73. Polyakov, A., C. Richter, A. Malhotra, D. Koulich, S. Borukhov, and S. A. Darst. 1998. Visualization of the binding site for the transcript cleavage factor GreB on Escherichia coli RNA polymerase. *J. Mol. Biol.* 281:465–473.

74. Reynolds, R., and M. J. Chamberlin. 1992. Parameters affecting transcription termination by Escherichia coli RNA. II. Construction and analysis of hybrid terminators. *J. Mol. Biol.* 224:53–63.

75. Rivetti, C., S. Codeluppi, G. Dieci, and C. Bustamante. 2003. Visualizing RNA extrusion and DNA wrapping in transcription elongation complexes of bacterial and eukaryotic RNA polymerases. *J. Mol. Biol.* 326:1413–1426.

76. Ross, W., A. Ernst, and R. L. Gourse. 2001. Fine structure of E. coli RNA polymerase-promoter interactions: alpha subunit binding to the UP element minor groove. *Genes Dev.* 15:491–506.

77. Ross, W., D. A. Schneider, B. J. Paul, A. Mertens, and R. L. Gourse. 2003. An intersubunit contact stimulating transcription initiation by E coli RNA polymerase: interaction of the alpha C-terminal domain and sigma region 4. *Genes Dev.* 17:1293–1307.

78. Saecker, R. M., O. V. Tsodikov, K. L. McQuade, P. E. Schlax, Jr., M. W. Capp, and M. T. Record, Jr. 2002. Kinetic studies and structural models of the association of E. coli sigma(70) RNA polymerase with the lambdaP(R) promoter: large scale conformational changes in forming the kinetically significant intermediates. *J. Mol. Biol.* 319:649–671.

79. Sasse-Dwight, S., and J. D. Gralla. 1990. Role of eukaryotic-type functional domains found in the prokaryotic enhancer receptor factor sigma 54. *Cell* 62:945–954.

80. Savery, N. J., G. S. Lloyd, S. J. Busby, M. S. Thomas, R. H. Ebright, and R. L. Gourse. 2002. Determinants of the C-terminal domain of the Escherichia coli RNA polymerase alpha subunit important for transcription at class I cyclic AMP receptor protein-dependent promoters. *J. Bacteriol.* 184:2273–2280.

81. Severinov, K. 2000. RNA polymerase structure-function: insights into points of transcriptional regulation. *Curr. Opin. Microbiol.* 3:118–125.

82. Severinov, K., and S. A. Darst. 1997. A mutant RNA polymerase that forms unusual open promoter complexes. *Proc. Natl. Acad. Sci. USA* 94:13481–13486.

83. Severinov, K., D. Fenyo, E. Severinova, A. Mustaev, B. T. Chait, A. Goldfarb, and S. A. Darst. 1994. The sigma subunit conserved region 3 is part of "5'-face" of active center of Escherichia coli RNA polymerase. *J. Biol. Chem.* 269:20826–20828.

84. Severinov, K., A. Mustaev, A. Kukarin, O. Muzzin, I. Bass, S. A. Darst, and A. Goldfarb. 1996. Structural modules of the large subunits of RNA polymerase. Introducing archaebac-

85. Severinova, E., K. Severinov, D. Fenyo, M. Marr, E. N. Brody, J. W. Roberts, B. T. Chait, and S. A. Darst. 1996. Domain organization of the Escherichia coli RNA polymerase sigma 70 subunit. *J. Mol. Biol.* 263:637–647.

86. Sharp, M. M., C. L. Chan, C. Z. Lu, M. T. Marr, S. Nechaev, E. W. Merritt, K. Severinov, J. W. Roberts, and C. A. Gross. 1999. The interface of sigma with core RNA polymerase is extensive, conserved, and functionally specialized. *Genes Dev.* 13:3015–3026.

87. Shimamoto, N., T. Kamigochi, and H. Utiyama. 1986. Release of the sigma subunit of Escherichia coli DNA-dependent RNA polymerase depends mainly on time elapsed after the start of initiation, not on length of product RNA. *J. Biol. Chem.* 261:11859–11865.

88. Sidorenkov, I., N. Komissarova, and M. Kashlev. 1998. Crucial role of the RNA:DNA hybrid in the processivity of transcription. *Mol. Cell* 2:55–64.

89. Siegele, D. A., J. C. Hu, W. A. Walter, and C. A. Gross. 1989. Altered promoter recognition by mutant forms of the sigma 70 subunit of Escherichia coli RNA polymerase. *J. Mol. Biol.* 206:591–603.

90. Sosunov, V., E. Sosunova, A. Mustaev, I. Bass, V. Nikiforov, and A. Goldfarb. 2003. Unified two-metal mechanism of RNA synthesis and degradation by RNA polymerase. *EMBO J.* 22:2234–2244.

91. Steitz, T. A., S. J. Smerdon, J. Jager, and C. M. Joyce. 1994. A unified polymerase mechanism for nonhomologous DNA and RNA polymerases. *Science* 266:2022–2025.

92. Svetlov, V., K. Nolan, and R. R. Burgess. 1998. Rpb3, stoichiometry and sequence determinants of the assembly into yeast RNA polymerase II in vivo. *J. Biol. Chem.* 273:10827–10830.

93. Sweetser, D., M. Nonet, and R. A. Young. 1987. Prokaryotic and eukaryotic RNA polymerases have homologous core subunits. *Proc. Natl. Acad. Sci. USA* 84:1192–1196.

94. Telesnitsky, A., and M. J. Chamberlin. 1989. Terminator-distal sequences determine the in vitro efficiency of the early terminators of bacteriophages T3 and T7. *Biochemistry* 28:5210–5218.

95. Tomsic, M., L. Tsujikawa, G. Panaghie, Y. Wang, J. Azok, and P. L. deHaseth. 2001. Different roles for basic and aromatic amino acids in conserved region 2 of Escherichia coli sigma(70) in the nucleation and maintenance of the single-stranded DNA bubble in open RNA polymerase-promoter complexes. *J. Biol. Chem.* 276:31891–31896.

96. Toulokhonov, I., I. Artsimovitch, and R. Landick. 2001. Allosteric control of RNA polymerase by a site that contacts nascent RNA hairpins. *Science* 292:730–733.

97. Vassylyev, D. G., S. Sekine, O. Laptenko, J. Lee, M. N. Vassylyeva, S. Borukhov, and S. Yokoyama. 2002. Crystal structure of a bacterial RNA polymerase holoenzyme at 2.6 Å resolution. *Nature* 417:712–719.

98. Vo, N. V., L. M. Hsu, C. M. Kane, and M. J. Chamberlin. 2003. In vitro studies of transcript initiation by Escherichia coli RNA polymerase. 3. Influences of individual DNA elements within the promoter recognition region on abortive initiation and promoter escape. *Biochemistry* 42:3798–3811.

99. Vogel, U., and K. F. Jensen. 1994. The RNA chain elongation rate in Escherichia coli depends on the growth rate. *J. Bacteriol.* 176:2807–2813.

100. Vuthoori, S., C. W. Bowers, A. McCracken, A. J. Dombroski, and D. M. Hinton. 2001. Domain 1.1 of the sigma(70) subunit of Escherichia coli RNA polymerase modulates the

formation of stable polymerase/promoter complexes. *J. Mol. Biol.* **309**:561–572.

101. **Wang, D., and R. Landick.** 1997. Nuclease cleavage of the upstream half of the nontemplate strand DNA in an Escherichia coli transcription elongation complex causes upstream translocation and transcriptional arrest. *J. Biol. Chem.* **272**:5989–5994.

102. **Wilson, C., and A. J. Dombroski.** 1997. Region 1 of sigma70 is required for efficient isomerization and initiation of transcription by Escherichia coli RNA polymerase. *J. Mol. Biol.* **267**:60–74.

103. **Wilson, K. S., C. R. Conant, and P. H. von Hippel.** 1999. Determinants of the stability of transcription elongation complexes: interactions of the nascent RNA with the DNA template and the RNA polymerase. *J. Mol. Biol.* **289**:1179–1194.

104. **Young, B. A., L. C. Anthony, T. M. Gruber, T. M. Arthur, E. Heyduk, C. Z. Lu, M. M. Sharp, T. Heyduk, R. R. Burgess, and C. A. Gross.** 2001. A coiled-coil from the RNA polymerase beta′ subunit allosterically induces selective nontemplate strand binding by sigma(70). *Cell* **105**:935–944.

105. **Young, B. A., T. M. Gruber, and C. A. Gross.** 2002. Views of transcription initiation. *Cell* **109**:417–420.

106. **Zakharova, N., B. J. Paster, I. Wesley, F. E. Dewhirst, D. E. Berg, and K. V. Severinov.** 1999. Fused and overlapping *rpoB* and *rpoC* genes in helicobacters, campylobacters, and related bacteria. *J. Bacteriol.* **181**:3857–3859.

107. **Zaychikov, E., L. Denissova, and H. Heumann.** 1995. Translocation of the Escherichia coli transcription complex observed in the registers 11 to 20: "jumping" of RNA polymerase and asymmetric expansion and contraction of the "transcription bubble." *Proc. Natl. Acad. Sci. USA* **92**:1739–1743.

108. **Zaychikov, E., L. Denissova, T. Meier, M. Gotte, and H. Heumann.** 1997. Influence of Mg2+ and temperature on formation of the transcription bubble. *J. Biol. Chem.* **272**:2259–2267.

109. **Zhang, G., E. A. Campbell, L. Minakhin, C. Richter, K. Severinov, and S. A. Darst.** 1999. Crystal structure of Thermus aquaticus core RNA polymerase at 3.3 Å resolution. *Cell* **98**:811–824.

The Bacterial Chromosome
Edited by N. Patrick Higgins
© 2005 ASM Press, Washington, D.C.

Chapter 16

How Transcription Initiation Can Be Regulated in Bacteria

SIMON L. DOVE AND ANN HOCHSCHILD

RNA polymerase (RNAP) in bacteria is a multisubunit entity. The RNAP core enzyme (subunit composition $\alpha_2\beta\beta'\omega$) contains all of the catalytic machinery required for the synthesis of RNA (20, 27). The RNAP holoenzyme consists of the core enzyme complexed with one or another σ factor, which confers on the holoenzyme its ability to recognize specific promoter sequences (32). The core enzyme is evolutionarily conserved from bacteria to humans, as has been strikingly confirmed by recent structural studies of both a prokaryotic and a eukaryotic enzyme (19, 98). Although the eukaryotic enzyme consists of a dozen or more subunits, subunits corresponding to β', β, α, and ω compose a structural scaffold that is remarkably congruent with the bacterial enzyme (20, 27, 69). Because of their simpler composition, the bacterial RNAPs provide valuable models for understanding the function of all multisubunit RNAPs.

The initiation of transcription by RNAP is a complex process consisting of multiple steps (reviewed in reference 66). In vitro studies indicate that the RNAP holoenzyme first recognizes and binds to duplex promoter DNA, forming what is called the closed complex. After this initial binding step, the RNAP-promoter complex undergoes a series of conformational changes. These conformational changes result ultimately in the formation of the transcriptionally active open complex in which the DNA is locally melted to expose the start site of transcription. Open complex formation can be described by a simplified two-step model (Fig. 1) (34, 66). Typically, the initial binding step is reversible and is characterized by an equilibrium binding constant, K_B. The conversion of the closed complex to the open complex (referred to as the isomerization step) is in most cases essentially irreversible, and is characterized by a forward rate constant, k_f. Once the open complex has been formed, RNAP typically directs the synthesis of short abortive RNA products while still maintaining contact with promoter elements (15; see also reference 43 and references therein). Finally, RNAP escapes from the promoter, permitting the production of a full-length transcript.

Most bacteria contain several different σ factors (*Escherichia coli* contains seven), with each σ factor specifying the recognition of a particular class of promoters (reviewed in reference 32). The primary σ factor in *E. coli* is σ^{70}, which is responsible for the bulk of the transcription that occurs in exponentially growing cells (48). The so-called alternative σ factors direct RNAP to specific sets of promoters under particular environmental or developmental conditions (reviewed in reference 32). When present in the RNAP holoenzyme, σ^{70} makes direct contact with two conserved promoter elements known as the -10 and -35 elements (TATAAT and TTGACA, respectively) (Fig. 2A) (32). While σ^{70}-dependent promoters typically require both a -10 and a -35 element, σ^{70}-containing RNAP can also initiate transcription from a special class of promoters defined by what is known as an extended -10 element (reviewed in reference 6). The extended -10 element is characterized by a TG dinucleotide located upstream of the -10 hexamer (typically separated from it by 1 bp). Additional contacts between σ^{70} and this TG dinucleotide can evidently compensate for the lack of a -35 element (4).

Primary σ factors share four regions of conserved sequence (regions 1-4) (60). Regions 2 and 4 contact the -10 and -35 elements of the promoter, respectively, when σ^{70} is complexed with the core enzyme (Fig. 2A). On its own, σ^{70} is unable to bind to promoters; however, removal of an N-terminal inhibitory domain (region 1.1) permits free σ^{70} to bind specifically to promoter sequences (22). The binding of σ^{70} to the core enzyme has been shown to trigger significant conformational changes in σ^{70}. These include a dramatic increase in the interdomain distance

Simon L. Dove • Division of Infectious Diseases, Children's Hospital, Harvard Medical School, Boston, MA 02115.　Ann Hochschild • Department of Microbiology and Molecular Genetics, Harvard Medical School, Boston, MA 02115.

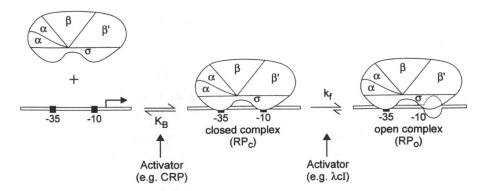

Figure 1. Two-step model for open complex formation. See text for details. Activators can influence open complex formation by exerting an effect at either step in the process.

between regions 2 and 4 to match the distance between the conserved promoter elements, as shown by luminescence resonance energy transfer measurements (13). A recent study indicates that this conformational change depends on an interaction between region 4 and a domain of the β subunit known as the β flexible flap domain (the β flap) (55). In fact, the β flap is required for RNAP holoenzyme containing σ^{70} to recognize typical −10/−35 promoters, presumably because it correctly positions region 4 relative to region 2 so that the −10 and −35 elements can be contacted simultaneously. In contrast, the β flap is not required for the σ^{70}-containing holoenzyme to recognize extended −10 promoters, consistent with the fact that region 4 itself is not required for the recognition of this class of promoters (55).

Interaction of region 2 of σ^{70} with the promoter −10 element is more complex than interaction of region 4 with the promoter −35 element. This stems from the fact that region 2 of σ^{70} plays an active role in the process of promoter melting, which initiates within the −10 element (reviewed in reference 32). In fact, region 2 of σ^{70} has both double-stranded and single-stranded DNA-binding activities. It can recognize the −10 element as duplex DNA (22). However, region 2 can also bind specifically to the nontemplate strand of the −10 element, an ability that was actually discovered in another context, during the course of studies on the mechanism of action of an antiterminator of transcription (82–84). Formation of the initiation-competent open complex presumably depends on the ability of core-bound σ^{70} to capture the nontemplate strand of the DNA and stabilize the initiation bubble.

The ability of σ^{70}-containing holoenzyme to bind selectively to the nontemplate strand of the −10 element can be assayed using single-stranded DNA oligonucleotides corresponding in sequence to either the template or the nontemplate strand (63). The use

of this assay has permitted the identification of a small (48-amino-acid) fragment of the β′ subunit of RNAP that suffices to confer on σ^{70} (or a σ^{70} fragment encompassing region 2) this selective single-stranded DNA-binding activity (97). This fragment of β′ consists of a coiled-coil motif, which evidently functions as an allosteric effector that induces in σ^{70} a conformational change required for it to bring about promoter melting.

Although the σ subunit of RNAP makes the critical contacts with promoter DNA, the α subunit of RNAP has also been found to play an important role in promoter recognition (reviewed in reference 30). The α subunit consists of two independently folded domains connected by a flexible linker region (Fig. 2A) (reviewed in reference 28). Whereas the α subunit N-terminal domain (α-NTD) mediates formation of the α dimer and serves as a scaffold for the assembly of the core enzyme (45), the α subunit C-terminal domain (α-CTD) is a DNA-binding domain that can contribute directly to promoter recognition (88). Specifically, the α-CTD binds to an A + T-rich sequence element (called the UP element) that is located just upstream of the −35 element of certain promoters (Fig. 2B) (reviewed in reference 30). Unlike the σ subunit and the core promoter elements, the α subunit can bind the UP element regardless of whether or not it is complexed with the other subunits of RNAP (88). The effect of the UP element can be very large: for example, in the case of the rRNA promoters, interactions between the α-CTD and the UP element can increase transcription 30-fold or more (87, 88). In fact, the α-CTD, though not strictly required for transcription in vitro, is essential for viability, suggesting that adequate expression of some number of essential genes depends on the UP element.

Because RNAP contains two copies of α, the question arises whether or not RNAP uses both of its

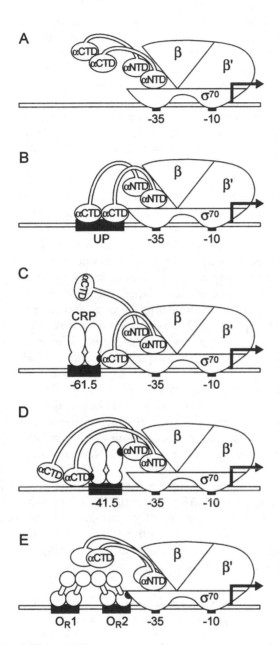

Figure 2. Transcription activation in bacteria. (A) RNAP (subunit composition $\alpha_2\beta\beta'\sigma^{70}$) bound to a σ^{70}-dependent promoter containing a -10 and a -35 element. The α subunits have been drawn to illustrate domain structure. αNTD designates the α N-terminal domain, and αCTD designates the α C-terminal domain. (B) RNAP bound to a σ^{70}-dependent promoter containing an UP element. (C) CRP-mediated transcription activation of a class I promoter. The activating region ARI of CRP (shaded black) is shown contacting the αCTD. (D) CRP-mediated transcription activation of a class II promoter. The activating regions ARI and ARII of CRP (shaded black) are shown contacting the αCTD and the αNTD, respectively. (E) λcI-mediated transcription activation from P$_{RM}$. λcI dimers are shown cooperatively bound to the operators O$_R$1 and O$_R$2. The activating region of λcI (shaded black) is shown contacting the σ^{70} subunit.

α-CTDs to contact the UP element. Detailed analysis has revealed that a so-called full UP element consists of two subsites, each of which binds an α-CTD (29). However, the two subsites can also function independently of one another, and in the case of a promoter that bears a proximal subsite only (located directly upstream of the -35 element), one α-CTD suffices to mediate full stimulation of transcription (29).

TRANSCRIPTION ACTIVATORS

Most activators of transcription initiation that affect σ^{70}-containing RNAP are sequence-specific DNA-binding proteins that bind to recognition sites located upstream of the core promoter elements (in some cases overlapping the -35 element). Generally, these DNA-bound activators make direct contact with one or more subunits of RNAP, and the α and σ^{70} subunits appear to be the preferred targets for these contacts (8, 39, 46). In particular, the DNA-binding domain of α (the α-CTD) and one of the DNA-binding domains of σ^{70} (region 4) provide contact surfaces for many DNA-bound activators. Activators that contact region 4 of σ^{70} generally bind to recognition sites that abut or partially overlap the promoter -35 element (61). On the other hand, activators that contact the α-CTD bind to recognition sites that are located at various positions upstream of the -35 element (47, 78). The surface of the activator that physically interacts with RNAP is often referred to as either a positive control surface or an activating region.

An Activator That Contacts the α-CTD: CRP

We begin by considering the mechanism of action of one of the most thoroughly characterized activators of σ^{70}-dependent transcription in *E. coli*, the cyclic AMP receptor protein (CRP), also known as the catabolite activator protein. CRP is a global regulator of transcription that activates the expression of more than 100 genes (10). It binds a cofactor, cyclic AMP (cAMP), that functions as an allosteric effector: only the cAMP-bound form of CRP is capable of sequence-specific DNA binding (53). When glucose is present in the growth medium, intracellular cAMP levels are low, whereas when glucose is absent, cAMP levels are elevated; thus, CRP-dependent gene expression is indirectly regulated by glucose levels.

CRP is a two-domain protein that binds its DNA recognition site as a dimer (91). The NTD mediates dimerization and also binds the cofactor cAMP; the CTD is the DNA-binding domain. At the well-studied *lac* promoter, the CRP recognition site is centered at position −61.5 relative to the start point of transcription (see reference 9). At this and other similarly arranged promoters, CRP uses a solvent-exposed patch in its CTD to contact the α-CTD (reviewed in reference 9). This patch is known as activating region I (ARI), and promoters at which the CRP recognition site is centered 61.5 bp or more upstream of the transcription start site are referred to as class I promoters. At the *lac* promoter, the protein-protein interaction between CRP and the α-CTD stabilizes the binding of the α-CTD to the DNA in the region between the CRP-recognition site and the promoter −35 element (Fig. 2C).

However, CRP also possesses a second activating region (ARII), which is located in the NTD. At so-called class II CRP-dependent promoters (such as the *gal* promoter), CRP binds to a recognition site that is centered at position −41.5 and makes two contacts with RNAP: it uses ARII to contact the α-NTD, and it uses ARI to contact the α-CTD (72). Under these circumstances, CRP stabilizes the binding of the α-CTD to the DNA just upstream of the CRP-recognition site, such that the DNA-bound CRP dimer is sandwiched between a DNA-bound α-CTD and the rest of RNAP (Fig. 2D) (see reference 9).

Studies of the kinetics of CRP-mediated activation at the *lac* promoter indicate that the interaction between ARI of CRP and the α-CTD stabilizes the initial binding of RNAP to the promoter (i.e., affects K_B) (62) (Fig. 1). Kinetic studies done at a class II promoter indicate that the interaction between ARI of CRP and the α-CTD similarly stabilizes the initial binding of RNAP to the promoter, but that the interaction between ARII of CRP and the α-NTD facilitates the isomerization step (i.e., affects k_f) (72, 79). Thus, whereas CRP affects only K_B at the *lac* promoter, CRP affects both K_B and k_f at class II promoters.

An Activator That Contacts Region 4 of σ⁷⁰: the λcI Protein

The cI protein of bacteriophage λ (λcI) provides an example of an activator that exerts its effect on transcription by contacting σ^{70} (39). Required for phage λ to grow lysogenically, λcI is both a repressor and an activator of transcription (75). In a λ lysogen, it represses transcription of the phage's early lytic genes and simultaneously activates transcription of its own gene (from promoter P_{RM}). λcI is also a two-domain protein that binds its DNA recognition site (the λ operator) as a dimer. The NTD is the DNA-binding domain, whereas the CTD mediates dimerization as well as higher-order oligomerization that results in cooperative binding to pairs of operator

sites (see references 38 and 75). λcI activates transcription from P_{RM} when bound to an operator site (O_R2) centered at position −42 relative to the start point of transcription (67). λcI mutants that are specifically defective for activation (positive-control mutants) bear amino acid substitutions within a solvent-exposed patch in the DNA-binding domain (12, 40), and λcI uses this positive-control surface to contact region 4 of σ^{70} (25, 54, 59) (Fig. 2E).

The identification of region 4 of σ^{70} as the contact site for λcI at P_{RM} was based on two lines of evidence. First, a σ^{70} mutant was isolated that restored the ability of a λcI positive-control mutant (λcI-D38N) to activate transcription; this mutant bore a single amino acid substitution in region 4 (R596H) (59). Second, several σ^{70} mutants were isolated that resulted in defects in λcI-stimulated transcription; these mutants also bore amino acid substitutions in region 4 (54). In accordance with this genetic evidence, molecular modeling suggests that the positive-control patch of O_R2-bound λcI is in close proximity to region 4 of σ^{70} when this region is bound to the −35 element (8). Subsequent experiments indicated that λcI can interact with a C-terminal fragment of σ^{70} encompassing region 4 (see below) (25).

Kinetic studies indicate that λcI working at P_{RM} stimulates only the rate of isomerization (i.e., affects k_f) (35). Moreover, positive-control mutant λcI-D38N stimulates the isomerization of RNAP containing σ^{70} with the R596H substitution (58). Unexpectedly, however, when wild-type λcI was tested in combination with RNAP containing the mutant σ^{70}, λcI was found to exert its effect primarily on the initial binding of RNAP (i.e., on K_B) (58). A possible explanation for this change in the kinetics of λcI-dependent activation is discussed below.

Using Artificial Activators To Probe the Mechanism of Transcription Activation

We now consider the question: how do protein-protein contacts between DNA-bound activators and RNAP stimulate transcription? Two alternative possibilities can be envisioned. On the one hand, the contact between a DNA-bound activator and RNAP could induce a conformational change in the enzyme that would facilitate a particular step in the initiation process. Implicit in this mechanism is the idea that activators contact special sites on RNAP to trigger the relevant conformational changes. Alternatively, a DNA-bound activator that makes an energetically favorable contact with RNAP could work by a cooperative binding mechanism, simply stabilizing the binding of RNAP to the promoter. In contrast to the first mechanism, this cooperative binding mechanism implies that activators could contact any accessible surface of the enzyme.

The use of artificial activators made it possible to subject the cooperative binding model to a critical test (24, 76). In particular, the following question was posed: can any sufficiently strong protein-protein contact between a DNA-bound protein and a subunit of RNAP activate transcription? Or, put another way: can the contact between a DNA-bound activator and RNAP be functionally replaced by a heterologous protein-protein interaction? Figure 3A illustrates the experimental strategy that was used to address this question. Two protein domains known to interact (X and Y) were fused, respectively, to a DNA-binding protein and a subunit of RNAP. In particular, domain X was fused to λcI (functioning simply as a DNA-binding protein in this context), and domain Y was fused to the end of the α linker, thereby replacing the α-CTD. It was then possible to ask whether or not the λcI-Y fusion protein could activate transcription from a suitably designed test promoter in cells containing the α-X chimera (Fig. 3A). The demonstration that such arbitrarily selected protein-protein interactions can activate transcription provided strong support for the cooperative binding mechanism for transcription activation (23, 26). Furthermore, as predicted by the model, a correlation was established between the strength of the engineered protein-protein interaction and the magnitude of the activation (26).

Further experiments indicated that transcription activation was also elicited when domain X was fused to another subunit of RNAP (ω), providing additional support for the idea that contact between a DNA-bound protein and any accessible surface of the enzyme can activate transcription (23). Finally, direct fusion of a DNA-binding protein to the α subunit of RNAP (or to ω) demonstrated that transcription can also be activated by a protein-DNA contact between the tethered protein and a cognate DNA recognition site suitably positioned upstream of the core promoter elements (23, 25, 52). Together, these observations establish that transcription can be activated by any interaction (protein-protein or protein-DNA) that helps tether RNAP to the promoter, presumably increasing the probability that the transcriptionally active open complex will form by decreasing the chance that RNAP will dissociate from the promoter.

These experiments with artificial activators suggest that at least some natural activators are likely to work by a simple cooperative binding mechanism. For example, in the case of CRP working at the *lac* promoter, it is likely that both the protein-protein interaction between CRP and the α-CTD and the

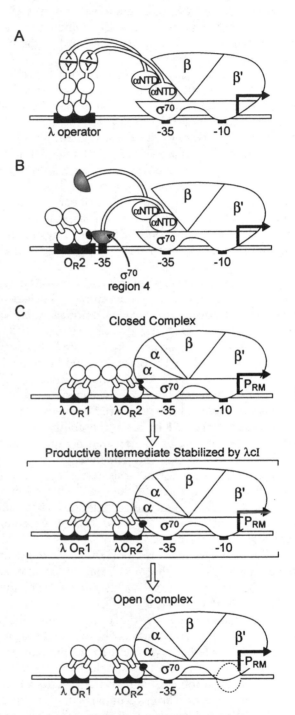

Figure 3. Use of artificial activators to probe activation mechanisms. (A) Interaction between protein domains X and Y can activate transcription. (B) Interaction between λcI and region 4 of σ^{70} tethered to the αNTD can activate transcription. The activating region of λcI (shaded black) is shown contacting the tethered σ^{70} moiety and stabilizing its binding to an ectopic −35 element. (C) Model for kinetic effect of λcI working at P_{RM}. Activating region of λcI (shaded black) and target surface on σ^{70} (shaded black) are misaligned in the closed complex, but come into alignment during the transition to the transcriptionally active open complex.

resulting protein-DNA interaction between the α-CTD and the DNA adjacent to the CRP-binding site are functioning simply to help tether RNAP to the promoter (8, 39).

How, then, does the interaction between λcI and region 4 of σ^{70} stimulate the rate of isomerization? An experimental strategy based on the previous findings with artificial activators was used to address this question. This strategy was designed specifically to test the idea that λcI can interact with region 4 of σ^{70} and stabilize its binding to an adjacently positioned −35 element (25). Accordingly, region 4 of σ^{70} was fused to the α-NTD and linker to create an α-σ^{70} chimera (Fig. 3B). In addition, a test promoter was created that bears a second −35 element upstream of the core promoter elements; this ectopic −35 element serves as a potential binding site for the tethered σ^{70} region 4 moiety. The test promoter also contained a λ operator just upstream of the ectopic −35 element, positioned so as to precisely recapitulate the relationship between O_R2 and the −35 element of P_{RM} (Fig. 3B). The previous work suggested that transcription from this test promoter would be activated by the binding of the tethered region 4 moiety to the ectopic −35 element. Moreover, the presence of the λ operator directly upstream of the ectopic −35 element made it possible to assay the ability of λcI to stabilize the binding of the tethered region 4 moiety to the ectopic −35 element. The finding that λcI did activate transcription from this test promoter specifically in the presence of the α-σ^{70} chimera indicated that λcI can make an energetically favorable interaction with region 4 of σ^{70} and stabilize its binding to a −35 element (25).

The simplest interpretation of the findings obtained with λcI and the α-σ^{70} chimera is that λcI functions analogously at P_{RM}, stabilizing the binding of region 4 of σ^{70} to the −35 element of P_{RM}. In support of this inference, the genetic requirements for λcI-dependent activation at the artificial test promoter mirrored those for λcI-dependent activation at P_{RM} (25). However, this simple picture presents an apparent paradox in light of the kinetics of λcI-dependent activation at P_{RM}. In particular, if λcI can stabilize the binding of region 4 of σ^{70} to the promoter −35 element, then why does it not stabilize the initial binding of RNAP to the promoter (i.e., affect K_B)? A possible answer (see reference 25) is illustrated in Fig. 3C: we have proposed that when RNAP forms a closed complex at P_{RM}, the positive control surface of λcI and its target site on σ^{70} may be misaligned so that no energetically favorable interaction can occur. We hypothesized further that these two surfaces come into alignment subsequently, during the transition from the closed to the

transcriptionally active open complex. By stabilizing the binding of region 4 of σ^{70} to the −35 element during this transition, λcI could stabilize a productive intermediate along the pathway to open complex formation, an effect that would manifest itself as an increase in the isomerization rate constant.

An implication of this view of λcI-mediated activation is that there need not necessarily be any fundamental difference between an activator that affects K_B and an activator that affects k_f (25, 39, 76, 89). In fact, as mentioned above, a single amino acid substitution in region 4 of σ^{70} changes the kinetic effect of wild-type λcI working at P_{RM} (58). Both types of activators (i.e., those that affect K_B and those that affect k_f) need only possess a surface that can interact with an accessible complementary surface on RNAP. Depending on when, during the initiation process, these surfaces can interact, one or the other kinetic parameter may be affected. In particular, a favorable interaction between a DNA-bound activator and RNAP that occurs when RNAP is in the closed complex would result in an effect on K_B, whereas a favorable interaction between a DNA-bound activator and RNAP that occurs only after the closed to open transition has begun would result in an effect on k_f. Thus, the critical factor that determines whether an activator affects K_B or k_f is not the detailed nature of the interaction between the activator and RNAP, but when during the initiation process the interaction can occur. In principle, therefore, the same protein-protein interaction between an activator and RNAP might produce an effect on K_B, k_f, or both, depending on the architecture of the promoter.

REPRESSORS THAT CONTACT RNAP

Many repressors apparently work by binding to DNA recognition sites that overlap with a promoter, thereby occluding the binding of RNAP. However, repressors can also work by contacting RNAP directly (39, 85, 89). In fact, in the case of the bacteriophage φ29 p4 protein, the same protein-protein contact between this regulator and RNAP can mediate either transcription activation or transcription repression (70, 71). Protein p4 activates transcription from the A3 promoter, whereas it represses transcription from the A2c promoter. In each case, p4 binds a specific recognition site located upstream of the core promoter elements (centered at position −82 of the A3 promoter and at position −71 of the A2c promoter) and interacts with the α-CTD (Fig. 4). Moreover, the same residues of p4 mediate this interaction in both cases (71).

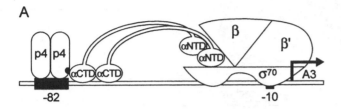

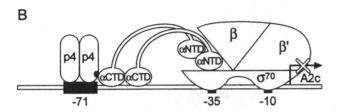

Figure 4. Transcription activation and repression by p4 of bacteriophage φ29. p4 activates transcription from the A3 promoter (A) and represses transcription from the A2c promoter (B). The same region of p4 (shaded black) contacts the αCTD to mediate both activation and repression.

Why then does p4 function as an activator at the A3 promoter and a repressor at the A2c promoter? It turns out that the effect of p4 depends on the characteristics of the target promoter itself (70). The A3 promoter is an inherently weak promoter that lacks a recognizable −35 element, whereas the A2c promoter is an inherently strong promoter that bears a near consensus −35 element. The introduction of a consensus −35 element appropriately positioned upstream of the A3 −10 element converted p4 from an activator to a repressor; conversely, inactivation of the −35 element upstream of the A2c −10 element converted p4 from a repressor to an activator (70). When functioning as an activator, p4 stabilizes the initial binding of RNAP to the promoter (73). However, when functioning as a repressor, p4 acts at a later step, inhibiting promoter clearance, the interaction between p4 and the α-CTD presumably overstabilizing an already stable open complex (70). Thus, the same protein-protein interaction between a DNA-bound regulator and RNAP can either facilitate promoter occupancy (when the promoter is poorly recognized by RNAP) or impede promoter clearance (when the promoter is efficiently recognized by RNAP). Again, it is not the nature of the protein-protein interaction between the DNA-bound regulator and RNAP that determines the regulatory outcome, but the context in which that interaction occurs.

The *gal* repressor (GalR) provides another example of a regulator that can repress transcription through a direct contact with RNAP (see reference 1). Ordinarily, GalR represses transcription from overlapping promoters P1 and P2 by a DNA-looping

mechanism, but when bound at a single operator site upstream of the promoters, GalR can repress transcription from P1. This repression involves a contact between the DNA-bound GalR molecule and the α-CTD, which is thought to reduce open complex formation by inhibiting isomerization (18).

ACTIVATION THROUGH CHANGES IN DNA STRUCTURE: THE MerR FAMILY

Thus far, we have considered activators that bind to DNA and affect the process of transcription initiation by making direct contacts with RNAP. We have not had to consider the precise details of their interactions with the DNA. However, members of the MerR family of transcription regulators induce distortions into the DNA that are essential to the activation process (37, 44, 93). MerR itself regulates the cell's response to mercury intoxication. It functions as both a repressor and an activator of transcription of the *merTPCAD* operon (encoding mercury

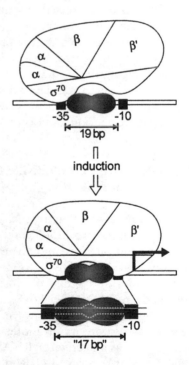

Figure 5. Transcription activation by MerR. MerR (shaded gray) is shown bound to its recognition site positioned between the −10 and −35 elements of its target promoter, which are separated by a noncanonical spacer of 19 bp. Under noninducing conditions, DNA-bound MerR stabilizes the formation of a transcriptionally inactive promoter complex (top). Upon induction, MerR distorts its recognition site, bringing the −10 and −35 elements of the target promoter closer together (effectively creating a canonical 17-bp spacer) so that they can be contacted simultaneously by RNAP (bottom).

detoxification genes). In the absence of mercury, MerR binds to a recognition site positioned between the −10 and −35 elements of the promoter, which are separated by a noncanonical 19-bp spacer (Fig. 5). Thus bound, MerR does not occlude the binding of RNAP, but rather stabilizes the formation of a transcriptionally inactive promoter complex in which RNAP contacts only the −35 element (2). Upon binding Hg(II), the DNA-bound MerR dimer distorts the spacer DNA so as to facilitate the formation of a transcriptionally active complex in which RNAP can engage with both the −10 and −35 elements of the promoter (2) (Fig. 5).

The detailed structural basis for this activator-induced promoter remodeling has recently been revealed by the crystal structure of another family member, the *Bacillus subtilis* BmrR protein, in complex with promoter DNA and a drug cofactor (36). When complexed with a coactivating ligand (the drug), BmrR activates transcription of a gene encoding a multidrug transporter. Like other members of the MerR family, BmrR binds between the −10 and −35 elements of a promoter that has a noncanonical 19-bp spacer. The structure of a BmrR-tetraphenyl phosphonium-DNA complex reveals a striking structural distortion in the center of the recognition site involving the disruption of two base pairs flanking the central axis of symmetry (36). This distortion has the effect of bringing the −10 and −35 elements of the promoter closer together and into proper alignment for simultaneous recognition by RNAP so that productive complex formation can occur.

ACTIVATOR-RNAP COMPLEXES: MarA, SoxS, AND Rob

In general, transcription activators that bind DNA and contact RNAP are thought to bind their specific DNA recognition sites and then, once appropriately positioned on the DNA, to interact with RNAP. Recent studies of a group of closely related transcription activators, MarA, SoxS, and Rob, suggest that in some cases the activator may be prebound to RNAP and locate its promoter-associated DNA recognition site as an activator-RNAP complex (31, 64) (Fig. 6). MarA, SoxS, and Rob activate a common set of genes referred to as the *mar/sox/rob* regulon (3, 74). The genes of this regulon help the cell defend against oxidative stress and a variety of toxins. The three proteins bind as monomers to DNA sites with the same degenerate consensus sequence (56, 77; see also reference 65), and estimates suggest that there are >10,000 of these sites on the *E. coli* chromosome (31, 64). The question thus arises: how

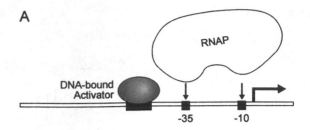

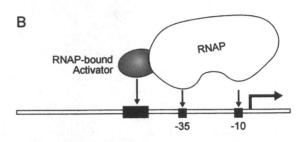

Figure 6. Transcription activation by an activator that is prebound to RNAP. (A) A classical activator of transcription that ordinarily binds to its specific recognition site on the DNA and then interacts with RNAP. (B) An activator such as MarA or SoxS that may ordinarily interact with RNAP prior to binding its specific recognition site on the DNA.

do these proteins distinguish between physiologically relevant sites and sites that are not associated with promoters? This problem is particularly relevant for both MarA and SoxS, because estimates suggest that either MarA- or SoxS-dependent activation of the regulon can occur when the number of MarA or SoxS molecules in the cell is considerably fewer than the number of potential binding sites. The finding that MarA, SoxS, and Rob can form complexes with RNAP (either core or holoenzyme) in the absence of DNA suggests a possible resolution to the apparent paradox (64; see also reference 31). An activator-RNAP complex would clearly be able to distinguish between promoter-associated and non-promoter-associated activator binding sites (and similarly, between promoters with and without an associated binding site for the activator). In effect, the RNAP-associated activator provides RNAP with an additional DNA-binding domain and the potential to form especially stable complexes with promoters of the *mar/sox/rob* regulon.

The proposal that MarA and SoxS bind to RNAP prior to binding specifically to the DNA leads to a testable genetic prediction, as pointed out by Griffith et al. (31). Activator mutants that are unable to bind to DNA should exert a dominant-negative effect when present in excess over wild-type activator, because they should compete with wild-type activator for binding to RNAP. On the other hand, positive-control mutants that are unable to bind to RNAP should behave as recessive mutants. Exactly the opposite is true when an activator that is pre-bound to the DNA functions to help recruit RNAP to the promoter. For example, the synthesis of high levels of a CRP positive-control mutant (but not a CRP DNA-binding mutant) in cells containing wild-type CRP inhibits CRP-dependent activation (J. K. Joung and A. Hochschild, unpublished observations).

ACTIVATORS THAT WORK FROM LARGE DISTANCES: THE NtrC FAMILY

Members of the NtrC family compose a group of activators that function quite differently from the σ^{70}-dependent activators we have considered so far. Activators in this group bind to sites that are typically located 100 bp or more upstream of a target promoter (96). Furthermore, members of the NtrC family activate transcription from promoters that are recognized by a form of RNAP containing the alternative σ factor σ^{54}, which appears to be evolutionarily unrelated to members of the σ^{70} family (see reference 7). NtrC itself activates transcription from the glnA promoter (pglnA) and other promoters that are induced under conditions of nitrogen limitation (see reference 100 and references therein). Unlike the σ^{70}-dependent promoters discussed above, pglnA binds σ^{54}-containing RNAP in a stable, but transcriptionally inactive, closed complex in the absence of the activator. Normally, an enhancer-bound NtrC oligomer contacts the promoter-bound RNAP through the formation of a DNA loop and catalyzes the formation of a transcriptionally active open complex in an ATP-dependent reaction (reviewed in reference 86). Interestingly, however, truncated and triple alanine-substituted mutants of NtrC that are incapable of binding to DNA can also activate transcription from pglnA, presumably because they can interact with preformed closed complexes directly from solution (72a; see also reference 43a). Evidence suggests that activators in this class specifically target σ^{54} (at least in part) and trigger a conformational switch that allows the prebound RNAP molecule to form a transcription-ready open complex (14, 16). Evidently, specific structural features of σ^{54} preclude the spontaneous formation of open complexes (33).

ACTIVATORS THAT DO NOT BIND TO SPECIFIC DNA RECOGNITION SITES

At least one σ^{70}-dependent activator has been described that can work without binding to DNA at all. The bacteriophage N4 single-stranded DNA-binding protein (N4 SSB) activates transcription of the phage's late genes by E. coli RNAP containing σ^{70} (17). In its role as activator, N4 SSB need not contact either single- or double-stranded DNA (68). This activation requires contact between a C-terminal region of N4 SSB and the β' subunit of RNAP (68).

Another bacteriophage-encoded activator that functions without binding to a specific DNA recognition site is the T4 Gp45 sliding clamp. This activator loads onto a specific enhancer site and then tracks along the DNA toward the promoter, where it activates late gene transcription by E. coli RNAP (94). In this case, the activation involves protein-protein contacts between Gp45 and two other phage-encoded proteins, σ factor Gp55 and coactivator Gp33 (90).

Specificity

We have discussed examples of activators that work only when bound to specific sites on the DNA, an activator that must be tethered to the DNA but remain mobile, and activators that can work directly from solution. Nevertheless, most prokaryotic activators are apparently designed to function when bound at specific sites on the DNA, thus providing an obvious mechanism to ensure that only the appropriate target genes are activated. It is perhaps not surprising, then, that the two currently known counter-examples, the N4 SSB and the T4 sliding clamp, are phage-encoded activators that are used in the context of a specific temporal program to activate a restricted set of promoters. The N4 SSB is specifically targeted to the N4 late promoters because they are the only N4 promoters that are recognized by E. coli RNAP, and the sliding clamp is targeted to the T4 late promoters by its interaction with the late-gene-specific σ factor Gp55. In the case of N4, the late promoters are presumably designed in such a way as to compete effectively with cellular promoters for RNAP, but also to be limited at some step that requires the action of the N4 SSB.

In the case of the NtrC family members, the mechanism of activation does not impose a requirement for the activator to bind DNA. However, to bypass the normal requirement for DNA binding, it is necessary to overproduce the activator. Thus, at normal physiological concentrations of the activator, binding to specific sites on the DNA increases the local concentration of the activator in the vicinity of a target promoter so that interaction between the activator and the DNA-bound RNAP can occur. Specificity is therefore maintained so long as the intracellular concentration of the activator does not rise above certain levels. We note that in the case of the MarA family members, their mechanism of action requires them to bind to the DNA regardless of whether or not they first associate with RNAP.

Complex Promoters with Multiple Activator-Binding Sites

Many bacterial promoters contain distinct binding sites for two or more regulators that can function synergistically to activate transcription (see, for example, references 21, 42, 92, and 99; for reviews, see references 41 and 78). Studies of a variety of artificial promoters first suggested that such synergistic effects can arise under circumstances that permit two (or more) DNA-bound activators to make direct contact with RNAP (11, 50, 51). For example, pairs of DNA-bound CRP dimers can activate transcription synergistically from derivatives of either a class I or a class II promoter (11, 51). That is, the promoter-proximal CRP-binding site can be positioned either 41.5 or 61.5 bp upstream of the transcription start site, and the promoter-distal CRP-binding site can function at a variety of different positions, provided the two sites are phased on approximately the same side of the DNA helix (5) (Fig. 7A). Various lines of genetic and biochemical evidence support the idea

A

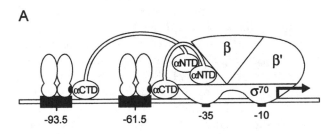

B

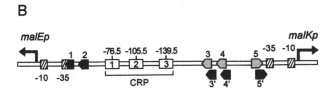

Figure 7. Transcription activator synergy. (A) Two DNA-bound CRP dimers activate transcription synergistically by contacting the αCTDs. The activating region of each CRP dimer (shaded black) is shown contacting an αCTD. (B) Regulatory region of *malEp* and *malKp*. Shown is the 271-bp regulatory region that mediates control of the divergent promoters *malEp* and *malKp* by MalT and CRP. Indicated are the −10 and −35 elements of the promoters (hatched boxes), the MalT recognition sites (pointed boxes), and the CRP recognition sites (open boxes). MalT sites 3/4/5 (shaded gray) are bound under repressing conditions, while MalT sites 1/2 and 3′/4′/5′ (shaded black) are bound under activating conditions. Adapted from reference 80.

that full activation in these cases involves a contact between each DNA-bound CRP dimer and an α-CTD (and, in the case of a CRP dimer bound at position −41.5, an additional contact with the α-NTD) (5, 11, 57). Similarly, heterologous activators can function synergistically by making simultaneous contact with RNAP (11, 50, 92). For example, λcI and CRP can work together synergistically to activate transcription from an artificial promoter bearing both a λ operator (centered at position −42) and a CRP-binding site (centered at position −93.5) (50). In this case, λcI evidently contacts the σ^{70} subunit, and the CRP dimer contacts an α-CTD.

A growing body of evidence suggests that many natural promoters are regulated by two (or more) DNA-bound activators, each of which interacts directly with RNAP (with at least one likely contacting an α-CTD). A particularly interesting example involves the *E. coli malE* promoter (*malEp*), which requires both the regulon-specific activator MalT and CRP for its activation (80). MalT and CRP also activate transcription from the divergently oriented *malK* promoter (*malKp*), and the activation of both promoters is a coupled process that depends on the formation of a higher-order nucleoprotein complex involving multiple MalT protomers and three CRP dimers. In particular, the 210-bp regulatory region spanning the two promoters consists of two MalT-binding sites (1/2, abutting *malEp*), three CRP-binding sites, and two overlapping sets of three MalT-binding sites (3/4/5 and 3′/4′/5′, abutting *malKp*) (Fig. 7B). In the activated state, the binding of CRP to its recognition sites promotes the formation of a higher-order complex in which MalT binds cooperatively to sites 1/2 and 3′/4′/5′, thereby activating transcription from *malEp* and *malKp*, respectively. Earlier work had demonstrated that CRP plays a primarily architectural role in mediating activation of *malKp*, facilitating the repositioning of MalT from high-affinity sites 3/4/5 to lower-affinity sites 1/2 and 3′/4′/5′ (81). A more recent study now demonstrates that CRP plays a more complex role in the events occurring at *malEp* (80). Not only does CRP function as an essential architectural element, but the CRP dimer bound at site 3 (centered at position −139.5 relative to the start site of *malE* transcription) also participates directly in the activation of *malEp* by interacting with the α-CTD. As is the case at simple class I CRP-dependent promoters, this CRP/α-CTD interaction depends on ARI of CRP (80).

The arabinose regulon provides other examples in which CRP works together with a regulon-specific activator (in this case, AraC), with both activators apparently contacting RNAP directly (49, 99). Interestingly, CRP is the promoter-proximal activator at one promoter (the *araFGH* promoter) and the promoter-distal activator at another (the *araBAD* promoter). Nevertheless, another recent study emphasizes that CRP can play many different roles at complex promoters, not necessarily functioning as either a direct activator of transcription or an architectural element (95). Transcription from the *melAB* promoter (p*melAB*) depends on both the melibiose-responsive activator MelR and CRP. The upstream promoter region consists of two promoter-proximal MelR-binding sites (centered at positions −42.5 and −62.5), a CRP-binding site (centered at position −81.5) and two promoter-distal MelR-binding sites (centered at positions −100.5 and −120.5). Here, direct protein-protein interactions between CRP and MelR result in cooperative DNA binding (in much the same way as the protein-protein interaction of adjacently bound λcI dimers [75] [Fig. 2E]), which in turn ensures the occupancy of the promoter-proximal MelR-binding site (95). Once bound at this site, MelR is thought to make direct contact with RNAP.

Regardless of mechanism, transcription activator synergy permits a complex promoter to respond to multiple regulatory inputs. Furthermore, when gene activation depends on multiple DNA-bound regulators, whether they are functioning in homotypic or heterotypic combinations, steep dose-response curves can be generated. This in turn provides a mechanistic basis for the function of sensitive biological switches.

ADDENDUM IN PROOF

We refer the reader to the following relevant publications that appeared after this chapter was written: A. Barnard, A. Wolfe, and S. Busby, *Curr. Opin. Microbiol.* 7:102–108, 2004; B. Benoff, H. Yang, C. L. Lawson, G. Parkinson, J. Liu, E. Blatter, Y. W. Ebright, H. M. Berman, and R. H. Ebright, *Science* 297:1562–1566, 2002; P. Bordes, S. R. Wigneshweraraj, J. Schumacher, X. Zhang, M. Chaney, and M. Buck, *Proc. Natl. Acad. Sci. USA* 100:2278–2283, 2003; B. Calles, M. Salas, and F. Rojo, *EMBO J.* 21:6185–6194, 2002; H. Chen, H. Tang, and R. H. Ebright, *Mol. Cell* 11:1621–1633, 2003; S. L. Dove, S. A. Darst, and A. Hochschild, *Mol. Microbiol.* 48:863–874, 2003; D. Jain, B. E. Nickels, L. Sun, A. Hochschild, and S. A. Darst, *Mol. Cell* 13:45–53, 2004; S. E. Kolesky, M. Ouhammouch, and E. P. Geiduschek, *J. Mol. Biol.* 321:767–784, 2002; K. S. Murakami, S. Masuda, E. A. Campbell, O. Muzzin, and S. A. Darst, *Science* 296:1285–1290, 2002; K. S. Murakami, S. Masuda, and S. A. Darst, *Science* 296:1280–1294, 2004; B. E. Nickels, S. L. Dove, K. S. Murakami, S. A. Darst, and A. Hochschild, *J. Mol. Biol.* 324:17–34, 2002; W. Ross, D. A. Schneider, B. J. Paul, A. Mertens, and R. L. Gourse, *Genes Dev.* 17:1293–1307, 2003; S. Semsey, M. Geanacopoulos, D. E. Lewis, and S. Adhya, *EMBO J.* 21:4349–4356, 2002; D. G. Vassylyev, S. Sekine, O. Laptenko, J. Lee, M. N. Vassylyeva, S. Borukhov, and S. Yokoyama, *Nature* 417:712–719, 2002; and B. A. Young, T. M. Gruber, and C. A. Gross, *Science* 303:1382–1384, 2004.

REFERENCES

1. Adhya, S., M. Geanacopoulos, D. E. Lewis, S. Roy, and T. Aki. 1998. Transcription regulation by repressosome and by RNA polymerase contact. *Cold Spring Harbor Symp. Quant. Biol.* 63:1–9.

2. Ansari, A. Z., J. E. Bradner, and T. V. O'Halloran. 1995. DNA-bend modulation in a repressor-to-activator switching mechanism. *Nature* 374:371–375.

3. Barbosa, T. M., and S. B. Levy. 2000. Differential expression of over 60 chromosomal genes in *Escherichia coli* by constitutive expression of MarA. *J. Bacteriol.* 182:3467–3474.

4. Barne, K. A., J. A. Bown, S. J. Busby, and S. D. Minchin. 1997. Region 2.5 of the *Escherichia coli* RNA polymerase σ⁷⁰ subunit is responsible for the recognition of the "extended −10" motif at promoters. *EMBO J.* 16:4034–4040.

5. Belyaeva, T. A., V. A. Rhodius, C. L. Webster, and S. J. Busby. 1998. Transcription activation at promoters carrying tandem DNA sites for the *Escherichia coli* cyclic AMP receptor protein: organisation of the RNA polymerase α subunits. *J. Mol. Biol.* 277:789–804.

6. Bown, J. A., K. A. Barne, S. D. Minchin, and S. J. W. Busby. 1997. Extended −10 promoters, p. 41–52. *In* F. Eckstein and D. M. J. Lilley (ed.), *Mechanisms of Transcription*, vol. 11. Springer-Verlag, Berlin, Germany.

7. Buck, M., M. T. Gallegos, D. J. Studholme, Y. Guo, and J. D. Gralla. 2000. The bacterial enhancer-dependent σ⁵⁴ (σᴺ) transcription factor. *J. Bacteriol.* 182:4129–4136.

8. Busby, S., and R. H. Ebright. 1994. Promoter structure, promoter recognition, and transcription activation in prokaryotes. *Cell* 79:743–746.

9. Busby, S., and R. H. Ebright. 1999. Transcription activation by catabolite activator protein (CAP). *J. Mol. Biol.* 293:199–213.

10. Busby, S., and A. Kolb. 1996. The CAP modulon, p. 255–279. *In* E. C. C. Lin and A. S. Lynch (ed.), *Regulation of Gene Expression in* Escherichia coli. RG Landes Co. Biomedical Publishers, Georgetown, Tex.

11. Busby, S., D. West, M. Lawes, C. Webster, A. Ishihama, and A. Kolb. 1994. Transcription activation by the *Escherichia coli* cyclic AMP receptor protein. Receptors bound in tandem at promoters can interact synergistically. *J. Mol. Biol.* 241:341–352.

12. Bushman, F. D., C. Shang, and M. Ptashne. 1989. A single glutamic acid residue plays a key role in the transcriptional activation function of lambda repressor. *Cell* 58:1163–1171.

13. Callaci, S., E. Heyduk, and T. Heyduk. 1999. Core RNA polymerase from *E. coli* induces a major change in the domain arrangement of the σ⁷⁰ subunit. *Mol. Cell* 3:229–238.

14. Cannon, W. V., M. T. Gallegos, and M. Buck. 2000. Isomerization of a binary sigma-promoter DNA complex by transcription activators. *Nat. Struct. Biol.* 7:594–601.

15. Carpousis, A. J., and J. D. Gralla. 1980. Cycling of ribonucleic acid polymerase to produce oligonucleotides during initiation in vitro at the lac UV5 promoter. *Biochemistry* 19:3245–3253.

16. Chaney, M., R. Grande, S. R. Wigneshweraraj, W. Cannon, P. Casaz, M. T. Gallegos, J. Schumacher, S. Jones, S. Elderkin, A. E. Dago, E. Morett, and M. Buck. 2001. Binding of transcriptional activators to σ⁵⁴ in the presence of the transition state analog ADP-aluminum fluoride: insights into activator mechanochemical action. *Genes Dev.* 15:2282–2294.

17. Cho, N. Y., M. Choi, and L. B. Rothman-Denes. 1995. The bacteriophage N4-coded single-stranded DNA-binding protein (N4SSB) is the transcriptional activator of *Escherichia coli* RNA polymerase at N4 late promoters. *J. Mol. Biol.* 246:461–471.

18. Choy, H. E., R. R. Hanger, T. Aki, M. Mahoney, K. Murakami, A. Ishihama, and S. Adhya. 1997. Repression and activation of promoter-bound RNA polymerase activity by Gal repressor. *J. Mol. Biol.* 272:293–300.

19. Cramer, P., D. A. Bushnell, J. Fu, A. L. Gnatt, B. Maier-Davis, N. E. Thompson, R. R. Burgess, A. M. Edwards, P. R. David, and R. D. Kornberg. 2000. Architecture of RNA polymerase II and implications for the transcription mechanism. *Science* 288:640–649.

20. Darst, S. A. 2001. Bacterial RNA polymerase. *Curr. Opin. Struct. Biol.* 11:155–162.

21. Darwin, A. J., E. C. Ziegelhoffer, P. J. Kiley, and V. Stewart. 1998. Fnr, NarP, and NarL regulation of *Escherichia coli* K-12 *napF* (periplasmic nitrate reductase) operon transcription in vitro. *J. Bacteriol.* 180:4192–4198.

22. Dombroski, A. J., W. A. Walter, M. T. Record, Jr., D. A. Siegele, and C. A. Gross. 1992. Polypeptides containing highly conserved regions of transcription initiation factor σ^{70} exhibit specificity of binding to promoter DNA. *Cell* 70:501–512.

23. Dove, S. L., and A. Hochschild. 1998. Conversion of the ω subunit of *Escherichia coli* RNA polymerase into a transcriptional activator or an activation target. *Genes Dev.* 12:745–754.

24. Dove, S. L., and A. Hochschild. 1998. Use of artificial activators to define a role for protein-protein and protein-DNA contacts in transcriptional activation. *Cold Spring Harbor Symp. Quant. Biol.* 63:173–180.

25. Dove, S. L., F. W. Huang, and A. Hochschild. 2000. Mechanism for a transcriptional activator that works at the isomerization step. *Proc. Natl. Acad. Sci. USA* 97:13215–13220.

26. Dove, S. L., J. K. Joung, and A. Hochschild. 1997. Activation of prokaryotic transcription through arbitrary protein-protein contacts. *Nature* 386:627–630.

27. Ebright, R. H. 2000. RNA polymerase: structural similarities between bacterial RNA polymerase and eukaryotic RNA polymerase II. *J. Mol. Biol.* 304:687–698.

28. Ebright, R. H., and S. Busby. 1995. The *Escherichia coli* RNA polymerase α subunit: structure and function. *Curr. Opin. Genet. Dev.* 5:197–203.

29. Estrem, S. T., W. Ross, T. Gaal, Z. W. Chen, W. Niu, R. H. Ebright, and R. L. Gourse. 1999. Bacterial promoter architecture: subsite structure of UP elements and interactions with the carboxy-terminal domain of the RNA polymerase α subunit. *Genes Dev.* 13:2134–2147.

30. Gourse, R. L., W. Ross, and T. Gaal. 2000. UPs and downs in bacterial transcription initiation: the role of the α subunit of RNA polymerase in promoter recognition. *Mol. Microbiol.* 37:687–695.

31. Griffith, K. L., I. M. Shah, T. E. Myers, M. C. O'Neill, and R. E. Wolf, Jr. 2002. Evidence for "pre-recruitment" as a new mechanism of transcription activation in *Escherichia coli*: the large excess of SoxS binding sites per cell relative to the number of SoxS molecules per cell. *Biochem. Biophys. Res. Commun.* 291:979–986.

32. Gross, C. A., C. Chan, A. Dombroski, T. Gruber, M. Sharp, J. Tupy, and B. Young. 1998. The functional and regulatory roles of sigma factors in transcription. *Cold Spring Harbor Symp. Quant. Biol.* 63:141–155.

33. Guo, Y., C. M. Lew, and J. D. Gralla. 2000. Promoter opening by σ^{54} and σ^{70} RNA polymerases: sigma factor-directed alterations in the mechanism and tightness of control. *Genes Dev.* 14:2242–2255.

34. Gussin, G. N. 1996. Kinetic analysis of RNA polymerase-promoter interactions. *Methods Enzymol.* 273:45–59.

35. Hawley, D. K., and W. R. McClure. 1982. Mechanism of activation of transcription initiation from the lambda P_{RM} promoter. *J. Mol. Biol.* 157:493–525.

36. Heldwein, E. E., and R. G. Brennan. 2001. Crystal structure of the transcription activator BmrR bound to DNA and a drug. *Nature* 409:378–382.

37. Hidalgo, E., and B. Demple. 1997. Spacing of promoter elements regulates the basal expression of the *soxS* gene and converts SoxR from a transcriptional activator into a repressor. *EMBO J.* 16:1056–1065.

38. Hochschild, A. 2002. The λ switch: cI closes the gap in autoregulation. *Curr. Biol.* 12:R87–R89.

39. Hochschild, A., and S. L. Dove. 1998. Protein-protein contacts that activate and repress prokaryotic transcription. *Cell* 92:597–600.

40. Hochschild, A., N. Irwin, and M. Ptashne. 1983. Repressor structure and the mechanism of positive control. *Cell* 32:319–325.

41. Hochschild, A., and J. K. Joung. 1997. Synergistic activation of transcription in E. coli, p. 101–114. *In* F. Eckstein and D. M. J. Lilley (ed.), *Mechanisms of Transcription*, vol. 11. Springer-Verlag, Berlin, Germany.

42. Holcroft, C. C., and S. M. Egan. 2000. Interdependence of activation at *rhaSR* by cyclic AMP receptor protein, the RNA polymerase α subunit C-terminal domain, and RhaR. *J. Bacteriol.* 182:6774–6782.

43. Hsu, L. M. 1996. Quantitative parameters for promoter clearance. *Methods Enzymol.* 273:59–71.

43a. Huala, E., and F. M. Ausubel. 1989. The central domain of *Rhizobium meliloti* NifA is sufficient to activate transcription from the *R. meliloti nifH* promoter. *J. Bacteriol.* 171:3354–3365.

44. Huffman, J. L., and R. G. Brennan. 2002. Prokaryotic transcription regulators: more than just the helix-turn-helix motif. *Curr. Opin. Struct. Biol.* 12:98–106.

45. Igarashi, K., and A. Ishihama. 1991. Bipartite functional map of the E. coli RNA polymerase α subunit: involvement of the C-terminal region in transcription activation by cAMP-CRP. *Cell* 65:1015–1022.

46. Ishihama, A. 1992. Role of the RNA polymerase α subunit in transcription activation. *Mol. Microbiol.* 6:3283–3288.

47. Ishihama, A. 1993. Protein-protein communication within the transcription apparatus. *J. Bacteriol.* 175:2483–2489.

48. Ishihama, A. 2000. Functional modulation of *Escherichia coli* RNA polymerase. *Annu. Rev. Microbiol.* 54:499–518.

49. Johnson, C. M., and R. F. Schleif. 2000. Cooperative action of the catabolite activator protein and AraC in vitro at the *araFGH* promoter. *J. Bacteriol.* 182:1995–2000.

50. Joung, J. K., D. M. Koepp, and A. Hochschild. 1994. Synergistic activation of transcription by bacteriophage λ cI protein and *Escherichia coli* cAMP receptor protein. *Science* 265:1863–1866.

51. Joung, J. K., L. U. Le, and A. Hochschild. 1993. Synergistic activation of transcription by *Escherichia coli* cAMP receptor protein. *Proc. Natl. Acad. Sci. USA* 90:3083–3087.

52. Joung, J. K., E. I. Ramm, and C. O. Pabo. 2000. A bacterial two-hybrid selection system for studying protein-DNA and protein-protein interactions. *Proc. Natl. Acad. Sci. USA* 97:7382–7387.

53. Kolb, A., S. Busby, H. Buc, S. Garges, and S. Adhya. 1993. Transcriptional regulation by cAMP and its receptor protein. *Annu. Rev. Biochem.* 62:749–795.

54. Kuldell, N., and A. Hochschild. 1994. Amino acid substitutions in the −35 recognition motif of σ[70] that result in defects in phage λ repressor-stimulated transcription. *J. Bacteriol.* **176:**2991–2998.

55. Kuznedelov, K., L. Minakhin, A. Niedziela-Majka, S. L. Dove, D. Rogulja, B. E. Nickels, A. Hochschild, T. Heyduk, and K. Severinov. 2002. A role for interaction of the RNA polymerase flap domain with the σ subunit in promoter recognition. *Science* **295:**855–857.

56. Kwon, H. J., M. H. Bennik, B. Demple, and T. Ellenberger. 2000. Crystal structure of the *Escherichia coli* Rob transcription factor in complex with DNA. *Nat. Struct. Biol.* **7:**424–430.

57. Langdon, R. C., and A. Hochschild. 1999. A genetic method for dissecting the mechanism of transcriptional activator synergy by identical activators. *Proc. Natl. Acad. Sci. USA* **96:**12673–12678.

58. Li, M., W. R. McClure, and M. M. Susskind. 1997. Changing the mechanism of transcriptional activation by phage λ repressor. *Proc. Natl. Acad. Sci. USA* **94:**3691–3696.

59. Li, M., H. Moyle, and M. M. Susskind. 1994. Target of the transcriptional activation function of phage λ CI protein. *Science* **263:**75–77.

60. Lonetto, M., M. Gribskov, and C. A. Gross. 1992. The σ[70] family: sequence conservation and evolutionary relationships. *J. Bacteriol.* **174:**3843–3849.

61. Lonetto, M. A., V. Rhodius, K. Lamberg, P. Kiley, S. Busby, and C. Gross. 1998. Identification of a contact site for different transcription activators in region 4 of the *Escherichia coli* RNA polymerase σ[70] subunit. *J. Mol. Biol.* **284:**1353–1365.

62. Malan, T. P., A. Kolb, H. Buc, and W. R. McClure. 1984. Mechanism of CRP-cAMP activation of *lac* operon transcription initiation activation of the P1 promoter. *J. Mol. Biol.* **180:**881–909.

63. Marr, M. T., and J. W. Roberts. 1997. Promoter recognition as measured by binding of polymerase to nontemplate strand oligonucleotide. *Science* **276:**1258–1260.

64. Martin, R. G., W. K. Gillette, N. I. Martin, and J. L. Rosner. 2002. Complex formation between activator and RNA polymerase as the basis for transcriptional activation by MarA and SoxS in *Escherichia coli*. *Mol. Microbiol.* **43:**355–370.

65. Martin, R. G., and J. L. Rosner. 2001. The AraC transcriptional activators. *Curr. Opin. Microbiol.* **4:**132–137.

66. McClure, W. R. 1985. Mechanism and control of transcription initiation in prokaryotes. *Annu. Rev. Biochem.* **54:**171–204.

67. Meyer, B. J., and M. Ptashne. 1980. Gene regulation at the right operator (O_R) of bacteriophage λ. III. λ repressor directly activates gene transcription. *J. Mol. Biol.* **139:**195–205.

68. Miller, A., D. Wood, R. H. Ebright, and L. B. Rothman-Denes. (1997). RNA polymerase β' subunit: a target of DNA binding-independent activation. *Science* **275:**1655–1657.

69. Minakhin, L., S. Bhagat, A. Brunning, E. A. Campbell, S. A. Darst, R. H. Ebright, and K. Severinov. 2001. Bacterial RNA polymerase subunit ω and eukaryotic RNA polymerase subunit RPB6 are sequence, structural, and functional homologs and promote RNA polymerase assembly. *Proc. Natl. Acad. Sci. USA* **98:**892–897.

70. Monsalve, M., B. Calles, M. Mencia, M. Salas, and F. Rojo. 1997. Transcription activation or repression by phage φ29 protein p4 depends on the strength of the RNA polymerase-promoter interactions. *Mol. Cell* **1:**99–107.

71. Monsalve, M., M. Mencia, F. Rojo, and M. Salas. 1996. Activation and repression of transcription at two different phage φ29 promoters are mediated by interaction of the same residues of regulatory protein p4 with RNA polymerase. *EMBO J.* **15:**383–391.

72. Niu, W., Y. Kim, G. Tau, T. Heyduk, and R. H. Ebright. 1996. Transcription activation at class II CAP-dependent promoters: two interactions between CAP and RNA polymerase. *Cell* **87:**1123–1134.

72a. North, A. K., and S. Kustu. 1997. Mutant forms of the enhancer-binding protein NtrC can activate transcription from solution. *J. Mol. Biol.* **267:**17–36.

73. Nuez, B., F. Rojo, and M. Salas. 1992. Phage φ29 regulatory protein p4 stabilizes the binding of the RNA polymerase to the late promoter in a process involving direct protein-protein contacts. *Proc. Natl. Acad. Sci. USA* **89:**11401–11405.

74. Pomposiello, P. J., M. H. Bennik, and B. Demple. 2001. Genome-wide transcriptional profiling of the *Escherichia coli* responses to superoxide stress and sodium salicylate. *J. Bacteriol.* **183:**3890–3902.

75. Ptashne, M. 1992. *A Genetic Switch, Phage Lambda and Higher Organisms.* Cell Press, Cambridge, Mass.

76. Ptashne, M., and A. Gann. 1997. Transcriptional activation by recruitment. *Nature* **386:**569–577.

77. Rhee, S., R. G. Martin, J. L. Rosner, and D. R. Davies. 1998. A novel DNA-binding motif in MarA: the first structure for an AraC family transcriptional activator. *Proc. Natl. Acad. Sci. USA* **95:**10413–10418.

78. Rhodius, V. A., and S. J. Busby. 1998. Positive activation of gene expression. *Curr. Opin. Microbiol.* **1:**152–159.

79. Rhodius, V. A., D. M. West, C. L. Webster, S. J. Busby, and N. J. Savery. 1997. Transcription activation at class II CRP-dependent promoters: the role of different activating regions. *Nucleic Acids Res.* **25:**326–332.

80. Richet, E. 2000. Synergistic transcription activation: a dual role for CRP in the activation of an *Escherichia coli* promoter depending on MalT and CRP. *EMBO J.* **19:**5222–5232.

81. Richet, E., D. Vidal-Ingigliardi, and O. Raibaud. 1991. A new mechanism for coactivation of transcription initiation: repositioning of an activator triggered by the binding of a second activator. *Cell* **66:**1185–1195.

82. Ring, B. Z., and J. W. Roberts. 1994. Function of a nontranscribed DNA strand site in transcription elongation. *Cell* **78:**317–324.

83. Ring, B. Z., W. S. Yarnell, and J. W. Roberts. 1996. Function of *E. coli* RNA polymerase sigma factor σ[70] in promoter-proximal pausing. *Cell* **86:**485–493.

84. Roberts, C. W., and J. W. Roberts. 1996. Base-specific recognition of the nontemplate strand of promoter DNA by *E. coli* RNA polymerase. *Cell* **86:**495–501.

85. Rojo, F. 2001. Mechanisms of transcriptional repression. *Curr. Opin. Microbiol.* **4:**145–151.

86. Rombel, I., A. North, I. Hwang, C. Wyman, and S. Kustu. 1998. The bacterial enhancer-binding protein NtrC as a molecular machine. *Cold Spring Harbor Symp. Quant. Biol.* **63:**157–166.

87. Ross, W., S. E. Aiyar, J. Salomon, and R. L. Gourse. 1998. *Escherichia coli* promoters with UP elements of different strengths: modular structure of bacterial promoters. *J. Bacteriol.* **180:**5375–5383.

88. Ross, W., K. K. Gosink, J. Salomon, K. Igarashi, C. Zou, A. Ishihama, K. Severinov, and R. L. Gourse. 1993. A third recognition element in bacterial promoters: DNA binding by the α subunit of RNA polymerase. *Science* **262:**1407–1413.

89. Roy, S., S. Garges, and S. Adhya. 1998. Activation and repression of transcription by differential contact: two sides of a coin. *J. Biol. Chem.* **273:**14059–14062.

90. Sanders, G. M., G. A. Kassavetis, and E. P. Geiduschek. 1997. Dual targets of a transcriptional activator that tracks on DNA. *EMBO J.* **16:**3124–3132.

91. Schultz, S. C., G. C. Shields, and T. A. Steitz. 1991. Crystal structure of a CAP-DNA complex: the DNA is bent by 90 degrees. *Science* **253:**1001–1007.

92. Scott, S., S. Busby, and I. Beacham. 1995. Transcriptional co-activation at the *ansB* promoters: involvement of the activating regions of CRP and FNR when bound in tandem. *Mol. Microbiol.* **18:**521–531.

93. Summers, A. O. 1992. Untwist and shout: a heavy metal-responsive transcriptional regulator. *J. Bacteriol.* **174:**3097–3101.

94. Tinker, R. L., K. P. Williams, G. A. Kassavetis, and E. P. Geiduschek. 1994. Transcriptional activation by a DNA-tracking protein: structural consequences of enhancement at the T4 late promoter. *Cell* **77:**225–237.

95. Wade, J. T., T. A. Belyaeva, E. I. Hyde, and S. J. Busby. 2001. A simple mechanism for co-dependence on two activators at an *Escherichia coli* promoter. *EMBO J.* **20:**7160–7167.

96. Weiss, D. S., K. E. Klose, T. R. Hoover, A. K. North, S. C. Porter, A. B. Wedel, and S. Kustu. 1992. Prokaryotic transcriptional enhancers, p. 667–694. *In* S. L. McKnight and K. R. Yamamoto (ed.), *Transcriptional Regulation,* vol. 2. Cold Spring Harbor Laboratory Press, Cold Spring Harbor, N.Y.

97. Young, B. A., L. C. Anthony, T. M. Gruber, T. M. Arthur, E. Heyduk, C. Z. Lu, M. M. Sharp, T. Heyduk, R. R. Burgess, and C. A. Gross. 2001. A coiled-coil from the RNA polymerase β' subunit allosterically induces selective nontemplate strand binding by σ70. *Cell* **105:**935–944.

98. Zhang, G. Y., E. A. Campbell, L. Minakhin, C. Richter, K. Severinov, and S. A. Darst. 1999. Crystal structure of *Thermus aquaticus* core RNA polymerase at 3.3 angstrom resolution. *Cell* **98:**811–824.

99. Zhang, X., and R. Schleif. 1998. Catabolite gene activator protein mutations affecting activity of the *araBAD* promoter. *J. Bacteriol.* **180:**195–200.

100. Zimmer, D. P., E. Soupene, H. L. Lee, V. F. Wendisch, A. B. Khodursky, B. J. Peter, R. A. Bender, and S. Kustu. 2000. Nitrogen regulatory protein C-controlled genes of *Escherichia coli*: scavenging as a defense against nitrogen limitation. *Proc. Natl. Acad. Sci. USA* **97:**14674–14679.

Chapter 17

Control of Transcription Termination and Antitermination

Irina Artsimovitch

In bacteria, a single RNA polymerase (RNAP) species is responsible for transcription of all genetic material, from the very short, single genes to long (up to 30-kb) operons, and from rapidly cotranslated proteins to the nontranslated tRNA and rRNA genes. This "relaxed specificity" of bacterial RNAP relies on multiple regulatory circuits that fine-tune the RNAP's response to the nucleic acid signals it encounters while scanning the DNA tape (63). The subjects of this chapter are the properties of RNAP and its extrinsic regulators (proteins and small molecules that bind to RNA, DNA, and RNAP) that balance two apparently contradictory abilities of RNAP: (i) to synthesize the nascent RNA chain at a high speed and processively at most positions along the template and yet (ii) to either halt transcription temporarily at pause sites or release the nascent RNA at terminators. To provide a framework for understanding these regulatory inputs, I briefly review the current structural and kinetic models for transcription elongation and termination. I then describe the regulatory molecules that are known to influence the elongation/termination decision by RNAP, with the emphasis on the most recent findings and on the mechanism of "active" regulators whose actions are not limited to changes in RNA folding. A number of recent reviews provide excellent, detailed descriptions of the structural (15, 47) and kinetic (20, 72) models of elongation and its different regulatory mechanisms (26, 36, 37, 69, 76).

INTRINSIC SIGNALS THAT MODULATE RNA CHAIN ELONGATION

The transcription elongation complex (TEC) forms when, upon making a nascent RNA chain of 8 to 12 nucleotides (nt), RNAP breaks all the contacts with the promoter elements and the initiation factors (escapes). The TEC is characterized by high resistance (22, 93) to high salt concentrations (1 M KCl), elevated temperature (60°C), and high pressure (up to 180 MPa). However, certain intrinsic (encoded within the DNA and the nascent RNA) signals instruct RNAP to halt transcription, either transiently (at pause sites) or indefinitely (at arrest sites), or to release the nascent RNA chain (at termination signals). A "minimal" model of transcription should account for the high thermodynamic stability and processivity of RNAP, as well as the ability of RNAP to recognize these signals.

At a pause site, RNAP stops temporarily, and then escapes upon addition of nucleoside monophosphate (NMP), either stochastically or following interaction with a regulatory molecule. Two classes of pause sites recognized by core RNAP have been described (4). The first critically depends on a nascent RNA hairpin whose interaction with the RNAP could allosterically affect catalysis (92). Signals of the second class do not involve hairpins; instead, they induce backtracking (reverse translocation) of RNAP along the RNA and DNA chains by 1 to 3 bp (46). Some pause signals have specific regulatory functions, such as attenuation control (52) and protein recruitment (5), whereas others appear to be randomly located along the template and cumulatively act to reduce the overall rate of transcription, thus facilitating coupling of transcription and translation and possibly altering the RNA folding pathways.

At an arrest site, the RNAP slides back along the template even further, replacing the 3′ OH of the nascent RNA in the active site with an internal phosphodiester bond. Restart requires cleavage of the extruding RNA fragment by the RNAP's active site in a reaction stimulated by the cleavage factors GreA and GreB or elevated pH (93).

Bacterial termination signals belong to either of two classes: factor independent (intrinsic) or factor dependent. Both induce dissociation of the TEC. At intrinsic terminators, RNA release is controlled

---

Irina Artsimovitch • Department of Microbiology, Ohio State University, 484 West 12th Ave., Columbus, OH 43210.

principally by an RNA signal consisting of a stable hairpin structure followed by U-rich RNA. Transcript release at such sites occurs 7 to 8 nt downstream from the base of the hairpin and does not require, but can be influenced by, general transcription factors such as NusA and NusG (93). In contrast, factor-dependent terminators critically depend on a regulatory protein (Alc, L4, Mfd, Nun, Rho; see below) to stop chain elongation and/or induce RNA release.

STRUCTURAL MODEL OF THE TEC

A combination of the extensive functional and cross-linking studies of the *Escherichia coli* TEC (see reference 48 and references within; 83), the X-ray structure of the *Thermus aquaticus* RNAP (112), and the model of the initiation complex (65) led to a functional model of the bacterial (hybrid) TEC (47). This model (Fig. 1) postulates that three separate interactions account for both the TEC's stability and its regulatory responses: the DNA-binding site (DBS), the front zip-lock (FZ), and the rear zip-lock (RZ); the last two sites compose the hybrid-binding site (HBS). DBS encircles the downstream DNA, whereas HBS accommodates the 8- to 9-bp RNA•DNA hybrid with the FZ that melts the DNA in front of the active site and the RZ that both maintains the upstream edge of the hybrid and actively displaces the nascent RNA from the hybrid, creating a double-stranded/single-stranded junction. This model provides the mechanistic explanations for the TEC's stability against dissociation, as well as the mechanisms for the formation of off-pathway intermediates: paused, arrested, and termination complexes.

TEC Stability

Interactions between RNAP and the nucleic acid chains, as well as the RNA:DNA pairing in the hybrid, all contribute to the extraordinary stability of the elongating TECs. In some studies, the RNA•DNA hybrid appears to be the critical determinant of the TEC's stability (88), whereas others also implicate the interactions between the 4- to 6-nt single-stranded RNA (ssRNA) segment upstream of the hybrid and the RNAP (94, 101). This discrepancy could be due in part to the different methods used to measure the half-life of the TEC. Reports concerning TEC at high ionic strength (which destabilizes the ionic contacts but strengthens the nonionic interactions, such as base pairing) implicate the hybrid as a key feature, while experiments performed under more physiolo-

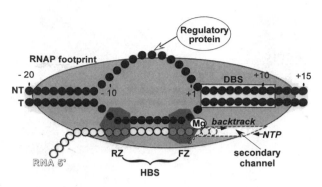

Figure 1. Schematic model of the TEC. RNAP (gray oval) is bound to the DNA duplex (black circles; T, template strand; NT, nontemplate strand) from ~ -20 to $\sim +15$ relative to the position of the 5' end of the encoded RNA (as judged from the protection against cleavage by various probes that it confers to the DNA). The 12 to 13 bp of the DNA duplex are melted in the transcription bubble. The nontemplate DNA strand is exposed on the surface of RNAP (48), where it becomes available for interactions with the regulatory proteins (5, 77).The nascent RNA (white circles) is annealed to the template strand to form 8 to 9 bp of the RNA•DNA hybrid (bound in the HBS) and is extruded from the TECs at ~ 14 nt from the 3' end. In the active TECs, the 3' end of the RNA is located in the active site. In the backtracked TECs, the 3' portion of the RNA (dashed circles) is threaded through the active site into the secondary channel, thus preventing the substrate NTP entry. Three principal interactions are distinguished (47): the DNA-binding site (DBS), the front zip-lock (FZ), and the rear zip-lock (RZ). Cross-linking analysis of the *E. coli* RNAP core ($\alpha_2\beta\beta'$ complex) bound to the RNA and DNA to form the TEC (see reference 48 and references therein) identifies the β' jaw/β lobe module as the DBS, the active site as the FZ, and a combination of β flap, β region D, and β' rudder as the RZ. Although originally the rudder was postulated to play a key role in RNA displacement, recent evidence indicates that the rudder stabilizes the TEC through direct contacts with RNA, but is not required for either RNA displacement or the maintenance of the transcription bubble (50).

gical conditions emphasize the ssRNA contribution. Analysis of these results is further complicated by the insufficient data about the nature of the ssRNA binding site, which could include a combination of ionic and nonionic interactions, some of which could even be sequence or structure specific. It is likely that both the RNA•DNA hybrid and the ssRNA contribute to the TEC's stability, together with the dsDNA at positions +3 to +11 (within the DBS). ssDNA at positions +1 to +2 and possibly the nontemplate DNA (see, however, reference 94) are also required for the maintenance of the stable and active TEC (49).

Lateral Oscillations of the TEC

The ability of RNAP to slide back and forth along the DNA and RNA chains is usually explained in terms of a thermodynamic "positional equilibrium" model (32), with the relative strength of the RNA•DNA hybrid determining the RNAP position. At positions with a particularly weak hybrid (such as rU:dA), RNAP is prone to backtrack to an upstream position (characterized by a stronger hybrid) by 1 to 3 bp at pause sites and up to 18 bp at arrest sites (93). Upon backtracking, the 3′ segment of RNA enters the secondary channel (Fig. 1) and blocks the NTP substrate's access to the active site, leading to pausing or arrest. Short-range backtracking is reversible in vitro and could be facilitated by protein factors such as NusG (4, 73) or by assisting mechanical force (24). In contrast, more extensive backtracking is irreversible except by the endonucleolytic cleavage of the extruding 3′ portion of the nascent transcript or by the action of Mfd protein (71). To provide for the lateral mobility of RNAP, DBS and HBS contacts must allow sliding of the RNAP along the DNA and RNA chains while maintaining both the constant length of the RNA•DNA hybrid "locked" between the FZ and RZ and the proper boundary between the hybrid and the extruded RNA. Irreversible arrest may be explained if extensive backward sliding allows a stable interaction between the 3′ RNA segment and the residues in the channel that prevents further sliding or triggers a conformational change in the TEC.

Transcription Termination

RNA release is triggered at sites where the nascent RNA folds into a stable, GC-rich hairpin followed by a stretch of the U-rich RNA. Both elements are essential for termination, but the size of the hairpin and the sequence of the 3′ RNA segment can be varied (the three U residues that abut the hairpin are the most conserved [17]). Termination occurs in two steps: pausing at the site of termination and RNA release. The terminator hairpin was originally thought, by analogy to pause hairpins, to induce pausing. Recent data suggest, however, that the pausing is induced by the U-rich RNA (34), whereas the role of RNA hairpin is limited to the destabilization of the hybrid and/or TEC. Termination can also be induced by oligonucleotides that pair to the nascent RNA to mimic the hairpin-duplex boundary if the chain extension by RNAP is slowed or halted (3, 107). The exact role of the termination hairpin is still debated. In one model, formation of the RNA hairpin disrupts critical interactions with the ssRNA-specific binding site in RNAP (34, 94). In another, formation of an RNA hairpin results in shortening of the RNA•DNA hybrid that is essential for TEC stability (96). The assumption that the hairpin directly competes with

and destroys a significant part of the RNA•DNA hybrid (106) is not supported by recent data: at many termination sites, the hairpin invades at most 1 nt into the 8-bp hybrid, and frequently destroys just the weakest rU:dA bp. However, the formation of the hairpin can destabilize the hybrid indirectly, either by disrupting the essential protein-nucleic acid contacts (34) or by altering the relative positions of the two double-stranded segments: the hairpin and the hybrid (45). Melting of the upstream portion of the hybrid that occurs during termination (34, 45) will trigger the collapse of the transcription bubble (reannealing of the two DNA strands) and, if the RZ module has to maintain its contacts to the newly established single-stranded/double-stranded junction, could also induce forward translocation of RNAP (107) and the loss of the FZ contacts to the 3′ end of the RNA. To leave the RNA completely behind, RNAP would have to forward-translocate by >8 bp. Alternatively, the loss of the RNA from the active site could trigger an allosteric change in the enzyme leading to the TEC dissociation and the simultaneous release of the RNA and DNA (51). Single-molecule studies of termination that can follow the release of the DNA and the nascent RNA from the individual TEC molecules should determine whether the pathway of nucleic acid release at a terminator is obligatory (and universal).

KINETIC MODELS OF TRANSCRIPTION

In contrast to the mechanistic models based on the analysis of TEC structure and biochemical analysis of the static complexes, kinetic models should allow quantitative predictions and identification of kinetic intermediates, which are likely too short-lived to be detectable by the currently utilized structural approaches. Over the years, the proposed models for elongation have assumed progressively more complex forms, reflecting the multiple choices of the TEC, which can follow several alternative pathways at any template position (96): elongation, termination, editing, arrest, and pausing. Transition to each of these pathways occurs at a characteristic rate, and the fate of the TEC can be determined by the relationships among these rates. Even at strong terminators, the fraction of TECs that actually release RNA is less than 1. Two classes of models were proposed to explain this heterogeneity. The first class, or "equilibrium" models, posits that, at any template position, the TEC chooses a pathway based on the relative heights of the activation energy barriers leading to each pathway (96). However, once RNAP moves to the next template position, the TEC is reprogrammed and the choice is again made on the basis of the relative energy barriers at a new site. Thus, an equilibrium between different states is established at each and every position. Alternative models (e.g., a "balanced branching" model) instead argue that some long-lived populations of TEC never pause or terminate, whereas others do so readily (35). These long-lived states do not equilibrate at every step, and in an extreme version of this model, can be programmed at the promoter to elongate or terminate. The principal distinction between these models (72) is that only one catalytically active state is allowed by the rapid equilibrium model, whereas several distinct catalytically competent states can exist in parallel for multiple rounds of the NTP addition within the framework of a "balanced branching" model.

An equilibrium "kinetic competition" model was developed by Yager and von Hippel (106) to quantify the propensity of the TEC to follow the elongation versus the termination path (considering only two alternatives at a time). This model used a number of assumptions to calculate the stability of the TEC ($\Delta G°_{TEC}$) as a sum of three components: the unfavorable standard free energy required to melt the 18 bp of the duplex DNA to form the transcription bubble ($\Delta G°_{bubble}$) balanced by two favorable terms: $\Delta G°_{hybrid}$, standard free energy of the 12-bp RNA·DNA hybrid, and $\Delta G°_{RNAP-NA}$, standard free energy of the interactions between RNAP and the nucleic acid strands (the latter was assumed to be constant). During elongation, TEC is stable and the barrier to termination is quite high, whereas at termination sites, formation of the RNA hairpin would destabilize the RNA·DNA hybrid (and consequently TEC) and lower the activation barrier toward termination. An attractive feature of this model is that small changes in barrier heights (2 to 4 kcal/mol) would be sufficient to shift the TEC population toward one or another state, allowing for modulation by regulatory proteins and other factors (96).

It is now known that the geometry of the TEC is somewhat different (the hybrid is only 8 to 9 bp, and the bubble is 12 to 13 bp long [Fig. 1]); therefore, formation of the termination RNA hairpin does not physically compete with the RNA·DNA hybrid. In addition, the assumption that $\Delta G°_{RNAP-NA}$ does not change during elongation (e.g., as a result of conformational changes at a pause or termination site) certainly is a simplification: the efficiency of different terminators changes differently in response to varying the concentrations of the mono- and divalent cations and Cl⁻, sequences of the not-yet-transcribed regions of the downstream DNA, and substitutions in RNAP, suggesting that changes in RNAP conformation and/

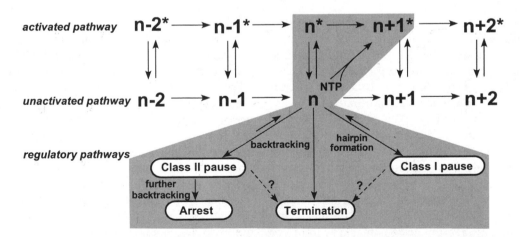

Figure 2. Branched regulatory pathways in transcript elongation. At least two TEC states that are competent for elongation can be distinguished (20). The activated pathway (asterisk) may require binding of the NTP in an "allosteric" site (25) or a conformational change induced by a substitution in RNAP. The unactivated pathway is characterized by a lower rate of NMP addition and represents a collection of intermediate states from which all the off-pathway states (pause, arrest, termination, and editing) are formed. This mechanism incorporates a previously proposed pathway (on a gray background) where certain nucleic acid signals trigger isomerization of TEC into a slow intermediate state (4). From the slow intermediate (n), the TEC can either slowly escape to the elongation pathway (upon nucleotide addition), or further isomerize into different types of pause, arrest, and termination complexes. Formation of an RNA hairpin would result in class I pausing, whereas backtracking would lead to a class II pause or arrest; termination could occur via either of these pathways. The principal difference between the slow intermediate and the parallel-path mechanisms is that the slow state is assumed to be short-lived in the former, and returns to the activated pathway upon NMP addition (escape) or NTP binding (reverse isomerization), but could persist for many rounds of catalysis in the latter. Isomerization into the termination and arrested states is irreversible and may proceed via formation of a significant kinetic intermediate (21, 24, 108).

or interactions with the nucleic acids also contribute to the termination/elongation decision (75). In fact, von Hippel and Yager (97) have also noted that the prediction of efficiency of class IV terminators requires that $\Delta G^{\circ}_{RNAP\text{-}NA}$ be increased by ~ 10 kcal/mol, implying that some of these interactions became significantly destabilized. The ability of the RNAP to terminate transcription on the ssDNA template, where the negative contribution of the bubble to the TEC stability does not exist (94), also supports models in which the RNA hairpin directly destabilizes a critical stability determinant within the TEC (either the RNA•DNA hybrid or the ssRNA contact) rather than offsets the energetic balance in the rapidly elongating TEC. Despite these limitations, the kinetic competition model correctly predicts the efficiencies for a collection of termination sites (97). This could indicate that the mechanism of termination is essentially identical at these sites (the same steps are rate-limiting) and, if RNAP contacts to the nucleic acids change at these terminators, they do so similarly.

The nonequilibrium, or "stable heterogeneity" (72), models postulate that TEC can exist in many states (Fig. 2), some of which do not equilibrate on the time scale of nucleotide addition. These models are supported by observations that RNAP possesses a memory of past and long-gone interactions (see reference 35 and references therein). Although most of these examples can be explained within the context of equilibrium models by assuming persistent interactions of the nascent RNA with the TEC, some cases remain unsolved. For example, termination protein Alc apparently "remembers" if RNAP has paused within 10 to 15 nt upstream from the site of Alc-induced termination (see below). The "stable heterogeneity" models can also explain the data from the single-molecule study in which the individual TEC molecules were observed (albeit still at low spatial resolution) to move at different rates (18) and the kinetic analyses (25, 70) that suggested the presence of more than one catalytically competent state of the TEC. An "alternative conformations" model (20) (Fig. 2) integrates these findings to suggest that elongating TEC exists in (at least) two distinct conformational states (activated and unactivated) that are catalytically active, do not equilibrate at each step of the transcription cycle, and are characterized by different rates of catalysis, responses to pause and termination sites, allosteric regulation by NTP, and elongation factors. This model also includes the off-pathway states, which are not in equilibrium with the elongation pathway(s), and provides possible targets for proteins that can affect elongation/termination decisions by the RNAP.

PROTEINS THAT INDUCE TRANSCRIPTION TERMINATION

Several protein factors are known to induce the nascent RNA release at template positions where the TEC is too stable to allow the spontaneous dissociation. These proteins either release the RNA directly, or halt RNAP until the arrival of a release factor, such as Mfd (see below).

Rho

Rho is the main termination protein in *E. coli*, where it is thought to control $\sim 50\%$ of all termination events (76). Unlike intrinsic termination sites, where all the absolutely required elements are clustered together, elements required for Rho-dependent termination are generally more complex. For Rho to terminate transcription, it must first attach to a 40-nt-long single-stranded nascent RNA segment with a high proportion of C residues, called the *rut* site (for Rho utilization). The attachment of Rho is inhibited by the ribosomes translating the nascent RNA; thus, Rho will terminate transcription efficiently only in the noncoding regions or when translation is inhibited. Once attached, Rho uses mechanical actions on the RNA (coupled to ATP hydrolysis) to track along the RNA. Finally, Rho catches up with an advancing RNAP in the *tsp* region (which can extend for 100 bp) where it releases the nascent RNA from the TEC. Positions at which Rho induces termination are determined by the efficiency of its loading at the *rut* site (114) and the sequence features that control the rate of elongation; there is good correlation between the distribution of the natural pause sites at which RNAP halts in the absence of Rho, and those at which Rho induces RNA release (64).

The low-resolution electron micrographic images of Rho reveal a ring-shaped sixfold symmetry structure, consistent with the notion that Rho functions as a hexamer, in which each subunit has an RNA-binding domain of a known structure (1, 10) and a putative ATP-binding domain whose structure is modeled on the F_1-ATPase (11). In a recently refined model for Rho action (76), two distinct sites for RNA binding exist in the hexamer: the C-rich *rut* RNA binds to the OB fold in a cleft that extends continuously across the six subunits, whereas the 3' RNA segment passes through the central hole and contacts the Q-loops of the ATP-binding domain. Capture of the 3' RNA fragment within the central hole would require the release of one or more subunits from the hexamer. Rebinding of the lost subunit would stabilize the hexamer and activate the RNA-dependent ATPase activity that is used by Rho for two types of motion. First, Rho can track along the RNA while possibly maintaining its original interaction with *rut*. Interaction of RNA with one subunit triggers ATP hydrolysis, followed by a conformational change in that subunit that transfers RNA to the next subunit, in the same direction upon each round of ATP hydrolysis, effectively translocating ssRNA in the 5′ to 3′ direction through the hole by a screwing motion at a rate of 1 nt per hydrolyzed ATP. Second, when Rho encounters the RNA:DNA duplex, it starts working as a helicase and unwinds the duplex. On naked nucleic acids, Rho moves along the ssRNA and the RNA:DNA duplex at the same speed of ~60 nt/s (estimated from the rate of ATP hydrolysis). At this rate, Rho could efficiently pull out the RNA from the TEC if RNAP moved slowly enough (76).

Rho can dissociate RNA by one of two mechanisms: it can either passively pull the RNA that has separated from the DNA as a result of breathing of the hybrid, or it can push RNAP forward (76), leading to the loss of the 3′-end position from the active site that could be accompanied by a conformational change (51). Rho action is probably the simplest among proteins that induce termination: Rho can release RNA from *E. coli*, T7, SP6 (73), and yeast polymerase II (pol II) (53), arguing against a specific contact with RNAP being required for termination. The observation that Rho cannot release RNA from stalled (and supposedly backtracked) TECs (73) suggests that Rho acts as an ATP-driven machine that is able to push RNAP easily, but only when lateral sliding is not inhibited by the 3′ segment of RNA in the secondary channel, and therefore acts similarly to the mechanical force that also cannot assist complexes that became arrested (24). The proposed mechanism of Rho action could explain the effect of NusG and NusA on RNA release. NusG, which is necessary in vivo for Rho-dependent termination at some sites (89), may assist Rho by preventing backtracking. In contrast, NusA inhibits Rho in vitro; this effect could be explained by stabilizing RNA secondary structures, facilitating backtracking, or denying Rho access to the TEC (NusA binds to the β flap, next to the emerging nascent RNA). In vivo, NusA could additionally inhibit Rho action by ensuring the close coupling between transcription and translation (113).

The universal mechanism of Rho action by destabilizing the RNA•DNA hybrid is consistent with the recently proposed universal mechanism of termination by pol II and *E. coli* RNAP (45) via shortening of the upstream part of the hybrid. However, unlike the experiments with nucleic acid duplexes in isolation, in TECs the hybrid is separated from the out-

side of RNAP by 4 to 6 nt of the ssRNA in the exit channel of RNAP, so that Rho does not have an easy access to the duplex, and the upstream boundary of the duplex may be stabilized by an RZ module (Fig. 1). Conceivably, to allow partial melting of the hybrid, a conformational change that destabilizes RZ (e.g., by lifting β flap at a pause site [51]) should occur first; if that were the case, the requirement for pausing prior to the Rho-dependent release would extend beyond a simple delay that allows Rho to catch up with the RNAP.

Mfd

Mfd is a transcription repair coupling factor that plays two key roles in the excision repair of the DNA lesions: it both removes transcription complexes stalled at these lesions and initiates the repair process by recruiting the DNA repair complex via contacts to UvrA (80). However, Mfd could have a more general role in transcription by removing RNAPs stalled on the template for any reason: Mfd also removes TECs stalled by a DNA-bound protein in vitro (81) and in vivo (109) and restores the register of the transcription complexes that have become arrested upon reverse translocation in vitro (71). Mfd dissociates TECs that are stalled in the absence of NTPs but activates them into elongating complexes when NTPs are present. The reactivation mechanism is unique: unlike GreB, Mfd does not induce the nascent RNA cleavage in the TEC, but induces instead a forward translocation of RNAP. The forward translocation has been proposed as a mechanism for both intrinsic and Rho-dependent termination. Interestingly enough, Mfd does not affect either intrinsic or Rho-dependent termination but rather induces release at sites of pausing (81). If Mfd action were slow and thus required stalling of RNAP, release of the nascent RNA at terminators could have occurred before Mfd had a chance to push RNAP forward. Both Mfd and Rho release TECs stalled by the binding of the cleavage-deficient EcoRI^{E111Q} (81). However, the pushing action of Mfd appears to be quite different: unlike Rho, Mfd does not act on a heterologous T7 RNAP and does not have a helicase activity on a variety of substrates (81), although it has helicase motifs and is d/rATP dependent. Most important, unlike Rho and mechanical force (24, 73), Mfd can rescue arrested complexes. The latter ability perhaps requires that Mfd induce a conformational change in the TEC that prohibits binding of the extruding 3′ segment of RNA in the arrested complexes within the secondary channel.

Like Rho, Mfd approaches the target TEC from behind: to act, it requires 25 to 30 bp of unobstructed (e.g., by σ in initiation complexes) upstream DNA, but the downstream DNA is not required. The Mfd binding site on RNAP has been mapped by the two-hybrid approach to the β^{1-142} segment, parts of which are located near the site where duplex DNA rewinds behind the transcription bubble (71). The differences in the fate of the TECs attacked by Mfd in the absence (dissociation) and in the presence (productive elongation) of NTP substrates may indicate that (Mfd-induced) forward translocation is catastrophic when the position of the 3′ end in the active site is not stabilized by the incoming NTP substrate. In the absence of the 3′ end in the active site, TEC may undergo a large-scale conformational change accompanied by opening of the TEC and release of the nucleic acid chains. In the presence of NTPs, the rate of elongation will be faster than the rate of Mfd-mediated release. Both Rho and Mfd seem to release the RNA at positions where TECs are not otherwise prone to dissociation in response to a regulatory signal—a lack of translation or a block to elongation, respectively.

Ribosomal Protein L4

Feedback control of the operons that encode ribosomal protein synthesis is commonly accomplished by autogenous regulation by one of the products. Among these operons, *E. coli* S10 is unique in that its regulator, the ribosomal protein L4 (when not bound to its 23S RNA target) not only represses translation of the first gene but also induces transcription termination in the leader region. Similar to many other attenuator regions, the S10 leader is predicted to encode five stem-loop structures (HA through HE), with the last one (HE) coding for the attenuator (terminator) hairpin. The first three hairpins appear to be dispensable for the L4 control mechanism (86), and the HD and HE structures tolerate a surprising number of alterations, including replacement by the heterologous hairpins (111), suggesting that L4 responds to multiple elements presented in the context of the secondary RNA structure. The attenuation mechanism critically depends on the NusA protein, which enhances intrinsic pausing at the attenuator site (87), likely through an interaction with the top part of attenuator hairpin. L4 additionally stabilizes the paused state (via interactions with the HD hairpin and the ascending part of HE), but cannot act in the absence of NusA even under conditions of NTP deprivation, where pausing is dramatically enhanced. The order-of-addition experiments revealed that NusA needs to act first for L4 to attenuate transcription: either L4 interacts with NusA directly or it only recognizes

TEC modified by NusA (87). The concerted action of NusA and L4 leads to the formation of a stable "pretermination" complex that releases the nascent RNA very slowly, at either a low or high UTP concentration in vitro: less than 50% of transcript is released after a 5-min incubation (87). In contrast, true termination occurs in vivo, suggesting a requirement for a release factor. The unusually high stability of these pretermination complexes could be due to the rather unusual structure of the attenuator hairpin, which contains many mismatches and only 50% G•C base pairs: the formation of the lower part of the hairpin would not be expected to compete efficiently with and disrupt the DNA•RNA hybrid. The essential role for NusA in this process would then be to stabilize the attenuator structure or its interaction with the β flap (92).

Transcription control by L4 likely involves interactions with RNA, RNAP, and/or NusA. In addition, L4 is a primary 23S rRNA-binding protein that binds to rRNA independently of the other proteins during early steps of 50S ribosomal subunit assembly and is thought to bring together distant regions of 23S RNA (105). Mutational analysis (55) implicates two distinct regions in RNA binding: rRNA (N-terminal) and mRNA (C-terminal). A recent structure (105) identifies two potential regions of L4 that could mediate rRNA and mRNA interactions at the molecular level. These α-helical regions are located at the opposite ends but on the same face of L4, whereas the other side could conceivably mediate interactions with the TEC. Although this side of the protein becomes buried within the ribosome, only free L4 is capable of transcriptional attenuation of S10 operon (110); thus, it remains possible that this region does indeed bind to TEC.

Bacteriophage HK022 Nun

Bacteriophage HK022 Nun protein is a member of the Arg-rich motif (ARM) family of proteins, which also includes λ N. Similarly to λ N, Nun binds to the boxB motif in the λ pL and pR operons (13, 16), but exerts the opposite effects on transcription: Nun induces termination shortly after BoxB, thus preventing expression of λ genes and arresting the λ lytic cycle (78). This effect is unique for the λ operons and, like the N action, is enhanced by the NusABEG proteins. Nun arrests transcription in the vicinity of a natural pause site, but can also act on TECs stalled by NTP deprivation if given sufficient time to bind (40, 41), suggesting that the pause simply allows Nun recruitment rather than shifts TEC into a Nun-receptive state. Moreover, the paused RNAP is somewhat resistant to the Nun action and can undergo multiple rounds of NMP incorporation prior to arrest. Once bound, Nun apparently inhibits translocation of RNAP and thus also inhibits the catalysis in both forward and reverse directions. A model for Nun action posits that Nun anchors RNAP to the template DNA via several simultaneous contacts: persistent contact to BoxB, Zn^{2+}-mediated binding to the RNAP (possibly to the putative Zn finger motif in the β′ N terminus), and intercalation of a critical penultimate Trp residue into the downstream DNA (99). In vitro, Nun arrests transcription, but no RNA release occurs (41); in fact, Nun additionally stabilizes the TEC at the site of release and during the template switching if the intercalating Trp residue is altered (99). In vivo, true termination occurs as a result of Mfd action (see reference 69 and references within).

Analysis of the Nun-N hybrids (38) and Nun interactions with the TEC and NusA (98), as well as the solution structure of Nun bound to BoxB (23), illuminated the differences between N and Nun. The similar N-terminal ARM motifs bind in the major groove of RNA hairpin, and the nonhomologous C-terminal regions confer the characteristic functions to N and Nun. The C-terminal portion of Nun is particularly information rich: it masks the N-terminal ARM unless NusA or RNAP is present, binds to RNAP and NusA, and contains the critical Trp^{109} residue; removal of 13 C-terminal Nun residues increases binding to BoxB but abolishes termination and RNAP binding (98). NusA alone inhibits Nun function by inhibiting its interactions with RNAP, but this effect is reversed by the addition of other Nus proteins (98). Differences between Nun and N interactions can also lie in the N-terminal domain, where, upon RNA binding, a hydrophobic ridge becomes exposed in Nun and can mediate interactions with TEC or host factors (23), whereas the corresponding surface of N is charged.

Bacteriophage T4 Alc

Like many other phages, T4 can selectively shut down the host transcription while allowing expression of the phage genome. Alc protein terminates transcription at several sites on a nonmodified host DNA, but not on the phage DNA that contains hydroxymethyl cytosine residues (43). Curiously, Alc terminates transcription by only rapidly transcribing complexes; it does not act on a slow RNAP mutant (β^{K1065A}) that pauses normally. Termination by Alc is maximal at high NTP concentration and, when only one NTP is limiting, is restricted to positions where this NMP has to be incorporated next. In addition, Alc termination is abolished if the TEC is halted within 10 to 15 nt from the point of release. These

observations led to the hypothesis that Alc is specific to a certain (activated?) state of RNAP which can be long-lived during elongation. Neither the mechanism of Alc-induced termination nor its inhibition by cytosine modifications is known. In vitro analysis demonstrated that Alc directly binds to the DNA in the absence of RNAP (43). During elongation, Alc might bind simultaneously to the DNA and to the region of RNAP that is positioned next to the downstream DNA: deletions in the β subunit SI1 region (residues 186 to 433 and 166 to 328) abrogate Alc function and alter contacts between the downstream DNA and β lobe (84), but do not impede elongation, suggesting that they identify the binding site for Alc (85). However, the observations that other mutations in RNAP, such as an active-site substitution β^{K1065A}, also interfere with Alc action but are not in contact with the downstream DNA suggest that conformational transitions in the TEC, perhaps accompanying the change in the downstream DNA interactions, are required for Alc binding/action. Upon binding, Alc could facilitate opening of the jaws and, consequently, dissociation of the TEC. Alternatively, Alc could anchor RNAP on the DNA (similarly to Nun) and inhibit forward translocation; however, unlike Nun, Alc does not appear to require a release factor in vitro (43). In vitro methylation of cytosine prevents Alc from releasing RNA at upstream or overlapping sites (-4 or -1; $+1$ sites relative to the modified C residue), suggesting that modifications either alter RNAP conformation or directly block Alc binding to the DNA. The second mechanism appears unlikely, as these modified residues are located within the DBS and are not expected to be accessible to Alc. On the other hand, pausing and termination are sensitive to changes in the downstream DNA sequences (54, 90); interactions with the (hydroxy)methyl C may induce a conformational change in RNAP that will make it Alc insensitive.

MECHANISMS THAT INHIBIT TERMINATION

Bacterial antitermination mechanisms can be divided into two groups (100). In one, formation of a terminator hairpin or binding of Rho (and thus RNA release) is inhibited by interactions with an antitermination factor (a regulatory protein, an upstream RNA segment, the ribosome, or a tRNA); as a result, RNAP transcribes through a single terminator (or the *tsp* region). The second group comprises "processive antitermination" mechanisms, in which the RNAP is modified to a termination-resistant form by a protein factor(s) or a nascent RNA structure; modified RNAP can transcribe through several consecutive terminators, becomes resistant to pause signals, and transcribes at a faster rate. In this case, the catalytic properties of the RNAP itself are altered: similar effects can be achieved by substitution(s) in the catalytic β and β' subunits (60, 107) and by interactions with the general elongation factors.

Attenuation control mechanisms, which adjust gene expression in response to a specific regulatory signal via modulation of transcription through the leader regions of different operons, exemplify the first group (26, 37). Most frequently, attenuation utilizes the intrinsic termination signals, likely because they can be confined to a short stretch of the transcript, in conjunction with the alternative RNA structures called antiterminators. The terminator and antiterminator structures are mutually exclusive; thus, by favoring one or the other structure, a regulatory factor can stimulate or inhibit termination, leading to a corresponding decrease or increase in the expression of the structural genes that follow the leader sequences. A more elaborate version of attenuation involves the formation of an antiantiterminator structure, which further stabilizes the termination hairpin by prohibiting the formation of antiterminator. Rho-attenuation mechanisms also exist; in the simplest case, the access of Rho to the RNA is denied by the stalled ribosome (27). Several types of nonprocessive mechanisms can be distinguished.

Translational Control

Amino acid biosynthetic operons regulate their expression in response to the supply of this same amino acid (e.g., histidine) by sensing the level of the corresponding tRNA charging. In this case, stalling of the ribosome at the regulatory position (e.g., at one of the seven tandem His codons in the leader region of the *Salmonella enterica* serovar Typhimurium *his* operon) signals a deficiency in the charged tRNA (52). Stalled ribosome allows the antiterminator hairpin to form, thus increasing the readthrough into the structural genes. Additional features within the leader regions can modulate the regulatory response. NusA-dependent pausing at the RNA hairpin preceding the antiterminator has been proposed to ensure the tight coupling of transcription and translation in the *E. coli trp* operon leader (52).

tRNA-Mediated Control

Regulation of amino acid metabolism is accomplished quite differently in gram-positive bacteria. Here the level of tRNA charging is sensed directly (in the absence of translation), via the interactions

between the tRNA and the leader transcript, which also encodes a terminator and an alternative antiterminator (29); the latter includes the conserved 14-nt sequence called the T-box. The expression of the T-box genes in *Bacillus subtilis* is induced by stabilization of an antiterminator element in the leader RNA by the cognate uncharged tRNA. The specificity of this response depends on a single codon in the leader that presumably pairs with the anticodon of the corresponding tRNA. A recent demonstration that the tRNA-leader RNA interaction can induce antitermination in vitro in the absence of accessory factors (31) indicates that this mechanism may represent a molecular fossil of the RNA world.

Transcription Control by Interaction of Small Metabolites with mRNA

Several recent reports demonstrated that the highly structured domains residing in the upstream noncoding RNA leader regions (termed riboswitches [103]) can directly sense the presence of small metabolites such as flavin mononucleotide, guanine, lysine, S-adenosylmethionine, and thiamin pyrophosphate (19, 30, 57, 59, 61, 102–104) in the absence of accessory protein factors. The riboswitches characterized thus far bind to their target molecules directly, with high affinity and selectivity. For example, the S-box leader RNA efficiently discriminates between S-adenosylmethionine, its natural signal, and S-adenosylhomocysteine, a close analogue that differs by a single methyl group (59, 104). Upon binding to their effectors, riboswitches, which encode various regulatory elements (terminator, antiterminator, and antiantiterminator RNA structures), undergo conformational changes that shift the equilibrium toward one of the mutually exclusive regulatory states, and thus determine whether transcription will terminate (e.g., if the termination hairpin formation is favored) or proceed into the structural genes. The variety of regulatory molecules discovered within the last year clearly indicates that this type of regulation is both essential and widespread in bacteria.

Antitermination by the Stalled Ribosome

Expression of the *tna* operon of *E. coli* allows bacteria to utilize tryptophan as a nitrogen or carbon source. Two structural genes of the operon, *tnaA* and *tnaB*, are preceded by a leader region that encodes a 24-residue TnaC. In the absence of Trp, Rho terminates transcription downstream of the *tnaC* stop codon. When present, Trp (or charged tRNA^{Trp}) competes with a release factor and inhibits the ribosome dissociation from the stop codon in response to a signal from a unique Trp^{12} residue located within the exit tunnel of the stalled ribosome (27). The stalled ribosome-leader peptide complex blocks the access of Rho to the overlapping *rut* sequences, thus preventing termination.

Antitermination by an RNA-Bound Protein

The *bgl* operon in *E. coli* is regulated in response to the availability of a substrate β-glucoside by BglG-induced antitermination. BglG binds to and stabilizes the antiterminator hairpin, preventing a formation of overlapping terminator (39). The RNA binding by BglG is regulated by BglF-mediated phosphorylation: in the absence of inducer, BglF phosphorylates BglG and inhibits its RNA binding activity (perhaps by inhibiting BglG dimerization); when β-glucosides are available, BglF dephosphorylates BglG, which now binds to its target and prevents transcription termination (2). A homologous system controls expression of the sugar utilization operons in *B. subtilis* (see reference 26 and references therein).

Termination by an RNA-Binding Protein

B. subtilis trp operon is regulated by another RNA-binding protein, TRAP. Binding of TRAP to the nascent leader RNA is activated by Trp and promotes termination of transcription before the *trp* biosynthetic genes by sequestering the upstream part of an antiterminator hairpin. TRAP (an 11-mer protein) binds to 11 tandem Trp codons separated by 2-nt linkers, wrapping the RNA around and melting the antiterminator structure, which is more stable than a terminator hairpin and thus predominates in the absence of Trp. RNA-binding activity of TRAP is fine-tuned to the charging ratio of $tRNA^{Trp}$: when levels of uncharged $tRNA^{Trp}$ increase, the T-box system induces expression of AT, an anti-TRAP protein that binds to TRAP and prevents its interaction with RNA (95), leading in turn to an increase in tryptophan biosynthesis. TRAP also controls the expression of an unlinked *trpG* gene by binding to the region overlapping the initiation codon and inhibiting translation (26).

Antitermination by RNA Chaperones That Melt Hairpins

A more global antitermination control is induced upon a shift of bacterial cells to a low temperature (6), and likely in response to other signals. The cold shock adaptation is accompanied by a dramatic increase in the expression of CspA, the major cold shock protein. CspA is an RNA chaperone that binds to and stabilizes the single-stranded RNAs, thereby destabilizing the RNA secondary structures (including the terminator hairpins). CspA (and its homologues CspC and E) induces expression of the *metY-rpsO* operon in vivo; at comparable concentrations, Csp proteins also inhibit pausing and antitermination in vitro (6). Although this mechanism could serve as an indiscriminate response to over-stabilization of many terminators upon cold shift, terminators that are affected more than others would be preferentially targeted. In addition, other Csp proteins are either constitutively produced at 37°C or induced upon entry into a stationary phase (see reference 6 and references therein), suggesting that their regulatory role is not limited to the cold shock.

Attenuation mechanisms operate within a particular regulatory region, and a separate act of recruitment would be necessary to modulate termination at each terminator. In contrast, the processive mechanisms utilize just one recruitment event to impose a lasting modification on the TEC upon binding of a primary regulator, which can be either protein or RNA. In the modified, fast state, RNAP becomes resistant to pause and termination signals, and thus continues transcription into the genes located downstream from the termination sequences. The nature of fast modification remains unknown. Conceivably, a regulator could bind to the TEC, switch it into a distinct state, and then dissociate again. If this switch does not involve a chemical modification, this mechanism would be supportive of the "stable heterogeneity" models. However, all well-characterized antitermination mechanisms have been shown (or proposed) to act on the TEC while maintaining a persistent contact with it.

Bacteriophage λ N Protein

Bacteriophage λ N protein regulates the transition between the early and late stages of phage transcription by allowing readthrough of transcriptional terminators (100). Although N can act alone (16), processive antitermination requires the assembly of a multipartite complex that is initiated by binding of N to the BoxB RNA hairpin within the NUT site and includes the cellular proteins NusA, -B, -G, and -E (62). N suppresses pausing and termination at both intrinsic and Rho-dependent sites. However, its kinetic effects on elongation are not sufficient to account for its antitermination effect, implying an additional role in stabilization of the TEC against dissociation (74, 107). A model for N action at the intrinsic sites posits that N, in conjunction with NusA, binds to the 5′ portion of the terminator RNA hairpin stem and precludes its formation (33). It is currently unclear how N inhibits Rho action; in this case the antipausing effect of N could conceivably allow RNAP to escape from Rho. Like HK022 Nun (see above), N can be functionally divided into two regions: an N-terminal RNA-binding arginine-rich motif that mediates N recruitment to TEC (via NUT) and a C-terminal region that likely mediates N-NusA and N-TEC contacts (69).

Bacteriophage λ Q Protein

Bacteriophage λ Q protein antiterminates transcription initiating from the late promoter $P_{R'}$. This specificity is determined by the mechanism of recruitment. Q initially binds a specific dsDNA target just upstream of the promoter and then transfers to the TEC that is stalled at a promoter-proximal site (+16) by σ^{70} contacts to a pseudo −10 element in the transcribed region (77). Q directly binds to region 4 of σ (that contacts the −35 hexamer at promoters) and repositions it with respect to other σ regions; the resulting conformational strain could explain why Q facilitates RNAP escape from the +16 pause site (67). After recruitment, Q modifies the TEC into a termination-resistant form (79), but is considerably more active toward TECs that are stabilized against backtracking, either by the sequence context around the pause site, or by the action of GreA and GreB proteins (58). In vitro, Q activity is stimulated by NusA; in vivo, Nus factor requirements are not characterized. Like N, Q apparently travels with RNAP over long distances (79). The model for Q action (107) considers two components of its antitermination activity: (i) Q may inhibit pausing in the region of RNA release and thus increase the kinetic barrier to termination, and (ii) like λ N, Q may directly interfere with the terminator hairpin formation (as it does with the invasion of oligos complementary to the nascent transcript). Both components likely contribute to Q function in vivo during rapid elongation.

Bacteriophage HK022 *put* RNAs

Bacteriophage HK022 *put* RNAs (for polymerase utilization), *putL* and *putR* are composed of two RNA stem-loop structures that, when present in the nascent RNA, modify RNAP into a state that bypasses the pause and termination sites in vitro and reads through downstream terminators located thousands of base pairs away in vivo (44). *put* mutations that disrupted the stems abolished antitermination, whereas mutations that restored the stems also restored function. The *put* activity was recapitulated

in a minimal in vitro system, indicating that the *put* transcript may itself modify RNAP (44). A direct interaction between the β′ Zn finger region of RNAP and *putL* has been suggested by the phenotypes of *rpoC* mutations that specifically affect *put* function (44), and detected by footprinting analysis in vitro (82). Antitermination requires that this contact is maintained during elongation, as the intervening RNA is looped out. The binding of the preformed *put* RNA to RNAP cannot be detected, suggesting that the stable association must be established while RNAP reads through *put* or requires very high local concentration of the RNA. In addition, changes in the structure of the in vitro prepared RNA cannot be excluded; in fact, some incongruity between the T7- and E. coli-derived *putL* RNAs has been detected by comparative in vitro footprinting (9). Although auxiliary proteins have not been implicated in *put* control, they might affect *put* RNA folding, either directly or via altering the rate of elongation; conceivably, RNAP itself could work as a chaperone for the bipartite *putL* secondary structure. It is also possible that some accessory factors are required to stabilize the *put*-RNAP interaction in vivo where the antitermination effect extends to the sites located as far as 5 kb downstream from *put*.

Bacteriophage Xp10 p7

Bacteriophage Xp10 p7 is a recently characterized regulatory protein from a lytic phage Xp10 of *Xanthomonas oryzae* (66) that inhibits intrinsic transcription termination by X. oryzae but not E. coli RNAP. Analysis of the E. coli/X. oryzae hybrid RNAPs implicates the β′ subunit as a binding site for p7. Unlike other antiterminator proteins, p7 does not appear to alter the rate of RNA chain elongation, and does not inhibit pausing at the site of RNA release. Interestingly, p7 also inhibits transcription initiation by preventing a conformational change in σ^{70}, and perhaps concomitant changes in core RNAP that accompany the open complex formation (12). Likewise, p7 could inhibit the hypothetical isomerization of the TEC into a state required for the RNA release, or simply inhibit the formation of a terminator hairpin. Persistent binding of p7 to the TEC and its effect on Rho termination have not been tested.

Bacteriophage P4 Psu

Bacteriophage P4 Psu inactivates Rho by an unknown mechanism (56) that may be nonprocessive: e.g., Psu may inhibit binding of Rho to the RNA.

Two cellular factors with the primary antitermination activity (as opposed to the enhancement of antitermination by NusA, NusG, etc.) have also been described: S4 and RfaH. Unlike most phage antitermination mechanisms that are able to challenge both intrinsic and Rho-dependent termination, cellular proteins are more specific to dealing with the Rho-dependent signals. The cellular systems may have evolved to allow transcription of the nontranslated, and thus vulnerable to Rho attack, *rrn* operons, as well as the unusually long operons, such as the 30-kb *tra*, which would be expected to contain a relatively large number of Rho-release sites.

Antitermination in the Ribosomal Operons

The operons that encode ribosomal operons (*rrn*) in *E. coli* are not transcribed and are, therefore, a potential target for Rho-mediated termination. However, these operons are not subject to Rho control. On one hand, rRNA transcripts fold into an extensive network of secondary structures that could mask the *rut* sites. However, this property alone does not account for Rho resistance: Rho-dependent terminators do exist within these operons. The rRNA antitermination mechanism with marked similarities to the λ N system ensures that RNAP is modified to a Rho-resistant state (14). With the exception of N, this mechanism utilizes the same nucleic acid (BoxA, -B, and -C) and protein determinants (NusA, -B, -G, and -E) to mediate antitermination. One of these proteins, NusE, is a ribosomal (r-) protein S10; its presence in the antitermination complexes can be either fortuitous (in the case of λ N) or can have a regulatory significance (e.g., in coregulation of the rRNA and r-proteins production). The second possibility is supported by a recent report that another r-protein, S4, acts as an antiterminator even in the absence of the rest of the proteins that together compose the ribosomal antitermination complex, thus being an analogue of λ N (91). Unlike N, however, S4 is specific toward Rho-dependent terminators, reflecting the specificity of the complete ribosomal antitermination complex. S4 directly and stably binds to RNAP in vitro, indicating that it could act on RNAP directly, rather than by affecting RNA structure. S4 has an additional role in rRNA synthesis: by binding to the 5′ portion of the emerging 16S RNA, it mediates its proper folding and assembly into the 30S subunit (68). Both mechanistic functions of S4 (as well as other r-proteins associated with the TEC) can be executed during antitermination: in addition to sustaining the unimpeded *rrn* operon transcription (so that the level of the downstream 23S RNA matches that of the 16S RNA), the

antitermination complex can directly deliver the r-proteins to their proper binding site (91).

RfaH

RfaH is a cellular antiterminator that controls expression of several virulence and fertility operons in gram-negative bacteria. RfaH action in vivo and in vitro depends on a 12-bp sequence called *ops* (operon polarity suppressor). RfaH increases expression of the downstream portions of *ops*-containing operons in vivo without affecting transcription initiation (8) and is therefore commonly described as an antiterminator. In vitro, RfaH significantly enhances the overall elongation rate and suppresses pausing, but only modestly inhibits termination. In contrast to λ N and Q, whose effects on chain elongation alone cannot account for their antitermination (74, 107), RfaH's effect on termination may be solely a consequence of decreased pausing. RfaH could lack the TEC stabilizing activity that is ascribed to λ N and Q; in this case, it would have a more profound effect at sites where termination is kinetically limited (including the Rho-dependent terminators).

RfaH is a paralogue of NusG, which also suppresses pausing (but enhances Rho-dependent termination; see above). Unlike NusG, RfaH requires a specific site for its recruitment to the TEC. RfaH does not bind strongly to either RNAP or nucleic acids in isolation, but recognizes a TEC that is paused at the *ops* site via a direct and specific contact to the nontemplate DNA segment exposed on the surface of RNAP (5). It is not known whether pausing is required for RfaH to bind; a pause could either provide a certain conformation of RNAP that RfaH binds to preferentially or simply increase the longevity of the target; unlike λ Q, RfaH delays RNAP escape from the *ops*. Upon recruitment, RfaH could induce conformational changes in RNAP (5) and, similarly to the phage regulators, could remain bound to the RNAP during transcription of the downstream genes (7, 28), thereby providing a persistent modification of the TEC. Although accessory proteins are not required for (and do not significantly alter) RfaH activity in vitro (5), they might modulate its action in vivo.

Bacteria and their phages use various regulatory factors to override the termination signals (Fig. 3). These factors can inhibit termination either directly, by interfering with the hairpin formation or Rho binding, or indirectly, by shifting the TEC into a termination-resistant state. While the direct antitermination mechanisms are well understood, the nature of the processive modification of the RNAP and the mechanism of RNA release remain unclear, chiefly because the different conformational states of

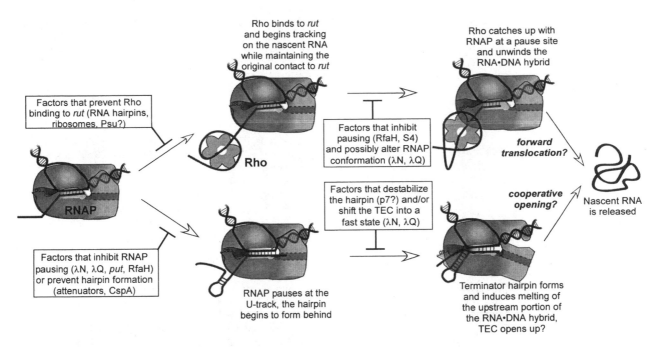

Figure 3. Antitermination mechanisms. Isomerization of a rapidly elongating TEC into a terminating complex proceeds through several steps at both Rho-dependent (top) and intrinsic (bottom) sites. Regulatory mechanisms that target one or more steps are known. As pausing precedes termination, RNA release can be inhibited by factors that also inhibit pausing (λ N, λ Q, and *put*). In contrast, other factors may preferentially inhibit hairpin formation at intrinsic terminators (alternative RNA structures, ssRNA-binding proteins, and possibly p7) or Rho access to the nascent RNA (the stalled ribosome, maybe Psu). Antitermination factors may also actively stabilize the TEC against dissociation (λ N and λ Q). Two alternative models for the mechanism of the nascent RNA release have been proposed. In a forward translocation model, RNAP slides forward (without necessarily changing its conformation), leaving the transcript behind (107). In an allosteric model, a regulatory signal (e.g., formation of a terminator hairpin) triggers a series of cooperative conformational changes in the TEC that lead to opening of a crab-claw-shaped TEC and the concomitant release of the nucleic acids (51).

preterminating TEC cannot be studied by conventional "slow" biochemical techniques that provide snapshots of the population-average mixture of TEC intermediates halted along the termination pathways or caught off-pathway (see references 42 and 93 for discussion). To further complicate this analysis, different terminators could be characterized by different rate-limiting steps, such that the conclusions from analysis of one site are not necessarily applicable to others. Thus, elucidation of the molecular mechanism(s) of termination and the exact roles of factors that modulate TEC behavior at terminators will require a combination of approaches that study the single TEC molecules in the real time.

Acknowledgments. I thank Vladimir Svetlov for discussions and helpful comments on the manuscript.

The research in my laboratory is supported by the NIH (GM067153) and the AHA (0265013B).

REFERENCES

1. **Allison, T. J., T. C. Wood, D. M. Briercheck, F. Rastinejad, J. P. Richardson, and G. S. Rule.** 1998. Crystal structure of the RNA-binding domain from transcription termination factor rho. *Nat. Struct. Biol.* **5:**352–356.

2. **Amster-Choder, O., and A. Wright.** 1997. BglG, the response regulator of the Escherichia coli bgl operon, is phosphorylated on a histidine residue. *J. Bacteriol.* **179:**5621–5624.

3. **Artsimovitch, I., and R. Landick.** 1998. Interaction of a nascent RNA structure with RNA polymerase is required for hairpin-dependent transcriptional pausing but not for transcript release. *Genes Dev.* **12:**3110–3122.

4. **Artsimovitch, I., and R. Landick.** 2000. Pausing by bacterial RNA polymerase is mediated by mechanistically distinct classes of signals. *Proc. Natl. Acad. Sci. USA* **97:**7090–7095.

5. **Artsimovitch, I., and R. Landick.** 2002. RfaH stimulates chain elongation by bacterial transcription complexes after recruitment by the exposed nontemplate DNA strand. *Cell* **109:**193–203.

6. **Bae, W., B. Xia, M. Inouye, and K. Severinov.** 2000. Escherichia coli CspA-family RNA chaperones are transcription antiterminators. *Proc. Natl. Acad. Sci. USA* **97:**7784–7789.

7. **Bailey, M. J., C. Hughes, and V. Koronakis.** 2000. In vitro recruitment of the RfaH regulatory protein into a specialised transcription complex, directed by the nucleic acid ops element. *Mol. Gen. Genet.* **262:**1052–1059.

8. **Bailey, M. J., C. Hughes, and V. Koronakis.** 1997. RfaH and the ops element, components of a novel system controlling bacterial transcription elongation. *Mol. Microbiol.* **26:**845–851.

9. Banik-Maiti, S., R. King, and R. Weisberg. 1997. The anti-terminator RNA of phage HK022. *J. Mol. Biol.* **272:**677–687.

10. Briercheck, D. M., T. C. Wood, T. J. Allison, J. P. Richardson, and G. S. Rule. 1998. The NMR structure of the RNA binding domain of E. coli rho factor suggests possible RNA-protein interactions. *Nat. Struct. Biol.* **5:**393–399.

11. Burgess, B. R., and J. P. Richardson. 2001. RNA passes through the hole of the protein hexamer in the complex with the Escherichia coli Rho factor. *J. Biol. Chem.* **276:**4182–4189.

12. Callaci, S., E. Heyduk, and T. Heyduk. 1998. Conformational changes of Escherichia coli RNA polymerase sigma70 factor induced by binding to the core enzyme. *J. Biol. Chem.* **273:**32995-32001.

13. Chattopadhyay, S., S. C. Hung, A. C. Stuart, A. G. Palmer III, J. Garcia-Mena, A. Das, and M. E. Gottesman. 1995. Interaction between the phage HK022 Nun protein and the nut RNA of phage lambda. *Proc. Natl. Acad. Sci. USA* **92:**12131–12135.

14. Condon, C., C. Squires, and C. L. Squires. 1995. Control of rRNA transcription in Escherichia coli. *Microbiol. Rev.* **59:** 623–645.

15. Cramer, P. 2002. Multisubunit RNA polymerases. *Curr. Opin. Struct. Biol.* **12:**89–97.

16. Das, A., M. Pal, J. Mena, W. Whalen, K. Wolska, R. Crossley, W. Rees, P. von Hippel, N. Costantino, D. Court, M. Mazzulla, A. Altieri, R. Byrd, S. Chattopadhyay, J. DeVito, and B. Ghosh. 1996. Components of multiprotein-RNA complex that controls transcription elongation in Escherichia coli phage lambda. *Methods Enzymol.* **274:**374–402.

17. d'Aubenton Carafa, Y., E. Brody, and C. Thermes. 1990. Prediction of Rho-independent *Escherichia coli* transcription terminators. A statistical analysis of their RNA stem-loop structures. *J. Mol. Biol.* **216:**835–858.

18. Davenport, R. J., G. J. Wuite, R. Landick, and C. Bustamante. 2000. Single-molecule study of transcriptional pausing and arrest by E. coli RNA polymerase. *Science* **287:**2497–2500.

19. Epshtein, V., A. S. Mironov, and E. Nudler. 2003. The riboswitch-mediated control of sulfur metabolism in bacteria. *Proc. Natl. Acad. Sci. USA* **100:**5052–5056.

20. Erie, D. 2002. The many conformational states of RNA polymerase elongation complexes and their roles in the regulation of transcription. *Biochim. Biophys. Acta* **1577:**224–239.

21. Erie, D. A., O. Hajiseyedjavadi, M. C. Young, and P. H. von Hippel. 1993. Multiple RNA polymerase conformations and GreA: control of fidelity of transcription. *Science* **262:**867–873.

22. Erijman, L., and R. M. Clegg. 1998. Reversible stalling of transcription elongation complexes by high pressure. *Biophys. J.* **75:**453–462.

23. Faber, C., M. Scharpf, T. Becker, H. Sticht, and P. Rosch. 2001. The structure of the coliphage HK022 Nun protein-lambda-phage boxB RNA complex. Implications for the mechanism of transcription termination. *J. Biol. Chem.* **276:** 32064–32070.

24. Forde, N. R., D. Izhaky, G. R. Woodcock, G. J. Wuite, and C. Bustamante. 2002. Using mechanical force to probe the mechanism of pausing and arrest during continuous elongation by Escherichia coli RNA polymerase. *Proc. Natl. Acad. Sci. USA* **99:**11682–11687.

25. Foster, J. E., S. F. Holmes, and D. A. Erie. 2001. Allosteric binding of nucleoside triphosphates to RNA polymerase regulates transcription elongation. *Cell* **106:**243–252.

26. Gollnick, P., and P. Babitzke. 2002. Transcription attenuation. *Biochim. Biophys. Acta* **1577:**240–250.

27. Gong, F., and C. Yanofsky. 2002. Instruction of translating ribosome by nascent peptide. *Science* **297:**1864–1867.

28. Grayhack, E. J., X. Yang, L. F. Lau, and J. W. Roberts. 1985. Phage lambda gene Q antiterminator recognizes RNA polymerase near the promoter and accelerates it through a pause site. *Cell* **42:**259–269.

29. Grundy, F. J., and T. M. Henkin. 1993. tRNA as a positive regulator of transcription antitermination in B. subtilis. *Cell* **74:**475–482.

30. Grundy, F. J., S. C. Lehman, and T. M. Henkin. 2003. The L box regulon: lysine sensing by leader RNAs of bacterial lysine biosynthesis genes. *Proc. Natl. Acad. Sci. USA* **100:** 12057–12062.

31. Grundy, F. J., W. C. Winkler, and T. M. Henkin. 2002. tRNA-mediated transcription antitermination in vitro: codon-anticodon pairing independent of the ribosome. *Proc. Natl. Acad. Sci. USA* **99:**11121–11126.

32. Guajardo, R., and R. Sousa. 1997. A model for the mechanism of polymerase translocation. *J. Mol. Biol.* **265:**8–19.

33. Gusarov, I., and E. Nudler. 2001. Control of intrinsic transcription termination by N and NusA: the basic mechanisms. *Cell* **107:**437–449.

34. Gusarov, I., and E. Nudler. 1999. The mechanism of intrinsic transcription termination. *Mol. Cell* **3:**495–504.

35. Harrington, K. J., R. B. Laughlin, and S. Liang. 2001. Balanced branching in transcription termination. *Proc. Natl. Acad. Sci. USA* **98:**5019–5024.

36. Henkin, T. M. 2000. Transcription termination control in bacteria. *Curr. Opin. Microbiol.* **3:**149–153.

37. Henkin, T. M., and C. Yanofsky. 2002. Regulation by transcription attenuation in bacteria: how RNA provides instructions for transcription termination/antitermination decisions. *Bioessays* **24:**700–707.

38. Henthorn, K. S., and D. I. Friedman. 1996. Identification of functional regions of the Nun transcription termination protein of phage HK022 and the N antitermination protein of phage gamma using hybrid nun-N genes. *J. Mol. Biol.* **257:**9–20.

39. Houman, F., M. R. Diaz-Torres, and A. Wright. 1990. Transcriptional antitermination in the bgl operon of E. coli is modulated by a specific RNA binding protein. *Cell* **62:**1153–1163.

40. Hung, S., and M. Gottesman. 1995. Phage HK022 Nun protein arrests transcription on phage lambda DNA in vitro and competes with the phage lambda N antitermination protein. *J. Mol. Biol.* **247:**428–442.

41. Hung, S. C., and M. E. Gottesman. 1997. The Nun protein of bacteriophage HK022 inhibits translocation of Escherichia coli RNA polymerase without abolishing its catalytic activities. *Genes Dev.* **11:**2670–2678.

42. Kashlev, M., and N. Komissarova. 2002. Transcription termination: primary intermediates and secondary adducts. *J. Biol. Chem.* **277:**14501–14508.

43. Kashlev, M., E. Nudler, A. Goldfarb, T. White, and E. Kutter. 1993. Bacteriophage T4 Alc protein: a transcription termination factor sensing local modification of DNA. *Cell* **75:**147–154.

44. King, R., S. Banik-Maiti, D. Jin, and R. Weisberg. 1996. Transcripts that increase the processivity and elongation rate of RNA polymerase. *Cell* **87:**893–903.

45. Komissarova, N., J. Becker, S. Solter, M. Kireeva, and M. Kashlev. 2002. Shortening of RNA:DNA hybrid in transcription elongation complex of RNA polymerase is a prerequisite for transcription termination. *Mol. Cell* **10:**1151–1162.

46. Komissarova, N., and M. Kashlev. 1997. RNA polymerase switches between inactivated and activated states by

translocating back and forth along the DNA and the RNA. *J. Biol. Chem.* **272:**15329–15338.

47. **Korzheva, N., and A. Mustaev.** 2001. Transcription elongation complex: structure and function. *Curr. Opin. Microbiol.* **4:**119–125.

48. **Korzheva, N., A. Mustaev, M. Kozlov, A. Malhotra, V. Nikiforov, A. Goldfarb, and S. A. Darst.** 2000. A structural model of transcription elongation. *Science* **289:**619–625.

49. **Korzheva, N., A. Mustaev, E. Nudler, V. Nikiforov, and A. Goldfarb.** 1998. Mechanistic model of the elongation complex of Escherichia coli RNA polymerase. *Cold Spring Harbor Symp. Quant. Biol.* **63:**337–345.

50. **Kuznedelov, K., N. Korzheva, A. Mustaev, and K. Severinov.** 2002. Structure-based analysis of RNA polymerase function: the largest subunit's rudder contributes critically to elongation complex stability and is not involved in the maintenance of RNA-DNA hybrid length. *EMBO J.* **21:**1369–1378.

51. **Landick, R.** 2001. RNA polymerase clamps down. *Cell* **105:**567–570.

52. **Landick, R., C. L. Turnbough, Jr., and C. Yanofsky.** 1996. Transcription attenuation, p. 1263–1286. *In* F. C. Neidhardt, R. Curtiss III, J. L. Ingraham, E. C. C. Lin, K. B. Low, B. Magasanik, W. S. Reznikoff, M. Riley, M. Schaechter, and H. E. Umbarger (ed.), Escherichia coli *and* Salmonella: *Cellular and Molecular Biology*, 2nd ed., vol. 1. ASM Press, Washington, D.C.

53. **Lang, W. H., T. Platt, and R. H. Reeder.** 1998. Escherichia coli rho factor induces release of yeast RNA polymerase II but not polymerase I or III. *Proc. Natl. Acad. Sci. USA* **95:**4900–4905.

54. **Lee, D. N., L. Phung, J. Stewart, and R. Landick.** 1990. Transcription pausing by *Escherichia coli* RNA polymerase is modulated by downstream DNA sequences. *J. Biol. Chem.* **265:**15145–15153.

55. **Li, X., L. Lindahl, and J. M. Zengel.** 1996. Ribosomal protein L4 from Escherichia coli utilizes nonidentical determinants for its structural and regulatory functions. *RNA* **2:**24–37.

56. **Linderoth, N. A., G. Tang, and R. Calendar.** 1997. In vivo and in vitro evidence for an anti-Rho activity induced by the phage P4 polarity suppressor protein Psu. *Virology* **227:**131–141.

57. **Mandal, M., B. Boese, J. E. Barrick, W. C. Winkler, and R. R. Breaker.** 2003. Riboswitches control fundamental biochemical pathways in Bacillus subtilis and other bacteria. *Cell* **113:**577–586.

58. **Marr, M. T., and J. W. Roberts.** 2000. Function of transcription cleavage factors GreA and GreB at a regulatory pause site. *Mol. Cell* **6:**1275–1285.

59. **McDaniel, B. A., F. J. Grundy, I. Artsimovitch, and T. M. Henkin.** 2003. Transcription termination control of the S box system: direct measurement of S-adenosylmethionine by the leader RNA. *Proc. Natl. Acad. Sci. USA* **100:**3083–3088.

60. **McDowell, J. C., J. W. Roberts, D. J. Jin, and C. Gross.** 1994. Determination of intrinsic transcription termination efficiency by RNA polymerase elongation rate. *Science* **266:**822–825.

61. **Mironov, A. S., I. Gusarov, R. Rafikov, L. E. Lopez, K. Shatalin, R. A. Kreneva, D. A. Perumov, and E. Nudler.** 2002. Sensing small molecules by nascent RNA: a mechanism to control transcription in bacteria. *Cell* **111:**747–756.

62. **Mogridge, J., T. Mah, and J. Greenblatt.** 1995. A protein-RNA interaction network facilitates the template-independent cooperative assembly on RNA polymerase of

a stable antitermination complex containing the lambda N protein. *Genes Dev.* **9:**2831–2845.

63. **Mooney, R. A., I. Artsimovitch, and R. Landick.** 1998. Information processing by RNA polymerase: recognition of regulatory signals during RNA chain elongation. *J. Bacteriol.* **180:**3265–3275.

64. **Morgan, W. D., D. G. Bear, and P. H. von Hippel.** 1984. Specificity of release by Escherichia coli transcription termination factor rho of nascent mRNA transcripts initiated at the lambda PR. *J. Biol. Chem.* **259:**8664–8671.

65. **Naryshkin, N., A. Revyakin, Y. Kim, V. Mekler, and R. H. Ebright.** 2000. Structural organization of the RNA polymerase-promoter open complex. *Cell* **101:**601–611.

66. **Nechaev, S., Y. Yuzenkova, A. Niedziela-Majka, T. Heyduk, and K. Severinov.** 2002. A novel bacteriophage-encoded RNA polymerase binding protein inhibits transcription initiation and abolishes transcription termination by host RNA polymerase. *J. Mol. Biol.* **320:**11–22.

67. **Nickels, B. E., C. W. Roberts, H. I. Sun, J. W. Roberts, and A. Hochschild.** 2002. The σ70 subunit of RNA polymerase is contacted by the λQ antiterminator during early elongation. *Mol. Cell* **10:**611–622.

68. **Noller, H., and M. Nomura.** 1996. Ribosomes, p. 167–186. *In* F. C. Neidhardt, R. Curtiss III, J. L. Ingraham, E. C. C. Lin, K. B. Low, B. Magasanik, W. S. Reznikoff, M. Riley, M. Schaechter, and H. E. Umbarger (ed.), Escherichia coli *and* Salmonella: *Cellular and Molecular Biology*, 2nd ed., vol. 1. ASM Press, Washington, D.C.

69. **Nudler, E., and M. E. Gottesman.** 2002. Transcription termination and anti-termination in E. coli. *Genes Cells* **7:**755–768.

70. **Palangat, M., and R. Landick.** 2001. Roles of RNA:DNA hybrid stability, RNA structure, and active site conformation in pausing by human RNA polymerase II. *J. Mol. Biol.* **311:**265–282.

71. **Park, J. S., M. T. Marr, and J. W. Roberts.** 2002. E. coli transcription repair coupling factor (Mfd protein) rescues arrested complexes by promoting forward translocation. *Cell* **109:**757–767.

72. **Pasman, Z., and P. von Hippel.** 2002. Active Escherichia coli transcription elongation complexes are functionally homogeneous. *J. Mol. Biol.* **322:**505.

73. **Pasman, Z., and P. H. von Hippel.** 2000. Regulation of rho-dependent transcription termination by NusG is specific to the Escherichia coli elongation complex. *Biochemistry* **39:**5573–5585.

74. **Rees, W. A., S. E. Weitzel, A. Das, and P. H. von Hippel.** 1997. Regulation of the elongation-termination decision at intrinsic terminators by antitermination protein N of phage lambda. *J. Mol. Biol.* **273:**797–813.

75. **Reynolds, R., R. M. Bermúdez-Cruz, and M. J. Chamberlin.** 1992. Parameters affecting transcription termination by *Escherichia coli* RNA polymerase. Analysis of 13 rho-independent terminators. *J. Mol. Biol.* **224:**31–51.

76. **Richardson, J.** 2002. Rho-dependent termination and ATPases in transcript termination. *Biochim. Biophys. Acta* **1577:**251–260.

77. **Ring, B., W. Yarnell, and J. Roberts.** 1996. Function of E. coli RNA polymerase σ factor σ70 in promoter-proximal pausing. *Cell* **86:**485–493.

78. **Robert, J., S. B. Sloan, R. A. Weisberg, M. E. Gottesman, R. Robledo, and D. Harbrecht.** 1987. The remarkable specificity of a new transcription termination factor suggests that the mechanisms of termination and antitermination are similar. *Cell* **51:**483–492.

79. Roberts, J. W., W. Yarnell, E. Bartlett, J. Guo, M. Marr, D. C. Ko, H. Sun, and C. W. Roberts. 1998. Antitermination by bacteriophage lambda Q protein. *Cold Spring Harbor Symp. Quant. Biol.* **63**:319–325.

80. Selby, C. P., and A. Sancar. 1994. Mechanisms of transcription-repair coupling and mutation frequency decline. *Microbiol. Rev.* **58**:317–329.

81. Selby, C. P., and A. Sancar. 1995. Structure and function of transcription-repair coupling factor. II. Catalytic properties. *J. Biol. Chem.* **270**:4890–4895.

82. Sen, R., R. A. King, and R. A. Weisberg. 2001. Modification of the properties of elongating RNA polymerase by persistent association with nascent antiterminator RNA. *Mol. Cell* **7**:993–1001.

83. Severinov, K. 2000. RNA polymerase structure-function: insights into points of transcriptional regulation. *Curr. Opin. Microbiol.* **3**:118–125.

84. Severinov, K., and S. A. Darst. 1997. A mutant RNA polymerase that forms unusual open promoter complexes. *Proc. Natl. Acad. Sci. USA* **94**:13481–13486.

85. Severinov, K., M. Kashlev, E. Severinova, I. Bass, K. McWilliams, E. Kutter, V. Nikiforov, L. Snyder, and A. Goldfarb. 1994. A non-essential domain of Escherichia coli RNA polymerase required for the action of the termination factor Alc. *J. Biol. Chem.* **269**:14254–14259.

86. Sha, Y., L. Lindahl, and J. M. Zengel. 1995. RNA determinants required for L4-mediated attenuation control of the S10 r-protein operon of Escherichia coli. *J. Mol. Biol.* **245**:486–498.

87. Sha, Y., L. Lindahl, and J. M. Zengel. 1995. Role of NusA in L4-mediated attenuation control of the S10 r-protein operon of Escherichia coli. *J. Mol. Biol.* **245**:474–485.

88. Sidorenkov, I., N. Komissarova, and M. Kashlev. 1998. Crucial role of the RNA:DNA hybrid in the processivity of transcription. *Mol. Cell* **2**:55–64.

89. Sullivan, S., and M. Gottesman. 1992. Requirement for E. coli NusG protein in factor-dependent transcription termination. *Cell* **68**:989–994.

90. Telesnitsky, A., and M. Chamberlin. 1989. Terminator-distal sequences determine the *in vitro* efficiency of the early terminators of bacteriophages T3 and T7. *Biochemistry* **28**:5210–5218.

91. Torres, M., C. Condon, J. M. Balada, C. Squires, and C. L. Squires. 2001. Ribosomal protein S4 is a transcription factor with properties remarkably similar to NusA, a protein involved in both non-ribosomal and ribosomal RNA antitermination. *EMBO J.* **20**:3811–3820.

92. Toulokhonov, I., I. Artsimovitch, and R. Landick. 2001. Allosteric control of RNA polymerase by a site that contacts nascent RNA hairpins. *Science* **292**:730–733.

93. Uptain, S., C. Kane, and M. Chamberlin. 1997. Basic mechanisms of transcript elongation and its regulation. *Annu. Rev. Biochem.* **66**:117–172.

94. Uptain, S. M., and M. J. Chamberlin. 1997. Escherichia coli RNA polymerase terminates transcription efficiently at rho-independent terminators on single-stranded DNA templates. *Proc. Natl. Acad. Sci. USA* **94**:13548–13553.

95. Valbuzzi, A., and C. Yanofsky. 2001. Inhibition of the B. subtilis regulatory protein TRAP by the TRAP-inhibitory protein, AT. *Science* **293**:2057–2059.

96. von Hippel, P. H. 1998. An integrated model of the transcription complex in elongation, termination, and editing. *Science* **281**:660–665.

97. von Hippel, P. H., and T. D. Yager. 1991. Transcript elongation and termination are competitive kinetic processes. *Proc. Natl. Acad. Sci. USA* **88**:2307–2311.

98. Watnick, R. S., and M. E. Gottesman. 1998. Escherichia coli NusA is required for efficient RNA binding by phage HK022 nun protein. *Proc. Natl. Acad. Sci. USA* **95**:1546–1551.

99. Watnick, R. S., S. C. Herring, A. G. Palmer III, and M. E. Gottesman. 2000. The carboxyl terminus of phage HK022 Nun includes a novel zinc-binding motif and a tryptophan required for transcription termination. *Genes Dev.* **14**:731–739.

100. Weisberg, R. A., and M. E. Gottesman. 1999. Processive antitermination. *J. Bacteriol.* **181**:359–367.

101. Wilson, K. S., C. R. Conant, and P. H. von Hippel. 1999. Determinants of the stability of transcription elongation complexes: interactions of the nascent RNA with the DNA template and the RNA polymerase. *J. Mol. Biol.* **289**:1179–1194.

102. Winkler, W., A. Nahvi, and R. R. Breaker. 2002. Thiamine derivatives bind messenger RNAs directly to regulate bacterial gene expression. *Nature* **419**:952–956.

103. Winkler, W. C., S. Cohen-Chalamish, and R. R. Breaker. 2002. An mRNA Structure that controls gene expression by binding FMN. *Proc. Natl. Acad. Sci. USA* **99**:15908–15913.

104. Winkler, W. C., A. Nahvi, N. Sudarsan, J. E. Barrick, and R. R. Breaker. 2003. An mRNA structure that controls gene expression by binding S-adenosylmethionine. *Nat. Struct. Biol.* **10**:701–707.

105. Worbs, M., R. Huber, and M. C. Wahl. 2000. Crystal structure of ribosomal protein L4 shows RNA-binding sites for ribosome incorporation and feedback control of the S10 operon. *EMBO J.* **19**:807–818.

106. Yager, T. D., and P. H. von Hippel. 1991. A thermodynamic analysis of RNA transcript elongation and termination in Escherichia coli. *Biochemistry* **30**:1097–1118.

107. Yarnell, W. S., and J. W. Roberts. 1999. Mechanism of intrinsic transcription termination and antitermination. *Science* **284**:611–615.

108. Yin, H., I. Artsimovitch, R. Landick, and J. Gelles. 1999. Nonequilibrium mechanism of transcription termination from observations of single RNA polymerase molecules. *Proc. Natl. Acad. Sci. USA* **96**:13124–13129.

109. Zalieckas, J. M., L. V. Wray, Jr., A. E. Ferson, and S. H. Fisher. 1998. Transcription-repair coupling factor is involved in carbon catabolite repression of the Bacillus subtilis hut and gnt operons. *Mol. Microbiol.* **27**:1031–1038.

110. Zengel, J. M., and L. Lindahl. 1993. Domain I of 23S rRNA competes with a paused transcription complex for ribosomal protein L4 of Escherichia coli. *Nucleic Acids Res.* **21**:2429–2435.

111. Zengel, J. M., Y. Sha, and L. Lindahl. 2002. Surprising flexibility of leader RNA determinants for r-protein L4-mediated transcription termination in the Escherichia coil S10 operon. *RNA* **8**:572–578.

112. Zhang, G., E. A. Campbell, L. Minakhin, C. Richter, K. Severinov, and S. A. Darst. 1999. Crystal structure of Thermus aquaticus core RNA polymerase at 3.3 A resolution. *Cell* **98**:811–824.

113. Zheng, C., and D. Friedman. 1994. Reduced Rho-dependent transcription termination permits NusA-independent growth of *Escherichia coli*. *Proc. Natl. Acad. Sci. USA* **91**:7543–7547.

114. Zhu, A. Q., and P. H. von Hippel. 1998. Rho-dependent termination within the trp t′ terminator. II. Effects of kinetic competition and rho processivity. *Biochemistry* **37**:11215–11222.

The Bacterial Chromosome
Edited by N. Patrick Higgins
© 2005 ASM Press, Washington, D.C.

Chapter 18

mRNA Decay and Processing

SIDNEY R. KUSHNER

In all biological systems, mRNAs serve as intermediates in the conversion of the genetic information contained within a cell's DNA into functional proteins. In prokaryotes, mRNAs are both synthesized and degraded rapidly, providing the organism with an excellent mechanism for both regulating gene expression and quickly adapting to changes in its environment. The half-lives of mRNAs can vary greatly (10 s to 20 min in *Escherichia coli*) but generally are never longer than the generation time of the bacterium. A unique feature in bacteria is the existence of polycistronic mRNAs. As discussed later in this chapter, many of these large transcripts are processed into smaller units as a means of controlling the expression of specific genes within each operon.

Furthermore, prokaryotic mRNAs can undergo posttranscriptional modification through the addition of poly(A) tails by either poly(A) polymerase or polynucleotide phosphorylase. Another feature that affects mRNA processing and stability is the coupling of transcription and translation. This coupling serves a number of functions, including preventing the formation of secondary structures within the mRNA as well as possibly directly protecting transcripts from either endo- or exonucleolytic degradation.

In *E. coli*, where mRNA decay and processing have been studied most extensively, considerable progress has been made over the past 30 years in identifying a variety of enzymes that degrade and process mRNA molecules either endo- or exonucleolytically. Furthermore, a series of structural features that help determine the overall stability of each transcript have also been characterized. Since there have been several extensive reviews of mRNA decay within the last several years (31, 70, 134, 150), this chapter focuses on issues that have not been completely resolved. These include the importance of RNA structural elements in mRNA decay, the existence and function of multiprotein mRNA decay complexes, the role of polyadenylation in mRNA decay, the regulation of

mRNA decay, the location of mRNA decay within the cell, whether *E. coli* is a suitable paradigm for mRNA processing and decay, the interrelationship between mRNA processing and decay, and whether all the proteins involved in mRNA decay and processing have been identified.

ENZYMES INVOLVED IN mRNA DECAY AND PROCESSING

E. coli and other bacteria contain a large number of RNases. For the sake of brevity, only those proteins that have been directly implicated in mRNA decay and processing are described in any detail (Table 1). However, as discussed below, there probably are additional enzymes involved.

Exonucleases

Exonucleases are defined as enzymes that degrade an RNA substrate from a terminus, one nucleotide at a time. They can act either hydrolytically, releasing nucleoside monophosphates, or phosphorolytically, employing inorganic phosphate and generating nucleoside monophosphates. Although they can start from either the 5' or 3' terminus, in *E. coli* there do not appear to be any 5' → 3' exonucleases (38). This is in contrast to *Saccharomyces cerevisiae*, in which most mRNA decay is carried out by the 5' → 3' Xrn exonuclease (24, 71).

PNPase

Polynucleotide phosphorylase (PNPase) degrades RNA by employing a phosphorolytic mechanism to generate nucleoside diphosphates (52, 145). The enzyme has been shown to be inhibited by secondary structures (147). Since the equilibrium constant of this reaction is close to 1, at low inorganic phosphate

---

Sidney R. Kushner • Department of Genetics, University of Georgia, Athens, GA 30602.

Table 1. Enzymes and proteins of *E. coli* that are involved in mRNA decay and processing

Enzyme	Gene	Subunit mol wt (10^3)	Substrate	Major characteristics and products
PNPase	*pnp*	86	mRNAs, poly(A) tails	$3' \rightarrow 5'$ exonuclease; phosphorolytic mechanism to release 5' nucleoside diphosphates in the presence of inorganic phosphate; reaction is reversible; does not degrade short oligonucleotides; highly conserved in both gram-negative and gram-positive bacteria; functions as a poly(A) polymerase in at least *E. coli* and photosynthetic bacteria
RNase II	*rnb*	73	mRNAs, poly(A) tails, untranslated RNAs	$3' \rightarrow 5'$ exonuclease; hydrolytic mechanism to release 5' nucleoside monophosphates; does not degrade short oligonucleotides; not present in many gram-positive bacteria such as *B. subtilis* and *S. aureus*; related to RNase R
RNase R	*rnr(vacB)*[a]	92	mRNAs?, rRNAs	$3' \rightarrow 5'$ exonuclease; hydrolytic mechanism to release 5' nucleoside monophosphates; does not degrade short oligonucleotides; homologues found in both gram-negative and gram-positive bacteria; related to RNase II; important for the degradation of nonfunctional rRNAs
Oligoribonuclease	*orn*	21	Short oligoribonucleotides	$3' \rightarrow 5'$ exonuclease; 5' nucleoside monophosphates; essential for cell viability; highly conserved in gram-negative bacteria; homologues not present in most gram-positive bacteria
RNase III	*rnc*	25	30s rRNA precursors; polycistronic mRNAs	Endonuclease; cleaves within the stem of specific stem-loop structures; found in both gram-negative and gram-positive bacteria; essential for cell viability in *B. subtilis* but not *E. coli*
RNase E	*rne(ams)*[b]	118	mRNAs, 9S and 16S rRNAs, tRNAs	Endonuclease; essential for cell viability; prefers RNA molecules with a 5' phosphomonoester terminus; cleaves within single-stranded AU-rich regions; highly conserved in gram-negative bacteria but found in only a limited number of gram-positive bacteria; amino terminus (through amino acid 468) significantly related to RNase G
RNase G	*rng(cafA)*[c]	51	5' end of 16S rRNA precursor; 9S rRNA; mRNAs	Endoribonuclease; not required for cell viability; prefers RNA molecules with a 5' phosphomonoester terminus; found in same species as RNase E; some bacteria have both RNase G and RNase E, others have only one
RNase P	*rnpA*[d] *rnpB*[e]	13	tRNA precursors; polycistronic mRNAs	Essential for cell viability; required for generating the mature 5' termini of tRNAs; contains a protein and a catalytic RNA subunit; highly conserved in all prokaryotes
Poly(A) polymerase I	*pcnB*	56	Any RNA with an available 3'-OH terminus	Generates poly(A) tails on mRNA molecules, usually after Rho-independent transcription terminators or on RNA degradation products that contain secondary structures at their 3' termini; highly conserved in both gram-negative and gram-positive organisms; related to tRNA nucleotidyltransferases
RhlB RNA helicase	*rhlB*	50	RNA with secondary structure	ATP-dependent RNA helicase; stimulated by RNase E; highly conserved in both gram-negative and gram-positive bacteria
Hfq	*Hfq*	11	Primarily an RNA-binding protein	Binds to AU-rich sequences which can also be cleavage sites for RNase E; has some effect on polyadenylation; highly conserved in gram-negative bacteria; more weakly conserved in gram-positive bacteria

[a]The structural gene for RNase R was originally called *vacB* (26).
[b]RNase E was independently identified as the *ams* locus (altered mRNA stability) (119).
[c]The structural gene for RNase G was originally identified as *cafA* (118).
[d]Encodes the protein subunit of RNase P.
[e]Encodes the M1 RNA subunit of RNase P.

concentrations the enzyme will synthesize RNA using nucleoside diphosphates as precursors (145). However, because the intracellular concentration of inorganic phosphate has been shown to be between 8 and 13 mM (109), it has generally been assumed that PNPase works exclusively as a degradative enzyme. Interestingly, it has now been demonstrated that PNPase can work biosynthetically in *E. coli*, serving as the backup poly(A) polymerase (104). In addition, it appears that in the cyanobacterium *Synechocystis* and in *Streptomyces coelicolor,* polyadenylation is carried out by PNPase and not poly(A) polymerase (139, 144). Genome-wide analysis indicates that PNPase is the major exonuclease involved in mRNA decay (106). Although *E. coli* contains a second phosphorolytic exonuclease (RNase PH) that is a functional homologue of PNPase, this enzyme does not seem to be involved in either polyadenylation (102) or mRNA decay (unpublished results). It has been shown to participate in tRNA maturation (67).

RNase II

RNase II is also a $3' \rightarrow 5'$ exonuclease, but it degrades RNA via a hydrolytic mechanism (146). This enzyme requires both Mg^{2+} and K^+ ions for full catalytic activity. RNase II is strongly inhibited by secondary structures (147). In addition, it has been shown that the enzyme can actually repress the degradation of RNA molecules by progressively removing nucleotides from the $3'$ end until it encounters a stable stem-loop structure. At this point it dissociates, but the molecule is now no longer a substrate for either RNase II or PNPase (28). Based on genomic analysis, it appears that RNase II actually protects a large number of mRNAs from degradation (106). Although mutations in either *pnp* (PNPase) or *rnb* (RNase II) do not affect cell viability, the double mutant is inviable and at the nonpermissive temperature accumulates partially degraded mRNA species (41).

RNase R

In *E. coli*, the RNase R enzyme is encoded by the *vacB/rnr* locus, a gene that was originally shown to be involved in virulence (26). In fact, RNase R has catalytic properties similar to those of RNase II and is significantly related at the amino acid level. Double mutants of PNPase and RNase R are inviable (26), a result that is similar to that observed with PNPase and RNase II multiple mutants (41). The loss of cell viability in the PNPase RNase R double mutant has now been attributed to a defect in the degradation of rRNAs (27). In contrast, RNase II RNase R double

mutants only show a small alteration in growth compared with either single mutant (unpublished results).

Oligoribonuclease

Oligoribonuclease is a $3' \rightarrow 5'$ exonuclease that, unlike the other enzymes described above, is specific for short oligonucleotides (116, 169). It has been shown that the enzyme is essential for cell viability and that oligonucleotides between 2 and 5 nucleotides (nt) in length accumulate at the nonpermissive temperature (50). Thus, oligoribonuclease is responsible for degrading the very short oligoribonucleotides that are no longer substrates for PNPase, RNase II, or RNase R. It is not clear, however, why an inability to degrade these oligonucleotides should be essential for cell viability.

Endonucleases

Endoribonucleases have traditionally been defined as enzymes that cleave phosphodiester bonds at a distance from a terminus.

RNase E

RNase E (*rne*) was first identified based on its role in the processing of a 9S rRNA precursor into a p5S form (6). Independently, the enzyme was discovered as a gene (*ams*) involved in the general degradation of mRNAs (7, 119). Subsequently, it has been shown that both loci encode the same protein, RNase E (9, 111, 153). After analysis of a number of RNase E cleavage sites, Ehretsmann et al. (44) proposed that the enzyme preferred a 5-nt AU-rich sequence that occurred in single-stranded regions of RNA molecules. However, other experiments have raised issues regarding the exact nature of the recognition site (64, 93, 94).

More important, the catalytic activity of the enzyme is stimulated significantly by its binding to the $5'$ terminus of an RNA molecule (89). In fact, it prefers a $5'$ monophosphate over a $5'$ triphosphate (148, 154). As discussed below in more detail, several additional proteins, including PNPase, copurify with RNase E (25, 128) (Fig. 1). Genetic studies have demonstrated that RNase E is essential for cell viability (6, 119).

RNase G/CafA

RNase G was first characterized as a protein (CafA) involved in the formation of cytoplasmic axial filaments (118). It was subsequently noted that

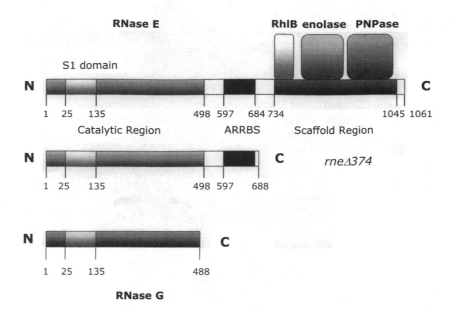

Figure 1. Physical relationship of RNase E and RNase G. The numbers below the horizontal rectangles represent the approximate domain boundaries as determined by McDowall and Cohen (95) and Vanzo et al. (156). ARRBS (rectangle) indicates the arginine-rich RNA binding site. The locations of the RhlB RNA helicase, enolase, and PNPase are as described by Vanzo et al. (156). The *rne*Δ374 allele retains the catalytic domain and the ARRBS and has been described by Ow et al. (121). The RNase G protein is 34% identical to the first 488 amino acids of RNase E.

there was considerable sequence identity (34.1% over the first 488 amino acids) between the CafA protein and the N terminus of RNase E (158) (Fig. 1). Subsequent experiments have demonstrated that *cafA* encodes a 5′ end-dependent endoribonuclease whose catalytic activity is similar to that of RNase E (77, 92, 154). It has now been shown that RNase G participates in the processing of the 5′ terminus of the 16S rRNA along with RNase E but that the two enzymes cleave at different sites (77, 154, 159). Interestingly, however, the inactivation of the RNase G protein does not lead to major phenotypic alterations in the cell (118, 158).

Following reports that RNase G alters the stability of the *adhE* and *eno* mRNAs (65, 155), microarray analysis of the *E. coli* transcriptosome showed that inactivation of RNase G led to an increase in the steady-state level of 11 mRNAs, including *adhE* and *eno* (73). In contrast, inactivation of RNase E affects the majority of *E. coli* transcripts (M. C. Ow and S. R. Kushner, unpublished results). Detailed genetic analysis has now shown that RNase G serves as a backup enzyme for RNase E in both the processing of 9S rRNA precursors and the decay of a variety of mRNAs, but does not participate in tRNA processing (124). However, there are conflicting data on whether RNase G can complement RNase E mutations (35, 73, 124, 158). What is clear is that overproduction of the native RNase G cannot complement RNase E mutants (35, 124).

RNase III

RNase III was first discovered as an enzyme that cleaves double-stranded RNA (137). It was subsequently shown to be involved in processing of the 30S rRNA precursor to help generate both 16S and 23S rRNA species (42). In vivo the enzyme recognizes specific stem-loop structures (Fig. 2C) and can cleave on either one or both sides of the stem. It initiates the decay of several mRNAs, including the polycistronic transcript that encodes PNPase (14, 126, 133). In general, most RNase III cleavage sites occur upstream or downstream of coding sequences. In *E. coli*, deletion of the *rnc* gene only results in a reduced growth rate (10, 152). In contrast, the gene seems to be essential for cell viability in *Bacillus subtilis* (58).

RNase P

RNase P is an essential enzyme that is involved in the processing of the 5′ ends of tRNA precursors (5). The enzyme is unusual in that it contains both a protein subunit and a catalytic RNA subunit (5). The RNA subunit is processed by RNase E (87). Although the enzyme was assumed to be primarily involved in tRNA processing, it was shown by Alifano et al. (3) that the polycistronic *his* mRNA of *Salmonella enterica* serovar Typhimurium was cleaved by RNase P. Subsequently, Li and Altman (76) used

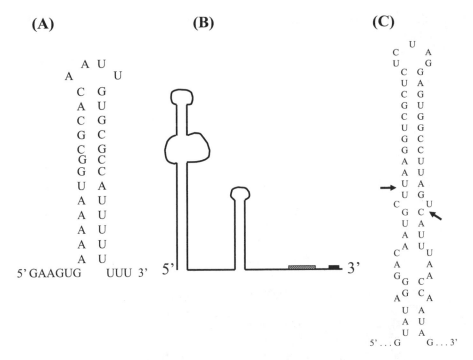

Figure 2. Secondary structures that affect mRNA decay. (A) Rho-independent transcription terminator at the 3′ terminus of the *E. coli lpp* mRNA. Of particular note is the very short region of single-stranded RNA that is present at the 3′ terminus. Poly(A) tails can be added at any of the three unpaired U's (22, 101). (B) Schematic view of the 5′ UTR of the *ompA* transcript as determined by Emory and Belasco (45). (C) Stem-loop structure recognized by RNase III. This is the R5 structure from the bacteriophage T7 early RNA (34). Arrows indicate the location of the RNase III cleavages.

microarray analysis in *E. coli* to demonstrate that RNase P can cleave in intercistronic regions in the *tna, secG, rbs,* and *his* operons. Cleavage at these locations leads to alteration in the half-lives of the downstream mRNAs (76).

Other Enzymes and Proteins Involved in mRNA Decay

RhlB

In *E. coli* there are a large number of potential RNA helicases, including at least five in the "DEAD" box family (80). One of these, encoded by the *rhlB* gene (66), has been implicated in mRNA decay because it copurifies with RNase E and PNPase (128). Although it is not clear that the isolated protein possesses helicase activity, Coburn et al. (32) have shown that in the presence of RNase E and ATP, a combination of PNPase and RhlB is sufficient to degrade a 147-nt *rpsT* degradation product that is known to contain stable secondary structures. Since many RNA molecules contain secondary structures that are known to inhibit 3′ → 5′ exonucleases, it makes considerable sense to think that some type of RNA helicase will be an important participant in mRNA decay.

Poly(A) polymerase I (PAP I)

In 1992 Cao and Sarkar (23) determined that the *pcnB* gene (84, 91) encoded poly(A) polymerase I (PAP I) of *E. coli*. Although PAP I was first identified in 1962 (8) and partially purified in the 1970s (100, 129, 149), not much attention was paid to the possibility that mRNAs in *E. coli* might contain poly(A) tails. However, along with the identification of the structural gene for poly(A) polymerase, experiments in that same year showed that *lpp* transcripts contained untemplated A residues after the Rho-independent transcription terminator (22) (Fig. 2A). Recent results have shown that in vitro PAP I can make mistakes, occasionally adding C, U, and G residues (164). However, in vivo, the enzyme appears to incorporate only A residues (104). An unusual feature of polyadenylation in *E. coli* is that the immature form of the 23S rRNA is the primary target of PAP I (101).

Because the level of polyadenylation in *E. coli* is rather low (estimated at 1 to 2% of total *E. coli* RNA [22, 101]), it was not clear whether other bacteria also polyadenylate their mRNAs. However, Yue et al. (165) showed that *E. coli* PAP I is part of a large superfamily of enzymes that also include tRNA nucleotidyltransferases. Raynal and Carpousis

(131) expanded on this work to demonstrate the presence of three motifs that are associated with the putative active-site E. coli of PAP I. Polyadenylated mRNAs have now been identified in S. coelicolor (18) and photosynthetic bacteria (139), although these species appear to be synthesized by PNPase (18, 139).

Hfq

Hfq was first identified based on its role in the replication of bacteriophage Qβ (49). Subsequently, it has been shown to be an abundant protein that appears to exist in vivo as a hexamer that forms a ring structure in both E. coli and Staphylococcus aureus (140, 142). The protein binds to both RNA and DNA but appears to prefer AU-rich sequences. In fact, Hfq is involved in controlling the levels of expression of a variety of mRNAs, such as rpoS (113) and ompA (157). Hfq has recently been shown to interact with a variety of small regulatory RNAs (108, 143, 166, 167). Its ability to bind to AU-rich sequences, some of which are RNase E cleavage sites, suggests a role for the protein in modulating both mRNA decay and tRNA processing (107, 167). In addition, Hfq has also been shown to affect the polyadenylation of the rpsO mRNA (55, 72). Based on in vitro experiments, it has been suggested that Hfq protects RNA, in particular, poly(A) tails, from exoribonucleolytic degradation (48). However, in vivo analysis indicates that Hfq serves to promote polyadenylation at specific locations on mRNA molecules rather than protecting poly(A) tails from degradation (B. K. Mohanty and S. R. Kushner, unpublished data).

Poly(A) binding proteins

Even though E. coli and other prokaryotes do not have homologues of the ubiquitous eukaryotic poly(A) binding proteins, there has been considerable interest in determining if other proteins serve a similar function. For example, an S1 RNA binding domain originally associated with ribosomal protein S1 (19) is found in a variety of RNases, including RNase E, RNase II, and PNPase. Since both PNPase and RNase II prefer poly(A) substrates, it would not be surprising if the S1 ribosomal protein also could bind to poly(A) tails. In fact, Feng et al. (47) demonstrated that in vitro S1 acts as a poly(A) binding protein. In addition, they also showed that the cold shock protein CspE not only bound to poly(A) sequences but inhibited the ability of PNPase to degrade such substrates (47).

IMPORTANT STRUCTURAL FEATURES OF mRNAs

3' Untranslated Regions

In E. coli and other bacteria, many transcripts are terminated by the formation of a Rho-independent stem-loop structure. Computer analysis suggests that almost 50% of the 2,592 annotated protein-encoding transcription units in E. coli are terminated in such a fashion (75). One of the interesting features of these structures is that they contain a very short 3' single-stranded region (Fig. 2A). Because most 3' → 5' exonucleases are not only inhibited by secondary structures but require a significant stretch of single-stranded RNA (>10 nt) to bind, Rho-independent terminators can provide an impediment to degradation. In addition, a second type of secondary structure, called repetitive extragenic elements (REP, an ~40-bp palindrome), was discovered to exist in the 3' untranslated regions of a number of RNAs (59, 151). It has been shown that 581 such sequences are located in 314 REP elements throughout the E. coli genome (16). These sequences are found either between genes or after the last gene in polycistronic operons (11). Some, but not all, of these REP sequences have been shown to have an effect on mRNA stability (115).

5' Untranslated Regions

Even though E. coli does not appear to have any 5' → 3' exonucleases (37), a variety of experiments have shown that 5' untranslated regions (5' UTR) can play a role in either stabilizing or destabilizing a particular transcript. In the case of ompA, the 133-nt 5' UTR forms a complex secondary structure, including a 5' proximal stem that leaves the terminus double stranded (46) (Fig. 2B). Alteration of the 5' UTR leads to dramatic changes in the half-life of the ompA transcript. In addition, when fused to the bla mRNA, the chimeric transcript is significantly stabilized (45, 46).

In contrast, the rne gene contains a 361-nt 5' UTR that is required, in part, for the autoregulation of the transcript (60). In this case, removal of the 5' UTR leads to the loss of autoregulation and the stabilization of the transcript. A series of phylogenetically conserved secondary structures within the 5' UTR is required for autoregulation to occur (40).

RNase III Processing Sites

RNase III recognizes specific stem-loop structures and cleaves them near the base of the stem (34).

Although many mRNAs contain a variety of stem-loop structures, it is difficult to accurately predict if they will be cleaved by RNase III and whether the enzyme will cut on one or both sides of the stem (Fig. 2C). For example, in the 5-kb early transcript of bacteriophage T7, there are five RNase III sites that are used to cleave it into smaller units (43). At four of the five sites, the enzyme introduces a single nick while at the fifth it cuts on both sides of the stem. Since RNA molecules contain large numbers of stem-loop structures and only a limited number of them are recognized by RNase III, considerable work has been carried out to establish the important features of a RNase III cleavage site. It now appears that double-helical RNA structures are protected from cleavage by the presence of specific base pairs that act as antideterminants (168). RNase III cleavage sites generally occur either in 5' UTRs (14) or inter-cistronic regions (132). Cleavages can result in the alteration of the stability of either the upstream or downstream mRNAs.

MULTIPROTEIN COMPLEXES ASSOCIATED WITH mRNA DECAY

Degradosomes

In 1994 several laboratories reported the copurification of RNase E and PNPase (25, 127). Subsequent work has shown that a multiprotein complex, called the degradosome, consists of RNase E, PNPase, the RhlB RNA helicase, and the glycolytic enzyme enolase (98, 128, 156) (Fig. 1). There have also been suggestions that several additional proteins, such as polyphosphate kinase, may weakly associate with the degradosome (17). With the discovery of the degradosome many investigators assumed that it was responsible for the bulk of mRNA decay (15, 130), even though all of the work establishing the existence of the degradosome employed in vitro techniques such as immunoprecipitation, cosedimentation, or copurification procedures. However, the model was supported by in vitro experiments that demonstrated that the degradosome is capable of decaying the *rpsT* transcript in a fashion comparable to that observed in vivo (30).

While it is not clear why the glycolytic enzyme enolase is found in the complex, the presence of the RhlB RNA helicase provides a mechanism for removing secondary structures from the RNA molecules being degraded. This is particularly important because PNPase, as well as other 3' → 5' exonucleases in *E. coli,* are inhibited by secondary structures

(28, 83, 96, 110, 147). Since many transcripts have stem-loop structures at their 3' termini, as noted above, degradation of transcripts by PNPase, RNase II, or RNase R will be impaired. Recently, Coburn et al. (32) have shown that under appropriate in vitro conditions, a complex containing RhlB, PNPase, and RNase E was sufficient to degrade a highly structured RNA fragment in an RNase E-independent fashion. In addition, it has also been suggested that RhlB can associate directly with PNPase in the absence of PNPase (82). These data indicate that under certain circumstances, the RhlB helicase in the presence of ATP can remove secondary structures from an RNA substrate, thereby improving the processivity of PNPase.

When taken together with the recent observations that RNase E is a 5' end-dependent endonuclease (89, 90, 148, 154), the degradosome is an attractive candidate to account for the very rapid degradation of full-length mRNAs without any detectable decay intermediates. In fact, Coburn and Mackie (31) have proposed a 5' tethering model to explain the action of the degradosome. As shown in Fig. 3, an alternative approach would have the simultaneous degradation of both ends of the transcript through the action of RNase E at the 5' end

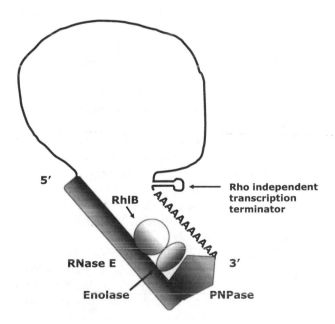

Figure 3. Model for mRNA decay involving degradosome attachment at both the 3' and 5' termini of a single mRNA. In this model, binding of PNPase to the 3' terminus of the poly(A) tail would bring along RNase E. RNase E would then bind to the 5' triphosphate terminus if this is a primary transcript. With such an arrangement, one could account for the rapid degradation of an individual mRNA.

and PNPase at the 3′ end. In this hypothesis, the degradosome could be brought to the mRNA by the association of PNPase with a poly(A) tail followed by the subsequent binding of RNase E to the 5′ terminus (89) or at internal entry sites (12). Because of the preferences of RNase E for a 5′ phosphomonoester (89), the initial binding by RNase E would be the rate-limiting step in this model. The RhlB helicase could remove any secondary structures that might impede the processivity of PNPase.

As attractive as these models may appear, there are still several issues that need to be addressed. In the first place, Deutscher and Reuven (36) have demonstrated that in extracts of *E. coli*, RNase II accounts for 90% of the degradative activity versus only 10% for PNPase. Taken at face value, this result would suggest that RNase II and not PNPase carries out most mRNA decay even though it has no association with the degradosome. While this observation at first glance might seem to cause a problem, it must be remembered that mRNAs only represent somewhere between 5 and 10% of the total cellular RNA at any given time. Furthermore, the Deutscher and Reuven (36) experiments used oligo(A) as a substrate, not naturally occurring RNA. In addition, it has now been shown that, in vivo, RNase II is far more effective in degrading poly(A) tails associated with rRNA than is PNPase (103). Finally, the results of genome-wide experiments examining steady-state RNA levels indicate that PNPase plays a much more significant role in mRNA decay than does RNase II (106). Thus it is still possible that RNase II provides the bulk of the degradative activity, but not that associated with the decay of most mRNAs.

A more serious issue relates to whether the degradosome actually exists in vivo, and if it does, if it is essential for mRNA decay. Recent work employing immunolocalization techniques has suggested that the degradosome exists in vivo and may be localized near the inner membrane (81). However, since *E. coli* is so small, immunolocalization techniques cannot unequivocally demonstrate that the four proteins colocalize inside the cell.

Of more concern are the experiments of Ow et al. (121). Taking advantage of a chromosomal deletion of the *rne* gene, they constructed a series of C-terminal truncation mutations that produced RNase E proteins that were defective in the assembly of the degradosome. Of particular interest is the *rne*Δ374 allele that retains both the catalytic region of the protein and the arginine-rich binding domain (ARRBS) but no longer contains the degradosome scaffold region (Fig. 1). Half-life measurements of seven individual mRNAs showed that mRNA decay was not significantly affected in the absence of de-

gradosome assembly (121). In addition, the growth rate of such a mutant was identical to the wild-type strain, within experimental error (121). It was not until both the ARRBS and the scaffold region were deleted that a significant defect in mRNA decay was observed (121). A similar result was obtained with a different *rne* allele that also was deleted for both the scaffold region and the ARRBS (86). Thus it would appear that failure to assemble the degradosome in *E. coli* is not essential for normal mRNA decay.

Another issue relating to the biological significance of the degradosome was the initial observation that it did not exist in a large number of bacteria. In particular, Kaberdin et al. (63) showed that a variety of bacteria, including *Haemophilus influenzae, Synechocystis, Porphyra purpurea,* and *Mycobacterium tuberculosis,* all contain homologues to RNase E. However, the similarities were limited to the amino-terminal 500 amino acids, i.e., that portion of the protein that contains the catalytic region (Fig. 1). In fact, the proteins from *Synechocystis* and *P. purpurea* are both less than 700 amino acids in length.

These conclusions were based on a limited sampling of bacterial genomes. With now over 160 bacterial genomes fully annotated, a slightly different story has emerged. The original observation that the most important sequence homology exists in the amino-terminal portion of the various RNase E homologues is still correct. In fact, proteins with greater than 50% identity over the first 500 amino acids vary in total length from 1,221 (*Yersinia pestis*) to 642 (*Burkholderia fungorum*) amino acids. BLAST searches using amino acids 734 to 1061 (degradosome scaffold region) of the *E. coli* RNase E protein (Fig. 1) yield relatively few homologues. Beyond certain species of *Shigella* and *Salmonella,* which show extensive homology, the next best alignments (*Vibrio, Pasteurella,* and *Haemophilus*) show sequence identities of under 30%. Thus the degradosome scaffolding region, based on the amino acid sequence found in *E. coli,* is not highly conserved. It is clearly missing in many organisms and where possibly present in others must utilize significantly different primary amino acid sequences.

Another complicating feature associated with genomic analysis is that many organisms have two proteins that show considerable homology to RNase E and/or RNase G. Since it is not clear at this time what features of the protein distinguish RNase E from RNase G, it is not possible to accurately assess how many of the putative RNase E homologues actually possess RNase E catalytic activity.

Despite this problem, it should be noted that all of the organisms that contain RNase E/G homologues also have RhlB and PNPase homologues. Thus it may be that in most prokaryotes a multiprotein complex, as defined by the *E. coli* degradosome, does not exist. It does not rule out, however, that mRNA decay still proceeds by a mechanism that involves the simultaneous utilization of RNase E, PNPase, and RhlB. It would also suggest that the proteins, if they do interact, might only do so after they are bound to an mRNA substrate. Furthermore, it indicates that the scaffolding region of *E. coli* RNase E protein may provide some functional advantage to the bacterium that is unrelated to mRNA decay.

Other Multiprotein Complexes

It was noted above that with the exception of RNase E, RNase G, RNase III, and possibly yet to be identified endonucleases, all the other RNases in *E. coli* initiate degradation of mRNAs at the 3' terminus. In eukaryotic organisms, a number of laboratories have demonstrated the existence of multiprotein complexes called exosomes (4, 99). Within these multiprotein complexes are a variety of 3' → 5' exonucleases that are homologous to *E. coli* RNase PH, RNase R, and RNase D. While these enzymes in *E. coli* seem to be exclusively involved in the processing of tRNAs (39), it would not be unreasonable to think that some type of bacterial exosome might exist to promote 3' → 5' mRNA decay.

Besides containing exonucleases, what other components might one expect to find in such a complex? One candidate that immediately comes to mind is PAP I. Coburn and Mackie (30) have already demonstrated that continuous cycles of polyadenylation promote the degradation of a 147-nt degradation fragment of the *rpsT* transcript by PNPase. In addition, Raynal and Carpousis (131) have argued, based on far-Western analysis, that PAP I interacts with RNase E in the same region of the scaffold where RhlB binds. Additional experiments have also indicated that PAP I, PNPase, and RNase II compete for 3' ends and that both PNPase and RNase II directly control the extent of polyadenylation by degrading newly synthesized tails (101, 103).

Thus it would not be surprising to find a multiprotein complex that contains at a minimum PNPase and PAP I. In fact, immunoprecipitation analysis has now shown that such a complex exists in wild-type *E. coli* (Mohanty and Kushner, unpublished). What would the expected function of such a multiprotein complex be? In the case of the 147-nt degradation product of the *rpsT* transcript (88), the binding of such a complex would lead to rapid exonucleolytic degradation through multiple polyadenylation/degradation steps. In fact, such a complex could deal with any mRNA decay product that contained secondary structures that were sufficiently strong to inhibit both RNase II and PNPase.

However, a complication with this idea is the potential for PNPase to act biosynthetically (104). Although Mohanty and Kushner (104) suggested that there could be transient changes in the intracellular phosphate concentration that would permit the enzyme to function biosynthetically, the fact that they isolated PNPase-synthesized tails over 125 nt in length indicates that biosynthesis can occur for extended periods of time. It is possible that some type of reversible modification might convert PNPase into a form that is no longer inhibited by P_i. Thus a PAP I/PNPase complex might function in either a biosynthetic or degradative capacity. Because of the large amount of PNPase in the cell, there would be sufficient enzyme present to form both degradosomes as well as PNPase/PAP I complexes.

Finally, it should be noted that some years ago, evidence was presented that suggested the existence of a multiprotein complex that contained RNase E, RNase III, and RNase P (61, 97). Although there is no indication that either RNase III or RNase P is associated with the degradosome (98, 128), since this purported complex was associated with the inner membrane, it might have been missed in using the standard RNase E purification procedure. The recent observations that RNase P can cleave certain polycistronic mRNAs (76) and that the degradosome is located close to the inner membrane (81) suggest that further study in this area is warranted.

THE ROLE OF POLYADENYLATION IN mRNA DECAY

Polyadenylation has been shown to be involved in the degradation of both mRNAs and untranslated RNAs (29, 54, 117, 162, 163). Some of these results led to the proposal that polyadenylation is a targeting mechanism for the initiation of mRNA decay (69, 117). However, the work of Coburn and Mackie (30) has shown that for the *rpsT* transcript, initial degradation is carried out by RNase E and is independent of polyadenylation. They argued that polyadenylation was important, but only for the degradation of a 3' 147-nt fragment that arose from RNase E cleavage (30). If all mRNAs behaved like *rpsT*, then clearly polyadenylation is not necessary to trigger initial degradation but rather to help PNPase and possibly other RNases, such as RNase R and RNase II, exonucleolytically degrade RNase E breakdown products that contain secondary structures.

However, this conclusion seems premature. In the first place, there are several mRNAs whose half-lives are definitely affected by deletion of the structural gene for PAP I. These include *lpp* (101, 117), *ompA* (101, 117), and *rpsO* (101). In addition, not all degraded *E. coli* transcripts appear to involve RNase E. For example, while *rpsT* and *rpsO* are clearly dependent of RNase E cleavage (29, 121), *trxA* and *lpp* are not (121). Thus it may be that if an mRNA is degraded primarily by an exonucleolytic mechanism and contains a Rho-independent transcription terminator, then polyadenylation is absolutely critical for initiation of decay. The *lpp* transcript would be a good example of this since it has a very strong Rho-independent stem-loop at its immediate 3′ terminus (Fig. 2A). Since a significant number of *E. coli* transcripts are terminated in a Rho-independent fashion (75), it would be expected that the decay of such full-length transcripts would be poly(A) dependent if they lacked endonucleolytic cleavage sites. Polyadenylation would be essential for the binding of any of the 3′ → 5′ exonucleases.

REGULATION OF mRNA DECAY AND PROCESSING

Over the past 15 years considerable evidence has been obtained to demonstrate that mRNA decay is not simply a constitutively expressed salvage pathway, but rather is highly regulated at several levels. In the first place, many of the genes involved in mRNA decay, including RNase III, PNPase, and RNase E, are controlled in part by autoregulation. For example, in the case of RNase III, the enzyme cleaves a stem-loop structure in the 5′ UTR of the transcript that is upstream of the ribosome binding site, leading to the rapid degradation of the transcript (14).

With PNPase, autoregulation of the PNPase transcript involves the association of PNPase with the 5′ end, in this case reducing the translation efficiency of the transcript (136). It has now been shown that PNPase is involved in the processing of the 5′ end of its own transcript (62). In addition, PNPase is part of a polycistronic mRNA that also includes the ribosomal protein S15 (*rpsO*). Cleavage of a stem-loop structure in the untranslated region between *rpsO* and *pnp* by RNase III leads to a five- to sevenfold reduction in the level of PNPase in the cell (132, 135).

In the case of RNase E, protein production is controlled in part by the interaction of the protein with the long 361-nt 5′ UTR (60). In fact, the half-life of the *rne* transcript increases significantly in an *rne-1* mutant at the nonpermissive temperature (56, 60). Stabilization of the transcript leads to a concomitant increase in the amount of RNase E protein

that is synthesized (112, 121). Although inactivation of an RNase E cleavage site within the 5′ UTR does not alter autoregulation (60), phylogenetic analysis of RNase E homologues in other bacteria has demonstrated the conservation of structural elements within the 5′ UTR that are important for autoregulation (40). Not only is the 5′ UTR required for autoregulation, but some portion of the 3′ coding region of the RNase E protein is also involved because carboxy-terminal deletions of the protein also lose their ability to autoregulate their own synthesis (40, 121).

The regulation of RNase E is even more complicated, as delineated by new studies of the gene's transcription. Specifically, it has now been demonstrated that the *rne* gene is transcribed from three independent promoters (123). Each promoter contributes significantly to the total amount of RNase E protein that is synthesized. In addition, the authors were able to show a direct correlation between in vivo RNase E levels and the half-lives of the *rpsT* and *rpsO* transcripts (123). Of more interest was the fact that deletion of one or more of the three promoters either reduced or abolished autoregulation (123).

Further evidence of the regulation of mRNA decay comes from experiments showing that in PNPase mutants, RNase II levels increase (170). Conversely, PNPase levels were also shown to increase in the absence of RNase II (170). In addition, it has recently been demonstrated that there is a protein in *E. coli*, called Gmr, that downregulates the synthesis of RNase II by a factor of 3 (20).

Besides the regulation associated with the various RNases, poly(A) polymerase I levels are also regulated. In the first place, they are kept relatively low by a combination of a weak promoter, poor translation initiation, and a UUG translation start codon (101). This downregulation of PAP I expression seems to be related to the fact that increased levels of the protein are toxic to the cell (101).

In addition, the levels of RNase E and PNPase appear to respond to the poly(A) content of the cell since increased levels of polyadenylation lead to the stabilization of both the *pnp* and *rne* mRNAs (101), leading to higher levels of each protein (103). Thus even though polyadenylation appears to lead to the destabilization of a large number of transcripts (54, 117, 162), in the case of the *rne* and *pnp* transcripts the effect is just the opposite. It has now been shown that in the absence of polyadenylation, the *pnp* and *rne* transcripts turn over more rapidly than under wild-type conditions (105). Thus the levels of two of the enzymes intimately involved in mRNA decay respond directly to the poly(A) content of the cell (105).

Recently, two other types of potential regulation have been reported. In the first case, it has been shown that the RNA binding protein Hfq, in binding to A/U in RNA molecules, can inhibit the ability of RNase E to cleave both mRNAs and tRNA precursors (107, 167). Thus in the case of mRNA decay, for some mRNAs RNase E cleavage will be blocked under conditions where there is sufficient Hfq to compete with the protein for AU-rich regions. Since there is estimated to be about 60,000 Hfq molecules/cell, it is reasonable to think that the presence of Hfq leads to at least some stabilization of some subset of *E. coli* transcripts.

The other new feature of regulation of mRNA decay relates to the identification of a new protein, called RraA, which appears to bind to RNase E and inhibit its endonucleolytic activity (74). Using microarray analysis, Lee et al. (74) demonstrated that the absence of RraA led to increased steady-state levels of almost 80 mRNAs. Taken together, it appears that the mRNA decay and processing capacity of the cells is tightly controlled by a variety of overlapping mechanisms.

LOCALIZATION OF mRNA DECAY WITHIN THE CELL

Results from several laboratories have provided some evidence that mRNA decay may actually take place at or near the inner membrane of the cell. The subcellular localization experiments of Liou et al. (81) demonstrated that over 91% of the RNase E protein was observed either on or within 100 nm of the cytoplasmic membrane. Since RNase E is probably involved with the decay of many *E. coli* transcripts, this would place the initial steps of degradation at or near the cytoplasmic membrane.

Further circumstantial evidence to support this hypothesis comes from the analysis of the MrsC/HflB protein. Originally identified by its ability to improve the lysogenic frequency of bacteriophage λ (13, 57), the gene was independently characterized as MrsC because of its effect on mRNA stability (53). Specifically, in the absence of the MrsC protein, the half-life of total pulse-labeled *E. coli* RNA increased from 2.9 to 5.9 min (53). In addition, the half-lives of specific transcripts also increased. In multiple mutants that carried both the *rne-1* and *mrsC505* alleles, mRNA half-lives increased even more (53). Subsequent experiments demonstrated that *mrsC505* is an allele of the *hflB* gene (160).

It has now been shown that the HflB protein is a membrane-bound ATP-dependent protease (1, 2, 68). The protein is inserted into the cytoplasmic membrane through its amino terminus while the protease domain extends into the cytoplasm. Since the HflB/MrsC protein is a protease, it is possible that there is another component of the mRNA decay pathway, presumably an RNase, that requires proteolytic processing to activate it. Because of the physical location of HflB, this activation would take place at the cytoplasmic membrane.

Alternatively, the effect of HflB/MrsC on mRNA decay could be through the inactivation of a protein that normally protects mRNAs from degradation. This is a highly speculative idea because at the present time there are only two known RNA binding proteins that contribute to the stability of specific mRNAs. The first is CsrA, first identified as a protein that controls the expression of genes involved in glycogen degradation and biosynthesis (85, 138). However, in this particular case, the CsrA protein binds to the mRNAs and stimulates their decay. Recently, it has been shown, however, that it can also stabilize the *flhDC* transcript by binding to its 5′ end (161).

A second RNA binding protein, Hfq, has also been shown to affect the stability of specific transcripts. In the case of the *rpoS* transcript, Hfq increases the stability of the transcript (113). In contrast, with the *ompA* transcript, Hfq binding inhibits translation initiation, resulting in a decreased half-life for the transcript (157). In fact, employing two-dimensional polyacrylamide gels, Muffler et al. (114) estimated that the levels of approximately 30 proteins were affected in Hfq-deficient strains. Clearly this would not be sufficient to cause the effects seen in the *mrsC/hflB* mutants.

Recent experiments have suggested that the Hfq protein increases poly(A) tail length by either improving the processivity of poly(A) polymerase I (55) or protecting the tail from exonucleolytic degradation (48). If in fact Hfq functions as a poly(A) binding protein, it could inhibit the degradation of the poly(A) tails by either RNase II or PNPase. For example, since degradation of the 174-nt *rpsT* fragment has been shown to be dependent on both poly(A) polymerase and PNPase (30), failure to turn over the Hfq protein could potentially lead to the stabilization of this fragment. Similarly, other transcripts that require either RNase II or PNPase for their decay would also be stabilized if the turnover of the Hfq protein were decreased. Thus reduced degradation of a general RNA binding protein could account for the significant stabilization of mRNA transcripts that was observed in the *mrsC/hflB* mutants (53, 160). However, *hfq* mutants do not show the large changes in mRNA stability (Mohanty and Kushner, unpublished) that are observed in *mrsC* strains (53).

IS *E. COLI* A GOOD PARADIGM FOR mRNA DECAY IN PROKARYOTES?

The bulk of the above discussion has dealt with our current knowledge of mRNA decay and processing in *E. coli*. In this organism it has been shown both genetically and biochemically that RNase E, RNase G, RNase III, RNase P, PNPase, RNase II, oligoribonuclease, RhlB, and PAP I are clearly involved in mRNA decay and/or processing. With the rapid proliferation of new genome sequencing data, what parallels can be drawn with other bacteria? In the case of PNPase, the level of sequence conservation among both gram-negative and gram-positive bacteria is quite remarkable. The *E. coli* protein is 734 amino acids in length. The homologues in other organisms only vary in length from 698 to 745 amino acids. Even in gram-positive bacteria, there is close to 50% sequence identity to *E. coli* PNPase. The enzyme has been found in such diverse organisms as *Yersinia enterocolitica, Pasteurella multocida, Xylella fastidiosa, Neisseria meningitidis, B. subtilis, S. aureus,* and *Rickettsia prowazekii.*

The RhlB RNA helicase also appears to be a highly conserved protein. Homologues with a high degree of sequence identity have been found in both gram-negative organisms, such as *Pseudomonas aeruginosa* and *Vibrio cholerae,* and gram-positive bacteria, such as *Streptococcus pyogenes* and *Lactococcus lactis.*

Although not as highly conserved at the amino acid level, RNase III homologues have now been identified in at least 30 different bacterial species, both gram negative and gram positive. In addition, these proteins also do not vary much in size, ranging from 212 to 239 amino acids in length.

With RNase II the situation is complicated by the fact that this protein is also homologous to RNase R. In particular, there are several domains within these two proteins that are nearly identical. This makes the determination of whether an organism has an RNase II or RNase R protein more difficult, particularly in light of the biochemical data that *B. subtilis* does not have RNase II activity (36) but, based on sequence homology, does contain an RNase R homologue. Since it is not clear how significant a role RNase R plays in *E. coli* mRNA decay (J. Gunnells-Ledford and S. R. Kushner, unpublished results), it is not certain what to make of the fact that most of the bacteria that have been sequenced have at least one member of this family.

Genomic analysis of RNase E homologues is particularly interesting. As noted above, the catalytic activity of the protein resides in the first 498 amino acids (95) (Fig. 2). If one carries out a search with just the first 498 amino acids of the *E. coli* protein, one finds a significantly large number of homologues, but they are almost exclusively found in gram-negative organisms. There are only a few gram-positive organisms, such as several species of *Streptomyces* and *Bacillus halodurans,* that appear to contain RNase E/G-type proteins. No homologues are present in *B. subtilis, S. aureus,* or *Streptococcus pneumoniae.* As noted earlier, however, it is not possible at this time to distinguish RNase E-type and RNase G-type proteins simply on the basis of sequence homology.

Thus even though an RNase E-based degradosome may not be present in certain species of bacteria, it seems clear that the features of mRNA decay, as currently understood in *E. coli*, i.e., endonucleolytic initiation of decay for most mRNAs followed by exonucleolytic degradation of the breakdown products, will probably be observed in most other bacteria. For example, in *B. subtilis*, which lacks RNase E, a homologue of the *Arabidopsis* tRNA processing enzyme RNase Z (141) has recently been identified (125). It has been shown that this enzyme cleaves polycistronic tRNAs (125) in a reaction similar to that seen for RNase E in *E. coli*. This protein appears to be essential for cell viability and may also be involved in mRNA decay.

RELATIONSHIP BETWEEN mRNA DECAY AND PROCESSING

It has already been noted that bacteria are unique in that many genes are transcribed as part of polycistronic operons. Processing of many of these large transcripts into smaller discrete units may actually occur before transcription has been completed. Some of the processing that takes place is probably the result of RNase III cleavages at stem-loop structures that occur in either the 5′ UTR of particular transcripts (the *rne era recO* operon) or within intercistronic regions (the *rpsO pnp* operon). In many cases, such processing decreases the stability of at least one of the reaction products (*rnc* and *pnp*). In fact, Gitelman and Apirion (51) suggested that RNase III processing could either increase or decrease the stability of a small percentage of transcripts based on two-dimensional polyacrylamide protein gels. In addition, other operons are processed by the action of RNase P (76). Thus processing of mRNAs can play a direct role in their stability.

It should also be noted that, while RNase E has been viewed primarily as an mRNA-degrading enzyme, it is involved in a number of rRNA processing reactions. For example, it has been shown to be required for both the processing of 9S rRNA into a p5s form (6) and the maturation of the 5′ end of the 16S rRNA (77, 154, 159). In addition, RNase E is also involved in the processing of the M1 RNA subunit of RNase P (87, 122) and tmRNA (79). More important, RNase E appears to initiate the maturation of tRNAs (78, 122). Thus it is not un-realistic to think that for certain mRNAs, RNase E cleavage may initially act as a processing reaction. As the action of this enzyme on more mRNAs is studied,

it is quite likely that this will in fact be found to be the case.

ADDITIONAL PROTEINS INVOLVED IN mRNA DECAY AND PROCESSING

Although the primary features of mRNA decay and processing in *E. coli* have probably already been

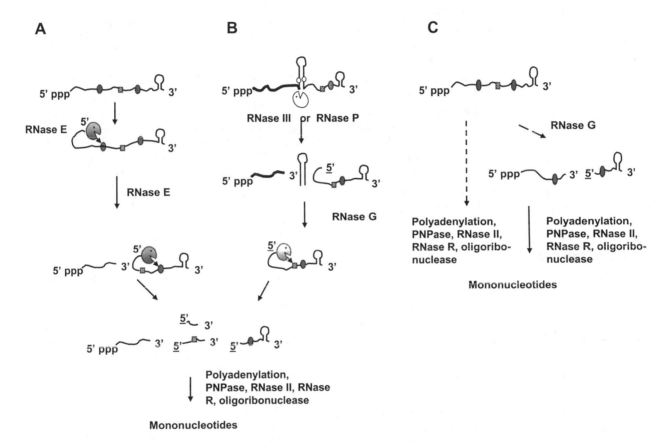

Figure 4. Current working model for mRNA decay in *E. coli*. (A) Initiation of mRNA decay by RNase E. Based on its catalytic properties (89, 90), RNase E, as part of the degradosome or independently, would first bind to an accessible 5′ end. The preference of RNase E for substrates that are 5′ monophosphorylated over those that contain 5′ triphosphates (89) suggests that this step will be the rate-limiting reaction in mRNA decay. Intermediates generated by the initial RNase E cleavage reaction can be further degraded endonucleolytically by either RNase E or RNase G or exonucleolytically by a combination of RNases. PNPase is probably the primary 3′ → 5′ exonuclease (106). Polyadenylation will be involved if an RNA fragment contains a stable stem-loop structure. Terminal degradation products (short oligonucleotides) will be degraded by oligoribonuclease (50). (B) Initiation of mRNA decay by either RNase III or RNase P. In a limited number of circumstances, such as observed with the *eno* or *his* mRNAs, either RNase III or RNase P, respectively, cleave within intercistronic regions of polycistronic mRNAs to generate a downstream fragment that would have a 5′ phosphomonoester, making it a better substrate for RNase G or RNase E. Unlike RNase E, which can cleave RNAs at internal sites without binding to a 5′ terminus (12), RNase G apparently cannot do this very efficiently. Once either RNase G or RNase E cleavages occur, the breakdown products would be susceptible to exonucleolytic degradation as in panel A. Some mRNAs may contain either potential RNase E or RNase G cleavage sites that are bypassed as shown in panels A and B, respectively. (C) mRNA decay in the absence of RNase E or for mRNAs that do not contain RNase E cleavage sites. In the absence of RNase E, decay of mRNAs dependent on this enzyme will proceed more slowly, either through RNase G cleavage or exonucleolytic degradation by PNPase and/or other exonucleases. For those mRNAs that do not contain any endonucleolytic cleavage sites, decay is probably initiated by polyadenylation of the 3′ terminus. Subsequently, the polyadenylated mRNA is degraded exonucleolytically as described for panel A. Ellipses, RNase E cleavage sites; squares, RNase G cleavage sites; circles, RNase III or RNase P cleavage sites. Heavy lines indicate gene 1 in a polycistronic mRNA. 5′ phosphomonoester termini are underlined; 5′ termini containing a triphosphate are not.

identified, it is clear that there must be additional gene products involved. There are several lines of evidence to support this conclusion. In the first place, even though mRNA decay is slowed by the inactivation of RNase E, PNPase, and RNase II (7), construction of a quadruple mutant deficient in RNase III, RNase E, PNPase, and RNase II actually led to an increase in mRNA decay rates (10). In addition, even in strains carrying multiple combinations of the genes described above (53, 117), degradation of many mRNAs is only moderately slowed.

Are there any candidates for other proteins that might participate in mRNA decay and processing? One possible enzyme is RNase I, a broad-specificity endoribonuclease that is predominantly found in the periplasm (21). However, a small amount may exist in the cytoplasm (21). Since analysis of mRNA decay in RNase I mutants has not been extensively studied, it is possible that it could serve some secondary role. In addition, new mutants have recently been identified that affect the decay of bacteriophage T4 mRNAs in *E. coli*. One called *std-2* significantly stabilizes the *soc* mRNA and was shown not to be an allele of any of the RNases discussed in this chapter (120). *E. coli* also has a homologue of the *B. subtilis* RNase Z (33) encoded by the *elaC* gene. Initial genomic analysis suggests that this enzyme may play a role in the decay of certain *E. coli* mRNAs (T. Perwez and S. R. Kushner, unpublished results). Only when all the participants in mRNA decay and processing are delineated will it be possible to fully understand this complex system.

FUTURE DIRECTIONS

The last 15 years have been extremely productive in the development of new insights into the mechanisms of mRNA decay in *E. coli*. Based on all of these studies, a current working model is described in Fig. 4. It is now clear that mRNA decay and processing play integral roles in the regulation of bacterial gene expression. In addition, the proliferation of new genomic sequencing data has shown that most of the enzymes found in *E. coli* have homologues in a wide variety of prokaryotes. However, there clearly is much more to learn about the set of overlapping pathways that control the processing and degradation of mRNAs in bacteria.

Acknowledgment. This work was supported in part by NIHGMS grant GM57220.

REFERENCES

1. Akiyama, Y., T. Yoshihisa, and K. Ito. 1995. FtsH, a membrane-bound ATPase, forms a complex in the cytoplasmic membrane of *Escherichia coli*. *J. Biol. Chem.* 270:23485–23490.

2. Akiyama, Y., A. Kihara, and K. Ito. 1996. Subunit a of proton ATPase F0 sector is a substrate of the FtsH protease in *Escherichia coli*. *FEBS Lett.* 399:26–28.

3. Alifano, P., F. Rivellini, C. Piscitelli, C. M. Arraiano, C. B. Bruni, and M. S. Carlomagno. 1994. Ribonuclease E provides substrates for ribonuclease P-dependent processing of a polycistronic mRNA. *Genes Dev.* 8:3021–3031.

4. Allmang, C., E. Petfalski, A. Podtelejnikov, M. Mann, D. Tollervey, and P. Mitchell. 1999. The yeast exosome and human PM-Scl are related complexes of 3′-5′ exonucleases. *Genes Dev.* 13:2148–2158.

5. Altman, S., L. Kirsebom, and S. Talbot. 1995. Recent studies of RNase P, p. 67–78. *In* D. Soll and U. L. RajBhandary (ed.), *tRNA: Structure, Biosynthesis, and Function.* ASM Press, Washington, D.C.

6. Apirion, D., and A. B. Lassar. 1978. A conditional lethal mutant of *Escherichia coli* which affects the processing of ribosomal RNA. *J. Biol. Chem.* 253:1738–1742.

7. Arraiano, C. M., S. D. Yancey, and S. R. Kushner. 1988. Stabilization of discrete mRNA breakdown products in *ams pnp rnb* multiple mutants of *Escherichia coli* K-12. *J. Bacteriol.* 170:4625–4633.

8. August, J., P. J. Ortiz, and J. Hurwitz. 1962. Ribonucleic acid-dependent ribonucleotide incorporation. I. Purification and properties of the enzyme. *J. Biol. Chem.* 237:3786–3793.

9. Babitzke, P., and S. R. Kushner. 1991. The Ams (altered mRNA stability) protein and ribonuclease E are encoded by the same structural gene of *Escherichia coli*. *Proc. Natl. Acad. Sci. USA* 88:1–5.

10. Babitzke, P., L. Granger, and S. R. Kushner. 1993. Analysis of mRNA decay and rRNA processing in *Escherichia coli* multiple mutants carrying a deletion in RNase III. *J. Bacteriol.* 175:229–239.

11. Bachellier, S., E. Gilson, M. Hofnung, and C. W. Hill. 1996. Repeated sequences, p. 2012–2040. *In* F. C. Neidhardt, R. Curtiss III, J. L. Ingraham, E. C. C. Lin, K. B. Low, B. Magasanik, W. S. Reznikoff, M. Riley, M. Schaechter, and H. E. Umbarger (ed.), *Escherichia coli and Salmonella: Cellular and Molecular Biology*, 2nd ed., vol. 2. ASM Press, Washington, D.C.

12. Baker, K. E., and G. A. Mackie. 2003. Ectopic RNase E sites promote bypass of 5′-end-dependent mRNA decay in *Escherichia coli*. *Mol. Microbiol.* 47:75–88.

13. Banuett, F., M. A. Hoyt, L. McFarlane, H. Echols, and I. Herskowitz. 1986. *hflB*, a new *Escherichia coli* locus regulating lysogeny and the level of bacteriophage lambda cII protein. *J. Mol. Biol.* 187:213–224.

14. Bardwell, J. C. A., P. Regnier, S.-M. Chen, Y. Nakamura, M. Grunberg-Manago, and D. L. Court. 1989. Autoregulation of RNase III operon by mRNA processing. *EMBO J.* 8:3401–3407.

15. Barlow, T., M. Berkmen, D. Georgellis, L. Bayr, S. Arvidson, and A. Von Gabain. 1998. RNase E, the major player in mRNA degradation, is down-regulated in *Escherichia coli* during a transient growth retardation (Diauxic lag). *Biol. Chem.* 379:33–38.

16. Blattner, F. R., G. Plunkett III, C. A. Bloch, N. T. Perna, V. Burland, M. Riley, J. Collado-Vides, J. D. Glasner, C. K. Rode, G. F. Mayhew, J. Gregor, N. W. Davis, H. A. Kirkpatrick, M. A. Goeden, D. J. Rose, B. Mau, and Y. Shao. 1997. The complete sequence of *Escherichia coli* K-12. *Science* 277:1453–1474.

17. Blum, E., B. Py, A. J. Carpousis, and C. F. Higgins. 1997. Polyphosphate kinase is a component of the *Escherichia coli* RNA degradosome. *Mol. Microbiol.* **26:**387–398.

18. Bralley, P., and G. H. Jones. 2001. Poly(A) polymerase activity and RNA polyadenylation in *Streptomyces coelicolor* A3. *Mol. Microbiol.* **40:**1155–1164.

19. Bycroft, M., T. J. P. Hubbard, M. Proctor, S. M. V. Freund, and A. G. Murzin. 1997. The solution structure of the S1 RNA binding domain: a member of an ancient nucleic acid-binding fold. *Cell* **88:**235–242.

20. Cairrao, F., A. Chora, R. Zilhao, A. J. Carpousis, and C. M. Arraiano. 2001. RNase II levels change according to the growth conditions: characterization of *gmr*, a new *Escherichia coli* gene involved in the modulation of RNase II. *Mol. Microbiol.* **39:**1550–1561.

21. Cannistraro, V. J., and D. Kennell. 1991. RNase I*, a form of RNase I, and mRNA degradation in *Escherichia coli*. *J. Bacteriol.* **173:**4653–4659.

22. Cao, G.-J., and N. Sarkar. 1992. Poly(A) RNA in *Escherichia coli*: nucleotide sequence at the junction of the *lpp* transcript and the polyadenylate moiety. *Proc. Natl. Acad. Sci. USA* **89:**7546–7550.

23. Cao, G.-J., and N. Sarkar. 1992. Identification of the gene for an *Escherichia coli* poly(A) polymerase. *Proc. Natl. Acad. Sci. USA* **89:**10380–10384.

24. Caponigro, G., and R. Parker. 1996. Mechanism and control of mRNA turnover in *Saccharomyces cerevisiae*. *Microbiol. Rev.* **60:**233–249.

25. Carpousis, A. J., G. Van Houwe, C. Ehretsmann, and H. M. Krisch. 1994. Copurification of *E. coli* RNAase E and PNPase: evidence for a specific association between two enzymes important in RNA processing and degradation. *Cell* **76:**889–900.

26. Cheng, Z. F., Y. Zuo, Z. Li, K. E. Rudd, and M. P. Deutscher. 1998. The *vacB* gene required for virulence in *Shigella flexneri* and *Escherichia coli* encodes the exoribonuclease RNase R. *J. Biol. Chem.* **273:**14077–14080.

27. Cheng, Z. F., and M. P. Deutscher. 2003. Quality control of ribosomal RNA mediated by polynucleotide phosphorylase and RNase R. *Proc. Natl. Acad. Sci. USA* **100:**6388–6393.

28. Coburn, G. A., and A. G. Mackie. 1996. Overexpression, purification and properties of *Escherichia coli* ribonuclease II. *J. Biol. Chem.* **271:**1048–1053.

29. Coburn, G. A., and G. A. Mackie. 1996. Differential sensitivities of portions of the mRNA for ribosomal protein S20 to 3′-exonucleases dependent on oligoadenylation and RNA secondary structure. *J. Biol. Chem.* **271:**15776–15781.

30. Coburn, G. A., and G. A. Mackie. 1998. Reconstitution of the degradation of the mRNA for ribosomal protein S20 with purified enzymes. *J. Mol. Biol.* **279:**1061–1074.

31. Coburn, G. A., and G. A. Mackie. 1999. Degradation of mRNA in *Escherichia coli*: an old problem with some new twists. *Prog. Nucleic Acid Res.* **62:**55–108.

32. Coburn, G. A., X. Miao, D. J. Briant, and G. A. Mackie. 1999. Reconstitution of a minimal RNA degradosome demonstrates functional coordination between a 3′ exonuclease and a DEAD-box RNA helicase. *Genes Dev.* **13:**2594–2603.

33. Condon, C. 2003. RNA processing and degradation in *Bacillus subtilis*. *Microbiol. Mol. Biol. Rev.* **67:**157–174.

34. Court, D. 1993. RNA processing and degradation by RNase III, p. 71–117. *In* J. Belasco and G. Brawerman (ed.), *Control of Messenger RNA Stability*. Academic Press, Inc., New York, N.Y.

35. Deana, A., and J. G. Belasco. 2004. The function of RNase G in *Escherichia coli* is constrained by its amino and carboxyl termini. *Mol. Microbiol.* **51:**1205–1217.

36. Deutscher, M. P., and N. B. Reuven. 1991. Enzymatic basis for hydrolytic versus phosphorolytic mRNA degradation in *Escherichia coli* and *Bacillus subtilis*. *Proc. Natl. Acad. Sci. USA* **88:**3277–3280.

37. Deutscher, M. P. 1993. Ribonuclease multiplicity, diversity and complexity. *J. Biol. Chem.* **268:**13011–13014.

38. Deutscher, M. P. 1993. Promiscuous exoribonucleases of *Escherichia coli*. *J. Bacteriol.* **175:**4577–4583.

39. Deutscher, M. P., and Z. Li. 2000. Exoribonucleases and their multiple roles in RNA metabolism. *Prog. Nucleic Acids Res.* **66:**67–105.

40. Diwa, A., A. L. Bricker, C. Jain, and J. G. Belasco. 2000. An evolutionarily conserved RNA stem-loop functions as a sensor that directs feedback regulation of RNase E gene expression. *Genes Dev.* **14:**1249–1260.

41. Donovan, W. P., and S. R. Kushner. 1986. Polynucleotide phosphorylase and ribonuclease II are required for cell viability and mRNA turnover in *Escherichia coli* K-12. *Proc. Natl. Acad. Sci. USA* **83:**120–124.

42. Dunn, J. J., and F. W. Studier. 1973. T7 early RNAs and *Escherichia coli* ribosomal RNAs are cut from large precursor RNAs in vivo by ribonuclease III. *Proc. Natl. Acad. Sci. USA* **70:**3296–3300.

43. Dunn, J. J., and F. W. Studier. 1983. Complete nucleotide sequence of bacteriophage T7 DNA and the locations of T7 genetic elements. *J. Mol. Biol.* **166:**477–535.

44. Ehretsmann, C. P., A. J. Carpousis, and H. M. Krisch. 1992. Specificity of *Escherichia coli* endoribonuclease RNase E: in vivo and in vitro analysis of mutants in a bacteriophage T4 mRNA processing site. *Genes Dev.* **6:**149–159.

45. Emory, S. A., and J. G. Belasco. 1990. The *ompA* 5′ untranslated RNA segment functions in *Escherichia coli* as a growth-rate-regulated mRNA stabilizer whose activity is unrelated to translational efficiency. *J. Bacteriol.* **172:**4472–4481.

46. Emory, S. A., P. Bouvet, and J. G. Belasco. 1992. A 5′-terminal stem-loop structure can stabilize mRNA in *Escherichia coli*. *Genes Dev.* **6:**135–148.

47. Feng, Y., H. Huang, J. Kiao, and S. N. Cohen. 2001. *Escherichia coli* poly(A) binding proteins that interact with components of degradosomes or impede RNA decay mediated by polynucleotide phosphorylase and RNase E. *J. Biol. Chem.* **276:**31651–31656.

48. Folichon, M., V. Arluison, O. Pellegrini, E. Huntzinger, P. Regnier, and E. Hajnsdorf. 2003. The poly(A) binding protein Hfq protects RNA from RNase E and exoribonucleolytic degradation. *Nucleic Acids Res.* **31:**7302–7310.

49. Franze de Fernandez, M. T., L. Eoyang, and T. L. August. 1968. Factor fraction required for the synthesis of bacteriophage Qbeta-RNA. *Nature* **219:**588–590.

50. Ghosh, S., and M. P. Deutscher. 1999. Oligoribonuclease is an essential component of the mRNA decay pathway. *Proc. Natl. Acad. Sci. USA* **96:**4372–4377.

51. Gitelman, D. R., and D. Apirion. 1980. The synthesis of some proteins is affected in RNA processing mutants of *Escherichia coli*. *Biochem. Biophys. Res. Commun.* **96:**1063–1070.

52. Godefroy-Colburn, T., and M. Grunberg-Manago. 1972. Polynucleotide phosphorylase, p. 533–574. *In* P. D. Boyer (ed.), *The Enzymes*, vol. 7. Academic Press, Inc., New York, N.Y.

53. Granger, L. L., E. B. O'Hara, R.-F. Wang, F. V. Meffen, K. Armstrong, S. D. Yancey, P. Babitzke, and S. R. Kushner. 1998. The *E. coli mrsC* gene is required for cell growth and mRNA decay. *J. Bacteriol.* **180:**1920–1928.

54. Hajnsdorf, E., F. Braun, J. Haugel-Nielsen, and P. Régnier. 1995. Polyadenylylation destabilizes the *rpsO* mRNA of *Escherichia coli*. *Proc. Natl. Acad. Sci. USA* **92:**3973–3977.

55. Hajnsdorf, E., and P. Régnier. 2000. Host factor Hfq of *Escherichia coli* stimulates elongation of poly(A) tails by poly(A) polymerase I. *Proc. Natl. Acad. Sci. USA* 97:1501–1505.

56. Henry, M., S. D. Yancey, and S. R. Kushner. 1992. The role of the heat-shock response in the stability of mRNA in *Escherichia coli* K-12. *J. Bacteriol.* 174:743–748.

57. Herman, C., T. Ogura, T. Tomoyasu, S. Hiraga, Y. Akiyama, K. Ito, R. Thomas, R. D'Ari, and P. Bouloc. 1993. Cell growth and λ phage development controlled by the same essential *Escherichia coli* gene, *ftsH/hflB*. *Proc. Natl. Acad. Sci. USA* 90:10861–10865.

58. Herskovitz, M. A., and D. H. Bechhofer. 2000. Endoribonuclease III is essential in *Bacillus subtilis*. *Mol. Microbiol.* 38:1027–1033.

59. Higgins, C. F., G. F.-L. Ames, W. M. Barnes, J. M. Clement, and M. Hofnung. 1982. A novel intercistronic regulatory element of prokaryotic operons. *Nature* 298:760–762.

60. Jain, C., and J. G. Belasco. 1995. RNase E autoregulates its synthesis by controlling the degradation rate of its own mRNA in *Escherichia coli*: unusual sensitivity of the *rne* transcript to RNase E activity. *Genes Dev.* 9:84–96.

61. Jain, S. K., B. Pragai, and D. Apirion. 1982. A possible complex containing RNA processing enzymes. *Biochem. Biophys. Res. Commun.* 106:768–778.

62. Jarrige, A.-C., N. Mathy, and C. Portier. 2001. PNPase autocontrols its expression by degrading a double-stranded structure in the *pnp* mRNA leader. *EMBO J.* 20:6845–6855.

63. Kaberdin, V. R., A. Miczak, J. S. Jakobsen, S. Lin-Chao, K. J. McDowall, and A. von Gabain. 1998. The endoribonucleolytic N-terminal half of *Escherichia coli* RNase E is evolutionarily conserved in *Synechocystis* sp. and other bacteria but not the C-terminal half, which is sufficient for degradosome assembly. *Proc. Natl. Acad. Sci. USA* 95:11637–11642.

64. Kaberdin, V. R. 2003. Probing the substrate specificity of *Escherichia coli* RNase E using a novel oligonucleotide-based assay. *Nucleic Acids Res.* 31:4710–4716.

65. Kaga, N., G. Umitsuki, K. Nagai, and M. Wachi. 2002. RNase G-dependent degradation of the *eno* mRNA encoding a glycolysis enzyme enolase in *Escherichia coli*. *Biosci. Biotechnol. Biochem.* 66:2216–2220.

66. Kalman, M., H. Murphy, and M. Cashel. 1991. *rhlB*, a new *Escherichia coli* K-12 gene with an RNA helicase-like protein sequence motif, one of at least five such possible genes in a prokaryote. *Nat. New Biol.* 3:886–895.

67. Kelly, K. O., N. B. Reuven, Z. Li, and M. P. Deutscher. 1992. RNase PH is essential for tRNA processing and viability in RNase-deficient *Escherichia coli* cells. *J. Biol. Chem.* 267:16015–16018.

68. Kihara, A., Y. Akiyama, and K. Ito. 1995. FtsH is required for proteolytic elimination of uncomplexed forms of SecY, an essential protein translocase subunit. *Proc. Natl. Acad. Sci. USA* 92:4532–4536.

69. Kushner, S. R. 1996. mRNA decay, p. 849–860. *In* F. C. Neidhardt, R. Curtiss III, J. L. Ingraham, E. C. C. Lin, K. B. Low, B. Magasanik, W. S. Reznikoff, M. Riley, M. Schaechter, and H. E. Umbarger (ed.), Escherichia coli *and* Salmonella: *Cellular and Molecular Biology*, 2nd ed., vol. 1. ASM Press, Washington, D.C.

70. Kushner, S. R. 2002. mRNA decay in *Escherichia coli* comes of age. *J. Bacteriol.* 184:4658–4665.

71. Larimer, F. W., C. L. Hsu, M. K. Maupin, and A. Stevens. 1992. Characterization of the XRN1 gene encoding a $5' \rightarrow 3'$ exoribonuclease: sequence data and analysis of disparate protein and mRNA levels of gene-disrupted yeast cells. *Gene* 120:51–57.

72. Le Derout, J., M. Folichon, F. Briani, G. Deho, P. Regnier, and E. Hajnsdorf. 2003. Hfq affects the length and the frequency of short oligo(A) tails at the $3'$ end of *Escherichia coli rpsO* mRNAs. *Nucleic Acids Res.* 31:4017–4023.

73. Lee, K., J. A. Bernstein, and S. N. Cohen. 2002. RNase G complementation of *rne* null mutation identified functional interrelationships with RNase E in *Escherichia coli*. *Mol. Microbiol.* 43:1445–1456.

74. Lee, K., X. Zhan, J. Gao, Y. Feng, R. Meganathan, S. N. Cohen, and G. Georgiou. 2003. RraA: a protein inhibitor of RNase E activity that globally modulates RNA abundance in *E. coli*. *Cell* 114:623–634.

75. Lesnik, E. A., R. Sampath, H. B. Levene, T. J. Henderson, J. A. McNeil, and D. J. Ecker. 2001. Prediction of rho-independent transcriptional terminators in *Escherichia coli*. *Nucleic Acids Res.* 29:3583–3594.

76. Li, Y., and S. Altman. 2003. A specific endoribonuclease, RNase P, affects gene expression of polycistronic operon mRNAs. *Proc. Natl. Acad. Sci. USA* 100:13213–13218.

77. Li, Z., S. Pandit, and M. P. Deutscher. 1999. RNase G (CafA protein) and RNase E are both required for the $5'$ maturation of 16S ribosomal RNA. *EMBO J.* 18:2878–2885.

78. Li, Z., and M. P. Deutscher. 2002. RNase E plays an essential role in the maturation of *Escherichia coli* tRNA precursors. *RNA* 8:97–109.

79. Lin-Chao, S., C.-L. Wei, and Y.-T. Lin. 1999. RNase E is required for the maturation of *ssrA* and normal *ssrA* RNA peptide-tagging activity. *Proc. Natl. Acad. Sci. USA* 96:12406–12411.

80. Linder, P., P. F. Lasko, M. Ashburner, P. Leroy, P. J. Nielsen, K. Nishi, J. Schnier, and P. P. Slonimski. 1989. Birth of the D-E-A-D box. *Nature* 340:246.

81. Liou, G.-G., W.-N. Jane, S. N. Cohen, N.-S. Lin, and S. Lin-Chao. 2001. RNA degradosomes exist *in vivo* in *Escherichia coli* as multicomponent complexes associated with the cytoplasmic membrane via the N-terminal region of ribonuclease E. *Proc. Natl. Acad. Sci. USA* 98:63–68.

82. Liou, G.-G., H.-Y. Chang, C.-S. Lin, and S. Lin-Chao. 2002. DEAD box RhlB RNA helicase physically associates with exoribonuclease PNPase to degrade double-stranded RNA independent of the degradosome-assembling region of RNase E. *J. Biol. Chem.* 277:41157–41162.

83. Littauer, U. Z., and H. Soreq. 1982. Polynucleotide phosphorylase, p. 517–553. *In* P. D. Boyer (ed.), *The Enzymes*, vol. 15. Academic Press, Inc., New York, N.Y.

84. Liu, J., and J. S. Parkinson. 1989. Genetics and sequence analysis of the *pcnB* locus, an *Escherichia coli* gene involved in plasmid copy number control. *J. Bacteriol.* 171:1254–1261.

85. Liu, M. A., and T. Romeo. 1997. The global regulator CsrA of *Escherichia coli* is a specific mRNA-binding protein. *J. Bacteriol.* 179:4639–4642.

86. Lopez, P. J., I. Marchand, S. A. Joyce, and M. Dreyfus. 1999. The C-terminal half of RNase E, which organizes the *Escherichia coli* degradosome, participates in mRNA degradation but not rRNA processing in vivo. *Mol. Microbiol.* 33:188–199.

87. Lundberg, U., and S. Altman. 1995. Processing of the precursor to the catalytic RNA subunit of RNase P from *Escherichia coli*. *RNA* 1:327–334.

88. Mackie, G. A. 1991. Specific endonucleolytic cleavage of the mRNA for ribosomal protein S20 of *Escherichia coli* requires the products of the *ams* gene in vivo and in vitro. *J. Bacteriol.* 173:2488–2497.

89. Mackie, G. A. 1998. Ribonuclease E is a $5'$-end-dependent endonuclease. *Nature* 395:720–723.

90. Mackie, G. A. 2000. Stabilization of circular *rpsT* mRNA demonstrates the 5'-end dependence of RNase E action in vivo. *J. Biol. Chem.* **275**:25069–25072.

91. March, J. B., M. D. Colloms, D. Hart-Davis, I. R. Oliver, and M. Masters. 1989. Cloning and characterization of an *Escherichia coli* gene, *pcnB*, affecting plasmid copy number. *Mol. Microbiol.* **3**:903–910.

92. Masaaki, W., U. Genryou, S. Miwa, T. Ayako, and N. Kazuo. 1999. *Escherichia coli cafA* gene encodes a novel RNase, designated as RNase G, involved in processing of the 5' end of 16S rRNA. *Biochem. Biophys. Res. Commun.* **259**:483–488.

93. McDowall, K. J., S. Lin-Chao, and S. N. Cohen. 1994. A + U content rather than a particular nucleotide order determines the specificity of RNase E cleavage. *J. Biol. Chem.* **269**:10790–10796.

94. McDowall, K. J., V. R. Kaberdin, S.-W. Wu, S. N. Cohen, and S. Lin-Chao. 1995. Site-specific RNase E cleavage of oligonucleotides and inhibition by stem-loops. *Nature* **374**:287–290.

95. McDowall, K. J., and S. N. Cohen. 1996. The N-terminal domain of the *rne* gene product has RNase E activity and is non-overlapping with the arginine-rich RNA-binding motif. *J. Mol. Biol.* **255**:349–355.

96. McLaren, R. S., S. F. Newbury, G. S. C. Dance, H. Causton, and C. F. Higgins. 1991. mRNA degradation by processive 3'-5' exonucleases in vitro and the implications for prokaryotic mRNA decay in vivo. *J. Mol. Biol.* **221**:81–95.

97. Miczak, A., R. A. K. Srivastava, and D. Apirion. 1991. Location of the RNA-processing enzymes RNase III, RNase E, and RNase P. *Mol. Microbiol.* **5**:1801–1810.

98. Miczak, A., V. R. Kaberdin, C.-L. Wei, and S. Lin-Chao. 1996. Proteins associated with RNase E in a multicomponent ribonucleolytic complex. *Proc. Natl. Acad. Sci. USA* **93**:3865–3869.

99. Mitchell, P., E. Petfalski, A. Shevchenko, M. Mann, and D. Tollervey. 1996. The exosome: a conserved eukaryotic RNA processing complex containing multiple 3'-5' exoribonucleases. *Cell* **91**:57–66.

100. Modak, M. J., and P. R. Srinivasan. 1973. Purification and properties of a ribonucleic acid primer independent polyriboadenylate polymerase from *Escherichia coli. J. Biol. Chem.* **248**:6904–6910.

101. Mohanty, B. K., and S. R. Kushner. 1999. Analysis of the function of *Escherichia coli* poly(A) polymerase I in RNA metabolism. *Mol. Microbiol.* **34**:1094–1108.

102. Mohanty, B. K., and S. R. Kushner. 1999. Residual polyadenylation in poly(A) polymerase I (*pcnB*) mutants of *Escherichia coli* does not result from the activity encoded by the *f310* gene. *Mol. Microbiol.* **34**:1109–1119.

103. Mohanty, B. K., and S. R. Kushner. 2000. Polynucleotide phosphorylase, RNase II and RNase E play different roles in the in vivo modulation of polyadenylation in *Escherichia coli. Mol. Microbiol.* **36**:982–994.

104. Mohanty, B. K., and S. R. Kushner. 2000. Polynucleotide phosphorylase functions both as a 3'-5' exonuclease and a poly(A) polymerase in *Escherichia coli. Proc. Natl. Acad. Sci. USA* **97**:11966–11971.

105. Mohanty, B. K., and S. R. Kushner. 2002. Polyadenylation of *Escherichia coli* transcripts plays an integral role in regulating intracellular levels of polynucleotide phosphorylase and RNase E. *Mol. Microbiol.* **45**:1315–1324.

106. Mohanty, B. K., and S. R. Kushner. 2003. Genomic analysis in *Escherichia coli* demonstrates differential roles for polynucleotide phosphorylase and RNase II in mRNA abundance and decay. *Mol. Microbiol.* **50**:645–658.

107. Moll, I., T. Afonyuskhin, O. Vytvytska, V. R. Kaberdin, and U. Blasi. 2003. Coincident Hfq binding and RNase E cleavage sites on mRNA and small regulator RNAs. *RNA* **9**:1308–1314.

108. Moller, T., T. Franch, P. Hojrup, D. R. Keene, H. P. Bachinger, R. G. Brennan, and P. Valentin-Hansen. 2002. Hfq: a bacterial Sm-like protein that mediates RNA-RNA interaction. *Mol. Cell* **9**:23–30.

109. Moreau, P. L., F. Gerard, N. W. Lutz, and P. Cozzone. 2001. Non-growing *Escherichia coli* cells starved for glucose or phosphate use different mechanisms to survive oxidative stress. *Mol. Microbiol.* **39**:1048–1060.

110. Mott, J. E., J. L. Galloway, and T. Platt. 1985. Maturation of *Escherichia coli* tryptophan operon mRNA: evidence for 3' exonucleolytic processing after rho-dependent termination. *EMBO J.* **4**:1887–1891.

111. Mudd, E. A., H. M. Krisch, and C. F. Higgins. 1990. RNase E, an endoribonuclease, has a general role in the chemical decay of *Escherichia coli* mRNA: evidence that *rne* and *ams* are the same genetic locus. *Mol. Microbiol.* **4**:2127–2135.

112. Mudd, E. A., and C. F. Higgins. 1993. *Escherichia coli* endoribonuclease RNase E: autoregulation of expression and site-specific cleavage of mRNA. *Mol. Microbiol.* **3**:557–568.

113. Muffler, A., D. Fischer, and R. Hengge-Aronis. 1996. The RNA-binding protein HF-1, known as a host factor for phage Qβ RNA replication, is essential for *rpoS* translation in *Escherichia coli. Genes Dev.* **10**:1143–1151.

114. Muffler, A., D. D. Traulsen, D. Fischer, R. Lange, and R. Hengge-Aronis. 1997. The RNA-binding protein HF-1 plays a global regulatory role which is largely, but not exclusively, due to its role in expression of the sigma S subunit of RNA polymerase in *Escherichia coli. J. Bacteriol.* **179**:297–300.

115. Newbury, S. F., N. H. Smith, E. C. Robinson, I. D. Hiles, and C. F. Higgins. 1987. Stabilization of translationally active mRNA by prokaryotic REP sequences. *Cell* **48**:297–310.

116. Niyogi, S. K., and A. K. Datta. 1975. A novel oligoribonuclease of *Escherichia coli*. I. Isolation and properties. *J. Biol. Chem.* **250**:7307–7312.

117. O'Hara, E. B., J. A. Chekanova, C. A. Ingle, Z. R. Kushner, E. Peters, and S. R. Kushner. 1995. Polyadenylylation helps regulate mRNA decay in *Escherichia coli. Proc. Natl. Acad. Sci. USA* **92**:1807–1811.

118. Okada, Y., M. Wachi, A. Hirata, K. Suzuki, K. Nagai, and M. Matsuhashi. 1994. Cytoplasmic axial filaments in *Escherichia coli* cells: possible function in the mechanism of chromosome segregation and cell division. *J. Bacteriol.* **176**:917–922.

119. Ono, M., and M. Kuwano. 1979. A conditional lethal mutation in an *Escherichia coli* strain with a longer chemical lifetime of mRNA. *J. Mol. Biol.* **129**:343–357.

120. Otsuka, Y., H. Ueno, and T. Yonesaki. 2003. *Escherichia coli* endoribonucleases involved in cleavage of bacteriophage T4 mRNAs. *J. Bacteriol.* **185**:983–990.

121. Ow, M. C., Q. Liu, and S. R. Kushner. 2000. Analysis of mRNA decay and rRNA processing in *Escherichia coli* in the absence of RNase E-based degradosome assembly. *Mol. Microbiol.* **38**:854–866.

122. Ow, M. C., and S. R. Kushner. 2002. Initiation of tRNA maturation by RNase E is essential for cell viability in *Escherichia coli. Genes Dev.* **16**:1102–1115.

123. Ow, M. C., Q. Liu, B. K. Mohanty, M. E. Andrew, V. F. Maples, and S. R. Kushner. 2002. RNase E levels in *Escherichia coli* are controlled by a complex regulatory system that involves transcription of the *rne* gene from three promoters. *Mol. Microbiol.* **43**:159–171.

124. Ow, M. C., T. Perwez, and S. R. Kushner. 2003. RNase G of *Escherichia coli* exhibits only limited functional overlap with its essential homologue, RNase E. *Mol. Microbiol.* **49:**607–622.

125. Pellegrini, O., J. Nezzar, A. Marchfelder, H. Putzer, and C. Condon. 2003. Endonucleolytic processing of CCA-less tRNA precursors by RNase E in *Bacillus subtilis*. *EMBO J.* **22:**4534–4543.

126. Portier, C., L. Dondon, M. Grunberg-Manago, and P. Regnier. 1987. The first step in the functional inactivation of the *Escherichia coli* polynucleotide phosphorylase messenger is ribonuclease III processing at the 5′ end. *EMBO J.* **6:**2165–2170.

127. Py, B., H. Causton, E. A. Mudd, and C. F. Higgins. 1994. A protein complex mediating mRNA degradation in *Escherichia coli*. *Mol. Microbiol.* **14:**717–729.

128. Py, B., C. F. Higgins, H. M. Krisch, and A. J. Carpousis. 1996. A DEAD-box RNA helicase in the *Escherichia coli* RNA degradosome. *Nature* **381:**169–172.

129. Ramanarayanan, M., and P. R. Srinivasan. 1976. Further studies on the isolation and properties of polyriboadenylate polymerase from *Escherichia coli* PR7 (RNase I⁻ *pnp*). *J. Biol. Chem.* **251:**6274–6286.

130. Rauhut, R., and G. Klug. 1999. mRNA degradation in bacteria. *FEMS Microbiol. Rev.* **23:**353–370.

131. Raynal, L. C., and A. J. Carpousis. 1999. Poly(A) polymerase I of *Escherichia coli*: characterization of the catalytic domain, an RNA binding site and regions for the interaction with proteins involved in mRNA degradation. *Mol. Microbiol.* **32:**765–775.

132. Regnier, P., and C. Portier. 1986. Initiation attenuation and RNase III processing of transcripts from the *Escherichia coli* operon encoding ribosomal protein S15 and polynucleotide phosphorylase. *J. Mol. Biol.* **187:**23–32.

133. Regnier, P., and M. Grunberg-Manago. 1989. Cleavage by RNase III in the transcripts of the *metY-nus-infB* operon of *Escherichia coli* releases the tRNA and initiates the decay of the downstream mRNA. *J. Mol. Biol.* **210:**293–302.

134. Regnier, P., and C. M. Arraiano. 2000. Degradation of mRNA in bacteria: emergence of ubiquitous features. *Bioessays* **22:**235–244.

135. Robert-Le Meur, M., and C. Portier. 1992. *Escherichia coli* polynucleotide phosphorylase expression is autoregulated through an RNase III-dependent mechanism. *EMBO J.* **11:**2633–2641.

136. Robert-Le Meur, M., and C. Portier. 1994. Polynucleotide phosphorylase of *Escherichia coli* induces the degradation of its RNase III processed messenger by preventing its translation. *Nucleic Acids Res.* **22:**397–403.

137. Robertson, H. D., R. E. Webster, and N. D. Zinder. 1967. A nuclease specific for double-stranded RNA. *Virology* **12:**718–719.

138. Romeo, T. 1996. Post-transcriptional regulation of bacterial carbohydrate metabolism: evidence that the gene product CsrA is a global mRNA decay factor. *Res. Microbiol.* **147:**505–512.

139. Rott, R., G. Zipor, V. Portnoy, V. Liveanu, and G. Schuster. 2003. RNA polyadenylation and degradation in cyanobacteria are similar to the chloroplast but different from *Escherichia coli*. *J. Biol. Chem.* **278:**15771–15777.

140. Sauter, C., J. Basquin, and D. Suck. 2003. Sm-like proteins in Eubacteria: the crystal structure of the Hfq protein from *Escherichia coli*. *Nucleic Acids Res.* **31:**4091–4098.

141. Schiffer, S., S. Rosch, and A. Marchfelder. 2002. Assigning a function to a conserved group of proteins: the tRNA 3′ processing enzymes. *EMBO J.* **21:**2769–2677.

142. Schumacher, M. A., R. F. Pearson, T. Moller, P. Valentin-Hansen, and R. G. Brennan. 2002. Structures of the pleiotropic translational regulator Hfq and an Hfq-RNA complex: a bacterial Sm-like protein. *EMBO J.* **21:**3546–3556.

143. Sledjeski, D. D., C. Whitman, and A. Zhang. 2001. Hfq is necessary for regulation by the untranslated RNA DsrA. *J. Bacteriol.* **183:**1997–2005.

144. Sohlberg, B., J. Huang, and S. N. Cohen. 2003. The *Streptomyces coelicolor* polynucleotide phosphorylase homologue, and not the putative poly(A) polymerase, can polyadenylate RNA. *J. Bacteriol.* **185:**7273–7278.

145. Soreq, H., and U. Z. Littauer. 1977. Purification and characterization of polynucleotide phosphorylase from *Escherichia coli*. *J. Biol. Chem.* **252:**6885–6888.

146. Spahr, P. F. 1964. Purification and properties of ribonuclease II from *Escherichia coli*. *J. Biol. Chem.* **239:**3716–3726.

147. Spickler, C., and G. A. Mackie. 2000. Action of RNase II and polynucleotide phosphorylase against RNAs containing stem-loops of defined structure. *J. Bacteriol.* **182:**2422–2427.

148. Spickler, C., V. Stronge, and G. A. Mackie. 2001. Preferential cleavage of degradative intermediates of *rpsT* mRNA by the *Escherichia coli* degradosome. *J. Bacteriol.* **183:**1106–1109.

149. Srinivasan, P. R., M. Ramanarayanan, and E. Rabbani. 1975. Presence of polyriboadenylate sequences in pulse-labeled RNA of *Escherichia coli*. *Proc. Natl. Acad. Sci. USA* **72:**2910–2914.

150. Steege, D. A. 2000. Emerging features of mRNA decay in bacteria. *RNA* **6:**1079–1090.

151. Stern, M. J., G. F.-L. Ames, N. H. Smith, E. C. Robinson, and C. F. Higgins. 1984. Repetitive extragenic palindromic sequences: a major component of the bacterial genome. *Cell* **37:**1015–1026.

152. Takiff, H. E., S. Chen, and D. L. Court. 1989. Genetic analysis of the *rnc* operon of *Escherichia coli*. *J. Bacteriol.* **171:**2581–2590.

153. Taraseviciene, L., A. Miczak, and D. Apirion. 1991. The gene specifying RNase E (*rne*) and a gene affecting mRNA stability (*ams*) are the same gene. *Mol. Microbiol.* **5:**851–855.

154. Tock, M. R., A. P. Walsh, G. Carroll, and K. J. McDowall. 2000. The CafA protein required for the 5′-maturation of 16 S rRNA is a 5′-end-dependent ribonuclease that has context-dependent broad sequence specificity. *J. Biol. Chem.* **275:**8726–8732.

155. Umitsuki, G., M. Wachi, A. Takada, T. Hikichi, and K. Nagia. 2001. Involvement of RNase G in *in vivo* mRNA metabolism in *Escherichia coli*. *Genes Cells* **6:**403–410.

156. Vanzo, N. F., Y. S. Li, B. Py, E. Blum, C. F. Higgins, L. C. Raynal, H. M. Krisch, and A. J. Carpousis. 1998. Ribonuclease E organizes the protein interactions in the *Escherichia coli* RNA degradosome. *Genes Dev.* **12:**2770–2781.

157. Vytvytska, O., I. Moll, V. R. Kaberdin, A. von Gabain, and U. Blasi. 2000. Hfq(HF1) stimulates *ompA* mRNA decay by interfering with ribosome binding. *Genes Dev.* **14:**1109–1118.

158. Wachi, M., G. Umitsuki, and K. Nagai. 1997. Functional relationship between *Escherichia coli* RNase E and the CafA protein. *Mol. Gen. Genet.* **253:**515–519.

159. Wachi, M., G. Umitsuki, M. Shimizu, A. Takada, and K. Nagai. 1999. *Escherichia coli cafA* gene encodes a novel RNase, designated as RNase G, involved in processing of the 5′ end of 16S rRNA. *Biochem. Biophys. Res. Commun.* **259:**483–488.

160. Wang, R.-F., E. B. O'Hara, M. Aldea, C. I. Bargmann, H. Gromley, and S. R. Kushner. 1998. *E. coli* MrsC is an

allele of HflB, a membrane associated ATPase and protease that is required for mRNA decay. *J. Bacteriol.* **180:**1929–1938.

161. **Wei, B. L., A. M. Brun-Zinkernagel, J. W. Simecka, B. M. Prub, P. Babitzke, and T. Romeo.** 2001. Positive regulation of motility and *flhDC* expression by the RNA-binding protein CsrA of *Escherichia coli. Mol. Microbiol.* **40:**245–256.

162. **Xu, F., S. Lin-Chao, and S. N. Cohen.** 1993. The *Escherichia coli pcnB* gene promotes adenylylation of antisense RNAI of ColE1-type plasmids *in vivo* and degradation of RNAI decay intermediates. *Proc. Natl. Acad. Sci. USA* **90:**6756–6760.

163. **Xu, F., and S. N. Cohen.** 1995. RNA degradation in *Escherichia coli* regulated by 3′ adenylation and 5′ phosphorylation. *Nature* **374:**180–183.

164. **Yehudai-Resheff, S., and G. Schuster.** 2000. Characterization of the *E. coli* poly(A) polymerase: nucleotide specificity, RNA-binding affinities and RNA structure dependence. *Nucleic Acids Res.* **28:**1139–1144.

165. **Yue, D., N. Maizels, and A. M. Weiner.** 1996. CCA-adding enzymes and poly(A) polymerases are all members of the same nucleotidyltransferase superfamily: characterization of the CCA-adding enzyme from the archaeal hyperthermophile *Sulfolobus shibatae. RNA* **2:**895–908.

166. **Zhang, A., S. Altuvia, A. Tiwari, L. Argaman, R. Hengge-Aronis, and G. Storz.** 1998. The *oxyS* regulatory RNA represses *rpoS* translation by binding Hfq (Hf-1) protein. *EMBO J.* **17:**6061–6068.

167. **Zhang, A., K. M. Wassarman, C. Rosenow, B. C. Tjaden, G. Storz, and S. Gottesman.** 2003. Global analysis of small RNA and mRNA targets of Hfq. *Mol. Microbiol.* **50:**1111–1124.

168. **Zhang, K., and A. W. Nicholson.** 1997. Regulation of ribonuclease III processing by double-helical sequence antideterminants. *Proc. Natl. Acad. Sci. USA* **94:**13437–13441.

169. **Zhang, X., L. Zhu, and M. P. Deutscher.** 1998. Oligoribonuclease is encoded by a highly conserved gene in the 3′-5′ exonuclease superfamily. *J. Bacteriol.* **180:**2779–2781.

170. **Zilhao, R., R. Cairrao, P. Régnier, and C. M. Arraiano.** 1996. PNPase modulates RNase II expression in *Escherichia coli*: implications for mRNA decay and cell metabolism. *Mol. Microbiol.* **20:**1033–1042.

IV. HOMOLOGOUS RECOMBINATION-REPAIR MACHINES

Chapter 19

Overview of Homologous Recombination and Repair Machines

Andrei Kuzminov and Franklin W. Stahl

GENETICS OF HOMOLOGOUS RECOMBINATION IN *ESCHERICHIA COLI*

Experimental Systems for Detection of Homologous Recombination in *E. coli*

Homologous recombination is exchange between homologous chromosomes which generates hybrid chromosomes when the parental chromosomes are genetically different. In eukaryotes, homologous recombination was detected genetically early in the 20th century (95), but in bacteria this happened only some 35 years later (66). Normally, bacterial cells are genetically haploid, although rapidly growing cells may contain several copies of the chromosome. Genetic haploidy is enforced by the lack of mechanisms for the preservation of diploidy, resulting in the segregation of homologous chromosomes into separate cells. This lack of diploidy requires that bacterial researchers adopt strategies for the detection of recombination that are different from those used traditionally by students of eukaryotes. To detect homologous recombination in bacteria, one must devise bacterial cells that carry additional DNA segments almost identical to, but genetically distinct from, the resident chromosome.

These additional chromosomal DNA segments can be introduced via conjugation (66), transformation (2), or phage-mediated transduction (138). During conjugation between an HFR strain and an F⁻ strain, a linear piece of variable length derived from the HFR chromosome enters the F⁻ cell, forming a merozygote (reviewed in reference 13). A fraction of these merozygotes give rise to cells in which genetically marked parts of the donor chromosome are found recombined into the recipient chromosome, replacing the corresponding alleles of the F⁻ parent. During generalized transduction, a random piece of the host chromosome is packaged in a phage capsid to be injected into a genetically distinct cell. The injected piece, or part of it, may then replace the homologous region of the chromosome using mechanisms like those involved in conjugational recombination (reviewed in reference 87). Some bacteria, such as *Bacillus subtilis* or *Haemophilus influenzae*, can be "transformed" by free DNA in the environment (reviewed in reference 78). They ingest DNA (probably for food [125]) and, if the DNA happens to be of the same species, incorporate it into the chromosome, replacing the corresponding resident segment.

All these approaches involve a transient merozygote, a cell containing an extra, nonreplicating DNA segment with homology to the chromosome (Fig. 1A). There are also ways to create stable merozygotes, either through the introduction of extrachromosomal elements (for example, plasmids carrying chromosomal segments [Fig. 1B]) or through incorporation of duplications into the chromosome (Fig. 1C). Upon culturing, these stable merozygotes produce recombinant progeny at a measurable rate. Finally, we may think of bacterial cells as short-lived zygotes of alien content when they are suffering mixed lytic-cycle infections by genetically different variants of the same bacteriophage.

In bacterial merozygotes, the ability of homologous chromosomes to recombine is almost proportional to the length of the segments of sequence identity; this ability drops precipitously below 30 to 50 nucleotides of identity, entering the realm of illegitimate recombination (52, 115, 117). The frequency of homologous recombination is also influenced by such factors as whether the recombining sequences are on the chromosome or on plasmids or phages, by the relative position of homologous sequences (if on the same chromosome), and by the type of replicon of the genetic element (reviewed in references 13 and 82).

Andrei Kuzminov • Department of Microbiology, University of Illinois, Urbana-Champaign, B103 C&LSL, 601 S. Goodwin Ave., Urbana, IL 61801-3709. **Franklin W. Stahl** • Institute of Molecular Biology, University of Oregon, Eugene, OR 97403-1229.

A: transient merozygote (conjugation, transduction)

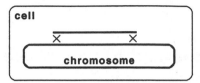

B: stable merozygote (an extrachromosomal element)

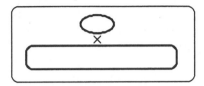

C: stable merozygote (a chromosomal repeat)

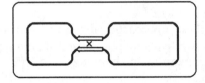

Figure 1. Basic types of merozygotes to detect homologous recombination in bacteria. See text for explanation.

Pathways of Homologous Recombination in *E. coli*

Recombination after conjugation or transduction in *E. coli* depends on *recA*, *recB*, and *recC* genes (reviewed in references 12 and 13). Phage lambda mutant for the phage's own recombination functions also requires *recABC* genes for homologous exchange (121). In *recA* mutants, homologous recombination is abolished, and the *recA* defect cannot be suppressed by further mutations. In contrast, recombination in *recB* and *recC* mutants is still detectable; this residual recombination is completely dependent on *recFOR* genes (76), revealing the existence of a secondary pathway. This *recFOR*-dependent recombination in *recBC* mutant cells is increased to wild-type (WT) levels by inactivation of exonuclease I due to *sbcB* mutations (130) and by simultaneous inactivation of the SbcCD nuclease (73). The robust recombination in *recBC sbcBC* (or *recBC sbcBD*) mutant cells is reduced by any of several other mutations, including *recG*, *recJ*, *recL*, *recN*, *recQ*, or *ruvABC* (reviewed in references 13

and 82). When introduced into the strains that are otherwise Rec$^+$, some of these new mutations induce hyperrecombination (*recL*) (1), some post no effect (*recJ* and *recQ*), and the rest cause a severalfold decrease in recombination (*recG*, *recN*, and *ruvABC*) (71), suggesting either modest participation in the WT recombination pathway or redundant functions. When these mutants were tested in pairwise combinations, only the *recG ruv* combinations proved to be fully recombination deficient (69), suggesting that *ruvABC* and *recG* genes (i) act at another stage of the recombination process distinct from the stage at which *recBCD* and *recFOR* genes act; (ii) act, at this stage, in different pathways from each other; and (iii) represent the only two such pathways.

The study of homologous recombination between plasmids, or between a plasmid and the chromosome, revealed that the RecFOR pathway is less of a poor cousin than first thought. Whereas plasmid recombination does still depend on *recA*, it is independent of *recBC* and requires the *recFOR* and *recJ* genes (reviewed in references 13 and 82). These differences between conjugational and transductional recombination versus plasmidic recombination are thought to reflect differences in the substrates: the RecBC pathway catalyzes exchanges when at least one of the participating chromosomes is linear (the linear piece brought in by conjugational transfer or by a transducing phage particle), whereas the RecFOR pathway in the WT cells catalyzes exchanges when both chromosomes are circular (e.g., two plasmids).

Since *recA* is the only gene whose inactivation completely blocks homologous recombination in *E. coli*, RecA protein must play the central role in this phenomenon. Analysis of epistatic interactions revealed that the products of *recBC* and *recFOR* genes work prior to RecA in the recombination pathways, while the products of *ruv* and *recG* genes work after the RecA-catalyzed step (reviewed in reference 62). These data were interpreted to signify the existence in *E. coli* of two early pathways of homologous recombination (RecBC for linear DNA and RecFOR for circular DNA) as well as two late pathways (*ruv* and *recG*) (Fig. 2) (15, 56).

The complete genetic requirements of homologous recombination in *E. coli* also include genes whose products participate in DNA replication and/or general DNA maintenance. Mutants in DNA replication genes rarely show up in screens for recombination-deficient mutants, because inactivation of these genes leads to gross perturbations in DNA replication. However, some partial loss-of-function mutants in DNA replication genes, while being relatively healthy, conspicuously affect homologous

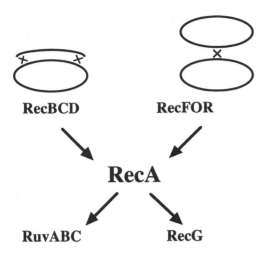

Figure 2. Genetic pathways for homologous recombination in *E. coli*. See text for explanation.

recombination. The DNA replication genes important for homologous recombination include *polA* and *lig* (28, 137), *ssb* (28, 34), *gyrAB* (26, 28, 44), *dnaE* (8), *dnaB* (10, 122), *priA* (53, 112), and *topA* (30). Inactivation of histonelike proteins, such as HU or OsmZ, also affects homologous recombination (26).

HOMOLOGOUS RECOMBINATION AND DNA DAMAGE

Stimulation of Homologous Recombination by DNA Damage, the SOS Response, and the Idea of Recombinational Repair

Soon after homologous recombination was described in *E. coli*, it was found that DNA-damaging treatments stimulate it (16, 51). The genetic requirements for DNA damage-induced recombination are the same as those for spontaneous recombination (45, 97), implying the same mechanisms. Most DNA lesions affect only one DNA strand at any point; such "one-strand" DNA lesions are efficiently removed by one of several excision repair systems (reviewed in reference 31). Since such repair would obviate any need for remedial interaction between homologues, the observed stimulation of homologous recombination by one-strand DNA damage might reasonably be attributed to induction of a cell-wide hyperrecombinogenic state.

It is well known that treatment of cells with one DNA-damaging agent, with subsequent incubation under growth conditions, makes them better prepared to repair DNA lesions induced later on by a different DNA-damaging treatment (86, 102). This

inducible (SOS) response to DNA damage is best demonstrated by the effect of split-dose DNA-damaging treatments: repair-proficient cells show better survival after a given damaging treatment if the damaging agent is delivered in several subdoses separated by short time intervals (27, 110). The SOS response (103) includes increased expression of about 30 genes in the *E. coli* chromosome and decreased expression of some 20 more genes and operons (23, 29). The products of induced genes participate in DNA repair and DNA damage tolerance; the products of suppressed genes participate in metabolism of various carbon sources as well as in cell division.

The idea that DNA damage induces homologous recombination exclusively via induction of a cell-wide hyperrecombinogenic state predicts that some hyperrecombination mutants in *E. coli* should be affected in the regulation of recombinational enzymes. Instead, all hyperrecombination mutants found to date seem to increase the level of endogenous DNA damage (54, 137), suggesting that, independently of whether recombination enzymes are overproduced or not, DNA lesions are required to stimulate recombination directly. Also, if damaged DNA is introduced into cells, the increased recombination is detected only between chromosomes that share homology with the damaged chromosome (35, 106), further arguing against the idea of indirect induction.

When the exquisite sensitivity to DNA damage of the first recombination-deficient mutants was found (14, 43), it became clear that homologous recombination might be the only way to repair certain DNA lesions. Still, it could be argued that homologous recombination happens independently of the DNA lesions, with the survivors being the lucky few that inherited, after recombination, only damage-free segments of the genome, as originally proposed in 1947 to explain the multiplicity reactivation phenomenon in bacteriophage (81). This idea predicts that the induction of homologous recombination by DNA damage should always be accompanied by a decrease in viability, with recombinants being more frequent only among the surviving chromosomes. However, doses of DNA-damaging agents that are lethal to recombination-deficient mutants, yet sublethal to recombination-proficient cells, do induce recombination. Thus, it seems certain that repair by homologous recombination results from an active mechanism rather than from a random redistribution of lesions.

Although the lesions requiring recombination for repair are still to be defined, the specificity of the inducible response provides a clue. When the original DNA damage affects only one strand of the DNA duplex, the inducible response to this damage requires

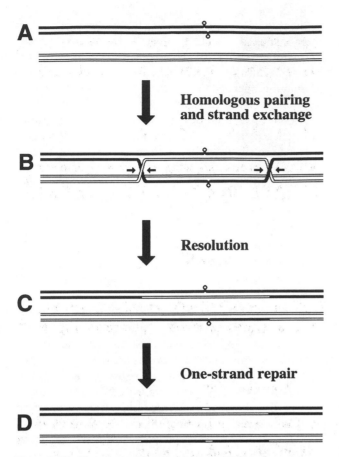

Figure 3. The idea of recombinational repair. (A) Two homologous chromosomes: the top one has a two-strand DNA lesion, and the bottom one is intact. (B) Homologous pairing and strand exchange between the two homologues, leading to conversion of the two-strand DNA lesion into a pair of one-strand DNA lesions in the hybrid DNA segments, bracketed by the double Holliday junction. One of two possible directions of the junction resolution is indicated by small arrows. (C) Holliday junction resolution breaks the joint molecule apart. (D) Excision repair removes the one-strand lesions. Open double line, the intact duplex; filled double line, the duplex with DNA lesions; lollipops in the filled strands, one-strand DNA lesions.

DNA replication (136). Similarly, induction of homologous recombination by treatments that affect only one strand of the DNA duplex requires replication of the damaged DNA (21, 68), suggesting that one-strand lesions spread to the opposite strands by replication, and the resulting two-strand lesions induce both the SOS response and homologous recombination. Accordingly, "recombinational repair" is thought to be recruited to repair two-strand lesions, when the individual DNA strands cannot be mended by one-strand repair (Fig. 3A). To make repair possible, the strands of an affected duplex are aligned by homology and then exchanged with the strands of an intact homologous duplex (Fig. 3B), as a result of which both lesions become single stranded and can

now be removed from the hybrid duplexes by one-strand repair (49).

Sensitivity to DNA Damage and Decreased Viability of *rec* Mutants

recA mutants, the first mutants in homologous recombination to be described, are exquisitely sensitive to DNA damage (14). Other *rec* mutants show varying degrees of sensitivity to DNA damage, their relative sensitivities being dependent on the DNA-damaging agent used (Table 1). UV light generates mostly pyrimidine dimers within the same DNA strand and is considered a typical one-strand DNA lesion-inducing agent (31). Nalidixic acid interferes with the function of DNA gyrase, which apparently leads to induction of double-strand DNA breaks (88, 90), so DNA gyrase can be viewed as an agent that induces predominantly two-strand DNA lesions. Ionizing radiation generates mostly one-strand DNA damage (oxidized bases, nicks), but also produces a measurable amount of direct double-strand breaks (31), so ionizing radiation can be viewed as an agent inducing both one-strand and two-strand DNA lesions. *recBC* and *recL* mutants show strong sensitivity to the same spectrum of DNA damage as *recA* mutants; *recFOR* and *ruv* mutants show strong sensitivity to UV irradiation and moderate (or no) sensitivity to ionizing radiation; *recN*, *recG*, and *recJ* mutants show weak sensitivity to UV irradiation, while *recG* mutants are moderately sensitive to ionizing radiation. Double *recBC recF* and *ruv recG* mutants are as sensitive to DNA damage as single *recA* mutants are (Table 1). Generally, the stronger the defect in homologous recombination, the higher the sensitivity to DNA damage. The only exception to this rule seems to be the *recL* defect, which makes cells quite sensitive to DNA damage but leaves them recombination proficient if all other *rec* genes are intact. *recL* is better known as *uvrD*, the gene for helicase II, which participates in various DNA transactions and is required to complete nucleotide excision repair (reviewed in reference 31).

The increased sensitivity of Rec⁻ mutants to artificially administered DNA-damaging treatments does not by itself prove an important adaptive role for recombination in DNA repair. However, the natural life cycle of *E. coli* alternates between periods of stable growth with little DNA damage per generation and periods of little growth with severe DNA damage (reviewed in reference 62), increasing the attractiveness of a repair role for homologous recombination. Furthermore, all mutants deficient in homologous recombination (*recA*, *recBC recF*, and *ruv recG*) have decreased viability even when grown in benign

Table 1. Viability of homologous recombination mutants and their sensitivity to different kinds of DNA-damaging treatments

Mutation	Sensitivity to DNA-damaging treatment[a]				Reference(s)
	UV[b]	NA[c]	IR[d]	Via[e]	
recA	<0.0001	0.005	0.002	0.5	37, 100, 134
recBC	0.003	0.005	0.01	0.3	37, 83
recD	1.0			1.0	74
recFOR	0.05	1.0		0.8–1.0	75, 77, 83
recG	0.2		0.1	0.8	72
recJ	0.3			0.9	83
recL	0.001	0.005		0.6	1, 70
recN	0.5		0.5	1.0	100
recQ	1.0			1.0	74
ruv	0.002		0.1	0.6	69, 99
priA	0.02		0.02	0.1[f]	53
xerCD or dif	1.0			0.6–0.9	93
recBC recF	0.00001	0.003		0.25	76, 77, 83
ruv recG	0.00001			0.25	69

[a]The doses of the DNA-damaging treatments are in all cases sublethal for the wild-type cells (survival is 80% or higher). The references were selected for consistency and, in many cases, are not original demonstrations.
[b]UV light sensitivity is shown as the ratio of survival of the mutant to the survival of the wild-type cells at 20 J/m^2. Cells were seeded on Luria-Bertani plates.
[c]Nalidixic acid (NA) sensitivity (88) is shown as the ratio of survival of the mutant to the survival of the wild-type cells after incubation in Luria broth with 100 μg/ml of the drug for 1 h at 37°C.
[d]Ionizing radiation (IR) sensitivity is shown as the ratio of survival of the mutant to the survival of the wild-type cells at 8 kilorads.
[e]Viability.
[f]Minimal media only.

laboratory conditions (Table 1), suggesting that DNA damage requiring recombinational repair results from normal cell metabolism. Thus, it is likely that homologous recombination ability is continuously selected for its role in DNA repair.

CHROMOSOMAL LESIONS AND THEIR REPAIR VIA HOMOLOGOUS STRAND EXCHANGE

Physical Detection of Chromosomal Lesions

Although one-strand DNA lesions compromise the accuracy of genetic information, they neither interfere with chromosomal replication and segregation nor threaten chromosome integrity. In contrast, two-strand DNA lesions (i) block replication of the chromosome, (ii) interfere with segregation of the daughter chromosomes, and (iii) compromise the integrity of the chromosome; therefore, they are regarded as "chromosomal lesions."

Treatments with ionizing radiation produce double-strand breaks, which promptly reveal themselves by chromosome fragmentation (7, 37) and render chromosomes susceptible to degradation. Treatments with bifunctional alkylating agents result in DNA that can be rapidly renatured after experimental denaturation, suggesting interstrand cross-links (104); such cross-links completely block DNA replication in their locality and prevent subsequent chromosomal

segregation. Because double-strand breaks and cross-links arise independently of DNA replication, they are classified as direct chromosomal lesions (Fig. 4).

The other group of chromosomal lesions forms at replication forks that are trying to traverse single-strand DNA lesions or difficult-to-replicate sequences, or when the replication machinery malfunctions (24, 92). In excision-repair-deficient mutants, single-strand gaps are found in DNA strands synthesized after UV irradiation (108). The average size of the gaps is half the length of Okazaki fragments (50), while the number of gaps is similar to the number of original UV lesions (108). The gaps are thought to form when replication forks traverse pyrimidine dimers in template DNA (Fig. 5B) (49). Replication forks apparently shed replisomes that are stalled at lesions (Fig. 5C). They then reinitiate downstream by recruiting new replisomes, leaving behind single-strand gaps opposite the lesions (Fig. 5D) (reviewed in reference 62). These so-called daughter-strand gaps preclude subsequent DNA replication, because both DNA strands now have lesions that would block a replisome.

Events after UV irradiation in uvr^+ cells, which efficiently excise UV-induced pyrimidine dimers, reveal another class of replication-dependent chromosomal lesions. In such cells, chromosomes break up; this fragmentation depends both on the ongoing excision repair and on DNA replication (reviewed in reference 62). It is proposed that replication forks in

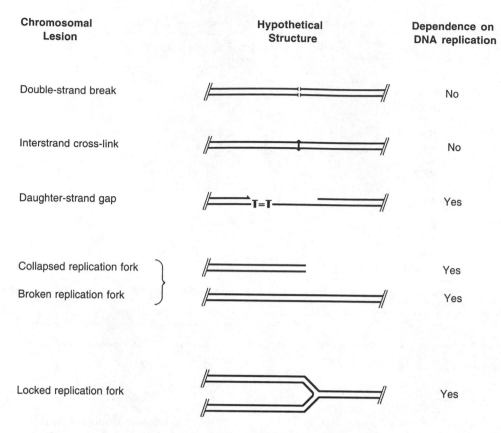

Figure 4. Configuration of chromosomal lesions.

these cells collapse when they run into transient excision gaps (Fig. 6) (42, 63). Collapse of a single replication fork in a theta-replicating circular chromosome should turn it into a sigma-replicating chromosome (Fig. 7A and B). Unless the eliminated replication fork is restored, the cell must be doomed, since it cannot segregate another circular chromosome (65). Collapse of the second replication fork is either good or bad news, depending on the DNA strand in which the second single-strand interruption happened. If the second single-strand interruption is in the same DNA strand as the first one, a linear subchromosomal fragment will be released and the circular chromosome will return to theta-replication (Fig. 7C to A). If, however, collapse of the second fork is due to a single-strand interruption in the other DNA strand, the chromosome becomes linear and will be degraded if not repaired (Fig. 7D) (47, 132).

Sometimes replication forks are stalled either because of replisome malfunction or because the template DNA is blocked by a tightly bound protein (48, 126) (Fig. 8B). A stalled replication fork is proposed to regress (roll back), extruding the newly synthesized DNA into a fourth arm, which emanates

from a Holliday junction (Fig. 8C) (32, 46) and has a double-strand end, accessible to exonucleases (79). Regressed replication forks have been recently detected in vivo in yeast mutants defective in cell cycle checkpoints (119). A regressed replication fork is hypothesized to be processed in any of three ways. One, "replication fork resetting," is imagined to be simply a reversal of the regression (Fig. 8D). Another, degradation of the open arm (Fig. 8E), should also result in the direct restoration of the replication fork structure (47, 60, 132). There is now genetic evidence substantiating the DNA degradation pathway of the replication fork restart (94, 114). The third way is Holliday junction resolution (Fig. 8F); this appears to be the least desirable way of processing because it breaks the replication fork (114). Such replication fork breakage creates a chromosomal lesion of the same type as the one generated by replication fork collapse (60) (compare Fig. 8G with Fig. 6D).

When a replication fork stalls while copying a palindromic sequence, yet another kind of chromosomal lesion might form. In this special case, regression of the fork may lead to template switching and,

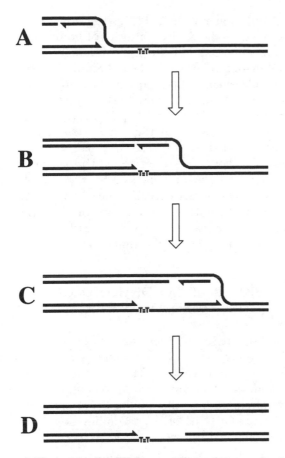

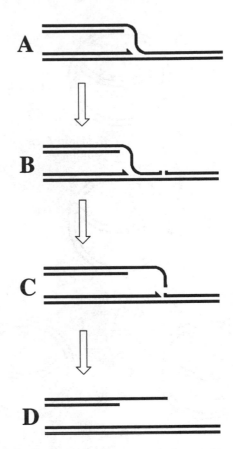

Figure 5. Formation of a daughter-strand gap during replication fork passage over a noncoding lesion. (A) A replication fork approaching a pyrimidine dimer. (B) The replication fork is traversing the pyrimidine dimer. (C) The stalled replisome is released, while the fork recruits a new replisome to reinitiate downstream from the lesion. (D) The replication fork moves away, leaving behind a daughter-strand gap. T=T, pyrimidine dimer (a noncoding lesion).

Figure 6. Replication fork collapse at a single-strand interruption in template DNA. (A) A replication fork. (B) The replication fork is approaching a single-strand interruption. (C) The replication fork has reached the interruption and come apart (collapsed). (D) The single-strand interruption in the full-length chromosome is repaired, while the detached double-strand end awaits its fate.

eventually, to replication fork "locking" (Fig. 9B to E). Plasmids with palindromes are known to form head-to-head dimers in *sbcCD* mutants (see "The SbcCD Enzyme" below), which is consistent with the template switching idea (5, 67, 84). The hypothetical way to process a locked replication fork is to continue to regress it in the same direction, so that the continuous strand of the new DNA pops out in a hairpin (Fig. 9F). When the single-strand loop of the hairpin is opened by the specialized endonuclease SbcCD ("The SbcCD Enzyme"), the cell can degrade the newly synthesized duplex, resetting the replication fork (Fig. 9G). In another hypothetical twist, if the locked replication fork is broken (as in Fig. 8F) before it is "unlocked," the loop on the double-strand end must be opened by SbcCD before the repair of the replication fork can continue.

The Inferred Pathways of Recombinational Repair

In *E. coli*, chromosomal lesions (two-strand DNA lesions) are repaired by homology-guided strand exchange between sister chromatids. The evidence in support of this notion comes in three forms (reviewed in reference 62). First, physical connections between parental and daughter strands, associated with lesion repair, can be detected. The formation of such hybrid strands is the most direct physical evidence of strand exchange; such evidence is available for daughter-strand gap repair (109) and for the repair of cross-links (18). Second, repair of chromosomal lesions is not observed in *recA* mutants. Such evidence is available for double-strand breaks (57), daughter-strand gaps (118), and cross-links (17), as well as for the yet to be physically confirmed lesions such as collapsed (58) or broken (60) replication forks. Third, DNA damage stimulates homologous recombination (see "Stimulation of Homologous Recombination by DNA

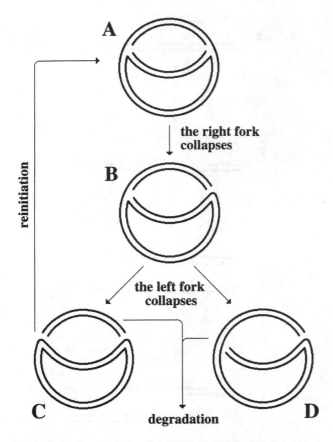

reinitiation

A

the right fork
collapses

B

the left fork
collapses

C

degradation

D

Figure 7. Replication fork collapse as a chromosomal lesion. (A) A theta-replicating chromosome. (B) As a result of collapse of the right replication fork, the chromosome starts replicating as a sigma-structure. (C) Collapse of the second replication fork terminates sigma-replication. (D) Collapse of the second replication fork linearizes the chromosome.

Damage, the SOS Response, and the Idea of Recombinational Repair" above), although the structure of chromosomal lesions in this case is unspecified. Also, the resolution of recombinational repair intermediates between sister duplexes in the *E. coli* chromosome is expected to generate sister chromatid exchanges, detected as dimeric chromosomes; formation of dimeric chromosomes is stimulated in *E. coli* mutants with increased DNA damage and is eliminated in single *recA* or in double *recB recF* mutants (123, 124).

The variety of chromosomal lesions notwithstanding, their final configuration belongs to one of the two basic types (Fig. 10): either an unfillable single-strand gap (a daughter-strand gap or the product of an attempt at excision repair of a crosslink) or a double-strand end (a pair of ends in the case of a direct double-strand break; a solitary end in the case of replication fork collapse or breakage). Double-strand-end repair is not observed in *recA* or

in *recBC* mutants (57, 113), suggesting that it is catalyzed by the RecBC pathway of homologous recombination (see "Pathways of Homologous Recombination in *E. coli*" above). Similarly, repair of unfillable single-strand gaps is blocked in *recA* or *recFOR* mutants (107, 118, 131), suggesting that it is catalyzed by the RecFOR pathway ("Pathways of Homologous Recombination in *E. coli*"). Plausible scenarios for these two types of repair are shown in Fig. 10. In both cases, the affected DNA region, with the help of RecA protein, invades the homologous region on the intact chromosome. After the resolution of DNA junctions and repair of one-strand lesions, the replication is restarted, either to fill the gaps or to resume a replication fork.

BIOCHEMISTRY OF HOMOLOGOUS RECOMBINATION IN *E. COLI*

Properties of various enzymes important in catalysis or regulation of homologous recombination are compiled in Table 2. Their biochemistry is covered in recent reviews (56, 105, 128, 133) and in several chapters of this book. The most important properties of some of the recombinational proteins are highlighted below.

SSB Protein

Single-stranded DNA-binding protein (SSB) complexes single-stranded DNA (ssDNA), facilitating its subsequent use in replication and in degradation and repair pathways of DNA metabolism (reviewed in reference 91). ssDNA wraps around SSB tetramers, reducing its contour length several times. In the high-cooperativity conformation, called SSB_{35}, the protein functions during DNA replication. In the low-cooperativity conformation, called SSB_{65}, the protein is thought to help RecA to polymerize on ssDNA (reviewed in reference 61). Without SSB, RecA filaments in vitro are unproductive unless assembled at low, nonphysiological concentrations of Mg^{2+}.

RecA Recombinase

RecA protein polymerizes into spiral filaments on ssDNA in the presence of SSB and high concentrations of Mg^{2+}. Since RecA prefers to polymerize in the $5' \rightarrow 3'$ direction, RecA filaments are more stable on the $3'$ ends of ssDNA than on the $5'$ ends. Such RecA filaments, which hydrolyze ATP, are the homologous pairing and strand exchange

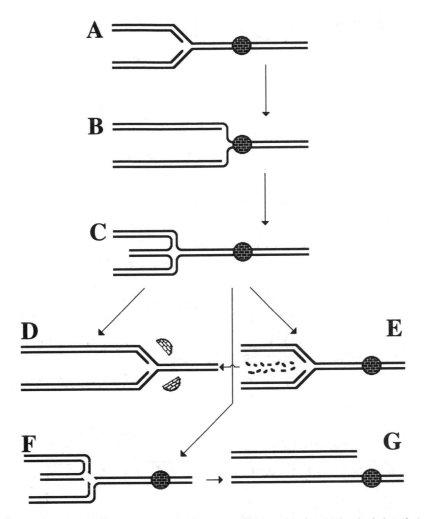

Figure 8. Regression of a stalled replication fork with subsequent resetting or breakage. The shaded circle indicates a protein tightly bound to the template DNA. (A) A replication fork approaches a block in the downstream template. (B) The replication fork stalls at the block. (C) The replication fork regresses from the block, forming a Holliday junction and extruding the newly replicated strands in a duplex of their own. (D) The regressed replication fork is reset and the block is removed. (E) A nuclease degrades the extruded fourth arm, recreating the replication fork structure. (F) Resolution of the Holliday junction leads to replication fork breakage. (G) Closure of the nicks completes the formation of a chromosomal lesion, in this case a double-strand end.

entities. They are also scaffolds for the binding and activation of autoproteases, such as the LexA repressor of the SOS response (see "Stimulation of Homologous Recombination by DNA Damage, the SOS Response, and the Idea of Recombinational Repair" above) or the UmuD subunit of the lesion bypass polymerase (see "Daughter-Strand Gap Repair" below).

Pairing and strand exchange is a critical process in recombinational repair and can be subdivided into three phases: (i) homology search, (ii) initial pairing, and (iii) strand exchange (reviewed in references 56, 62, and 105). SSB helps RecA at all phases of this process. The final products of the RecA-catalyzed

reactions in vitro (when RecA is not saturating) are joint molecules, with the region of strand exchange bracketed by DNA junctions. In vivo, RecA polymerization and depolymerization are likely to be aided by other Rec proteins (see "RecBCD Enzyme," "RecF, RecO, and RecR Proteins," and "RuvABC Resolvasome and RecG Helicase" below). For a detailed discussion of the RecA recombinase, see chapter 20.

RecBCD Enzyme

RecBCD enzyme binds tightly to double-strand DNA ends. The enzyme has several activities in vitro: (i) DNA end-specific helicase, (ii) ATP-dependent

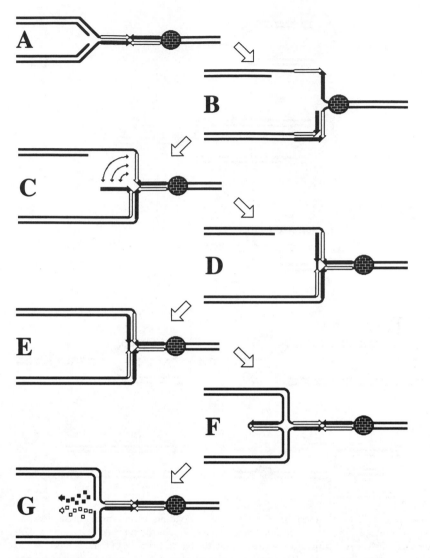

Figure 9. Locking and unlocking of a replication fork stalled at a small palindrome: a hypothesis. (A) A replication fork approaches a block in the downstream template, which happens to be near a small palindrome. (B) The replication fork is stalled at the block; one strand of the palindrome is replicated, while the opposite strand is complexed with SSB and remains single stranded. (C) The replication fork regresses from the block, extruding the newly replicated strand. The possibility of homologous pairing between the single-strand regions is shown by arrows. (D) Template switching due to the annealing of the complementary strands. (E) DNA synthesis, primed by the switched end, locks the replication fork. (F) Further regression of the locked replication fork extrudes the palindrome into a hairpin. (G) Hairpin degradation by SbcCD regenerates a replication fork structure. Shaded circle, a protein tightly bound to the template DNA; black and white arrows, palindrome (a black arrow forms a duplex with a codirectional white arrow).

double-stranded DNA exonuclease, (iii) ATP-dependent ssDNA exonuclease, (iv) ATP-stimulated ssDNA endonuclease, (v) Chi recognition, and (vi) RecA filament targeting to a particular DNA strand. Circular duplex DNA or SSB-complexed circular ssDNA are completely resistant to the action of RecBCD. The recombination-relevant subunits are RecB and RecC; the recombination-relevant functions are DNA helicase and RecA filament targeting. The RecD subunit and the other functions are im-

portant for linear DNA degradation (whose importance for *E. coli* is still unclear, but see "Double-Strand-End Repair" below) and for turning the enzyme from a degradase into a recombinase at Chi sites in the *E. coli* chromosome (reviewed in references 56 and 98). RecBCD binds at the ends of a linear duplex and processes them to generate 3' overhangs. Then RecBCD promotes RecA polymerization on these overhangs in the presence of SSB. The mechanism of DNA unwinding by RecBCD has been

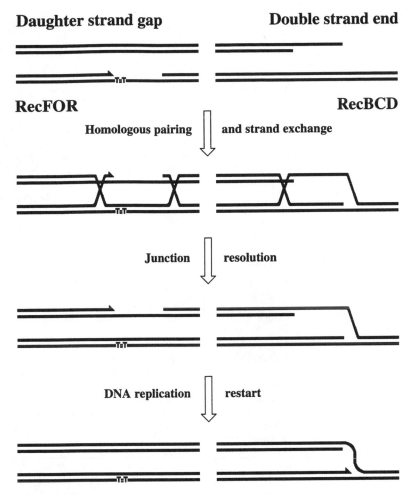

Figure 10. The two hypothetical pathways of recombinational repair. A scheme for daughter-strand gap repair, catalyzed by RecFOR and RecA, is shown on the left; a scheme for double-strand-end repair, catalyzed by RecBCD and RecA, is shown on the right. T = T, pyrimidine dimer (a noncoding lesion).

the subject of recent exciting experiments (6, 25, 120, 129). For a detailed discussion of the RecBCD enzyme, see chapter 21.

RecF, RecO, and RecR Proteins

RecF, RecO, and RecR proteins in vitro can bind ssDNA as well as duplex DNA. RecF protein binds ATP but does not hydrolyze it. The ATP-exchange factor for RecF has yet to be identified. No other activity is known for this protein. RecO and RecR proteins stimulate RecA polymerization on SSB-complexed ssDNA. Apparently, RecO and RecR could help RecA to polymerize on single-strand DNA gaps complexed by SSB in vivo, but they themselves would require targeting by gap-associated proteins (reviewed in reference 62). RecFOR-promoted polymerization of RecA on SSB-complexed gapped DNA was recently demonstrated (96).

RuvABC Resolvasome and RecG Helicase

RuvA in vitro binds completely double-stranded four-way DNA junctions (Holliday junctions) and isomerizes them from the folded conformation (difficult to branch-migrate) to the square planar conformation (easy to branch-migrate) (Fig. 11A and B). RuvB in vitro forms hexameric doughnuts around duplex DNA, interacts with RuvA, and, while holding on to RuvA, can "pump" duplex DNA through the central cavity of the hexameric ring (Fig. 11D). RuvC also binds Holliday junctions, changing their conformation to the planar one, and then symmetrically cleaves a pair of opposite strands at consensus sequences to "resolve" the junctions (Fig. 11E and F). RuvABC, working as a single complex called resolvasome, is believed to bind, isomerize, protect, branch-migrate, and resolve Holliday junctions (reviewed in reference 133). RecG helicase binds on one

Table 2. The pageant of homologous recombination enzymes of *E. coli*

Protein	Mol wt	Monomer copy no. (regular/SOS)	Active form	Interaction with DNA
DnaB	52,224	?	Hexamer	DNA unwinding; the main helicase of the replication fork
DnaE	129,686	20	DNA pol III	The catalytic subunit of DNA pol III
GyrA	96,767	4,000 (?)	Heterotetramer with GyrB	DNA supercoiling through transient double-strand breakage
GyrB	89,784	4,000 (?)	Heterotetramer with GyrA	DNA-dependent ATPase, DNA supercoiling
LexA	22,213	1,000	Dimer	Binding to SOS boxes (SOS repressor)
LigA	73,426	?	Monomer (?)	DNA ligase, closes nicks in duplex DNA
PolA	102,918	400	Monomer	DNA polymerization, $5' \rightarrow 3'$ and $3' \rightarrow 5'$ exonuclease
PriA	81,473	?	?	Primosome assembly at replication fork structures
RecA	37,816	1,000/100,000	Polymer	Homologous pairing, strand exchange
RecB	133,739	10	RecBCD	dsDNA unwinding and degradation, RecA filament assembly
RecC	128,634	10	RecBCD	dsDNA unwinding and degradation, RecA filament assembly
RecD	66,726	10	RecBCD	dsDNA degradation
RecE	96,171	?	Tetramer	Exo VIII, dsDNA degradation, $5' \rightarrow 3'$ strand only
RecF	40,356	190	?	dsDNA, ssDNA binding
RecG	76,249	10 (?)	Monomer	DNA junction processing
RecJ	63,216	?	?	ssDNA degradation ($5' \rightarrow 3'$)
RecL(UvrD)	81,804	5,000/25,000	Dimer	dsDNA unwinding from nicks
RecN	61,199	?/?	?	?
RecO	27,241	?	?	ssDNA binding, RecA filament assembly
RecQ	68,045	?	Hexamer (?)	DNA helicase
RecR	21,817	?	?	dsDNA, ssDNA binding, RecA filament assembly
RecT	29,570	?	Polymer	ssDNA binding, strand annealing
RuvA	21,940	700/5,600	Tetramer	Holliday junction binding
RuvB	37,018	200/1,600	Hexamer	dsDNA binding, RuvA-dependent DNA pumping
RuvC	18,604	?	Dimer	Holliday junction resolution
SbcB	54,334	?	Monomer (?)	Exo I, ssDNA degradation ($3' \rightarrow 5'$)
SbcC	118,515	?	$SbcC_6\ SbcD_{12}$ (?)	DNA hairpin degradation
SbcD	44,554	?	$SbcC_6\ SbcD_{12}$ (?)	DNA hairpin degradation
SSB	18,832	6,000	Tetramer	ssDNA complexing
TopA	97,152	?	Monomer	DNA relaxation through transient nicking

side of three-way or four-way DNA junctions and branch-migrates them to "squeeze out" the DNA on the opposite side of the junction (reviewed in reference 89). Genetic data suggest that RecG, like RuvABC, participates in DNA junction removal (69), but mechanisms of its action are less well understood. For a detailed discussion of the Holliday junction-processing enzymes, see chapter 22.

PriABC and DnaT

PriA has two major activities: one is the unwinding of DNA, which may be important for stabilization of inhibited replication forks (discussed in reference 59). The other, better-understood activity of PriA is recruitment of replisomes to replication fork structures. PriB, PriC, and DnaT are auxiliary factors required in vitro for this replisome recruitment. By attracting replisomes, PriABC + DnaT are hypothesized to complete the repair of replication forks (reviewed in references 85 and 111). Recently, replication fork structure assembly by RecA +

RecBCD with subsequent PriA-dependent initiation of DNA synthesis has been achieved in vitro (135). For more on reinitiation of DNA replication, see chapter 12.

The SbcCD Enzyme

sbcCD mutations allow normally unstable DNA palindromes to be stably propagated in *E. coli* (11, 33). Palindromes are assumed to form stem-loop structures in vivo, which, during DNA replication, may lead to palindrome deletion by strand slippage (101) or to the inability to complete replication of the palindrome-carrying chromosomes. In vitro, SbcCD enzyme is an ATP-dependent dsDNA exonuclease (19), as well as an ATP-stimulated endonuclease specific for single-strand loops of stem-loop structures (20). *sbcCD* mutants are recombination proficient; *sbcCD* mutations, together with *sbcB* mutations, are required in *recBC* mutant cells for the full level of recombination proficiency (by the RecFOR pathway) (33, 73).

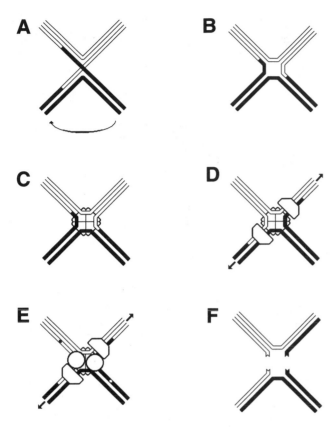

Figure 11. Holliday junction processing by the RuvABC resolvasome. (A) A Holliday junction in the folded conformation (difficult to process but preferred in physiological conditions). The arrow indicates the 180° rotation required to convert the folded conformation into the square planar one. (B) A Holliday junction in the square planar conformation (easy to process, observed in the absence of Mg^{2+} ions). (C) RuvA tetramers bind Holliday junctions under physiological conditions and isomerize them into the square planar conformation. (D) RuvB hexamers are shown as washers on opposite arms of the Holliday junction; they interact with RuvA and "pump" DNA through their central openings (the direction of DNA movement is shown by arrows). At this stage of the Holliday junction processing, two RuvA tetramers assemble around the junction in a turtle-shell configuration (not shown). (E) One of the RuvA tetramers is replaced with a RuvC dimer (two circles), while the RuvC consensus resolution sequences (diamonds) are drawn into the junction by RuvB pumping. (F) RuvC symmetrically cuts at the resolution consensus sequences, resolving the joint molecule (RuvABC proteins are not shown). The two original duplexes, forming a joint molecule, are shown as either open or filled double lines.

XerCD and *dif*

XerC and XerD are a pair of site-specific recombinases, working at the 28-bp *dif* site, situated in the middle of the replication terminus on the *E. coli* chromosome (reviewed in references 62 and 116). This site-specific resolution system is not required for homologous recombination, but helps complete recombinational repair by monomerizing dimeric chromosomes, which occasionally arise as a result of such

repair (see "The Inferred Pathways of Recombinational Repair" above). In fact, chromosomal dimerization in *E. coli* itself creates a chromosomal lesion, because it prevents segregation of the replicated chromosomes into daughter cells. For a detailed discussion of the resolution of chromosome dimers, see chapter 28.

THE HYPOTHETICAL MECHANISMS OF RECOMBINATIONAL REPAIR

Double-Strand-End Repair

A disintegrated replication fork, whether starting as a collapsed, broken, or locked replication fork, has the same final structure: an intact chromosome and a subchromosomal branch ending with an open double-strand end (Fig. 10). In the case of a theta-replicating circular chromosome of *E. coli*, a disintegrated replication fork translates into a sigma-replicating chromosome (Fig. 12B). To reattach the free end in a homology-guided process, the cell must (i) unwind the duplex end, because RecA cannot catalyze strand exchange between completely double-stranded DNAs; (ii) eliminate the 5'-ending strand, so that the RecA filament is stable at the DNA end, while the subsequent homology search is not poisoned by the presence of the complementary strand; and (iii) help RecA to polymerize on this end in the presence of SSB, which is always the first protein to bind to ssDNA. Based on its biochemical activities, the RecBCD enzyme is thought to catalyze all three tasks (reviewed in references 56 and 62). In particular, the enzyme binds in vitro to duplex DNA ends and degrades them until its degradase activity is inactivated at Chi sites, after which the enzyme generates a single-strand 3' overhang and promotes RecA polymerization on this overhang. The switch in RecBCD from degradase to recombinase at Chi requires RecA polymerization; in the absence of RecA, RecBCD continues as a degradase, disregarding Chi sites (64, 65). This could be physiologically important: when the replication fork cannot be restored because of a shortage of RecA, the linear tail could be degraded, returning the chromosome to the circular form (Fig. 12D).

RecA-mediated, homology-guided single-strand-end invasion instantly regenerates a replication fork structure (Fig. 12C), but, to resume replication, the fork must attract a replisome. Judging from in vitro activities, PriA is involved in this reaction. Proposed steps include (i) PriA binding on the displaced strand near the invading 3' end, (ii) PriBC and DnaT binding to PriA, (iii) DnaB helicase recruitment onto the displaced strand from the DnaB-DnaC complex, and (iv)

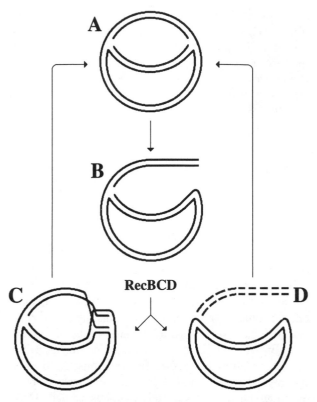

Figure 12. The two ways of restoring theta-replication to a circular chromosome that suffered replication fork collapse. (A) A theta-replicating chromosome. (B) The right fork of the replication bubble has collapsed, shifting chromosome replication into sigma-mode. (C) RecBCD- and RecA-catalyzed strand exchange restores replication fork structure, returning the chromosome to theta-replication. (D) RecBCD-catalyzed degradation of the linear tail makes theta-replication possible again without repairing the collapsed replication fork.

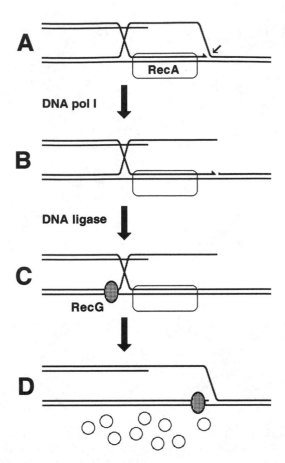

Figure 13. A hypothetical, RecG-dependent way of Holliday junction processing. (A) An end-invasion intermediate, with the 3' end being extended by DNA pol I. (B) DNA pol I cleaves the displaced strand and starts nick-translating. (C) DNA ligase seals the nick, while RecG helicase binds to the side of the Holliday junction opposite the side bound by the RecA filament. (D) RecG translocates the Holliday junction toward the DNA end, dispersing the RecA filament and restoring replication fork structure. Rectangle with rounded corners, RecA filament; gray oval, RecG; open circles, RecA monomers; small arrow, the position of DNA strand cleavage; one-sided arrow, the 3' end used by DNA pol I.

DnaG primase recruitment onto DnaB at the replication fork (reviewed in reference 85). DnaB starts unwinding the fork, while DnaG deposits primers for both leading and lagging strand synthesis.

Now the cell needs to clean up the repair mess behind the replication fork, which involves resolving the DNA junction and removing the spent RecA filament from the hybrid duplex. The in vitro activities of RuvA, RuvB, and RuvC qualify them for the role (reviewed in references 62 and 133). RuvAB complex can strip RecA filaments from duplex DNA and from recombination intermediates. Addition of RuvC to the RuvAB complex turns it into a resolvasome, able to translocate DNA junctions to the RuvC consensus sequences and to resolve the junctions once the RuvC consensus sequence is positioned properly within the junction. RuvABC can resolve both completely double-stranded four-way junctions (Holliday junctions) and four-way junctions with some arms single stranded.

There is a scenario for the processing of strand invasion intermediates that would make Holliday junction resolution by RuvABC detrimental for double-strand-end repair. The 3' end of the invading strand might be used as a primer by DNA polymerase I, which would then cleave the displaced strand with its 5' → 3' exonuclease activity (which can also act as an endonuclease [80]) and nick-translate until DNA ligase seals the nick (Fig. 13A to C). This connects the invading strand with the resident strand but eliminates the replication fork structure. Moreover, if RuvABC now resolves the Holliday junction (in Fig. 13C), the prerepair situation will be reached. This conundrum is avoided if, rather than being resolved, the Holliday junction is translocated toward the double-strand end, recreating a replication fork structure without the need for RecA-catalyzed reinvasion (Fig. 13C and D). The biochemical properties of the RecG helicase make it a good candidate for catalysis of this step in such a hypothetical processing pathway (reviewed in reference 62). In doing so, RecG will also disperse the RecA filament from the hybrid portion of the duplex.

Daughter-Strand Gap Repair

A daughter-strand gap arising at a replication fork may have the following structure: (i) a length of about 800 nucleotides (about half the length of an Okazaki fragment), all complexed with SSB; (ii) the 3' end blocked at the noncoding lesion, with the DNA polymerase still idling there; and (iii) the 5' end terminating with the RNA primer of the next Okazaki fragment (62). How does the cell distinguish an unfillable single-strand gap from a slowly replicating Okazaki fragment? How does the cell catalyze RecA polymerization on SSB-complexed ssDNA without an end if RecBCD has an absolute requirement for a DNA end? The suppression of the *recFOR* mutant phenotype by certain *recA* mutations (reviewed in reference 62) and the recent demonstration of RecFOR-promoted RecA polymerization on gapped DNA in the presence of SSB (96) implicate RecFOR in helping RecA to polymerize on daughter-strand DNA gaps in vivo.

After strand invasion, the single-strand gap is filled using the intact template strand, and there is again a need to remove the two DNA junctions and the spent RecA filament from DNA. As in double-strand-end repair (see "Double-Strand-End Repair" above), RuvABC resolvasome and RecG helicase are good candidates for this role. The higher UV sensitivity of *ruv* mutants as compared with *recG* mutants (Table 1) implies that RuvABC is used preferentially to complete the repair of daughter-strand gaps, which are likely to be the predominant chromosomal lesion after UV irradiation.

To deal with occasional daughter-strand gaps lacking an intact homologous partner for repair, *E. coli* cells employ a different strategy: they use RecA filaments to induce synthesis of a nonprocessive translesion DNA polymerase, UmuCD'$_2$ (DNA pol V), which is then targeted to the 3' ends associated with RecA filaments (reviewed in reference 36). The auxiliary subunit UmuD' helps to dissociate the now useless RecA filament, while the UmuCD'$_2$ complex adds a few nucleotides across the lesion. Once past the lesion, UmuCD'$_2$ is replaced with DNA pol I or DNA pol III, which fills in the gap. UmuCD'$_2$ is a sloppy copier, but the low processivity of the enzyme

limits the mutagenesis associated with this translesion DNA synthesis.

PERSPECTIVES

The complexity of recombinational repair notwithstanding, we have a satisfying core of ideas to test (summarized in reference 62). One notable exception is the paucity of insights regarding the central stage of recombinational repair—the homology search. Recent experimental results suggest that (i) the precision and efficiency of the homology search are the product of several readily reversible steps (3, 4), and (ii) A:T base pairing is more important for homology recognition than G:C base pairing (40, 41). More experiments are needed to understand the structure-function relationship within the RecA filament and to learn the mechanistic basis of the homology search reaction.

Our understanding of the formation of replication-dependent chromosomal lesions is still primitive. There is one in vivo study on the structure of stalled replication forks (39), a report documenting replication fork reversal in vivo (119), as well as a few reports of replication fork reversal in vitro, likely to be an artifact of DNA isolation. There are also two reports confirming the possibility of replication fork collapse at single-strand interruptions (63, 127). In the case of daughter-strand gaps, old findings should be revisited, if only to reaffirm the original interpretations in the face of recently proposed alternatives (22, 38).

The bulk of chromosomal lesions result from endogenous DNA damage whose level, in its own turn, is determined by the general metabolism of the cell. The metabolic pathways leading to DNA damage that, through DNA replication, eventually translates into chromosomal lesions are mostly uncharted. In this respect, genetic and metabolic characterization of hyperrecombinational and recombination-dependent mutants is expected to be illuminating (9, 55).

Acknowledgments. Studies in the laboratory of F.W.S. are supported by grant MCB-0109809 from the National Science Foundation. Studies in the laboratory of A.K. are supported by grant MCB-0196020 from the National Science Foundation.

REFERENCES

1. **Arthur, H. M., and R. G. Lloyd.** 1980. Hyper-recombination in *uvrD* mutants of *Escherichia coli* K-12. *Mol. Gen. Genet.* **180:**185–191.
2. **Avery, O. T., C. M. Macleod, and M. McCarty.** 1944. Studies on the chemical nature of the substance inducing transformation of pneumococcal types. I. Induction of

transformation by a desoxyribonucleic acid fraction isolated from Pneumococcus type III. *J. Exp. Med.* **79**:137–158.

3. Bazemore, L. R., E. Folta-Stogniew, M. Takahashi, and C. M. Radding. 1997. RecA tests homology at both pairing and strand exchange. *Proc. Natl. Acad. Sci. USA* **94**:11863–11868.

4. Bazemore, L. R., M. Takahashi, and C. M. Radding. 1997. Kinetic analysis of pairing and strand exchange catalyzed by RecA. Detection by fluorescence energy transfer. *J. Biol. Chem.* **272**:14672–14682.

5. Bi, X., and L. F. Liu. 1996. DNA rearrangements mediated by inverted repeats. *Proc. Natl. Acad. Sci. USA* **93**:819–823.

6. Bianco, P. R., and S. C. Kowalczykowski. 2000. Translocation step size and mechanism of the RecBC DNA helicase. *Nature* **405**:368–372.

7. Billen, D., and R. Hewitt. 1967. Concerning the dynamics of chromosome replication and degradation in a bacterial population exposed to X-rays. *Biochim. Biophys. Acta* **138**:587–595.

8. Blinkowa, A. 1976. The role of polymerase III in conjugation between *E. coli* K12 donor and recipient strains carrying *dna*Ets mutation. *Acta Microbiol. Pol.* **25**:95–108.

9. Bradshaw, J. S., and A. Kuzminov. 2003. RdgB acts to avoid chromosome fragmentation in *Escherichia coli*. *Mol. Microbiol.* **48**:1711–1725.

10. Bresler, S. E., V. A. Lanzov, and A. A. Lukjaniec-Blinkova. 1968. On the mechanism of conjugation in *Escherichia coli* K12. *Mol. Gen. Genet.* **102**:269–284.

11. Chalker, A. F., D. R. F. Leach, and R. G. Lloyd. 1988. *Escherichia coli sbcC* mutants permit stable propagation of DNA replicons containing a long palindrome. *Gene* **71**:201–205.

12. Clark, A. J. 1971. Toward a metabolic interpretation of genetic recombination of *E. coli* and its phages. *Annu. Rev. Microbiol.* **25**:437–464.

13. Clark, A. J., and K. B. Low. 1988. Pathways and systems of homologous recombination in *Escherichia coli*, p. 155–215. *In* K. B. Low (ed.), *The Recombination of Genetic Material*. Academic Press, Inc., San Diego, Calif.

14. Clark, A. J., and A. D. Margulies. 1965. Isolation and characterization of recombination-deficient mutants of *Escherichia coli* K-12. *Proc. Natl. Acad. Sci. USA* **53**:451–459.

15. Clark, A. J., and S. J. Sandler. 1994. Homologous recombination: the pieces begin to fall into place. *Crit. Rev. Microbiol.* **20**:125–142.

16. Clark, J. B., F. Haas, W. S. Stone, and O. Wyss. 1950. The stimulation of gene recombination in *Escherichia coli*. *J. Bacteriol.* **59**:375–379.

17. Cole, R. S. 1971. Inactivation of *Escherichia coli*, F′ episomes at transfer, and bacteriophage lambda by psoralen plus 360-nm light: significance of deoxyribonucleic acid cross-links. *J. Bacteriol.* **107**:846–852.

18. Cole, R. S. 1973. Repair of DNA containing interstrand crosslinks in *Escherichia coli*: sequential excision and recombination. *Proc. Natl. Acad. Sci. USA* **70**:1064–1068.

19. Connelly, J. C., E. S. de Leau, E. A. Okely, and D. R. F. Leach. 1997. Overexpression, purification, and characterization of the SbcCD protein from *Escherichia coli*. *J. Biol. Chem.* **272**:19819–19826.

20. Connelly, J. C., L. A. Kirkham, and D. R. F. Leach. 1998. The SbcCD nuclease of *Escherichia coli* is a structural maintenance of chromosomes (SMC) family protein that cleaves hairpin DNA. *Proc. Natl. Acad. Sci. USA* **95**:7969–7974.

21. Cordone, L., R. M. Sperandeo-Mineo, and S. Mannino. 1975. UV-induced enhancement of recombination among

lambda bacteriophages: relation with replication of irradiated DNA. *Nucleic Acids Res.* **2**:1129–1142.

22. Courcelle, J., A. K. Ganesan, and P. C. Hanawalt. 2001. Therefore, what are recombination proteins there for? *Bioessays* **23**:463–470.

23. Courcelle, J., A. Khodursky, B. Peter, P. O. Brown, and P. C. Hanawalt. 2001. Comparative gene expression profiles following UV exposure in wild-type and SOS-deficient *Escherichia coli*. *Genetics* **158**:41–64.

24. Cox, M. M. 1998. A broadening view of recombinational DNA repair in bacteria. *Genes Cells* **3**:65–78.

25. Dillingham, M. S., M. Spies, and S. C. Kowalczykowski. 2003. RecBCD enzyme is a bipolar DNA helicase. *Nature* **423**:893–897.

26. Dri, A.-M., P. L. Moreau, and J. Rouviere-Yaniv. 1992. Role of the histone-like proteins OsmZ and HU in homologous recombination. *Gene* **120**:11–16.

27. Dzidic, S., E. Salaj-Smic, and Z. Trgovcevic. 1986. The relationship between survival and mutagenesis in *Escherichia coli* after fractionated ultraviolet irradiation. *Mutat. Res.* **173**:89–91.

28. Ennis, D. G., S. K. Amundsen, and G. R. Smith. 1987. Genetic functions promoting homologous recombination in *Escherichia coli*: a study of inversion in phage λ. *Genetics* **115**:11–24.

29. Fernández de Henestrosa, A. R., T. Ogi, S. Aoyagi, D. Chafin, J. J. Hayes, H. Ohmori, and R. Woodgate. 2000. Identification of additional genes belonging to the LexA regulon in *Escherichia coli*. *Mol. Microbiol.* **35**:1560–1572.

30. Fishel, R. A., and R. Kolodner. 1984. *Escherichia coli* strains containing mutations in the structural gene for topoisomerase I are recombination deficient. *J. Bacteriol.* **160**:1168–1170.

31. Friedberg, E. C., G. C. Walker, and W. Siede. 1995. *DNA Repair and Mutagenesis*. ASM Press, Washington, D.C.

32. Fujiwara, Y., and M. Tatsumi. 1976. Replicative bypass repair of ultraviolet damage to DNA of mammalian cells: caffeine sensitive and caffeine resistant mechanisms. *Mutat. Res.* **37**:91–110.

33. Gibson, F. P., D. R. F. Leach, and R. G. Lloyd. 1992. Identification of *sbcD* mutations as cosuppressors of *recBC* that allow propagation of DNA palindromes in *Escherichia coli* K-12. *J. Bacteriol.* **174**:1222–1228.

34. Glassberg, J., R. R. Meyer, and A. Kornberg. 1979. Mutant single-strand binding protein of *Escherichia coli*: genetic and physiological characterization. *J. Bacteriol.* **140**:14–19.

35. Golub, E. I., and K. B. Low. 1983. Indirect stimulation of genetic recombination. *Proc. Natl. Acad. Sci. USA* **80**:1401–1405.

36. Goodman, M. F. 2000. Coping with replication "train wrecks" in *Escherichia coli* using Pol V, Pol II and RecA proteins. *Trends Biochem. Sci.* **25**:189–195.

37. Gray, W. J. H., M. H. L. Green, and B. A. Bridges. 1972. DNA synthesis in gamma-irradiated recombination deficient strains of *Escherichia coli*. *J. Gen. Microb.* **71**:359–366.

38. Gregg, A. V., P. McGlynn, R. P. Jaktaji, and R. G. Lloyd. 2002. Direct rescue of stalled DNA replication forks via the combined action of PriA and RecG helicase activities. *Mol. Cell* **9**:241–251.

39. Gruber, M., R. E. Wellinger, and J. M. Sogo. 2000. Architecture of the replication fork stalled at the 3′ end of yeast ribosomal genes. *Mol. Cell. Biol.* **20**:5777–5787.

40. Gupta, R. C., E. Folta-Stogniew, S. O'Malley, M. Takahashi, and C. M. Radding. 1999. Rapid exchange of A:T base pairs is essential for recognition of DNA homology by human Rad51 recombination protein. *Mol. Cell* **4**:705–714.

41. Gupta, R. C., E. Folta-Stogniew, and C. M. Radding. 1999. Human Rad51 protein can form homologous joints in the absence of net strand exchange. *J. Biol. Chem.* **274:**1248–1256.

42. Hanawalt, P. C. 1966. The U.V. sensitivity of bacteria: its relation to the DNA replication cycle. *Photochem. Photobiol.* **5:**1–12.

43. Harm, W. 1964. On the control of UV-sensitivity of phage T4 by the gene *x*. *Mutat. Res.* **1:**344–354.

44. Hays, J. B., and S. Boehmer. 1978. Antagonists of DNA gyrase inhibit repair and recombination of UV-irradiated phage λ. *Proc. Natl. Acad. Sci. USA* **75:**4125–4129.

45. Hays, J. B., B. K. Duncan, and S. Boehmer. 1981. Recombination of uracil-containing lambda bacteriophages. *J. Bacteriol.* **145:**306–320.

46. Higgins, N. P., K. Kato, and B. Strauss. 1976. A model for replication repair in mammalian cells. *J. Mol. Biol.* **101:**417–425.

47. Horiuchi, T., and Y. Fujimura. 1995. Recombinational rescue of the stalled DNA replication fork: a model based on analysis of an *Escherichia coli* strain with a chromosomal region difficult to replicate. *J. Bacteriol.* **177:**783–791.

48. Horiuchi, T., Y. Fujimura, H. Nishitani, T. Kobayashi, and M. Hidaka. 1994. The DNA replication fork blocked at the *Ter* site may be an entrance for the RecBCD enzyme into duplex DNA. *J. Bacteriol.* **176:**4656–4663.

49. Howard-Flanders, P. 1973. DNA repair and recombination. *Br. Med. Bull.* **29:**226–235.

50. Iyer, V. N., and W. D. Rupp. 1971. Usefulness of benzoylated naphthoylated DEAE-cellulose to distinguish and fractionate double-stranded DNA bearing different extents of single-stranded regions. *Biochim. Biophys. Acta* **228:**117–126.

51. Jacob, F., and E. L. Wollman. 1955. Étude génétique d'un bactériophage tempéré d'*Escherichia coli*. III. Effet du rayonnement ultraviolet sur la recombinaison génétique. *Ann. Inst. Pasteur* **88:**724–749.

52. King, S. R., and J. P. Richardson. 1986. Role of homology and pathway specificity for recombination between plasmids and bacteriophage λ. *Mol. Gen. Genet.* **204:**141–147.

53. Kogoma, T., G. W. Cadwell, K. G. Barnard, and T. Asai. 1996. The DNA replication priming protein, PriA, is required for homologous recombination and double-strand break repair. *J. Bacteriol.* **178:**1258–1264.

54. Konrad, E. B. 1977. Method for the isolation of *Escherichia coli* mutants with enhanced recombination between chromosomal duplications. *J. Bacteriol.* **130:**167–172.

55. Kouzminova, E. A., and A. Kuzminov. 2004. Chromosomal fragmentation in dUTPase-deficient mutants of *Escherichia coli* and its recombinational repair. *Mol. Microbiol.* **51:**1279–1295.

56. Kowalczykowski, S. C., D. A. Dixon, A. K. Eggleston, S. D. Lauder, and W. M. Rehrauer. 1994. Biochemistry of homologous recombination in *Escherichia coli*. *Microbiol. Rev.* **58:**401–465.

57. Krasin, F., and F. Hutchinson. 1977. Repair of DNA double-strand breaks in *Escherichia coli*, which requires *recA* function and the presence of a duplicate genome. *J. Mol. Biol.* **116:**81–98.

58. Kuzminov, A. 1995. Collapse and repair of replication forks in *Escherichia coli*. *Mol. Microbiol.* **16:**373–384.

59. Kuzminov, A. 2001. DNA replication meets genetic exchange: chromosomal damage and its repair by homologous recombination. *Proc. Natl. Acad. Sci. USA* **98:**8461–8468.

60. Kuzminov, A. 1995. Instability of inhibited replication forks in *E. coli*. *Bioessays* **17:**733–741.

61. Kuzminov, A. 1995. A mechanism for induction of the SOS response in *E. coli*: insights into the regulation of reversible protein polymerization *in vivo*. *J. Theor. Biol.* **177:**29–43.

62. Kuzminov, A. 1999. Recombinational repair of DNA damage in *Escherichia coli* and bacteriophage λ. *Microbiol. Mol. Biol. Rev.* **63:**751–813.

63. Kuzminov, A. 2001. Single-strand interruptions in replicating chromosomes cause double-strand breaks. *Proc. Natl. Acad. Sci. USA* **98:**8241–8246.

64. Kuzminov, A., E. Schabtach, and F. W. Stahl. 1994. χ-sites in combination with RecA protein increase the survival of linear DNA in *E. coli* by inactivating ExoV activity of RecBCD nuclease. *EMBO J.* **13:**2764–2776.

65. Kuzminov, A., and F. W. Stahl. 1997. Stability of linear DNA in *recA* mutant *Escherichia coli* cells reflects ongoing chromosomal DNA degradation. *J. Bacteriol.* **179:**880–888.

66. Lederberg, J. 1947. Gene recombination and linked segregations in *Escherichia coli*. *Genetics* **32:**505–525.

67. Lin, C.-T., Y. L. Lyu, and L. F. Liu. 1997. A cruciform-dumbbell model for inverted dimer formation mediated by inverted repeats. *Nucleic Acids Res.* **25:**3009–3016.

68. Lin, P.-F., and P. Howard-Flanders. 1976. Genetic exchanges caused by ultraviolet photoproducts in phage λ DNA molecules: the role of DNA replication. *Mol. Gen. Genet.* **146:**107–115.

69. Lloyd, R. G. 1991. Conjugational recombination in resolvase-deficient *ruvC* mutants of *Escherichia coli* K-12 depends on *recG*. *J. Bacteriol.* **173:**5414–5418.

70. Lloyd, R. G. 1983. *lexA* dependent recombination in *uvrD* strains of *Escherichia coli*. *Mol. Gen. Genet.* **189:**157–161.

71. Lloyd, R. G., and C. Buckman. 1995. Conjugational recombination in *Escherichia coli*: genetic analysis of recombinant formation in Hfr x F⁻ crosses. *Genetics* **139:**1123–1148.

72. Lloyd, R. G., and C. Buckman. 1991. Genetic analysis of the *recG* locus of *Escherichia coli* K-12 and of its role in recombination and DNA repair. *J. Bacteriol.* **173:**1004–1011.

73. Lloyd, R. G., and C. Buckman. 1985. Identification and genetic analysis of *sbcC* mutations in commonly used *recBC sbcB* strains of *Escherichia coli* K-12. *J. Bacteriol.* **164:**836–844.

74. Lloyd, R. G., and C. Buckman. 1991. Overlapping functions of *recD*, *recJ* and *recN* provide evidence of three epistatic groups of genes in *Escherichia coli* recombination and DNA repair. *Biochimie* **73:**313–320.

75. Lloyd, R. G., C. Buckman, and F. E. Benson. 1987. Genetic analysis of conjugational recombination in *Escherichia coli* K12 strains deficient in RecBCD enzyme. *J. Gen. Microbiol.* **133:**2531–2538.

76. Lloyd, R. G., N. P. Evans, and C. Buckman. 1987. Formation of recombinant *lacZ⁺* DNA in conjugational crosses with a *recB* mutant of *Escherichia coli* K12 depends on *recF*, *recJ* and *recO*. *Mol. Gen. Genet.* **209:**135–141.

77. Lloyd, R. G., M. C. Porton, and C. Buckman. 1988. Effect of *recF*, *recJ*, *recN*, *recO* and *ruv* mutations on ultraviolet survival and genetic recombination in a *recD* strain of *Escherichia coli* K12. *Mol. Gen. Genet.* **212:**317–324.

78. Lorenz, M. G., and W. Wackernagel. 1994. Bacterial gene transfer by natural genetic transformation in the environment. *Microbiol. Rev.* **58:**563–602.

79. Louarn, J.-M., J. Louarn, V. François, and J. Patte. 1991. Analysis and possible role of hyperrecombination in the termination region of the *Escherichia coli* chromosome. *J. Bacteriol.* **173:**5097–5104.

80. Lundquist, R. C., and B. M. Olivera. 1982. Transient generation of displaced single-stranded DNA during nick translation. *Cell* **31:**53–60.

81. Luria, S. E. 1947. Reactivation of irradiated bacteriophage by transfer of self-reproducing units. *Proc. Natl. Acad. Sci. USA* **33**:253–264.

82. Mahajan, S. K. 1988. Pathways of homologous recombination in *Escherichia coli*, p. 87–140. *In* R. Kucherlapati and G. R. Smith (ed.), *Genetic Recombination.* American Society for Microbiology, Washington, D.C.

83. Mahdi, A. A., and R. G. Lloyd. 1989. Identification of the *recR* locus of *Escherichia coli* K-12 and analysis of its role in recombination and DNA repair. *Mol. Gen. Genet.* **216**:503–510.

84. Malagón, F., and A. Aguilera. 1998. Genetic stability and DNA rearrangements associated with a 2 x 1.1-Kb perfect palindrome in *Escherichia coli*. *Mol. Gen. Genet.* **259**:639–644.

85. Marians, K. J. 2000. PriA-directed replication fork restart in *Escherichia coli*. *Trends Biochem. Sci.* **25**:185–189.

86. Martignoni, K. D. 1978. Inhibition of x-ray-induced protection of *Escherichia coli* K-12 cells against the lethal effects of ultra-violet light by nitrofurantoin. *Int. J. Radiat. Biol.* **33**:577–585.

87. Masters, M. 1996. Generalized transduction, p. 2421–2441. *In* F. C. Neidhardt, R. Curtiss III, J. L. Ingraham, E. C. C. Lin, K. B. Low, B. Magasanik, W. S. Reznikoff, M. Riley, M. Schaechter, and H. E. Umbarger (ed.), Escherichia coli *and* Salmonella: *Cellular and Molecular Biology*, 2nd ed., vol. 2. ASM Press, Washington, D.C.

88. McDaniel, L. S., L. H. Rogers, and W. E. Hill. 1978. Survival of recombination-deficient mutants of *Escherichia coli* during incubation with nalidixic acid. *J. Bacteriol.* **134**:1195–1198.

89. McGlynn, P., and R. G. Lloyd. 2002. Genome stability and the processing of damaged replication forks by RecG. *Trends Genet.* **18**:413–419.

90. McPartland, A., L. Green, and H. Echols. 1980. Control of *recA* gene RNA in *E. coli*: regulatory and signal genes. *Cell* **20**:731–737.

91. Meyer, R. R., and P. S. Laine. 1990. The single-stranded DNA-binding protein of *Escherichia coli*. *Microbiol. Rev.* **54**:342–380.

92. Michel, B. 2000. Replication fork arrest and DNA recombination. *Trends Biochem. Sci.* **25**:173–178.

93. Michel, B., G. D. Recchia, M. Penel-Colin, S. D. Ehrlich, and D. J. Sherratt. 2000. Resolution of Holliday junctions by RuvABC prevents dimer formation in *rep* mutants and UV-irradiated cells. *Mol. Microbiol.* **37**:180–191.

94. Miranda, A., and A. Kuzminov. 2003. Chromosomal lesion suppression and removal in *Escherichia coli* via linear DNA degradation. *Genetics* **163**:1255–1271.

95. Morgan, T. H., and E. Cattell. 1912. Data for the study of sex-linked inheritance in Drosophila. *J. Exp. Zool.* **13**:79–101.

96. Morimatsu, K., and S. C. Kowalczykowski. 2003. RecFOR proteins load RecA protein onto gapped DNA to accelerate DNA strand exchange: a universal step of recombinational repair. *Mol. Cell* **11**:1337–1347.

97. Mudgett, J. S., M. Buckholt, and W. D. Taylor. 1991. Ultraviolet light-induced plasmid-chromosome recombination in *Escherichia coli*: the role of *recB* and *recF*. *Gene* **97**:131–136.

98. Myers, R. S., and F. W. Stahl. 1994. χ and the RecBCD enzyme of *Escherichia coli*. *Annu. Rev. Genet.* **28**:49–70.

99. Otsuji, N., H. Iyehara, and Y. Hideshima. 1974. Isolation and characterization of *Escherichia coli ruv* mutant which forms nonseptate filaments after low doses of ultraviolet light irradiation. *J. Bacteriol.* **117**:337–344.

100. Picksley, S. M., P. V. Attfield, and R. G. Lloyd. 1984. Repair of DNA double-strand breaks in *Escherichia coli* K12 requires a functional *recN* product. *Mol. Gen. Genet.* **195**:267–274.

101. Pinder, D. J., C. E. Blake, J. C. Lindsey, and D. R. F. Leach. 1998. Replication strand preference for deletions associated with DNA palindromes. *Mol. Microbiol.* **28**:719–727.

102. Pollard, E. C., and J. K. J. Fugate. 1978. Relative rates of repair of single-strand breaks and postirradiation DNA degradation in normal and induced cells of *Escherichia coli*. *Biophys. J.* **24**:429–437.

103. Radman, M. 1975. SOS repair hypothesis: phenomenology of an inducible DNA repair which is accompanied by mutagenesis, p. 355–367. *In* P. C. Hanawalt and R. B. Setlow (ed.), *Molecular Mechanisms for Repair of DNA*, vol. A. Plenum Press, New York, N.Y.

104. Roberts, J. J. 1978. The repair of DNA modified by cytotoxic, mutagenic, and carcinogenic chemicals. *Adv. Radiat. Biol.* **7**:211–436.

105. Roca, A. I., and M. M. Cox. 1997. RecA protein: structure, function, and role in recombinational DNA repair. *Prog. Nucleic Acid Res. Mol. Biol.* **56**:129–223.

106. Ross, P., and P. Howard-Flanders. 1977. Initiation of $recA^+$-dependent recombination in *Escherichia coli* (λ). II. Specificity in the induction of recombination and strand cutting in undamaged covalent circular bacteriophage 186 and lambda DNA molecules in phage-infected cells. *J. Mol. Biol.* **117**:159–174.

107. Rothman, R. H., T. Kato, and A. J. Clark. 1975. The beginning of an investigation of the role of *recF* in the pathways of metabolism of ultraviolet-irradiated DNA in *Escherichia coli*, p. 283–291. *In* P. C. Hanawalt and R. B. Setlow (ed.), *Molecular Mechanisms for Repair of DNA*, vol. A. Plenum Press, New York, N.Y.

108. Rupp, W. D., and P. Howard-Flanders. 1968. Discontinuities in the DNA synthesized in an excision-defective strain of *Escherichia coli* following ultraviolet irradiation. *J. Mol. Biol.* **31**:291–304.

109. Rupp, W. D., C. E. Wilde III, D. L. Reno, and P. Howard-Flanders. 1971. Exchanges between DNA strands in ultraviolet-irradiated *Escherichia coli*. *J. Mol. Biol.* **61**:25–44.

110. Salaj-Smic, E., S. Dzidic, and Z. Trgovcevic. 1985. The effect of a split UV dose on survival, division delay and mutagenesis in *Escherichia coli*. *Mutat. Res.* **144**:127–130.

111. Sandler, S. J., and K. J. Marians. 2000. Role of PriA in replication fork reactivation in *Escherichia coli*. *J. Bacteriol.* **182**:9–13.

112. Sandler, S. J., H. S. Samra, and A. J. Clark. 1996. Differential suppression of *priA2::kan* phenotypes in *Escherichia coli* K-12 by mutations in *priA*, *lexA*, and *dnaC*. *Genetics* **143**:5–13.

113. Sargentini, N. J., and K. C. Smith. 1986. Quantitation of the involvement of the *recA*, *recB*, *recC*, *recF*, *recJ*, *recN*, *lexA*, *radA*, *radB*, *uvrD*, and *umuC* genes in the repair of X-ray-induced DNA double-strand breaks in *Escherichia coli*. *Radiat. Res.* **107**:58–72.

114. Seigneur, M., V. Bidnenko, S. D. Ehrlich, and B. Michel. 1998. RuvAB acts at arrested replication forks. *Cell* **95**:419–430.

115. Shen, P., and H. V. Huang. 1986. Homologous recombination in *Escherichia coli*: dependence on substrate length and homology. *Genetics* **112**:441–457.

116. Sherratt, D. J., L. K. Arciszewska, G. Blakely, S. Colloms, K. Grant, N. Leslie, and R. McCulloch. 1995. Site-specific recombination and circular chromosome segregation. *Phil. Trans. R. Soc. Lond.* **347**:37–42.

117. **Singer, B. S., L. Gold, P. Gauss, and D. H. Doherty.** 1982. Determination of the amount of homology required for recombination in bacteriophage T4. *Cell* **31:**25–33.

118. **Smith, K. C., and D. H. C. Meun.** 1970. Repair of radiation-induced damage in *Escherichia coli*. I. Effect of *rec* mutations on post-replication repair of damage due to ultraviolet radiation. *J. Mol. Biol.* **51:**459–472.

119. **Sogo, J. M., M. Lopes, and M. Foiani.** 2002. Fork reversal and ssDNA accumulation at stalled replication forks owing to checkpoint defects. *Science* **297:**599–602.

120. **Spies, M., P. R. Bianco, M. S. Dillingham, N. Handa, R. J. Baskin, and S. C. Kowalczykowski.** 2003. A molecular throttle: the recombination hotspot χ controls DNA translocation by the RecBCD helicase. *Cell* **114:**647–654.

121. **Stahl, F. W., and M. M. Stahl.** 1977. Recombination pathway specificity of Chi. *Genetics* **86:**715–725.

122. **Stallions, D. R., and R. Curtiss III.** 1971. Chromosome transfer and recombinant formation with deoxyribonucleic acid temperature-sensitive strains of *Escherichia coli*. *J. Bacteriol.* **105:**886–895.

123. **Steiner, W. W., and P. L. Kuempel.** 1998. Cell division is required for resolution of dimer chromosomes at the *dif* locus of *Escherichia coli*. *Mol. Microbiol.* **27:**257–268.

124. **Steiner, W. W., and P. L. Kuempel.** 1998. Sister chromatid exchange frequencies in *Escherichia coli* analyzed by recombination at the *dif* resolvase site. *J. Bacteriol.* **180:**6269–6275.

125. **Stewart, G. J., and C. A. Carlson.** 1986. The biology of natural transformation. *Annu. Rev. Microbiol.* **40:**211–235.

126. **Strauss, B. S.** 1972. The relationship of repair mechanisms to the induction of chromosome aberrations in eukaryotic cells, p. 151–171. *In* H. Altmann (ed.), *DNA-Repair Mechanisms.* F.K. Schattauer, Stuttgart, Germany.

127. **Strumberg, D., A. A. Pilon, M. Smith, R. Hickey, L. Malkas, and Y. Pommier.** 2000. Conversion of topoisomerase I cleavage complexes on the leading strand of ribosomal DNA into 5′-phosphorylated DNA double-strand breaks by replication runoff. *Mol. Cell. Biol.* **20:**3977–3987.

128. **Taylor, A. F.** 1988. RecBCD enzyme of *Escherichia coli*, p. 231–263. *In* R. Kucherlapati and G. R. Smith (ed.), *Genetic Recombination.* American Society for Microbiology, Washington, D.C.

129. **Taylor, A. F., and G. R. Smith.** 2003. RecBCD enzyme is a DNA helicase with fast and slow motors of opposite polarity. *Nature* **423:**889–893.

130. **Templin, A., S. R. Kushner, and A. J. Clark.** 1972. Genetic analysis of mutations indirectly suppressing *recB* and *recC* mutations. *Genetics* **72:**205–215.

131. **Tseng, Y.-C., J.-L. Hung, and T.-C. V. Wang.** 1994. Involvement of the RecF pathway recombination genes in postreplication repair in UV-irradiated *Escherichia coli* cells. *Mutat. Res.* **315:**1–9.

132. **Uzest, M., S. D. Ehrlich, and B. Michel.** 1995. Lethality of *rep recB* and *rep recC* double mutants of *Escherichia coli*. *Mol. Microbiol.* **17:**1177–1188.

133. **West, S. C.** 1997. Processing of recombination intermediates by the RuvABC proteins. *Annu. Rev. Genet.* **31:**213–244.

134. **Willetts, N. S., and A. J. Clark.** 1969. Characteristics of some multiply recombination-deficient strains of *Escherichia coli*. *J. Bacteriol.* **100:**231–239.

135. **Xu, L., and K. J. Marians.** 2003. PriA mediates DNA replication pathway choice at recombination intermediates. *Mol. Cell* **11:**817–826.

136. **Yarmolinsky, M. B., and E. Stevens.** 1983. Replication-control functions block the induction of an SOS response by a damaged P1 bacteriophage. *Mol. Gen. Genet.* **192:**140–148.

137. **Zieg, J., V. F. Maples, and S. R. Kushner.** 1978. Recombination levels of *Escherichia coli* K-12 mutants deficient in various replication, recombination, or repair genes. *J. Bacteriol.* **134:**958–966.

138. **Zinder, N. D., and J. Lederberg.** 1952. Genetic exchange in *Salmonella*. *J. Bacteriol.* **64:**679–699.

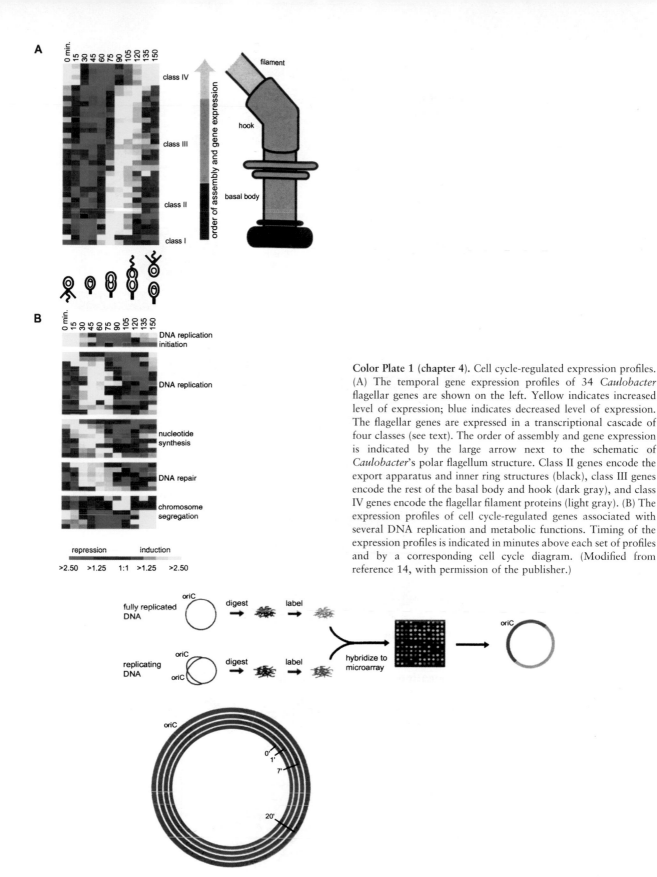

Color Plate 1 (chapter 4). Cell cycle-regulated expression profiles. (A) The temporal gene expression profiles of 34 *Caulobacter* flagellar genes are shown on the left. Yellow indicates increased level of expression; blue indicates decreased level of expression. The flagellar genes are expressed in a transcriptional cascade of four classes (see text). The order of assembly and gene expression is indicated by the large arrow next to the schematic of *Caulobacter*'s polar flagellum structure. Class II genes encode the export apparatus and inner ring structures (black), class III genes encode the rest of the basal body and hook (dark gray), and class IV genes encode the flagellar filament proteins (light gray). (B) The expression profiles of cell cycle-regulated genes associated with several DNA replication and metabolic functions. Timing of the expression profiles is indicated in minutes above each set of profiles and by a corresponding cell cycle diagram. (Modified from reference 14, with permission of the publisher.)

Color Plate 2 (chapter 4). Tracking DNA replication fork progression in *E. coli* using microarrays. Replication was synchronized using a temperature-sensitive *dnaC* initiation mutant strain. Shifting this strain to the restrictive temperature for 90 min allowed completion of replication, but with no additional initiations. Shifting this strain back to the permissive temperature then produced synchronized initiations. Samples of partially replicated DNA and fully replicated DNA were collected, labeled, and competitively hybridized on a whole-genome microarray (Fig. 1). Spots for genes that have been replicated are expected to have fluorescence ratios near 2, compared with a ratio of 1 for unreplicated genes. This information could be mapped back to the *E. coli* chromosome, drawn as a gray-and-red circle, with red indicating genes that have been replicated. The limits of the red area thus indicate the replication fork location. The concentric circles at the bottom show how executing this assay at different time points—0, 1, 7, and 20 min—allowed measurement of replication fork progression. (Modified from reference 13, with permission of the publisher.)

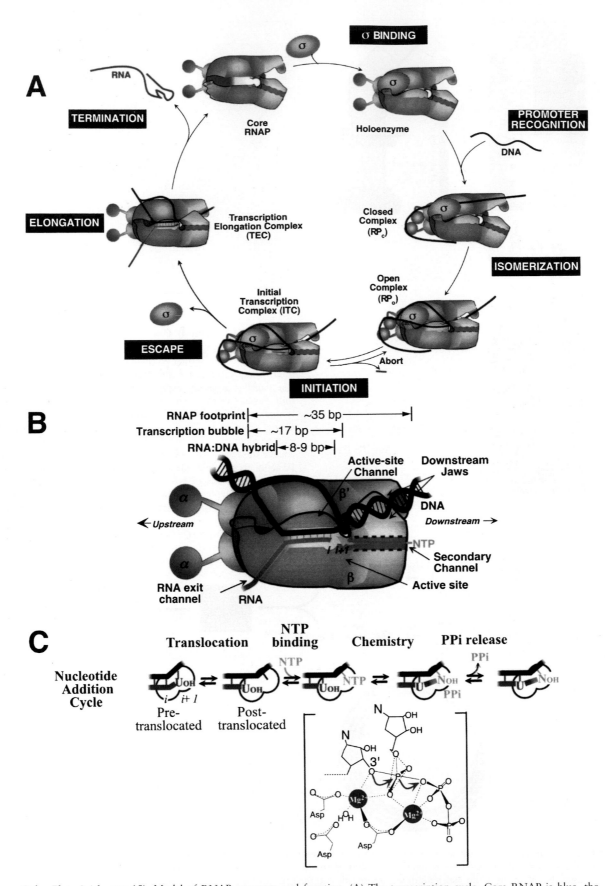

Color Plate 3 (chapter 15). Model of RNAP structure and function. (A) The transcription cycle. Core RNAP is blue, the initiation factor σ is orange, the DNA template is black, and the nascent RNA transcript is red. (B) Model of the TEC. The incoming NTP and an arrow showing its path into the TEC through the secondary channel are green. (C) The nucleotide addition cycle. The *i* and *i* + *1* sites are shown as two joined circles. (Inset) Two Mg^{2+} ions (red circles) catalyze phosphodiester bond formation between the 3′ end of the nascent transcript and an incoming NTP.

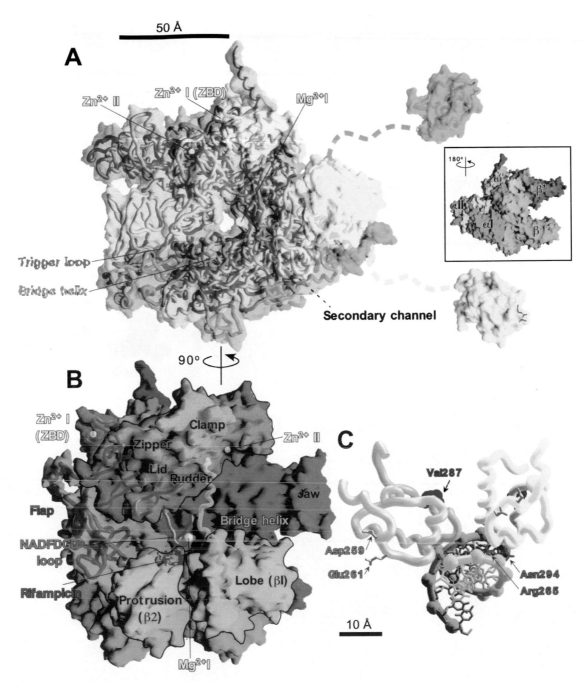

Color Plate 4 (chapter 15). Structure of core RNAP. (A) The downstream face of core RNAP. The model is based on the coordinates of the *Thermus thermophilus* holoenzyme (PDB [Protein Data Bank] ID 1IW7 [97]), with the σ subunit and a nonconserved region not present in *Escherichia coli* β′ (amino acids 164 to 448) removed and the RNAP conformation adjusted to that observed in the core *Thermus aquaticus* RNAP (PDB ID 1I6V [109]) by movement of RNAP mobile modules (23). Sequence insertions present in *E. coli* (23) are not depicted. The path of the secondary channel is illustrated by a dashed line. The α-carbon backbone is shown as a worm inside a semitransparent surface. Subunits are color-coded as follows: β′, pink; β, cyan; αI, green; αII, yellow; ω, gray. The β′ bridge helix and trigger loops are depicted as green and orange worms, respectively. Zn^{2+} and Mg^{2+} atoms are depicted as yellow balls. The αCTDs are shown in arbitrary positions 43 Å from the core. They are shown as isolated domains but may be present as a dimer in RNAP (7). The boxed inset depicts the upstream face of core RNAP, illustrating the "crab-claw" shape of the enzyme. (B) The active-site channel. The RNAP model in panel A is shown rotated 90° to the right. The subunits are shown as solid surfaces except that the ZBD, β′ Mg^{2+}-binding loop, rudder, lid, and zipper are shown as pink worms in a semitransparent surface, and the β flap domain is shown as a dark blue worm in a semitransparent surface. The β′ bridge helix is depicted as a green worm. The antibiotic rifampin (labeled Rifampicin in the figure) is depicted in red (β is rendered semitransparent in front of rifampin to reveal the antibiotic nestled in its binding pocket.) The clamp, protrusion, and lobe are outlined in black. (C) Two αCTDs bound to UP element DNA. This model is based on the crystal structure of αCTD, DNA, and catabolite activator protein (PDB ID 1LB2 [6]). Nontemplate DNA is light green; template DNA is dark green. Two residues involved in recognition of the UP element (Arg-265 and Asn-294) are shown as red sticks. The CAP interaction determinant is indicated by the blue space-fill valine residue at position 287. Asp-259 and Glu-261, two residues that interact with σ region 4, are shown as orange sticks. Only one of each symmetric pair of residues is labeled.

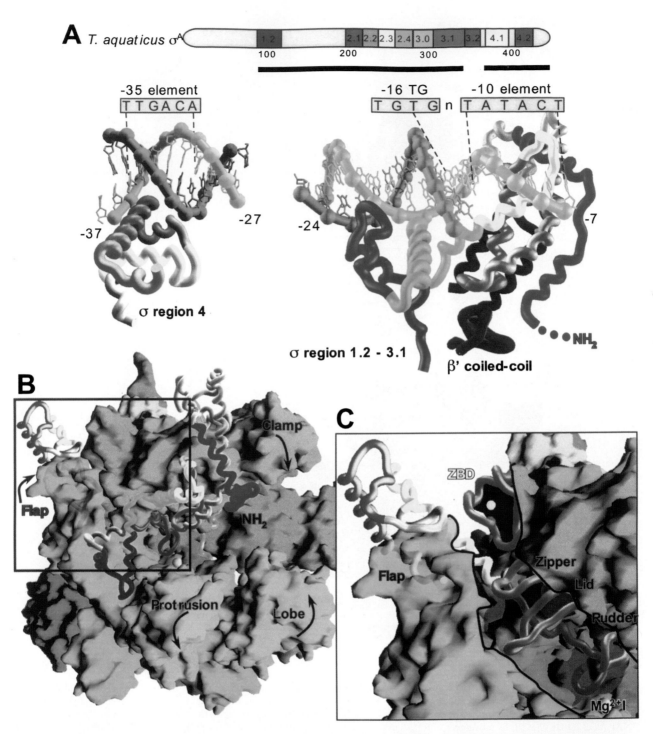

Color Plate 5 (chapter 15). Structure of σ and holoenzyme. (A) The structure of the σ subunit. *T. aquaticus* σ^A is depicted as a gray rod with the regions of sequence conservation (1.2 to 4.2) in different colored boxes. Black bars beneath σ represent the segments of σ crystallized. The σ region 4-DNA cocrystal is shown (PDB ID 1KU7 [19]). The σ region 1.2 to 3.1 crystallized fragment (19) is modeled with partially single-stranded promoter DNA and the coiled-coil domain from β' with which it interacts in holoenzyme (shown in dark gray), based on the fork junction RNAP structure (PDB ID 19LZ [65]). DNA is colored as in Color Plate 4. The regions of σ are colored as depicted in the schematic shown above the structures. The sequence of promoter elements is shown in gray boxes above the DNA, and selected bases are indicated with dotted lines. An arbitrary sequence has been modeled into the DNA of the region 1.2 to 3.1 fragment, rather than the promoter sequence. **(B)** A model of the structure of holoenzyme. The model is based on the *T. thermophilus* holoenzyme crystal structure (PDB ID 1IW7 [97]) with a missing segment of σ region 4 modeled based on the *T. aquaticus* holoenzyme structure (PDB ID 1L9U [66]). The view of RNAP is similar to that in Color Plate 4B. Arrows indicate the movement of the clamp, flap, lobe, and protrusion domains in holoenzyme relative to core RNAP (66). Core subunits are colored as in Color Plate 4. σ is colored as in panel A. The box indicates the area magnified in panel C. **(C)** The path of the σ 3.2 linker. Parts of the surface from β' and β and σ 2.3 to 3.1 have been cut away to make the path of the linker through the RNA-exit channel visible.

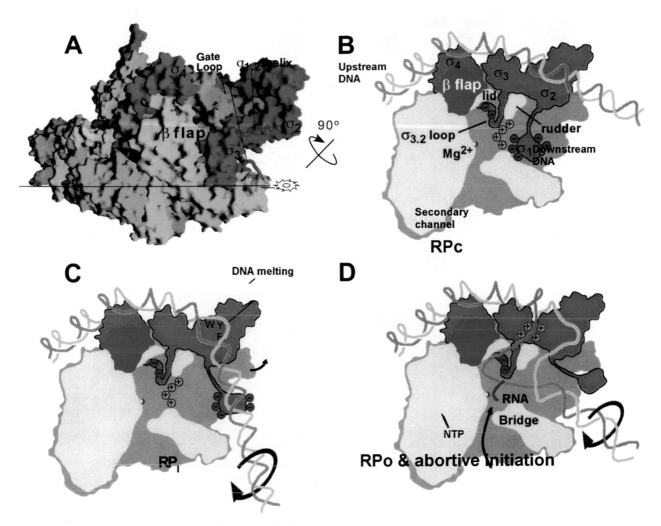

Color Plate 6 (chapter 15). A model for RP_o formation. (A) The holoenzyme depicted as in Color Plate 5B, rotated to show the upstream face (the same as in the inset to Color Plate 4A) and with σ shown as an orange surface. The gate loop is shown as a blue worm, and σ 1.2 helix is red (it lies behind σ_2 in this view). The line across the RNAP depicts the plane at which RNAP would be cut to generate the view shown in panels B through D. (B) A cutaway view of RP_c. The core is gray, σ is orange, and the flap is blue. The DNA is dark and light green and lies along the surface of the upstream face of RNAP. The negatively charged σ region 1.1 lies in the positively charged downstream DNA channel. The β lobe and gate loop would lie above the plane of the page, protruding into the downstream DNA channel above σ region 1.2. (C) RP_i, a possible intermediate in RP_o formation. A kink forms in the DNA within the −10 element. Aromatic residues in σ region 2.3 interact with the nontemplate strand to assist DNA opening. The DNA moves into the downstream DNA channel, replacing region 1.1. The β gate loop prevents entry of double-stranded DNA into the active site and may assist in unwinding the DNA. (D) RP_o. The downstream DNA is inserted in the downstream DNA channel, and melting has extended from the −10 element to the transcription start site. The template strand is guided into the active-site channel by positive residues in σ regions 2.4 and 3.0. NTPs entering through the secondary channel can be incorporated into a nascent transcript (red), extension of which is blocked by the σ 3.2 loop, resulting in abortive initiation. Adapted from Fig. 3 of reference 64, with permission.

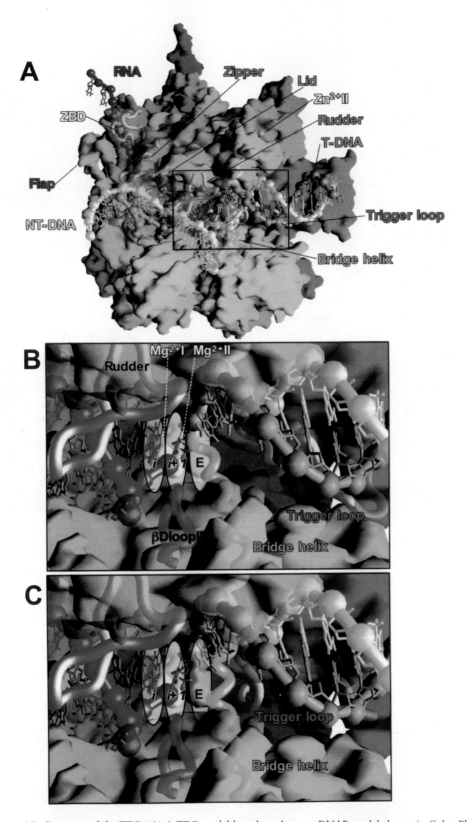

Color Plate 7 (chapter 15). Structure of the TEC. (A) A TEC model based on the core RNAP model shown in Color Plate 4 with mobile modules (23) adjusted to the conformation of the *Saccharomyces cerevisiae* RNA II TEC (PDB ID 1I6H [34]). The RNA-DNA hybrid and downstream DNA positions are those observed in the *S. cerevisiae* RNAP II TEC (34) with the scaffold dimensions and upstream DNA as modeled by Korzheva et al. (45). Subunits and DNA are colored as in Color Plate 4. RNA is red. Active-site Mg^{2+} ions are yellow. The trigger loop is depicted as an orange worm. The βD loop II is shown as a dark blue worm. The box encloses the portion of the active-site channel magnified in panels B and C. (B) The conformation of the active site in *T. thermophilus* holoenzyme (97). A portion of the nontemplate strand of the DNA has been removed to allow a clearer view of the active center. Ovals represent the *i, i + 1*, and E sites. The RNA 3′ end is depicted in the *i + 1* site; however, the kinked bridge helix in this structure would sterically clash with the base in the *i + 1* site. (C) The conformation of the active site in the yeast RNAP II TEC (34). The view is the same as that in panel B. In this conformation, the bridge helix is straight and there is no steric clash between the bases in the *i + 1* site and residues from the helix.

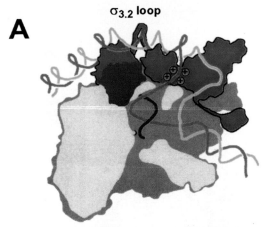

A

σ₃.₂ loop

end of abortive initiation

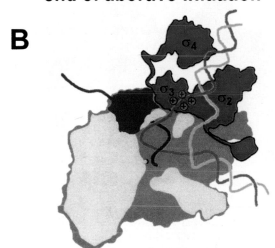

B

σ₄

σ₃

σ₂

Promoter clearance

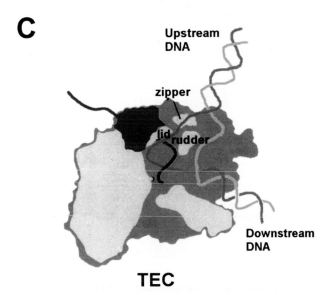

C

Upstream
DNA

zipper

lid rudder

Downstream
DNA

TEC

Color Plate 8 (chapter 15). Postinitiation events in transcription. The color code and orientation are the same as in Color Plate 6. (A) The end of abortive initiation. Once the RNA transcript exceeds 8 nt, it has fully displaced the σ 3.2 linker from the RNA-exit channel. This weakens the interaction between σ and the promoter, allowing promoter clearance. (B) Promoter clearance. Removal of the 3.2 linker from the RNA-exit channel destabilizes the interaction between the flap and region 4, and consequently the interaction with the −35 element. (C) The TEC. After promoter clearance, the remaining contact between σ and the core is weakened, and σ is stochastically lost from the complex. Adapted from Fig. 3 of reference 64, with permission.

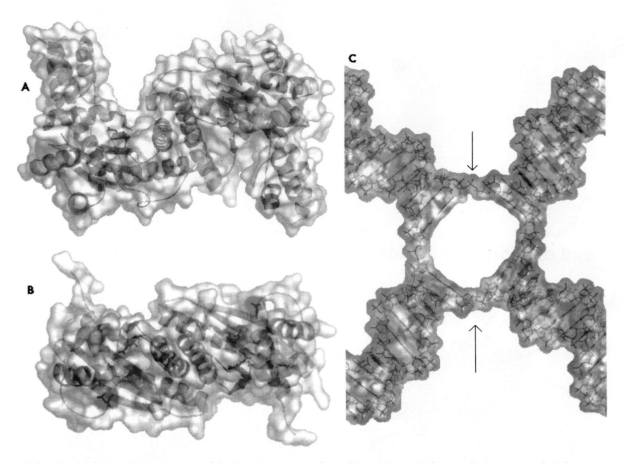

Color Plate 9 (chapter 22). Comparison of the dimeric structure of *E. coli* RuvC (B) and *Schizosaccharomyces pombe* Ydc2 (A), showing a solvent-accessible surface and secondary structure cartoon. The four conserved acidic residues that form the catalytic metal-binding pockets in each enzyme are highlighted in red. (C) Structure of the Holliday junction. The junction is presented in the open, fourfold symmetric form seen in the absence of metal ions. This corresponds closely to the global conformation of the junction when bound by RuvC, RuvA, or Cce1/Ydc2, and the positions of the paired nicks in the phosphodiester backbone introduced by these enzymes are indicated by arrows. Some distortion of the center of the junction from the idealized B-form DNA represented is expected when it is complexed with these proteins.

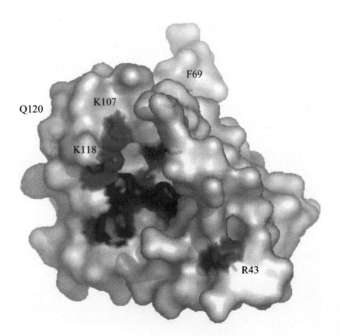

Color Plate 10 (chapter 22). A monomer of RuvC is shown, with conserved residues highlighted and labeled. The four acidic residues (pink) constitute the metal-binding site of the enzyme. Lys-107 and Lys-118 (blue) may play a role in stabilization of the transition state. The possible functions of the other residues are discussed in the text.

The Bacterial Chromosome
Edited by N. Patrick Higgins
© 2005 ASM Press, Washington, D.C.

Chapter 20

The RecA Protein

Michael M. Cox

The RecA protein of *Escherichia coli* has established the paradigm for a nearly universal class of proteins that facilitate some of the central steps in recombination. After the discovery of the *recA* gene in 1965 by Clark and Margulies (26), *recA* mutants were found to have a complex pleiotropic phenotype that has fascinated and challenged researchers ever since. Many aspects of RecA action are still not understood despite over 3 decades of intensive research.

The heart of RecA function, and the one activity shared by all proteins in this class, is DNA strand exchange. In the simplest and most common version of these reactions, RecA protein and its homologues bind to single-stranded DNA, forming a helical nucleoprotein filament. A homologous duplex DNA is then taken up, and one strand of the duplex is transferred from the duplex to the complementary single strand originally bound. The other strand of the duplex, the one identical to the strand originally bound, is displaced. This general reaction plays a central role in recombinational DNA repair, recombination during conjugation and transduction in bacteria, meiotic recombination in eukaryotes, and many other processes.

The RecA protein is found in essentially all bacteria (17, 178). The only apparent exceptions involve bacteria (*Buchnera* sp.) that have a greatly reduced genome size, possibly arising from their endosymbiotic relationship with aphids (142, 227). Bacterial RecA proteins are highly homologous to each other, having similar sequences, sizes, and domain structures (17, 178). Every bacterial RecA protein in a survey of 64 examples had at least a 49% sequence identity to the *E. coli* RecA protein (178). In eukaryotes, the important RecA homologues are the Rad51 and Dmc1 proteins (9, 16, 71, 123, 196, 206). In archaea, the RadA proteins provide the same recombination functions (186). This chapter focuses on the bacterial RecA proteins, which have at least three major roles.

The first function involves a direct participation in the central steps of recombination, via the DNA strand exchange activity already described. These reactions have a classical role in the recombination that occurs during conjugation, but current evidence indicates that RecA did not evolve to promote the generation of genetic diversity. In bacteria, the most important role of recombination proteins, including RecA, is in the nonmutagenic repair of stalled replication forks. Under normal aerobic growth conditions, a large fraction of bacterial replication forks, perhaps most of them, are halted by encounters with DNA damage or other barriers (30, 31, 33, 36, 89, 102, 103). Recombination represents the primary path to repair of these events. An encounter with a strand break can create a double-strand break in which one branch of the fork is detached (Fig. 1A). An encounter with an unrepaired DNA lesion can result in a stalled fork in which the lesion is left in a single-strand gap (Fig. 1B). There are probably numerous other structures that can result from a collapsed fork. Classical paths of DNA repair, such as excision repair, rely on the presence of an undamaged DNA strand to repair against, something which does not exist in either of the situations in Fig. 1. Recombination provides a nonmutagenic path to repair of these structures (30, 31, 33, 36, 89, 102, 103), and the RecA protein plays a central role in these processes. From this standpoint, the function of RecA and other recombination proteins in recombination during conjugation and transduction is quite incidental to bacterial DNA metabolism.

Second, RecA protein has a role in regulation. As a regulatory function, the RecA protein facilitates the autocatalytic cleavage of the LexA repressor and certain other proteins to induce the SOS response to DNA damage (113, 114, 154, 194). The LexA repressor regulates the expression of over 40 genes encoding enzymes and proteins involved in DNA repair and control of cell division. A current view of

Michael M. Cox • Department of Biochemistry, University of Wisconsin—Madison, 433 Babcock Dr., Madison, WI 53706-1544.

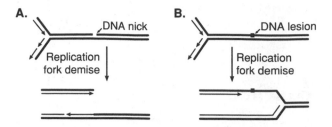

Figure 1. Replication fork demise at the sites of DNA damage. (A) Encounter with a strand break leads to a double-strand break and dissociation of one arm of the fork. (B) Encounter with an unrepaired DNA lesion can result in the creation of a gap at a stalled fork. In either case, repair pathways involve recombination.

SOS induction can be summarized briefly. When the cell is subjected to high levels of DNA damage, single-strand gaps form in the DNA, leading to the binding and activation of RecA protein to the DNA. In this bound form, RecA filaments interact with the LexA repressor, leading to the autocatalytic cleavage and inactivation. This activity of RecA is often called a coprotease function. As the levels of active LexA repressor drop, the SOS genes are induced.

Finally, the RecA protein participates in yet another type of repair process. Late in the SOS response, especially when DNA damage levels are particularly high and nonmutagenic DNA repair is insufficient to get replication restarted, a need arises to restart replication via lesion bypass. This is something of a desperation measure for the cell, since it inevitably leads to mutation. The RecA protein participates directly in the replicational bypass of DNA lesions (48, 61, 209). Mutagenic lesion bypass is mediated by a specialized class of DNA polymerases, represented in *E. coli* by DNA polymerases IV and V. The RecA protein participates directly in the lesion bypass reactions promoted by DNA polymerase V by a mechanism that is still under intensive investigation.

The known biochemical activities of the RecA protein parallel these cellular roles. These include binding to DNA, ATP hydrolysis, filament formation, DNA strand exchange, and the coprotease activity. In this chapter, I try to provide a fairly thorough and current overview. A series of recent reviews can be consulted for details (15, 32, 92, 178). Each of these activities is replete with complexity that has made the protein a focus of over 2 decades of research. Several themes wind through the continuing discussion. RecA protein functions as a filament in all of its known activities, and this filament provides an excellent platform for the central process of aligning homologous DNA sequences and switching DNA strands. The hydrolysis of ATP confers properties on RecA protein that suggest the presence of a motor

function. Also, the RecA homologues found in other classes of organisms do not have all of the RecA functions, and even the shared functions can be very different in their details. This is making it increasingly clear that each of these proteins is highly adapted to a very different cellular milieu. The Rad51, Dmc1, and RadA proteins will be compared and contrasted with RecA protein wherever such comparisons may provide some insight.

STRUCTURE

The RecA protein of *E. coli* is a polypeptide chain with 352 amino acid residues with a combined molecular weight of 37,842 (37,711 after the initiating methionine is removed). This is quite average for a bacterial RecA protein. Molecular weights of other bacterial RecAs are similar, ranging from 34,344 (318 amino acid residues including the initiating methionine) for the RecA of *Bacteroides fragilis* (67) to 41,949 (388 amino acid residues including the initiating Met) for the RecA of *Streptococcus pneumoniae* (122). Bacterial strains engineered to produce the RecA protein at very high levels are widely available. Purification is relatively simple (28, 39, 69, 99, 192), and good protocols have been developed for the purification of many different mutant RecA proteins as well as the RecA proteins from a variety of bacterial species. The bacterial RecA proteins are generally quite soluble in aqueous buffers, and are stable for many months if stored at −70°C. The pI of the purified *E. coli* RecA protein has been measured in at least six studies, giving values from 5.0 (198) to 6.2 (70). Averaging all the measurements (21, 65, 70, 94, 156, 198) gives a representative pI of 5.6. In solution, the protein has a tendency to aggregate into oligomers, filaments, and bundles of filaments (19, 20, 94, 164, 210, 230). The aggregation state of a given solution of the protein is affected by pH, salt concentration, and temperature. Concentrated solutions of the protein often appear opalescent as a result of light scattering by filamentous structures.

The structure of RecA protein without DNA bound has been determined at 2.3-Å resolution (204, 205). The crystal packing features a continuous spiral filament, with six monomers per right-hand helical turn (Fig. 2). These results are consistent with other studies showing that RecA protein forms a helical nucleoprotein filament on DNA, also with six monomers per helical turn. In the electron microscope, the contiguous filaments have a striated appearance (Fig. 2). The filament exhibits a deep helical groove, where DNA binding and DNA pairing take place. The structure was solved in two forms, one with

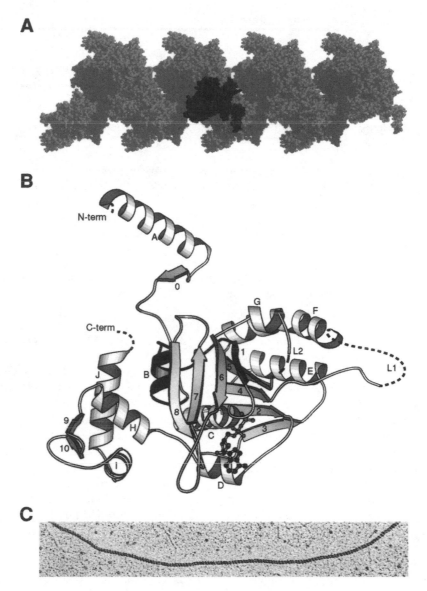

Figure 2. RecA protein structure. (A) A RecA filament is shown, with 24 RecA monomers, based on the 1992 structure by Story and Steitz (204) (see text). One monomer is colored in a darker gray. (B) A RecA monomer, with bound ADP. (C) An electron micrograph of one segment of a RecA filament formed on DNA.

bound ADP (204) and the other with no nucleotide (205). No crystal form with RecA bound to DNA has been reported. The filament seen in the crystal is not as extended as that observed on DNA in the presence of ATP, and hence the structure may reflect an inactive form of the protein (53, 54, 59, 157, 237).

In each monomer, there is a central core domain flanked by two smaller subdomains at the N and C termini. Many parts of the core domain are highly conserved among the bacterial RecAs, the eukaryotic Rad51 and Dmc1 proteins, and the archaeal RadA protein (17, 178). In contrast, the C-terminal subdomain of RecA is unique in this class of proteins (Fig. 3). Conversely, the N-terminal subdomains of

Rad51, Dmc1, and RadA are not found in the bacterial RecA proteins (Fig. 3). The 20 to 30 amino acids at the C terminus of most bacterial RecA proteins, including that of *E. coli*, feature a high concentration of acidic residues. This segment of the protein is otherwise among the most variable regions of the bacterial *recA* sequences (178). The N-terminal domains of Rad51, Dmc1, and RadA proteins also feature stretches of acidic amino acid residues, suggesting that these subdomains might have the same (currently unknown) function as the RecA C-terminal subdomain (17, 178).

Many sequences of *recA* genes and the genes of *recA* homologues are available. This has led to

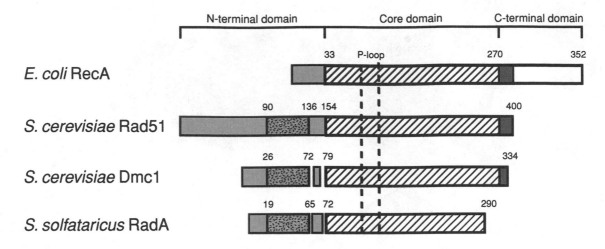

Figure 3. Domain structures of RecA protein and its homologues in *Saccharomyces cerevisiae* and *Sulfolobus solfataricus*. The open box is the core domain shared by all four proteins. The RecA C-terminal domain is unique to bacterial RecA proteins. Homologies among other domain elements are indicated by shading patterns.

extensive sequence alignment efforts and detailed structural analyses (17, 82, 177, 178). The RecA proteins from bacteria are all closely related, with none having less than a 49% amino acid sequence identity to the *E. coli* RecA protein. In a comparison of 64 bacterial *recA* genes (178), there were 59 invariant residues and another 100 that could be classified as chemically conserved. The protein has a binding site for ATP, and less well defined binding sites for up to three strands of DNA within the helical groove. Many of the conserved residues are involved in the binding of DNA or ATP, or are at the monomer-monomer filament interface. In general, the Rad51, Dmc1, and RadA sequences are more similar to each other than any of them is to a bacterial RecA sequence (17, 178).

The RecA protein of certain mycobacteria is expressed as a larger precursor containing a large, self-splicing intein (64, 98, 140, 181, 195). The intein is found in about 20% of mycobacterial species. While initially thought to be restricted to pathogenic species, such as *Mycobacterium tuberculosis* and *Mycobacterium leprae*, evidence now suggests a more general invasion of the mycobacteria with the inteins (181). The inteins are comparable in size to the RecA proteins themselves (98). The inteins are not spliced when the mycobacterial genes are cloned into *E. coli*, but are spliced when cloned into mycobacterial species that otherwise lack the *recA* intein (64).

Most of the central core domain of RecA protein, from residues 61 to 246, exhibits an intriguing structural similarity to the F1-ATPase, even though the two proteins share very little sequence similarity (239). The same structural folding pattern is seen in hexameric helicases and a number of other proteins, most involved in nucleic acid metabolism (4, 14, 51, 239).

DNA BINDING

RecA protein binding to DNA is a multistep process. The first step is nucleation, involving the binding of a monomer or small oligomer. Additional protomers are added in an extension phase to create and lengthen a filament. RecA protein filaments both assemble and disassemble in a unidirectional and end-dependent manner, with protomers added at one end and subtracted from the other (Fig. 4) (110, 187). This is true on both single-stranded DNA (ssDNA) and double-stranded DNA (dsDNA). The oligomeric species that is added and subtracted has not been defined for *E. coli* RecA protein. However, work with the RecA protein from *Thermus thermophilus* indicates that monomeric RecA protein is an intermediate in filament assembly on ssDNA (124).

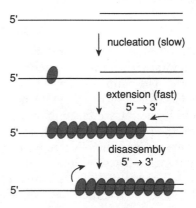

Figure 4. Assembly and disassembly pathways for RecA filaments. After nucleation, filament extension proceeds 5' to 3' relative to the ssDNA, and can encompass adjacent duplex regions. RecA dissociation occurs from the opposite filament end, and proceeds also 5' to 3'.

Without nucleotide cofactor or with ADP, RecA protein binds to ssDNA in a compact form called a "collapsed" filament, with a helical pitch of 64 Å and an axial rise of 2.1 Å per nucleotide (74, 202). An extended right-hand helical filament is formed when ATP or dATP is added. Alternatively, the ATP analogue ATPγS [adenosine 5′-O-(3-thio)triphosphate] or ADP·AlF$_4^-$ can support extended filament formation. Both analogues are bound by RecA protein, but not readily hydrolyzed. The extended filament has 6 to 6.1 RecA monomers per turn, a pitch of 95 Å, and a diameter of 100 Å (45, 55, 155, 240). The single strand of DNA binds deep within the filament groove and near the filament axis (57). The DNA is bound along the ribose-phosphate backbone, with the DNA bases displayed in the filament groove (46, 47, 105). The same extended filament can be observed when RecA is bound to dsDNA. In this case the DNA is underwound to approximately 18 bp per turn, removing either 43% (in the presence of ATPγS [201]) or 39.6% (in the presence of ATP [166]) of the helical turns in the DNA.

The structure of RecA-bound DNA has been studied by nuclear magnetic resonance. The experiments revealed a novel deoxyribose-base stacking in which the 2′-methylene moiety of each deoxyribose was positioned above the base of the neighboring residue (152, 153). Adjacent bases were not stacked as in DNA in solution. The deoxyribose-base stacking led to a base spacing of about 5 Å, as expected from other measurements. This structure has been proposed to facilitate base flipping (152).

The nucleation of RecA protein on ssDNA is slowed considerably if the DNA is bound by the *E. coli* single-strand DNA-binding protein SSB (120, 187). This inhibition is overcome in at least some contexts by other proteins. The RecO and RecR proteins form a complex that greatly facilitates RecA nucleation on SSB-bound ssDNA (219). Also, the RecBCD protein prepares duplex DNAs for RecA binding by trimming the 5′-ending strand more than the 3′-ending strand to generate a 3′ single-strand extension. The RecBCD protein also directly loads RecA protein onto the prepared DNA substrate (6, 25).

RecA binding to ssDNA is generally nonspecific with respect to sequence, although the protein exhibits an enhanced affinity to certain homopolymers such as poly(dT) (5, 22, 128). RecA also exhibits a higher affinity for a range of GT-enriched sequences, among them the recombination hotspot chi (216, 217). Optimal binding of wild-type RecA protein to poly(dT) generally occurs only with polymers more than 50 nucleotides in length (18), although somewhat shorter polymers can be employed in concert with the addition of ATPγS.

RecA protein binding to ssDNA is cooperative (92, 134). Cooperativity is also manifested in the ATP hydrolytic activity, as the binding of ATP or ATPγS to individual filament subunits can stimulate nucleoside triphosphate (NTP) hydrolysis in neighboring subunits (106, 136). Binding to ATP induces a conformation change in which the RecA protein is activated and bound DNA is extended and (if duplex) underwound. Monomer-monomer interactions help to maintain the entire filament in an extended active state.

In the assembly process, nucleation of filament formation is rate-limiting (91, 163, 165). Nucleation is much faster on ssDNA than on dsDNA. Slow binding of RecA to dsDNA at neutral pH and above is due to slow nucleation (91, 163, 165). Faster nucleation on dsDNA is seen at pHs below 6.5 (163, 165). Nucleation on dsDNA is facilitated by DNA treatments that tend to unwind the DNA. The most effective nucleation site is a ssDNA gap (91, 163, 165), although nucleation is accelerated if the DNA is supercoiled, linearized, or nicked (163). Nucleation is not enhanced by DNA damage as has been reported (86, 87, 118, 225), but by the perturbations in DNA structure (in particular DNA unwinding) that are caused by DNA damage (163).

The assembly of an active RecA filament requires the binding of ATP, but not hydrolysis. Extended filaments are formed in the presence of ATPγS. On ssDNA, the filament is extended uniquely 5′ to 3′ (168, 187). In a ssDNA gap, the extension of the filament continues into the adjoining dsDNA, continuing until the available contiguous DNA or RecA protein is depleted.

End-dependent RecA filament disassembly occurs on the end opposite to that at which assembly occurs (110, 187), and is coupled to ATP hydrolysis. ATP is hydrolyzed by RecA monomers uniformly throughout RecA filaments (18), and there is no evidence for enhanced rates at either filament end. The hydrolysis of ATP by interior monomers does not generally result in dissociation, and under some conditions ATP hydrolysis can proceed with no evident dissociation of RecA monomers (151, 187, 190). Thus, ATP hydrolysis occurs everywhere, apparently resulting in dissociation only for monomers at the disassembling end (7). Dissociation at that end occurs with a probability that is some function of reaction conditions. Some exchange of RecA protein between free and bound forms can be detected in the interior of RecA filaments under some conditions (190). This is most easily explained by end-dependent dissociation or association processes at filament discontinuities, although some dissociation of fully integrated interior monomers under some conditions has not been ruled out. On ssDNA, the maximum observed rate of end-dependent disassembly (seen above pH 7.5) is 60 to 65 monomers per end per min (7).

Filament dynamics can also be studied with the aid of mixed filaments of wild-type RecA protein and a mutant (190). Mixed filaments are formed with wild-type RecA protein and the mutant protein RecA K72R, which binds but does not hydrolyze dATP (190). When the proteins are added together, proportions of mutant and wild-type proteins in the resulting mixed filaments are the same as the proportions in the overall experiment. The presence of a mutant affects the observed levels of dATP hydrolysis. If the mutant protein is added after wild-type filaments are formed on ssDNA, little exchange of mutant protein into the filaments is observed. On dsDNA, there is a rather facile exchange of mutant protein into the filaments. In general, the filaments on dsDNA are more dynamic with respect to monomer exchange than are those on ssDNA (190).

In a RecA filament, there is one RecA monomer per three bound nucleotides or base pairs of DNA. A few titration experiments have provided a stoichiometry twice as high (104, 133, 135, 141). The apparent higher site size reflects the binding of a second DNA strand to another DNA binding site within the filament (56, 57, 157, 203, 244). RecA filaments can actually bind up to three DNA strands within the filament groove, with three distinct DNA strand-binding sites (95, 100, 101, 211, 212). The sites are usually called simply I, II, and III (95, 100, 101, 211, 212). Site I has the highest affinity for a DNA strand and is generally the site at which ssDNA is bound to RecA protein. Site II is the site generally occupied by a strand complementary to that in site I, although heterologous sequences are tolerated. Site III would then be the site occupied by a DNA strand displaced during DNA strand exchange. The numbering of the sites varies somewhat from research group to research group. When strands of identical ssDNA are added successively to a RecA filament, site III actually binds to the DNA with an affinity higher than that of site II (when sites are identified by the system used here), and it is therefore sometimes referred to as site II.

In a competing formalism, the filament groove is described by some workers to have two DNA binding sites, each of which can bind to either a single- or double-stranded DNA (125, 127, 145, 170). Since no more than three total strands can be bound at one time, the two sites as defined this way presumably overlap.

On dsDNA, the two strands of the DNA are bound asymmetrically. The strand occupying the binding site normally occupied by ssDNA (site I) is protected from nuclease digestion 2 to 3 times better than its complement (site II) (24, 111). The strand bound in site I determines the orientation of the entire filament, with assembly processes proceeding 5' to 3' relative to that strand.

The RecA protein filament has been the paradigm for DNA strand exchange proteins, and similar filaments are formed by the eukaryotic Rad51 protein (157) and the archaeal RadA protein (186). The extent of structural similarity in these filaments, and the similar extents to which DNA is extended within them (50), is remarkable. Notably, the eukaryotic Dmc1 protein has not formed filaments in any experiment reported to date, and appears to function as an octameric ring (158). The mechanism by which such a structure can promote even the relatively weak DNA strand exchange reactions so far observed for Dmc1 protein provides a fascinating experimental challenge.

ATP HYDROLYSIS

RecA-mediated ATP hydrolysis is almost entirely DNA dependent. The measured monomer k_{cat} is about 30 min^{-1} on ssDNA, decreasing to 16 to 20 min^{-1} on dsDNA. The maximum reported value for k_{cat}/K_m at 37°C is about 2×10^4 M^{-1} s^{-1}, well below the diffusion-controlled limit of 10^8 to 10^9 M^{-1} s^{-1}. The kinetics are complicated by cooperativity, and the K_m is more accurately reported as $S_{0.5}$. RecA-mediated ATP hydrolysis in the absence of DNA is slow, with a maximum k_{cat} of 0.1 min^{-1} observed at about pH 6.0 (226). The rate is stimulated markedly by high (1.5 to 2 M) concentrations of a wide range of salts (164). The salt appears to induce a conformation change that to some degree mimics that induced by the binding of DNA to RecA protein (96, 234).

A variety of rNTPs and dNTPs are hydrolyzed by RecA protein, including dATP, UTP, PTP (purine ribonucleoside triphosphate), ITP, CTP, dCTP, GTP, and dUTP (90, 137, 226). Only ATP, dATP, and PTP (those with a measured $S_{0.5}$ below 100 μM) serve as cofactors in the DNA strand exchange reaction (90, 137, 138, 226).

The DNA-dependent ATPase activity exhibits no dependence on pH between 5.5 and 9.0 (226). The effects of temperature between 25 and 45°C on ssDNA-dependent ATP hydrolysis produce a linear Arrhenius plot, from which an Arrhenius activation energy of 11.8 ± 0.3 kcal mol^{-1} can be derived (13) for the reaction. The ATP hydrolysis occurs uniformly throughout the RecA filament, with no enhancement at filament ends (18, 91, 182). When longer DNAs are used as cofactors, the relationship between ATP hydrolysis and DNA binding is sufficiently constant that ATP hydrolysis can be used as an indirect measure of DNA binding under many conditions (7, 187). However, this method of estimating DNA binding levels is obviously indirect and should not be used without first determining the k_{cat} for ATP hydrolysis and confirming that the k_{cat} is constant under all conditions of a given experiment.

ATP is hydrolyzed to ADP and P_i. ADP acts as a competitive inhibitor of ATP hydrolysis. As already noted, the ATP analogue ATPγS, which is bound but not appreciably hydrolyzed by RecA protein, also acts as a potent competitive inhibitor. The inhibition patterns in both cases are greatly complicated by additional effects of these two nucleotides (106, 107). Mixtures of ATP and ADP result in collapse and dissociation of RecA filaments from both ssDNA and dsDNA when the ADP/ATP ratio approaches 50% (41, 106, 107). ATP and ADP tend to cooperatively stabilize different conformations of a RecA filament, and these forms are not directly interconvertible via ATP hydrolysis (107, 166, 240). Very low concentrations of ATPγS can stimulate the ATPase activity under some conditions (106).

ATP hydrolysis does not proceed to completion. RecA filaments tend to come apart when the ADP/ATP ratio reaches about 1.0 (41, 106), effectively halting ATP hydrolysis. ADP and ATPγS are antagonistic inhibitors that evidently stabilize different conformations of the protein (106). When the concentration of ADP is quite high, ATPγS can partially relieve the inhibition due to ADP (106).

The hydrolysis of dATP tends to enhance some RecA reactions. With dATP, hydrolytic rates are increased about 20% relative to ATP hydrolysis. RecA filament stability and the rates of DNA strand exchange are also enhanced (133, 190). As already noted, the RecA mutant RecA K72R will promote a limited DNA strand exchange reaction only if dATP is used as the nucleotide cofactor (169, 191). The use of dATP also prevents an end-dependent disassembly of filaments from ssDNA, as described below (187).

DNA STRAND EXCHANGE

Alignment of DNA and Strand Transfer: Reactions without ATP Hydrolysis

The fundamental process of DNA pairing and strand transfer occurs within the filament groove, an arrangement originally proposed by Howard-Flanders (77). As promoted by RecA protein, DNA strand exchange can involve either three or four DNA strands (Fig. 5). RecA protein filaments first form on

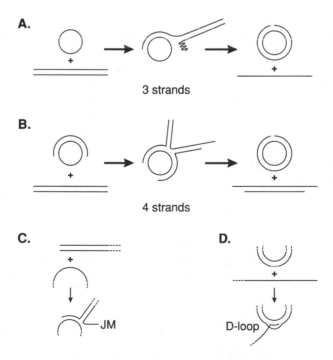

Figure 5. RecA protein-mediated DNA strand exchange reactions. (A) The most common three-strand reaction used in many studies. (B) The standard four-strand reaction. RecA protein filaments are nucleated in the gap of the circular DNA substrate, and DNA pairing is also initiated in this gap. (C) Initiation of DNA strand exchange at a free duplex DNA end. (D) Initiation of DNA strand exchange at a free single-strand (RecA-bound) DNA end.

the single-strand DNA, or on the gapped DNA substrate of a four-strand exchange reaction. The second DNA duplex is then drawn into the filament, and aligned with the first. Alignment triggers the strand switch. ATP hydrolysis is not required for the alignment or strand switch, so these basic pairing steps are presumably catalyzed by binding interactions within the filament groove (38, 83, 84, 93, 132, 169, 191). The pairing and strand switch can occur in a more or less concerted way along a region of several hundred base pairs. Thus, for reactions involving short oligonucleotides, this represents the entire process, and no ATP hydrolysis is required for any part of the reaction. When long (>1,000 bp) DNA substrates are employed, the product of these reactions is often referred to as a joint molecule (Fig. 5), a branched structure that must be extended to complete a strand exchange reaction. ATP hydrolysis is involved in certain aspects of the extended strand exchange reactions seen with long DNA substrates, and it is considered separately below.

These reactions proceed in a prescribed order, in which a single strand is bound first and the homologous duplex is drawn in second. This sequence is the same for the reactions reported for RecA and all of the homologues for which this has been examined to date. However, an active site constructed to promote this reaction should in principle be able to promote the same process in the opposite direction. An inverse reaction has been reported for the *E. coli* RecA protein, in which the duplex DNA is bound first and the single strand is drawn in (241). The reaction has been demonstrated only for quite short duplex DNA substrates (63 bp) in which RecA protein filaments are stabilized by the presence of long circular single strands contiguous with the duplex. The reaction promoted is effectively the reverse of the normal reaction, although there are some different reaction requirements that suggest it may have some unique properties (241). The DNA binding affinities of the RecA protein make it difficult to set up and observe such a "duplex-first" reaction.

The mechanism by which RecA protein aligns the bound single strand with the duplex has received a great deal of attention. Analysis of the kinetics of DNA pairing with short oligonucleotides reveals at least two steps: a rapid second-order alignment of the two homologous DNA molecules, followed by a slow first-order process (11, 12, 236). The initial alignment may involve base flipping by some bases in the duplex (especially A and T bases) to align the DNAs via Watson-Crick pairing, as revealed for the human Rad51 protein (72). In this mechanism, the slow first-order process may be a completion of the strand switch to include all remaining base pairs over an extended region. The homologous alignment does not involve a sliding of the RecA-ssDNA complex along a duplex DNA, but instead is best explained by random collision (2).

In principle, the duplex could approach the ssDNA within the filament via either its major or minor groove. Within the major-groove-first path, a novel triplex DNA intermediate, formed prior to a strand switch or base flipping, has been proposed. This triplex has been called R-form DNA (85, 243). In the alternative minor-groove-first path, homologous alignment would involve only the standard Watson-Crick base pairing provided by base flipping. As the duplex bound, it would be extended and underwound such that its bases would be free to rotate and "sample" the bound ssDNA for complementarity. The RecA and Rad51 proteins both promote DNA strand exchange readily with DNA substrates heavily substituted with base analogues that disrupt many of the interactions proposed to stabilize the R-form triplex as it has been presented (79, 173, 174). Overall, the weight of the evidence currently supports a minor-groove-first pathway for DNA pairing that excludes an R-form triplex intermediate (8, 62, 72, 97, 160, 161, 242), although arguments for a major-groove-first path have also appeared (121). The RecA filament appears to stabilize the products of DNA strand exchange, using binding energy to promote the strand switch (3).

The "search for homology" involves the transient underwinding of the duplex DNA substrate, an effect that can be observed with heterologous DNA substrates (180). In addition, the binding of a heterologous duplex DNA to a RecA-ssDNA complex "activates" the duplex for a strand exchange reaction with a homologous single strand introduced external to the filament (126). This strand exchange "in *trans*" is a much weaker reaction than the conventional strand exchange, and has been demonstrated only for short oligonucleotides. At a minimum, interesting structural changes are conferred on a duplex DNA when it binds to a RecA nucleoprotein filament, and the alterations are likely to facilitate strand exchange reactions in general.

A duplex DNA can initiate pairing at any point along the RecA-ssDNA complex. In fact, the formation of paired complexes is readily detected even when both the single-stranded and homologous duplex DNAs are circular (112, 183, 235). However, for a stable net strand exchange to occur, topology considerations dictate that at least one of the two DNA substrates undergoing DNA strand exchange must have a free end. Thus, a reaction can occur between a circular single strand and a linear duplex, or a linear single strand and a circular duplex (Fig. 5C and D).

The latter reaction is sometimes called D-loop formation or DNA strand invasion, and provides a good model for RecA protein's likely role in the repair of double-strand breaks.

If the initial DNA pairing interaction occurs away from a DNA end, the joint must migrate until a free end is encountered. This can also happen within the RecA filament and does not require ATP hydrolysis, or dissociation movement of any part of the RecA filament. Migration of such a joint can be halted by secondary DNA pairing interactions, or by other nonproductive interactions of the unreacted duplex DNA with the RecA-ssDNA complex (Fig. 6). Such interactions can limit the extent of strand exchange reactions of RecA (191), as well as Rad51 protein (197). In effect, a DNA pairing process that is too facile can restrict the extension of joints. This may be one function of ATP hydrolysis in the RecA system (191).

The four-strand reactions require a significant single-strand gap in one of the two duplex DNAs. The gap serves two purposes. It is a nucleation site for the formation of a RecA filament, which can then encompass the adjacent duplex regions. It is also the location where the DNA pairing is initiated; i.e. four-strand exchanges must be initiated as three-strand reactions (23, 27, 112). Thus, the linear duplex DNA substrate must include sequences at one or both ends that are homologous with those in the single-strand gap of the other substrate.

The four-strand reactions do not occur without ATP hydrolysis. When a gapped DNA is paired with a linear duplex in situations where ATP or an analogue is bound, but not hydrolyzed, then DNA strand exchange is initiated in the gap. However, the exchange halts at the boundary of the gap and does not extend into the region of duplex-duplex (four-strand) exchange (84, 191). Some early proposals suggested that RecA might promote the alignment of two homologous duplex DNAs, and a quadruplex recombination intermediate was entertained as a hypothesis (58, 60, 77, 92, 129, 228, 231). However, many data now indicate that four DNA strands (two duplexes) cannot be accommodated in the filament interior (29, 84, 95, 100, 101, 112, 145, 188, 191, 211, 212). Inherent ambiguities in the DNA pairing results that led to the original duplex-duplex pairing proposals have been pointed out (29, 112, 188, 233).

Reactions That Require ATP Hydrolysis

When ATP is hydrolyzed, the course of a RecA-mediated DNA strand exchange reaction with long DNA substrates follows an established pattern. The reaction begins by binding RecA protein to ssDNA, forming the extended helical filament already described. The nucleoprotein filament hydrolyzes ATP with the monomer k_{cat} of up to 30 min^{-1}. Filament assembly is cooperative, but individual monomers hydrolyze ATP independently (7). When a homologous duplex DNA is added to initiate strand exchange, a rapid decrease in ATP hydrolysis (by 30%) is observed, to a rate characteristic of filaments on

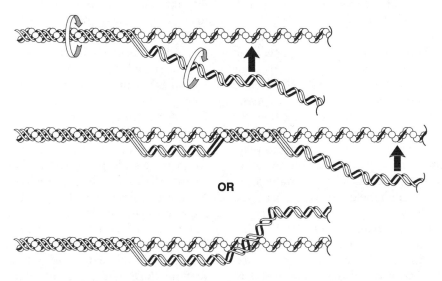

Figure 6. Unproductive complexes that can stall DNA strand exchange reactions. The formation of a joint molecule involves the uptake of a duplex DNA into the RecA filament and its alignment with the previously bound single-stranded DNA. Extension of the region of paired DNA requires a continued spooling of the duplex into the filament, as shown in the top panel. This extension can be blocked by a secondary DNA pairing involving another part of the duplex (leaving an external loop of DNA [middle panel]) or any unproductive interaction of the duplex and the filament (bottom panel) that halts the needed spooling process.

dsDNA (182). During DNA strand exchange, the RecA filaments are in a state analogous to that observed when filaments are formed on dsDNA (189). The filaments become more dynamic and monomer exchange is observed between free and bound forms, as is seen with filaments on dsDNA (189). This induced exchange between bound and free RecA monomers occurs only when a complementary strand is already paired with the ssDNA in the filament; i.e., strand exchange must be complete. It is not seen when the duplex DNA is heterologous (189), and thus has no role in the "search for homology" that must occur prior to homologous alignment of the two DNAs.

Once strand exchange is initiated, the joint molecules are extended unidirectionally (37, 81, 229), uniquely $5'$ to $3'$ relative to the ssDNA to which RecA protein initially binds (Fig. 5). DNA strand exchange is directed at a measured rate (380 ± 20 bp min^{-1} at 37°C) through the available homologous DNA sequences (13). The four-strand exchange reactions are also unidirectional, proceeding $5'$ to $3'$ relative to the single strand in the gap where the reaction is initiated (23, 188).

For the *E. coli* RecA protein, there are three properties of the DNA strand exchange reaction that require ATP hydrolysis, and a fourth that is affected by it. The three that require hydrolysis are as follows. (i) The movement of DNA branches during the late stages of strand exchange is rendered unidirectional (78, 191), with the polarity already noted. (ii) ATP hydrolysis allows the bypass of substantial DNA structural barriers (83, 179, 191). These structural barriers can include a heterologous insertion of 100 bp or more in the duplex DNA substrate, which is unwound as it is bypassed by RecA protein-mediated strand exchange. The bypass involves a generation of torsional stress, since a nick in the heterologous region blocks the bypass (80). (iii) ATP hydrolysis is required for any DNA strand exchange involving two duplex DNAs (the four-strand reaction [Fig. 5]) (84, 191). The fourth function that is affected by ATP hydrolysis is the overall efficiency of the reaction. ATP hydrolysis allows the reaction to go to completion, generating hybrid DNAs over many thousands of base pairs in high yield. As expanded upon below, this effect does not apply to the eukaryotic Rad51 proteins, and the requirement for ATP hydrolysis is only partial even in the case of RecA. In addition to these direct roles in DNA strand exchange, ATP hydrolysis is coupled to end-dependent dissociation of RecA protein from a filament on DNA, as already noted.

None of the four roles for ATP hydrolysis in RecA-mediated DNA strand exchange are observed for any of the eukaryotic or archaeal RecA homologues, and the utilization of ATP is one area where the RecA and its homologues differ greatly. The rates of Rad51-mediated ATP hydrolysis are as much as 2 orders of magnitude less than that observed for RecA protein (206), and the ATP that is hydrolyzed appears to have little effect on DNA strand exchange. The eukaryotic Dmc1 protein and the archaeal RadA protein also hydrolyze ATP more slowly than RecA by an order of magnitude or more (108, 186). The Rad51 protein does not require ATP hydrolysis to promote a robust DNA strand exchange with long DNA substrates (208). In addition, the Rad51 protein cannot promote DNA strand exchange through significant heterologous insertions (149, 207) or promote four-strand exchange reactions (P. Sung, personal communication), whether or not ATP is hydrolyzed. Like RecA, some initial reports provided evidence that Rad51-mediated DNA strand exchange was unidirectional (10, 73, 207, 208). However, some of these reports conflicted as to the direction of strand exchange, and additional reports indicated that no unique polarity to the reactions existed (147–149). The issue appears to have been a function of the structure of the DNA ends of the linear duplex DNA substrate. A single-strand overhang at the end is necessary for an efficient initiation of strand exchange by Rad51, and an overhang at either end can lead to initiation. Initiation at either end can lead experimentally to either polarity of the ensuing strand exchange as it propagates from the site of initiation (147–149). In general, there is no evidence that ATP hydrolysis plays a significant role in DNA strand exchange for any proteins in this class, other than the bacterial RecA proteins.

ATP hydrolysis is not always required for extensive strand exchange reactions to generate long hybrid DNA products. In the case of RecA protein, complete products from the reaction diagrammed in Fig. 5A are generated if the concentration of Mg ion is maintained at levels similar to the ATP concentration (191). The products are generated much more slowly (on a time scale measured in hours) and to lower extents than when ATP is hydrolyzed. The lowered Mg ion also slows the basic DNA pairing processes of the RecA filament (191). Excess Mg ion prevents the generation of complete nicked circular products. These results suggest that secondary DNA pairing events limit the extent of DNA strand exchange without ATP hydrolysis at high Mg ion concentrations, and that full product formation requires conditions that minimize these unproductive interactions between the duplex DNA substrate and the RecA-ssDNA complex. Although the Rad51 protein does not require ATP hydrolysis to promote a strong and extensive DNA strand exchange reaction (208), it does require solution additives that weaken the basic DNA pairing process and/or eliminate other types of unproductive duplex-filament interactions (Fig. 6) (174, 197).

The capacity to promote uniquely unidirectional DNA strand exchange reactions, to bypass significant structural barriers, and to promote four-strand exchange reactions is so far unique to the bacterial RecA proteins, and all of these processes require ATP hydrolysis. These functions may also be associated with the elevated (relative to RecA homologues) levels of ATP hydrolysis seen with RecA. For Rad51, Dmc1, and RadA proteins, the slow ATP hydrolysis observed may play a role in the assembly and disassembly of functional complexes with DNA, but this has not yet been explored. These unique functions of RecA indicate that this protein, unlike the Rad51, Dmc1, or RadA proteins, clearly couples ATP hydrolysis directly to DNA strand exchange. Rates of DNA strand exchange are linked to the rates of ATP hydrolysis (13, 150), and potential mechanisms for this coupling can now be described.

Models for Coupling ATP Hydrolysis to RecA Protein-Mediated DNA Strand Exchange

Certain early models for RecA-mediated DNA strand exchange proposed that RecA filament disassembly and/or reassembly on the displaced single strand was coupled directly to the movement of the DNA branch in the late stages of DNA strand exchange (77, 88, 143). However, under optimal conditions, there is no net dissociation of RecA protein during DNA strand exchange, and the filament is found on the hybrid DNA product rather than on the displaced single strand (218). Rates of end-dependent filament assembly and/or disassembly are affected greatly by changes in reaction conditions (pH, dilution, addition of very small amounts of ATPγS) that have little or no effect on the rates of DNA strand exchange (110). No attempt has been made to explain how the proposed assembly and disassembly mechanisms would permit the bypass of DNA lesions and heterologous insertions during DNA strand exchange. Finally, most of these models envision that two duplex DNAs are both inside the filament during a four-strand exchange reaction, conferring on the RecA filament interior a binding capacity it does not appear to have. For these and other reasons, models that employ filament assembly or disassembly as the enabling process in ATP-dependent DNA strand exchange have been largely discarded. Two models remain under active investigation.

The RecA redistribution model

In the RecA redistribution model, it is proposed that RecA filaments are discontinuous, that DNA strand exchange halts at the discontinuities when ATP is not hydrolyzed, and that ATP hydrolysis serves to recycle RecA protein so as to fill in the discontinuities (93, 132, 169). This idea is subtly different from the ideas considered above, since the filament is seen to stay largely intact and bound to the hybrid DNA duplex at the end of a reaction, as consistent with observation. Also, the recycling of RecA protein need not occur at the migrating branch point of a DNA strand exchange reaction. The model can explain how strand exchange might bypass a heterologous insertion in the duplex DNA substrate (15). In brief, the RecA protein underwinds bound DNA. Also, when filaments form on circular ssDNA substrates, there are no free filament ends with the potential for free rotation. If the dsDNA substrate is bound within the filament on both sides of the insertion, then a modest dissociation of RecA protein from the filament might release underwound DNA. In the absence of free ends to dissipate the torsional stress, the released underwinding could be translated into unwinding of the DNA in the heterologous insertion to allow bypass. The exchange of RecA between free and bound forms during strand exchange (189) is generally consistent with this idea. However, two predictions of this model, that heterology bypass should be blocked in exchanges where both DNA substrates are linear, and that excess RecA should fill in the discontinuities and allow more extensive strand exchange, have not been borne out by experiment (119, 191). In addition, conditions exist in which ATP is hydrolyzed and an exchange of RecA between free and bound forms is observed, but where the presence of small amounts of a mutant RecA protein in a mixed filament completely blocks DNA strand exchange (189). Thus, RecA protein redistribution is not sufficient in itself to explain the role of ATP hydrolysis in strand exchange reactions. There has been no attempt to explain how four-strand exchange reactions might proceed with the RecA redistribution model, or why ATP hydrolysis might be required for them.

The facilitated DNA rotation model

The facilitated DNA rotation model assigns a motor function to the RecA protein (34, 35, 40, 178, 191). The filament disassembly role of ATP hydrolysis then becomes a separate function. The model proposes that without ATP hydrolysis, DNA strand exchange is halted prematurely because a discontinuous DNA

Figure 7. Model for RecA protein-mediated rotation of DNA to effect DNA strand exchange. Any duplex DNA external to the filament, perhaps as a result of the formation of external DNA loops as shown in Fig. 6, would be rotated around the outside of the filament in a reaction coupled to ATP hydrolysis. Rotation in the direction of the curved arrows will result in branch movement in the direction of the black arrows.

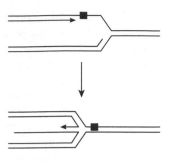

Figure 8. Fork regression as might occur at a stalled replication fork. The product of this reaction is sometimes called a chicken foot.

pairing intermediate is created by secondary DNA pairing events (Fig. 6). The discontinuities take the form of DNA loops external to the filament. The contiguous region of strand exchange that occurs in the absence of ATP hydrolysis is limited to the region between the end of the duplex DNA and the first such external loop. When ATP is hydrolyzed, the loops are resolved by rotating this external DNA around the outside of the filament, keeping the DNA itself more or less parallel to the filament axis (Fig. 7). Unidirectional rotation will bring about a unidirectional migration of the loop, with DNA spooling into the filament on one end and out on the other. The rotation is driven by an ATPase motor.

Quantitative and qualitative predictions of the facilitated DNA rotation model in three-strand exchanges have been tested, and the results conform to expectations based on the model (7, 13, 119). The model provides a basis for the requirement for ATP hydrolysis in four-strand exchanges. If four DNA strands cannot be bound within the RecA filament, then an ATP-facilitated rotation of one duplex around another (with only one bound within the filament) would allow a four-strand exchange to proceed. The four-strand exchanges also exhibit properties that conform to model predictions (188). A rotational motor working as shown in Fig. 7 would provide the energy for the bypass (and unwinding) of heterologous inserts in the duplex DNA substrate, and would render the strand exchange unidirectional. The torsional stress used to unwind the heterologous DNA would be generated by the ATP-driven rotation.

A motor function is also consistent with the structure of the core domain of RecA protein. This domain is a close structural homologue of the F1-ATPase, a family of hexameric helicases, and several other proteins that may act as motors or undergo large conformation changes in the course of their catalytic activity (1, 51, 52, 66, 215, 239).

A motor function could play a role in the recombinational DNA repair of stalled replication

forks. Some of the most prominent models for the repair of stalled forks begin with a reaction called fork regression (Fig. 8) (31, 33, 36, 76, 130, 139). The replication fork is simply forced backwards, with the original template strands being re-paired and the newly synthesized strands extruded to form a "chicken foot" structure (162). There are several pathways for fork regression (33), one of which involves catalysis by the RecA protein (185). RecA protein promotes an efficient fork regression in vitro, using model DNA molecules designed to mimic the structure of the stalled fork shown in Fig. 1B (176). The reannealing of the template strands at such a fork requires the winding of one about the other to recreate the DNA helix, and such a reaction would be consistent with the proposed motor function of the RecA filament. RecA-mediated fork regression is completely dependent on ATP hydrolysis (176).

REGULATION OF THE SOS RESPONSE

In 1974, a series of earlier observations suggesting the existence of an elaborate pathway for DNA repair and mutagenesis induced by DNA damage (43, 144) were unified under the SOS hypothesis by Radman (167). RecA was implicated in the regulation of SOS, and the first activity with which RecA protein was associated in vitro was the cleavage of the bacteriophage λ repressor as a biochemical step in SOS induction (175). The LexA repressor, responsible for the in vivo repression of the bacterial SOS genes, was also soon found to be cleaved by RecA protein (115). This work led to the first purification of RecA protein (175).

A great deal is now known about the SOS induction pathway and the role of RecA protein in it (44, 63, 115, 221, 222). High levels of DNA damage cause an interruption of replication. The single-stranded

DNA created as one result of the replication halt is the primary signal for SOS induction (75). RecA protein binds to the ssDNA, assuming the extended filament conformation in the presence of ATP. However, the RecA protein is not a protease. Instead, the LexA repressor, the bacteriophage λ repressor, the UmuD protein, and a few other proteins bind to the extended RecA filament. When thus bound, these proteins undergo an autocatalytic cleavage (113, 114, 154, 194). The RecA filament simply facilitates this integrated cleavage function of the other proteins. Thus, this function of RecA protein has been dubbed the "coprotease" activity (113, 114). The cleavage of the LexA repressor results in the induction of over 40 SOS genes, including the RecA protein itself.

Facilitating the autocatalytic cleavage of LexA protein involves the binding of LexA repressor deep within the groove of a RecA filament (238), situated so as to span two adjacent RecA monomers. The RecA species involved in the coprotease activities is almost certainly identical to the extended RecA filament involved in recombinational processes (238). The propensity of RecA to bind ssDNA almost exclusively under physiological conditions serves to target RecA protein to ssDNA gaps created by the halt in DNA replication. Some RecA protein mutants bind more readily to dsDNA, leading to more indiscriminate binding in the cell. These mutants often produce an SOS constitutive phenotype (117, 214).

SOS MUTAGENESIS

Heavy DNA damage leads to a complete cessation of DNA replication. After 30 to 45 min, replication resumes. The resumption is generally referred to as replication restart (49). The SOS system induced by the replication halt is actually a staged accumulation of enzymes involved in DNA repair (42, 103, 193, 199, 200, 224). The stages reflect a measured response to different levels of damage.

There are two general paths to replication restart. Nonmutagenic replication fork repair provides one way for a fork to bypass a lesion, as already discussed. The enzymes needed for nonmutagenic fork repair are expressed under normal growth conditions, but many (including RecA protein) are also induced to higher levels in the early stages of the SOS response. The second path to restart involves a specialized system for mutagenic DNA lesion bypass that is induced in the late stages of the SOS response (44, 116, 146, 223). Induction of the enzymes needed for each of these is thus organized temporally during SOS so that nonmutagenic repair processes are induced early, followed by mutagenic processes (42, 103).

The mutagenic pathway inevitably produces some lethal mutations, and a rationale is needed for the evolution of such a system. DNA damage may simply reach levels that are beyond the capacity of the nonmutagenic systems, including recombinational replication fork repair, to handle. Thus, there may be an advantage at some point to accepting the lethality associated with mutagenic replicational bypass in order to get replication going again and rescue at least some cells in the population. Another possibility has been described by Sedgwick (184, 232): a high level of damage produces DNA lesions closely placed on opposite template strands, leading to the generation of overlapping daughter-strand gaps. With such a scenario, it may be impossible for bypass to take place via the nonmutagenic recombinational DNA repair paths. The mutagenic replicational bypass would then represent a last-ditch effort by the "stuck" cell to survive.

Mutagenic replicational bypass of DNA lesions is mediated by a special class of error-prone DNA polymerases (68, 109). In *E. coli*, this class is represented by DNA polymerases IV and V (131, 213, 220). DNA polymerase V is the enzyme primarily responsible for SOS mutagenesis. It consists of the products of the *umuC* and *umuD* genes, with the UmuD protein processed by a RecA-dependent autocatalytic cleavage to generate the active UmuD' (154, 194). Both genes are under SOS control and expressed late in the SOS process. The DNA polymerase V is a heterotrimeric UmuD'$_2$C complex. Also required for optimal activity is the βγ complex from DNA polymerase III, the SSB protein, and the RecA protein (68, 159, 171, 172, 213). The role of RecA in this reaction may be the least-understood aspect of RecA function, although it is clear that RecA greatly activates the bypass process. The lesion bypass activity of DNA polymerase V appears to require the establishment of a RecA filament in the gap downstream from the primer that DNA polymerase V will extend in the lesion bypass process (159, 172). Such a filament would have to be displaced by the polymerase as it moves while at the same time maintaining contact with it. This displacement would take place at the 3'-proximal end of the filament, where filament dissociation generally does not occur. In addition, a filament forming in a DNA gap would extend into the adjacent duplex DNA, and thus block access to the free 3' end that the polymerase would require for use as a primer. It has not been determined how DNA polymerase V is directed to the sites where it must act in this environment.

Acknowledgments. Work in my laboratory is supported by grants GM32335 and GM52725 from the National Institutes of Health.

REFERENCES

1. **Abrahams, J. P., A. G. Leslie, R. Lutter, and J. E. Walker.** 1994. Structure at 2.8 A resolution of F1-ATPase from bovine heart mitochondria. *Nature* **370:**621–628.

2. **Adzuma, K.** 1998. No sliding during homology search by RecA protein. *J. Biol. Chem.* **273:**31565–31573.

3. **Adzuma, K.** 1992. Stable synapsis of homologous DNA molecules mediated by the Escherichia coli RecA protein involves local exchange of DNA strands. *Genes Dev.* **6:**1679–1694.

4. **Amano, T., M. Yoshida, Y. Matsuo, and K. Nishikawa.** 1994. Structural model of the ATP-binding domain of the F1-beta subunit based on analogy to the RecA protein. *FEBS Lett.* **351:**1–5.

5. **Amaratunga, M., and A. S. Benight.** 1988. DNA sequence dependence of ATP hydrolysis by RecA protein. *Biochem. Biophys. Res. Commun.* **157:**127–133.

6. **Anderson, D. G., and S. C. Kowalczykowski.** 1997. The translocating RecBCD enzyme stimulates recombination by directing RecA protein onto ssDNA in a chi-regulated manner. *Cell* **90:**77–86.

7. **Arenson, T. A., O. V. Tsodikov, and M. M. Cox.** 1999. Quantitative analysis of the kinetics of end-dependent disassembly of RecA filaments from ssDNA. *J. Mol. Biol.* **288:**391–401.

8. **Baliga, R., J. W. Singleton, and P. B. Dervan.** 1995. RecA oligonucleotide filaments bind in the minor groove of double-stranded DNA. *Proc. Natl. Acad. Sci. USA* **92:**10393–10397.

9. **Baumann, P., F. E. Benson, and S. C. West.** 1996. Human Rad51 protein promotes ATP-dependent homologous pairing and strand transfer reactions in vitro. *Cell* **87:**757–766.

10. **Baumann, P., and S. C. West.** 1997. The human Rad51 protein: polarity of strand transfer and stimulation by Hrp-A. *EMBO J.* **16:**5198–5206.

11. **Bazemore, L. R., E. Foltastogniew, M. Takahashi, and C. M. Radding.** 1997. RecA tests homology at both pairing and strand exchange. *Proc. Natl. Acad. Sci. USA* **94:**11863–11868.

12. **Bazemore, L. R., M. Takahashi, and C. M. Radding.** 1997. Kinetic analysis of pairing and strand exchange catalyzed by RecA. Detection by fluorescence energy transfer. *J. Biol. Chem.* **272:**14672–14682.

13. **Bedale, W. A., and M. Cox.** 1996. Evidence for the coupling of ATP hydrolysis to the final (extension) phase of RecA protein-mediated DNA strand exchange. *J. Biol. Chem.* **271:**5725–5732.

14. **Bianchet, M. A., Y. H. Ko, L. M. Amzel, and P. L. Pedersen.** 1997. Modeling of nucleotide binding domains of ABC transporter proteins based on a F1-ATPase/recA topology: structural model of the nucleotide binding domains of the cystic fibrosis transmembrane conductance regulator (CFTR). *J. Bioenerg. Biomembr.* **29:**503–524.

15. **Bianco, P. R., R. B. Tracy, and S. C. Kowalczykowski.** 1998. DNA strand exchange proteins: a biochemical and physical comparison. *Front. Biosci.* **3:**560–603.

16. **Bishop, D. K., D. Park, L. Xu, and N. Kleckner.** 1992. DMC1: a meiosis-specific yeast homolog of E. coli recA required for recombination, synaptonemal complex formation, and cell cycle progression. *Cell* **69:**439–456.

17. **Brendel, V., L. Brocchieri, S. J. Sandler, A. J. Clark, and S. Karlin.** 1997. Evolutionary comparisons of RecA-like proteins across all major kingdoms of living organisms. *J. Mol. Evol.* **44:**528–541.

18. **Brenner, S. L., R. S. Mitchell, S. W. Morrical, S. K. Neuendorf, B. C. Schutte, and M. M. Cox.** 1987. RecA protein-promoted ATP hydrolysis occurs throughout RecA nucleoprotein filaments. *J. Biol. Chem.* **262:**4011–4016.

19. **Brenner, S. L., A. Zlotnick, and J. D. Griffith.** 1988. RecA protein self-assembly. Multiple discrete aggregation states. *J. Mol. Biol.* **204:**959–972.

20. **Brenner, S. L., A. Zlotnick, and W. F. Stafford.** 1990. RecA protein self-assembly. 2. Analytical equilibrium ultracentrifugation studies of the entropy-driven self-association of RecA. *J. Mol. Biol.* **216:**949–964.

21. **Cazaux, C., F. Larminat, G. Villani, N. P. Johnson, M. Schnarr, and M. Defais.** 1994. Purification and biochemical characterization of Escherichia coli RecA proteins mutated in the putative DNA binding site. *J. Biol. Chem.* **269:**8246–8254.

22. **Cazenave, C., M. Chabbert, J. J. Toulme, and C. Helene.** 1984. Absorption and fluorescence studies of the binding of the *recA* gene product from *E. coli* to single-stranded and double-stranded DNA. Ionic strength dependence. *Biochim. Biophys. Acta* **781:**7–13.

23. **Chow, S. A., S. K. Chiu, and B. C. Wong.** 1992. RecA protein-promoted homologous pairing and strand exchange between intact and partially single-stranded duplex DNA. *J. Mol. Biol.* **223:**79–93.

24. **Chow, S. A., S. M. Honigberg, and C. M. Radding.** 1988. DNase protection by RecA protein during strand exchange. Asymmetric protection of the Holliday structure. *J. Biol. Chem.* **263:**3335–3347.

25. **Churchill, J. J., D. G. Anderson, and S. C. Kowalczykowski.** 1999. The RecBC enzyme loads RecA protein onto ssDNA asymmetrically and independently of chi, resulting in constitutive recombination activation. *Genes Dev.* **13:**901–911.

26. **Clark, A. J., and A. D. Margulies.** 1965. Isolation and characterization of recombination-deficient mutants of *Escherichia coli* K12. *Proc. Natl. Acad. Sci. USA* **53:**451–459.

27. **Conley, E. C., and S. C. West.** 1990. Underwinding of DNA associated with duplex-duplex pairing by RecA protein. *J. Biol. Chem.* **265:**10156–10163.

28. **Cotterill, S. M., A. C. Satterthwait, and A. R. Fersht.** 1982. RecA protein from *Escherichia coli*. A very rapid and simple purification procedure: binding of adenosine 5′-triphosphate and adenosine 5′-diphosphate by the homogeneous protein. *Biochemistry* **21:**4332–4337.

29. **Cox, M. M.** 1995. Alignment of three (but not four) DNA strands in a RecA protein filament. *J. Biol. Chem.* **270:**26021–26024.

30. **Cox, M. M.** 1998. A broadening view of recombinational DNA repair in bacteria. *Genes Cells* **3:**65–78.

31. **Cox, M. M.** 2001. Historical overview: searching for replication help in all the *rec* places. *Proc. Natl. Acad. Sci. USA* **98:**8173–8180.

32. **Cox, M. M.** 1999. Recombinational DNA repair in bacteria and the RecA protein. *Prog. Nucleic Acid Res. Mol. Biol.* **63:**310–366.

33. **Cox, M. M.** 2001. Recombinational DNA repair of damaged replication forks in Escherichia coli: questions. *Annu. Rev. Genet.* **35:**53–82.

34. **Cox, M. M.** 1989. The role of RecA protein in homologous genetic recombination, p. 43–70. *In* K. W. Adolph (ed.), *Molecular Biology of Chromosome Function.* Springer-Verlag, New York, N.Y.

35. **Cox, M. M.** 1994. Why does RecA protein hydrolyze ATP. *Trends Biochem. Sci.* **19:**217–222.

36. Cox, M. M., M. F. Goodman, K. N. Kreuzer, D. J. Sherratt, S. J. Sandler, and K. J. Marians. 2000. The importance of repairing stalled replication forks. *Nature* 404:37–41.

37. Cox, M. M., and I. R. Lehman. 1981. Directionality and polarity in RecA protein-promoted branch migration. *Proc. Natl. Acad. Sci. USA* 78:6018–6022.

38. Cox, M. M., and I. R. Lehman. 1981. RecA protein of *Escherichia coli* promotes branch migration, a kinetically distinct phase of DNA strand exchange. *Proc. Natl. Acad. Sci. USA* 78:3433–3437.

39. Cox, M. M., K. McEntee, and I. R. Lehman. 1981. A simple and rapid procedure for the large scale purification of the RecA protein of *Escherichia coli*. *J. Biol. Chem.* 256:4676–4678.

40. Cox, M. M., B. F. Pugh, B. C. Schutte, J. E. Lindsley, J. Lee, and S. W. Morrical. 1987. On the mechanism of RecA protein-promoted DNA branch migration, p. 597–607. *In* R. McMacken and T. J. Kelly (ed.), *DNA Replication and Recombination*. Alan R. Liss, Inc., New York, N.Y.

41. Cox, M. M., D. A. Soltis, I. R. Lehman, C. DeBrosse, and S. J. Benkovic. 1983. ADP-mediated dissociation of stable complexes of RecA protein and single-stranded DNA. *J. Biol. Chem.* 258:2586–2592.

42. Defais, M., and R. Devoret. 2000. *SOS Response*, vol. 2001, 1st ed. Macmillan Publishers Ltd., Basingstoke, Hampshire, England.

43. Defais, M., P. Fauquet, M. Radman, and M. Errera. 1971. Ultraviolet reactivation and ultraviolet mutagenesis of lambda in different genetic systems. *Virology* 43:495–503.

44. Devoret, R. 1992. Les fonctions SOS ou comment les bactéries survivent aux lésions de leur ADN. *Ann. Inst. Pasteur Actual.* 1:11–20.

45. Di Capua, E., A. Engel, A. Stasiak, and T. Koller. 1982. Characterization of complexes between RecA protein and duplex DNA by electron microscopy. *J. Mol. Biol.* 157:87–103.

46. Di Capua, E., and B. Müller. 1987. The accessibility of DNA to dimethylsulfate in complexes with RecA protein. *EMBO J.* 6:2493–2498.

47. Dombroski, D. F., D. G. Scraba, R. D. Bradley, and A. R. Morgan. 1983. Studies of the interaction of RecA protein with DNA. *Nucleic Acids Res.* 11:7487–7504.

48. Dutreix, M., B. Burnett, A. Bailone, C. M. Radding, and R. Devoret. 1992. A partially deficient mutant, recA1730, that fails to form normal nucleoprotein filaments. *Mol. Gen. Genet.* 232:489–497.

49. Echols, H., and M. F. Goodman. 1991. Fidelity mechanisms in DNA replication. *Annu. Rev. Biochem.* 60:477–511.

50. Egelman, E. 2001. Does a stretched DNA structure dictate the helical geometry of RecA-like filaments? *J. Mol. Biol.* 309:539–602.

51. Egelman, E. 2000. A ubiquitous structural core. *Trends Biochem.* 25:183–184.

52. Egelman, E. H. 1998. Bacterial helicases. *J. Struct. Biol.* 124:123–128.

53. Egelman, E. H. 1993. What do x-ray crystallographic and electron microscopic structural studies of the RecA protein tell us about recombination? *Curr. Opin. Struct. Biol.* 3:189–197.

54. Egelman, E. H., and A. Stasiak. 1993. Electron microscopy of RecA-DNA complexes: two different states, their functional significance and relation to the solved crystal structure. *Micron* 24:309–324.

55. Egelman, E. H., and A. Stasiak. 1986. Structure of helical RecA-DNA complexes. Complexes formed in the presence of ATP-γ-S or ATP. *J. Mol. Biol.* 191:677–697.

56. Egelman, E. H., and A. Stasiak. 1988. Structure of helical RecA-DNA complexes. II. Local conformational changes visualized in bundles of RecA-ATP-γ-S filaments. *J. Mol. Biol.* 200:329–349.

57. Egelman, E. H., and X. Yu. 1989. The location of DNA in RecA-DNA helical filaments. *Science* 245:404–407.

58. Eggleston, A. K., A. H. Mitchell, and S. C. West. 1997. In vitro reconstitution of the late steps of genetic recombination in E. coli. *Cell* 89:607–617.

59. Ellouze, C., M. Takahashi, P. Wittung, K. Mortensen, M. Schnarr, and B. Norden. 1995. Evidence for elongation of the helical pitch of the RecA filament upon ATP and ADP binding using small-angle neutron scattering. *Eur. J. Biochem.* 233:579–583.

60. Fishel, R. A., and A. Rich. 1988. The role of left-handed Z-DNA in general genetic recombination, p. 23–32. *In* E. C. Friedberg and P. C. Hanawalt (ed.), *Mechanisms and Consequences of DNA Damage Processing*. Alan R. Liss, Inc., New York, N.Y.

61. Frank, E. G., J. Hauser, A. S. Levine, and R. Woodgate. 1993. Targeting of the UmuD, UmuD′, and MucA′ mutagenesis proteins to DNA by RecA protein. *Proc. Natl. Acad. Sci. USA* 90:8169–8173.

62. Frank-Kamenetskii, M. D., and S. M. Mirkin. 1995. Triplex DNA structures. *Annu. Rev. Biochem.* 64:65–95.

63. Friedberg, E. C., G. C. Walker, and W. Siede. 1995. *DNA Repair and Mutagenesis*. ASM Press, Washington, D.C.

64. Frischkorn, K., B. Springer, E. C. Bottger, E. O. Davis, M. J. Colston, and P. Sander. 2000. In vivo splicing and functional characterization of Mycobacterium leprae RecA. *J. Bacteriol.* 182:3590–3592.

65. Garvey, N., A. C. St. John, and E. M. Witkin. 1985. Evidence for RecA protein association with the cell membrane and for changes in the levels of major outer membrane proteins in SOS-induced Escherichia coli cells. *J. Bacteriol.* 163:870–876.

66. Gomis-Ruth, F. X., G. Moncalian, R. Perez-Luque, A. Gonzalez, E. Cabezon, F. de la Cruz, and M. Coll. 2001. The bacterial conjugation protein TrwB resembles ring helicases and F1-ATPase. *Nature* 409:637–641.

67. Goodman, H. J. K., and D. R. Woods. 1990. Molecular analysis of the *Bacteroides fragilis recA* gene. *Gene* 94:77–82.

68. Goodman, M. F., and B. Tippin. 2000. The expanding polymerase universe. *Nat. Rev. Mol. Cell. Biol.* 1:101–109.

69. Griffith, J., and C. G. Shores. 1985. RecA protein rapidly crystallizes in the presence of spermidine: a valuable step in its purification and physical characterization. *Biochemistry* 24:158–162.

70. Gudas, L. J., and D. W. Mount. 1977. Identification of the *recA* (*tif*) gene product of *Escherichia coli*. *Proc. Natl. Acad. Sci. USA* 74:5280–5284.

71. Gupta, R. C., L. R. Bazemore, E. I. Golub, and C. M. Radding. 1997. Activities of human recombination protein Rad51. *Proc. Natl. Acad. Sci. USA* 94:463–468.

72. Gupta, R. C., E. Folta-Stogniew, S. O'Malley, M. Takahashi, and C. M. Radding. 1999. Rapid exchange of A:T base pairs is essential for recognition of DNA homology by human Rad51 recombination protein. *Mol. Cell* 4:705–714.

73. Gupta, R. C., E. I. Golub, M. S. Wold, and C. M. Radding. 1998. Polarity of DNA strand exchange promoted by recombination proteins of the RecA family. *Proc. Natl. Acad. Sci. USA* 95:9843–9848.

74. Heuser, J., and J. Griffith. 1989. Visualization of RecA protein and its complexes with DNA by quick-freeze/deep-etch electron microscopy. *J. Mol. Biol.* 210:473–484.

75. Higashitani, N., A. Higashitani, A. Roth, and K. Horiuchi. 1992. SOS induction in Escherichia coli by infection with mutant filamentous phage that are defective in initiation of complementary-strand DNA synthesis. *J. Bacteriol.* **174:** 1612–1618.

76. Higgins, N. P., K. Kato, and B. Strauss. 1976. A model for replication repair in mammalian cells. *J. Mol. Biol.* **101:**417–425.

77. Howard-Flanders, P., S. C. West, and A. Stasiak. 1984. Role of RecA protein spiral filaments in genetic recombination. *Nature* **309:**215–219.

78. Jain, S. K., M. M. Cox, and R. B. Inman. 1994. On the role of ATP hydrolysis in RecA protein-mediated DNA strand exchange. III. Unidirectional branch migration and extensive hybrid DNA formation. *J. Biol. Chem.* **269:**20653–20661.

79. Jain, S. K., R. B. Inman, and M. M. Cox. 1992. Putative 3-stranded DNA pairing intermediate in RecA protein-mediated DNA strand exchange: no role for guanine N-7. *J. Biol. Chem.* **267:**4215–4222.

80. Jwang, B., and C. M. Radding. 1992. Torsional stress generated by RecA protein during DNA strand exchange separates strands of a heterologous insert. *Proc. Natl. Acad. Sci. USA* **89:**7596–7600.

81. Kahn, R., R. P. Cunningham, C. Das Gupta, and C. M. Radding. 1981. Polarity of heteroduplex formation promoted by *Escherichia coli* RecA protein. *Proc. Natl. Acad. Sci. USA* **78:**4786–4790.

82. Karlin, S., and L. Brocchieri. 1996. Evolutionary conservation of RecA genes in relation to protein structure and function. *J. Bacteriol.* **178:**1881–1894.

83. Kim, J. I., M. M. Cox, and R. B. Inman. 1992. On the role of ATP hydrolysis in RecA protein-mediated DNA strand exchange. I. Bypassing a short heterologous insert in one DNA substrate. *J. Biol. Chem.* **267:**16438–16443.

84. Kim, J. I., M. M. Cox, and R. B. Inman. 1992. On the role of ATP hydrolysis in RecA protein-mediated DNA strand exchange. II. Four-strand exchanges. *J. Biol. Chem.* **267:** 16444–16449.

85. Kim, M. G., V. B. Zhurkin, R. L. Jernigan, and R. D. Camerini-Otero. 1995. Probing the structure of a putative intermediate in homologous recombination: the third strand in the parallel DNA triplex is in contact with the major groove of the duplex. *J. Mol. Biol.* **247:**874–889.

86. Kim, S. K., M. Takahashi, B. Jernstrom, and B. Norden. 1993. Enhancement of binding rate of RecA protein to DNA by carcinogenic benzo[a]pyrene derivatives and selective change of adduct conformation. *Carcinogenesis* **14:**311–313.

87. Kojima, M., M. Suzuki, T. Morita, T. Ogawa, H. Ogawa, and M. Tada. 1990. Interaction of RecA protein with pBR322 DNA modified by N-hydroxy-2-acetylaminofluorene and 4-hydroxyaminoquinoline 1-oxide. *Nucleic Acids Res.* **18:**2707–2714.

88. Konforti, B. B., and R. W. Davis. 1992. ATP hydrolysis and the displaced strand are two factors that determine the polarity of RecA-promoted DNA strand exchange. *J. Mol. Biol.* **227:**38–53.

89. Kowalczykowski, S. C. 2000. Initiation of genetic recombination and recombination-dependent replication. *Trends Biochem. Sci.* **25:**156–165.

90. Kowalczykowski, S. C. 1986. Interaction of RecA protein with a photoaffinity analogue of ATP, 8-azido-ATP: determination of nucleotide cofactor binding parameters and of the relationship between ATP binding and ATP hydrolysis. *Biochemistry* **25:**5872–5881.

91. Kowalczykowski, S. C., J. Clow, and R. A. Krupp. 1987. Properties of the duplex DNA-dependent ATPase activity of *Escherichia coli* RecA protein and its role in branch migration. *Proc. Natl. Acad. Sci. USA* **84:**3127–3131.

92. Kowalczykowski, S. C., and A. K. Eggleston. 1994. Homologous pairing and DNA strand-exchange proteins. *Annu. Rev. Biochem.* **63:**991–1043.

93. Kowalczykowski, S. C., and R. A. Krupp. 1995. DNA-strand exchange promoted by RecA protein in the absence of ATP: implications for the mechanism of energy transduction in protein-promoted nucleic acid transactions. *Proc. Natl. Acad. Sci. USA* **92:**3478–3482.

94. Krueger, J. H., and G. C. Walker. 1984. groEL and dnaK genes of Escherichia coli are induced by UV irradiation and nalidixic acid in an htpR+-dependent fashion. *Proc. Natl. Acad. Sci. USA* **81:**1499–1503.

95. Kubista, M., T. Simonson, R. Sjöback, H. Widlund, and A. Johansson. 1996. Towards an understanding of the mechanism of DNA strand exchange mediated by RecA protein, p. 49–59. *In* R. H. Sarma and M. H. Sarma (ed.), *Biological Structure and Dynamics: Proceedings of the Ninth Conversation*, vol. 1. Adenine Press, Schenectady, N.Y.

96. Kumar, K. A., S. Mahalakshmi, and K. Muniyappa. 1993. DNA-induced conformational changes in RecA protein. Evidence for structural heterogeneity among nucleoprotein filaments and implications for homologous pairing. *J. Biol. Chem.* **268:**26162–26170.

97. Kumar, K. A., and K. Muniyappa. 1992. Use of structure-directed DNA ligands to probe the binding of recA protein to narrow and wide grooves of DNA and on its ability to promote homologous pairing. *J. Biol. Chem.* **267:**24824–24832.

98. Kumar, R. A., M. B. Vaze, N. R. Chandra, M. Vijayan, and K. Muniyappa. 1996. Functional characterization of the precursor and spliced forms of RecA protein of *Mycobacterium tuberculosis. Biochemistry* **35:**1793–1802.

99. Kuramitsu, S., K. Hamaguchi, T. Ogawa, and H. Ogawa. 1981. A large-scale preparation and some physicochemical properties of recA protein. *J. Biochem.* **90:**1033–1045.

100. Kurumizaka, H., B. J. Rao, T. Ogawa, C. M. Radding, and T. Shibata. 1994. A chimeric Rec-A protein that implicates non-Watson-Crick interactions in homologous pairing. *Nucleic Acids Res.* **22:**3387–3391.

101. Kurumizaka, H., and T. Shibata. 1996. Homologous recognition by RecA protein using non-equivalent three DNA-strand-binding sites. *J. Biochem.* **119:**216–223.

102. Kuzminov, A. 2001. DNA replication meets genetic exchange: chromosomal damage and its repair by homologous recombination. *Proc. Natl. Acad. Sci. USA* **98:**8461–8468.

103. Kuzminov, A. 1999. Recombinational repair of DNA damage in Escherichia coli and bacteriophage lambda. *Microbiol. Mol. Biol. Rev.* **63:**751–813.

104. Lauder, S. D., and S. C. Kowalczykowski. 1991. Asymmetry in the RecA protein-DNA filament. *J. Biol. Chem.* **266:**5450–5458.

105. Leahy, M. C., and C. M. Radding. 1986. Topography of the interaction of RecA protein with single-stranded deoxyoligonucleotides. *J. Biol. Chem.* **261:**6954–6960.

106. Lee, J. W., and M. M. Cox. 1990. Inhibition of RecA protein-promoted ATP hydrolysis. I. ATPγS and ADP are antagonistic inhibitors. *Biochemistry* **29:**7666–7676.

107. Lee, J. W., and M. M. Cox. 1990. Inhibition of RecA protein-promoted ATP hydrolysis. II. Longitudinal assembly and disassembly of RecA protein filaments mediated by ATP and ADP. *Biochemistry* **29:**7677–7683.

108. Li, Z. F., E. I. Golub, R. Gupta, and C. M. Radding. 1997. Recombination activities of Hsdmc1 protein, the meiotic human homolog of RecA protein. *Proc. Natl. Acad. Sci. USA* **94:**11221–11226.

109. Lindahl, T., and R. D. Wood. 1999. Quality control by DNA repair. *Science* 286:1897–1905.

110. Lindsley, J. E., and M. M. Cox. 1990. Assembly and disassembly of RecA protein filaments occurs at opposite filament ends: relationship to DNA strand exchange. *J. Biol. Chem.* 265:9043–9054.

111. Lindsley, J. E., and M. M. Cox. 1989. Dissociation pathway for RecA nucleoprotein filaments formed on linear duplex DNA. *J. Mol. Biol.* 205:695–711.

112. Lindsley, J. E., and M. M. Cox. 1990. On RecA protein-mediated homologous alignment of 2 DNA molecules: 3 strands versus 4 strands. *J. Biol. Chem.* 265:10164–10171.

113. Little, J. W. 1984. Autodigestion of LexA and phage lambda repressors. *Proc. Natl. Acad. Sci. USA* 81:1375–1379.

114. Little, J. W. 1991. Mechanism of specific LexA cleavage: autodigestion and the role of RecA coprotease. *Biochimie* 73:411–422.

115. Little, J. W., S. H. Edmiston, L. Z. Pacelli, and D. W. Mount. 1980. Cleavage of the *Escherichia coli* LexA protein by the RecA protease. *Proc. Natl. Acad. Sci. USA* 77:3225–3229.

116. Livneh, Z., F. O. Cohen, R. Skaliter, and T. Elizur. 1993. Replication of damaged DNA and the molecular mechanism of ultraviolet light mutagenesis. *Crit. Rev. Biochem. Mol. Biol.* 28:465–513.

117. Lu, C., and H. Echols. 1987. RecA protein and SOS. Correlation of mutagenesis phenotype with binding of mutant RecA proteins to duplex DNA and LexA cleavage. *J. Mol. Biol.* 196:497–504.

118. Lu, C., R. H. Scheuermann, and H. Echols. 1986. Capacity of RecA protein to bind preferentially to UV lesions and inhibit the editing subunit (ε) of DNA polymerase III: a possible mechanism for SOS-induced targeted mutagenesis. *Proc. Natl. Acad. Sci. USA* 83:619–623.

119. MacFarland, K. J., Q. Shan, R. B. Inman, and M. M. Cox. 1997. RecA as a motor protein. Testing models for the role of ATP hydrolysis in DNA strand exchange. *J. Biol. Chem.* 272:17675–17685.

120. Madiraju, M. V., P. E. Lavery, S. C. Kowalczykowski, and A. J. Clark. 1992. Enzymatic properties of the RecA803 protein, a partial suppressor of recF mutations. *Biochemistry* 31:10529–10535.

121. Malkov, V. A., I. G. Panyutin, R. D. Neumann, V. B. Zhurkin, and R. D. Camerini-Otero. 2000. Radio-probing of a RecA-three-stranded DNA complex with iodine 125: evidence for recognition of homology in the major groove of the target duplex. *J. Mol. Biol.* 299:629–640.

122. Martin, B., J. M. Ruellan, J. F. Angulo, R. Devoret, and J. P. Claverys. 1992. Identification of the recA gene of Streptococcus pneumoniae. *Nucleic Acids Res.* 20:6412.

123. Masson, J. Y., A. A. Davies, N. Hajibagheri, E. Van Dyck, F. E. Benson, A. Z. Stasiak, A. Stasiak, and S. C. West. 1999. The meiosis-specific recombinase hDmc1 forms ring structures and interacts with hRad51. *EMBO J.* 18:6552–6560.

124. Masui, R., T. Mikawa, R. Kato, and S. Kuramitsu. 1998. Characterization of the oligomeric states of RecA protein: monomeric RecA protein can form a nucleoprotein filament. *Biochemistry* 37:14788–14797.

125. Mazin, A. V., and S. C. Kowalczykowski. 1998. The function of the secondary DNA binding site of RecA protein during DNA strand exchange. *EMBO J.* 17:1161–1168.

126. Mazin, A. V., and S. C. Kowalczykowski. 1999. A novel property of the RecA nucleoprotein filament: activation of double-stranded DNA for strand exchange in trans. *Genes Dev.* 13:2005–2016.

127. Mazin, A. V., and S. C. Kowalczykowski. 1996. The specificity of the secondary DNA binding site of RecA protein

128. McEntee, K., G. M. Weinstock, and I. R. Lehman. 1981. Binding of the recA protein of Escherichia coli to single- and double-stranded DNA. *J. Biol. Chem.* 256:8835–8844.

129. McGavin, S. 1971. Models of specifically paired like (homologous) nucleic acid structures. *J. Mol. Biol.* 55:293–298.

130. McGlynn, P., R. G. Lloyd, and K. J. Marians. 2001. Formation of Holliday junctions by regression of nascent DNA in intermediates containing stalled replication forks: RecG stimulates regression even when the DNA is negatively supercoiled. *Proc. Natl. Acad. Sci. USA* 98:8235–8240.

131. McKenzie, G. J., P. L. Lee, M. J. Lombardo, P. J. Hastings, and S. M. Rosenberg. 2001. SOS mutator DNA polymerase IV functions in adaptive mutation and not adaptive amplification. *Mol. Cell* 7:571–579.

132. Menetski, J. P., D. G. Bear, and S. C. Kowalczykowski. 1990. Stable DNA heteroduplex formation catalyzed by the *Escherichia coli* RecA protein in the absence of ATP hydrolysis. *Proc. Natl. Acad. Sci. USA* 87:21–25.

133. Menetski, J. P., and S. C. Kowalczykowski. 1989. Enhancement of *Escherichia coli* RecA protein enzymatic function by dATP. *Biochemistry* 28:5871–5881.

134. Menetski, J. P., and S. C. Kowalczykowski. 1985. Interaction of RecA protein with single-stranded DNA. Quantitative aspects of binding affinity modulation by nucleotide cofactors. *J. Mol. Biol.* 181:281–295.

135. Menetski, J. P., and S. C. Kowalczykowski. 1987. Transfer of RecA protein from one polynucleotide to another. Kinetic evidence for a ternary intermediate during the transfer reaction. *J. Biol. Chem.* 262:2085–2092.

136. Menge, K. L., and F. R. Bryant. 1988. ATP-stimulated hydrolysis of GTP by RecA protein: kinetic consequences of cooperative RecA protein-ATP interactions. *Biochemistry* 27:2635–2640.

137. Menge, K. L., and F. R. Bryant. 1992. Effect of nucleotide cofactor structure on recA protein-promoted DNA pairing. 1. Three-strand exchange reaction. *Biochemistry* 31:5151–5157.

138. Menge, K. L., and F. R. Bryant. 1992. Effect of nucleotide cofactor structure on recA protein-promoted DNA pairing. 2. DNA renaturation reaction. *Biochemistry* 31:5158–5165.

139. Michel, B. 2000. Replication fork arrest and DNA recombination. *Trends Biochem. Sci.* 25:173–178.

140. Mills, K. V., and H. Paulus. 2001. Reversible inhibition of protein splicing by zinc ion. *J. Biol. Chem.* 276:10832–10838.

141. Mitchell, R. S., A. Zlotnick, and S. L. Brenner. 1988. Direct evidence for two ssDNA-binding sites in a RecA nucleoprotein filament. *Biophys. J.* 53:220a.

142. Moran, N. A., and P. Baumann. 2000. Bacterial endosymbionts in animals. *Curr. Opin. Microbiol.* 3:270–275.

143. Morel, P., A. Stasiak, S. D. Ehrlich, and E. Cassuto. 1994. Effect of length and location of heterologous sequences on RecA-mediated strand exchange. *J. Biol. Chem.* 269:19830–19835.

144. Mount, D. W., K. B. Low, and S. J. Edmiston. 1972. Dominant mutations (lex) in *Escherichia coli* K-12 which affect radiation sensitivity and frequency of ultraviolet light-induced mutations. *J. Bacteriol.* 112:886–893.

145. Müller, B., T. Koller, and A. Stasiak. 1990. Characterization of the DNA binding activity of stable RecA-DNA complexes: interaction between the two DNA binding sites within RecA helical filaments. *J. Mol. Biol.* 212:97–112.

146. Murli, S., and G. C. Walker. 1993. SOS mutagenesis. *Curr. Opin. Genet. Dev.* 3:719–725.

147. Namsaraev, E., and P. Berg. 1997. Characterization of strand exchange activity of yeast Rad51 protein. *Mol. Cell. Biol.* 17:5359–5368.

148. Namsaraev, E. A., and P. Berg. 1998. Branch migration during Rad51-promoted strand exchange proceeds in either direction. *Proc. Natl. Acad. Sci. USA* 95:10477–10481.

149. Namsaraev, E. A., and P. Berg. 2000. Rad51 uses one mechanism to drive DNA strand exchange in both directions. *J. Biol. Chem.* 275:3970–3976.

150. Nayak, S., and F. R. Bryant. 1999. Differential rates of NTP hydrolysis by the mutant [S69G]RecA protein. Evidence for a coupling of NTP turnover to DNA strand exchange. *J. Biol. Chem.* 274:25979–25982.

151. Neuendorf, S. K., and M. M. Cox. 1986. Exchange of RecA protein between adjacent RecA protein-single-stranded DNA complexes. *J. Biol. Chem.* 261:8276–8282.

152. Nishinaka, T., Y. Ito, S. Yokoyama, and T. Shibata. 1997. An extended DNA structure through deoxyribose-base stacking induced by RecA protein. *Proc. Natl. Acad. Sci. USA* 94:6623–6628.

153. Nishinaka, T., A. Shinohara, Y. Ito, S. Yokoyama, and T. Shibata. 1998. Base pair switching by interconversion of sugar puckers in DNA extended by proteins of RecA-family: a model for homology search in homologous genetic recombination. *Proc. Natl. Acad. Sci. USA* 95:11071–11076.

154. Nohmi, T., J. R. Battista, L. A. Dodson, and G. C. Walker. 1988. RecA-mediated cleavage activates UmuD for mutagenesis: mechanistic relationship between transcriptional derepression and posttranslational activation. *Proc. Natl. Acad. Sci. USA* 85:1816–1820.

155. Nordén, B., C. Elvingson, M. Kubista, B. Sjoberg, H. Ryberg, M. Ryberg, K. Mortensen, and M. Takahashi. 1992. Structure of RecA-DNA complexes studied by combination of linear dichroism and small-angle neutron scattering measurements on flow-oriented samples. *J. Mol. Biol.* 226:1175–1191.

156. Ogawa, T., H. Wabiko, T. Tsurimoto, T. Horii, H. Masukata, and H. Ogawa. 1979. Characteristics of purified recA protein and the regulation of its synthesis in vivo. *Cold Spring Harbor Symp. Quant. Biol.* 2:909–915.

157. Ogawa, T., X. Yu, A. Shinohara, and E. H. Egelman. 1993. Similarity of the yeast RAD51 filament to the bacterial RecA filament. *Science* 259:1896–1899.

158. Passy, S. I., X. Yu, Z. F. Li, C. M. Radding, J. Y. Masson, S. C. West, and E. H. Egelman. 1999. Human Dmc1 protein binds DNA as an octameric ring. *Proc. Natl. Acad. Sci. USA* 96:10684–10688.

159. Pham, P., J. G. Bertram, M. O'Donnell, R. Woodgate, and M. F. Goodman. 2001. A model for SOS-lesion-targeted mutations in Escherichia coli. *Nature* 409:366–370.

160. Podyminogin, M. A., R. B. Meyer, and H. B. Gamper. 1996. RecA-catalyzed, sequence-specific alkylation of DNA by crosslinking oligonucleotides. Effects of length and nonhomologous base substitution. *Biochemistry* 35:7267–7274.

161. Podyminogin, M. A., R. B. Meyer, and H. B. Gamper. 1995. Sequence-specific covalent modification of DNA by crosslinking oligonucleotides. Catalysis by RecA and implication for the mechanism of synaptic joint formation. *Biochemistry* 34:13098–13108.

162. Postow, L., C. Ullsperger, R. W. Keller, C. Bustamante, A. V. Vologodskii, and N. R. Cozzarelli. 2001. Positive torsional strain causes the formation of a four-way junction at replication forks. *J. Biol. Chem.* 267:2790–2796.

163. Pugh, B. F., and M. M. Cox. 1988. General mechanism for RecA protein binding to duplex DNA. *J. Mol. Biol.* 203:479–493.

164. Pugh, B. F., and M. M. Cox. 1988. High salt activation of RecA protein ATPase in the absence of DNA. *J. Biol. Chem.* 263:76–83.

165. Pugh, B. F., and M. M. Cox. 1987. Stable binding of RecA protein to duplex DNA. Unraveling a paradox. *J. Biol. Chem.* 262:1326–1336.

166. Pugh, B. F., B. C. Schutte, and M. M. Cox. 1989. Extent of duplex DNA underwinding induced by RecA protein binding in the presence of ATP. *J. Mol. Biol.* 205:487–492.

167. Radman, M. 1974. Phenomenology of an inducible mutagenic DNA repair pathway in *Escherichia coli*: SOS repair hypothesis, p. 128–142. *In* L. Prakash, F. Sherman, M. Miller, C. Lawrence, and H. W. Tabor (ed.), *Molecular and Environmental Aspects of Mutagenesis.* Charles C Thomas Publisher, Springfield, Ill.

168. Register, J. C., III, and J. Griffith. 1985. The direction of RecA protein assembly onto single strand DNA is the same as the direction of strand assimilation during strand exchange. *J. Biol. Chem.* 260:12308–12312.

169. Rehrauer, W. M., and S. C. Kowalczykowski. 1993. Alteration of the nucleoside triphosphate (NTP) catalytic domain within Escherichia coli recA protein attenuates NTP hydrolysis but not joint molecule formation. *J. Biol. Chem.* 268:1292–1297.

170. Rehrauer, W. M., and S. C. Kowalczykowski. 1996. The DNA binding site(s) of the Escherichia coli RecA protein. *J. Biol. Chem.* 271:11996–12002.

171. Reuven, N. B., G. Arad, A. Maor-Shoshani, and Z. Livneh. 1999. The mutagenesis protein UmuC is a DNA polymerase activated by UmuD′, RecA, and SSB and is specialized for translesion replication. *J. Biol. Chem.* 274:31763–31766.

172. Reuven, N. B., G. Arad, A. Z. Stasiak, A. Stasiak, and Z. Livneh. 2001. Lesion bypass by the Escherichia coli DNA polymerase V requires assembly of a RecA nucleoprotein filament. *J. Biol. Chem.* 276:5511–5517.

173. Rice, K. P., J. C. Chaput, M. M. Cox, and C. Y. Switzer. 2000. RecA protein promotes DNA strand exchange with substrates containing isoguanine and 5-methyl isocytosine. *Biochemistry* 39:10177–10188.

174. Rice, K. P., A. L. Eggler, P. Sung, and M. M. Cox. 2001. DNA pairing and strand exchange by the *Escherichia coli* RecA and yeast Rad51 proteins without ATP hydrolysis: on the importance of not getting stuck. *J. Biol. Chem.* 276:38570–38581.

175. Roberts, J. W., C. W. Roberts, and N. L. Craig. 1978. Escherichia coli recA gene product inactivates phage lambda repressor. *Proc. Natl. Acad. Sci. USA* 75:4714–4718.

176. Robu, M. E., R. B. Inman, and M. M. Cox. 2001. RecA protein promotes the regression of stalled replication forks in vitro. *Proc. Natl. Acad. Sci. USA* 98:8211–8218.

177. Roca, A. I., and M. M. Cox. 1990. The RecA protein: structure and function. *Crit. Rev. Biochem. Mol. Biol.* 25:415–456.

178. Roca, A. I., and M. M. Cox. 1997. RecA protein: structure, function, and role in recombinational DNA repair. *Prog. Nucleic Acid Res. Mol. Biol.* 56:129–223.

179. Rosselli, W., and A. Stasiak. 1991. The ATPase activity of RecA is needed to push the DNA strand exchange through heterologous regions. *EMBO J.* 10:4391–4396.

180. Rould, E., K. Muniyappa, and C. M. Radding. 1992. Unwinding of heterologous DNA by RecA protein during the search for homologous sequences. *J. Mol. Biol.* 226:127–139.

181. Saves, I., M. A. Laneelle, M. Daffe, and J. M. Masson. 2000. Inteins invading mycobacterial RecA proteins. *FEBS Lett.* 480:221–225.

182. Schutte, B. C., and M. M. Cox. 1987. Homology-dependent changes in adenosine 5'-triphosphate hydrolysis during RecA protein promoted DNA strand exchange: evidence for long paranemic complexes. *Biochemistry* **26:**5616–5625.

183. Schutte, B. C., and M. M. Cox. 1988. Homology-dependent underwinding of duplex DNA in RecA protein generated paranemic complexes. *Biochemistry* **27:**7886–7894.

184. Sedgwick, S. G. 1976. Misrepair of overlapping daughter strand gaps as a possible mechanism for UV induced mutagenesis in UVR strains of Escherichia coli: a general model for induced mutagenesis by misrepair (SOS repair) of closely spaced DNA lesions. *Mutat. Res.* **41:**185–200.

185. Seigneur, M., S. D. Ehrlich, and B. Michel. 2000. RuvABC-dependent double-strand breaks in dnaBts mutants require RecA. *Mol. Microbiol.* **38:**565–574.

186. Seitz, E. M., J. P. Brockman, S. J. Sandler, A. J. Clark, and S. C. Kowalczykowski. 1998. RadA protein is an archaeal RecA protein homolog that catalyzes DNA strand exchange. *Genes Dev.* **12:**1248–1253.

187. Shan, Q., J. M. Bork, B. L. Webb, R. B. Inman, and M. M. Cox. 1997. RecA protein filaments: end-dependent dissociation from ssDNA and stabilization by RecO and RecR proteins. *J. Mol. Biol.* **265:**519–540.

188. Shan, Q., and M. M. Cox. 1998. On the mechanism of RecA-mediated repair of double-strand breaks: no role for four-strand DNA pairing intermediates. *Mol. Cell* **1:**309–317.

189. Shan, Q., and M. M. Cox. 1997. RecA filament dynamics during DNA strand exchange reactions. *J. Biol. Chem.* **272:**11063–11073.

190. Shan, Q., and M. M. Cox. 1996. RecA protein dynamics in the interior of RecA nucleoprotein filaments. *J. Mol. Biol.* **257:**756–774.

191. Shan, Q., M. M. Cox, and R. B. Inman. 1996. DNA strand exchange promoted by RecA K72R. Two reaction phases with different Mg2+ requirements. *J. Biol. Chem.* **271:**5712–5724.

192. Shibata, T., R. P. Cunningham, and C. M. Radding. 1981. Homologous pairing in genetic recombination. Purification and characterization of *Escherichia coli* RecA protein. *J. Biol. Chem.* **256:**7557–7564.

193. Shinagawa, H. 1996. SOS response as an adaptive response to DNA damage in prokaryotes. *EXS* **77:**221–235.

194. Shinagawa, H., H. Iwasaki, T. Kato, and A. Nakata. 1988. RecA protein-dependent cleavage of UmuD protein and SOS mutagenesis. *Proc. Natl. Acad. Sci. USA* **85:**1806–1810.

195. Shingledecker, K., S. Jiang, and H. Paulus. 2000. Reactivity of the cysteine residues in the protein splicing active center of the Mycobacterium tuberculosis RecA intein. *Arch. Biochem. Biophys.* **375:**138–144.

196. Shinohara, A., H. Ogawa, and T. Ogawa. 1992. Rad51 protein involved in repair and recombination in S. cerevisiae is a RecA-like protein. *Cell* **69:**457–470.

197. Sigurdsson, S., K. Trujillo, B. W. Song, S. Stratton, and P. Sung. 2001. Basis for avid homologous DNA strand exchange by human Rad51 and RPA. *J. Biol. Chem.* **276:**8798–8806.

198. Simonson, T., M. Kubista, R. Sjoback, H. Ryberg, and M. Takahashi. 1994. Properties of RecA-oligonucleotide complexes. *J. Mol. Recogn.* **7:**199–206.

199. Smith, B. T., and G. C. Walker. 1998. Mutagenesis and more: umuDC and the Escherichia coli SOS response. *Genetics* **148:**1599–1610.

200. Sommer, S., F. Boudsocq, R. Devoret, and A. Bailone. 1998. Specific RecA amino acid changes affect RecA-UmuD'C interaction. *Mol. Microbiol.* **28:**281–291.

201. Stasiak, A., and E. Di Capua. 1982. The helicity of DNA in complexes with RecA protein. *Nature* **299:**185–186.

202. Stasiak, A., and E. H. Egelman. 1988. Visualization of recombination reactions, p. 265–307. *In* R. Kucherlapati and G. R. Smith (ed.), *Genetic Recombination*. American Society for Microbiology, Washington, D.C.

203. Stasiak, A., E. H. Egelman, and P. Howard-Flanders. 1988. Structure of helical RecA-DNA complexes. III. The structural polarity of RecA filaments and functional polarity in the RecA-mediated strand exchange reaction. *J. Mol. Biol.* **202:**659–662.

204. Story, R. M., and T. A. Steitz. 1992. Structure of the RecA protein-ADP complex. *Nature* **355:**374–376.

205. Story, R. M., I. T. Weber, and T. A. Steitz. 1992. The structure of the E. coli RecA protein monomer and polymer. *Nature* **355:**318–325.

206. Sung, P. 1994. Catalysis of ATP-dependent homologous DNA pairing and strand exchange by yeast RAD51 protein. *Science* **265:**1241–1243.

207. Sung, P., and D. L. Robberson. 1995. DNA strand exchange mediated by a RAD51-ssDNA nucleoprotein filament with polarity opposite to that of RecA. *Cell* **82:**453–461.

208. Sung, P., and S. A. Stratton. 1996. Yeast Rad51 recombinase mediates polar DNA strand exchange in the absence of ATP hydrolysis. *J. Biol. Chem.* **271:**27983–27986.

209. Sweasy, J. B., E. M. Witkin, N. Sinha, and V. Roegnermaniscalco. 1990. RecA protein of *Escherichia coli* has a 3rd essential role in SOS mutator activity. *J. Bacteriol.* **172:**3030–3036.

210. Takahashi, M. 1989. Analysis of DNA-RecA protein interactions involving the protein self-association reaction. *J. Biol. Chem.* **264:**288–295.

211. Takahashi, M., M. Kubista, and B. Nordén. 1991. Coordination of multiple DNA molecules in RecA fiber evidenced by linear dichroism spectroscopy. *Biochimie* **73:**219–226.

212. Takahashi, M., and B. Nordén. 1994. Structure of RecA-DNA complex and mechanism of DNA strand exchange reaction in homologous recombination. *Adv. Biophys.* **30:**1–35.

213. Tang, M. J., X. Shen, E. G. Frank, M. O'Donnell, R. Woodgate, and M. F. Goodman. 1999. UmuD'$_2$C is an error-prone DNA polymerase, Escherichia coli pol V. *Proc. Natl. Acad. Sci. USA* **96:**8919–8924.

214. Tateishi, S., T. Horii, T. Ogawa, and H. Ogawa. 1992. C-terminal truncated Escherichia coli RecA protein RecA5327 has enhanced binding affinities to single- and double-stranded DNAs. *J. Mol. Biol.* **223:**115–129.

215. Thompson, T. B., M. G. Thomas, J. C. Escalante-Semerena, and I. Rayment. 1999. Three-dimensional structure of adenosylcobinamide kinase/adenosylcobinamide phosphate guanylyltransferase (CobU) complexed with GMP: evidence for a substrate-induced transferase active site. *Biochemistry* **38:**12995–13005.

216. Tracy, R. B., F. Chedin, and S. C. Kowalczykowski. 1997. The recombination hot spot chi is embedded within islands of preferred DNA pairing sequences in the E. coli genome. *Cell* **90:**205–206.

217. Tracy, R. B., and S. C. Kowalczykowski. 1996. In vitro selection of preferred DNA pairing sequences by the Escherichia coli RecA protein. *Genes Dev.* **10:**1890–1903.

218. Ullsperger, C. J., and M. M. Cox. 1995. Quantitative RecA protein binding to the hybrid duplex product of DNA strand exchange. *Biochemistry* **34:**10859–10866.

219. Umezu, K., and R. D. Kolodner. 1994. Protein interactions in genetic recombination in Escherichia coli. Interactions involving RecO and RecR overcome the inhibition of RecA

by single-stranded DNA-binding protein. *J. Biol. Chem.* **269:**30005–30013.

220. **Wagner, J., and T. Nohmi.** 2000. *Escherichia coli* DNA polymerase IV mutator activity: genetic requirements and mutational specificity. *J. Bacteriol.* **182:**4587–4595.

221. **Walker, G. C.** 1985. Inducible DNA repair systems. *Annu. Rev. Biochem.* **54:**425–457.

222. **Walker, G. C.** 1987. The SOS response of *Escherichia coli*, p. 1346–1357. *In* F. C. Neidhardt, J. L. Ingraham, K. B. Low, B. Magasanik, M. Schaechter, and H. E. Umbarger (ed.), Escherichia coli *and* Salmonella typhimurium: *Cellular and Molecular Biology*, vol. 2. American Society for Microbiology, Washington, D.C.

223. **Walker, G. C.** 1995. SOS-regulated proteins in translesion DNA synthesis and mutagenesis. *Trends Biochem. Sci.* **20:**416–420.

224. **Walker, G. C., B. T. Smith, and M. D. Sutton.** 2000. The SOS response to DNA damage, p. 131–144. *In* G. Storz and R. Hengge-Aronis (ed.), *Bacterial Stress Responses.* American Society for Microbiology, Washington, D.C.

225. **Wang, Y. H., C. D. Bortner, and J. Griffith.** 1993. RecA binding to bulge- and mismatch-containing DNAs. Certain single base mismatches provide strong signals for RecA binding equal to multiple base bulges. *J. Biol. Chem.* **268:**17571–17577.

226. **Weinstock, G. M., K. McEntee, and I. R. Lehman.** 1981. Hydrolysis of nucleoside triphosphates catalyzed by the recA protein of Escherichia coli. Characterization of ATP hydrolysis. *J. Biol. Chem.* **256:**8829–8834.

227. **Wernegreen, J. J., H. Ochman, I. B. Jones, and N. A. Moran.** 2000. Decoupling of genome size and sequence divergence in a symbiotic bacterium. *J. Bacteriol.* **182:**3867–3869.

228. **West, S. C.** 1992. Enzymes and molecular mechanisms of genetic recombination. *Annu. Rev. Biochem.* **61:**603–640.

229. **West, S. C., E. Cassuto, and P. Howard-Flanders.** 1981. Heteroduplex formation by RecA protein: polarity of strand exchanges. *Proc. Natl. Acad. Sci. USA* **78:**6149–6153.

230. **Wilson, D. H., and A. S. Benight.** 1990. Kinetic analysis of the pre-equilibrium steps in the self-assembly of RecA protein from *Escherichia coli. J. Biol. Chem.* **265:**7351–7359.

231. **Wilson, J. H.** 1979. Nick-free formation of reciprocal heteroduplexes: a simple solution to the topological problem. *Proc. Natl. Acad. Sci. USA* **76:**3641–3645.

232. **Witkin, E. M.** 1976. Ultraviolet mutagenesis and inducible DNA repair in *Escherichia coli. Bacteriol. Rev.* **40:**869–907.

233. **Wittung, P., B. Nordén, S. K. Kim, and M. Takahashi.** 1994. Interactions between DNA molecules bound to RecA filament. Effects of base complementarity. *J. Biol. Chem.* **269:**5799–5803.

234. **Wittung, P., B. Norden, and M. Takahashi.** 1995. Secondary structure of RecA in solution. The effects of cofactor, DNA and ionic conditions. *Eur. J. Biochem.* **228:**149–154.

235. **Wu, A. M., M. Bianchi, C. Das Gupta, and C. M. Radding.** 1983. Unwinding associated with synapsis of DNA molecules by RecA protein. *Proc. Natl. Acad. Sci. USA* **80:**1256–1260.

236. **Yancey-Wrona, J. E., and R. D. Camerini-Otero.** 1995. The search for DNA homology does not limit stable homologous pairing promoted by RecA protein. *Curr. Biol.* **5:**1149–1158.

237. **Yu, X., and E. H. Egelman.** 1992. Direct visualization of dynamics and co-operative conformational changes within RecA filaments that appear to be associated with the hydrolysis of adenosine 5′-O-(3-thiotriphosphate). *J. Mol. Biol.* **225:**193–216.

238. **Yu, X., and E. H. Egelman.** 1993. The LexA repressor binds within the deep helical groove of the activated RecA filament. *J. Mol. Biol.* **231:**29–40.

239. **Yu, X., and E. H. Egelman.** 1997. The RecA hexamer is a structural homologue of ring helicases. *Nat. Struct. Biol.* **4:**101–104.

240. **Yu, X., and E. H. Egelman.** 1992. Structural data suggest that the active and inactive forms of the RecA filament are not simply interconvertible. *J. Mol. Biol.* **227:**334–346.

241. **Zaitsev, E. N., and S. C. Kowalczykowski.** 2000. A novel pairing process promoted by Escherichia coli RecA protein: inverse DNA and RNA strand exchange. *Genes Dev.* **14:**740–749.

242. **Zhou, X., and K. Adzuma.** 1997. DNA strand exchange mediated by the Escherichia coli RecA protein initiates in the minor groove of double-stranded DNA. *Biochemistry* **36:**4650–4661.

243. **Zhurkin, V. B., G. Raghunathan, N. B. Ulyanov, O. R. Camerini, and R. L. Jernigan.** 1994. A parallel DNA triplex as a model for the intermediate in homologous recombination. *J. Mol. Biol.* **239:**181–200.

244. **Zlotnick, A., R. S. Mitchell, R. K. Steed, and S. L. Brenner.** 1993. Analysis of two distinct single-stranded DNA binding sites on the recA nucleoprotein filament. *J. Biol. Chem.* **268:**22525–22530.

Chapter 21

Homologous Recombination by the RecBCD and RecF Pathways

MARIA SPIES AND STEPHEN C. KOWALCZYKOWSKI

RECOMBINATIONAL REPAIR OF DNA DAMAGE

Homologous, or general, recombination is a crucial biological process that involves the paring and transfer of strands between DNA molecules that share a region of significant sequence homology. After its discovery in 1946 by Lederberg and Tatum (48), homologous recombination in bacteria was associated with the sexual process of conjugation and was viewed as an evolutionary mechanism both for shuffling the genome and for spreading favorable alleles. However, more recently, a more immediate function of homologous recombination has been recognized: namely, it is a mechanism for the maintenance of chromosomal integrity that acts to repair DNA lesions, both double-strand DNA breaks and single-strand DNA gaps, generated during the course of DNA replication. The intimate connection between the processes of replication and recombination was initially appreciated in the life cycle of bacteriophage T4 (56) and then later recognized as an important determinant of viability in bacteria (39, 45). In T4 phage, recombination is linked to replication to produce a high yield of phage DNA; in *Escherichia coli*, recombination is linked to replication to permit its completion when interrupted by DNA damage, and also to initiate DNA replication in the absence of origin function. This view of recombination as an integral part of efficient chromosome duplication also reconciled the high level of inviability (up to 95%) of recombination-deficient cells (12).

GENERATION OF A DSB IN DNA

A significant fraction of DNA damage affects only one strand of the DNA duplex. Such lesions can often be repaired by one of the repair systems that are specific for a particular DNA lesion (see reference 28 for a review). These repair systems use the intact complementary strand as a template to restore the damaged DNA molecule to its original state. Occasionally both strands of the double-stranded DNA (dsDNA) can be broken opposite to each other, resulting in a double-strand break (DSB). DSBs can be produced, for example, as a direct consequence of ionizing radiation (Fig. 1A). However, the bulk of the DSBs in bacteria are generated indirectly as the result of DNA replication through an unrepaired break in just a single strand of DNA (Fig. 1B). Replication of DNA containing a single-strand nick or a gap in the leading strand results in the dissociation of the DNA polymerase holoenzyme complex and the generation of one blunt DSB (Fig. 1B), whereas replication of DNA with a nick in the lagging strand produces a DSB with a 3'-terminated single-stranded DNA (ssDNA) tail.

A DSB can also be produced when the DNA replication process is halted by anything that might block the progress of the replisome. Upon dissociation of the replisome, the stalled replication fork can regress to produce a Holliday junction. Such structure contains both a dsDNA end and an intermediate of recombination and, hence, attracts recombination machinery in a manner similar to that of the simple DSB. In addition, the Holliday junction can be cleaved to provide yet another path for DSB formation (Fig. 2). Left unrepaired, DSBs are lethal, and *E. coli* can repair only a few DSBs per chromosome without dying (43). Thus, DNA replication is a major source of endogenous DSBs that, in turn, are repaired by homologous recombination. This relationship between DNA replication and recombinational repair of DSBs is summarized in Fig. 1.

Recombinational repair, however, requires a homologous DNA molecule to be used as a template from which to restore, by DNA synthesis, the genetic

Maria Spies and Stephen C. Kowalczykowski • Sections of Microbiology and of Molecular and Cellular Biology, Center for Genetics and Development, University of California, Davis, CA 95616-8665.

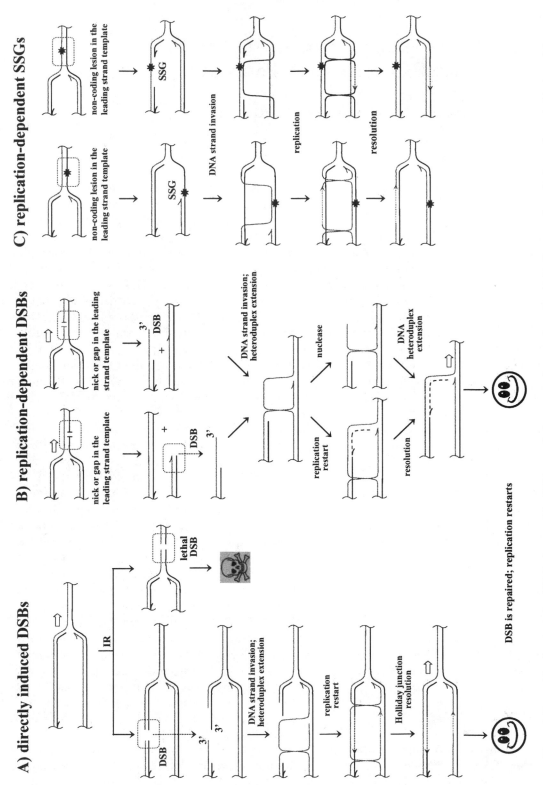

Figure 1. Recombinational repair of DNA damage. The DNA strands that are used as templates for the leading strand synthesis are shown in black, and the strands used as templates for the lagging strand synthesis are shown in gray. Arrows indicate the direction of DNA synthesis. (A) Recombinational repair of directly induced DSBs. A DSB, which occurs in newly synthesized DNA, can be repaired by completion of the following steps. First, the DSB is processed to produce 3'-terminated ssDNA. Then, one of the ssDNA tails can invade the homologous dsDNA daughter, displacing one of the resident strands to form a D-loop. This structure can be used as a template for DNA synthesis, ultimately resulting in the formation of a Holliday junction. Upon resolution of the Holliday junction, the replication fork is restored to its original form. When a DSB occurs in a part of the chromosome that is not yet replicated, there is no homologous DNA to serve as a template, and such a DSB can be lethal. (B) Recombinational repair of replication-dependent DSBs. DSBs can be produced by

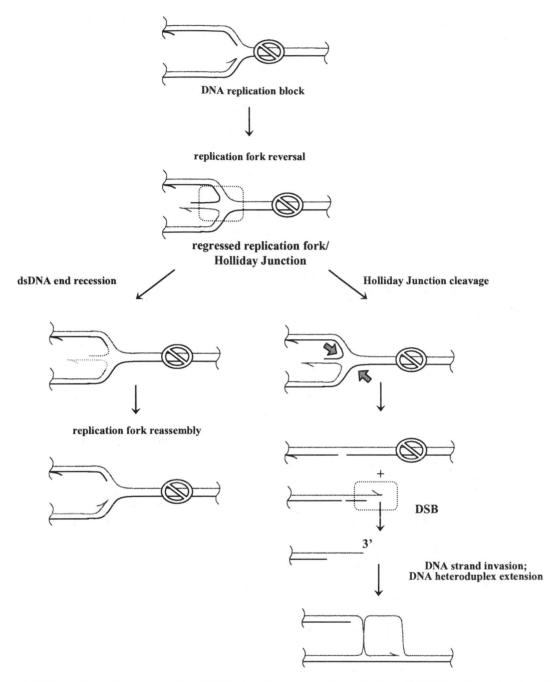

Figure 2. DSBs can be produced as a result of Holliday junction cleavage. Reversal of a stalled DNA replication fork results in the formation of a regressed replication fork, which is a four-way (Holliday) junction that contains a dsDNA end. There are two potential fates for this regressed replication fork: it can be degraded by the recombination machinery to produce a three-way junction that resembles a replication fork, or the Holliday junction can be cleaved to produce a DSB that is repaired by a DSB repair pathway.

replication through a single-strand DNA break. The source of the replication-dependent DSBs is a nick or an ssDNA gap in one of the strands of the replicated DNA molecule. Replication through the strand discontinuity results in the formation of one intact and one broken DNA molecule. The end of the broken chromosome is processed to form a 3′-terminated ssDNA tail, which invades the intact homologous DNA to form a D-loop, which can then be used to restore a normal replication fork. (C) Recombinational repair of replication-dependent SSGs. SSGs can be produced when synthesis of only one DNA strand is halted by an encounter of a noncoding lesion in that DNA strand. After DNA strand exchange of the ssDNA in the gap with a strand in the intact daughter homologue, the displaced DNA strand can be used as a template for DNA synthesis, resulting in the restoration of the replication fork. Upon completion of replication, one of the DNA molecules will still contain the original DNA damage. If this damage is not repaired by the appropriate repair system, then an SSG will be formed again by the next round of DNA replication.

content of the damaged DNA molecule. These homologous DNA sequences are generated by DNA replication, and, therefore, the meridiploid character of bacterial chromosomes provides the necessary templates for recombinational repair.

GENERATION OF A SINGLE-STRAND GAP IN DNA

When synthesis of a DNA strand is blocked by a noncoding lesion (for example, an abasic site or an intrastrand cross-link, such as a thymine dimer), continued replication of the flawless strand beyond the lesion produces a single-strand DNA gap (SSG) (Fig. 1C) (42, 54). SSGs are also known as daughter-strand gaps, since they appear on one of the newly synthesized daughter strands during semiconservative DNA replication.

The interstrand DNA cross-link is a special class of chemical damage to DNA, since it blocks DNA replication completely by preventing DNA strand separation (Fig. 3) (28). Repair of this special class of DNA damage is absolutely dependent on homologous recombination (65). Even a single cross-link is lethal for recombination-deficient cells, while wild-type *E. coli* cells can tolerate up to 70 cross-links per chromosome (65). DNA cross-links are repaired through an incisional-recombinational mechanism. An incision is made at both sides of the cross-link on one DNA strand, and displacement of the incised oligonucleotide results in an SSG that can be repaired through an SSG repair pathway. Subsequent incision on the second strand releases the cross-linked strands, producing a second SSG that is repaired by a second round of recombinational repair.

For all SSGs, if the ssDNA in the gap is cleaved before the completion of SSG repair, then the lesion

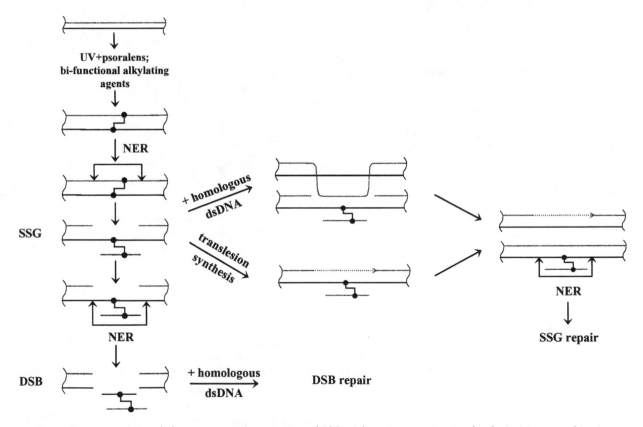

Figure 3. Interstrand cross-links are converted into DSBs and SSGs. Schematic representation for the incision-recombination mechanism of interstrand DNA cross-link repair. The repair of an interstrand cross-link depends on both the nucleotide excision repair (NER) and recombination machineries. The NER enzymes make an incision on either side of the lesion on one DNA strand and also displace the cross-link-containing oligonucleotide to produce an SSG; this SSG can be repaired by an SSG repair pathway. However, the incised oligonucleotide remains cross-linked to the second DNA strand. If this lesion is recognized by NER machinery prior to the completion of SSG repair of the top strand, then the subsequent incisions on the second strand convert the interstrand cross-link into a DSB, which is then repaired by a DSB repair pathway.

is converted into a DSB. Thus, some fraction of SSGs are converted into DSBs and are repaired through the DSB repair pathway.

CONJUGATION, TRANSDUCTION, AND TRANSFORMATION CREATE DSBs AND SSGs

DSBs arise not only as a consequence of DNA damage; in fact, DSBs are formed as a natural step in several normal cellular processes. The molecular feature common to each of these processes is the cellular acquisition of a linear segment of dsDNA. The end of this linear DNA molecule is seen as a DSB, and the recombinational repair of the DSB is initiated. Bacteria can acquire linear DNA by any of three different routes. First, during conjugation, a copy of a chromosome is transferred from one bacterium, which possesses a fertility factor (the F' plasmid), to another. Genetic studies of this process led to isolation of the first recombination-deficient mutants (21). Transformation is a second route by which some bacteria take up a segment of DNA from the environment; unrelated in process, but not in the form of the DNA involved, artificial transformation is widely used as a laboratory technique to introduce foreign dsDNA into cells. Finally, infection by bacteriophage also results in the introduction of linear DNA into bacteria. Note, however, that in addition to DSBs, the processes of conjugation and natural transformation also produce SSGs due to incomplete replication of the DNA intermediates that form during these processes.

HOMOLOGOUS PAIRING AND DNA STRAND EXCHANGE MEDIATED BY RecA PROTEIN

A step common to all pathways of recombinational repair discussed below is the homologous pairing of DNA (Fig. 1). In bacteria, the ubiquitous RecA protein catalyzes an invasion of ssDNA into homologous duplex DNA and the exchange of DNA strands. Regardless of whether the DNA lesion is a DSB or an SSG, pathway-specific processing of the break produces an extensive region of ssDNA that serves as the substrate for assembly of the RecA nucleoprotein filament. The process of finding DNA sequence homology and exchanging DNA strands occurs in three defined stages: (i) presynapsis, during which the RecA nucleoprotein is assembled; (ii) synapsis, during which the homology search and ex-

change of DNA strands occur; and (iii) postsynapsis, during which branch migration can occur. Afterward, resolution of the resulting recombination intermediate produces a recombinant molecule.

In *E. coli*, *recA* null mutations reduce conjugational recombination by 100,000-fold (21). The active form of RecA protein is a nucleoprotein filament formed by the cooperative binding of RecA protein to the ssDNA tails of the processed DSB, or to the ssDNA in the SSG (see reference 6 for a review). Thus, ssDNA is an essential, although transient, intermediate in the process of homologous recombination. Most of the SSGs, whose average length is minimally 200 nucleotides (78) (but can be as large as 800 nucleotides [37], or half the average size of an Okazaki fragment [66]), are sufficiently large for RecA nucleoprotein filament assembly; however, DSBs are not the normal direct targets of RecA protein binding. Therefore, DSBs and short SSGs are processed to produce longer regions of ssDNA. This processing requires either a nuclease, a helicase, or both (see below).

However, production of ssDNA is not sufficient for RecA nucleoprotein filament assembly in vivo. Within the cell, ssDNA rapidly forms a complex with the ssDNA-binding (SSB) protein. SSB protein binding has the beneficial consequences of protecting ssDNA from degradation by nucleases and of disrupting inhibitory DNA secondary structures. Unfortunately, SSB protein blocks assembly of the RecA nucleoprotein filament by competing directly with RecA protein for ssDNA binding. Thus, the SSB protein-ssDNA complex is a kinetic barrier for the assembly of a RecA-ssDNA nucleoprotein filament, and its formation is inhibitory for the subsequent steps of homologous recombination. However, to counteract this inhibitory effect of SSB protein, there is a class of the recombination/replication mediator proteins whose function is to facilitate assembly of RecA protein (and its homologues in other organisms) on ssDNA, thereby alleviating the kinetic barrier imposed by SSB protein (the role of replication mediator proteins is reviewed in reference 5). Thus, in addition to the production of a single-stranded region within dsDNA, recombination requires the "loading" of RecA protein onto this ssDNA.

In wild-type *E. coli*, the processing of a broken DNA molecule, and the subsequent delivery of RecA protein to this ssDNA, occurs by either of two pathways: RecBCD or RecF. The pathway names reflect critical and unique enzymes acting in each. The RecBCD pathway is used primarily to initiate recombination at a DSB, whereas the RecF pathway is used for recombinational repair at SSGs. The concept of recombination pathways was initially postulated by Clark (20). In both pathways, the Holliday junctions that result are resolved by the RuvABC enzyme complex (see reference 46 for a review) into the recombinant progeny.

DSB REPAIR BY RECOMBINATION: THE RecBCD PATHWAY

The RecBCD pathway comprises the RecBCD, SSB, RecA, and RuvABC proteins; in addition, a specific DNA locus called χ (Chi [crossover hot spot instigator], 5'-GCTGGTGG-3') is required in the dsDNA that is broken.

In wild-type *E. coli*, RecBCD enzyme is required for approximately 99% of the recombination events associated with conjugation and transduction, i.e., the processes that involve linear DNA and, hence, a DSB. Genetic studies revealed that the products of *recB* and *recC* genes are necessary for the repair of DSBs and for conjugational recombination. Deletion of either *recB* or *recC* genes reduces the levels of sexual recombination 100- to 1,000-fold and increases sensitivity to DNA-damaging agents, such as UV irradiation and mitomycin C (26, 36, 84). This effect, however, is significantly smaller then the 100,000-fold decrease in homologous recombination in the *recA* mutant cells. In contrast, cells deleted for *recD* are not only proficient in homologous recombination and DSB repair, but they also catalyze homologous recombination at high rates (1).

The RecBCD enzyme is a heterotrimer consisting of three nonidentical polypeptides, RecB (134 kDa), RecC (129 kDa), and RecD (67 kDa) (1, 58). The three subunits compose a complex, multifunctional enzyme that possesses a number of seemingly disparate catalytic activities, including DNA-dependent ATPase, DNA helicase, ssDNA endo- and exonuclease, and dsDNA exonuclease. In vivo, these activities enable RecBCD enzyme to carry out a highly coordinated set of biochemical reactions resulting in conversion of the broken dsDNA into the active species in homologous recombination: the RecA nucleoprotein filament assembled on the ssDNA containing χ.

Among DNA helicases, an unusual feature of the RecBCD enzyme is its ability to initiate unwinding at a blunt or nearly blunt dsDNA end. RecBCD enzyme binds to blunt DNA ends with a very high affinity (~ 1 nM), forming an initiation complex (30) that has a footprint of about 20 or 21 nucleotides on the 5'-terminated strand and 16 to 17 nucleotides on the 3'-terminated strand.

RecB protein has a modular organization. The N-terminal domain of the RecB subunit contains motifs characteristic for the superfamily 1 (SF1) DNA helicases. Moreover, the purified RecB protein is an ssDNA-dependent ATPase and a DNA helicase (10). Its behavior is related to the well-characterized Rep, UvrD, and PcrA helicases. Similar to these enzymes, RecB protein displays $3' \rightarrow 5'$ helicase activity, requiring a 3'-terminated ssDNA region flanking the duplex DNA that is to be unwound. The C-terminal domain of this same subunit contains a nuclease motif resembling that of other nucleases, such as FokI and BamHI (77, 85) and λ exonuclease (2c). Besides being responsible for the complicated nuclease activity of RecBCD enzyme, the C-terminal domain of the RecB subunit has another crucial function: it harbors a site for interaction with RecA protein (19).

The amino acid sequence of RecC protein provides no clue as to its role in the holoenzyme. However, the existence of the RecC mutants that enable the holoenzyme to recognize an altered χ sequence suggests a significant role in χ recognition (3). Interaction with the RecC subunit greatly stimulates the weak nuclease activity of the RecB helicase, and increases its affinity for dsDNA ends (61). The resulting RecBC enzyme is a fast, processive helicase that can initiate homologous recombination. However, RecBC enzyme displays negligible nuclease activity. As a result, homologous recombination in the *recD* mutant cells strongly depends on the function of RecJ nuclease (51, 51a).

The RecD subunit also contains SF1 helicase motifs. The purified RecD protein is a DNA-dependent ATPase that was recently shown to be a $5' \rightarrow 3'$ helicase (23) similar to the closely related TraI and Dda helicases. Like these other helicases, RecD protein unwinds substrates containing 5'-terminated ssDNA flanking the duplex DNA. Interaction with the RecD subunit further stimulates both the helicase activity and the dsDNA end-binding affinity of RecBC enzyme. Given that the RecB subunit is bound to the 3'-terminated strand, the RecD subunit is bound to the 5'-terminated strand (27), and each motor subunit moves with an opposite polarity, the resulting bipolar RecBCD enzyme translocates with each motor subunit moving in the same direction relative to the DSB. In addition, the resulting RecBCD holoenzyme now manifests its vigorous nuclease activity, implying that the RecD subunit also activates the nuclease contained within the RecB subunit (41).

The activities described above permit the following description of the mechanism of action of the RecBCD helicase/nuclease (Fig. 4). Upon binding to the DNA end (Fig. 4a), the enzyme uses the energy of ATP hydrolysis to translocate along and unwind the dsDNA molecule, consuming approximately two

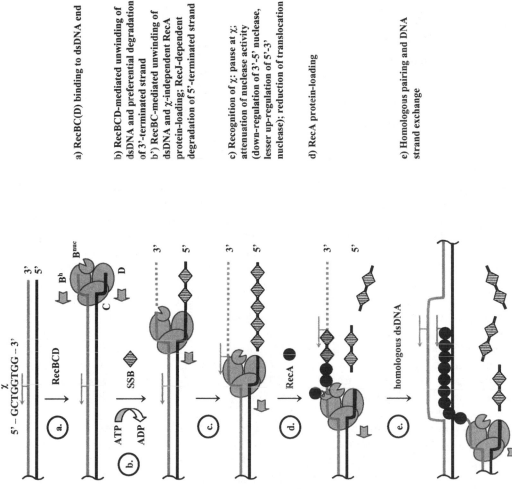
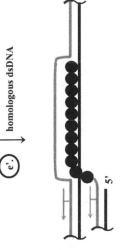
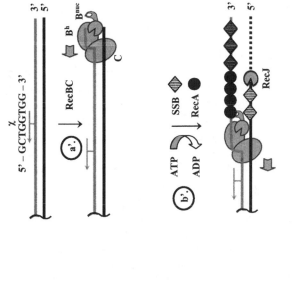

a) RecBC(D) binding to dsDNA end

b) RecBCD-mediated unwinding of dsDNA and preferential degradation of 3'-terminated strand

b') RecBC-mediated unwinding of dsDNA and χ-independent RecA protein-loading; RecJ-dependent degradation of 5'-terminated strand

c) Recognition of χ: pause at χ; attenuation of nuclease activity (down-regulation of 3'–5' nuclease, lesser up-regulation of 5'–3' nuclease); reduction of translocation

d) RecA protein-loading

e) Homologous pairing and DNA strand exchange

Figure 4. Initiation of homologous recombination at DSBs by the RecBCD pathway. Schematic representation of the early enzymatic steps of RecBCD pathway of recombinational repair. (a and a') RecBCD (or RecBC) enzyme binds to the blunt or nearly blunt dsDNA end. (b) RecBCD enzyme uses the energy of ATP hydrolysis to translocate along and to unwind the dsDNA. The associated nuclease activity degrades the newly produced ssDNA. (c) Interaction with χ results in the attenuation of the RecBCD nuclease activity and production of ssDNA terminated with the χ sequence at its 3' end. (d) The χ-modified enzyme loads RecA protein onto the χ-containing ssDNA to the exclusion of SSB protein, forming a RecA nucleoprotein filament. (e and e') The RecA-ssDNA nucleoprotein filament invades homologous dsDNA. (b') The RecBC enzyme (i.e., lacking the RecD subunit) behaves as a constitutively χ-modified RecBCD enzyme that can load RecA protein without the need for χ interaction. The RecBC enzyme lacks nuclease activity that is compensated for by the action of the RecJ nuclease.

molecules of ATP per base pair unwound (63) (Fig. 4b). RecBCD enzyme unwinds, on average, 30,000 bp of dsDNA per binding event (62) at a rate of approximately 1,000 to 1,300 bp/s (at 37°C). DNA unwinding by RecBCD enzyme is accompanied by endonucleolytic cleavage of the newly produced ssDNA. The nuclease activity of RecBCD enzyme is asymmetric, with digestion occurring preferentially on the 3′-terminated strand relative to the DSB (25).

THE RECOMBINATION HOT SPOT, χ

Originally, χ was identified as a *cis*-acting mutation in bacteriophage λ that allowed its more efficient growth in *E. coli* by stimulating the host's recombination system. Stimulation of recombination is approximately 10-fold (47), and it is confined to loci that are downstream of χ (70): stimulation is highest at χ and then decreases exponentially (17, 47). The χ sequence is overrepresented in *E. coli*: 1,009 χ sequences are found in the 4.6-Mb genome of the MG1655 strain (9). Furthermore, over 60% of χ sequences are oriented toward replication origin. Such an orientation would facilitate RecBCD-mediated recombination repair of DSBs created during DNA replication (11).

Biochemically, the activities of RecBCD enzyme are altered upon recognition of χ. Alteration of RecBCD enzyme activity is manifested only when the enzyme approaches χ, 5′-GCTGGTGG-3′, from its 3′ side. In the schematic depiction (Fig. 4), χ, which is the sequence in the "top strand" (7), is recognized by an enzyme moving only from right to left. In vivo, interaction with χ results in the stimulation of homologous recombination downstream of χ (69, 71). In vitro, recognition of the χ sequence causes RecBCD enzyme to switch the polarity of its nuclease activity (Fig. 4c): upon interaction with χ, degradation of the 3′-terminated strand is downregulated (24, 25), while degradation of the 5′-terminated strand is upregulated (2a). Consequently, the enzyme produces a lengthy ssDNA tail with χ at the 3′-terminated end. As determined from production of the χ-specific ssDNA fragments, the probability of recognizing a single χ is about 30 to 40% (3, 72). Interaction with χ also affects the helicase activity of RecBCD enzyme. Recognition of χ causes the enzyme to pause briefly (typically, a few seconds) at χ and to resume translocation after the χ site, but at a rate that is reduced by approximately twofold (68). Another consequence of χ modification is that the RecBCD enzyme gains the ability to "load" RecA protein onto the newly produced ssDNA (2b) (Fig. 4d). Thus, in response to χ, the RecBCD enzyme accomplishes both tasks

essential for initiation of homologous recombination: (i) it recesses the DSB to produce an ssDNA-tailed duplex DNA with χ at its terminus, and (ii) it catalyzes formation of the RecA nucleoprotein filament on the ssDNA produced. This RecA nucleoprotein filament now can search for homology, promote invasion of the homologous recipient, and exchange the DNA strands (Fig. 4e).

The exact molecular mechanism by which χ recognition is translated into the observed changes in the activities of the RecBCD enzyme remains unknown. Most models propose either dissociation or inactivation of the RecD subunit (2, 19, 57, 73). Indeed, as mentioned previously, the RecBC enzyme (lacking the RecD subunit) is recombinationally proficient both in vivo (13) and in vitro (18). The RecBC enzyme is a processive helicase but with little or no nuclease activity (41), which is in contrast to the χ-activated form of RecBCD enzyme (Fig. 4a′); however, this distinction is consistent with the requirement for RecJ nuclease activity in *recD* mutant cells in vivo (Fig. 4b′) (51). Similar to RecBCD enzyme, RecBC helicase facilitates asymmetric assembly of RecA protein only onto ssDNA that is 3′ terminal at the enzyme entry site (18) (Fig. 4e′). The RecBC-mediated loading of RecA protein is constitutive and is independent of χ, consistent with the phenotypic behavior of *recD* mutant cells in vivo.

REGULATED HELICASES / NUCLEASES IN OTHER BACTERIAL SPECIES

For decades, the interaction between RecBCD enzyme and χ was known to exist only in the species of enteric bacteria closely related to *E. coli*. But relatively recently, short (5 to 8 bp) sequences, which protect linear dsDNA from degradation by attenuation of the nuclease activity of RecBCD-like helicase/ nuclease enzymes, were found in several distantly related bacteria (8, 16, 67). This finding indicates that, although apparently not universal, the regulation of recombinational helicases/nucleases by specific DNA sequences is widely spread among prokaryotes. While some bacteria possess clear homologues of RecBCD enzyme, other species contain its functional equivalent, the AddAB enzyme (reviewed in reference 15). AddAB helicase/nuclease comprises two subunits encoded by *addA* and *addB* genes. The sole motor subunit of AddAB enzyme, AddA protein, contains an SF1 helicase and a nuclease domain, which display a high degree of similarity to those in RecB protein (32). Also similar to RecB protein, AddA is a 3′ → 5′ helicase, and its 3′ → 5′ nuclease activity is downregulated upon interaction with the cognate recombination hot spot of *Bacillus subtilis*, χ_{Bs} (5'-AGCGG-3') (14, 16). The AddB subunit has no substantial similarity to either the RecC or the RecD subunit, but it does contain a putative ATPase motif and a second nuclease site similar to that of the AddA protein. The AddB subunit is responsible for the degradation of the 5'-terminated strand. Despite the limited sequence similarity to the RecBCD enzyme, AddAB enzyme is functional in *E. coli*: its expression overcomes the recombination and repair defects of *recBC*-deficient cells (40). Similar to RecBCD enzyme, AddAB enzyme binds to blunt-ended dsDNA, and uses the energy of ATP hydrolysis to translocate along and unwind dsDNA. However, whereas RecBCD enzyme degrades the dsDNA asymmetrically, the AddAB enzyme degrades both strands of the DNA duplex equally. Interaction with a correctly oriented χ_{Bs} results in downregulation of only the 3' → 5' nuclease activity of the translocating AddAB enzyme. The outcome, therefore, is the same as that occurring for the *E. coli* enzyme, namely, the production of ssDNA-tailed dsDNA with χ at the 3' terminus. Homologues of AddAB enzyme are found in 12 different species of gram-positive bacteria, and χ homologues were identified in several bacterial species (reviewed in reference 15).

SINGLE-STRAND GAP REPAIR BY RECOMBINATION: THE RecF PATHWAY

The conjugal recombination deficiency of *recB* or *recC* mutants can be overcome by the combined effect of two extragenic suppressor mutations: *sbcB* and either *sbcC* or *sbcD* (suppressor of *recBC*). The *sbcB* mutation disables the nuclease activity of exonuclease I (44), while *sbcC* (31) disables one of the two subunits of the SbcCD nuclease, which, in wild-type *E. coli*, cleaves DNA hairpin and cruciform structures formed during replication of palindromic sequences (22, 49). The combined effect of these mutations is the full activation of an alternative pathway of sexual homologous recombination, referred to as the RecF pathway. Interestingly, the efficiency of conjugational and transductional recombination by the RecF pathway in the *recBC sbcBC* cells is similar to that of the RecBCD pathway in wild-type cells, showing that the machinery of this pathway can be as productive as that of the RecBCD pathway. Moreover, some bacterial species whose survival depends on homologous recombination (such as *Deinococcus radiodurans*) do not possess obvious RecBCD or AddAB enzymes, implying that a RecF-like pathway is the wild-type pathway in those bacteria.

Homologous recombination in the *recBC sbcB sbcC* mutant background depends on RecF, RecJ, RecN, RecO, RecQ, RecR, and SSB proteins. The processing of a DSB is likely achieved by the combined action of the RecQ helicase and RecJ nuclease. Although RecQ helicase is responsible for about 75% of conjugal recombination events occurring in *recBC sbcB sbcC* mutant cells, the remaining 25% require either UvrD (helicase II) or HelD (helicase IV) (53). The RecJ protein is an exonuclease that degrades ssDNA in the 5' → 3' direction (Fig. 5a to c). RecQ protein is an SF2 DNA helicase with a 3' → 5' polarity. Similar to RecBCD enzyme, RecQ helicase can unwind blunt-ended dsDNA (76). The helicase activity of RecQ protein is not limited to blunt dsDNA ends: the enzyme can also unwind an ssDNA-dsDNA junction with a 3'-ssDNA overhang (76), and even internal regions of dsDNA (33). It is hypothesized that DNA unwinding by RecQ helicase is coupled to the degradation of the 5'-terminated strand by RecJ nuclease, resulting in the production of the 3'-terminated ssDNA overhang, which can then be used as a substrate for RecA nucleoprotein assembly (Fig. 5c). In contrast to RecBCD enzyme, RecQ does not facilitate RecA nucleoprotein filament assembly. Therefore, the ssDNA produced by RecQ is bound by SSB protein and must be protected from degradation by nucleases. This explains the requirement for the *sbcB* mutation, since the preferred substrate for exonuclease I is SSB-complexed ssDNA.

Intriguingly, two major classes of mutations in the structural gene for exonuclease I were found. One class of mutations, referred to as *sbcB*, restores both recombination and the UV resistance of *recBC* cells. In contrast, the other class, known as *xonA* mutants, suppresses only the UV sensitivity but not the recombination deficiency of the *recBC* mutant bacteria (44). Exonuclease I activity is significantly reduced in both classes of mutants; moreover, UV sensitivity directly correlates with the amount of residual activity of the enzyme (60). The *sbcB* mutations are not the same as null mutations and, therefore, are likely gain-of-function mutations. The molecular mechanism distinguishing the two types of mutations still remains to be elucidated, but protection of the 3' end of ssDNA by the *sbcB* mutations is envisioned.

The loading of RecA protein is an essential aspect of recombination in the RecBCD pathway (4). Not unexpectedly, the RecF pathway provides a RecA-loading activity in the form of the RecFOR complex. The genetic data had suggested that the loading of RecA protein onto SSB-coated ssDNA depends on the concerted action of RecF, RecO, and RecR proteins. First, mutant RecA proteins that

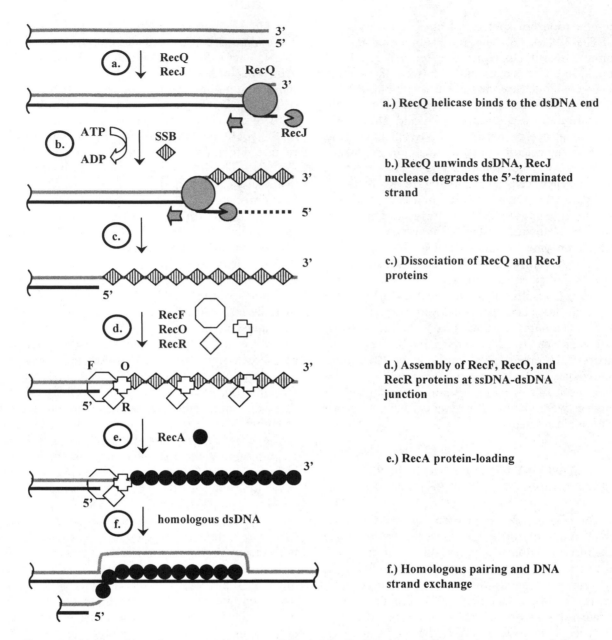

Figure 5. Initiation of homologous recombination at DSBs by the RecF pathway. Schematic representation of the early enzymatic steps of the RecF pathway. (a, b, and c) The combined action of RecQ helicase and RecJ nuclease converts a DSB into an ssDNA-dsDNA junction with 3'-terminated ssDNA overhang; this ssDNA is complexed with SSB protein. (d and e) The RecF, RecO, and RecR proteins form a complex at the junction and facilitate RecA nucleoprotein filament assembly on the SSB-coated ssDNA. (f) The RecA-ssDNA nucleoprotein filament invades homologous dsDNA.

suppress the UV sensitivity of *recF* mutations (such as the RecA803 protein) displace SSB protein from ssDNA much faster and more extensively than does the wild-type RecA protein (52), implying that RecF protein plays a role in SSB displacement. Second, a number of mutant *recA* alleles cosuppress mutations in *recF*, *recO*, and *recR* genes (79). Finally, this suppression is dependent on *recJ* function, suggesting that RecF, RecO, and RecR proteins function together as a complex (79). This view is strongly sup-

ported by biochemical observations showing that these proteins interact with one another (34, 74).

The RecFOR proteins form a number of complexes with different activities: RecO protein interacts with both RecR and SSB proteins (34) and facilitates RecA nucleoprotein filament assembly on SSB-coated ssDNA (75). RecO protein also promotes the annealing of ssDNA and of SSB-ssDNA complexes (38). Finally, RecR protein interacts with RecF protein to form a complex (34, 81) that will bind to an ssDNA-dsDNA junction (55). Biochemical analysis revealed that RecF protein binds preferentially to the ssDNA-dsDNA junction (35), and that DNA binding by RecF protein is controlled by ATP hydrolysis (80). The RecFOR complex will bind to an ssDNA-dsDNA junction with a base-paired 5′ terminus at the junction region, and it will facilitate assembly of RecA protein onto ssDNA adjacent to the junction (55). RecF protein (or RecFR complex) recognizes an ssDNA-dsDNA junction with a 5′ end, which is the structure that should be produced by RecQ and RecJ proteins. The RecOR complex (or RecO protein) binds to the DNA-RecF(R) complex, which alters the SSB-ssDNA complex nearby and allows RecA protein nucleation; subsequent nucleoprotein filament extension permits assembly of RecA protein on the entire ssDNA tail.

Another component of the RecF pathway, RecN protein, has not yet been assigned any biochemical function. However, the slight recombination deficiency and mild UV sensitivity of the *recJ recN* double mutant, combined with the severe recombination defect (50- to 100-fold reduction) of the *recD recJ recN* mutant, suggests that RecN protein might be a functional equivalent of the RecJ nuclease (50).

Despite its apparent complexity, the enzymatic machinery of the RecF pathway is as functional in DSB repair as the RecBCD pathway. Moreover, the components of the RecF pathway have functional homologues or paralogues in all organisms, from bacteriophage to human (5).

SINGLE-STRAND GAP REPAIR

Because conjugation (and transduction) involves a DSB as the initiating site for recombination, it should not be surprising that the genes that emerged from a screen for mutants defective in sexual recombination are essential for the repair of DSBs. However, the consequence of this nearly singular focus on conjugational recombination events led to the erroneous conclusion that the RecF pathway serves only a minor function in recombination in wild-type cells. Also incorrect was the belief that the discovery of the RecF pathway as the set of genes that permitted recombination in the absence of the primary RecBCD pathway implied that the RecF pathway was a cryptic recombination pathway that could be activated to compensate for the loss of RecBCD enzyme function. Rather, in wild-type cells, the RecF pathway is responsible for the repair of all SSGs. This conclusion emerged from many studies, but was made clearest from recombination assays that did not employ sexual events.

A genetic assay system that observes recombination between direct repeats of a chromosomal segment allows following sister chromosome exchanges required for the repair of DSBs (29). This assay was developed to approximate the function of homologous recombination in the repair of the DNA lesions produced during replication. The recombination events occurring between direct chromosomal repeats are detected in the colony-sectoring recombination assay, since the detected recombination events eliminate the joint-point markers located between the repeats. Recombination events detected in this assay are absolutely dependent on RecA protein function. The *recF* and *recJ* mutants display a *rec*+ phenotype, *recB* mutants show only a slight defect, and *recB recJ* double mutants are capable of supporting duplication segregation. On the other hand, *recB recF* double mutants are deficient in recombination between chromosomal direct repeats, suggesting that both RecBCD and RecF pathways play major roles in recombination.

Initiation of homologous recombination on SSGs is presented in Fig. 6. First, the RecFOR complex assembles at the ssDNA-dsDNA junction of an SSG containing a base-paired 5′ end (Fig. 6a) and facilitates RecA nucleoprotein assembly in the ssDNA region of the gap (55) (Fig. 6e). The RecFR complex can bind to the 3′-containing end of SSG, limiting RecA filament extension into the dsDNA region (82) (Fig. 6f). The resulting RecA nucleoprotein filament formed on the SSG can then invade the homologous dsDNA molecule (Fig. 6g). DNA strand exchange followed by heteroduplex extension results in the formation of two Holliday junctions. To complete the repair, RecA protein must be removed from the junctions, and the Holliday junctions themselves need to be resolved. The processing of a Holliday junction into mature recombinant molecules is achieved by the RuvA, RuvB, and RuvC proteins (see references 64 and 83 for reviews). The RuvA tetramer is a four-way junction-specific recognition protein that binds to the Holliday junction and induces the square planar conformation of this junction. The RuvB protein is a specialized translocation protein that can "pump" or move DNA through its circular hexameric active form. To catalyze branch migration, two hexameric rings of RuvB protein bind to opposing arms of the Holliday junction that was recognized by the RuvA proteins, and then they translocate the dsDNA outward through the center of each ring, resulting in the relative movement of the junction (Fig. 6h). RuvC protein is a four-way junction-specific endonuclease that resolves a Holliday junction by symmetrically cleaving the opposite arm of the junction to

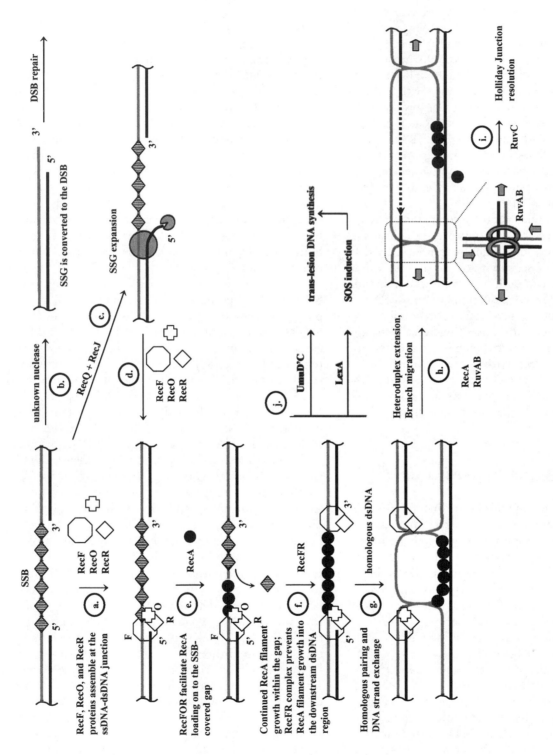

Figure 6. SSG repair by the RecF pathway. The SSG has several potential fates. The RecFOR proteins can bind to the 5′ end of the ssDNA-dsDNA junction (a), the SSG can be cleaved to produce a DSB (b), or the ssDNA region can be expanded by the combined activities of the RecQ helicase and RecJ nuclease (c and d). (e) RecFOR proteins facilitate RecA protein loading onto the SSB-coated ssDNA. (f) Growth of the RecA nucleoprotein filament beyond the ssDNA region is prevented by the RecFR complex bound to the ssDNA-dsDNA junction containing a free 3′ end. (g) The RecA nucleoprotein filament invades homologous dsDNA and catalyzes DNA strand exchange. (h) DNA heteroduplex expansion results in the formation of two Holliday junctions. (i) RuvABC proteins facilitate branch migration and Holliday junction resolution to produce repaired recombinant molecules. (j) Translesion DNA synthesis by DNA polymerase (UmuD′C) can also repair the SSG due to direct interaction of the polymerase with the RecA nucleoprotein filament.

produce the recombinant DNA product molecules (Fig. 6i).

Alternatively, an ssDNA endonuclease can convert the SSG into a DSB, which can be repaired through a DSB repair mechanism (Fig. 6b). If the region of ssDNA in the gap is too small, then it can be expanded by the combined action of RecQ helicase and RecJ nuclease. Similar to the DSB repair situation, RecQ helicase function can be substituted by UvrD helicase or helicase IV (53) (Fig. 6c). The SSG, on which the RecA nucleoprotein filament is assembled, is not necessarily repaired only by homologous recombination. The error-prone UmuD′C DNA polymerase can be attracted to such a RecA filament assembled on the SSG to catalyze translesion DNA synthesis (59) (Fig. 6j).

CONCLUSION

Homologous recombination can be initiated at either DSBs or SSGs in duplex DNA. Two major pathways are responsible for homologous recombination in wild-type *E. coli*: the RecBCD and RecF pathways. The RecBCD pathway is specific for the recombinational repair of DSBs, and in the wild-type cells, the RecF pathway is primarily used for recombination that initiates at SSGs. However, with appropriate suppressor mutations in *E. coli*, and presumably in bacteria that lack a RecBCD pathway, the RecF pathway can efficiently act at DSBs as well. Despite the different initiating lesions, both pathways have the same subsequent step: conversion of the broken DNA molecule into a central intermediate of recombination, which is the RecA protein nucleoprotein filament assembled along the ssDNA. In the RecBCD pathway, this process is carried out by the combined helicase/nuclease activity of RecBCD enzyme, and depends on the presence of the recombination hot spot, χ. In the RecF pathway, the combined efforts of RecQ helicase and RecJ nuclease are needed in combination with the RecFOR complex. In the RecBCD pathway, RecBCD enzyme facilitates assembly of the RecA protein onto SSB-coated ssDNA; in the RecF pathway, this task is accomplished by RecF, RecO, and RecR proteins. The RecA nucleoprotein filament can then initiate invasion of ssDNA into homologous dsDNA, progressing into the final stage of homologous recombination, which is resolution by RuvABC proteins.

REFERENCES

1. Amundsen, S. K., A. F. Taylor, A. M. Chaudhury, and G. R. Smith. 1986. *recD*: the gene for an essential third subunit of exonuclease V. *Proc. Natl. Acad. Sci. USA* 83:5558–5562.

2. Anderson, D. G., J. J. Churchill, and S. C. Kowalczykowski. 1997. Chi-activated RecBCD enzyme possesses 5′ → 3′ nucleolytic activity, but RecBC enzyme does not: evidence suggesting that the alteration induced by Chi is not simply ejection of the RecD subunit. *Genes Cells* 2:117–128.

2a. Anderson, D. G., and S. C. Kowalczykowski. 1997. The recombination hot spot χ is a regulatory element that switches the polarity of DNA degradation by the RecBCD enzyme. *Genes Dev.* 11:571–581.

2b. Anderson, D. G., and S. C. Kowalczykowski. 1997. The translocating RecBCD enzyme stimulates recombination by directing RecA protein onto ssDNA in a χ-regulated manner. *Cell* 90:77–86.

2c. Aravind, L., K. S. Makarova, and E. V. Koonin. 2000. Holliday junction resolvases and related nucleases: identification of new families, phyletic distribution and evolutionary trajectories. *Nucleic Acids Res.* 28:3417–3432.

3. Arnold, D. A., P. R. Bianco, and S. C. Kowalczykowski. 1998. The reduced levels of χ recognition exhibited by the RecBC^{1004}D enzyme reflect its recombination defect in vivo. *J. Biol. Chem.* 273:16476–16486.

4. Arnold, D. A., and S. C. Kowalczykowski. 2000. Facilitated loading of RecA protein is essential to recombination by RecBCD enzyme. *J. Biol. Chem.* 275:12261–12265.

5. Beernink, H. T., and S. W. Morrical. 1999. RMPs: recombination/replication mediator proteins. *Trends Biochem. Sci.* 24:385–389.

6. Bianco, P. R., and S. C. Kowalczykowski. 1999. RecA protein. In *Encyclopedia of Life Sciences*. [Online.] http://www.els.net. Nature Publishing Group, London, England.

7. Bianco, P. R., and S. C. Kowalczykowski. 1997. The recombination hotspot Chi is recognized by the translocating RecBCD enzyme as the single strand of DNA containing the sequence 5′-GCTGGTGG-3′. *Proc. Natl. Acad. Sci. USA* 94:6706–6711.

8. Biswas, I., E. Maguin, S. D. Ehrlich, and A. Gruss. 1995. A 7-base-pair sequence protects DNA from exonucleolytic degradation in Lactococcus lactis. *Proc. Natl. Acad. Sci. USA* 92:2244–2248.

9. Blattner, F. R., G. Plunkett III, C. A. Bloch, N. T. Perna, V. Burland, M. Riley, J. Collado-Vides, J. D. Glasner, C. K. Rode, G. F. Mayhew, J. Gregor, N. W. Davis, H. A. Kirkpatrick, M. A. Goeden, D. J. Rose, B. Mau, and Y. Shao. 1997. The complete genome sequence of Escherichia coli K-12. *Science* 277:1453–1474.

10. Boehmer, P. E., and P. T. Emmerson. 1992. The RecB subunit of the Escherichia coli RecBCD enzyme couples ATP hydrolysis to DNA unwinding. *J. Biol. Chem.* 267:4981–4987.

11. Burland, V., G. Plunkett III, D. L. Daniels, and F. R. Blattner. 1993. DNA sequence and analysis of 136 kilobases of the Escherichia coli genome: organizational symmetry around the origin of replication. *Genomics* 16:551–561.

12. Capaldo-Kimball, F., and S. D. Barbour. 1971. Involvement of recombination genes in growth and viability of *Escherichia coli* K-12. *J. Bacteriol.* 106:204–212.

13. Chaudhury, A. M., and G. R. Smith. 1984. A new class of Escherichia coli *recBC* mutants: implications for the role of recBC enzyme in homologous recombination. *Proc. Natl. Acad. Sci. USA* 81:7850–7854.

14. Chédin, F., S. D. Ehrlich, and S. C. Kowalczykowski. 2000. The Bacillus subtilis AddAB helicase/nuclease is regulated by its cognate Chi sequence in vitro. *J. Mol. Biol.* 298:7–20.

15. Chédin, F., and S. C. Kowalczykowski. 2002. A novel family of regulated helicases/nucleases from Gram-positive bacteria: insights into the initiation of DNA recombination. *Mol. Microbiol.* **43:**823–834.

16. Chédin, F., P. Noirot, V. Biaudet, and S. D. Ehrlich. 1998. A five-nucleotide sequence protects DNA from exonucleolytic degradation by AddAB, the RecBCD analogue of Bacillus subtilis. *Mol. Microbiol.* **29:**1369–1377.

17. Cheng, K. C., and G. R. Smith. 1989. Distribution of Chi-stimulated recombinational exchanges and heteroduplex endpoints in phage lambda. *Genetics* **123:**5–17.

18. Churchill, J. J., D. G. Anderson, and S. C. Kowalczykowski. 1999. The RecBC enzyme loads RecA protein onto ssDNA asymmetrically and independently of Chi, resulting in constitutive recombination activation. *Genes Dev.* **13:**901–911.

19. Churchill, J. J., and S. C. Kowalczykowski. 2000. Identification of the RecA protein-loading domain of RecBCD enzyme. *J. Mol. Biol.* **297:**537–542.

20. Clark, A. J. 1973. Recombination deficient mutants of E. coli and other bacteria. *Annu. Rev. Genet.* **7:**67–86.

21. Clark, A. J., and A. D. Margulies. 1965. Isolation and characterization of recombination-deficient mutants of *Escherichia coli* K12. *Proc. Natl. Acad. Sci. USA* **53:**451–459.

22. Connelly, J. C., L. A. Kirkham, and D. R. Leach. 1998. The SbcCD nuclease of Escherichia coli is a structural maintenance of chromosomes (SMC) family protein that cleaves hairpin DNA. *Proc. Natl. Acad. Sci. USA* **95:**7969–7974.

23. Dillingham, M. S., M. Spies, and S. C. Kowalczykowski. 2003. RecBCD enzyme is a bipolar DNA helicase. *Nature* **423:**893–897.

24. Dixon, D. A., J. J. Churchill, and S. C. Kowalczykowski. 1994. Reversible inactivation of the *Escherichia coli* RecBCD enzyme by the recombination hotspot χ in vitro: evidence for functional inactivation or loss of the RecD subunit. *Proc. Natl. Acad. Sci. USA* **91:**2980–2984.

25. Dixon, D. A., and S. C. Kowalczykowski. 1993. The recombination hotspot χ is a regulatory sequence that acts by attenuating the nuclease activity of the E. coli RecBCD enzyme. *Cell* **73:**87–96.

26. Emmerson, P. T. 1968. Recombination deficient mutants of *Escherichia coli* K12 that map between thyA and argA. *Genetics* **60:**19–30.

27. Farah, J. A., and G. R. Smith. 1997. The RecBCD enzyme initiation complex for DNA unwinding: enzyme positioning and DNA opening. *J. Mol. Biol.* **272:**699–715.

28. Friedberg, E. C., G. C. Walker, and W. Siede. 1995. *DNA Repair and Mutagenesis.* ASM Press, Washington, D.C.

29. Galitski, T., and J. R. Roth. 1997. Pathways for homologous recombination between chromosomal direct repeats in Salmonella typhimurium. *Genetics* **146:**751–767.

30. Ganesan, S., and G. R. Smith. 1993. Strand-specific binding to duplex DNA ends by the subunits of Escherichia coli recBCD enzyme. *J. Mol. Biol.* **229:**67–78.

31. Gibson, F. P., D. R. Leach, and R. G. Lloyd. 1992. Identification of sbcD mutations as cosuppressors of recBC that allow propagation of DNA palindromes in *Escherichia coli* K-12. *J. Bacteriol.* **174:**1222–1228.

32. Haijema, B. J., G. Venema, and J. Kooistra. 1996. The C terminus of the AddA subunit of the *Bacillus subtilis* ATP-dependent DNase is required for the ATP-dependent exonuclease activity but not for the helicase activity. *J. Bacteriol.* **178:**5086–5091.

33. Harmon, F. G., and S. C. Kowalczykowski. 2001. Biochemical characterization of the DNA helicase activity of the Escherichia coli RecQ helicase. *J. Biol. Chem.* **276:**232–243.

34. Hegde, S. P., M. H. Qin, X. H. Li, M. A. Atkinson, A. J. Clark, M. Rajagopalan, and M. V. Madiraju. 1996. Interactions of RecF protein with RecO, RecR, and single-stranded DNA binding proteins reveal roles for the RecF-RecO-RecR complex in DNA repair and recombination. *Proc. Natl. Acad. Sci. USA* **93:**14468–14473.

35. Hegde, S. P., M. Rajagopalan, and M. V. V. S. Madiraju. 1996. Preferential binding of *Escherichia coli* RecF protein to gapped DNA in the presence of adenosine (γ-thio) triphosphate. *J. Bacteriol.* **178:**184–190.

36. Howard-Flanders, P., and L. Theriot. 1966. Mutants of Escherichia coli K-12 defective in DNA repair and in genetic recombination. *Genetics* **53:**1137–1150.

37. Iyer, V. N., and W. D. Rupp. 1971. Usefulness of benzoylated naphthoylated DEAE-cellulose to distinguish and fractionate double-stranded DNA bearing different extents of single-stranded regions. *Biochim. Biophys. Acta* **228:**117–126.

38. Kantake, N., M. V. Madiraju, T. Sugiyama, and S. C. Kowalczykowski. 2002. Escherichia coli RecO protein anneals ssDNA complexed with its cognate ssDNA-binding protein: a common step in genetic recombination. *Proc. Natl. Acad. Sci. USA* **99:**15327–15332.

39. Kogoma, T. 1997. Stable DNA replication: interplay between DNA replication, homologous recombination, and transcription. *Microbiol. Mol. Biol. Rev.* **61:**212–238.

40. Kooistra, J., B. J. Haijema, and G. Venema. 1993. The Bacillus subtilis addAB genes are fully functional in Escherichia coli. *Mol. Microbiol.* **7:**915–923.

41. Korangy, F., and D. A. Julin. 1993. Kinetics and processivity of ATP hydrolysis and DNA unwinding by the RecBC enzyme from Escherichia coli. *Biochemistry* **32:**4873–4880.

42. Kowalczykowski, S. C. 2000. Initiation of genetic recombination and recombination-dependent replication. *Trends Biochem. Sci.* **25:**156–165.

43. Krasin, F., and F. Hutchinson. 1977. Repair of DNA double-strand breaks in *Escherichia coli*, which requires recA function and the presence of a duplicate genome. *J. Mol. Biol.* **116:**81–98.

44. Kushner, S. R., H. Nagaishi, and A. J. Clark. 1972. Indirect suppression of recB and recC mutations by exonuclease I deficiency. *Proc. Natl. Acad. Sci. USA* **69:**1366–1370.

45. Kuzminov, A. 1995. Collapse and repair of replication forks in Escherichia coli. *Mol. Microbiol.* **16:**373–384.

46. Kuzminov, A. 1999. Recombinational repair of DNA damage in *Escherichia coli* and bacteriophage λ. *Microbiol. Mol. Biol. Rev.* **63:**751–813.

47. Lam, S. T., M. M. Stahl, K. D. McMilin, and F. W. Stahl. 1974. Rec-mediated recombinational hot spot activity in bacteriophage lambda. II. A mutation which causes hot spot activity. *Genetics* **77:**425–433.

48. Lederberg, J., and E. L. Tatum. 1953. Sex in bacteria; genetic studies, 1945–1952. *Science* **118:**169–175.

49. Lloyd, R. G., and C. Buckman. 1985. Identification and genetic analysis of sbcC mutations in commonly used recBC sbcB strains of *Escherichia coli* K-12. *J. Bacteriol.* **164:**836–844.

50. Lloyd, R. G., and C. Buckman. 1991. Overlapping functions of recD, recJ and recN provide evidence of three epistatic groups of genes in Escherichia coli recombination and DNA repair. *Biochimie* **73:**313–320.

51. Lloyd, R. G., M. C. Porton, and C. Buckman. 1988. Effect of recF, recJ, recN, recO and ruv mutations on ultraviolet survival and genetic recombination in a recD strain of Escherichia coli K12. *Mol. Gen. Genet.* **212:**317–324.

51a. Lovett, S. T., C. Luisi-DeLuca, and R. D. Kolodner. 1988. The genetic dependence of recombination in *recD* mutants of *Escherichia coli*. *Genetics* 120:37–45.

52. Madiraju, M. V. V. S., A. Templin, and A. J. Clark. 1988. Properties of a mutant recA-encoded protein reveal a possible role for Escherichia coli recF-encoded protein in genetic recombination. *Proc. Natl. Acad. Sci. USA* 85:6592–6596.

53. Mendonca, V. M., H. D. Klepin, and S. W. Matson. 1995. DNA helicases in recombination and repair: construction of a Δ*uvrD* Δ*helD* Δ*recQ* mutant deficient in recombination and repair. *J. Bacteriol.* 177:1326–1335.

54. Michel, B., S. D. Ehrlich, and M. Uzest. 1997. DNA double-strand breaks caused by replication arrest. *EMBO J.* 16:430–438.

55. Morimatsu, K., and S. C. Kowalczykowski. 2003. RecFOR proteins load RecA protein onto gapped DNA to accelerate DNA strand exchange: a universal step of recombinational repair. *Mol. Cell* 11:1337–1347.

56. Mosig, G. 1987. The essential role of recombination in phage T4 growth. *Annu. Rev. Genet.* 21:347–371.

57. Myers, R. S., A. Kuzminov, and F. W. Stahl. 1995. The recombination hot spot χ activates RecBCD recombination by converting *Escherichia coli* to a *recD* mutant phenocopy. *Proc. Natl. Acad. Sci. USA* 92:6244–6248.

58. Myers, R. S., and F. W. Stahl. 1994. Chi and the RecBCD enzyme of Escherichia coli. *Annu. Rev. Genet.* 28:49–70.

59. Pham, P., E. M. Seitz, S. Saveliev, X. Shen, R. Woodgate, M. M. Cox, and M. F. Goodman. 2002. Two distinct modes of RecA action are required for DNA polymerase V-catalyzed translesion synthesis. *Proc. Natl. Acad. Sci. USA* 99:11061–11066.

60. Phillips, G. J., D. C. Prasher, and S. R. Kushner. 1988. Physical and biochemical characterization of cloned *sbcB* and *xonA* mutations from *Escherichia coli* K-12. *J. Bacteriol.* 170:2089–2094.

61. Phillips, R. J., D. C. Hickleton, P. E. Boehmer, and P. T. Emmerson. 1997. The RecB protein of Escherichia coli translocates along single-stranded DNA in the 3′ to 5′ direction: a proposed ratchet mechanism. *Mol. Gen. Genet.* 254:319–329.

62. Roman, L. J., A. K. Eggleston, and S. C. Kowalczykowski. 1992. Processivity of the DNA helicase activity of *Escherichia coli* recBCD enzyme. *J. Biol. Chem.* 267:4207–4214.

63. Roman, L. J., and S. C. Kowalczykowski. 1989. Characterization of the adenosinetriphosphatase activity of the *Escherichia coli* RecBCD enzyme: relationship of ATP hydrolysis to the unwinding of duplex DNA. *Biochemistry* 28:2873–2881.

64. Shinagawa, H., and H. Iwasaki. 1995. Molecular mechanisms of Holliday junction processing in Escherichia coli. *Adv. Biophys.* 31:49–65.

65. Sinden, R. R., and R. S. Cole. 1978. Repair of cross-linked DNA and survival of *Escherichia coli* treated with psoralen and light: effects of mutations influencing genetic recombination and DNA metabolism. *J. Bacteriol.* 136:538–547.

66. Smith, C. L., J. G. Econome, A. Schutt, S. Klco, and C. R. Cantor. 1987. A physical map of the Escherichia coli K12 genome. *Science* 236:1448–1453.

67. Sourice, S., V. Biaudet, M. El Karoui, S. D. Ehrlich, and A. Gruss. 1998. Identification of the Chi site of Haemophilus influenzae as several sequences related to the Escherichia coli Chi site. *Mol. Microbiol.* 27:1021–1029.

68. Spies, M., P. R. Bianco, M. S. Dillingham, N. Handa, R. J. Baskin, and S. C. Kowalczykowski. 2003. A molecular throttle: the recombination hotspot, χ, controls DNA translocation by the RecBCD helicase. *Cell* 114:647–654.

69. Stahl, F. W., J. M. Crasemann, and M. M. Stahl. 1975. Rec-mediated recombinational hot spot activity in bacteriophage lambda. III. Chi mutations are site-mutations stimulating rec-mediated recombination. *J. Mol. Biol.* 94:203–212.

70. Stahl, F. W., K. D. McMilin, M. M. Stahl, J. M. Crasemann, and S. Lam. 1974. The distribution of crossovers along unreplicated lambda bacteriophage chromosomes. *Genetics* 77:395–408.

71. Stahl, F. W., and M. M. Stahl. 1975. Rec-mediated recombinational hot spot activity in bacteriophage lambda. IV. Effect of heterology on Chi-stimulated crossing over. *Mol. Gen. Genet.* 140:29–37.

72. Taylor, A. F., D. W. Schultz, A. S. Ponticelli, and G. R. Smith. 1985. RecBC enzyme nicking at Chi sites during DNA unwinding: location and orientation-dependence of the cutting. *Cell* 41:153–163.

73. Taylor, A. F., and G. R. Smith. 1999. Regulation of homologous recombination: Chi inactivates RecBCD enzyme by disassembly of the three subunits. *Genes Dev.* 13:890–900.

74. Umezu, K., N. W. Chi, and R. D. Kolodner. 1993. Biochemical interaction of the Escherichia coli RecF, RecO, and RecR proteins with RecA protein and single-stranded DNA binding protein. *Proc. Natl. Acad. Sci. USA* 90:3875–3879.

75. Umezu, K., and R. D. Kolodner. 1994. Protein interactions in genetic recombination in Escherichia coli. Interactions involving RecO and RecR overcome the inhibition of RecA by single-stranded DNA-binding protein. *J. Biol. Chem.* 269:30005–30013.

76. Umezu, K., K. Nakayama, and H. Nakayama. 1990. Escherichia coli RecQ protein is a DNA helicase. *Proc. Natl. Acad. Sci. USA* 87:5363–5367.

77. Wang, J., R. Chen, and D. A. Julin. 2000. A single nuclease active site of the Escherichia coli RecBCD enzyme catalyzes single-stranded DNA degradation in both directions. *J. Biol. Chem.* 275:507–513.

78. Wang, T. C., and S. H. Chen. 1992. Similar-sized daughter-strand gaps are produced in the leading and lagging strands of DNA in UV-irradiated E. coli uvrA cells. *Biochem. Biophys. Res. Commun.* 184:1496–1503.

79. Wang, T.-C. V., H. Y. Chang, and J. L. Hung. 1993. Co-suppression of *recF*, *recR* and *recO* mutations by mutant *recA* alleles in *Escherichia coli* cells. *Mutat. Res.* 294:157–166.

80. Webb, B. L., M. M. Cox, and R. B. Inman. 1999. ATP hydrolysis and DNA binding by the Escherichia coli RecF protein. *J. Biol. Chem.* 274:15367–15374.

81. Webb, B. L., M. M. Cox, and R. B. Inman. 1995. An interaction between the Escherichia coli RecF and RecR proteins dependent on ATP and double-stranded DNA. *J. Biol. Chem.* 270:31397–31404.

82. Webb, B. L., M. M. Cox, and R. B. Inman. 1997. Recombinational DNA repair: the RecF and RecR proteins limit the extension of RecA filaments beyond single-strand DNA gaps. *Cell* 91:347–356.

83. West, S. C. 1996. The RuvABC proteins and Holliday junction processing in *Escherichia coli*. *J. Bacteriol.* 178:1237–1241.

84. Willetts, N. S., and D. W. Mount. 1969. Genetic analysis of recombination-deficient mutants of *Escherichia coli* K-12 carrying *rec* mutations cotransducible with *thyA*. *J. Bacteriol.* 100:923–934.

85. Yu, M., J. Souaya, and D. A. Julin. 1998. Identification of the nuclease active site in the multifunctional RecBCD enzyme by creation of a chimeric enzyme. *J. Mol. Biol.* 283:797–808.

Chapter 22

Recombination Machinery: Holliday Junction-Resolving Enzymes

MALCOLM F. WHITE

The Holliday junction (or four-way DNA junction) was predicted 40 years ago, as an elegant solution to the puzzle of how organisms exchange genetic information and reorganize their genomes by homologous recombination (22). Experimental confirmation of the existence of the Holliday junction lagged somewhat behind the theory, though we now understand the structure and dynamics of four-way DNA junctions at a molecular level (reviewed in reference 31). Holliday junctions are branch points linking homologous DNA duplexes, allowing physical exchange of genetic information (homologous recombination). Holliday junctions are also formed during the repair of double-strand breaks and the rescue of stalled or collapsed DNA replication forks. Holliday junctions are resolved into recombinant duplex DNA species by a class of structure-specific endonucleases known as the Holliday junction-resolving enzymes. The primary cellular resolving enzyme in bacteria is RuvC, which is the main focus of this chapter. I also discuss the RusA protein, which may act as an alternative to RuvC in some bacterial species, and I attempt to place RuvC in a wider context based on our knowledge of other junction-resolving enzymes.

IDENTIFICATION AND DISTRIBUTION OF BACTERIAL JUNCTION-RESOLVING ENZYMES

As with so many other proteins involved in the manipulation of DNA, the first junction-resolving enzymes identified were of bacteriophage origin. DNA packaging defects due to the accumulation of branched DNA species were correlated to mutations in gene 49 of bacteriophage T4 and gene 3 of bacteriophage T7 (27, 28, 52). Both of these gene products were sub-sequently cloned and shown to encode Holliday junction-resolving enzymes: the somewhat confusingly named T4 endonuclease VII (39) and T7 endonuclease I (13). These enzymes are relatively aggressive nucleases with general specificity for branched DNA and play a dual role both in recombination and in debranching phage DNA to allow packaging (reviewed in references 26 and 57). Both enzymes have a limited distribution among related bacteriophage.

The first cellular Holliday junction-resolving enzyme identified was RuvC from *Escherichia coli*. Initial genetic data suggesting a role for the *ruvC* gene product in recombination (34, 46) were confirmed by the identification of the RuvC protein as a nuclease specific for four-way DNA junctions (9, 10, 14, 24, 49, 51). RuvC can be regarded as the major cellular resolving enzyme in bacteria. It is widely distributed across the major bacterial phyla, including the proteobacteria, cyanobacteria, firmicutes, fusobacteria, chlamydiales, spirochaetales, thermotogae, and deinococci (Fig. 1). Many bacteriophage also encode homologues of RuvC (45); however, it is not universally present, with notable absences in the bacilli, aquifex, mycoplasmas, and lactococci.

The nonuniversality of RuvC is partially explained by the distribution of another junction-resolving enzyme, RusA. RusA was originally identified in screens for suppressors of the repair-deficient phenotypes of *ruv* mutants (36). In *E. coli*, RusA is encoded by a defective prophage and is normally transcriptionally silent; there are seven copies of phage-encoded RusA genes scattered throughout the *E. coli* O157 genome. RusA is widely distributed among bacteriophage, but also appears to be the main cellular resolving enzyme in *Bacillus subtilis*, *Legionella pneumoniae*, and *Aquifex aeolicus* (47). The *rap* gene in bacteriophage λ occupies the position equivalent to *rusA* in gene order,

Malcolm F. White • Centre for Biomolecular Sciences, University of St. Andrews, North Haugh, St. Andrews, Fife KY16 9ST, United Kingdom.

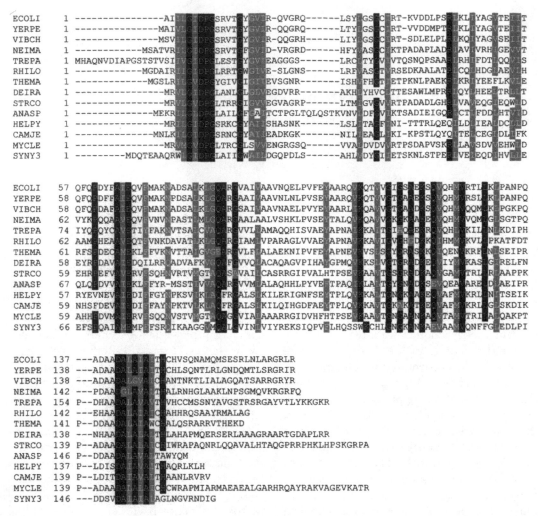

Figure 1. Sequence alignment of a diverse selection of RuvC homologues: ECOLI (*E. coli*, Swiss-Prot database database accession number P24239), YERPE (*Yersinia pestis*, Q8ZEU7), VIBCH (*Vibrio cholerae*, Q9KR00), NEIMA (*Neisseria meningitidis*, Q9JTU3), TREPA (*Treponema pallidum*, O83530), RHILO (*Rhizobium loti*, Q98F72), THEMA (*Thermotoga maritima*, Q9WZ45), DEIRA (*Deinococcus radiodurans*, Q9RX75), STRCO (*Streptomyces coelicolor*, Q9L289), ANASP (*Anabaena* sp., O52751), HELPY (*Helicobacter pylori*, O25544), CAMJE (*Campylobacter jejuni*, Q9PLU8), MYCLE (*Mycobacterium leprae*, P40834), and SYNY3 (*Synechocystis* sp., Q55506).

and has been shown to have nuclease activity against branched DNA species, including four-way junctions; however, the two enzymes share no sequence similarity (reviewed in reference 45).

RESOLVING ENZYME SEQUENCES AND STRUCTURES

Structure of RuvC

E. coli RuvC is a dimeric enzyme with 173 residues per subunit and a pI of 9.6. RuvC was the first resolving enzyme structure solved by X-ray crystallography (2), which revealed an α/β-type structure with an unexpected similarity to the folds of RNase

HI and the human immunodeficiency virus and Mu integrases (2, 25, 40). Approximately 60 sequence homologues of RuvC are currently deposited in the public data banks. A selection covering a diverse cross section of the bacterial sequences is shown aligned in Fig. 1. It is immediately apparent that very few residues are absolutely conserved in all sequences, and a corollary of this is that conserved residues have a fundamental role in either the structure or function of the protein. The most commonly conserved residues are glycines, which are necessary to confer flexibility in the protein backbone, e.g., in turns. Of the remainder, it is striking that four acidic residues (Asp-7, Glu-66, Asp-138, and Asp-142) are highly conserved. These four residues cluster together in the crystal structure of RuvC, defining the active-site pocket (2)

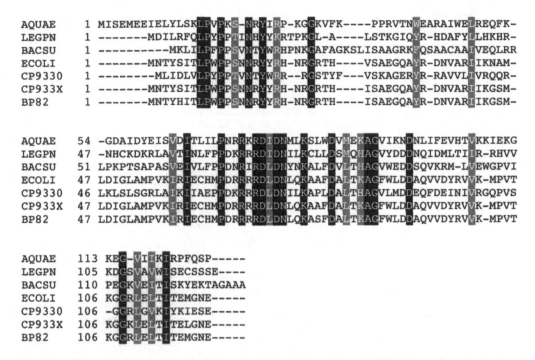

Figure 2. Sequence alignment of a selection of cellular and viral RusA homologues: AQUAE (*A. aeolicus*, Swiss-Prot database accession number O67766), LEGPN (*L. pneumoniae*, Q9AKY6), BACSU (*B. subtilis*, Q8X556), ECOLI (*E. coli*, P40116), CP9330 (phage CP-933O from *E. coli* O157:H7, AAL89445), CP933X (phage CP-933X from *E. coli* O157:H7, Q8X707), and BP82 (phage BP-82, Q37873).

(see Color Plate 9 [color insert]). Mutagenesis studies have confirmed that each of these residues is essential for catalytic activity (41). Three of these acidic residues have structural equivalents in RNase HI, suggesting that the similarity in gross fold extends to the proteins' active sites. This family of related proteins has been termed the "integrase superfamily" (33).

Primary Structure of RusA

E. coli RusA is a dimer of 120 amino acid monomers with a calculated pI of 9.6. A selection of six RusA homologues from bacterial and viral sources is shown aligned in Fig. 2. RusA bears no detectable sequence identity to any other resolving enzyme or to any other protein of known function. It is assumed to represent a separate class of resolving enzyme, but definitive confirmation of this should await the solution of the crystal structure. As with RuvC, four acidic residues are absolutely conserved, and these are likely to constitute the metal ligands in the active site. Three of these aspartate residues have been shown by site-directed mutagenesis to be essential for catalytic activity (7, 20).

Distant Cousins Reunited

Homologous recombination is ubiquitous among cellular life forms and many prokaryotic and eu-

karyotic viruses, and wherever Holliday junctions are formed, junction-resolving enzymes can be confidently expected. While this ubiquity suggests a fundamental, and therefore ancient, process, the sheer variety of Holliday junction-resolving enzymes was until recently a source of some disquiet. The sequences of the bacteriophage T7 endonuclease I, T4 endonuclease VII, RusA, bacterial RuvC, and fungal mitochondrial Cce1 resolving enzymes appeared so divergent that it was difficult to predict a homologous relationship between any two of them. While it is recognized that viruses often scavenge cellular enzymes and adapt them for their own purposes, the lack of similarity between the sequences of the bacterial and mitochondrial resolving enzymes was puzzling, given the bacterial ancestry of the mitochondrion. Suspicions were raised from a comparison of their very similar biochemical characteristics (see below, and reviewed in reference 57), and the smoking gun was uncovered when two groups independently identified the resolving enzyme from eukaryotic poxviruses (19, 33).

Poxviruses replicate in the cell cytosol and therefore must encode all the enzymes necessary for DNA replication and viral assembly. They express a junction-resolving enzyme that probably functions both to catalyze viral DNA recombination and to resolve linear concatemers of the viral genome, which are linked by a telomeric hairpin sequence that is essentially a four-way DNA junction (50). Sensitive sequence analyses were used to match an uncharacterized viral gene product with the fungal (33) and bacterial (1) resolving enzymes. The viral protein was cloned and shown to possess clear junction-resolving activity (19). Thus the identification of a poxviral "halfway house" sequence allowed the unification of the mitochondrial and bacterial resolving enzymes in the integrase superfamily. The enzymes have diverged to the point where they share extremely limited sequence similarity—essentially only the four acidic metal binding residues in the active site and one or two structurally important residues close to them. However, the evolutionary link was placed beyond doubt by the recent solution of the crystal structure of the fission yeast mitochondrial resolving enzyme Ydc2, which shares the RuvC fold (8) (Color Plate 9).

MOLECULAR RECOGNITION AND MANIPULATION OF THE HOLLIDAY JUNCTION

Resolving enzymes recognize the branched structure of the Holliday junction and introduce paired phosphodiester bond cleavages in opposing strands to collapse the junction, releasing nicked duplex DNA products. Because they have evolved to deal with a common substrate, all the resolving enzymes seem to have certain features in common: they are dimeric, have a pI >9, and have conserved acidic residues that bind the catalytic metal ions (magnesium, manganese, or cobalt). The structures of diverse resolving enzymes reveal a fairly flat, highly basic binding surface with two acidic active-site pockets positioned to attack the phosphodiester backbone on either side of the junction. The chemistry of phosphodiester bond cleavage is also likely to be conserved, with a metal-activated hydroxyl ion hydrolyzing the scissile bond to generate (ligatable) 5'-phosphate and 3'-hydroxyl termini (reviewed in reference 32).

RuvC binds Holliday junctions and immobile synthetic four-way junctions of diverse sequence with an equilibrium dissociation constant (K_d) of approximately 1 nM (16). This structure- (but not sequence-)specific binding is one of the two main mechanisms by which RuvC gains its substrate specificity. Other branched DNA species, such as three-way junctions, are also bound by RuvC, with a K_d in the range of 10 to 20 nM (16). The enzyme may sense the branched nature of its substrate by "measuring" the angle between DNA duplex arms, probing the junction center, or most likely using a combination of both mechanisms. The lack of sequence

specificity for binding suggests that the majority of contacts are via the invariant phosphodiester backbone or the sugar moieties.

The global conformation of four-way junctions bound by RuvC has been analyzed both by comparative gel electrophoresis (6) and by fluorescence resonance energy transfer measurements in solution (17). While a free four-way junction adopts a "stacked X" structure under most physiological conditions, both methods indicate that the junction is held by RuvC in an open conformation similar to that of the free junction in the presence of EDTA and the junction bound by the RuvA tetramer (see above). This type of distortion is also imposed by the mitochondrial resolving enzymes Cce1 (59) and Ydc2 (58) and the archaeal resolving enzyme Hjc (17, 30). Permanganate probing studies suggest local distortion of the base pairs flanking the cleavage site (61), as has been observed for Cce1 (12) and Hjc (30).

CATALYSIS OF JUNCTION RESOLUTION BY RuvC

Sequence-Dependent Junction Cleavage

Soon after the discovery of RuvC, it was realized that, as well as possessing a structural specificity for substrate binding, the enzyme cleaved DNA in a sequence-specific fashion. Cleavage of a synthetic mobile junction was observed only 3' of thymines, even though all junction sequences were bound equally well (4). The cleavage sequence preference of RuvC was further refined by using large recombination intermediates that allowed 1,500 bp of sequence to be scanned by the enzyme, resulting in the definition of a $5'\text{-}^A/_T TT\downarrow^G/_C$ consensus sequence (43). This consensus was confirmed by Fogg et al., who varied the consensus sequence systematically in a fixed junction, and showed that any change from the central TT dinucleotide resulted in at least a 100-fold decrease in cleavage rate by RuvC (16). This sequence is located preferentially at the point of strand exchange (5), though a consensus sequence positioned 1 nucleotide 3' of the junction center is also cut, at a 10-fold reduced rate (16). The same types of observations have been made for the yeast mitochondrial enzyme Cce1 (42), whereas *E. coli* RusA cuts 5' of GG sequences (36).

Given that these enzymes exhibit sequence-dependent junction cleavage, the question arises, why should this be so? One clue comes from studies of cleavage of three-way junctions. Both RuvC and Cce1 can cut three-way junctions quite efficiently (16; M. J. Schofield et al., unpublished data) as long as the recognition sequence is correctly positioned with respect to the point of strand exchange. The point is that three-way junctions, along with other distorted and branched DNA structures, are fixed with respect to the DNA sequence, and would present as substrates to RuvC only if the consensus cleavage sequence were correctly positioned by chance. Holliday junctions, on the other hand, are mobile branch points within homologous DNA, and can therefore scan a large amount of sequence space. Sooner or later a consensus sequence will arrive at the correct position, and the junction will be resolved. The sequence requirement of RuvC may therefore act as a second filter on the endonuclease activity, restricting it to bona fide Holliday junctions (42). The archaeal enzymes Hjc and Hje do not cleave three-way junctions even though they lack sequence specificity for Holliday junction cleavage. They are evolutionarily unrelated to RuvC/Cce1, and appear to have achieved the required level of specificity by another route that is presumably exercised at the level of substrate-specific binding (30).

Mechanism of Phosphodiester Bond Cleavage

As we have seen, all resolving enzymes probably use a two-metal-ion, hydrolytic mechanism for phosphodiester bond cleavage via an activated water molecule. This mechanism is utilized by a great many nucleases and is typified by the exonuclease domain of DNA polymerase I (3). Cce1 has been shown to bind two magnesium or manganese ions by isothermal titration calorimetry and electron paramagnetic resonance (29), and RNase HI has been crystallized with two manganese ions present at the active site (21). We know that the four acidic residues conserved in RuvC are all essential for activity and are likely to act as ligands for the metal ions (41). Other than the acidic residues and structurally important residues such as glycines, five residues that map to the junction-binding surface of RuvC are almost absolutely conserved, and may therefore play important roles in catalysis or junction recognition (Color Plate 10 [see color insert]). Arg-43 is the most distant of these residues from the active site and is probably important for DNA binding rather than catalysis. An R43L mutant has a significantly reduced affinity for four-way junctions but retains some catalytic activity (23). Phe-69 is thought to form a stacking interaction with the DNA near the point of cleavage, with mutations to residues other than Tyr resulting in loss of catalytic activity and altered binding characteristics (61). Lys-107 and Lys-118 are both important for catalytic activity, and may help stabilize the developing negative charge in the transition state (60). Last, Gln-120 has not been studied to the best of my knowledge, but

it may have a structural role in the conserved area around the two catalytic lysines, or may conceivably interact with the DNA near the point of strand cleavage.

Coordinating Junction Resolution by Bilateral Strand Scission

Like all junction-resolving enzymes, RuvC is dimeric and can resolve junctions only by introducing paired symmetric cleavages in opposing DNA strands. Under such circumstances, it is reasonable to assume some mechanism of coordination to ensure dual incision leading to a productive resolution event. Simultaneous paired cleavage events are not an essential feature of RuvC's activity, as heterodimers with one active and one inactive subunit can still nick DNA junctions in one strand (18), and junctions with one cleavable and one uncleavable (phosphorothiolate-containing) strand are nicked by wild-type RuvC (44). Nevertheless, Lilley and coworkers have demonstrated that nicking of the junction at one active site stimulates subsequent cleavage at the opposing site by approximately 150-fold. This is particularly striking when the second site does not conform to the consensus -TT- sequence, where up to 500-fold rate enhancements are observed (18). At present it is unclear how RuvC achieves this second-site enhancement, which has also been observed to a less dramatic extent for Cce1 (18). Although the reason for this rate enhancement is still unclear, it may occur because a nicked junction, with inherently greater conformational flexibility, is easier to deform toward the transition state, resulting in a lower free energy barrier for the second cleavage event (18). In any case, the result is that junctions are almost always resolved by paired cleavages. This may be particularly important where junctions are being processed rapidly by branch migration (see below), or in patches of heteroduplex DNA where paired consensus sites may not always occur.

INTERACTION WITH RuvAB: A "RESOLVASOME" MACHINE

Part of the attraction of the bacterial RuvABC system as a model for Holliday junction resolution arises from the idea that the three proteins form a molecular machine or "resolvasome." Clearly, there is a very tight molecular and genetic association between the RuvA and RuvB proteins: they are part of the same operon, and we have a good understanding of how they function together to promote branch migration (see above). The best genetic evidence that

RuvC might form the third component of a tripartite resolvasome machine comes from the observation that the alternative branch migration motor RecG cannot substitute for RuvAB in *ruvAB* mutants unless RusA is also expressed (35–37, 48, 56). In other words, RuvC appears to require the presence of RuvAB for normal activity in vivo, at least in *E. coli*. The converse does not appear to be true, as organisms such as *B. subtilis* and *A. aeolicus* utilize RuvAB with RusA rather than RuvC.

The first evidence for a functional interaction of RuvAB and RuvC came from the observation that RuvA and RuvC can simultaneously bind a four-way DNA junction (55), suggesting that RuvC can recognize the open face of a junction already bound on one side by the RuvA tetramer. Direct physical interactions between RuvC and RuvB have also been observed (15, 54), and RuvC can even partially substitute for RuvA in targeting RuvB to Holliday junctions (54). The tight functional association of RuvAB and RuvC has fundamental ramifications, as one consequence is that the choice of strands cleaved by RuvC is dictated by the orientation of binding of the RuvB hexamers (53). Biases in strand cleavage influence the ratio of crossover (splice) to noncrossover (patch) recombination events, and there is evidence that *E. coli* controls this ratio to generate favorable outcomes—for example, in reducing the number of chromosome dimers generated by recombination in response to UV damage (38). Intriguingly, the (as yet unidentified) junction-resolving activity detected in mammalian cell extracts by the West laboratory initially copurifies with a branch migration activity (11), raising the tantalizing prospect that eukaryotes utilize a similar resolvasome machine to that found in bacteria.

CONCLUSIONS AND FUTURE PROSPECTS

The study of homologous recombination and the Holliday junction was for many years the realm of geneticists. Today, through a powerful combination of molecular and structural biology and biochemistry, combined with genetics, we are close to understanding the mechanisms of Holliday junction migration and resolution at a molecular level. This work has largely been driven by studies of the *E. coli* RuvABC resolvasome, emphasizing the continuing utility of bacteria as a model system to study some of the most interesting problems in biology. The recent realization (described elsewhere in this book) that homologous recombination plays a major role in the restart of stalled or collapsed replication forks has also provided a spur to this research field. Within the next

few years we can look forward to the identification of the eukaryotic Holliday junction-resolving enzyme, which may yet turn out to be related to RuvC. We still do not understand fully how resolving enzymes recognize their junction substrates, nor do we understand the mechanism of junction resolution at a molecular level. These goals probably require the solution of a cocrystal structure of RuvC bound to a four-way DNA junction, or ultimately the structure of the RuvABC:junction quaternary complex.

REFERENCES

1. Aravind, L., K. S. Makarova, and E. V. Koonin. 2000. Survey and summary: Holliday junction resolvases and related nucleases: identification of new families, phyletic distribution and evolutionary trajectories. *Nucleic Acids Res.* 28:3417–3432.
2. Ariyoshi, M., D. G. Vassylyev, H. Iwasaki, H. Nakamura, H. Shinagawa, and K. Morikawa. 1994. Atomic structure of the RuvC resolvase: a Holliday junction-specific endonuclease from *E. coli*. *Cell* 78:1063–1072.
3. Beese, L. S., and T. A. Steitz. 1991. Structural basis for the 3′-5′ exonuclease activity of *Escherichia coli* DNA polymerase I: a two metal ion mechanism. *EMBO J.* 10:25–33.
4. Bennett, R. J., H. J. Dunderdale, and S. C. West. 1993. Resolution of Holliday junctions by RuvC resolvase: cleavage specificity and DNA distortion. *Cell* 74:1021–1031.
5. Bennett, R. J., and S. C. West. 1996. Resolution of Holliday junctions in genetic recombination: RuvC protein nicks DNA at the point of strand exchange. *Proc. Natl. Acad. Sci. USA* 93:12217–12222.
6. Bennett, R. J., and S. C. West. 1995. Structural analysis of the RuvC-Holliday junction complex reveals an unfolded junction. *J. Mol. Biol.* 252:213–226.
7. Bolt, E. L., G. J. Sharples, and R. G. Lloyd. 1999. Identification of three aspartic acid residues essential for catalysis by the RusA Holliday junction resolvase. *J. Mol. Biol.* 286:403–415.
8. Ceschini, S., A. Keeley, M. S. McAlister, M. Oram, J. Phelan, L. H. Pearl, I. R. Tsaneva, and T. E. Barrett. 2001. Crystal structure of the fission yeast mitochondrial Holliday junction resolvase Ydc2. *EMBO J.* 20:6601–6611.
9. Connolly, B., C. Parsons, F. Benson, H. Dunderdale, R. Lloyd, and S. West. 1991. Resolution of Holliday junctions in vitro requires the *Escherichia coli ruvC* gene product. *Proc. Natl. Acad. Sci. USA* 88:6063–6067.
10. Connolly, B., and S. C. West. 1990. Genetic recombination in *Escherichia coli*: Holliday junctions made by RecA protein are resolved by fractionated cell-free extracts. *Proc. Natl. Acad. Sci. USA* 87:8476–8480.
11. Constantinou, A., A. A. Davies, and S. C. West. 2001. Branch migration and Holliday junction resolution catalyzed by activities from mammalian cells. *Cell* 104:259–268.
12. Declais, A. C., and D. M. Lilley. 2000. Extensive central disruption of a four-way junction on binding CCE1 resolving enzyme. *J. Mol. Biol.* 296:421–433.
13. de Massey, B., R. A. Weisberg, and F. W. Studier. 1987. Gene 3 endonuclease of bacteriophage T7 resolves conformationally branched structures in double-stranded DNA. *J. Mol. Biol.* 193:359–376.
14. Dunderdale, H. J., F. E. Benson, C. A. Parsons, G. J. Sharples, R. G. Lloyd, and S. C. West. 1991. Formation and resolution

of recombination intermediates by *E. coli* RecA and RuvC proteins. *Nature* **354:**506–510.

15. Eggleston, A. K., A. H. Mitchell, and S. C. West. 1997. In vitro reconstitution of the late steps of genetic recombination in *E. coli. Cell* **89:**607–617.

16. Fogg, J., M. J. Schofield, M. F. White, and D. M. J. Lilley. 1999. Sequence and functional-group specificity for cleavage of DNA junctions by RuvC of *Escherichia coli. Biochemistry* **38:**11349–11358.

17. Fogg, J. M., M. Kvaratskhelia, M. F. White, and D. M. Lilley. 2001. Distortion of DNA junctions imposed by the binding of resolving enzymes: a fluorescence study. *J. Mol. Biol.* **313:**751–764.

18. Fogg, J. M., and D. M. Lilley. 2000. Ensuring productive resolution by the junction-resolving enzyme RuvC: large enhancement of the second-strand cleavage rate. *Biochemistry* **39:**16125–16134.

19. Garcia, A. D., L. Aravind, E. Koonin, and B. Moss. 2000. Bacterial-type DNA Holliday junction resolvases in eukaryotic viruses. *Proc. Natl. Acad. Sci. USA* **97:**8926–8931.

20. Giraud-Panis, M. J., and D. M. Lilley. 1998. Structural recognition and distortion by the DNA junction-resolving enzyme RusA. *J. Mol. Biol.* **278:**117–133.

21. Goedken, E. R., and S. Marqusee. 2001. Co-crystal of *Escherichia coli* RNase HI with Mn^{2+} ions reveals two divalent metals bound in the active site. *J. Biol. Chem.* **276:**7266–7271.

22. Holliday, R. 1964. A mechanism for gene conversion in fungi. *Genet. Res.* **5:**282–304.

23. Ichiyanagi, K., H. Iwasaki, T. Hishida, and H. Shinagawa. 1998. Mutational analysis on structure-function relationship of a Holliday junction specific endonuclease RuvC. *Genes Cells* **3:**575–586.

24. Iwasaki, H., M. Takahagi, T. Shiba, A. Nakata, and H. Shinagawa. 1991. *Escherichia coli* RuvC protein is an endonuclease that resolves the Holliday structure. *EMBO J.* **10:**4381–4389.

25. Katayanagi, K., M. Miyagawa, M. Matsuchima, M. Ishikawa, S. Kanaya, H. Nakamura, M. Ikehara, T. Matsuzaki, and K. Morikawa. 1992. Structural details of ribonuclease H from *Escherichia coli* as refined to an atomic resolution. *J. Mol. Biol.* **223:**1029–1052.

26. Kemper, B. 1997. Branched DNA resolving enzymes (X-solvases), p. 179–204. *In* J. A. Nickoloff and M. Hoekstra (ed.), *DNA Damage and Repair: Biochemistry, Genetics and Cell Biology.* Humana Press, Ottawa, Ontario, Canada.

27. Kemper, B., and D. T. Brown. 1976. Function of gene 49 of bacteriophage T4. II. Analysis of intracellular development and the structure of very fast sedimenting DNA. *J. Virol.* **18:**1000–1015.

28. Kemper, B., and E. Janz. 1976. Function of gene 49 of bacteriophage T4. 1. Isolation and biochemical charcterisation of very fast sedimenting DNA. *J. Virol.* **18:**992–999.

29. Kvaratskhelia, M., S. George, A. Cooper, and M. F. White. 1999. Quantitation of binding of metal ions and DNA junctions by the Holliday junction resolving enzyme Cce1. *Biochemistry* **38:**16613–16619.

30. Kvaratskhelia, M., B. N. Wardleworth, D. G. Norman, and M. F. White. 2000. A conserved nuclease domain in the archaeal Holliday junction resolving enzyme Hjc. *J. Biol. Chem.* **275:**25540–25546.

31. Lilley, D. M., and D. G. Norman. 1999. The Holliday junction is finally seen with crystal clarity. *Nat. Struct. Biol.* **6:**897–899.

32. Lilley, D. M. J., and M. F. White. 2001. The junction-resolving enzymes. *Nat. Rev. Mol. Cell. Biol.* **2:**433–443.

33. Lilley, D. M. J., and M. F. White. 2000. Resolving the relationships of resolving enzymes. *Proc. Natl. Acad. Sci. USA* **97:**9351–9353.

34. Lloyd, R. G., F. E. Benson, and C. E. Shurvington. 1984. Effect of ruv mutations on recombination and DNA repair in Escherichia coli K12. *Mol. Gen. Genet.* **194:**303–309.

35. Lloyd, R. G., and G. J. Sharples. 1993. Dissociation of synthetic Holliday junctions by E. coli RecG protein. *EMBO J.* **12:**17–22.

36. Mahdi, A. A., G. J. Sharples, T. N. Mandal, and R. G. Lloyd. 1996. Holliday junction resolvases encoded by homologous rusA genes in Escherichia coli K-12 and phage 82. *J. Mol. Biol.* **257:**561–573.

37. Mandal, T., A. Mahdi, G. Sharples, and R. Lloyd. 1993. Resolution of Holliday intermediates in recombination and DNA repair: indirect suppression of *ruvA*, *ruvB*, and *ruvC* mutations. *J. Bacteriol.* **175:**4325–4334.

38. Michel, B., G. D. Recchia, M. Penel-Colin, S. D. Ehrlich, and D. J. Sherratt. 2000. Resolution of Holliday junctions by RuvABC prevents dimer formation in rep mutants and UV-irradiated cells. *Mol. Microbiol.* **37:**180–191.

39. Nishimoto, H., M. Takayama, and T. Minagawa. 1979. Purification and some properties of a deoxyribonuclease whose synthesis is controlled by gene 49 of bacteriophage T4. *J. Biochem.* **100:**433–440.

40. Rice, P., and K. Mizuuchi. 1995. Structure of the bacteriophage Mu transposase core: a common structural motif for DNA transposition and retroviral integration. *Cell* **82:**209–220.

41. Saito, A., H. Iwasaki, M. Ariyoshi, K. Morikawa, and H. Shinagawa. 1995. Identification of four acidic amino acids that constitute the catalytic centre of the RuvC Holliday junction resolvase. *Proc. Natl. Acad. Sci. USA* **92:**7470–7474.

42. Schofield, M. J., D. M. J. Lilley, and M. F. White. 1998. Dissection of the sequence specificity of the Holliday junction endonuclease CCE1. *Biochemistry* **37:**7733–7740.

43. Shah, R., R. J. Bennett, and S. C. West. 1994. Genetic recombination in E. coli: RuvC protein cleaves Holliday junctions at resolution hotspots in vitro. *Cell* **79:**853–864.

44. Shah, R., R. Cosstick, and S. C. West. 1997. The RuvC protein dimer resolves Holliday junctions by a dual incision mechanism that involves base-specific contacts. *EMBO J.* **16:**1464–1472.

45. Sharples, G. J. 2001. The X philes: structure-specific endonucleases that resolve Holliday junctions. *Mol. Microbiol.* **39:**823–834.

46. Sharples, G. J., F. E. Benson, G. T. Iling, and R. G. Lloyd. 1990. Molecular and functional analysis of the ruv region of Escherichia coli K-12 reveals three genes involved in DNA repair and recombination. *Mol. Gen. Genet.* **221:**219–226.

47. Sharples, G. J., E. L. Bolt, and R. G. Lloyd. 2002. RusA proteins from the extreme thermophile Aquifex aeolicus and lactococcal phage r1t resolve Holliday junctions. *Mol. Microbiol.* **44:**549–559.

48. Sharples, G. J., S. N. Chan, A. A. Mahdi, M. C. Whitby, and R. G. Lloyd. 1994. Processing of intermediates in recombination and DNA repair: identification of a new endonuclease that specifically cleaves Holliday junctions. *EMBO J.* **13:**6133–6142.

49. Sharples, G. J., and R. G. Lloyd. 1991. Resolution of Holliday junctions in *Escherichia coli*: identification of the *ruvC* gene product as a 19-kilodalton protein. *J. Bacteriol.* **173:**7711–7715.

50. Stuart, D., K. Ellison, K. Graham, and G. McFadden. 1992. In vitro resolution of poxvirus replicative intermediates into

linear minichromosomes with hairpin termini by a virally induced Holliday junction endonuclease. *J. Virol.* **66:**1551–1563.

51. **Takahagi, M., H. Iwasaki, A. Nakata, and H. Shinagawa.** 1991. Molecular analysis of the *Escherichia coli ruvC* gene, which encodes a Holliday junction-specific endonuclease. *J. Bacteriol.* **173:**5747–5753.

52. **Tsujimoto, Y., and H. Ogawa.** 1978. Intermediates in genetic recombination of bacteriophage T7 DNA. Biological activity and the roles of gene 3 and gene 5. *J. Mol. Biol.* **125:**255–273.

53. **van Gool, A. J., N. M. Hajibagheri, A. Stasiak, and S. C. West.** 1999. Assembly of the Escherichia coli RuvABC resolvasome directs the orientation of Holliday junction resolution. *Genes Dev.* **13:**1861–1870.

54. **van Gool, A. J., R. Shah, C. Mezard, and S. C. West.** 1998. Functional interactions between the Holliday junction resolvase and the branch migration motor of Escherichia coli. *EMBO J.* **17:**1838–1845.

55. **Whitby, M. C., E. L. Bolt, S. N. Chan, and R. G. Lloyd.** 1996. Interactions between RuvA and RuvC at Holliday junctions: inhibition of junction cleavage and formation of a RuvA-RuvC-DNA complex. *J. Mol. Biol.* **264:**878–890.

56. **Whitby, M. C., L. Ryder, and R. G. Lloyd.** 1993. Reverse branch migration of Holliday junctions by RecG protein—a new mechanism for resolution of intermediates in recombination and DNA repair. *Cell* **75:**341–350.

57. **White, M. F., M.-J. E. Giraud-Panis, J. R. G. Pohler, and D. M. J. Lilley.** 1997. Recognition and manipulation of branched DNA structures by junction resolving enzymes. *J. Mol. Biol.* **269:**647–664.

58. **White, M. F., and D. M. Lilley.** 1998. Interaction of the resolving enzyme YDC2 with the four-way DNA junction. *Nucleic Acids Res.* **26:**5609–5616.

59. **White, M. F., and D. M. Lilley.** 1997. The resolving enzyme CCE1 of yeast opens the structure of the four-way DNA junction. *J. Mol. Biol.* **266:**122–134.

60. **Yoshikawa, M., H. Iwasaki, K. Kinoshita, and H. Shinagawa.** 2000. Two basic residues, Lys-107 and Lys-118, of RuvC resolvase are involved in critical contacts with the Holliday junction for its resolution. *Genes Cells* **5:**803–813.

61. **Yoshikawa, M., H. Iwasaki, and H. Shinagawa.** 2001. Evidence that phenylalanine 69 in Escherichia coli RuvC resolvase forms a stacking interaction during binding and destabilization of a Holliday junction DNA substrate. *J. Biol. Chem.* **276:**10432–10436.

The Bacterial Chromosome
Edited by N. Patrick Higgins
© 2005 ASM Press, Washington, D.C.

Chapter 23

Dr. Jekyll and Mr. Hyde: How the MutSLH Repair System Kills the Cell

M. G. Marinus

The association of mismatch repair defects with sporadic and hereditary nonpolyposis colon carcinoma in humans (10, 23, 44) drew heavily upon the large body of information about mismatch repair in bacteria, particularly *Pneumococcus* (15) and *Escherichia coli*. Indeed, the complete biochemical reconstitution of the system has been achieved only with *E. coli*, and our understanding of the repair process is most developed in this organism (56). The basic biochemical steps in *E. coli* and humans are remarkably similar, and the two key proteins, MutS and MutL, are each part of a highly conserved superfamily. In this chapter I concentrate on the repair mechanism in *E. coli*, but the lessons learned in this organism should also apply to analogous systems in other organisms. Although there are several distinct DNA mismatch repair systems, in this chapter the term is used to denote the MutSLH system.

MutSL mismatch repair proteins have been detected, or inferred from genome sequencing, in various bacteria, plants, yeasts, and mammals. However, the distribution of these proteins is not universal; among sequenced microbial genomes, the mycobacteria and mycoplasmas conspicuously lack these proteins. Perhaps another system substitutes for MutSL in these organisms. The MutH protein is found only in bacteria, and its distribution is restricted to the enterobacteria, *Haemophilus influenzae*, and *Pasteurella multocida*. In organisms lacking MutH, it is not known what mechanism substitutes for MutH to introduce nicks or gaps in the newly synthesized DNA chain to initiate the mismatch repair process (see below).

Because of the relationship between mismatch repair deficiency and human disease, there have been many reviews on mismatch repair over the past few years, and there are certainly more to come (8, 10,

28, 30). To differentiate this review from others, I include studies on *dam* DNA methylation, the relationship of *dam* DNA methylation and mismatch repair to recombination, as well as how mismatch repair kills cells exposed to certain experimental and chemotherapeutic agents. Some primary experimental data are included as well as speculative models. The action of mismatch repair in preventing mutations and recombination between related DNA sequences can be viewed as beneficial in contrast to its capacity to kill normal cells exposed to certain drugs and to elicit drug resistance in tumor cells when inactivated. The title of this chapter reflects this duality.

Dam METHYLATION

In *E. coli* DNA almost all adenines in GATC sequences are methylated to form N^6-methyladenine by the DNA adenine methyltransferase (Dam) encoded by the *dam* gene (48). This enzyme is present at about 130 molecules per cell (9) and is barely able to keep up with progression of the replication fork during chromosome replication; if it does not, it is excluded by the replisome. Consequently, there is a delay in the methylation of newly replicated DNA resulting in the transient formation of hemimethylated DNA trailing the replication fork. It is this hemimethylated DNA that is the substrate for MutSLH repair, described below. After conversion of the hemimethylated DNA to fully methylated DNA, mismatch repair is inhibited because it cannot use fully methylated DNA as a substrate (56, 60).

Hemimethylated DNA also occurs at the origin of replication (*oriC*) for chromosome duplication for much of the cell cycle (11). This appears to be due to the binding of the SeqA protein to this site to prevent

M. G. Marinus • Biochemistry and Molecular Pharmacology, University of Massachusetts Medical School, 55 Lake Ave., Worcester, MA 01655.

methylation by Dam and its subsequent release late in the cell cycle. *oriC* must be released from SeqA to allow Dam methylation, as only fully methylated *oriC*'s are optimal for initiation. The sequestration of the *oriC* region by SeqA is important to prevent re-initiation of chromosome replication by DnaA, so that initiation occurs only once per cell cycle. Hemimethylated and fully unmethylated GATC sequences occur at various locations on the chromosome, presumably due to protein binding that prevents Dam action (63, 74).

Inactivation of Dam by mutation leads to cells in which DNA is completely unmethylated at GATC sequences (65). Such *dam* mutants display a variety of phenotypes, some of which are listed in Table 1. Several of these phenotypes are considered in detail in this chapter. At this point, it is clear that the pleiotropic phenotypes in a *dam* mutant are consistent with the role of Dam methyltransferase and *dam* methylation in mismatch repair, modulation of gene transcription, chromosome initiation, and synchronous initiation of chromosome duplication.

Dam is also present in *Salmonella*, and *dam* derivatives display the same phenotypes as those in *E. coli* (12, 73) with one significant exception. *Salmonella* Dam mutant infection protects mice from subsequent challenge with virulent wild-type strains (26, 29). The basis for this attenuation of virulence in

dam mutants is likely gene expression at inappropriate times during infection.

THE MutSLH SYSTEM

Mismatch repair in *E. coli* requires MutS, MutL, and MutH proteins. The process is shown in Fig. 1 (56). MutS recognizes mismatches in DNA and, by a process that is still unclear, forms a complex with MutL and MutH. In the ternary complex, MutL activates MutH's latent endonuclease activity to cleave at the phosphodiester bond 5' to the G in GATC sequences in the unmethylated strand of hemimethylated DNA. In *E. coli*, nearly all these sequences are methylated by Dam, but as the Dam concentration is limiting, methylation of newly synthesized DNA chains is delayed, forming a region of hemimethylated DNA behind the replication fork. The MutH endonuclease shows high activity on hemimethylated but not methylated DNA, thereby restricting mismatch correction to the replication fork. After incision of the newly synthesized strand by MutH, the ternary complex must dissociate. The subsequent steps of the pathway involve helicase (UvrD) translocation in either the 5'-3' or 3'-5' direction followed by the appropriate exonucleases. DNA polymerase III, the replicative enzyme, catalyzes resynthesis of nucleotides and ligation followed by Dam methylation to complete the process. Note that in mismatch repair it is the replicative polymerase III holoenzyme that synthesizes the repair patches and not polymerase I.

The biochemical evidence for this model has been presented in several recent reviews and will not be covered here (31, 56, 80). In a later section of this chapter, the role of mismatch repair in antirecombination will be discussed briefly because a detailed review of this topic has been published (28). In the past 2 years the atomic structures of the Mut proteins have been determined by X-ray crystallography; the implications of these structures for biological function will be covered briefly here. More detailed information about the structures can be found in recent reviews (31, 33, 68, 80).

The *E. coli* and *Thermus aquaticus* MutS cocrystal structures were solved using oligonucleotides with a G-T mismatch or a single T deletion/insertion, respectively (42, 57). Both structures are remarkably similar, which is not surprising given the strong amino acid sequence conservation among members of the MutS superfamily. The protein in the crystals is a dimer and has two large channels, but only one of these binds the oligonucleotide. The complex can be visualized as a pair of hands held

Table 1. Altered physiological properties in a *dam* mutant

Reduced Dam activity in vitro and in vivo, leading to a reduction of N^6-methyladenine in GATC sequences in DNA

A mutator phenotype

A hyperrecombination phenotype

Alleviation of EcoK restriction

Increased number of DNA single-strand breaks

Increased sensitivity to certain chemical agents, such as cisplatin and methylating agents

Increased drug-induced mutagenesis

Derepression of certain genes in the SOS regulon

Increased spontaneous induction of lysogenic phages

Inviability of *dam* mutant cells with *recA*, *recBC*, *lexA*, *polA*, *lig*, and *ruvABC*

Increased precise excision and transposition of Tn*10* and other transposons

Altered expression of certain chromosomal and nonchromosomal genes, such as *trpS*, *sulA*, *glnS*, *mom*, *dnaA*, *pap*, and *tsp*

Suppression of some *dam* phenotypes by second-site mutation in *mutS*, *mutL*, and *mutH*

Deficient for very-short-patch (VSP) repair

Reduced *dam* P1 phage DNA packaging into virions

Asynchronous initiation of chromosome DNA replication

Failure to support the growth of plasmids containing the *E. coli* origin (*oriC*) of chromosome replication, or the phage P1 *ori*, or those dependent upon the RepI protein

Inefficient transformation with Dam-methylated plasmids

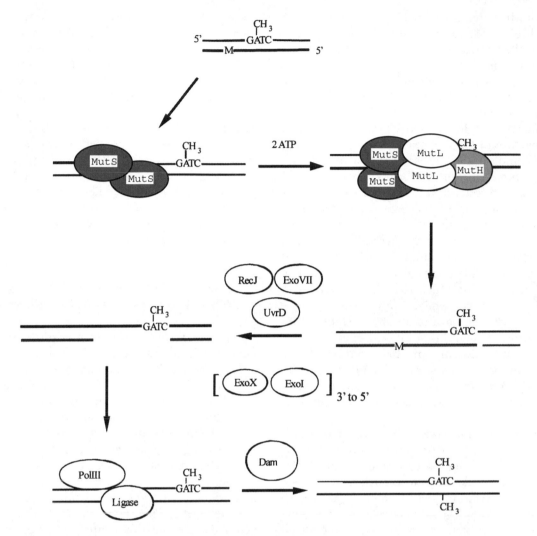

Figure 1. MutSLH DNA repair in *E. coli*. One arm of a replication fork is shown at the top of the figure with methylated and unmethylated GATC (*dam*) sequences and a replication mistake (M) generating a base-base or deletion/insertion mismatch. The mismatch is bound by MutS, and, in an ATP-dependent reaction, a ternary complex is formed with MutS, MutL, and MutH proteins. Incision by activated MutH occurs in the newly synthesized strand at an unmethylated GATC sequence. The nick is extended into a gap by excision in either the 3' to 5' or the 5' to 3' direction. Only the 5' to 3' direction is depicted in the figure. The gap is formed by the action of exonucleases, including exonuclease I (ExoI), ExoVII, ExoX, and RecJ, and the direction of excision is determined by the UvrD helicase. Resynthesis is accomplished by DNA polymerase III holoenzyme, and the nick is sealed by DNA ligase. Subsequent methylation by Dam completes the process.

together as if in prayer and the DNA passing through the gap formed between the finger tips and the thumbs. Of particular interest is that the MutS homodimer behaves like a heterodimer, in that only one of the subunits (the tip of one thumb) contacts the mismatch. This asymmetry of binding explains why in higher organisms active MutS is always a heterodimer. The binding of MutS distorts the DNA conformation to induce a 60° bend, which is stabilized by the insertion into the DNA helix of the highly conserved Phe-39 residue. ATP binding and hydrolysis are important to MutS function (the protein has weak ATPase activity), but the role of this cofactor is controversial and the crystal structures do not supply an answer. In the MutS crystal structure, however, a molecule of ADP, which stabilizes DNA binding, was present.

An unresolved problem is how mismatch recognition by MutS translates into a signal that allows MutH to nick the DNA at sites thousands of bases away. It may be that ATP-induced conformational changes in MutS after mismatch binding induce conformational changes in MutL and in turn MutH. MutS and MutL have been shown to interact in vitro and a large fraction of MutS isolated under gentle conditions from *E. coli* is bound to MutL (79).

Binding of MutS to mismatched DNA in the presence of MutL leads to the formation of a large complex at the site of the mismatch, as determined by DNase I footprinting (71). It may be that studies employing only MutS in DNA binding studies do not accurately reflect the situation in vivo, where both MutS and MutL may be bound at the mismatched site. MutSL can stimulate MutH-mediated cleavage even when the mismatch and GATC cleavage sites are on separate DNA molecules, suggesting that MutL can act as a bridge between MutS and MutH (34). MutS has also been shown to bind to the beta-clamp loader of the DNA polymerase III holoenzyme, thereby suggesting a mechanism for MutS recruitment in the vicinity of the replication machinery (47).

The crystal structure of a large N-terminal fragment of *E. coli* MutL has been solved (4). The N-terminal region of MutL is highly conserved in members of this superfamily. The protein has weak ATPase activity in a motif found in the DNA gyrase family of proteins, and this motif is distinct from that found in MutS. Binding of ATP converts MutL monomers into dimers, and these interact to activate MutH. The interaction has been modeled at the atomic level, but currently no structural data are available (80).

MutL also interacts with other proteins, including UvrD and Vsr. The interaction with UvrD is thought to recruit this helicase to the site of the MutH-induced strand break to begin the excision process. The C-terminal region of MutL interacts with both UvrD and MutH, although it seems likely that only one or the other protein can be bound by the MutS-MutL complex (27). Still unresolved is how the mismatch repair complex instructs the UvrD protein to unwind the DNA in a $5'$ to $3'$ direction or the reverse.

MutL interacts with Vsr, a component of the very-short-patch (Vsp) repair pathway that removes thymine (arising from deamination of 5-methylcytosine) from T-G mismatches (45). Efficient Vsp repair requires the *mutS*, *mutH*, and *mutL* gene products, although the molecular basis for this requirement is unknown (45). Vsr is the strand-specific endonuclease that incises the strand containing thymine, and it has been shown to interact with MutL in a catalytic manner (20).

The MutH protein structure (free, not bound) has been determined at the atomic level (5). The MutH protein resembles a large clamp with a cleft in which the DNA substrate is thought to bind. The movement of the clamp arms appears to be modulated by the position of the N-terminal helix, which can be likened to a lever. MutL alone, in the presence of ATP, can activate the latent single-strand endo-

nuclease activity of MutH. An attractive feature of the structural model is that MutL binding allows movement of the lever to close the clamp holding the bound DNA. This forces the DNA into the proximity of the catalytic residues at the bottom of the cleft and on one of the arms (5, 46). Mutational analysis has confirmed that the catalytic residues do affect catalysis and not DNA binding (78).

The next step to be solved in this saga is the atomic structure of the MutSLH complex bound to DNA with a mismatch. The use of MutH variants defective in incision may help to realize this goal. Such a structure should help to identify the binding faces of the proteins and the conformational changes that probably occur in the ternary complex.

SOME BACKGROUND INFORMATION ON DNA REPAIR AND RECOMBINATION

DNA damage induces transcription of about 40 LexA-regulated SOS genes (17). The sensor of DNA damage is the RecA protein, which binds single-stranded DNA (such as single-stranded gaps produced by Uvr DNA repair enzymes) in the presence of ATP. In the single-strand bound form, RecA promotes LexA repressor inactivation by autoproteolysis. Among the genes induced by this response is the *lexA* gene itself, as well as genes for DNA repair (the *uvr* genes), cell cycle arrest (*umuDC*), cell division inhibition (*sulA*), translesion synthesis (*umuDC*), and recombination (*recA*, *ruvA*, *ruvB*). Table 2 lists the genes included in this chapter together with their functions. As the damaged DNA is repaired, there are fewer gaps and less activated RecA, leading to an increase in LexA concentration and the reestablishment of basal levels of transcription of the SOS genes. The *lexA3* mutation produces a LexA protein that is resistant to RecA coprotease cleavage. Consequently, cells with this mutation cannot induce the SOS response (Ind⁻) and are very sensitive to DNA-damaging agents.

A brief description of DNA repair-associated recombination follows; details can be found in recent reviews (7, 18, 19, 40, 41, 55, 64). Daughter-strand gap (DSG) repair occurs at noncoding (replication-blocking) lesions such as cisplatin adducts, as shown in Fig. 2A. DSGs are produced when the DNA replication machinery stalls at the lesion, dissociates from the template strand containing the lesion, and reassociates downstream (Fig. 2A, step 1). Recombinational repair is initiated by the RecA protein, which synapses the $3'$ end of the invading strand with its complement on the lesion-containing strand (Fig. 2A,

Table 2. Proteins encoded by some *E. coli* DNA repair and recombination genes

Gene	Description
ada	O⁶-Methylguanine methyltransferase, inducible
alkA	3-Methyladenine DNA glycosylase II
alkB	AlkB binds to single-stranded alkylated DNA
dam	DNA adenine methyltransferase
dnaA	DnaA binds to the replicative origin
fpg (mutM)	Formamidopyrimidine glycosylase
gyr	DNA gyrase
lexA	Repressor for SOS regulon
lig	DNA ligase
mutHLS	Binding of base mismatches and endonuclease action
nei	Endonuclease VIII
nth	Endonuclease III
nfo	Endonuclease IV
ogt	O⁶-methylguanine methyltransferase, constitutive
polA	DNA polymerase I
priABC	Primosome assembly
radA	Sensitivity to gamma radiation
recA	RecA protein strand transferase and coprotease
recBCD	Exonuclease V
recFOR	Helps to load RecA at daughter-strand gaps
recG	RecG binds to Holliday junctions
recJ	5′-3′ single-stranded exonuclease
recN	Needed for DSB repair
rep	Helicase
ruvABC	Bind and cleave Holliday junctions
sbcC, sbcD	Exonucleases active on recombination intermediates
soxRS	Oxidative stress regulon
ssb	Single-strand binding protein
tag	3-Methyladenine glycosylase
topA	DNA topoisomerase I
umuDC	DNA polymerase V and damage checkpoint signal
ung	Uracil glycosylase
uvrABC	Encodes proteins for bulky adduct recognition and nicking
uvrD(mutU)	Helicase II
xthA	Exonuclease III

step 2). The loading of the RecA protein at DSGs appears to be facilitated by the RecFOR proteins. After formation of a Holliday junction (Fig. 2A, step 3), the resulting gap is filled by a polymerase (Pol I or Pol III), and the junction is translocated by either the RuvAB proteins or the RecG protein (Fig. 2A, step 4). Resolution of the junction by RuvC restores the normal structure behind the fork and allows for removal of the adduct by nucleotide excision repair.

Figure 2B shows that when a replication fork encounters a gap opposite a DNA lesion, the replication fork collapses, resulting in the dislocation of one arm of the fork. Formally, this arm results from

the formation of a double-strand DNA break (DSB), whereas the other arm can undergo DSG repair (Fig. 2B, step 6). The RecBCD enzyme loads onto the end of the linear molecule and uses its helicase function to separate strands and its exonuclease function to degrade the strand with a 3′ end. When RecBCD encounters a chi site (GCTGGTGG), however, a conformational change occurs and exonuclease activity is now predominantly on the 5′-ended strand (Fig. 2B, step 7). This action leaves a single-stranded "tail" with a 3′ end. The RecA protein can bind to this tail and synapse it with the homologous strand of the intact chromosome to form a D-loop (Fig. 2B, step 8). Recent data suggest that RecA prefers to act at chi sites through relief of inhibition by the RecD subunit (2, 13). Strand assimilation leads to the formation of a Holliday junction; the action of the RuvABC and RecG proteins is as described above for DSG repair (Fig. 2B, steps 9 and 10). Both DSG and DSB repair also require housekeeping enzymes such as DNA ligase, topoisomerase, and single-strand binding protein, which are encoded by the *lig*, *gyr*, and *ssb* genes, respectively.

RECOMBINATION IS ESSENTIAL FOR *dam* MUTANTS

Bacteria that are mutated in *dam* and that contain the *lexA3* mutation cannot be constructed, because this combination of mutations results in inviable cells (51). This result suggests that one or more of the LexA-regulated SOS gene products is required for *dam* bacteria to survive. One of these required SOS gene products is RecA, as *dam recA* double mutants are inviable (51). However, since *dam lexA3* mutants cannot be constructed even when RecA is supplied from a multicopy plasmid, additional SOS gene products are needed. The experimental data for the statements above are shown in Table 3 (49). When a *dam*::Kan Hfr is mated with a *lexA3* recipient (DE407), no authentic *dam lexA3* recombinants are recovered (Table 3, line 1). A few colonies do appear on the selection plates, but these are *dam lex⁺* due to transfer of the wild-type *lexA* gene from the donor. This result indicates that transfer and recombination can occur during the mating, and the failure to recover *dam lexA3* recombinants is not due to the lack of these processes. Inclusion of a multicopy plasmid with the *recA* gene in the recipient does not enhance the recombination frequency (Table 3, line 3).

A systematic search was undertaken over a number of years to identify the SOS gene products

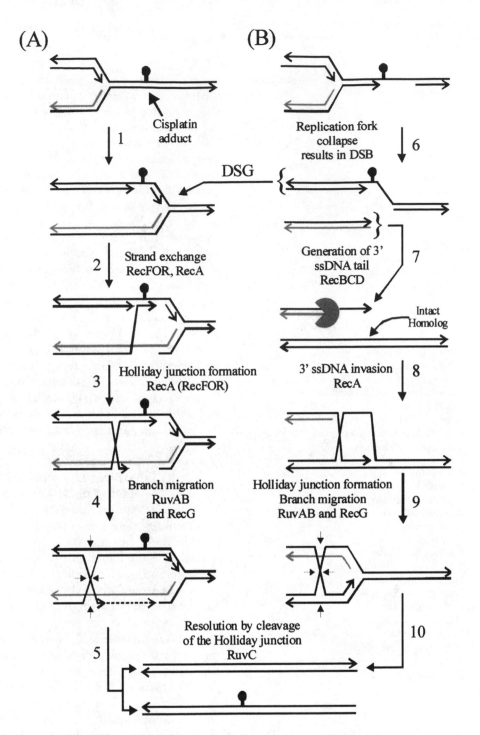

Figure 2. Daughter-strand gap (DSG) and double-strand break (DSB) repair of platinated DNA. (A) DSG pathway. Step 1: persistent cisplatin DNA adducts are encountered by the replication complex. Stalled replication results in the formation of a DSG opposite the adduct. Step 2: interactions between the proteins of the RecFOR pathway and the replication fork initiate RecA nucleation and strand exchange. Step 3: the ensuing RecA-catalyzed strand exchange (with the aid of the RecFOR accessory proteins) results in the formation of a Holliday junction. Step 4: branch migration of the Holliday junction catalyzed by the RuvAB or RecG proteins results in the repair of the DSG and restoration of the replication fork. Step 5: resolution of the Holliday junction by RuvC restores two double-stranded DNA molecules. This could be a mechanism of damage tolerance, as the cisplatin adduct is bypassed by recombinational repair and persists in the DNA for subsequent removal by nucleotide excision repair. (B) DSB pathway. Step 6: the replication complex encounters an unrepaired DSG or a nick opposite the adduct. Collapse of the replication fork forms a DSB and a DSG; the DSG portion of the collapsed replication fork is processed by the DSG pathway using a complementary strand from a homologous chromosome. Step 7: the RecBCD complex (sectored circle) binds the free end of the DSB and generates single-stranded DNA that is a substrate for RecA nucleation. Step 8: RecA nucleoprotein filaments catalyze the invasion of the RecBCD-generated single-strand tail into the homologous duplex. Step 9: RecA-catalyzed strand exchange and branch migration result in the formation of a Holliday junction and restoration of the replication fork. Step 10: resolution of the Holliday junction by RuvC yields two intact duplexes (only one molecule is shown). Reprinted from *Chemistry and Biology* (82) with permission of the publisher.

Table 3. Effect of *ruvAB* and *recA* plasmids on recombination frequency in Hfr *dam16*::Kan Str^S × F$^-$ *lexA3* Str^R crosses[a]

Recipient strain	Recombination frequency
DE407 (*lexA3*)	0 (5)
p*ruvAB*/DE407	0 (5)
p*recA*/DE407	0 (7)
p*recA*/p*ruvAB*/DE407	1,000
GM7362 (DE407 *mutS453*)	260
GM7363 (DE407 *mutL451*)	300
p*recA*/p*ruvAB*/GM7362	1,300
p*recA*/p*ruvAB*/GM7363	1,300

[a]Logarithmic phase donor and recipient strains were mated at a ratio of 1:10, respectively, and plated to select for kanamycin (Kan)- and streptomycin (Str)-resistant recombinants. The recombination frequency is the number of recombinants in 50 µl of mating mixture. The numbers in parentheses indicate recombinants that are Lex$^+$ due to transfer of this marker during mating. p*recA* denotes a multicopy plasmid encoding RecA; p*ruvAB* denotes a multicopy plasmid encoding RuvB.

required for viability of *dam* bacteria (58, 59). The identification of these genes is shown in Table 3. When the recipient strain contains multicopy plasmids expressing RecA, RuvA, and RuvB, *dam lexA3* recombinants are recovered at high frequency (Table 3, line 4). Expression of RuvA and RuvB alone, however, is not sufficient to enhance the recombination frequency (Table 3, line 2). Therefore, of all the SOS genes, only *recA* and *ruvAB* need to be expressed constitutively to allow for the survival of a *dam* mutant in a *lexA3* background.

Inactivation of the mismatch repair system by mutation in *mutS* or *mutL* greatly enhances recovery of *dam lexA3* recombinants (Table 3, lines 5 and 6), indicating that mismatch repair is ultimately responsible for the lethality of this combination. The recombination frequency can be increased to a high level by supplying RecA, RuvA, and RuvB in *trans* (Table 3, lines 7 and 8). This suggests that the concentration of these proteins may be limiting during recombination in the *lexA3* strain despite the inactivation of the mismatch repair system.

In addition to the requirement for expression of the SOS-regulated *recA* and *ruvAB* genes for viability in a *dam* mutant, expression of non-SOS genes is also required. These are *ruvC*, *recB*, *recC*, *polA*, and *lig*, which when mutated cannot coexist in a cell with a *dam* mutation. Since all these SOS and non-SOS gene products are involved in recombination, it was concluded that recombination is essential in a *dam* mutant.

Is a specific recombination pathway required for DSB repair in *dam* mutants? The answer is based on data obtained with the wild type and with *recBC sbcB* mutants (58). In the wild type, recombination occurs almost completely by the RecBCD pathway, which includes the repair and recombination gene products

that we have been discussing so far (14). In *recBC sbcB* bacteria, a different recombination pathway operates, designated the RecF pathway (14). RecF-dependent recombination requires the products of the *recFOR*, *recN*, and *recQ* gene products. *E. coli* strains that are *dam recBC sbcB* are viable, but *recFOR*, *recN*, and *recQ* derivatives of these strains are not. The conclusion from these data is that in *dam* mutants the RecBCD pathway provides recombinational repair but in its absence the RecF pathway can substitute to effect repair. Thus any recombination pathway capable of DSB repair will prevent *dam* lethality. To further bolster this argument, the phage lambda *red* recombination system can prevent *dam* inviability in the absence of the RecBCD pathway (unpublished data).

The requirement for recombination in *dam* strains must be associated with the production of MutH-catalyzed single-strand breaks because inactivation of mismatch repair abrogates the necessity for recombinational repair (Table 3). The DNA of a *dam* mutant contains single-strand nicks produced by the action of MutH (51). The protein prefers hemimethylated DNA as a substrate but it is also active on unmethylated DNA. On both substrates it produces a nick 5′ to the G at -GATC- sequences (76). The nicks result from the initial binding of MutS to some as yet unidentified DNA lesion or mismatch in nonreplicating DNA. It was proposed that these single-strand breaks were converted to DSBs and that recombination was required to repair the DSBs (49). The single-strand breaks can be converted to DSBs by two separate mechanisms (Fig. 3). In the first mechanism, MutH makes two nicks at the same -GATC- sequence on each complementary strand (3) (Fig. 3A). This is equivalent to a DSB and requires the action of DSB-specific recombination proteins (e.g., RecA, RecBCD, RuvABC) for repair. In the second mechanism, a DSB is formed by a replication fork approaching a MutH strand break, resulting in replication fork collapse and requiring DSB-specific recombination proteins to restore the fork. A key difference between the two mechanisms is that DNA replication is required in one but not the other. A second difference between the models is the requirement for the PriA protein (see below) in the replication fork collapse model. The requirement for DNA replication to produce DSBs in a *dam* mutant is currently being tested.

The presence of single- and double-strand breaks in *dam* bacteria DNA should induce the SOS regulon and, indeed, this is observed ("subinduction") (59). It is expected, therefore, that in *dam mut* bacteria the level of transcription of these genes should be at wild-type basal levels, as the need for DSB repair should be reduced. This is not the case,

A.

B.

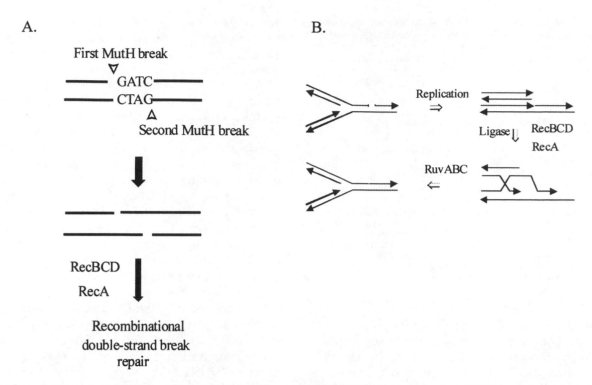

Figure 3. Generation of DSBs in a *dam* mutant. (A) DSBs produced by MutH nicking at the same GATC on complementary strands produces a substrate for DSB repair. (B) Replication through DNA with a nick or gap leads to replication fork collapse, effectively producing a DSB of one fork arm, which becomes a substrate for RecBCD action in DSB repair. Ligation of the nick by DNA ligase produces an intact duplex on the complementary strand.

however, and the level of subinduction of the SOS regulon is the same in *dam* and *dam mut* bacteria (58). This result suggests that there is more than one SOS signal being generated in *dam* cells and that one or more is *mut* independent and due to a nonlethal DNA substrate. The identity of this substrate remains unknown. Alternatively, it could result from the direct effects of Dam methylation on transcription of these SOS genes.

It is believed that only about three-quarters of the replication forks initiating at *oriC* complete their traverse to the termination (Ter) region of the chromosome (67). The rest are thought to stall at lesions or sequences in the chromosome or to collapse. Such stalling can, for example, lead to isomerization of the fork to a cruciform structure (Fig. 4). Such a cruciform structure is equivalent to a Holliday junction and a substrate for RuvC resolvase which cleaves the cruciform to form a DSB (Fig. 4) with concomitant dissociation of the polymerase III holoenzyme from its substrate. Such dissociation also occurs at collapsed replication forks. Recombinational DSB repair leads to restoration of the fork, but now the DNA helicases plus polymerase III holoenzyme need to reassociate at the point of interruption. This requires a poorly understood phenomenon referred to

as "stable DNA replication," which is an SOS- and RecA-dependent process requiring the PriA protein (37, 38). The PriA protein binds to the D-loop formed during recombination and recruits the DnaB helicase and the DnaG primase and then the polymerase holoenzyme complex (66). The RecJ exonuclease and RecQ helicase proteins may also be involved in remodeling the replication fork after encountering DNA damage before PriA action (16). After resolution of the Holliday junction, the replication fork is restored and replication can continue. Genetic and in vitro data support such a recruitment model by the PriA protein. The above scenario for restoration of the replication fork after stalling or collapse suggests that the role of cellular recombination proteins is primarily in this kind of repair. This may account for the poor viability of various recombination mutants, such as the *recBC* strains.

If DSBs in *dam* mutants are formed by replication fork collapse (Fig. 3B), then PriA protein should be required for restoration of the fork. If so, then *dam priA* double mutants should be inviable. For the MutH-induced DSB (Fig. 3A), there should be no requirement for PriA, and *dam priA* mutants should be viable. The construction of such mutants is being attempted.

A.

B.

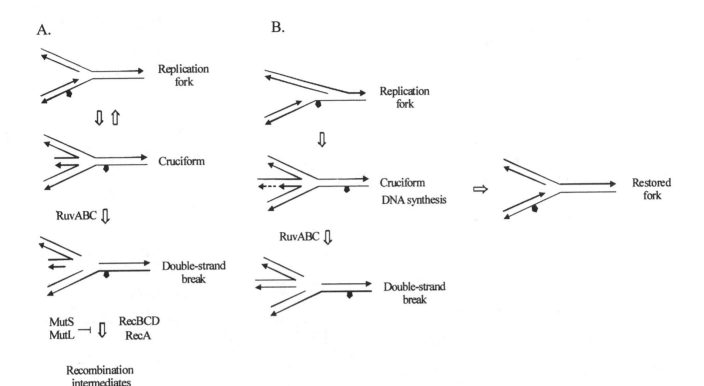

Recombination
intermediates

Figure 4. RuvC-induced replication fork DSB. (A) The pentagon represents a miscoding lesion subject to mismatch repair. After replication of this substrate, the fork stalls owing to the need for mismatch repair. The stalled fork isomerizes into a cruciform and becomes a substrate for RuvC cleavage. DSB repair is required to restore the fork, and MutSL is predicted to inhibit this repair. Reprinted from the *Journal of Bacteriology* (49) with permission of the publisher. (B) The pentagon represents a DNA polymerase-blocking lesion which, when encountered by a replication fork, leads to unequal replication of template strands. The fork isomerizes and DNA synthesis extends the 3′ end to produce a flush-ended molecule that can revert to its normal conformation, yielding a replication fork in which the lesion has been bypassed. Alternatively, the cruciform becomes a substrate for RuvC cleavage that requires DSB repair.

MISMATCH REPAIR
AND DRUG RESISTANCE

Bacteria with mutations in the *dam* gene are more sensitive to methylating agents and cisplatin than wild-type bacteria (Fig. 5). Double mutants with *dam* and either *mutS* or *mutL* mutations are as resistant to these agents as the wild type (Fig. 5). Therefore, mismatch repair sensitizes *dam* bacteria to these agents (24, 36). Remarkably, human cells behave like *dam* mutants; they are also sensitive to these agents but resistant if a mutation preventing mismatch repair is present (35). We showed that in *dam* bacteria, O⁶-methylguanine (OMG) was the base responsible for the mismatch repair effect produced by methylating agents, and we postulated that OMG paired with either C or T is a substrate for mismatch repair recognition (36). That is, all possible OMG base pairs are subject to mismatch repair. Consequently, upon replication of the OMG-containing strand, a futile cycle of mismatch repair ensues (Fig. 6). As the replicative polymerase III holoenzyme syn-thesizes mismatch repair tracts, this event causes polymerase III holoenzyme stalling and eventual cell death. In the absence of mismatch repair, there is no stalling of the polymerase because there is no mismatch repair and no sensitivity to the cytotoxic agents. The stalled replication fork could, in theory, isomerize to a cruciform structure and become a substrate for RuvC cleavage (Fig. 4). If this were the case, however, no DSB would be formed in the absence of RuvC, and a *ruvC* mutant strain should be as resistant as the wild type to N-methyl-N′-nitro-N-nitrosoguanidine (MNNG). *RuvABC* mutants, however, are sensitive to MNNG, making RuvC cleavage unlikely unless the Ruv proteins are required at a subsequent step in recombinational repair (see Fig. 8). Overall, the futile cycling model is compatible with all the known experimental facts in *E. coli* exposed to methylating agents.

The situation with regard to cisplatin, however, is more complex. In this case, unlike OMG, cisplatin adducts are replication-blocking lesions which form DSGs behind the replication fork (54). It is, therefore,

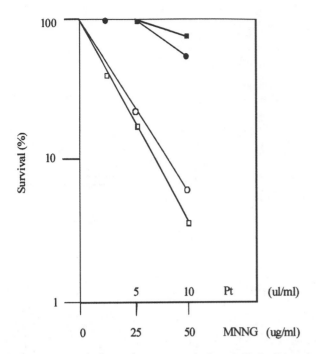

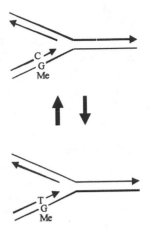

Figure 6. Futile cycling at the replication fork. DNA polymerase III holoenzyme has placed a C opposite an O^6-methylguanine (G-Me) template residue. This base pair is recognized and acted upon by the mismatch repair system. Replacement of the C with a T again results in a mismatched base pair subject to repair. Because DNA polymerase III holoenzyme is required for repair synthesis, the replication fork stalls at the mismatch. In the absence of mismatch repair, no stalling of the fork occurs.

Figure 5. Survival of *E. coli* strains exposed to cisplatin (Pt) and MNNG. *E. coli dam* bacteria in the exponential phase of growth were exposed to cisplatin (open squares) or MNNG (open circles), and survival was measured as a function of drug dose. The survival of *dam mutS* cells exposed to cisplatin (closed squares) or MNNG (closed circles) was measured the same way. The wild-type strain gave the same results as the *dam mutS* strain.

difficult to invoke a futile cycling model as for the methylating agents. However, in both cases replication fork stalling or collapse occurs and this may be the common denominator in the cytotoxic action by these agents. An alternative to the futile cycling model based on DSB recombinational repair is described below to explain how mismatch repair sensitizes *E. coli dam* mutants (and human cells) to methylating agents and cisplatin.

In *dam* mutants there is constant repair of DSBs, and the recombinational capacity of the cell is probably near its maximum. This conclusion is based on the higher basal level of transcription of certain SOS genes in *dam* cells, suggesting that one or more of the RecA or RuvA or RuvB proteins is limiting. If additional DSBs are introduced in the chromosome of *dam* cells by exposure to cisplatin or methylating agents, then the recombinational capacity would not be sufficient to repair them and the cells would die. Inactivation of mismatch repair by mutation in *mutS* or *mutL* in *dam* cells (Table 2), however, would dramatically decrease the number of endogenous DSBs, thereby allowing increased repair of drug-induced DSBs and increasing resistance to methylating agents or cisplatin.

The recombination model above predicts that recombination-deficient cells should be more sensitive to cisplatin and methylating agents than wild-type cells. This prediction was tested with cisplatin and *dam*⁺ recombination-deficient strains. (*dam rec* strains cannot be used because they are inviable. However, the results for wild-type cells should also apply to *dam* mutants.) It was found that mutations affecting recombination (*recBCD*, *recFOR*, *ruvABC*, and *recG*) were differentially sensitive to cisplatin, indicating that, indeed, recombinational repair of DSBs and DSGs is crucially important for repairing cisplatin damage (82). Until this study, it was believed that only nucleotide excision repair (NER) was important for survival of cells exposed to this drug (54). Figure 7 shows that recombination-deficient *recBC* cells are as sensitive to cisplatin as the NER-deficient *uvrA* bacteria, indicating that recombinational repair is as important as NER. The striking result, however, is the extreme sensitivity of the double mutant (*recBCD uvrA*) lacking both NER and recombination repair to cisplatin, confirming the independent nature of the two repair pathways (Fig. 7) (82).

Preliminary results with recombination-deficient *dam*⁺ *ruvABC* bacteria also indicate an increased sensitivity to methylating agents (Fig. 8) (unpublished data). Coupled with the previously known sensitivity of *recA* strains to MNNG (25), this suggests that recombinational repair will also be important for damage inflicted to DNA by methylating agents. The sensitivity of the *recA* strain can also be interpreted as a requirement for SOS-induced transcription of *ruvAB* in addition to *recA* itself. It should also be

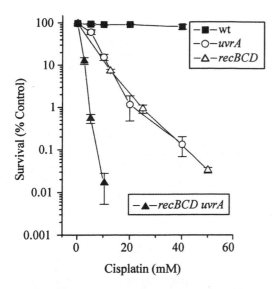

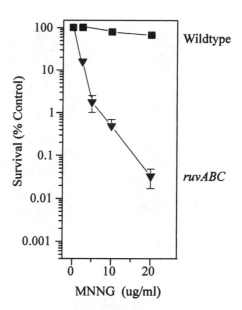

Figure 7. Survival of *uvrA* and *recBCD E. coli* strains exposed to cisplatin. Exponential-phase *E. coli* cells growing in broth were exposed to various doses of cisplatin and then plated to determine survival. Reprinted from *Chemistry and Biology* (82) with permission of the publisher.

Figure 8. Survival of wild-type and *ruvABC* cells exposed to MNNG. Exponential-phase *E. coli* cells growing in broth were exposed to various doses of MNNG for 15 min and then plated to determine survival.

noted that base excision repair enzymes such as AlkA and Tag as well as the methyl-abstracting proteins Ada and Ogt are important for repair of methylated bases (see Table 2 for protein functions) (25).

Given the sensitivity of strains affecting recombination to cisplatin and methylating agents, the idea that these agents induce the formation of DSBs is reasonable. Therefore, the hypothesis that *dam* bacteria are sensitive to these agents because of inability to repair all DSBs is quite plausible. An important common theme is the requirement for replication forks to stall or collapse at lesions.

DRUG-INDUCED RECOMBINATION

One of the phenotypic traits of *dam* mutants is a hyperrecombination phenotype (Table 1) (50). The necessity for DSB recombination functions in these mutants correlates nicely with current ideas that DSBs can initiate recombination (72). If there are more DSBs than in the wild type, then there should be an increased level of recombination in *dam* cells, thereby explaining the hyperrecombination phenotype due to increased DSBs.

If cisplatin and methylating agents induce the formation of DSBs, then they should also increase the spontaneous recombination frequency. To test this hypothesis, we used a genetic assay designed by Konrad (39) in which two inactive *lac* operons on the chromosome can recombine to form a functional

lac operon. Recombination can be scored easily by measuring the frequency of Lac$^+$ recombinants. A typical result using this assay is shown in Fig. 9 (82). MacConkey plates were spread with an aliquot of an overnight culture of wild-type cells (GM7330), and filter paper disks were placed on the plates. Various concentrations of cisplatin, *trans*-DDP (the biologically inactive isomer of cisplatin), and MNNG were added to the disks. Figure 9A shows that a dose-dependent increase in recombination and cytotoxicity was elicited by cisplatin. The *trans*-DDP isomer showed an increase in recombination only at the highest doses tested (Fig. 9B). MNNG also elicited a dose-dependent increase in recombination and cytotoxicity (Fig. 9C). These data support the hypothesis that cisplatin and MNNG induce the formation of DSBs.

AN ASIDE: A GENERAL MECHANISM FOR DRUG-INDUCED CYTOTOXICITY

In the previous section, evidence was presented that recombination is important for wild-type and *dam* cells to repair damage caused by cisplatin and methylating agents such as MNNG. Further, it seems probable that repair of DSBs is critical for survival. Could it be that the common denominator by which cells are killed by drugs is failure to repair DSBs induced by DNA replication?

Nitric oxide (NO) produces deaminated and oxidized bases in the DNA of bacterial cells exposed

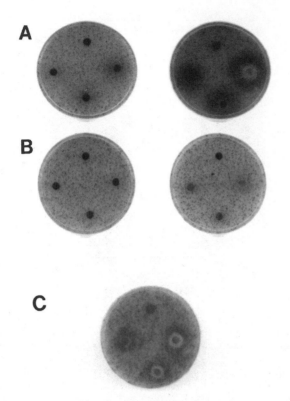

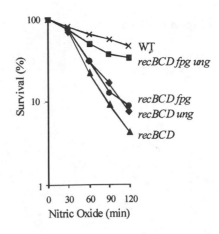

Figure 10. Survival of *E. coli* strains exposed to nitric oxide (NO). Exponential-phase *E. coli* cells growing in broth were exposed to various doses of NO gas and then plated to determine survival. Reprinted from the *Journal of Bacteriology* (69) with permission of the publisher.

Figure 9. Drug-induced recombination: (A) cisplatin, (B) *trans*-DDP, and (C) MNNG. Wild-type cells (GM7330) containing duplicate inactive *lac* operons were spread on MacConkey agar plates, and filter paper disks containing various amounts of drug were added. After incubation, the plates were placed on a scanner, and a digitized image was obtained. Recombination was measured by the formation of Lac⁺ colonies (which appear as dark dots on the plates) on a background of Lac⁻ bacteria. Drug concentration increases in a counterclockwise manner on each plate except for the disks without drug, which are at the top of each plate.

to it. These damaged bases are removed by glycosylases to produce abasic (AP) sites, which in turn are recognized, and the DNA chain containing them is cleaved by AP endonucleases. The resulting nicks are entry sites for DNA polymerase I, which excises nucleotides and replaces the AP sites with the correct base sequence. The final step is the action of DNA ligase to seal the nick. The major AP endonucleases in *E. coli* are exonuclease III (Xth) and endonuclease IV (Nfo) (see Table 2); double mutants lacking AP endonuclease activity are also more sensitive to NO than the wild type (70). Recombination-deficient *E. coli* strains (e.g., *recBCD* [Fig. 10]) are much more sensitive to NO than wild-type strains or strains deficient in nucleotide excision repair (70). The NO sensitivity of the *recBCD* mutants is abrogated if mutations in both *ung* and *fpg* are present (Fig. 10) (69). These genes encode glycosylases that remove uracil and ring-opened and oxidized bases (Table 2). Further, NO is recombinogenic in wild-type cells (70).

The most reasonable hypothesis to explain these data is that the Ung and Fpg glycosylases act on NO-damaged DNA to produce lesions that are acted upon by AP endonucleases, and that DNA replication converts the nicks or gaps into DSBs. These DSBs require RecBCD in order to be repaired and in addition are responsible for the increased recombinogenicity. This scenario is made more plausible by the observation that NO-induced recombination is decreased in *ung fpg* double mutants (69).

In the examples described above, DSBs are formed in the DNA of cells treated with agents that produce very different DNA modifications that are repaired by nonoverlapping repair systems. They all, however, provoke recombinational repair, and this is the common thread among them, presumably owing to the formation of DSBs. It is tempting to extrapolate these results to other agents to make the case that, in general, unrepaired DSBs are the basis for cytotoxicity. Agents such as bleomycin and gamma radiation produce DSBs directly, and therefore DNA replication is not needed to produce them. A small number of unrepaired DSBs might also explain the sensitivity of *E. coli* to UV light, given that DSBs have been detected in *uvr recBC* cells at low doses (75).

MISMATCH REPAIR AND ANTIRECOMBINATION

As discussed above, one of the functions of mismatch repair is to rectify errors introduced into the newly replicated DNA chain. Another function of mismatch repair is to inhibit recombination when the DNA substrates are similar but not identical (homeologous). In contrast, the inhibitory action of mismatch repair on homologous recombination is small. This feature of mismatch repair was first described in conjugational crosses between *E. coli* and its close relative *Salmonella enterica* serovar Typhimurium, which are about 15% divergent in their genomic sequences (62). Crosses between these two species are sterile; however, if the recipient is defective in mismatch repair, recombinants are formed at a frequency that is about 1/100th to 1/1,000th of that for the corresponding cross between *E. coli* donors and recipients. The interspecies recombinants are mosaics containing large patches of serovar Typhimurium DNA embedded in the *E. coli* genome. It was postulated that during interspecific crosses between wild-type cells, recombinant DNA molecules are formed that have a large number of mismatches (homeologous DNA), and that mismatch repair acts to abort recombination on such substrates (antirecombination). Inactivation of mismatch repair by mutation in the *mutS* and *mutL* genes was most effective in generating recombinants in interspecific crosses. Induction of the SOS system stimulated interspecific recombination (53) and, in the model proposed here, this could occur by increasing production of RecA and RuvAB.

A biochemical demonstration of the inhibition of homeologous recombination by MutS and MutL was developed using the closely related phages fd and M13 (3% difference in nucleotide sequence) in a RecA strand-transfer reaction (Fig. 11) (77). In the presence of MutS and MutL, there was a strong inhibition of strand transfer compared with when they were absent. In addition, there was no inhibition of strand transfer by the Mut proteins with homologous (M13 × M13, fd × fd) substrates. The extent of Mut protein inhibition was proportional to both the number of DNA mismatches (of which at least 3 to 4 per 100 bases were needed) and the length of the DNA.

It was shown previously that inclusion of the RuvAB proteins in the assay for homologous recombination stimulates RecA-catalyzed strand exchange (1). However, MutS and MutL inhibit this

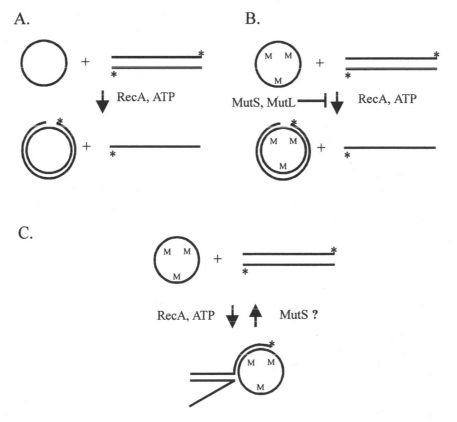

Figure 11. RecA-catalyzed strand-transfer reaction. (A) The RecA protein, in the presence of ATP, catalyzes the transfer of a strand from a linear duplex to the complementary single-stranded circular substrate to form a nicked circle. The asterisks indicate end labels. MutSL proteins have no effect on this reaction. (B) The same reaction is shown, except that the circular DNA substrate is either methylated or platinated. MutSL proteins are predicted to inhibit the strand-transfer reaction. (C) Proposed reversal of RecA action by MutS. A recombination intermediate is shown resulting from the action of RecA. It is postulated that MutS might reverse this reaction.

stimulation in the absence of RecA by inhibiting branch migration (22). The antirecombinogenic effect of mismatch repair on homeologous DNA recombination intermediates, therefore, appears to occur in two distinct steps.

ANTIRECOMBINATION AND DRUG RESISTANCE

Mismatch repair sensitizes *dam E. coli* (and human cells) to the toxic effects of cisplatin and methylating agents (Fig. 5). If it is assumed that the drugs act by producing lesions in DNA that overwhelm the recombinational repair capacity of the cell, then how can mismatch repair elicit drug sensitivity? The model proposed here is that mismatch repair inhibits recombinational repair by a mechanism similar to that for antirecombination (Fig. 11). It assumes that cisplatin-adducted or methylated DNA contains drug-modified mismatches recognized by MutS as homeologous DNA when these are formed during recombinational repair. By reducing the efficiency of the recombinational repair processes, mismatch repair sensitizes cells to the drugs. In cells defective for mismatch repair, no inhibition of recombinational repair occurs, and they are as resistant to these agents as the wild type.

It is already known that MutS recognizes and binds to O^6-methylguanine paired with either thymine or cytosine (61). However, the binding affinity is quite weak for the OMG-cytosine pair, which is probably the most important one in vivo; it is much less than that for normal base mismatches. The effect of MutS binding to multiple OMG mismatches on long (about 150-bp) oligonucleotides has not been tested. When such oligonucleotides are modified with cisplatin, there is little detectable binding of MutS when single adducts are present, but binding becomes cooperative when the number of adducts increases (Fig. 12) (81). Such cooperative DNA binding might also occur with OMG-containing DNA, which would result in a higher affinity for these lesions by MutS.

Cisplatin forms three types of adducts in DNA: (i) two intrastrand cross-links between either adjacent guanine-guanine (GpG) or guanine-adenine (GpA) residues (the 1,2-adduct) or (ii) guanines separated by a base (1,3-adduct), and (iii) an interstrand cross-link between guanine residues. The GpG intrastrand cross-link makes up 65% of all adducts, the GpA makes up 25%, and the 1,3-adduct makes up 5 to 10% (54). Less frequent lesions include interstrand cross-links and monoadducts. Of these adducts, only the 1,2-GpG modification is specifically bound by

MutS with an affinity equal to that of a G-T mismatch, which in turn has the highest affinity among base-base mismatches (Z. Z. Zdraveski, J. A. Mello, and M. G. Marinus, unpublished data).

All three types of cisplatin cross-links are recognized by the NER system, but the rate of removal of the 1,3-adduct is about 50 times greater than that of the 1,2-adduct (54). Therefore, the 1,2-adduct should persist longer in modified DNA, and replication forks should have a greater chance of encountering this replication-blocking lesion. The possible fates of this encounter are diagrammed in Fig. 2 (82). In the presence of a nick or gap ahead of the fork, DSB repair will occur (Fig. 2B), but if no interruption is present, DSG repair will ensue as a consequence of replication blockage (Fig. 2A). In both pathways of repair, RecA-mediated strand exchange must occur, followed by branch migration of the Holliday junction (Fig. 2, steps 2 to 4 and 8 and 9). It is during these steps that the antirecombination effect of mismatch repair of 1,2-adducts is predicted to occur, because the cisplatin-modified DNA is perceived by MutS as homeologous DNA, and it will bind to the cisplatin 1,2-adducted mismatches as they are formed by branch migration. Together with MutL, MutS then aborts the recombination process as it does for mismatches during antirecombination.

As noted above, MutS and MutL block the stimulation of strand exchange produced by RuvAB proteins on homeologous substrates by inhibiting branch migration. With platinated recombination intermediates, this mechanism may also operate. However, it has been reported that RuvAB can reverse the action of RecA-mediated strand transfer on homeologous DNA (32). Therefore, an alternative possibility is that on platinated substrates, RuvAB simply reverses the action of RecA rather than inhibiting branch migration.

The antirecombination action of mismatch repair as modeled above would effectively block recombinational DNA repair of platinated DNA, thereby leaving unrepaired lesions in the DNA and triggering pathways leading to cell death. The same model would also apply to cells containing methylated DNA produced by alkylating agents, although in this case there may be additional inhibition by futile cycling. Cells in which mismatch repair is defective will be resistant to the drugs because the inhibition of recombinational repair is no longer possible. MutS also binds to cisplatin or methylated mismatches in normal *E. coli* chromosomal DNA, but because the DNA is fully methylated, no incision of DNA occurs and there is also no initiation of recombinational repair. The predictions of this model are currently being tested.

A.

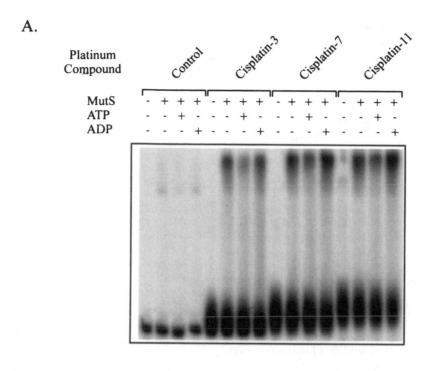

B.

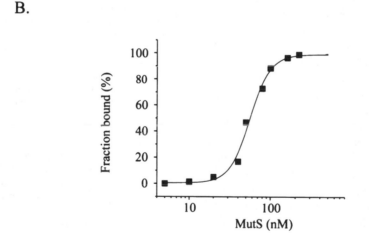

Figure 12. Binding isotherm of MutS and cisplatin-modified DNA. (A) MutS protein was titrated into binding reactions with or without ADP or ATP containing radiolabeled 162-bp probes with various levels of cisplatin adducts. (B) The fraction of bound cisplatin-7 probe as a function of MutS concentration. Reprinted from the *Journal of Biological Chemistry* (81) with permission of the publisher.

It is tempting to speculate that MutS binding to DNA mismatches may be fundamentally different during repair of premutational lesions at the replication fork and during antirecombination. Near the replication fork, high-affinity recognition of single mismatches by MutS in short DNA sequences would seem optimal. In contrast, during antirecombination it seems necessary for MutS to recognize multiple mismatches over a long stretch of DNA. One possible model is that the MutS active form is a dimer in mutation avoidance and a tetramer in antirecombination. Dimers of MutS bound to widely separated mismatches during antirecombination could interact by protein-protein interaction to generate tetramers and form a looped DNA-protein structure. Such a structure may be required for the antirecombinogenic effect of MutS in reversing RecA action.

The MutS protein efficiently binds single-stranded DNA, and ATP hydrolysis occurs when this nucleotide is present (56). Is it possible that MutS catalyzes the reverse reaction to that of RecA? The MutS crystal structure indicates that two channels exist between the subunits, one of which is the double-stranded DNA binding site (42, 57). Could the other serve as a binding site for single-stranded DNA? The channel is about 30×20 Å, large enough for single-stranded DNA to penetrate. If so, this might explain how MutS could reverse the action of RecA, as shown in Fig. 11.

PHENOTYPES OF *dam* MUTANTS

The realization that recombination plays a vital role in the life of *dam* mutants has helped to explain some of the phenotypes associated with these mutants. The hyperrecombination phenotype is explained by the increased number of DSBs leading to increased initiation of recombination. The restriction of "foreign" DNA requires the action of restriction endonucleases followed by RecBCD to degrade the fragments. The reduction in restriction ability of *dam* mutants (21, 52) can be explained by the limiting concentration of RecBCD due to increased chromosomal DSB repair. The increased sensitivity to various drugs is due to interference of recombination repair by mismatch repair, and the resistance of *dam mut* bacteria to these agents is due to a reduction in the number of single-strand breaks being converted to DSBs and by lack of mismatch repair inhibition of recombination. The subinduction of the SOS regulon is, in part, a consequence of single-strand gaps in DNA which can activate RecA for coprotease activity. Increased DSB formation is likely the reason for lysogenic phages to be induced and to abandon their *dam* host. The synthetic lethality of *dam* with *rec*, *lexA*, *lig*, *pol*, and *ruv* genes is clearly related to the requirement for DSB repair. The necessity for Dam in Vsp repair remains puzzling (6). Together with the roles for Dam methylation in controlling transcription initiation and its role in regulating initiation of chromosome replication and its synchronization, almost all the phenotypic properties can now be explained.

Acknowledgments. I thank my collaborators without whom this review would not have been possible: Bevin Engleward, John Essigmann, Jill Mello, Erik Spek, and Zoran Zdraveski. The manuscript was much improved by comments from Kenan Murphy, Tony Poteete, Erik Spek, and Mike Volkert.

REFERENCES

1. **Adams, D. E., and S. C. West.** 1996. Bypass of DNA heterologies during RuvAB-mediated three- and four-stranded branch migration. *J. Mol. Biol.* **263:**582–596.
2. **Amundsen, S. K., A. F. Taylor, and G. R. Smith.** 2000. The RecD subunit of the *Escherichia coli* RecBCD enzyme inhibits RecA loading, homologous recombination, and DNA repair. *Proc. Natl. Acad. Sci. USA* **97:**7399–7404.
3. **Au, K. G., K. Welsh, and P. Modrich.** 1992. Initiation of methyl-directed mismatch repair. *J. Biol. Chem.* **267:**12142–12148.
4. **Ban, C., and W. Yang.** 1998. Crystal structure and ATPase activity of MutL: implications for DNA repair and mutagenesis. *Cell* **95:**541–552.
5. **Ban, C., and W. Yang.** 1998. Structural basis for MutH activation in *E. coli* mismatch repair and relationship of MutH to restriction endonucleases. *EMBO J.* **17:**1526-1534.
6. **Bell, D. C., and C. G. Cupples.** 2001. Very-short-patch repair in *Escherichia coli* requires the *dam* adenine methylase. *J. Bacteriol.* **183:**3631–3635.
7. **Bianco, P. R., R. B. Tracy, and S. C. Kowalczykowski.** 1998. DNA strand exchange proteins: a biochemical and physical comparison. *Front. Biosci.* **3:**D570–D603.
8. **Borts, R. H., S. R. Chambers, and M. F. Abdullah.** 2000. The many faces of mismatch repair in meiosis. *Mutat. Res.* **451:**129–150.
9. **Boye, E., M. G. Marinus, and A. Lobner-Olesen.** 1992. Quantitation of Dam methyltransferase in *Escherichia coli*. *J. Bacteriol.* **174:**1682–1685.
10. **Buermeyer, A. B., S. M. Deschenes, S. M. Baker, and R. M. Liskay.** 1999. Mammalian DNA mismatch repair. *Annu. Rev. Genet.* **33:**533–564.
11. **Campbell, J. L., and N. Kleckner.** 1990. *E. coli oriC* and the *dnaA* gene promoter are sequestered from *dam* methyltransferase following the passage of the chromosomal replication fork. *Cell* **62:**967–979.
12. **Casadesus, J., and J. Torreblanca.** 1996. Methylation-related epigenetic signals in bacterial DNA, p. 141–153. *In* R. E. A. Russo, R. A. Marteinssen, and A. D. Riggs (ed.), *Epigenetic Mechanisms of Gene Regulation.* Cold Spring Harbor Laboratory Press, Cold Spring Harbor, N.Y.
13. **Churchill, J. J., and S. C. Kowalczykowski.** 2000. Identification of the RecA protein-loading domain of RecBCD enzyme. *J. Mol. Biol.* **297:**537–542.
14. **Clark, A. J., and K. B. Low.** 1988. Pathways and systems of homologous recombination in *Escherichia coli*, p. 155–215. *In* K. B. Low (ed.), *The Recombination of Genetic Material.* Academic Press, Inc., San Diego, Calif.
15. **Claverys, J. P., and S. A. Lacks.** 1986. Heteroduplex deoxyribonucleic acid base mismatch repair in bacteria. *Microbiol. Rev.* **50:**133–165.
16. **Courcelle, J., C. Carswell-Crumpton, and P. C. Hanawalt.** 1997. *recF* and *recR* are required for the resumption of replication at DNA replication forks in *Escherichia coli. Proc. Natl. Acad. Sci. USA* **94:**3714–3719.
17. **Courcelle, J., A. Khodursky, B. Peter, P. O. Brown, and P. C. Hanawalt.** 2001. Comparative gene expression profiles following UV exposure in wild-type and SOS-deficient *Escherichia coli. Genetics* **158:**41–64.
18. **Cox, M. M.** 1999. Recombinational DNA repair in bacteria and the RecA protein. *Prog. Nucleic Acid Res. Mol. Biol.* **63:**311–366.
19. **Cox, M. M., M. F. Goodman, K. N. Kreuzer, D. J. Sherratt, S. J. Sandler, and K. J. Marians.** 2000. The importance of repairing stalled replication forks. *Nature* **404:**37–41.
20. **Drotschmann, K., A. Aronshtam, H. J. Fritz, and M. G. Marinus.** 1998. The *Escherichia coli* MutL protein stimulates binding of Vsr and MutS to heteroduplex DNA. *Nucleic Acids Res.* **26:**948–953.
21. **Efimova, E. P., E. P. Delver, and A. A. Belogurov.** 1988. Alleviation of type I restriction in adenine methylase (*dam*) mutants of *Escherichia coli. Mol. Gen. Genet.* **214:**313–316.

22. Fabisiewicz, A., and L. Worth, Jr. 2001. *Escherichia coli* MutS, L modulate RuvAB-dependent branch migration between diverged DNA. *J. Biol. Chem.* **276:**9413–9420.

23. Fishel, R., M. K. Lescoe, M. R. Rao, N. G. Copeland, N. A. Jenkins, J. Garber, M. Kane, and R. Kolodner. 1993. The human mutator gene homolog MSH2 and its association with hereditary nonpolyposis colon cancer. *Cell* **75:**1027–1038.

24. Fram, R. J., P. S. Cusick, J. M. Wilson, and M. G. Marinus. 1985. Mismatch repair of cis-diamminedichloroplatinum(II)-induced DNA damage. *Mol. Pharmacol.* **28:**51–55.

25. Friedberg, E. C., G. C. Walker, and W. Siede. 1995. *DNA Repair and Mutagenesis.* ASM Press, Washington D.C.

26. Garcia-Del Portillo, F., M. G. Pucciarelli, and J. Casadesus. 1999. DNA adenine methylase mutants of *Salmonella typhimurium* show defects in protein secretion, cell invasion, and M cell cytotoxicity. *Proc. Natl. Acad. Sci. USA* **96:**11578–11583.

27. Hall, M. C., and S. W. Matson. 1999. The *Escherichia coli* MutL protein physically interacts with MutH and stimulates the MutH-associated endonuclease activity. *J. Biol. Chem.* **274:**1306–1312.

28. Harfe, B. D., and S. Jinks-Robertson. 2000. DNA mismatch repair and genetic instability. *Annu. Rev. Genet.* **34:**359–399.

29. Heithoff, D. M., R. L. Sinsheimer, D. A. Low, and M. J. Mahan. 1999. An essential role for DNA adenine methylation in bacterial virulence. *Science* **284:**967–970.

30. Hoeijmakers, J. H. 2001. Genome maintenance mechanisms for preventing cancer. *Nature* **411:**366–374.

31. Hsieh, P. 2001. Molecular mechanisms of DNA mismatch repair. *Mutat. Res.* **486:**71–87.

32. Iype, L. E., R. B. Inman, and M. M. Cox. 1995. Blocked RecA protein-mediated DNA strand exchange reactions are reversed by the RuvA and RuvB proteins. *J. Biol. Chem.* **270:**19473–19480.

33. Jiricny, J. 2000. Mismatch repair: the praying hands of fidelity. *Curr. Biol.* **10:**R788–R790.

34. Junop, M. S., G. Obmolova, K. Rausch, P. Hsieh, and W. Yang. 2001. Composite active site of an ABC ATPase: MutS uses ATP to verify mismatch recognition and authorize DNA repair. *Mol. Cell* **7:**1–12.

35. Karran, P., and M. Bignami. 1994. DNA damage tolerance, mismatch repair and genome instability. *Bioessays* **16:**833–839.

36. Karran, P., and M. G. Marinus. 1982. Mismatch correction at O^6-methylguanine residues in *E. coli* DNA. *Nature* **296:**868–869.

37. Kogoma, T. 1997. Stable DNA replication: interplay between DNA replication, homologous recombination, and transcription. *Microbiol. Mol. Biol. Rev.* **61:**212–238.

38. Kogoma, T., G. W. Cadwell, K. G. Barnard, and T. Asai. 1996. The DNA replication priming protein, PriA, is required for homologous recombination and double-strand break repair. *J. Bacteriol.* **178:**1258–1264.

39. Konrad, E. B. 1977. Method for isolation of *Escherichia coli* mutants with enhanced recombination between chromosomal duplications. *J. Bacteriol.* **130:**167–172.

40. Kuzminov, A. 1999. Recombinational repair of DNA damage in *Escherichia coli* and bacteriophage lambda. *Microbiol. Mol. Biol. Rev.* **63:**751–813.

41. Kuzminov, A. 2001. DNA replication meets genetic exchange: chromosomal damage and its repair by homologous recombination. *Proc. Natl. Acad. Sci. USA* **98:**8461–8468.

42. Lamers, M. H., A. Perrakis, J. H. Enzlin, H. H. Winterwerp, N. de Wind, and T. K. Sixma. 2000. The crystal structure of DNA mismatch repair protein MutS binding to a G × T mismatch. *Nature* **407:**711–717.

43. Reference deleted.

44. Leach, F. S., N. C. Nicolaides, N. Papadopoulos, B. Liu, J. Jen, R. Parsons, P. Peltomaki, P. Sistonen, L. A. Aaltonen, M. Nystrom-Lahti, et al. 1993. Mutations of a mutS homolog in hereditary nonpolyposis colorectal cancer. *Cell* **75:**1215–1225.

45. Lieb, M., and A. S. Bhagwat. 1996. Very short patch repair: reducing the cost of cytosine methylation. *Mol. Microbiol.* **20:**467–473.

46. Loh, T., K. C. Murphy, and M. G. Marinus. 2001. Mutational analysis of the MutH protein from *Escherichia coli.* *J. Biol. Chem.* **276:**12113–12119.

47. Lopez de Saro, F. J., and M. O'Donnell. 2001. Interaction of the beta sliding clamp with MutS, ligase, and DNA polymerase I. *Proc. Natl. Acad. Sci. USA* **98:**8376–8380.

48. Marinus, M. G. 1996. Methylation of DNA, p. 782–791. *In* F. C. Neidhardt, R. Curtiss III, J. L. Ingraham, E. C. C. Lin, K. B. Low, B. Magasanik, W. S. Reznikoff, M. Riley, M. Schaechter, and H. E. Umbarger (ed.), Escherichia coli *and* Salmonella: *Cellular and Molecular Biology*, 2nd ed., vol. 1. ASM Press, Washington, D.C.

49. Marinus, M. G. 2000. Recombination is essential for viability of an *Escherichia coli dam* (DNA adenine methyltransferase) mutant. *J. Bacteriol.* **182:**463–468.

50. Marinus, M. G., and E. B. Konrad. 1976. Hyper-recombination in *dam* mutants of *Escherichia coli* K-12. *Mol. Gen. Genet.* **149:**273–277.

51. Marinus, M. G., and N. R. Morris. 1974. Biological function for 6-methyladenine residues in the DNA of *Escherichia coli* K12. *J. Mol. Biol.* **85:**309–322.

52. Marinus, M. G., and N. R. Morris. 1975. Pleiotropic effects of a DNA adenine methylation mutation (*dam-3*) in *Escherichia coli* K12. *Mutat. Res.* **28:**15–26.

53. Matic, I., F. Taddei, and M. Radman. 1996. Genetic barriers among bacteria. *Trends Microbiol.* **4:**69–72.

54. Mello, J. A., E. E. Trimmer, M. Kartalou, and J. M. Essigmann. 1998. Conflicting roles of mismatch and nucleotide excision repair in cellular susceptibility to anticancer drugs, p. 249–274. *In* F. Eckstein and D. J. M. Liley (ed.), *Nucleic Acids and Molecular Biology*. Springer-Verlag, Heidelberg, Germany.

55. Michel, B. 2000. Replication fork arrest and DNA recombination. *Trends Biochem. Sci.* **25:**173–178.

56. Modrich, P., and R. Lahue. 1996. Mismatch repair in replication fidelity, genetic recombination, and cancer biology. *Annu. Rev. Biochem.* **65:**101–133.

57. Obmolova, G., C. Ban, P. Hsieh, and W. Yang. 2000. Crystal structures of mismatch repair protein MutS and its complex with a substrate DNA. *Nature* **407:**703–710.

58. Peterson, K. R., and D. W. Mount. 1993. Analysis of the genetic requirements for viability of *Escherichia coli* K-12 DNA adenine methylase (*dam*) mutants. *J. Bacteriol.* **175:**7505–7508.

59. Peterson, K. R., K. F. Wertman, D. W. Mount, and M. G. Marinus. 1985. Viability of *Escherichia coli* K-12 DNA adenine methylase (*dam*) mutants requires increased expression of specific genes in the SOS regulon. *Mol. Gen. Genet.* **201:**14–19.

60. Pukkila, P. J., J. Peterson, G. Herman, P. Modrich, and M. Meselson. 1983. Effects of high levels of DNA adenine methylation on methyl-directed mismatch repair in *Escherichia coli.* *Genetics* **104:**571–582.

61. Rasmussen, L. J., and L. Samson. 1996. The *Escherichia coli* MutS DNA mismatch binding protein specifically binds O(6)-methylguanine DNA lesions. *Carcinogenesis* **17:**2085–2088.

62. Rayssiguier, C., D. S. Thaler, and M. Radman. 1989. The barrier to recombination between *Escherichia coli* and

Salmonella typhimurium is disrupted in mismatch-repair mutants. *Nature* **342:**396–401.

63. **Ringquist, S., and C. L. Smith.** 1992. The *Escherichia coli* chromosome contains specific, unmethylated *dam* and *dcm* sites. *Proc. Natl. Acad. Sci. USA* **89:**4539–4543.

64. **Rothstein, R., B. Michel, and S. Gangloff.** 2000. Replication fork pausing and recombination or "gimme a break." *Genes Dev.* **14:**1–10.

65. **Russell, D. W., and R. K. Hirata.** 1989. The detection of extremely rare DNA modifications. Methylation in *dam⁻* and *hsd⁻ Escherichia coli* strains. *J. Biol. Chem.* **264:**10787–10794.

66. **Sandler, S. J., and K. J. Marians.** 2000. Role of PriA in replication fork reactivation in *Escherichia coli. J. Bacteriol.* **182:**9–13.

67. **Seigneur, M., V. Bidnenko, S. D. Ehrlich, and B. Michel.** 1998. RuvAB acts at arrested replication forks. *Cell* **95:**419–430.

68. **Sixma, T. K.** 2001. DNA mismatch repair: MutS structures bound to mismatches. *Curr. Opin. Struct. Biol.* **11:**47–52.

69. **Spek, E. J., L. N. Vuong, T. Matsuguchi, M. G. Marinus, and B. P. Engelward.** 2002. Nitric oxide-induced homologous recombination in *Escherichia coli* is promoted by DNA glycosylases. *J. Bacteriol.* **184:**3501–3507.

70. **Spek, E. J., T. L. Wright, M. S. Stitt, N. R. Taghizadeh, S. R. Tannenbaum, M. G. Marinus, and B. P. Engelward.** 2001. Recombinational repair is critical for survival of *Escherichia coli* exposed to nitric oxide. *J. Bacteriol.* **183:**131–138.

71. **Su, S. S., and P. Modrich.** 1986. *Escherichia coli* mutS-encoded protein binds to mismatched DNA base pairs. *Proc. Natl. Acad. Sci. USA* **83:**5057–5061.

72. **Szostak, J. W., T. L. Orr-Weaver, R. J. Rothstein, and F. W. Stahl.** 1983. The double-strand-break repair model for recombination. *Cell* **33:**25–35.

73. **Torreblanca, J., and J. Casadesus.** 1996. DNA adenine methylase mutants of *Salmonella typhimurium* and a novel *dam*-regulated locus. *Genetics* **144:**15–26.

74. **Wang, M. X., and G. M. Church.** 1992. A whole genome approach to in vivo DNA-protein interactions in *E. coli. Nature* **360:**606–610.

75. **Wang, T. C., and K. C. Smith.** 1986. Postreplicational formation and repair of DNA double-strand breaks in UV-irradiated *Escherichia coli uvrB* cells. *Mutat. Res.* **165:**39–44.

76. **Welsh, K. M., A. L. Lu, S. Clark, and P. Modrich.** 1987. Isolation and characterization of the *Escherichia coli mutH* gene product. *J. Biol. Chem.* **262:**15624–15629.

77. **Worth, L., S. R. M. Clark, and P. Modrich.** 1994. Mismatch repair proteins MutS and MutL inhibit RecA-catalyzed strand transfer between diverged DNAs. *Proc. Natl. Acad. Sci. USA* **91:**3238–3241.

78. **Wu, T. H., T. Loh, and M. G. Marinus.** 2002. The function of Asp70, Glu77 and Lys79 in the *Escherichia coli* MutH protein. *Nucleic Acids Res.* **30:**818–822.

79. **Wu, T. H., and M. G. Marinus.** 1999. Deletion mutation analysis of the *mutS* gene in *Escherichia coli. J. Biol. Chem.* **274:**5948–5952.

80. **Yang, W.** 2000. Structure and function of mismatch repair proteins. *Mutat. Res.* **460:**245–256.

81. **Zdraveski, Z. Z., J. A. Mello, C. K. Farinelli, J. M. Essigmann, and M. G. Marinus.** 2002. MutS preferentially recognizes cisplatin- over oxaliplatin-modified DNA. *J. Biol. Chem.* **277:**1255–1260.

82. **Zdraveski, Z. Z., J. A. Mello, M. G. Marinus, and J. M. Essigmann.** 2000. Multiple pathways of recombination define cellular responses to cisplatin. *Chem. Biol.* **7:**39–50.

The Bacterial Chromosome
Edited by N. Patrick Higgins
© 2005 ASM Press, Washington, D.C.

Chapter 24

Excision Repair and Bypass

Bernard S. Strauss

The introduction of isotopes into biological research in the second half of the 20th century permitted investigators to study the turnover of compounds in the absence of net increases or decreases. These early investigations led to what appeared to be a most satisfying conclusion: that proteins and ribonucleic acids "turned over" rapidly but that, in contrast, the DNA was stable and aloof from the metabolic busyness of the cell (60). Only during replication did metabolic events intrude upon this aloof behavior, a behavior then thought to be appropriate to a storehouse of the cell's information. More recent studies have turned this concept around. A multitude of enzymes and metabolic reactions are now understood to be needed to protect and reconstitute the genetic material, which is continually being altered as a result of external and internal insults (57). The DNA exists (as do all biopolymers) in an environment which is almost 55.6 M water and is thus subject to the mass action pressure of hydrolysis on its phosphodiester, N-glycosyl, and amino bonds. Cells are exposed to oxidizing agents both directly and as a result of the generation of metabolic intermediates, and they are exposed to environmental radiation both ionizing and in the ultraviolet region. All of these agents result in alterations of portions of the DNA structure. A major result of ionizing radiation is the production of double-strand breaks (DSBs) in the DNA. A major result of UV irradiation is the formation of intrastrand pyrimidine-pyrimidine dimers of two main types, cyclobutane dimers and 6-4 dimers (Fig. 1). Ionizing radiation, acting in part via free radical mechanisms and oxidizing, alkylating, and other agents both environmental and endogenous, produces a variety of base damages in the DNA (Fig. 2). Many of these changes alter the coding properties of the particular bases and, in ad-

dition, many serve as blocks to either or both RNA polymerase and the replicative DNA polymerase. The number of such damaged sites in DNA can be very large. An estimate can be made for *Escherichia coli* based on a rationale developed by Holmquist (36) by using an estimate of the levels of some of the more common oxidation and alkylation products in human tissues (59). Even excluding the most frequent change (hydroxymethyluracil), there are a total of about 100 of these per 10^7 bases or about 100 per *E. coli* genome. Lesions are produced and repaired, and the rate of generation (L_g) is related to the equilibrium frequency (L_{eq}) found in the DNA by the relationship $L_g = L_{eq} \times \ln2/t_{1/2}$, where $t_{1/2}$ is the half-life for repair (36). If there is a 30-min half-life for repair, there must be $100 \times 0.693/30$, or about 2 to 3 lesions per min, generated in each cell. In addition, enzymes whose function it is to repair damaged DNA do attack normal DNA at a significant rate. The *E. coli alkA* gene product removes guanine from DNA, albeit at a rate 2,100 times slower than that of its supposed substrate, 3-methyladenine, and the other bases are removed at a lower rate still (5). The UV excision repair enzyme attacks normal DNA at a significant rate (7). Damaged bases, abasic sites, and breaks are being produced constantly. It is therefore important that organisms deal with such lesions and, in fact, numerous biological mechanisms have developed for this purpose. Over the past few decades it has been shown that almost all classes of organism have such repair mechanisms and that the general mechanisms by which these repair pathways act is remarkably similar in all classes from bacteria to humans (or higher plants) (21). The protein machinery does differ in detail; for example, in nucleotide excision repair (see below), *E. coli* does with five proteins what eukaryotes do with about 15, but

Bernard S. Strauss • Department of Molecular Genetics and Cell Biology, The University of Chicago, 920 E. 58th St., Chicago, IL 60637.

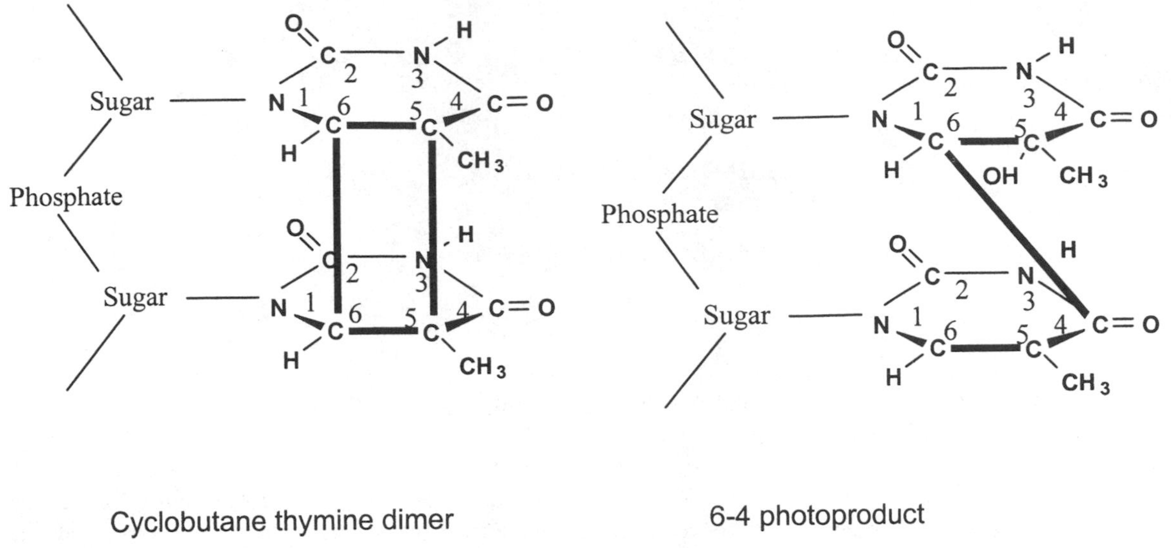

Cyclobutane thymine dimer   6-4 photoproduct

Figure 1. Pyrimidine dimers.

the reaction pathways are remarkably similar (3). For mismatch repair in which organisms deal with mistakes in replication resulting in mismatched bases, the proteins themselves have retained homology going from prokaryotes to eukaryotes (20). The structure of some of the polymerases which appear to be designed to deal with damaged DNA has retained homology across biological kingdoms. The *dinB* gene product, *E. coli* DNA polymerase IV (Pol IV), has human homologues (27). This chapter deals primarily with *E. coli*, but it remains clear that what is true for *E. coli* is (more or less) true for elephants!

The early studies in both prokaryotes and eukaryotes dealt mainly with the details of the pathways involved. Almost from the beginning there was the finding, in *E. coli*, of an inducible mechanism by which damage induced the formation of the enzymatic machinery needed to deal with that damage (78, 108). There is indeed a mechanism restricted to bacteria (21) which permits response to the environment. Analogous mechanisms, different in the way they solve the problem, permit higher organisms to respond to damage. What is important is that cells can sense damage in their DNA, halt replication, and induce systems which can deal with the problem. Indeed, one of the major problems faced by cells is the determination of what constitutes damage. Not all changes in DNA are recognized as damage: the substitution of bromouracil for thymine, for example, does not provoke a damage response. On the other hand, change of a cytosine to a thymine within DNA producing a G:T pair is recognized as damage by a particular glycosylase which specifically removes the T from the site (see below). In general, damage is

"defined" by the cell as an alteration in DNA structure that blocks or severely retards replication or transcription. The mechanisms used by cells to deal with such damage are the subject of this chapter. In particular, the chapter deals with three questions: (i) How does *E. coli* recognize damage to its DNA—how is the DNA "scanned"? (ii) How do cells deal with damage once it has been recognized? (iii) Are there preexisting molecular machines ready to deal with damage, or is the damage itself the signal for the assembly of such machines?

GENERAL CELL STRATEGY

UV-induced damage is the paradigm for damage recognition and handling in *E. coli*. This concentration on UV-induced damage does not represent an estimate of the quantitatively most prevalent type of damage, which is probably that due to oxidation (55), but rather has to do with the historical development of the study (26). The first step in the cells' response is damage recognition. At least four options are available to *E. coli* once damage has been recognized (Fig. 3): (i) the damage may be removed by special mechanisms which recognize specific lesions and restore DNA to its undamaged state without cleaving the phosphodiesterase bonds; (ii) the damage may be removed by the "cut and patch" process of excision repair; (iii) the damage may be bypassed by DNA synthesis utilizing processes related to those of genetic recombination; or (iv) the damage may be bypassed by the use of special DNA polymerases which can synthesize past the lesion, so-called "translesion

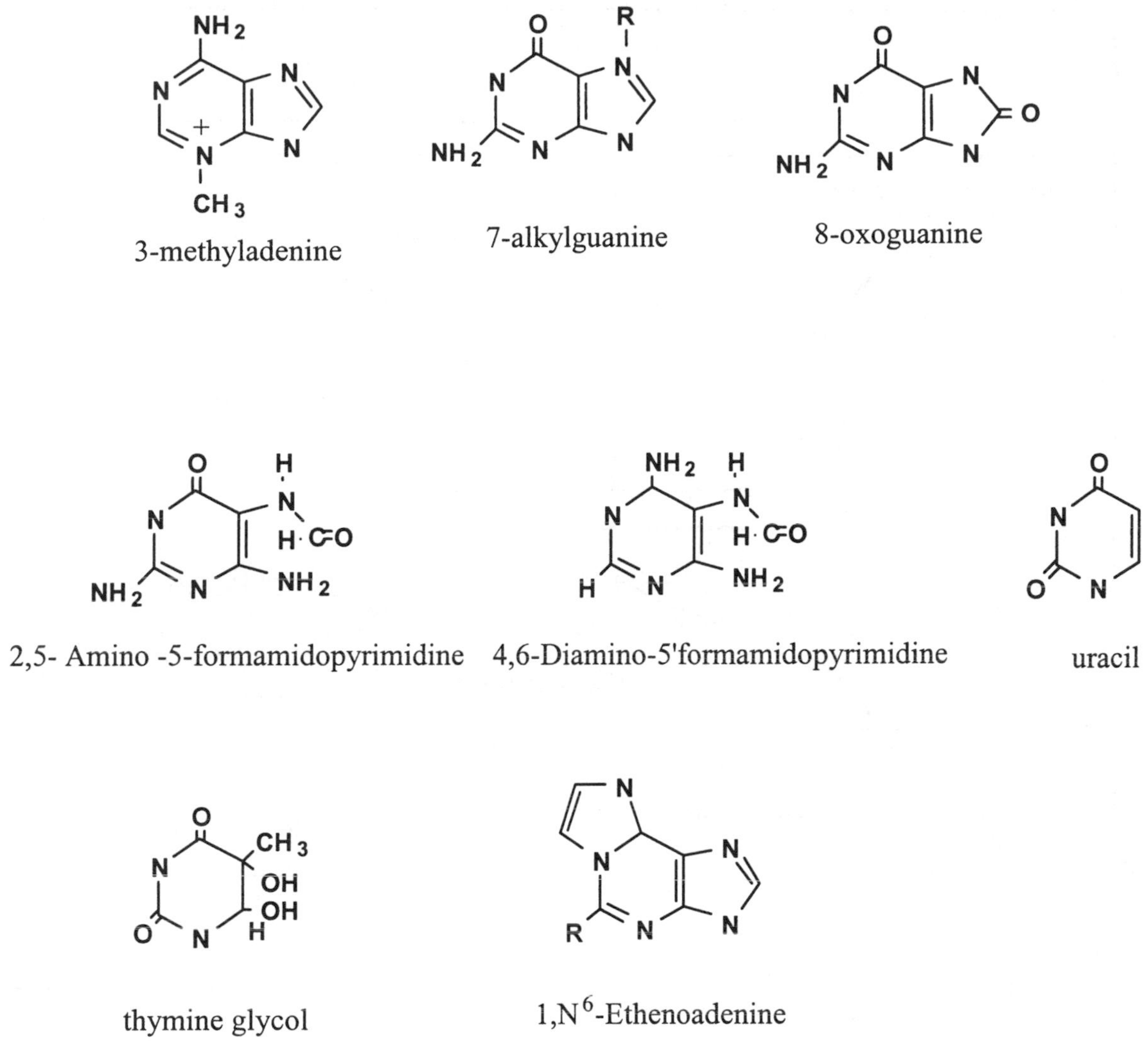

Figure 2. Some bases recognized by DNA glycosylases as abnormal.

synthesis." This last option is available only at the cost of increasing mutation rate (74). Processes involving recombination-like processes or special polymerases have not been reported as participating in the recovery of RNA synthesis. However, there is a variant of the excision repair system which is effective on transcriptionally active DNA and which is specific for the transcribed strand (99)!

The choice of which damage repair pathway to use may be determined by the particular stage of the cell cycle (or DNA replication cycle) in which the damage is recognized. Cells in which neither transcription nor replication is occurring have the opportunity to scan their DNA over a relatively long time. In cells in which a replication fork or a transcription bubble has just encountered a lesion which is a barrier to progression, there needs to be an immediate response.

One of the consequences of damage in E. coli is the activation of an inducible system, the SOS repair system. An active replication fork is required for induction. This system is activated by interaction of the RecA protein with single-stranded DNA, accumulating at replication forks stalled as a result of damage (33, 87). The interaction with single-stranded DNA alters the product of the RecA gene so that it activates the autodigestion of the LexA repressor. Inactivation of LexA induces at least 26 and probably 43 gene products, including additional production of the excision repair enzymes and three inducible DNA polymerases (15). Modulation of the activity of repair enzymes provides an additional level of control. An early idea of the relative importance of the excision repair and recombination-function-dependent pathways was given by Howard-Flanders and Boyce at the time of the original recognition that there might be

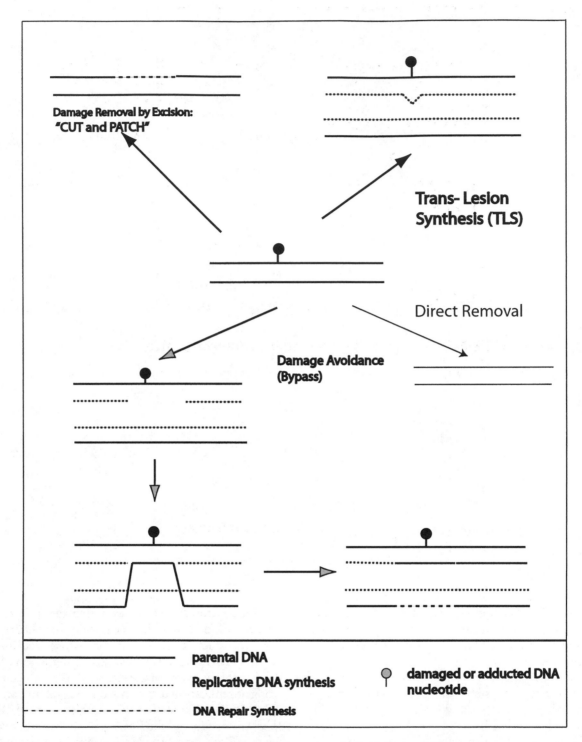

Figure 3. Four methods by which organisms deal with damaged DNA.

separate repair pathways (37). These data (Table 1) were interpreted to mean that there are at least two independent repair pathways and that a single pyrimidine dimer is sufficient to inactivate a bacterium in the absence of both. However, mutation in either *uvr* or *recA* genes results in major hypersensitivity and illustrates that for most cells, survival requires that both genes be functioning (11). The two pathways may therefore not be functionally independent. The RecA-dependent pathway actually involves at least two separable methods for damage avoidance and the RecA gene product itself carries out at least

Table 1. Dose of radiation at which 37% of bacteria form colonies[a]

Genetic characteristics		Dimers per 10^7 bases	Dimers per haploid genome
rec	uvrA		
rec-13	uvrA-6	1.5	1.4
rec-13	+	22	20.4
+	uvrA-6	60	55.7
+	+	3,700	3,433

[a]Dimers per 10^7 bases were calculated on the assumption that the yield of pyrimidine dimers is 3×10^{-6} per thymine per erg per mm^2 (3×10^{-5} per thymine per J/m^2) (37).

two separable and mutually reinforcing functions (see below).

DIRECT REPAIR MECHANISMS

Specialized mechanisms exist which remove particular kinds of damage from DNA and restore the damaged bases without breaking the phosphodiester bonds of the nucleic acid backbone. The first reaction of this type to be discovered is photoreactivation, in which the flavin chromophore of an E. coli enzyme uses light energy to directly split the cyclic bonds in cyclobutane thymine dimers (Fig. 1) to regenerate undamaged thymines (31). This enzyme is encoded by phrB. There is a suggestion that a second enzyme, encoded by phrA, has specificity for the 6-4 photoproduct (19), but although such an enzyme has been obtained from other organisms (97), there are no reports of its isolation from E. coli. E. coli adapts to treatment with methylating agents by induction of the ada operon (56). ada encodes the protein O^6-methylguanine DNA methyltransferase. This protein transfers a methyl group from O^6-methylguanine in DNA (produced by reaction with methylating agents) to one of its own cysteines. O^6-methylguanine DNA methyltransferase is irreversibly inactivated by this transfer, but unmodified guanine is regenerated. The ada operon includes alkA, which encodes 3-methyladenine DNA glycosylase, and alkB, whose product was not known until recently. Within the past 2 years it has been discovered that alkB encodes a special sort of dioxygenase dependent on both iron (Fe^{2+}) and alpha-ketoglutarate (22, 100). The enzyme couples the oxidative decarboxylation of ketoglutarate to hydroxylation of the methylated bases in 1-methyladenine and 3-methycytosine, releasing formaldehyde in the process and regenerating the native bases. The enzyme is widely distributed in prokaryotes and eukaryotes. This oxidative demethylation

reaction acts on altered RNA, as indicated by an ability to reactivate the RNA bacteriophage MS2 (1).

EXCISION REPAIR

In excision repair, damaged DNA is recognized as altered and the damage is cut out. The excised section is replaced by polymerase action using the undamaged strand as a template. How this occurs depends on the nature of the damage and where it is located—certain types of damage located on the transcribed strand of an active gene are handled by a variant of the general system. Two types of excision repair, each with important variations, can be distinguished (57). The first, base excision repair (BER), uses particular enzymes, the DNA N-glycosylases to sense specific damaged bases. The second, nucleotide excision repair (NER; sometimes called "global NER") is a multiprotein system which recognizes generalized deformation in the DNA. A variant of the process, transcription-coupled repair (TCR), copes with damage in the transcribed strand of active genes.

Base Excision Repair

A specialized class of enzyme, the DNA N-glycosylases, recognizes and removes altered bases, leaving an abasic site which is then cut by an endonuclease. Particular glycosylases have an associated AP lyase activity which results in chain scission (Table 2). The resulting DNA is then processed to make the DNA a polymerase substrate. A gap of one nucleotide (major pathway) or about 15 nucleotides (minor pathway) is then repaired by polymerase and ligase action (61). (Most recent studies on gap size have been done with eukaryotes. It is not clear what the exact gap size is in E. coli, whether one nucleotide, several, or an indeterminate but small number.) Base excision repair depends on the ability of glycosylases to sense altered bases. Although many of the glycosylases, particularly the formamide pyrimidine glycosylase (MutM/Fpg, the product of the mutM [synonym fpg] gene), sense multiple substrates, the substrates are related in structure (Table 3) at least as compared with the NER system. Furthermore, in contrast to NER, the substrates do not distort the double-helical structure. The glycosylases are relatively small (~ 30 kDa) monomeric proteins which are able to gain access to, and remove, bases from the DNA. Structural studies with uracil glycosylase give a picture of what seems to be a general mechanism for both eukaryotic and prokaryotic glycosylases. The description of the structural events is taken from Krokan et al. (47). The three-dimensional structures

Table 2. Glycosylases of *E. coli*

Glycosylase	Substrate(s)[a]	β-Lyase activity
Uracil DNA glycosylase	Uracil, 5-hydroxyuracil, fluorouracil, isodialuric acid	No
Alkyl base-DNA glycosylases		No
AlkA (3-methyladenine glycosylase II)	3-Methyladenine, 7-methyladenine, 3-methylguanine, 7-methylguanine, O$_2$-alkylthymine, O$_2$-alkylcytosine, 5-formyluracil, hypoxanthine, 1,N6-ethenoadenine, 3,N4-ethenocytosine	
Tag	3-Methyladenine, 3-methylguanine	
Formamide pyrimidine glycosylase (Fpg/MutM)	8-Oxoguanine	Yes
	2,5-Amino-5-formamidopyrimidine, 4,6-diamino-5-formamidopyrimidine, 2,6-diamino-4-hydroxy-5-formamidopyrimidine	
8-oxoguanine DNA glycosylase (OGG-1)		
MutY adenine glycosylase Endo III (Nth)	A in G/A	
	5-Hydroxycytosine, 5,6-dihydrothymine, 5-hydroxy-5,6-dihydrothymine, thymine glycol, uracil glycol, alloxan, 5,6-dihydrouracil, 5-hydroxy-5,6-dihydrouracil, 5-hydroxyuracil, urea, methyltartronylurea, 5-hydroxyhydantoin	Yes
Endo VIII	5,6-Dihydrothymine, thymine glycol, 8-oxoguanine	Yes
Endo IX	Urea	

[a]See reference 47.

of human and *E. coli* uracil glycosylases are very similar (47). A positively charged groove running along the uracil glycosylase surface orients the enzyme active site along the DNA. There is a conserved amino acid residue located above the uracil binding pocket involved in scanning of the minor groove and expulsion of the dUMP residue from the double-stranded DNA base stack via the major groove. The DNA backbone phosphates flanking the uracil are compressed, "pushing" the uracil out of the double helix and stabilizing the extrahelical configuration. The active site of the enzyme is able to recognize

uracil in this configuration and cleave the base. This rotation of the uracil and the deoxyribose phosphate is described as "nucleotide flipping" and appears to be a characteristic mode of action of the glycosylases.

Once this feat of molecular legerdemain has been completed, BER can proceed by one of two related but biochemically different pathways. Very likely the pathway chosen is determined by the glycosylase that has been involved (25). The result of glycosylase action can be the production of either an abasic site or a cleaved oligonucleotide strand, depending on whether the glycosylase has associated

Table 3. Increased amounts of transcript following UV irradiation

Gene	Gene product	Maximum increase in amt of transcript[a]	Time of maximum increase[a] (min)	Induction ratio[b]	
				32°C	42°C
umuC	Pol V	39	40	28	17
umuD	Pol V	29	20		
polB	Pol II	2.1	60	5.2	9.3
dinB	Pol IV	7.1	20	7.3	9.5
recA	RecA	9.3	20	11	13
uvrA	UvrA	2	5	3.4	6.2
uvrB	UvrB	5	40	3.6	3.7
ruvA	RuvA	4.6	10		
ruvB	RuvB	3.2	5		
ruvC	RuvC	1.01	5		

[a]Reference 15. Values represent the fold increase in transcript level over that at the beginning of the experiment.
[b]Reference 73. Values are ratios of β-galactosidase activity (Miller units) of a *lexA* strain to that of a wild-type *lexA*$^+$ strain.

lyase activity and whether such activity is functional in vivo. The abasic site is cleaved, but the manner of cleavage determines which enzymes are necessary to "clean up" the damaged site. Cleavage by an AP endonuclease results in a 3' hydroxynucleotide which is a substrate for DNA polymerase. A 5' deoxyribose phosphate terminus is also produced, and this residue needs to be removed for repair to continue. Removal is likely carried out by a deoxyribophosphodiesterase (dRpase) activity which is a property of both *fpg*(*mutM*) and *recJ* gene products. Exonuclease I also has dRpase activity. *fpg* and *recJ* mutants can both still repair abasic sites, but strains deficient in all three dRpase activities are inviable (84), indicating both an interchangeability and a requirement for at least one of the enzymes.

In a quantitative sense, BER probably accounts for the majority of repair events required to maintain the DNA. Hydrolytic and oxidative damage, exacerbated by free radical production resulting from ionizing and nonionizing radiation along with inadvertent and inappropriate methylation by *S*-adenosylmethionine, results in lethal and mutagenic damage which is mainly repaired by the BER pathway. Although thymine glycols induced by ionizing radiation and oxidizing agents are repaired more readily on transcribed DNA, at least in human cells, most BER-susceptible lesions are probably kept in check by constant surveillance with glycosylase proteins. There have been no major interactions reported between the different proteins of the BER pathway, although AP endonucleases often do stimulate glycosylase activity (30). This lack of demonstrated physical interaction implies the absence of a preexisting machine. There must be sufficient numbers of the different glycosylase molecules as well as time to continually test the DNA for damage. The findings that significant numbers of normal bases are removed and that overexpression of the *alkA* glycosylase increases spontaneous mutation rate (5) imply that the system is adjusted to tolerate normal base turnover in the interest of removing damage.

Nucleotide Excision Repair

The second mode of excision repair, NER, uses a single multiprotein complex to deal with a wide range of substrates with specificity rules that are yet to be completely defined (see below). NER has two major branches. Global excision repair detects and removes damage throughout the genome. TCR detects damage on strands being transcribed and repairs the transcribed strand. All NER substrates appear to both disrupt the duplex and be chemically altered nucleotides (32). Three *E. coli* gene products

are involved in the removal of a section of DNA containing a recognizable lesion. These are the UvrA, UvrB, and UvrC proteins (54). The UvrA protein forms a dimer which combines with the UvrB product to form a $UvrA_2B$ complex (Fig. 4). This complex recognizes the damage (70, 71). Both proteins are ATPases (29), and the result of encountering recognizable damage is the binding of the UvrB protein to the DNA, forming a preincision complex and the release of the UvrA protein. The key step is the recognition of damage. Although the UvrA protein binds preferentially to damaged DNA, the actual search for damage is by the $UvrA_2B$ complex. This complex separates the two DNA strands at the expense of ATP hydrolysis (112), and the $UvrA_2B$ protein is deposited at the site of damage so that the DNA is wrapped around the protein in an asymmetric manner imposed by the UvrA protein (103). DNA damage is recognized in a sequence of two reactions (111, 112). In the first, the helix distortion is recognized, and there is an initial local separation of the strands based on the helicase activity of the $UvrA_2B$ complex. The UvrA then dissociates. The open structure formed is accessible to the other repair proteins and permits discrimination between altered and unaltered bases. UvrC combines with the UvrB-DNA complex and the DNA is then cut both 3' (at the fourth or fifth phosphodiester bond) and 5' (at the eighth phosphodiester bond) (52, 53, 102). The catalytic sites for both incisions are located in the UvrC protein (102). A helicase (UvrD) then removes the damaged sequence (which has been nicked at both sides). An oligonucleotide of 12 to 13 nucleotides is released. DNA Pol I fills in the resulting gap, which is then sealed with ligase (Fig. 4). The $UvrA_2B$ complex is presumably preformed and, it might be supposed, continually scans the DNA for damage. Damage must be located before it can be excised, and the larger the number of scanning structures, the better the chance of locating damage. On the other hand, there is the danger that nondamaged bases will be removed (7). The SOS-inducible system in *E. coli* is arranged to increase the probability that damage will be removed. The key genes in damage recognition, *uvrA* and *uvrB*, include LexA boxes in their promoter regions, which means these genes are normally (partially) repressed by the LexA protein. Recognition that a lesion is present by its property of inhibiting the movement of the DNA replication fork induces the SOS functions and increases the amount of transcript made, thereby permitting the location and removal of (other) damaged sites.

Lesions in the DNA can not only block replication but can perhaps do more immediate damage by blocking transcription. Lesions which block transcription

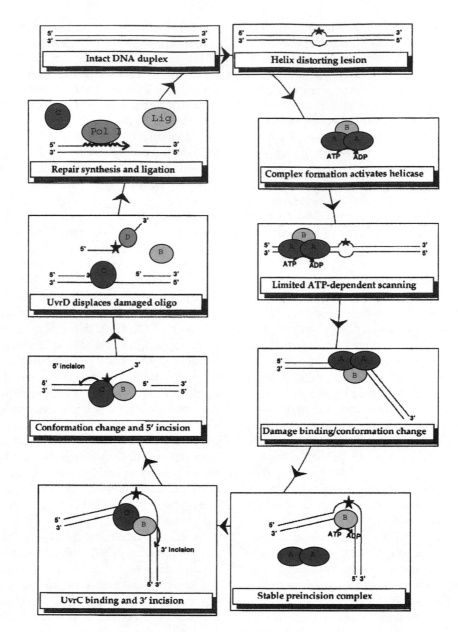

Figure 4. Model for the nucleotide excision repair of nontranscribed DNA in *E. coli*. Reproduced from reference 3 (copyright 2000) with permission from Elsevier and R. D. Wood.

are repaired not only by global NER but by the special process of TCR. There is good but not perfect correlation between interference by a lesion of RNA polymerase progression and the extent of its TCR (99). The phenomenon of TCR is found in both eukaryotes and prokaryotes, but in *E. coli* the mechanism hinges on a protein, the transcription repair coupling factor (TCRF), the product of the *mfd* gene (90). TCRF is a 130-kDa protein with helicase motifs that binds to DNA, to RNA polymerase, and to the UvrA protein. This protein displaces *E. coli* RNA polymerase and facilitates the delivery of the UvrA₂B complex to the lesion site (91) so that NER can take place. The transcription complex stalls on encountering a lesion in the DNA, and the stalled complex undergoes reversible backtracking (2) in which the 3′ OH of the growing chain dissociates from the RNA polymerase. After repair, the TCRF realigns the 3′ OH of the transcript with the active site of RNA polymerase, presumably by translocating the enzyme forward or by pulling the DNA back from under the enzyme (72). The RNA polymerase structure of eukaryotes is more complex than that of prokaryotic cells, and TCR in eukaryotes seems also to be a more complex process. There is a human TCRF, but the ternary complex of stalled RNA polymerase is not disrupted by this factor (89), indicating that some other mechanism must be operative.

The specificities of the NER system are so broad that it is perhaps not surprising that normal DNA should serve as a substrate. Branum et al. (7) estimate that undamaged DNA is excised at a rate of about 1% of that of the 6-4 photoproduct in human extracts. The investigators also detected excision products from ostensibly undamaged DNA treated with the *E. coli* UvrABC nuclease (7). Since it is practically impossible to obtain DNA absolutely free of endogenous damage (55), this may be an overestimate, but it probably mirrors what happens in vivo. The presence of a sensitive repair system apparently requires that the organism tolerate a certain percentage of turnover (with its risk of mutation) in the DNA. Extensive turnover is often tolerated to prevent error. For example, approximately 13% of dATP, 10% of dTTP, and 5 to 6% of dGTP and dCTP are wastefully hydrolyzed by the DNA Pol III proofreader in vitro (24), apparently to minimize the chance of incorporating an incorrect base. Over the long run, organisms can adjust the mutator activity they are willing to tolerate by controlling the level of proofreading (86): the level of mutations an organism is willing to tolerate needs to be balanced against the energy cost of proofreading. Mutation rate can also be regulated by changes in the size of the nucleotide pools. In yeasts (but possibly also in prokaryotes) DNA damage leads to an increase in deoxynucleoside triphosphate (dNTP) levels because of a relaxation in dATP-mediated feedback inhibition of ribonucleoside reductase (9). Survival is increased, presumably because the mass action of the increased dNTPs permits DNA synthesis past the lesions. The cost, however, is a doubling in mutation rate, presumably because of the forced misincorporation of dNTPs present in exceptionally high concentration.

REPLICATION, RECOMBINATION, AND REPAIR

What happens to the replication complex when it encounters damaged DNA? Time is clearly an element in the analysis. At the moment the cell is damaged, there are between 10 and 20 DNA polymerase III alpha subunits in the cell (110) (of which presumably at least four are involved in the movement of the bidirectional replication fork), 30 to 50 molecules of DNA Pol II (77), 250 molecules of DNA Pol IV (45), and nondetectable levels of either *umuC* or *umuD'* products (i.e., fewer than 20 molecules of

Pol V) (109). About 180 molecules of the nonactivated UmuD product were detected per uninduced cell. Pol V is the product of two closely linked genes, *umuC* and *umuD*, which are tightly controlled by the *lexA* operator (109), but the polymerase activity actually resides in the UmuC protein (81, 96). The UmuD protein is transformed in SOS-induced cells to UmuD' by a proteolytic cleavage between residues Cys-24 and Gly-25 (8, 68). Two molecules of the large C-terminal fragment (UmuD') combine with the UmuC product to form a $UmuD'_2C$ complex (23), which has polymerase activity (96). After SOS induction there are 210 to 350 molecules of Pol II, 200 molecules of Pol V (actually, 200 molecules of the UmuC product and 1,900 of the UmuD' product) (109), and about 2,500 molecules of DNA Pol IV. The time course with which these products (or their transcripts) increase differs from gene to gene (Table 3). UvrA, the damage recognition protein, increases the fastest, and later the Pol V proteins increase, and to a greater extent. Given the speed of traverse of a replication fork of nucleotides, about 1,000 nucleotides per s, if there are as many as 60 lesions present (induced, for example, by UV exposure [see Table 1]), then it is likely that any cell in which replication is occurring will encounter a newly induced lesion within a minute. (With a genome of 4.64×10^6 bp [9.3 Mb], a generation time of 20 to 40 min, and bidirectional replication with leading and lagging strands, there need to be at least four growing points moving at $(9.3 \times 10^6)/(4 \times 20)$, or 1.16×10^5 bases/min or from 1,883 nucleotides/s [20-min generation] to 940 nucleotides per s [40-min generation]. This compares with the figure of 750 nucleotides per s for chain elongation by Pol III holoenzyme in vitro [43].) The cell will then have to deal with the consequentially stalled replication fork with what protein apparatus is immediately available.

DNA synthesis in wild-type *E. coli* is inhibited by UV irradiation but returns to a normal rate within 10 to 15 min (79) with recovery beginning as early as 30 s after irradiation (75). Mutants deficient in either Pol II (79) or PriA delay restart of synthesis for 50 min (85). Since UV lesions do block the progression of DNA polymerases (66), it is evident that some process(es) must occur to permit replication. The obvious explanation, that the lesions are excised, is not always correct. Recovery of DNA synthesis occurs in *uvrA* mutants (44) although the formation of long chains is delayed (82). It was shown early on that the RecA product is involved in the restoration of synthesis, and a recent compilation (Table 4) indicates at least 23 different gene products that may be involved. There are arguments that this list is not complete (14). A classic explanation suggests that once a replication

Table 4. Proteins that may participate in replication fork reactivation[a]

Protein	Main activity	Role in fork reaction
RecA	Strand pairing/exchange	Formation of joint molecules
RecBCD	Chi site-modulated nuclease	Initiation of repair at DSB
RecF	Binds single- and double-stranded DNA	Limits extension of RecA filament
RecO	Binds SSB[b]-coated DNA	Facilitates RecA loading to SSB-coated DNA
RecR	Binds SSB-coated DNA	Facilitates RecA loading to SSB-coated DNA
RuvAB	Branch migration DNA helicase	Resolution of joint molecules
RecG	Branch migration DNA helicase	Modulate structure of stalled fork?
RuvC	Holliday structure endonuclease	Resolution of joint molecules
RusA	Holliday structure endonuclease	Resolution of joint molecules
PriA	$3' \rightarrow 5'$ DNA helicase, binds bent DNA	Initiates replication fork assembly
PriB	Facilitates complex formation between PriA and DnaT	Replication fork assembly
PriC	Primosome assembly	Replication fork assembly
DnaT	Primosome assembly	Replication fork assembly
DnaB	$5' \rightarrow 3'$ helicase	Replication fork helicase
DnaC	Binds and loads DnaB to DNA	Replication fork assembly
DnaG	Primase	Okazaki fragment primase
Rep	$3' \rightarrow 5'$ helicase	Processing of stalled fork?
DNA Pol III holoenzyme	DNA polymerase	Replicative polymerase
DNA Pol II	DNA polymerase	Replication restart at template lesions
DNA Pol I	DNA polymerase, $5' \rightarrow 3'$ exonuclease	Gap sealing
DNA ligase	Ligase	Gap sealing
SSB	Single-stranded DNA-binding protein	Coats single-stranded DNA
XerCD	Site-specific resolvase	Resolves dimeric chromosome

[a]Reprinted from reference 17 with permission.
[b]SSB, single-strand binding protein.

block is encountered, DNA synthesis continues along one strand, using the nondamaged strand as a template; then, at some point downstream of the lesion, synthesis is resumed, leaving a single-strand gap. The model is supported by finding that after UV irradiation, newly synthesized DNA is in shorter pieces (82) and that single-stranded DNA can be detected in UV-irradiated cells (38). The most convincing evidence came from experiments in which cells allowed to replicate in density-labeled medium were transferred to light medium, irradiated, and incubated with thymidine (83). The heavy strand was isolated by alkaline CsCl centrifugation, and its density was determined after shearing. The shearing of DNA from irradiated but not from unirradiated cells resulted in a density shift of radioactivity toward lighter densities. This shift is what would be expected if newly synthesized DNA were covalently linked to parental DNA, which in turn is what would be expected if the daughter-strand gap had been filled by recombination. A recent reevaluation of this experiment (13) implies that the results are not as definitive as they appear. The experiments that demonstrated a covalent linkage of old and new strands indicative of recombination (83) employed excision repair mutants that had been irradiated with a dose of UV light sufficient to reduce survival to about 0.5%; i.e., 99.5% of the cells were not viable (37)! Thus, the finding that parental and daughter strands are joined may or may not be informative as to the mechanisms of survival used by the majority of cells (14). In an experiment with *Saccharomyces cerevisiae*, it was shown that UV irradiation in the G$_1$ phase stimulated sister chromatid recombination in an excision-defective diploid. Sister chromatids are not present until after replication, and the result implies that the unexcised UV lesions induced recombination during or after replication (42).

A more recent approach considers what happens to the replicative DNA complex which starts at *oriC* and moves processively when it encounters a damaged site. In such studies the type of damage is important, since an unexcised pyrimidine dimer leaves the replication fork intact whereas the presence of a single-strand nick in the DNA (49) leads to "collapse" of the replication fork and the production of what is essentially a double-strand break (Fig. 5). Recent models which indicate how cells can overcome this fork collapse involve a recombinational restart mechanism (16), since the double-strand break that is formed when replication encounters a single-strand break would appear to require recombination to patch the broken parts back into the DNA (Fig. 5). The clear requirement for the RecA protein is explained by the need to support the strand exchanges required for recombination and for the filling of gaps

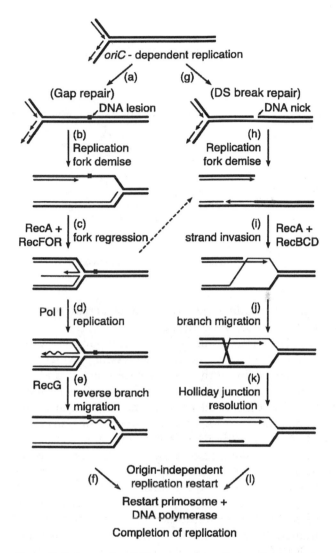

Figure 5. Pathways for DNA repair of a stalled replication fork. Reproduced from reference 16 (copyright 2001 National Academy of Sciences) with permission.

left by the blockage of DNA synthesis along one strand only at the site of pyrimidine dimer or other lesion (Fig. 5). Furthermore, the production and movement of a Holliday junction indicates the need for the *ruvABC* gene products. There is another way of looking at the data, particularly the data involving DNA pausing at lesions which are not immediately translated into strand breaks. It has been pointed out that recombination gene functions play a role in processes other than recombination. In particular, it has been suggested that regression of the growing point (48) forming a Holliday junction and resulting in the "extrusion of the newly synthesized DNA in a duplex which may include single-stranded DNA could return the lesion at the block to a double stranded form in which excision repair could take place" (14). A similar model was proposed for eu-

karyotes based on the observation of increased repair at DNA growing points (88). An extension of this model (34) suggests a mechanism for synthesis of DNA past the blocking lesion (Fig. 5). Direct evidence for the proposed reversed replication fork intermediate in *E. coli* has been obtained by two-dimensional gel electrophoresis. The intermediate is stabilized by RecA and RecF proteins and is degraded in their absence (12). Reversal of the replication fork in DNA replication is in some ways similar to the reversal of the transcription complex in TCR (72), illustrating the conservatism in the manner in which biochemical problems are solved.

Related to replication restart is a kind of NER which results in the insertion of long patches (1.5 to 9 kb or more) into the DNA (long-path excision repair). The process requires *uvr* gene products (hence the use of the phrase "excision repair") and also *recA* and *recF* (10). These long patches may in fact occur at the DNA growing point (14). The nucleotide excision may be due to growing point regression (Fig. 5), which returns the lesion to a double-stranded state where excision is possible. This excision and repair in turn permits progression of the growing point, producing what appears to be a long patch as part of the replication restart phenomenon (14).

Most of the experiments to date bearing on the question as to how replication "gets past" the blocking lesion involve the demonstration of the importance of recombination functions and the existence of Holliday junctions as part of the process of reinitiation (see, e.g., references 62 and 64). However, the Rupp experiment (83) remains the major evidence for actual recombination as part of the process in *E. coli*.

THE ROLE OF AUXILIARY DNA POLYMERASES

The *E. coli* genome codes for at least five different DNA polymerases (28). The replicative polymerase (Pol III) subunit alpha, along with nine other subunits (43), constitutes the replicating holoenzyme responsible for chromosome replication. DNA Pol I, the first to be discovered, is critical for excision repair, both in the substitution of DNA for the RNA used to prime Okazaki pieces and in excision repair. Deletions of Pol I are lethal when *E. coli* cells are grown on a complex medium but not when grown on minimal medium (41), indicating that other polymerases can carry out its functions. The three other polymerases, DNA Pol II, Pol IV, and Pol V, all have LexA binding sites in their promoters and are inducible to varying degrees. Pol V is the most stringently controlled, with nondetectable amounts in noninduced cells, whereas cells not induced for SOS function do produce detectable amounts of both Pol II (77) and Pol IV (45) (see above). Polymerase II functions in the restart of replication at regions other than *oriC* (74, 79) and apparently does this by replacing the Pol III catalytic unit in the holoenzyme complex (6, 80). At some point after the restart, Pol II is displaced by Pol III, which then completes chromosomal replication. How this displacement occurs is unknown. The RecA protein stimulates Pol V activity by 4 orders of magnitude. Both Pol V and Pol IV (105) are greatly stimulated by interaction with the beta processivity clamp. In vitro and in vivo experiments suggest that the UmuD$'_2$C product, Pol V, can carry out translesion synthesis past both major UV lesions (the 6-4 photoproduct and the cyclobutane dimer). In the presence of the RecA protein, beta sliding clamp, single-strand binding protein, and gamma clamp loader, Pol V can synthesize DNA using a template with either a pyrimidine dimer or an abasic site (81, 95) present. In vitro experiments indicating the specificity of (mis)-incorporation opposite the 6-4 T-T photoproduct mimic the in vivo findings. G insertion is favored opposite the 3′ T, whereas A is favored opposite the 5′ T (95), supporting the conclusion that this polymerase is responsible for SOS-induced mutagenesis. RecA protein is required, targeting and elongating by different mechanisms (76).

Other enzymes are used to deal with some chemical damage, particularly the addition of large adducts to DNA bases (67). After SOS induction, a replication complex confronted with DNA that has been modified by the addition of acetylaminofluorene to guanine residues in the midst of a 5′GGC GCC3′ sequence can replicate past the lesion without error but is also likely to make −2 frameshifts. In the absence of *umuDC*, error-free translesion synthesis does not occur; in the absence of Pol II, −2 frameshifts do not occur (4). The *umuDC* product Pol V is therefore required for an accurate translesion synthesis. In the presence of Pol II (the *polB* product) synthesis can occur, but presumably the chain cannot be extended if the adducted template base remains in place. However, template slippage aligns a base able to complement at the growing point and provides a structure which can be extended by Pol II. Becherel and Fuchs (4) suppose (Fig. 6) that the adducted DNA with repetitive sequences can exist in either slipped or nonslipped configurations, which must be in some equilibrium with one another. Since extension can occur without mutagenesis in strains induced for Pol V, and since extension occurs with mutagenesis only when Pol II is available, it is concluded that Pol II extends the slipped and distorted end. Pol V in this case extends the nonslipped mol-

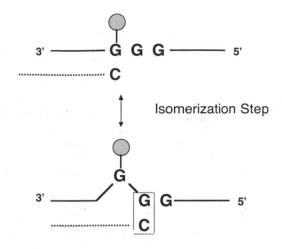

Figure 6. The isomerization step in lesion avoidance (4).

ecules by adding the correct base across from an acetylaminofluorene adduct.

The explanation is further complicated by the finding that the selection of a successful polymerase is determined by both the sequence context and the nature of the lesion involved (67, 104). Furthermore, Pol V is considered error prone in the generation of mutations arising as a consequence of translesion synthesis of UV-induced lesions, but in the case of the acetylaminofluorene in a particularly repetitious sequence, it carries out translesion synthesis in an error-free manner (4). The role of DNA Pol IV (the *dinB* polymerase) in bypass is not clear. Overexpression of this polymerase leads to base substitutions and frameshifts, but these are exclusively −1 frameshifts (106). When investigated in a strain with normal polymerase activity but deficient in mismatch repair activity, spontaneous +1 frameshifts outnumber −1 frameshifts by a factor of at least 3 when using a particular tester sequence (18). In a different context, a mismatch-deficient strain overexpressing DNA Pol IV is reported to make mainly −1 frameshifts (45), implying that under normal circumstances Pol IV (*dinB*) is not involved in most of the spontaneous mutations. Mutation in the *dinB* operon results in modestly lowered spontaneous mutation rate (63, 93), but this may be due to the neighboring genes in the operon. Pol IV itself, the *dinB* product, is responsible for stationary-phase mutation (98).

DNA polymerase action involves a number of steps in which a base is positioned, put in place, edited, and extended in a kinetically complex series of interrelated steps (39, 65). In addition, different proteins need to be assembled to make up the replication fork (35, 92). Any mechanistic picture of bypass will require a detailed understanding of the replicating machine discussed elsewhere in this book.

The dynamics of the bypass process involving the different polymerases can be thought of as follows. The complete replication system is in place and DNA is being replicated, either by progression of the complex along a chromosome, as usually thought, or, as has been suggested for *Bacillus subtilis*, by the chromosomal DNA moving through a fixed replication site (50). The DNA is held to the replication complex by the beta clamp, which ensures processivity. At some point, because of either a lesion or a transient secondary structure in the DNA, the process stalls and the polymerase subunit and the growing DNA strand (the 3′ OH end) dissociate. It is possible that the polymerase subunit dissociates by swiveling on the beta clamp (35) so that the complex remains intact even though stalled. It is probably an important clue that all five *E. coli* DNA polymerases have been reported to interact with the beta clamp (6, 43, 51, 58, 94, 105), although there is no evidence indicating that several polymerases can be simultaneously bound to the same sliding clamp molecule(s). (The MutS protein also interacts with the beta clamp [58]. It is generally assumed that the mismatch repair complex acts separately from the replication machinery, but in eukaryotes, mismatch repair proteins are associated with the replication complex and may play a structural role [40, 101, 107] by interacting with the eukaryotic analogue of the beta clamp, PCNA.)

There is then a competition for the free 3′-OH end of the DNA in which one or the other of the polymerases fixes the growing point of the DNA. For *E. coli*, there are five possible choices. Pol II and Pol IV are available only in limited quantities before SOS induction. Pol V is available only after SOS induction, and increased numbers of Pol II and Pol IV molecules are available after induction. Although the RecA protein is not (necessarily) involved in normal DNA replication, it is required for translesion DNA synthesis (76). This requirement is in addition to its role in the induction of the SOS system. One of the available polymerases must attack the now free 3′-OH group and attempt to add and extend this DNA end. Given the short time lag for the replication restart phenomenon, the recombination gene products may be catalyzing a backing up of the growing point or the formation of a recombination intermediate at essentially the same time. The time constraints (restart within a matter of seconds) imply that the DNA and the replication machinery stay together. This may be particularly important for cases in which a break would lead to collapse of a replication fork! UV-induced mutagenesis is dependent on *umuD′C* products that do not accumulate until almost 40 min after irradiation (15). Either the replication complex dissociates and then reassociates with Pol V, or at

least some replication forks can remain stably stalled for the extended period of time required for Pol V formation. Either before or after extension, the auxiliary DNA polymerase dissociates. Depending on the processivity of the successful complex, synthesis continues for a period, but at some point the polymerase subunit dissociates and is replaced by the Pol III alpha polymerase subunit.

SUMMARY

E. coli has a variety of enzymes which cooperate to detect and remove damage from the DNA. The operation of these mechanisms is dependent on the location of the lesions with respect to DNA growing points. Lesions far from such growing points are detected either by small DNA glycosylases that continually test the DNA for aberrant bases or by the UvrA$_2$UvrB protein complex, which detects helix distortions and searches the distorted helix for altered bases. Lesions which block DNA synthesis lead to the accumulation of single-stranded DNA, possibly by regression of the growing point with strand displacement, which activates the SOS set of inducible reactions. Growing point regression either permits excision repair of the lesion or error-free replication past the lesion by a copy-choice mechanism. Where single-strand breaks have occurred ahead of the replication complex, the replication fork collapse may be reversed by the formation of recombination intermediates which, with the aid of the recombination proteins, restore a replication fork. In some cases, replication forks may remain stalled for extended periods of time until the synthesis of polymerases induced by SOS signals permits error-prone synthesis directly past the lesion. The events at the growing point involve the cooperation of at least 23 proteins as well as those of the replication complex itself.

These considerations illustrate the complex nature of the reactions possible at the DNA growing point, a set of reactions influenced by the state of the cell (the concentration of SOS-inducible products) as well as by the physical state of the DNA, which in turn must be influenced by DNA-binding proteins. The response to DNA damage in a rapidly replicating cell depends critically upon the replication complex machinery and very likely on the beta sliding clamp, which plays a role in *E. coli* similar to that of PCNA in eukaryotic organisms. The *E. coli* cell does not appear to have preassembled a set of repair "machines" to deal with damage, but rather assembles the necessary proteins in an ordered sequence most likely governed by protein-protein interactions (46).

Not included in this discussion, but clearly important in determining the in vivo response and the eventual viability of damaged cells, is the question of the role of the repair proteins as cell cycle control proteins, a role recently ascribed to the UmuDC proteins (69).

Acknowledgments. This chapter was prepared with support provided in part by a grant from the National Cancer Institute, National Institutes of Health (CA 32436).

I am grateful to Douglas Bishop for a critical reading of an earlier version of the manuscript.

REFERENCES

1. Aas, P. A., M. Otterlei, P. O. Falnes, C. B. Vagbo, F. Skorpen, M. Akbari, O. Sundheim, M. Bjoras, G. Slupphaug, E. Seeberg, and H. E. Krokan. 2003. Human and bacterial oxidative demethylases repair alkylation damage in both RNA and DNA. *Nature* **421:**859–863.

2. Artsimovitch, I., and R. Landick. 2000. Pausing by bacterial RNA polymerase is mediated by mechanistically distinct classes of signals. *Proc. Natl. Acad. Sci. USA* **97:**7090–7095.

3. Batty, D. P., and R. D. Wood. 2000. Damage recognition in nucleotide excision repair of DNA. *Gene* **241:**193–204.

4. Becherel, O., and R. P. Fuchs. 2001. Mechanism of DNA polymerase II-mediated frameshift mutagenesis. *Proc. Natl. Acad. Sci. USA* **98:**8566–8571.

5. Berdal, K. G., R. F. Johansen, and E. Seeberg. 1998. Release of normal bases from intact DNA by a native DNA repair enzyme. *EMBO J.* **17:**363–367.

6. Bonner, C. A., P. T. Stukenberg, M. Rajagopalan, R. Eritja, M. O'Donnell, K. McEntee, H. Echols, and M. F. Goodman. 1992. Processive DNA synthesis by DNA polymerase II mediated by DNA polymerase III accessory proteins. *J. Biol. Chem.* **267:**11431–11438.

7. Branum, M. E., J. T. Reardon, and A. Sancar. 2001. DNA repair excision nuclease attacks undamaged DNA. A potential source of spontaneous mutations. *J. Biol. Chem.* **276:**25421–25426.

8. Burckhardt, S. E., R. Woodgate, R. H. Scheuermann, and H. Echols. 1988. UmuD mutagenesis protein of *Escherichia coli:* overproduction, purification, and cleavage by RecA. *Proc. Natl. Acad. Sci. USA* **85:**1811–1815.

9. Chabes, A., B. Georgieva, V. Domkin, X. Zhao, R. Rothstein, and L. Thelander. 2003. Survival of DNA damage in yeast directly depends on increased dNTP levels allowed by relaxed feedback inhibition of ribonucleotide reductase. *Cell* **112:**391–401.

10. Cooper, P. K. 1982. Characterization of long patch excision repair of DNA in ultraviolet-irradiated *Escherichia coli:* an inducible function under rec-lex control. *Mol. Gen. Genet.* **185:**189–197.

11. Courcelle, J., D. J. Crowley, and P. C. Hanawalt. 1999. Recovery of DNA replication in UV-irradiated *Escherichia coli* requires both excision repair and recF protein function. *J. Bacteriol.* **181:**916–922.

12. Courcelle, J., J. R. Donaldson, K. H. Chow, and C. T. Courcelle. 2003. DNA damage-induced replication fork regression and processing in *Escherichia coli*. *Science* **299:**1064–1067.

13. Courcelle, J., A. K. Ganesan, and P. C. Hanawalt. 2001. Therefore, what are recombination proteins there for? *Bioessays* **23:**463–470.

14. Courcelle, J., and P. C. Hanawalt. 2001. Participation of recombination proteins in rescue of arrested replication forks in UV-irradiated *Escherichia coli* need not involve recombination. *Proc. Natl. Acad. Sci. USA* **98:**8196–8202.

15. Courcelle, J., A. Khodursky, B. Peter, P. O. Brown, and P. C. Hanawalt. 2001. Comparative gene expression profiles following UV exposure in wild-type and SOS-deficient *Escherichia coli*. *Genetics* **158:**41–64.

16. Cox, M. M. 2001. Historical overview: searching for replication help in all of the rec places. *Proc. Natl. Acad. Sci. USA* **98:**8173–8180.

17. Cox, M. M., M. F. Goodman, K. N. Kreuzer, D. J. Sherratt, S. J. Sandler, and K. J. Marians. 2000. The importance of repairing stalled replication forks. *Nature* **404:**37–41.

18. Cupples, C. G., M. Cabrera, C. Cruz, and J. H. Miller. 1990. A set of lacZ mutations in *Escherichia coli* that allow rapid detection of specific frameshift mutations. *Genetics* **125:**275–280.

19. Dorrell, N., D. J. Davies, and S. H. Moss. 1995. Evidence of photoenzymatic repair due to the phrA gene in a phrB mutant of *Escherichia coli* K-12. *J. Photochem. Photobiol. B* **28:**87–92.

20. Eisen, J. A. 1998. A phylogenomic study of the MutS family of proteins. *Nucleic Acids Res.* **26:**4291–4300.

21. Eisen, J. A., and P. C. Hanawalt. 1999. A phylogenomic study of DNA repair genes, proteins, and processes. *Mutat. Res.* **435:**171–213.

22. Falnes, P. O., R. F. Johansen, and E. Seeberg. 2002. AlkB-mediated oxidative demethylation reverses DNA damage in *Escherichia coli*. *Nature* **419:**178–182.

23. Ferentz, A. E., G. C. Walker, and G. Wagner. 2001. Converting a DNA damage checkpoint effector (UmuD(2)C) into a lesion bypass polymerase (UmuD'(2)C). *EMBO J.* **20:**4287–4298.

24. Fersht, A. R., J. W. Knill-Jones, and W. C. Tsui. 1982. Kinetic basis of spontaneous mutation. Misinsertion frequencies, proofreading specificities and cost of proofreading by DNA polymerases of Escherichia coli. *J. Mol. Biol.* **156:**37–51.

25. Fortini, P., E. Parlanti, O. M. Sidorkina, J. Laval, and E. Dogliotti. 1999. The type of DNA glycosylase determines the base excision repair pathway in mammalian cells. *J. Biol. Chem.* **274:**15230–15236.

26. Friedberg, E. 1997. *Correcting the Blueprint of Life. An Historical Account of the Discovery of DNA Repair Mechanisms.* Cold Spring Harbor Laboratory Press, Plainview, N.Y.

27. Gerlach, V. L., L. Aravind, G. Gotway, R. A. Schultz, E. V. Koonin, and E. C. Friedberg. 1999. Human and mouse homologs of *Escherichia coli* DinB (DNA polymerase IV), members of the UmuC/DinB superfamily. *Proc. Natl. Acad. Sci. USA* **96:**11922–11927.

28. Goodman, M. F. 2002. Error-prone repair DNA polymerases in prokaryotes and eukaryotes. *Annu. Rev. Biochem.* **71:**17–50.

29. Goosen, N., and G. Moolenar. 2001. Role of ATP hydrolysis by UvrA and UvrB during nucleotide excision repair. *Res. Microbiol.* **152:**401–409.

30. Hang, B., and B. Singer. 2003. Protein-protein interactions involving DNA glycosylases. *Chem. Res. Toxicol.* **16:**1181–1195.

31. Heelis, P. F., S. T. Kim, T. Okamura, and A. Sancar. 1993. The photo repair of pyrimidine dimers by DNA photolyase and model systems. *J. Photochem. Photobiol. B* **17:**219–228.

32. Hess, M. T., U. Schwitter, M. Petretta, B. Giese, and H. Naegeli. 1997. Bipartite substrate discrimination by human

nucleotide excision repair. *Proc. Natl. Acad. Sci. USA* **94**:6664–6669.

33. Higashitani, N., A. Higashitani, and K. Horiuchi. 1995. SOS induction in *Escherichia coli* by single-stranded DNA of mutant filamentous phage: monitoring by cleavage of LexA repressor. *J. Bacteriol.* **177**:3610–3612.

34. Higgins, N. P., K. Kato, and B. Strauss. 1976. A model for replication repair in mammalian cells. *J. Mol. Biol.* **101**:417–425.

35. Hingorani, M. M., and M. O'Donnell. 2000. Sliding clamps: a (tail)ored fit. *Curr. Biol.* **10**:R25–R29.

36. Holmquist, G. P. 1998. Endogenous lesions, S-phase-independent spontaneous mutations, and evolutionary strategies for base excision repair. *Mutat. Res.* **400**:59–68.

37. Howard-Flanders, P., and R. P. Boyce. 1966. DNA repair and genetic recombination: studies on mutants of *Escherichia coli* defective in these processes. *Radiat. Res. Suppl.* **6**:156–184.

38. Iyer, V. N., and W. D. Rupp. 1971. Usefulness of benzoylated naphthoylated DEAE-cellulose to distinguish and fractionate double-stranded DNA bearing different extents of single-stranded regions. *Biochim. Biophys. Acta* **228**:117–126.

39. Johnson, K. A. 1993. Conformational coupling in DNA polymerase fidelity. *Annu. Rev. Biochem.* **62**:685–713.

40. Johnson, R. E., G. K. Kovvali, S. N. Guzder, N. S. Amin, C. Holm, Y. Habraken, P. Sung, L. Prakash, and S. Prakash. 1996. Evidence for involvement of yeast proliferating cell nuclear antigen in DNA mismatch repair. *J. Biol. Chem.* **271**:27987–27990.

41. Joyce, C. M., and N. D. Grindley. 1984. Method for determining whether a gene of *Escherichia coli* is essential: application to the *polA* gene. *J. Bacteriol.* **158**:636–643.

42. Kadyk, L. C., and L. H. Hartwell. 1993. Replication-dependent sister chromatid recombination in *rad1* mutants of *Saccharomyces cerevisiae*. *Genetics* **133**:469–487.

43. Kelman, Z., and M. O'Donnell. 1995. DNA polymerase III holoenzyme: structure and function of a chromosomal replicating machine. *Annu. Rev. Biochem.* **64**:171–200.

44. Khidhir, M. A., S. Casaregola, and I. B. Holland. 1985. Mechanism of transient inhibition of DNA synthesis in ultraviolet-irradiated *E. coli*: inhibition is independent of *recA* whilst recovery requires RecA protein itself and an additional, inducible SOS function. *Mol. Gen. Genet.* **199**:133–140.

45. Kim, S., K. Matsui, M. Yamada, P. Gruz, and T. Nohmi. 2001. Role of chromosomal and episomal *dinB* genes encoding DNA pol IV in targeted and untargeted mutagenesis in *Escherichia coli*. *Mol. Genet. Genom.* **266**:207–215.

46. Kowalczykowski, S. C. 2000. Some assembly required. *Nat. Struct. Biol.* **7**:1087–1089.

47. Krokan, H. E., R. Standal, and G. Slupphaug. 1997. DNA glycosylases in the base excision repair of DNA. *Biochem. J.* **325**:1–16.

48. Kuzminov, A. 2001. DNA replication meets genetic exchange: chromosomal damage and its repair by homologous recombination. *Proc. Natl. Acad. Sci. USA* **98**:8461–8468.

49. Kuzminov, A. 1999. Recombinational repair of DNA damage in *Escherichia coli* and bacteriophage λ. *Microbiol. Mol. Biol. Rev.* **63**:751–813.

50. Lemon, K. P., and A. D. Grossman. 2000. Movement of replicating DNA through a stationary replisome. *Mol. Cell* **6**:1321–1330.

51. Lenne-Samuel, N., J. Wagner, H. Etienne, and R. P. Fuchs. 2002. The processivity factor beta controls DNA polymerase IV traffic during spontaneous mutagenesis and translesion synthesis in vivo. *EMBO Rep.* **3**:45–49.

52. Lin, J. J., A. M. Phillips, J. E. Hearst, and A. Sancar. 1992. Active site of (A)BC excinuclease. II. Binding, bending, and catalysis mutants of UvrB reveal a direct role in 3′ and an indirect role in 5′ incision. *J. Biol. Chem.* **267**:17693–17700.

53. Lin, J. J., and A. Sancar. 1992. Active site of (A)BC excinuclease. I. Evidence for 5′ incision by UvrC through a catalytic site involving Asp399, Asp438, Asp466, and His538 residues. *J. Biol. Chem.* **267**:17688–17692.

54. Lin, J. J., and A. Sancar. 1992. (A)BC excinuclease: the *Escherichia coli* nucleotide excision repair enzyme. *Mol. Microbiol.* **6**:2219–2224.

55. Lindahl, T. 1996. The Croonian Lecture, 1996: endogenous damage to DNA. *Philos. Trans. R. Soc. Lond. B* **351**:1529–1538.

56. Lindahl, T., B. Sedgwick, M. Sekiguchi, and Y. Nakabeppu. 1988. Regulation and expression of the adaptive response to alkylating agents. *Annu. Rev. Biochem.* **57**:133–157.

57. Lindahl, T., and R. D. Wood. 1999. Quality control by DNA repair. *Science* **286**:1897–1905.

58. Lopez De Saro, F. J., and M. O'Donnell. 2001. Interaction of the beta sliding clamp with MutS, ligase, and DNA polymerase I. *Proc. Natl. Acad. Sci. USA* **98**:8376–8380.

59. Marnett, L. J., and P. C. Burcham. 1993. Endogenous DNA adducts: potential and paradox. *Chem. Res. Toxicol.* **6**:771–785.

60. Mazia, D. 1952. Physiology of the cell nucleus, p. 77–122. *In* E. S. G. Barron (ed.), *Modern Trends in Physiology and Biochemistry*. Academic Press, Inc., New York, N.Y.

61. McCullough, A. K., M. L. Dodson, and R. S. Lloyd. 1999. Initiation of base excision repair: glycosylase mechanisms and structures. *Annu. Rev. Biochem.* **68**:255–285.

62. McGlynn, P., R. G. Lloyd, and K. J. Marians. 2001. Formation of Holliday junctions by regression of nascent DNA in intermediates containing stalled replication forks: RecG stimulates regression even when the DNA is negatively supercoiled. *Proc. Natl. Acad. Sci. USA* **98**:8235–8240.

63. McKenzie, G. J., D. B. Magner, P. L. Lee, and S. M. Rosenberg. 2003. The *dinB* operon and spontaneous mutation in *Escherichia coli*. *J. Bacteriol.* **185**:3972–3977.

64. Michel, B., M. J. Flores, E. Viguera, G. Grompone, M. Seigneur, and V. Bidnenko. 2001. Rescue of arrested replication forks by homologous recombination. *Proc. Natl. Acad. Sci. USA* **98**:8181–8188.

65. Mizrahi, V., and S. J. Benkovic. 1988. The dynamics of DNA polymerase-catalyzed reactions. *Adv. Enzymol. Relat. Areas Mol. Biol.* **61**:437–457.

66. Moore, P., and B. S. Strauss. 1979. Sites of inhibition of in vitro DNA synthesis in carcinogen- and UV-treated phi X174 DNA. *Nature* **278**:664–666.

67. Napolitano, R., R. Janel-Bintz, J. Wagner, and R. P. Fuchs. 2000. All three SOS-inducible DNA polymerases (Pol II, Pol IV and Pol V) are involved in induced mutagenesis. *EMBO J.* **19**:6259–6265.

68. Nohmi, T., J. R. Battista, L. A. Dodson, and G. C. Walker. 1988. RecA-mediated cleavage activates UmuD for mutagenesis: mechanistic relationship between transcriptional derepression and posttranslational activation. *Proc. Natl. Acad. Sci. USA* **85**:1816–1820.

69. Opperman, T., S. Murli, B. T. Smith, and G. C. Walker. 1999. A model for a *umuDC*-dependent prokaryotic DNA damage checkpoint. *Proc. Natl. Acad. Sci. USA* **96**:9218–9223.

70. Orren, D. K., and A. Sancar. 1989. The (A)BC excinuclease of *Escherichia coli* has only the UvrB and UvrC subunits in the incision complex. *Proc. Natl. Acad. Sci. USA* **86**:5237–5241.

71. Orren, D. K., and A. Sancar. 1990. Formation and enzymatic properties of the UvrB.DNA complex. *J. Biol. Chem.* **265:** 15796–15803.

72. Park, J. S., M. T. Marr, and J. W. Roberts. 2002. *E. coli* transcription repair coupling factor (Mfd protein) rescues arrested complexes by promoting forward translocation. *Cell* **109:**757–767.

73. Peterson, K. R., and D. W. Mount. 1987. Differential repression of SOS genes by unstable LexA41 (tsl-1) protein causes a "split-phenotype" in *Escherichia coli* K-12. *J. Mol. Biol.* **193:**27–40.

74. Pham, P., J. G. Bertram, M. O'Donnell, R. Woodgate, and M. F. Goodman. 2001. A model for SOS-lesion-targeted mutations in *Escherichia coli*. *Nature* **409:**366–370.

75. Pham, P., S. Rangarajan, R. Woodgate, and M. F. Goodman. 2001. Roles of DNA polymerases V and II in SOS-induced error-prone and error-free repair in *Escherichia coli*. *Proc. Natl. Acad. Sci. USA* **98:**8350–8354.

76. Pham, P., E. M. Seitz, S. Saveliev, X. Shen, R. Woodgate, M. M. Cox, and M. F. Goodman. 2002. Two distinct modes of RecA action are required for DNA polymerase V-catalyzed translesion synthesis. *Proc. Natl. Acad. Sci. USA* **99:**11061–11066.

77. Qiu, Z., and M. F. Goodman. 1997. The *Escherichia coli polB* locus is identical to *dinA*, the structural gene for DNA polymerase II. Characterization of Pol II purified from a polB mutant. *J. Biol. Chem.* **272:**8611–8617.

78. Radman, M. 1974. Phenomenology of an inducible mutagenic DNA repair pathway in *Escherichia coli:* SOS repair hypothesis, p. 128–142. *In* L. Prakash, F. Sherman, M. Miller, C. Lawrence, and H. W. Tabor (ed.), *Molecular and Environmental Aspects of Mutagenesis.* Charles C Thomas, Springfield, Ill.

79. Rangarajan, S., R. Woodgate, and M. F. Goodman. 1999. A phenotype for enigmatic DNA polymerase II: a pivotal role for pol II in replication restart in UV-irradiated *Escherichia coli*. *Proc. Natl. Acad. Sci. USA* **96:**9224–9229.

80. Rangarajan, S., R. Woodgate, and M. F. Goodman. 2002. Replication restart in UV-irradiated *Escherichia coli* involving Pols II, III, V, PriA, RecA and RecFOR proteins. *Mol. Microbiol.* **43:**617–628.

81. Reuven, N. B., G. Arad, A. Maor-Shoshani, and Z. Livneh. 1999. The mutagenesis protein UmuC is a DNA polymerase activated by UmuD', RecA, and SSB and is specialized for translesion replication. *J. Biol. Chem.* **274:**31763–31766.

82. Rupp, W. D., and P. Howard-Flanders. 1968. Discontinuities in the DNA synthesized in an excision-defective strain of *Escherichia coli* following ultraviolet irradiation. *J. Mol. Biol.* **31:**291–304.

83. Rupp, W. D., C. E. Wilde III, D. L. Reno, and P. Howard-Flanders. 1971. Exchanges between DNA strands in ultraviolet-irradiated *Escherichia coli*. *J. Mol. Biol.* **61:**25–44.

84. Sandigursky, M., G. A. Freyer, and W. A. Franklin. 1998. The post-incision steps of the DNA base excision repair pathway in *Escherichia coli:* studies with a closed circular DNA substrate containing a single U:G base pair. *Nucleic Acids Res.* **26:**1282–1287.

85. Sandler, S. J., and K. J. Marians. 2000. Role of PriA in replication fork reactivation in *Escherichia coli*. *J. Bacteriol.* **182:**9–13.

86. Santos, M. E., and J. W. Drake. 1994. Rates of spontaneous mutation in bacteriophage T4 are independent of host fidelity determinants. *Genetics* **138:**553–564.

87. Sassanfar, M., and J. W. Roberts. 1990. Nature of the SOS-inducing signal in *Escherichia coli*. The involvement of DNA replication. *J. Mol. Biol.* **212:**79–96.

88. Scudiero, D., and B. Strauss. 1976. Increased repair in DNA growing point regions after treatment of human lymphoma cells with N-methyl-N'-nitro-N-nitrosoguanidine. *Mutat. Res.* **35:**311–324.

89. Selby, C. P., and A. Sancar. 1997. Human transcription-repair coupling factor CSB/ERCC6 is a DNA-stimulated ATPase but is not a helicase and does not disrupt the ternary transcription complex of stalled RNA polymerase II. *J. Biol. Chem.* **272:**1885–1890.

90. Selby, C. P., and A. Sancar. 1993. Molecular mechanism of transcription-repair coupling. *Science* **260:**53–58.

91. Selby, C. P., and A. Sancar. 1995. Structure and function of transcription-repair coupling factor. II. Catalytic properties. *J. Biol. Chem.* **270:**4890–4895.

92. Sexton, D. J., A. J. Berdis, and S. J. Benkovic. 1997. Assembly and disassembly of DNA polymerase holoenzyme. *Curr. Opin. Chem. Biol.* **1:**316–322.

93. Strauss, B. S., R. Roberts, L. Francis, and P. Pouryazdanparast. 2000. Role of the *dinB* gene product in spontaneous mutation in *Escherichia coli* with an impaired replicative polymerase. *J. Bacteriol.* **182:**6742–6750.

94. Sutton, M. D., M. F. Farrow, B. M. Burton, and G. C. Walker. 2001. Genetic interactions between the *Escherichia coli umuDC* gene products and the beta processivity clamp of the replicative DNA polymerase. *J. Bacteriol.* **183:**2897–2909.

95. Tang, M., P. Pham, X. Shen, J. S. Taylor, M. O'Donnell, R. Woodgate, and M. F. Goodman. 2000. Roles of *E. coli* DNA polymerases IV and V in lesion-targeted and untargeted SOS mutagenesis. *Nature* **404:**1014–1018.

96. Tang, M., X. Shen, E. G. Frank, M. O'Donnell, R. Woodgate, and M. F. Goodman. 1999. UmuD'(2)C is an error-prone DNA polymerase, *Escherichia coli* pol V. *Proc. Natl. Acad. Sci. USA* **96:**8919–8924.

97. Todo, T., H. Takemori, H. Ryo, M. Ihara, T. Matsunaga, O. Nikaido, K. Sato, and T. Nomura. 1993. A new photoreactivating enzyme that specifically repairs ultraviolet light-induced (6-4)photoproducts. *Nature* **361:**371–374.

98. Tompkins, J. D., J. L. Nelson, J. C. Hazel, S. L. Leugers, J. D. Stumpf, and P. L. Foster. 2003. Error-prone polymerase, DNA polymerase IV, is responsible for transient hypermutation during adaptive mutation in *Escherichia coli*. *J. Bacteriol.* **185:**3469–3472.

99. Tornaletti, S., and P. C. Hanawalt. 1999. Effect of DNA lesions on transcription elongation. *Biochimie* **81:** 139–146.

100. Trewick, S. C., T. F. Henshaw, R. P. Hausinger, T. Lindahl, and B. Sedgwick. 2002. Oxidative demethylation by *Escherichia coli* AlkB directly reverts DNA base damage. *Nature* **419:**174–178.

101. Umar, A., A. B. Buermeyer, J. A. Simon, D. C. Thomas, A. B. Clark, R. M. Liskay, and T. A. Kunkel. 1996. Requirement for PCNA in DNA mismatch repair at a step preceding DNA resynthesis. *Cell* **87:**65–73.

102. Verhoeven, E. E., M. van Kesteren, G. F. Moolenaar, R. Visse, and N. Goosen. 2000. Catalytic sites for 3' and 5' incision of *Escherichia coli* nucleotide excision repair are both located in UvrC. *J. Biol. Chem.* **275:**5120–5123.

103. Verhoeven, E. E., C. Wyman, G. F. Moolenaar, J. H. Hoeijmakers, and N. Goosen. 2001. Architecture of nucleotide excision repair complexes: DNA is wrapped by UvrB before and after damage recognition. *EMBO J.* **20:** 601–611.

104. Wagner, J., H. Etienne, R. Janel-Bintz, and R. P. Fuchs. 2002. Genetics of mutagenesis in E. coli: various combinations of translesion polymerases (Pol II, IV and V) deal

with lesion/sequence context diversity. *DNA Repair* **1:**159–167.

105. **Wagner, J., S. Fujii, P. Gruz, T. Nohmi, and R. P. Fuchs.** 2000. The beta clamp targets DNA polymerase IV to DNA and strongly increases its processivity. *EMBO Rep.* **1:**484–488.

106. **Wagner, J., and T. Nohmi.** 2000. *Escherichia coli* DNA polymerase IV mutator activity: genetic requirements and mutational specificity. *J. Bacteriol.* **182:**4587–4595.

107. **Wang, Y., D. Cortez, P. Yazdi, N. Neff, S. J. Elledge, and J. Qin.** 2000. BASC, a super complex of BRCA1-associated proteins involved in the recognition and repair of aberrant DNA structures. *Genes Dev.* **14:**927–939.

108. **Witkin, E. M.** 1974. Thermal enhancement of ultraviolet mutability in a *tif-1 uvrA* derivative of *Escherichia coli* B-r: evidence that ultraviolet mutagenesis depends upon an inducible function. *Proc. Natl. Acad. Sci. USA* **71:**1930–1934.

109. **Woodgate, R., and D. G. Ennis.** 1991. Levels of chromosomally encoded Umu proteins and requirements for in vivo UmuD cleavage. *Mol. Gen. Genet.* **229:**10–16.

110. **Wu, Y. H., M. A. Franden, J. R. Hawker, Jr., and C. S. McHenry.** 1984. Monoclonal antibodies specific for the alpha subunit of the *Escherichia coli* DNA polymerase III holoenzyme. *J. Biol. Chem.* **259:**12117–12122.

111. **Zou, Y., C. Luo, and N. E. Geacintov.** 2001. Hierarchy of DNA damage recognition in *Escherichia coli* nucleotide excision repair. *Biochemistry* **40:**2923–2931.

112. **Zou, Y., and B. Van Houten.** 1999. Strand opening by the UvrA(2)B complex allows dynamic recognition of DNA damage. *EMBO J.* **18:**4889–4901.

Chapter 25

Misalignment-Mediated Mutations and Genetic Rearrangements at Repetitive DNA Sequences

Susan T. Lovett

Strand mispairing interactions between repeated DNA sequences provoke a host of mutations and genetic rearrangements in bacteria. These processes provide a source of genetic diversity in bacterial populations and serve to shape the bacterial chromosome. Repeated DNA sequences in either direct or inverted orientation can promote misalignment-mediated mutations and rearrangements by very similar mechanisms.

Rearrangements at direct repeats generate duplications and deletions within the chromosome; those at inverted repeats can cause inversions of genomic segments. If repeats are large enough (>100 to 200 nucleotides), homologous recombination pathways, dependent on the RecA protein of bacteria, can catalyze rearrangements (see chapter 21). However, genetic rearrangements can also occur independently of the RecA strand exchange protein, and hence the general homologous recombination pathways of bacteria. Rearrangements between short sequence homologies (<100 bp) appear to be mediated almost exclusively by RecA-independent means.

Strand mispairing at repetitive sequences drives these RecA-independent rearrangements; related mechanisms of transient strand misalignment can generate mutations with or without accompanying rearrangements. This chapter catalogs several types of misalignment-mediated genetic changes in bacteria, with an emphasis on the mechanisms by which they occur. Systematic study of misalignment-mediated mutation and genetic rearrangements has revealed many elements of the mechanisms of these processes, including the integral role for DNA replication. The impact of these misalignment processes on bacterial physiology and genomic evolution is also discussed.

FRAMESHIFT MUTATIONS AND THE "REPLICATION SLIPPAGE" MODEL

Frameshift mutations are a common source of genetic change that arises from slipped mispairing of DNA. From his mutational studies of the lysozyme gene of bacteriophage T4, George Streisinger was the first to apply a slipped misalignment model to frameshift mutagenesis (103). Streisinger discovered that frameshift mutations do not arise at random but predominate in repetitive DNA sequence arrays such as mononucleotide or dinucleotide runs. Streisinger proposed the "replication slippage" model, in which, during DNA replication, frameshift mutations are produced by the misalignment of the nascent strand at the repetitive sequence on its template (Fig. 1). This slippage phenomenon appears to be a problem intrinsic to DNA replication and can be seen with many different polymerases in vitro (18).

The result of frameshift strand slippage is a looped structure with unpaired bases on the template or on the nascent strand (Fig. 1). These unpaired loops, intermediates of frameshift mutagenesis, are very efficiently recognized by the bacterial mismatch repair system. Frameshift mutations are therefore among those most highly elevated by mismatch repair deficiency (26, 97). The *Escherichia coli* MutHLS mismatch repair pathway recognizes loops of 1 to 3 bases as substrates for repair. There is, however, a limit to the size of the loop that can be detected as a mispair: loops of four bases or greater are not efficiently bound by the MutS protein and therefore escape detection (80).

Frameshifts at simple sequence repeats are among the strongest of mutational hot spots. In the *E. coli lacI* gene, frameshifts at a tetranucleotide repeat, repeated three times, account for over 65% of the

Susan T. Lovett • Rosenstiel Center, Brandeis University, Waltham, MA 02454-9110.

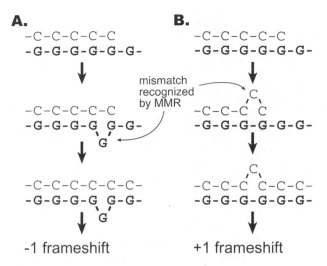

Figure 1. Frameshifts by slipped misalignment. "Slippage" of the nascent strand with its template leading to −1 frameshift (A) or +1 frameshift (B) mutations is shown. The one base loop in the slippage intermediate is efficiently recognized by the mismatch repair (MMR) system, leading to excision of the nascent strand and removal of the potential frameshift.

mutations that inactivate the gene (33). A run of eight G residues within the *E. coli xylB* gene is a hot spot for frameshift mutation evident in mismatch repair mutants (39).

FRAMESHIFT MUTATIONS AND BACTERIAL DIVERSIFICATION

Although frameshift mutations usually cause a severe loss-of-function of affected genes and are therefore deleterious, loss of gene expression can occasionally be beneficial to bacteria. For example, pathogenic organisms profit from loss of surface antigens that are targets of the host immune system. Not surprisingly, then, bacteria capitalize on naturally occurring slipped misalignment as a mechanism to generate phenotypic diversity. As a mechanism for generating genetic variation, an increase or a decrease in the number of tandem sequence repeats by slipped misalignment has the advantageous properties that it (i) can occur at reasonably high frequency in the population and (ii) is reversible in nature. These changes can therefore turn on or off the expression of genes by shifts in the reading frame, thereby producing "contingency" loci whose expression is hypervariable in the population (5). Variable expression of antigenic determinants or factors specifying host interaction is desirable for pathogenic bacteria to ensure that some members of the population escape destruction by the immune system or can colonize new locations in their hosts.

Frameshifts at monotonic and other simple repetitive arrays are commonly the source for variation in genes of pathogenic bacteria. One well-studied example of this mechanism is from *Neisseria gonorrhoeae*, in which the *opa* genes, which control tissue attachment, carry repetitive pentanucleotide sequences within their coding regions (65). Slipped misalignment at these repeats produces frameshifts that alter the genes' expression. Likewise, within *N. gonorrhoeae*, genetic variation of *pilC* pilin expression is controlled by frameshifts in a monotonic G run (49). When genomic sequence information became available, computer searches for short tandem repeats within *Haemophilus influenzae* genes revealed nine new examples of loci likely to be regulated by frameshifting (45). These genes carry tetranucleotide repeat arrays, repeated from 6 to 36 times, and show homology to known factors involved in lipopolysaccharide synthesis, adhesion, iron binding, or restriction/modification. The presence of simple sequence repeats can therefore be useful as a criterion to identify new virulence genes in bacterial genomes.

Slipped misalignment at simple sequence repeats as a mechanism of genetic control is probably widespread in pathogenic bacteria (for reviews, see references 111 and 112). Genomes of pathogenic organisms are strikingly enriched for genes carrying simple, 1- to 8-nucleotide sequence repeats (45, 93). The prevalence of the repeat elements that are 4 nucleotides and greater probably reflects the fact that these larger frameshift errors escape detection by the mismatch repair system (80) and so may occur at higher frequencies in mismatch repair-proficient strains. It is not known whether the sequence repeats involved in contingency loci or the sequence context in which they exist makes them especially prone to misalignment.

DISLOCATION AND QUASI-DIRECT REPEAT MUTAGENESIS: TRANSIENT SLIP-TEMPLATED MUTATIONS

Whereas frameshift mutations are believed to be generated from a stable slipped misalignment of nascent strand and template, transient mispairing may also allow the templating of mutational changes that can become fixed in the genome. In this proposed mechanism, the nascent strand dislocates along its template, promoted to mispair by short complementary sequences at its 3′ end, where it primes a small amount of DNA synthesis before returning to its previous position (Fig. 2). Although "dislocation mutagenesis" was first demonstrated for eukaryotic DNA polymerases in vitro (50, 51), evidence from

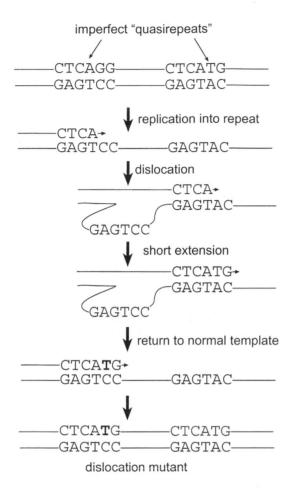

imperfect "quasirepeats"

————CTCAGG———— CTCATG————
————GAGTCC———— GAGTAC————

↓ replication into repeat

————CTCA→
————GAGTCC————————GAGTAC————

↓ dislocation

————————————————CTCA→
————————————————GAGTAC————
————GAGTCC—

↓ short extension

————————————————CTCATG→
————————————————GAGTAC————
————GAGTCC—

↓ return to normal template

————CTCATG→
————GAGTCC————————GAGTAC————

↓

————CTCATG————————CTCATG————
————GAGTCC————————GAGTAC————

dislocation mutant

Figure 2. Dislocation mutagenesis by transient slipped misalignment. Transient slipped misalignment between imperfect or quasirepeats can cause a mutation to be templated. A second strand switch restores normal alignment, leaving a mismatched heteroduplex intermediate.

mutational spectra indicate that the process also occurs in *E. coli* in vivo (95, 96). This type of error is especially evident in the mutational spectra from *E. coli* DNA polymerase III mutants lacking proofreading activity (38, 95). Because in many instances the templated errors are only a single nucleotide base, this class of mutation can be difficult to recognize; these events are more easily discernible in the case of complex mutations, where two or more mutational changes in a region are detected and a suitable donor sequences lies nearby. Frameshift mutations that do not occur in nucleotide runs may also be suspected to occur by this mechanism (88). In vitro, dislocation mutagenesis is more common adjacent to T runs (51). For bacteriophage T4, frameshift mutagenesis at quasirepeats (dislocation mutagenesis involving imperfect direct repeats) is strongly enhanced by DNA polymerase mutants with impaired processivity (90). The influences on dislocation mutagenesis, including the effects of repeat length, proximity and context, remains unknown.

LONG-RANGE REPLICATION SLIPPAGE AND TANDEM REPEAT REARRANGEMENTS

In a similar fashion to frameshift mutations, which occur over very short genetic distances of just a few nucleotides, slipped misalignment also can provoke rearrangements over much larger distances, as long as a kilobase or more. Larger DNA sequences (from hundreds to thousands of nucleotides), directly repeated in tandem, suffer extraordinarily frequent deletions and expansions, in a manner independent of *recA* (7, 29, 58, 60). Even short repeats of 10 bases or fewer dispersed within a gene are vulnerable mutational sites. In *E. coli*, deletions between dispersed repeats of 8 or more nucleotides are found as hot spots of mutation in *lac* (1, 33). These long-range interactions must contribute significantly to genetic change in bacteria and may serve to streamline the bacterial chromosome by deletion of unnecessary genetic information.

The rate of such longer-range rearrangements is dependent on the size and perfection of homology between the repeats. Deletion of large repeats, >100 nucleotides in length, is quite prevalent, at frequencies approaching 10^{-3} in the population (7, 29, 58, 60). For these large repeats, the efficiency of RecA-independent deletion rivals that of RecA-dependent recombination. For direct repeats of 10 to 20 bases within a gene, deletion occurs at detectable rates, equivalent to or higher than those of point mutations (1, 7, 68). Size is not the only determinant of rearrangement efficiency: the "perfection" of the repeats plays an important role. For example, we find that as few as four mismatches between 100-bp tandem repeats reduces deletion formation by almost 3 orders of magnitude (59), and even one base mismatch in 100 can significantly reduce deletion efficiency (V. V. Feschenko and S. T. Lovett, unpublished results). This depression of deletion by mismatches is dependent on a functional Dam MutHLS UvrD-dependent mismatch repair pathway (59). A likely explanation is that sequence mispairs present in the heteroduplex slippage intermediate elicit mismatch repair, leading to destruction of the slipped intermediate and abortion of the deletion process. In addition, mutational inactivation of single-stranded DNA exonucleases of *E. coli*, particularly the major 3' to 5' exonuclease I, elevates deletion between mismatched repeats, in a manner independent of mismatch repair (V. V. Feschenko, L. A. Rajman, and S. T. Lovett,

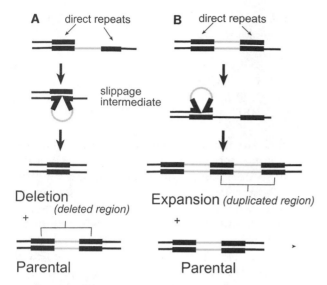

Figure 3. Long-range slipped misalignment and tandem repeat deletion or expansion. Slipped misalignment can occur between direct repeats, leading to deletion of one repeat and any intervening material (A) or duplication of one repeat and the intervening sequence (B). The long-range slippage intermediate consists of a large single-stranded loop, which is not recognized by the mismatch repair system. However, if sequence heterologies exist between the two direct repeats, mismatches in the heteroduplex region of the slipped intermediate elicit mismatch repair.

unpublished results). In this case, transient unwinding of 3′ DNA ends may elicit 3′-end degradation, thereby destroying slippage intermediates.

LONG-RANGE SLIPPED MISALIGNMENT AND DNA REPLICATION

For these longer-range rearrangements, the replication slippage model can account for both expansion in the number of tandem repeats and their loss by deletion (Fig. 3). Because expansion is relatively difficult to select, most systematic studies have addressed the formation of deletion. One system designed to select for expansion of a repeat from two copies to three or more concluded that the repeat expansion occurred at a rate similar to deletion of the same repeats and shared many of the same genetic effects (74).

One prediction of the replication slippage model is that this class of rearrangements should be limited to those in which the repeated sequences coexist in the single-stranded region of the replication fork. Therefore, one might imagine the limit for RecA-independent deletion events to be several thousands of bases, the size of an Okazaki fragment (104). Indeed, observed RecA-independent deletion between large repeats is exquisitely sensitive to the proximity of the repeats, with an exponential decrease in deletion rate

or frequency as the distance between the repeats is increased (7, 22, 60). This contrasts to RecA-dependent recombination of tandem repeats that is blind to the relative proximity of repeats (7). (Note, however, that RecA-dependent processes contribute to rearrangements of repeats only if they exceed 200 bp or so in length [7]).

DNA replication plays an integral role in the mechanism of RecA-independent tandem repeat rearrangements. Studies of the effects of mismatch repair on imperfectly homologous repeats shows that deletion between tandem repeats occurs during or shortly after DNA replication in the context of hemimethylated DNA in *E. coli* (59). A variety of replication mutants exhibit an elevated deletion rate, with rearrangements partially or completely independent of RecA in *E. coli* and *Bacillus subtilis* (12, 14, 94). An explanation for these effects is that accumulation of single-stranded DNA in the fork and defects in processivity of replication may promote slipped misalignment in vivo. In vitro, slippage potential of polymerases is indeed correlated with their lack of processivity (18, 113), and deletion endpoints correlate with replication pause sites (79). The dissociation of the DNA polymerase, as well as the disassembly of part of the replication fork complex, may be required to allow the nascent strand to find a mispairing partner sequence within the fork. Several lines of evidence support the notion that replication frequently arrests and the fork is repaired and then restarted (see reference 24); molecular events during this arrest/repair/restart process may actively contribute to long-range slipped mispairing. We have suggested that the misalignment process could be initiated by unwinding of the nascent strand from its template during repair of blocked replication forks, allowing the opportunity for slipped misalignments (58).

NATURAL EXAMPLES OF LONG TANDEM REPEATS IN BACTERIA

Longer-sequence tandem repeats are somewhat more rare than simple sequence repeats in bacterial replicons, possibly because of their intrinsic instability. However, one example comes from group B streptococci, in which a 246-bp segment of the gene encoding the alpha C protein is repeated in tandem nine times. RecA-independent rearrangements at this sequence occur during infection and alter the immunogenicity of the molecule (85). Other cases of larger repeats (36 to 561 bp in length) regulate protein translation or function in various bacterial pathogens (cataloged references in 111 and 112).

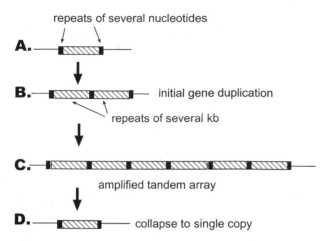

Figure 4. Gene amplification through tandem direct-repeat arrays. Short repeats in DNA (A) promote a gene duplication (B) through RecA-dependent recombination or RecA-independent slippage. Formation of the initial gene duplication is the rate-limiting step. Under selection for high gene expression, the duplication expands to higher copy, usually through RecA-dependent recombination at the large direct repeats (C). In the absence of selection, the amplicon collapses to a single-copy locus (D), again usually by RecA-dependent homologous recombination.

Naturally occurring tandem arrays can also be found associated with genes whose amplified expression has been selected (3, 31, 70, 107). The mechanism of tandem amplification (Fig. 4) is believed to begin with formation of a duplicated DNA segment, either by RecA-dependent recombination at long spurious repeats (such as insertion elements [47, 71, 81]) or by RecA-independent duplication at short random sequence repeats (31). This RecA-independent sequence duplication probably occurs via slipped misalignment during replication, as described above. Once this duplication has been established, further expansion of the array to larger copy number occurs under selection. Gene amplification also accounts for some (43) or all (44) of so-called "adaptive" mutations in F' *lac* that occur in nondividing cells under selection. In the absence of selection, amplified arrays tend to collapse to a single copy.

Most studied cases of amplification and deamplification have been deduced to occur by RecA-dependent recombination (3, 4, 21, 40, 47, 67, 91, 115). Given a RecA-independent slippage mechanism, RecA-independent amplification and deamplification should, in theory, also occur. Indeed, a few instances of RecA-independent amplification have been reported (4, 67). The relative inefficiency of RecA-independent amplification is understandable since most cases of amplified arrays involve repeats of 10 kb or greater, which places the repeats out of the optimum proximity for RecA-independent slippage (see above).

CROSS-FORK SLIPPAGE: SCE-ASSOCIATED MISALIGNMENT EVENTS

The previous examples of "simple slippage" errors involve either stable or transient intramolecular misalignment between a nascent strand and its template at repetitive sequences. Misalignment at repeated sequences may also occur intermolecularly, between strands across a replication fork. If the repeats are direct in orientation, these misalignments would produce deletions and duplications, products identical to the simple intramolecular slippage events; however, cross-fork slippage can be recognized genetically because it can lead to crossing-over between sister chromosomes. The hallmark of such sister chromosome exchange (SCE) between circular bacterial replicons is the formation of dimeric replicon products.

SCE-associated misalignment has been studied only on plasmid replicons. When *recA*-independent deletions or expansions between tandem repeats are selected on plasmids, concomitant replicon dimerization is often seen (9, 29, 58, 68, 74, 118). These dimers appear to be formed from crossing-over between sister chromosomes and not from plasmid-by-plasmid interactions (58). SCE associated with RecA-independent repeat rearrangements can be reciprocal: that is, for a plasmid bearing a duplication, the product is a plasmid dimer with both a triplication and a single-repeat locus (58, 74). To explain this phenomenon, two models have been proposed that invoke misalignment between the strands of replicating sister chromosomes (8, 58). Experimental data employing genetically marked repeats support the first model (34). This proposed mechanism (Fig. 5) is similar in many respects to simple replication slippage: instead of misalignment of the nascent strand with its template, the two nascent strands misalign with each other across the replication fork (58, 74). The intermediate of the cross-fork mispairing is a Holliday junction. In agreement with this, mutations in the RuvAB Holliday junction helicase alter the character of deletion products (58).

We favor the view that RecA-independent crossing-over may be the result of a repair reaction initiated by the displacement of the 3' nascent strands when polymerization has become blocked (58, 74). Mutations affecting the 3' exonuclease, exonuclease I, and *dnaE*, the polymerase subunit of DNA polymerase III, specifically increase SCE-associated RecA-independent deletions (17). Although this mechanism should involve unwinding and annealing of the 3' nascent strands, no factor involved in this mechanism has yet been identified. Therefore, its role in replication fork repair on the bacterial chromosome remains

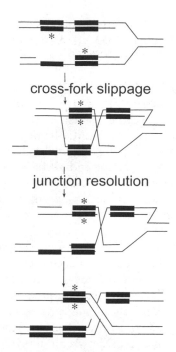

cross-fork slippage

↓

junction resolution

↓

↓

Figure 5. Deletion associated with SCE by cross-fork slipped misalignment. In a blocked replication fork, slipped misalignment can occur between the nascent strands at direct repeat sequences (asterisk). After processing of the Holliday junction branched intermediate, crossovers between the sister chromosomes can occur concomitantly with a reciprocal rearrangement (formation of both deletion and triplication product). Such crossing-over can be recognized since it leads to replicon dimerization.

reciprocal crossover product

unclear. Nevertheless, this type of *recA*-independent crossing-over between low-copy-number plasmid sister chromosomes is detected at relatively high frequencies (74). The rate of SCE-associated slippage relative to simple intramolecular slippage is difficult to estimate since dimeric plasmids have a competitive advantage over monomers, especially for high-copy-number plasmids (10, 69), in which virtually all deletion products are dimeric (9, 29, 68). For low-copy-number plasmid pBR322, the frequency of dimeric deletion products is approximately the same whether arising from a monomeric plasmid parent or isolated by transformation into a plasmid-free host (V. V. Feschenko, M. Bzymek, and S. T. Lovett, unpublished results). This allows us to estimate that SCE-associated deletion occurs at 10^{-5} to 10^{-4} per cell generation for 100- to 800-bp tandem direct repeats, roughly 10 to 40% of all RecA-independent deletion events. Presumably, the juxtaposition of the interacting strands and the single-strand nature of the replication fork facilitate this SCE reaction and eliminate the need for a protein that promotes strand invasion, such as RecA.

INVERTED REPEAT-STIMULATED REPLICATION SLIPPAGE AT DIRECT REPEATS

Slipped misalignment between repeats found either in tandem or dispersed can be stimulated by the presence of an inverted repeat sequence nearby. Spontaneous deletion hot spots often include short direct repeats flanking an inverted repeat sequence. For example, a naturally occurring hot spot for deletion between short repeats in *E. coli lacI* is associated with an inverted repeat with the potential of forming a stem-loop structure containing 18 bp (1). By promoting strand-pairing within dispersed inverted repeat sequences, these structures can also increase the local proximity and subsequent slipped misalignment of otherwise dispersed direct repeats (Fig. 6). However, inverted repeats stimulate deletion even of tandemly juxtaposed direct repeats, where no proximity benefit is applicable. Systematic study of deletion between tandem repeats borne on plasmids or carried by bacteriophage shows that inverted repeats stimulate deletion by several orders of magnitude (2, 15, 16, 83, 101, 116). In several studies, inverted repeats have been deduced to stimulate deletion specifically on the lagging strand (15, 84, 92, 109). Presumably, the lagging strand template has sufficient single-strand character to allow the inverted repeats to assume a hairpin secondary structure. These hairpins block DNA polymerase (53), presumably causing its dissociation, which then allows the nascent strand to misalign at repeats in the vicinity. In vitro slippage associated with inverted repeats is inversely correlated to the propensity of various polymerases to proceed through the hairpin structure (18) and is accompanied by pausing and dissociation of the polymerase (113). In addition to their role in promoting replication slippage, inverted repeats also stimulate deletion via an additional mechanism, single-strand annealing (see below).

DELETION BY SINGLE-STRAND ANNEALING: SbcCD CLEAVAGE AT INVERTED REPEATS

In theory, deletions between direct repeats can occur, not only by slipped misalignment, but also by a single-strand annealing (SSA) mechanism initiated by a double-strand break between the repeats. In this mechanism, resection of one strand of the double-strand ends eventually reveals complementarity that allows the exposed strands to anneal. After trimming and religation, SSA produces a deletion of one repeat and any intervening DNA (Fig. 7). In contrast to

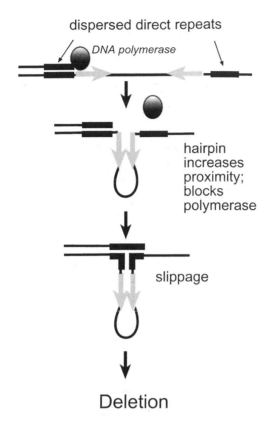

dispersed direct repeats

DNA polymerase

hairpin
increases
proximity;
blocks
polymerase

slippage

Deletion

Figure 6. Inverted repeat stimulation of deletion at direct repeats. Inverted repeat sequences stimulate slipped misalignment at flanking direct repeats, facilitating deletion of the inverted repeats and one direct repeat. The stimulatory effect is due to formation of a hairpin structure by the inverted repeats, which can both block replication and increase the local proximity of the direct repeats.

replication slippage at direct repeats that generates both expansions and deletions, the SSA mechanism is necessarily subtractive and generates only deletions.

SSA contributes as a major pathway to rearrangements in eukaryotes (35, 57, 66) and in bacteriophage-infected *E. coli* (102, 108) but may occur efficiently only under restricted circumstances in normal bacterial growth. Tandem repeat deletion assayed on plasmids in *E. coli* has properties consistent with replication slippage and is not compatible with an SSA mechanism (16). SSA's role in prokaryotes is limited by the RecBCD or AddAB class of ATP-dependent double-stranded DNA exonucleases. In *E. coli*, linear DNA is rapidly degraded unless it contains multiple and properly oriented 8-nucleotide chi sites, which attenuate RecBCD's nuclease activity in vivo (27, 52) and in vitro (30). Naturally occurring plasmids such as those used in many assays for tandem repeat rearrangements lack chi sites, which makes them vulnerable to destruction if they are broken. Healing of linear plasmid DNA bearing terminal homologies in *recA* mutant strains

is a measure of SSA activity in vivo and is efficient in *E. coli* only in mutants with mutationally attenuated RecBCD activity (16).

On the *E. coli* chromosome, SSA may contribute to deletions at microhomologies that occur at very low rates in the population. One experimental system that has been studied extensively is the formation of lambda bio transducing phage that excise from the chromosome by deletion between dispersed microhomologies (48, 117). The frequency of such deletions is enhanced by UV irradiation and other genotoxic agents (99, 117), although such events are still infrequent. This form of illegitimate recombination requires the function of the RecJ 5' to 3' exonuclease (110), is independent of the RecA strand transfer protein (48), and is inhibited by the RecQ helicase protein (41). RecJ may be required to resect the ends to reveal single-strand complementary ends; RecQ may act at a late step to unwind annealed but unligated intermediates. An SSA mechanism rather than replication slippage is suggested by the fact that levels of DNA ligase in vivo are correlated with the efficiency of illegitimate recombination at microhomologies (78). Illegitimate recombination in this assay requires ongoing DNA replication (42), which may provide a source of chromosome breaks or deliver other processing enzymes that enhance deletion formation. Deletions at microhomologies, induced by Tus/Ter stalled replication forks on *E. coli* plasmids, are also inferred to occur by SSA since their formation is stimulated 10-fold by the inactivation of RecBCD nuclease (11).

SSA has also been implicated in efficient RecA-independent deletion that is associated with secondary structure in *E. coli* (15, 16). SbcCD-mediated cleavage of cruciform structures, formed by perfect palindromic DNA, appears to initiate SSA between direct repeats flanking the site of the break. Deletion between 101-bp direct repeats separated by a 50-bp palindrome is reasonably efficient, at 10^{-3} to 10^{-2} per cell generation. SbcCD possesses both endonuclease and exonuclease activities and will specifically cleave hairpin structures in vitro (23) and produce breaks in the chromosome in vivo (25, 55). Its role in SSA may be to initiate the break, to resect and potentially protect the broken ends (Fig. 7). Since SbcCD-dependent deletion appears to be sensitive to the cruciform-forming potential of inverted repeats in duplex DNA rather than hairpin-forming potential in single-stranded DNA (15), it is the formation of cruciform structures in duplex DNA that attracts SbcCD to initiate SSA. Since SbcCD-dependent SSA is not inhibited by RecBCD even when the replicon contains no chi sites to attenuate RecBCD (16), SbcCD may preclude RecBCD's access to the broken

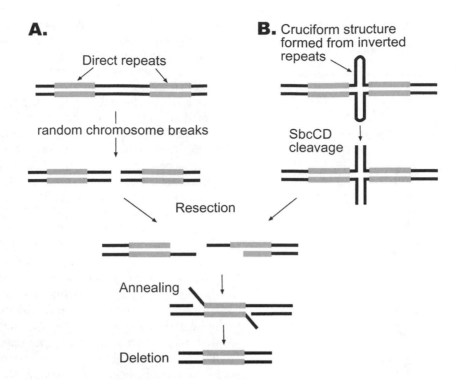

Figure 7. SSA mechanism for deletion formation. Breakage of the chromosome, followed by degradation of one of the two strands at the broken end, will reveal complementary sequences at direct repeats flanking the break. These complementary sequences anneal and are processed to form a deletion between the direct repeats. The initial break may be caused by random breakage due to spontaneous DNA damage (A) or may be induced more efficiently by endonucleolytic cleavage by SbcCD at cruciform structures formed from inverted repeats (B).

end. The relative efficiency of SSA in this assay versus that of the microhomology systems described above may be due to the increased length of the homology, the more efficient cleavage reaction, or other protective activity afforded by the participation of SbcCD.

TRANSPOSON EXCISION: A SPECIAL CASE OF INVERTED REPEAT-ASSOCIATED DELETION AT SHORT DIRECT REPEATS

Transposition produces a short duplication (<10 bp) of the host target sequence that flanks the transposable element. Since many transposons possess inverted repeats at their ends, these inverted repeats can stimulate deletion at the flanking short target-site direct repeats. Transposon "precise excision" is therefore a special, well-studied case of the inverted repeat-associated deletion discussed above (Fig. 8). Tn10 also possesses direct repeats within its inverted ends, and deletion at these sites, so-called "nearly precise" excision, can occur. Both of these excision reactions have outcomes advantageous to the bacterium: precise excision can remove a deleterious insertion entirely, and nearly precise excision can relieve the transposon's polar effect on transcription.

Precise excision and nearly precise excision in *E. coli* appear to be mechanistically similar (37) and do not require transposase function (32). The inverted repeats of the element are essential for its deletion and presumably interact to form a stem-loop structure that in some way is stimulatory to excision (Fig. 8). It has been suggested that precise excision and nearly precise excision occur by a replication slippage mechanism (32, 37, 105) because it is stimulated when the transposon becomes single stranded, such as during single-strand DNA phage replication (28) or conjugal transfer (106). Nevertheless, SSA is also a viable mechanism. These mechanisms need not be exclusive: spontaneous excision may occur by a combination of means.

Mutations that stimulate precise and nearly precise excision of Tn10 or Tn5 (Table 1), the so-called "Tex" phenotype (46, 61, 62, 87), provide some insights into the mechanisms of the deletion process. The *tex* mutations have been classified into two types: type I *tex* mutations include those that stimulate excision of both intact and "mini" versions of Tn10, and type II *tex* mutations stimulate excision of wild-type Tn10 but not mini-Tn10 (86). Wild-type Tn10 and mini-Tn10 differ in that the inverted repeats of wild-type Tn10 are not completely identical,

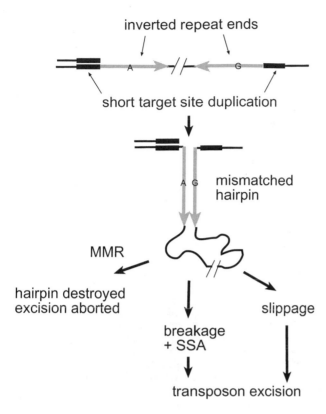

inverted repeat ends

short target site duplication

mismatched hairpin

MMR

hairpin destroyed excision aborted

breakage + SSA

slippage

transposon excision

Figure 8. Transposon excision. After transposition, short target-site duplications flank large inverted repeats at the transposon ends. Formation of a hairpin structure at the inverted repeats can promote deletion at the flanking direct repeats by stimulation of replication slippage. Breakage followed by SSA can also promote excision. For wild-type Tn*10*, mismatches within the inverted ends prompt mismatch repair to destroy the hairpin, thereby reducing the efficiency of transposon excision.

whereas those of mini-Tn*10* are perfect repeats. Therefore, the stem structure potentially formed by the imperfect wild-type Tn*10* inverted ends would contain mispairs at several sites. Action of the mismatch repair system on such mispairs should destabilize a hairpin structure formed from the inverted repeats (Fig. 8). Accordingly, the type II *tex* mutations affect all components of the mismatch repair pathway: *dam*, *mutHLS*, and *uvrD*. Type I *tex* mutations include *ssb* (single-strand DNA-binding protein) (62, 86), *polA* (DNA polymerase I) (64, 86), *topA* (DNA topoisomerase I) (86), *dnaQ* (DNA polymerase III proofreading exonuclease) (79), *uup* (putative ATPase) (46, 86, 87), *dnaE* (DNA polymerase III polymerase subunit), and *dnaB* (replicative helicase) (76).

In addition to their stimulation of transposon excision, mutations in some of the type I *tex* genes also stimulate other types of deletion events not involving inverted repeats (12, 86, 94). This may be because they have general stimulatory effects on replication slippage. Mutations in *polA* and *ssb* may

stimulate slippage by the accumulation of single-stranded DNA; this may also facilitate hairpin formation at the inverted repeats. Impaired processivity conferred by mutations in the replicative apparatus (*dnaE*, *dnaQ*, and *dnaB*) may also favor slipped misalignment of direct repeats.

Other *tex* mutants suggest that transposon excision can occur by mechanisms other than replication slippage. Special nonnull mutations in the *recB* and *recC* subunits of the RecBCD nuclease were isolated by their Tex phenotype (63). Since RecBCD nuclease activity in vivo appears to require a double-strand end, this suggests that transposon excision, at least to some extent, can occur from double-strand-break processing, perhaps by an SSA mechanism.

Treatment of *E. coli* with UV light also stimulates precision excision of Tn*10*, which occurs by a mechanism genetically distinct from that for spontaneous excision. This stimulated level of excision requires several homologous recombination factors—*recN*, *ruvAB*, and *ruvC* (19, 77)—that do not affect spontaneous excision (Table 1). The *recA* gene is also required (20) although it is not yet clear whether RecA plays a direct role or promotes excision indirectly by its role in induction of the SOS response (56). (Both the *recN* and *ruvAB* loci are part of the SOS regulon.) Replication fork reversal in the response to template lesions that is mediated by RuvAB helicase, followed by cleavage of the retrograde fork by RuvC endonuclease (98), is a potential initiating event for UV light-stimulated transposon excision. A candidate mechanism is RecA-independent SSA promoted by this breakage of the replication fork. RecN is known to be required for efficient double-strand-break repair in *E. coli* (82), but its biochemical properties are not known. RecN may conceivably participate in protection or processing of this break to provoke SSA at the short direct repeats.

TRANSIENT STRAND SWITCHING AT INVERTED REPEATS AND QUASIPALINDROME-TEMPLATED MUTATIONS

Similar to the fashion in which strands dislocate and mispair at direct repeats to generate deletion and duplication slippage products, nascent strands can dislocate, mispair, and prime DNA synthesis at inverted repeats. However, such strand switching at inverted repeats reverses the direction of replication and therefore confounds recovery of such products. Hence, most illustrations of inverted repeat-associated mutation involve transient dislocation or mispairing and two template switch reactions. One

Table 1. Genes affecting transposon excision in *E. coli*

Gene	Function	Effect of mutation[a]	Reference(s)
dam, mutHLS, uvrD	Methyl-directed mismatch repair	+, Tn*10* PE and NPE 0, mini-Tn*10* PE	61, 86
recBC	Helicase/nuclease	+, Tn*10* PE and NPE	63
dnaQ	DNA Pol III proofreading exonuclease	+, Tn*10* PE +, tandem repeat deletion	61, 94
polA	DNA polymerase I	+, Tn*10* PE and NPE, mini-Tn*10* PE +, deletion at MH	36, 75, 86
ihfAB	IHF DNA bending protein	−, Mu, Tn5, Tn*10* PE −, deletion at MH	72, 100
uup	Predicted ATPase subunit of ABC transporter	+, Tn5, Tn*10*, mini-Tn*10*, PE +, Tn*10* NPE +, tandem repeat deletion	46, 86
ssb	Single-strand DNA-binding protein	+, Tn*10* and mini-Tn*10* PE +, Tn*10* NPE +, tandem repeat deletion	58, 61, 75, 86, 87
dnaB	Replication fork helicase	+, Tn*10* and mini-Tn*10* PE +, tandem repeat deletion	76, 94
topA	Topoisomerase I	+, Tn*10* and mini-Tn*10* PE +, IR-associated tandem repeat deletion	86, 101
dnaE	Polymerase subunit of DNA polymerase III	+, Tn*10* and mini-Tn*10* PE +, tandem repeat deletion (SCE associated) +, IR-associated tandem repeat deletion	12, 15, 76, 94
recN	Unknown role in DSB repair	−, UV- and MMC-induced Tn*10* PE	19
ruvAB	Holliday junction helicase, replication fork regression	−, UV- and MMC-induced Tn*10* PE	77
recG	Holliday junction helicase, replication fork regression	−, MMC-induced Tn*10* PE	77

[a]A minus (−) refers to a reduction of the specified events by mutation of the denoted gene; +, an elevation; 0, no effect. PE, precise excision; NPE, nearly precise excision; MMC, mitomycin C; SCE, sister chromosome exchange; MH microhomology; IR, inverted repeat.

template switch occurs by mispairing of the nascent strand at an inverted repeat, and the second restores the original direction of replication.

Inverted repeats that contain slight sequence differences between the two repeats are known as imperfect or "quasipalindromes." Transient misalignment of nascent strands between the two repeats allows one repeat to template sequence changes into the other repeat (Fig. 9), a mechanism first proposed by Lynn Ripley (89). The simplest case, intramolecular quasipalindrome-templated mutations, can arise after DNA synthesis has passed the center of symmetry between the inverted repeats. Because of the inverted symmetry, the nascent strand can snap into a hairpin structure, allowing the first strand of the hairpin to template the other. A second template switch, in which the newly synthesized second repeat aligns with its normal parental template, restores the normal direction of replication. The end result of this reaction is to convert an "imperfect" inverted repeat into a more "perfect" one.

Quasipalindrome-associated mutations have been found in many organisms, including *E. coli* and its bacteriophages (reviewed in references 13 and 88).

They are often recognized as complex mutations, in which several sequence changes are introduced in a single mutational event, or as frameshift mutations that are not associated with nucleotide runs. In a mutational spectrum in the *E. coli rpsL* gene, quasipalindrome-associated complex mutations were elevated more than 200,000-fold by a mutator allele of DNA polymerase III, *dnaE173*, and constitute 8% of all detected mutations in this strain (73). (These complex mutations were not detectable in the wild-type control spectrum.) An extraordinarily potent mutational hot spot of *E. coli* is found within the *thyA* gene, which encodes thymidylate synthase. The hot spot mutation, an AT to TA transversion, can be accounted for by a simple intramolecular strand switch at a natural quasipalindrome (114). These inverted repeats could assume a hairpin stabilized by 17 bp with three imperfections within the stem. This hot spot, internal to the *thyA* coding region, accounts for over 50% of the mutations that inactivate *thyA* in wild-type *E. coli*. Inactivation of the major 3′ single-strand DNA exonucleases, ExoI and ExoVII, strongly increases the frequency of this quasipalindrome-associated mutation (114). Presumably, the

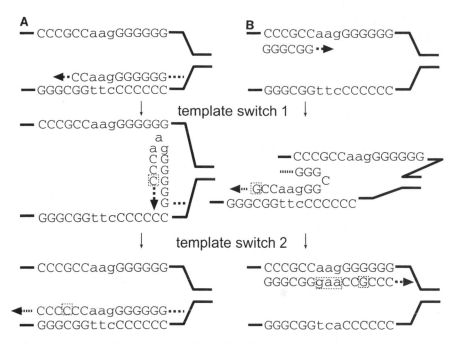

Figure 9. Quasipalindrome-associated mutation. Strand-switch replication leads to the templating of mutational changes at imperfect or quasipalindrome sequences. Two strand-switch reactions are required to recover products. (A) Simple intramolecular pairing. (B) Intermolecular cross-fork mispairing. Note that this strand switch produces an inversion of the sequence lying between the direct repeats. The product of both intramolecular and intermolecular strand switching is a more perfect (and potentially longer) inverted repeat.

exonucleases normally abort these mutations by degrading the 3' end of single strands that are displaced during the two template switch reactions. Influence of the mismatch repair machinery supports the conclusion that the majority of these mutations arise during normal chromosomal replication rather than during more localized repair DNA synthesis. The *thyA* hot spot mutation occurs during lagging strand replication, in which dissociation and reassociation cycling of the lagging strand polymerase and the single-stranded nature of the template may facilitate hairpin formation. During in vitro DNA polymerization, quasipalindrome-associated mutagenesis was highly correlated with replication pause sites (79).

In her original proposal, Ripley also realized that intermolecular strand-switching reactions could also account for some fraction of quasipalindrome-associated mutations (88, 89). In this mechanism, one strand mispairs with its inverted repeat across the replication fork, templating a short amount of synthesis before a second strand switch that restores normal replication. If the switch occurs during replication of the first repeat of an inverted pair, it can cause an inversion of any intervening sequence between the two inverted repeats. Rosche et al. designed a clever assay to detect such intermolecular strand switches by selecting inversion of the sequences between two 17-bp inverted repeats, con-

comitant with correction of a quasipalindrome (92). Such strand-switch events were found to arise more commonly from the leading strand. The rationale for this specificity is that the intermolecular mispairing target for a nascent leading strand would be on the lagging strand and would be more likely to be single stranded and therefore available for mispairing. One might imagine that such intermolecular strand switching is more infrequent than simple fold-back intramolecular strand switching; there is some support for this from in vitro reactions since central region inversions are rarely detected (79).

REPLICON INVERTED DUPLICATION CAUSED BY RECIPROCAL STRAND SWITCHING AT SHORT INVERTED REPEATS

For small replicons, intermolecular strand switching at short inverted repeat sequences can lead to formation of a replicon dimer in which virtually the entire replicon has been converted to an inverted duplication. An assay designed to detect inversion between 353-bp inverted repeats carried on plasmid pBR322 showed that inversion occurred at high frequencies (ca. 10^{-3}) independent of RecA (6). Like direct repeat rearrangements, the RecA-independent inversion reaction was highly sensitive to the proximity of the inverted repeats. However, when the inversion products were examined, they consisted of complex head-to-head plasmid dimers. Rather than a simple inversion restricted to the repeats, inverted dimeric plasmid products were exclusively recovered. A reciprocal strand-switching mechanism initiated by misalignment at the inverted repeats (Fig. 10) can explain these events. (Both nascent strands cross the fork and mispair in a manner similar but not identical to that of the SCE-associated deletion mechanism described above. Both mechanisms create Holliday junction intermediates in the fork.) Since large inverted repeats are poorly viable (54), these products are recoverable only on small plasmids or when a subsequent rearrangement limits the extent of the inverted repeat produced by this strand-switching mechanism. Nevertheless, it illustrates the potential

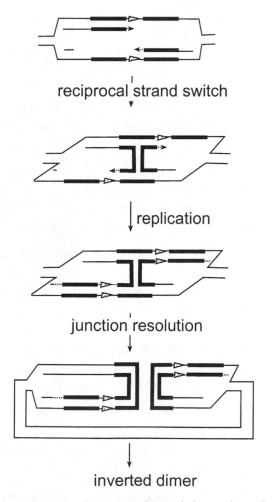

Figure 10. Inversion by reciprocal cross-fork strand switching. Strand switching of both nascent strands across the fork at inverted repeat sequences causes an inversion of the interval between the inverted repeats (marked with a triangle) and is associated with an inverted duplication of the entire replicon.

for strand switching at inverted repeats to extend, potentially in a dramatic way, the inverted repeat length.

GENOMIC CONSEQUENCES OF INVERTED REPEAT STRAND SWITCHING

The high mutation rate at the *thyA* natural quasipalindrome demonstrates that strand switching at inverted repeats can account for a significant fraction of mutational burden at certain chromosomal sites. How many other similarly mutable targets exist in bacterial genomes is not known. These studies also suggest that, once an inverted repeat sequence of sufficient size and stability is formed, the strand-switching mechanism should drive the formation of longer and more perfect palindromes. Inverted repeats can therefore be considered "selfish" elements that persist because their rate of formation by strand-switching replication exceeds that of their destruction by mutation (114). Creation of short inverted repeats may also confer selective advantage if they form favorable sites of binding for nucleic acid regulatory proteins. However, there is a limit to the tolerance for inverted repeats in bacterial genomes, even if they do not influence protein coding potential in a negative way. Large palindromes, ca. 200 bp, are known to be selectively lost because they impede replication (54). Large inverted repeats may also negatively influence mRNA structure or function.

SUMMARY

The process of strand misalignment dynamically molds bacterial chromosomes. Short spurious repeats can be sites for deletion, duplication, or inversion; these processes create larger repeats that can lead to even higher rates of rearrangements. These rearrangements can occur independently of the bacterial homologous recombination pathways and are dependent on the length and perfection of the repeats, as well as their proximity. The rate of rearrangement between long (>100-bp) tandem repeats can be quite high (10^{-5} to 10^{-4} per cell generation), rivaling the rates of homologous recombination. Mechanisms involving strand misalignment during replication (replication slippage) can account for many features of these rearrangements. Even transient misalignments during replication can cause base sequence changes, sometimes at very high frequencies, as in the case of certain quasipalindrome-associated mutations. Other chromosomal rearrangements appear to arise from mispairing at processed broken chromosomes, by the SSA mechanism. Cleavage of cruciform structures by SbcCD nuclease or replication of damaged DNA templates may cause chromosomal breakage, initiating these deletion events.

Although many rearrangements are deleterious, some can lead to selective advantage. Bacteria can use such rearrangements to generate "contingency" loci that are expressed in a subfraction of the population or to generate high-copy amplicons to provide increased expression of a gene under selection. Generation of inverted repeats can create new regulatory sites. Deletion rearrangements can remove deleterious or superfluous DNA (including transposable elements and prophage), thereby streamlining the bacterial chromosome.

Although the mechanisms promoting rearrangements or mutations at DNA sequence repeats have been loosely defined, much remains to be discovered regarding the cellular control of such misalignment processes.

REFERENCES

1. **Albertini, A. M., M. Hofer, M. P. Calos, and J. H. Miller.** 1982. On the formation of spontaneous deletions: the importance of short sequence homologies in the generation of large deletions. *Cell* **29**:319–328.
2. **Albertini, A. M., M. Hofer, M. P. Calos, T. D. Tlsty, and J. H. Miller.** 1983. Analysis of spontaneous deletions and gene amplification in the *lac* region of *Escherichia coli*. *Cold Spring Harbor Symp. Quant. Biol.* **47**:841–850.
3. **Anderson, R. P., and J. R. Roth.** 1978. Tandem genetic duplications in *Salmonella typhimurium*: amplification of the histidine operon. *J. Mol. Biol.* **126**:53–71.
4. **Barten, R., and T. F. Meyer.** 2001. DNA circle formation in *Neisseria gonorrhoeae*: a possible intermediate in diverse genomic recombination processes. *Mol. Gen. Genet.* **264**: 691–701.
5. **Bayliss, C. D., D. Field, and E. R. Moxon.** 2001. The simple sequence contingency loci of *Haemophilus influenzae* and *Neisseria meningitidis*. *J. Clin. Investig.* **107**:657–662.
6. **Bi, X., and L. F. Liu.** 1996. DNA rearrangement mediated by inverted repeats. *Proc. Natl. Acad. Sci. USA* **93**:819–823.
7. **Bi, X., and L. F. Liu.** 1994. *recA*-independent and *recA*-dependent intramolecular plasmid recombination. Differential homology requirement and distance effect. *J. Mol. Biol.* **235**:414–423.
8. **Bi, X., and L. F. Liu.** 1996. A replicational model for DNA recombination between direct repeats. *J. Mol. Biol.* **256**:849–858.
9. **Bi, X., Y. L. Lyu, and L. F. Liu.** 1995. Specific stimulation of *recA*-independent plasmid recombination by a DNA sequence at a distance. *J. Mol. Biol.* **247**:890–902.
10. **Bierne, H., S. D. Ehrlich, and B. Michel.** 1995. Competition between parental and recombinant plasmids affects the measure of recombination frequencies. *Plasmid* **33**:101–112.
11. **Bierne, H., S. D. Ehrlich, and B. Michel.** 1997. Deletions at stalled replication forks occur by two different pathways. *EMBO J.* **16**:3332–3340.
12. **Bierne, H., D. Vilette, S. D. Ehrlich, and B. Michel.** 1997. Isolation of a *dnaE* mutation which enhances RecA-inde-

13. **Bissler, J. J.** 1998. DNA inverted repeats and human disease. *Front. Biosci.* **3**:d408–d418.
14. **Bruand, C., V. Bidnenko, and S. D. Ehrlich.** 2001. Replication mutations differentially enhance RecA-dependent and RecA-independent recombination between tandem repeats in *Bacillus subtilis*. *Mol. Microbiol.* **39**:1248–1258.
15. **Bzymek, M., and S. T. Lovett.** 2001. Evidence for two mechanisms of palindrome-stimulated deletion in *Escherichia coli*: single-strand annealing and replication slipped mispairing. *Genetics* **158**:527–540.
16. **Bzymek, M., and S. T. Lovett.** 2001. Instability of repetitive DNA sequences: the role of replication in multiple mechanisms. *Proc. Natl. Acad. Sci. USA* **98**:8319–8325.
17. **Bzymek, M., C. J. Saveson, V. V. Feschenko, and S. T. Lovett.** 1999. Slipped misalignment mechanisms of deletion formation: in vivo susceptibility to nucleases. *J. Bacteriol.* **181**:477–482.
18. **Canceill, D., E. Viguera, and S. D. Ehrlich.** 1999. Replication slippage of different DNA polymerases is inversely related to their strand displacement efficiency. *J. Biol. Chem.* **274**: 27481–27490.
19. **Chan, A., M. S. Levy, and R. Nagel.** 1994. RecN SOS gene and induced precise excision of Tn*10* in *Escherichia coli*. *Mutat. Res.* **325**:75–79.
20. **Chan, A., and R. Nagel.** 1997. Involvement of *recA* and *recF* in the induced precise excision of Tn*10* in *Escherichia coli*. *Mutat. Res.* **381**:111–115.
21. **Chandler, M., and D. J. Galas.** 1983. IS1-mediated tandem duplication of plasmid pBR322. Dependence on *recA* and on DNA polymerase I. *J. Mol. Biol.* **165**:183–190.
22. **Chedin, F., E. Dervyn, R. Dervyn, S. D. Ehrlich, and P. Noirot.** 1994. Frequency of deletion formation decreases exponentially with distance between short direct repeats. *Mol. Microbiol.* **12**:561–569.
23. **Connelly, J. C., E. S. de Leau, E. A. Okely, and D. R. Leach.** 1997. Overexpression, purification, and characterization of the SbcCD protein from *Escherichia coli*. *J. Biol. Chem.* **272**:19819–19826.
24. **Cox, M. M., M. F. Goodman, K. N. Kreuzer, D. J. Sherratt, S. J. Sandler, and K. J. Marians.** 2000. The importance of repairing stalled replication forks. *Nature* **404**:37–41.
25. **Cromie, G. A., C. B. Millar, K. H. Schmidt, and D. R. Leach.** 2000. Palindromes as substrates for multiple pathways of recombination in *Escherichia coli*. *Genetics* **154**:513–522.
26. **Cupples, C. G., M. Cabrera, C. Cruz, and J. H. Miller.** 1990. A set of *lacZ* mutations in *Escherichia coli* that allow rapid detection of specific frameshift mutations. *Genetics* **125**:275–280.
27. **Dabert, P., S. D. Ehrlich, and A. Gruss.** 1992. Chi sequence protects against RecBCD degradation of DNA in vivo. *Proc. Natl. Acad. Sci. USA* **89**:12073–12077.
28. **d'Alençon, E., M. Petranovic, B. Michel, P. Noirot, A. Aucouturier, M. Uzest, and S. D. Ehrlich.** 1994. Copy-choice illegitimate DNA recombination revisited. *EMBO J.* **13**: 2725–2734.
29. **Dianov, G. L., A. V. Kuzminov, A. V. Mazin, and R. I. Salganik.** 1991. Molecular mechanisms of deletion formation in *Escherichia coli* plasmids. I. Deletion formation mediated by long direct repeats. *Mol. Gen. Genet.* **228**:153–159.
30. **Dixon, D. A., and S. C. Kowalczykowski.** 1993. The recombination hotspot chi is a regulatory sequence that acts by attenuating the nuclease activity of the E. coli RecBCD enzyme. *Cell* **73**:87–96.

31. Edlund, T., and S. Normark. 1981. Recombination between short DNA homologies causes tandem duplication. *Nature* **292:**269–271.

32. Egner, C., and D. E. Berg. 1981. Excision of transposon Tn*5* is dependent on the inverted repeats but not on the transposase function of Tn*5*. *Proc. Natl. Acad. Sci. USA* **78:**459–463.

33. Farabaugh, P. J., U. Schmeissner, M. Hofer, and J. H. Miller. 1978. Genetic studies of the *lac* repressor. VII. On the molecular nature of spontaneous hotspots in the *lacI* gene of *Escherichia coli. J. Mol. Biol.* **126:**847–857.

34. Feschenko, V. V., and S. T. Lovett. 1998. Slipped misalignment mechanisms of deletion formation: analysis of deletion endpoints. *J. Mol. Biol.* **276:**559–569.

35. Fishman-Lobell, J., N. Rudin, and J. E. Haber. 1992. Two alternative pathways of double-strand break repair that are kinetically separable and independently modulated. *Mol. Cell. Biol.* **12:**1292–1303.

36. Fix, D. F., P. A. Burns, and B. W. Glickman. 1987. DNA sequence analysis of spontaneous mutation in a PolA1 strain of *Escherichia coli* indicates sequence-specific effects. *Mol. Gen. Genet.* **207:**267–272.

37. Foster, T. J., V. Lundblad, S. Hanley-Way, S. M. Halling, and N. Kleckner. 1981. Three Tn*10*-associated excision events: relationship to transposition and role of direct and inverted repeats. *Cell* **23:**215–227.

38. Fowler, R. G., R. M. Schaaper, and B. W. Glickman. 1986. Characterization of mutational specificity within the *lacI* gene for a *mutD5* mutator strain of *Escherichia coli* defective in 3′ → 5′ exonuclease (proofreading) activity. *J. Bacteriol.* **167:**130–137.

39. Funchain, P., A. Yeung, J. L. Stewart, R. Lin, M. M. Slupska, and J. H. Miller. 2000. The consequences of growth of a mutator strain of *Escherichia coli* as measured by loss of function among multiple gene targets and loss of fitness. *Genetics* **154:**959–970.

40. Goldberg, I., and J. J. Mekalanos. 1986. Effect of a *recA* mutation on cholera toxin gene amplification and deletion events. *J. Bacteriol.* **165:**723–731.

41. Hanada, K., T. Ukita, Y. Kohno, K. Saito, J. Kato, and H. Ikeda. 1997. RecQ DNA helicase is a suppressor of illegitimate recombination in *Escherichia coli. Proc. Natl. Acad. Sci. USA* **94:**3860–3865.

42. Hanada, K., T. Yamashita, Y. Shobuike, and H. Ikeda. 2001. Role of DnaB helicase in UV-induced illegitimate recombination in *Escherichia coli. J. Bacteriol.* **183:**4964–4969.

43. Hastings, P. J., H. J. Bull, J. R. Klump, and S. M. Rosenberg. 2000. Adaptive amplification: an inducible chromosomal instability mechanism. *Cell* **103:**723–731.

44. Hendrickson, H., E. S. Slechta, U. Bergthorsson, D. I. Andersson, and J. R. Roth. 2002. Amplification-mutagenesis: evidence that "directed" adaptive mutation and general hypermutability result from growth with a selected gene amplification. *Proc. Natl. Acad. Sci. USA* **99:**2164–2169.

45. Hood, D. W., M. E. Deadman, M. P. Jennings, M. Bisercic, R. D. Fleischmann, J. C. Venter, and E. R. Moxon. 1996. DNA repeats identify novel virulence genes in *Haemophilus influenzae. Proc. Natl. Acad. Sci. USA* **93:**11121–11125.

46. Hopkins, J. D., M. Clements, and M. Syvanen. 1983. New class of mutations in *Escherichia coli* (*uup*) that affect precise excision of insertion elements and bacteriophage Mu growth. *J. Bacteriol.* **153:**384–389.

47. Huffman, G. A., and R. H. Rownd. 1984. Transition of deletion mutants of the composite resistance plasmid NR1 in *Escherichia coli* and *Salmonella typhimurium. J. Bacteriol.* **159:**488–498.

48. Ikeda, H., H. Shimizu, T. Ukita, and M. Kumagai. 1995. A novel assay for illegitimate recombination in *Escherichia coli*: stimulation of lambda bio transducing phage formation by ultra-violet light and its independence from RecA function. *Adv. Biophys.* **31:**197–208.

49. Jonsson, A. B., G. Nyberg, and S. Normark. 1991. Phase variation of gonococcal pili by frameshift mutation in *pilC*, a novel gene for pilus assembly. *EMBO J.* **10:**477–488.

50. Kunkel, T. A. 1990. Misalignment-mediated DNA synthesis errors. *Biochemistry* **29:**8003–8011.

51. Kunkel, T. A., and A. Soni. 1988. Mutagenesis by transient misalignment. *J. Biol. Chem.* **263:**14784–14789.

52. Kuzminov, A., E. Schabtach, and F. W. Stahl. 1994. Chi sites in combination with RecA protein increase the survival of linear DNA in *Escherichia coli* by inactivating exoV activity of RecBCD nuclease. *EMBO J.* **13:**2764–2776.

53. LaDuca, R. J., P. J. Fay, C. Chuang, C. S. McHenry, and R. A. Bambara. 1983. Site-specific pausing of deoxyribonucleic acid synthesis catalyzed by four forms of *Escherichia coli* DNA polymerase III. *Biochemistry* **22:**5177–5188.

54. Leach, D. R. 1994. Long DNA palindromes, cruciform structures, genetic instability and secondary structure repair. *Bioessays* **16:**893–900.

55. Leach, D. R., E. A. Okely, and D. J. Pinder. 1997. Repair by recombination of DNA containing a palindromic sequence. *Mol. Microbiol.* **26:**597–606.

56. Levy, M. S., E. Balbinder, and R. Nagel. 1993. Effect of mutations in SOS genes on UV-induced precise excision of Tn*10* in Escherichia coli. *Mutat. Res.* **293:**241–247.

57. Lin, F. L., K. Sperle, and N. Sternberg. 1990. Repair of double-stranded DNA breaks by homologous DNA fragments during transfer of DNA into mouse L cells. *Mol. Cell. Biol.* **10:**113–119.

58. Lovett, S. T., P. T. Drapkin, V. A. Sutera, Jr., and T. J. Gluckman-Peskind. 1993. A sister-strand exchange mechanism for *recA*-independent deletion of repeated DNA sequences in *Escherichia coli. Genetics* **135:**631–642.

59. Lovett, S. T., and V. V. Feschenko. 1996. Stabilization of diverged tandem repeats by mismatch repair: evidence for deletion formation via a misaligned replication intermediate. *Proc. Natl. Acad. Sci. USA* **93:**7120–7124.

60. Lovett, S. T., T. J. Gluckman, P. J. Simon, V. A. Sutera, Jr., and P. T. Drapkin. 1994. Recombination between repeats in *Escherichia coli* by a *recA*-independent, proximity-sensitive mechanism. *Mol. Gen. Genet.* **245:**294–300.

61. Lundblad, V., and N. Kleckner. 1985. Mismatch repair mutants of *Escherichia coli* K12 enhance transposon excision. *Genetics* **109:**3–19.

62. Lundblad, V., and N. Kleckner. 1982. Mutants of *Escherichia coli* K12 which affect excision, p. 245–258. *In* J. F. Lemontt and W. M. Generoso (ed.), *Molecular and Cellular Mechanisms of Mutagenesis*. Academic Press, Inc., New York, N.Y.

63. Lundblad, V., A. F. Taylor, G. R. Smith, and N. Kleckner. 1984. Unusual alleles of *recB* and *recC* stimulate excision of inverted repeat transposons Tn*10* and Tn*5*. *Proc. Natl. Acad. Sci. USA* **81:**824–828.

64. MacPhee, D. G., and L. M. Hafner. 1988. Antimutagenic effects of chemicals on mutagenesis resulting from excision of a transposon in *Salmonella typhimurium. Mutat. Res.* **207:**99–105.

65. Makino, S., J. P. van Putten, and T. F. Meyer. 1991. Phase variation of the opacity outer membrane protein controls invasion by *Neisseria gonorrhoeae* into human epithelial cells. *EMBO J.* **10:**1307–1315.

66. Maryon, E., and D. Carroll. 1991. Characterization of recombination intermediates from DNA injected into *Xenopus laevis* oocytes: evidence for a nonconservative mechanism of homologous recombination. *Mol. Cell. Biol.* 11:3278–3287.

67. Mattes, R., H. J. Burkardt, and R. Schmitt. 1979. Repetition of tetracycline resistance determinant genes on R plasmid pRSD1 in *Escherichia coli. Mol. Gen. Genet.* 168:173–184.

68. Mazin, A. V., A. V. Kuzminov, G. L. Dianov, and R. I. Salganik. 1991. Mechanisms of deletion formation in *Escherichia coli* plasmids. II. Deletions mediated by short direct repeats. *Mol. Gen. Genet.* 228:209–214.

69. Mazin, A. V., T. V. Timchenko, M. K. Saparbaev, and O. M. Mazina. 1996. Dimerization of plasmid DNA accelerates selection for antibiotic resistance. *Mol. Microbiol.* 20:101–108.

70. Mekalanos, J. J. 1983. Duplication and amplification of toxin genes in *Vibrio cholerae. Cell* 35:253–263.

71. Meyer, J., and S. Iida. 1979. Amplification of chloramphenicol resistance transposons carried by phage P1Cm in *Escherichia coli. Mol. Gen. Genet.* 176:209–219.

72. Miller, H. I., and D. I. Friedman. 1980. An *E. coli* gene product required for lambda site-specific recombination. *Cell* 20:711–719.

73. Mo, J. Y., H. Maki, and M. Sekiguchi. 1991. Mutational specificity of the *dnaE173* mutator associated with a defect in the catalytic subunit of DNA polymerase III of *Escherichia coli. J. Mol. Biol.* 222:925–936.

74. Morag, A. S., C. J. Saveson, and S. T. Lovett. 1999. Expansion of DNA repeats in *Escherichia coli*: effects of recombination and replication functions. *J. Mol. Biol.* 289:21–27.

75. Mukaihara, T., and M. Enomoto. 1997. Deletion formation between the two *Salmonella typhimurium* flagellin genes encoded on the mini F plasmid: *Escherichia coli ssb* alleles enhance deletion rates and change hot-spot preference for deletion endpoints. *Genetics* 145:563–572.

76. Nagel, R., and A. Chan. 2000. Enhanced Tn*10* and mini-Tn*10* precise excision in DNA replication mutants of *Escherichia coli* K12. *Mutat. Res.* 459:275–284.

77. Nagel, R., A. Chan, and E. Rosen. 1994. *ruv* and *recG* genes and the induced precise excision of Tn*10* in *Escherichia coli. Mutat. Res.* 311:103–109.

78. Onda, M., J. Yamaguchi, K. Hanada, Y. Asami, and H. Ikeda. 2001. Role of DNA ligase in the illegitimate recombination that generates lambdabio-transducing phages in *Escherichia coli. Genetics* 158:29–39.

79. Papanicolaou, C., and L. S. Ripley. 1991. An in vitro approach to identifying specificity determinants of mutagenesis mediated by DNA misalignments. *J. Mol. Biol.* 221:805–821.

80. Parker, B. O., and M. G. Marinus. 1992. Repair of DNA heteroduplexes containing small heterologous sequences in *Escherichia coli. Proc. Natl. Acad. Sci. USA* 89:1730–1734.

81. Peterson, B. C., and R. H. Rownd. 1983. Homologous sequences other than insertion elements can serve as recombination sites in plasmid drug resistance gene amplification. *J. Bacteriol.* 156:177–185.

82. Picksley, S. M., P. V. Attfield, and R. G. Lloyd. 1984. Repair of DNA double-strand breaks in *Escherichia coli* K12 requires a functional *recN* product. *Mol. Gen. Genet.* 195:267–274.

83. Pierce, J. C., D. Kong, and W. Masker. 1991. The effect of the length of direct repeats and the presence of palindromes on deletion between directly repeated DNA sequences in bacteriophage T7. *Nucleic Acids Res.* 19:3901–3905.

84. Pinder, D. J., C. E. Blake, J. C. Lindsey, and D. R. Leach. 1998. Replication strand preference for deletions associated with DNA palindromes. *Mol. Microbiol.* 28:719–727.

85. Puopolo, K. M., S. K. Hollingshead, V. J. Carey, and L. C. Madoff. 2001. Tandem repeat deletion in the alpha C protein of group B streptococcus is *recA* independent. *Infect. Immun.* 69:5037–5045.

86. Reddy, M., and J. Gowrishankar. 2000. Characterization of the *uup* locus and its role in transposon excisions and tandem repeat deletions in *Escherichia coli. J. Bacteriol.* 182:1978–1986.

87. Reddy, M., and J. Gowrishankar. 1997. Identification and characterization of *ssb* and *uup* mutants with increased frequency of precise excision of transposon Tn*10* derivatives: nucleotide sequence of *uup* in *Escherichia coli. J. Bacteriol.* 179:2892–2899.

88. Ripley, L. S. 1990. Frameshift mutation: determinants of specificity. *Annu. Rev. Genet.* 24:189–213.

89. Ripley, L. S. 1982. Model for the participation of quasi-palindromic DNA sequences in frameshift mutation. *Proc. Natl. Acad. Sci. USA* 79:4128–4132.

90. Ripley, L. S., B. W. Glickman, and N. B. Shoemaker. 1983. Mutator versus antimutator activity of a T4 DNA polymerase mutant distinguishes two different frameshifting mechanisms. *Mol. Gen. Genet.* 189:113–117.

91. Romero, D., J. Martinez-Salazar, L. Girard, S. Brom, G. Davilla, R. Palacios, M. Flores, and C. Rodriguez. 1995. Discrete amplifiable regions (amplicons) in the symbiotic plasmid of *Rhizobium etli* CFN42. *J. Bacteriol.* 177:973–980.

92. Rosche, W. A., T. Q. Trinh, and R. R. Sinden. 1995. Differential DNA secondary structure-mediated deletion mutation in the leading and lagging strands. *J. Bacteriol.* 177:4385–4391.

93. Saunders, N. J., A. C. Jeffries, J. F. Peden, D. W. Hood, H. Tettelin, R. Rappuoli, and E. R. Moxon. 2000. Repeat-associated phase variable genes in the complete genome sequence of *Neisseria meningitidis* strain MC58. *Mol. Microbiol.* 37:207–215.

94. Saveson, C. J., and S. T. Lovett. 1997. Enhanced deletion formation by aberrant DNA replication in *Escherichia coli. Genetics* 146:457–470.

95. Schaaper, R. M. 1988. Mechanisms of mutagenesis in the *Escherichia coli* mutator *mutD5*: role of DNA mismatch repair. *Proc. Natl. Acad. Sci. USA* 85:8126–8130.

96. Schaaper, R. M., B. N. Danforth, and B. W. Glickman. 1986. Mechanisms of spontaneous mutagenesis: an analysis of the spectrum of spontaneous mutation in the *Escherichia coli lacI* gene. *J. Mol. Biol.* 189:273–284.

97. Schaaper, R. M., and R. L. Dunn. 1987. Spectra of spontaneous mutations in *Escherichia coli* strains defective in mismatch correction: the nature of in vivo DNA replication errors. *Proc. Natl. Acad. Sci. USA* 84:6220–6224.

98. Seigneur, M., V. Bidnenko, S. D. Ehrlich, and B. Michel. 1998. RuvAB acts at arrested replication forks. *Cell* 95:419–430.

99. Shanado, Y., K. Hanada, and H. Ikeda. 2001. Suppression of gamma ray-induced illegitimate recombination in *Escherichia coli* by the DNA-binding protein H-NS. *Mol. Genet. Genom.* 265:242–248.

100. Shanado, Y., J. Kato, and H. Ikeda. 1997. Fis is required for illegitimate recombination during formation of lambda *bio* transducing phage. *J. Bacteriol.* 179:4239–4245.

101. Sinden, R. R., G. X. Zheng, R. G. Brankamp, and K. N. Allen. 1991. On the deletion of inverted repeated DNA in

Escherichia coli: effects of length, thermal stability, and cruciform formation in vivo. *Genetics* **129**:991–1005.

102. Stahl, M. M., L. Thomason, A. R. Poteete, T. Tarkowski, A. Kuzminov, and F. W. Stahl. 1997. Annealing vs. invasion in phage lambda recombination. *Genetics* **147**:961–977.

103. Streisinger, G., Y. Okada, J. Emrich, J. Newton, A. Tsugita, E. Terzaghi, and I. Inouye. 1966. Frameshift mutations and the genetic code. *Cold Spring Harbor Symp. Quant. Biol.* **31**:77–86.

104. Sugimoto, K., T. Okazaki, and R. Okazaki. 1968. Mechanism of DNA chain growth, II. Accumulation of newly synthesized short chains in *E. coli* infected with ligase-defective T4 phages. *Proc. Natl. Acad. Sci. USA* **60**:1356–1362.

105. Syvanen, M. 1988. Bacterial insertion sequences, p. 331–356. *In* R. Kucherlapati and G. R. Smith (ed.), *Genetic Recombination*. American Society for Microbiology, Washington, D.C.

106. Syvanen, M., J. D. Hopkins, T. J. Griffin IV, T. Y. Liang, K. Ippen-Ihler, and R. Kolodner. 1986. Stimulation of precise excision and recombination by conjugal proficient F′ plasmids. *Mol. Gen. Genet.* **203**:1–7.

107. Tlsty, T. D., A. M. Albertini, and J. H. Miller. 1984. Gene amplification in the *lac* region of *E. coli*. *Cell* **37**:217–224.

108. Tomso, D. J., and K. N. Kreuzer. 2000. Double-strand break repair in tandem repeats during bacteriophage T4 infection. *Genetics* **155**:1493–1504.

109. Trinh, T. Q., and R. R. Sinden. 1991. Preferential DNA secondary structure mutagenesis in the lagging strand of replication in *E. coli*. *Nature* **352**:544–547.

110. Ukita, T., and H. Ikeda. 1996. Role of the *recJ* gene product in UV-induced illegitimate recombination at the hotspot. *J. Bacteriol.* **178**:2362–2367.

111. van Belkum, A., S. Scherer, L. van Alphen, and H. Verbrugh. 1998. Short-sequence DNA repeats in prokaryotic genomes. *Microbiol. Mol. Biol. Rev.* **62**:275–293.

112. van Belkum, A., W. van Leeuwen, S. Scherer, and H. Verbrugh. 1999. Occurrence and structure-function relationship of pentameric short sequence repeats in microbial genomes. *Res. Microbiol.* **150**:617–626.

113. Viguera, E., D. Canceill, and S. D. Ehrlich. 2001. Replication slippage involves DNA polymerase pausing and dissociation. *EMBO J.* **20**:2587–2595.

114. Viswanathan, M., J. J. Lacirignola, R. L. Hurley, and S. T. Lovett. 2000. A novel mutational hotspot in a natural quasipalindrome in *Escherichia coli*. *J. Mol. Biol.* **302**:553–564.

115. Volff, J. N., and J. Altenbuchner. 1998. Genetic instability of the *Streptomyces* chromosome. *Mol. Microbiol.* **27**:239–246.

116. Weston-Hafer, K., and D. E. Berg. 1989. Palindromy and the location of deletion endpoints in *Escherichia coli*. *Genetics* **121**:651–658.

117. Yamaguchi, H., T. Yamashita, H. Shimizu, and H. Ikeda. 1995. A hotspot of spontaneous and UV-induced illegitimate recombination during formation of lambda *bio* transducing phage. *Mol. Gen. Genet.* **248**:637–643.

118. Yi, T. M., D. Stearns, and B. Demple. 1988. Illegitimate recombination in an *Escherichia coli* plasmid: modulation by DNA damage and a new bacterial gene. *J. Bacteriol.* **170**:2898–2903.

V. NONHOMOLOGOUS RECOMBINATION

Chapter 26

DNA Transposons: Different Proteins and Mechanisms but Similar Rearrangements

KEITH M. DERBYSHIRE AND NIGEL D. F. GRINDLEY

She moves in mysterious ways.

Bono, U2

For many years this is how we felt about transposon movement. However, over the last decade the detailed studies of several in vitro transposition systems, the elucidation of the three-dimensional structures of transposases, and clever genetic experiments have enabled us to understand, or at least predict, the many different ways these segments of DNA can translocate. In this chapter we define four different families of transposable DNA elements and describe how transposon-mediated recombination molds the organization of the bacterial chromosome. We do not attempt to describe the movement of transposable elements in exquisite molecular and biochemical detail, but instead refer the reader to the comprehensive discussions in this book, in *Mobile DNA II*, and in other reviews (12, 37, 39, 41a, 65, 69, 102). Our goals are to define the families of transposable elements, highlighting and contrasting their different features and the various types of rearrangements they generate. This will provide a framework for understanding how these fascinating elements play such an important role in bacterial evolution and chromosome structure.

DEFINING TRANSPOSABLE ELEMENTS

Before we discuss the ins and outs of transposition, we will reiterate an early definition of these mobile genetic elements (91): "Prokaryotic transposable elements [or transposons] are defined genetic entities which are capable of inserting as discrete, non-permuted DNA segments at many different sites in prokaryotic genomes." Although not stated explicitly, this definition, while including certain bacteriophage genomes (for example, Mu and D108), was clearly intended to exclude others, such as λ. Integration of phage λ was known to occur by a mechanism of "conservative site-specific recombination" that could not account for the properties of various transposons that were under investigation at that time (118). In addition to other site-specific recombination systems (e.g., dimer resolution), this definition also precludes group I intron homing, which requires DNA homology to drive a gene conversion event requiring the host recombination machinery (11). For purely historical reasons, certain transposon characteristics, including the presence of terminal inverted repeats and the duplication of a short target sequence, became accepted as additional defining hallmarks. However, it is now clear that many transposons do not have inverted repeats and do not generate a target duplication yet are capable of translocation to many different sites. So we shall return to the broader definition above, which includes not only the classical bacterial transposons, but also the rolling circle transposons and two families with transposase proteins related to conservative site-specific recombinases. Transposons from each of these families meet the criteria of our definition, but quite different transposase proteins using distinct recombination mechanisms mediate their movement (Fig. 1). Although outside the scope of this chapter, our definition would include the long terminal repeat retrotransposons and retroviruses (12, 39). These all transpose via an RNA intermediate that is reverse-transcribed into a cDNA copy, which is then inserted into many different target sites.

Keith M. Derbyshire • Division of Infectious Disease, Wadsworth Center, New York State Department of Health, and Department of Biomedical Sciences, State University of New York at Albany, Albany, NY 12201-2002. **Nigel D. F. Grindley** • Department of Molecular Biophysics and Biochemistry, Yale University, New Haven, CT 06520-8114.

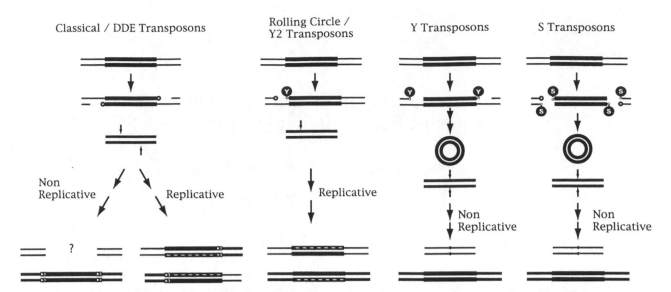

Figure 1. Comparison of the mechanisms of transposition utilized by the four transposon families. The figure highlights the major differences between transposition pathways; for specific details, see the text, Table 1, and Fig. 3, 7, 9, and 10. Heavy bar, transposon DNA; thin line, flanking donor DNA; thick line, target DNA; vertical arrows, sites of target cleavage; dashed bar, newly replicated DNA; boxed arrowhead, target duplication; circled Y or S, covalent phosphotyrosine or phosphoserine linkages of recombinase to DNA; open circles, exposed 3'-OH nucleophile. The fate of the donor DNA is not fully determined for nonreplicative transposition and is indicated by a question mark. Major differences to note are initial strand-cleavage events, transposase-DNA covalent linkages, role of DNA replication, circular intermediates, target-site duplications, and fate of the donor DNA.

Defining Transposon Families

To facilitate our discussion, we have divided the transposable elements into four families based on the proteins they encode for their mobility: (i) the DDE transposons, which include the majority of the classical bacterial elements such as IS3, IS50 (Tn5), IS10 (Tn10), Tn3, and phage Mu; (ii) the rolling circle transposons, which we call Y2-transposons and include IS91; (iii) the Y-transposons, which include the conjugative transposon Tn916; and (iv) the S-transposons, a newly recognized family that includes IS1535, IS607, and the mobilizable transposon Tn4451. The defining feature of each family is the catalytic motif (or nucleophile) encoded by the transposase protein, since this largely dictates the mechanism of transposition; our name for each family derives from this motif. To highlight the differences between families and similarities within a family, the basic transposition steps are compared in Fig. 1 and listed in Table 1. Thus, distinguishing features of the DDE transposons include transposase-mediated cleavage at the transposon ends by hydrolysis and the absence of covalent protein-DNA intermediates. Staggered DNA cleavages at the site of insertion also ensure that the new insertion is flanked by directly repeated sequences. By contrast, S-, Y-, and Y2-transposases form covalent protein-DNA intermediates, and target

duplications are not generated on insertion. Transposition of the Y2 elements requires replication of the entire transposon, while transposition of S- and Y-transposons does not involve any replication: it is a conservative process with the energy of all cleaved phosphodiester bonds maintained by the transient formation of covalent protein-DNA intermediates.

DDE transposons

The majority of bacterial transposons described to date encode a protein that contains a triad of acidic amino acids that are intimately involved in catalysis (29, 69, 93, 131). These three residues, collectively known as the DDE motif (Asp, Asp, Glu), coordinate divalent cations that play a key role in catalysis of the various phosphoryl transfer reactions during transposition. The DDE family of transposases also includes the integrases encoded by retroviruses and long terminal repeat retrotransposons (134) and transposases encoded by many other eukaryotic DNA transposons, such as Tc, Mariner, and P elements (39). It is clear from the wealth of biochemical data generated from work with members of this family that the chemistry of transposition is essentially identical, while the different products made by these elements are due to subtle variations that occur during strand cleavage. We discuss these in more detail below.

Table 1. Distinguishing characteristics of the four transposon families

Characteristic	Classical or DDE transposons		Y-transposons	S-transposons	Rolling circle or Y2-transposons
Active-site residues	DDE		Y	S[a]	YY
Related proteins	Retroviral integrases, RNase H, RuvC		λ-like tyrosine integrases	Serine, site-specific resolvases and invertases	φX174 A protein and Rep proteins of rolling circle plasmids
Examples	Mu, Tn3	IS10, IS50, Tn7	Tn916	IS607, Tn5397, Tn4451	IS91, IS801, IS1294
Is insertion associated with transposon replication?	Replicative integration	Replicative excision[b]	Nonreplicative	Nonreplicative	Replicative
Role of DNA replication	Copies element during insertion	Replication generates a circular intermediate; gap repair in new target	Resolution of mismatches in coupling sequences[c]	None	Displaces transposon strand in donor and copies transferred transposon strand in recipient
Does transposition involve formation of a protein-DNA covalent linkage?	No	No	Yes, to 3' P	Yes, to 5' P	Yes, to 5' P
Does transposition involve a circular intermediate?	No	Yes	Yes	Yes	Perhaps a single-stranded circle
Does transposition generate a target duplication?	Yes	Yes	No, but flanking coupling sequences are transferred to one side of target	No	No
Target specificity	Low	Low; exception is Tn7	Preferred sites but not site specific	Inserts at dinucleotides identical to that at transposon flank	Strong specificity for 5'-CTTG-3' and 5'-GTTC-3'
Initial reaction step involves:	Concerted nicks at 3' ends of transposon	Single nick at 3' end	Concerted nicks on one strand at each transposon end	Concerted double-strand break at each end	Single nick at 5' end of transposon

[a] No biochemical data exist for this family of elements. However, based on the resolvase/invertase family of recombinases, we would predict that this family would have the properties listed.

[b] Replication is not associated with insertion of these elements but is required for transposon excision.

[c] Mismatches are corrected by the host replication or mismatch repair machinery and are therefore not an active part of the transposition process.

Y2-transposons

Those elements that transpose via a rolling circle mechanism constitute the second transposon family (53). These elements encode a transposase that is related to a family of proteins that mediate replication of several phage and plasmids by a rolling circle mechanism. The transposon and phage-encoded proteins contain a pair of absolutely conserved tyrosines, which have been shown in protein A of ϕX174 to catalyze the initial strand cleavage event via a phosphotyrosine linkage (67). Genetic and biochemical data support a similar role for these two tyrosine residues in this family of transposases, which we call Y2-transposases (54). The Y2-transposons do not generate a target duplication, but it is clear that replication plays an intrinsic role in their transposition.

Y-transposons

The Y-transposons are those elements whose mobility depends on a tyrosine recombinase, that is, one related to the phage λ integrase. The Y-transposons, named for the tyrosine nucleophile, are mobilized by a two-step process. The initial excision is a true site-specific recombination (involving cleavage and joining of the two transposon ends), but the subsequent integration shows target-site promiscuity. In each step, phosphodiester bond energy is conserved during DNA cleavages by the formation of recombinase-DNA covalent linkages via a phosphotyrosine. The promiscuity at the integration step is in marked contrast to the high degree of specificity exhibited by a classical site-specific tyrosine recombinase, such as λ Int, in both excision and integration.

We note that the similarities of both site specificity and recombination mechanism of this group of elements to λ integration create a nomenclature problem: are they transposons or site-specific integrating elements? Here, we particularly wish to distinguish between those elements (e.g., Tn916) that behave as true transposons by inserting into many different sites, and those (e.g., Tn554) that, like λ, are highly target specific and thus do not meet our criteria of a transposon. We suggest "specifically integrating element" (spinel) as a more appropriate term for the latter elements.

S-transposons

S-transposons, like the Y-transposons, have adopted a site-specific recombinase for their transposase. The S-transposases all have a catalytic domain related to that of the serine recombinases, a family of recombinases exemplified by $\gamma\delta$ (or Tn3)

resolvase, or the Hin DNA invertase (58, 80). Like the Y-transposons, S-transposons move by a two-step (excision-integration) process, using an authentic site-specific recombination for excision, but exhibiting target-site degeneracy for integration. Both steps are anticipated to be wholly conservative, with recombinase-DNA phosphoserine linkages preserving phosphodiester bond energy during DNA cleavages.

The mechanisms of transposition for each of these four families are very different, although the end products may often look quite similar. Below we summarize their features, and illustrate the different ways a transposon can move from one site to another and the different types of chromosomal rearrangement they can create. Inevitably, there will be a strong bias toward the DDE family, as so much elegant work has been carried out with these transposons, but we have tried to highlight what we believe are the novel features of the other families, as they too play a critical role in chromosomal evolution.

THE DDE TRANSPOSASE FAMILY

The DDE motif was originally noted as a triad of amino acids conserved between the integrase protein of retroviruses and the transposases encoded by the IS3 family of elements, but it has since been identified in the majority of transposases described to date (29, 69, 93, 131). These include the transposases from not only the well-characterized elements (IS3 family, IS10, IS50, IS903, Tn7, Tn552, and Mu), but also the less well characterized insertion sequences, many of which have been described only by sequence analysis (see Table 1 in reference 29). The recent identification of a DDE-like motif in the Rag 1 protein, required for VDJ recombination, further emphasizes the role that this class of protein plays in genome rearrangements in both prokaryotes and eukaryotes (52, 88, 94).

The transposase protein recognizes specific sequences at the transposon ends, which generally take the form of short inverted repeats of between 10 and 40 bp. Some elements, such as Mu and Tn7, have multiple transposase binding sites at their ends, but their purpose is still the same: to facilitate the interaction of the two transposon ends to form a synaptic complex that ensures the coordination of the transposition reaction. These nucleoprotein complexes, or transpososomes, have been described for a number of elements (42, 113, 145, 182). Assembly of the transpososome plays a critical role in regulating the transposition reaction, bringing together all the key proteins and DNA substrates to promote a productive transposition event. A hallmark of DDE transposons is that insertions are flanked by short, directly repeated sequences of the target DNA; as we will see below, this is a direct consequence of the transposition reaction.

The DDE transposases (and retroviral integrases, an unfortunate and confusing terminology) are not closely related at the primary sequence level, and the three invariant acidic residues are found in small, noncontiguous patches of amino acids, which are weakly conserved within the different DDE transposon subfamilies. Their importance was originally demonstrated by site-specific mutagenesis, which abolished catalytic activity of both transposase (8, 17, 89, 120) and integrase proteins (48, 93, 177). This led to the suggestion that the DDE residues might coordinate a pair of divalent metal ions that are essential for catalysis, as described for other transesterification reactions (82, 112a). However, it was the X-ray crystal structures of transposases Mu and IS50 (42, 132) and retroviral integrases avian sarcoma virus and human immunodeficiency virus (20, 47) that clarified their role in transposition (133). Although each of the DDE residues is located on distinct structural elements in the protein, these structures are folded such that the three residues are clustered together to form an active site. Most remarkably, despite almost nonexistent amino acid sequence homologies between the proteins, the core structures of these proteins are almost superimposable around the active-site residues (133). Furthermore, the presence of a divalent cation coordinated by the two aspartate residues in several of the structures supports the model that the DDE motif is responsible for coordination of metal ions required for catalysis (18, 19, 99).

How Do DDE Transposons Move?

Until relatively recently, it was thought that DDE transposons moved by one of two alternative pathways, either nonreplicative to produce simple insertions, or replicative to form cointegrates (Fig. 1 and 2). IS10 was the prototypical nonreplicative (or cut-and-paste) transposon, while Tn3 and phage Mu (during its lytic growth phase) provided the replicative paradigm. A few elements, such as Mu and

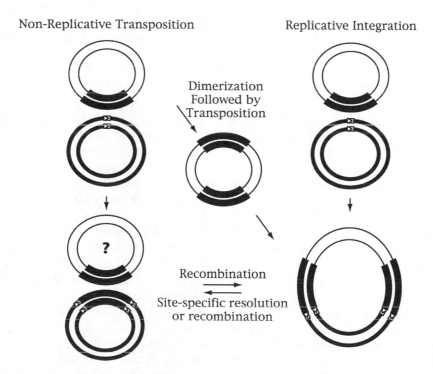

Figure 2. Intermolecular transposition can result in a simple insertion or a cointegrate, irrespective of the transposition pathway. Shown vertically are the expected transposition products for cut-and-paste and replicative integration. Each results in a target duplication, but only the replicative pathway duplicates the transposon and fuses the donor and target replicons. The question mark in the upper product for cut-and-paste transposition indicates that the fate of the donor DNA is unclear; the double-strand gap may be repaired, or the DNA may be degraded (see text and Fig. 6d for more details). Cointegrates can be reduced to form the target with a simple insertion and the initial donor, either by site-specific resolution or by homologous recombination between each transposon copy. Homologous recombination between donor and simple insertion product can also generate cointegrates. Dimerization of the donor plasmid followed by cut-and-paste transposition of the two transposons and intervening donor DNA (a composite transposon) will also result in a cointegrate. Symbols are as described in the legend to Fig. 1.

Tn*903*, were able to exploit both pathways (70, 183). A third transposition pathway has since been revealed by studies of members of the IS*3* family (141); these elements form simple insertions by a process that very likely involves insertional sequence (IS) replication and, in contrast to both IS*10* and Mu, certainly requires the intervention of host replication and repair activities in advance of the insertion into target DNA. Thus, we include the IS*3* family of elements in the category of those that require DNA replication (Table 1; also see below).

The analysis of transposition products in vivo resulted in the generation of many models for the actual mechanism of transposition (46, 60). Unfortunately, the effects of homologous recombination, multicopy plasmids, and dimerization tended to complicate these analyses and their interpretation. It was the advent of in vitro transposition systems starting in 1983 with Mu (111) that allowed the molecular details of transposition to be determined; this work has since been augmented by the establishment of similar systems for IS*10*, IS*50*, Tn*7*, and IS*911* (36, 68, 130, 141). The outcome from a particular transposition event depends on the location of the target with respect to the transposon ends and the transposition pathway. We will discuss how each influences the outcome in detail later, but first we will briefly review the biochemical steps that occur during transposition and why some transposons generate simple insertions, while others form cointegrates.

Although the actions of different DDE transposases differ in detail, all perform the following pair of critical biochemical processes: (i) they act as a hydrolytic endonuclease to cleave the transposon 3′ ends, creating a 3′-terminal OH, and (ii) they join these ends to new DNA in a single-step strand-transfer reaction using the 3′ OH as a nucleophile to directly attack a target phosphodiester (Fig. 3). An additional common feature is that a short 5′ stagger separates the targeted phosphates; replication or repair of this segment generates the short target duplication that flanks the transposon, the length of which is characteristic for each transposon. The major distinguishing feature between nonreplicative and replicative transposition is whether additional cleavages at the transposon 5′ ends sever all the connections of the transposon to donor DNA and therefore prevent its replication.

Transposition Involving DNA Replication

The Mu paradigm

The first biochemical step in replicative transposition of Mu (and presumably of Tn*3*, Tn*552*, and

similar replicative transposons) is the transposase-mediated cleavage of both 3′ ends (Fig. 3). The two 3′ OHs then act as nucleophiles to attack the opposing strands of a target, resulting in the fusion of donor and target DNAs joined by a single copy of the transposon. This strand-transfer intermediate was first postulated in transposition models in 1979 (5, 154) and was subsequently identified and characterized in a series of elegant biochemical studies in the early 1980s with the transposon Mu (40, 111, 112). The Y-shaped (or branched) structures at each end of the transposon are identical in structure to replication forks and can assemble the host replication machinery (26, 117). Subsequent replication of the entire transposon (from one or both ends) resolves the intermediate into a cointegrate—a fusion of the donor and target replicons with a copy of the transposon at each junction (Fig. 2 and 3).

The IS*911* paradigm: one-ended transposition and a circular IS intermediate

As revealed by studies of IS*911* and IS*2*, members of the IS*3* family of elements use an alternative pathway of transposition—one that is very likely replicative but does not form cointegrates (97, 126, 152). Although the principal product of transposition is a simple insertion, the transposase alone is incapable of excising the IS and instead requires the early intervention of the host replication (or possibly repair) machinery to generate a novel transposition intermediate: an IS circle (Fig. 3).

In contrast to other transposons, the first biochemical step is asymmetric and involves a one-ended transposition event. A nick is made at one of the transposon ends to generate a free 3′ OH. This acts as a nucleophile in a strand-transfer event, in which the target is the other end of the element. Strand transfer occurs just outside the transposon (3 nucleotides for IS*911* and IS*3*, and 1 nucleotide for IS*2*) on the same DNA strand (Fig. 3). This generates a figure-eight molecule, in which one strand of the transposon is a covalently closed circle with the two transposon ends separated by a few nucleotides. The rest of the strand remains unsealed. The figure-eight intermediate is processed by the host to release a double-stranded, covalently closed IS circle. This probably occurs by replication (either initiated at the free 3′ end at one branch point, or as a result of passage of a chromosomal replication fork through the branched structure), although it is possible that resolution may also occur by a repair process (for example, by a Holliday junction resolvase). A linear transposon is generated from the IS circle by transposase nicking at each transposon end. The resulting exposed 3′ OHs attack

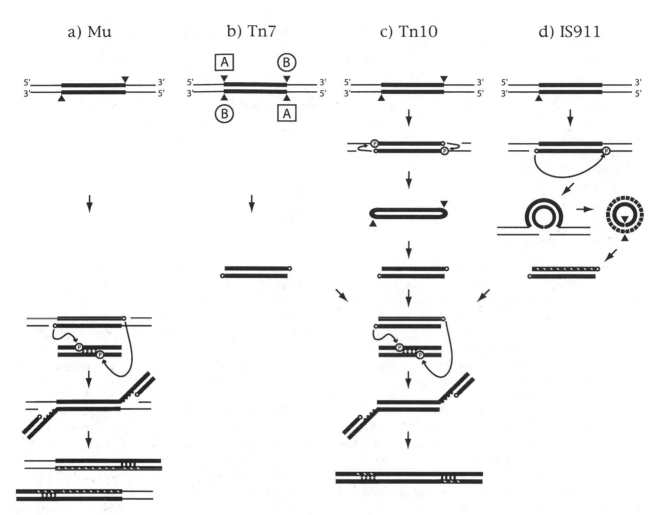

a) Mu   b) Tn7   c) Tn10   d) IS911

Figure 3. Comparison of DDE transposition mechanisms (see text for details). Although, for simplicity, transposons are represented as straight lines, all transposase-mediated processes occur within a complex of two transposon ends and transposase. Cleavage at the ends of Tn7 (but not of the other elements shown) also requires the presence of the target DNA. Symbols are as described in Fig. 1. A and B for the Tn7 pathway indicate the TnsA and TnsB proteins required for cleavage at the 5′ and 3′ ends of the transposon, respectively; black triangles indicate the sites of transposase-mediated hydrolysis; scissile phosphates (P in a circle) are the sites of concerted cleavage and strand joining by the indicated 3′-OH nucleophile (connecting arrows). Note that for IS911 transposition, the initial strand-transfer event generates a figure-eight molecule, which is probably resolved by replication (not shown) to regenerate the parent molecule and a transposon circle with left and right ends separated by a few nucleotides. Nicking at each transposon end then linearizes the transposon circle. Adapted from reference 69.

a target DNA to give a simple insertion product. Although the IS911 integration step is nonreplicative, replication plays a key role by processing the initial strand transfer to form the substrate for integration. By contrast, Mu and Tn3 employ the replication apparatus during the integration step of transposition, i.e., replicative integration.

Although evidence for the IS911 mode of transposition has only been obtained with IS911, IS2, and IS3 (141), there are several descriptions of non-IS3 family transposons that form a transpositionally hyperactive junction of their left and right ends (usually observed as a head-to-tail tandem repeat of the IS).

These transposons include the prototype members of the IS21, IS30, and IS256 families (29). Since such structures are readily formed from plasmid dimers by one-ended transposition following the IS911 model (see Fig. 6), it seems likely that these transposons will move by a similar mechanism.

Nonreplicative (Cut-and-Paste) Transposition

In nonreplicative transposition, the transposase cleaves not only the 3′ transposon ends but also both 5′ ends. This releases the element from the donor DNA to generate an excised transposon (174). The transposon is then integrated into a target site by the concerted action of the 3' OHs at the ends of the transposon on a suitable target DNA. The biochemistry of the strand-transfer reaction is identical to that for replicative transposition. However, the role of replication is limited to the repair of the short terminal gaps that are generated by the staggered cleavage of the target DNA. One of the most interesting aspects of nonreplicative transposition is the different ways that elements have evolved to sever the 5' flanking DNA (Fig. 3).

The IS*10* paradigm: terminal hairpin intermediates

Cleavage at the 5' ends of IS*10* and IS*50* is achieved via a hairpin intermediate in a series of transposase-catalyzed steps that follow, and are dependent on, the initial 3'-end cleavages (Fig. 3) (14, 85). The free terminal 3' OHs directly attack the backbone at the 5' ends of the transposon. This is mechanistically equivalent to the strand-transfer step of integration, except that the target is the complementary strand. Hairpin formation simultaneously releases the flanking DNA and excises the transposon with a 3'-5' phosphodiester hairpin structure at each end. The transposase resolves the hairpins by nicking at the 3' ends of the transposon, a repeat of the first hydrolysis step, to generate free 3' OH ends, in preparation for target attack. Thus, transposition of IS*10* and IS*50* involves a pair of reiterated steps: hydrolysis, transesterification, hydrolysis, and transesterification (86). Elucidation of this mechanism also explained earlier results, which demonstrated that a single IS*10* transposase monomer carried out all the strand cleavage and transfer steps at each transposon end (17). The issue at hand here was how a single active site could cleave two DNA strands with opposite polarities. The solution lies in the formation of the hairpin, whose formation allows release of the flanking DNA, but, more important, allows hydrolysis at the same 3' end by the same active-site residues to generate an excised transposon.

Hairpin intermediates have also been described in VDJ recombination (108, 178). In fact, there are many parallels between VDJ recombination and transposition (38, 139). The recombinase proteins RAG1 and RAG2 cleave via a hairpin intermediate; however, the polarity is different from that seen with IS*10* and IS*50*, as the nick occurs at the 5' ends of the "transposon," and therefore the hairpin is formed on the flanking DNA. More recent work has identified a DDE-like motif in RAG1 and has shown that mutation of these residues prevents recombination (52, 88, 94).

The Tn*7* paradigm: a 5' endonuclease

Tn*7* is a rather unusual transposon that encodes five proteins involved in transposition (TnsA, B, C, D, and E) (36). Unlike other transposons, Tn*7* uses two different proteins, TnsA and B, to carry out the cleavage and joining steps of transposition. TnsB contains a DDE motif and, like other transposases, is the subunit responsible for terminal binding, 3'-end cleavage, and strand transfer (15, 148). Transposon excision is achieved by cleavage of the 5' flanking DNA by the TnsA subunit; however, TnsA is not a DDE transposase. The crystal structure of TnsA has revealed that it is related to type II restriction endonucleases (71). Under nonstandard in vitro conditions, TnsAB will carry out cleavage and strand-transfer steps without the other added Tns proteins (15), confirming that together they form a heterodimeric transposase (148). However, transposition in vivo and under standard in vitro conditions requires all the Tns proteins and target, indicating that transposition occurs in a coordinated fashion, presumably within a large nucleoprotein complex (148).

The difference between replicative and nonreplicative transposition is actually a rather subtle one, as illustrated by the effect of mutations in both Tn*7* and IS*903*. Endonuclease-defective mutants of TnsA converted Tn*7* from a nonreplicative to a replicative transposon (107). The presence of the defective TnsA protein allowed 3'-end cleavage and strand transfer by TnsB, but the TnsA mutations prevented 5'-end cleavage and therefore transposon excision (Fig. 4). This resulted in the formation of classical branched strand-transfer (or Shapiro) intermediates, which were identified and characterized in vitro. Furthermore, when these TnsA mutants were used in vivo, cointegrates were formed, suggesting that these intermediates were processed in the same way as other replicative transposons. Thus a single amino acid change quite dramatically altered the type of transposition product generated. A similar switch between pathways has been described for IS*903* (167, 168). IS*903* transposes predominantly by a simple insertion pathway, but amino acid substitutions of residues that map around the putative active site of the transposase increased cointegrate formation. In addition, mutation of either the terminal nucleotide of the transposon or the flanking nucleotide increased replicative transposition. These results led to the suggestion that the DNA and protein mutants were allowing normal 3'-end cleavage but delaying cleavage of the 5' flanking DNA, presumably by disturbing the conformation of the active site and substrate. It was proposed that

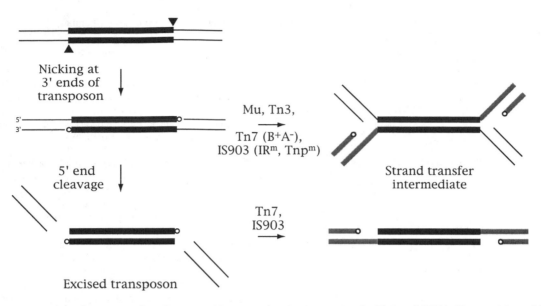

Figure 4. Switching from cut-and-paste transposition to replicative integration by Tn7 and IS903. Transposition of both elements is consistent with efficient cleavage of the 5' ends of the transposon to generate an excised transposon (shown on the left), which is then integrated to form a simple insertion. TnsA mutants (B$^+$ and A$^-$) of Tn7 fail to cleave at the 5' ends of Tn7 and form strand-transfer intermediates. Mutations in IS903 located either at the transposon termini (IR^m) or close to the transposase active site (Tnp^m) result in elevated levels of cointegrate formation consistent with formation of the strand-transfer intermediate.

the delay in 5'-end cleavage increased the chance of strand transfer by the longer-lived 3' nicked intermediate, resulting in replicative transposition.

What Types of Rearrangements Are Made by DDE Transposons?

Intermolecular transposition

The product of intermolecular replicative transposition mediated by Mu or members of the Tn3 family is a cointegrate, in which donor and target replicons are fused by two copies of the transposon (Fig. 2). By contrast, simple insertions are generated by both the IS3-like elements and those that transpose via a nonreplicative pathway (Fig. 2 and 3). However, in vivo each pathway can lead to the formation of both cointegrate and simple insertion products (Fig. 2)! Although Mu forms stable cointegrates, Tn3 and Tn552 encode a resolvase (a site-specific recombinase), which is able to reduce a cointegrate into its two constituents: the donor replicon and the target now with a transposon insertion (58). Conversely, if a plasmid containing a nonreplicative element such as IS10 dimerizes, a composite transposon is formed by the two copies of the IS. Transposition of this composite transposon (containing a copy of the entire donor plasmid) into the target will result in a cointegrate.

Intramolecular transposition

Transposition events are normally depicted as intermolecular hops (Fig. 2). However, transposons also mediate intramolecular rearrangements, and this is particularly relevant when considering how transposons influence chromosome structure (Table 2). Replicative transposition mediated by Mu and the Tn3 family of elements generates adjacent deletions and adjacent inversions when an intramolecular target is selected (Fig. 5); the outcome simply depends on transposon-target strand connections. If each 3' end of the transposon is joined to the same DNA strand at the target, a deletion results; if they are joined to the opposite strand, an inversion is generated. In vivo both of these events can be detected; however, for an adjacent deletion, only one of the two circular products will contain a plasmid or chromosomal origin and be retained by the cell.

The situation with elements that integrate in a nonreplicative fashion (IS3 family and IS10) is more complicated. Nonreplicative transposons (IS10 and IS50) can rearrange adjacent DNA to make deletions and inversions, which are readily detected in vitro, only if the target is located within the transposon. This is simply because a double-strand break at each transposon end excises the element and therefore prevents it from rearranging adjacent DNA. For

Table 2. Comparison of the intramolecular products generated by the transposon families

Transposon family	Mode of transposition	Example(s)	Adjacent deletions	Adjacent inversions	Comments
DDE	Replicative integration	Mu, Tn3	Yes	Yes	Products depend on strand connections. No loss of DNA with inversions. Only deletion circles containing *oriV* are retained.
DDE	IS3 replicative resolution	IS3, IS911	Yes	Yes	Requires interaction of ends from two different transposons: a composite transposon. Inversions (and deletions) are accompanied by deletion of the DNA segment between the two interacting ends.
DDE	Nonreplicative	IS10, IS50	Yes	Yes	Requires interaction of ends from two different transposons, e.g., a composite transposon, or perhaps between opposite ends of IS copies on sister chromatids. If originating from a composite Tn, inversions (and deletions) are accompanied by deletion of the DNA segment between the two interacting ends.
Y2	Rolling circle	IS91	Yes	No	Predict adjacent deletions possible for one-ended events. Inversions are not topologically viable.
Y	Conservative	Tn916	Yes	Yes	Predict adjacent deletions and inversions possible if a pseudo-end, present on the same chromosome, participates in recombination. Potentially, deletions could either leave the Tn in place, or delete it, and inversions may or may not be accompanied by Tn inversion.
S	Conservative	IS607	Yes	Yes	Predict adjacent deletions and inversions possible (see above).

these elements to make biologically meaningful rearrangements (closed circular DNAs with an origin of replication), ends from two separate elements must interact and select a target within the composite transposon. This allows one end of each element to maintain the connection to the host DNA (Fig. 6a). Excision of the composite element separates the transposon DNA from the adjacent flank, which, as a linear DNA, is degraded and lost from the cell. Insertion of the ends of the excised composite transposon into an internal site will generate either an adjacent deletion or a deletion-inversion. As most bacterial chromosomes are circular and often contain multiple copies of the same element, they can be imagined as containing many composite transposons, each of which is capable of undergoing a rearrangement. However, unlike replicative events, there is always loss of some flanking DNA (the "e f" segment in Fig. 6a) and this will obviously limit the extent, and number, of rearrangements that the cell can sustain before deletion of essential genes occurs.

In a further twist, in vitro studies with IS10 have shown that IS10 ends on two different DNA

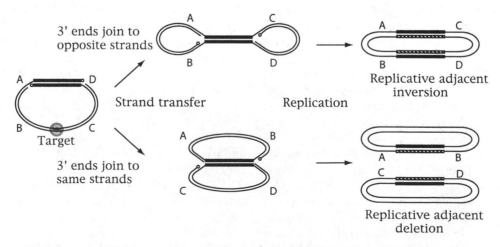

Figure 5. Intramolecular transposition by elements that integrate replicatively generates adjacent deletions or adjacent inversions. The outcome of this event is determined simply by transposon-target strand connections. Symbols are as for Fig. 1. A, B, C, and D are four hypothetical genes used to depict the inversion event.

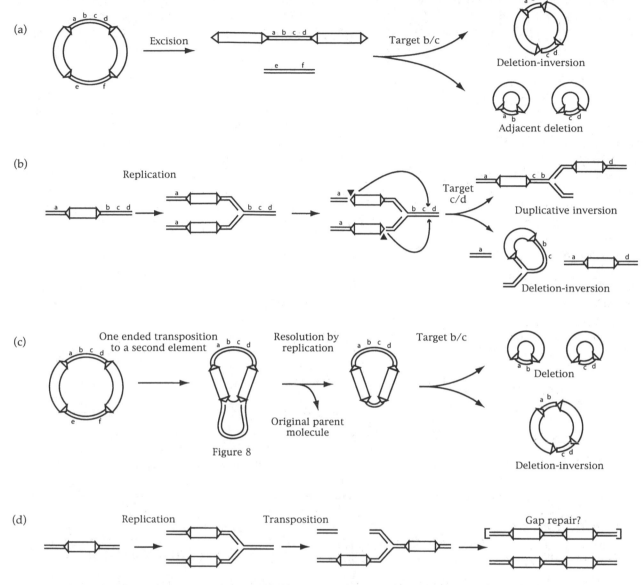

Figure 6. Intramolecular transposition. The figure shows how nonreplicative (a and b) and IS*911*-like (c) elements can generate adjacent deletions and inversions even though each forms an excised, linear transposon intermediate. In each case, transposition involves a composite transposon, or ends from two transposons in the sister chromosome pathway (b). Note that duplicative inversion, as shown in panel b, is not associated with any DNA loss, in contrast to deletion-inversions (a and c). Cut-and-paste transposons can also transpose to intramolecular targets if transposition is associated with replication (d). The fate of the donor is unclear (indicated by brackets). The chromosome may be degraded, or it may be rescued by gap repair using the sister chromosome as a template.

molecules can synapse and transpose to a new site. This suggests that two opposite ends of IS*10* copies from sister chromosomes may also be able to transpose in an intramolecular fashion (28). The product from such events can be either an adjacent deletion or an inversion, depending on which strand connections are made (Fig. 6b). These in vitro experiments provided the first molecular insight that might explain the rare adjacent deletions observed in vivo for IS*10* and IS*50* (79, 137).

At first glance, the formation of the excised circular intermediate observed with the IS*3* family of transposons would seem to preclude them from forming intramolecular rearrangements. However, as for the IS elements described above, if more than one IS*3* element resides on a chromosome, then the same repertoire of rearrangements can be generated (175). In this case, the 3′ OH created by the asymmetric cleavage at one end of the element would be joined not to its other end, but rather to the end of a second element (Fig. 6c). This would result in two abutted elements with ends positioned correctly for nicking to form the linear excised transposon, which could then insert at a target within the element to generate a viable product.

Target-Site Selection

Although DDE transposons generally insert into multiple sites, many exhibit varying degrees of target-site preference. In vitro assays have shown that target choice is primarily determined by the transposase protein (57, 83, 114). The one notable exception is Tn7, which encodes two additional proteins that determine very different types of insertion site (36). In addition to the specific items discussed below, evidence is accumulating that many other host factors and processes can influence transposon insertion specificity. These include cellular processes such as DNA replication, transcription (43), and supercoiling. For example, DNA gyrase mutants affect the level of IS50 transposition into plasmids (98), and conjugation influences targeting of both Tn7 and IS903 (76, 187).

Some Transposons Prefer To Insert at Specific DNA Sites

The ability of transposons to insert into different sites without any obvious sequence preference makes them ideal for use as insertional mutagens. However, careful inspection of many independent insertion sites (obtained both in vivo and in vitro) shows that most elements exhibit some sequence preferences, although the preferences are often rather weak. A hallmark of DDE transposons is that they generate a target duplication of specific length on insertion (Fig. 3). Compilation of many of these duplications has allowed degenerate consensus target sequences to be generated for some elements. For example, IS10 prefers to insert at 5' NGCTNAGCN 3' (62), IS50 prefers 5' AGNTYWRANCT 3', especially when it is embedded within related sequences (57), and Mu prefers 5' CYG/CRG 3' (63, 114). Each of these sequences is palindromic, reflecting the symmetry of the two transposase-bound ends. In addition, more recent genetic and biochemical studies have shown that the target recognized by some transposases is significantly longer than the sequence duplicated and that DNA structure is likely to play a role (63, 76, 128).

In vitro studies with IS10 transposase target-DNA complexes have shown that the transposase contacts about 24 bp, including the insertion site (128). By using different chemical footprinting probes, it was inferred that the transposase made base-specific

contacts to the central 9 bp of the target (before duplication), while flanking sequence contacts were primarily to the phosphate backbone, and lacked sequence specificity. In addition, two sites of enhanced sensitivity were identified on either side of the consensus core, indicative of deformation in the DNA structure. It was proposed that these deformations would facilitate transposase binding to both the central core and the flanking DNA, and may also promote strand transfer by activating the phosphate backbone for nucleophilic attack.

Compilation and statistical analysis of many insertion sites for IS903, Mu, and Tn552 has identified symmetric target sites and suggests that flanking sequences contribute to targeting efficiency (63, 76, unpublished results). More comprehensive genetic analyses with IS903 have shown that the preferred target is 21 bp long, palindromic, and can be dissected into three components: the flanking sequence, the target duplication, and the dinucleotide pair that is cleaved on transposon insertion (77). In contrast to the studies with IS10, this analysis showed that the contributions of the flanking sequence to targeting were more significant than those within the 9-bp target duplication. The requirement of the preferred sequence on either side of the 9-bp target is also consistent with target selection's being mediated by a transposase dimer, or multimer, making symmetrical interactions to both sides of the target. Alteration of the dinucleotide pair that is cleaved on transposon insertion also affected insertion efficiency, perhaps by allowing adoption of a structure that is more conducive to cleavage and insertion.

Proteins That Influence Target-Site Selection

Two transposon-encoded proteins select different targets for Tn7 insertion

Although it is an exceptional case, Tn7 provides the best example of how transposon-encoded proteins influence target-site selection. Tn7 transposes by two alternative pathways, depending on which subset of transposition proteins are used (36). In combination with TnsAB (the transposase) and TnsC (the transposase-activating protein), TnsD targets Tn7 to the unique attTn7 site, while TnsE facilitates transposition to non-attTn7 sites.

In the TnsD-dependent pathway, TnsD binds specifically to a site on the bacterial chromosome adjacent to attTn7 (7), inducing a DNA distortion that is recognized and bound by TnsC (92). TnsC, bound to the minor-groove side of the DNA, is thought to form a "platform" that promotes binding of TnsAB and the paired transposon ends to the opposite face of the insertion site, and activates transposition (92).

The TnsE-dependent pathway is significantly less efficient (and less well understood) than the TnsD pathway. It directs insertions of Tn7 preferentially into conjugative plasmids, specifically in recipient (rather than donor) cells and always in the same orientation. A variety of experiments suggest that TnsE, a DNA binding protein, targets structures formed during lagging strand DNA synthesis, perhaps the 3' ends of the extending strands (123). Following conjugal transfer, the single-stranded transferred DNA is converted to a double-stranded form exclusively by lagging strand DNA synthesis in the recipient cell (50). Thus, the presence of multiple 3' ends and the replication apparatus on the transferred DNA could provide recognition sites for the TnsE protein (123). This hypothesis is also consistent with regions of the chromosome associated with double-strand breaks being targeted via the TnsE pathway, since the breaks will also be recognized by the repair and replication apparatus (124).

Selection of a suitable target for Tn7 is clearly a highly evolved process, both ensuring the stable maintenance of Tn7 via integration into the chromosome and enhancing its dissemination to different bacterial hosts by targeting conjugating DNAs. *attTn7* is highly conserved (at the 3' end of the *glmS* gene) and is therefore present in a diverse set of organisms (35, 36). Thus, targeting to *attTn7* via the TnsD pathway provides a "safe haven" for the transposon and has no obvious deleterious effect on the host since insertion does not disrupt expression of *glmS*. We note that targeting of Tn7 is in stark contrast to that of other transposons: Tn7 appears to have evolved to minimize disruption of host genes, while the others will insert almost anywhere and deal with the consequences later.

Immunity

Another way for transposition proteins to influence target use is via the process of target immunity—the phenomenon in which a target, containing a copy of the transposon or even just a single transposon end, is refractory to further insertions of the same transposon. Immunity affects only targets containing the transposon: the overall level of transposition to other DNA molecules or replicons is unaffected. Only certain transposons exhibit immunity: e.g., the Mu, Tn3, and Tn7 families (3, 4). It is thought that immunity plays a key role in protecting a transposon from the damaging effects of its own transposition. The molecular details for immunity were first described for the transposon Mu (2, 3) and have since been shown to occur in a similar fashion for Tn7 (164).

Efficient transposition of the bacteriophage Mu is mediated by the MuA and MuB proteins (26). MuA is the DDE transposase, while MuB is an activator of MuA that binds DNA nonspecifically in the presence of ATP. A DNA bound by MuB is the preferred target for Mu insertion. If two targets are available, one with and one without MuA binding sites (i.e., Mu ends), then in the presence of MuA, MuB accumulates on the target without an end. This is because MuA binds to the Mu end and, by direct interaction with MuB, causes the latter's rapid dissociation from that DNA, making it a poor target for Mu insertions. Immunity in Tn7 parallels that of Mu, with TnsB and TnsC playing the roles of MuA and MuB. Although the TnsC and MuB proteins are not related at the amino acid level, they do share similar features. Both are ATP-dependent, nonspecific DNA binding proteins, and ATP hydrolysis stimulated by their cognate transposases causes them to dissociate from DNA.

The levels of immunity exhibited by Mu and Tn7 differ quite dramatically. Mu immunity occurs over a region up to 25 kb from a Mu end (103). This is sufficient to prevent insertion of Mu into itself during the lytic cycle (when multiple transposition events occur), since immunity extends from each end of the 39-kb Mu genome. By contrast, immunity exhibited by Tn7 occurs over much greater distances. A single Tn7 element confers immunity to further insertion in the 65-kb plasmid pOX38. Even more dramatically, a Tn7 end 190 kb from an *attTn7* site on the bacterial chromosome can inhibit Tn7 insertion threefold (44). Although not absolute, it shows that protein-protein interactions that mediate Tn7 immunity can occur at great distances along the bacterial chromosome. Given the two pathways Tn7 has evolved to select very different and specific targets (see above), it is not immediately clear why it should exert immunity for distances that are so much bigger than the length of Tn7 itself. Perhaps Tn7 immunity is designed to protect the host in addition to preventing self-inactivation. If a chromosome, or plasmid, were to receive a second copy of Tn7, then the intervening DNA would be subject to the same rearrangements mediated by cut-and-paste transposons (Fig. 6). For transposons of the Tn3 family, with very efficient cointegrate resolution systems, immunity may well function primarily to protect the host (and the transposon) from the potentially disastrous consequences of resolvase-mediated recombination between two transposon copies.

Chromosomal Consequences of Transposition

What is the fate of the double-strand break generated by excision of the cut-and-paste transposons? This question still has not been rigorously explored. There are two likely scenarios: (i) loss of the broken chromosome by degradation or (ii) gap repair. However, since bacterial replicons (plasmids and chromosomes) are generally present in multiple copies in actively growing cells, degradation of a broken chromosome is unlikely to be as deadly as it may seem. Gap repair of a broken DNA chromosome has been thoroughly documented in eukaryotes, where the sister chromosome can be used as a template for gene conversion (49, 125). Although double-strand-break repair (using a second copy of the genomic locus as template) is thought to be a rare event in *Escherichia coli*, such events have been described for Tn7 by using a *recD* mutant host to reduce exonucleolytic degradation (64) (Fig. 6d).

In both the repair and chromosome loss scenarios, coordination of transposition with DNA replication would enhance cell survival by ensuring the presence of a second intact chromosome for viability or gap repair. Consistent with this, elegant genetic studies with IS10, IS903, and IS50 demonstrated that their transposition is coupled to DNA replication via methylation of *dam* (GATC) sites within the transposon. When these sites are in their hemimethylated form (i.e., for a brief period following replication), transposition is activated (136). In these cases the transposition process per se is not replicative, but a new copy of the element is maintained at the old site and so the overall process is duplicative (Fig. 6d). The IS911 mode of transposition also generates an excised linear transposon (Fig. 3); however, it does not generate a double-strand break at the original donor site. This is because replication of the early figure-eight intermediate regenerates the original donor molecule, in addition to releasing an excised circular transposon.

Irrespective of the mode of transposition, intermolecular events (depicted in Fig. 1 and 2) clearly provide an opportunity for transposon dissemination via hops onto bacteriophage and transmissible plasmids. Although most events are likely to be harmful or neutral, some will clearly be beneficial to the host. For example, insertion of a transposon upstream of a gene can lead to its activation if it carries an outward reading promoter. In addition, there are several reports of transposon movement occurring at a higher frequency in colonies growing under carbon source limitation, suggesting a role for transposons in facilitating host survival under stressful conditions (104, 129, 153, 155). Transposition also plays a

significant role in the transfer of nontransposon DNA in bacteria (22). For example, in a donor strain carrying coexisting plasmids, transposition that results in a cointegrate between a conjugative plasmid and a smaller nonconjugative plasmid can result in mobilization of the nonconjugative plasmid to a recipient cell (62). Perhaps most significantly, transposition between a conjugative plasmid and the chromosome to form a cointegrate would create an Hfr strain capable of transfer of chromosomal DNA into a recipient strain. Indeed, such events have been clearly documented for IS21, which mediates cointegrate formation between the conjugative plasmid R68.45 and the host chromosome (186).

Replicative transposition can also facilitate acquisition of new plasmid or phage DNA genes in a recipient cell (transconjugant or transductant). While simple insertion of a transposon would result in integration of only the genes carried by that element, replicative transposition would result in complete integration of the entire plasmid or phage into the host genome. This would be most important when the incoming plasmid or phage cannot replicate in that particular host: unless integrated into a viable replicon, the genetic information carried on the incoming DNA would be lost.

Intramolecular events play a major role in defining the organization of the bacterial chromosome, but their regulation is also critical to the host, which must prevent excessive transposition and, in particular, adjacent deletions of potentially essential genes. Of the two events, intramolecular inversions are clearly less deleterious, as less genetic material is likely to be lost. In addition, they can also be construed as more advantageous to the host as each new inversion product forms a new composite transposon (Fig. 6a). Interestingly, integration host factor has been shown to bias intramolecular transposition of IS10 toward inversion events (27, 160), which suggests that the host has evolved mechanisms to regulate the type, and level, of transposition that occurs in the cell.

Finally, all transposons can also facilitate chromosomal rearrangements in the absence of transposition, when they act as substrates for RecA-mediated homologous recombination. The presence of multiple transposon copies on chromosomes and plasmids can lead to integration, deletion, and inversion events (140). Therefore, even defective elements can be responsible for DNA rearrangements.

Y2-TRANSPOSONS

The Y2-transposons, typified by IS91, IS1294, and IS801, share some unique and unusual characteristics

(Fig. 1 and Table 1) (53). These elements exhibit a very strong preference to insert 3′ of the sequences 5′-CTTG-3′ or 5′-GTTC-3′, with the same end of the transposon always inserting adjacent to the target. Furthermore, there is no target duplication. As might be expected for the insertion orientation bias, the two transposon ends are dissimilar and play very different roles in transposition (109). These rather unusual characteristics, along with the fact that their transposases are not related to DDE transposases, imply that these elements move by a novel mechanism.

The amino acid sequence similarities of the transposase proteins to the Rep proteins of single-stranded DNA (ssDNA) phages and plasmids that replicate via a rolling circle mechanism provided the first clue as to the likely mechanism of transposition (110). There are five shared sequence motifs among these proteins; one, in particular, contains a pair of conserved tyrosines, which are absolutely conserved in both the transposases and the ϕX174 gene A protein; the plasmid Rep proteins appear to be more distantly related, as they share only a single conserved tyrosine (45). In the gene A protein of ϕX174, these two residues are directly involved in both the initial strand-cleavage event, in which one of the tyrosines is covalently joined to the 5′ end of the nick site, and in the termination of rolling circle replication, when the single-stranded phage DNA is circularized (67). Substitution of these residues in the IS91 transposase abolishes transposition and indicates they are likely to play a catalytic role in transposition (54).

The DNA sequences at the ends of these transposons are also atypical. They are not inverted repeat sequences, nor do they contain related multiple transposase binding sites. Instead, each end contains a series of unrelated inverted repeats, suggesting that the transposase recognizes each in a unique fashion. Most strikingly, the end inserted adjacent to the target site bears a strong resemblance to the origin of ssDNA phage and rolling circle plasmids at both the sequence and secondary structure levels, further supporting a rolling circle mechanism for transposition. The non-ori-like end is not essential for transposition. However, in one-ended transposition the products generated are not discrete insertions, but instead extend into the flanking DNA, ending at CTTG or GTTC sequences found in the adjacent vector DNA. This led to the suggestion that this end behaves more like a terminus of transposition, and was termed ter (109).

Models for Y2 Transposition

Two models have been proposed for Y2 transposition. These are based on the similarities to ssDNA phage and rolling circle plasmid replication systems and the transposition products observed in vivo (see below).

In a model proposed by de la Cruz et al. (53, 109), each active-site tyrosine acts as a nucleophile in a transesterification reaction that nicks a DNA strand, resulting in a covalent linkage between transposase and DNA (Fig. 7). Transposition is initiated by transposase-mediated cleavage at the ori end of the IS, resulting in a 5′ phosphotyrosine linkage and exposing a 3′ OH in the flanking donor DNA (Fig. 7b). A similar reaction joins the second tyrosine to the 5′ side of the insertion site; i.e., donor and target are cleaved in a concerted fashion. The free 3′ OH in the target attacks the phosphotyrosine linkage at the 5′ end of the transposon, joining one transposon strand to the target and releasing the tyrosine (Fig. 7c). Replication initiated from the 3′ OH in the donor flank displaces the target-linked transposon strand (Fig. 7d and e). Once the replication-transposase complex reaches the ter end, a reversal of the first pair of cleavages occurs. The free tyrosine cleaves at the 3′ end of ter after the terminal CTCG (for IS91), linking to the 5′ side of the nick (Fig. 7f). The exposed 3′ OH then attacks the phosphotyrosine junction at the target site, releasing the tyrosine and covalently joining the 3′ OH of the transposon to the 5′ end of the target (Fig. 7f and g). The looped-out transposon is then replicated by the host machinery.

The model proposed by Tavakoli et al. (166) uses the same premise but differs in key steps (Fig. 7h to m). In this model, the cleavage of the donor and the cleavage of the target are not coupled, and a free single-stranded transposon intermediate is formed. After nicking at the ori end, DNA replication displaces the transposon strand (Fig. 7h and i). On reaching ter, a transposase-mediated, ssDNA endonuclease activity is hypothesized to cut the DNA precisely after the CTTG sequence (of IS1294). This generates a free 3′ OH at the 3′ end of the transposon and no phosphotyrosine linkage to the donor flank (in contrast to the first model) (Fig. 7j). A single strand of the transposon is released, presumably held together as a non-covalent circle by the transposase. The free 3′ OH of the released transposon strand is proposed to attack a target DNA in a transesterification reaction joining the 3′ end of the transposon to the 5′ end of the target (Fig. 7k). The newly exposed 3′ OH of the target in turn attacks the phosphotyrosine linkage at the 5′ end of the transposon, resulting in a single-stranded insertion (Fig. 7l). Thus, this model also differs in how the transposon is joined to the target: by a one-step transesterification as opposed to a two-step mechanism involving a phosphotyrosine linkage.

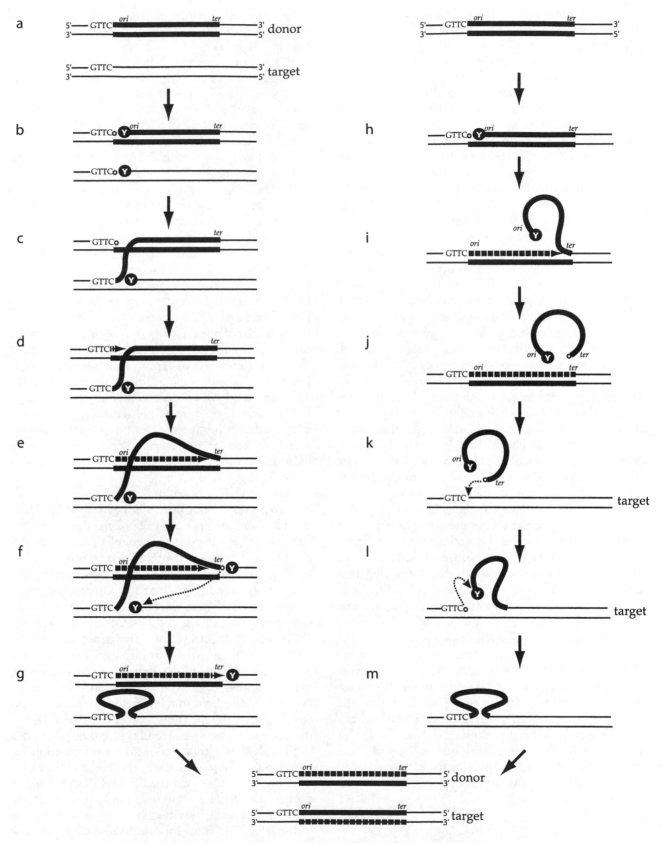

Figure 7. Two models for rolling circle transposition. Left, the model of Mendiola et al. (109); right, the model of Tavakoli et al. (166). See the text for details. The Y2-transposase, which binds to the *ori* end of the transposon, cleaves 3′ to the sequence GTTC. *ter* is the second site of transposase cleavage and defines the 3′ end of the ssDNA form of the transposon.

One unsatisfactory aspect of this second model is that it fails to explain why two tyrosines are so highly conserved and are essential for transposition of IS91 (54). A modified version combines aspects of both models, and fulfills the requirement for two tyrosines by utilizing the second tyrosine for both *ter* and target cleavage. In addition, it provides a satisfactory explanation for the close proximity (four residues apart) of these two tyrosines in one active site: to hold the different DNAs together and to coordinate transposition. In this model, transposase mediates cleavage at *ter* via the second active-site tyrosine to release the 3' end of the transposon (instead of using a hypothetical ssDNA endonuclease activity [Fig. 7i and j]). The flank-linked tyrosine must be released from the donor DNA to allow the transposase-ssDNA transposon complex to seek a target. Tyrosine release is presumably achieved by attack of the Y-covalent linkage by the 3' OH of the newly replicated transposon strand. Target cleavage and insertion could occur either via the free 3' OH (Fig. 7k to m) or, as depicted in Fig. 7b, by using the free tyrosine as the attacking nucleophile, followed by transposon insertion, as shown in Fig. 7c and f (but without a transposon-donor connection). Adding weight to this model, single-strand circles of IS91 with a unique polarity have been detected in vivo, and production of these circles depended on the critical tyrosine residues of the transposase (54). However, the ssDNA circles observed were covalently closed, and no evidence was presented to show they were true intermediates in transposition.

Clearly, reconciliation of these models will require development of an in vitro system to determine which of the predicted key intermediates are formed. However, each model is consistent with one key aspect of Y2 transposition: the target site is not duplicated on insertion because only a single strand of the target is cleaved.

Y2 Transposition Products Observed In Vivo

Intermolecular transposition

Three types of intermolecular product are observed in vivo. The predominant product is a simple insertion. This is predicted from each model and requires precise cleavage at each end of the transposon. Replicon fusions of the donor and target occur at a lower frequency than simple insertions for each of the Y2-transposons (from 20% to fewer than 1% of simple insertions). The fusions are of two forms, either standard cointegrates (Fig. 2) or multiple tandem insertions of the donor plasmid into the target. The rolling circle model satisfactorily explains for-

mation of each fusion if the recognition of *ter* is incomplete. If *ter* is not recognized, replication and strand displacement of the entire donor replicon would occur; termination on reaching *ter* for the second time would generate a cointegrate, while tandem insertions would occur if *ter* was recognized after several rounds of rolling circle replication. In contrast to cointegrates generated by DDE transposons, where replication is limited to transposon DNA, replication of the entire donor plasmid would be required to form this product. Finally, the models can explain the products observed in the absence of a *ter* end (so-called one-ended transposition), in which DNA flanking the "non-*ori*" end of the transposon is also transferred to the target site. In these cases, strand displacement occurs until a *ter*-like cleavage sequence (CTTG or GTTC) is detected in the flanking DNA, and then cleavage and strand transfer occur (109). The ability of this family of elements to ignore their normal *ter* sequence, but still use pseudo-*ter* sites in adjacent DNA, provides them with a unique opportunity to acquire new functions flanked by new *ter* ends. Thus, they can constantly evolve and, in contrast to most other transposons, are not constrained by defined ends. Intriguingly, a new family of rolling circle-like transposons, called Helitrons, have recently been described in eukaryotes (84). These elements encode a transposase that has both the conserved tyrosine motif and, at its C-terminal end, a helicase motif, likely acquired by the above mechanism.

As an intriguing aside, we note that the Y2-transposases are distantly related to relaxases of conjugative plasmids, which become attached to the 5' end of *oriT* via a highly conserved tyrosine and then mediate the transfer of a single strand into the recipient (96). Is it possible that in a recipient cell following DNA transfer, instead of recircularizing with the 3' end of *oriT* to regenerate a circular plasmid, the transferred strand could be inserted into the host chromosome and the newly inserted DNA could thus create a novel rolling circle transposon (Fig. 7k to m)?

Intramolecular transposition

Theoretically, intramolecular transposition of Y2 elements should be able to occur without DNA rearrangements, as the precise translocation of a single strand to a second location via a replicative process precludes the deletion, or inversion, of the intervening DNA that is observed with DDE transposons. In fact, the products would look like simple insertions from a second replicon, making it difficult to deduce the origin of such events. This was also the conclusion from an analysis of secondary insertions into a transposon-containing plasmid (13).

Closer inspection of the models allows us to predict that Y2 elements are also capable of making adjacent deletions but not inversions (Table 2). These deletions would be of DNA adjacent to the *ori* end of the transposon, and would be formed by one-ended (i.e., *ori*-mediated) transposition events. In the first model, following transfer of the 5′ transposon end to an intramolecular target on the same strand, replacement strand synthesis from the 3′ OH would simply displace a strand of DNA circularized between *ori* and the target. Complementary strand synthesis of this circle would generate a product that would form a viable adjacent deletion if it contained the plasmid origin of replication. Inversions are not possible, as opposite strand joining would generate a nonviable product. Similarly, in the Tavakoli model (166), once a free, circular strand is generated (Fig. 7j), an adjacent deletion would be formed if the 3′ OH attacked the phosphotyrosine linkage to covalently seal the single strand rather than attack a target. We note that these are analogous to the ssDNA circles observed in vivo by de la Cruz and colleagues (54). In the Tavakoli model, the absence of the second strand in the intermediate precludes inversions. A carefully established genetic selection should allow detection of these products and provide further genetic evidence to support the models presented.

Y-TRANSPOSONS

Y-transposons consist of a broad, rather heterogeneous group of elements that have co-opted a "site-specific" recombinase for their mobility (for a sampling, see Table 3; for a comprehensive review, see reference 32). The defining feature of these transposons is that they encode a tyrosine recombinase, essential for transposon insertion and excision. The prototypes of this group of elements, Tn916 and the very closely related Tn1545, behaved as true transposons, with multiple insertion sites. Thus, the subsequent discovery that the Tn1545 transposase, essential for its excision and integration, was a homologue of λ integrase (127) came as a considerable surprise. Furthermore, a second Tn1545-encoded gene product that appeared to be weakly homologous to several phage-encoded Xis proteins strongly stimulated excision (127). These similarities to the λ phage integration system provided a simple explanation for several uncharacteristic features of these transposons: their lack of terminal inverted repeats, their failure to duplicate a short target sequence upon

Table 3. A sampling of Y-transposons

Element	Original host	Description[a]	Transposition genes	Target specificity	Reference(s)
Tn916	*Enterococcus faecalis*	Conjugative, Tet[r]	*int, xis*	Many sites (tracts of A's and T's preferred), hot spot in *Clostridium difficile*	32, 33
Tn1545	*Streptococcus pneumoniae*	Conjugative, Tet[r], Erm[r], Kan[r]	*int, xis*	Many sites in *E. coli* (tracts of A's and T's preferred)	23, 127
CTnDOT	*Bacteroides* spp.	Conjugative, Tet[r], Erm[r]	*int, exc (orf4),* and regulatory genes *rteA, B, C*	Relatively site specific: ~7 sites in *Bacteroides*	30, 31
Tn4555	*Bacteroides vulgatus*	Mobilizable, Cfx[r]	*int, xis, tnpA, tnpC*	A primary target (80% of insertions) and many secondary targets (20%) in *Bacteroides fragilis.* Primary site targeting requires TnpA (compare TnsD in Tn7)	171, 172
NBU1	*Bacteroides* sp.	Nonreplicating *Bacteroides* unit, mobilizable, cryptic	*int,* several other genes needed for excision	Single site in *Bacteroides,* several sites in *E. coli*	158, 159
SXT[b]	*Vibrio cholerae* O139	Conjugative, Sul[r], Cam[r], Trm[r], Str[r]	*int, xis*	Single site in *Vibrio* and *E. coli*	74
Tn554[b]	*Staphylococcus aureus*	Spc[r], Erm[r]	*tnpA, tnpB* (both *int* homologues), *tnpC* (affects frequency and orientation of insertion)	Single site in *S. aureus;* insertions at secondary sites are very rare (compare λ phage)	10

[a] Tet, tetracycline; Erm, erythromycin; Kan, kanamycin; Cfx, cefoxitin; Sul, sulfamethoxazole; Cam, chloramphenicol; Trm, trimethoprim; Str, streptomycin; Spc, spectinomycin.
[b] As these are site-specific integrating elements, they would be classed as spinels according to our nomenclature.

insertion, and their ability to excise precisely from a target, not only regenerating the original target sequence but also forming a circular transposon (23, 32, 55).

Most Y-transposons are from gram-positive bacteria but are functional in *E. coli*. One particularly interesting property of these transposons is that many of them contain genes for their conjugative transfer, or mobilization by other conjugative systems (32). Thus, these elements have acquired the ability to transfer not only to different DNA sites within a cell, but also between cells. The Y-transposons exhibit a wide variety of target preferences that range from relatively nonspecific to highly specific, although, as with phage λ (184), even the highly specific transposons occasionally insert into secondary targets. The high target specificity of some (e.g., SXT and Tn554 [Table 3]) suggests that not all should be considered transposons (after all, we do not call phage λ a transposon), and thus we propose the term "site-specific integrating element" (spinel) to describe these elements. Nevertheless, the promiscuous target selectivity of others, particularly Tn916 and Tn1545, lead us to treat at least these examples as transposons. Since Tn916 and Tn1545 are also the most studied, we focus on these two in describing the behavior and properties of Y-transposons. It seems likely that most other members of the family will behave in a rather similar manner, differing in details but not in general processes.

The Proteins and Sites for Y-Transposon Movement

Tn916 and Tn1545 contain two genes needed for transposition (32, 127, 165). The *int* gene encodes the tyrosine recombinase (or Y-transposase), and the adjacent (and upstream) *xis* gene encodes a small basic protein needed for efficient excision. Int and Xis are the only proteins required for the excision reaction in vitro, although the low efficiency of this reaction may indicate that other (possibly host-encoded) proteins might also play a stimulatory role (142). Xis appears not to be important for integration (105, 165).

Tn916 Y-transposase, like phage λ Int (but in contrast to the Cre or Flp tyrosine recombinases), is a bivalent protein with two independent domains that bind distinct DNA sequences (100). The C-terminal domain, which contains the catalytic active site, binds the sites of recombination—the junctions between the ends of the transposon and the flanking DNA. The N-terminal domain binds to short repeated sequences within the element located about 150 bp from the left end and 90 bp from the right end (Fig. 8) (100). These sets of sequences can be considered analogous to the "core"- and "arm"-type sequences found in the λ *att* sites that are bound, respectively, by the C- and N-terminal domains of λ Int (6, 115). Tn916 Xis is a DNA bending protein that binds to sites close to the two transposon ends (Fig. 8) (32, 142). The Xis site near the left end, which is needed for efficient excision, is between the arm- and core-type sites and probably facilitates simultaneous occupancy of pairs of arm- and core-type sites by single Int subunits. Surprisingly, the Xis site near the right end is further from the transposon end than the arm-type sites (Fig. 8); here, Xis appears to play a regulatory role by competing with Int binding to the adjacent arm-type site (32, 72).

How Do Y-Transposons Move?

It was recognized that transposition of Tn916 differed from classical transposition in significant ways well before the realization that it encoded a tyrosine recombinase. Conjugative transfer of a plasmid carrying Tn916 caused the transposon to excise at a high frequency, restoring the function of a plasmid gene inactivated by the original insertion. In many, but not all, cases, excision was accompanied by insertion of Tn916 into new target sites. This behavior (and its similarity to the zygotic induction of phage λ) led Gawron-Burke and Clewell (55) to propose that transposition occurred by a two-step process involving precise excision of the transposon (resealing the target DNA and perhaps forming a transposon circle), followed by insertion into a new target site, conjugative transfer to a new host, or loss.

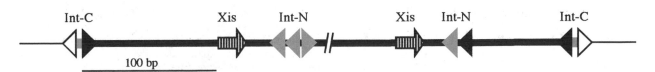

Figure 8. The ends of Tn916, showing binding sites for the transposition proteins. The heavy line is Tn916; the thin lines are the flanking DNA. Binding sites for the two domains of Int (the Y-transposase) are shown as arrowheads: black and white indicate sites for the C-terminal catalytic domain (the white arrowheads are the variable sites acquired from each target DNA); gray indicates sites for the N-terminal arm-site binding domain. Barred arrows indicate the Xis binding sites.

Several years later, the free circular form of Tn916 was detected in vivo, and purified circles were shown to retain the ability to transpose, consistent with their proposed function as a transposition intermediate (151). The excision step has now been recapitulated in an in vitro reaction (144).

All the data obtained thus far, from studies of the process and consequences of transposition of Tn916 or Tn1545 (32), are consistent with a strand exchange mechanism essentially identical to that of a prototypical tyrosine recombinase. This process, elucidated by thorough biochemical analysis of phage λ integration (6, 95, 119), has been dramatically illustrated and confirmed by the elegant structural studies of the tyrosine recombinase, Cre, in a series of complexes with loxP that represent the various stages of a recombination cycle (176). The anticipated mechanism for Tn916 excision is shown in Fig. 9. It involves two sequential cycles of single-strand cleav-

age, switching of the free ends, and strand rejoining. Cleavages are carried out by direct attack of the recombinase active-site tyrosine on the scissile phosphodiester. The sites of cleavage on the two strands of each duplex are generally separated by a 6-bp segment called the overlap region or the coupling sequence. Rejoining is accomplished by attack of the 5'-OH DNA termini on the phosphotyrosine recombinase-DNA joints. Each recombinant product contains a 6-bp coupling sequence that is heteroduplex, meaning that its two (complementary) strands are derived from different parental duplexes. (Here and throughout this chapter, the term heteroduplex simply indicates that the two strands of a duplex have different hereditary origins; it does not imply that the duplex contains one or more mismatches, although mismatches may occur.) The first recombination cycle (Fig. 9, top row) forms a Holliday junction. This is then resolved by the second cycle (bottom row) to

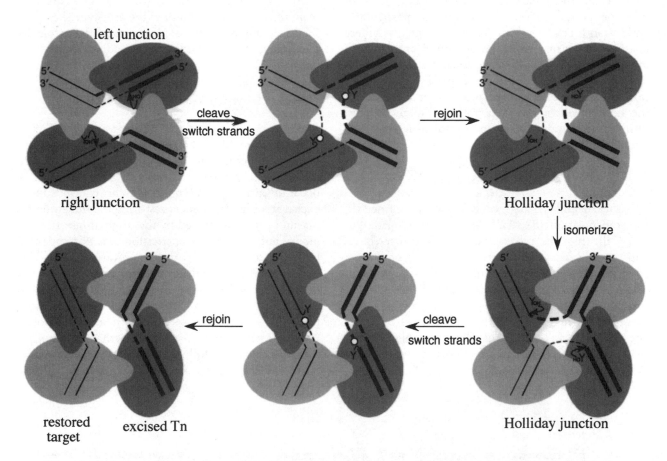

Figure 9. Mechanism of Y-transposon excision. The heavy lines are the transposon ends, and the thin lines are the flanking donor DNA. The dashed portions represent the coupling sequences—the 6-bp duplex segment between the cleavage sites. Y_{OH}, the free tyrosine nucleophiles; Y-, the covalent tyrosine-DNA linkages; O, free 5' OHs positioned to attack the phosphotyrosine covalent linkages. Curved arrows show the nucleophilic attack of the Y_{OH} on the DNA cleavage site. The shaded asymmetric shapes represent the Y-transposase; dark subunits are in the active conformation, light subunits are inactive. Note the switch of both subunit activities and DNA conformations at the isomerization step. Transposon insertion involves synapsis of an excised transposon and a target site and occurs by reversal of the entire process. Adapted from reference 176.

form the recombinant products, the excised transposon circle, and the restored integration site. Insertion of the transposon involves the same overall process in reverse, and is initiated by the capture of a suitable target DNA in a synaptic complex with the joined transposon ends.

While the illustrated mechanism is well suited to the site-specific processes of Cre-mediated recombination, its adaptation for transposition to many target sites raises several significant questions. Among these are (i) how are sequence mismatches within the heteroduplex segment at the center of the crossover site dealt with, and (ii) how are target sites recognized and captured, without unacceptable levels of recombination between two targets?

The mismatch problem

In typical reactions performed by tyrosine recombinases, the recombinant product contains a 6- to 8-bp heteroduplex consisting of one strand from one partner and the complementary strand from the other. The formation of this does not pose a problem for the site-specific recombinases, since the sequences of both partners are identical. However, in the case of λ recombination, it has been shown that artificially introduced mismatches are inhibitory to strand exchange, and altered sequences in this region can result in new target specificity (90, 121, 184). In the Cre-*loxP* structures, the 5′ ends of the crossing strands that form the Holliday junction base pair with the complementary strand of the partner duplex; mismatches would inhibit this base pairing and would be likely to alter the position of the attacking 5′ OH, preventing the joining reaction (176). A similar problem would arise if mismatches were encountered during the Holliday junction resolution step.

Analyses of Tn*916* and Tn*1545* targets and junctions indicate that the Y-transposase cleavage sites usually are separated by a 6-base 5′ stagger (143, 173). This conclusion was supported by using purified Int protein in vitro to map the cleavage sites, although some variation of the site of cleavage was observed at the right end (169). Excision or integration of these elements should thus form a 6-bp heteroduplex and, indeed, such a heteroduplex has been detected in unreplicated Tn*916* circles (24). However, neither excision nor integration frequencies appear to depend on identity between the two flanking coupling sequences (105, 173). Furthermore, Trieu-Cuot et al. (173) have shown examples of multiple insertions into a single target in which both transposon orientations are equally represented, even though one orientation provided perfect cou-

pling sequence complementarity, while the other was completely mismatched. If the coupling sequence is indeed 6 bp, the only possible explanation is that base pairing of the attacking 5′ base is simply not important for strand joining. This is a rather remarkable conclusion, given what we know from the in vitro studies of other tyrosine recombinases. If, however, there is truly some variation allowed in the length of the coupling sequence (to 7 or 8 bp), as implied by the biochemical results of Taylor and Churchward (169), then base pairing of at least one 5′ end remains a possibility.

Target choice

Tn*916* does not insert randomly and in some cases shows pronounced site specificity; for example, there is a single hot spot in *Clostridium difficile* at which >90% of insertions occur (181). Nonetheless, Tn*916* is capable of inserting into many different sites (e.g., four different insertions into the *cat* gene of pIP501 have been detected [150; see also reference 173]). A target-site consensus sequence that strongly resembles the sequence of the joint between the right and left ends formed upon Tn*916* excision has been derived. This consists of [TTTTTN$_{(5-6)}$AAAAA] but known insertion sites may differ from this consensus at several positions (75, 150). Target specificity appears to be determined by Int recognition of the portion of the target that flanks the central 6-bp coupling sequence and appears not to be affected by the actual sequence of the latter (173).

The similarity of target sites to the joint formed by transposon excision raises the following potential problem. If the free Y-transposase protein were to recognize targets of transposition as recombinase binding sites, transposase-mediated synapsis between two target sites (or between a target and one transposon end) potentially could initiate recombination, resulting in deletion or inversion of the intervening DNA segment—a highly undesirable consequence. To avoid this, it seems likely that targets are recognized efficiently only by the complex formed when transposase binds to the R-end/L-end joint that is formed upon transposon excision.

A precedent for such behavior is provided by phage λ. *attB*, the equivalent of the target site for λ integration, is not recognized as a recombinational partner either by another *attB* or by *attL* or *attR*; rather, *attB* appears to be captured only by the *attP*-Int presynaptic (intasome) complex (135). The structure of the *attP* intasome is determined by the presence of multiple Int binding sites, the ability of Int to bridge arm and core sites, and the influence of DNA bending proteins such as Xis and integration host factor (6).

Capture of *attB* is presumably achieved by the unoccupied C-terminal domains of a pair of arm-bound Int protomers that extend from the intasome. The recent demonstration that the affinity of the core-binding C-terminal domain is substantially increased by interaction of the N-terminal domain (in *cis*) with an arm site provides an explanation for the ability of the intasome, but not free Int, to capture *attB* (146).

It seems likely that Tn916 employs an analogous regulatory process, since, like phage λ Int, the Tn916 Y-transposase is a bivalent recombinase, able to bind both core and distinct arm sequences within the element (100). The requirement for arm-type sequences for excision has been demonstrated, but their role in integration has yet to be explored.

Anticipated Genetic Consequences of Integrated Y-Transposons

Integrated Y-transposons would be expected to influence local chromosome dynamics in a variety of ways, resulting both from actions of their recombinase activities at a transposon end and from the conjugation functions (or mobilizability) associated with most of these elements.

Site-specific recombination between two sites on the same DNA molecule results in either deletion or inversion of the intervening segment, the outcome depending on the orientation of the sites. For a simple monovalent tyrosine recombinase such as Cre or Flp, the recognition of a recombination site is essentially twofold symmetrical and does not involve the central overlap region. Thus, synapsis between two identical sites can occur in either orientation, but is productive only in the orientation that results in complementary overlap heteroduplexes. If the sequence of the overlap region is palindromic, both deletion and inversion can occur.

The principal purpose of the interaction of a Y-transposase with its transposon ends is to precisely excise the transposon in preparation for its transfer and subsequent integration. Thus, it is likely that the bivalent nature of the Tn916 transposase, and the presence of the arm-type binding sites within the ends of the element, ensures that excision is the predominant reaction. Nevertheless, since the Y-transposase of Tn916/Tn1545 appears to ignore the complementarity of the partner coupling sequences during recombination, it is not particularly surprising that inversion of a Tn916 insertion can occur (122). The efficiency of inversion was low, but it is not known if the efficiency of excision was any higher. The observed inversion suggests that these controlling signals can be ignored or overridden, raising the possibility that a transposon end might interact productively with a target site rather than with a second end (Table 2). Such interactions would result in inversions or deletions and, even if relatively infrequent, could have significant effects on chromosome organization.

Many of the Y-transposons encode their own conjugative functions. One would expect an integrated conjugative element to mobilize gene transfer from its integration site just as an integrated F plasmid gives polarized transfer of the *E. coli* chromosome in an Hfr strain. This is precisely what is observed with the spinel SXT (73). Surprisingly, Tn916 appears not to mobilize adjacent genes (51). It appears that such mobilization is prevented by an elegant control mechanism. Expression of the transfer genes depends on a promoter at the opposite end of the transposon that directs transcription outward. Following excision, transcription from this promoter reads across the newly joined recombination site into the *tra* genes (25). These authors also suggested that insertion of Tn916 in the appropriate orientation downstream of a promoter might uncouple *tra* gene expression from transposon excision.

S-TRANSPOSONS

The S-transposons are a group of elements that use a serine recombinase to mediate their mobility. The prototypical serine site-specific recombinases, γδ resolvase and the Hin invertase, are both small, two-domain proteins of about 190 amino acid residues with an N-terminal catalytic domain (about 130 residues) and a C-terminal DNA binding domain. The S-transposon-encoded transposases differ from these recombinases, falling into at least two new groups with alternative domain structures.

The prototypes of the first group, Tn4451 and Tn5397 (Table 4), were originally isolated from *Clostridium* spp. (1, 116). Both transposons encode a type of serine recombinase that has become known as a "large resolvase" (TnpX of Tn4451 is 707 residues, and TndX of Tn5397 is 533 residues). In each case, the N-terminal portion contains an easily recognized serine recombinase catalytic domain of the standard size. However, the roles of the oversized C-terminal domain(s) and the location of the DNA binding domain remain unknown. This is also true for other "large resolvase" proteins, including the enzymes for integration and excision of several phages (e.g., φC31 and R4) (106, 170), and SpoIVCA, the recombinase responsible for excision of the SKIN element during *Bacillus subtilis* sporulation (149).

The second group of S-transposons contains several elements initially identified from genomic

Table 4. A sampling of S-transposons

Element	Original host	Description	Transposition gene	Transposase size (amino acids)	Target specificity	Reference
Tn4451	*Clostridium perfringens*	6.3 kb, Cam^r, mobilizable	*tnpX*	707	Several sites in *E. coli*	9
Tn5397	*Clostridium difficile*	~20 kb, Tet^r, conjugative	*tndX*	533	Single site in *C. difficile*, multiple sites in *B. subtilis*	180
IS607	*Helicobacter pylori*	2.0 kb, only 2 ORFs^a	*orfA*	217	No obvious specificity	87
IS1535	*Mycobacterium tuberculosis*,	2.3 kb, only 2 ORFs	Rv0921	193	Unknown	34
	Methanococcus jannaschii		MJ0014	213	Unknown	21
ISC1904	*Sulfolobus solfataricus*	1.9 kb, only 2 ORFs	*orf1*, SSO3171	192	Unknown	156

^{a}ORF, open reading frame.

sequences (Table 4). Of these, only the *Helicobacter pylori* element IS607 has been demonstrated to transpose (87). The S-transposases of this group, although similar in size to the prototypical serine recombinases, have a reversed domain structure—an N-terminal DNA binding domain (about 55 amino acids) and a typical C-terminal catalytic domain.

Target-site information for the S-transposons is rather limited. Nevertheless, despite clear site selectivity on the part of Tn4451 and Tn5397, these and IS607 exhibit sufficient promiscuity to be considered "true" transposons.

Sites, Proteins, and the Mechanism of S Transposition

The S-transposase is the only transposon-encoded protein needed for transposition, being sufficient for both the excision and insertion steps (41, 87, 101, 179). As predicted, mutation of several catalytically crucial residues conserved throughout the serine recombinase family (including the serine nucleophile) eliminates transposase activity. In contrast to the Y-transposons, there is no requirement for a specific cofactor for excision. The extent of the sites needed for transposition remains to be determined.

As expected for elements that have adopted a site-specific recombinase, S-transposons (like Y-transposons) transpose by a two-step excision-insertion pathway, forming a circular transposon as an intermediate. Precise excision of the transposon is readily detected, forming a fully restored target site and a circular transposon with left and right ends joined (9, 41, 179; H. Ickes and N. D. F. Grindley, unpublished results). The preformed circular transposon also integrates at elevated frequencies, consistent with its anticipated role as a transposition intermediate.

All experimental data with the three prototype S-transposons are fully compatible with a standard serine recombinase-promoted strand exchange process, as shown in Fig. 10 (58, 59, 162). Excision of the transposon would be initiated by synapsis of the left and right ends, mediated by a dimer (at least) of the S-transposase at each end. Synapsis activates a cycle of double-strand DNA cleavage, switching of ends, and religation. Concerted cleavages are carried out by direct attack of the catalytic serine residues on the scissile phosphodiester bonds, and generate double-strand breaks with a 2-base 3' extension and with the serine covalently attached to the recessed 5' ends. After strand switching, the breaks are sealed by attack of the 3'-OH ends on the phosphoserine linkages. A single cycle is sufficient for transposon excision and target restoration. Integration involves repeating the catalytic cycle within a newly formed synaptic complex between the left end-right end junction of the excised transposon and a new target site.

Again, adaptation of a site-specific recombination process for transposition to many target sites raises the questions: (i) how are potential mismatches within recombinant heteroduplex regions dealt with, and (ii) how are target sites recognized and captured without promoting target-by-target recombination?

The mismatch problem

Serine recombinase strand exchange creates a 2-bp heteroduplex at the center of the crossover site, with one strand derived from each parent. The artificial introduction of mismatches into this region inhibits recombination by resolvases or DNA invertases (78, 81, 163). Not surprisingly, therefore, the one feature of S-transposon insertion targets that has always been conserved is the sequence of the

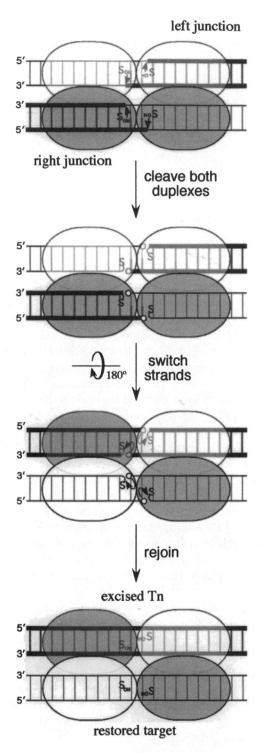

Figure 10. Mechanism of S-transposon excision. The heavy lines are the transposon ends, and the thin lines are the flanking donor DNA. The cartoon shows the two DNA duplexes separated by a tetramer of S-transposase catalytic domains, as proposed in a recent model for synapsis by $\gamma\delta$ resolvase (147). The shaded subunits are bound to the junction at the transposon's right end. S_{OH}, the free serine nucleophiles; S-, the covalent serine-DNA linkages; O, free 3' OHs. Curved arrows show the coordinated nucleophilic attacks of S_{OH} on the DNA cleavage sites (top panel) or 3' OHs on the phosphoserines (third panel). Note that in this

flanking dinucleotide—for Tn*4451* and Tn*5397* it is GA, while for IS*607* it is GG.

Crellin and Rood (41) have examined the effect on Tn*4451* excision of mutating one of the two flanking dinucleotides. In most cases, a potential mismatch reduced excision and stabilized the insert. The one exception was a change of GA to TA, which increased excision, suggesting that the TA sequence enhanced TnpX activity and more than compensated for the inhibitory effect of the mismatch. This was confirmed by showing that Tn*4451* with two flanking TA dinucleotides was excised much more frequently than the GA-flanked wild-type element. Thus, in contrast to the Y-transposases, the S-transposases are sensitive to the sequence at the center of the crossover site. Further analyses using a mismatched pair of recombination sites showed that the excision products had either of the flanking sequences, suggesting that the initial product contained the expected heteroduplex.

Another important role for the central dinucleotide of the S-transposons may be to promote excision rather than inversion in situ. The actual crossover sites of serine recombinases are functionally symmetric and can synapse with another site in either of the two possible orientations. Recombination would then produce a mixture of deletions (i.e., excision) and inversions unless other features of the system cause a specific bias. In resolution (*res*) sites, a bias toward a deletion-specific synapse results from the effect of resolvase bound to the adjacent accessory sites; in inversion sites, the bias toward inversion depends on the interaction of the enhancer-bound accessory protein FIS (for reviews, see references 58, 80, and 161). It is possible that transposon excision (rather than inversion) may also be promoted by as yet undetected extra interactions of the transposases with the transposon ends, or by additional factors provided by the host. Alternatively, the nonsymmetric nature of the flanking sequences may well limit productive recombination to the deletion reaction, since an attempt at inversion would result in a mismatched heteroduplex inhibitory to the joining step.

Target choice and avoiding the undesirable consequences of relaxed target recognition

There is very little information concerning the target recognition by S-transposons. A limited analysis

model, strand switching is accompanied by switching of the recombinase catalytic domains, since the two are covalently linked. Transposon insertion involves synapsis of an excised transposon and a target site, and occurs by reversal of the entire process. For further details, see the text.

of target sites for Tn4451 and Tn5397 suggests that their targets resemble the joint formed between the transposon ends upon its excision (41, 181). This, together with the rather pronounced specificity of Tn4451 and Tn5397, is consistent with the idea that target recognition is similar to transposon end recognition, but more relaxed. The extent of the sequences required for transposition is also unknown for Tn4451 and Tn5397, so we cannot tell whether these transposons employ a bivalent S-transposase (like the Y-transposase of Tn916) and arm and core binding sites to regulate the capture of target sites and minimize recombination between pairs of target sites or a transposon end and a target site. It may be relevant that the "integrase" of phage φC31, which is also a "large" serine recombinase (but is site specific), is the sole protein required for phage integration, and that the recombination sites, attP and attB, are both small (39 and 34 bp, respectively) (61, 170). This suggests that a tetramer of the recombinase (a dimer at each site) may be sufficient for synapsis and recombination.

For IS607 and its much smaller S-transposase, the situation is a little clearer. The size and domain structure of the S-transposase essentially eliminate the possibility that it recognizes more than one recognition sequence. Moreover, preliminary deletion analyses of the extent of the sites needed for transposition indicate that the excised transposon with only 40 bp from each transposon end integrates with undiminished efficiency (Ickes and Grindley, unpublished). Despite its adoption of a site-specific recombinase as the transposase, IS607 exhibits no obvious target specificity. Ten independent insertions into the E. coli F plasmid are at different sites, and comparison of these sequences (or the 20 half sites they represent) shows no readily discernible conservation, or similarity to the IS607 ends (87).

How might the S-transposases avoid recombining pairs of target sites? φC31 may again provide a clue, at least for Tn4451 and Tn5397. Although the φC31 recombinase binds all of its attachment sites with similar affinities, it is only able to recombine the attP × attB combination. Thorpe and Smith (170) propose that the attP and attB sequences each impose a conformation on the recombinase that allows this pair (and no others) to synapse and recombine. IS607 seems to present a somewhat different problem, since target sites do not seem to look like the joined transposon ends; how does the transposase recognize the target at all? One intriguing possibility is that a transposase tetramer assembles on the end-end joint of the excised IS607, and this is then able to capture a relatively random target sequence using a very much relaxed recognition process (that may or

may not employ the unoccupied pair of DNA binding domains). Clearly, much more research is needed before we can hope to answer these questions and provide an accurate picture of the actual processes of recombination-site synapsis and strand exchange.

Consequences of Integrated S-Transposons

Site-specific recombinases are likely to have evolved to be highly precise in their DNA transactions. However, an element that has evolved to insert with relaxed specificity (perhaps for the purpose of survival in a broad range of host strains) may have to pay the price of reduced precision. We have found with IS607 in E. coli that excision is occasionally imprecise, joining one correct end of the transposon to an illegitimate site, resulting in a deletion in one of the products (and presumably a reciprocal insertion in the other [Table 2]) (N. D. F. Grindley, unpublished observations). Inversions have not been detected (nor have they been sought), but since the transposition system shows no topological specificity (i.e., it can promote both excision and integration), there is no a priori reason not to expect the formation of inversions as well. It also seems probable that analogous recombinational events will be seen at a low frequency with other S-transposons.

In addition to their transposition functions, some S-transposons, like many Y-transposons, carry genes for conjugation or mobilization (Table 4), and thus can ensure their rapid dissemination via horizontal transfer. Such transposons also have the potential to mobilize the flanking DNA sequences, although it is not yet known if they do so, or whether they, too, encode regulatory mechanisms that inhibit such activity.

WHAT TYPES OF NOVEL TRANSPOSASES MIGHT BE DISCOVERED IN THE FUTURE?

In this chapter we have compared and contrasted four families of transposable elements found in bacteria. Each family has developed distinct ways of translocating defined segments of DNA, although there are some unifying themes and many of the end products look identical. All of the transposase proteins recognize specific sites at the transposon ends, pair ends together in synaptic complexes, and cleave and join multiple DNA strands in highly coordinated ways. They have also evolved ways to hold onto the cleaved transposon ends in large nucleoprotein complexes either noncovalently or by protein-DNA covalent linkages (Y2- and S- and Y-transposases).

We anticipate that yet other novel transposons will inevitably emerge both from experimental research and from analyses of genome sequence, and that these novel elements will move in new and mysterious ways. Regardless, their transposases will have to coordinate strand cleavages and strand transfer in order to maintain chromosome and transposon integrity. Thus they will need to bind specific terminal sequences, cleave DNA without letting go, and capture a target for insertion. Can we anticipate their form? A clue might come from examining potential ancestors of the known transposases, and then identifying other proteins that could similarly evolve into transposases. The DDE transposases are structurally and mechanistically related to RNase H and RuvC, suggesting they may have evolved from phosphoryl transferases (47, 134). The Y2-transposases are likely to have evolved from the replication initiation proteins of single-stranded bacteriophages (53). In addition, not only are the S- and Y-transposases modified site-specific recombinases, but the Y-recombinases, as a group, are also homologues of type 1 topoisomerases (157).

Clearly, enzymes involved in DNA metabolism are the most likely candidates to evolve into a transposase. Restriction endonucleases would satisfy at least some of the above criteria, as these proteins recognize and cleave specific sites. However, to evolve into a transposase, a typical endonuclease would have to cleave outside its binding site so that it can keep hold of the free DNA ends. Furthermore, to allow coordinated transposition of a segment of DNA, the endonuclease would have to recognize and cleave at two distinct DNA sites in a coordinated fashion. A type IIs endonuclease would satisfy the first of these criteria, as these enzymes cleave at a distance from their binding site, e.g., FokI (138). The type IIf endonucleases bind and cleave two sites concurrently and thus satisfy the requirement of coordinating cleavage at two sites, e.g., SfiI (16, 185). Most interestingly, BspMI has characteristics of both families (56). This enzyme can bind and synapse two recognition sites and cleaves outside this binding sequence, thus coordinating the first step of transposition: excision. Perhaps one day . . . ?

Acknowledgments. We thank Bryan Swingle for all his efforts with multiple drafts of figures. We gratefully acknowledge the valuable comments and discussions from all our colleagues in the field of transposition, and we look forward to more fun in the future.

Research in our laboratories was supported by NIH grants GM50699 (K.M.D.) and GM24870 (N.D.F.G.).

ADDENDUM IN PROOF

We refer readers to reference 41a for additional discussions on transposon rearrangements.

REFERENCES

1. **Abraham, L. J., and J. I. Rood.** 1987. Identification of Tn4451 and Tn4452, chloramphenicol resistance transposons from *Clostridium perfringens. J. Bacteriol.* **169:**1579–1584.
2. **Adzuma, K., and K. Mizuuchi.** 1989. Interaction of proteins located at a distance along DNA: mechanism of target immunity in the Mu DNA strand-transfer reaction. *Cell* **57:**41–47.
3. **Adzuma, K., and K. Mizuuchi.** 1988. Target immunity of Mu transposition reflects a differential distribution of Mu B protein. *Cell* **53:**257–266.
4. **Arciszewska, L. K., D. Drake, and N. L. Craig.** 1989. Transposon Tn7. cis-acting sequences in transposition and transposition immunity. *J. Mol. Biol.* **207:**35–52.
5. **Arthur, A., and D. Sherratt.** 1979. Dissection of the transposition process: a transposon-encoded site-specific recombination system. *Mol. Gen. Genet.* **175:**267–274.
6. **Azaro, M. A., and A. Landy.** 2002. λ integrase and the λ Int family, p. 118–148. *In* N. L. Craig, R. Craigie, M. Gellert, and A. M. Lambowitz (ed.), *Mobile DNA II.* ASM Press, Washington, D.C.
7. **Bainton, R. J., K. M. Kubo, J. N. Feng, and N. L. Craig.** 1993. Tn7 transposition: target DNA recognition is mediated by multiple Tn7-encoded proteins in a purified in vitro system. *Cell* **72:**931–943.
8. **Baker, T. A., and L. Luo.** 1994. Identification of residues in the Mu transposase essential for catalysis. *Proc. Natl. Acad. Sci. USA* **91:**6654–6658.
9. **Bannam, T. L., P. K. Crellin, and J. I. Rood.** 1995. Molecular genetics of the chloramphenicol-resistance transposon Tn4451 from *Clostridium perfringens*: the TnpX site-specific recombinase excises a circular transposon molecule. *Mol. Microbiol.* **16:**535–551.
10. **Bastos, M. C., and E. Murphy.** 1988. Transposon Tn554 encodes three products required for transposition. *EMBO J.* **7:**2935–2941.
11. **Belfort, M., V. Derbyshire, M. M. Parker, B. Cousineau, and A. M. Lambowitz.** 2002. Mobile introns: pathway and proteins, p. 761–783. *In* N. L. Craig, R. Craigie, M. Gellert, and A. M. Lambowitz (ed.), *Mobile DNA II.* ASM Press, Washington, D.C.
12. **Berg, D. E., and M. M. Howe (ed.).** 1989. *Mobile DNA.* American Society for Microbiology, Washington, D.C.
13. **Bernales, I., M. V. Mendiola, and F. de la Cruz.** 1999. Intramolecular transposition of insertion sequence IS91 results in second-site simple insertions. *Mol. Microbiol.* **33:**223–234.
14. **Bhasin, A., I. Y. Goryshin, and W. S. Reznikoff.** 1999. Hairpin formation in Tn5 transposition. *J. Biol. Chem.* **274:**37021–37029.
15. **Biery, M. C., F. J. Stewart, A. E. Stellwagen, E. A. Raleigh, and N. L. Craig.** 2000. A simple in vitro Tn7-based transposition system with low target site selectivity for genome and gene analysis. *Nucleic Acids Res.* **28:**1067–1077.
16. **Bilcock, D. T., and S. E. Halford.** 1999. DNA restriction dependent on two recognition sites: activities of the SfiI restriction-modification system in Escherichia coli. *Mol. Microbiol.* **31:**1243–1254.
17. **Bolland, S., and N. Kleckner.** 1996. The three chemical steps of Tn10/IS10 transposition involve repeated utilization of a single active site. *Cell* **84:**223–233.
18. **Bujacz, G., J. Alexandratos, A. Wlodawer, G. Merkel, M. Andrake, R. A. Katz, and A. M. Skalka.** 1997. Binding of

different divalent cations to the active site of avian sarcoma virus integrase and their effects on enzymatic activity. *J. Biol. Chem.* **272**:18161–18168.

19. **Bujacz, G., M. Jaskolski, J. Alexandratos, A. Wlodawer, G. Merkel, R. A. Katz, and A. M. Skalka.** 1996. The catalytic domain of avian sarcoma virus integrase: conformation of the active-site residues in the presence of divalent cations. *Structure* **4**:89–96.

20. **Bujacz, G., M. Jaskolski, J. Alexandratos, A. Wlodawer, G. Merkel, R. A. Katz, and A. M. Skalka.** 1995. High-resolution structure of the catalytic domain of avian sarcoma virus integrase. *J. Mol. Biol.* **253**:333–346.

21. **Bult, C. J., O. White, G. J. Olsen, L. Zhou, R. D. Fleischmann, G. G. Sutton, J. A. Blake, L. M. FitzGerald, R. A. Clayton, J. D. Gocayne, A. R. Kerlavage, B. A. Dougherty, J. F. Tomb, M. D. Adams, C. I. Reich, R. Overbeek, E. F. Kirkness, K. G. Weinstock, J. M. Merrick, A. Glodek, J. L. Scott, N. S. Geoghagen, and J. C. Venter.** 1996. Complete genome sequence of the methanogenic archaeon, Methanococcus jannaschii. *Science* **273**:1058–1073.

22. **Bushman, F.** 2002. *Lateral DNA Transfer: Mechanisms and Consequences.* Cold Spring Harbor Laboratory Press, Cold Spring Harbor, N.Y.

23. **Caillaud, F., and P. Courvalin.** 1987. Nucleotide sequence of the ends of the conjugative shuttle transposon Tn1545. *Mol. Gen. Genet.* **209**:110–115.

24. **Caparon, M. G., and J. R. Scott.** 1989. Excision and insertion of the conjugative transposon Tn916 involves a novel recombination mechanism. *Cell* **59**:1027–1034.

25. **Celli, J., and P. Trieu-Cuot.** 1998. Circularization of Tn916 is required for expression of the transposon-encoded transfer functions: characterization of long tetracycline-inducible transcripts reading through the attachment site. *Mol. Microbiol.* **28**:103–117.

26. **Chaconas, G., and R. M. Harshey.** 2002. Transposition of phage Mu DNA, p. 384–402. *In* N. L. Craig, R. Craigie, M. Gellert, and A. M. Lambowitz (ed.), *Mobile DNA II.* ASM Press, Washington, D.C.

27. **Chalmers, R., A. Guhathakurta, H. Benjamin, and N. Kleckner.** 1998. IHF modulation of Tn10 transposition: sensory transduction of supercoiling status via a proposed protein/DNA molecular spring. *Cell* **93**:897–908.

28. **Chalmers, R. M., and N. Kleckner.** 1996. IS10/Tn10 transposition efficiently accommodates diverse transposon end configurations. *EMBO J.* **15**:5112–5122.

29. **Chandler, M., and J. Mahillon.** 2002. Insertion sequences revisited, p. 305–366. *In* N. L. Craig, R. Craigie, M. Gellert, and A. M. Lambowitz (ed.), *Mobile DNA II.* ASM Press, Washington D.C.

30. **Cheng, Q., B. J. Paszkiet, N. B. Shoemaker, J. F. Gardner, and A. A. Salyers.** 2000. Integration and excision of a *Bacteroides* conjugative transposon, CTnDOT. *J. Bacteriol.* **182**:4035–4043.

31. **Cheng, Q., Y. Sutanto, N. B. Shoemaker, J. F. Gardner, and A. A. Salyers.** 2001. Identification of genes required for excision of CTnDOT, a Bacteroides conjugative transposon. *Mol. Microbiol.* **41**:625–632.

32. **Churchward, G.** 2002. Conjugative transposons and related mobile elements, p. 177–191. *In* N. L. Craig, R. Craigie, M. Gellert, and A. M. Lambowitz (ed.), *Mobile DNA II.* ASM Press, Washington, D.C.

33. **Clewell, D. B., S. E. Flannagan, and D. D. Jaworski.** 1995. Unconstrained bacterial promiscuity: the Tn916-Tn1545 family of conjugative transposons. *Trends Microbiol.* **3**:229–236.

34. **Cole, S. T., R. Brosch, J. Parkhill, T. Garnier, C. Churcher, D. Harris, S. V. Gordon, K. Eiglmeier, S. Gas, C. E. Barry III,** F. Tekaia, K. Badcock, D. Basham, D. Brown, T. Chillingworth, R. Connor, R. Davies, K. Devlin, T. Feltwell, S. Gentles, N. Hamlin, S. Holroyd, T. Hornsby, K. Jagels, B. G. Barrell, et al. 1998. Deciphering the biology of Mycobacterium tuberculosis from the complete genome sequence. *Nature* **393**:537–544.

35. **Craig, N. L.** 1997. Target site selection in transposition. *Annu. Rev. Biochem.* **66**:437–474.

36. **Craig, N. L.** 2002. Tn7, p. 423–456. *In* N. L. Craig, R. Craigie, M. Gellert, and A. M. Lambowitz (ed.), *Mobile DNA II.* ASM Press, Washington, D.C.

37. **Craig, N. L.** 1996. Transposition, p. 2339–2362. *In* F. C. Neidhardt, R. Curtiss III, J. L. Ingraham, E. C. C. Lin, K. B. Low, B. Magasanik, W. S. Reznikoff, M. Riley, M. Schaechter, and H. E. Umbarger (ed.), Escherichia coli *and* Salmonella: *Cellular and Molecular Biology*, 2nd ed., vol. 2. ASM Press, Washington, D.C.

38. **Craig, N. L.** 1996. V(D)J recombination and transposition: closer than expected. *Science* **271**:1512.

39. **Craig, N. L., R. Craigie, M. Gellert, and A. M. Lambowitz.** 2002. *Mobile DNA II.* ASM Press, Washington, D.C.

40. **Craigie, R., and K. Mizuuchi.** 1985. Mechanism of transposition of bacteriophage Mu: structure of a transposition intermediate. *Cell* **41**:867–876.

41. **Crellin, P. K., and J. I. Rood.** 1997. The resolvase/invertase domain of the site-specific recombinase TnpX is functional and recognizes a target sequence that resembles the junction of the circular form of the *Clostridium perfringens* transposon Tn4451. *J. Bacteriol.* **179**:5148–5156.

41a. **Curcio, M. J., and K. Derbyshire.** 2003. The ins and outs of transposition: from Mu to Kangaroo. *Nat. Rev. Mol. Cell. Biol.* **4**:865–877.

42. **Davies, D. R., I. Y. Goryshin, W. S. Reznikoff, and I. Rayment.** 2000. The three-dimensional structure of the Tn5 synaptic complex intermediate. *Science* **289**:77–85.

43. **DeBoy, R. T., and N. L. Craig.** 2000. Target site selection by Tn7: *att*Tn7 transcription and target activity. *J. Bacteriol.* **182**:3310–3313.

44. **DeBoy, R. T., and N. L. Craig.** 1996. Tn7 transposition as a probe of *cis* interactions between widely separated (190 kilobases apart) DNA sites in the *Escherichia coli* chromosome. *J. Bacteriol.* **178**:6184–6191.

45. **del Solar, G., R. Giraldo, M. J. Ruiz-Echevarria, M. Espinosa, and R. Diaz-Orejas.** 1998. Replication and control of circular bacterial plasmids. *Microbiol. Mol. Biol. Rev.* **62**:434–464.

46. **Derbyshire, K. M., and N. D. Grindley.** 1986. Replicative and conservative transposition in bacteria. *Cell* **47**:325–327.

47. **Dyda, F., A. B. Hickman, T. M. Jenkins, A. Engelman, R. Craigie, and D. R. Davies.** 1994. Crystal structure of the catalytic domain of HIV-1 integrase: similarity to other polynucleotidyl transferases. *Science* **266**:1981–1986.

48. **Engelman, A., and R. Craigie.** 1992. Identification of conserved amino acid residues critical for human immunodeficiency virus type 1 integrase function in vitro. *J. Virol.* **66**:6361–6369.

49. **Engels, W. R., D. M. Johnson-Schlitz, W. B. Eggleston, and J. Sved.** 1990. High-frequency P element loss in Drosophila is homolog dependent. *Cell* **62**:515–525.

50. **Firth, N., K. Ippen-Ihler, and R. A. Skurray.** 1996. Structure and function of the F factor and mechanism of conjugation, p. 2377–2401. *In* F. C. Neidhardt, R. Curtiss III, J. L. Ingraham, E. C. C. Lin, K. B. Low, B. Magasanik, W. S. Reznikoff, M. Riley, M. Schaechter, and H. E. Umbarger (ed.), Escherichia coli *and* Salmonella: *Cellular and Molecular Biology*, 2nd ed., vol. 2. ASM Press, Washington, D.C.

51. Flannagan, S. E., and D. B. Clewell. 1991. Conjugative transfer of Tn916 in *Enterococcus faecalis*: *trans* activation of homologous transposons. *J. Bacteriol.* 173:7136–7141.

52. Fugmann, S. D., I. J. Villey, L. M. Ptaszek, and D. G. Schatz. 2000. Identification of two catalytic residues in RAG1 that define a single active site within the RAG1/RAG2 protein complex. *Mol. Cell* 5:97–107.

53. Garcillan-Barcia, M. P., I. Bernales, M. V. Mendiola, and F. de la Cruz. 2002. IS91 rolling-circle transposition, p. 891–904. *In* N. L. Craig, R. Craigie, M. Gellert, and A. M. Lambowitz (ed.), *Mobile DNA II*. ASM Press, Washington, D.C.

54. Garcillan-Barcia, M. P., I. Bernales, M. V. Mendiola, and F. de la Cruz. 2001. Single-stranded DNA intermediates in IS91 rolling-circle transposition. *Mol. Microbiol.* 39:494–501.

55. Gawron-Burke, C., and D. B. Clewell. 1982. A transposon in Streptococcus faecalis with fertility properties. *Nature* 300:281–284.

56. Gormley, N. A., A. L. Hillberg, and S. E. Halford. 2001. The type IIs restriction endonuclease BspMI is a tetramer that acts concertedly at two copies of its recognition sequence. *J. Biol. Chem.* 29:29.

57. Goryshin, I. Y., J. A. Miller, Y. V. Kil, V. A. Lanzov, and W. S. Reznikoff. 1998. Tn5/IS50 target recognition. *Proc. Natl. Acad. Sci. USA* 95:10716–10721.

58. Grindley, N. D. F. 2002. The movement of Tn3-like elements: transposition and cointegrate resolution, p. 272–302. *In* N. L. Craig, R. Craigie, M. Gellert, and A. Lambowitz (ed.), *Mobile DNA II*. ASM Press, Washington, D.C.

59. Grindley, N. D. F. 1994. Resolvase-mediated site-specific recombination, p. 236–267. *In* F. Eckstein and D. M. J. Lilley (ed.), *Nucleic Acids and Molecular Biology*, vol. 8. Springer-Verlag, Berlin, Germany.

60. Grindley, N. D. F., and R. R. Reed. 1985. Transpositional recombination in prokaryotes. *Annu. Rev. Biochem.* 54:863–896.

61. Groth, A. C., E. C. Olivares, B. Thyagarajan, and M. P. Calos. 2000. A phage integrase directs efficient site-specific integration in human cells. *Proc. Natl. Acad. Sci. USA* 97:5995–6000.

62. Guyer, M. S. 1978. The gamma delta sequence of F is an insertion sequence. *J. Mol. Biol.* 126:347–365.

63. Haapa-Paananen, S., H. Rita, and H. Savilahti. 2002. DNA transposition of bacteriophage Mu: a quantitative analysis of target site selection in vitro. *J. Biol. Chem.* 277:2843–2851.

64. Hagemann, A. T., and N. L. Craig. 1993. Tn7 transposition creates a hotspot for homologous recombination at the transposon donor site. *Genetics* 133:9–16.

65. Hallet, B., and D. J. Sherratt. 1997. Transposition and site-specific recombination: adapting DNA cut-and-paste mechanisms to a variety of genetic rearrangements. *FEMS Microbiol. Rev.* 21:157–178.

66. Halling, S. M., and N. Kleckner. 1982. A symmetrical six-basepair target site sequence determines Tn10 insertion specificity. *Cell* 28:155–163.

67. Hanai, R., and J. C. Wang. 1993. The mechanism of sequence-specific DNA cleavage and strand transfer by phi X174 gene A* protein. *J. Biol. Chem.* 268:23830–23836.

68. Haniford, D. B. 2002. Transposon Tn10, p. 457–483. *In* N. L. Craig, R. Craigie, M. Gellert, and A. M. Lambowitz (ed.), *Mobile DNA II*. ASM Press, Washington, D.C.

69. Haren, L., B. Ton-Hoang, and M. Chandler. 1999. Integrating DNA: transposases and retroviral integrases. *Annu. Rev. Microbiol.* 53:245–281.

70. Harshey, R. M. 1984. Transposition without duplication of infecting bacteriophage Mu DNA. *Nature* 311:580–581.

71. Hickman, A. B., Y. Li, S. V. Mathew, E. W. May, N. L. Craig, and F. Dyda. 2000. Unexpected structural diversity in DNA recombination: the restriction endonuclease connection. *Mol. Cell* 5:1025–1034.

72. Hinerfeld, D., and G. Churchward. 2001. Xis protein of the conjugative transposon Tn916 plays dual opposing roles in transposon excision. *Mol. Microbiol.* 41:1459–1467.

73. Hochhut, B., J. Marrero, and M. K. Waldor. 2000. Mobilization of plasmids and chromosomal DNA mediated by the SXT element, a constin found in *Vibrio cholerae* O139. *J. Bacteriol.* 182:2043–2047.

74. Hochhut, B., and M. K. Waldor. 1999. Site-specific integration of the conjugal *Vibrio cholerae* SXT element into prfC. *Mol. Microbiol.* 32:99–110.

75. Hosking, S. L., M. E. Deadman, E. R. Moxon, J. F. Peden, N. J. Saunders, and N. J. High. 1998. An in silico evaluation of Tn916 as a tool for generalized mutagenesis in *Haemophilus influenzae* Rd. *Microbiology* 144:2525–2530.

76. Hu, W.-Y., and K. M. Derbyshire. 1998. Target choice and orientation preference of the insertion sequence IS903. *J. Bacteriol.* 180:3039–3048.

77. Hu, W. Y., W. Thompson, C. E. Lawrence, and K. M. Derbyshire. 2001. Anatomy of a preferred target site for the bacterial insertion sequence IS903. *J. Mol. Biol.* 306:403–416.

78. Iida, S., and R. Hiestand-Nauer. 1986. Localized conversion at the crossover sequences in the site-specific DNA inversion system of bacteriophage P1. *Cell* 45:71–79.

79. Jilk, R. A., J. C. Makris, L. Borchardt, and W. S. Reznikoff. 1993. Implications of Tn5-associated adjacent deletions. *J. Bacteriol.* 175:1264–1271.

80. Johnson, R. C. 2002. Bacterial site-specific DNA inversion systems, p. 230–271. *In* N. L. Craig, R. Craigie, M. Gellert, and A. M. Lambowitz (ed.), *Mobile DNA II*. ASM Press, Washington, D.C.

81. Johnson, R. C., and M. F. Bruist. 1989. Intermediates in Hin-mediated DNA inversion: a role for Fis and the recombinational enhancer in the strand exchange reaction. *EMBO J.* 8:1581–1590.

82. Joyce, C. M., and T. A. Steitz. 1994. Function and structure relationships in DNA polymerases. *Annu. Rev. Biochem.* 63:777–822.

83. Junop, M. S., and D. B. Haniford. 1997. Factors responsible for target site selection in Tn10 transposition: a role for the DDE motif in target DNA capture. *EMBO J.* 16:2646–2655.

84. Kapitonov, V. V., and J. Jurka. 2001. Rolling-circle transposons in eukaryotes. *Proc. Natl. Acad. Sci. USA* 98:8714–8719.

85. Kennedy, A. K., A. Guhathakurta, N. Kleckner, and D. B. Haniford. 1998. Tn10 transposition via a DNA hairpin intermediate. *Cell* 95:125–134.

86. Kennedy, A. K., D. B. Haniford, and K. Mizuuchi. 2000. Single active site catalysis of the successive phosphoryl transfer steps by DNA transposases: insights from phosphorothioate stereoselectivity. *Cell* 101:295–305.

87. Kersulyte, D., A. K. Mukhopadhyay, M. Shirai, T. Nakazawa, and D. E. Berg. 2000. Functional organization and insertion specificity of IS607, a chimeric element of *Helicobacter pylori*. *J. Bacteriol.* 182:5300–5308.

88. Kim, D. R., Y. Dai, C. L. Mundy, W. Yang, and M. A. Oettinger. 1999. Mutations of acidic residues in RAG1 define the active site of the V(D)J recombinase. *Genes Dev.* 13:3070–3080.

89. Kim, K., S. Y. Namgoong, M. Jayaram, and R. M. Harshey. 1995. Step-arrest mutants of phage Mu transposase. Implications in DNA-protein assembly, Mu end cleavage, and strand transfer. *J. Biol. Chem.* 270:1472–1479.

90. Kitts, P. A., and H. A. Nash. 1987. Homology-dependent interactions in phage lambda site-specific recombination. *Nature* **329**:346–348.

91. Kleckner, N. 1981. Transposable elements in prokaryotes. *Annu. Rev. Genet.* **15**:341–404.

92. Kuduvalli, P. N., J. E. Rao, and N. L. Craig. 2001. Target DNA structure plays a critical role in Tn7 transposition. *EMBO J.* **20**:924–932.

93. Kulkosky, J., K. S. Jones, R. A. Katz, J. P. Mack, and A. M. Skalka. 1992. Residues critical for retroviral integrative recombination in a region that is highly conserved among retroviral/retrotransposon integrases and bacterial insertion sequence transposases. *Mol. Cell. Biol.* **12**:2331–2338.

94. Landree, M. A., J. A. Wibbenmeyer, and D. B. Roth. 1999. Mutational analysis of RAG1 and RAG2 identifies three catalytic amino acids in RAG1 critical for both cleavage steps of V(D)J recombination. *Genes Dev.* **13**:3059–3069.

95. Landy, A. 1989. Dynamic, structural, and regulatory aspects of lambda site-specific recombination. *Annu. Rev. Biochem.* **58**:913–949.

96. Lanka, E., and B. M. Wilkins. 1995. DNA processing reactions in bacterial conjugation. *Annu. Rev. Biochem.* **64**:141–169.

97. Lewis, L. A., and N. D. Grindley. 1997. Two abundant intramolecular transposition products, resulting from reactions initiated at a single end, suggest that IS2 transposes by an unconventional pathway. *Mol. Microbiol.* **25**:517–529.

98. Lodge, J. K., and D. E. Berg. 1990. Mutations that affect Tn5 insertion into pBR322: importance of local DNA supercoiling. *J. Bacteriol.* **172**:5956–5960.

99. Lovell, S., I. Y. Goryshin, W. R. Reznikoff, and I. Rayment. 2002. Two-metal active site binding of a Tn5 transposase synaptic complex. *Nat. Struct. Biol.* **9**:278–281.

100. Lu, F., and G. Churchward. 1994. Conjugative transposition: Tn916 integrase contains two independent DNA binding domains that recognize different DNA sequences. *EMBO J.* **13**:1541–1548.

101. Lyras, D., and J. I. Rood. 2000. Transposition of Tn4451 and Tn4453 involves a circular intermediate that forms a promoter for the large resolvase, TnpX. *Mol. Microbiol.* **38**:588–601.

102. Mahillon, J., and M. Chandler. 1998. Insertion sequences. *Microbiol. Mol. Biol. Rev.* **62**:725–774.

103. Manna, D., and N. P. Higgins. 1999. Phage Mu transposition immunity reflects supercoil domain structure of the chromosome. *Mol. Microbiol.* **32**:595–606.

104. Manna, D., X. Wang, and N. P. Higgins. 2001. Mu and IS1 transpositions exhibit strong orientation bias at the *Escherichia coli bgl* locus. *J. Bacteriol.* **183**:3328–3335.

105. Marra, D., and J. R. Scott. 1999. Regulation of excision of the conjugative transposon Tn916. *Mol. Microbiol.* **31**:609–621.

106. Matsuura, M., T. Noguchi, D. Yamaguchi, T. Aida, M. Asayama, H. Takahashi, and M. Shirai. 1996. The *sre* gene (ORF469) encodes a site-specific recombinase responsible for integration of the R4 phage genome. *J. Bacteriol.* **178**:3374–3376.

107. May, E. W., and N. L. Craig. 1996. Switching from cut-and-paste to replicative Tn7 transposition. *Science* **272**:401–404.

108. McBlane, J. F., D. C. van Gent, D. A. Ramsden, C. Romeo, C. A. Cuomo, M. Gellert, M. A. Oettinger, G. Bujacz, J. Alexandratos, A. Wlodawer, G. Merkel, M. Andrake, R. A. Katz, and A. M. Skalka. 1995. Cleavage at a V(D)J recombination signal requires only RAG1 and RAG2 proteins and occurs in two steps. *Cell* **83**:387–395.

109. Mendiola, M. V., I. Bernales, and F. de la Cruz. 1994. Differential roles of the transposon termini in IS91 transposition. *Proc. Natl. Acad. Sci. USA* **91**:1922–1926.

110. Mendiola, M. V., and F. de la Cruz. 1992. IS91 transposase is related to the rolling-circle-type replication proteins of the pUB110 family of plasmids. *Nucleic Acids Res.* **20**:3521.

111. Mizuuchi, K. 1983. In vitro transposition of bacteriophage Mu: a biochemical approach to a novel replication reaction. *Cell* **35**:785–794.

112. Mizuuchi, K. 1984. Mechanism of transposition of bacteriophage Mu: polarity of the strand transfer reaction at the initiation of transposition. *Cell* **39**:395–404.

112a. Mizuuchi, K., and T. A. Baker. 2002. Chemical mechanisms for mobilizing DNA, p. 12–23. *In* N. L. Craig, R. Craigie, M. Gellert, and A. M. Lambowitz (ed.), *Mobile DNA II*. ASM Press, Washington, D.C.

113. Mizuuchi, M., T. A. Baker, and K. Mizuuchi. 1992. Assembly of the active form of the transposase-Mu DNA complex: a critical control point in Mu transposition. *Cell* **70**:303–311.

114. Mizuuchi, M., and K. Mizuuchi. 1993. Target site selection in transposition of phage Mu. *Cold Spring Harbor Symp. Quant. Biol.* **58**:515–523.

115. Moitoso de Vargas, L., C. A. Pargellis, N. M. Hasan, E. W. Bushman, and A. Landy. 1988. Autonomous DNA binding domains of λ integrase recognize two different sequence families. *Cell* **54**:923–929.

116. Mullany, P., M. Wilks, I. Lamb, C. Clayton, B. Wren, and S. Tabaqchali. 1990. Genetic analysis of a tetracycline resistance element from *Clostridium difficile* and its conjugal transfer to and from *Bacillus subtilis*. *J. Gen. Microbiol.* **136**:1343–1349.

117. Nakai, H., and R. Kruklitis. 1995. Disassembly of the bacteriophage Mu transposase for the initiation of Mu DNA replication. *J. Biol. Chem.* **270**:19591–19598.

118. Nash, H. A. 1981. Integration and excision of bacteriophage λ: the mechanism of conservative site specific recombination. *Annu. Rev. Genet.* **15**:143–167.

119. Nash, H. A. 1996. Site-specific recombination: integration, excision, resolution, and inversion of defined DNA segments, p. 2363–2376. *In* F. C. Neidhardt, R. Curtiss III, J. L. Ingraham, E. C. C. Lin, K. B. Low, B. Magasanik, W. S. Reznikoff, M. Riley, M. Schaechter, and H. E. Umbarger (ed.), Escherichia coli *and* Salmonella: *Cellular and Molecular Biology*, 2nd ed., vol. 2. ASM Press, Washington, D.C.

120. Naumann, T. A., and W. S. Reznikoff. 2000. Trans catalysis in Tn5 transposition. *Proc. Natl. Acad. Sci. USA* **97**:8944–8949.

121. Nunes-Duby, S. E., D. Yu, and A. Landy. 1997. Sensing homology at the strand-swapping step in lambda excisive recombination. *J. Mol. Biol.* **272**:493–508.

122. O'Keeffe, T., C. Hill, and R. P. Ross. 1999. In situ inversion of the conjugative transposon Tn916 in Enterococcus faecium DPC3675. *FEMS Microbiol. Lett.* **173**:265–271.

123. Peters, J. E., and N. L. Craig. 2001. Tn7 recognizes transposition target structures associated with DNA replication using the DNA-binding protein TnsE. *Genes Dev.* **15**:737–747.

124. Peters, J. E., and N. L. Craig. 2000. Tn7 transposes proximal to DNA double-strand breaks and into regions where chromosomal DNA replication terminates. *Mol. Cell* **6**:573–582.

125. Plasterk, R. H., and J. T. Groenen. 1992. Targeted alterations of the Caenorhabditis elegans genome by transgene instructed DNA double strand break repair following Tc1 excision. *EMBO J.* **11**:287–290.

126. Polard, P., B. Ton-Hoang, L. Haren, M. Betermier, R. Walczak, and M. Chandler. 1996. IS911-mediated transpositional recombination in vitro. *J. Mol. Biol.* **264**:68–81.

127. Poyart-Salmeron, C., P. Trieu-Cuot, C. Carlier, and P. Courvalin. 1989. Molecular characterization of two proteins involved in the excision of the conjugative transposon Tn1545: homologies with other site-specific recombinases. *EMBO J.* **8**:2425–2433.

128. Pribil, P. A., and D. B. Haniford. 2000. Substrate recognition and induced DNA deformation by transposase at the target-capture stage of Tn10 transposition. *J. Mol. Biol.* **303**:145–159.

129. Reynolds, A. E., J. Felton, and A. Wright. 1981. Insertion of DNA activates the cryptic bgl operon in E. coli K12. *Nature* **293**:625–629.

130. Reznikoff, W. S. 2002. Tn5 transposition, p. 403–422. *In* N. L. Craig, R. Craigie, M. Gellert, and A. M. Lambowitz (ed.), *Mobile DNA II.* ASM Press, Washington, D.C.

131. Rezsohazy, R., B. Hallet, J. Delcour, and J. Mahillon. 1993. The IS4 family of insertion sequences: evidence for a conserved transposase motif. *Mol. Microbiol.* **9**:1283–1295.

132. Rice, P., and K. Mizuuchi. 1995. Structure of the bacteriophage Mu transposase core: a common structural motif for DNA transposition and retroviral integration. *Cell* **82**:209–220.

133. Rice, P. A., and T. A. Baker. 2001. Comparative architecture of transposase and integrase complexes. *Nat. Struct. Biol.* **8**:302–307.

134. Rice, P. A., R. Craigie, and D. R. Davies. 1996. Retroviral integrases and their cousins. *Curr. Opin. Struct. Biol.* **6**:76–83.

135. Richet, E., P. Abcarian, and H. A. Nash. 1988. Synapsis of attachment sites during lambda integrative recombination involves capture of a naked DNA by a protein-DNA complex. *Cell* **52**:9–17.

136. Roberts, D., B. C. Hoopes, W. R. McClure, and N. Kleckner. 1985. IS10 transposition is regulated by DNA adenine methylation. *Cell* **43**:117–130.

137. Roberts, D. E., D. Ascherman, and N. Kleckner. 1991. IS10 promotes adjacent deletions at low frequency. *Genetics* **128**:37–43.

138. Roberts, R. J., and D. Macelis. 2001. REBASE—restriction enzymes and methylases. *Nucleic Acids Res.* **29**:268–269.

139. Roth, D. B., and N. L. Craig. 1998. VDJ recombination: a transposase goes to work. *Cell* **94**:411–414.

140. Roth, J. R., N. Benson, T. Galitski, K. Haack, J. G. Lawrence, and L. Miesel. 1996. Rearrangements of the bacterial chromosome: formation and applications, p. 2256–2276. *In* F. C. Neidhardt, R. Curtiss III, J. L. Ingraham, E. C. C. Lin, K. B. Low, B. Magasanik, W. S. Reznikoff, M. Riley, M. Schaechter, and H. E. Umbarger (ed.), Escherichia coli *and* Salmonella: *Cellular and Molecular Biology*, 2nd ed., vol. 2. ASM Press, Washington, D.C.

141. Rousseau, P., C. Normand, C. Loot, C. Turlan, R. Alazard, G. Duval-Valentin, and M. Chandler. 2002. Transposition of IS911, p. 367–383. *In* N. L. Craig, R. Craigie, M. Gellert, and A. M. Lambowitz (ed.), *Mobile DNA II.* ASM Press, Washington, D.C.

142. Rudy, C., K. L. Taylor, D. Hinerfeld, J. R. Scott, and G. Churchward. 1997. Excision of a conjugative transposon in vitro by the Int and Xis proteins of Tn916. *Nucleic Acids Res.* **25**:4061–4066.

143. Rudy, C. K., and J. R. Scott. 1994. Length of the coupling sequence of Tn916. *J. Bacteriol.* **176**:3386–3388.

144. Rudy, C. K., J. R. Scott, and G. Churchward. 1997. DNA binding by the Xis protein of the conjugative transposon Tn916. *J. Bacteriol.* **179**:2567–2572.

145. Sakai, J., R. M. Chalmers, and N. Kleckner. 1995. Identification and characterization of a pre-cleavage synaptic complex that is an early intermediate in Tn10 transposition. *EMBO J.* **14**:4374–4383.

146. Sarkar, D., M. Radman-Livaja, and A. Landy. 2001. The small DNA binding domain of lambda integrase is a context-sensitive modulator of recombinase functions. *EMBO J.* **20**:1203–1212.

147. Sarkis, G. J., L. L. Murley, A. E. Leschziner, M. R. Boocock, W. M. Stark, and N. D. Grindley. 2001. A model for the gamma delta resolvase synaptic complex. *Mol. Cell* **8**:623–631.

148. Sarnovsky, R. J., E. W. May, and N. L. Craig. 1996. The Tn7 transposase is a heteromeric complex in which DNA breakage and joining activities are distributed between different gene products. *EMBO J.* **15**:6348–6361.

149. Sato, T., Y. Samori, and Y. Kobayashi. 1990. The *cisA* cistron of *Bacillus subtilis* sporulation gene *spoIVC* encodes a protein homologous to a site-specific recombinase. *J. Bacteriol.* **172**:1092–1098.

150. Scott, J. R., F. Bringel, D. Marra, G. Van Alstine, and C. K. Rudy. 1994. Conjugative transposition of Tn916: preferred targets and evidence for conjugative transfer of a single strand and for a double-stranded circular intermediate. *Mol. Microbiol.* **11**:1099–1108.

151. Scott, J. R., P. A. Kirchman, and M. G. Caparon. 1988. An intermediate in transposition of the conjugative transposon Tn916. *Proc. Natl. Acad. Sci. USA* **85**:4809–4813.

152. Sekine, Y., K. Aihara, and E. Ohtsubo. 1999. Linearization and transposition of circular molecules of insertion sequence IS3. *J. Mol. Biol.* **294**:21–34.

153. Shapiro, J. A. 1997. Genome organization, natural genetic engineering and adaptive mutation. *Trends Genet.* **13**:98–104.

154. Shapiro, J. A. 1979. Molecular model for the transposition and replication of bacteriophage Mu and other transposable elements. *Proc. Natl. Acad. Sci. USA* **76**:1933–1937.

155. Shapiro, J. A., and N. P. Higgins. 1989. Differential activity of a transposable element in *Escherichia coli* colonies. *J. Bacteriol.* **171**:5975–5986.

156. She, Q., R. K. Singh, F. Confalonieri, Y. Zivanovic, G. Allard, M. J. Awayez, C. C. Chan-Weiher, I. G. Clausen, B. A. Curtis, A. De Moors, G. Erauso, C. Fletcher, P. M. Gordon, I. Heikamp-de Jong, A. C. Jeffries, C. J. Kozera, N. Medina, X. Peng, H. P. Thi-Ngoc, P. Redder, M. E. Schenk, C. Theriault, N. Tolstrup, R. L. Charlebois, W. F. Doolittle, M. Duguet, T. Gaasterland, R. A. Garrett, M. A. Ragan, C. W. Sensen, and J. Van der Oost. 2001. The complete genome of the crenarchaeon *Sulfolobus solfataricus* P2. *Proc. Natl. Acad. Sci. USA* **98**:7835–7840.

157. Sherratt, D. J., and D. B. Wigley. 1998. Conserved themes but novel activities in recombinases and topoisomerases. *Cell* **93**:149–152.

158. Shoemaker, N. B., G. R. Wang, and A. A. Salyers. 2000. Multiple gene products and sequences required for excision of the mobilizable integrated *Bacteroides* element NBU1. *J. Bacteriol.* **182**:928–936.

159. Shoemaker, N. B., G. R. Wang, and A. A. Salyers. 1996. NBU1, a mobilizable site-specific integrated element from *Bacteroides* spp., can integrate nonspecifically in *Escherichia coli. J. Bacteriol.* **178**:3601–3607.

160. Signon, L., and N. Kleckner. 1995. Negative and positive regulation of Tn10/IS10-promoted recombination by IHF: two distinguishable processes inhibit transposition off of multicopy plasmid replicons and activate chromosomal events that favor evolution of new transposons. *Genes Dev.* **9**:1123–1136.

161. Stark, W. M., and M. R. Boocock. 1995. Topological selectivity in site-specific recombination, p. 101–129. *In* D. J. Sherratt (ed.), *Mobile Genetic Elements*. Oxford University Press, Oxford, United Kingdom.

162. Stark, W. M., M. R. Boocock, and D. J. Sherratt. 1992. Catalysis by site-specific recombinases. *Trends Genet.* 8:432–439.

163. Stark, W. M., N. D. Grindley, G. F. Hatfull, and M. R. Boocock. 1991. Resolvase-catalysed reactions between res sites differing in the central dinucleotide of subsite I. *EMBO J.* 10:3541–3548.

164. Stellwagen, A. E., and N. L. Craig. 1997. Avoiding self: two Tn7-encoded proteins mediate target immunity in Tn7 transposition. *EMBO J.* 16:6823–6834.

165. Storrs, M. J., C. Poyart-Salmeron, P. Trieu-Cuot, and P. Courvalin. 1991. Conjugative transposition of Tn*916* requires the excisive and integrative activities of the transposon-encoded integrase. *J. Bacteriol.* 173:4347–4352.

166. Tavakoli, N., A. Comanducci, H. M. Dodd, M. C. Lett, B. Albiger, and P. Bennett. 2000. IS1294, a DNA element that transposes by RC transposition. *Plasmid* 44:66–84.

167. Tavakoli, N. P., and K. M. Derbyshire. 1999. IS903 transposase mutants that suppress defective inverted repeats. *Mol. Microbiol.* 31:1183–1195.

168. Tavakoli, N. P., and K. M. Derbyshire. 2001. Tipping the balance between replicative and simple transposition. *EMBO J.* 20:2923–2930.

169. Taylor, K. L., and G. Churchward. 1997. Specific DNA cleavage mediated by the integrase of conjugative transposon Tn916. *J. Bacteriol.* 179:1117–1125.

170. Thorpe, H. M., and M. C. Smith. 1998. In vitro site-specific integration of bacteriophage DNA catalyzed by a recombinase of the resolvase/invertase family. *Proc. Natl. Acad. Sci. USA* 95:5505–5510.

171. Tribble, G. D., A. C. Parker, and C. J. Smith. 1997. The *Bacteroides* mobilizable transposon Tn*4555* integrates by a site-specific recombination mechanism similar to that of the gram-positive bacterial element Tn*916*. *J. Bacteriol.* 179:2731–2739.

172. Tribble, G. D., A. C. Parker, and C. J. Smith. 1999. Transposition genes of the Bacteroides mobilizable transposon Tn4555: role of a novel targeting gene. *Mol. Microbiol.* 34:385–394.

173. Trieu-Cuot, P., C. Poyart-Salmeron, C. Carlier, and P. Courvalin. 1993. Sequence requirements for target activity in site-specific recombination mediated by the Int protein of transposon Tn1545. *Mol. Microbiol.* 8:179–185.

174. Turlan, C., and M. Chandler. 2000. Playing second fiddle: second-strand processing and liberation of transposable elements from donor DNA. *Trends Microbiol.* 8:268–274.

175. Turlan, C., B. Ton-Hoang, and M. Chandler. 2000. The role of tandem IS dimers in IS911 transposition. *Mol. Microbiol.* 35:1312–1325.

176. Van Duyne, G. D. 2001. A structural view of Cre-*loxP* site-specific recombination. *Annu. Rev. Biophys. Biomol. Struct.* 30:87–104.

177. van Gent, D. C., A. A. Groeneger, and R. H. Plasterk. 1992. Mutational analysis of the integrase protein of human immunodeficiency virus type 2. *Proc. Natl. Acad. Sci. USA* 89:9598–9602.

178. van Gent, D. C., J. F. McBlane, D. A. Ramsden, M. J. Sadofsky, J. E. Hesse, and M. Gellert. 1995. Initiation of V(D)J recombination in a cell-free system. *Cell* 81:925–934.

179. Wang, H., and P. Mullany. 2000. The large resolvase TndX is required and sufficient for integration and excision of derivatives of the novel conjugative transposon Tn*5397*. *J. Bacteriol.* 182:6577–6583.

180. Wang, H., A. P. Roberts, D. Lyras, J. I. Rood, M. Wilks, and P. Mullany. 2000. Characterization of the ends and target sites of the novel conjugative transposon Tn*5397* from *Clostridium difficile*: excision and circularization is mediated by the large resolvase, TndX. *J. Bacteriol.* 182:3775–3783.

181. Wang, H., A. P. Roberts, and P. Mullany. 2000. DNA sequence of the insertional hot spot of Tn916 in the *Clostridium difficile* genome and discovery of a Tn916-like element in an environmental isolate integrated in the same hot spot. *FEMS Microbiol. Lett.* 192:15–20.

182. Watson, M. A., and G. Chaconas. 1996. Three-site synapsis during Mu DNA transposition: a critical intermediate preceding engagement of the active site. *Cell* 85:435–445.

183. Weinert, T. A., K. M. Derbyshire, F. M. Hughson, and N. D. Grindley. 1984. Replicative and conservative transpositional recombination of insertion sequences. *Cold Spring Harbor Symp. Quant. Biol.* 49:251–260.

184. Weisberg, R. A., and A. Landy. 1983. Site-specific recombination in phage lambda, p. 211–250. *In* R. W. Hendrix, J. W. Roberts, F. W. Stahl, and R. A. Weisberg (ed.), *Lambda II*. Cold Spring Harbor Laboratory, Cold Spring Harbor, N.Y.

185. Wentzell, L. M., T. J. Nobbs, and S. E. Halford. 1995. The SfiI restriction endonuclease makes a four-strand DNA break at two copies of its recognition sequence. *J. Mol. Biol.* 248:581–595.

186. Willetts, N. S., C. Crowther, and B. W. Holloway. 1981. The insertion sequence IS21 of R68.45 and the molecular basis for mobilization of the bacterial chromosome. *Plasmid* 6:30–52.

187. Wolkow, C. A., R. T. DeBoy, and N. L. Craig. 1996. Conjugating plasmids are preferred targets for Tn7. *Genes Dev.* 10:2145–2157.

The Bacterial Chromosome
Edited by N. Patrick Higgins
© 2005 ASM Press, Washington, D.C.

Chapter 27

Potential Mechanisms for Linking Phage Mu Transposition with Cell Physiology

STELLA H. NORTH AND HIROSHI NAKAI

INTRODUCTION

Transposable elements are ubiquitous in cellular chromosomes and have undoubtedly played a major role in shaping the genome. The interactions between transposable elements and their respective hosts represent a very successful host-parasite relationship. For transposons to effectively exploit the host for their propagation, the frequency of transposition must be maintained at low levels within the cell to minimize lethal alterations to the host chromosome. For example, the accumulation of defective retroelements in eukaryotic genomes is one strategy for reducing transposition to very low levels. The reverse transcriptase that replicates these elements is error prone, limiting the number of active transposing elements in the genome. Alternatively, active elements can be tightly repressed, and the mechanisms that derepress transposition can serve as an important regulatory switch for increasing transposon copy number in the cell or disseminating copies to other cells. Derepression of transposition can potentially benefit the host under conditions of stress, and these mechanisms can be part of the cellular stress response. However, the mechanisms that trigger transposition and how this process is linked to physiological conditions such as nutritional status remain a mystery.

Bacteriophage Mu is a model of regulated transposition, for it functions within its host as a fully active transposon as well as a virus. Transposition of the Mu genome into the host chromosome establishes lysogeny and replicates Mu DNA for lytic development (for a review see reference 103). Following injection into a host cell, the phage DNA is inserted at a random site of the host chromosome (104) without replicating Mu DNA to form a simple insert (31).

This can lead to the establishment of lysogeny, requiring the Mu repressor encoded by gene *c* (Fig. 1). Alternatively, Mu DNA may then continue to transpose replicatively to form cointegrates and related transposition products. In replicative transposition, Mu DNA is replicated each time it is inserted into a new site until 100 to 200 copies are produced. As a protein that regulates the decision between lysogeny and lytic development, the Mu repressor must be responsive to the changing physiological conditions of the host. Mechanisms that lead to the inactivation of the repressor can provide insights as to how transposition can be linked to host physiology.

The Role of the Phage-Encoded MuA and MuB Proteins

The Mu genome is 36,717 bp in length (75), encoding proteins and *cis*-acting elements that direct transposition. The key sequences required for transposition include the coding sequences for the transposition proteins MuA and MuB, the operator that regulates their expression, and the Mu DNA end sequences that comprise the binding sites for MuA and the attachment sites to host DNA (Fig. 1). MuA transposase catalyzes cleavage and joining through two sequential reactions (72). First, a hydrolysis reaction introduces a nick at the two attachment sites of the transposon to produce Mu 3'-OH ends. Then the Mu ends are integrated into target DNA by transesterification. In this reaction each Mu 3'-OH end makes a nucleophilic attack at phosphodiester linkages located on opposite strands of target DNA and separated by 5 bp. This strand exchange produces a forked DNA structure (16, 92) at each Mu end (Fig. 2a and b). One Mu strand remains attached to the

---

Stella H. North and Hiroshi Nakai • Department of Biochemistry and Molecular Biology, Georgetown University Medical Center, 3900 Reservoir Rd. NW, Washington, DC 20007.

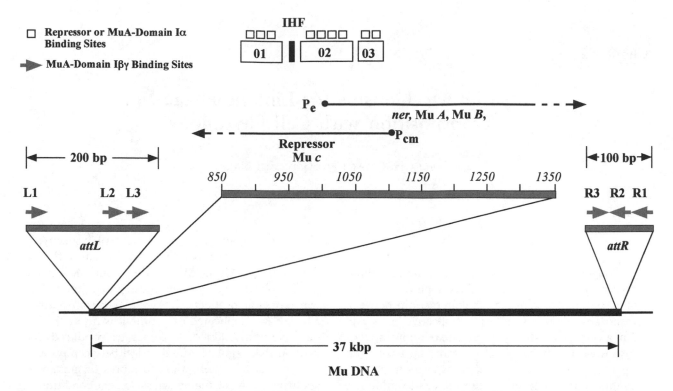

Figure 1. Bacteriophage Mu genome. Regions for Mu DNA transposition include a 200-bp left-end (*attL*) sequence and a 100-bp right-end (*attR*) sequence, which contain binding sites for domain Iβγ of MuA and make up attachment sites to host DNA. The operator sequences (O1, O2, and O3) are contained within a 200-bp region and regulate the P_e and P_{cm} promoters. They contain binding sites for the Mu repressor, encoded by the *c* gene. O1 and O2 also include an internal transpositional enhancer that is recognized by domain Iα of MuA and required for the assembly of transposase under physiological conditions.

former Mu vector and makes up the lagging strand arm of the potential replication fork. The opposite Mu strand attaches to the target of transposition and becomes the leading strand arm. Replication initiates at one of the two forks and duplicates Mu to form the final transposition product, the cointegrate.

The MuB protein is an ATPase and DNA binding protein (11, 68) and plays an accessory role to the MuA transposase. In vivo, Mu transposes at a low frequency and cannot replicate without the MuB function (18, 81, 112). MuB significantly increases the rate of strand exchange by capturing target DNA (68, 77) and by allosterically activating MuA transposase (6, 100, 101). While MuB's ability to activate MuA transposase appears to be sufficient for its role in enhancing nonreplicative Mu transposition (10), its function in binding and capturing target DNA is necessary for promoting Mu DNA replication (88).

Organization of the Mu Genome as a Transposable Element

MuA and MuB transposition functions are regulated by the 196-residue Mu repressor protein that binds cooperatively to the Mu operator sequences (Fig. 1). These sequences are found in three discrete segments (O1, O2, and O3) contained within a 200-bp region (50, 108). The segments contain a total of nine nonsymmetric, 11- to 14-bp consensus sequences that likely make up the individual binding sites of the repressor (50, 108). Repressor molecules bind sites in O1, O2, and O3 to form discrete nucleoprotein complexes (89) that shut down the P_e and P_{cm} promoters. The P_e promoter drives expression of transposition functions MuA, MuB, and *ner*, encoding another repressor that negatively regulates transcription from both the P_{cm} and P_e promoters (see reference 28 for a review). In the absence of the Mu *c* repressor, an integration host factor binding site located between O1 and O2 enhances transcription directed by the P_e promoter contained within O2 while simultaneously decreasing transcription directed by the P_{cm} promoter contained within O3 (50). In contrast, in the presence of repressor, integration host factor facilitates the shutdown of the P_e promoter by stabilizing repressor binding to the operator (2, 7, 23), presumably by promoting formation of a looped, repressor-operator complex that cross-links

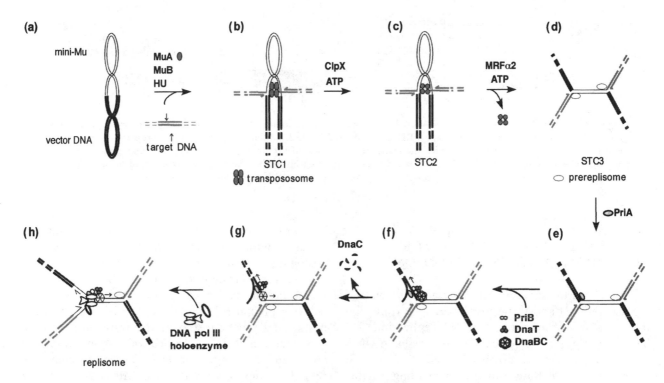

Figure 2. Mu replicative transposition. The Mu transposition reaction catalyzed in vitro is depicted. (a) MuA transposase binds to sites on the left and right ends of mini-Mu DNA, which is harbored on a supercoiled plasmid vector (thick black lines). In the presence of the histone-like protein HU, MuA is assembled into a stable tetramer that holds the two Mu ends together in a synaptic complex. MuB plays an accessory role, functioning in target DNA capture and activating transposase activity. (b) MuA introduces nicks at the Mu ends and transfers them to phosphodiester linkages that are 5 bp apart (indicated by full arrows) on target DNA (thick gray lines) to form STC1. (c) Strand exchange produces a fork at each Mu end, the target DNA providing 3'-OH ends (indicated by half arrows) that can be potentially used as primer for leading strand synthesis. The molecular chaperone ClpX then destabilizes the quaternary interactions of the MuA transpososome to convert STC1 to STC2. (d) Factors present in a host extract (MRFα2) displace the transpososome and form the prereplisome STC3. PriA binds to one of the forked structures formed by strand exchange (e), and this initiates the assembly of the primosome by bringing PriB, DnaT, and DnaBC to the fork (f and g). (f) The 3' to 5' helicase of PriA can function to unwind the lagging strand arm of the fork to create a binding site for the DnaB helicase. DnaC disassembles from the DnaBC complex as the DnaB is loaded onto the lagging strand template (g), and DNA polymerase III holoenzyme is bound to the fork to complete the formation of the replisome (h).

O1 and O2 (89, 108). The P$_{cm}$ promoter, which drives the expression of the *c* gene (50), is also suppressed by the Mu repressor but requires much higher repressor concentrations, which may cause O3 to be a part of the overall nucleoprotein complex. This mechanism may account for autoregulation of the repressor gene (26).

In addition to the transcriptional regulation of factors required for Mu transposition, the binding of the repressor to the operator suppresses transposition by a second means. The Mu repressor and the MuA transposase both have DNA binding domains (DBDs) at their N termini. The repressor DBD (Fig. 3), which binds operator sequences, is made up of approximately the first 80 N-terminal residues (39, 108). The MuA DBD, found in domain I at the MuA N terminus, is made up of the first 243 out of 663 total

residues (80). The N-terminal portion of this domain, domain Iα, is homologous to the Mu repressor DBD (32). These homologous domains bind to the same sequences within the Mu operator and adopt a winged helix-turn-helix structure (13, 39, 115). The C-terminal portion of the MuA DBD, domain Iβγ, binds to the Mu end sequences. As a binding site for MuA, the operator sequences O1 and O2 act as an internal enhancer (57, 74) (Fig. 1) that increases transposition rate (58, 102) by facilitating transpososome assembly (73, 99). Catalysis of Mu strand exchange requires the assembly of inactive MuA monomers (53) into an active oligomeric transpososome (5, 56, 73, 98) that tightly binds to both Mu ends. Thus, by competing for MuA binding sites in the operator, the repressor can directly inhibit the strand exchange reaction.

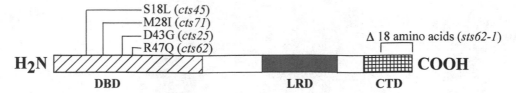

Figure 3. Domains of the Mu repressor. Depicted are the DNA binding domain (DBD), which makes up approximately 80 N-terminal residues; the leucine-rich domain (LRD [L121 to L162]), which is thought to function in repressor oligomerization; and the C-terminal domain (CTD), which modulates repressor degradation by ClpXP protease and DNA binding (I170 to V196). The indicated mutations in the DBD result in temperature-sensitive DNA binding (*cts*). Deletion of the last 18 amino acids (*sts62-1*) from the C terminus suppresses the *cts* DBD mutations and confers dominance over *vir*.

HOW Mu EXPLOITS CONSERVED CELLULAR FUNCTIONS FOR REPLICATIVE TRANSPOSITION

Mechanisms involved in Mu transposition and replication (12) and those involved in the transition from Mu strand exchange to DNA replication (78) have recently been reviewed. The processes in DNA replication relevant to potential mechanisms in Mu derepression and those properties providing insights about Mu's relationship with its host are summarized here.

The ClpX Molecular Chaperone

In addition to the homologous DNA binding domains at the N terminus, another common feature shared by MuA transposase and the repressor is that they are both substrates for the *Escherichia coli* molecular chaperone ClpX (24, 52, 59, 70, 111). ClpX promotes both the degradation of the Mu repressor for derepression and the remodeling of the MuA transpososome for replicative transposition. ClpX, a member of the Clp/Hsp100 family of ATPases, possesses two modes of function: (i) recognition and unfolding of protein substrates and (ii) delivery of protein substrates to the peptidase active sites of the ClpP protease (29, 110, 116). The ClpP protease consists of two stacked heptameric rings which contain serine active sites within the inner chambers. The hexameric molecular chaperones, ClpX and ClpA, each with distinct substrate specificity, can form a stacked complex at either end of the ClpP barrel for delivery of the substrate into the protease chamber (30, 47, 83, 109). Recognition determinants for ClpX have been identified at the N or C terminus of ClpX substrates (21, 27, 62); three recurring motifs at the N terminus and two at the C terminus have been found among *E. coli* proteins recognized by ClpXP (22). ClpX binding is especially dependent upon the presence of substrate recognition motifs; ClpX cannot stably bind even unfolded substrates if they lack such

a motif (97). In contrast, while ClpA is unable to interact with native proteins lacking ClpA-specific recognition signals, ClpA can bind to those same proteins in their denatured state (35).

The molecular chaperone function of ClpX plays a pivotal role in Mu replicative transposition, catalyzing the first step in the transition from recombination to DNA replication. Upon completion of strand exchange, the transpososome remains in a tight complex with the two Mu ends in what is known as the type II transpososome or the strand-transfer complex (STC) (Fig. 2b) (98), posing as an impediment to the assembly of a replisome (51). The action of ClpX converts the transpososome into an altered, more fragile form called the type III transpososome, or STC2 (Fig. 2c), which can be readily disassembled by moderate NaCl concentrations or heparin challenge, conditions that cannot disassemble the type II transpososome (52). The conversion does not necessarily require recognition of every protomer in the transpososome (8). When a transpososome is assembled using oligonucleotides containing the Mu right end, recognition of a single protomer destabilizes the entire core tetramer. The type III transpososome is essential in the transition to DNA synthesis, permitting replication by a specific pathway dependent on *E. coli* replication proteins PriA, PriB, DnaT, DnaB, DnaC, and the DNA polymerase III holoenzyme (Fig. 2d to h) and inhibiting DNA synthesis that can be nonspecifically initiated on a deproteinized DNA template (44, 45, 51, 52, 79).

The presence of ClpP does not promote degradation of the MuA protomers that make up the type III transpososome (45). Even after the action of ClpXP, the transpososome continues to maintain the two Mu ends in a synaptic complex, albeit less stably. The transpososome is eventually disassembled for the purpose of DNA synthesis but left undegraded in its oligomeric form (52, 79). The last 10 residues of the MuA C terminus contain the ClpX recognition determinant, which promotes ClpXP-dependent degradation of phage P22 Arc repressor when attached to its C terminus (62). Thus, some property of MuA transpososome, distinct from its recognition determinant, renders it resistant to proteolysis. This demonstrates the ability of Mu to employ ClpX to remodel protein complexes without destroying them so that they can perform additional functions. Although ClpP is not needed for Mu DNA replication, ClpX must work as part of a chaperone-linked protease, ClpXP, to promote Vir-induced Mu derepression (45).

Primosome Assembly Proteins

The requirement of PriA, PriB, and DnaT for Mu DNA replication and how they are exploited for replisome assembly at the Mu fork (42–44) indicate a distinct strategy for exploiting the host apparatus. These proteins play a vital role in host chromosomal replication by promoting restart of DNA replication when replication forks stall (14, 15, 66). Homologous recombination functions are involved in repairing lesions created by arrested forks and preparing the template for resumption of DNA synthesis. Binding of PriA to the DNA template leads to the assembly of a multiprotein complex known as the primosome (66), which includes primase and DnaB helicase needed for the propagation of the replication fork. PriA binds to forked DNA structures (69) found in recombination intermediates and arrested replication forks. Since replication forks can potentially stall at any site on the chromosome, the signal for initiation depends on the DNA structure at the arrested fork or D loop rather than an origin sequence to which initiator proteins such as DnaA protein bind. PriA is thus able to assemble a primosome at the D loop (63) and Mu fork (43), binding at a Mu end being dependent upon the presence of a forked DNA structure and not a DNA sequence (41).

For Mu DNA replication, PriA functions much like an initiator protein at an origin of DNA replication. First, it binds to the DNA template at which DNA synthesis initiates. Second, PriA's helicase activity can play a vital role at the Mu fork in opening the DNA duplex for loading of the major replicative helicase DnaB on the lagging strand template of the fork (Fig. 2e to g) (41, 42). Third, PriA promotes assembly of the replisome by initiating the primosome assembly. Unlike many bacteriophages, Mu appears to lack an origin like oriC with a specific DNA sequence recognized by an initiator protein. For the initiation of Mu replication, specific DNA sequences are apparently required only for the strand exchange reaction catalyzed by MuA transposase. Mu seems to engage the host replication apparatus in

the same way that host replication proteins are re-engaged for the restart of replication.

Viruses that exploit the host replication apparatus for their own propagation generally have evolved DNA elements such as origins or proteins that specifically interact with the host replication proteins. Lacking origin sequences, Mu apparently exploits the branched DNA structure created by strand exchange to promote the assembly of the replisome. However, there are factors involved in transition from transpososome to replisome yet to be defined. These factors (Mu replication factors α2 [MRFα2]) promote transpososome disassembly and form a prereplisome in preparation for the assembly of the primosome (52, 78) (Fig. 2c to h). Assembly of the prereplisome at the Mu fork is dependent on the presence of the transpososome or a DNA structure maintained by transpososome binding (79), but whether a specific sequence from the Mu ends is required remains to be determined. The data so far indicate that Mu exploits for its own replication a process vital in all cells for chromosomal replication—the mechanism for restarting arrested forks. This minimizes specific interactions between Mu-encoded DNA elements or proteins with the host replication proteins it exploits. Such a strategy may allow Mu to be more versatile in exploiting the replication apparatus of a wide range of hosts than viruses that encode DNA elements or proteins interacting directly with the host apparatus. In a similar fashion, Mu may potentially exploit the cellular mechanism for rescuing stalled ribosomes as a way of linking transposition with cell physiology while minimizing specific host-virus interactions necessary to produce this link. We speculate below that aborted translation products from repressor mRNA might be exploited as inducers of transposition, a mechanism in which a basic cellular function produces signal molecules from a Mu-encoded mRNA.

CHARACTERISTICS OF THE Mu REPRESSOR AND THE FUNCTIONS OF ITS C-TERMINAL DOMAIN AND DBD

Two types of repressor mutants which induce lytic development in Mu lysogens have provided insight as to how Mu derepression may be triggered. Proteins with a temperature-sensitive mutation in the DBD (cts45, S18L; cts71, M28I; cts25, D43G; cts62, R47Q [Fig. 3]) bind to operator DNA at 30°C but not at 42°C, allowing thermoinduction of lysogens (37, 108). The second class of mutants are dominant-negative forms of repressor (Vir) having a virulent phenotype (106). The vir3060 and vir3061 (or

```
                170        180        190    196
Rep       ILSKYGIHEQESVVVPSQEPQEVKKAV
Vir3060   ILSKYGIHEQESVVVPFRNHRR
                         *****
Vir3051   IFCQSMGFMNRKVL
          ************
```

Figure 4. Dominant-negative forms of repressor. The sequence within the Rep CTD is shown. Vir3060 and Vir3051 are produced by frameshift mutations that alter the last 11 to 26 residues of the C terminus. The residues of the Vir proteins that differ from Rep are marked with an asterisk.

vir3051) alleles are produced by frameshift mutations that alter the last 11 to 26 residues at the C terminus (Fig. 4) (25). Although the wild-type repressor (Rep) is stable in vivo, the Vir proteins are rapidly degraded by ClpXP protease and cause Rep to be degraded as well (25, 70, 111). The altered sequence at the C terminus of Vir contains a determinant for ClpX recognition. When attached to the C terminus of two heterologous proteins, CcdA and CcdB encoded by the F plasmid, the last seven C-terminal residues of the Vir3061 protein converted CcdA, but not CcdB, into a ClpXP protease substrate (54). The results indicate that these residues are necessary but not sufficient for recognition and degradation by ClpXP protease.

The C-terminal domain (CTD) of Rep plays an important role not only in eliciting thermolability of *cts* DBD mutations present in *cis* but also in promoting Rep degradation induced by Vir expressed in *trans* (107). Deletions of 1 to 18 residues from the C terminus of repressor suppress temperature sensitivity of *cts* mutations in the DBD. Moreover, these deletion mutants promote gain of dominance over the *vir* alleles (76, 107). Thus, these *sts* (survival of temperature shifts or suppression of temperature sensitivity) mutant proteins, such as the *sts62-1*Rep missing 18 residues at the C terminus (Fig. 3), are not rapidly degraded in the *vir* genetic background. These results indicate that the CTD functions as a negative modulator, reducing DNA binding affinity as well as providing a determinant that promotes derepression through repressor degradation. DNA binding studies have indicated that truncation of the CTD by 18 residues (the *sts62-1* mutation) stabilizes the binding of *cts62* repressor to operator DNA at elevated temperatures (107). Vogel et al. (107) proposed that the CTD performs this function by destabilizing the folding of the DBD or altering the multimeric structure of the repressor such that the DNA binding or necessary repressor subunit interactions are sterically hindered. They also proposed that the CTD's role in transmitting protease sensi-

tivity may be a result of its ability to stimulate subunit exchange that facilitates entry of Vir into Rep oligomeric complexes, thereby introducing determinants for protease recognition.

Fluorescence resonance energy transfer experiments examining energy transfer from two tryptophan residues in the DBD to a fluorescent probe attached at the C terminus have indicated movement of the repressor CTD as a result of DNA binding (84). When the repressor is not interacting with DNA, the CTD is found in close proximity to the DBD, but upon interaction with DNA, the CTD moves away. At 42°C this movement is prevented in the *cts62* repressor but not the wild-type repressor. These results are consistent with the hypothesis that the CTD hinders interaction between the DBD and DNA. Alternatively, the CTD, in close proximity to the DBD, may hinder oligomerization and consequently destabilize the repressor protomers. Because the repressor can be found in large oligomeric complexes in the absence of DNA (3, 89), conditions under which the CTD is found in close proximity to the DBD (84), we favor the hypothesis that CTD can sterically block DNA binding rather than obstruct repressor oligomerization. Truncation of the CTD may favor stable binding of the lower-affinity *cts62* repressor to DNA by eliminating the need to move the CTD out of the way. The presence of the CTD in the repressor molecule creates a trypsin-hypersensitive site within the DBD (R70), most likely produced by the interaction of the CTD with the DBD (84). This interaction is apparently more stable in the *cts62* repressor, for this site is even more sensitive to trypsin than the analogous site on wild-type repressor. Thus, the CTD's interaction with the DBD may play an important role in modulating DNA binding activity of the repressor.

A second function of the CTD's movement is to influence the repressor's susceptibility to degradation by ClpXP. Rep is a ClpXP protease substrate and is degraded with essentially the same catalytic rate constant as Vir (111). But Vir is degraded more rapidly with a Michaelis constant (K_m) 20 times lower than that of Rep. In vivo, when Rep interacts with DNA (even nonspecific DNA with no operator DNA sequences), Rep essentially becomes resistant to proteolysis by ClpXP. In contrast, Vir remains susceptible even when it is part of a repressor-operator complex. Vir induces Rep to become sensitive to ClpXP even in the presence of DNA and accelerates degradation of Rep molecules by reducing the K_m to the same value as for its own degradation.

Vir induces a conformational change in Rep, resulting in the exposure of the Rep CTD. When the environmentally sensitive fluorescent probe MIANS

(2-(4'-maleimidylanilino) naphthalene-6-sulfonic acid) is attached to the C terminus of Rep, the emission spectrum shows a hydrophobic environment, suggesting that the CTD is buried in the tertiary structure of the protein (84). In contrast, the emission spectrum of the MIANS bound to the C terminus of Vir3060 displays a much more polar environment. Moreover, the rate of attachment of the MIANS probe to the Vir C terminus is 10-fold greater than to the Rep C terminus, indicating that the Vir CTD is more accessible than the Rep CTD. Other lines of evidence implicate Vir as having an active role in inducing Rep CTD to become accessible to the protease ClpXP (67). In the presence of Vir, the MIANS attached to the Rep C terminus enters a polar environment comparable to the environment of MIANS bound to the C terminus of Vir. The presence of Vir accelerates the rate of attachment of MIANS to the Rep C terminus. When a cyclic AMP-dependent protein kinase motif is attached to the C terminus of Rep, it is phosphorylated at a relatively low rate. Phosphorylation of the Rep C terminus is accelerated by the presence of Vir. In contrast, Vir cannot accelerate the phosphorylation of the kinase motif placed at the Rep N terminus, which is also phosphorylated at a relatively low rate. These results indicate that Vir can induce movement of Rep CTD which results in the accelerated exposure of ClpX determinants.

These results have led to the hypothesis that Rep may have its own ClpX recognition determinant at its C terminus (111) and that this determinant is essential for Rep degradation induced by Vir. The last five residues of Rep (VKKAV), which resemble a motif at the MuA C terminus, are capable of acting as a ClpX recognition determinant when attached to a heterologous protein (16a). Although both the altered C-terminal sequence of Vir and the analogous wild-type sequence may be recognized with high affinity by ClpX, the Vir C-terminal sequence is more exposed and therefore more accessible. Vir may promote Rep degradation by interacting with Rep molecules and inducing a conformational change that exposes the ClpX determinants within the Rep CTD. In vitro, heat denaturation increases the rate of Rep degradation by ClpXP similar to the effect of Vir on native Rep (76). In contrast, *sts62-1*Rep and mutants with certain single amino acid replacements in the CTD not only exhibit dominance over *vir* but also are resistant to ClpXP even after heat denaturation. These observations are consistent with the hypothesis that the exposure of the Rep CTD promotes efficient Rep degradation.

In addition to the repressor DBD and the CTD, a third domain plays a critical role in the establishment of repression. A leucine-rich domain (LRD) spanning L121 to L162 is believed to function in repressor oligomerization (108). Several lines of evidence indicate that this domain is needed for high-affinity DNA binding. The DBD domain by itself binds specifically to operator sequences, but the DBD peptides bind with lower affinity (dissociation constant, K_d, of 18 μM to individual repressor binding sites) compared with the binding affinity of full-length repressor (K_d of 3 to 5 nM to operator DNA) (39, 108). DBD peptides expressed in vivo have no repressor activity (82, 105), but when fused with the C-terminal fragment of β-galactosidase, which contains a domain for tetramerization, repressor activity is observed (108). The *cts4* mutation, which encodes an L129V alteration within the LRD, alters the oligomeric structure of the repressor protein as detected by protein cross-linking and destabilizes DNA binding at elevated temperatures (90). Thus, the LRD is likely to play an important role in formation of a higher-order complex of Rep and operator DNA, and it can potentially play an important role in linking Mu transposition with cell physiology.

PHYSIOLOGICAL CONDITIONS THAT PROMOTE Mu DEREPRESSION

The classic example of how cellular signals can lead to viral development is the derepression of phage λ triggered by the SOS-inducing signal in *E. coli*. Arrested replication forks resulting from DNA damage by agents such as UV light and mitomycin C lead to the activation of RecA, which promotes cleavage of the lambda *c*I immunity repressor and subsequently leads to phage lytic development (86, 91). However, the cellular signals that lead to Mu transposition are distinct from those that trigger λ development; neither UV light nor other DNA-damaging agents will induce Mu lytic development (36). In fact, identifying physiological conditions that induce Mu transposition has been especially challenging for two reasons: derepression triggered by specific physiological conditions is not necessarily sufficient to trigger lytic development or transposition, and no known physiological condition of wild-type Mu lysogens (no known chemical or physical treatment) can induce the majority of cells to undergo lytic development. The regulation of Mu transposition has been previously reviewed (105), and we focus here on current studies that may suggest the type of signals involved in triggering Mu derepression.

Two known physiological conditions that promote Mu derepression are carbon starvation and entry into stationary phase (55). Derepression has been termed S derepression to reflect a mostly stochastic (i.e., taking place in only a fraction of the cell culture) process that is induced by starvation and stationary-phase conditions. Under these conditions, the derepression of the P_e promoter, regulated by *cts62* at noninducing temperatures, becomes dependent upon *clpXP* and *lon* protease functions and *rpoS*, which encodes σ^s, the RNA polymerase subunit for regulating stationary-phase and stress response genes.

The link between starvation and stationary phase conditions and Mu transposition was first demonstrated in studies of *araB-lacZ* fusions promoted by a *cts62* prophage. Employing a system designed by Casadaban (9) to generate fusions between selected promoters and the *lacZ* gene, fusion formation was examined in an *E. coli* strain with a Mu*cts62* prophage inserted between the *araBAD* promoter and a promoterless *lacZ* gene. Thermal induction resulted in the imprecise excision of Mu (64), a process that placed *lacZ* under the control of the *araBAD* promoter. Shapiro and his colleagues (55, 64, 65, 93, 94, 96) further examined the physiological conditions and the host functions required for Mu-induced fusion to take place at noninducing temperatures. Fusion does not arise during logarithmic growth but does so upon prolonged incubation of the lysogens under aerobic carbon starvation conditions (65, 71, 93). Fusion formation requires the MuA transposase (96), and sequencing of independent fusions have indicated the presence of rearranged Mu right-end sequences at all junctions between the *araB* and *lacZ* segments. This is consistent with fusions arising from the standard STC intermediate that undergoes an aborted replication or processing event (64). These results indicate that prolonged incubation conditions trigger derepression to promote MuA expression and STC formation, but complete replication of Mu DNA template is somehow prevented, yielding an aborted transposition product.

Using a reporter strain with *lacZ* driven by the P_e promoter and regulated by the *cts62* repressor, Lamrani et al. (55) further delineated the requirements for S derepression and the remaining events leading to fusion formation. Although derepression is essential, it is not sufficient to promote *araB-lacZ* fusions. When Mu*cts62* prophages are thermally induced, fusion formation occurs at an accelerated rate but still requires starvation conditions to complete fusions. These results indicate that host physiology influences not only derepression but also the mechanisms that carry out strand exchange and the processing of the STC.

The link between derepression and growth phase has also been shown by the transposition of a *cts62*

mini-Mu element, which contains a promoterless *lacZ* gene, into sites adjacent to active promoters in growing *E. coli* colonies (95). Transposition is indicated by the expression of β-galactosidase, induced by the integration of mini-Mu elements adjacent to active promoters. Bacterial colonies display concentric rings of transposition, indicating synchronous derepression of Mu at specific stages of colony growth. These patterns of Mu transposition in growing colonies and the Mu-induced *araB-lacZ* fusions formed in starving cells are both blocked in both *clpX* and *clpP* mutants, suggesting that derepression relies on the degradation of the repressor by ClpXP protease.

Recent evidence implicates a role for the Mu repressor CTD in S derepression and regulation of transposition. As previously discussed, the repressor lacking 18 residues at the C terminus (*sts62-1* mutation) is stable in vivo even in the presence of Vir and has been found to be a superrepressor in vivo, being able to shut down the P_e promoter approximately sixfold more effectively than the wild-type repressor (107). This most likely reflects the very low level of spontaneous induction permitted by the *sts* repressor, induction that may be triggered by cellular signals when the repressor has an intact CTD. As further discussed below, S derepression is prevented when truncated repressor molecules missing the CTD accumulate in host cells with *ssrA* knockout mutations (85).

Spontaneous Mu lytic development may take place predominantly in exponentially growing cells, but Mu derepression levels are significantly higher in populations of cells under starvation or stationary-phase conditions (55, 85). Although phage development would not be expected to occur at optimal levels under such conditions, it is speculated that the Mu-induced genetic rearrangements triggered by conditions of stress may be able to confer upon the host a selective advantage for adapting to stressful environmental changes (105). Other transposable elements have displayed the potential for promoting adaptation by the host. For example, insertion sequence IS*50*R confers upon *E. coli* K-12 a growth rate advantage proportional to the number of elements it harbors (33). Characterization of heat shock-resistant *Saccharomyces cerevisiae* mutants has shown that multistress resistance may be achieved through insertion of retrotransposon Ty into adenylate cyclase gene CYR1 (38). In addition, under growth-limiting conditions such as glucose deprivation, P1, P2, and Mu lysogens display greater fitness relative to their isogenic counterparts (17). In view of what is known of the Mu repressor with its CTD function and the variable physiological conditions that influence Mu transposition, an important question emerges as to how the Mu repressor can be an effective sensor for host physiology under conditions of stress.

POTENTIAL SIGNAL MOLECULES THAT PROMOTE Mu DEREPRESSION

The dominant-negative properties of Vir and how it promotes Rep degradation suggest a general mechanism by which Mu derepression might be triggered. During conditions of stress, Vir-like molecules may be formed, and they in turn may promote the degradation of a larger, unmodified repressor population. Since frameshift mutations give rise to *vir* alleles, it is conceivable that translational frameshifting could give rise to Vir-like molecules that induce derepression. However, there is no evidence that such frameshifting occurs and no apparent reason why it should occur at high frequency under conditions for S derepression. An attractive alternative to frameshifting is tagging by the host *ssrA* function, which can convert repressor peptides into ClpXP substrates. Unlike frameshifting, tagging would be expected to occur at increased frequencies under starvation and stationary-phase conditions.

The *ssrA* gene encodes a small RNA molecule with properties of both an alanine tRNA and mRNA (19, 20, 46, 49, 113). Stalling of ribosomes on mRNA (4, 40) can be caused by the ribosomes encountering a stop codon, the end of a truncated mRNA, or lesions in mRNA (1, 34, 87). The *ssrA* gene product functions in two ways to alleviate this dysfunction. First, it rescues the stalled ribosomes by releasing them from the mRNA. Second, it modifies the partially translated peptide with an 11-amino-acid degradation tag (AANDENYALAA). This is accomplished by addition of the alanine from the charged *ssrA* RNA to the stalled peptide. The ribosome is then transferred to an open reading frame on *ssrA* RNA in order to add another 10 residues (46). The ribosome-associated protein SspB, which binds *ssrA*-tagged peptides and enhances their degradation by ClpXP (60), is elevated by conditions of carbon, nitrogen, and phosphate starvation (114), a process that may be designed to cope with higher levels of *ssrA*-tagged peptides under starvation conditions.

Ranquet et al. (85) demonstrated that when the Mu *c* gene is expressed in an *ssrA* knockout strain, repressor molecules that are truncated at the C terminus accumulate. These molecules are like the *sts* repressors, having a deletion in the CTD and being no longer capable of negatively modulating DNA binding and promoting repressor degradation. In this strain, S derepression is blocked (85), confirming the importance of the CTD in linking derepression with host physiology and also implicating the *ssrA* function as a key factor in preventing truncated, *sts*-like repressor molecules from accumulating in the cell. A high incidence of ribosome stalling may promote accumulation of truncated repressor molecules analogous to the *sts* repressors. Ranquet et al. (85) have suggested that the higher concentration of *ssrA* RNA present in stationary phase versus the logarithmic phase (48) may be partly responsible for stimulating S derepression. The *ssrA* tagging system prevents the accumulation of truncated repressor and increases the proportion of full-length repressor through its ability to rescue stalled ribosomes. That is, the repressor population would be most responsive to physiological signals that promote its inactivation when *sts*-like repressor molecules are present at minimal levels. We propose a second pathway for S derepression in which the attachment of *ssrA* degradation tags to repressor peptides generates a positive signal for Mu derepression.

Repressor peptides bearing an *ssrA* tag are able to confer a dominant-negative phenotype like Vir (82). When nested deletions of Rep were generated starting at the C terminus and the tag coding sequence was added at the deletion endpoint, the resulting tagged repressor peptides were generally found to be potent inducers of Mu derepression. The most potent inducers are tagged DBD peptides, in which the tag is attached near or at the C-terminal boundary of the DBD. These peptides promote *clpP*-dependent reduction in cellular Rep levels. However, tagged peptides with considerable deletions of the DBD are virtually deficient in their ability to promote derepression. Tagged peptides with an intact DBD but without the LRD are unable to shut down either the P_e or P_{cm} promoters, but they can still physically interact with Rep to transmit protease sensitivity from the peptide to Rep. These results indicate that repressor peptides with a degradation tag can transmit protease sensitivity to Rep even when they lack the LRD, which is thought to be involved in repressor oligomerization. Thus, tagged repressor peptides can potentially act as signal molecules that trigger Mu derepression. An increased incidence of ribosome stalling on repressor mRNA may lead to the formation of tagged repressor peptides that induce repressor degradation.

The CTD's influence on DNA binding as well as repressor degradation represents two potential pathways by which derepression may be triggered. These two distinct functions of the CTD may be coupled when the repressor is targeted by the ClpXP protease. When Rep is in solution without DNA, it is degraded at a high K_m, whereas it is essentially resistant to

ClpXP when bound to DNA. That is, Rep more readily assumes the protease-sensitive conformation when it is not bound to DNA and is in the "closed conformation," in which the CTD is in close proximity to the DBD. Thus, when tagged DBD peptides trigger Rep degradation, they may be inducing a conformation similar to the closed conformation. In fact, the untagged DBD peptide (comprising residues M1 to G78) can promote high levels of derepression even though it does not promote Rep degradation, suggesting that it may inhibit formation of the Rep-operator complex (82). However, it does not seem to be simply competing with Rep for operator sequences since it cannot promote repression by itself and it cannot prevent repression established by the sts62-1 repressor. One possibility is that the DBD is inducing the closed conformation of Rep.

Such a dual mode of Rep inactivation, through induction of a conformation that has low affinity for DNA and a conformation that is susceptible to proteolytic degradation, would be consistent with the different requirements for derepression by various tagged repressor peptides (82). Tagged repressor peptides that have no LRD and no repressor activity generally retain a fair amount of inducing activity in a ClpP⁻ cell or when they bear an $ssrA^{DD}$ tag, a mutant tag whose ClpX recognition determinant has been inactivated by replacing the C-terminal pair of alanines with aspartates (21). In contrast, repressor molecules with intact repressor activity require both the clpX and clpP functions as well as a wild-type ssrA tag to promote derepression. We speculate that under conditions of S derepression, the prevailing population of tagged repressor peptides is capable of inducing a closed or protease-sensitive conformation of Rep. While degradation signals on inducing molecules may facilitate Rep degradation, such signals may not be essential if the inducing molecules themselves have no repressor activity. Even protease-resistant peptides may be capable of inducing Rep to adopt the closed or protease-sensitive conformations. In a heterogeneous population of tagged repressor peptides produced under conditions of S derepression, not every peptide may be able to induce both conformations with equal facility. Thus, both mechanisms for inactivating Rep may be essential for derepression triggered by such peptides.

The mechanism by which Vir proteins or ssrA-tagged repressor peptides promote Rep degradation apparently depends upon their ability to confer conformational changes that expose ClpX recognition determinants in Rep. In vitro, Vir accelerates the degradation of a much larger population of Rep molecules (111). One explanation for this observation is that the conformational changes in Rep in-duced by Vir are allosterically communicated to other Rep protomers. Alternatively, conformational changes may be catalytically disseminated from Vir to Rep molecules. In fact, ClpX can increase Vir-induced exposure of the Rep CTD (67). ClpX may promote protomer exchange needed to turn over Rep-Vir complexes to catalytically convert Rep to the ClpXP-sensitive conformation. Alternatively, the unfolding of repressor protomers within oligomeric structures may serve to expose ClpX determinants in adjacent protomers.

STRATEGY IN LINKING HOST AND TRANSPOSON FUNCTIONS

Mu derepression can have at least two potential outcomes, one that serves to propagate Mu as phage and transposon and another that may benefit the host, both regulated with respect to host physiology. We hypothesize that the incidence of ribosomes stalled on repressor mRNA can be a measure of cell physiology, and the formation of ssrA-tagged repressor peptides above a certain threshold may trigger Mu transposition. Much like Mu's general strategy for exploiting the host replication apparatus, regulation of transposition would be achieved without relying heavily upon specific interactions between the repressor and host-encoded signal molecules and relying instead upon the self-associating properties of the repressor molecule to disseminate a degradation signal. Mu derepression and replication require the Mu repressor and the MuA transposase, respectively, to be recognized by ClpX, which has substrate binding domains that are highly conserved among the Clp/Hsp100 family of molecular chaperones (61). Aside from the role of ClpX in these two processes, no other specific interactions between host and Mu factors may be involved in linking host and viral functions. For the link between recombination and replication, Mu strand exchange creates a DNA structure recognized by the host system for rescuing stalled replication forks. For the link between transposition and cell physiology, the host system for rescuing stalled ribosomes may be exploited to produce tagged peptides of the repressor under starvation conditions, peptides that interact with the functional Rep population to induce and propagate a conformational change.

Acknowledgments. We thank Mick Chandler for critically reading the manuscript.

Work in the Nakai laboratory has been supported by grants from the National Institutes of Health (GM58265 and GM49649). S.H.N. is a recipient of a predoctoral training grant from the Department of Defense Breast Cancer Research Program (DAMD 17-98-1-8090).

REFERENCES

1. Abo, T., T. Inada, K. Ogawa, and H. Aiba. 2000. SsrA-mediated tagging and proteolysis of LacI and its role in the regulation of lac operon. *EMBO J.* **19:**3762–3769.

2. Alazard, R., M. Bétermier, and M. Chandler. 1992. *Escherichia coli* integration host factor stabilizes bacteriophage Mu repressor interactions with operator DNA in vitro. *Mol. Microbiol.* **6:**1707–1714.

3. Alazard, R., C. Ebel, V. Venien-Bryan, L. Mourey, J. P. Samama, and M. Chandler. 1998. Oligomeric structure of the repressor of the bacteriophage Mu early operon. *Eur. J. Biochem.* **252:**408–415.

4. Atkins, J. F., and R. F. Gesteland. 1996. A case for *trans* translation. *Nature* **379:**769–771.

5. Baker, T. A., and K. Mizuuchi. 1992. DNA-promoted assembly of the active tetramer of the Mu transposase. *Genes Dev.* **6:**2221–2232.

6. Baker, T. A., M. Mizuuchi, and K. Mizuuchi. 1991. MuB protein allosterically activates strand transfer by the transposase of phage Mu. *Cell* **65:**1003–1013.

7. Bétermier, M., P. Rousseau, R. Alazard, and M. Chandler. 1995. Mutual stabilisation of bacteriophage Mu repressor and histone-like proteins in a nucleoprotein structure. *J. Mol. Biol.* **249:**332–341.

8. Burton, B. M., T. L. Williams, and T. A. Baker. 2001. ClpX-mediated remodeling of Mu transpososomes: selective unfolding of subunits destabilizes the entire complex. *Mol. Cell* **8:**449–454.

9. Casadaban, M. J. 1976. Transposition and fusion of the lac genes to selected promoters in Escherichia coli using bacteriophage lambda and Mu. *J. Mol. Biol.* **104:**541–555.

10. Chaconas, G., E. B. Giddens, J. L. Miller, and G. Gloor. 1985. A truncated form of the bacteriophage Mu B protein promotes conservative integration, but not replicative transposition, of Mu DNA. *Cell* **41:**857–865.

11. Chaconas, G., G. Gloor, and J. L. Miller. 1985. Amplification and purification of the bacteriophage Mu encoded B transposition protein. *J. Biol. Chem.* **260:**2662–2669.

12. Chaconas, G., and R. M. Harshey. 2002. Transposition of phage Mu DNA, p. 384–402. *In* N. L. Craig, R. Craigie, M. Gellert, and A. M. Lambowitz (ed.), *Mobile DNA II.* ASM Press, Washington, D.C.

13. Clubb, R. T., J. G. Omichinski, H. Savilahti, K. Mizuuchi, A. M. Gronenborn, and G. M. Clore. 1994. A novel class of winged helix-turn-helix protein: the DNA binding domain of Mu transposase. *Structure* **2:**1041–1048.

14. Cox, M. M. 1998. A broadening view of recombinational DNA repair in bacteria. *Genes Cells* **3:**65–78.

15. Cox, M. M., M. F. Goodman, K. N. Kreuzer, D. J. Sherratt, S. J. Sandler, and K. J. Marians. 2000. The importance of repairing stalled replication forks. *Nature* **404:**37–41.

16. Craigie, R., and K. Mizuuchi. 1985. Mechanism of transposition of bacteriophage Mu: structure of a transposition intermediate. *Cell* **41:**867–876.

16a. Defenbaugh, D. A., and H. Nakai. 2003. A context-dependent ClpX recognition determinant located at the C terminus of phage Mu repressor. *J. Biol. Chem.* **278:**52333–52339.

17. Edlin, G., L. Lin, and R. Bitner. 1977. Reproductive fitness of P1, P2, and Mu lysogens of *Escherichia coli. J. Virol.* **21:**560–564.

18. Faelen, M., and A. Toussaint. 1978. Stimulation of deletions in the *Escherichia coli* chromosome by partially induced Mucts62 prophages. *J. Bacteriol.* **136:**477–483.

19. Felden, B., K. Hanawa, J. F. Atkins, H. Himeno, A. Muto, R. F. Gesteland, J. A. McCloskey, and P. F. Crain. 1998. Presence and location of modified nucleotides in Escherichia coli tmRNA: structural mimicry with tRNA acceptor branches. *EMBO J.* **17:**3188–3196.

20. Felden, B., H. Himeno, A. Muto, J. P. McCutcheon, J. F. Atkins, and R. F. Gesteland. 1997. Probing the structure of the Escherichia coli 10Sa RNA (tmRNA). *RNA* **3:**89–103.

21. Flynn, J. M., I. Levchenko, M. Seidel, S. H. Wickner, R. T. Sauer, and T. A. Baker. 2001. Overlapping recognition determinants within the ssrA degradation tag allow modulation of proteolysis. *Proc. Natl. Acad. Sci. USA* **98:**10584–10589.

22. Flynn, J. M., S. B. Neher, Y. I. Kim, R. T. Sauer, and T. A. Baker. 2003. Proteomic discovery of cellular substrates of the ClpXP protease reveals five classes of ClpX-recognition signals. *Mol. Cell* **11:**671–683.

23. Gama, M. J., A. Toussaint, and N. P. Higgins. 1992. Stabilization of bacteriophage Mu repressor-operator complexes by the *Escherichia coli* integration host factor protein. *Mol. Microbiol.* **6:**1715–1722.

24. Geuskens, V., A. Mhammedi-Alaoui, L. Desmet, and A. Toussaint. 1992. Virulence in bacteriophage Mu: a case of *trans*-dominant proteolysis by the *Escherichia coli* Clp serine protease. *EMBO J.* **11:**5121–5127.

25. Geuskens, V., J. L. Vogel, R. Grimaud, L. Desmet, N. P. Higgins, and A. Toussaint. 1991. Frameshift mutations in the bacteriophage Mu repressor gene can confer a *trans*-dominant virulent phenotype to the phage. *J. Bacteriol.* **173:**6578–6585.

26. Giphart-Gassler, M., J. Reeve, and P. van de Putte. 1981. Polypeptides encoded by the early region of bacteriophage Mu synthesized in minicells of Escherichia coli. *J. Mol. Biol.* **145:**165–191.

27. Gonciarz-Swiatek, M., A. Wawrzynow, S. J. Um, B. A. Learn, R. McMacken, W. L. Kelley, C. Georgopoulos, O. Sliekers, and M. Zylicz. 1999. Recognition, targeting, and hydrolysis of the lambda O replication protein by the ClpP/ClpX protease. *J. Biol. Chem.* **274:**13999–14005.

28. Goosen, N., and P. van de Putte. 1987. Regulation of transcription, p. 41–52. *In* N. Symonds, A. Toussaint, P. van de Putte, and M. M. Howe (ed.), *Phage Mu.* Cold Spring Harbor Laboratory, Cold Spring Harbor, N.Y.

29. Gottesman, S., W. P. Clark, V. de Crécy-Lagard, and M. R. Maurizi. 1993. ClpX, an alternative subunit for the ATP-dependent Clp protease of *Escherichia coli*: sequence and *in vivo* activities. *J. Biol. Chem.* **268:**22618–22626.

30. Grimaud, R., M. Kessel, F. Beuron, A. C. Steven, and M. R. Maurizi. 1998. Enzymatic and structural similarities between the Escherichia coli ATP-dependent proteases, ClpXP and ClpAP. *J. Biol. Chem.* **273:**12476–12481.

31. Harshey, R. M. 1984. Transposition without duplication of infecting bacteriophage Mu DNA. *Nature* **311:**580–581.

32. Harshey, R. M., E. D. Getzoff, D. L. Baldwin, J. L. Miller, and G. Chaconas. 1985. Primary structure of phage Mu transposase: homology to Mu repressor. *Proc. Natl. Acad. Sci. USA* **82:**7676–7680.

33. Hartl, D. L., D. E. Dykhuizen, R. D. Miller, L. Green, and J. de Framond. 1983. Transposable element IS50 improves growth rate of E. coli cells without transposition. *Cell* **35:**503–510.

34. Hayes, C. S., B. Bose, and R. T. Sauer. 2002. Stop codons preceded by rare arginine codons are efficient determinants of SsrA tagging in *Escherichia coli. Proc. Natl. Acad. Sci. USA* **99:**3440–3445.

35. Hoskins, J. R., S. K. Singh, M. R. Maurizi, and S. Wickner. 2000. Protein binding and unfolding by the chaperone ClpA

and degradation by the protease ClpAP. *Proc. Natl. Acad. Sci. USA* **97:**8892–8897.

36. Howe, M. M. 1987. Phage Mu: an overview, p. 25–39. *In* N. Symonds, A. Toussaint, P. van de Putte, and M. M. Howe (ed.), *Phage Mu.* Cold Spring Harbor Laboratory, Cold Spring Harbor, N.Y.

37. Howe, M. M. 1973. Prophage deletion mapping of bacteriophage Mu-1. *Virology* **54:**93–101.

38. Iida, H. 1988. Multistress resistance of *Saccharomyces cerevisiae* is generated by insertion of retrotransposon Ty into the 5′ coding region of the adenylate cyclase gene. *Mol. Cell. Biol.* **8:**5555–5560.

39. Ilangovan, U., J. M. Wojciak, K. M. Connolly, and R. T. Clubb. 1999. NMR structure and functional studies of the Mu repressor DNA-binding domain. *Biochemistry* **38:**8367–8376.

40. Jentsch, S. 1996. When proteins receive deadly messages at birth. *Science* **271:**955–956.

41. Jones, J. M., and H. Nakai. 1999. Duplex opening by primosome protein PriA for replisome assembly on a recombination intermediate. *J. Mol. Biol.* **289:**503–515.

42. Jones, J. M., and H. Nakai. 2001. *Escherichia coli* PriA helicase: synergism between fork binding and helicase activity stimulates unwinding of arrested forks. *J. Mol. Biol.* **312:**935–947.

43. Jones, J. M., and H. Nakai. 1997. The φX174-type primosome promotes replisome assembly at the site of recombination in bacteriophage Mu transposition. *EMBO J.* **16:**6886–6895.

44. Jones, J. M., and H. Nakai. 2000. PriA and T4 gp59: factors that promote DNA replication on forked DNA substrates. *Mol. Microbiol.* **36:**519–527.

45. Jones, J. M., D. J. Welty, and H. Nakai. 1998. Versatile action of *Escherichia coli* ClpXP as protease or molecular chaperone for bacteriophage Mu transposition. *J. Biol. Chem.* **273:**459–465.

46. Keiler, K. C., P. R. H. Waller, and R. T. Sauer. 1996. Role of a peptide tagging system in degradation of proteins synthesized from damaged messenger RNA. *Science* **271:**990–993.

47. Kim, Y. I., R. E. Burton, B. M. Burton, R. T. Sauer, and T. A. Baker. 2000. Dynamics of substrate denaturation and translocation by the ClpXP degradation machine. *Mol. Cell* **5:**639–648.

48. Komine, Y., M. Kitabatake, and H. Inokuchi. 1996. 10Sa RNA is associated with 70S ribosome particles in Escherichia coli. *J. Biochem.* (Tokyo) **119:**463–467.

49. Komine, Y., M. Kitabatake, T. Yokogawa, K. Nishikawa, and H. Inokuchi. 1994. A tRNA-like structure is present in 10Sa RNA, a small stable RNA from Escherichia coli. *Proc. Natl. Acad. Sci. USA* **91:**9223–9227.

50. Krause, H. M., and N. P. Higgins. 1986. Positive and negative regulation of the Mu operator by Mu repressor and *Escherichia coli* integration host factor. *J. Biol. Chem.* **261:**3744–3752.

51. Kruklitis, R., and H. Nakai. 1994. Participation of bacteriophage Mu A protein and host factors in initiation of Mu DNA synthesis in vitro. *J. Biol. Chem.* **269:**16469–16477.

52. Kruklitis, R., D. J. Welty, and H. Nakai. 1996. ClpX protein of *Escherichia coli* activates bacteriophage Mu transposase in the strand transfer complex for initiation of Mu DNA synthesis. *EMBO J.* **15:**935–944.

53. Kuo, C.-F., A. Zou, M. Jayaram, E. Getzoff, and R. Harshey. 1991. DNA-protein complexes during attachment-site synapsis in Mu DNA transposition. *EMBO J.* **10:**1585–1591.

54. Laachouch, J. E., L. Desmet, V. Geuskens, R. Grimaud, and A. Toussaint. 1996. Bacteriophage Mu repressor as a target

for the *Escherichia coli* ATP-dependent Clp protease. *EMBO J.* **15:**437–444.

55. Lamrani, S., C. Ranquet, M. J. Gama, H. Nakai, J. A. Shapiro, A. Toussaint, and G. Maenhaut-Michel. 1999. Starvation-induced Mucts62-mediated coding sequence fusion: a role for ClpXP, Lon, RpoS and Crp. *Mol. Microbiol.* **32:**327–343.

56. Lavoie, B. D., B. S. Chan, R. G. Allison, and G. Chaconas. 1991. Structural aspects of a higher order nucleoprotein complex: induction of an altered DNA structure at the Mu-host junction of the Mu Type 1 transpososome. *EMBO J.* **10:**3051–3059.

57. Leung, D. W., F. Chen, and D. V. Goeddel. 1989. A method for random mutagenesis of a defined DNA segment using a modified polymerase chain reaction. *Technique* **1:**11–15.

58. Leung, P. C., D. B. Teplow, and R. M. Harshey. 1989. Interaction of distinct domains in Mu transposase with Mu DNA ends and an internal transpositional enhancer. *Nature* **338:**656–658.

59. Levchenko, I., L. Luo, and T. A. Baker. 1995. Disassembly of the Mu transposase tetramer by the ClpX chaperone. *Genes Dev.* **9:**2399–2408.

60. Levchenko, I., M. Seidel, R. T. Sauer, and T. A. Baker. 2000. A specificity-enhancing factor for the ClpXP degradation machine. *Science* **289:**2354–2356.

61. Levchenko, I., C. K. Smith, N. P. Walsh, R. T. Sauer, and T. A. Baker. 1997. PDZ-like domains mediate binding specificity in the Clp/Hsp100 family of chaperones and protease regulatory subunits. *Cell* **91:**939–947.

62. Levchenko, I., M. Yamauchi, and T. A. Baker. 1997. ClpX and MuB interact with overlapping regions of Mu transposase: implications for control of the transposition pathway. *Genes Dev.* **11:**1561–1572.

63. Liu, J., and K. J. Marians. 1999. PriA-directed assembly of a primosome on D loop DNA. *J. Biol. Chem.* **274:**25033–25041.

64. Maenhaut-Michel, G., C. E. Blake, D. R. Leach, and J. A. Shapiro. 1997. Different structures of selected and unselected *araB-lacZ* fusions. *Mol. Microbiol.* **23:**1133–1145.

65. Maenhaut-Michel, G., and J. A. Shapiro. 1994. The roles of starvation and selective substrates in the emergence of *araB-lacZ* fusion clones. *EMBO J.* **13:**5229–5239.

66. Marians, K. J. 2000. PriA-directed replication fork restart in Escherichia coli. *Trends Biochem. Sci.* **25:**185–189.

67. Marshall-Batty, K., and H. Nakai. 2003. Trans-targeting of the phage Mu repressor is promoted by conformational changes that expose its ClpX recognition determinant. *J. Biol. Chem.* **278:**1612–1617.

68. Maxwell, A., R. Craigie, and K. Mizuuchi. 1987. B protein of bacteriophage Mu is an ATPase that preferentially stimulates intermolecular DNA strand transfer. *Proc. Natl. Acad. Sci. USA* **84:**699–703.

69. McGlynn, P., A. A. Al-Deib, J. Liu, K. J. Marians, and R. G. Lloyd. 1997. The DNA replication protein PriA and the recombination protein RecG bind D-loops. *J. Mol. Biol.* **270:**212–221.

70. Mhammedi-Alaoui, A., M. Pato, M.-J. Gama, and A. Toussaint. 1994. A new component of bacteriophage Mu replicative transposition machinery: the *Escherichia coli* ClpX protein. *Mol. Microbiol.* **11:**1109–1116.

71. Mittler, J., and R. E. Lenski. 1990. Further experiments on excisions of Mu from *Escherichia coli* MCS2 cast doubt on directed mutation hypothesis. *Nature* **344:**173–175.

72. Mizuuchi, K., and K. Adzuma. 1991. Inversion of the phosphate chirality at the target site of Mu DNA strand transfer: evidence for a one-step transesterification mechanism. *Cell* **66:**129–140.

73. Mizuuchi, M., T. A. Baker, and K. Mizuuchi. 1992. Assembly of the active form of the transposase-Mu DNA complex: a critical control point in Mu transposition. *Cell* **70:**303–311.

74. Mizuuchi, M., and K. Mizuuchi. 1989. Efficient Mu transposition requires interaction of transposase with a DNA sequence at the Mu operator: implications for regulation. *Cell* **58:**399–408.

75. Morgan, G. J., G. F. Hatfull, S. Casjens, and R. W. Hendrix. 2002. Bacteriophage Mu genome sequence: analysis and comparison with Mu-like prophages in Haemophilus, Neisseria and Deinococcus. *J. Mol. Biol.* **317:**337–359.

76. Mukhopadhyay, B., K. R. Marshall-Batty, B. D. Kim, D. O'Handley, and H. Nakai. 2003. Modulation of phage Mu repressor DNA binding and degradation by distinct determinants in its C-terminal domain. *Mol. Microbiol.* **47:**171–182.

77. Naigamwalla, D. Z., and G. Chaconas. 1997. A new set of Mu DNA transposition intermediates: alternate pathways of target capture preceding strand transfer. *EMBO J.* **16:**5227–5234.

78. Nakai, H., V. Doseeva, and J. M. Jones. 2001. Handoff from recombinase to replisome: insights from transposition. *Proc. Natl. Acad. Sci. USA* **98:**8247–8254.

79. Nakai, H., and R. Kruklitis. 1995. Disassembly of the bacteriophage Mu transposase for the initiation of Mu DNA replication. *J. Biol. Chem.* **270:**19591–19598.

80. Nakayama, C., D. B. Teplow, and R. M. Harshey. 1987. Structural domains in phage Mu transposase: identification of the site-specific DNA-binding domain. *Proc. Natl. Acad. Sci. USA* **84:**1809–1813.

81. O'Day, K. J., D. W. Schultz, and M. M. Howe. 1978. Search for integration-deficient mutants of bacteriophage Mu, p. 48–51. *In* D. Schlessinger (ed.), *Microbiology—1978.* American Society for Microbiology, Washington, D.C.

82. O'Handley, D., and H. Nakai. 2002. Derepression of bacteriophage Mu transposition functions by truncated forms of the immunity repressor. *J. Mol. Biol.* **322:**311–324.

83. Ortega, J., S. K. Singh, T. Ishikawa, M. R. Maurizi, and A. C. Steven. 2000. Visualization of substrate binding and translocation by the ATP-dependent protease, ClpXP. *Mol. Cell* **6:**1515–1521.

84. Rai, S. S., D. O'Handley, and H. Nakai. 2001. Conformational dynamics of a transposition repressor in modulating DNA binding. *J. Mol. Biol.* **312:**311–322.

85. Ranquet, C., J. Geiselmann, and A. Toussaint. 2001. The tRNA function of SsrA contributes to controlling repression of bacteriophage Mu prophage. *Proc. Natl. Acad. Sci. USA* **98:**10220–10225.

86. Roberts, J. W., and C. W. Roberts. 1975. Proteolytic cleavage of bacteriophage lambda repressor in induction. *Proc. Natl. Acad. Sci. USA* **72:**147–151.

87. Roche, E. D., and R. T. Sauer. 2001. Identification of endogenous SsrA-tagged proteins reveals tagging at positions corresponding to stop codons. *J. Biol. Chem.* **276:**28509–28515.

88. Roldan, L. A., and T. A. Baker. 2001. Differential role of the Mu B protein in phage Mu integration vs. replication: mechanistic insights into two transposition pathways. *Mol. Microbiol.* **40:**141–155.

89. Rousseau, P., M. Bétermier, M. Chandler, and R. Alazard. 1996. Interactions between the repressor and the early operator region of bacteriophage Mu. *J. Biol. Chem.* **271:**9739–9745.

90. Rousseau, P., J. E. Laachouch, M. Chandler, and A. Toussaint. 2002. Characterization of the cts4 repressor mutation

91. Sassanfar, M., and J. W. Roberts. 1990. Nature of the SOS-inducing signal in *Escherichia coli*: the involvement of DNA replication. *J. Mol. Biol.* **212:**79–96.

92. Shapiro, J. A. 1979. Molecular model for the transposition and replication of bacteriophage Mu and other transposable elements. *Proc. Natl. Acad. Sci. USA* **76:**1933–1937.

93. Shapiro, J. A. 1984. Observations on the formation of clones containing *araB-lacZ* cistron fusions. *Mol. Gen. Genet.* **194:**79–90.

94. Shapiro, J. A. 1993. A role for the Clp protease in activating Mu-mediated DNA rearrangements. *J. Bacteriol.* **175:**2625–2631.

95. Shapiro, J. A., and N. P. Higgins. 1989. Differential activity of a transposable element in *Escherichia coli* colonies. *J. Bacteriol.* **171:**5975–5986.

96. Shapiro, J. A., and D. Leach. 1990. Action of a transposable element in coding sequence fusions. *Genetics* **126:**293–299.

97. Singh, S. K., R. Grimaud, J. R. Hoskins, S. Wickner, and M. R. Maurizi. 2000. Unfolding and internalization of proteins by the ATP-dependent proteases ClpXP and ClpAP. *Proc. Natl. Acad. Sci. USA* **97:**8898–8903.

98. Surette, M. G., S. J. Buch, and G. Chaconas. 1987. Transpososomes: stable protein-DNA complexes involved in the in vitro transposition of bacteriophage Mu DNA. *Cell* **49:**253–262.

99. Surette, M. G., and G. Chaconas. 1992. The Mu transpositional enhancer can function in trans: requirement of the enhancer for synapsis but not strand cleavage. *Cell* **68:**1101–1108.

100. Surette, M. G., and G. Chaconas. 1991. Stimulation of the Mu DNA strand cleavage and intramolecular strand transfer reactions by the Mu B protein is independent of stable binding of Mu B protein to DNA. *J. Biol. Chem.* **266:**17306–17313.

101. Surette, M. G., T. Harkness, and G. Chaconas. 1991. Stimulation of the Mu A protein-mediated strand cleavage reaction by the Mu B protein, and the requirement of DNA nicking for stable Type 1 transpososome formation. *J. Biol. Chem.* **266:**3118–3124.

102. Surette, M. G., B. D. Lavoie, and G. Chaconas. 1989. Action at a distance in Mu DNA transposition: an enhancer-like element is the site of action of supercoiling relief activity by integration host factor (IHF). *EMBO J.* **8:**3483–3489.

103. Symonds, N., A. Toussaint, P. van de Putte, and M. M. Howe (ed.). 1987. *Phage Mu.* Cold Spring Harbor Laboratory, Cold Spring Harbor, N.Y.

104. Taylor, A. L. 1963. Bacteriophage-induced mutation in *E. coli. Proc. Natl. Acad. Sci. USA* **50:**1043–1051.

105. Toussaint, A., M. J. Gama, J. Laachouch, G. Maenhaut-Michel, and A. Mhammedi-Alaoui. 1994. Regulation of bacteriophage Mu transposition. *Genetica* **93:**27–39.

106. van Vliet, F., M. Couturier, L. Desmet, M. Faelen, and A. Toussaint. 1978. Virulent mutants of temperate phage Mu-1. *Mol. Gen. Genet.* **160:**195–202.

107. Vogel, J. L., V. Geuskens, L. Desmet, N. P. Higgins, and A. Toussaint. 1996. C-terminal deletions can suppress temperature-sensitive mutations and change dominance in the phage Mu repressor. *Genetics* **142:**661–672.

108. Vogel, J. L., Z. J. Li, M. M. Howe, A. Toussaint, and N. P. Higgins. 1991. Temperature-sensitive mutations in bacteriophage Mu *c* repressor locate a 63-amino-acid DNA-binding domain. *J. Bacteriol.* **173:**6568–6577.

109. Wang, J., J. A. Hartling, and J. M. Flanagan. 1997. The structure of ClpP at 2.3 Å resolution suggests a model for ATP-dependent proteolysis. *Cell* **91:**447–456.

in transposable bacteriophage Mu. *Res. Microbiol.* **153:**511–518.

110. **Wawrzynow, A., D. Wojtkowiak, J. Marszalek, B. Banecki, M. Jonsen, B. Graves, C. Georgopoulos, and M. Zylicz.** 1995. The ClpX heat-shock protein of *Escherichia coli*, the ATP-dependent substrate specificity component of the ClpP-ClpX protease, is a novel molecular chaperone. *EMBO J.* **14:**1867–1877.

111. **Welty, D. J., J. M. Jones, and H. Nakai.** 1997. Communication of ClpXP protease hypersensitivity to bacteriophage Mu repressor isoforms. *J. Mol. Biol.* **272:**31–41.

112. **Wijffelman, C., and B. Lotterman.** 1977. Kinetics of Mu DNA synthesis. *Mol. Gen. Genet.* **151:**169–174.

113. **Williams, K. P., and D. P. Bartel.** 1996. Phylogenetic analysis of tmRNA secondary structure. *RNA* **2:**1306–1310.

114. **Williams, M. D., T. X. Ouyang, and M. C. Flickinger.** 1994. Starvation-induced expression of SspA and SspB: the effects of a null mutation in *sspA* on Escherichia coli protein synthesis and survival during growth and prolonged starvation. *Mol. Microbiol.* **11:**1029–1043.

115. **Wojciak, J. M., J. Iwahara, and R. T. Clubb.** 2001. The Mu repressor-DNA complex contains an immobilized "wing" within the minor groove. *Nat. Struct. Biol.* **8:**84–90.

116. **Wojtkowiak, D., C. Georgopoulos, and M. Zylicz.** 1993. Isolation and characterization of ClpX, a new ATP-dependent specificity component of the Clp protease of *Escherichia coli*. *J. Biol. Chem.* **268:**22609–22617.

The Bacterial Chromosome
Edited by N. Patrick Higgins
© 2005 ASM Press, Washington, D.C.

Chapter 28

Chromosome Dimer Resolution

FRANÇOIS-XAVIER BARRE AND DAVID J. SHERRATT

Faithful inheritance of the genetic material requires that the replication of chromosomes be accurate and complete, and that newly replicated chromosomes be separated prior to their segregation to daughter cells at cell division.

Crossing over by homologous recombination can lead to sister chromosome exchanges and hence to the dimerization of the two newly replicated molecules in bacteria containing circular chromosomes. These dimeric chromosomes must be converted to monomers if normal chromosome segregation is to occur. In *Escherichia coli*, Xer site-specific recombination converts dimers to monomers; homologues of this system are present in most eubacteria with circular chromosomes. Similarly, circular plasmids and the plasmid forms of temperate viruses can form dimers by recombinational processes and need to be converted to monomers if their stable inheritance at cell division is to be ensured. As a consequence, they also have evolved dedicated dimer resolution systems.

Dimerization can also happen when the replicons are linear. Indeed, in some bacteria with linear replicons, such as *Borrelia burgdorferi*, circular chromosome dimers are formed during each round of replication as a direct consequence of the structure of the replicon and need to be resolved into two new monomeric linear replicons with hairpin ends by dedicated chromosome dimer resolution machinery.

In this chapter, we present the processes that can lead to the dimerization of replicons and discuss the mechanisms that ensure their resolution. Additionally, we discuss how chromosome dimer resolution is integrated into other aspects of DNA processing during the bacterial cell cycle.

CHROMOSOME DIMER FORMATION

Linear Chromosomes with Hairpin Ends

Although knowledge of chromosome biology in *Borrelia*, the causative agent of Lyme disease, is relatively recent, it provides a convenient starting point for discussion of chromosome dimer formation and resolution, since dimers form during each and every replication cycle. The chromosomes and linear plasmids of *Borrelia* are linear with hairpin ends (Fig. 1A). Bidirectional replication is initiated internally and, when complete, generates a circular dimer of the original linear replicon (19, 59). In this circular dimer, the hairpin sites of the parental DNA are converted into palindromic "telomere" sites, which are used for chromosome dimer resolution. Specialized resolution machinery acts at these sites to convert the dimeric products of replication to linear molecules with hairpin ends (45).

The DNA of the poxviruses is also packaged as a linear duplex with hairpin ends. Upon infection of cells, initiation of viral replication occurs in the telomeric regions. Replication is believed to be unidirectional, giving rise to linear dimers and to higher multimeric forms by additional rounds of replication (35, 44). The temperate *E. coli* bacteriophage N15 is maintained as a linear prophage with hairpin ends; although the replication mechanism is unknown, the presence of a functional telomere resolvase suggests that replication generates multimeric DNA that is resolved to hairpin monomers by a mechanism similar to that in *Borrelia* (31).

François-Xavier Barre • Laboratoire de Microbiologie et de Génétique moléculaire du CNRS, 118 route de Narbonne, 31062 Toulouse Cedex, France. David J. Sherratt • Division of Molecular Genetics, Department of Biochemistry, University of Oxford, South Parks Rd., Oxford OX1 3QU, United Kingdom.

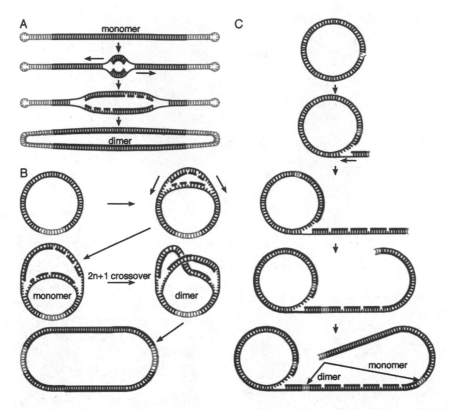

Figure 1. Chromosome dimer formation. The DNA backbone and the base pairing between two DNA strands in a duplex DNA molecule are schematically represented by "ladders." Parental strands are represented as thin lines, and replicated daughter strands are represented as thick lines to help in the visualization of strand exchanges. The leading strand is shown as a continuous line, whereas the lagging strand is shown first as a dashed line representing the Okazaki fragments, which later become continuous strands. Arrows depict directions of replication. (A) *Borrelia*'s replication strategy for linear replicons with covalently closed hairpin ends. The so-called "telomeres" are shown in light gray. Initiation of bidirectional replication occurs internally within the chromosome. Completion of replication produces a dimeric chromosome with two palindromic inverted repeats of the "telomere" sites. (B) Bidirectional replication of circular replicons. Replication will produce two catenated sister chromosomes in the absence of crossing over. An odd number of crossovers between the sister chromosomes generates a dimeric replicon. The region opposite the origin of replication, where chromosome dimer resolution occurs, is shown in light gray. (C) Rolling circle replication of circular replicons. Replication is initiated at a nick and is unidirectional. After one round of replication, the replication fork can continue to displace one of the sister chromosomes, leading to the formation of a multimeric linear concatemer of sister chromosomes that can be processed back into circular monomers or circular multimers by recombination.

Plasmid and Circular Virus Chromosomes

Dimeric and multimeric forms of plasmids occur during crossing over by homologous recombination (Fig. 1B; see below). In *E. coli*, such multimers form readily in Rec$^+$ strains, whereas multimer formation is largely absent in *recA* strains. Multimers can also form during conjugal rolling circle replication (Fig. 1C) (78) and in vegetative cells under special circumstances, for example, when either the RecF or the RecET pathway is activated in *E. coli* (47, 66). Multimer formation is detrimental to cells not only because it leads to segregational instability, but because multimeric plasmids appear to have a replication advantage over their monomeric parents (74). Similarly, in the circular forms of viruses, crossing

over by homologous recombination can lead to multimer formation; in the plasmid prophage state of P1 this is thought to lead to segregational instability (8). In contrast, in bacteriophage lambda, which has no dedicated multimer resolution system, multimer formation by sister chromosome homologous exchanges is essential for viral DNA packaging into virions, when rolling circle replication during a lytic cycle is prevented (76).

Circular Bacterial Chromosomes

In contrast to plasmid circular dimers, which can be visualized directly by physical methods such as agarose gel electrophoresis, the formation of chromosome dimers in bacteria remains to be demonstrated directly. Indeed, Xer site-specific recombination, which is responsible for chromosome dimer resolution, was initially discovered through its role in converting multimers of ColE1-related multicopy plasmids to monomers, and hence in ensuring their stable inheritance within *E. coli* (75). However, it was soon realized that Xer recombination was also required for normal *E. coli* chromosome segregation at cell division, apparently by converting chromosome dimers to monomers through its action at a chromosomal recombination site, *dif*, located in the replication terminus region (13, 22, 46). Xer recombination seems to be a conserved feature of most eubacteria that contain circular chromosomes, and presumably functions ubiquitously in converting chromosomal dimers to monomers (61).

Chromosome dimer formation has been estimated to occur about once in every six generations in normally growing laboratory strains of *E. coli* (58, 69). This frequency can be modulated both genetically and by changes in environment (52, 57, 69). Chromosome dimers form mostly when an odd number of crossovers has been added by homologous recombination during recombinational repair events. It is believed that most such recombination occurs between newly replicated sister chromosomes, before the homologous regions are separated and segregated to daughter cells (Fig. 1B). It is generally accepted that these homologous recombination repair events occur most frequently at stalled or broken replication forks (30). Many intracellular events can lead to replication fork stalling or breakage. For example, the replication machinery may encounter strand discontinuities that arise within the parental DNA through DNA damage and repair, or through DNA lesions, which prevent polymerase passage. Homologous recombination between the newly replicated sister chromosomes is therefore necessary to ensure faithful and efficient replication.

CONVERSION OF CIRCULAR DIMERS TO MONOMERS

The *Borrelia* and Other Linear Hairpin Chromosomes

In *Borrelia* and in bacteriophage N15 of *E. coli*, resolution of chromosome dimers is due to the action of a single enzyme, ResT or TelN, respectively. Those proteins share relatively little homology but are clearly related to the tyrosine site-specific recombinases and to the type Ib topoisomerases of eukaryotes and eukaryotic viruses (31, 32, 45). Correspondingly, their

mechanism of action is expected to be a two-step transesterification using a 3′-phosphotyrosyl protein-DNA intermediate (Fig. 2A). Indeed, the apparent mechanism is very similar to that of the tyrosine site-specific recombinases, except that two resolvase molecules bound to a single resolution site act in concert, each forming a protein-DNA intermediate which is then acted on by the free 5′ OH located 3 bp away on the opposite strand to form the required hairpin (Fig. 2A) (19). Similar reactions to form hairpins can be mediated by the tyrosine recombinases in vitro (21). This activity is normally precluded in vivo because tyrosine recombinases require cyclic interactions inside a tetramer of molecules so that adjacent molecules bound to a single duplex cannot be simultaneously active (Fig. 2B) (see below).

The proposed mechanism of hairpin formation during chromosome dimer resolution by ResT or TelN is different from that proposed for the resolution of circular dimers in poxviruses, where it has been suggested that the palindromic telomere site is extruded into a Holliday junction (HJ), which is then cleaved either by an enzyme structurally related to the RuvC bacterial HJ-resolvase (38, 39), or by the vaccinia topoisomerase, which is known to cleave extruded HJs at the telomere palindromic resolution site and rejoin the DNA fragments into hairpins in vitro (56, 63). It is not yet clear which of these two enzymes is responsible for the resolution activity that produces hairpins in vivo.

Plasmid and Circular Virus Chromosomes

Although more complex than the ResT and TelN systems, the systems dedicated to the resolution of plasmid and circular virus dimers have been characterized in greater detail (Fig. 2B). Dimer resolution is performed by site-specific recombination using tyrosine recombinases. The two best-studied dimer resolution systems are Cre/*loxP* from phage P1 (1, 2) and XerCD recombination of *E. coli*, which functions on a number of sites from many different plasmids, notably on the *cer* and *psi* recombination sites from plasmids ColE1 and pSC101, respectively (14, 26, 28). The Flp/*frt* system from *Saccharomyces cerevisiae* 2μm plasmids may also function in multimer resolution in addition to its role in controlling plasmid copy number (62).

During tyrosine recombinase-mediated site-specific recombination, two tyrosine recombinase molecules bind cooperatively to ~30-bp specific core recombination sites in the DNA (Fig. 2B). Recombinase-recombinase interactions then synapse distant sites together, forming a nucleoprotein complex containing four recombinases and two DNA

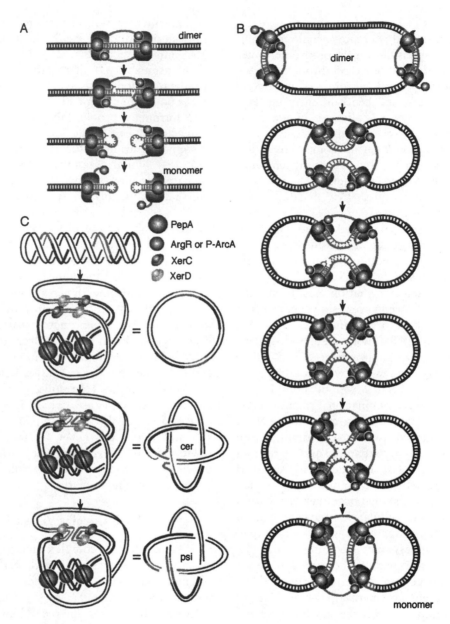

Figure 2. Chromosome dimer resolution. (A) Dimer resolution of *Borrelia* and bacteriophage N15. DNA is shown as in Fig. 1. Two ResT (or two TelN) enzymes bind to each of the two "telomere" sequences present as an inverted repeat in the dimeric replicon. Although no structural data are available, they are depicted as interacting with each other by an extension protruding from one molecule, which fits into a socket of the other molecule. (B) The mechanism of dimer resolution by tyrosine site-specific recombinases. DNA is shown as in Fig. 1. The C-terminal domain and the N-terminal domain of the recombinase monomers are represented by a large ellipse and small ellipses shaded with a gradient. These two domains form a C-shaped clamp that encircles half of the recombination site. The extreme C-terminal extension from each monomer is depicted by a small shaded circle. In Cre, this C-terminal extension fits into a socket in the C-terminal domain of a neighbor recombinase. DNA strand exchanges are performed successively by one pair of diagonally opposite recombinases in the complex and then the other. A complete cleavage-rejoining reaction by one of the recombinases proceeds in four steps identical to the ones performed by type IB topoisomerases: the initial protein-DNA complex is converted into a stable covalent enzyme-DNA adduct involving a 3′ phosphotyrosine linkage at the active site, before completion of the reaction by rejoining of the DNA. The phosphodiester linkage of the DNA substrate backbone which is attacked in this reaction lies 3 bp (*loxP*, *frt*, and *psi*) or 4 bp (*cer*) away from the center of the recombination site toward the bound recombinase. (C) Xer recombination at *cer* and *psi*. The *cer* and *psi* recombination sites are indicated by a thicker line on one of their DNA strands. The core recombination site is shown in gray. XerC binds to the half-site proximal to the accessory sequences. Three negative supercoils are entrapped by the accessory proteins and sequences. Synapsis of the core recombination sites is in antiparallel, with the C-terminal domain of all four recombinases facing the accessory sequences and proteins complex. The nucleoprotein complex structure is not planar but slightly bent, with the four arms of the DNA strands coming from the side of the C-terminal domains of the recombinases when they enter the nucleoprotein complex.

sites. Within this complex, two pairs of DNA strand exchanges are performed sequentially by two sets of partner recombinases. The first pair of strand exchanges create an HJ intermediate that is resolved into a crossover product by a second pair of strand exchanges, 6 to 8 bp away (Fig. 2B). It therefore follows that, at any one time, two recombinase molecules are active and two are inactive, there being a reciprocal switch between activity and inactivity at the HJ intermediate stage.

Tyrosine recombinases form a C-shaped clamp around the DNA, with the N-terminal domain on one side of the DNA duplex and the C-terminal domain on the other side (21, 42, 73). The crystal structures of Cre and FLP recombinases bound to their DNA substrates show that the nucleoprotein complex is formed around two DNA recombination sites aligned in an antiparallel configuration (21, 40, 42). The N-terminal domains of the four recombinases of the nucleoprotein complex are all on the same side of the complex (Fig. 2B and C), and the C-terminal domains are all on the other side. With Cre, each recombinase interacts through a C-terminal extension with a pocket in the C-terminal domain of one of its neighbors in a cyclic arrangement, thereby allowing communication between the four recombinase molecules and coordination of catalysis.

The C-terminal domain contains all of the catalytic residues and the residues required for sequence-specific DNA binding to the outer region of the core recombination site, while the N-terminal domain contacts the inner region of the recombinase-binding site close to the scissile phosphate. Both the C- and N-terminal domains are implicated in coordinating catalysis (4, 37, 43). In addition to the tyrosine nucleophile, five other residues, forming a characteristic RKHRH signature, are implicated in the strand exchange reaction; they are presumed to participate directly in acid-base catalysis or in the stabilization of the transition state (21).

Xer site-specific recombination differs from the Cre/loxP and Flp/frt systems because it requires two separate tyrosine recombinases, XerC and XerD, which are invariably encoded by different regions of the host genome (14, 26) and which are responsible for the first and second separate DNA pairs of strands exchanged, respectively, during recombination at psi (24, 25). The use of two recombinases rather than one would appear to provide additional levels of control of the recombination reaction. During recombination at cer, XerC initiates recombination to form HJ intermediates, but these appear not to be acted on by XerD to complete the reaction. The process which resolves these HJ intermediates remains unknown; it is not linked to the catalytic ac-

tivity of XerC or XerD, or apparently to the cellular HJ resolvases such as RuvC or RusA. One possibility is that these figure-eight molecules are converted into one dimer and two monomers by a new round of replication.

Xer recombination at cer and psi depends on the presence of accessory sequences and accessory proteins. For example, in addition to XerC and XerD, recombination at psi or cer depends on two other host-encoded proteins: PepA and ArgR at cer (70, 71) and PepA and phosphorylated ArcA at psi (23, 28). All three proteins are known transcription factors with specific DNA binding properties. The complete cer and psi recombination sites consist of a canonical site-specific recombination core site of ~30 bp and approximately 178 bp (cer) or 158 bp (psi) of accessory sequence adjacent to the XerC binding site. ArgR binds as a hexamer to a single ArgR box within the accessory sequences of cer (16), while PepA binds as a hexamer to synapsed cer accessory sequences on both sides of the ArgR binding site (3). ArcA-P binds to the psi accessory sequences at a place similar to that bound by ArgR in cer (23). The accessory sequences and proteins direct the formation of a productive synapse that has a precise geometry, which is reflected in the precise topology of the products of recombination at cer and psi (Fig. 2C) (24). The psi recombination product consists of two right-handed catenated circles containing four interlinks, and with the psi sites oriented antiparallel with respect to each other (Fig. 2C [psi]). The cer recombination product is an HJ in which the recombined strands form catenated rings with an identical topology to that of the psi product, while the unrecombined strand remains as an unknotted circle (Fig. 2C [cer]). Based on this topological information and on the known binding of the accessory proteins to the accessory sequences, a model for the synaptic structure was proposed (Fig. 2C). This model was refined with the help of the PepA crystal structure (72): two PepA hexamers are sandwiched around a hexamer of ArgR and the recombination sites interwrap around accessory proteins as a right-handed superhelix that contains 3-negative superhelical turns. As the recombination sites are believed to be aligned in an antiparallel arrangement by the Xer recombinases, this translates into a synaptic structure in which 4-negative supercoils are entrapped, three being bound by accessory proteins and the fourth being "free." A similar complex is thought to form between the psi accessory sequences, PepA and ArcA-P.

Two roles have been identified for the accessory elements. First, the accessory sequences and proteins compensate for the weak stability of synapses formed between cer or psi core sites, and enable the formation and stabilization of the synapse when the interacting *cer* or *psi* sites are directly repeated. This absolute requirement for proteins and sequences that impose a synapse with a defined topology provides a "topological filter" that ensures resolution selectivity, because synapses of the observed defined topology can only readily form between directly repeated sites on the same molecules (67). In addition, the accessory factors also have roles in determining the order of strand exchange and in controlling progression of the recombination reaction at *psi*. For example, it is possible to reverse the order of strand exchanges at *psi* by placing the accessory sequences adjacent to the XerD binding site rather than the XerC binding site of the core site (15). Furthermore, in the absence of PepA, supercoiling, or catenation, XerCD converts the HJ intermediate of *psi* reactions back to substrate through a second round of strand exchanges by XerC (M. Robertson and D. J. Sherratt, unpublished data), thereby demonstrating that activation of the second pair of strand exchanges catalyzed by XerD requires a precise accessory factor-mediated synapse geometry. Thus, the catenated HJ intermediate formed by XerC on supercoiled *psi*-containing substrates in vitro requires PepA and presumably the associated synapse geometry in order to promote the conformational change required for catalysis by XerD. These observations are in agreement with those that show that synthetic HJ substrates are recombined much more efficiently by XerC than by XerD (5, 6). Thus, the accessory sequences and proteins can now be viewed as regulatory elements which (i) act during synapsis to ensure resolution selectivity by allowing a productive synaptic complex of precise geometry to assemble, and (ii) enable successive sequential activation of each of the two partner recombinases when the correct substrate has been encountered.

Resolution of Bacterial Chromosome Dimers and Integration with the Cell Cycle

The XerC and XerD site-specific recombinases function in chromosome dimer resolution by adding a single crossover at *dif*, a specific 28-bp core site located in the region of termination of replication of the chromosome. Since a functional *dif* site contains no accessory sequences, it was surprising to discover eventually that chromosome dimer resolution by Xer recombination at *dif* does require an additional accessory protein, FtsK, which interacts with DNA in a DNA sequence-independent manner (60, 68). Just as Xer recombination within plasmids needs to be preferentially intramolecular, the reaction at chromosomal *dif* must be restricted so that no Xer-dependent crossover is added if the two replicated sister chromosomes are monomeric. Indeed, no crossovers can be detected at *dif* in strains that cannot form chromosome dimers by homologous recombination (69). FtsK plays a key role in this selectivity for intramolecular resolution of chromosomal dimers, although it is not required for the resolution of plasmid dimers at *cer* and *psi*.

Any region of a sister chromosome can come into contact with its homologous counterpart on the other sister chromosome as judged by the efficiency of homologous recombination in the repair of DNA damage. Correspondingly, HJs can be formed between *dif* sites whatever their location on the chromosome (9). However, there is a higher proportion of HJs formed between *dif* sites located in the terminus region of the chromosome (9). This is likely to be because homologous newly replicated sister terminus regions remain in close association for longer than other newly replicated regions of the chromosome, perhaps because of the inevitable catenation that will occur in the replication terminus region.

Although HJs can form efficiently between sister *dif* sites inserted throughout the 200-kb terminus region, a complete crossover reaction is restricted to a 50-kb zone within this region, the "*dif* activity zone" (DAZ) (27, 58). This is due to the requirement for FtsK during Xer recombination at *dif*, and its association with the division septum. Overexpressing full-length FtsK so that it becomes distributed throughout the cellular inner membrane, or an FtsK derivative with a diffuse cytoplasmic distribution, overcomes the DAZ constraint and allows for crossing over between *dif* sites anywhere on the chromosome. However, chromosome dimer resolution is not restored, probably because the system can no longer distinguish between *dif* sites carried by a chromosome dimer or two monomers (9, 57).

One explanation for Xer selectivity in chromosome dimer resolution is that the DAZ is "marked" as the region of the nucleoid that will be trapped in the closing septum if a chromosome dimer has been formed; this will bring the Xer synaptic complex into contact with the FtsK protein (Fig. 3D). In agreement with this model, it has been shown that the terminus region, which contains the DAZ, is the last region to be segregated away from mid cell before cell division (41, 54, 55). The DNA sequence of the chromosome is polarized along the two replichore arms from the origin of replication to the *dif* site in the terminus region. It has been proposed that this polarization is partly responsible for the cellular localization of the DAZ (17). There is some genetic evidence that the FtsK protein itself could be implicated in detecting the polarization of the chromosome (29).

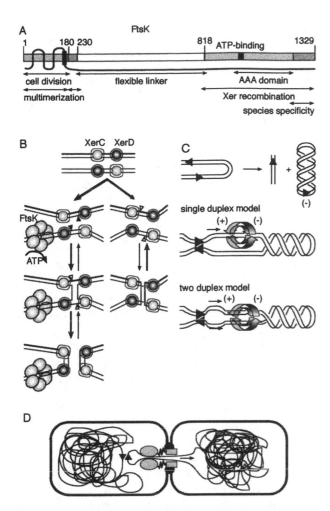

FtsK Links Chromosome Segregation and Cell Division

FtsK is a large, 1,329-amino-acid (aa) integral membrane protein that is involved in cell division, chromosome segregation, and chromosome dimer resolution (Fig. 3A) (12, 49, 60, 68, 81). FtsK contains three distinct domains. The N-terminal domain of FtsK (aa 1 to 180) is required for septum formation and cell division (34, 77, 80). FtsK is recruited to the septum in an FtsZ- and FtsA-dependent manner (77, 80), and is necessary for the later recruitment of other septal proteins (20). The N-terminal domain of FtsK contains four transmembrane helices, one of them being twice the size of classical membrane-spanning helices (33). The cell division defect of *ftsK* mutant cells is partially suppressed by mutations in genes involved in peptidoglycan synthesis, making it likely that the N-terminal domain of FtsK is involved in completing septum closure (12). Indeed, SpoIIIE, the bacterial gram-positive homologue of FtsK in *Bacillus subtilis*, has been reported to have a role in membrane fusion (64). The largest domain of FtsK (aa 230 to 818) is a proline- and glutamine-rich region of low complexity. This region is highly divergent in length and sequence among the FtsK homologues that have been found in other bacterial species and is likely to serve as a flexible linker. Observation of the intracellular distribution of several FtsK derivatives tagged with a green fluorescent

Figure 3. FtsK and the control of chromosome dimer resolution. (A) Scheme of the FtsK protein, showing the different domains that have been identified and the roles that they have been assigned. The N-terminal domain is shown by a shaded box, and the four transmembrane regions it contains are represented as black lines. A thicker line indicates the long transmembrane helix found at the end of the N-terminal domain. The extreme N terminus points toward the cytoplasm as well as the C-terminal domain of the protein. A dark box at the junction between the N-terminal domain and the linker region of the protein indicates the 50 amino acid residues potentially implicated in multimerization of the *E. coli* FtsK protein. The ATP binding site is shown as a darker box inside the C-terminal domain. (B) FtsK-dependent and independent pathways of Xer recombination at *dif*. In the absence of FtsK, the Xer synaptic complex adopts a conformation suitable for XerC-mediated strand exchanges, depicted by a kink at the XerD binding site. FtsK can use the energy of ATP to switch the Xer synaptic complex to a conformation suitable for XerD-strand exchanges, depicted by a kink at the XerC binding site. The intensity of the arrows reflects the probability of recombinational events. XerC and XerD cleavage sites are shown by white and black triangles, respectively. (C) The DNA translocase activity of FtsK can influence the topological outcome of the Xer recombination. Xer recombination between two directly repeated *dif* sites (black triangles) on a linear duplex creates one linear and one circular duplex with single *dif* sites. The circular product traps negative supercoils (−) preferentially. This is linked to the DNA translocation activity of FtsK, which creates positive supercoils in front of the advancing protein and negative supercoils in its wake.

We propose that FtsK translocates along the DNA toward the synaptic complex to contact the recombinases and activate crossover formation by introducing positive writhe and twist onto the complex. Thus, positive supercoils are created between FtsK and the Xer complex, and negative supercoils are created on the other side of FtsK. On the substrate shown, FtsK should load most frequently between the repeated sites. To explain the preferential global negative supercoiling of the circular substrate, we propose two models: (i) the FtsK protein encircles only one duplex DNA; the Xer synaptic complex and/or an additional contact of DNA with the outside of FtsK prevent(s) negative supercoils from diffusing at the ends of the substrate, but allow(s) positive supercoils to diffuse away; (ii) the FtsK protein encircles two duplexes, thus preventing the negative supercoils from diffusing away; the Xer synaptic complex does not prevent the positive supercoils from diffusing away. (D) FtsK plays a safeguard role in DNA segregation when cellular events such as chromosome dimer formation have delayed separation and migration of the two sister chromosomes into the two daughter cells. The two replicated terminus regions remain associated asymmetrically in one of the daughter cells. We propose that FtsK forms an oriented pore through which the two duplex strands of a chromosome can pass. Its directional translocation activity would pump DNA when necessary, while its interaction with Xer-*dif* (triangles) can lead to chromosome dimer resolution. FtsK$_C$ is represented by ovoids, the linker is represented by a zigzag, and the N-terminal part of the protein is represented by rectangles.

protein molecule at their N terminus has led us to believe that the linker interacts with other components of the septum machinery (F.-X. Barre, unpublished results).

The C-terminal domain of FtsK (FtsK$_C$; aa 818 to 1329) is the most conserved feature of the protein. It belongs to the AAA (ATPase associated with various activities) superfamily (50, 53) and defines, along with the C-terminal domain of SpoIIIE, a class of proteins which are involved in DNA translocation (36). SpoIIIE is thought to translocate chromosomal DNA from the mother cell into the prespore during the asymmetric cell division of *B. subtilis* sporulation (79). Consistent with this role, a truncated version of SpoIIIE has been shown to modulate the topology of a DNA molecule in vitro (11). A purified derivative of FtsK with an intact C-terminal domain has biochemical activities similar to those of SpoIIIE (7). Thus FtsK$_C$, in addition to its role in chromosome dimer resolution, is likely to be involved in DNA translocation through the closing septum in *E. coli* (see later).

The C-Terminal Domain of FtsK Activates Xer Recombination at *dif*

The initial observation that synapses between *dif* sites occurred independently of FtsK$_C$ in vitro and in vivo, and that overexpression of only the C-terminal domain of the protein activated Xer recombination between *dif* sites independently of the DAZ or the presence of chromosome dimers, led us to believe that FtsK was directly implicated in the recombination reaction (9). Using a purified derivative of FtsK, which is active in DNA translocation in vitro, we have been able to reconstitute a complete Xer recombination reaction between *dif* sites in vitro (7).

In vivo and in vitro studies show that in the absence of FtsK, HJs are created and resolved back to the original substrate in cycles of XerC-mediated strand exchanges (Fig. 3B). FtsK promotes a complete Xer recombination reaction at *dif* by reversing the catalytic state of XerC and XerD in the synaptic complex, presumably by modifying the conformation of the recombinational complex. XerD then mediates a first pair of strand exchanges, creating a new HJ intermediate which is resolved into crossing over by XerC-strand exchanges (Fig. 3B). FtsK can activate a complete recombination reaction by XerC and XerD at a core *psi* site in a similar way (H. Capiaux, D. J. Sherratt, and F.-X. Barre, unpublished results).

Because FtsK is a DNA translocase, it would be reasonable to believe that activation of Xer recombination is a consequence of the topological changes that translocation could impose on the DNA globally.

However, FtsK is able to promote Xer recombination between a small 34-bp linear duplex containing the 28-bp *dif* site and a longer linear duplex of 200 bp (49a). Therefore, FtsK does not require translocation of the DNA to activate Xer recombination to switch the catalytic activity at Xer-*dif*. It can act locally in close vicinity to the nucleoprotein complex. Indeed, a functional study of FtsK$_C$ in which domains were swapped between *Haemophilus influenzae* and *E. coli* FtsK has revealed a species-specific interaction between FtsK$_C$ and the Xer recombinases, which involves the last 100 aa of FtsK (79a). This result suggests that activation of XerD occurs only when FtsK$_C$ interacts specifically with the synaptic nucleoprotein complex. We propose that FtsK$_C$ imposes writhe or twist on the XerCD-*dif* complex by using the same ATP-dependent mechanism as the one used for translocation; this leads to a switch in conformation of the complex so that XerD, rather than XerC, initiates catalysis.

Other Possible Biological Roles of FtsK$_C$

There is now substantial evidence that FtsK$_C$ is involved in other processes important for chromosome segregation in addition to chromosome dimer resolution. For example, FtsK$_C$ has ATP-dependent DNA translocation activity in vitro that is independent of Xer-*dif*. This translocation activity could be used in vivo during the pumping of residual DNA away from the division septum, with the polarity of the two replichores around *dif* dictating an orderly translocation process; this could explain the observation that DAZ functions only when it is correctly located in the replication termination region (see chapter 13).

The molecular mechanism of this translocation remains to be determined, although we can envisage two different models (Fig. 3C). In the first, FtsK$_C$ translocates a single duplex. If this were the case, the observed partition of preferentially negative supercoils into the excised circles formed during Xer recombination at two directly repeated *dif* sites in a linear substrate would arise because there is a barrier to the free diffusion of the local negative supercoils generated behind the translocating complex, but there is no barrier to the free diffusions of the positive supercoils generated in front of the complex. In the second model, FtsK$_C$ translocates two duplexes simultaneously; consistent with this, electron microscopy images of FtsK$_{50C}$ with DNA frequently show two duplexes apparently entering and leaving the hole in the ring, which has a diameter ($\sim$100 Å) just sufficient to accommodate two plectonemically interwound duplexes (L. Aussel, A. Staziak, and

D. J. Sherratt, unpublished results). If this is the case, the FtsK multimeric complex would itself trap the negative and positive supercoils it generates on either side of the complex. Such an activity could be relevant to the cellular role of FtsK, since in a cell there may be a need to translocate the two duplex arms of a circular chromosome through the septum from one cellular compartment to the other (Fig. 3D). Also consistent with a global role in DNA translocation in vivo is the observation that Cre/*loxP* recombination can partially substitute for the XerCD-*dif* in chromosome dimer resolution in *E. coli* (48). Although Cre/*loxP* recombination is independent of FtsK, the ability to substitute it for XerCD/*dif* in chromosome dimer resolution still depends on FtsK, perhaps because of the requirement for the DNA translocation activity of FtsK in passing the resolved DNA through the closing septum (18).

Any global role of $FtsK_C$ in DNA translocation in vivo has not been revealed from studies of $FtsK_C^-$ mutants, which have a similar phenotype to Xer$^-$ mutants, and whose only obvious defect is in chromosome dimer resolution (60). Perhaps, the DNA translocation activity of $FtsK_C$ is only necessary when DNA is trapped within the septum at a late stage of cell division, which happens almost exclusively when chromosome dimers are present. At this later stage, the septum is almost completely closed, and FtsK is required not only as a DNA motor protein but also to make a pore in the septum through which DNA passes. Although $FtsK_C$ may use its translocation activity in the absence of dimers, the absence of active $FtsK_C$ could be compensated for by the segregation activities of MukB and the DNA replication process, which together will tend to pull the DNA away from the division septum (65). Indeed, $FtsK_C$ and MukB are likely to play complementary and in part overlapping roles in chromosome segregation, consistent with the observation that mutations in either gene are viable, but the combination is synthetically lethal (81). The DNA translocation activity of $FtsK_C$ could also lead to local DNA condensation through the differential activity of topoisomerases on the positive and negative supercoiled domains created during translocation; for example, if DNA gyrase acts preferentially on positively supercoiled domains, the result will be an overall increase in negative supercoiling.

In addition to its translocation roles, a further possible role of $FtsK_C$ is revealed by the observation that the products of a reaction between two directly repeated *dif* sites carried by a supercoiled plasmid are free circles in vitro (7), in contrast to reactions performed on the same substrate by *H. influenzae* XerC and *E. coli* XerD in the absence of FtsK and by Cre (43a). Such selectivity in vivo could act to minimize sister chromosome entanglement and catenation during dimer resolution. $FtsK_C$ may also stimulate decatenation by topoisomerase IV by interacting with ParE, one of the topoisomerase IV subunits (36a).

By simultaneously monitoring the position of the region of the origin of replication and of replication termination in live cells, using arrays of repeated *lacO* sequences and *tetO* sequences, which can be bound by LacI-YFP and TetR-CFP molecules, respectively (47a, 65), we have observed that the two replicated terminus regions frequently stay on the same side of the septum as a single focus until just before cell division; immediately prior to division, the two foci separate and one passes through the septum. This could reflect the DNA translocation activity of FtsK and/or a role of $FtsK_C$ in breaking the cohesion present in the terminus region, for example, by activating decatenation. Furthermore, the process that generates this asymmetry of newly replicated termini could also serve to insert FtsK asymmetrically into the septum (Fig. 3D).

PERSPECTIVE

Although Barbara McClintock noted that crossing over by homologous recombination could generate dimeric circular chromosomes from circular monomers (51), it was not until the 1990s that molecular biologists realized the importance of this phenomenon for bacterial chromosome segregation. It was shown that Xer site-specific recombination functions in chromosome dimer resolution as well as plasmid multimer resolution (13, 22, 46). Since then, much effort has been devoted to understanding how the Xer site-specific recombination is controlled to ensure that monomers are made from dimers and not dimers from monomers. The realization that Xer recombination uses different strategies to ensure resolution selectivity during plasmid and chromosome dimer resolution demonstrates the sophistication that has developed during bacterial evolution. Furthermore, the involvement of FtsK in chromosome dimer resolution and in cell division suggests it may play an important part in coordinating cell division with DNA replication and chromosome segregation; indeed, it seems likely that FtsK and XerCD-*dif* have coevolved in the eubacterial lineage. How archaea process their circular chromosomes, and whether dimers occur, remains to be determined; tyrosine recombinases are certainly present, although we have not yet been able to find convincing evidence for either FtsK-like proteins or the presence of both XerC and XerD homologues. We believe that DNA replication is a major driving force in bacterial chromosome segregation, with Muk and FtsK$_C$ playing important and complementary roles in different regions of the cell (65). The demonstration that tyrosine recombinases and mechanistically related enzymes function in circular dimer resolution and in the production of linear hairpin chromosomes from the primary, dimeric replication products in organisms like *Borrelia* suggests a very early and fundamental evolutionary role of tyrosine nucleophile topoisomerase-like enzymes in chromosome processing. Future work will further understanding of how replication and its initiation are controlled, of how "replication factories" are organized and their activities are controlled, of how replication and recombinational processes are linked, of how positional information is used to place highly condensed and organized chromosomes and their processing machinery at specific positions in a cell, and of how DNA processing is coordinated with the cell cycle.

Acknowledgments. F.-X.B. was supported by an EMBO fellowship, and research in the laboratory of D.J.S. was supported by the Wellcome Trust.

We thank Lidia Arciszewska (Oxford, United Kingdom), Sergio Filipe (Oxford), Francois Cornet (Toulouse, France), and Jean-Michel Louarn (Toulouse) for valuable discussions and comments on the manuscript.

REFERENCES

1. **Abremski, K., and R. Hoess.** 1985. Phage P1 Cre-*loxP* site-specific recombination. Effects of DNA supercoiling on catenation and knotting of recombinant products. *J. Mol. Biol.* **184:**211–220.

2. **Abremski, K., R. Hoess, and N. Sternberg.** 1983. Studies on the properties of P1 site-specific recombination: evidence for topologically unlinked products following recombination. *Cell* **32:**1301–1311.

3. **Alen, C., D. J. Sherratt, and S. D. Colloms.** 1997. Direct interaction of aminopeptidase A with recombination site DNA in Xer site-specific recombination. *EMBO J.* **16:**5188–5197.

4. **Arciszewska, L. K., R. A. Baker, B. Hallet, and D. J. Sherratt.** 2000. Coordinated control of XerC and XerD catalytic activities during Holliday junction resolution. *J. Mol. Biol.* **299:**391–403.

5. **Arciszewska, L. K., I. Grainge, and D. J. Sherratt.** 1997. Action of site-specific recombinases XerC and XerD on tethered Holliday junctions. *EMBO J.* **16:**3731–3743.

6. **Arciszewska, L. K., and D. J. Sherratt.** 1995. Xer site-specific recombination in vitro. *EMBO J.* **14:**2112–2120.

7. **Aussel, L., F. X. Barre, M. Aroyo, A. Stasiak, A. Z. Stasiak, and D. Sherratt.** 2002. FtsK is a DNA motor protein that activates chromosome dimer resolution by switching the catalytic state of the XerC and XerD recombinases. *Cell* **108:**195–205.

8. **Austin, S., M. Ziese, and N. Sternberg.** 1981. A novel role for site-specific recombination in maintenance of bacterial replicons. *Cell* **25:**729–736.

9. **Barre, F. X., M. Aroyo, S. D. Colloms, A. Helfrich, F. Cornet, and D. J. Sherratt.** 2000. FtsK functions in the processing of a Holliday junction intermediate during bacterial chromosome segregation. *Genes Dev.* **14:**2976–2988.

10. **Barre, F. X., B. Soballe, B. Michel, M. Aroyo, M. Robertson, and D. Sherratt.** 2001. Circles: the replication-recombination-chromosome segregation connection. *Proc. Natl. Acad. Sci. USA* **98:**8189–8195.

11. **Bath, J., L. J. Wu, J. Errington, and J. C. Wang.** 2000. Role of *Bacillus subtilis* SpoIIIE in DNA transport across the mother cell-prespore division septum. *Science* **290:**995–997.

12. **Begg, K. J., S. J. Dewar, and W. D. Donachie.** 1995. A new *Escherichia coli* cell division gene, *ftsK. J. Bacteriol.* **177:**6211–6222.

13. **Blakely, G., S. Colloms, G. May, M. Burke, and D. Sherratt.** 1991. *Escherichia coli* XerC recombinase is required for chromosomal segregation at cell division. *New Biol.* **3:**789–798.

14. **Blakely, G., G. May, R. McCulloch, L. K. Arciszewska, M. Burke, S. T. Lovett, and D. J. Sherratt.** 1993. Two related recombinases are required for site-specific recombination at *dif* and *cer* in *E. coli* K12. *Cell* **75:**351–361.

15. **Bregu, M., D. J. Sherratt, and S. D. Colloms.** 2002. Accessory factors determine the order of strand exchange in Xer recombination at psi. *EMBO J.* **21:**3888–3897.

16. **Burke, M., A. F. Merican, and D. J. Sherratt.** 1994. Mutant *Escherichia coli* arginine repressor proteins that fail to bind L-arginine, yet retain the ability to bind their normal DNA-binding sites. *Mol. Microbiol.* **13:**609–618.

17. **Capiaux, H., F. Cornet, J. Corre, M. Guijo, K. Perals, J. E. Rebollo, and J. Louarn.** 2001. Polarization of the *Escherichia coli* chromosome. A view from the terminus. *Biochimie* **83:**161–170.

18. **Capiaux, H., C. Lesterlin, K. Perals, J. M. Louarn, and F. Cornet.** 2002. A dual role for the FtsK protein in *Escherichia coli* chromosome segregation. *EMBO Rep.* **3:**532–536.

19. **Chaconas, G., P. E. Stewart, K. Tilly, J. L. Bono, and P. Rosa.** 2001. Telomere resolution in the Lyme disease spirochete. *EMBO J.* **20:**3229–3237.

20. **Chen, J. C., and J. Beckwith.** 2001. FtsQ, FtsL and FtsI require FtsK, but not FtsN, for co-localization with FtsZ during *Escherichia coli* cell division. *Mol. Microbiol.* **42:**395–413.

21. **Chen, Y., U. Narendra, L. E. Iype, M. M. Cox, and P. A. Rice.** 2000. Crystal structure of a Flp recombinase-Holliday junction complex: assembly of an active oligomer by helix swapping. *Mol. Cell* **6:**885–897.

22. **Clerget, M.** 1991. Site-specific recombination promoted by a short DNA segment of plasmid R1 and by a homologous segment in the terminus region of the *Escherichia coli* chromosome. *New Biol.* **3:**780–788.

23. **Colloms, S. D., C. Alen, and D. J. Sherratt.** 1998. The ArcA/ArcB two-component regulatory system of *Escherichia coli* is essential for Xer site-specific recombination at psi. *Mol. Microbiol.* **28:**521–530.

24. **Colloms, S. D., J. Bath, and D. J. Sherratt.** 1997. Topological selectivity in Xer site-specific recombination. *Cell* **88:**855–864.

25. **Colloms, S. D., R. McCulloch, K. Grant, L. Neilson, and D. J. Sherratt.** 1996. Xer-mediated site-specific recombination in vitro. *EMBO J.* **15:**1172–1181.

26. **Colloms, S. D., P. Sykora, G. Szatmari, and D. J. Sherratt.** 1990. Recombination at ColE1 *cer* requires the *Escherichia coli xerC* gene product, a member of the lambda integrase family of site-specific recombinases. *J. Bacteriol.* **172:**6973–6980.

27. **Cornet, F., J. Louarn, J. Patte, and J. M. Louarn.** 1996. Restriction of the activity of the recombination site *dif* to a small zone of the *Escherichia coli* chromosome. *Genes Dev.* **10:**1152–1161.

28. Cornet, F., I. Mortier, J. Patte, and J.-M. Louarn. 1994. Plasmid pSC101 harbors a recombination site, *psi*, which is able to resolve plasmid multimers and to substitute for the analogous chromosomal *Escherichia coli* site *dif*. *J. Bacteriol.* **176:**3188–3195.

29. Corre, J., and J.-M. Louarn. 2002. Evidence from terminal recombination gradients that FtsK uses replichore polarity to control chromosome terminus positioning at division in *Escherichia coli*. *J. Bacteriol.* **184:**3801–3807.

30. Cox, M. M., M. F. Goodman, K. N. Kreuzer, D. J. Sherratt, S. J. Sandler, and K. J. Marians. 2000. The importance of repairing stalled replication forks. *Nature* **404:**37–41.

31. Deneke, J., G. Ziegelin, R. Lurz, and E. Lanka. 2000. The protelomerase of temperate *Escherichia coli* phage N15 has cleaving-joining activity. *Proc. Natl. Acad. Sci. USA* **97:** 7721–7726.

32. Deneke, J., G. Ziegelin, R. Lurz, and E. Lanka. 2002. Phage N15 telomere resolution. Target requirements for recognition and processing by the protelomerase. *J. Biol. Chem.* **277:**10410–10419.

33. Dorazi, R., and S. J. Dewar. 2000. Membrane topology of the N-terminus of the *Escherichia coli* FtsK division protein. *FEBS Lett.* **478:**13–18.

34. Draper, G. C., N. McLennan, K. Begg, M. Masters, and W. D. Donachie. 1998. Only the N-terminal domain of FtsK functions in cell division. *J. Bacteriol.* **180:**4621–4627.

35. Du, S., and P. Traktman. 1996. Vaccinia virus DNA replication: two hundred base pairs of telomeric sequence confer optimal replication efficiency on minichromosome templates. *Proc. Natl. Acad. Sci. USA* **93:**9693–9698.

36. Errington, J., J. Bath, and L. J. Wu. 2001. DNA transport in bacteria. *Nat. Rev. Mol. Cell. Biol.* **2:**538–545.

36a. Espeli, O., C. Lee, and K. J. Marians. 2003. A physical and functional interaction between *Escherichia coli* FtsK and topoisomerase IV. *J. Biol. Chem.* **278:**44639–44644.

37. Ferreira, H., D. Sherratt, and L. Arciszewska. 2001. Switching catalytic activity in the XerCD site-specific recombination machine. *J. Mol. Biol.* **312:**45–57.

38. Garcia, A. D., L. Aravind, E. V. Koonin, and B. Moss. 2000. Bacterial-type DNA Holliday junction resolvases in eukaryotic viruses. *Proc. Natl. Acad. Sci. USA* **97:**8926–8931.

39. Garcia, A. D., and B. Moss. 2001. Repression of vaccinia virus Holliday junction resolvase inhibits processing of viral DNA into unit-length genomes. *J. Virol.* **75:**6460–6471.

40. Gopaul, D. N., F. Guo, and G. D. Van Duyne. 1998. Structure of the Holliday junction intermediate in Cre-*loxP* site-specific recombination. *EMBO J.* **17:**4175–4187.

41. Gordon, G. S., and A. Wright. 1998. DNA segregation: putting chromosomes in their place. *Curr. Biol.* **8:**R925–R927.

42. Guo, F., D. N. Gopaul, and G. D. van Duyne. 1997. Structure of Cre recombinase complexed with DNA in a site-specific recombination synapse. *Nature* **389:**40–46.

43. Hallet, B., L. K. Arciszewska, and D. J. Sherratt. 1999. Reciprocal control of catalysis by the tyrosine recombinases XerC and XerD: an enzymatic switch in site-specific recombination. *Mol. Cell* **4:**949–959.

43a. Ip, S. Y. P., M. Bregu, F.-X. Barre, and D. J. Sherratt. 2003. Decatenation of DNA circles by FtsK-dependent Xer site-specific recombination. *EMBO J.* **22:**6399–6407.

44. Kobryn, K., and G. Chaconas. 2001. The circle is broken: telomere resolution in linear replicons. *Curr. Opin. Microbiol.* **4:**558–564.

45. Kobryn, K., and G. Chaconas. 2002. ResT, a telomere resolvase encoded by the Lyme disease spirochete. *Mol. Cell* **9:**195–201.

46. Kuempel, P. L., J. M. Henson, L. Dircks, M. Tecklenburg, and D. F. Lim. 1991. *dif*, a recA-independent recombination site in the terminus region of the chromosome of *Escherichia coli*. *New Biol.* **3:**799–811.

47. Kusano, K., K. Nakayama, and H. Nakayama. 1989. Plasmid-mediated lethality and plasmid multimer formation in an *Escherichia coli recBC sbcBC* mutant. Involvement of RecF recombination pathway genes. *J. Mol. Biol.* **209:**623–634.

47a. Lau, I. F., S. R. Filipe, B. Søballe, O.-A. Økstad, F.-X. Barre, and D. J. Sherratt. 2003. Spatial and temporal organisation of replicating *Escherichia coli* chromosomes. *Mol. Microbiol.* **49:**731–743.

48. Leslie, N. R., and D. J. Sherratt. 1995. Site-specific recombination in the replication terminus region of *Escherichia coli*: functional replacement of *dif*. *EMBO J.* **14:**1561–1570.

49. Liu, G., G. C. Draper, and W. D. Donachie. 1998. FtsK is a bifunctional protein involved in cell division and chromosome localization in *Escherichia coli*. *Mol. Microbiol.* **29:** 893–903.

49a. Massey, T., L. Aussel, F.-X. Barre, and D. J. Sherratt. 2004. Asymmetric activation of Xer site-specific recombination by FtsK. *EMBO Rep.* **5:**399–404.

50. Maurizi, M. R., and C. C. Li. 2001. AAA proteins: in search of a common molecular basis: International meeting on cellular functions of AAA proteins. *EMBO Rep.* **2:**980–985.

51. McClintock, B. 1932. A correlation of ring-shaped chromosomes with variegation in Zea mays. *Proc. Natl. Acad. Sci. USA* **18:**677–681.

52. Michel, B., G. D. Recchia, M. Penel-Colin, S. D. Ehrlich, and D. J. Sherratt. 2000. Resolution of Holliday junctions by RuvABC prevents dimer formation in *rep* mutants and UV irradiated cells. *Mol. Microbiol.* **37:**181–191.

53. Neuwald, A. F., L. Aravind, J. L. Spouge, and E. V. Koonin. 1999. AAA+: a class of chaperone-like ATPases associated with the assembly, operation, and disassembly of protein complexes. *Genome Res.* **9:**27–43.

54. Niki, H., and S. Hiraga. 1998. Polar localization of the replication origin and terminus in *Escherichia coli* nucleoids during chromosome partitioning. *Genes Dev.* **12:**1036–1045.

55. Niki, H., Y. Yamaichi, and S. Hiraga. 2000. Dynamic organization of chromosomal DNA in *Escherichia coli*. *Genes Dev.* **14:**212–223.

56. Palaniyar, N., E. Gerasimopoulos, and D. H. Evans. 1999. Shope fibroma virus DNA topoisomerase catalyses Holliday junction resolution and hairpin formation in vitro. *J. Mol. Biol.* **287:**9–20.

57. Perals, K., H. Capiaux, J. B. Vincourt, J. M. Louarn, D. J. Sherratt, and F. Cornet. 2001. Interplay between recombination, cell division and chromosome structure during chromosome dimer resolution in *Escherichia coli*. *Mol. Microbiol.* **39:**904–913.

58. Perals, K., F. Cornet, Y. Merlet, I. Delon, and J. M. Louarn. 2000. Functional polarization of the *Escherichia coli* chromosome terminus: the *dif* site acts in chromosome dimer resolution only when located between long stretches of opposite polarity. *Mol. Microbiol.* **36:**33–43.

59. Picardeau, M., J. R. Lobry, and B. J. Hinnebusch. 1999. Physical mapping of an origin of bidirectional replication at the centre of the *Borrelia burgdorferi* linear chromosome. *Mol. Microbiol.* **32:**437–445.

60. Recchia, G. D., M. Aroyo, D. Wolf, G. Blakely, and D. J. Sherratt. 1999. FtsK-dependent and -independent pathways of Xer site-specific recombination. *EMBO J.* **18:**5724–5734.

61. Recchia, G. D., and D. J. Sherratt. 1999. Conservation of Xer site-specific recombination genes in bacteria. *Mol. Microbiol.* **34:**1146–1148.

62. Sadowski, P. D. 1995. The Flp recombinase of the 2-microns plasmid of *Saccharomyces cerevisiae*. *Prog. Nucleic Acid Res. Mol. Biol.* **51:**53–91.

63. Sekiguchi, J., C. Cheng, and S. Shuman. 2000. Resolution of a Holliday junction by vaccinia topoisomerase requires a spacer DNA segment 3′ of the CCCTT/ cleavage sites. *Nucleic Acids Res.* **28:**2658–2663.

64. Sharp, M. D., and K. Pogliano. 1999. An in vivo membrane fusion assay implicates SpoIIIE in the final stages of engulfment during Bacillus subtilis sporulation. *Proc. Natl. Acad. Sci. USA* **96:**14553–14558.

65. Sherratt, D. J., I. F. Lau, and F. X. Barre. 2001. Chromosome segregation. *Curr. Opin. Microbiol.* **4:**653–659.

66. Silberstein, Z., S. Maor, I. Berger, and A. Cohen. 1990. Lambda Red-mediated synthesis of plasmid linear multimers in Escherichia coli K12. *Mol. Gen. Genet.* **223:**496–507.

67. Stark, W. M., and M. R. Boocock. 1995. Topological selectivity in site-specific recombination, p. 101–129. *In* D. J. Sherratt (ed.), *Mobile Genetic Elements*. IRL Press, Oxford, United Kingdom.

68. Steiner, W., G. Liu, W. D. Donachie, and P. Kuempel. 1999. The cytoplasmic domain of FtsK protein is required for resolution of chromosome dimers. *Mol. Microbiol.* **31:**579–583.

69. Steiner, W. W., and P. L. Kuempel. 1998. Sister chromatid exchange frequencies in *Escherichia coli* analyzed by recombination at the *dif* resolvase site. *J. Bacteriol.* **180:**6269–6275.

70. Stirling, C. J., S. D. Colloms, J. F. Collins, G. Szatmari, and D. J. Sherratt. 1989. *xerB*, an *Escherichia coli* gene required for plasmid ColE1 site-specific recombination, is identical to *pepA*, encoding aminopeptidase A, a protein with substantial similarity to bovine lens leucine aminopeptidase. *EMBO J.* **8:**1623–1627.

71. Stirling, C. J., G. Szatmari, G. Stewart, M. C. Smith, and D. J. Sherratt. 1988. The arginine repressor is essential for plasmid-stabilizing site-specific recombination at the ColE1 *cer* locus. *EMBO J.* **7:**4389–4395.

72. Strater, N., D. J. Sherratt, and S. D. Colloms. 1999. X-ray structure of aminopeptidase A from *Escherichia coli* and a model for the nucleoprotein complex in Xer site-specific recombination. *EMBO J.* **18:**4513–4522.

73. Subramanya, H. S., L. K. Arciszewska, R. A. Baker, L. E. Bird, D. J. Sherratt, and D. B. Wigley. 1997. Crystal structure of the site-specific recombinase, XerD. *EMBO J.* **16:**5178–5187.

74. Summers, D. K., C. W. Beton, and H. L. Withers. 1993. Multicopy plasmid instability: the dimer catastrophe hypothesis. *Mol. Microbiol.* **8:**1031–1038.

75. Summers, D. K., and D. J. Sherratt. 1984. Multimerization of high copy number plasmids causes instability: ColE1 encodes a determinant essential for plasmid monomerization and stability. *Cell* **36:**1097–1103.

76. Thaler, D. S., and F. W. Stahl. 1988. DNA double-chain breaks in recombination of phage lambda and of yeast. *Annu. Rev. Genet.* **22:**169–197.

77. Wang, L., and J. Lutkenhaus. 1998. FtsK is an essential cell division protein that is localized to the septum and induced as part of the SOS response. *Mol. Microbiol.* **29:**731–740.

78. Warren, G. J., and A. J. Clark. 1980. Sequence-specific recombination of plasmid ColE1. *Proc. Natl. Acad. Sci. USA* **77:**6724–6728.

79. Wu, L. J., P. J. Lewis, R. Allmansberger, P. M. Hauser, and J. Errington. 1995. A conjugation-like mechanism for prespore chromosome partitioning during sporulation in *Bacillus subtilis*. *Genes Dev.* **9:**1316–1326.

79a. Yates, J., M. Aroyo, D. J. Sherratt, and F.-X. Barre. 2003. Species specificity in the activation of Xer recombination at *dif* by FtsK. *Mol. Microbiol.* **49:**241–249.

80. Yu, X. C., A. H. Tran, Q. Sun, and W. Margolin. 1998. Localization of cell division protein FtsK to the *Escherichia coli* septum and identification of a potential N-terminal targeting domain. *J. Bacteriol.* **180:**1296–1304.

81. Yu, X. C., E. K. Weihe, and W. Margolin. 1998. Role of the C terminus of FtsK in *Escherichia coli* chromosome segregation. *J. Bacteriol.* **180:**6424–6428.

Chapter 29

Replication of Linear Bacterial Chromosomes: No Longer Going Around in Circles

GEORGE CHACONAS AND CARTON W. CHEN

One of the ubiquitous features of almost all bacterial chromosomes is their circular structure. However, an increasing number of notable exceptions to this rule exist (26, 51, 75, 92, 111). The first linear bacterial chromosome was identified in the genus *Borrelia* about 15 years ago (36). Linear chromosomes and plasmids have also been noted in *Streptomyces* species (26, 71) and other actinomycetes (93), as well as in *Coxiella burnetii* (114) and a biovar of *Agrobacterium tumefaciens* (2, 41, 42, 115).

The structural deviation of linear replicons from the traditional circular molecules has profound implications in the way in which these molecules must be replicated (20, 26, 62, 74). The purpose of this chapter is to focus on the structure and associated functional features of the linear chromosomes of *Borrelia* and *Streptomyces* species. In particular, the unique aspects of the replication process resulting from the structure of these exceptional molecules and their strategies to deal with the problem of replicating the 3′ ends of the linear template molecules (113) are discussed.

THE *BORRELIA* GENOME

The genus *Borrelia* contains spirochetes causing Lyme disease (also known as Lyme borreliosis) and relapsing fever (7, 12, 79, 98). The genome of the prototype Lyme disease spirochete, *Borrelia burgdorferi* strain B31, has been recently sequenced (19, 22, 39) and displays several fascinating features. The genome is segmented and consists of about 22 independent replicons peacefully coexisting under one roof— more replicons than found in any other bacterial species.

The problems associated with replication and faithful partitioning of such a large complement of DNA molecules are interesting and heretofore unencountered. Proper maintenance of the *B. burgdorferi* replicons is an important feature for the organism since loss of certain plasmids is known to correlate with loss of infectivity in mammals (66, 90, 97).

Other intriguing features of the *Borrelia* genome are outside the scope of this chapter but have been recently summarized (19, 22). They include extensive and apparently recent DNA rearrangements as well as a large number of nonfunctional pseudogenes. The vast majority of open reading frames on *B. burgdorferi* plasmids have no similarity to nonborrelial genes, suggesting specialized functions.

Further information on the *B. burgdorferi* genome can be found at http://www.tigr.org/tigr-scripts/CMR2/GenomePage3.spl?database=gbb.

LINEAR REPLICONS IN *BORRELIA*

The 911-kb chromosome from *B. burgdorferi*, as well as 12 of the plasmids (5 to 56 kb), are linear; the remaining nine replicons are circular. Most of the circular molecules appear to be phages or remnants of phage genomes (see reference 35) and carry antigens expressed during mammalian infection (3, 17, 46, 105, 106).

The linear replicons are all terminated by covalently closed hairpin ends or "telomeres" (9, 20, 21, 49, 50). This type of telomeric structure in a bacterial chromosome was unique in *Borrelia* until very recently, when the second chromosome in *A. tumefaciens* was shown to have a similar structure

George Chaconas • Department of Biochemistry and Molecular Biology and Department of Microbiology and Infectious Diseases, University of Calgary, Calgary, Alberta T2N 4N1, Canada.   Carton W. Chen • Institute of Genetics, National Yang-Ming University, Shih-Pai, Taipei 112, Taiwan.

(41). The temperate *Escherichia coli* prophage N15 is also maintained as a linear molecule and has covalently closed hairpin ends or telomeres (95), as does the *Yersinia enterocolitica* prophage PY54 (EMBL locus BPY348844, accession number AJ348844.1) and the phage K02 from *Klebsiella oxytoca* (S. Casjens, W. Huang, G. Hatfull, and R. Hendrix, personal communication).

The sequence of a number of *Borrelia* telomeres has now been reported (20, 21), although only the left and right of the 17-kb linear plasmid (lp17) have been sequenced around the hairpins (49). Nonetheless, all the sequenced *Borrelia* telomeres are clearly related (Fig. 1) and are likely to interact with a single enzyme involved in generating all the telomeres (to be discussed below).

The conformational state of the linear *Borrelia* DNA molecules in vivo is at this point an open question. *B. burgdorferi* encodes topoisomerase I (Topo I), DNA gyrase, and Topo IV (39). (It does not carry a gene for Topo III, which is believed to be involved in RecA-mediated recombination [122].) It is possible that tethering of each linear molecule to two or more fixed points could result in DNA supercoiling; however, no evidence for or against this exists at present. This scenario would be more difficult to accomplish with the smaller linear replicons. It is also noteworthy that a recent study of the nucleoids from *B. burgdorferi* and *Borrelia hermsii* revealed a loose network of DNA strands without nodes, in contrast to the *E. coli* nucleoid, which contains some 20 to 50 loops emanating from a dense node of DNA (47).

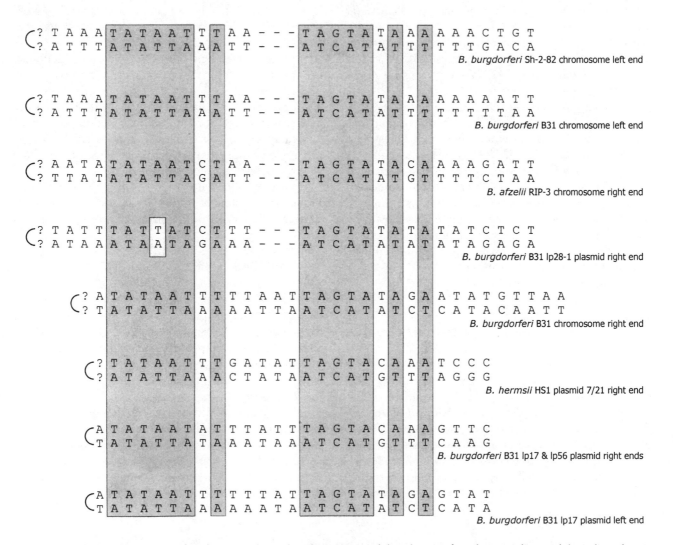

Figure 1. Alignment of *Borrelia* telomeres. The nucleotide sequences of the telomeres from linear replicons of the indicated species is shown (20, 21, 39, 49, 50, 59, 121). Conserved telomeric regions are boxed and shaded. The question marks indicate that DNA sequencing around the hairpin was not performed, so there is some uncertainty as to the exact sequence of the turnaround.

The difference in nucleoid structure may well be a simple reflection of an altered packaging of the linear DNA in *Borrelia*.

REPLICATION OF THE LINEAR *BORRELIA* CHROMOSOME

Replication of the linear *B. burgdorferi* chromosome is initiated from an origin near the center of the chromosome. Nascent strand analysis localized the bidirectional origin to within a 2-kb region (88). There is also a dramatic switch in CG skew, which occurs within the 240-bp *dnaA-dnaN* intergenic region, suggesting that the origin lies within this region (88). However, unlike other bacteria, no clearly identifiable DnaA boxes have been identified in the region where the origin has been mapped. The most likely explanation for this is that *Borrelia* DnaA recognizes a sequence distinct from previously reported DnaA boxes.

B. burgdorferi does not contain the accessory factors involved in the initiation of replication in many bacteria. IHF, HU, and Fis have been replaced by Hbb (64, 108) and Gac (60, 61) in *Borrelia*. The Hbb protein displays properties of both HU and IHF as well as sequence similarity to these proteins but recognizes a site unrelated to the IHF consensus (64). An Hbb binding site has been reported between *dnaA* and *dnaN* and is flanked on both sides by 13-mer repeats. It is tempting to speculate that this region is part of the origin and is involved in the assembly of the open complex mediated by Hbb-directed DNA bending. The Gac protein (the independently expressed C-terminal domain of the gyrase A subunit) may also play a role in the assembly of replication or other higher-order complexes in *B. burgdorferi*. This protein, although unrelated by DNA sequence, can efficiently substitute for the HU protein in the Mu DNA transposition reaction in vitro and complement an *E. coli* HU mutant (60, 61).

Following the initiation of bidirectional chromosomal replication, the reaction is believed to proceed to generate a circular head-to-head, tail-to-tail dimer as shown in Fig. 2. Such dimeric intermediates have not been observed, but are inferred to exist because of the demonstration of the telomere resolution step, whereby dimer junctions are processed into linear monomers through a DNA breakage-and-reunion reaction. It is noteworthy that a linear head-to-head plasmid dimer has been reported in *B. burgdorferi* and was likely derived by mutation of one of the telomeres to block the telomere resolution step (73).

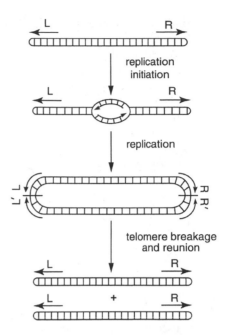

Figure 2. The replication strategy for linear replicons with covalently closed hairpin ends in *B. burgdorferi*. The arrows labeled L and R indicate the inverted repeats at the left and right ends, respectively. The line bisecting the head-to-head (L'-L) and tail-to-tail (R-R') telomere junctions in the replication intermediate is an axis of 180° rotational symmetry. The telomere breakage and reunion reaction is referred to as telomere resolution. Reprinted from reference 63 with permission from Elsevier.

TELOMERE RESOLUTION

A telomere resolution step in *B. burgdorferi* has been demonstrated both in vivo and in vitro. A telomere resolution substrate of 70 or 140 bp consisting of the replicated left-end telomere from lp17 (L'-L in Fig. 2) was shown to promote the in vivo formation of a covalently closed hairpin end at an internal position of lp17 (23). The same replicated telomere was capable of converting a small *B. burgdorferi* circular plasmid to a linear form in vivo. More recently, the *B. burgdorferi* telomere resolvase protein (ResT) was purified and shown to efficiently perform the telomere resolution step in vitro (63). ResT is encoded by the *resT* gene (BBB03) carried on the circular 26-kb plasmid, cp26. This plasmid is therefore expected to be an essential replicon in *Borrelia* perhaps better termed a minichromosome, as previously suggested (8). ResT is likely the only telomere resolvase encoded by *B. burgdorferi* since no other proteins with sequence homology are present in the genome. Moreover, the similarity in sequence found in all *Borrelia* telomeres suggests that they are recognized by a single enzyme (Fig. 1).

Telomere resolvases are a new class of DNA breakage-and-reunion enzyme that performs a unique reaction: breakage of two phosphodiester bonds in a single DNA duplex (one in each strand) and joining of each end with the opposite DNA strand to form a covalently closed hairpin telomere. The reaction is more complex than that promoted by a type I topoisomerase (breakage and reunion of a single DNA strand to alter linking number) but simpler than that promoted by a site-specific recombinase (breakage of four DNA strands and joining of each set to a new site to generate a recombinant product). Telomere resolvases have also been referred to as protelomerases, for prokaryotic or proteic telomerases (95). However, the term protelomerase is somewhat confusing since these enzymes do not display telomerase activity as observed in eukaryotic telomerases. In contrast, the term telomere resolvase is clear, biochemically correct, and applicable to eukaryotic systems such as the poxviruses, where the term originated (83, 99). We have therefore suggested the use of the term telomere resolvase rather than protelomerase (62).

Telomere resolvases have now been purified from *E. coli* phage N15 (31, 32), from *B. burgdorferi* (63), and from *K. oxytoca* phage K02 (W. M. Huang, personal communication). A putative telomere resolvase also exists in the *Y. enterocolitica* phage PY54 (EMBL locus BPY348844, accession number AJ348844.1) and in the genome of *A. tumefaciens* (accession number AAK88254 [41]). These telomere resolvases all display limited sequence homology to the tyrosine recombinases (the λ integrase family of proteins) as first predicted for the N15 enzyme (95). The tyrosine recombinases, like the type Ib topoisomerases, perform a two-step transesterification through a 3′ phosphotyrosine-linked protein-DNA intermediate (43–45, 102). *B. burgdorferi* ResT was recently shown to perform telomere resolution through a two-step transesterification (Fig. 3) and to cleave its substrate 3 nucleotides away from the axis of symmetry, leaving 6-nucleotide 5′ overhangs (63). ResT performs the telomere resolution reaction in the absence of divalent metal ions, high-energy cofactors, and DNA accessory proteins.

An intriguing feature of the ResT reaction is that it does not occur on supercoiled DNA because of the spontaneous extrusion of the replicated telomere as a cruciform (Fig. 3) (63). This raises the question of how and when ResT functions in vivo. Complete DNA replication in vivo to generate a circular dimeric intermediate might also be expected to result in cruciform extrusion. However, the high AT content of *Borrelia* DNA might well result in DNA unwinding at positions other than at the telomeres, in contrast to when telomeres are carried on a relatively GC-rich *E. coli* plasmid. Alternatively, telomere resolution in vivo might occur at some time before all the nicks are

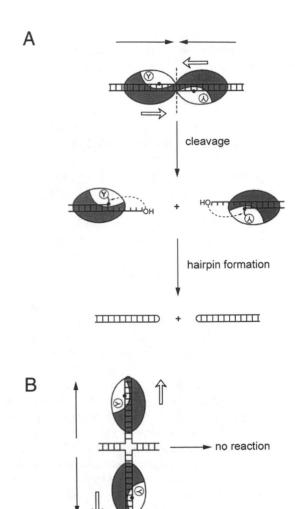

Figure 3. Mechanism of action of ResT. (A) In a relaxed or linearized plasmid the telomere junction is presented as lineform DNA with a head-to-head structure for the inverted repeat (thin arrows). The scissile phosphates are noted with black dots and are 6 nucleotides apart on opposite strands, placing them on the same face of the DNA double helix. The shaded ovals represent ResT protomers, and the unshaded portions denote the active site with its putative tyrosine nucleophile (Y). The open arrows indicate the orientation of the ResT protomers. For simplicity, the reaction is drawn with active-site function in *cis*, although whether catalysis actually occurs in *cis* or in *trans* is not yet known. DNA cleavage is effected through nucleophilic attack by an active-site tyrosine residue which makes a covalent intermediate with the DNA through a 3′ phosphotyrosine linkage. The 5′ hydroxyl groups are brought into proximity with the phosphotyrosine linkage for transesterification by a conformational change in the complex or by simple dissociation, with joining of the bottom strand to the top strand to produce the DNA hairpin. (B) In a supercoiled plasmid the telomere junction is presented as cruciform DNA with the inverted repeats in the opposite orientation to that found in the lineform DNA. This structure would block interaction of ResT protomers by reversing their relative orientation. They would also be separated in space on the long arms of the extruded cruciform. Additionally, the cleavage sites are moved far from the strand to which they need to be joined for hairpin formation. Reprinted from reference 63 with permission from Elsevier.

sealed, so that the replicated molecules are not yet supercoiled. Finally, ResT might act upon supercoiled DNA molecules with the assistance of some cellular protein to lower DNA supercoiling or remove the extruded cruciforms.

REPLICATION OF LINEAR *BORRELIA* PLASMIDS

The replication origins of the linear plasmids in *Borrelia* have not yet been mapped. Plasmid copy number in *B. burgdorferi* appears to be tightly coupled with that of the linear chromosome (48). Analysis of the CG skew of the plasmids does not result in a picture as dramatic as that observed for the chromosome, but nonetheless points to internal origins that map near a *parA*-like partition function in most of the plasmids (22, 40, 87). Based upon the similarity of the telomeres on the chromosome to those on the linear plasmids (21), replication of the linear plasmids is also expected to be bidirectional from an internal origin. This would generate circular dimeric intermediates that are processed by ResT. However, initiation of replication on the linear plasmids may differ significantly from that of the chromosome.

A set of paralogous gene families (PF) believed to be involved in replication and partitioning of *B. burgdorferi* plasmids has been described (22, 40, 87). Representative members of each of these families (32, 49, 50 and the 57/62 superfamily) are found on most of the 21 plasmids in *B. burgdorferi* B31. Incomplete sets have been found only on the smaller replicons: lp5, lp17, lp21, and cp9. The precise function of each of these gene families remains unknown, with the exception of the family 32 genes, which likely encode partitioning proteins related to the ParA/SopA/Soj family.

A 3.3-kb DNA fragment carrying three paralogous gene family members (49, 50 and 57/62) in cp9 was shown to be sufficient for autonomous replication of a small circular cp9 derivative in *B. burgdorferi*; all three family members were required for autonomous replication (107). This is the first direct evidence of a requirement for these paralogous families in the DNA replication process. More recently (34), it has been shown that a 4-kb fragment containing a family 32, 49, 50, and 57 gene from one of the cp32 plasmids is capable of autonomous replication. The minimal region required for replication was 2 kb in size and included all of the PF57 gene and the entire intergenic region between the PF57 and PF161 genes. Replication of this construct may, however, require the contribution of other cp32 replication proteins in

trans. The intergenic region contained six DnaA-like boxes using relaxed homology requirements, suggesting a possible role for DnaA in cp32 replication. Finally, transformation of *B. burgdorferi* with the deleted construct resulted in selective deletion of cp32-3, which contained the most closely related PF32 protein. Hence, PF32 may govern compatibility of the cp32 plasmids. These results are in keeping with the proposal (22) for a series of distinct but related replication and partition functions as a mechanism for the peaceful coexistence (replication, compatibility, and partitioning) of the large complement of replicons in *B. burgdorferi*.

WHY LINEAR REPLICONS IN *BORRELIA*?

The philosophical question of why *Borrelia* should have a linear chromosome and linear plasmids remains an enigma. The argument for linearity as a promoter of recombination for sequences located near the ends of molecules has been previously discussed (20), and stimulation of recombination by telomeric localization in *B. burgdorferi* has recently been demonstrated (23). In particular, the recombinational events underlying the switching of information from silent to expressed loci in the antigenic variation process in *Borrelia hermsii* (59, 89) and *B. burgdorferi* (121) may be stimulated by the location of the expression loci adjacent to telomeric regions. But this reason alone is not entirely satisfying, as most linear replicons in *Borrelia* do not have any apparent need for enhanced recombinational tactics. It also does not help to explain the covalently closed hairpin telomeres observed in *Agrobacterium* and in the phage systems noted earlier.

Another interesting question regarding the topological state of the replicons in *Borrelia* is why there is a persistence of two different forms. Is there some fundamental difference between the circular and linear replicons? Both circular and linear replicons seem to rigidly maintain their topological state. Only one example of the conversion of a linear to a circular plasmid (*B. hermsii*) has been reported (37), and no naturally occurring transitions from circular to linear form have been observed. A circular derivative of cp9 (107) has been converted to a linear form through the insertion of a telomere resolution substrate (23). However, the resulting linear replicon displayed a copy number reduced fivefold from the circular parental plasmid. This suggests the possibility of some fundamental difference between circular and linear replicons in *Borrelia*, but additional examples will need to be found to establish this idea as a general tenet.

THE *STREPTOMYCES* GENOMES

The genetics of the gram-positive genus *Streptomyces* has been studied for half a century (see reference 53 for a review), mostly with the model species, *S. coelicolor* A3(2) and, to a lesser extent, *S. lividans* 66, from both of which much of the discussion below is drawn. Based on these studies, circular genetic maps of the *Streptomyces* chromosomes have been established (52, 55). It is thus a very significant yet unexpected discovery that these chromosomes are linear (71). Although this is also true in *Borrelia*, several basic features are different. Among eubacteria, *Borrelia* chromosomes are almost the smallest in size (910 kb) and the lowest in G+C content (910 kb and 28.6% for *B. burgdorferi* [39]), while *Streptomyces* chromosomes are at the other end of the spectrum (8.7 Mb and 72.1% G+C for *S. coelicolor* [http://www.sanger.ac.uk/Projects/S_coelicolor]). The large size of *Streptomyces* chromosomes reflects the complex life cycles and high capacity for secondary metabolite production of these free-living soil microbes, while the small size of *Borrelia* chromosomes signifies their simpler parasitic life style. *Borrelia* telomeres are hairpin loops, while *Streptomyces* chromosomes have proteins covalently attached to the 5′ ends of the linear DNA (71, 116). *Borrelia* chromosomes contain short (26 bp in *B. burgdorferi* [39]) and imperfect "terminal inverted repeats" (TIRs), whereas the TIR of *Streptomyces* chromosomes are generally very long, ranging from tens to hundreds of kilobases (68, 70, 71, 84). The *Streptomyces avermitilis* chromosome, however, has an exceptionally short (174-bp) TIR (56b, 81).

Chromosome linearity is not unique to *Streptomyces*, nor is it universal for all members of the actinomycetes. It appears that those species with more sophisticated structures and life cycles, such as *Streptomyces* spp. (71), *Saccharopolyspora erythraea* (94), *Actinoplanes philippinensis*, *Micromonospora chalcea*, *Nocardia asteroides*, *Streptoverticillium abikoense*, and *Streptoverticillium cinnamoneus* (28, 93), have larger (7.5 to 9.4 Mb) linear chromosomes, whereas those with simpler life styles, such as *Corynebacterium*, *Mycobacterium*, and *Rhodococcus*, have smaller (3.0 to 6.5 Mb) circular chromosomes (93). The extra genes harbored by the *Streptomyces* chromosomes include multitudes of gene clusters for secondary metabolites.

Linear plasmids are widespread among actinomycetes, regardless of the topology of the host chromosome. These linear DNA molecules have the same structural features as the linear host chromosomes—internal replication origins, TIRs, and covalently bound terminal proteins (TPs). The size of these plasmids varies greatly, from just over 10 kb to nearly 1 Mb, and so does that of their TIR, from 41 bp in SLP2 (27) to 80 kb in SCP1 (12a, 58).

REPLICATION OF *STREPTOMYCES* CHROMOSOMES

As in other gram-positive bacteria, an *oriC* is found between the *dnaA* and *dnaN* genes (18, 119) located at the center of the *Streptomyces* chromosome. Compared with the initiation system for replication of *E. coli* chromosomes, the *Streptomyces* system appears to be more complex: a longer DNA sequence, a larger number of DnaA boxes, and a larger DnaA protein. The *oriC* of *Streptomyces* spp. contains two clusters of 19 "DnaA boxes" (consensus: TTGTCCACA) in relatively conserved positions (118, 119). DnaA proteins, which control initiation of chromosome replication, bind both clusters of DnaA boxes and bend the spacer DNA to form a functional loop through cooperative protein-protein interactions (57, 72). Such dimerization of *Streptomyces* DnaA on binding DnaA boxes has not been seen in *E. coli* DnaA.

A temperature-sensitive mutation in *dnaA* of *S. lividans* was created by site-directed mutagenesis in the conserved C terminus of DnaA (69), and synchronization of the mutant culture was achieved by temperature manipulation. Measurements of thymidine incorporation gave a doubling time (generation time) of about 105 min at 30° and 95 min at 39°, and a time for chromosome replication of 90 to 100 min. These results suggest the lack of overlapping rounds of replication in *Streptomyces* even at (or near) the maximum growth rate, consistent with the conclusion of Shahab et al. (101) based on analysis of macromolecular compositions and synthesis. Considering the size of the *S. lividans* chromosome (about 8 Mb [68a, 71]) and the number of replication forks (supposedly two), this would give a chain growth rate of polymerization of about 670 to 740 nucleotides (nt) per s at 39°. This is comparable to the DNA chain growth rate of 560 nt/s for *E. coli* growing at a doubling time of 100 min at 30° (14).

Once initiation at *oriC* has started, the cell must prevent reinitiation at the newly replicated *oriC*. Neither the *dam* methylation system used in *E. coli* for discriminating old (fully methylated) from new (semimethylated) origins nor a homologue of the sequestration protein, SeqA, which binds the methylated GATC sequences in *oriC*, is present in *S. coelicolor* or other gram-positive bacteria such as *Bacillus subtilis* (65). One or more alternative systems must be in play in these bacteria. Boye et al. (15)

suggested that in gram-positive bacteria the location of *oriC* next to *dnaA* provides a high-capacity sink for DnaA, thereby reducing the level of free DnaA.

In many bacteria, such as *E. coli*, *B. subtilis*, and *B. burgdorferi*, GC skew displays a minus-to-plus sign inversion around *oriC* on the chromosomes, indicating a preference for G over C in the leading strands during replication (77). A sign inversion of GC skew at *oriC* is also seen in the *S. coelicolor* chromosome but in the opposite (plus to minus) polarity, i.e., a preference for C over G in the leading strand (S. Bentley and D. A. Hopwood, personal communication). The reason for this is not clear.

TERMINAL PATCHING AT *STREPTOMYCES* TELOMERES

When replication reaches the ends of the chromosome, a stretch at the 3' ends will remain single-stranded until it is filled by a patching reaction, which certainly involves the TP. The structure of the exposed 3' overhangs is supposedly crucial for patching.

The telomere DNA from many *Streptomyces* chromosomes and linear plasmids has been cloned and sequenced. The majority of these sequences form a very conserved group that contains at least seven tightly packed palindromic sequences in the first 166 to 168 nt (56). The 174-bp TIR of the *S. avermitilis* chromosome also contains such a conserved sequence (56b, 81).

The 3' overhangs containing these terminal palindromic sequences exhibit complex secondary structures with few unpaired nucleotides as predicted by energy minimization (Fig. 4). Interestingly, some of the structural features are also present in the 3' ends of autonomous parvovirus genomes (30, 56), such as hairpin loops and "bulges" that contain non-Watson-Crick purine:purine sheared pairings and confer resistance to single-strand-specific nucleases. Both features are important for replication of autonomous parvovirus genomes (4).

TPs from two species, *Streptomyces rochei* (6) and *S. coelicolor* (116), have been purified, and their amino acid sequences have been determined. TP coding sequences (designated *tpg*) lie about 100 kb

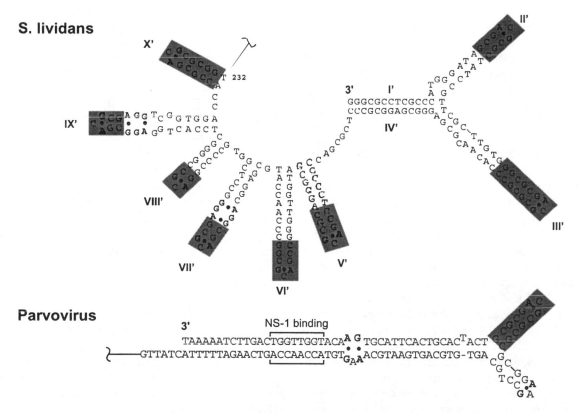

Figure 4. Predicted secondary structure of the 3' end of the *S. lividans* chromosome. Roman numerals indicate palindrome numbers. The "rabbit ears" secondary structure of the terminal 119 nt based on autonomous parvovirus (30) is included for comparison. The hairpin sequences that are homologous to a hairpin of the parvoviral genome are shaded. The putative Pu:Pu sheared pairings are indicated by black dots. Palindromes I' to VII' are conserved in most *Streptomyces* chromosomes. Palindromes VIII' to X' are absent from some. Modified from references 13 and 56.

from one end of the chromosomes of *S. lividans* and *S. coelicolor* (6, 116), and near the left end of the 50-kb linear plasmid, SLP2, of *S. lividans* (56a, 116). In addition, a putative pseudogene was identified in the homologous sequence shared by the right arm of SLP2 and the TIR of the *S. lividans* chromosome. Therefore, *S. lividans* possesses five copies (three species) of *tpg* homologues. Similarly, at least three copies of *tpg* homologues were identified in the genome of *S. rochei* (6; H. Kinashi, personal communication). The genome of *S. coelicolor*, however, contains only one *tpg* homologue.

The amino acid sequences of the Tpg homologues, whether experimentally determined or conceptually translated, are very conserved in length (184 or 185 amino acids [aa]) and sequence (except for the pseudogene), and contain a putative helix-turn-helix domain that shows limited similarity to the DNA-binding "thumb" domain of human immunodeficiency virus reverse transcriptase (6, 116) and putative amphiphilic beta-sheets. The latter may participate in protein-protein or protein-membrane interactions, as has been observed in adenoviruses and phage ϕ29 (16, 82, 96). Interactions between the *Streptomyces* TPs also appear to occur spontaneously under routine laboratory conditions (for example, see reference 71) and have been inferred from genetic (112) and cytological (117) studies. Such intramolecular end-to-end interactions would force a bias toward an even number of crossovers during recombination between two linear chromosomes, thus leading automatically to a circular genetic map (103, 104).

TERMINAL PATCHING MODELS

Patching of the 3' single-strand gaps at the termini of *Streptomyces* chromosomes during replication represents a novel situation, to which no known molecular models may be readily applied. The well-studied TP-capped genomes of phage ϕ29 and adenoviruses do not use such terminal patching; instead, TP-primed replication is initiated at the double-strand ends. The TIRs in these genomes are very short (e.g., 6 bp in ϕ29 and 103 bp in adenovirus type 2) and lack the complicated sequence patterns found in *Streptomyces* replicons, perhaps reflecting the absence of exposed 3' ends during replication and thus the need to form specific secondary structures.

Several models for terminal patching of the linear *Streptomyces* replicons have been proposed (Fig. 5) (26), and partially tested (6). These models may be divided into two general groups, in which the TP acts either as a primer or as a nickase. In both

cases, the TP becomes covalently bound to the 5' ends of the telomere DNA. In the first group (TP as primer [Fig. 5, left]), the substrate is either the complex secondary structures formed by the 3' strand (TP priming model), or the TP-capped double-stranded DNA (recombination model). In the former (Fig. 5, upper left), the TP and DNA polymerase act in concert to initiate synthesis directly from the 3' end using the TP as primer. In the latter (Fig. 5, lower left), TP-primed synthesis displaces the TP-capped 5' parental strand, which pairs with the protruding 3' strand. Displacement and strand exchange continue until the whole gap is filled, and the half Holliday junction is resolved by homologous recombination.

The recombination model suggests that homologous recombination would be essential for chromosomal replication and thus viability. The difficulties encountered by several laboratories (29, 76, 78) in constructing null mutations of the *recA* gene in *Streptomyces* seemed to support the model. However, Vierling et al. (109) created a null *recA* mutation through a two-step procedure during which a second unknown mutation (affecting sporulation) had been acquired. It is possible that this secondary mutation suppressed the lethality of the *recA* deficiency by providing a new *recA*-independent recombination pathway. An essential *recA* gene, accompanied by a nonessential one, was also identified in *Myxococcus xanthus* (80).

Thus, the feasibility of the recombination model is still an open question. Nevertheless, it is noteworthy that the recombination model in theory is blind to the sophisticated secondary structures at the 3' overhangs during replication (except for their possible roles in structural integrity of the intermediates). In contrast, the complex terminal structure is an important feature of the TP priming model. In support of the model, integrity of most of the terminal palindromic sequences (including the highly conserved 13-nt palindrome I) was essential for replication of the linear plasmid pSLA2 (91) and a minichromosome (C.-H. Huang, H.-H. Tsai, and C. W. Chen, unpublished results).

In the second group of models (TP as nickase), the classical rolling hairpin model (Fig. 5, upper right), based on helper-dependent parvovirus replication, proposes that the 3' overhang folds back onto itself and provides a primer for DNA synthesis to fill the gap. This loop is resolved by the TP nickase at a site facing the original 3' end, and a new round of DNA synthesis primed by the new 3' end at the nick restores the TP-capped duplex DNA. This model predicts a flip-flop of the terminal palindromic sequences with each round of replication, and thus

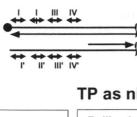

TP as primer

TP-primed patching

Recombinational patching

TP as nickase

Rolling hairpin

Modified Rolling hairpin

Figure 5. Four models for terminal patching of *Streptomyces* linear replicons. The intermediate telomere structure with the 3′ single-stranded overhang, which contains the conserved terminal palindromes (I′ to IV′ are indicated), is shown on the top. The terminal protein (TP) is indicated by the filled circle. New TP is represented by open circles, DNA polymerase is represented by open arrows, and newly synthesized DNA is represented by dashed lines. See the text for details about the models. Modified from reference 26.

the existence of a mixed population of chromosomes with two opposite orientations of the terminal palindromes. Such flip-flopped terminal sequences have not been detected among populations of terminal sequences analyzed (56, 91).

The second model in this group, the modified rolling hairpin model (Fig. 5, lower right) (unpublished observations), is based on the asymmetric resolution model for autonomous parvovirus replication, which employs hairpinned termini as an alternative (and recycled) form of telomere, thus avoiding flip-flops of the terminal palindromes (4, 5). This model predicts equal presence of the TP-capped ends ("extended form") and the hairpinned ends ("turn-

around form") of *Streptomyces* chromosomes. This model is also disfavored, because C.-H. Huang and C. W. Chen (unpublished results), using a two-dimensional gel electrophoresis technique, detected only the extended form but no turn-around form of the *Streptomyces* telomeres.

In other TP-capped genomes (such as those of adenoviruses and φ29), the DNA polymerases involved in TP-primed replication are homologues of DNA polymerase PolII (α type). No PolII homologue, however, could be detected in the *S. coelicolor* genome. This suggests that a different type of polymerase catalyzes terminal patching—not surprising, in view of the contrast in the two reactions (see above). Among enzymes encoded by the *S. coelicolor* genome, candidates for the terminal patching polymerase are a PolA (DNA polymerase I) homologue, a PolA-like protein, and a small (212-aa) conserved "DNA polymerase-related" protein.

A *dnaE* gene, encoding the catalytic α subunit of PolIII, was identified as an essential component in chromosome replication (38). The *S. coelicolor* chromosome sequence, however, contains an open reading frame encoding another α subunit (1,185 aa) of PolIII. There are also two copies of β and γ/τ subunits and three of ε subunits. The meaning of this multiplicity is not clear, but it is unlikely that an extra PolIII enzyme functions in terminal patching, because the DNA strand synthesized in terminal patching is expected to be relatively short (24), not necessitating the high processivity of such enzyme.

REPLICATION OF LINEAR *STREPTOMYCES* PLASMIDS

Linear plasmids of *Streptomyces* also replicate bidirectionally, starting from internal origins (24, 25) and leaving about 280-nt 3′ overhangs at the two ends (24), which are expected to be patched by the same mechanism as that for the chromosomes. The telomere sequences of many linear plasmids are homologous to those of the chromosomes, but a few are more divergent, being conserved only in the first few palindromes (56). Identical or highly similar terminal sequences may be capped by the same TP. Shorter derivatives of linear plasmids lacking a *tpg* gene may be capped by TPs encoded by a larger plasmid or the chromosome (6). The right end of the SLP2 plasmid in *S. lividans*, being identical to the ends of the chromosome, probably shares the same TP (116). SCP1, the sole linear plasmid in *S. coelicolor*, has the most divergent terminal sequence and lacks a *tpg* homologue. It is likely that a very different TP is involved in capping its telomeres.

CIRCULARIZATION OF *STREPTOMYCES* CHROMOSOMES

An exceptionally high incidence of structural instability of the chromosomes has been a hallmark of *Streptomyces* genetics for many years. Deletions of very long stretches of terminal DNA (up to about 1 Mb [91a]), which are often accompanied by tandem amplifications of nearby sequences, occur at frequencies of 10^{-4} to 10^{-2} per spore (recently reviewed in references 67 and 110). Often these deletions lead to chromosome circularization (71). While some mu-

tants exhibit certain defects (such as nonsporulation), many grow as vigorously as the wild-type strains with linear chromosomes.

In theory, circularity may pose a segregation problem that requires the assistance of a type II topoisomerase, as in other bacteria. In *E. coli*, the primary enzyme involved in decatenation of the daughter chromosomes and circular plasmids is Topo IV (1, 120). The two subunits of Topo IV, ParC and ParE, are highly homologous to the subunits of gyrase: GyrA and GyrB, respectively. The genome of *S. coelicolor* contains both *gyrA/gyrB* and a corresponding homologous pair, SCO5836 (*gyrA*-like)/SCO5822 (*gyrB*-like). The latter probably constitute the Topo IV decatenase, and its existence is intriguing, because this enzyme would not be needed for linear chromosomes. Interestingly, the linear chromosome of *B. burgdorferi* also has both sets of type II topoisomerases (39).

In contrast, the absence of homologues of the site-specific resolution enzyme, XerCD, in the *Streptomyces* genome does not seem to affect partitioning of circular chromosomes during sporulation (H.-H. Lee and C. W. Chen, unpublished fluorescence microscopy observation). This may be because (i) homologous recombination between daughter sister chromosomes is rare; (ii) resolution of the Holliday junctions by RuvABC is strongly biased toward noncrossovers with only exchanges of single strands (patch) instead of crossover of double strands (splice) (11) as during eukaryotic mitosis (85); or (iii) formation of anucleated spores is suppressed by an unknown mechanism. (See reference 26a for a discussion of evolutionary significance.)

CONJUGAL TRANSFER OF *STREPTOMYCES* CHROMOSOMES AND PLASMIDS

Essentially all natural isolates of plasmids, circular or linear, in *Streptomyces* appear to be conjugation competent, being able to transfer themselves (at up to 100% efficiency per recipient) and the host chromosomes (at lower efficiency). In circular plasmid-mediated conjugation, the recombinant chromosomes contain only short internal stretches of the donor chromosomes, as deduced by Hopwood (54) from purely genetic data and later confirmed physically by Wang et al. (112). This is consistent with the classical *E. coli* F plasmid model, i.e., transfer of the chromosome initiated internally from the origin of transfer, *oriT*, of the integrated plasmid. Such an *oriT* may correspond to a *cis*-acting element identified on plasmids pIJ101 (33, 86) and pJV1 (100), which is needed in *cis* for transfer of the plasmid but not for transfer of the host chromosome.

Transfer of linear *Streptomyces* replicons, however, cannot follow any established model. Surprisingly, the recombinant chromosomes produced in linear plasmid-mediated interspecies conjugation inherit essentially complete donor sequences, with only short stretches of the recipient chromosomes containing the selected marker (112). While this is consistent with the end-first transfer model of Chen (26), whereby the transfer of linear *Streptomyces* replicons is initiated at the telomeres (led by the TP-capped 5′ strands that are displaced by TP-primed synthesis), the prevalence of the donor chromosome and the departure of the recipient chromosome from the recombinants was totally unexpected. One possible "aggressor" may be the terminally located transposable elements on the donor chromosome, which, on entering the recipient mycelium, might be released from a repressed state and assault the defenseless recipient chromosome in packs. This is analogous to "hybrid dysgenesis" in *Drosophila* induced by introduction of P elements on sperm chromosomes into the P elementless eggs.

The best support for the end-first transfer model came from the observation (H.-H. Lee, Y.-L. Lin, H.-H. Tsai, I.-T. Hsieh, and C. W. Chen, unpublished results) that circular chromosomes can be mobilized by circular plasmids but not by linear plasmids, and that linear chromosomes can be mobilized by both. The presence of telomeres appears to be essential for linear plasmid-mediated transfer.

CONCLUDING REMARKS

The linear replicons of *Borrelia* and *Streptomyces*, despite their identical topology, appear to be highly diversified in their structures and modes of replication. These two organisms use varied but highly effective mechanisms to ensure that the ends of their linear DNA molecules are faithfully replicated, thereby eliminating the need for circularity of their DNA. The reasons for the existence of such diverse approaches for dealing with the end replication problem are not currently understood.

Acknowledgments. We thank David A. Hopwood (John Innes Centre, Norwich, United Kingdom), Stephen Bentley (Sanger Centre, Cambridge, United Kingdom), H. Kinashi (University of Hiroshima, Hiroshima, Japan), Sherwood Casjens (University of Utah), and Wai Mun Huang (University of Utah) for communication of research results prior to publication. The contributions of Yvonne Tourand in generating Fig. 1 and Chi Chen for assistance in bioinformatic analysis are gratefully acknowledged. We also thank Cecile Beaurepaire, David Hopwood, Kerri Kobryn, Scott Samuels, and Yvonne Tourand for critical review of parts of the manuscript.

G.C. is supported by research grants and a Distinguished Scientist Award from the Canadian Institutes of Health Research. C.W.C. is supported by research grants from the National Science Council (NSC88-2312-B010-002, NSC89-2312-B010-002, and NSC90-2312-B010-002) and the Ministry of Education (89-B-FA22-2-4) of Taiwan.

ADDENDUM IN PROOF

Since the submission of the manuscript, several papers related to *Borrelia* and *Streptomyces* DNA replication have been published or accepted for publication.

Those related to *Borrelia* DNA replication are as follows: T. Bankhead and G. Chaconas, *Proc. Natl. Acad. Sci. USA*, in press; R. Byram, P. E. Stewart, and P. Rosa, *J. Bacteriol.* **186:**3561–3569, 2004; S. R. Casjens, E. B. Gilcrease, W. M. Huang, K. L. Bunny, M. L. Pedulla, M. E. Ford, J. M. Houtz, G. F. Hatfull, and R. W. Hendrix, *J. Bacteriol.* **186:**1818–1832, 2004; S. Hertwig, I. Klein, R. Lurz, E. Lanka, and B. Appel, *Mol. Microbiol.* **48:**989–1003, 2003; W. M. Huang, L. Joss, T. Hsieh, and S. Casjens, *J. Mol. Biol.* **337:**77–92, 2004; N. V. Ravin, *FEMS Microbiol. Lett.* **221:**1–6, 2003; N. V. Ravin, V. V. Kuprianov, E. B. Gilcrease, and S. R. Casjens, *Nucleic Acids Res.* **31:**6552–6560, 2003; P. Stewart, P. A. Rosa, and K. Tilly, p. 291–301, *in* B. E. Funnell and G. J. Phillips (ed.), *Plasmid Biology* (ASM Press, Washington, D.C.), 2004; P. E. Stewart, G. Chaconas, and P. Rosa, *J. Bacteriol.* **185:**3202–3209, 2003; and Y. Tourand, K. Kobryn, and G. Chaconas, *Mol. Microbiol.* **48:**901–911, 2003.

Those related to *Streptomyces* DNA replication are as follows: K. Bao and S. N. Cohen, *Genes Dev.* **17:**774–785, 2003; S. D. Bentley, S. Brown, L. D. Murphy, D. E. Harris, M. A. Quail, J. Parkhill, B. G. Barrell, J. R. McCormick, R. I. Santamaria, R. Losick, M. Yamasaki, H. Kinashi, C. W. Chen, G. Chandra, D. Jakimowicz, H. M. Kieser, T. Kieser, and K. F. Chater, *Mol. Microbiol.* **51:**1615–1628, 2004; S. D. Bentley, K. F. Chater, A.-M. Cerdeño-Tárraga, G. L. Challis, N. R. Thomson, K. D. James, D. E. Harris, M. A. Quail, H. Kieser, D. Harper, A. Bateman, S. Brown, G. Chandra, C. W. Chen, M. Collins, A. Cronin, A. Fraser, A. Goble, J. Hidalgo, T. Hornsby, S. Howarth, C. H. Huang, T. Kieser, L. Larke, L. Murphy, K. Oliver, S. O'Neil, E. Rabbinowitsch, M. A. Rajandream, K. Rutherford, S. Rutter, K. Seeger, D. Saunders, S. Sharp, R. Squares, S. Squares, K. Taylor, T. Warren, A. Wietzorrek, J. Woodward, B. G. Barrell, J. Parkhill, and D. A. Hopwood, *Nature* **417:**141–147, 2002; C. W. Chen, C.-H. Huang, H.-H. Lee, H.-H. Tsai, and R. Kirby, *Trends Genet.* **18:**522–529, 2002; C. H. Huang, C. Y. Chen, H. H. Tsai, C. Chen, Y. S. Lin, and C. W. Chen, *Mol. Microbiol.* **47:**1563–1576, 2003; H. Ikeda, J. Ishikawa, A. Hanamoto, M. Shinose, H. Kikuchi, T. Shiba, Y. Sakaki, M. Hattori, and S. Omura, *Nat. Biotechnol.* **21:**526–531, 2003; and Z. Qin, M. Shen, and S. N. Cohen, *J. Bacteriol.* **185:**6575–6582, 2003.

REFERENCES

1. **Adams, D. E., E. M. Shekhtman, E. L. Zechiedrich, M. B. Schmid, and N. R. Cozzarelli.** 1992. The role of topoisomerase IV in partitioning bacterial replicons and the structure of catenated intermediates in DNA replication. *Cell* **71:**277–288.

2. **Allardet-Servent, A., S. Michaux-Charachon, E. Jumas-Bilak, L. Karayan, and M. Ramuz.** 1993. Presence of one linear and one circular chromosome in the *Agrobacterium tumefaciens* C58 genome. *J. Bacteriol.* **175:**7869–7874.

3. Anguita, J., S. Samanta, B. Revilla, K. Suk, S. Das, S. W. Barthold, and E. Fikrig. 2000. *Borrelia burgdorferi* gene expression in vivo and spirochete pathogenicity. *Infect. Immun.* **68:**1222–1230.

4. Astell, C. R., M. B. Chow, and D. C. Ward. 1985. Sequence analysis of the termini of virion and replicative forms of minute virus of mice suggests a modified rolling hairpin model for autonomous parvovirus DNA replication. *J. Virol.* **54:**171–177.

5. Astell, C. R., M. Thomson, M. B. Chow, and D. C. Ward. 1983. Structure and replication of minute virus of mice DNA. *Cold Spring Harbor Symp. Quant. Biol.* **47:**751–762.

6. Bao, K., and S. N. Cohen. 2001. Terminal proteins essential for the replication of linear plasmids and chromosomes in *Streptomyces. Genes Dev.* **15:**1518–1527.

7. Barbour, A. G. 2001. *Borrelia:* a diverse and ubiquitous genus of tick-borne pathogens, p. 153–173. *In* M. W. Scheld, W. A. Craig, and J. M. Hughes (ed.), *Emerging Infections 5.* ASM Press, Washington, D.C.

8. Barbour, A. G., and D. Fish. 1993. The biological and social phenomenon of Lyme disease. *Science* **260:**1610–1616.

9. Barbour, A. G., and C. F. Garon. 1987. Linear plasmids of the bacterium *Borrelia burgdorferi* have covalently closed ends. *Science* **237:**409–411.

10. Baril, C., C. Richaud, G. Baranton, and I. S. Saint Girons. 1989. Linear chromosome of *Borrelia burgdorferi. Res. Microbiol.* **140:**507–516.

11. Barre, F. X., B. Soballe, B. Michel, M. Aroyo, M. Robertson, and D. Sherratt. 2001. Circles: the replication-recombination-chromosome segregation connection. *Proc. Natl. Acad. Sci. USA* **98:**8189–8195.

12. Barthold, S. W. 2000. Lyme borreliosis, p. 281–304. *In* J. P. Nataro, M. J. Blaser, and S. Cunningham-Rundles (ed.), *Persistent Bacterial Infections.* ASM Press, Washington, D.C.

12a. Bentley, S. D., S. Brown, L. D. Murphy, D. E. Harris, M. A. Quail, J. Parkhill, B. G. Barrell, J. R. McCormick, R. I. Santamaria, R. Losick, M. Yamasaki, H. Kinashi, C. W. Chen, G. Chandra, D. Jakimowicz, H. M. Kieser, T. Kieser, and K. F. Chater. 2004. SCP1, a 356,023 base pair linear plasmid adapted to the ecology and developmental biology of its host, *Streptomyces coelicolor* A3(2). *Mol. Microbiol.* **51:**1615–1628.

13. Bey, S. J., M. F. Tsou, C. H. Huang, C. C. Yang, and C. W. Chen. 2000. The homologous terminal sequence of the *Streptomyces lividans* chromosome and SLP2 plasmid. *Microbiology* **146:**911–922.

14. Bipatnath, M., P. P. Dennis, and H. Bremer. 1998. Initiation and velocity of chromosome replication in *Escherichia coli* B/r and K-12. *J. Bacteriol.* **180:**265–273.

15. Boye, E., A. Lobner-Olesen, and K. Skarstad. 2000. Limiting DNA replication to once and only once. *EMBO Rep.* **1:**479–483.

16. Bravo, A., and M. Salas. 1997. Initiation of bacteriophage f29 DNA replication in vivo: assembly of a membrane-associated multiprotein complex. *J. Mol. Biol.* **269:**102–112.

17. Caimano, M. J., X. Yang, T. G. Popova, M. L. Clawson, D. R. Akins, M. V. Norgard, and J. D. Radolf. 2000. Molecular and evolutionary characterization of the cp32/18 family of supercoiled plasmids in *Borrelia burgdorferi* 297. *Infect. Immun.* **68:**1574–1586.

18. Calcutt, M. J., and F. J. Schmidt. 1992. Conserved gene arrangement in the origin region of the *Streptomyces coelicolor* chromosome. *J. Bacteriol.* **174:**3220–3226.

19. Casjens, S. 2000. *Borrelia* genomes in the year 2000. *J. Mol. Microbiol. Biotechnol.* **2:**401–410.

20. Casjens, S. 1999. Evolution of the linear DNA replicons of the *Borrelia* spirochetes. *Curr. Opin. Microbiol.* **2:**529–534.

21. Casjens, S., M. Murphy, M. DeLange, L. Sampson, R. van Vugt, and W. M. Huang. 1997. Telomeres of the linear chromosomes of Lyme disease spirochaetes: nucleotide sequence and possible exchange with linear plasmid telomeres. *Mol. Microbiol.* **26:**581–596.

22. Casjens, S., N. Palmer, R. Van Vugt, W. H. Huang, B. Stevenson, P. Rosa, R. Lathigra, G. Sutton, J. Peterson, R. J. Dodson, D. Haft, E. Hickey, M. Gwinn, O. White, and C. M. Fraser. 2000. A bacterial genome in flux: the twelve linear and nine circular extrachromosomal DNAs in an infectious isolate of the Lyme disease spirochete *Borrelia burgdorferi. Mol. Microbiol.* **35:**490–516.

23. Chaconas, G., P. E. Stewart, K. Tilly, J. L. Bono, and P. Rosa. 2001. Telomere resolution in the Lyme disease spirochete. *EMBO J.* **20:**3229–3237.

24. Chang, P. C., and S. N. Cohen. 1994. Bidirectional replication from an internal origin in a linear *Streptomyces* plasmid. *Science* **265:**952–954.

25. Chang, P. C., E. S. Kim, and S. N. Cohen. 1996. *Streptomyces* linear plasmids that contain a phage-like, centrally located, replication origin. *Mol. Microbiol.* **22:**789–800.

26. Chen, C. W. 1996. Complications and implications of linear bacterial chromosomes. *Trends Genet.* **12:**192–196.

26a. Chen, C. W., C.-H. Huang, H.-H. Lee, H.-H. Tsai, and R. Kirby. 2002. Once the circle has been broken: dynamics and evolution of *Streptomyces* chromosomes. *Trends Genet.* **18:**522–529.

27. Chen, C. W., T.-W. Yu, Y. S. Lin, H. M. Kieser, and D. A. Hopwood. 1993. The conjugative plasmid SLP2 of *Streptomyces lividans* is a 50 kb linear molecule. *Mol. Microbiol.* **7:**925–932.

28. Chen, Y.-T. 2000. Linear chromosomes and linear plasmids in Actinomycetes. M.S. thesis. National Yang-Ming University, Taipei, Taiwan.

29. Cheng, A.-J. 1995. Construction and characterization of *recA* mutants of *Streptomyces lividans.* M.S. thesis. National Yang-Ming University, Taipei, Taiwan.

30. Chou, S.-H., L. Zhu, and B. R. Reid. 1997. Sheared purine-purine pairing in biology. *J. Mol. Biol.* **267:**1055–1067.

31. Deneke, J., G. Ziegelin, R. Lurz, and E. Lanka. 2002. Phage N15 telomere resolution: target requirements for recognition and processing by the protelomerase. *J. Biol. Chem.* **277:**10410–10419.

32. Deneke, J., G. Ziegelin, R. Lurz, and E. Lanka. 2000. The protelomerase of temperate *Escherichia coli* phage N15 has cleaving-joining activity. *Proc. Natl. Acad. Sci. USA* **97:**7721–7726.

33. Ducote, M. J., S. Prakash, and G. S. Pettis. 2000. Minimal and contributing sequence determinants of the *cis*-acting locus of transfer (*clt*) of streptomycete plasmid pIJ101 occur within an intrinsically curved plasmid region. *J. Bacteriol.* **182:**6834–6841.

34. Eggers, C. H., M. J. Caimano, M. L. Clawson, W. G. Miller, D. S. Samuels, and J. D. Radolf. 2002. Identification of loci critical for replication and compatibility of a *Borrelia burgdorferi* cp32 plasmid and use of a cp32-based shuttle vector for the expression of fluorescent reporters in the Lyme disease spirochaete. *Mol. Microbiol.* **43:**281–295.

35. Eggers, C. H., S. Casjens, S. F. Hayes, C. F. Garon, C. J. Damman, D. B. Oliver, and D. S. Samuels. 2000. Bacteriophages of spirochetes. *J. Mol. Microbiol. Biotechnol.* **2:**365–373.

36. Ferdows, M. S., and A. G. Barbour. 1989. Megabase-sized linear DNA in the bacterium *Borrelia burgdorferi,* the Lyme disease agent. *Proc. Natl. Acad. Sci. USA* **86:**5969–5973.

37. Ferdows, M. S., P. Serwer, G. A. Griess, S. J. Norris, and A. G. Barbour. 1996. Conversion of a linear to a circular plasmid in the relapsing fever agent *Borrelia hermsii*. *J. Bacteriol.* **178:**793–800.

38. Flett, F., D. de Mello Jungmann-Campello, V. Mersinias, S. L. Koh, R. Godden, and C. P. Smith. 1999. A "gram-negative-type" DNA polymerase III is essential for replication of the linear chromosome of *Streptomyces coelicolor* A3(2). *Mol. Microbiol.* **31:**949–958.

39. Fraser, C. M., S. Casjens, W. M. Huang, G. G. Sutton, R. Clayton, R. Lathigra, O. White, K. A. Ketchum, R. Dodson, E. K. Hickey, M. Gwinn, B. Dougherty, J. F. Tomb, R. D. Fleischmann, D. Richardson, J. Peterson, A. R. Kerlavage, J. Quackenbush, S. Salzberg, M. Hanson, R. van Vugt, N. Palmer, M. D. Adams, J. Gocayne, J. Weidman, T. Utterback, L. Watthey, L. McDonald, P. Artiach, C. Bowman, S. Garland, C. Fujii, M. D. Cotton, K. Horst, K. Roberts, B. Hatch, H. O. Smith, and J. C. Venter. 1997. Genomic sequence of a Lyme disease spirochaete, *Borrelia burgdorferi*. *Nature* **390:**580–586.

40. Garcia-Lara, J., M. Picardeau, B. J. Hinnebusch, W. M. Huang, and S. Casjens. 2000. The role of genomics in approaching the study of *Borrelia* DNA replication. *J. Mol. Microbiol. Biotechnol.* **2:**447–454.

41. Goodner, B., G. Hinkle, S. Gattung, N. Miller, M. Blanchard, B. Qurollo, B. S. Goldman, Y. Cao, M. Askenazi, C. Halling, L. Mullin, K. Houmiel, J. Gordon, M. Vaudin, O. Iartchouk, A. Epp, F. Liu, C. Wollam, M. Allinger, D. Doughty, C. Scott, C. Lappas, B. Markelz, C. Flanagan, C. Crowell, J. Gurson, C. Lomo, C. Sear, G. Strub, C. Cielo, and S. Slater. 2001. Genome sequence of the plant pathogen and biotechnology agent *Agrobacterium tumefaciens* C58. *Science* **294:**2323–2328.

42. Goodner, B. W., B. P. Markelz, M. C. Flanagan, C. B. Crowell, J. L. Racette, B. A. Schilling, L. M. Halfon, J. S. Mellors, and G. Grabowski. 1999. Combined genetic and physical map of the complex genome of *Agrobacterium tumefaciens*. *J. Bacteriol.* **181:**5160–5166.

43. Gopaul, D. N., and G. D. Duyne. 1999. Structure and mechanism in site-specific recombination. *Curr. Opin. Struct. Biol.* **9:**14–20.

44. Grainge, I., and M. Jayaram. 1999. The integrase family of recombinase: organization and function of the active site. *Mol. Microbiol.* **33:**449–456.

45. Hallet, B., and D. J. Sherratt. 1997. Transposition and site-specific recombination: adapting DNA cut-and-paste mechanisms to a variety of genetic rearrangements. *FEMS Microbiol. Rev.* **21:**157–178.

46. Hefty, P. S., S. E. Jolliff, M. J. Caimano, S. K. Wikel, J. D. Radolf, and D. R. Akins. 2001. Regulation of OspE-related, OspF-related, and Elp lipoproteins of *Borrelia burgdorferi* strain 297 by mammalian host-specific signals. *Infect. Immun.* **69:**3618–3627.

47. Hinnebusch, B. J., and A. J. Bendich. 1997. The bacterial nucleoid visualized by fluorescence microscopy of cells lysed within agarose: comparison of *Escherichia coli* and spirochetes of the genus *Borrelia*. *J. Bacteriol.* **179:**2228–2237.

48. Hinnebusch, J., and A. G. Barbour. 1992. Linear- and circular-plasmid copy numbers in *Borrelia burgdorferi*. *J. Bacteriol.* **174:**5251–5257.

49. Hinnebusch, J., and A. G. Barbour. 1991. Linear plasmids of *Borrelia burgdorferi* have a telomeric structure and sequence similar to those of a eukaryotic virus. *J. Bacteriol.* **173:**7233–7239.

50. Hinnebusch, J., S. Bergstrom, and A. G. Barbour. 1990. Cloning and sequence analysis of linear plasmid telomeres of the bacterium *Borrelia burgdorferi*. *Mol. Microbiol.* **4:**811–820.

51. Hinnebusch, J., and K. Tilly. 1993. Linear plasmids and chromosomes in bacteria. *Mol. Microbiol.* **10:**917–922.

52. Hopwood, D. A. 1965. A circular linkage map in the actinomycete *Streptomyces coelicolor*. *J. Mol. Biol.* **12:**514–516.

53. Hopwood, D. A. 1999. Forty years of genetics with *Streptomyces*: from in vivo through in vitro to in silico. *Microbiology* **145:**2183–2202.

54. Hopwood, D. A. 1967. Genetic analysis and genome structure in *Streptomyces coelicolor*. *Bacteriol. Rev.* **31:**373–403.

55. Hopwood, D. A. 1966. Lack of constant genome ends in *Streptomyces coelicolor*. *Genetics* **54:**1177–1184.

56. Huang, C.-H., Y.-S. Lin, Y.-L. Yang, S.-W. Huang, and C. W. Chen. 1998. The telomeres of *Streptomyces* chromosomes contain conserved palindromic sequences with potential to form complex secondary structures. *Mol. Microbiol.* **28:**905–926.

56a. Huang, C. H., C. Y. Chen, H. H. Tsai, C. Chen, Y. S. Lin, and C. W. Chen. 2003. Linear plasmid SLP2 of *Streptomyces lividans* is a composite replicon. *Mol. Microbiol.* **47:**1563–1576.

56b. Ikeda, H., J. Ishikawa, A. Hanamoto, M. Shinose, H. Kikuchi, T. Shiba, Y. Sakaki, M. Hattori, and S. Omura. 2003. Complete genome sequence and comparative analysis of the industrial microorganism *Streptomyces avermitilis*. *Nat. Biotechnol.* **21:**526–531.

57. Jakimowicz, D., J. Majkadagger, G. Konopa, G. Wegrzyn, W. Messer, H. Schrempf, and J. Zakrzewska-Czerwinska. 2000. Architecture of the *Streptomyces lividans* DnaA protein-replication origin complexes. *J. Mol. Biol.* **298:**351–364.

58. Kinashi, H., and M. Shimaji-Murayama. 1991. Physical characterization of SCP1, a giant linear plasmid from *Streptomyces coelicolor*. *J. Bacteriol.* **173:**1523–1529.

59. Kitten, T., and A. G. Barbour. 1990. Juxtaposition of expressed variable antigen genes with a conserved telomere in the bacterium *Borrelia hermsii*. *Proc. Natl. Acad. Sci. USA* **87:**6077–6081.

60. Knight, S. W., B. J. Kimmel, C. H. Eggers, and D. S. Samuels. 2000. Disruption of the *Borrelia burgdorferi gac* gene, encoding the naturally synthesized GyrA C-terminal domain. *J. Bacteriol.* **182:**2048–2051.

61. Knight, S. W., and D. S. Samuels. 1999. Natural synthesis of a DNA-binding protein from the C-terminal domain of DNA gyrase A in *Borrelia burgdorferi*. *EMBO J.* **18:**4875–4881.

62. Kobryn, K., and G. Chaconas. 2001. The circle is broken: telomere resolution in linear replicons. *Curr. Opin. Microbiol.* **4:**558–564.

63. Kobryn, K., and G. Chaconas. 2002. ResT, a telomere resolvase encoded by the Lyme disease spirochete. *Mol. Cell* **9:**195–201.

64. Kobryn, K., D. Z. Naigamwalla, and G. Chaconas. 2000. Site-specific DNA binding and bending by the *Borrelia burgdorferi* Hbb protein. *Mol. Microbiol.* **37:**145–155.

65. Kunst, F., N. Ogasawara, I. Moszer, A. M. Albertini, G. Alloni, V. Azevedo, M. G. Bertero, P. Bessieres, A. Bolotin, S. Borchert, R. Borriss, L. Boursier, A. Brans, M. Braun, S. C. Brignell, S. Bron, S. Brouillet, C. V. Bruschi, B. Caldwell, V. Capuano, N. M. Carter, S. K. Choi, J. J. Codani, I. F. Connerton, N. J. Cummings, R. A. Daniel, F. Denizot, K. M. Devine, A. Düsterhöft, S. D. Ehrlich, P. T. Emmerson, K. D. Entian, J. Errington, C. Fabret, E. Ferrari, D. Foulger, C. Fritz, M. Fujita, Y. Fujita, S. Fuma, A. Galizzi, N. Galleron, S.-Y. Ghim, P. Glaser, A. Goffeau, E. J. Golightly, G. Grandi, G. Guiseppi, B. J. Guy, K. Haga, J. Haiech, C. R. Harwood, A. Hénaut, H. Hilbert, S. Holsappel, S. Hosono, M.-F.

Hullo, M. Itaya, L. Jones, B. Joris, D. Karamata, Y. Kasahara, M. Klaerr-Blanchard, C. Klein, Y. Kobayashi, P. Koetter, G. Koningstein, S. Krogh, M. Kumano, K. Kurita, A. Lapidus, S. Lardinois, J. Lauber, V. Lazarevic, S.-M. Lee, A. Levine, H. Liu, S. Masuda, C. Mauël, C. Médigue, N. Medina, R. P. Mellado, M. Mizuno, D. Moestl, S. Nakai, M. Noback, D. Noone, M. O'Reilly, K. Ogawa, A. Ogiwara, B. Oudega, S.-H. Park, V. Parro, T. M. Pohl, D. Portetelle, S. Porwollik, A. M. Prescott, E. Presecan, P. Pujic, B. Purnelle, G. Rapoport, M. Rey, S. Reynolds, M. Rieger, C. Rivolta, E. Rocha, B. Roche, M. Rose, Y. Sadaie, T. Sato, E. Scanlan, S. Schleich, R. Schroeter, F. Scoffone, J. Sekiguchi, A. Sekowska, S. J. Seror, P. Serror, B.-S. Shin, B. Soldo, A. Sorokin, E. Tacconi, T. Takagi, H. Takahashi, K. Takemaru, M. Takeuchi, A. Tamakoshi, T. Tanaka, P. Terpstra, A. Tognoni, V. Tosato, S. Uchiyama, M. Vandenbol, F. Vannier, A. Vassarotti, A. Viari, R. Wambutt, E. Wedler, H. Wedler, T. Weitzenegger, P. Winters, A. Wipat, H. Yamamoto, K. Yamane, K. Yasumoto, K. Yata, K. Yoshida, H.-F. Yoshikawa, E. Zumstein, H. Yoshikawa, and A. Danchin. 1997. The complete genome sequence of the gram-positive bacterium *Bacillus subtilis*. *Nature* 390:249–256.

66. Labandeira-Rey, M., and J. T. Skare. 2001. Decreased infectivity in *Borrelia burgdorferi* strain B31 is associated with loss of linear plasmid 25 or 28-1. *Infect. Immun.* 69:446–455.

67. Leblond, P., and B. Decaris. 1994. New insights into the genetic instability of *Streptomyces*. *FEMS Microbiol. Lett.* 123:225–232.

68. Leblond, P., G. Fischer, F. Francou, F. Berger, M. Guerineau, and B. Decaris. 1996. The unstable region of *Streptomyces ambofaciens* includes 210-kb terminal inverted repeats flanking the extremities of the linear chromosomal DNA. *Mol. Microbiol.* 19:261–271.

68a. Leblond, P., M. Redenbach, and J. Cullum. 1993. Physical map of the *Streptomyces lividans* 66 genome and comparison with that of the related strain *Streptomyces coelicolor* A3(2). *J. Bacteriol.* 175:3422–3429.

69. Lee, L. F., S. H. Yeh, and C. W. Chen. 2002. Construction and synchronization of *dnaA* temperature-sensitive mutants of *Streptomyces*. *J. Bacteriol.* 184:1214–1218.

70. Lezhava, A., T. Mizukami, T. Kajitani, D. Kameoka, M. Redenbach, H. Shinkawa, O. Nimi, and H. Kinashi. 1995. Physical map of the linear chromosome of *Streptomyces griseus*. *J. Bacteriol.* 177:6492–6498.

71. Lin, Y.-S., H. M. Kieser, D. A. Hopwood, and C. W. Chen. 1993. The chromosomal DNA of *Streptomyces lividans* 66 is linear. *Mol. Microbiol.* 10:923–933.

72. Majka, J., J. Zakrzewska-Czerwinska, and W. Messer. 2001. Sequence recognition, cooperative interaction, and dimerization of the initiator protein DnaA of *Streptomyces*. *J. Biol. Chem.* 276:6243–6252.

73. Marconi, R. T., S. Casjens, U. G. Munderloh, and D. S. Samuels. 1996. Analysis of linear plasmid dimers in *Borrelia burgdorferi* sensu lato isolates: implications concerning the potential mechanism of linear plasmid replication. *J. Bacteriol.* 178:3357–3361.

74. Meijer, W. J. J., J. A. Horcajadas, and M. Salas. 2001. ϕ29 family of phages. *Microbiol. Mol. Biol. Rev.* 65:261–287.

75. Meinhardt, F., R. Schaffrath, and M. Larsen. 1997. Microbial linear plasmids. *Appl. Microbiol. Biotechnol.* 47:329–336.

76. Mikoc, A., I. Ahel, and V. Gamulin. 2000. Construction and characterization of a *Streptomyces rimosus recA* mutant: the RecA-deficient strain remains viable. *Mol. Gen. Genet.* 264:227–232.

77. Mrazek, J., and S. Karlin. 1998. Strand compositional asymmetry in bacterial and large viral genomes. *Proc. Natl. Acad. Sci. USA* 95:3720–3725.

78. Muth, G., D. Frese, A. Kleber, and W. Wohlleben. 1997. Mutational analysis of the *Streptomyces lividans recA* gene suggests that only mutants with residual activity remain viable. *Mol. Gen. Genet.* 255:420–428.

79. Nordstrand, A., A. G. Barbour, and S. Bergstrom. 2000. *Borrelia* pathogenesis research in the post-genomic and post-vaccine era. *Curr. Opin. Microbiol.* 3:86–92.

80. Norioka, N., M. Y. Hsu, S. Inouye, and M. Inouye. 1995. Two *recA* genes in *Myxococcus xanthus*. *J. Bacteriol.* 177:4179–4182.

81. Omura, S., H. Ikeda, J. Ishikawa, A. Hanamoto, C. Takahashi, M. Shinose, Y. Takahashi, H. Horikawa, H. Nakazawa, T. Osonoe, H. Kikuchi, T. Shiba, Y. Sakaki, and M. Hattori. 2001. Genome sequence of an industrial microorganism *Streptomyces avermitilis*: deducing the ability of producing secondary metabolites. *Proc. Natl. Acad. Sci. USA* 98:12215–12220.

82. Ortin, J., E. Vinuela, M. Salas, and C. Vasquez. 1971. DNA-protein complex in circular DNA from phage f29. *Nat. New Biol.* 234:275–277.

83. Palaniyar, N., E. Gerasimopoulos, and D. H. Evans. 1999. Shope fibroma virus DNA topoisomerase catalyses Holliday junction resolution and hairpin formation in vitro. *J. Mol. Biol.* 287:9–20.

84. Pandza, K., G. Pfalzer, J. Cullum, and D. Hranueli. 1997. Physical mapping shows that the unstable oxytetracycline gene cluster of *Streptomyces rimosus* lies close to one end of the linear chromosome. *Microbiology* 143:1493–1501.

85. Paques, F., and J. E. Haber. 1999. Multiple pathways of recombination induced by double-strand breaks in *Saccharomyces cerevisiae*. *Microbiol. Mol. Biol. Rev.* 63:349–404.

86. Pettis, G. S., and S. N. Cohen. 1994. Transfer of the pIJ101 plasmid in *Streptomyces lividans* requires a *cis*-acting function dispensable for chromosomal gene transfer. *Mol. Microbiol.* 13:955–964.

87. Picardeau, M., J. R. Lobry, and B. J. Hinnebusch. 2000. Analyzing DNA strand compositional asymmetry to identify candidate replication origins of *Borrelia burgdorferi* linear and circular plasmids. *Genome Res.* 10:1594–1604.

88. Picardeau, M., J. R. Lobry, and B. J. Hinnebusch. 1999. Physical mapping of an origin of bidirectional replication at the centre of the *Borrelia burgdorferi* linear chromosome. *Mol. Microbiol.* 32:437–445.

89. Plasterk, R. H., M. I. Simon, and A. G. Barbour. 1985. Transposition of structural genes to an expression sequence on a linear plasmid causes antigenic variation in the bacterium *Borrelia hermsii*. *Nature* 318:257–263.

90. Purser, J. E., and S. J. Norris. 2000. Correlation between plasmid content and infectivity in *Borrelia burgdorferi*. *Proc. Natl. Acad. Sci. USA* 97:13865–13870.

91. Qin, Z., and S. N. Cohen. 1998. Replication at the telomeres of the *Streptomyces* linear plasmid pSLA2. *Mol. Microbiol.* 28:893–904.

91a. Redenbach, M., F. Flett, W. Piendl, I. Glocker, U. Rauland, O. Wafzig, P. Leblond, and J. Cullum. 1993. The *Streptomyces lividans* 66 chromosome contains a 1 Mb deletogenic region flanked by two amplifiable regions. *Mol. Gen. Genet.* 241:255–262.

92. Redenbach, M., E. Kleinert, and A. Stoll. 2000. Identification of DNA amplifications near the center of the *Streptomyces coelicolor* M145 chromosome. *FEMS Microbiol. Lett.* 191:123–129.

93. Redenbach, M., J. Scheel, and U. Schmidt. 2000. Chromosome topology and genome size of selected actinomycetes species. *Antonie Leeuwenhoek* 78:227–235.

94. Reeves, A. R., D. A. Post, and T. J. Vanden Boom. 1998. Physical-genetic map of the erythromycin-producing organism *Saccharopolyspora erythraea*. *Microbiology* 144:2151–2159.

95. Rybchin, V. N., and A. N. Svarchevsky. 1999. The plasmid prophage N15: a linear DNA with covalently closed ends. *Mol. Microbiol.* 33:895–903.

96. Schaack, J., W. Y. Ho, P. Freimuth, and T. Shenk. 1990. Adenovirus terminal protein mediates both nuclear matrix association and efficient transcription of adenovirus DNA. *Genes Dev.* 4:1197–1208.

97. Schwan, T. G., W. Burgdorfer, and C. F. Garon. 1988. Changes in infectivity and plasmid profile of the Lyme disease spirochete, *Borrelia burgdorferi*, as a result of in vitro cultivation. *Infect. Immun.* 56:1831–1836.

98. Schwan, T. G., W. Burgdorfer, and P. A. Rosa. 1999. *Borrelia*, p. 746–758. *In* P. R. Murray, E. J. Baron, M. A. Pfaller, F. C. Tenover, and R. H. Yolken (ed.), *Manual of Clinical Microbiology*, 7th ed. ASM Press, Washington, D.C.

99. Sekiguchi, J., C. Cheng, and S. Shuman. 2000. Resolution of a Holliday junction by vaccinia topoisomerase requires a spacer DNA segment 3' of the CCCTT cleavage sites. *Nucleic Acids Res.* 28:2658–2663.

100. Servin-Gonzalez, L. 1996. Identification and properties of a novel *clt* locus in the *Streptomyces phaeochromogenes* plasmid pJV1. *J. Bacteriol.* 178:4323–4326.

101. Shahab, N., F. Flett, S. G. Oliver, and P. R. Butler. 1996. Growth rate control of protein and nucleic acid content in *Streptomyces coelicolor* A3(2) and *Escherichia coli* B/r. *Microbiology* 142(pt. 8):1927–1935.

102. Shuman, S. 1998. Vaccinia virus DNA topoisomerase: a model eukaryotic type IB enzyme. *Biochim. Biophys. Acta* 1400:321–337.

103. Stahl, F. W. 1967. Circular genetic maps. *J. Cell. Physiol.* 70:1–12.

104. Stahl, F. W., and C. M. Steinberg. 1964. The theory of formal phage genetics for circular maps. *Genetics* 50:531–538.

105. Stevenson, B., S. F. Porcella, K. L. Oie, C. A. Fitzpatrick, S. J. Raffel, L. Lubke, M. E. Schrumpf, and T. G. Schwan. 2000. The relapsing fever spirochete *Borrelia hermsii* contains multiple, antigen-encoding circular plasmids that are homologous to the cp32 plasmids of Lyme disease spirochetes. *Infect. Immun.* 68:3900–3908.

106. Stevenson, B., W. R. Zuckert, and D. R. Akins. 2000. Repetition, conservation, and variation: the multiple cp32 plasmids of *Borrelia* species. *J. Mol. Microbiol. Biotechnol.* 2:411–422.

107. Stewart, P. E., R. Thalken, J. L. Bono, and P. Rosa. 2001. Isolation of a circular plasmid region sufficient for autonomous replication and transformation of infectious *Borrelia burgdorferi*. *Mol. Microbiol.* 39:714–721.

108. Tilly, K., J. Fuhrman, J. Campbell, and D. S. Samuels. 1996. Isolation of *Borrelia burgdorferi* genes encoding homologues of DNA-binding protein HU and ribosomal protein S20. *Microbiology* 142:2471–2479.

109. Vierling, S., T. Weber, W. Wohlleben, and G. Muth. 2001. Evidence that an additional mutation is required to tolerate insertional inactivation of the *Streptomyces lividans recA* gene. *J. Bacteriol.* 183:4374–4381.

110. Volff, J. N., and J. Altenbuchner. 1998. Genetic instability of the *Streptomyces* chromosome. *Mol. Microbiol.* 27:239–246.

111. Volff, J. N., and J. Altenbuchner. 2000. A new beginning with new ends: linearisation of circular chromosomes during bacterial evolution. *FEMS Microbiol. Lett.* 186:143–150.

112. Wang, S.-J., H.-M. Chang, Y.-S. Lin, C.-H. Huang, and C. W. Chen. 1999. *Streptomyces* genomes: circular genetic maps from the linear chromosomes. *Microbiology* 145:2209–2220.

113. Watson, J. D. 1972. Origin of concatemeric T7 DNA. *Nat. New Biol.* 239:197–201.

114. Willems, H., C. Jager, and G. Baljer. 1998. Physical and genetic map of the obligate intracellular bacterium *Coxiella burnetii*. *J. Bacteriol.* 180:3816–3822.

115. Wood, D. W., J. C. Setubal, R. Kaul, D. E. Monks, J. P. Kitajima, V. K. Okura, Y. Zhou, L. Chen, G. E. Wood, N. F. Almeida, Jr., L. Woo, Y. Chen, I. T. Paulsen, J. A. Eisen, P. D. Karp, D. Bovee, Sr., P. Chapman, J. Clendenning, G. Deatherage, W. Gillet, C. Grant, T. Kutyavin, R. Levy, M. J. Li, E. McClelland, A. Palmieri, C. Raymond, G. Rouse, C. Saenphimmachak, Z. Wu, P. Romero, D. Gordon, S. Zhang, H. Yoo, Y. Tao, P. Biddle, M. Jung, W. Krespan, M. Perry, B. Gordon-Kamm, L. Liao, S. Kim, C. Hendrick, Z. Y. Zhao, M. Dolan, F. Chumley, S. V. Tingey, J. F. Tomb, M. P. Gordon, M. V. Olson, and E. W. Nester. 2001. The genome of the natural genetic engineer *Agrobacterium tumefaciens* C58. *Science* 294:2317–2323.

116. Yang, C.-C., C.-H. Huang, C.-Y. Li, Y.-G. Tsay, S.-C. Lee, and C. W. Chen. 2002. The terminal proteins of linear *Streptomyces* chromosomes and plasmids: a novel class of replication priming proteins. *Mol. Microbiol.* 43:297–305.

117. Yang, M. C., and R. Losick. 2001. Cytological evidence for association of the ends of the linear chromosome in *Streptomyces coelicolor*. *J. Bacteriol.* 183:5180–5186.

118. Zakrzewska-Czerwinska, J., D. Jakimowicz, J. Majka, W. Messer, and H. Schrempf. 2000. Initiation of the *Streptomyces* chromosome replication. *Antonie Leeuwenhoek* 78:211–221.

119. Zakrzewska-Czerwinska, J., and H. Schrempf. 1992. Characterization of an autonomously replicating region from the *Streptomyces lividans* chromosome. *J. Bacteriol.* 147:2688–2693.

120. Zechiedrich, E. L., and N. R. Cozzarelli. 1995. Roles of topoisomerase IV and DNA gyrase in DNA unlinking during replication in *Escherichia coli*. *Genes Dev.* 9:2859–2869.

121. Zhang, J. R., J. M. Hardham, A. G. Barbour, and S. J. Norris. 1997. Antigenic variation in Lyme disease borreliae by promiscuous recombination of VMP-like sequence cassettes. *Cell* 89:275–285.

122. Zhu, Q., P. Pongpech, and R. J. DiGate. 2001. Type I topoisomerase activity is required for proper chromosomal segregation in *Escherichia coli*. *Proc. Natl. Acad. Sci. USA* 98:9766–9771.

INDEX